One-Stop Internet Resources

Log on to

fl.pre-algebra.com

Florida Online Book

- Complete Student Edition
- Links to Online Study Tools

Florida Online Study Tools

- Extra Examples
- Self-Check Quizzes
- Vocabulary Review
- Chapter Test Practice
- FCAT Practice

Online Activities

- WebQuest Projects
- USA TODAY Activities
- Career Links
- Data Updates

Parent & Student Study Guide Workbook

- Printable Worksheets

Graphing Calculator Keystrokes

- Calculator Keystrokes for other calculators

For more information on these resources, see page 1.

Pre-Algebra

Contents

Florida Teacher Advisory Board .ii
Florida Sunshine State Standards and Grade 8 Expectations iii
How to Prepare for FCAT FL2
Countdown to FCAT FL4

New York, New York Columbus, Ohio Chicago, Illinois Peoria, Illinois Woodland Hills, California

ISBN: 0-07-860368-4 (*Florida Student Edition*)

Florida Teacher Advisory Board

Sunshine State Standards and Grade 8 Expectations

Florida

Strand A NUMBER SENSE, CONCEPTS, AND OPERATIONS

Standard 1: *The student understands the different ways numbers are represented and used in the real world.*

Benchmark MA.A.1.3.1: The student associates verbal names, written word names, and standard numerals with integers, fractions, decimals; numbers expressed as percents; numbers with exponents; numbers in scientific notation; radicals; absolute value; and ratios.

1	knows word names and standard numerals for integers, fractions, decimals, numbers expressed as percents, numbers with exponents, numbers expressed in scientific notation, absolute value, radicals, and ratios.

Benchmark MA.A.1.3.2: The student understands the relative size of integers, fractions, and decimals; numbers expressed as percents; numbers with exponents; numbers in scientific notation; radicals; absolute value; and ratios.

1	compares and orders fractions, decimals, integers, and radicals using graphic models, number lines, and symbols.
2	compares and orders numbers expressed in absolute value, scientific notation, integers, percents, numbers with exponents, fractions, decimals, radicals, and ratios.

Benchmark MA.A.1.3.3: The student understands concrete and symbolic representations of rational numbers and irrational numbers in real-world situations.

1	knows examples of rational and irrational numbers in real-world situations.
2	describes the meanings of rational and irrational numbers using physical or graphical displays.
3	constructs models to represent rational and irrational numbers.

Benchmark MA.A.1.3.4: The student understands that numbers can be represented in a variety of equivalent forms, including integers, fractions, decimals, percents, scientific notation, exponents, radicals, and absolute value.

1	knows the relationships among fractions, decimals, and percents given a real-world context.
2	simplifies expressions using integers, exponents, and radicals.
3	knows equivalent forms of large and small numbers in scientific and standard notation.
4	identifies and explains the absolute value of a number.

Standard 2: *The student understands number systems.*

Benchmark MA.A.2.3.1: The student understands and uses exponential and scientific notation.

1	expresses rational numbers in exponential notation including negative exponents (for example, $2^{-3} = \frac{1}{2^3} = \frac{1}{8}$).
2	expresses numbers in scientific or standard notation including decimals between 0 and 1.
3	evaluates numerical or algebraic expressions that contain exponential notation.

Benchmark MA.A.2.3.2: The student understands the structure of number systems other than the decimal number system.

1	expresses base ten numbers as equivalent numbers in different bases, such as base two, base five, and base eight.
2	discusses the application of the binary (base two) number system in computer technology.
3	expresses non-base ten numbers as equivalent numbers in base ten.

Standard 3: *The student understands the effects of operations on numbers and the relationships among these operations, selects appropriate operations, and computes for problem solving.*

Benchmark MA.A.3.3.1: The student understands and explains the effects of addition, subtraction, multiplication, and division on whole numbers, fractions, including mixed numbers, and decimals, including the inverse relationships of positive and negative numbers.

1	knows the effects of the four basic operations on whole numbers, fractions, mixed numbers, decimals, and integers.
2	knows the inverse relationship of positive and negative numbers.
3	applies the properties of real numbers to solve problems (commutative, associative, distributive, identity, equality, inverse, and closure).

Benchmark MA.A.3.3.2: The student selects the appropriate operation to solve problems involving addition, subtraction, multiplication, and division of rational numbers, ratios, proportions, and percents, including the appropriate application of the algebraic order of operations.

1	knows the appropriate operations to solve real-world problems involving integers, ratios, rates, proportions, numbers expressed as percents, decimals, and fractions.
2	solves real-world problems involving integers, ratios, proportions, numbers expressed as percents, decimals, and fractions in two- or three-step problems.
3	solves real-world problems involving percents including percents greater than 100% (for example percent of change, commission).
4	writes and simplifies expressions from real-world situations using the order of operations.

Benchmark MA.A.3.3.3: The student adds, subtracts, multiplies, and divides whole numbers, decimals, and fractions, including mixed numbers, to solve real-world problems, using appropriate methods of computing, such as mental mathematics, paper and pencil, and calculator.

1	solves multi-step real-world problems involving fractions, decimals, and integers using appropriate methods of computation, such as mental computation, paper and pencil, and calculator.

Standard 4: *The student uses estimation in problem solving and computation.*

Benchmark MA.A.4.3.1: The student uses estimation strategies to predict results and to check the reasonableness of results.

1	knows appropriate estimation techniques for a given situation using real numbers.
2	estimates to predict results and to check reasonableness of results.

Standard 5: *The student understands and applies theories related to numbers.*

Benchmark MA.A.5.3.1: The student uses concepts about numbers, including primes, factors, and multiples, to build number sequences.

1	knows if numbers are relatively prime.
2	applies number theory concepts to determine the terms in a real number sequence.
3	applies number theory concepts, including divisibility rules, to solve real-world or mathematical problems.

Strand B MEASUREMENT

Standard 1: *The student measures quantities in the real world and uses the measures to solve problems.*

Benchmark MA.B.1.3.1: The student uses concrete and graphic models to derive formulas for finding perimeter, area, surface area, circumference, and volume of two- and three-dimensional shapes, including rectangular solids and cylinders.

1	uses concrete and graphic models to explore and derive formulas for surface area and volume of three-dimensional regular shapes, including pyramids, prisms, and cones.
2	solves and explains real-world problems involving surface area and volume of three-dimensional shapes.

Benchmark MA.B.1.3.2: The student uses concrete and graphic models to derive formulas for finding rates, distance, time, and angle measures.

1	applies formulas for finding rates, distance, time and angle measures.
2	describes and uses rates of change (for example, temperature as it changes throughout the day, or speed as the rate of change in distance over time) and other derived measures.

Benchmark MA.B.1.3.3: The student understands and describes how the change of a figure in such dimensions as length, width, height, or radius affects its other measurements such as perimeter, area, surface area, and volume.

1	knows how a change in a figure's dimensions affects its perimeter, area, circumference, surface area, or volume.
2	knows how changes in the volume, surface area, area, or perimeter of a figure affect the dimensions of the figure.
3	solves real-world or mathematical problems involving the effects of changes either to the dimensions of a figure or to the volume, surface area, area, perimeter, or circumference of figures.

Benchmark MA.B.1.3.4: The student constructs, interprets, and uses scale drawings such as those based on number lines and maps to solve real-world problems.

1	interprets and applies various scales including those based on number lines, graphs, models, and maps. (Scale may include rational numbers.)
2	constructs and uses scale drawings to recreate a given situation.

Standard 2: *The student compares, contrasts, and converts within systems of measurement (both standard/nonstandard and metric/customary).*

Benchmark MA.B.2.3.1: The student uses direct (measured) and indirect (not measured) measures to compare a given characteristic in either metric or customary units.

1	finds measures of length, weight or mass, and capacity or volume using proportional relationships and properties of similar geometric figures.

Benchmark MA.B.2.3.2: The student solves problems involving units of measure and converts answers to a larger or smaller unit within either the metric or customary system.

1	solves problems using mixed units within each system, such as feet and inches, hours and minutes.
2	solves problems using the conversion of measurements within the customary system.

3	solves problems using the conversions of measurement within the metric system.

Standard 3: *The student estimates measurements in real-world problem situations.*

Benchmark MA.B.3.3.1: The student solves real-world and mathematical problems involving estimates of measurements including length, time, weight/mass, temperature, money, perimeter, area, and volume, in either customary or metric units.

1	knows a variety of strategies to estimate, describe, make comparisons, and solve real-world and mathematical problems involving measurements.

Standard 4: *The student selects and uses appropriate units and instruments for measurement to achieve the degree of precision and accuracy required in real-world situations.*

Benchmark MA.B.4.3.1: The student selects appropriate units of measurement and determines and applies significant digits in a real-world context. (Significant digits should relate to both instrument precision and to the least precise unit of measurement).

1	selects the appropriate unit of measure for a given situation.
2	knows the precision of different measuring instruments.
3	determines the appropriate precision unit for a given situation.
4	identifies the number of significant digits as it relates to the least precise unit of measure.
5	determines the greatest possible error of a given measurement and the possible actual measurements of an object.

Benchmark MA.B.4.3.2: The student selects and uses appropriate instruments, technology, and techniques to measure quantities in order to achieve specified degrees of accuracy in a problem situation.

1	applies significant digits in the real-world context.
2	selects and uses appropriate instruments, technology, and techniques to measure quantities and dimensions to a specified degree of accuracy.

Strand C GEOMETRY AND SPATIAL SENSE

Standard 1: *The student describes, draws, identifies, and analyzes two- and three-dimensional shapes.*

Benchmark MA.C.1.3.1: The student understands the basic properties of, and relationships pertaining to, regular and irregular geometric shapes in two- and three-dimensions.

1	determines and justifies the measures of various types of angles based upon geometric relationships in two- and three-dimensional shapes.
2	compares regular and irregular polygons and two- and three-dimensional shapes.
3	draws and builds three-dimensional figures from various perspectives (for example, flat patterns, isometric drawings, nets).
4	knows the properties of two- and three-dimensional figures.

Standard 2: *The student visualizes and illustrates ways in which shapes can be combined, subdivided, and changed.*

Benchmark MA.C.2.3.1: The student understands the geometric concepts of symmetry, reflections, congruency, similarity, perpendicularity, parallelism, and transformations, including flips, slides, turns, and enlargements.

1	use the properties of parallelism, perpendicularity, and symmetry in solving real-world problems.
2	identifies congruent and similar figures in real-world situations and justifies the identification.
3	identifies and performs the various transformations (reflection, translation, rotation, dilation) of a given figure on a coordinate plane.

Benchmark MA.C.2.3.2: The student predicts and verifies patterns involving tessellations (a covering of a plane with congruent copies of the same pattern with no holes and no overlaps, like floor tiles).

1	continues a tessellation pattern using the needed transformations.
2	creates an original tessellating tile and tessellation pattern using a combination of transformations.

Standard 3: *The student uses coordinate geometry to locate objects in both two- and three-dimensions and to describe objects algebraically.*

Benchmark MA.C.3.3.1: The student represents and applies geometric properties and relationships to solve real-world and mathematical problems.

1	observes, explains, makes and tests conjectures regarding geometric properties and relationships (among regular and irregular shapes of two and three dimensions).
2	applies the Pythagorean Theorem in real-world problems (for example, finds the relationship among sides in 45°-45° and 30°-60° right triangles).

Benchmark MA.C.3.3.2: The student identifies and plots ordered pairs in all four quadrants of a rectangular coordinate system (graph) and applies simple properties of lines.

1	given an equation or its graph, finds ordered-pair solutions (for example, $y = 2x$).
2	given the graph of a line, identifies the slope of the line (including the slope of vertical and horizontal lines).
3	given the graph of a linear relationship, applies and explains the simple properties of lines on a graph, including parallelism, perpendicularity, and identifying the *x*- and *y*-intercepts, the midpoint of a horizontal or vertical line segment, and the intersection point of two lines.

Strand D ALGEBRAIC THINKING

Standard 1: *The student describes, analyzes, and generalizes a wide variety of patterns, relations, and functions.*

Benchmark MA.D.1.3.1: The student describes a wide variety of patterns, relationships, and functions through models, such as manipulatives, tables, graphs, expressions, equations, and inequalities.

1	reads, analyzes, and describes graphs of linear relationships.
2	uses variables to represent unknown quantities in real-world problems.
3	uses the information provided in a table, graph, or rule to determine if a function is linear and justifies reasoning.
4	finds a function rule to describe tables of related input-output variables.
5	predicts outcomes based upon function rules.

Benchmark MA.D.1.3.2: The student creates and interprets tables, graphs, equations, and verbal descriptions to explain cause-and-effect relationships.

1	interprets and creates tables and graphs (function tables).

2	writes equations and inequalities to express relationships.
3	graphs equations and inequalities to explain cause-and-effect relationships.
4	interprets the meaning of the slope of a line from a graph depicting a real-world situation.

Standard 2: *The student uses expressions, equations, inequalities, graphs, and formulas to represent and interpret situations.*

Benchmark MA.D.2.3.1: The student represents and solves real-world problems graphically, with algebraic expressions, equations, and inequalities.

1	translates verbal expressions and sentences into algebraic expressions, equations, and inequalities.
2	translates algebraic expressions, equations, or inequalities representing real-world relationships into verbal expressions or sentences.
3	solves single- and multiple-step linear equations and inequalities in concrete or abstract form.
4	graphs linear equations on the coordinate plane using tables of values.
5	graphically displays real-world situations represented by algebraic equations or inequalities.
6	evaluates algebraic expressions, equations, and inequalities by substituting integral values for variables and simplifying the results.
7	simplifies algebraic expressions that represent real-world situations by combining like terms and applying the properties of real numbers.

Benchmark MA.D.2.3.2: The student uses algebraic problem-solving strategies to solve real-world problems involving linear equations and inequalities.

1	simplifies algebraic expressions with a maximum of two variables.
2	solves single- and multi-step linear equations and inequalities that represent real-world situations.

Strand E DATA ANALYSIS AND PROBABILITY

Standard 1: *The student understands and uses the tools of data analysis for managing information.*

Benchmark MA.E.1.3.1: The student collects, organizes, and displays data in a variety of forms, including tables, line graphs, charts, bar graphs, to determine how different ways of presenting data can lead to different interpretations.

1	reads and interprets data displayed in a variety of forms including histograms.
2	constructs and interprets displays of data, (including circle, line, bar, and box-and-whisker graphs) and explains how different displays of data can lead to different interpretations.

Benchmark MA.E.1.3.2: The student understands and applies the concepts of range and central tendency (mean, median, and mode).

1	finds the mean, median, and mode of a set of data using raw data, tables, charts, or graphs.
2	interprets measures of dispersion (range) and of central tendency.
3	determines appropriate measures of central tendency for a given situation or set of data.

Benchmark MA.E.1.3.3: The student analyzes real-world data by applying appropriate formulas for measures of central tendency and organizing data in a quality display, using appropriate technology, including calculators and computers.

1	determines the mean, median, mode, and range of a set of real-world data using appropriate technology.
2	organizes, graphs and analyzes a set of real-world data using appropriate technology.

Standard 2: *The student identifies patterns and makes predictions from an orderly display of data using concepts of probability and statistics.*

Benchmark MA.E.2.3.1: The student compares experimental results with mathematical expectations of probabilities.

1	compares and explains the results of an experiment with the mathematically expected outcomes.
2	calculates simple mathematical probabilities for independent and dependent events.

Benchmark MA.E.2.3.2: The student determines odds for and odds against a given situation.

1	predicts the mathematical odds for and against a specified outcome in a given real-world situation.

Standard 3: *The student uses statistical methods to make inferences and valid arguments about real-world situations.*

Benchmark MA.E.3.3.1: The student formulates hypotheses, designs experiments, collects and interprets data, and evaluates hypotheses by making inferences and drawing conclusions based on statistics (range, mean, median, and mode) and tables, graphs, and charts.

1	formulates a hypothesis and designs an experiment.
2	performs the experiment and collects, organizes, and displays the data.
3	evaluates the hypothesis by making inferences and drawing conclusions based on statistical results.

Benchmark MA.E.3.3.2: The student identifies the common uses and misuses of probability or statistical analysis in the everyday world.

1	knows appropriate uses of statistics and probability in real-world situations.
2	knows when statistics and probability are used in misleading ways.
3	identifies and uses different types of sampling techniques (for example, random, systematic, stratified).
4	knows whether a sample is biased.

Image Credits: **i** Digital Vision; **ii, FL1** PhotoDisc

How To...

Prepare for FCAT

Countdown To FCAT

Pages FL4–FL24 of this text include a section called **Countdown to FCAT**. Each page contains 7 problems that are just like those on FCAT. You should plan to complete one page each week to help you prepare for the test.

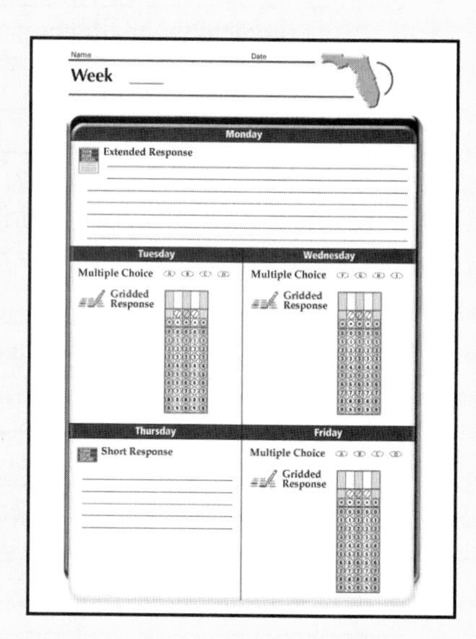

Plan to spend a few minutes each day working on the FCAT problem(s) for that day unless your teacher asks you to do otherwise. Each day of the week has the same type of problem(s). If you have difficulty with any problem, you can refer to the lesson that is referenced in parentheses after the problem.

Monday Extended Response
Tuesday Multiple Choice, Gridded Response
Wednesday Multiple Choice, Gridded Response
Thursday Short Response
Friday Multiple Choice or Gridded Response

Your teacher can provide you with an answer sheet to record your work and your answers for each week. A printable worksheet is also available at fl.pre-alg.com. At the end of the week, your teacher may want you to turn in the answer sheet.

FCAT Practice and Sample Test Workbook

The **FCAT Practice and Sample Test Workbook, Grade 8,** contains a diagnostic test, practice for each of the Sunshine State Standards, and a sample test.

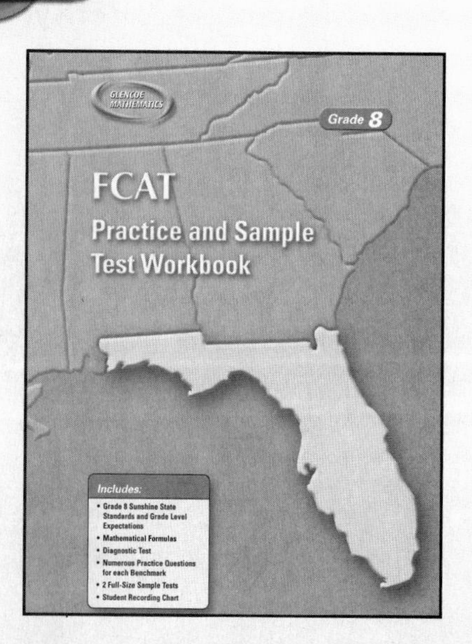

As you practice and master each objective, you can record your progress in the Student Recording Chart in your workbook.

Your teacher may also ask you to take a sample test at various points throughout the year to see if you're ready to take the real FCAT.

Your Textbook

Your textbook contains many opportunities for you to get ready for FCAT every day. Take advantage of these so you don't need to cram before the test.

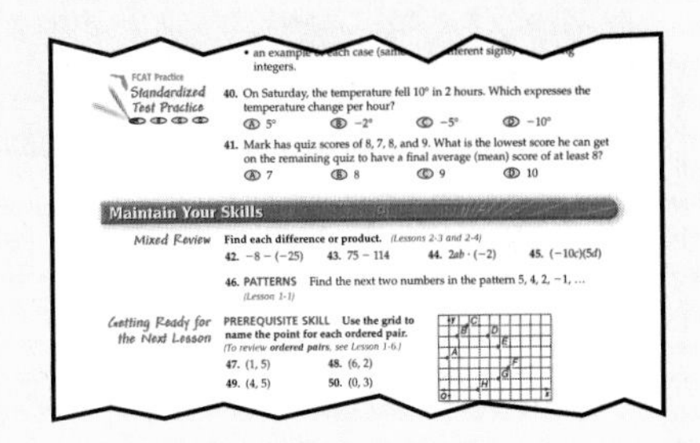
FCAT Practice

Standardized Test Practice

40. On Saturday, the temperature fell 10° in 2 hours. Which expresses the temperature change per hour?

Ⓐ 5° Ⓑ −2° Ⓒ −5° Ⓓ −10°

41. Mark has quiz scores of 8, 7, 8, and 9. What is the lowest score he can get on the remaining quiz to have a final average (mean) score of at least 8?

Ⓐ 7 Ⓑ 8 Ⓒ 9 Ⓓ 10

Maintain Your Skills

Mixed Review Find each difference or product. (Lessons 2-3 and 2-4)

42. −8 − (−25) 43. 75 − 114 44. 2ab · (−2) 45. (−10c)(5d)

46. PATTERNS Find the next two numbers in the pattern 5, 4, 2, −1, ... (Lesson 1-1)

Getting Ready for the Next Lesson PREREQUISITE SKILL Use the grid to name the point for each ordered pair. (To review ordered pairs, see Lesson 1-6.)

47. (1, 5) 48. (6, 2)

49. (4, 5) 50. (0, 3)

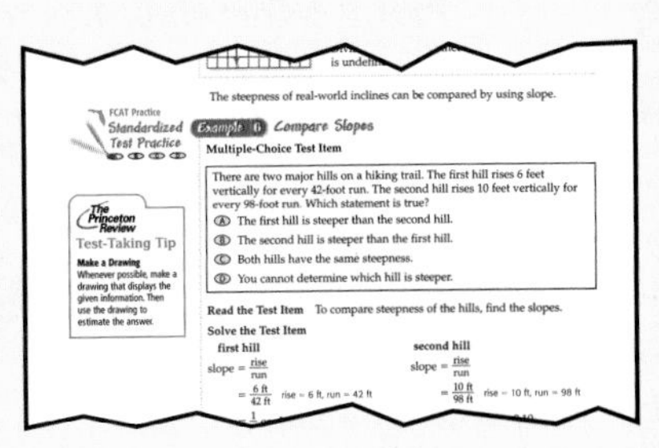
The steepness of real-world inclines can be compared by using slope.

FCAT Practice

Standardized Test Practice

Example 6 Compare Slopes

Multiple-Choice Test Item

There are two major hills on a hiking trail. The first hill rises 6 feet vertically for every 42-foot run. The second hill rises 10 feet vertically for every 98-foot run. Which statement is true?

Ⓐ The first hill is steeper than the second hill.

Ⓑ The second hill is steeper than the first hill.

Ⓒ Both hills have the same steepness.

Ⓓ You cannot determine which hill is steeper.

The Princeton Review Test-Taking Tip

Make a Drawing Whenever possible, make a drawing that displays the given information. Then use the drawing to estimate the answer.

Read the Test Item To compare steepness of the hills, find the slopes.

Solve the Test Item

first hill slope = $\frac{\text{rise}}{\text{run}}$ = $\frac{6 \text{ ft}}{42 \text{ ft}}$ rise = 6 ft, run = 42 ft

second hill slope = $\frac{\text{rise}}{\text{run}}$ = $\frac{10 \text{ ft}}{98 \text{ ft}}$ rise = 10 ft, run = 98 ft

- **Each lesson** contains at least two FCAT practice problems. You can use these problems every day to keep your FCAT skills sharp. The **Chapter Practice Test** also includes FCAT practice problems.
- **Worked-out examples** in each chapter show you step-by-step solutions of FCAT problems. Just like the practice problems, these include multiple choice, gridded response, short response, and extended response. **Test-Taking Tips** are also included.
- Two pages of **FCAT Practice** are included at the end of each chapter. These problems may cover any of the content up to and including the chapter they follow.

Test-Taking Tips

- ✓ Go to bed early the night before the test. You will think more clearly after a good night's rest.
- ✓ Read each problem carefully and think about ways to solve the problem before you try to answer the question.
- ✓ Relax. Most people get nervous when taking a test. It's natural. Just do your best.
- ✓ Answer questions you are sure about first. If you do not know the answer to a question, skip it and go back to that question later.
- ✓ Think positively. Some problems may seem hard to you, but you may be able to figure out what to do if you read each question carefully.
- ✓ If no figure is provided, draw one. If one is furnished, mark it up to help you solve the problem.
- ✓ When you have finished each problem, reread it to make sure your answer is reasonable.
- ✓ Become familiar with common formulas and when they should be used. Use the FCAT Reference sheet found at the back of this book.
- ✓ Make sure that the number of the question on the answer sheet matches the number of the question on which you are working in your test booklet.

Countdown to FCAT

Week 1

Monday

Extended Response Mr. Gray surveyed his eighth grade students about their favorite beverage. *(Lesson 1-1)*

Part A How many students were surveyed?

Part B What three beverages make up exactly 50% of those surveyed? Explain how you determined your answer.

Beverage	Number of Students
water	63
milk	36
orange juice	24
soda	15
other	12

Tuesday

Multiple Choice Sara went to the mall to spend her birthday money. She bought a shirt for $24.95 and a pair of pants for $39.95. How much did she spend for her new outfit? *(Prerequisite Skill)*

Ⓐ $53.80 Ⓑ $64.90
Ⓒ $54.90 Ⓓ $63.80

Gridded Response Alicia ran the 50-yard dash in 5.7 seconds. Taryn ran it in 6.2 seconds. How much faster in seconds was Alicia? *(Prerequisite Skill)*

Wednesday

Multiple Choice Evaluate $\frac{2(6 + 6 \cdot 4)}{27 + 3}$. *(Lesson 1-2)*

Ⓕ 0.5 Ⓖ 2.4
Ⓗ 3.2 Ⓘ 2.0

Gridded Response What is the next term in the pattern? *(Lesson 1-1)*

2, 8, 14, 20, 26, . . .

Thursday

Short Response The state of Florida has a total area of 58,560 square miles. If the total water area is 4,308 square miles, what is the total land area? Show your work. *(Prerequisite Skill)*

Friday

Multiple Choice Which algebraic expression represents the relationship in the table? *(Lesson 1-3)*

Number of Movie tickets	Total Cost
1	$5.75
2	$11.50
3	$17.25
4	$23.00
n	

Ⓐ $\$5.75 + n$ Ⓑ $n \div \$5.75$
Ⓒ $\$5.75n$ Ⓓ $\$5.75 \div n$

Week 2

Monday

Extended Response The scatter plot at the right shows the average monthly high temperatures for Clearwater, Florida. *(Lesson 1-7)*

Part A Does the scatter plot show a positive, negative, or no relationship from January to August? Explain.

Part B Describe the relationship between the month and average temperature from August to December.

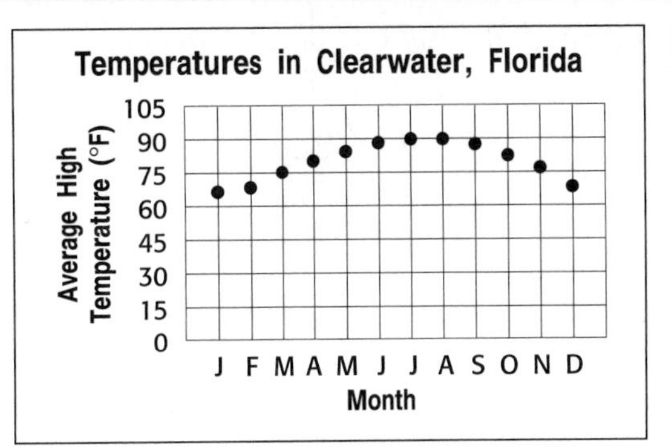

Tuesday

Multiple Choice Which ordered pair represents point P? *(Lesson 1-6)*

(A) (3, 1)
(B) (−3, 1)
(C) (−1, 3)
(D) (3, −1)

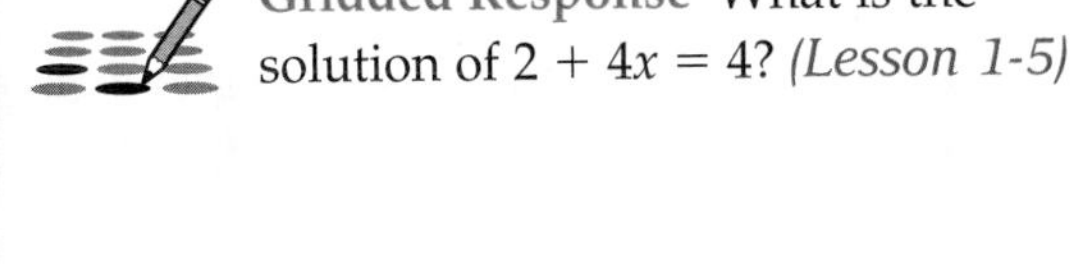

Gridded Response What is the solution of $2 + 4x = 4$? *(Lesson 1-5)*

Wednesday

Multiple Choice Trey is going to fence in his garden. How much fencing does he need if his garden measures 20 feet by 10 feet? *(Prerequisite Skill)*

(F) 200 feet (G) 60 feet
(H) 40 feet (I) 30 feet

20 ft

10 ft

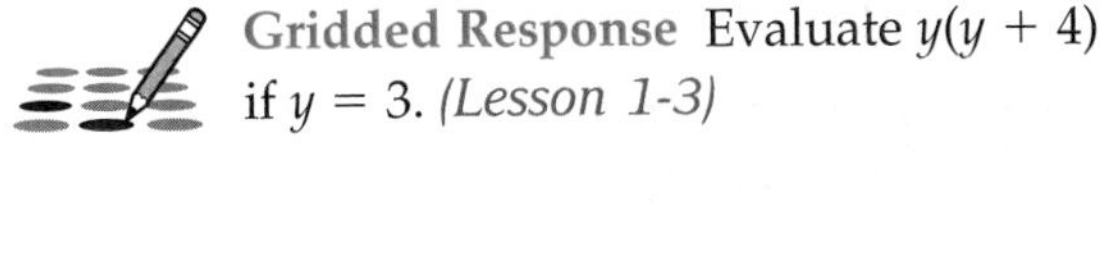

Gridded Response Evaluate $y(y + 4)$ if $y = 3$. *(Lesson 1-3)*

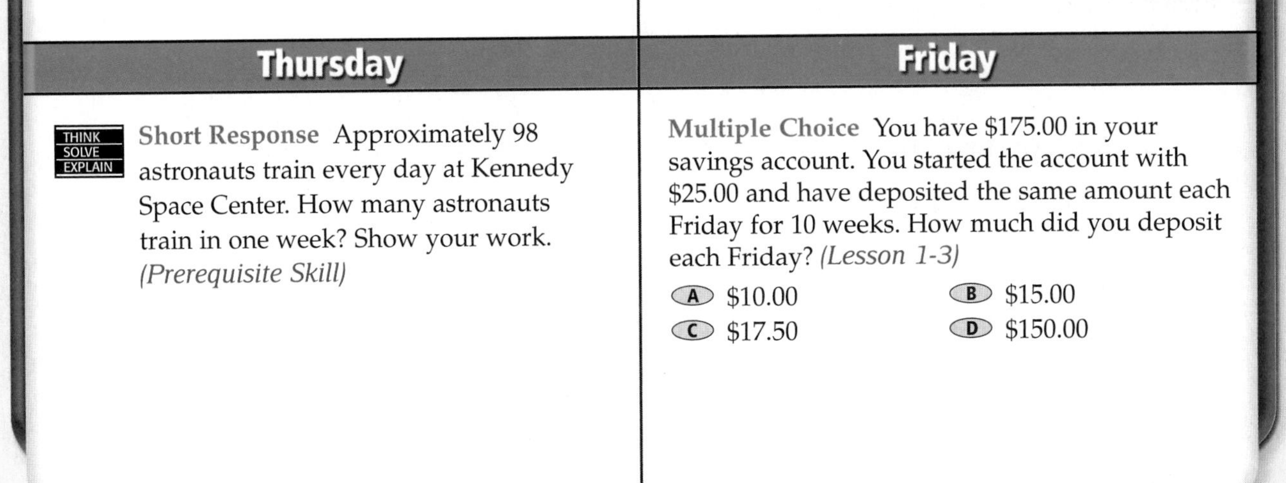

Thursday

Short Response Approximately 98 astronauts train every day at Kennedy Space Center. How many astronauts train in one week? Show your work. *(Prerequisite Skill)*

Friday

Multiple Choice You have $175.00 in your savings account. You started the account with $25.00 and have deposited the same amount each Friday for 10 weeks. How much did you deposit each Friday? *(Lesson 1-3)*

(A) $10.00 (B) $15.00
(C) $17.50 (D) $150.00

Week 3

Monday

THINK SOLVE EXPLAIN

Extended Response Jaron wants to join the soccer team at Miami Beach Middle School. His weekly workout times are shown below. *(Lesson 1-6)*

Week	1	2	3	4	5	6	7	8
Time (min)	60	85	90	90	100	105	110	120

Part A Write a set of ordered pairs for the each of the weekly workouts.

Part B Graph the data on a coordinate system.

Part C Use the graph to predict the **total** number of minutes of Jaron's workouts in Week 9. Explain.

Tuesday

Multiple Choice There are 24 hours in a day on Earth. On Venus, a day lasts eight Earth months. How long is a day on Venus in Earth hours? (Assume there are 30 days in a month.) *(Lesson 1-4)*

Ⓐ 5,760 h Ⓑ 720 h
Ⓒ 240 h Ⓓ 192 h

Gridded Response What is the missing value in the table? *(Lesson 1-1)*

x	0	−1	−2	−3
y	3.5	4.5	?	6.5

Wednesday

Multiple Choice Which statement is true? *(Lesson 2-1)*

Ⓕ $|-6| - |-3| = 6$
Ⓖ $|-6| - |3| = -9$
Ⓗ $|-6| + |-3| = -9$
Ⓘ $|-6| - |-3| = 3$

Gridded Response Evaluate $|a| - |b|$ if $a = -4$ and $b = 3$. *(Lesson 2-1)*

Thursday

THINK SOLVE EXPLAIN

Short Response If the pattern shown in the graph continues, what would you expect enrollment to be in 2004? Explain your reasoning. *(Lesson 1-1)*

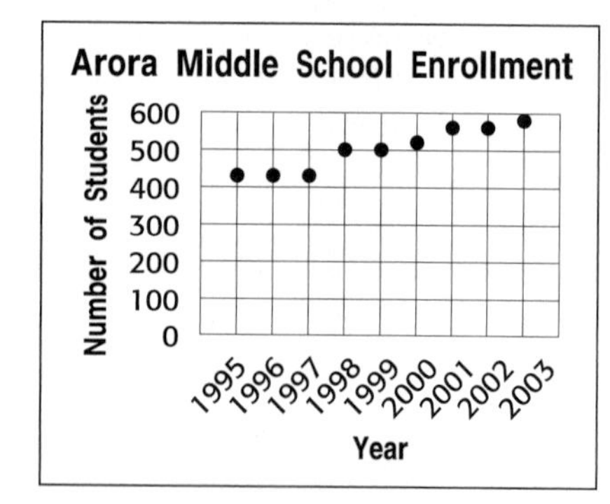

Friday

Multiple Choice Evaluate $x - (-y)$ if $x = 10$ and $y = -4$. *(Lesson 2-3)*

Ⓐ −14 Ⓑ −6
Ⓒ 6 Ⓓ 14

Week 4

Monday

Extended Response Mrs. Schaefer has her students practice integer operations by using playing cards. The students work in pairs and turn over one card at the same time. The red cards represent negative values, and the black cards positive values. Face cards are worth ten and aces are worth eleven. *(Lesson 2-4)*

Part A If they are multiplying the values of their cards and they both turn over a red card, describe the result.

Part B If they are adding the values of their cards and they both turn over a red card, what will be the result?

Part C Give an example of a draw whose outcome value would depend on the color of the cards.

Tuesday

Multiple Choice Order 12, -10, -12, 15, and 0 from **least to greatest**. *(Lesson 2-1)*

- (A) 15, 12, 0, -12, -10
- (B) 0, -10, -12, 12, 15
- (C) -12, -10, 0, 12, 15
- (D) 15, 12, 0, -10, -12

Gridded Response Carlos wants to purchase chicken for dinner. The chicken is on sale for $2.49 per pound, and he needs four pounds for the recipe he is using. How much will the chicken cost in dollars? *(Prerequisite Skill)*

Wednesday

Multiple Choice Evaluate $-7x$ if $x = -3$. *(Lesson 2-4)*

- (F) -21
- (G) -10
- (H) 4
- (I) 21

Gridded Response Orange County had an estimated population of 883,308 in the 2000 Florida census. Pinellas County had an estimated population of 903,502. What is the difference in population between the two counties? *(Lesson 2-3)*

Thursday

Short Response The distance from Pensacola to Key West is 792 miles. The longest river in Florida, the St. Johns, is 273 miles long. How many times farther is the distance from Pensacola to Key West than the length of the St. John's River? Round to the nearest whole number. *(Lesson 2-5)*

Friday

Gridded Response What is the mean low temperature for the week in degrees Fahrenheit? *(Lesson 2-3)*

Day	Low Temperature (°F)
Monday	47
Tuesday	50
Wednesday	60
Thursday	64
Friday	53
Saturday	35
Sunday	41

Week 5

Monday

Extended Response Mr. King sells music equipment. His weekly salary is $350, plus a bonus of $50 for each guitar, g, that he sells. *(Lesson 3-6)*

Part A Write an expression to represent the total amount of money that Mr. King earns in a week.

Part B If he sells three guitars in one week, what are his total earnings for the week?

Part C How much would he make in four weeks if he sold a total of 15 guitars? Show your work.

Tuesday

Multiple Choice What is the area of the rectangle? *(Lesson 3-7)*

25 m

15 m

Ⓐ 40 square meters
Ⓑ 80 square meters
Ⓒ 300 square meters
Ⓓ 375 square meters

Gridded Response Evaluate $(ab)(c)$ if $a = -3$, $b = 55$, and $c = -1$. *(Lesson 2-4)*

Wednesday

Multiple Choice A plumber charges $55 for a service call plus $26 per hour. Which expression represents the plumber's total cost for h hours of work? *(Lesson 3-2)*

Ⓕ $\$26 + h$
Ⓖ $\$55 + \$26h$
Ⓗ $\$81h$
Ⓘ $\$26h$

Gridded Respone It takes two feet of ribbon to make a large bow. Jeannette has 250 feet of ribbon. How many bows can she make? *(Lesson 3-4)*

Thursday

Short Response The longest river in Florida, the St. Johns, is 273 miles long. Write and solve an equation to find the number of hours, n, that it would take to boat the entire length of the river at an average speed of 20 miles per hour. *(Lesson 3-4)*

Friday

Multiple Choice Swim practice was m minutes long on Monday. Each day after that practice was 5 minutes longer than the previous day. Write a simplified expression to represent the total swim practice times from Monday through Friday. *(Lesson 3-2)*

Ⓐ $5m + 50$
Ⓑ $25m + 50$
Ⓒ $5m + 5$
Ⓓ $5m + 25$

Week 6

Monday

Extended Response The population of Florida is approximately 13 million people. Suppose 132 thousand people move to the state each year. *(Lesson 3-6)*

Part A Write an equation that represents the population in y years.

Part B What will be the population in five years?

Part C In how many years will the population be 14 million?

Tuesday

Multiple Choice In which quadrant is point P? *(Lesson 2-6)*

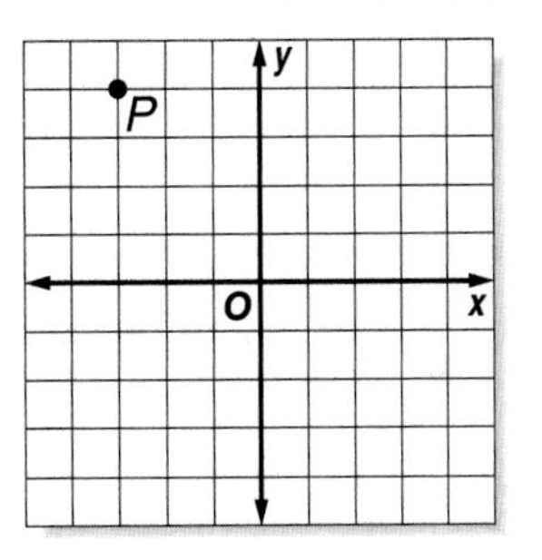

Ⓐ quadrant I
Ⓑ quadrant II
Ⓒ quadrant III
Ⓓ quadrant IV

Gridded Response Anna is 3 years older than her sister Sue. The sum of their ages is 33. How old is Sue? *(Lesson 3-6)*

Wednesday

Multiple Choice The difference of a number and 12 is −10. What is the number? *(Lesson 3-3)*

Ⓕ −22
Ⓖ −2
Ⓗ 2
Ⓘ 22

Gridded Response Solve $\frac{w}{4} - 10 = 2$. *(Lesson 3-5)*

Thursday

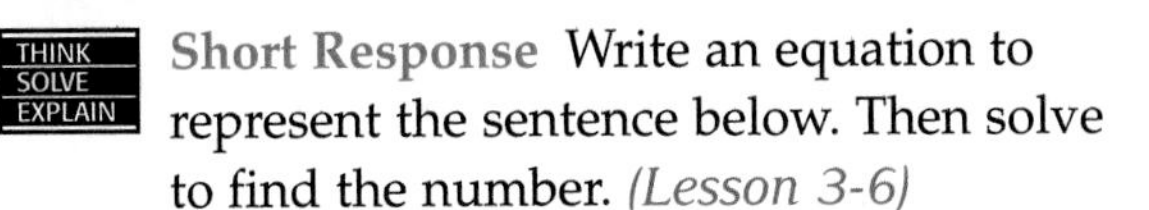

Short Response Write an equation to represent the sentence below. Then solve to find the number. *(Lesson 3-6)*

Thirteen more than three times a number is 25.

Friday

Multiple Choice A cell phone company charges \$0.15 for the first minute and \$0.03 for each additional minute, as shown in the table. How much would a 10-minute call cost? *(Lesson 3-5)*

Ⓐ \$1.50
Ⓑ \$0.42
Ⓒ \$0.33
Ⓓ \$0.30

Time (min)	Cost ($)
1	0.15
2	0.18
3	0.21
4	0.24
5	0.27
6	0.30

Week 7

Monday

Extended Response Mr. Ling wants to put in a lap pool. The pool will be 10 feet from the property line on the three sides and 15 feet from his house, which is 90 feet by 90 feet. His house sits 15 feet from the property line. *(Lesson 3-7)*

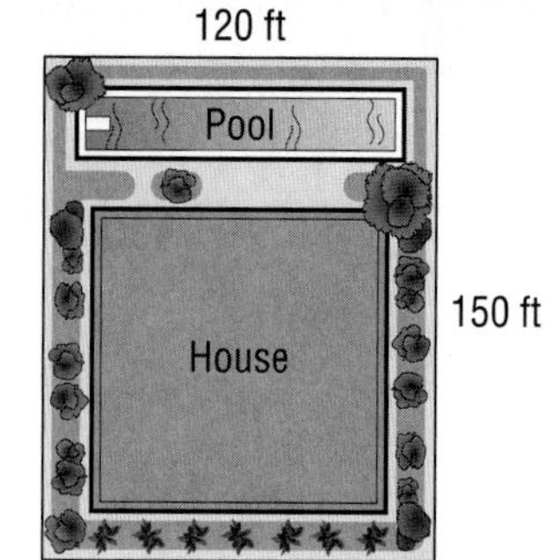

Part A Write and solve an equation to find the total length of the pool.

Part B Write and solve an equation to find the total width of the pool.

Part C What will be the area of Mr. Ling's pool? Show your work.

Tuesday

Multiple Choice Which of the following is a monomial? *(Lesson 4-1)*

Ⓐ $6(a + b)$ Ⓑ $x + 4$
Ⓒ $-6a$ Ⓓ $2c - d$

Gridded Response The distance across Florida from the Atlantic Ocean to the Perdido River is 361 miles. How long would it take to drive this distance at a rate of 50 miles per hour? Round to the nearest hour. *(Lesson 3-7)*

Wednesday

Multiple Choice Evaluate $2(x + y)^3$ if $x = 4$ and $y = 1$. *(Lesson 4-2)*

Ⓕ 10 Ⓖ 30
Ⓗ 250 Ⓘ 320

Gridded Response Logan is training for a marathon. In one week, she ran a total of 105 miles. If she ran the same distance each day, how many miles did she run in one day? *(Lesson 2-5)*

Thursday

Short Response The area of a square playground is 8,100 square feet. What are the dimensions of the playground? *(Lesson 3-7)*

Friday

Multiple Choice Florida has an average annual rainfall of 53 inches. If it rained that much every year, what would be the total rainfall after 7 years? *(Lesson 2-4)*

Ⓐ 371 inches Ⓑ 350 inches
Ⓒ 60 inches Ⓓ 46 inches

Week 8

Monday

Extended Response On a map that Joshua is using, $\frac{1}{4}$ inch represents 150 miles. On the map, he measured $1\frac{1}{4}$ inches from his home to his grandparents' home. *(Lesson 3-4)*

Part A Describe how he could find the number of miles to his grandparents' home.

Part B Find the distance.

Tuesday

Multiple Choice Write $\frac{12c^3}{3c^2}$ in simplest form. *(Lesson 4-5)*

(A) $36c^5$ (B) $4c$
(C) $4c^5$ (D) $9c$

Gridded Response If $x = 10$ and $y = -4$, what is the value of $(x - y) - (x + y)$? *(Lesson 2-3)*

Wednesday

Multiple Choice What is the prime factorization of 375? *(Lesson 4-3)*

(F) $5 \cdot 75$ (G) $5^2 \cdot 15$
(H) $3 \cdot 5^3$ (I) $3^5 \cdot 3$

Gridded Response Find the GCF of 108 and 324. *(Lesson 4-4)*

Thursday

THINK SOLVE EXPLAIN

Short Response Describe how to multiply $6xy^4$ and $2x^2y^2$. Then find the product. *(Lesson 4-6)*

Friday

Gridded Response Jack Russell Stadium is the spring training site for the Philadelphia Phillies and Al Lang Stadium is the spring training site for the Tampa Bay Devil Rays. If each seat sells for \$8.50, how much more money in dollars can St. Petersburg make if all seats sell out for both stadiums? *(Lesson 3-5)*

Stadium	Location	Number of Seats
Jack Russell	Clearwater	4,700
Al Lang	St. Petersburg	7,000

Countdown to FCAT

Week 9

Monday

Extended Response A high school student surveyed his friends about their favorite "hang out" place. *(Lesson 5-1)*

Part A Of those surveyed, $\frac{2}{5}$ chose which place as their favorite?

Part B Of those surveyed, $\frac{3}{100}$ chose which place as their favorite?

Part C Make a bar graph to represent the data.

Favorite Place to Hang Out

Place	Percent
park	40% = 0.40
mall	23% = 0.23
bedroom	18% = 0.18
movies	16% = 0.16
library	3% = 0.03

Tuesday

Multiple Choice A pair of houseflies can produce 5,000,000,000,000 offspring in one season. Which is the scientific notation for this number? *(Lesson 4-8)*

Ⓐ 5×10^{12} Ⓑ 5×10^{4}

Ⓒ 5×10^{-4} Ⓓ 5×10^{-12}

Gridded Response The coldest temperature ever recorded in Florida was −2°F. The hottest temperature ever recorded was 109°F. What is the difference in degrees Fahrenheit between the hottest and coldest temperatures? *(Lesson 2-3)*

Wednesday

Multiple Choice Some human hair is $\frac{1}{1,500}$ of an inch in diameter. What is this measure in scientific notation? *(Lesson 4-8)*

Ⓕ 1.5×10^{3} in. Ⓖ 2×10^{3} in.

Ⓗ 6.7×10^{-4} in. Ⓘ 2.2×10^{-3} in.

Gridded Response Most Americans eat three times a day. Crocodiles eat only 50 times a year. What decimal compares the crocodile's meals to the human's dinners in one year? (Assume there are 365 days in a year.) Round to the nearest hundredth. *(Lesson 5-1)*

Thursday

THINK SOLVE EXPLAIN

Short Response Explain the difference between the sets of natural numbers and whole numbers. *(Lesson 5-2)*

Friday

Gridded Response Evaluate $(xy)^{-2}$ if $x = -2$ and $y = -1$. Write as a fraction. *(Lesson 4-7)*

Week 10

Monday

Extended Response Jose is making a quilt. Each piece of blue and red will be $\frac{1}{4}$ of a square yard. Each piece of green will be $\frac{1}{3}$ of a square yard. Each piece of yellow will be $\frac{1}{6}$ of a square yard. *(Lessons 5-5, 5-6, and 5-7)*

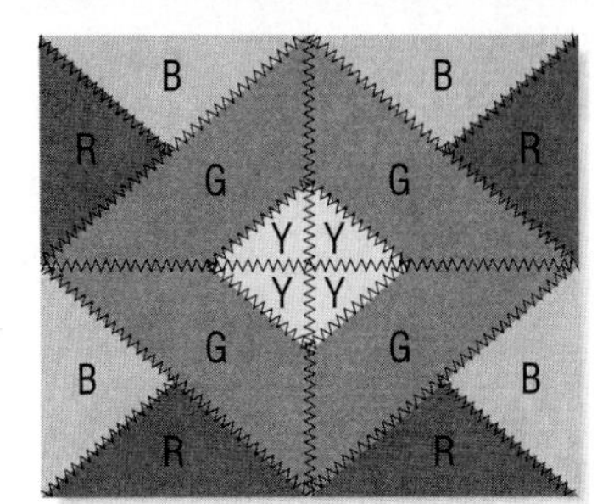

Part A How many square yards of each color will he need? Show all your work.

Part B What is the total number of square yards of all the pieces?

Tuesday

Multiple Choice The average length of a dwarf goby male fish is 0.339 of a inch. Which is the length written in scientific notation? *(Lesson 4-8)*

Ⓐ 3.39×10^{-1} in.
Ⓑ 3.39×10 in.
Ⓒ 0.339×10^{-1} in.
Ⓓ 339×10^{-3} in.

Gridded Response Miss Okuma is covering her kitchen table with tiles. A box of tiles covers 600 square inches. If her square table measures $40\frac{3}{4}$ inches on a side, how many boxes of tiles will she need? *(Lessons 5-3 and 5-4)*

Wednesday

Multiple Choice What is the LCD of $\frac{2}{ab}$ and $\frac{3}{a^2}$? *(Lesson 5-6)*

Ⓕ a^2b
Ⓖ ab
Ⓗ a^2
Ⓘ $2ab$

Gridded Response Evaluate $x + y$ if $x = 2\frac{10}{15}$ and $y = 2\frac{1}{3}$. *(Lesson 5-7)*

Thursday

Short Response Students in English class are to write a short story using $1\frac{1}{4}$-inch margins on both sides and $1\frac{3}{4}$-inch margins on the top and bottom. If a normal sheet of paper measures $8\frac{1}{2}$ by 11 inches, what are the dimensions of the short story? *(Lesson 5-9)*

Friday

Gridded Response In four consecutive swim meets, Andrew dropped three tenths of a second in his butterfly stroke. If his time at the first meet was 56.4 seconds, what was his time in the fifth meet in seconds? *(Lesson 5-10)*

Countdown to FCAT

Week 11

Monday

Extended Response The seven longest bridges in Florida are listed. *(Lesson 5-8)*

Part A What is the mean of the set of data?

Part B What is the mode of the set of data?

Part C What is the median of the set of data?

Part D How does the mean length compare to the median length?

Florida Bridges

Bridge	Length (ft)
Seven Mile	35,716
Sunshine Skyway	21,872
Mid Bay	19,257
Garcon Point	18,420
Henry H. Buckman, Sr.	16,296
Howard Franklin, South	15,896
Howard Franklin, North	15,868

Tuesday

Multiple Choice Find $\frac{2a}{xy} \div \frac{2a}{x}$. *(Lesson 5-4)*

Ⓐ $\frac{4a}{xy}$ Ⓑ $\frac{2}{xy}$

Ⓒ $\frac{2a}{y}$ Ⓓ $\frac{1}{y}$

Gridded Response In 1999, approximately 32 million tourists came to Florida by air and 26 million by auto. Write the ratio of auto travelers to air travelers as a fraction in simplest form. *(Lesson 6-1)*

Wednesday

Multiple Choice The cost of 24 cans of soda is $6.99. At this rate, how much would 6 cans of soda cost? *(Lesson 6-2)*

Ⓕ $0.29 Ⓖ $0.40

Ⓗ $1.17 Ⓘ $1.75

Gridded Response The model tractor had a scale factor of 1:32. What is the actual height of the tractor in feet if the model height is $\frac{1}{2}$ foot? *(Lesson 6-3)*

Thursday

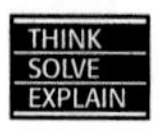

Short Response The correct mixture of oil and gas is critical in lawn equipment usage. The manual for a gas edger states that 2.6 ounces of oil should be added for every gallon of gas. Write and solve a proportion to find how much oil would be needed for 2.5 gallons of gas. *(Lesson 6-2)*

Friday

Multiple Choice Dimercus was making a poster of Florida's first operating railroad, the St. Joseph Railroad. It was mule-powered and began in 1836. His scale factor was 0.5 inch to 2 miles. If he drew a length of two inches to represent the railroad, what was the actual length of the St. Joseph Railroad? *(Lesson 6-3)*

Ⓐ 8 miles Ⓑ 2 miles

Ⓒ 4 miles Ⓓ 1 mile

Week 12

Monday

Extended Response The Disney World Marathon held each January is 26.2 miles long. *(Lesson 6-1 and 6-2)*

Part A If there are approximately 3.28 feet in one meter, how many meters is the marathon? Write and solve a proportion. Round to the nearest meter.

Part B Approximately how many kilometers is this?

Tuesday

Multiple Choice Find the value of x that makes $\frac{6.9}{x} = \frac{27.6}{68.4}$ a proportion. *(Lesson 6-2)*

(A) 1.71 (B) 4.71
(C) 47.7 (D) 17.1

Gridded Response A Peregrine falcon can reach speeds of 185 miles per hour when diving. What is this speed in feet per second? Round to the nearest tenth. *(Lesson 6-1)*

Wednesday

Multiple Choice What is the scale factor of a drawing if it has a scale of 3 inches = 5 feet? *(Lesson 6-3)*

(F) $\frac{3}{5}$ (G) $\frac{5}{3}$
(H) $\frac{1}{20}$ (I) $\frac{1}{5}$

Gridded Response What number is 22.3% of 32? *(Lesson 6-5)*

Thursday

THINK SOLVE EXPLAIN

Short Response Which investor earns more simple interest after 2 years? How much more? *(Lesson 6-7)*

Investor	Principal	Interest Rate
A	$12,000	6.5%
B	$18,000	5.0%

Friday

Multiple Choice About 29.5% of Florida's 65,755 square miles is covered by wetlands. About how many square miles in Florida are covered by wetlands? *(Lesson 6-7)*

(A) 19,398 sq mi
(B) 32,878 sq mi
(C) 193,977 sq mi
(D) 1,939,773 sq mi

Countdown to FCAT

Week 13

Monday

Extended Response Two fair number cubes are rolled. *(Lesson 6-9)*

Part A What is the total number of outcomes?

Part B What is the probability of rolling a sum of 7?

Part C What is the probability of NOT rolling a sum of seven?

Tuesday

Multiple Choice At the 1896 Olympics in Athens, the winning time for the 400-meter race was 54.2 seconds. In the 1996 Olympics in Atlanta, the winning time was 43.49 seconds. What was the percent of decrease from the 1896 Olympics time to the 1996 time? *(Lesson 6-8)*

(A) 4.9% (B) 10.7%
(C) 12.3% (D) 19.8%

Gridded Response Solve $3(3x - 4) = 4(2x + 9)$. *(Lesson 7-2)*

Wednesday

Multiple Choice Sabrina earns $7 per hour working at the library. Which inequality can be used to find how many hours she must work to earn more than $100? *(Lesson 7-5)*

(F) $7x > 100$ (G) $7x \geq 100$
(H) $7x < 100$ (I) $7x \leq 100$

Gridded Response Solve $-4s + 90 = -3s + 50$. *(Lesson 7-1)*

Thursday

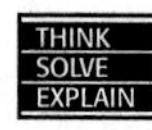

Short Response A new gym has two membership plans as shown below.

Membership Plans		
Plan	Membership Fee	Cost Per Month
1	$100	$25
2	$150	$15

In how many months would the total cost of the two plans be the same? Show the equation that you used to solve the problem. *(Lesson 7-1)*

Friday

Multiple Choice The state bird of Florida is the mockingbird. It has a wingspan of 15 inches and measures 10 inches in length. What is the ratio of length to the wingspan? *(Lesson 6-1)*

(A) 5:2 (B) 2:5
(C) 3:2 (D) 2:3

Countdown to FCAT

Week 14

Monday

Extended Response The band at Tarpon Springs High School is selling gift wrap as a fund-raiser. They make a 50% profit for each item sold. They need to make at least \$1,500 in order to have enough to travel to state competition. *(Lesson 7-3)*

Part A If the gift wrap sells for \$10 each, write an inequality that represents how many of each they need to sell.

Part B How many would they need to sell in order to go to the competition?

Tuesday

Multiple Choice There are at least 50 more public schools in Jacksonville than in Tampa. If Jacksonville has 233 public schools, which inequality represents the number of public schools in Tampa? *(Lesson 7-4)*

(A) $x \leq 183$ (B) $x \leq 283$
(C) $x \geq 183$ (D) $x \geq 283$

Gridded Response Solve $\frac{1}{3}(x + 4) = \frac{1}{2}(x - 3)$. *(Lesson 7-2)*

Wednesday

Multiple Choice Which inequality is graphed? *(Lesson 7-3)*

−4 −3 −2 −1 0 1 2 3 4

(F) $x > -2$ (G) $x \geq -2$
(H) $x < -2$ (I) $x \leq -2$

Gridded Response The perimeter of a rectangle is 60 inches. The length is 6 inches greater than twice the width. Find the length in inches. *(Lesson 7-2)*

Thursday

THINK SOLVE EXPLAIN

Short Response There are 4 black, 10 white, 2 red, and 8 blue chips in a bag. What is the probability of randomly selecting a black or red chip? *(Lesson 6-9)*

Friday

Multiple Choice Tamika has \$5 to buy a salad, an apple, and milk in the school cafeteria. A salad costs \$2.50, and an apple costs one-fifth that amount. If milk costs one-half the salad, which inequality represents the situation? *(Lesson 7-5)*

(A) $\$2.50 + \frac{1}{5} + \frac{1}{2} \leq \$5.00$

(B) $\$2.50 + \frac{1}{5}(\$2.50) + \frac{1}{2}(\$2.50) \leq \$5.00$

(C) $\$2.50 + \frac{1}{5} + \frac{1}{2} \geq \$5.00$

(D) $\$2.50 + 1\frac{1}{5}(\$2.50) + \frac{1}{2}(\$2.50) \geq \$5.00$

Week 15

Monday

Extended Response A restaurant sold 1,693 appetizers in one month. *(Lesson 6-7)*

Part A About how many quesadillas were sold if they were 23% of total?

Part B The least number of appetizers sold was oysters at 3%. How many is this?

Part C The greatest number of appetizers sold was cheese sticks at 45%. About how much money did the restaurant earn on sales of cheese sticks?

Part D What is the buffalo shrimp and chips/salsa total number?

Appetizer	Price
cheese sticks	$4.99
chips/salsa	$1.99
buffalo shrimp	$7.99
steamed oysters	$8.99
quesadillas	$5.99

Tuesday

Multiple Choice The relation {(−2, 4), (6, 0), (4, 7), (0,5)} is NOT a function when which ordered pair is added to the set? *(Lesson 8-1)*

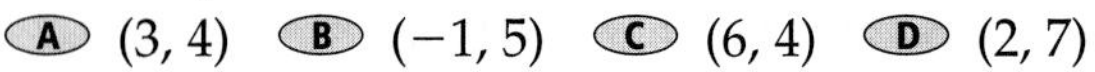

A (3, 4) B (−1, 5) C (6, 4) D (2, 7)

Gridded Response In 1944, there were 2,372,292 active United States Air Force personnel. In 2001, there were 351,935. What is the percent of decrease of the United States Air Force? Round to the nearest whole percent. *(Lesson 6-8)*

Wednesday

Multiple Choice Which ordered pair is a solution to the graph? *(Lesson 8-2)*

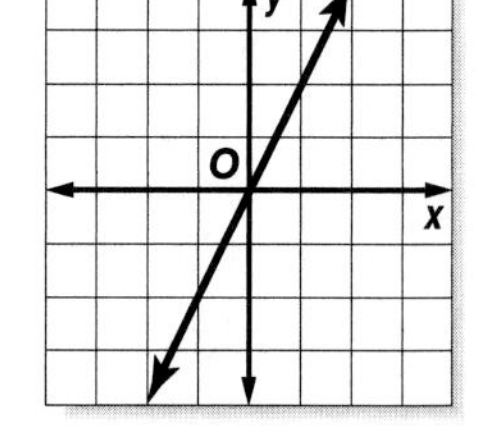

F (0, −1) G (1, 2)
H (−1, 1) I (2, 1)

Gridded Response In the linear equation $3x - 4y = -10$, if $x = 2$ what is the value of y? *(Lesson 8-2)*

Thursday

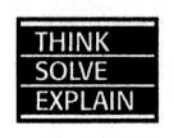

Short Response An independent computer analyst charges $125 for an initial consulting fee, plus $85 per hour. The total cost can be represented by $y = 125 + 85x$, as shown in the graph below. Find the y-intercept and explain what it means. *(Lesson 8-3)*

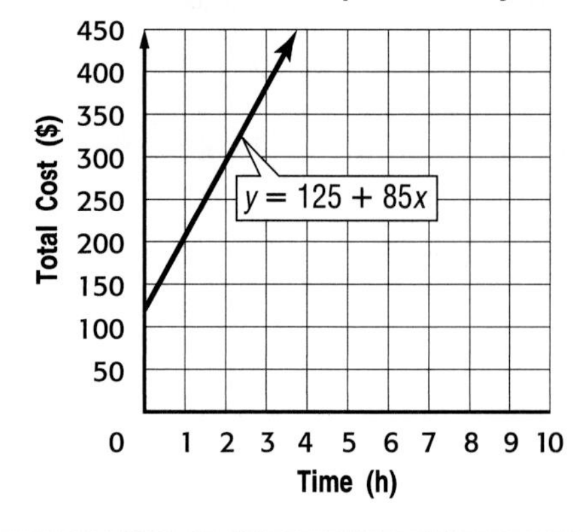

Friday

Multiple Choice In 1980, the population of Volusia County, Florida, was 258,762. In 1990, there was an increase in population of 43%. Approximately how many people lived in Volusia County in 1990? *(Lesson 6-8)*

A 200,000 B 370,000
C 450,000 D 1,800,000

Week 16

Monday

Extended Response Use the table to answer the following questions. *(Lesson 8-3)*

Part A Graph the line containing the points.

Part B Determine the slope.

Part C Determine the x-intercept and the y-intercept.

Part D Write an equation of the line in slope-intercept form.

x	y
6	4
3	1
0	−2
−3	−5

Tuesday

Multiple Choice Which describes the line graphed at the right? *(Lesson 8-4)*

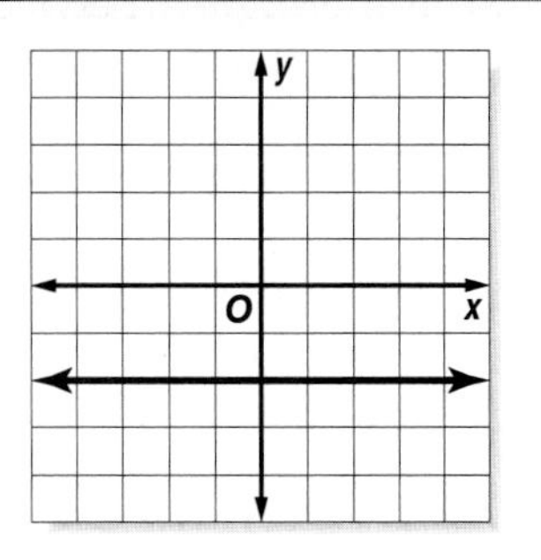

Ⓐ positive slope

Ⓑ negative slope

Ⓒ undefined slope

Ⓓ zero slope

Gridded Response What is the slope of the graph of $2x - 3y = 12$? *(Lesson 8-6)*

Wednesday

Multiple Choice If y varies directly with x, which equation relates x and y if $y = 10$ when $x = 28$? *(Lesson 8-5)*

Ⓕ $x = \frac{5}{14}y$

Ⓖ $y = \frac{5}{14}x$

Ⓗ $y = \frac{14}{5}x$

Ⓘ $x = \frac{14}{5}y$

Gridded Response What is the y-intercept of the graph of $5x + 4y = 15$? *(Lesson 8-6)*

Thursday

Short Response Miguel is printing Buccaneer 2003 Super Bowl T-shirts at the rate shown in the table. Write an equation that represents the number of T-shirts printed in h hours. *(Lesson 8-5)*

Time (h)	Number of T-shirts
0	0
1	225
2	450
3	675

Friday

Multiple Choice In Florida's Everglade National Park, a red mangrove tree had a recorded height of 75 feet and a circumference of 77 inches. What is the ratio of the circumference of the mangrove to its height? *(Lesson 6-1)*

Ⓐ $\frac{77}{900}$

Ⓑ $\frac{75}{77}$

Ⓒ $\frac{77}{75}$

Ⓓ $\frac{900}{77}$

Week 17

Monday

Extended Response Julian and Erin live $4\frac{1}{4}$ miles apart. It is $3\frac{1}{2}$ miles from the library to Erin's house. *(Lesson 9-5)*

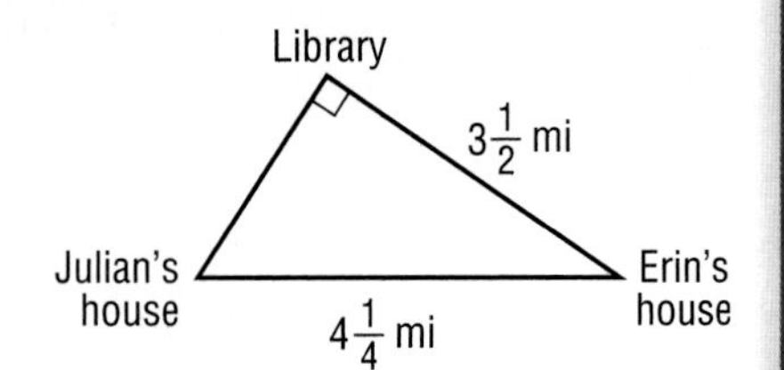

Part A Explain how to find the distance from the library to Julian's house. Then find the distance to the nearest tenth.

Part B Erin rode from her house to the library then to Julian's. How far did she ride?

Tuesday

Multiple Choice Complete 43 ___?___ $6\frac{11}{20}$ to make a true statement. *(Lesson 9-2)*

Ⓐ = Ⓑ <
Ⓒ > Ⓓ ≤

Gridded Response Find $m\angle XYV$ in degrees. *(Lesson 9-3)*

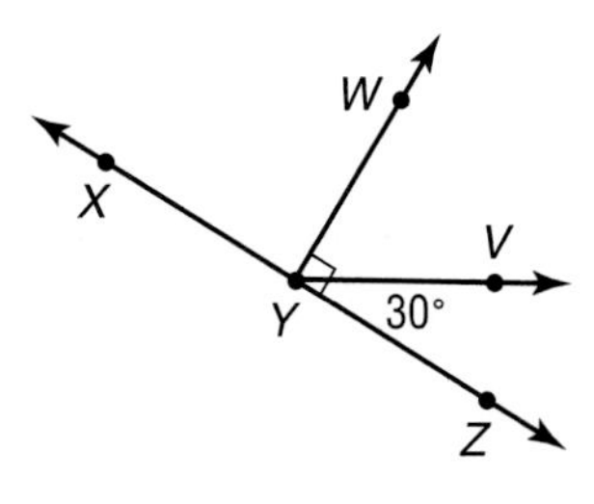

Wednesday

Multiple Choice Between which two numbers is $\sqrt{61}$ on a number line? *(Lesson 9-1)*

Ⓕ 64, 81 Ⓖ 49, 64
Ⓗ 7, 8 Ⓘ 8, 9

Gridded Response Solve $y^2 = 152$. Round to the nearest tenth if necessary. *(Lesson 9-2)*

Thursday

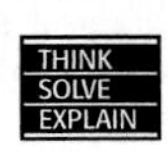

Short Response A weather reporter shows that a hurricane is forming in the Gulf of Mexico 175 miles from Tampa. How far is the hurricane from Fort Walton Beach? Round to the nearest tenth of a mile. *(Lesson 9-5)*

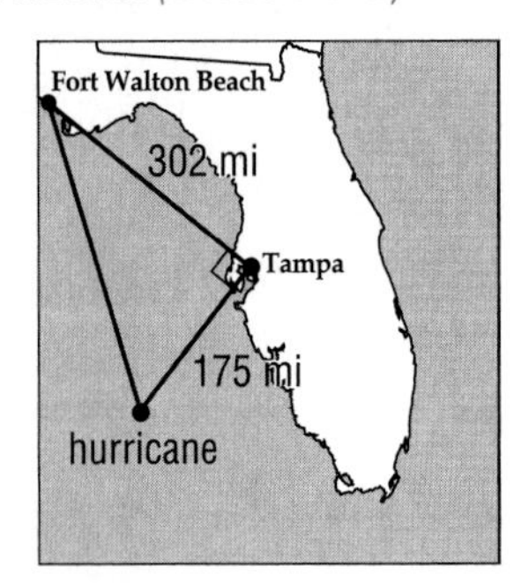

Friday

Multiple Choice What is the distance between M and N? *(Lesson 9-6)*

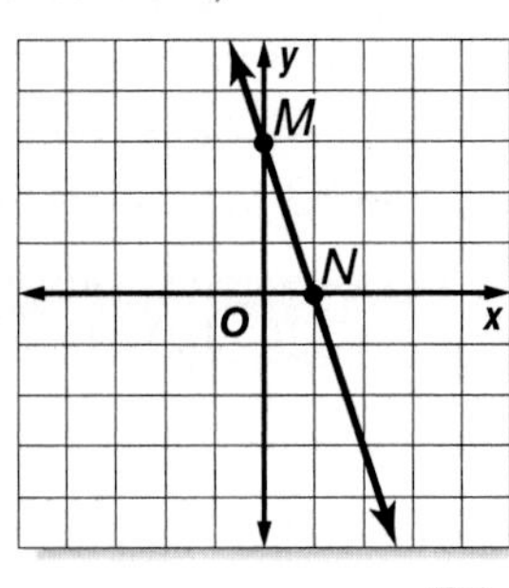

Ⓐ 16 Ⓑ $\sqrt{10}$
Ⓒ 3 Ⓓ 2

Week 18

Monday

Extended Response An architect is designing a new outdoor mall. The center of the mall will be a park with a playground and benches, as shown at the right. *(Lesson 10-5)*

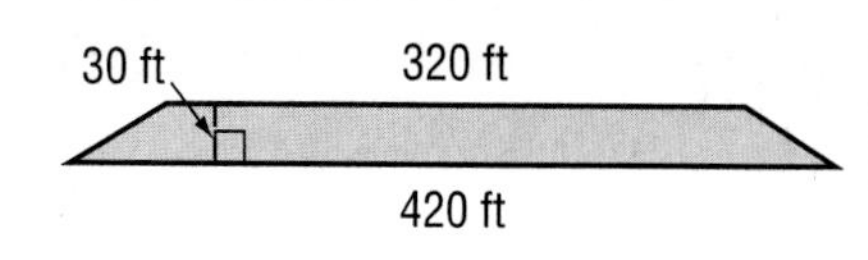

Part A What is the area of the park?

Part B The developer wants to change the base measures but keep the park area approximately the same. What does the sum of the bases need to be if the height stays the same?

Part C Give one example of the new base measures with approximately the same park area.

Tuesday

Multiple Choice What is the perimeter of the Pentagon in Washington, D.C., if each side of this regular pentagon measures 921 feet? *(Lesson 10-6)*

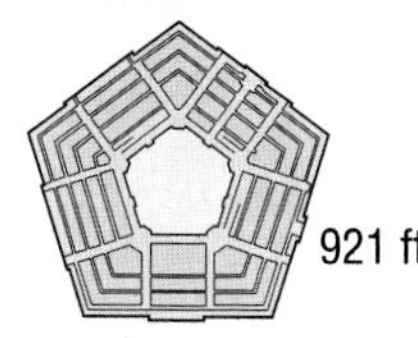

(A) 3,684 feet (B) 4,605 feet
(C) 5,526 feet (D) 9,210 feet

Gridded Response What is the measure in degrees of an interior angle of a regular dodecagon? *(Lesson 10-6)*

Wednesday

Multiple Choice What is the sum of the measures of the interior angles of a 20-gon? *(Lesson 10-6)*

(F) 360° (G) 3,240°
(H) 3,600° (I) 3,960°

Gridded Response According to the 2001 Florida Almanac, a camphor tree in Hardee County has a circumference of 368 feet. What is the diameter of the tree? Round to the nearest foot. *(Lesson 10-7)*

Thursday

Short Response Find the value of x if $\triangle ABC \cong \triangle EDC$. Round to the nearest tenth. Explain in words how you determined your answer. *(Lesson 10-2)*

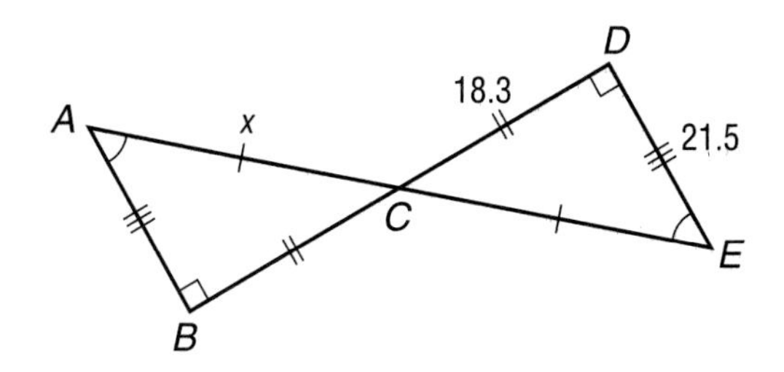

Friday

Multiple Choice Find the area of the irregular figure. *(Lesson 10-8)*

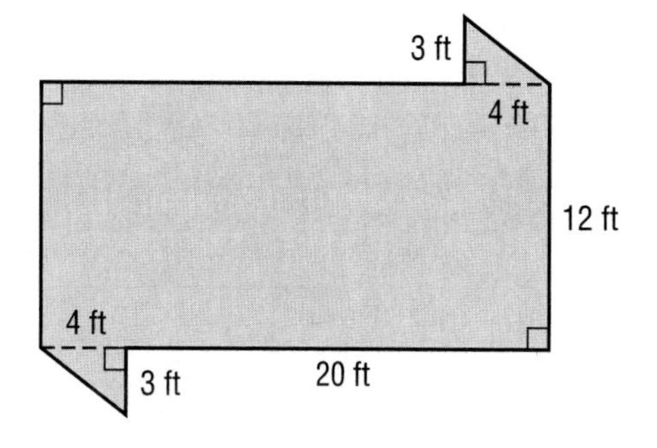

(A) 252 square feet (B) 300 square feet
(C) 326 square feet (D) 720 square feet

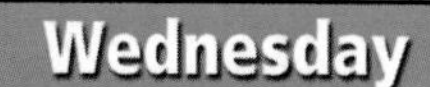

Week 19

Monday

Extended Response The figure at the right is a rectangular prism. *(Lesson 11-2)*

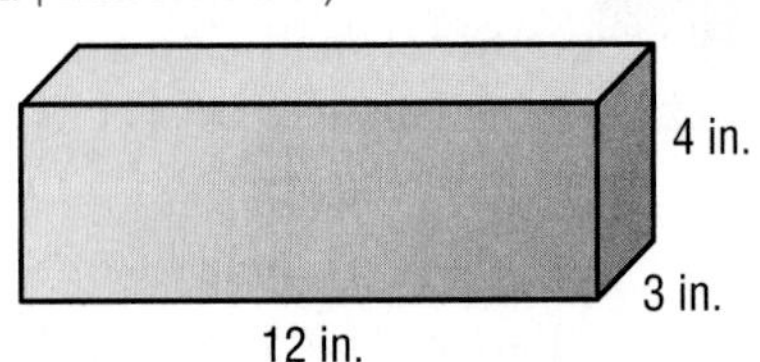

Part A What is the volume of the figure?

Part B What is the volume if all of the sides are doubled?

Part C By how many times has the volume increased?

Part D What can you generalize about the volume of a rectangular prism if the side lengths are doubled?

Tuesday

Multiple Choice Florida A & M University has approximately 12,000 students. Sixteen percent are from out of state. How many students are from Florida? *(Lesson 6-7)*

Ⓐ 720　Ⓑ 1,500
Ⓒ 1,920　Ⓓ 10,080

Gridded Response What is the missing measure for the pair of similar solids, in inches? *(Lesson 11-6)*

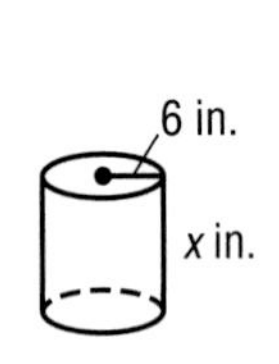

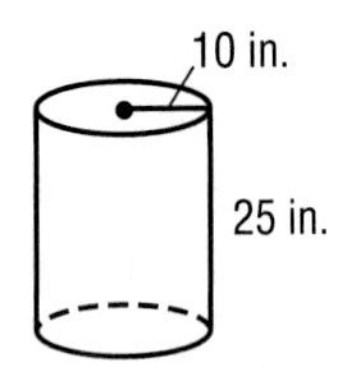

Wednesday

Multiple Choice What is the surface area of the figure shown at the right? *(Lesson 11-4)*

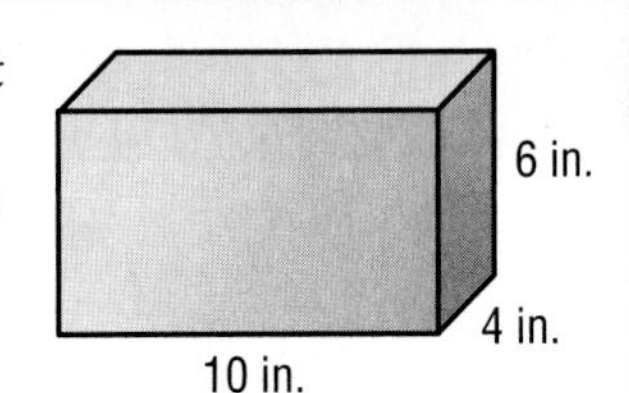

Ⓕ 120 square inches　Ⓖ 240 square inches
Ⓗ 248 square inches　Ⓘ 960 square inches

Gridded Response Find the height of a cylinder if the surface area is 2,488 square millimeters and the diameter is 12 millimeters. Use 3.14 for π and round to the nearest millimeter. *(Lesson 11-4)*

Thursday

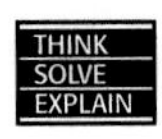

Short Response The table shows record 24-hour rainfalls in Florida. Use the correct number of significant digits to find the mean. *(Lesson 11-7)*

Record Rainfall		
Year	**Location**	**Rainfall (in.)**
1941	Trento	30
1950	Yankeetown	38.7
1950	Cedar Key	34
1969	Fernandino Beach	22
1980	Key West	23.3

Friday

Multiple Choice According to the 2001 *Guinness Book of World Records*, the narrowest optical fibers for communication measure only 0.000000002 of an inch. What is this number written in scientific notation? *(Lesson 4-8)*

Ⓐ 2.0×10^{8}　Ⓑ 2.0×10^{-9}
Ⓒ 2.0×10^{9}　Ⓓ 2.0×10^{-8}

Week 20

Monday

Extended Response The table shows enrollment data for seven universities in Florida. *(Lesson 12-3)*

Florida Universities, 2000–2001							
University	UF	FSU	USF	UCF	FIU	UNF	FAU
Enrollment	45,114	33,951	36,015	33,713	31,945	12,550	21,229

Part A What is the mean of the enrollments?

Part B What is the median?

Part C Draw a box-and-whisker plot for the data.

Tuesday

Multiple Choice What is the range of the men's springboard diving scores? *(Lesson 12-2)*

Olympic Men's Springboard Diving	
Year	Score
1984	710.91
1988	638.61
1992	677.31
1996	692.34
2000	724.53

Ⓐ 688.74 Ⓑ 85.92
Ⓒ 72.30 Ⓓ 13.62

Gridded Response What is the interquartile range for the data set {32, 16, 27, 45, 12, 2}? *(Lesson 12-2)*

Wednesday

Multiple Choice Determine the number of significant digits in 0.00081. *(Lesson 11-7)*

Ⓕ 2 Ⓖ 3
Ⓗ 5 Ⓘ 6

Gridded Response The Indosat Telkom Tower in Jakarta, Indonesia, is 1,831 feet tall. If it casts a shadow 1,221 feet long and at the same time a tree casts a shadow 10 feet long, how tall is the tree to the nearest foot? *(Lesson 9-7)*

Thursday

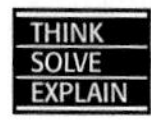

Short Response A bowl of fruit contains 3 oranges, 5 grapefruit, and 7 bananas. Find the probability of randomly choosing an orange and then a banana. Assume the first fruit is not replaced. Write as a fraction. *(Lesson 12-9)*

Friday

Multiple Choice A code uses only the vowels a, e, i, o, u and the digits 0 through 9. How many different passwords are possible if they contain a vowel and a digit? *(Lesson 12-6)*

Ⓐ 10 Ⓑ 15
Ⓒ 45 Ⓓ 50

Week 21

Monday

Extended Response Refer to the figures below. *(Lesson 10-6)*

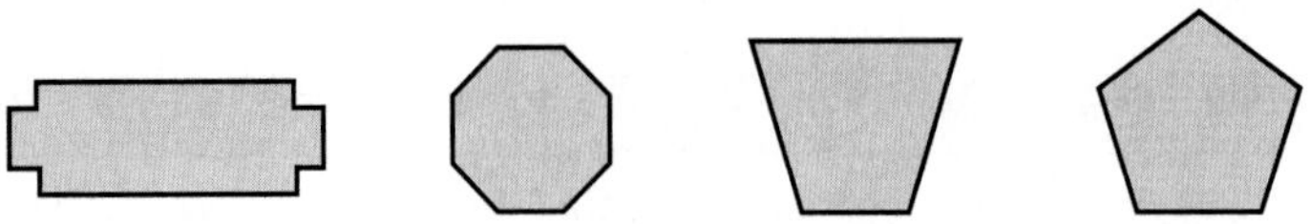

Figure 1

Figure 2

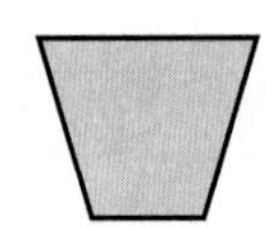

Figure 3

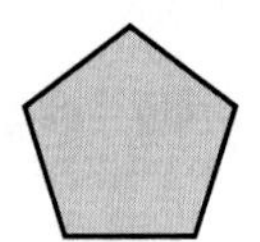

Figure 4

Part A Which of the figures will tessellate by themselves?

Part B Which will be able to tessellate if another shape is added?

Part C Draw your own tessellation.

Tuesday

Multiple Choice The perimeter of the pentagon is $8x + 24$ units. Find the length of the missing side. *(Lesson 13-3)*

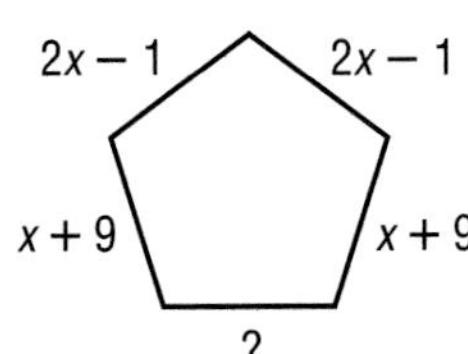

(A) $6x - 8$ units (B) $2x + 8$ units

(C) $2x + 16$ units (D) $6x + 16$ units

Gridded Response Evaluate $(2a + 2b) + (6a - 6b)$ if $a = 4$ and $b = -0.5$. *(Lesson 13-2)*

Wednesday

Multiple Choice Which rule describes a linear equation? *(Lesson 13-5)*

(F) $y = 4x(x - 9)$ (G) $y = 4x^2 + 9$

(H) $y = 4x^3 + 9$ (I) $y = 4x + 9$

Gridded Response A bag of chips contains 3 white chips, 6 red chips, and 5 blue chips. Once a chip has been drawn, it does not get replaced. What is the probability of drawing a red chip then blue chip? Write as a percent to the nearest hundredth. *(Lesson 12-9)*

Thursday

THINK SOLVE EXPLAIN

Short Response Interstate 95 runs along the east coast of Florida. Use a geometric term to describe the relationship between the interstate and the coast. *(Lesson 10-1)*

Friday

Multiple Choice Simplify $4ab(2a - 4ab + b)$. *(Lesson 13-4)*

(A) $6ab - 8ab + 4ab$

(B) $8a^2b - 16a^2b^2 + 4ab^2$

(C) $6a^2b - 8a^2b^2 + 4ab^2$

(D) $8ab - 16ab + 4ab$

GLENCOE MATHEMATICS

Florida Edition

Pre-Algebra

Malloy
Price
Willard
Sloan

New York, New York
Columbus, Ohio
Chicago, Illinois
Peoria, Illinois
Woodland Hills, California

The McGraw-Hill Companies

The Standardized Test Practice features in this book were aligned and verified by The Princeton Review, the nation's leader in test preparation. Through its association with McGraw-Hill, The Princeton Review offers the best way to help students excel on standardized assessments.

The Princeton Review is not affiliated with Princeton University or Educational Testing Service.

Send all inquiries to:
Glencoe/McGraw-Hill
8787 Orion Place
Columbus, OH 43240-4027

ISBN: 0-07-860368-4

Printed in the United States of America.

1 2 3 4 5 6 7 8 9 10 027/043 12 11 10 09 08 07 06 05 04 03

Contents in Brief

Unit 1 Algebra and Integers

Chapter 1 The Tools of Algebra **4**

Chapter 2 Integers **54**

Chapter 3 Equations **96**

Unit 2 Algebra and Rational Numbers

Chapter 4 Factors and Fractions **146**

Chapter 5 Rational Numbers **198**

Chapter 6 Ratio, Proportion, and Percent **262**

Unit 3 Linear Equations, Inequalities, and Functions

Chapter 7 Equations and Inequalities **326**

Chapter 8 Functions and Graphing **366**

Unit 4 Applying Algebra to Geometry

Chapter 9 Real Numbers and Right Triangles **434**

Chapter 10 Two-Dimensional Figures **490**

Chapter 11 Three-Dimensional Figures **552**

Unit 5 Extending Algebra to Statistics and Polynomials

Chapter 12 More Statistics and Probability **604**

Chapter 13 Polynomials and Nonlinear Functions **666**

Authors

Carol Malloy, Ph.D.
Associate Professor of Mathematics Education
University of North Carolina at Chapel Hill
Chapel Hill, North Carolina

Jack Price, Ed.D.
Professor Emeritus, Mathematics Education
California State Polytechnic University
Pomona, California

Teri Willard, Ed.D.
Mathematics Consultant
Belgrade, Montana

Leon L. "Butch" Sloan, Ed.D.
Secondary Mathematics Coordinator
Garland ISD
Garland, Texas

Contributing Authors

USA TODAY

The USA TODAY Snapshots®, created by USA TODAY®, help students make the connection between real life and mathematics.

Dinah Zike
Educational Consultant
Dinah-Might Activities, Inc.
San Antonio, Texas

Consultants

Content Consultants

Each of the Content Consultants reviewed every chapter and gave suggestions for improving the effectiveness of the mathematics instruction.

Mathematics Consultants

Rhonda Bailey
Mathematics Consultant
Mathematics by Design
DeSoto, Texas

Gunnar Carlsson, Ph.D.
Professor of Mathematics
Stanford University
Stanford, California

Ralph Cohen, Ph.D.
Professor of Mathematics
Stanford University
Stanford, California

William Leschensky
Former Mathematics Teacher
Glenbard South High School
College of DuPage
Glen Ellyn, Illinois

Yuria Orihuela
Mathematics Supervisor
Miami-Dade County Public Schools
Miami, Florida

Reading Consultant

Nancy Klores Welday
Language Arts Chairperson and
Reading Resource Teacher
Hialeah-Miami Senior High School
Hialeah, Florida

Teacher Reviewers

Teacher Reviewers

Each Teacher Reviewer reviewed at least two chapters of the Student Edition, giving feedback and suggestions for improving the effectiveness of the mathematics instruction.

Steven L. Arnofsky
Assistant Principal, Supervision
George W. Wingate High School
Brooklyn, New York

James M. Barr, Jr.
Teacher
Sells Middle School
Dublin, Ohio

Fay Bonacorsi
High School Mathematics Teacher
Lafayette High School
Brooklyn, New York

Rose H. Boothe
Subject Area Leader
A.J. Ferrell MST
Tampa, Florida

Diana L. Boyle
Mathematics Teacher, 6–8
Judson Middle School
Salem, Oregon

Beverly Burke
7th Grade Mathematics Teacher
USD 362
LaCygne, Kansas

Barbara A. Cain
Mathematics Teacher
Thomas Jefferson Middle School
Merritt Island, Florida

Rusty Campbell
Mathematics Instructor/Chairperson
North Marion High School
Farmington, West Virginia

Carol Caroff
Mathematics Department Chair/Teacher
Solon High School
Solon, Ohio

Vincent Ciraulo
Supervisor of Mathematics
J.P. Stevens High School
Edison, New Jersey

Lisa Cook
Mathematics Teacher
Kaysville Junior High School
Kaysville, Utah

Dianne Coppa
Mathematics Supervisor
Linden School District
Linden, New Jersey

Andrea L. Ellyson
Teacher/Department Chairperson
Great Bridge Middle School
Chesapeake, Virginia

James E. Ewing
7th Grade Pre-Algebra
Hiawatha Middle School
Hiawatha, Kansas

Eve Fingerett
Mathematics Teacher
Mountain Brook Junior High School
Mountain Brook, Alabama

Larry T. Gathers
Mathematics Teacher
Springfield South High School
Springfield, Ohio

Pamela M. Huskey
Mathematics Teacher
Buckingham County Middle School
Dillwyn, Virginia

Donald T. Jobbins
Supervisor of Mathematics
Edison Township Public Schools
Edison, New Jersey

Bonnie Nalesnik Johnston
Academically Gifted Program Facilitator
Valley Springs Middle School
Arden, North Carolina

Michael J. Klein
Math and Science Curriculum Consultant
Warren Consolidated Schools
Warren, Michigan

Thomas Massa
Math Teacher
Los Altos Middle School
Camarillo, California

Jenny L. Miller
Math Teacher
Tuttle Middle School
Crawfordsville, Indiana

Aletha T. Paskett
Mathematics Teacher
Indian Hills Middle School
Sandy, Utah

Mary D. Pistor
Collaborative Peer Teacher
Osborn School District
Phoenix, Arizona

Debra K. Prowse
Mathematics Teacher
Beloit Memorial High School
Beloit, Wisconsin

Carol Read
8th Grade Mathematics Teacher
Legg Middle School
Coldwater, Michigan

Vicki Rentz
Mathematics Teacher
Burnett Middle School
Seffner, Florida

Cherie Rhoades
Mathematics Department Chairperson
Davidson Middle School
Crestview, Florida

Sherri Roberti
TEA/Math Mentor
Beaverton Schools
Beaverton, Oregon

Jack F. Rose, Jr.
Teacher/Department Chairperson
John Winthrop Middle School
Deep River, Connecticut

Sandy Schoff
Math Curriculum Coordinator K–12
Anchorage School District
Anchorage, Alaska

Carol A. Spice
Teacher of Mathematics
Woodlawn Beach Middle School
Gulf Breeze, Florida

Paula Allen Tibbs
Middle School Math
Narrows Elementary/Middle School
Narrows, Virginia

Christine Waddell
Mathematics Department Chair/Teacher
Albion Middle School
Sandy, Utah

Field Test Schools

Glencoe/McGraw-Hill wishes to thank the following schools that field-tested pre-publication manuscript during the 2001–2002 school year. They were instrumental in providing feedback and verifying the effectiveness of this program.

Burnett Middle School
Seffner, Florida

Carwise Middle School
Palm Harbor, Florida

Ft. Zumwalt Middle School
O'Fallon, Missouri

Graham Middle School
Bluefield, Virginia

John F. Kennedy Middle School
Bethpage, New York

McLane Middle School
Brandon, Florida

Martin Middle School
Raleigh, North Carolina

Parkway Southwest Middle School
Ballwin, Missouri

Safety Harbor Middle School
Safety Harbor, Florida

Table of Contents

Algebra and Integers

Chapter 1 The Tools of Algebra 4

Getting Started ... 5
1-1 Using a Problem-Solving Plan ... 6
Reading Mathematics: Translating Expressions Into Words ... 11
1-2 Numbers and Expressions ... 12
1-3 Variables and Expressions ... 17
Practice Quiz 1: Lessons 1-1 through 1-3 ... 21
Spreadsheet Investigation: Expressions and Spreadsheets ... 22
1-4 Properties ... 23
1-5 Variables and Equations ... 28
Practice Quiz 2: Lessons 1-4 and 1-5 ... 32
1-6 Ordered Pairs and Relations ... 33
Algebra Activity: Scatter Plots ... 39
1-7 Scatter Plots ... 40
Graphing Calculator Investigation: Scatter Plots ... 45
Study Guide and Review ... 47
Practice Test ... 51
Standardized Test Practice ... 52

- Introduction 3
- Follow-Ups 43, 79, 135
- Culmination 136

Lesson 1-2, page 15

Prerequisite Skills

- Getting Started 5
- Getting Ready for the Next Lesson 10, 16, 21, 27, 32, 38

Study Organizer 5

Reading and Writing Mathematics

- Translating Expressions into Words 11
- Reading Math Tips 17, 23, 24, 29
- Writing in Math 10, 16, 21, 27, 32, 37, 44

Standardized Test Practice

- Multiple Choice 10, 16, 21, 27, 29, 30, 32, 38, 44, 51, 52
- Short Response/Grid In 53
- Open Ended 53

Snapshots 3, 8, 16

Chapter 2 Integers 54

Getting Started 55

2-1 Integers and Absolute Value 56

Algebra Activity: Adding Integers 62

2-2 Adding Integers 64

Reading Mathematics: Learning Mathematics Vocabulary 69

2-3 Subtracting Integers 70

Practice Quiz 1: Lessons 2-1 through 2-3 74

2-4 Multiplying Integers 75

2-5 Dividing Integers 80

Practice Quiz 2: Lessons 2-4 and 2-5 84

2-6 The Coordinate System 85

Study Guide and Review 90

Practice Test 93

Standardized Test Practice 94

Prerequisite Skills

- Getting Started 55
- Getting Ready for the Next Lesson 61, 68, 74, 79, 84

FOLDABLES™

Study Organizer 55

Reading and Writing Mathematics

- Learning Mathematics Vocabulary 69
- Reading Math Tips 56, 57, 64, 75, 80, 88
- Writing in Math 61, 68, 74, 79, 84, 89

Standardized Test Practice

- Multiple Choice 61, 68, 74, 76, 77, 79, 84, 89, 93, 94
- Short Response/Grid In 95
- Open Ended 95

USA TODAY **Snapshots** 60

Lesson 2-4, page 78

Chapter 3 Equations 96

Getting Started 97

3-1 The Distributive Property 98

3-2 Simplifying Algebraic Expressions....103

Practice Quiz 1: Lessons 3-1 and 3-2 107

Algebra Activity: Solving Equations Using Algebra Tiles....108

3-3 Solving Equations by Adding or Subtracting....110

3-4 Solving Equations by Multiplying or Dividing....115

3-5 Solving Two-Step Equations....120

Reading Mathematics: Translating Verbal Problems into Equations....125

3-6 Writing Two-Step Equations....126

Practice Quiz 2: Lessons 3-3 through 3-6....130

3-7 Using Formulas 131

Spreadsheet Investigation: Perimeter and Area....137

Study Guide and Review....138

Practice Test 141

Standardized Test Practice 142

Lesson 3-6, page 127

Prerequisite Skills

- Getting Started 97
- Getting Ready for the Next Lesson 102, 107, 114, 119, 124, 130

FOLDABLES™ Study Organizer 97

Reading and Writing Mathematics

- Translating Verbal Problems into Equations 125
- Reading Math Tips 98, 103
- Writing in Math 101, 106, 114, 119, 123, 130, 136

Standardized Test Practice

- Multiple Choice 102, 107, 112, 113, 114, 119, 124, 130, 136, 141, 142
- Short Response/Grid In 124, 143
- Open Ended 143

Snapshots 101

UNIT 2

Algebra and Rational Numbers

Chapter 4 Factors and Fractions 146

	Getting Started	147
4-1	Factors and Monomials	148
4-2	Powers and Exponents	153
	Algebra Activity: Base 2	158
4-3	Prime Factorization	159
	Practice Quiz 1: Lessons 4-1 through 4-3	163
4-4	Greatest Common Factor (GCF)	164
4-5	Simplifying Algebraic Fractions	169
	Reading Mathematics: Powers	174
4-6	Multiplying and Dividing Monomials	175
	Algebra Activity: A Half-Life Simulation	180
4-7	Negative Exponents	181
	Practice Quiz 2: Lessons 4-4 through 4-7	185
4-8	Scientific Notation	186
	Study Guide and Review	191
	Practice Test	195
	Standardized Test Practice	196

Lesson 4-8, page 189

- Introduction 145
- Follow-Ups 173, 242, 301
- Culmination 314

Prerequisite Skills
- Getting Started 147
- Getting Ready for the Next Lesson 152, 157, 163, 168, 173, 179, 185

FOLDABLES™

Study Organizer 147

Reading and Writing Mathematics
- Powers 174
- Reading Math Tips 148, 149, 150, 159, 177
- Writing in Math 152, 157, 162, 168, 173, 179, 184, 190

Standardized Test Practice
- Multiple Choice 152, 157, 163, 168, 171, 173, 179, 184, 190, 195, 196
- Short Response/Grid In 197
- Open Ended 197

Snapshots 145, 156

Chapter 5 Rational Numbers 198

Getting Started 199

5-1 Writing Fractions as Decimals 200

5-2 Rational Numbers 205

5-3 Multiplying Rational Numbers 210

5-4 Dividing Rational Numbers 215

5-5 Adding and Subtracting Like Fractions 220

Practice Quiz 1: Lessons 5-1 through 5-5 224

Reading Mathematics: Factors and Multiples 225

5-6 Least Common Multiple 226

Algebra Activity: Juniper Green 231

5-7 Adding and Subtracting Unlike Fractions 232

Algebra Activity: Analyzing Data 237

5-8 Measures of Central Tendency 238

Graphing Calculator Investigation: Mean and Median 243

5-9 Solving Equations with Rational Numbers 244

Practice Quiz 2: Lessons 5-6 through 5-9 248

5-10 Arithmetic and Geometric Sequences 249

Algebra Activity: Fibonacci Sequence 253

Study Guide and Review 254

Practice Test 259

Standardized Test Practice 260

Algebra Activity, page 253

Prerequisite Skills

- Getting Started 199
- Getting Ready for the Next Lesson 204, 209, 214, 219, 224, 230, 236, 242, 248

 Study Organizer 199

Reading and Writing Mathematics

- Factors and Multiples 225
- Reading Math Tips 200, 205, 206, 215
- Writing in Math 204, 209, 214, 219, 223, 230, 236, 242, 247, 251

Standardized Test Practice

- Multiple Choice 204, 209, 214, 219, 224, 230, 236, 240, 241, 242, 247, 252, 259, 260
- Short Response/Grid In 240, 261
- Open Ended 261

 Snapshots 203, 213

Chapter 6 Ratio, Proportion, and Percent 262

Getting Started 263

6-1 Ratios and Rates 264

Reading Mathematics: Making Comparisons 269

6-2 Using Proportions 270

Algebra Activity: Capture-Recapture 275

6-3 Scale Drawings and Models 276

6-4 Fractions, Decimals, and Percents 281

Algebra Activity: Using a Percent Model 286

6-5 Using the Percent Proportion 288

Practice Quiz 1: Lessons 6-1 through 6-5 292

6-6 Finding Percents Mentally 293

6-7 Using Percent Equations 298

Spreadsheet Investigation: Compound Interest....303

6-8 Percent of Change 304

Practice Quiz 2: Lessons 6-6 through 6-8 308

Algebra Activity: Taking a Survey 309

6-9 Probability and Predictions 310

Graphing Calculator Activity: Probability Simulation 315

Study Guide and Review 316

Practice Test 321

Standardized Test Practice 322

Lesson 6-7, page 299

Prerequisite Skills

- Getting Started 263
- Getting Ready for the Next Lesson 268, 274, 280, 285, 292, 297, 302, 308

FOLDABLES™ Study Organizer 263

Reading and Writing Mathematics

- Making Comparisons 269
- Reading Math Tips 281, 300, 311
- Writing in Math 268, 274, 280, 285, 292, 297, 302, 307, 314

Standardized Test Practice

- Multiple Choice 268, 274, 280, 285, 292, 297, 302, 305, 306, 308, 314, 321, 322
- Short Response/Grid In 323
- Open Ended 323

Snapshots 289, 290, 312, 314

UNIT 3

Linear Equations, Inequalities, and Functions

Chapter 7 Equations and Inequalities 326

Getting Started 327

Algebra Activity: Equations with Variables on Each Side 328

7-1 Solving Equations with Variables on Each Side 330

7-2 Solving Equations with Grouping Symbols 334

Practice Quiz 1: Lessons 7-1 and 7-2 338

Reading Mathematics: Meanings of *At Most* and *At Least* 339

7-3 Inequalities 340

7-4 Solving Inequalities by Adding or Subtracting 345

7-5 Solving Inequalities by Multiplying or Dividing 350

Practice Quiz 2: Lessons 7-3 through 7-5 354

7-6 Solving Multi-Step Inequalities 355

Study Guide and Review 360

Practice Test 363

Standardized Test Practice 364

WebQuest Internet Project
- Introduction 325
- Follow-Ups 333, 411
- Culmination 422

Prerequisite Skills
- Getting Started 327
- Getting Ready for the Next Lesson 333, 338, 344, 349, 354

FOLDABLES™
Study Organizer 327

Reading and Writing Mathematics
- Meanings of *At Most* and *At Least* 339
- Reading Math Tips 341
- Writing in Math 333, 338, 344, 349, 354, 359

Standardized Test Practice
- Multiple Choice 333, 338, 344, 349, 351, 353, 354, 359, 363, 364
- Short Response/Grid In 354, 365
- Open Ended 365

Snapshots 325, 343

Lesson 7-5, page 354

Chapter 8 Functions and Graphing 366

Getting Started 367

Algebra Activity: Input and Output 368

8-1 Functions 369

Graphing Calculator Investigation: Function Tables 374

8-2 Linear Equations in Two Variables 375

Reading Mathematics: Language of Functions 380

8-3 Graphing Linear Equations Using Intercepts 381

Algebra Activity: It's All Downhill 386

8-4 Slope 387

Algebra Activity: Slope and Rate of Change 392

8-5 Rate of Change 393

Practice Quiz 1: Lessons 8-1 through 8-5 397

8-6 Slope-Intercept Form 398

Graphing Calculator Investigation: Families of Graphs 402

8-7 Writing Linear Equations 404

8-8 Best-Fit Lines 409

8-9 Solving Systems of Equations 414

Practice Quiz 2: Lessons 8-6 through 8-9 418

8-10 Graphing Inequalities 419

Graphing Calculator Investigation: Graphing Inequalities 423

Study Guide and Review 424

Practice Test 429

Standardized Test Practice 430

Prerequisite Skills

- Getting Started 367
- Getting Ready for the Next Lesson 373, 379, 385, 391, 397, 401, 408, 413, 418

FOLDABLES™

Study Organizer 367

Reading and Writing Mathematics

- Language of Functions 380
- Reading Math Tips 370, 381, 383
- Writing in Math 373, 379, 385, 391, 397, 401, 408, 412, 418, 422

Standardized Test Practice

- Multiple Choice 373, 379, 385, 389, 390, 391, 397, 401, 408, 413, 418, 422, 429, 430
- Short Response/Grid In 431
- Open Ended 429, 431

Lesson 8-9, page 417

UNIT 4

Applying Algebra to Geometry

Chapter 9 Real Numbers and Right Triangles 434

Getting Started 435

9-1 Squares and Square Roots 436

9-2 The Real Number System 441

Reading Mathematics: Learning Geometry Vocabulary 446

9-3 Angles 447

Practice Quiz 1: Lessons 9-1 through 9-3 451

Spreadsheet Investigation: Circle Graphs and Spreadsheets 452

9-4 Triangles 453

Algebra Activity: The Pythagorean Theorem 458

9-5 The Pythagorean Theorem 460

Algebra Activity: Graphing Irrational Numbers 465

9-6 The Distance and Midpoint Formulas 466

Practice Quiz 2: Lessons 9-4 through 9-6 470

9-7 Similar Triangles and Indirect Measurement 471

Algebra Activity: Ratios in Right Triangles 476

9-8 Sine, Cosine, and Tangent Ratios 477

Graphing Calculator Investigation: Finding Angles of a Right Triangle 482

Study Guide and Review 483

Practice Test 487

Standardized Test Practice 488

WebQuest Internet Project

- Introduction 433
- Follow-Ups 481, 542, 571
- Culmination 594

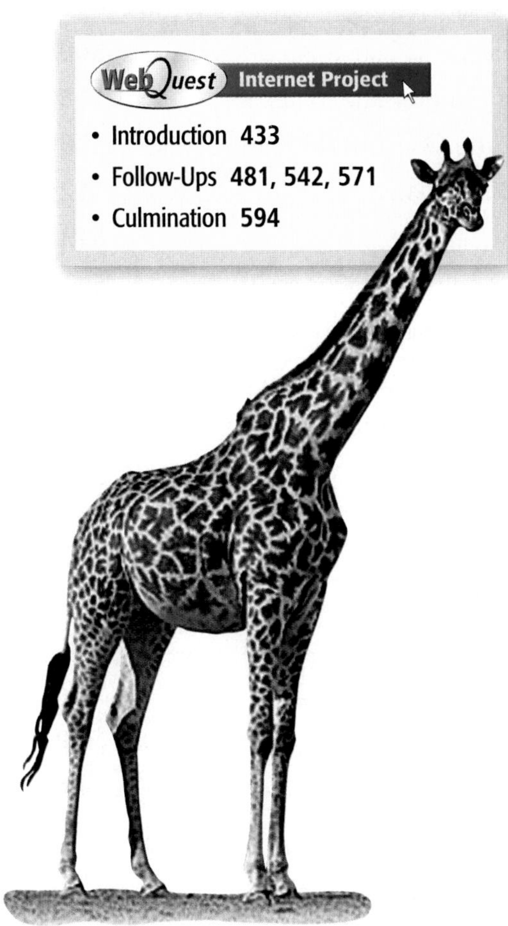

Lesson 9-7, page 474

Prerequisite Skills

- Getting Started 435
- Getting Ready for the Next Lesson 440, 445, 451, 457, 464, 470, 475

FOLDABLES™ Study Organizer 435

Reading and Writing Mathematics

- Learning Geometry Vocabulary 446
- Reading Math Tips 437, 448, 453, 455, 471, 472, 477
- Writing in Math 440, 445, 451, 457, 464, 469, 475, 481

Standardized Test Practice

- Multiple Choice 440, 445, 451, 457, 461, 462, 464, 470, 475, 481, 487, 488
- Short Response/Grid In 489
- Open Ended 489

Snapshots 433, 450

Chapter 10 Two-Dimensional Figures 490

Getting Started ... 491

10-1 Line and Angle Relationships ... 492

Algebra Activity: Constructions ... 498

10-2 Congruent Triangles ... 500

Algebra Activity: Symmetry ... 505

10-3 Transformations on the Coordinate Plane ... 506

Algebra Activity: Dilations ... 512

10-4 Quadrilaterals ... 513

Practice Quiz 1: Lessons 10-1 through 10-4 ... 517

Algebra Activity: Area and Geoboards ... 518

10-5 Area: Parallelograms, Triangles, and Trapezoids ... 520

Reading Mathematics: Learning Mathematics Prefixes ... 526

10-6 Polygons ... 527

Algebra Activity: Tessellations ... 532

10-7 Circumference and Area: Circles ... 533

Practice Quiz 2: Lessons 10-5 through 10-7 ... 538

10-8 Area: Irregular Figures ... 539

Study Guide and Review ... 544

Practice Test ... 549

Standardized Test Practice ... 550

Prerequisite Skills

- Getting Started 491
- Getting Ready for the Next Lesson 497, 504, 511, 517, 525, 531, 538

FOLDABLES™

Study Organizer 491

Reading and Writing Mathematics

- Learning Mathematics Prefixes 526
- Reading Math Tips 493, 500, 508
- Writing in Math 497, 504, 511, 517, 525, 531, 537, 543

Standardized Test Practice

- Multiple Choice 494, 495, 497, 504, 511, 517, 525, 531, 537, 543, 549, 550
- Short Response/Grid In 517, 531, 537, 551
- Open Ended 551

Snapshots 537

Lesson 10-5, page 524

Chapter 11 Three-Dimensional Figures 552

Getting Started 553

Geometry Activity: Building Three-Dimensional Figures 554

11-1 Three-Dimensional Figures 556

Geometry Activity: Volume 562

11-2 Volume: Prisms and Cylinders 563

11-3 Volume: Pyramids and Cones 568

Practice Quiz 1: Lessons 11-1 through 11-3 572

11-4 Surface Area: Prisms and Cylinders 573

11-5 Surface Area: Pyramids and Cones 578

Geometry Activity: Similar Solids 583

11-6 Similar Solids 584

Practice Quiz 2: Lessons 11-4 through 11-6 588

Reading Mathematics: Precision and Accuracy 589

11-7 Precision and Significant Digits 590

Study Guide and Review 595

Practice Test 599

Standardized Test Practice 600

Lesson 11-2, page 564

Prerequisite Skills

- Getting Started **553**
- Getting Ready for the Next Lesson **561, 567, 572, 577, 582, 588**

FOLDABLES™ **Study Organizer 553**

Reading and Writing Mathematics

- Precision and Accuracy **589**
- Writing in Math **561, 567, 571, 577, 582, 588, 594**

Standardized Test Practice

- Multiple Choice **561, 564, 566, 567, 572, 577, 582, 588, 594, 599, 600**
- Short Response/Grid In **601**
- Open Ended **601**

Snapshots 593

UNIT 5

Extending Algebra to Statistics and Polynomials

Chapter 12 More Statistics and Probability 604

Getting Started 605

12-1 Stem-and-Leaf Plots 606

12-2 Measures of Variation 612

12-3 Box-and-Whisker Plots 617

Graphing Calculator Investigation: Box-and-Whisker Plots 622

12-4 Histograms 623

Practice Quiz 1: Lessons 12-1 through 12-4 628

Graphing Calculator Investigation: Histograms 629

12-5 Misleading Statistics 630

Reading Mathematics: Dealing with Bias 634

12-6 Counting Outcomes 635

Algebra Activity: Probability and Pascal's Triangle 640

12-7 Permutations and Combinations 641

Practice Quiz 2: Lessons 12-5 through 12-7 645

12-8 Odds 646

12-9 Probability of Compound Events 650

Algebra Activity: Simulations 656

Study Guide and Review 658

Practice Test 663

Standardized Test Practice 664

- Introduction 603
- Follow-Ups 626, 690
- Culmination 696

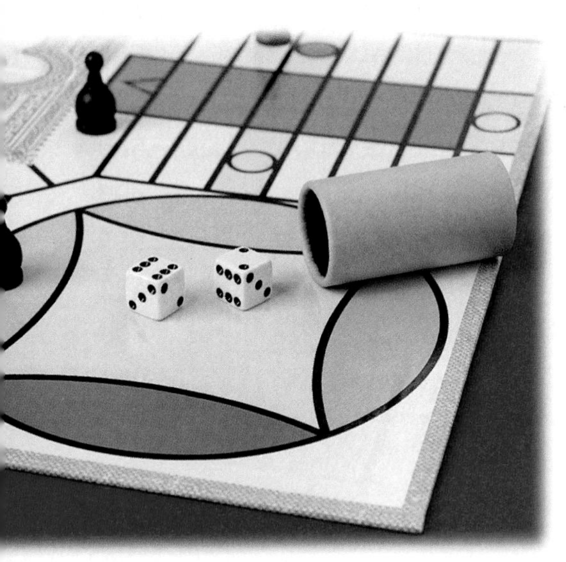

Lesson 12-9, page 651

Prerequisite Skills

- Getting Started 605
- Getting Ready for the Next Lesson 611, 616, 621, 628, 633, 639, 645, 649

Reading and Writing Mathematics

- Dealing with Bias 634
- Reading Math Tips 624, 641, 642, 643, 647, 650
- Writing in Math 610, 616, 621, 627, 633, 639, 645, 649, 654

Study Organizer 605

Standardized Test Practice

- Multiple Choice 611, 616, 621, 627, 633, 639, 645, 647, 648, 649, 655, 663, 664
- Short Response/Grid In 665
- Open Ended 665

Snapshots 603, 610, 649, 654

Chapter 13 Polynomials and Nonlinear Functions 666

Getting Started ... 667

Reading Mathematics: Prefixes and Polynomials ... 668

13-1 Polynomials ... 669

Algebra Activity: Modeling Polynomials with Algebra Tiles ... 673

13-2 Adding Polynomials ... 674

13-3 Subtracting Polynomials ... 678

Practice Quiz 1: Lessons 13-1 through 13-3 ... 681

Algebra Activity: Modeling Multiplication ... 682

13-4 Multiplying a Polynomial by a Monomial ... 683

13-5 Linear and Nonlinear Functions ... 687

Practice Quiz 2: Lessons 13-4 and 13-5 ... 691

13-6 Graphing Quadratic and Cubic Functions ... 692

Graphing Calculator Investigation: Families of Quadratic Functions ... 697

Study Guide and Review ... 698

Practice Test ... 701

Standardized Test Practice ... 702

Prerequisite Skills

- Getting Started 667
- Getting Ready for the Next Lesson 672, 677, 681, 686, 691

FOLDABLES™

Study Organizer 667

Reading and Writing Mathematics

- Prefixes and Polynomials 668
- Reading Math Tips 678, 684, 688
- Writing in Math 672, 677, 681, 686, 691, 695

Standardized Test Practice

- Multiple Choice 672, 677, 681, 686, 689, 691, 695, 696, 701, 702
- Short Response/Grid In 703
- Open Ended 703

Snapshots 690

Student Handbook

Skills

Prerequisite Skills ... 706

Extra Practice ... 724

Mixed Problem Solving ... 758

Reference

English-Spanish Glossary ... R1

Selected Answers ... R19

Photo Credits ... R51

Index ... R52

Lesson 13-2, page 675

One-Stop Internet Resources

Need extra help or information? Log on to math.glencoe.com or any of the Web addresses to learn more.

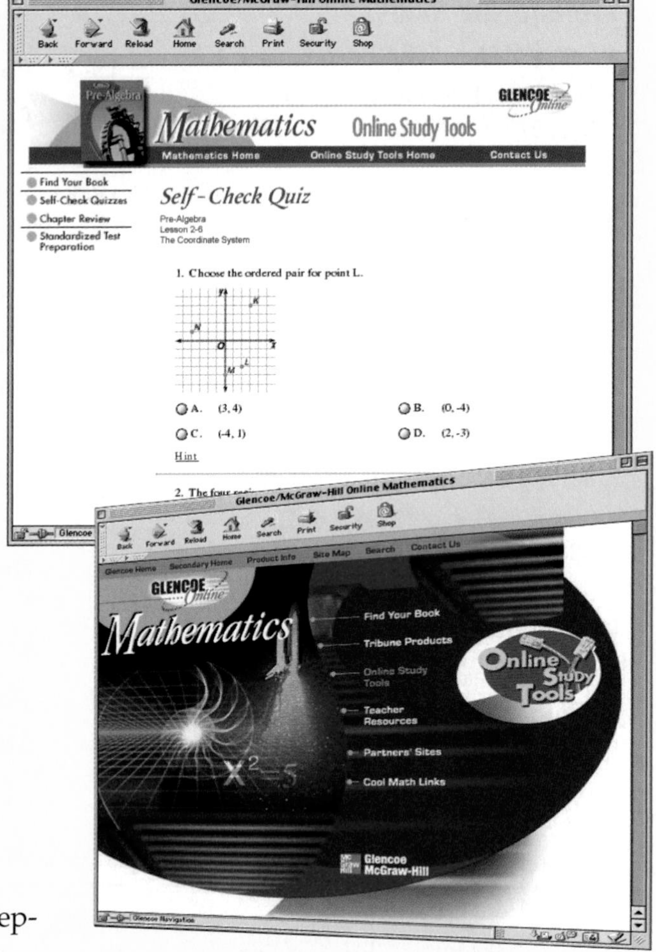

Online Study Tools

- www.pre-alg.com/extra_examples shows you additional worked-out examples that mimic the ones in your book.
- www.pre-alg.com/self_check_quiz provides you with a self-checking practice quiz for each lesson.
- www.pre-alg.com/vocabulary_review lets you check your understanding of the terms and definitions used in each chapter.
- www.pre-alg.com/chapter_test allows you to take a self-checking test before the actual test.
- www.pre-alg.com/standardized_test is another way to brush up on your standardized test-taking skills.

Research Options

- www.pre-alg.com/webquest walks you step-by-step through a long-term project using the Web. One WebQuest for each unit is explored using the mathematics from that unit.
- www.pre-alg.com/usa_today provides activities related to the concept of the lesson as well as up-to-date Snapshot data.
- www.pre-alg.com/careers links you to additional information about interesting careers.
- www.pre-alg.com/data_update links you to the most current data available for subjects such as basketball and family.

Calculator Help

- www.pre-alg.com/other_calculator_keystrokes provides you with keystrokes other than the TI-83 Plus used in your textbook.

FOLDABLES™

A Handy Way to Help You Study

As easy as 1-2-3! Just fold and you are ready to go! Each chapter provides you with a different Foldable that's easy to create. It's a fun way to organize what you learn and a great study tool.

Make the Foldable below and add definitions and examples to it as you learn new properties throughout the year. Begin with eight half-sheets of plain paper.

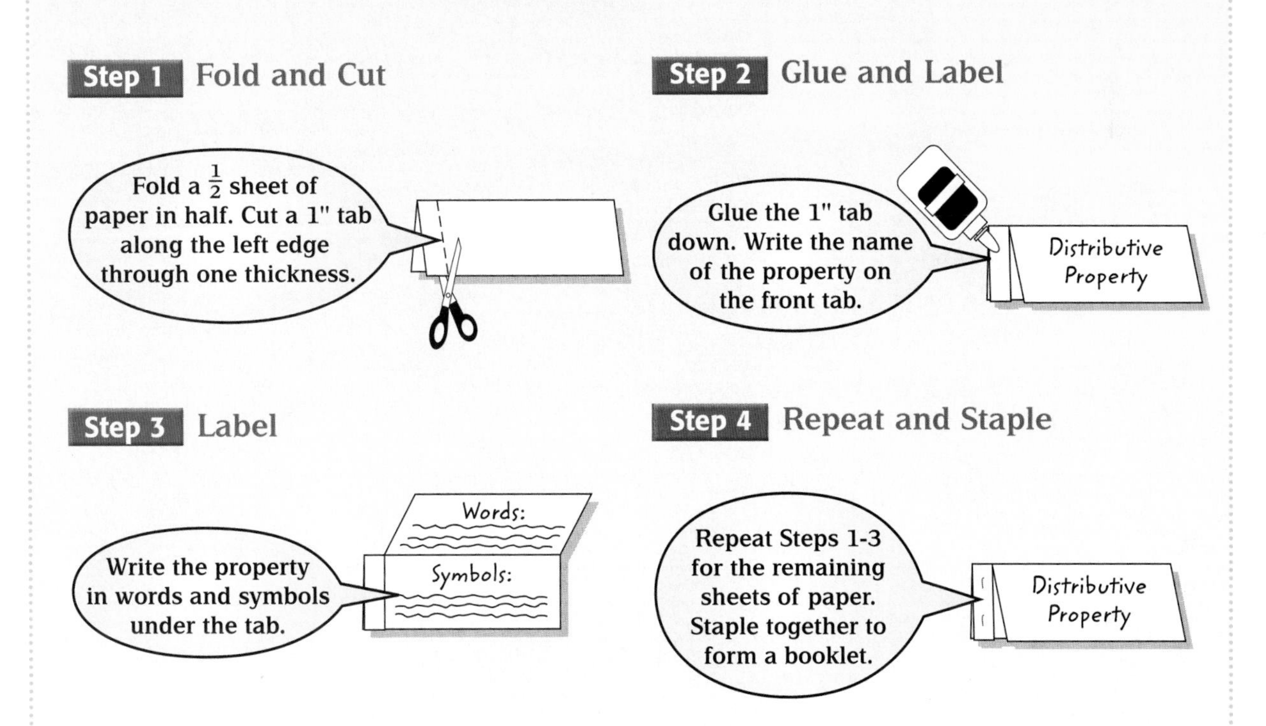

Step 1 Fold and Cut
Fold a $\frac{1}{2}$ sheet of paper in half. Cut a 1" tab along the left edge through one thickness.
Step 2 Glue and Label
Glue the 1" tab down. Write the name of the property on the front tab.
Distributive Property
Step 3 Label
Write the property in words and symbols under the tab.
Words:
Symbols:
Step 4 Repeat and Staple
Repeat Steps 1-3 for the remaining sheets of paper. Staple together to form a booklet.
Distributive Property

FOLDABLES can be found on the following pages: 5, 55, 97, 147, 199, 263, 327, 367, 435, 491, 553, 605, and 667.

UNIT 1

Algebra and Integers

The word *algebra* comes from the Arabic word *al-jebr*, which was part of the title of a book about equations and how to solve them. In this unit, you will lay the foundation for your study of algebra by learning about the language of algebra, its properties, and methods of solving equations.

Chapter 1
The Tools of Algebra

Chapter 2
Integers

Chapter 3
Equations

WebQuest Internet Project

Vacation Travelers Include More Families

"Taking the kids with you is increasingly popular among Americans, according to a travel report that predicts an expanding era of kid-friendly attractions and services." **Source:** *USA TODAY*, November 17, 1999

In this project, you will be exploring how graphs and formulas can help you plan a family vacation.

Log on to www.pre-alg.com/webquest. Begin your WebQuest by reading the Task.

Then continue working on your WebQuest as you study Unit 1.

Lesson	1-7	2-4	3-7
Page	43	79	135

USA TODAY Snapshots®

Spouses are top travel partners

Spouses	58%
Children/grandchildren	34%
Friends	18%
Other family members	14%
Solo	13%
Group tour	8%

Source: Travel Industry Association of America

By Cindy Hall and Sam Ward, USA TODAY

1 The Tools of Algebra

What You'll Learn

- **Lesson 1-1** Use a four-step plan to solve problems and choose the appropriate method of computation.
- **Lessons 1-2 and 1-3** Translate verbal phrases into numerical expressions and evaluate expressions.
- **Lesson 1-4** Identify and use properties of addition and multiplication.
- **Lesson 1-5** Write and solve simple equations.
- **Lesson 1-6** Locate points and represent relations.
- **Lesson 1-7** Construct and interpret scatter plots.

Key Vocabulary

- order of operations (p. 12)
- variable (p. 17)
- algebraic expression (p. 17)
- ordered pair (p. 33)
- relation (p. 35)

Why It's Important

Algebra is important because it can be used to show relationships among variables and numbers. You can use algebra to describe how fast something grows. For example, the growth rate of bamboo can be described using variables. *You will find the growth rate of bamboo in Lesson 1-6.*

Getting Started

Prerequisite Skills To be successful in this chapter, you'll need to master these skills and be able to apply them in problem-solving situations. Review these skills before beginning Chapter 1.

For Lesson 1-1 — Add and Subtract Decimals

Find each sum or difference. *(For review, see page 713.)*

1. $6.6 + 8.2$ **2.** $4.7 + 8.5$ **3.** $5.4 - 2.3$

4. $8.6 - 4.9$ **5.** $2.65 + 0.3$ **6.** $1.08 + 1.2$

7. $4.25 - 0.7$ **8.** $4.3 - 2.89$ **9.** $9.06 - 1.18$

For Lessons 1-1 through 1-5 — Estimate with Whole Numbers

Estimate each sum, difference, product, or quotient.

10. $1800 + 285$ **11.** $328 + 879$ **12.** $22{,}431 - 13{,}183$

13. $659 - 536$ **14.** 68×12 **15.** 189×89

16. $3845 \div 82$ **17.** $21{,}789 \div 97$ **18.** $\$1951 \div 49$

For Lessons 1-1 through 1-5 — Estimate with Decimals

Estimate each sum, difference, product, or quotient. *(For review, see pages 712 and 714.)*

19. $8.8 + 5.3$ **20.** $47.2 + 9.75$ **21.** $\$7.34 - \$2.16$

22. $83.6 - 75.32$ **23.** 4.2×29.3 **24.** $18.8(5.3)$

25. $7.8 \div 2.3$ **26.** $54 \div 9.1$ **27.** $21.3 \div 1.7$

Make this Foldable to help you organize your strategies for solving problems. Begin with a sheet of unlined paper.

Step 1 Fold

Fold the short sides so they meet in the middle.

Step 2 Fold Again

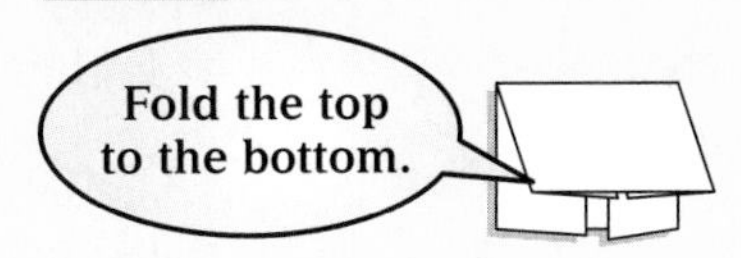

Step 3 Cut

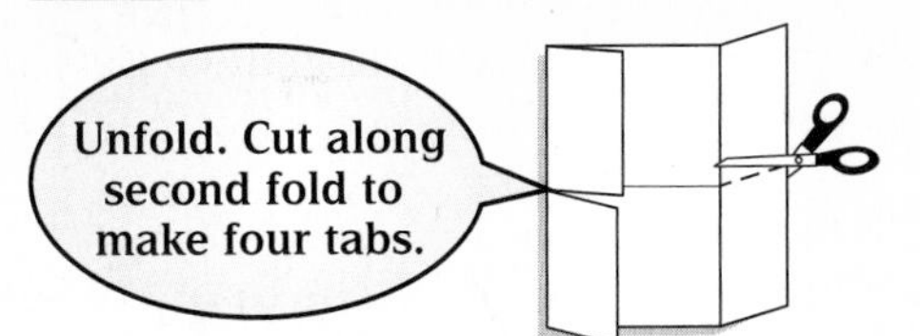

Step 4 Label

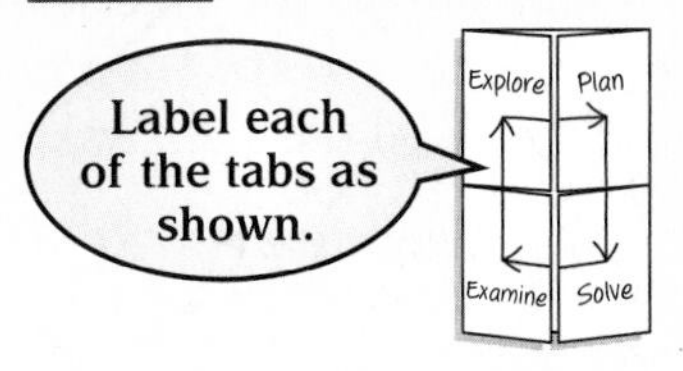

Label each of the tabs as shown.
Explore
Plan
Examine
Solve

Reading and Writing As you read and study the chapter, you can write examples of each problem-solving step under the tabs.

1-1 Using a Problem-Solving Plan

Sunshine State Standards
MA.A.3.3.2-1, MA.A.3.3.2-2, MA.A.3.3.3-1, MA.A.4.3.1-2

Vocabulary

- conjecture
- inductive reasoning

What You'll Learn

- Use a four-step plan to solve problems.
- Choose an appropriate method of computation.

Why is it helpful to use a problem-solving plan to solve problems?

The table shows the first-class mail rates in 2001.

Weight (oz)	Cost
1	$0.34
2	$0.55
3	$0.76
4	$0.97
5	$1.18

Source: www.ups.com

a. Find a pattern in the costs.

b. How can you determine the cost to mail a 6-ounce letter?

c. Suppose you were asked to find the cost of mailing a letter that weighs 8 ounces. What steps would you take to solve the problem?

FOUR-STEP PROBLEM-SOLVING PLAN It is often helpful to have an organized plan to solve math problems. The following four steps can be used to solve any math problem.

1. Explore
- Read the problem quickly to gain a general understanding of it.
- Ask yourself, "What facts do I know?" and "What do I need to find out?"
- Ask, "Is there enough information to solve the problem? Is there extra information?"

2. Plan
- Reread the problem to identify relevant facts.
- Determine how the facts relate to each other.
- Make a plan to solve the problem.
- Estimate the answer.

3. Solve
- Use your plan to solve the problem.
- If your plan does not work, revise it or make a new plan. Ask, "What did I do wrong?"

4. Examine
- Reread the problem.
- Ask, "Is my answer reasonable and close to my estimate?"
- Ask, "Does my answer make sense?"
- If not, solve the problem another way.

Study Tip

Problem-Solving Strategies
Here are a few strategies you will use to solve problems in this book.
- Look for a pattern.
- Solve a simpler problem.
- Guess and check.
- Draw a diagram.
- Make a table or chart.
- Work backward.
- Make a list.

Concept Check Which step involves estimating the answer?

Example 1 Use the Four-Step Problem-Solving Plan

POSTAL SERVICE Refer to page 6. How much would it cost to mail a 9-ounce letter first-class?

Explore The table shows the weight of a letter and the respective cost to mail it first-class. We need to find how much it will cost to mail a 9-ounce letter.

Plan Use the information in the table to solve the problem. Look for a pattern in the costs. Extend the pattern to find the cost for a 9-ounce letter.

Solve First, find the pattern.

Weight (oz)	1	2	3	4	5
Cost	\$0.34	\$0.55	\$0.76	\$0.97	\$1.18

+ 0.21 + 0.21 + 0.21 + 0.21

Each consecutive cost increases by \$0.21. Next, extend the pattern.

Weight (oz)	5	6	7	8	9
Cost	\$1.18	\$1.39	\$1.60	\$1.81	\$2.02

+ 0.21 + 0.21 + 0.21 + 0.21

It would cost \$2.02 to mail a 9-ounce letter.

Examine It costs \$0.34 for the first ounce and \$0.21 for each additional ounce. To mail a 9-ounce letter, it would cost \$0.34 for the first ounce and $8 \times \$0.21$ or \$1.68 for the eight additional ounces. Since $\$0.34 + \$1.68 = \$2.02$, the answer is correct.

Study Tip

Reasonableness Always check to be sure your answer is reasonable. If the answer seems unreasonable, solve the problem again.

A **conjecture** is an educated guess. When you make a conjecture based on a pattern of examples or past events, you are using **inductive reasoning**. In mathematics, you will use inductive reasoning to solve problems.

Example 2 Use Inductive Reasoning

a. Find the next term in 1, 3, 9, 27, 81, ...

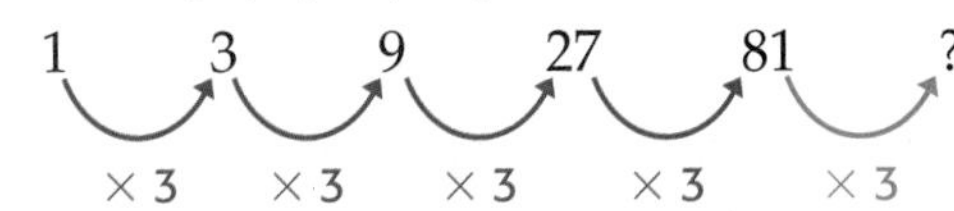

Assuming the pattern continues, the next term is 81×3 or 243.

b. Draw the next figure in the pattern.

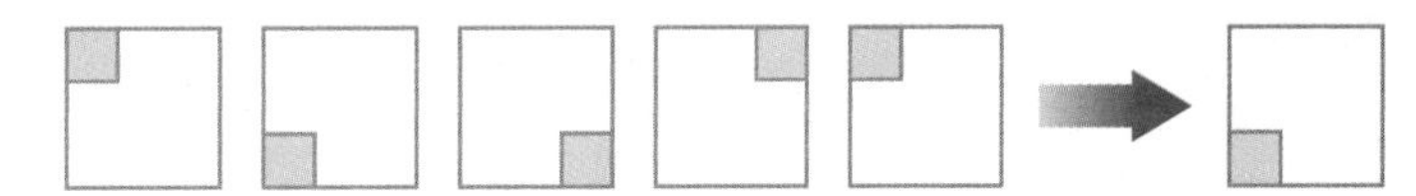

In the pattern, the shaded square moves counterclockwise. Assuming the pattern continues, the shaded square will be positioned at the bottom left of the figure.

Concept Check What type of reasoning is used when you make a conclusion based on a pattern?

CHOOSE THE METHOD OF COMPUTATION Choosing the method of computation is also an important step in solving problems. Use the diagram below to help you decide which method is most appropriate.

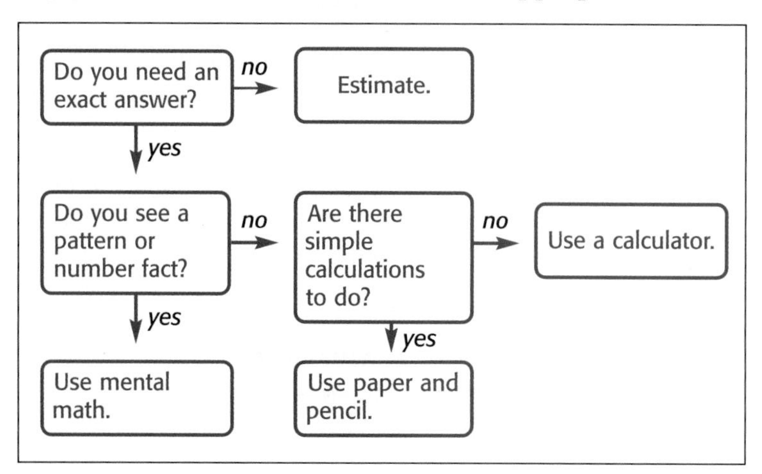

Do you need an exact answer?
no
Estimate.
yes
Do you see a pattern or number fact?
no
Are there simple calculations to do?
no
Use a calculator.
yes
yes
Use mental math.
Use paper and pencil.

Example 3 Choose the Method of Computation

Log on for:
- Updated data
- More activities on Using a Problem-Solving Plan

www.pre-alg.com/usa_today

TRAVEL **The graph shows the seating capacity of certain baseball stadiums in the United States. About how many more seats does Comerica Park have than Fenway Park?**

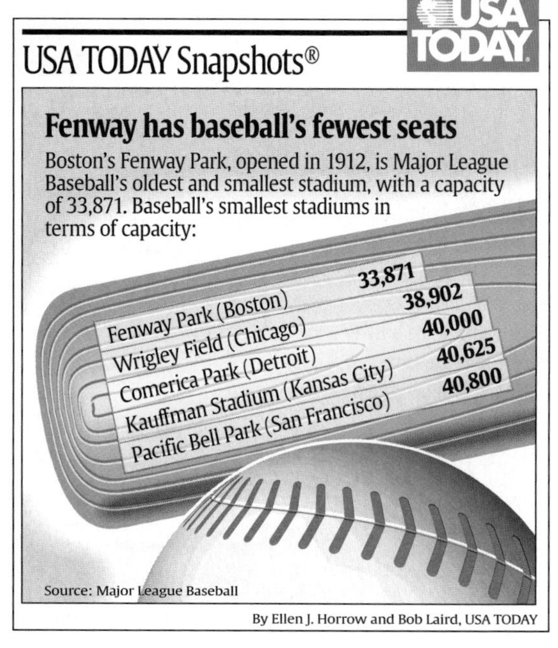

USA TODAY
USA TODAY Snapshots®
Fenway has baseball's fewest seats
Boston's Fenway Park, opened in 1912, is Major League Baseball's oldest and smallest stadium, with a capacity of 33,871. Baseball's smallest stadiums in terms of capacity:
Fenway Park (Boston) 33,871
Wrigley Field (Chicago) 38,902
Comerica Park (Detroit) 40,000
Kauffman Stadium (Kansas City) 40,625
Pacific Bell Park (San Francisco) 40,800
Source: Major League Baseball
By Ellen J. Horrow and Bob Laird, USA TODAY

Explore You know the seating capacities of Comerica Park and Fenway Park. You need to find how many more seats Comerica Park has than Fenway Park.

Plan The question uses the word *about*, so an exact answer is not needed. We can solve the problem using estimation. Estimate the amount of seats for each park. Then subtract.

Solve Comerica Park: $40{,}000 \rightarrow 40{,}000$

Fenway Park: $33{,}871 \rightarrow 34{,}000$ Round to the nearest thousand.

$40{,}000 - 34{,}000 = 6000$ Subtract 34,000 from 40,000.

So, Comerica Park has about 6000 more seats than Fenway Park.

Examine Since $34{,}000 + 6000 = 40{,}000$, the answer makes sense.

Check for Understanding

Concept Check

1. **Tell** when it is appropriate to solve a problem using estimation.
2. **OPEN ENDED** Write a list of numbers in which four is added to get each succeeding term.

Guided Practice

3. **TRAVEL** The ferry schedule at the right shows that the ferry departs at regular intervals. Use the four-step plan to find the earliest time a passenger can catch the ferry if he/she cannot leave until 1:30 P.M.

South Bass Island Ferry Schedule

Departures	Arrivals
8:45 A.M.	9:
9:33 A.M.	1
10:21 A.M.	
11:09 A.M.	

Find the next term in each list.

4. 10, 20, 30, 40, 50, …
5. 37, 33, 29, 25, 21, …
6. 12, 17, 22, 27, 32, …
7. 3, 12, 48, 192, 768, …

Application

8. **MONEY** In 1999, the average U.S. household spent $12,057 on housing, $1891 on entertainment, $5031 on food, and $7011 on transportation. How much was spent on food each month? Round to the nearest cent.
Source: Bureau of Labor Statistics

Practice and Apply

Homework Help

For Exercises	See Examples
9, 10	1
11–20	2
21–26	3

Extra Practice
See page 724.

HEALTH For Exercises 9 and 10, use the table that gives the approximate heart rate a person should maintain while exercising at 85% intensity.

Age	20	25	30	35	40	45
Heart Rate (beats/min)	174	170	166	162	158	154

9. Assume the pattern continues. Use the four-step plan to find the heart rate a 15-year-old should maintain while exercising at this intensity.
10. What heart rate should a 55-year old maintain while exercising at this intensity?

Find the next term in each list.

11. 2, 5, 8, 11, 14, …
12. 4, 8, 12, 16, 20, …
13. 0, 5, 10, 15, 20, …
14. 2, 6, 18, 54, 162, …
15. 54, 50, 46, 42, 38, …
16. 67, 61, 55, 49, 43, …
17. 2, 5, 9, 14, 20, …
18. 3, 5, 9, 15, 23, …

GEOMETRY Draw the next figure in each pattern.

19. 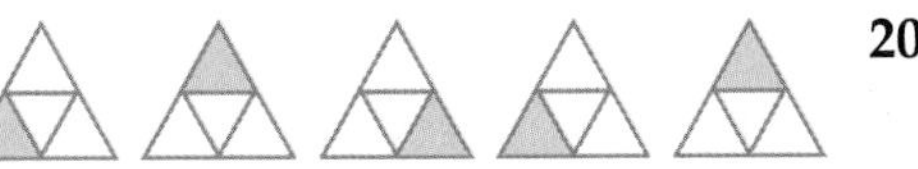

20.  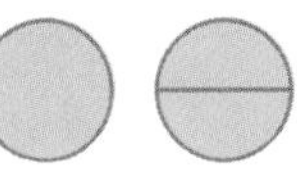 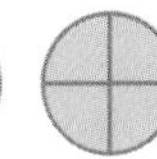 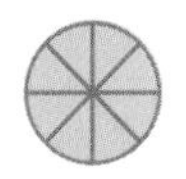

21. **MONEY** Ryan needs to save $125 for a ski trip. He has $68 in his bank. He receives $15 for an allowance and earns $20 delivering newspapers and $16 shoveling snow. Does he have enough money for the trip? Explain.

22. MONEY Using eight coins, how can you make change for 65 cents that will not make change for a quarter?

23. TRANSPORTATION A car traveled 280 miles at 55 mph. About how many hours did it take for the car to reach its destination?

24. CANDY A gourmet jelly bean company can produce 100,000 pounds of jelly beans a day. One ounce of these jelly beans contains 100 Calories. If there are 800 jelly beans in a pound, how many jelly beans can be produced in a day?

More About. . .

Candy

In 1981, $3\frac{1}{2}$ tons of red, blue, and white jelly beans were sent to the Presidential Inaugural Ceremonies for Ronald Reagan.

Source: www.jellybelly.com

25. MEDICINE The number of different types of transplants that were performed in the United States in 1999 are shown in the table. About how many transplants were performed?

Transplant	Number
heart	2185
liver	4698
kidney	12,483
heart-lung	49
lung	885
pancreas	363
intestine	70
kidney-pancreas	946

Source: *The World Almanac*

26. COMMUNICATION A telephone tree is set up so that every person calls three other people. Anita needs to tell her co-workers about a time change for a meeting. Suppose it takes 2 minutes to call 3 people. In 10 minutes, how many people will know about the change of time?

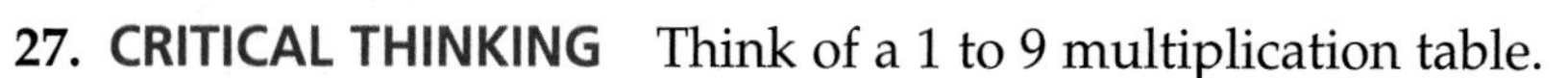

27. CRITICAL THINKING Think of a 1 to 9 multiplication table.

a. Are there more odd or more even products? How can you determine the answer without counting?

b. Is this different from a 1 to 9 addition facts table?

28. WRITING IN MATH Answer the question that was posed at the beginning of the lesson.

Why is it helpful to use a problem-solving plan to solve problems?

Include the following in your answer:

- an explanation of the importance of performing each step of the four-step problem-solving plan, and
- an explanation of why it is beneficial to estimate the answer in the *Plan* step.

FCAT Practice
Standardized Test Practice

A B C D

29. Find the next figure in the pattern shown below.

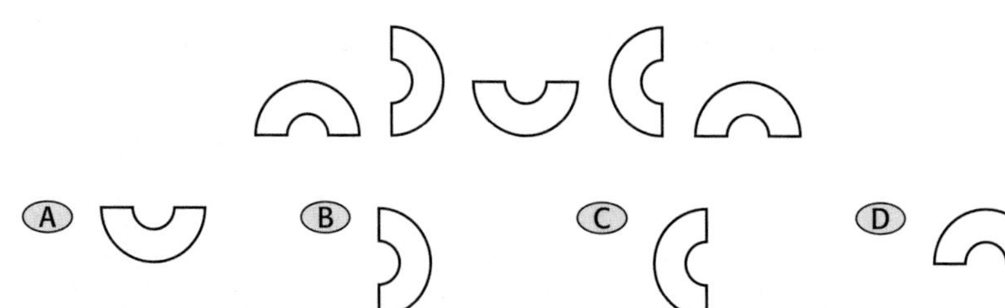

30. A wagon manufacturing plant in Chicago, Illinois, can produce 8000 wagons a day at top production. Which of the following is a reasonable amount of wagons that can be produced in a year?

(A) 24,000 (B) 240,000 (C) 2,400,000 (D) 240,000,000

Getting Ready for the Next Lesson

BASIC SKILL **Round each number to the nearest whole number.**

31. 2.8 **32.** 5.2 **33.** 35.4

34. 49.6 **35.** 109.3 **36.** 999.9

Reading Mathematics

Sunshine State Standards
M.A.A.1.3.1-1, MA.D.1.3.1-1, MA.D.1.3.1-2, MA.D.1.3.1-5

Translating Expressions Into Words

Translating numerical expressions into verbal phrases is an important skill in algebra. Key words and phrases play an essential role in this skill.

The following table lists some words and phrases that suggest addition, subtraction, multiplication, and division.

Addition	Subtraction	Multiplication	Division
plus	minus	times	divided
sum	difference	product	quotient
more than	less than	multiplied	per
increased by	subtract	each	rate
in all	decreased by	of	ratio
	less	factors	separate

A few examples of how to write an expression as a verbal phrase are shown.

Expression	Key Word	Verbal Phrase
5×8	times	5 times 8
$2 + 4$	sum	the sum of 2 and 4
$16 \div 2$	quotient	the quotient of 16 and 2
$8 - 6$	less than	6 less than 8
2×5	product	the product of 2 and 5
$5 - 2$	less	5 less 2

Reading to Learn

1. Refer to the table above. Write a different verbal phrase for each expression.

Choose the letter of the phrase that best matches each expression.

2. $9 - 3$

3. $3 \div 9$

4. $9 \cdot 3$

5. $3 + 9$

6. $9 \div 3$

a. the sum of 3 and 9

b. the quotient of 9 and 3

c. 3 less than 9

d. 9 multiplied by 3

e. 3 divided by 9

Write two verbal phrases for each expression.

7. $5 + 1$

8. $8 + 6$

9. 9×5

10. $2(4)$

11. $12 \div 3$

12. $\frac{20}{4}$

13. $8 - 7$

14. $11 - 5$

1-2 Numbers and Expressions

What You'll Learn

- Use the order of operations to evaluate expressions.
- Translate verbal phrases into numerical expressions.

Sunshine State Standards
MA.A.3.3.2-3

Why do we need to agree on an order of operations?

Vocabulary

- numerical expression
- evaluate
- order of operations

Scientific calculators are programmed to find the value of an expression in a certain order.

Expression	$1 + 2 \times 5$	$8 - 4 \div 2$	$10 \div 5 + 14 \times 2$
Value	11	6	30

a. Study the expressions and their respective values. For each expression, tell the order in which the calculator performed the operations.

b. For each expression, does the calculator perform the operations in order from left to right?

c. Based on your answer to parts **a** and **b**, find the value of each expression below. Check your answer with a scientific calculator.

$12 - 3 \times 2$ $\qquad$ $16 \div 4 - 2$ $\qquad$ $18 + 6 - 8 \div 2 \times 3$

d. **Make a conjecture** as to the order in which a scientific calculator performs operations.

ORDER OF OPERATIONS Expressions like $1 + 2 \times 5$ and $10 \div 5 + 14 \div 2$ are **numerical expressions**. Numerical expressions contain a combination of numbers and operations such as addition, subtraction, multiplication, and division.

When you **evaluate** an expression, you find its numerical value. To avoid confusion, mathematicians have agreed upon the following **order of operations**.

Study Tip

Grouping Symbols
Grouping symbols include:
- parentheses ( ),
- brackets [ ], and
- fraction bars, as in $\frac{6+4}{2}$, which means $(6 + 4) \div 2$.

Concept Summary — *Order of Operations*

Step 1 Simplify the expressions inside grouping symbols.
Step 2 Do all multiplications and/or divisions from left to right.
Step 3 Do all additions and/or subtractions from left to right.

Numerical expressions have only one value. Consider $6 + 4 \times 3$.

$$6 + 4 \times 3 = 6 + 12 = 18$$ (Multiply, then add.)

~~$6 + 4 \times 3 = 10 \times 3 = 30$~~ (Add, then multiply.)

Which is the correct value, 18 or 30? Using the order of operations, the correct value of $6 + 4 \times 3$ is 18.

Concept Check Which operation would you perform first to evaluate $10 - 2 + 3$?

Example 1 Evaluate Expressions

Find the value of each expression.

a. $3 + 4 \times 5$

$3 + 4 \times 5 = 3 + 20$ Multiply 4 and 5.
$= 23$ Add 3 and 20.

b. $18 \div 3 \times 2$

$18 \div 3 \times 2 = 6 \times 2$ Divide 18 by 3.
$= 12$ Multiply 6 and 2.

c. $6(2 + 9) - 3 \cdot 8$

$6(2 + 9) - 3 \cdot 8 = 6(11) - 3 \cdot 8$ Evaluate (2 + 9) first.
$= 66 - 3 \cdot 8$ 6(11) means 6×11.
$= 66 - 24$ $3 \cdot 8$ means 3 times 8.
$= 42$ Subtract 24 from 66.

d. $4[(15 - 9) + 8(2)]$

$4[(15 - 9) + 8(2)] = 4[6 + 8(2)]$ Evaluate (15 − 9).
$= 4(6 + 16)$ Multiply 8 and 2.
$= 4(22)$ Add 6 and 16.
$= 88$ Multiply 4 and 22.

e. $\frac{53 + 15}{17 - 13}$

$\frac{53 + 15}{17 - 13} = (53 + 15) \div (17 - 13)$ Rewrite as a division expression.
$= 68 \div 4$ Evaluate 53 + 15 and 17 − 13.
$= 17$ Divide 68 by 4.

Study Tip

Multiplication and Division Notation
A raised dot or parentheses represents multiplication. A fraction bar represents division.

TRANSLATE VERBAL PHRASES INTO NUMERICAL EXPRESSIONS

You have learned to translate numerical expressions into verbal phrases. It is often necessary to translate verbal phrases into numerical expressions.

Example 2 Translate Phrases into Expressions

Write a numerical expression for each verbal phrase.

a. the product of eight and seven

Phrase the product of eight and seven
Key Word product
Expression 8×7

b. the difference of nine and three

Phrase the difference of nine and three
Key Word difference
Expression $9 - 3$

Study Tip

Differences and Quotients
In this book, *the difference of 9 and 3* means to start with 9 and subtract 3, so the expression is $9 - 3$. Similarly, *the quotient of 9 and 3* means to start with 9 and divide by 3, so the expression is $9 \div 3$.

Example 3 Use an Expression to Solve a Problem

TRANSPORTATION A taxicab company charges a fare of $4 for the first mile and $2 for each additional mile. Write and then evaluate an expression to find the fare for a 10-mile trip.

Words	$4 for the first mile	and	$2 for each additional mile
Expression	4	+	2 × 9

$4 + 2 \times 9 = 4 + 18$ Multiply.
$= 22$ Add.

The fare for a 10-mile trip is $22.

Check for Understanding

Concept Check

1. **OPEN ENDED** Give an example of an expression involving multiplication and subtraction, in which you would subtract first.
2. **Tell** whether $2 \times 4 + 3$ and $2 \times (4 + 3)$ have the same value. Explain.
3. **FIND THE ERROR** Emily and Marcus are evaluating $24 \div 2 \times 3$.

Emily	Marcus
$24 \div 2 \times 3 = 12 \times 3$	$24 \div 2 \times 3 = 24 \div 6$
$= 36$	$= 4$

Who is correct? Explain your reasoning.

Guided Practice

Name the operation that should be performed first. Then find the value of each expression.

4. $3 \cdot 6 - 4$
5. $32 - 24 \div 2$
6. $5(8) + 7$
7. $6(15 - 4)$
8. $\frac{10 - 4}{1 + 2}$
9. $11 + 56 \div (2 \cdot 7)$

Write a numerical expression for each verbal phrase.

10. the quotient of fifteen and five
11. the difference of twelve and nine

Application

12. **MUSIC** Hector purchased 3 CDs for $13 each and 2 cassette tapes for $9 each. Write and then evaluate an expression for the total cost of the merchandise.

Practice and Apply

Homework Help

For Exercises	See Examples
13–28	1
31–38	2
39–42, 47, 48	3

Extra Practice

See page 724.

Find the value of each expression.

13. $2 \cdot 6 - 8$
14. $12 - 3 \times 3$
15. $12 \div 3 + 21$
16. $9 + 18 \div 3$
17. $8 + 5(6)$
18. $4(7) - 11$
19. $\frac{15 + 9}{32 - 20}$
20. $\frac{45 - 18}{9 \div 3}$
21. $11(6 - 1)$
22. $(9 - 7) \cdot 13$
23. $56 \div (7 \cdot 2) \times 6$
24. $75 \div (7 + 8) - 3$
25. $2[5(11 - 3)] - 16$
26. $5[4 + (12 - 4) \div 2]$
27. $9[(22 - 17) + 5(1 + 2)]$
28. $10[9(2 + 4) - 6 \cdot 2]$

29. Find the value of *six added to the product of four and eleven.*

30. What is the value of *sixty divided by the sum of two and ten*?

Write a numerical expression for each verbal phrase.

31. six minus three

32. seven increased by two

33. nine multiplied by five

34. eleven more than fifteen

35. twenty-four divided by six

36. four less than eighteen

37. the cost of 3 notebooks at $6 each

38. the total amount of CDs if Erika has 4 and Roberto has 5

GARDENING **For Exercises 39 and 40, use the following information.**
A bag of potting soil sells for $2, and a bag of fertilizer sells for $13.

39. Write an expression for the total cost of 4 bags of soil and 2 bags of fertilizer.

40. What is the total cost of the gardening supplies?

TRAVEL **For Exercises 41 and 42, use the following information.**
Miko is packing for a trip. The total weight of her luggage cannot exceed 200 pounds. She has 3 suitcases that weigh 57 pounds each and 2 sport bags that weigh 12 pounds each.

41. Write an expression for the total weight of the luggage.

42. Is Miko's luggage within the 200-pound limit? Explain.

Copy each sentence. Then insert parentheses to make each sentence true.

43. $61 - 15 + 3 = 43$

44. $12 \times 3 \div 1 + 2 = 12$

45. $56 \div 2 + 6 - 4 = 3$

46. $5 + 2 \cdot 9 - 3 = 42$

FOOTBALL **For Exercises 47 and 48, use the table and the following information.**
A national poll ranks college football teams using votes from sports reporters. Each vote is worth a certain number of points. Suppose the University of Oklahoma receives 50 first-place votes, 7 second-place votes, 4 fourth-place votes, and 3 tenth-place votes.

Number of Points for Each Vote

Vote	Points
1st place	25
2nd place	24
3rd place	23
4th place	22
5th place	21
⋮	⋮
25th place	1

47. Write an expression for the number of points that the University of Oklahoma receives.

48. Find the total number of points.

More About. . .

Football

The University of Oklahoma Sooners ended the 2000 season ranked No. 1 in NCAA Division I-A college football.

Source: www.espn.com

PUBLISHING **For Exercises 49 and 50, use the following information.**
An ISBN number is used to identify a published book. To determine if an ISBN number is correct, multiply each of the numbers in order by 10, 9, 8, 7, and so on. If the sum of the products can be divided by 11, with no remainder, the number is correct.

49. Find the ISBN number on the back cover of this book.

50. Is the number correct? Explain why or why not.

51. **CRITICAL THINKING** Suppose only the 1, [+], [−], [×], [÷], [(], [)], and [ENTER] keys on a calculator are working. How can you get a result of 75 if you are only allowed to push these keys fewer than 20 times?

52. **WRITING IN MATH** Answer the question that was posed at the beginning of the lesson.

Why do we need to agree on an order of operations?

Include the following in your answer:

- an explanation of how the order operations are performed, and
- an explanation of what will happen to the value of an expression if the order of operations are not followed.

53. Which expression has a value of 18?

Ⓐ $2[2(6 - 3)] + 5$ Ⓑ $27 \div 3 + (12 - 4)$

Ⓒ $(9 \times 3) - 63 \div 7$ Ⓓ $6(3 + 2) \div (9 - 7)$

54. Identify the expression that represents *the quotient of ten and two.*

Ⓐ $2 \div 10$ Ⓑ $\frac{10}{2}$ Ⓒ 10×2 Ⓓ $10 - 2$

Maintain Your Skills

Mixed Review

Find the next term in each list. *(Lesson 1-1)*

55. 2, 4, 8, 16, 32, …

56. 45, 42, 39, 36, 33, …

57. 1, 3, 6, 10, 15, 21, …

58. 15, 18, 22, 25, 29, …

Solve each problem. *(Lesson 1-1)*

59. **BUSINESS** Mrs. Lewis is a sales associate for a computer company. She receives a salary, plus a bonus for any computer package she sells. Find Mrs. Lewis' bonus if she sells 16 computer packages.

Packages	Bonus
2	$100
4	$125
6	$150
8	$175

60. **TRAVEL** The graph shows the projected number of travelers for 2020. How many more people will travel to the United States than to Spain?

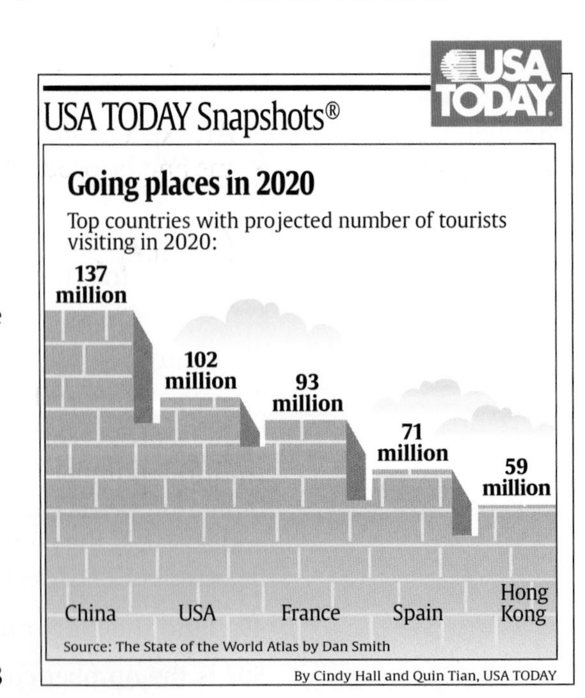

61. **SPACE SHUTTLE** The space shuttle can carry a payload of about 65,000 pounds. If a compact car weighs about 2450 pounds, about how many compact cars can the space shuttle carry?

Getting Ready for the Next Lesson

BASIC SKILL **Find each sum.**

62. $18 + 34$

63. $85 + 41$

64. $342 + 50$

65. $535 + 28$

1-3 Variables and Expressions

What You'll Learn

- Evaluate expressions containing variables.
- Translate verbal phrases into algebraic expressions.

Sunshine State Standards
MA.A.3.3.2-4, MA.D.1.3.1-2, MA.D.2.3.1-1, MA.D.2.3.1-2, MA.D.2.3.1-6

Vocabulary

- variable
- algebraic expression
- defining a variable

How are variables used to show relationships?

A baby-sitter earns \$5 per hour. The table shows several possibilities for number of hours and earnings.

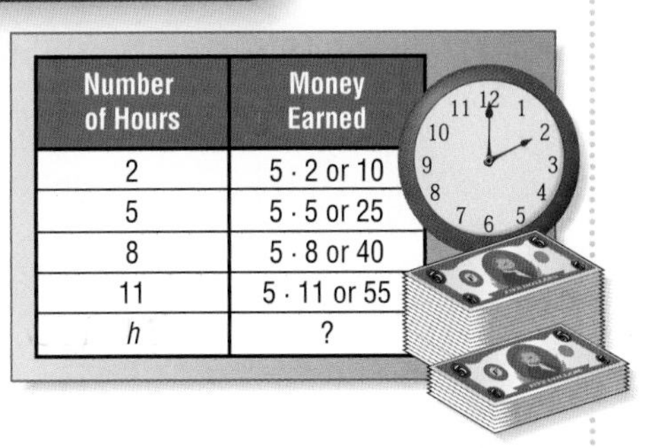

Number of Hours	Money Earned
2	5 · 2 or 10
5	5 · 5 or 25
8	5 · 8 or 40
11	5 · 11 or 55
h	?

a. Suppose the baby-sitter worked 10 hours. How much would he or she earn?

b. What is the relationship between the number of hours and the money earned?

c. If h represents *any number of hours*, what expression could you write to represent the amount of money earned?

Reading Math

Variable
Root Word: Vary
The word *variable* means *likely to change or vary.*

EVALUATE EXPRESSIONS Algebra is a language of symbols. One symbol that is frequently used is a variable. A **variable** is a placeholder for any value. As shown above, h represents some *unknown number of hours.*

Any letter can be used as a variable. Notice the special notation for multiplication and division with variables.

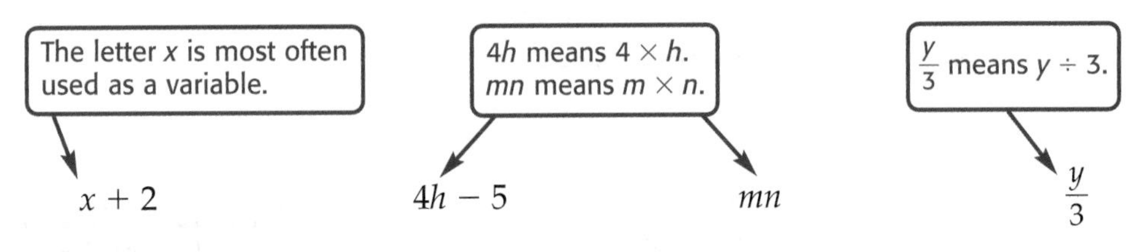

The letter x is most often used as a variable.
4h means 4 × h.
mn means m × n.
y/3 means y ÷ 3.
x + 2
4h − 5
mn
y/3

An expression like $x + 2$ is an **algebraic expression** because it contains sums and/or products of variables and numbers.

Concept Check *True* or *false:* 2x is an example of an algebraic expression. Explain your reasoning.

To evaluate an algebraic expression, replace the variable or variables with known values and then use the order of operations.

Example 1 Evaluate Expressions

Evaluate $x + y - 9$ if $x = 15$ and $y = 26$.

$x + y - 9 = 15 + 26 - 9$ Replace x with 15 and y with 26.
$= 41 - 9$ Add 15 and 26.
$= 32$ Subtract 9 from 41.

Replacing a variable with a number demonstrates the **Substitution Property of Equality**.

Key Concept — *Substitution Property of Equality*

- **Words** If two quantities are equal, then one quantity can be replaced by the other.
- **Symbols** For all numbers a and b, if $a = b$, then a may be replaced by b.

Example 2 Evaluate Expressions

Evaluate each expression if $k = 2$, $m = 7$, and $n = 4$.

a. $6m - 3k$

$6m - 3k = 6(7) - 3(2)$ Replace m with 7 and k with 2.
$= 42 - 6$ Multiply.
$= 36$ Subtract.

b. $\frac{mn}{2}$

$\frac{mn}{2} = mn \div 2$ Rewrite as a division expression.
$= (7 \cdot 4) \div 2$ Replace m with 7 and n with 4.
$= 28 \div 2$ Multiply.
$= 14$ Divide.

c. $n + (k + 5m)$

$n + (k + 5m) = 4 + (2 + 5 \cdot 7)$ Replace n with 4, k with 2, and m with 7.
$= 4 + (2 + 35)$ Multiply 5 and 7.
$= 4 + 37$ Add 2 and 35.
$= 41$ Add 4 and 37.

TRANSLATE VERBAL PHRASES The first step in translating verbal phrases into algebraic expressions is to choose a variable and a quantity for the variable to represent. This is called **defining a variable**.

Example 3 Translate Verbal Phrases into Expressions

Study Tip

Look Back
To review **key words and phrases**, see p. 11.

Translate each phrase into an algebraic expression.

a. twelve points more than the Dolphins scored

Words twelve points more than the Dolphins scored
Variable Let p represent the points the Dolphins scored.

	twelve points	more than	the Dolphins scored	
Expression	12	+	p	The expression is $p + 12$.

b. four times a number decreased by 6

Words four times a number decreased by 6
Variable Let n represent the number.

	four times a number	decreased by	six	
Expression	$4n$	−	6	The expression is $4n - 6$.

More About. . .

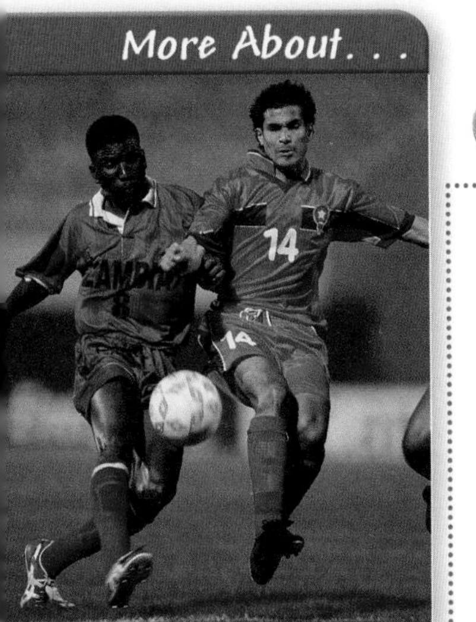

Soccer

Soccer is the fastest growing and most popular sport in the world. It is estimated that more than 100,000,000 people in more than 150 countries play soccer.

Source: *The World Almanac for Kids*

Algebraic expressions can be used to represent real-world situations.

Example 4 Use an Expression to Solve a Problem

SOCCER The Johnstown Soccer League ranks each team in their league using points. A team gets three points for a win and one point for a tie.

a. Write an expression that can be used to find the total number of points a team receives.

Words three points for a win and one point for a tie

Variables Let w = number of wins and t = number of ties.

Expression $\underbrace{3w}_{\text{three points for a win}} \underbrace{+}_{\text{and}} \underbrace{1t}_{\text{one point for a tie}}$

The expression $3w + 1t$ can be used to find the total number of points a team will receive.

b. Suppose in one season, the North Rockets had 17 wins and 4 ties. How many points did they receive?

$3w + 1t = 3(17) + 1(4)$ Replace w with 17 and t with 4.

$= 51 + 4$ Multiply.

$= 55$ Add.

The North Rockets received 55 points.

Check for Understanding

Concept Check

1. **OPEN ENDED** Give two examples of an algebraic expression and two examples of expressions that are *not* algebraic.
2. **Define** *variable.*
3. **Write** an expression that is the same as $4cd$.

Guided Practice

ALGEBRA Evaluate each expression if $a = 5$, $b = 12$, and $c = 4$.

4. $b + 6$
5. $18 - 3c$
6. $\frac{2b}{8}$
7. $5a - (b - c)$

ALGEBRA Translate each phrase into an algebraic expression.

8. eight more than the amount Kira saved
9. five goals less than the Pirates scored
10. the quotient of a number and four, minus five
11. seven increased by the quotient of a number and eight

Application

12. **SPACE** Due to gravity, objects weigh three times as much on Earth as they do on Mercury.
    a. Suppose the weight of an object on Mercury is w. Write an expression for the object's weight on Earth.
    b. How much would an object weigh on Earth if it weighs 25 pounds on Mercury?

Practice and Apply

Homework Help

For Exercises	See Examples
13–32, 43, 44	1, 2
33–42	3
48–50	4

Extra Practice
See page 724.

ALGEBRA **Evaluate each expression if $x = 7$, $y = 3$, and $z = 9$.**

13. $z + 2$ **14.** $5 + x$ **15.** $2 + 4z$

16. $15 - 2x$ **17.** $\frac{6y}{z}$ **18.** $\frac{9x}{y}$

19. $\frac{xy}{3} + 2$ **20.** $10 - \frac{xz}{9}$ **21.** $4z - 3y$

22. $3x - 2y$ **23.** $2x + 3z + 5y$ **24.** $5z - 3x - 2y$

25. $7z - (y + x)$ **26.** $(8y + 5) - 2z$ **27.** $3y + (7z - 4x)$

28. $6x - (z - 2y) + 15$ **29.** $2x + (4z - 13) - 5$ **30.** $(9 - 3y) + 4z - 5$

SCIENCE **For Exercises 31 and 32, use the following information.**
The number of times a cricket chirps can be used to estimate the temperature in degrees Fahrenheit. Use $c \div 4 + 37$ where c is the number of chirps in one minute.

31. Find the approximate temperature if a cricket chirps 136 times in a minute.

32. What is the temperature if a cricket chirps 100 times in a minute?

ALGEBRA **Translate each phrase into an algebraic expression.**

33. Mark's salary plus a $200 bonus

34. three more than the number of cakes baked

35. six feet shorter than the mountain's height

36. two seconds faster than Sarah's time

37. five times a number, minus four

38. seven less than a number times eight

39. nine more than a number divided by six

40. the quotient of eight and twice a number

41. the difference of seventeen and four times a number

42. three times the product of twenty-five and a number

43. Evaluate $\frac{10mn}{3p - 3}$ if $m = 6$, $n = 3$, and $p = 7$.

44. What is the value of $\frac{3(4a - 3b)}{b - 4}$ if $a = 6$ and $b = 7$?

ALGEBRA **Write an algebraic expression that represents the relationship in each table.**

45.

Age Now	Age in Three Years
10	13
12	15
15	18
20	23
x	■

46.

Number of Items	Total Cost
5	25
6	30
8	40
10	50
n	■

47.

Regular Price	Sale Price
$12	8
$15	11
$18	14
$24	20
$$p$	■

48. BUSINESS Cornet Cable charges \$32.50 a month for basic cable television. Each premium channel selected costs an additional \$4.95 per month. Write an expression to find the cost of a month of cable service.

SALES For Exercises 49 and 50, use the following information.
The selling price of a sweater is the cost plus the markup minus the discount.

49. Write an expression to show the selling price s of a sweater. Use c for cost, m for markup, and d for discount.

50. Suppose the cost of a sweater is \$25, the markup is \$20, and the discount is \$6. What is the selling price of the sweater?

51. CRITICAL THINKING What value of t makes the expressions $6t$, $t + 5$, and $2t + 4$ equal?

52. WRITING IN MATH Answer the question that was posed at the beginning of the lesson.

How are variables used to show relationships?

Include the following in your answer:

- an explanation of variables and what they represent, and
- an example showing how variables are used to show relationships.

FCAT Practice
Standardized Test Practice

53. If the value of $c + 5$ is 18, what is the value of c?

Ⓐ 3 Ⓑ 8 Ⓒ 7 Ⓓ 13

54. Which expression represents *four less than twice a number*?

Ⓐ $4n - 2$ Ⓑ $2n - 4$ Ⓒ $4(2 + n)$ Ⓓ $2n + 4$

Maintain Your Skills

Mixed Review

Find the value of each expression. *(Lesson 1-2)*

55. $3 + (6 \times 2) - 8$ **56.** $5(16 - 5 \times 3)$ **57.** $36 \div (9 \cdot 2) + 7$

58. FOOD The table shows the amount in pounds of certain types of pasta sold in a recent year. About how many million pounds of these types of pasta were sold? *(Lesson 1-1)*

Pasta	Amount (millions)
Spaghetti	308
Elbow	121
Noodles	70
Twirl	52
Penne	51
Lasagna	35
Fettuccine	24

Source: *National Pasta Association*

Getting Ready for the Next Lesson

BASIC SKILL Find each difference.

59. $53 - 17$ **60.** $97 - 28$ **61.** $104 - 82$ **62.** $152 - 123$

Practice Quiz 1

Lessons 1-1 through 1-3

1. What is the next term in the list 4, 5, 7, 10, ...?

Find the value of each expression. *(Lesson 1-2)*

2. $28 \div 4 \times 2$ **3.** $7(3 + 10) - 2 \cdot 6$ **4.** $3[6(12 - 3)] - 17$

5. Evaluate $7x - 3y$ if $x = 4$ and $y = 2$. *(Lesson 1-3)*

Spreadsheet Investigation

A Follow-Up of Lesson 1-3

Expressions and Spreadsheets

One of the most common computer applications is a spreadsheet program. A **spreadsheet** is a table that performs calculations. It is organized into boxes called **cells**, which are named by a letter and a number. In the spreadsheet below, cell B1 is highlighted.

An advantage of using a spreadsheet is that values in the spreadsheet are recalculated when a number is changed. You can use a spreadsheet to investigate patterns in data.

Example

Here's a mind-reading trick! Think of a number. Then double it, add six, divide by two, and subtract the original number. What is the result?

You can use a spreadsheet to test different numbers. Suppose we start with the number 10.

Mind-Reading Trick

	A	B	C
1	Think of a number.	10	10
2	Double it.	2*B1	20
3	Add 6.	B2+6	26
4	Divide by 2.	B3/2	13
5	Subtract the original number.	B4-B1	3
6			

Sheet1 Sheet2 Sheet3

Ready

The spreadsheet takes the value in B1, doubles it, and enters the value in B2. Note the * is the symbol for multiplication.

The spreadsheet takes the value in B3, divides by 2, and enters the value in B4. Note that / is the symbol for division.

The result is 3.

Exercises

To change information in a spreadsheet, move the cursor to the cell you want to access and click the mouse. Then type in the information and press Enter. Find the result when each value is entered in B1.

1. 6 **2.** 8 **3.** 25 **4.** 100 **5.** 1500

Make a Conjecture

6. What is the result if a decimal is entered in B1? a negative number?

7. Explain why the result is always 3.

8. Make up your own mind-reading trick. Enter it into a spreadsheet to show that it works.

1-4 Properties

What You'll Learn

- Identify and use properties of addition and multiplication.
- Use properties of addition and multiplication to simplify algebraic expressions.

Sunshine State Standards
MA.A.3.3.1-3

Vocabulary

- properties
- counterexample
- simplify
- deductive reasoning

How are real-life situations commutative?

Abraham Lincoln delivered the Gettysburg Address more than 130 years ago. The table lists the number of words in certain historic documents.

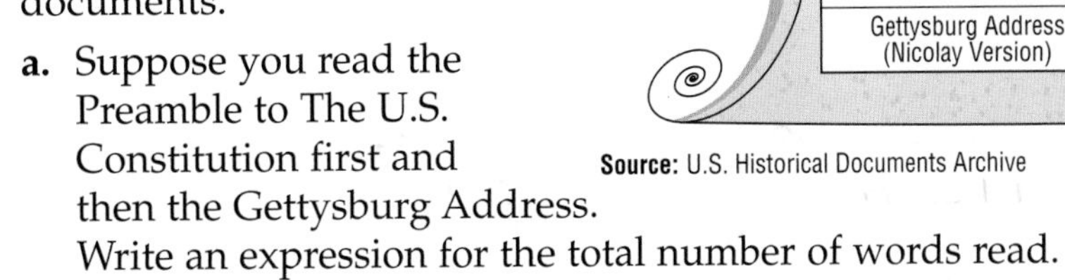

Historical Document	Words
Preamble to The U.S.Constitution	52
Mayflower Compact	196
Atlantic Charter	375
Gettysburg Address (Nicolay Version)	238

Source: U.S. Historical Documents Archive

a. Suppose you read the Preamble to The U.S. Constitution first and then the Gettysburg Address. Write an expression for the total number of words read.

b. Suppose you read the Gettysburg Address first and then the Preamble to the U.S. Constitution. Write an expression for the total number of words read.

c. Find the value of each expression. What do you observe?

d. Does it matter in which order you add any two numbers? Why or why not?

PROPERTIES OF ADDITION AND MULTIPLICATION In algebra, **properties** are statements that are true for any numbers. For example, the expressions $3 + 8$ and $8 + 3$ have the same value, 11. This illustrates the **Commutative Property of Addition**. Likewise, $3 \cdot 8$ and $8 \cdot 3$ have the same value, 24. This illustrates the **Commutative Property of Multiplication**.

Reading Math

Commutative

Root Word: Commute

The everyday meaning of the word *commute* means *to change or exchange.*

Key Concept — Commutative Property of Addition

- **Words** The order in which numbers are added does not change the sum.
- **Symbols** For any numbers a and b, $a + b = b + a$.
- **Example** $2 + 3 = 3 + 2$
  $5 = 5$

Commutative Property of Multiplication

- **Words** The order in which numbers are multiplied does not change the product.
- **Symbols** For any numbers a and b, $a \cdot b = b \cdot a$.
- **Example** $2 \cdot 3 = 3 \cdot 2$
  $6 = 6$

Concept Check Write an example that shows the Commutative Property of Multiplication.

Reading Math

Associative

Root Word: Associate

The word *associate* means *to join together, connect, or combine.*

When evaluating expressions, it is often helpful to group or *associate* the numbers. The **Associative Property** says that the way in which numbers are grouped when added or multiplied does not change the sum or the product.

The Associative Property also holds true when multiplying numbers.

Key Concept — *Associative Property of Addition*

- **Words** The way in which numbers are grouped when added does not change the sum.
- **Symbols** For any numbers a, b, and c, $(a + b) + c = a + (b + c)$.
- **Example** $(5 + 8) + 2 = 5 + (8 + 2)$
  $13 + 2 = 5 + 10$
  $15 = 15$

Associative Property of Multiplication

- **Words** The way in which numbers are grouped when multiplied does not change the product.
- **Symbols** For any numbers a, b, and c, $(a \cdot b) \cdot c = a \cdot (b \cdot c)$.
- **Example** $(4 \cdot 6) \cdot 3 = 4 \cdot (6 \cdot 3)$
  $24 \cdot 3 = 4 \cdot 18$
  $72 = 72$

Concept Check Write an example showing the Associative Property of Addition.

The following properties are also true.

Key Concept — *Properties of Numbers*

Property	Words	Symbols	Examples
Additive Identity	When 0 is added to any number, the sum is the number.	For any number a, $a + 0 = 0 + a = a$.	$5 + 0 = 5$ $0 + 9 = 9$
Multiplicative Identity	When any number is multiplied by 1, the product is the number.	For any number a, $a \cdot 1 = 1 \cdot a = a$.	$7 \cdot 1 = 7$ $1 \cdot 6 = 6$
Multiplicative Property of Zero	When any number is multiplied by 0, the product is 0.	For any number a, $a \cdot 0 = 0 \cdot a = 0$.	$4 \cdot 0 = 0$ $0 \cdot 2 = 0$

Example 1 *Identify Properties*

Name the property shown by each statement.

a. $3 + 7 + 9 = 7 + 3 + 9$

The order of the numbers changed. This is the Commutative Property of Addition.

b. $(a \cdot 6) \cdot 5 = a \cdot (6 \cdot 5)$

The grouping of the numbers and variables changed. This is the Associative Property of Multiplication.

c. $0 \cdot 12 = 0$

The number was multiplied by zero. This is the Multiplicative Property of Zero.

You can use the properties of numbers to find sums and products mentally. Look for sums or products that end in zero.

Example 2 Mental Math

Find $4 \cdot (25 \cdot 11)$ mentally.

Group 4 and 25 together because $4 \cdot 25 = 100$. It is easy to multiply by 100 mentally.

$4 \cdot (25 \cdot 11) = (4 \cdot 25) \cdot 11$	Associative Property of Addition
$= 100 \cdot 11$	Multiply 4 and 25 mentally.
$= 1100$	Multiply 100 and 11 mentally.

Study Tip

Counterexample
You can disprove a statement by finding only one counterexample.

You may wonder whether these properties apply to subtraction. One way to find out is to look for a counterexample. A **counterexample** is an example that shows a conjecture is not true.

Example 3 Find a Counterexample

State whether the following conjecture is *true* or *false*. If false, provide a counterexample.

Subtraction of whole numbers is associative.

Write two subtraction expressions using the Associative Property, and then check to see whether they are equal.

$9 - (5 - 3) \stackrel{?}{=} (9 - 5) - 3$	State the conjecture.
$9 - 2 \stackrel{?}{=} 4 - 3$	Simplify within the parentheses.
$7 \neq 1$	Subtract.

We found a counterexample. That is, $9 - (5 - 3) \neq (9 - 5) - 3$. So, subtraction is *not* associative. The conjecture is false.

ALGEBRA CONNECTION

SIMPLIFY ALGEBRAIC EXPRESSIONS To **simplify** algebraic expressions means to write them in a simpler form. You can use the Associative or Commutative Properties to simplify expressions.

Example 4 Simplify Algebraic Expressions

Simplify each expression.

a. $(k + 2) + 7$

$(k + 2) + 7 = k + (2 + 7)$	Associative Property of Addition
$= k + 9$	Substitution Property of Equality; $2 + 7 = 9$

b. $5 \cdot (d \cdot 9)$

$5 \cdot (d \cdot 9) = 5 \cdot (9 \cdot d)$	Commutative Property of Multiplication
$= (5 \cdot 9)d$	Associative Property of Multiplication
$= 45d$	Substitution Property of Equality; $5 \cdot 9 = 45$

Study Tip

Inductive Reasoning
In inductive reasoning, conclusions are made based on past events or patterns.

Notice that each step in Example 4 was justified by a property. The process of using facts, properties, or rules to justify reasoning or reach valid conclusions is called **deductive reasoning**.

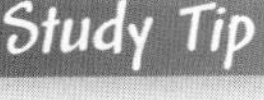

Check for Understanding

Concept Check

1. **OPEN ENDED** Write a numerical sentence that illustrates the Commutative Property of Multiplication.
2. **Tell** the difference between the Commutative and Associative Properties.
3. **FIND THE ERROR** Kimberly and Carlos are using the Associative Properties of Addition and Multiplication to rewrite expressions.

Kimberly	Carlos
$(4 + 3) + 6 = 4 + (3 + 6)$	$(2 + 7) \cdot 5 = 2 + (7 \cdot 5)$

Who is correct? Explain your reasoning.

Guided Practice

Name the property shown by each statement.

4. $7 + 5 = 5 + 7$
5. $8 + 0 = 8$
6. $8 \cdot 4 \cdot 13 = 4 \cdot 8 \cdot 13$

Find each sum or product mentally.

7. $13 + 8 + 7$
8. $6 \cdot 9 \cdot 5$
9. $8 + 11 + 22 + 4$

10. State whether the conjecture *division of whole numbers is commutative* is *true* or *false*. If false, provide a counterexample.

ALGEBRA Simplify each expression.

11. $6 + (n + 7)$
12. $(3 \cdot w) \cdot 9$

Application

13. **SHOPPING** Denyce purchased a pair of jeans for \$26, a T-shirt for \$12, and a pair of socks for \$4. What is the total cost of the items? Explain how the Commutative Property of Addition can be used to find the total.

Practice and Apply

Homework Help

For Exercises	See Examples
14–25	1
26–34	2
35–37	3
39–47	4

Extra Practice

See page 725.

Name the property shown by each statement.

14. $5 \cdot 3 = 3 \cdot 5$
15. $1 \cdot 4 = 4$
16. $6 \cdot 2 \cdot 0 = 0$
17. $12 \cdot 8 = 8 \cdot 12$
18. $0 + 13 = 13 + 0$
19. $(4 + 5) + 15 = 4 + (5 + 15)$
20. $1h = h$
21. $7k + 0 = 7k$
22. $(5 + x) + 6 = 5 + (x + 6)$
23. $4(mn) = (4m)(n)$
24. $9(gh) = (9g)h$
25. $(3a + b) + 2c = 2c + (3a + b)$

Find each sum or product mentally.

26. $11 + 8 + 19$
27. $17 + 5 + 33$
28. $15 \cdot 0 \cdot 2$
29. $5 + 18 + 15 + 2$
30. $2 \cdot 7 \cdot 30$
31. $11 \cdot 9 \cdot 10$
32. $23 + 3 + 17 + 7$
33. $125 \cdot 4 \cdot 0$
34. $16 + 57 + 94 + 33$

State whether each conjecture is *true* or *false*. If false, provide a counterexample.

35. Division of whole numbers is associative.
36. The sum of two whole numbers is always greater than either addend.
37. Subtraction of whole numbers is commutative.

38. **SCIENCE** In chemistry, water is used to dilute acid. Since pouring water into acid could cause spattering and burns, it's important to pour the acid into the water. Is combining acid and water commutative? Explain.

More About. . .

Science

A water molecule is composed of two atoms of hydrogen and one atom of oxygen. Thus, in chemistry, the formula for water is H_2O.

Source: *Merrill Chemistry*

ALGEBRA Simplify each expression.

39. $(m + 8) + 4$
40. $(17 + p) + 9$
41. $15 + (12 + a)$
42. $21 + (k + 16)$
43. $6 \cdot (y \cdot 2)$
44. $7 \cdot (d \cdot 4)$
45. $(6 \cdot c) \cdot 8$
46. $(3 \cdot w) \cdot 5$
47. $25s(3)$

48. **CRITICAL THINKING** The **Closure Property** states that because the sum or product of two whole numbers (0, 1, 2, 3, ...) is also a whole number, the set of whole numbers is *closed* under addition and multiplication. Tell whether the set of whole numbers is closed under subtraction and division. If not, give counterexamples.

49. **WRITING IN MATH** Answer the question that was posed at the beginning of the lesson.

How are real-life situations commutative?

Include the following in your answer:

- an example of a real-life situation that is commutative,
- an example of a real-life situation that is not commutative, and
- an explanation of why each situation is or is not commutative.

FCAT Practice
Standardized Test Practice

50. The statement $e + (f + g) = (f + g) + e$ is an example of which property of addition?

(A) Commutative
(B) Associative
(C) Identity
(D) Substitution

51. Rewrite the expression $(7 \cdot m) \cdot 8$ using the Associative Property.

(A) $(8 \cdot 7) \cdot m$
(B) $7 \cdot (m \cdot 8)$
(C) $8 \cdot (7 \cdot m)$
(D) $7 \cdot m \cdot 8$

3, 4, 18 21
6

Maintain Your Skills

Mixed Review **ALGEBRA Evaluate each expression if $a = 6$, $b = 4$, and $c = 5$.** *(Lesson 1-3)*

52. $a + c - b$
53. $8a - 3b$
54. $4a - (b + c)$

55. Translate the phrase *the difference of w and 12* into an algebraic expression. *(Lesson 1-3)*

Find the value of each expression. *(Lesson 1-2)*

56. $7 - 2 \times 3$
57. $21 \div 3 \times 5$
58. $4 \cdot (8 + 9) + 6$

59. Find the next two terms in the list 0, 1, 3, 6, 10, ... *(Lesson 1-1)*

Getting Ready for the Next Lesson **BASIC SKILL Find each product.**

60. 48×5
61. 8×37
62. 16×12
63. 25×42
64. 106×13
65. 59×127

1-5 Variables and Equations

Sunshine State Standards
MA.D.1.3.1-2, MA.D.1.3.2-2, MA.D.2.3.1-1

Vocabulary

- equation
- open sentence
- solution
- solving the equation

What You'll Learn

- Identify and solve open sentences.
- Translate verbal sentences into equations.

How is solving an open sentence similar to evaluating an expression?

Emilio is seven years older than his sister Rebecca.

a. If Rebecca is x years old, what expression represents Emilio's age?

Suppose Emilio is 19 years old. You can write a mathematical sentence that shows two expressions are equal.

Words Emilio's age is 19.

Symbols $x + 7 = 19$

b. What two expressions are equal?

c. If Emilio is 19, how old is Rebecca?

EQUATIONS AND OPEN SENTENCES A mathematical sentence that contains an equals sign (=) is called an **equation**. A few examples are shown.

$5 + 9 = 14$ $\qquad 2(6) - 3 = 9$ $\qquad x + 7 = 19$ $\qquad 2m - 1 = 13$

An equation that contains a variable is an **open sentence**. An open sentence is neither true nor false. When the variable in an open sentence is replaced with a number, you can determine whether the sentence is true or false.

$x + 7 = 19$

$11 + 7 \stackrel{?}{=} 19$ Replace x with 11.

$18 \neq 19$ false

When $x = 11$, this sentence is false.

$x + 7 = 19$

$12 + 7 \stackrel{?}{=} 19$ Replace x with 12.

$19 = 19$ true

When $x = 12$, this sentence is true.

Study Tip

Symbols
The symbol $\neq$ means *is not equal to*.

A value for the variable that makes an equation true is called a **solution**. For $x + 7 = 19$, the solution is 12. The process of finding a solution is called **solving the equation**.

Example 1 Solve an Equation

Find the solution of $12 - m = 8$. Is it 2, 4, or 7?

Replace m with each value.

Value for m	$12 - m = 8$	True or False?
2	$12 - 2 \stackrel{?}{=} 8$	false
4	$12 - 4 \stackrel{?}{=} 8$	true ✓
7	$12 - 7 \stackrel{?}{=} 8$	false

Therefore, the solution of $12 - m = 8$ is 4.

Most standardized tests include questions that ask you to solve equations.

Example 2 Solve an Equation

Multiple-Choice Test Item

Which value is the solution of $2x + 1 = 7$?

Ⓐ 6 Ⓑ 5 Ⓒ 4 Ⓓ 3

Read the Test Item

The *solution* is the value that makes the equation true.

Solve the Test Item Test each value.

$2x + 1 = 7$
$2(6) + 1 = 7$ Replace x with 6.
$13 \neq 7$

$2x + 1 = 7$
$2(5) + 1 = 7$ Replace x with 5.
$11 \neq 7$

$2x + 1 = 7$
$2(4) + 1 = 7$ Replace x with 4.
$9 \neq 7$

$2x + 1 = 7$
$2(3) + 1 = 7$ Replace x with 3.
$7 = 7$ ✓

Since 3 makes the equation true, the answer is D.

The Princeton Review

Test-Taking Tip

The strategy of testing each value is called *backsolving*. You can also use this strategy with complex equations.

Example 3 Solve Simple Equations Mentally

Solve each equation mentally.

a. $5x = 30$

$5 \cdot 6 = 30$ Think: What number times 5 is 30?
$x = 6$ The solution is 6.

b. $\frac{72}{d} = 8$

$\frac{72}{9} = 8$ Think: 72 divided by what number is 8?
$d = 9$ The solution is 9.

In Lesson 1-4, you learned that certain properties are true for any number. Two properties of equality are shown below.

Reading Math

Symmetric
Root Word: Symmetry
The word *symmetry* means *similarity of form or arrangement on either side.*

Key Concept — Properties of Equality

Property	Words	Symbols	Example
Symmetric	If one quantity equals a second quantity, then the second quantity also equals the first.	For any numbers a and b, if $a = b$, then $b = a$.	If $10 = 4 + 6$, then $4 + 6 = 10$.
Transitive	If one quantity equals a second quantity and the second quantity equals a third quantity, then the first equals the third.	For any numbers a, b, and c, if $a = b$ and $b = c$, then $a = c$.	If $3 + 5 = 8$ and $8 = 2(4)$, then $3 + 5 = 2(4)$.

Example 4 Identify Properties of Equality

Name the property of equality shown by each statement.

a. If $5 = x + 2$, then $x + 2 = 5$.

If $a = b$, then $b = a$. This is the Symmetric Property of Equality.

b. If $y + 8 = 15$ and $15 = 7 + 8$, then $y + 8 = 7 + 8$.

If $a = b$ and $b = c$, then $a = c$. This is the Transitive Property of Equality.

TRANSLATE VERBAL SENTENCES INTO EQUATIONS Just as verbal phrases can be translated into algebraic expressions, verbal sentences can be translated into equations and then solved.

Example 5 Translate Sentences Into Equations

The difference of a number and ten is seventeen. Find the number.

Words The difference of a number and ten is seventeen.

Variables Let n = the number. Define the variable.

Equation

The difference of a number and ten	is	seventeen.
$n - 10$	$=$	17

$n - 10 = 17$ Write the equation.

$27 - 10 = 17$ Think: What number minus 10 is 17?

$n = 27$ The solution is 27.

Check for Understanding

Concept Check

1. **OPEN ENDED** Write two different equations whose solutions are 5.
2. **Tell** what it means to *solve an equation.*

Guided Practice

ALGEBRA Find the solution of each equation from the list given.

3. $h + 15 = 21$; 5, 6, 7
4. $13 - m = 4$; 7, 8, 9

ALGEBRA Solve each equation mentally.

5. $a + 8 = 13$
6. $12 - d = 9$
7. $3x = 18$
8. $4 = \frac{36}{t}$

Name the property of equality shown by each statement.

9. If $x + 4 = 9$, then $9 = x + 4$.
10. If $5 + 7 = 12$ and $12 = 3 \cdot 4$, then $5 + 7 = 3 \cdot 4$.

ALGEBRA Define a variable. Then write an equation and solve.

11. A number increased by 8 is 23.
12. Twenty-five is 10 less than a number.

FCAT Practice

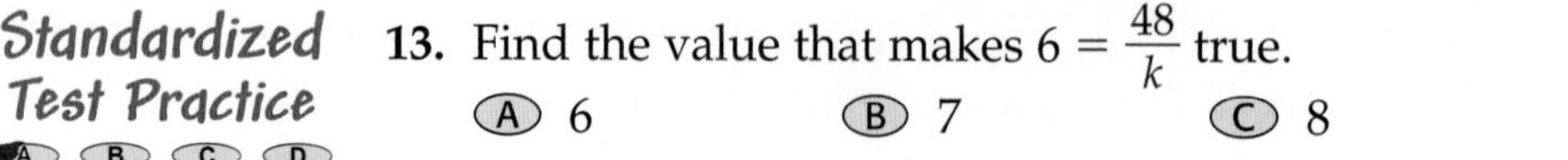

Standardized Test Practice

13. Find the value that makes $6 = \frac{48}{k}$ true.

Ⓐ 6 Ⓑ 7 Ⓒ 8 Ⓓ 12

Practice and Apply

Homework Help

For Exercises	See Examples
14–23	1
26–41	3
42–49, 54, 55	5
50–53	4

Extra Practice

See page 725.

ALGEBRA **Find the solution of each equation from the list given.**

14. $c + 12 = 30$; 8, 16, 18

15. $g + 17 = 28$; 9, 11, 13

16. $23 - m = 14$; 7, 9, 11

17. $18 - k = 6$; 8, 10, 12

18. $14k = 42$; 2, 3, 4

19. $75 = 15n$; 3, 4, 5

20. $\frac{51}{z} = 3$; 15, 16, 17

21. $\frac{60}{p} = 4$; 15, 16, 17

22. What is the solution of $3n + 13 = 25$; 2, 3, 4?

23. Find the solution of $7 = 4w - 29$. Is it 8, 9, or 10?

Tell whether each sentence is *sometimes, always,* or *never* true.

24. An equation is an open sentence.

25. An open sentence contains a variable.

ALGEBRA **Solve each equation mentally.**

26. $d + 7 = 12$

27. $19 = 4 + y$

28. $8 + j = 27$

29. $22 + b = 22$

30. $20 - p = 11$

31. $15 - m = 0$

32. $16 = x - 7$

33. $12 = y - 5$

34. $7s = 49$

35. $8c = 88$

36. $63 = 9h$

37. $72 = 8w$

38. $\frac{30}{r} = 3$

39. $\frac{24}{y} = 8$

40. $12 = \frac{36}{p}$

41. $14 = \frac{56}{d}$

ALGEBRA **Define a variable. Then write an equation and solve.**

42. The sum of 7 and a number is 23.

43. A number minus 10 is 27.

44. Twenty-four is the product of 8 and a number.

45. The sum of 9 and a number is 36.

46. The difference of a number and 12 is 54.

47. A number times 3 is 45.

More About. . .

Movie Industry

In 1990, the total number of indoor movie screens was about 22,000. Today, there are over 37,000 indoor movie screens and the number keeps rising.

Source: National Association of Theatre Owners

MOVIE INDUSTRY **For Exercises 48 and 49, use the following information.**
Megan purchased movie tickets for herself and two friends. The cost was $24.

48. Define a variable. Then write an equation that can be used to find how much Megan paid for each ticket.

49. What was the cost of each ticket?

Name the property of equality shown by each statement.

50. If $2 + 3 = 5$ and $5 = 1 + 4$, then $2 + 3 = 1 + 4$.

51. If $3 + 4 = 7$ then $7 = 3 + 4$.

52. If $(1 + 2) + 6 = 9$, then $9 = (1 + 2) + 6$.

53. If $m + n = p$, then $p = m + n$.

HEIGHT **For Exercises 54 and 55, use the following information.**
Sean grew from a height of 65 inches to a height of 68 inches.

54. Define a variable. Then write an equation that can be used to find the increase in height.

55. How many inches did Sean grow?

56. **CRITICAL THINKING** Write three different equations in which there is no solution that is a whole number.

57. **WRITING IN MATH** Answer the question that was posed at the beginning of the lesson.

How is solving an open sentence similar to evaluating an expression?

Include the following in your answer:

- an explanation of how to evaluate an expression, and
- an explanation of what makes an open sentence true.

58. Find the solution of $9m = 54$.

Ⓐ 4 Ⓑ 7 Ⓒ 5 Ⓓ 6

59. Which value satisfies $2n - 5 = 19$?

Ⓐ 11 Ⓑ 12 Ⓒ 13 Ⓓ 14

Extending the Lesson

60. The table shows equations that have one variable or two variables.

One Variable	Two Variables
$4 + x = 7$	$z + y = 7$
$3t = 24$	$ab = 24$
$s - 5 = 2$	$m - n = 2$

a. Find as many whole number solutions as you can for each equation.

b. **Make a conjecture** about the relationship between the number of variables in equations like the ones above and the number of solutions.

Maintain Your Skills

Mixed Review

Simplify each expression. *(Lesson 1-4)*

61. $16 + (7 + d)$

62. $(4 \cdot p) \cdot 6$

ALGEBRA Translate each phrase into an algebraic expression. *(Lesson 1-3)*

63. ten decreased by a number

64. the sum of three times a number and four

Find the value of each expression. *(Lesson 1-2)*

65. $3 \cdot 7 - 2(1 + 4)$

66. $3[(17 - 7) - 2(3)]$

67. What is the next term in 67, 62, 57, 52, 47, …? *(Lesson 1-1)*

Getting Ready for the Next Lesson

PREREQUISITE SKILL Evaluate each expression for the given value.

*(To review **evaluating expressions**, see Lesson 1-5.)*

68. $4x; x = 3$

69. $3m; m = 6$

70. $2d; d = 8$

71. $5c; c = 10$

72. $8a; a = 9$

73. $6y; y = 15$

Practice Quiz 2

Lessons 1-4 and 1-5

Name the property shown by each statement. *(Lesson 1-4)*

1. $6 \cdot 1 = 6$

2. $9 + 6 = 6 + 9$

3. Simplify $8 \cdot (h \cdot 3)$. *(Lesson 1-4)*

4. Find the solution of $2w - 6 = 14$. Is it 8, 10, or 12? *(Lesson 1-4)*

5. Solve $72 = 9x$ mentally. *(Lesson 1-5)*

1-6 Ordered Pairs and Relations

Sunshine State Standards
MA.C.3.3.2-1, MA.D.1.3.2-1, MA.D.2.3.1-1

Vocabulary

- coordinate system
- *y*-axis
- coordinate plane
- origin
- *x*-axis
- ordered pair
- *x*-coordinate
- *y*-coordinate
- graph
- relation
- domain
- range

What You'll Learn

- Use ordered pairs to locate points.
- Use tables and graphs to represent relations.

How are ordered pairs used to graph real-life data?

Maria and Hiroshi are playing a game. The player who gets four Xs or Os in a row wins.

1st move Maria places an X at 1 over and 3 up.
2nd move Hiroshi places an O at 2 over and 2 up.
3rd move Maria places an X at 1 over and 1 up.
4th move Hiroshi places an O at 1 over and 2 up.

Starting Position

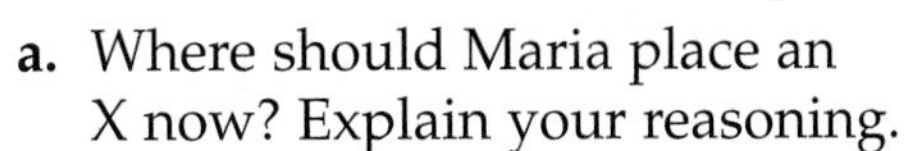

a. Where should Maria place an X now? Explain your reasoning.

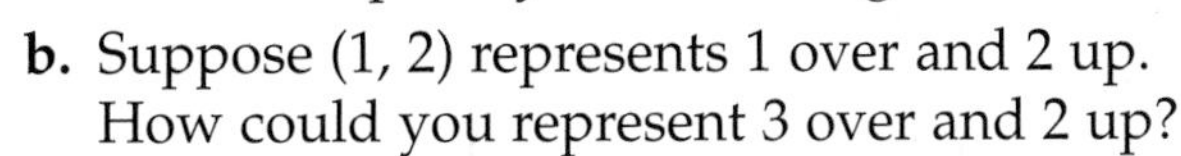

b. Suppose (1, 2) represents 1 over and 2 up. How could you represent 3 over and 2 up?
c. How are (5, 1) and (1, 5) different?
d. Where is a good place to put the next O?
e. Work with a partner to finish the game.

ORDERED PAIRS In mathematics, a **coordinate system** is used to locate points. The coordinate system is formed by the intersection of two number lines that meet at right angles at their zero points.

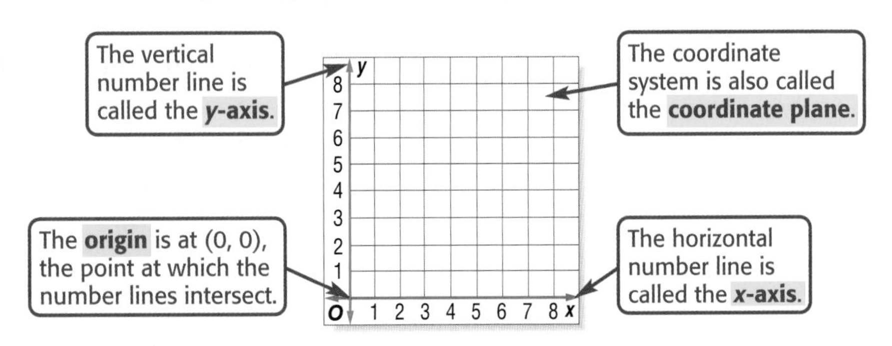

The vertical number line is called the *y*-axis.
The coordinate system is also called the coordinate plane.
The origin is at (0, 0), the point at which the number lines intersect.
The horizontal number line is called the *x*-axis.
y
8
7
6
5
4
3
2
1
O
1 2 3 4 5 6 7 8
x

An **ordered pair** of numbers is used to locate any point on a coordinate plane. The first number is called the ***x*-coordinate**. The second number is called the ***y*-coordinate**.

The *x*-coordinate corresponds to a number on the *x*-axis. (3, 2) The *y*-coordinate corresponds to a number on the *y*-axis.

To **graph** an ordered pair, draw a dot at the point that corresponds to the ordered pair. The coordinates are your directions to locate the point.

Study Tip

Coordinate System
You can assume that each unit on the *x*- and *y*-axis represents 1 unit. *Axes* is the plural of *axis*.

Example 1 Graph Ordered Pairs

Graph each ordered pair on a coordinate system.

a. (4, 1)

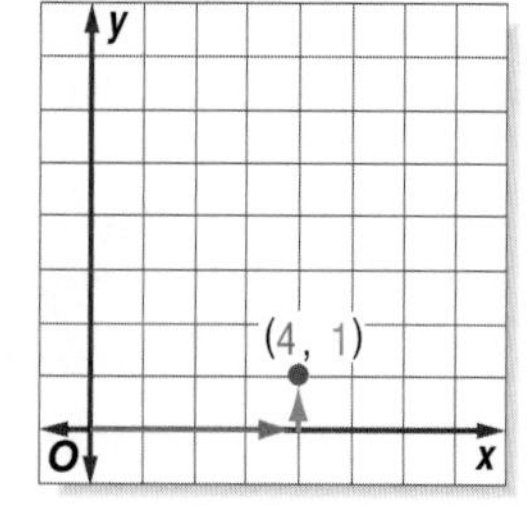

Step 1 Start at the origin.

Step 2 Since the x-coordinate is 4, move 4 units to the right.

Step 3 Since the y-coordinate is 1, move 1 unit up. Draw a dot.

b. (3, 0)

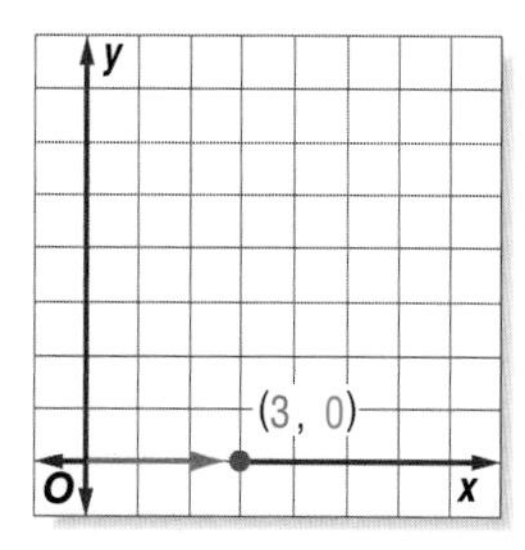

Step 1 Start at the origin.

Step 2 The x-coordinate is 3. So, move 3 units to the right.

Step 3 Since the y-coordinate is 0, you will not need to move up. Place the dot on the axis.

Concept Check Where is the graph of (0, 4) located?

Sometimes a point on a graph is named by using a letter. To identify its location, you can write the ordered pair that represents the point.

Example 2 Identify Ordered Pairs

Write the ordered pair that names each point.

a. M

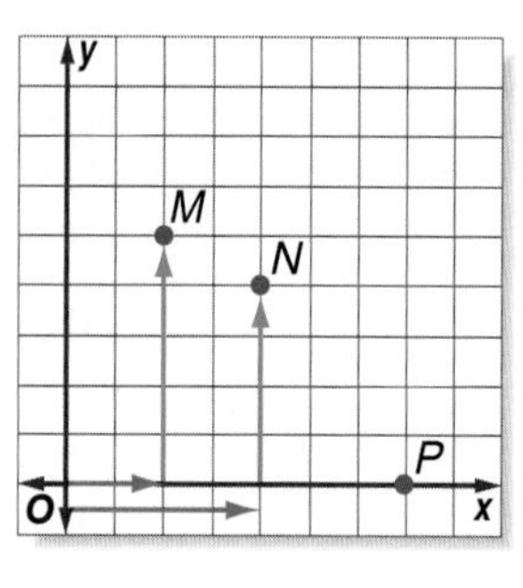

Step 1 Start at the origin.

Step 2 Move right on the x-axis to find the x-coordinate of point M, which is 2.

Step 3 Move up the y-axis to find the y-coordinate, which is 5.

The ordered pair for point M is (2, 5).

b. N

The x-coordinate of N is 4, and the y-coordinate is 4.

The ordered pair for point N is (4, 4).

c. P

The x-coordinate of P is 7, and the y-coordinate is 0.

The ordered pair for point P is (7, 0).

RELATIONS A set of ordered pairs such as {(1, 2), (2, 4), (3, 0), (4, 5)} is a **relation**. The **domain** of the relation is the set of x-coordinates. The **range** of the relation is the set of y-coordinates.

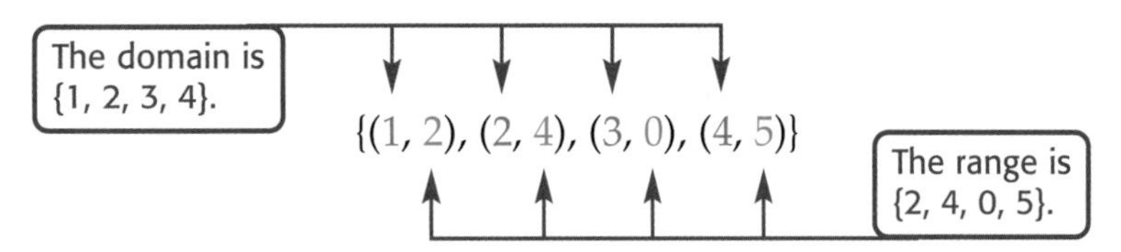

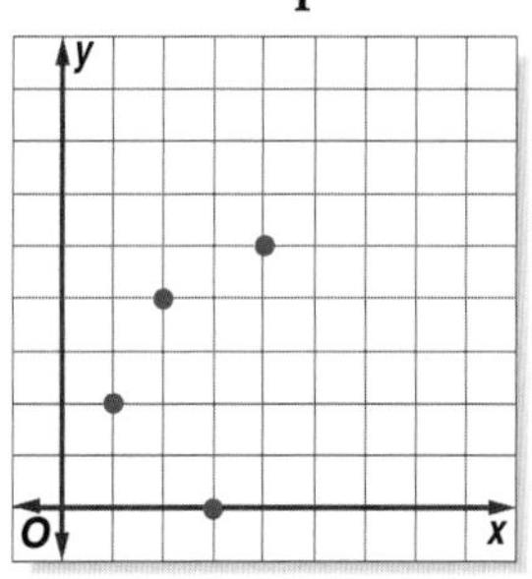

A relation can be shown in several ways.

Ordered Pairs

(1, 2)

(2, 4)

(3, 0)

(4, 5)

Table

x	y
1	2
2	4
3	0
4	5

Graph

Example 3 Relations as Tables and Graphs

Express the relation {(0, 0), (2, 1), (1, 3), (5, 2)} as a table and as a graph. Then determine the domain and range.

x	y
0	0
2	1
1	3
5	2

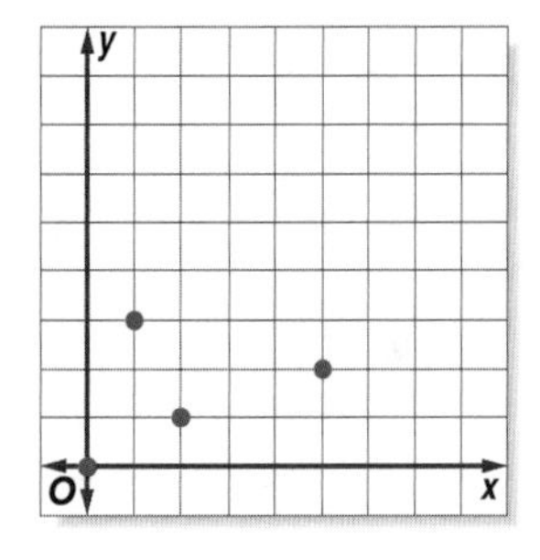

The domain is {0, 2, 1, 5}, and the range is {0, 1, 3, 2}.

Example 4 Apply Relations

PLANTS Some species of bamboo grow 3 feet in one day.

a. Make a table of ordered pairs in which the x-coordinate represents the number of days and the y-coordinate represents the amount of growth for 1, 2, 3, and 4 days.

x	y	(x, y)
1	3	(1, 3)
2	6	(2, 6)
3	9	(3, 9)
4	12	(4, 12)

b. Graph the ordered pairs.

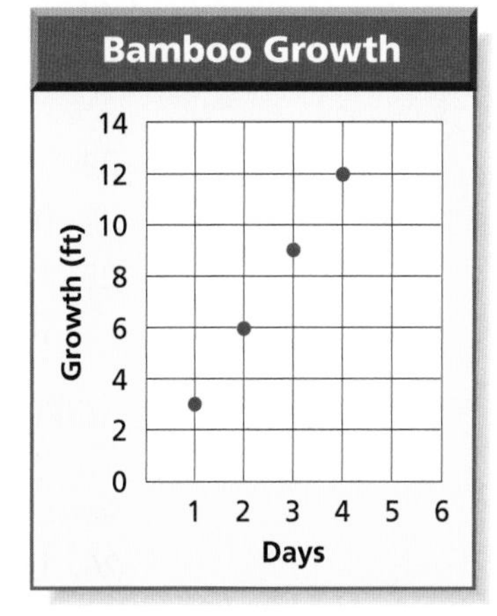

c. Describe the graph.

The points appear to fall in a line.

More About. . .

Plants

Bamboo is a type of grass. It can vary in height from one-foot dwarf plants to 100-foot giant timber plants.

Source: American Bamboo Society

Check for Understanding

Concept Check

1. **OPEN ENDED** Give an example of an ordered pair, and identify the x- and y-coordinate.
2. **Name** three ways to represent a relation.
3. **Define** *domain* and *range.*

Guided Practice

Graph each point on a coordinate system.

4. $H(5, 3)$
5. $D(6, 0)$

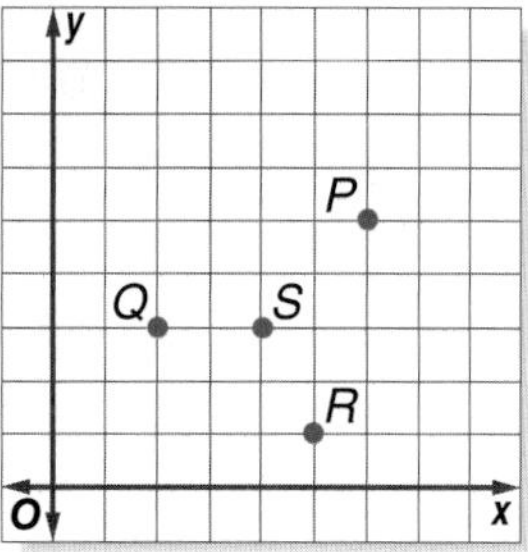

Refer to the coordinate system shown at the right. Write the ordered pair that names each point.

6. Q
7. P

Express each relation as a table and as a graph. Then determine the domain and range.

8. $\{(2, 5), (0, 2), (5, 5)\}$
9. $\{(1, 6), (6, 4), (0, 2), (3, 1)\}$

Application

ENTERTAINMENT **For Exercises 10 and 11, use the following information.** It costs $4 to buy a student ticket to the movies.

10. Make a table of ordered pairs in which the x-coordinate represents the number of student tickets and the y-coordinate represents the cost for 2, 4, and 5 tickets.
11. Graph the ordered pairs (number of tickets, cost).

Practice and Apply

Homework Help

For Exercises	See Examples
12–17	1
18–23	2
26–30, 37–43	4
31–36	3

Extra Practice

See page 725.

Graph each point on a coordinate system.

12. $A(3, 3)$
13. $D(1, 8)$
14. $G(2, 7)$
15. $X(7, 2)$
16. $P(0, 6)$
17. $N(4, 0)$

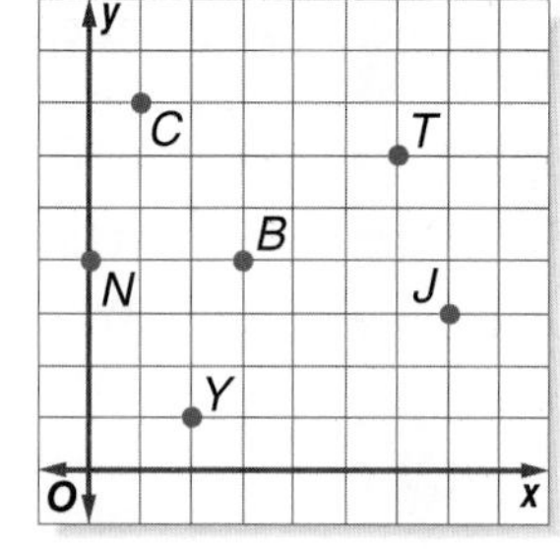

Refer to the coordinate system shown at the right. Write the ordered pair that names each point.

18. C
19. J
20. N
21. T
22. Y
23. B
24. What point lies on both the x-axis and y-axis?
25. Where are all of the possible locations for the graph of (x, y) if $y = 0$?, if $x = 0$?

SCIENCE **For Exercises 26 and 27, use the following information.** The average speed of a house mouse is 12 feet per second.
Source: *Natural History Magazine*

26. Find the distance traveled in 3, 5, and 7 seconds.
27. Graph the ordered pairs (time, distance).

SCIENCE For Exercises 28–30, use the following information.
Keyson is conducting a physics experiment. He drops a tennis ball from a height of 100 centimeters and then records the height after each bounce. The results are shown in the table.

Bounce	0	1	2	3	4
Height (cm)	100	50	25	13	6

28. Write a set of ordered pairs for the data.

29. Graph the data.

30. How high do you think the ball will bounce on the fifth bounce? Explain.

Express each relation as a table and as a graph. Then determine the domain and range.

31. {(4, 5), (5, 2), (1, 6)}

32. {(6, 8), (2, 9), (0, 1)}

33. {(7, 0), (3, 2), (4, 4), (5, 1)}

34. {(2, 4), (1, 3), (5, 6), (1, 1)}

35. {(0, 1), (0, 3), (0, 5), (2, 0)}

36. {(4, 3), (3, 4), (1, 2), (2, 1)}

AIR PRESSURE For Exercises 37–39, use the table and the following information.
The air pressure decreases as the distance from Earth increases. The table shows the air pressure for certain distances.

Height (mi)	Pressure (lb/in^2)
sea level	14.7
1	10.2
2	6.4
3	4.3
4	2.7
5	1.6

37. Write a set of ordered pairs for the data.

38. Graph the data.

39. State the domain and the range of the relation.

SCIENCE For Exercises 40–43, use the following information and the information at the left.
Water boils at sea level at 100°C. The boiling point of water decreases about 5°C for every mile above sea level.

40. Make a table that shows the boiling point at sea level and at 1, 2, 3, 4, and 5 miles above sea level.

41. Show the data as a set of ordered pairs.

42. Graph the ordered pairs.

43. At about what temperature does water boil in Albuquerque, New Mexico? in Alpine, Texas? (*Hint:* 1 mile = 5280 feet)

More About. . .

Science

Albuquerque, New Mexico, is at 7200 feet above sea level. Alpine, Texas, is at 4490 feet above sea level.

Source: *The World Almanac*

44. **CRITICAL THINKING** Where are all of the possible locations for the graph of (x, y) if $x = 4$?

45. WRITING IN MATH Answer the question that was posed at the beginning of the lesson.

How are ordered pairs used to graph real-life data?

Include the following in your answer:
- an explanation of how an ordered pair identifies a specific point on a graph, and
- an example of a situation where ordered pairs are used to graph data.

46. **Graph each relation on a coordinate system. Then find the coordinates of another point that follows the pattern in the graph.**

a.

x	1	3	5	7
y	2	4	6	8

b.

x	0	2	4	6
y	10	8	6	4

47. State the domain of the relation shown in the graph.

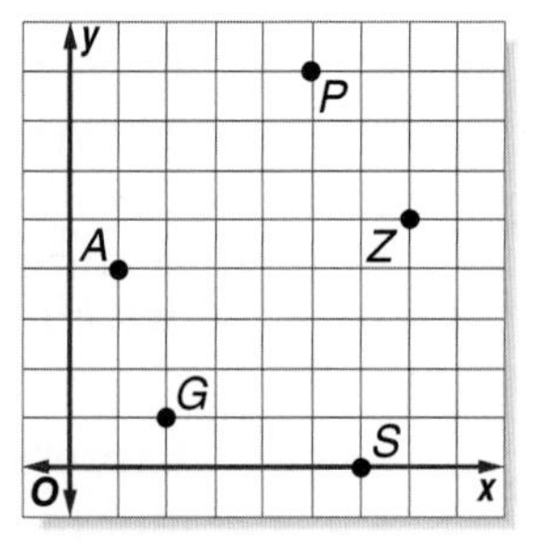

Ⓐ {0, 1, 4, 5, 8}

Ⓑ {A, G, P, S, Z}

Ⓒ {0, 1, 2, 4, 5}

Ⓓ {1, 2, 5, 6, 7}

48. What relationship exists between the x- and y-coordinates of each of the data points shown on the graph?

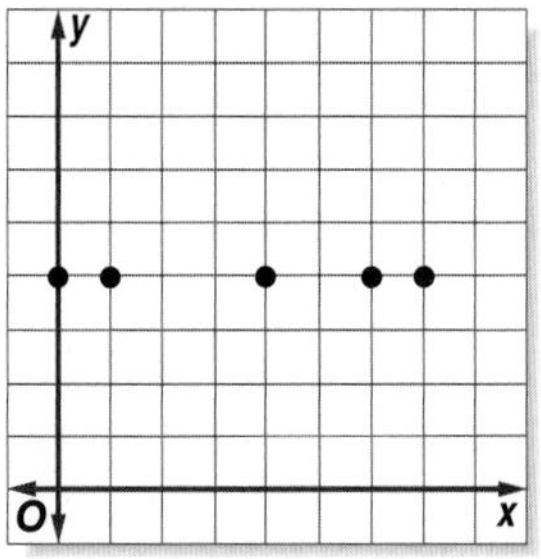

Ⓐ The y-coordinate varies, and the x-coordinate is always 4.

Ⓑ The y-coordinate is 4 more than the x-coordinate.

Ⓒ The sum of the x- and y-coordinate is always 4.

Ⓓ The x-coordinate varies, and the y-coordinate is always 4.

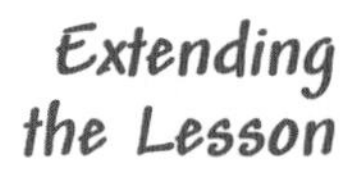

49. **Draw a coordinate grid.**

a. Graph (2, 1), (2, 4), and (5, 1).

b. Connect the points with line segments. Describe the figure formed.

c. Multiply each coordinate in the set of ordered pairs by 2.

d. Graph the new ordered pairs. Connect the points with line segments. What figure is formed?

e. **MAKE A CONJECTURE** How do the figures compare? Write a sentence explaining the similarities and differences of the figures.

Maintain Your Skills

Mixed Review

ALGEBRA **Solve each equation mentally.** *(Lesson 1-5)*

50. $a + 6 = 17$

51. $7t = 42$

52. $\frac{54}{n} = 6$

53. Name the property shown by $4 \cdot 1 = 4$. *(Lesson 1-4)*

ALGEBRA **Evaluate each expression if $a = 5$, $b = 1$, and $c = 3$.** *(Lesson 1-3)*

54. $ca - cb$

55. $5a - 6c$

Write a numerical expression for each verbal phrase. *(Lesson 1-2)*

56. fifteen less than twenty-one

57. the product of ten and thirty

Getting Ready for the Next Lesson

BASIC SKILL **Find each quotient.**

58. $74 \div 2$

59. $96 \div 8$

60. $102 \div 3$

61. $112 \div 4$

62. $80 \div 16$

63. $91 \div 13$

64. $132 \div 22$

65. $153 \div 17$

Algebra Activity

A Preview of Lesson 1-7

Sunshine State Standards
MA.E.1.3.1-1, MA.E.1.3.1-2

Scatter Plots

Sometimes, it is difficult to determine whether a relationship exists between two sets of data by simply looking at them. To determine whether a relationship exists, we can write the data as a set of ordered pairs and then graph the ordered pairs on a coordinate system.

Collect the Data

Let's investigate whether a relationship exists between height and arm span.

Step 1 Work with a partner. Use a centimeter ruler to measure the length of your partner's height and arm span to the nearest centimeter. Record the data in a table like the one shown.

Name	Height (cm)	Arm Span (cm)

Step 2 Extend the table. Combine your data with that of your classmates.

Step 3 Make a list of ordered pairs in which the x-coordinate represents height and the y-coordinate represents arm span.

Step 4 Draw a coordinate grid like the one shown and graph the ordered pairs (height, arm span).

Arm Span (cm)

y

O

Height (cm)

x

Analyze the Data

1. Does there appear to be a trend in the data? If so, describe the trend.

Make a Conjecture

2. Estimate the arm span of a person whose height is 60 inches. 72 inches.

3. How does a person's arm span compare to his or her height?

4. Suppose the variable x represents height, and the variable y represents arm span. Write an expression for arm span.

Extend the Activity

5. Collect and graph data to determine whether a relationship exists between height and shoe length. Explain your results.

1-7 Scatter Plots

What You'll Learn

- Construct scatter plots.
- Interpret scatter plots.

Sunshine State Standards
MA.E.1.3.1-1, MA.E.3.3.1-3

Vocabulary

- scatter plot

How can scatter plots help spot trends?

Suppose you work in the video department of a home entertainment store. The number of movies on videocassettes you have sold in a five-year period is shown in the graph.

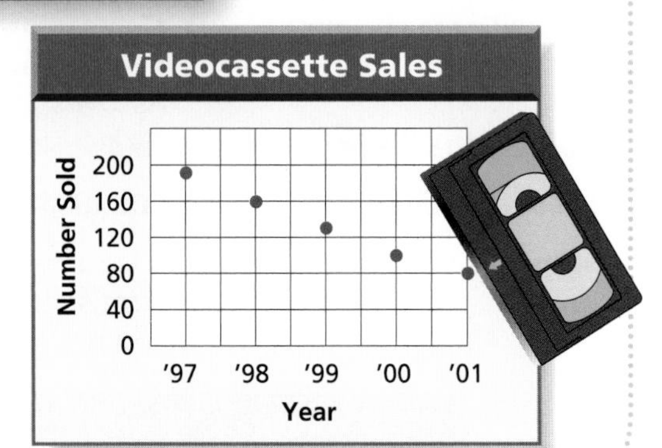

a. What appears to be the trend in sales of movies on videocassette?

b. Estimate the number of movies on videocassette sold for 2003.

CONSTRUCT SCATTER PLOTS A **scatter plot** is a graph that shows the relationship between two sets of data. In a scatter plot, two sets of data are graphed as ordered pairs on a coordinate system.

Example 1 *Construct a Scatter Plot*

TEST SCORES The table shows the average SAT math scores from 1990–2000. Make a scatter plot of the data.

Year	'90	'91	'92	'93	'94	'95	'96	'97	'98	'99	'00
Score	501	500	501	503	504	506	508	511	512	511	514

Source: *The World Almanac*

Let the horizontal axis, or x-axis, represent the year. Let the vertical axis, or y-axis, represent the score. Then graph ordered pairs (year, score).

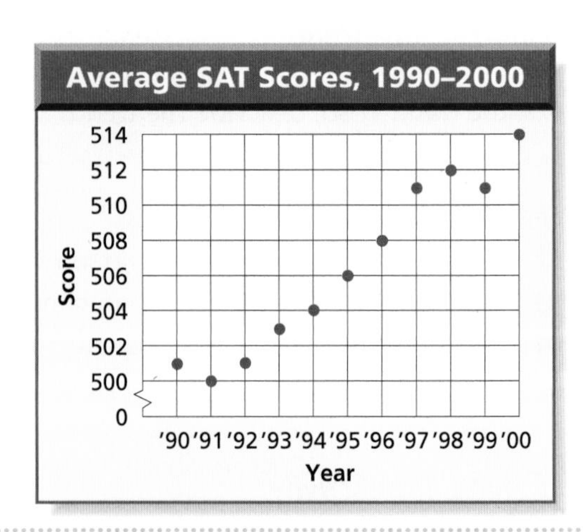

Concept Check *True* or *false*: A scatter plot represents one set of data. Explain.

INTERPRET SCATTER PLOTS The following scatter plots show the types of relationships or patterns of two sets of data.

Study Tip

Scatter Plots
Data that appear to go uphill from left to right show a positive relationship. Data that appear to go downhill from left to right show a negative relationship.

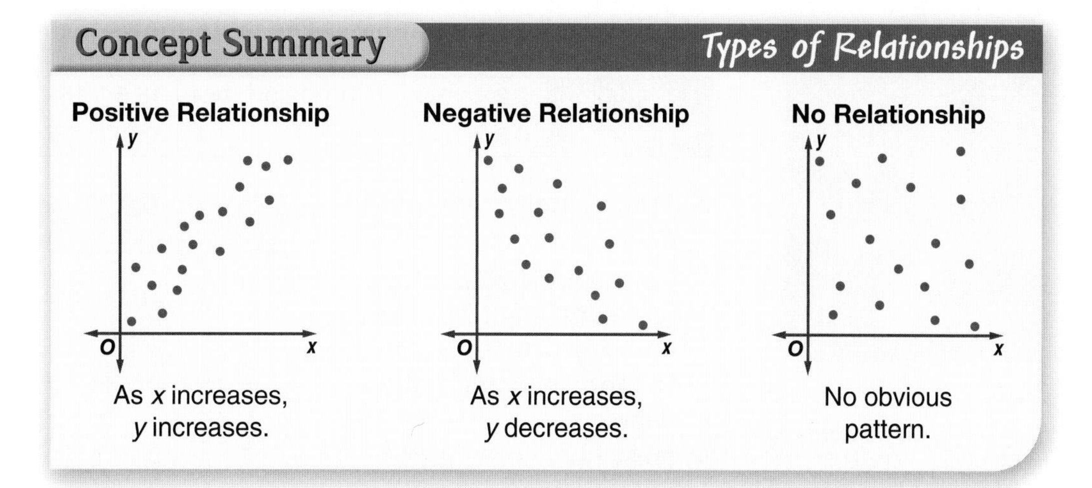

Concept Summary
Types of Relationships
Positive Relationship
As x increases, y increases.
Negative Relationship
As x increases, y decreases.
No Relationship
No obvious pattern.

Concept Check What type of relationship is shown on a graph that shows as the values of x increase, the values of y decrease?

Example 2 Interpret Scatter Plots

Determine whether a scatter plot of the data for the following might show a *positive, negative,* or *no* relationship. Explain your answer.

a. age of car and value of car

As the age of a car increases, the value of the car decreases. So, a scatter plot of the data would show a negative relationship.

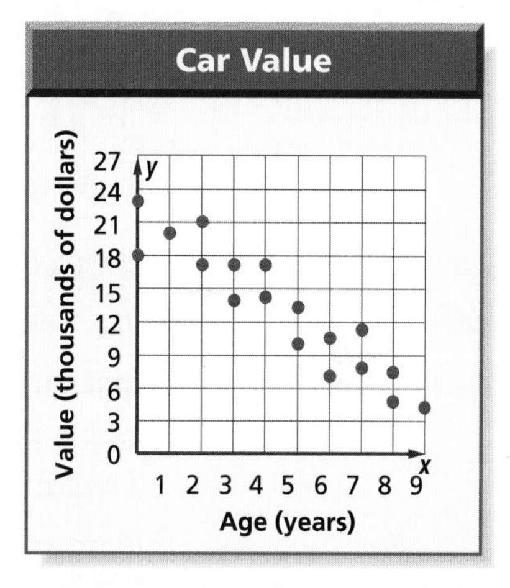

b. birth month and birth weight

A person's birth weight is not affected by their birth month. Therefore, a scatter plot of the data would show no relationship.

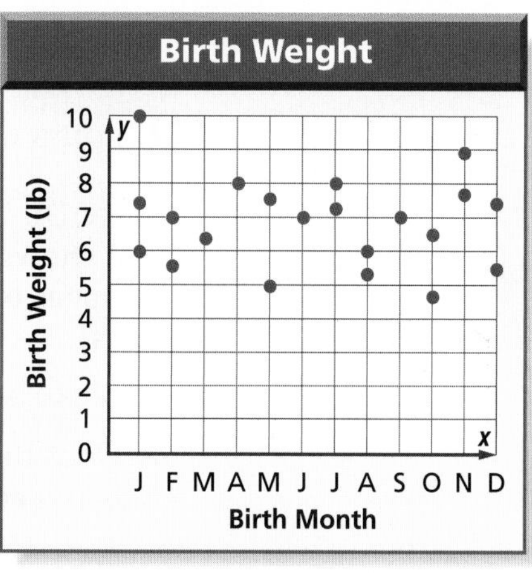

You can also use scatter plots to spot trends, draw conclusions, and make predictions about the data.

Career Choices

Biologist

Wildlife biologists work in the field of fish and wildlife conservation. Duties may include studying animal populations and monitoring trends of migrating animals.

Online Research For information about a career as a wildlife biologist, visit: www.pre-alg.com/careers

Example 3 Use Scatter Plots to Make Predictions

BIOLOGY A wildlife biologist is recording the lengths and weights of a sampling of largemouth bass. The table shows the results.

Length (in.)	9.2	10.9	12.3	12.0	14.1	15.5	16.4	16.9	17.7	18.4	19.8
Weight (lb)	0.5	0.8	0.9	1.3	1.7	2.2	2.5	3.2	3.6	4.1	4.8

a. Make a scatter plot of the data.

Let the horizontal axis represent length, and let the vertical axis represent weight. Then graph the data.

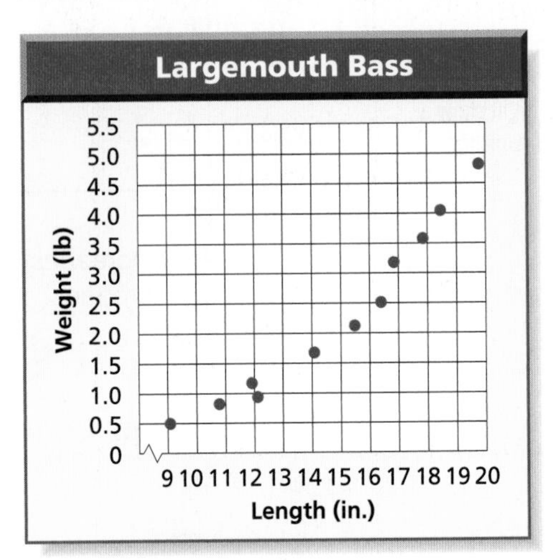

b. Does the scatter plot show a relationship between the length and weight of a largemouth bass? Explain.

As the length of the bass increases, so does its weight. So, the scatter plot shows a positive relationship.

c. Predict the weight of a bass that measures 22 inches.

By looking at the pattern in the graph, we can predict that the weight of a bass measuring 22 inches would be between 5 and 6 pounds.

Check for Understanding

Concept Check

1. **List** three ways a scatter plot can be used.
2. **OPEN ENDED** Draw a scatter plot with ten ordered pairs that show a negative relationship.
3. **Name** the three types of relationships shown by scatter plots.

Guided Practice

Determine whether a scatter plot of the data for the following might show a *positive*, *negative*, or *no* relationship. Explain your answer.

4. hours worked and earnings
5. hair color and height

Application

SCHOOL For Exercises 6 and 7, use the table that shows the heights and grade point averages of the students in Mrs. Stanley's class.

6. Make a scatter plot of the data.
7. Does there appear to be a relationship between the scores? Explain.

Name	Height (in.)	GPA
Jenna	66	3.6
Michael	61	3.2
Laura	59	3.9
Simon	64	2.8
Marcus	61	3.8
Timothy	65	3.1
Brandon	70	2.6
Emily	64	2.2
Eduardo	65	4.0

Practice and Apply

Homework Help

For Exercises	See Examples
8–13	2
14–20	1, 3

Extra Practice
See page 726.

Determine whether a scatter plot of the data for the following might show a *positive, negative,* or *no* relationship. Explain your answer.

8. size of household and amount of water bill
9. number of songs on a CD and cost of a CD
10. size of a car's engine and miles per gallon
11. speed and distance traveled
12. outside temperature and amount of heating bill
13. size of a television screen and the number of channels it receives

ANIMALS **For Exercises 14–16, use the scatter plot shown.**

14. Do the data show a *positive, negative,* or *no* relationship between the year and the number of bald eagle hatchlings?
15. What appears to be the trend in the number of hatchlings between 1965 and 1972?
16. What appears to be the trend between 1972 and 1985?

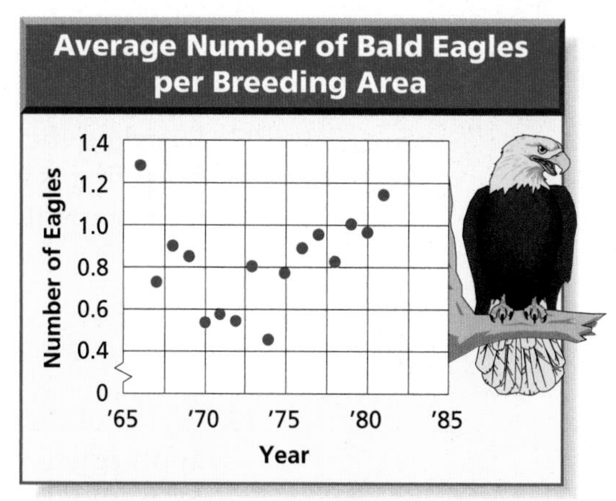

Source: *CHANCE*

Online Research **Data Update** How has the total number of bald eagle pairs in the United States changed since 1980? Visit www.pre-alg.com/data_update to learn more.

The high and low temperatures for your vacation destinations can be shown in a scatter plot. Visit www.pre-alg.com/webquest to continue work on your WebQuest project.

BASKETBALL **For Exercises 17–19, use the following information.**
The number of minutes played and the number of field goal attempts for certain players of the Indiana Pacers for the 1999–2000 season is shown below.

Player	Minutes Played	Field Goal Attempts	Player	Minutes Played	Field Goal Attempts
Rose	2978	1196	Best	1691	561
Miller	2987	1041	Jackson	2190	570
Smits	1852	890	Perkins	1620	441
Croshere	1885	653	Mullin	582	187
Davis	2127	602	McKey	634	108

17. Make a scatter plot of the data.
18. Does the scatter plot show any relationship? If so, is it positive or negative? Explain your reasoning.
19. Suppose a player played 2500 minutes. Predict the number of field goal attempts for that player.
20. **RESEARCH** Use the Internet or another source to find two sets of sports statistics that can be shown in a scatter plot. Identify any trends in the data.

21. CRITICAL THINKING Refer to Example 1 on page 40. Do you think the trend in the test scores would continue in the years to come? Explain your reasoning.

22. WRITING IN MATH Answer the question that was posed at the beginning of the lesson.

How can scatter plots help us spot trends?

Include the following in your answer:

- definitions of positive relationship, negative relationship, and no relationship, and
- examples of real-life situations that would represent each type of relationship.

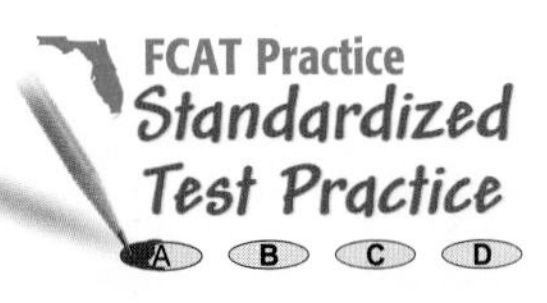

The scatter plot shows the study time and test scores for the students in Mr. Mock's history class.

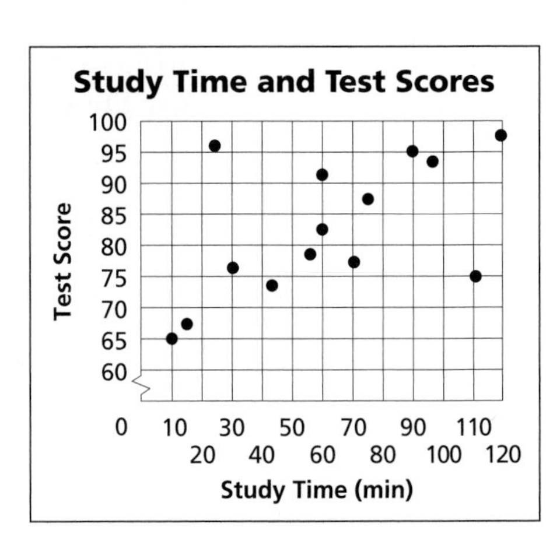

23. Based on the results, which of the following is an appropriate score for a student who studies for 1 hour?

Ⓐ 68 Ⓑ 98

Ⓒ 87 Ⓓ 72

24. Which of the following is an appropriate score for a student who studies for 1.5 hours?

Ⓐ 78 Ⓑ 92

Ⓒ 81 Ⓓ 74

Maintain Your Skills

Mixed Review **Graph each ordered pair on a coordinate system.** *(Lesson 1-6)*

25. $M(3, 2)$ **26.** $X(5, 0)$ **27.** $K(0, 2)$

Write the ordered pair that names each point. *(Lesson 1-6)*

28.

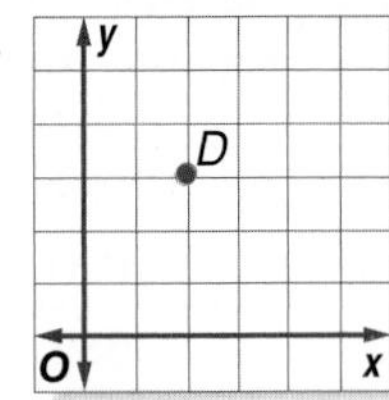

29.

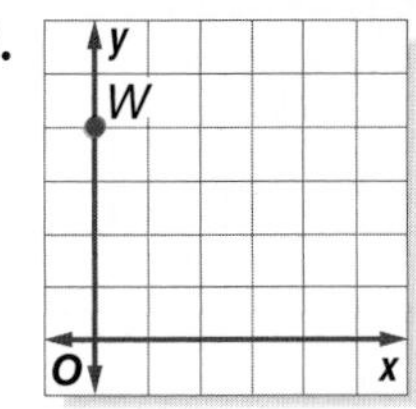

30.

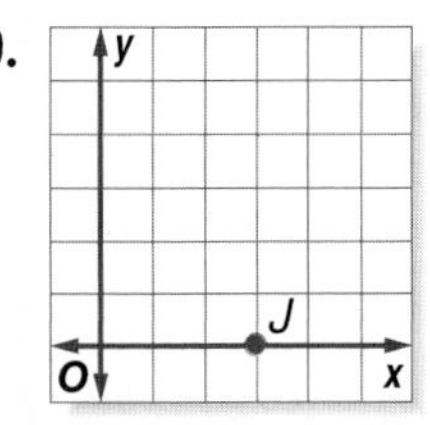

31. Determine the domain and range of the relation {(0, 9), (4, 8), (2, 3), (6, 1)}. *(Lesson 1-6)*

ALGEBRA Solve each equation mentally. *(Lesson 1-5)*

32. $3c = 81$ **33.** $15 - x = 8$ **34.** $8 = \frac{32}{m}$

35. ALGEBRA Simplify $15 + (b + 3)$. *(Lesson 1-4)*

ALGEBRA Evaluate each expression if $m = 8$ and $y = 6$. *(Lesson 1-3)*

36. $(2m + 3y) - m$ **37.** $3m + (y - 2) + 3$

Graphing Calculator Investigation

A Follow-Up of Lesson 1-7

Sunshine State Standards
MA.E.1.3.1-1, MA.E.1.3.1-2

Scatter Plots

You have learned that graphing ordered pairs as a scatter plot on a coordinate plane is one way to make it easier to "see" if there is a relationship. You can use a TI-83 Plus graphing calculator to create scatter plots.

SCIENCE A zoologist studied extinction times (in years) of island birds. The zoologist wanted to see if there was a relationship between the average number of nests and the time needed for each bird to become extinct on the islands. Use the table of data below to make a scatter plot.

Bird Name	Bird Size	Average Number of Nests	Extinction Time
Buzzard	Large	2.0	5.5
Quail	Large	1.0	1.5
Curlew	Large	2.8	3.1
Cuckoo	Large	1.4	2.5
Magpie	Large	4.5	10.0
Swallow	Small	3.8	2.6
Robin	Small	3.3	4.0
Stonechat	Small	3.6	2.4
Blackbird	Small	4.7	3.3
Tree-sparrow	Small	2.2	1.9

Step 1 *Enter the data.*

- Clear any existing lists.

KEYSTROKES: [STAT] [ENTER] [▲] [CLEAR] [ENTER]

- Enter the average number of nests as L1 and extinction times as L2.

KEYSTROKES: [STAT] [ENTER] 2 [ENTER] 1 [ENTER] … 2.2 [ENTER] [▶] 5.5 [ENTER] 1.5 [ENTER] … 1.9 [ENTER]

The first data pair is (2, 5.5).

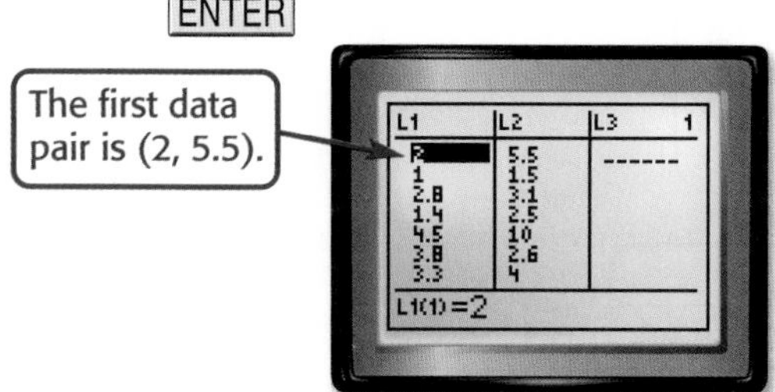

Step 2 *Format the graph.*

- Turn on the statistical plot.

KEYSTROKES: [2nd] [STAT PLOT] [ENTER] [ENTER]

- Select the scatter plot, L1 as the Xlist and L2 as the Ylist.

KEYSTROKES: [▼] [ENTER] [▼] [2nd] [L1] [ENTER] [2nd] [L2] [ENTER]

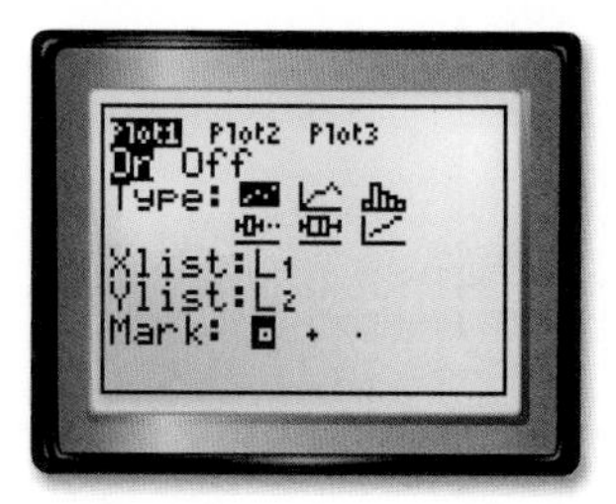

www.pre-alg.com/other_calculator_keystrokes

Step 3 *Graph the data.*

- Display the scatter plot.

  KEYSTROKES: ZOOM 9

- Use the TRACE feature and the left and right arrow keys to move from one point to another.

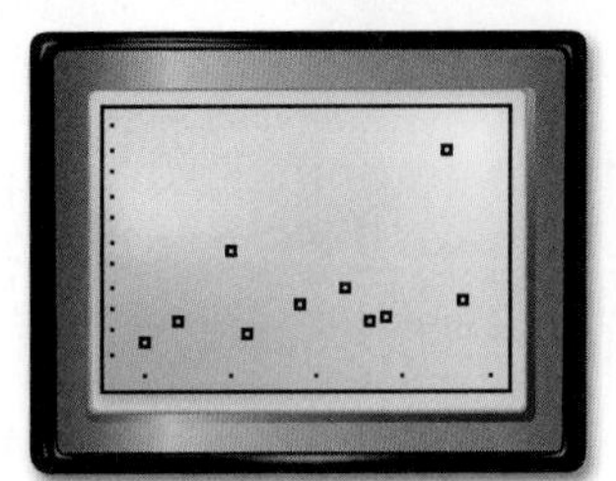

Exercises

1. Press TRACE. Use the left and right arrow keys to move from one point to another. What do the coordinates of each data point represent?
2. Describe the scatter plot.
3. Is there a relationship between the average number of nests and extinction times? If so, write a sentence or two that describes the relationship.
4. Are there any differences between the extinction times of large birds versus small birds?
5. Separate the data by bird size. Enter average number of nests and extinction times for large birds as lists L1 and L2 and for small birds as lists L3 and L4. Use the graphing calculator to make two scatter plots with different marks for large and small birds. Does your scatter plot agree with your answer in Exercise 4? Explain.

For Exercises 6–8, make a scatter plot for each set of data and describe the relationship, if any, between the x- and y-values.

6.

x	y
70	323
80	342
40	244
50	221
30	121
80	399
60	230
60	200
50	215
40	170

7.

x	y
8	89
5	32
9	30
10	18
3	26
4	72
10	51
7	34
6	82
7	60

8.

x	5.2	5.8	6.3	6.7	7.4	7.6	8.4	8.5	9.1
y	12.1	11.9	11.5	9.8	10.2	9.6	8.8	9.1	8.5

9. **RESEARCH** Find two sets of data on your own. Then determine whether a relationship exists between the data.

Chapter 1

Study Guide and Review

Vocabulary and Concept Check

algebraic expression (p. 17)
conjecture (p. 7)
coordinate plane (p. 33)
coordinate system (p. 33)
counterexample (p. 25)
deductive reasoning (p. 25)
defining a variable (p. 18)
domain (p. 35)
equation (p. 28)
evaluate (p. 12)
graph (p. 34)
inductive reasoning (p. 7)
numerical expression (p. 12)
open sentence (p. 28)
ordered pair (p. 33)
order of operations (p. 12)
origin (p. 33)
properties (p. 23)
range (p. 35)
relation (p. 35)
scatter plot (p. 40)
simplify (p. 25)
solution (p. 28)
solving the equation (p. 28)
variable (p. 17)
x-axis (p. 33)
x-coordinate (p. 33)
y-axis (p. 33)
y-coordinate (p. 33)

Choose the letter of the term that best matches each statement or phrase. Use each letter once.

1. $m + 3n - 4$
2. to find the value of a numerical expression
3. the set of all *y*-coordinates of a relation
4. $20 + 12 \div 4 - 1 \times 2$
5. the set of all *x*-coordinates of a relation

a. numerical expression
b. evaluate
c. domain
d. algebraic expression
e. range

Lesson-by-Lesson Review

1-1 Using a Problem-Solving Plan

See pages 6–10.

Concept Summary

- The four steps of the four-step problem-solving plan are *explore, plan, solve,* and *examine.*
- Some problems can be solved using inductive reasoning.

Example **What is the next term in the list 1, 5, 9, 13, 17, …?**

Explore We know the first five terms. We need to find the next term.

Plan Use inductive reasoning to determine the next term.

Solve Each term is 4 more than the previous term.

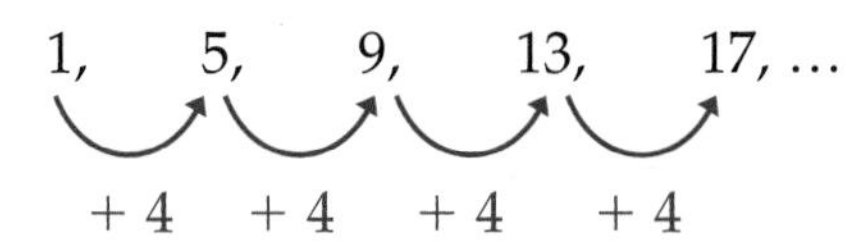

By continuing the pattern, the next term is 17 + 4 or 21.

Examine Subtract 4 from each term. $21 - 4 = 17$, $17 - 4 = 13$, $13 - 4 = 9$, $9 - 4 = 5$, and $5 - 4 = 1$. So, the answer is correct.

Exercises **Find the next term in each list.** *See Example 2 on page 7.*

6. 2, 4, 6, 8, 10, …
7. 5, 8, 11, 14, 17, …
8. 2, 6, 18, 54, 162, …
9. 1, 2, 4, 7, 11, 16, …

10. **FOOD** The table below shows the cost of various-sized hams. How much will it cost to buy a ham that weighs 7 pounds? *See Examples 1 and 3 on pages 7 and 8.*

Weight (lb)	1	2	3	4	5
Cost	$4.38	$8.76	$13.14	$17.52	$21.90

1-2 Numbers and Expressions

See pages 12–16.

Concept Summary

- When evaluating an expression, follow the order of operations.
  - **Step 1** Simplify the expressions inside grouping symbols.
  - **Step 2** Do all multiplications and/or divisions from left to right.
  - **Step 3** Do all additions and/or subtractions from left to right.

Example **Find the value of $3[(10 - 7) + 2]$.**

$3[(10 - 7) + 2] = 3[3 + 2]$ Evaluate $(10 - 7)$.
$= 3[5]$ Add 3 and 2.
$= 15$ Multiply 3 and 5.

Exercises **Find the value of each expression.** *See Example 1 on page 13.*

11. $7 + 3 \cdot 5$
12. $36 \div 9 - 3$
13. $5 \cdot (7 - 2) - 9$
14. $\frac{2(17 + 4)}{3}$
15. $18 \div (7 - 4) + 6$
16. $4[9 + (1 \cdot 16) - 8]$

1-3 Variables and Expressions

See pages 17–21.

Concept Summary

- To evaluate an algebraic expression, replace each variable with its known value, and then use the order of operations.

Example **Evaluate $5a + 2$ if $a = 7$.**

$5a + 2 = 5(7) + 2$ Replace a with 7.
$= 35 + 2$ Multiply 5 and 7.
$= 37$ Add 35 and 2.

Exercises **ALGEBRA** **Evaluate each expression if $x = 3$, $y = 8$, and $z = 5$.**
See Examples 1 and 2 on pages 17 and 18.

17. $y + 6$
18. $17 - 2x$
19. $z - 3 + y$
20. $6x - 2z + 7$
21. $\frac{6y}{x} + 9$
22. $9x - (y + z)$

1-4 Properties

See pages 23–27.

Concept Summary

For any numbers a, b, and c:

- $a + b = b + a$ Commutative Property of Addition
- $(a + b) + c = a + (b + c)$ Associative Property of Addition
- $a \cdot b = b \cdot a$ Commutative Property of Multiplication
- $(a \cdot b) \cdot c = a \cdot (b \cdot c)$ Associative Property of Multiplication
- $a + 0 = 0 + a = a$ Additive Identity
- $a \cdot 0 = 0 \cdot a = 0$ Multiplicative Property of Zero
- $a \cdot 1 = 1 \cdot a = a$ Multiplicative Identity

Example **Name the property shown by each statement.**

$8 \cdot 1 = 8$ Multiplicative Identity

$(2 + 3) + 6 = 2 + (3 + 6)$ Associative Property of Addition

$1 \cdot 6 \cdot 9 = 6 \cdot 1 \cdot 9$ Commutative Property of Multiplication

Exercises **Name the property shown by each statement.**
See Example 1 on page 24.

23. $1 + 9 = 9 + 1$
24. $6 + 0 = 6$
25. $15 \times 0 = 0$
26. $(x \cdot 8) \cdot 2 = x \cdot (8 \cdot 2)$

1-5 Variables and Equations

See pages 28–32.

Concept Summary

- To solve an equation, find the value for the variable that makes the equation true.

Example **Find the solution of $26 = 33 - w$. Is it 5, 6, or 7?**

Replace w with each value.

Value for w	$26 = 33 - w$	True or False?
5	$26 \stackrel{?}{=} 33 - 5$	false
6	$26 \stackrel{?}{=} 33 - 6$	false
7	$26 \stackrel{?}{=} 33 - 7$	true ✓

Therefore, the solution of $26 = 33 - w$ is 7.

ALGEBRA **Solve each equation mentally.** *See Example 3 on page 29.*

27. $n + 3 = 13$
28. $9 = k - 6$
29. $24 = 7 + g$
30. $6x = 48$
31. $54 = 9h$
32. $\frac{56}{a} = 14$

Chapter 1 **For More ...**

- Extra Practice, see pages 724–726.
- Mixed Problem Solving, see page 758.

1-6 Ordered Pairs and Relations

See pages 33–38.

Concept Summary

- Ordered pairs are used to graph a point on a coordinate system.
- A relation is a set of ordered pairs. The set of x-coordinates is the domain, and the set of y-coordinates is the range.

Example **Express the relation {(1, 4), (3, 2), (4, 3), (0, 5)} as a table and as a graph. Then determine the domain and range.**

x	y
1	4
3	2
4	3
0	5

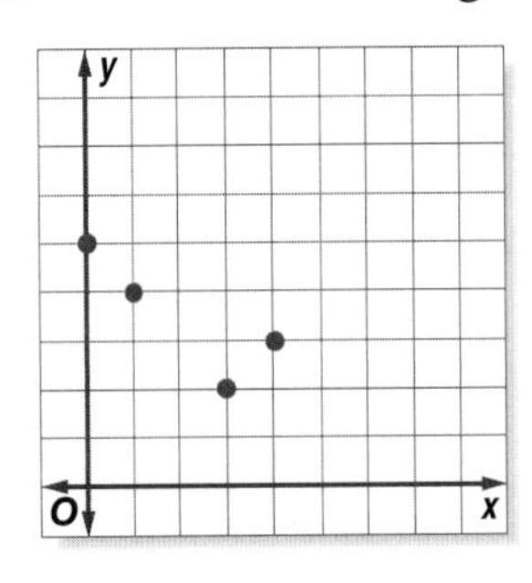

The domain is {1, 3, 4, 0}, and the range is {4, 2, 3, 5}.

Exercises **Express each relation as a table and as a graph. Then determine the domain and range.** *See Example 3 on page 35.*

33. {(2, 3), (6, 1), (7, 5)}

34. {(0, 2), (1, 7), (5, 2), (6, 5)}

1-7 Scatter Plots

See pages 40–44.

Concept Summary

- A scatter plot is a graph that shows the relationship between two sets of data.

Example **The scatter plot shows the approximate heights and circumferences of various giant sequoia trees.**

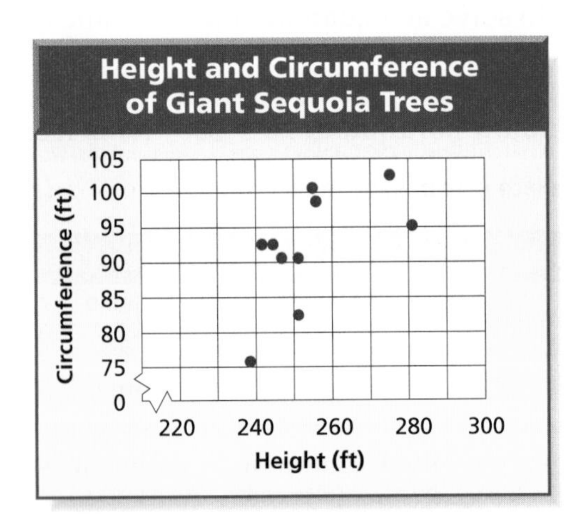

Exercises **Refer to the scatter plot.** *See Example 3 on page 42.*

35. Does the scatter plot show a *positive, negative,* or *no* relationship? Explain.

36. Predict the circumference of a 245-foot sequoia. Explain your reasoning.

Chapter 1 Practice Test

Vocabulary and Concepts

1. **Write** the steps of the four-step problem-solving plan.
2. **List** the order of operations used to find the value of a numerical expression.

Skills and Applications

Find the value of each expression.

3. $24 - 8 \div 2 \cdot 3$
4. $16 \div 4 + 3(9 - 7)$
5. $3[18 - 5(7 - 5 + 1)]$

Write a numerical expression for each verbal phrase.

6. three less than fifteen
7. twelve increased by seven
8. the quotient of twelve and six

ALGEBRA Evaluate each expression if $a = 7$, $b = 3$, and $c = 5$.

9. $4a - 3c$
10. $42 \div [a(c - b)]$
11. $5c + (a + 2b) - 8$

12. What property is shown by $(5 \cdot 6) \cdot 8 = 5 \cdot (6 \cdot 8)$?

ALGEBRA Simplify each expression.

13. $9 + (p + 3)$
14. $6 \cdot (7 \cdot k)$

ALGEBRA Solve each equation mentally.

15. $4m = 20$
16. $16 - a = 9$

17. Graph $A(2, 5)$ on a coordinate system.

Refer to the coordinate system shown at the right. Write the ordered pair that names each point.

18. C
19. D

20. Express $\{(8, 5), (4, 3), (2, 2), (6, 1)\}$ as a table and as a graph. Then determine the domain and range.

Determine whether a scatter plot of the data for the following might show a *positive*, *negative*, or *no* relationship. Explain your answer.

21. outside temperature and air conditioning bill
22. number of siblings and height
23. Find the next three terms in the list 3, 5, 9, 15, ….
24. **MONEY** Mrs. Adams rents a car for a week and pays \$79 for the first day and \$49 for each additional day. Mr. Lowe rents a car for \$350 a week. Which was the better deal? Explain.

25. **STANDARDIZED TEST PRACTICE** Katie purchased 6 loaves of bread at the grocery store and paid a total of \$12. Which equation can be used to find how much Katie paid for each loaf of bread?

   (A) $x + 6 = 12$ (B) $6x = 12$ (C) $x - 6 = 12$ (D) $x \div 6 = 12$

Chapter 1 Standardized Test Practice

FCAT Practice

Part 1 Multiple Choice

Record your answers on the answer sheet provided by your teacher or on a sheet of paper.

1. Find the next two terms in the pattern 4, 12, 36, 108, … . (Lesson 1-1)

Ⓐ 116 and 124 Ⓑ 116 and 140

Ⓒ 324 and 648 Ⓓ 324 and 972

2. Evaluate $2(15 - 3 \cdot 4)$. (Lesson 1-2)

Ⓐ 6 Ⓑ 16 Ⓒ 18 Ⓓ 96

3. Of the six books in a mystery series, four have 200 pages and two have 300 pages. Which expression represents the total number of pages in the series? (Lesson 1-3)

Ⓐ 200 + 300 Ⓑ 6(200 + 300)

Ⓒ 4(200) + 2(300) Ⓓ 6(200) + 6(300)

4. The postage for a first-class letter is \$0.34 for the first ounce and \$0.21 for each additional ounce. Which expression best represents the cost of postage for a letter that weighs 5 ounces? (Lesson 1-3)

Ⓐ 0.34 + 0.21(5) Ⓑ 0.21 + 0.34(4)

Ⓒ 0.34(5) Ⓓ 0.34 + 0.21(4)

5. Which property is represented by the equation below? (Lesson 1-4)

$$8 \cdot (5 \cdot 3) = (8 \cdot 5) \cdot 3$$

Ⓐ Commutative Property of Addition

Ⓑ Commutative Property of Multiplication

Ⓒ Associative Property of Multiplication

Ⓓ Identity Property of Multiplication

6. Which number is the solution of the equation $17 - 2x = 9$? (Lesson 1-5)

Ⓐ 2 Ⓑ 4 Ⓒ 6 Ⓓ 8

> **The Princeton Review Test-Taking Tip**
>
> **Question 6** To solve an equation, you can replace the variable in the equation with the values given in each answer choice. The answer choice that results in a true statement is the correct answer.

7. Which sentence does the equation $n + 9 = 15$ represent? (Lesson 1-5)

Ⓐ A number is the sum of 9 and 15.

Ⓑ A number decreased by 9 is 15.

Ⓒ The product of a number and 9 is 15.

Ⓓ Nine more than a number is 15.

8. What are the coordinates of point P? (Lesson 1-6)

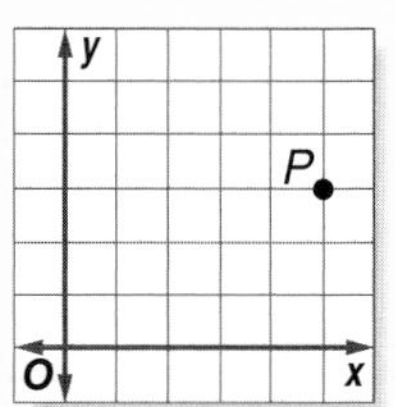

Ⓐ (3, 5)

Ⓑ (5, 3)

Ⓒ (3, 3)

Ⓓ (5, 5)

9. Which table shows the set of ordered pairs that represents the points graphed on the grid below? (Lesson 1-6)

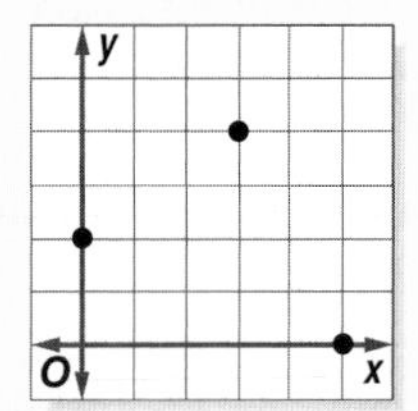

Ⓐ

x	y
2	0
3	4
0	5

Ⓑ

x	y
0	2
4	3
5	0

Ⓒ

x	y
0	2
3	4
5	0

Ⓓ

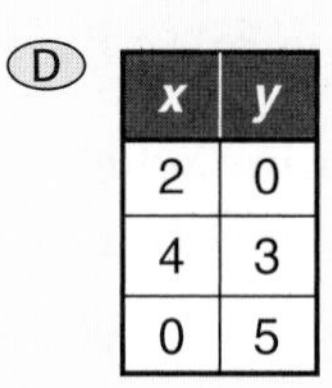

x	y
2	0
4	3
0	5

10. What type of relationship does the scatter plot below show? (Lesson 1-7)

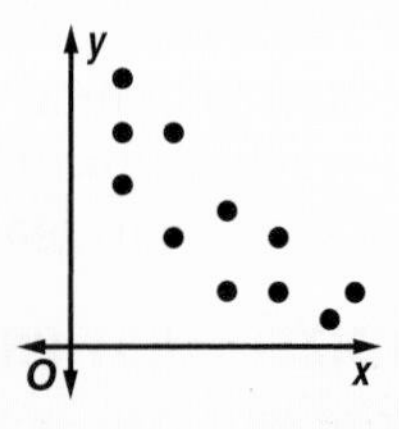

Ⓐ positive Ⓑ negative

Ⓒ associative Ⓓ none

Part 2 Short Response/Grid In

Record your answers on the answer sheet provided by your teacher or on a sheet of paper.

11. The graph below shows the number of students absent from school each day of one week. On what day were the fewest students absent? (Prerequisite Skill, p. 722)

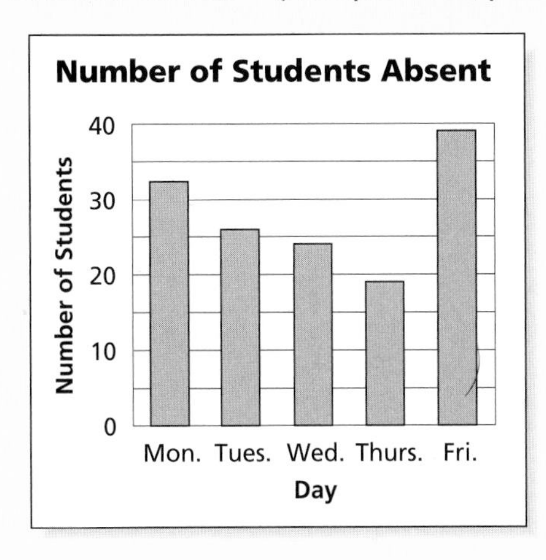

12. The number of Olympic events for women is shown. About how many more events for women were held in 2000 than in 1980? (Prerequisite Skill, p. 722)

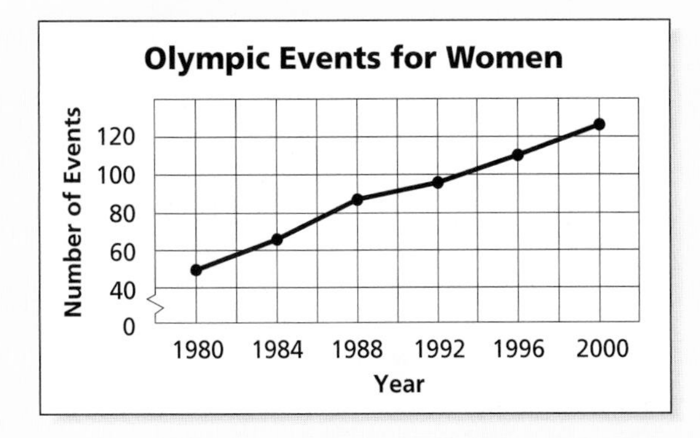

13. Six tables positioned in a row will be used to display science projects. Each table is 8 feet long. How many yards of fabric are needed to make a banner that will extend from one end of the row of tables to the other? (Lesson 1-1)

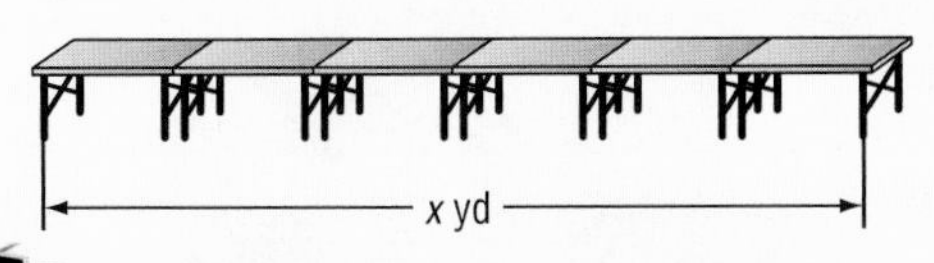

14. What is the value of the expression $5 + 4 \times 6 \div 3$? (Lesson 1-2)

15. Evaluate $x(xy + 3)$ if $x = 5$ and $y = 2$. (Lesson 1-3)

16. Write *14 is 12 less than twice the value of x* as an equation. (Lesson 1-5)

Part 3 Open Ended

Record your answers on a sheet of paper. Show your work.

17. Kenneth is recording the time it takes him to run various distances. The results are shown. (Lesson 1-6)

Distance (mi)	2	3	5	7	9
Time (min)	13	20	35	53	72

a. Write a set of ordered pairs for the data.
b. Graph the data.
c. How many minutes do you think it will take Kenneth to run 4 miles? Explain.
d. Predict how far Kenneth will run if he runs for 1 hour.

18. The table below shows the results of a survey about the average time that individual students spend studying on weekday evenings. (Lesson 1-7)

Grade	Time (min)	Grade	Time (min)
2	20	6	60
2	15	6	45
2	20	6	55
4	30	6	60
4	20	8	70
4	25	8	80
4	40	8	75
4	30	8	60

a. Make a scatter plot of the data.
b. What are the coordinates of the point that represents the longest time spent on homework?
c. Does a relationship exist between grade level and time spent studying? If so, write a sentence to describe the relationship. If not, explain why not.

Chapter 2

Integers

What You'll Learn

- **Lesson 2-1** Compare and order integers, and find the absolute value of an expression.
- **Lessons 2-2 through 2-5** Add, subtract, multiply, and divide integers.
- **Lessons 2-3 and 2-4** Evaluate and simplify algebraic expressions.
- **Lesson 2-5** Find the average of a set of data.
- **Lesson 2-6** Graph points, and show algebraic relationships on a coordinate plane.

Key Vocabulary

- **integer** (p. 56)
- **inequality** (p. 57)
- **absolute value** (p. 58)
- **additive inverse** (p. 66)
- **quadrants** (p. 86)

Why It's Important

In both mathematics and everyday life, there are many situations where integers are used. Some examples include temperatures, sports such as golf and football, and measuring the elevation of points on Earth or the depth below sea level. *You will represent real-world situations with integers in Lesson 2-1.*

Getting Started

Prerequisite Skills To be successful in this chapter, you'll need to master these skills and be able to apply them in problem-solving situations. Review these skills before beginning Chapter 2.

For Lesson 2-1 — Evaluate Expressions

Evaluate each expression if $a = 4$, $b = 10$, and $c = 8$. *(For review, see Lesson 1-3.)*

1. $a + b + c$ **2.** $bc - ab$ **3.** $b + ac$

4. $4c + 3b$ **5.** $2b - (a + c)$ **6.** $2c - b + a$

For Lesson 2-3 — Patterns

Find the next term in each list. *(For review, see Lesson 1-1.)*

7. 34, 28, 22, 16, 10, ... **8.** 120, 105, 90, 75, ...

For Lesson 2-6 — Graph Points

Use the grid to name the point for each ordered pair. *(For review, see Lesson 1-6.)*

9. (1, 3) **10.** (5, 2) **11.** (5, 5)

12. (3, 4) **13.** (0, 2) **14.** (6, 1)

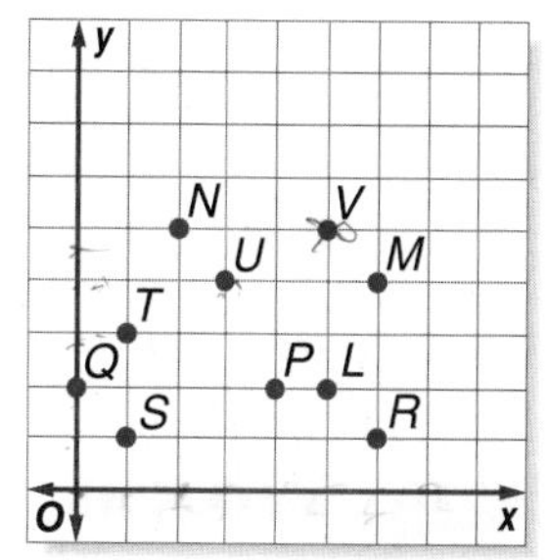

Make this Foldable to help you organize your notes on operations with integers. Begin with a piece of graph paper.

Step 1 Fold in Half

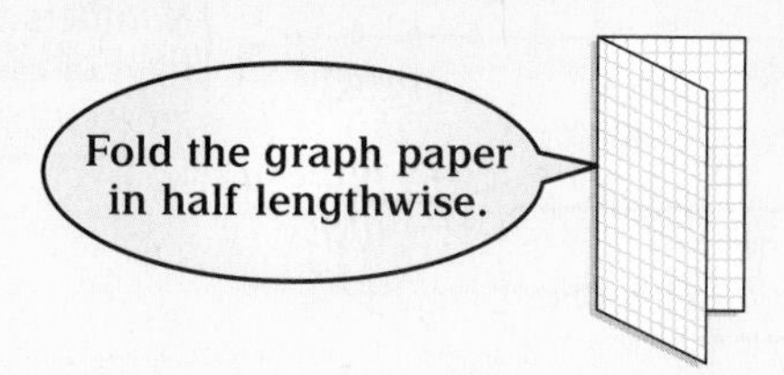

Fold the graph paper in half lengthwise.

Step 2 Fold Again in Fourths

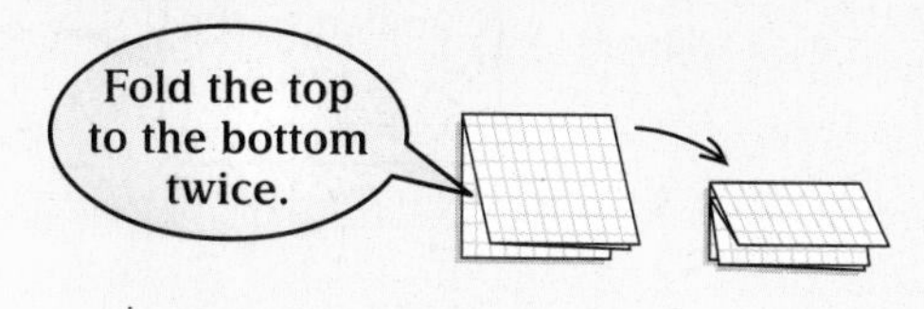

Fold the top to the bottom twice.

Step 3 Cut

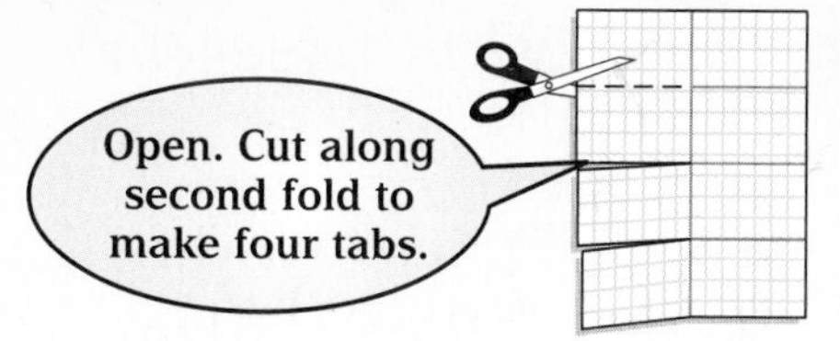

Open. Cut along second fold to make four tabs.

Step 4 Label

Fold lengthwise. Draw a number line on the outside. Label each tab as shown.

+ − × ÷

Reading and Writing As you read and study the chapter, write rules and examples for each integer operation under the tabs.

2-1 Integers and Absolute Value

Sunshine State Standards
MA.A.1.3.1-1, MA.A.1.3.2-1, MA.A.1.3.2-2, MA.A.1.3.4-4

Vocabulary

- negative number
- integers
- coordinate
- inequality
- absolute value

What You'll Learn

- Compare and order integers.
- Find the absolute value of an expression.

How are integers used to model real-world situations?

The summer of 1999 was unusually dry in parts of the United States. In the graph, a value of -8 represents 8 inches below the normal rainfall.

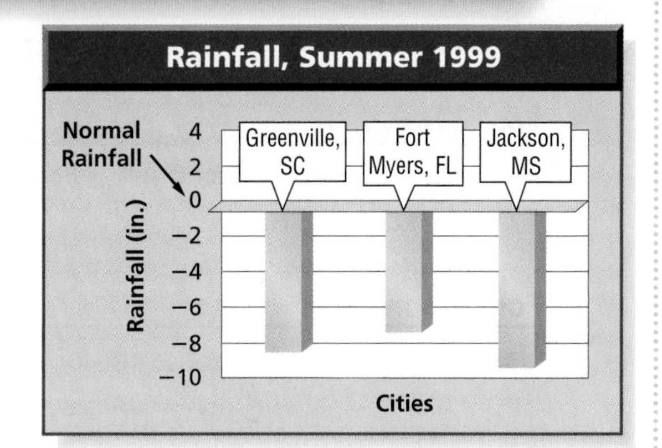

a. What does a value of -7 represent?

b. Which city was farthest from its normal rainfall?

c. How could you represent 5 inches above normal rainfall?

Reading Math

Integers

Read -8 as *negative 8.* A positive integer like 6 can be written as $+6$. It is usually written without the $+$ sign, as 6.

COMPARE AND ORDER INTEGERS With normal rainfall as the starting point of 0, you can express 8 inches below normal as $0 - 8$, or -8. A **negative number** is a number less than zero.

Negative numbers like -8, positive numbers like $+6$, and zero are members of the set of **integers**. Integers can be represented as points on a number line.

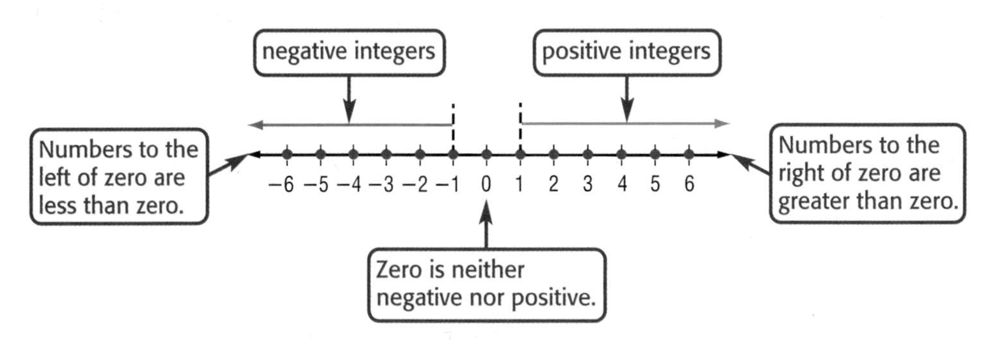

This set of integers can be written $\{\ldots, -3, -2, -1, 0, 1, 2, 3, \ldots\}$ where … means continues indefinitely.

Example 1 Write Integers for Real-World Situations

Write an integer for each situation.

a. 500 feet below sea level The integer is -500.

b. a temperature increase of 12° The integer is $+12$.

c. a loss of $240 The integer is -240.

Concept Check Which integer is neither positive nor negative?

To graph integers, locate the points named by the integers on a number line. The number that corresponds to a point is called the **coordinate** of that point.

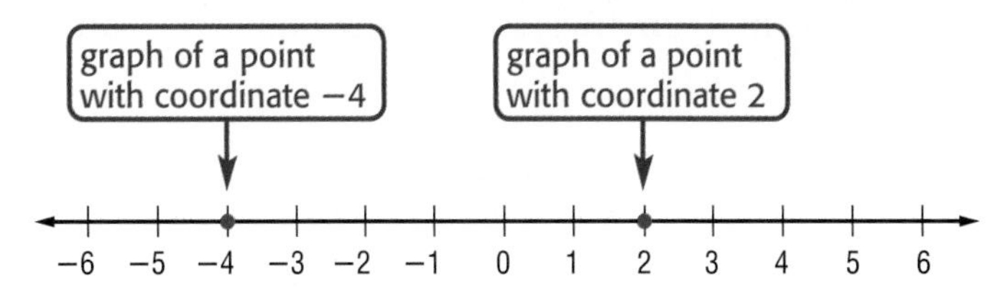

Notice that the numbers on a number line increase as you move from left to right. This can help you determine which of two numbers is greater.

Reading Math

Inequality Symbols
Read the symbol < as *is less than*. Read the symbol > as *is greater than*.

Words -4 is less than 2. OR 2 is greater than -4.

Symbols $-4 < 2$ $2 > -4$

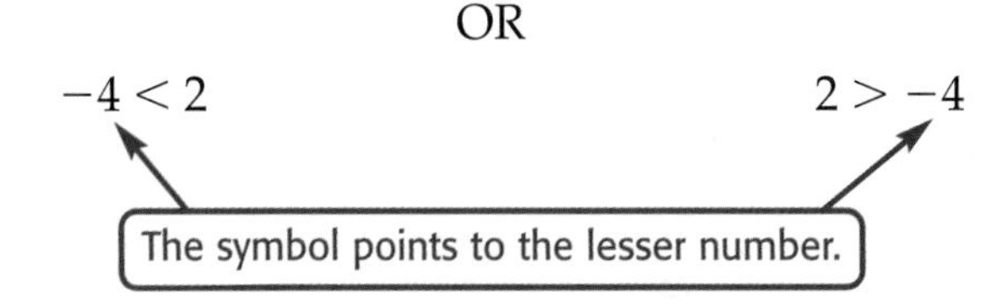

Any mathematical sentence containing < or > is called an inequality. An **inequality** compares numbers or quantities.

Example 2 Compare Two Integers

Use the integers graphed on the number line below.

−6 −5 −4 −3 −2 −1 0 1 2 3 4 5 6

a. **Write two inequalities involving −3 and 4.**
Since -3 is to the left of 4, write $-3 < 4$.
Since 4 is to the right of -3, write $4 > -3$.

b. **Replace the ● with < or > in −5 ● −1 to make a true sentence.**
-1 is greater since it lies to the right of -5. So write $-5 < -1$.

More About. . .

Golf

Karrie Webb finished the 2000 World Championship at 2 under par. In 2000, she was the LPGA's leading money winner at $1.8 million.

Source: www.LPGA.com

Integers are used to compare numbers in many real-world situations.

Example 3 Order Integers

GOLF The final round scores of the top ten finishers in the 2000 World Championship LPGA tournament were −4, −14, −1, +1, +2, +5, 0, +3, −10, and −2. Order the scores from least to greatest.

Graph each integer on a number line.

−14 −12 −10 −8 −6 −4 −2 0 2 4 6

Write the numbers as they appear from left to right.

The scores $-14, -10, -4, -2, -1, 0, +1, +2, +3, +5$ are in order from least to greatest.

Concept Check Why is the sentence $5 > 2$ an inequality?

ABSOLUTE VALUE On the number line, notice that -5 and 5 are on opposite sides of zero, and they are the same distance from zero. In mathematics, we say they have the same **absolute value**, 5.

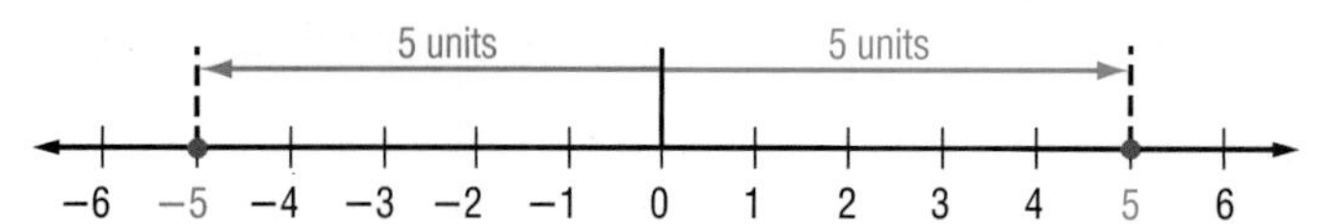

The symbol for absolute value is two vertical bars on either side of the number.

$|5| = 5$ The absolute value of 5 is 5.
$|-5| = 5$ The absolute value of -5 is 5.

Key Concept — *Absolute Value*

- **Words** The absolute value of a number is the distance the number is from zero on the number line. The absolute value of a number is always greater than or equal to zero.
- **Examples** $|5| = 5$ $|-5| = 5$

Example 4 *Expressions with Absolute Value*

Evaluate each expression.

a. $|-8|$

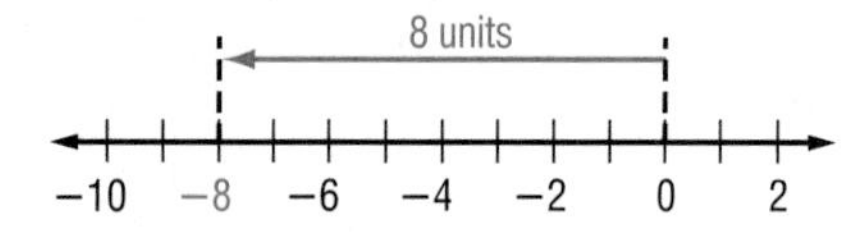

$|-8| = 8$ The graph of -8 is 8 units from 0.

b. $|9| + |-7|$ The absolute value of 9 is 9.

$|9| + |-7| = 9 + 7$ The absolute value of -7 is 7.

$= 16$ Simplify.

c. $|-4| - |3|$

$|-4| - |3| = 4 - 3$ $|-4| = 4$, $|3| = 3$

$= 1$ Simplify.

Study Tip

Common Misconception
It is not always true that the absolute value of a number is the opposite of the number. Remember that absolute value is always positive or zero.

ALGEBRA CONNECTION

Since variables represent numbers, you can use absolute value notation with algebraic expressions involving variables.

Example 5 *Algebraic Expressions with Absolute Value*

ALGEBRA **Evaluate $|x| - 3$ if $x = -5$.**

$|x| - 3 = |-5| - 3$ Replace x with -5.

$= 5 - 3$ The absolute value of -5 is 5.

$= 2$ Simplify.

Check for Understanding

Concept Check

1. **Explain** how you would graph −4 on a number line.
2. **OPEN ENDED** Write two inequalities using integers.
3. **Define** *absolute value.*

Guided Practice

Write an integer for each situation. Then graph on a number line.

4. 8° below zero
5. a 15-yard gain
6. Graph the set of integers {0, −3, 6} on a number line.

Write two inequalities using the numbers in each sentence. Use the symbols < or >.

7. −4° is colder than 2°.
8. −6 is greater than −10.

Replace each ● with <, >, or = to make a true sentence.

9. −18 ● −8
10. 0 ● −3
11. 9 ● −9
12. Order the integers {28, −6, 0, −2, 5, −52, 115} from least to greatest.

Evaluate each expression.

13. $|-10|$
14. $|10| - |-4|$
15. $|16| + |-5|$

ALGEBRA **Evaluate each expression if $a = -8$ and $b = 5$.**

16. $9 + |a|$
17. $|a| - b$
18. $2|a|$

Application

19. **WEATHER** The table shows the record low temperatures in °F for selected states. Order the temperatures from least to greatest.

State	AL	CA	FL	IN	KY	NY	NC	OK	OR
Temperature	−27	−45	−2	−36	−37	−52	−34	−27	−54

Practice and Apply

Homework Help

For Exercises	See Examples
20–25, 66	1
26–43	2
44–47, 67–70	3
48–59	4
60–65	5

Extra Practice

See page 726.

Write an integer for each situation. Then graph on a number line.

20. a bank withdrawal of $100
21. a loss of 6 pounds
22. a salary increase of $250
23. a gain of 9 yards
24. 12° above zero
25. 5 seconds before liftoff

Graph each set of integers on a number line.

26. {0, −2, 4}
27. {−3, 1, 2, 5}
28. {−2, −4, −5, −8}
29. {−4, 0, 6, −7, −1}

Write two inequalities using the numbers in each sentence. Use the symbols < or >.

30. 3 meters is taller than 2 meters.
31. A temperature of −5°F is warmer than a temperature of −10°F.
32. 55 miles per hour is slower than 65 miles per hour.

Write two inequalities using the numbers in each sentence. Use the symbols < or >.

33. Yesterday's pollen count was 248. Today's count is 425.

34. Yesterday's low temperature was -2°F. The high temperature was 23°F.

35. Water boils at 212°F, and it freezes at 32°F.

Replace each ● with <, >, or = to make a true sentence.

36. $-6 \bullet -2$	**37.** $-10 \bullet -13$	**38.** $0 \bullet -9$	**39.** $14 \bullet 0$
40. $-18 \bullet 8$	**41.** $5 \bullet -23$	**42.** $\lvert 9 \rvert \bullet \lvert -9 \rvert$	**43.** $\lvert -20 \rvert \bullet \lvert -4 \rvert$

Order the integers in each set from least to greatest.

44. {5, 0, -8}

45. {-15, -1, -2, -4}

46. {24, 5, -46, 9, 0, -3}

47. {98, -57, -60, 38, 188}

Evaluate each expression.

48. $\lvert -15 \rvert$	**49.** $\lvert 46 \rvert$	**50.** $-\lvert 20 \rvert$	**51.** $-\lvert 5 \rvert$
52. $\lvert 0 \rvert$	**53.** $\lvert 7 \rvert$	**54.** $\lvert -5 \rvert + \lvert 4 \rvert$	**55.** $\lvert 0 \rvert + \lvert -2 \rvert$
56. $\lvert 15 \rvert - \lvert -1 \rvert$	**57.** $\lvert 0 + 9 \rvert$	**58.** $-\lvert -24 \rvert$	**59.** $-\lvert \lvert -6 \rvert + \lvert 14 \rvert \rvert$

ALGEBRA **Evaluate each expression if $a = 0$, $b = 3$, and $c = -4$.**

60. $14 + \lvert b \rvert$	**61.** $\lvert c \rvert - a$	**62.** $a + b + \lvert c \rvert$
63. $ab + \lvert -40 \rvert$	**64.** $\lvert c \rvert - b$	**65.** $\lvert ab \rvert + b$

66. **GEOGRAPHY** The Caribbean Sea has an average depth of 8685 feet below sea level. Use an integer to express this depth.

WEATHER **For Exercises 67–70, use the graphic.**

67. Graph the temperatures on a number line.

68. Compare the lowest temperature in the United States and the lowest temperature east of the Mississippi using the < symbol.

69. Compare the lowest temperatures of the contiguous 48 states and east of the Mississippi using the > symbol.

70. Write the temperatures in order from greatest to least.

71. How many units apart are -4 and 3 on a number line?

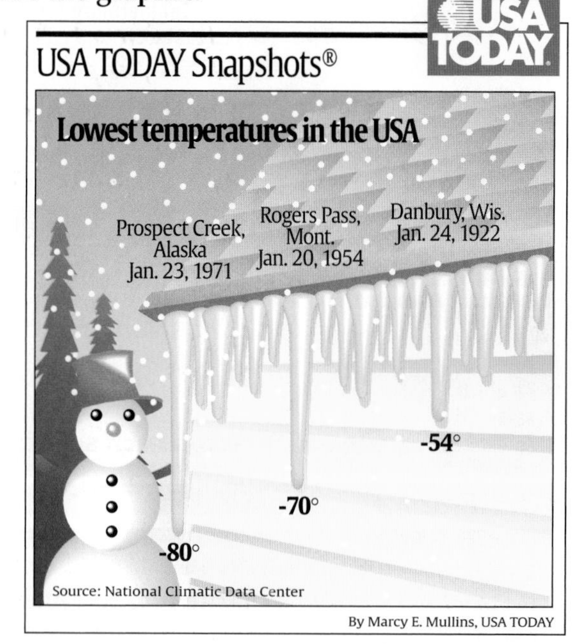

Study Tip

Contiguous States
Contiguous states are those states that touch each other. Alaska and Hawaii are *not* contiguous states.

72. **CRITICAL THINKING** Consider any two points on the number line where $X > Y$. Is it *always*, *sometimes*, or *never* true that $\lvert X \rvert > \lvert Y \rvert$? Explain.

73. **CRITICAL THINKING** Consider two numbers A and B on a number line. Is it *always, sometimes,* or *never* true that the distance between A and B equals the distance between $|A|$ and $|B|$? Explain.

74. **WRITING IN MATH** Answer the question that was posed at the beginning of the lesson.

How are integers used to model real-world situations?

Include the following in your answer:
- an explanation of how integers are used to describe rainfall, and
- some situations in the real world where negative numbers are used.

75. Which of the following describes the absolute value of $-2°$?

Ⓐ It is the distance from -2 to 2 on a thermometer.

Ⓑ It is the distance from -2 to 0 on a thermometer.

Ⓒ It is the actual temperature outside when a thermometer reads $-2°$.

Ⓓ None of these describes the absolute value of $-2°$.

76. What is the temperature shown on the thermometer at the right?

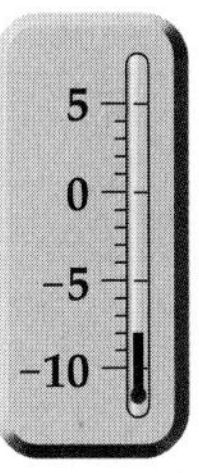

Ⓐ 8

Ⓑ 7

Ⓒ -7

Ⓓ -8

Maintain Your Skills

Mixed Review

Determine whether a scatter plot of the data for the following might show a *positive, negative,* or *no* relationship. Explain your answer. *(Lesson 1-7)*

77. height and arm length

78. birth month and weight

Express each relation as a table and as a list of ordered pairs. *(Lesson 1-6)*

79. 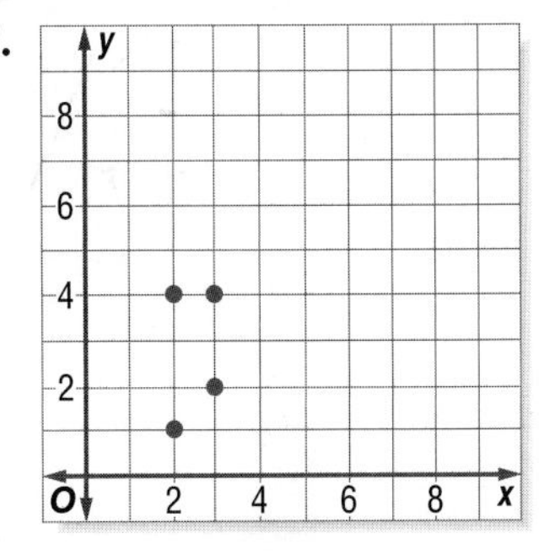

80. 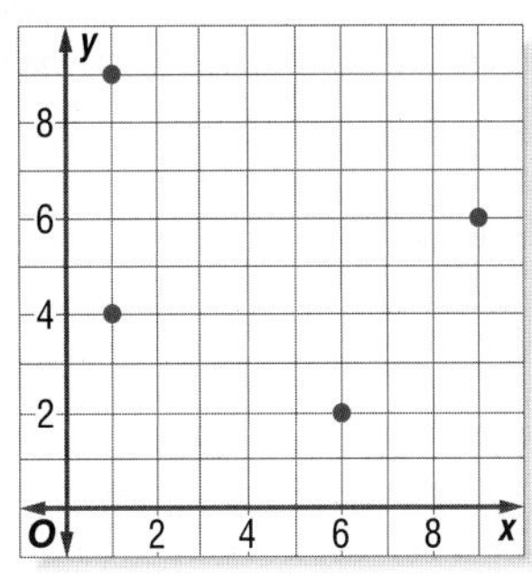

Name the property shown by each statement. *(Lesson 1-4)*

81. $20 \cdot 18 = 18 \cdot 20$

82. $9 \cdot 8 \cdot 0 = 0$

83. $3ab = 3ba$

Getting Ready for the Next Lesson

BASIC SKILL Find each sum or difference.

84. $18 + 29 + 46$

85. $232 + 156$

86. $451 + 629 + 1027$

87. $36 - 19$

88. $479 - 281$

89. $2011 - 962$

Algebra Activity

A Preview of Lesson 2-2

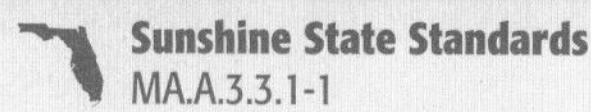

Sunshine State Standards
MA.A.3.3.1-1

Adding Integers

In a set of algebra tiles, [1] represents the integer 1, and [−1] represents the integer −1. You can use algebra tiles and an integer mat to model operations with integers.

Activity 1

The following example shows how to find the sum $-3 + (-2)$ using algebra tiles. Remember that addition means *combining*. $-3 + (-2)$ tells you to combine a set of 3 negative tiles with a set of 2 negative tiles.

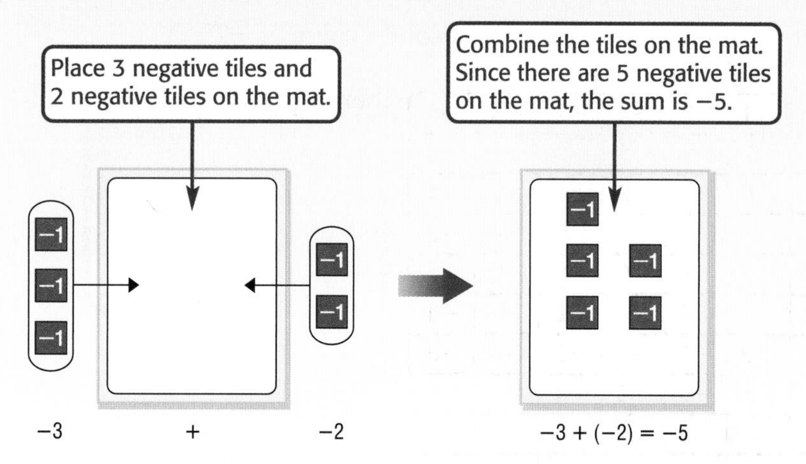

Place 3 negative tiles and 2 negative tiles on the mat.
Combine the tiles on the mat. Since there are 5 negative tiles on the mat, the sum is −5.
−1
−1
−1
−1
−1
−3
+
−2
−3 + (−2) = −5

Therefore, $-3 + (-2) = -5$.

There are two important properties to keep in mind when you model operations with integers.

- When one positive tile is paired with one negative tile, the result is called a **zero pair**.
- You can add or remove zero pairs from a mat because removing or adding zero does not change the value of the tiles on the mat.

The following example shows how to find the sum $-4 + 3$.

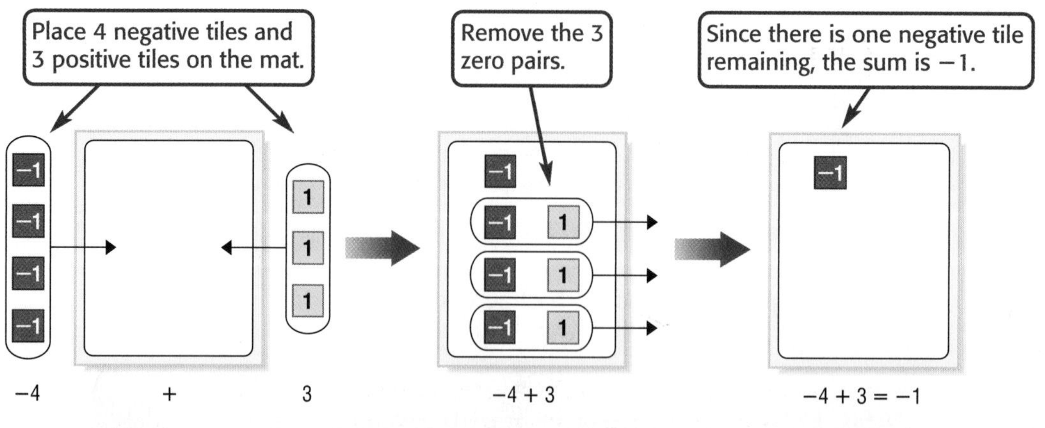

Place 4 negative tiles and 3 positive tiles on the mat.
Remove the 3 zero pairs.
Since there is one negative tile remaining, the sum is −1.
−1
1
−4
+
3
−4 + 3
−4 + 3 = −1

Therefore, $-4 + 3 = -1$.

Model

Use algebra tiles to model and find each sum.

1. $-2 + (-4)$ **2.** $-3 + (-5)$ **3.** $-6 + (-1)$ **4.** $-4 + (-5)$

5. $-4 + 2$ **6.** $2 + (-5)$ **7.** $-1 + 6$ **8.** $4 + (-4)$

Activity 2

The Addition Table was completed using algebra tiles. In the highlighted portion of the table, the addends are -3 and 1, and the sum is -2. So, $-3 + 1 = -2$. You can use the patterns in the Addition Table to learn more about integers.

Addition Table

+	4	3	2	1	0	−1	−2	−3	−4
4	8	7	6	5	4	3	2	1	0
3	7	6	5	4	3	2	1	0	−1
2	6	5	4	3	2	1	0	−1	−2
1	5	4	3	2	1	0	−1	−2	−3
0	4	3	2	1	0	−1	−2	−3	−4
−1	3	2	1	0	−1	−2	−3	−4	−5
−2	2	1	0	−1	−2	−3	−4	−5	−6
−3	1	0	−1	−2	−3	−4	−5	−6	−7
−4	0	−1	−2	−3	−4	−5	−6	−7	−8

← addends (top row); sums (body); addends (left column)

Make a Conjecture

9. Locate all of the positive sums in the table. Describe the addends that result in a positive sum.

10. Locate all of the negative sums in the table. Describe the addends that result in a negative sum.

11. Locate all of the sums that are zero. Describe the addends that result in a sum of zero.

12. The Identity Property says that when zero is added to any number, the sum is the number. Does it appear that this property is true for addition of integers? If so, write two examples that illustrate the property. If not, give a counterexample.

13. The Commutative Property says that the order in which numbers are added does not change the sum. Does it appear that this property is true for addition of integers? If so, write two examples that illustrate the property. If not, give a counterexample.

14. The Associative Property says that the way numbers are grouped when added does not change the sum. Is this property true for addition of integers? If so, write two examples that illustrate the property. If not, give a counterexample.

2-2 Adding Integers

What You'll Learn

- Add two integers.
- Add more than two integers.

Sunshine State Standards
MA.A.3.3.1-1, MA.A.3.3.1-2, MA.A.3.3.1-3, MA.A.3.3.2-1, MA.A.3.3.2-2

Vocabulary

- opposites
- additive inverse

How can a number line help you add integers?

In football, forward progress is represented by a positive integer. Being pushed back is represented by a negative integer. Suppose on the first play a team loses 5 yards and on the second play they lose 2 yards.

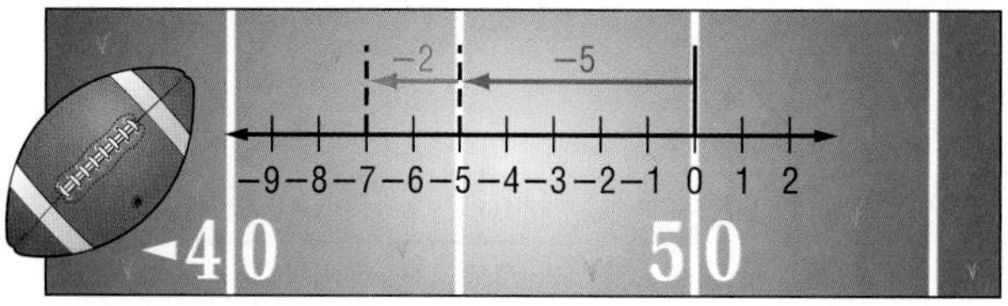

a. What integer represents the total yardage on the two plays?

b. Write an addition sentence that describes this situation.

Reading Math

Addends and Sums
Recall that the numbers you add are called *addends*. The result is called the *sum*.

ADD INTEGERS The equation $-5 + (-2) = -7$ is an example of adding two integers with the same sign. Notice that the sign of the sum is the same as the sign of the addends.

Example 1 Add Integers on a Number Line

Find $-2 + (-3)$.

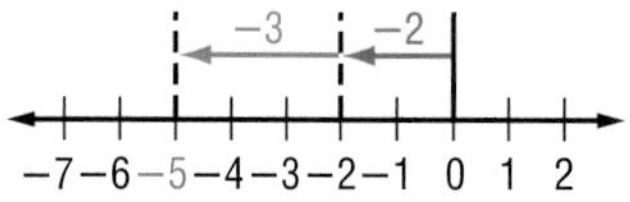

Start at zero.
Move 2 units to the left.
From there, move 3 more units to the left.

$-2 + (-3) = -5$

This example suggests a rule for adding integers with the same sign.

Key Concept — Adding Integers with the Same Sign

- **Words** To add integers with the same sign, add their absolute values. Give the result the same sign as the integers.
- **Examples** $-5 + (-2) = -7$ $\quad 6 + 3 = 9$

Example 2 Add Integers with the Same Sign

Find $-4 + (-5)$.

$-4 + (-5) = -9$ Add $|-4|$ and $|-5|$. Both numbers are negative, so the sum is negative.

A number line can also help you understand how to add integers with different signs.

Example 3 Add Integers on a Number Line

Find each sum.

a. 7 + (−4)

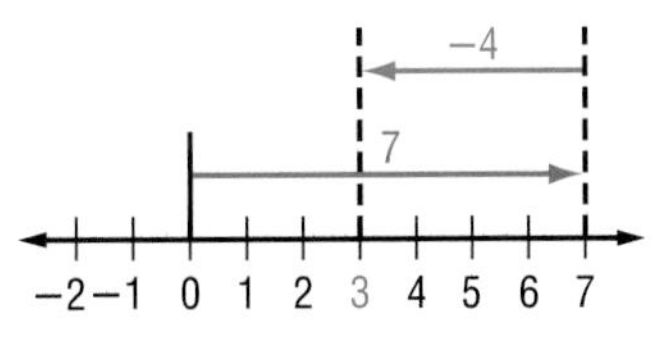

Start at zero.
Move 7 units to the right.
From there, move 4 units to the left.

$7 + (-4) = 3$

Study Tip

Adding Integers on a Number Line
Always start at zero. Move right to model a positive integer. Move left to model a negative integer.

b. 2 + (−3)

−3
2
−4 −3 −2 −1 0 1 2 3 4 5

Start at zero.
Move 2 units to the right.
From there, move 3 units to the left.

$2 + (-3) = -1$

Notice how the sums in Example 3 relate to the addends.

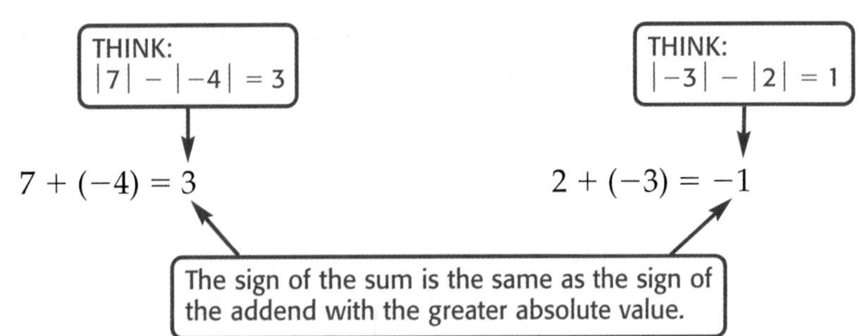

Key Concept — *Adding Integers with Different Signs*

- **Words** To add integers with different signs, subtract their absolute values. Give the result the same sign as the integer with the greater absolute value.
- **Examples** $7 + (-2) = 5$ $\quad -7 + 2 = -5$

Example 4 Add Integers with Different Signs

Find each sum.

a. −8 + 3

$-8 + 3 = -5$ To find $-8 + 3$, subtract $|3|$ from $|-8|$.
The sum is negative because $|-8| > |3|$.

b. 10 + (−4)

$10 + (-4) = 6$ To find $10 + |-4|$, subtract $|-4|$ from $|10|$.
The sum is positive because $|10| > |-4|$.

Example 5 Use Integers to Solve a Problem

ASTRONOMY **During the night, the average temperature on the moon is $-140°C$. By noon, the average temperature has risen $252°C$. What is the average temperature on the moon at noon?**

Words The temperature at night is $-140°C$. It increases $252°C$ by noon. What is the temperature at noon?

Variables Let x = the temperature at noon.

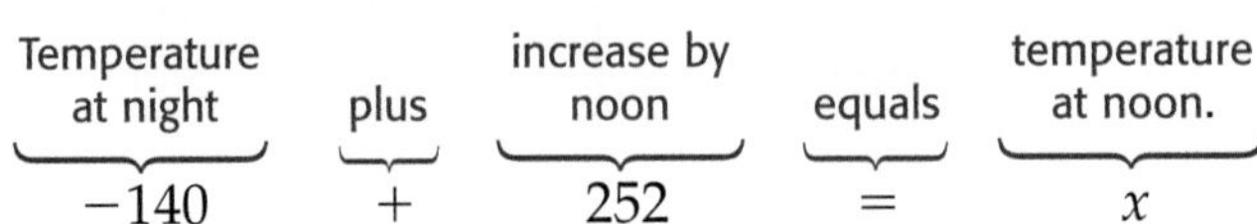

	Temperature at night	plus	increase by noon	equals	temperature at noon.
Equation	-140	$+$	252	$=$	x

Solve the equation.

$-140 + 252 = x$ To find the sum, subtract $|-140|$ from 252.

$112 = x$ The sum is positive because $|252| > |-140|$.

The average temperature at noon is 112°C.

More About. . .

Astronomy

The temperatures on the moon are so extreme because the moon does not have any atmosphere to trap heat.

ADD MORE THAN TWO INTEGERS Two numbers with the same absolute value but different signs are called **opposites**. For example, -4 and 4 are opposites. An integer and its opposite are also called **additive inverses**.

Key Concept — Additive Inverse Property

- **Words** The sum of any number and its additive inverse is zero.
- **Symbols** $x + (-x) = 0$
- **Example** $6 + (-6) = 0$

Concept Check What is the additive inverse of 2? What is the additive inverse of -6?

The commutative, associative, and identity properties also apply to integers. These properties can help you add more than two integers.

Example 6 Add Three or More Integers

Find each sum.

a. $9 + (-3) + (-9)$

$9 + (-3) + (-9) = 9 + (-9) + (-3)$ Commutative Property

$= 0 + (-3)$ Additive Inverse Property

$= -3$ Identity Property of Addition

b. $-4 + 6 + (-3) + 9$

$-4 + 6 + (-3) + 9 = -4 + (-3) + 6 + 9$ Commutative Property

$= [-4 + (-3)] + (6 + 9)$ Associative Property

$= -7 + 15$ or 8 Simplify.

Study Tip

Adding Mentally One way to add mentally is to group the positive addends together and the negative addends together. Then add to find the sum. You should also look for addends that are opposites. You can always add in order from left to right.

Check for Understanding

Concept Check

1. **State** whether each sum is positive or negative. Explain your reasoning.
   a. $-4 + (-5)$ b. $12 + (-2)$
   c. $-11 + 9$ d. $15 + 10$

2. **OPEN ENDED** Give an example of two integers that are additive inverses.

Guided Practice

Find each sum.

3. $-2 + (-4)$
4. $-10 + (-5)$
5. $7 + (-2)$
6. $11 + (-3)$
7. $8 + (-5)$
8. $9 + (-12)$
9. $8 + (-6) + 2$
10. $-6 + 5 + (-10)$

Application

11. **FOOTBALL** A team gained 4 yards on one play. On the next play, they lost 5 yards. Write an addition sentence to find the change in yardage.

Practice and Apply

Homework Help

For Exercises	See Examples
12–21	1, 2
22–29	3, 4
32–39	6
40, 41	5

Extra Practice

See page 726.

Find each sum.

12. $-4 + (-1)$
13. $-5 + (-2)$
14. $-4 + (-6)$
15. $-3 + (-8)$
16. $-7 + (-8)$
17. $-12 + (-4)$
18. $-9 + (-14)$
19. $-15 + (-6)$
20. $-11 + (-15)$
21. $-23 + (-43)$
22. $8 + (-5)$
23. $6 + (-4)$
24. $3 + (-7)$
25. $4 + (-6)$
26. $-15 + 6$
27. $-5 + 11$
28. $18 + (-32)$
29. $-45 + 19$

30. What is the additive inverse of 14?
31. What is the additive inverse of -21?

Find each sum.

32. $6 + (-9) + 9$
33. $7 + (-13) + 4$
34. $-9 + 16 + (-10)$
35. $-12 + 18 + (-12)$
36. $14 + (-9) + 6$
37. $28 + (-35) + 4$
38. $-41 + 25 + (-10)$
39. $-18 + 35 + (-17)$

40. **ACCOUNTING** The starting balance in a checking account was \$50. What was the balance after checks were written for \$25 and for \$32?

41. **GOLF** A score of 0 is called *even par*. Two under par is written as -2. Two over par is written as $+2$. Suppose a player shot 4 under par, 2 over par, even par, and 3 under par in four rounds of a tournament. What was the player's final score?

Find each sum.

42. $|18 + (-13)|$
43. $|-27 + 19|$
44. $|-25 + (-12)|$
45. $|-28 + (-12)|$

POPULATION For Exercises 46 and 47, use the table below that shows the change in population of several cities from 1990 to 2000.

City	1990 Population	Change as of 2000
Dallas, TX	1,006,877	+181,703
Honolulu, HI	365,272	+6385
Jackson, MS	196,637	−12,381
Philadelphia, PA	1,585,577	−68,027

46. What was the population in each city in 2000?

47. What was the total change in population of these cities?

Online Research **Data Update** How have the populations of other cities changed since 2000? Visit www.pre-alg.com/data_update to learn more.

48. CRITICAL THINKING *True* or *false*: $-n$ always names a negative number. If false, give a counterexample.

49. WRITING IN MATH Answer the question that was posed at the beginning of the lesson.

How can a number line help you add integers?

Include the following in your answer:

- an example showing the sum of a positive and a negative integer, and
- an example showing the sum of two negative integers.

FCAT Practice
Standardized Test Practice

50. What is the sum of $-32 + 20$?

Ⓐ -52 Ⓑ -18 Ⓒ -12 Ⓓ 12

51. What is the value of $-|-2 + 8|$?

Ⓐ -10 Ⓑ 10 Ⓒ 6 Ⓓ -6

Maintain Your Skills

Mixed Review

52. CHEMISTRY The freezing point of oxygen is 219 degrees below zero on the Celsius scale. Use an integer to express this temperature. *(Lesson 2-1)*

Order the integers in each set from least to greatest. *(Lesson 2-1)*

53. $\{14, -12, -8, 3, -9, 0\}$

54. $\{-242, 35, -158, 99, -24\}$

Determine whether a scatter plot of the data for the following might show a *positive*, *negative*, or *no* relationship. *(Lesson 1-7)*

55. age and family size

56. temperature and sales of mittens

Identify the solution of each equation from the list given. *(Lesson 1-5)*

57. $18 - n = 12; 6, 16, 30$

58. $25 = 16 + x; 9, 11, 41$

59. $\frac{x}{2} = 10; 5, 12, 20$

60. $7a = 49; 7, 42, 343$

Getting Ready for the Next Lesson

PREREQUISITE SKILL Evaluate each expression if $a = 6$, $b = 10$, and $c = 3$.
*(To review **evaluating expressions**, see Lesson 1-3.)*

61. $a + 19$

62. $2b - 6$

63. $ab - ac$

64. $3a - (b + c)$

65. $5b + 5c$

66. $\frac{6b}{c}$

Reading Mathematics

Learning Mathematics Vocabulary

Some words used in mathematics are also used in English and have similar meanings. For example, in mathematics *add* means *to combine*. The meaning in English is *to join or unite*.

Some words are used only in mathematics. For example, *addend* means *a number to be added to another*.

Some words have more than one mathematical meaning. For example, an *inverse* operation *undoes the effect of another operation*, and an additive *inverse* is *a number that when added to a given number gives zero*.

The list below shows some of the mathematics vocabulary used in Chapters 1 and 2.

Vocabulary	Meaning	Examples
algebraic expression	an expression that contains at least one variable and at least one mathematical operation	$2 + x, \frac{4}{c}, 3b$
evaluate	to find the value of an expression	$2 + 5 = 7$
simplify	to find a simpler form of an expression	$3b + 2b = 5b$
integer	a whole number, its inverse, or zero	$-3, 0, 2$
factor	a number that is multiplied by another number	$3(4) = 12$ 3 and 4 are factors.
product	the result of multiplying	$3(4) = 12$ ⟵ product
quotient	the result of dividing two numbers	$\frac{12}{4} = 3$ ⟵ quotient
dividend	the number being divided	$\frac{12}{4} = 3$ dividend (12)
divisor	the number being divided into another number	$\frac{12}{4} = 3$ divisor (4)
coordinate	a number that locates a point	(5, 2)

Reading to Learn

1. Name two of the words above that are also used in everyday English. Use the Internet, a dictionary, or another reference to find their everyday definition. How do the everyday definitions relate to the mathematical definitions?
2. Name two words above that are used only in mathematics.
3. Name two words above that have more than one mathematical meaning. List their meanings.

2-3 Subtracting Integers

Sunshine State Standards
MA.A.3.3.1-1, MA.A.3.3.1-2, MA.A.3.3.1-3, MA.A.3.3.2-1, MA.A.3.3.2-2, MA.D.2.3.1-6

What You'll Learn

- Subtract integers.
- Evaluate expressions containing variables.

How are addition and subtraction of integers related?

You can use a number line to subtract integers. The model below shows how to find $6 - 8$.

Step 1 Start at 0. Move 6 units right to show positive 6.

Step 2 From there, move 8 units left to subtract positive 8.

−8
+6
−5 −4 −3 −2 −1 0 1 2 3 4 5 6 7

a. What is $6 - 8$?

b. What direction do you move to indicate subtracting a positive integer?

c. What addition sentence is also modeled by the number line above?

SUBTRACT INTEGERS When you subtract $6 - 8$, as shown on the number line above, the result is the same as adding $6 + (-8)$. When you subtract $-3 - 5$, the result is the same as adding $-3 + (-5)$.

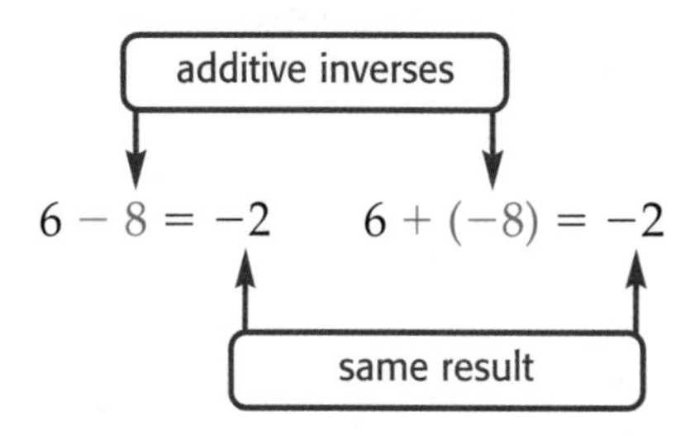

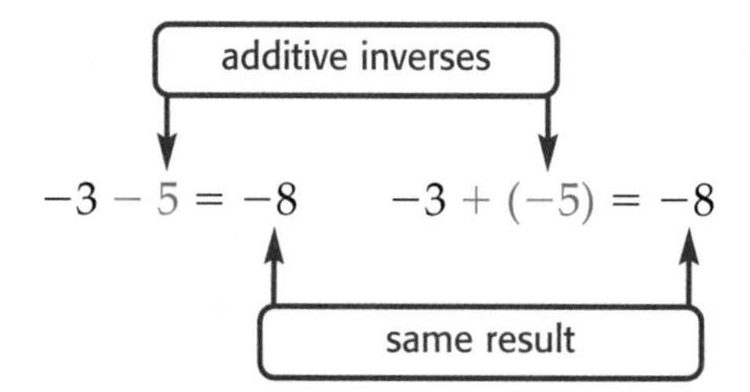

These examples suggest a method for subtracting integers.

Key Concept — Subtracting Integers

- **Words** To subtract an integer, add its additive inverse.
- **Symbols** $a - b = a + (-b)$
- **Examples** $5 - 9 = 5 + (-9)$ or -4 $-2 - 7 = -2 + (-7)$ or -9

Study Tip

Subtracting a Positive Integer
To subtract a positive integer, think about moving left on a number line from the starting integer. In Example 1a, start at 8, then move left 13. You'll end at −5. In Example 1b, start at −4, then move left 10. You'll end at −14.

Example 1 Subtract a Positive Integer

Find each difference.

a. 8 − 13

$8 - 13 = 8 + (-13)$ To subtract 13, add −13.

$= -5$ Simplify.

b. −4 − 10

$-4 - 10 = -4 + (-10)$ To subtract 10, add −10.

$= -14$ Simplify.

In Example 1, you subtracted a positive integer by adding its additive inverse. Use inductive reasoning to see if the method also applies to subtracting a negative integer.

Study Tip

Look Back
To review **inductive reasoning**, see Lesson 1-1.

Subtracting an Integer	↔	**Adding Its Additive Inverse**
$2 - 2 = 0$		$2 + (-2) = 0$
$2 - 1 = 1$		$2 + (-1) = 1$
$2 - 0 = 2$		$2 + 0 = 2$
$2 - (-1) = ?$		$2 + 1 = 3$

Continuing the pattern in the first column, $2 - (-1) = 3$. The result is the same as when you add the additive inverse. This suggests that the method also works for subtracting a negative integer.

Example 2 Subtract a Negative Integer

Find each difference.

a. 7 − (−3)

$7 - (-3) = 7 + 3$ To subtract −3, add 3.

$= 10$

b. −2 − (−4)

$-2 - (-4) = -2 + 4$ To subtract −4, add 4.

$= 2$

Concept Check How do you find the difference $9 - (-16)$?

Example 3 Subtract Integers to Solve a Problem

WEATHER The table shows the record high and low temperatures recorded in selected states through 1999. What is the range, or difference between the highest and lowest temperatures, for Virginia?

State	Lowest Temp. °F	Highest Temp. °F
Utah	−69	117
Vermont	−50	105
Virginia	−30	110
Washington	−48	118
West Virginia	−37	112
Wisconsin	−54	114
Wyoming	−66	114

Source: *The World Almanac*

Explore You know the highest and lowest temperatures. You need to find the range for Virginia's temperatures.

Plan To find the range, or difference, subtract the lowest temperature from the highest temperature.

Solve $110 - (-30) = 110 + 30$ To subtract −30, add 30.

$= 140$ Add 110 and 30.

The range for Virginia is 140°.

Examine Think of a thermometer. The difference between 110° above zero and 30° below zero must be 110 + 30 or 140°. The answer appears to be reasonable.

ALGEBRA CONNECTION

EVALUATE EXPRESSIONS You can use the rule for subtracting integers to evaluate expressions.

Example 4 Evaluate Algebraic Expressions

a. Evaluate $x - (-6)$ if $x = 12$.

$x - (-6) = 12 - (-6)$ Write the expression. Replace x with 12.

$= 12 + 6$ To subtract -6, add its additive inverse, 6.

$= 18$ Add 12 and 6.

b. Evaluate $s - t$ if $s = -9$ and $t = -3$.

$s - t = -9 - (-3)$ Replace s with -9 and t with -3.

$= -9 + 3$ To subtract -3, add 3.

$= -6$ Add -9 and 3.

c. Evaluate $a - b + c$ if $a = 15$, $b = 5$, and $c = -8$.

$a - b + c = 15 - 5 + (-8)$ Replace a with 15, b with 5, and c with -8.

$= 10 + (-8)$ Order of operations

$= 2$ Add 10 and -8.

Concept Check How do you subtract integers using additive inverses?

Check for Understanding

Concept Check

1. **OPEN ENDED** Write examples of a positive and a negative integer and their additive inverses.

2. **FIND THE ERROR** José and Reiko are finding $8 - (-2)$.

José	Reiko
$8 - (-2) = 8 + 2$	$8 - (-2) = 8 + (-2)$
$= 10$	$= 6$

Who is correct? Explain your reasoning.

Guided Practice

Find each difference.

3. $8 - 11$
4. $-9 - 3$
5. $5 - (-4)$
6. $7 - (-10)$
7. $-6 - (-4)$
8. $-2 - (-8)$

ALGEBRA **Evaluate each expression if $x = 10$, $y = -4$, and $z = -15$.**

9. $x - (-10)$
10. $y - x$
11. $x + y - z$

Application

WEATHER **For Exercises 12 and 13, use the table in Example 3 on page 71.**

12. Find the range in temperature for Vermont.
13. Name a state that has a greater range than Vermont's.

Practice and Apply

Homework Help

For Exercises	See Examples
14–21, 30–33	1
22–29, 34–37	2
38, 39	3
40–51	4

Extra Practice

See page 727.

Find each difference.

14. $3 - 8$
15. $4 - 5$
16. $2 - 9$
17. $9 - 12$
18. $-3 - 1$
19. $-5 - 4$
20. $-6 - 7$
21. $-4 - 8$
22. $6 - (-8)$
23. $4 - (-6)$
24. $7 - (-4)$
25. $9 - (-3)$
26. $-9 - (-7)$
27. $-7 - (-10)$
28. $-11 - (-12)$
29. $-16 - (-7)$
30. $10 - 24$
31. $45 - 59$
32. $-27 - 14$
33. $-16 - 12$
34. $48 - (-50)$
35. $125 - (-114)$
36. $-320 - (-106)$
37. $-2200 - (-3500)$

38. **WEATHER** During January, the normal high temperature in Duluth, Minnesota, is 16°F, and the normal low temperature is −2°F. Find the difference between the temperatures.

39. **GEOGRAPHY** The highest point in California is Mount Whitney, with an elevation of 14,494 feet. The lowest point is Death Valley, elevation −282 feet. Find the difference in the elevations.

ALGEBRA **Evaluate each expression if $x = -3$, $y = 8$, and $z = -12$.**

40. $y - 10$
41. $12 - z$
42. $3 - x$
43. $z - 24$
44. $x - y$
45. $z - x$
46. $y - z$
47. $z - y$
48. $x + y - z$
49. $z - y + x$
50. $x - y - z$
51. $z - y - x$

Career Choices

Veterinarian

Veterinarians work with animals to diagnose, treat, and prevent disease, disorders, and injuries.

Online Research For information about a career as a veterinarian, visit: www.pre-alg.com/careers

PETS **For Exercises 52 and 53, use the following table.**

52. Describe the change in the number of dogs of each breed registered from Year 1 to Year 2.

53. What was the total change in the number of dogs of these breeds registered from Year 1 to Year 2?

Registration in American Kennel Club

Breed	Year 1	Year 2
Airedale Terrier	2891	2950
Beagle	53,322	49,080
Chinese Shar-Pei	8614	6845
Chow Chow	6241	4342
Labrador Retriever	157,936	154,897
Pug	21,487	21,555

Source: www.akc.org

54. **BUSINESS** The formula $P = I - E$ is used to find the profit (P) when income (I) and expenses (E) are known. One month a small business has income of $19,592 and expenses of $20,345.
    a. What is the profit for the month?
    b. What does a negative profit mean?

55. **CRITICAL THINKING** Determine whether each statement is *true* or *false*. If false, give a counterexample.
    a. Subtraction of integers is commutative.
    b. Subtraction of integers is associative.

56. **WRITING IN MATH** Answer the question that was posed at the beginning of the lesson.

How are addition and subtraction of integers related?

Include the following in your answer:
- a model that shows how to find the difference $4 - 10$, and
- the expression $4 - 10$ rewritten as an addition expression using the additive inverse.

57. The terms in a pattern are given in the table. What is the value of the 5th term?

Term	1	2	3	4	5
Value	13	8	3	−2	?

(A) −7 (B) −5
(C) 7 (D) 5

58. When 5 is subtracted from a number, the result is −4. What is the number?

(A) 9 (B) 1 (C) −1 (D) −9

Maintain Your Skills

Mixed Review

59. **OCEANOGRAPHY** A submarine at 1300 meters below sea level descends an additional 1150 meters. What integer represents the submarine's position with respect to sea level? *(Lesson 2-2)*

60. **ALGEBRA** Evaluate $|b| - |a|$ if $a = 2$ and $b = -4$. *(Lesson 2-1)*

ALGEBRA Solve each equation mentally. *(Lesson 1-5)*

61. $x + 9 = 12$ 62. $18 = w - 2$ 63. $5a = 35$ 64. $\frac{64}{b} = 8$

ALGEBRA Translate each phrase into an algebraic expression. *(Lesson 1-3)*

65. a number divided by 5
66. the sum of t and 9
67. the quotient of eighty-six and b
68. s decreased by 8

Find the value of each expression. *(Lesson 1-2)*

69. $2 \times (5 + 8) - 6$
70. $96 \div (6 \times 8) \div 2$

Getting Ready for the Next Lesson

BASIC SKILL Find each product.

71. $5 \cdot 15$ 72. $8 \cdot 12$ 73. $3 \cdot 5 \cdot 8$ 74. $2 \cdot 7 \cdot 5 \cdot 9$

Practice Quiz 1 — *Lessons 2-1 through 2-3*

1. **WEATHER** The three states with the lowest recorded temperatures are Alaska at −80°F, Utah at −69°F, and Montana at −70°F. Order the temperatures from least to greatest. *(Lesson 2-1)*

Find each sum. *(Lesson 2-2)*

2. $-5 + (-15)$
3. $-5 + 11$
4. $-6 + 9 + (-8)$

Find each difference. *(Lesson 2-3)*

5. $16 - 23$
6. $-15 - 8$
7. $25 - (-7)$

ALGEBRA Evaluate each expression if $x = 5$, $y = -2$, and $z = -3$. *(Lesson 2-3)*

8. $x - y$
9. $z - 6$
10. $x - y - z$

2-4 Multiplying Integers

Sunshine State Standards
MA.A.1.3.4-2, MA.A.3.3.1-1, MA.A.3.3.1-3, MA.A.3.3.2-1, MA.A.3.3.2-2, MA.D.2.3.1-7, MA.D.2.3.2-1

What You'll Learn

- Multiply integers.
- Simplify algebraic expressions.

How are the signs of factors and products related?

The temperature drops 7°C for each 1 kilometer increase in altitude. A drop of 7°C is represented by −7. So, the temperature change equals the altitude times −7. The table shows the change in temperature for several altitudes.

Altitude (km)	Altitude × Rate of Change	Temperature Change (°C)
1	1(−7)	−7
2	2(−7)	−14
3	3(−7)	−21
...	...	...
11	11(−7)	−77

a. Suppose the altitude is 4 kilometers. Write an expression to find the temperature change.

b. Use the pattern in the table to find 4(−7).

Reading Math

Parentheses
Recall that a product can be written using parentheses. Read 3(−7) as *3 times negative 7.*

MULTIPLY INTEGERS Multiplication is repeated addition. So, 3(−7) means that −7 is used as an addend 3 times.

$3(-7) = (-7) + (-7) + (-7)$
$= -21$

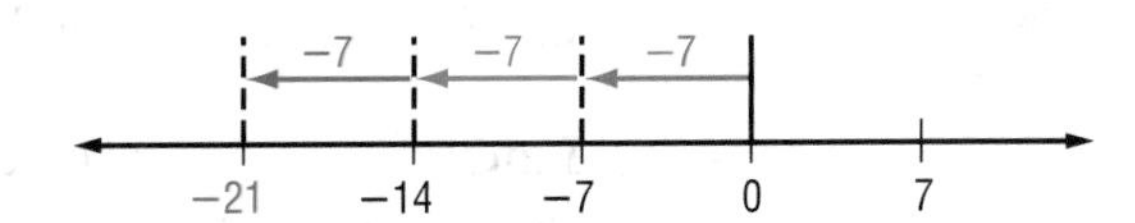

By the Commutative Property of Multiplication, 3(−7) = −7(3).

This example suggests the following rule.

Key Concept *Multiplying Two Integers with Different Signs*

- **Words** The product of two integers with different signs is negative.
- **Examples** 4(−3) = −12 −3(4) = −12

Example 1 Multiply Integers with Different Signs

Find each product.

a. 5(−6)

5(−6) = −30 The factors have different signs. The product is negative.

b. −4(16)

−4(16) = −64 The factors have different signs. The product is negative.

The product of two positive integers is positive. What is the sign of the product of two negative integers? Use a pattern to find $(-4)(-2)$.

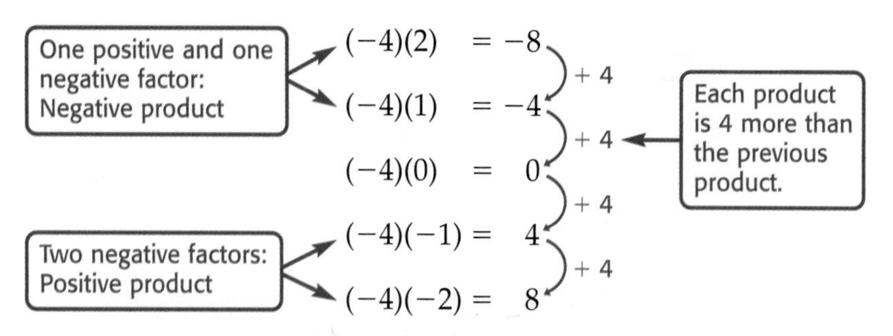

This example suggests the following rule.

Key Concept *Multiplying Two Integers with the Same Sign*

- **Words** The product of two integers with the same sign is positive.
- **Examples** $4(3) = 12$ $-4(-3) = 12$

Example 2 Multiply Integers with the Same Sign

Find $-6(-12)$.

$-6(-12) = 72$ The two factors have the same sign. The product is positive.

Example 3 Multiply More Than Two Integers

Find $-4(-5)(-8)$.

$-4(-5)(-8) = [(-4)(-5)](-8)$ Associative Property

$= 20(-8)$ $(-4)(-5) = 20$

$= -160$ $20(-8) = -160$

Study Tip

Look Back
To review the **Associative Property**, see Lesson 1-4.

Example 4 Use Integers to Solve a Problem

Multiple-Choice Test Item

A glacier was receding at a rate of 300 feet per day. What is the glacier's movement in 5 days?

(A) 305 feet (B) -1500 feet (C) -300 feet (D) -60 feet

Read the Test Item

The word *receding* means moving backward, so the rate per day is represented by -300. Multiply 5 times -300 to find the movement in 5 days.

Solve the Test Item

$5(-300) = -1500$ The product is negative.

The answer is B.

The Princeton Review

Test-Taking Tip

Read the problem. Try to picture the situation. Look for words that suggest mathematical concepts.

ALGEBRA CONNECTION

ALGEBRAIC EXPRESSIONS You can use the rules for multiplying integers to simplify and evaluate algebraic expressions.

Example 5 Simplify and Evaluate Algebraic Expressions

a. Simplify $-4(9x)$.

$-4(9x) = (-4 \cdot 9)x$ Associative Property of Multiplication

$= -36x$ Simplify.

b. Simplify $-2x(3y)$.

$-2x(3y) = (-2)(x)(3)(y)$ $-2x = (-2)(x)$, $3y = (3)(y)$

$= (-2 \cdot 3)(x \cdot y)$ Commutative Property of Multiplication

$= -6xy$ $-2 \cdot 3 = -6$, $x \cdot y = xy$

c. Evaluate $4ab$ if $a = 3$ and $b = -5$.

$4ab = 4(3)(-5)$ Replace a with 3 and b with -5.

$= [4(3)](-5)$ Associative Property of Multiplication

$= 12(-5)$ The product of 4 and 3 is positive.

$= -60$ The product of 12 and -5 is negative.

Check for Understanding

Concept Check

1. **Write** the product that is modeled on the number line below.

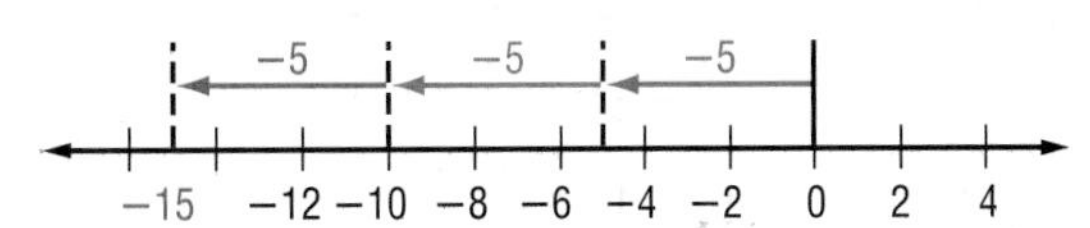

2. **State** whether each product is positive or negative.

   a. $-5 \cdot 8$ b. $6(-4)$ c. $8 \cdot 24$

   d. $-9(-7)$ e. $-2(9)(-3)$ f. $-7(-5)(-11)$

3. **OPEN ENDED** Give an example of three integers whose product is negative.

Guided Practice

Find each product.

4. $-3 \cdot 8$ 5. $5(-8)$ 6. $4 \cdot 30$

7. $-7(-4)$ 8. $-4(2)(-6)$ 9. $-5(-9)(-12)$

ALGEBRA **Simplify each expression.**

10. $-4 \cdot 3x$ 11. $7(-3y)$ 12. $-8a(-3b)$

ALGEBRA **Evaluate each expression.**

13. $-6h$, if $h = -20$ 14. $-4st$, if $s = -9$ and $t = 3$

FCAT Practice

Standardized Test Practice

15. The research submarine *Alvin*, used to locate the wreck of the *Titanic*, descends at a rate of about 100 feet per minute. Which integer describes the distance *Alvin* travels in 5 minutes?

   Ⓐ -500 ft Ⓑ -100 ft Ⓒ -20 ft Ⓓ 100 ft

Practice and Apply

Homework Help

For Exercises	See Examples
16–21	1
22–25	2
26–33	3
34, 35, 54, 55	4
36–53	5

Extra Practice

See page 727.

Find each product.

16. $-3 \cdot 4$
17. $-7 \cdot 6$
18. $4(-8)$
19. $9 \cdot (-8)$
20. $-12 \cdot 3$
21. $14(-5)$
22. $6 \cdot 19$
23. $4(32)$
24. $-8(-11)$
25. $-15(-3)$
26. $-5(-4)(6)$
27. $5(-13)(-2)$
28. $-7(-8)(-3)$
29. $-11(-4)(-7)$
30. $-12(-9)(6)$
31. $-6(-8)(11)$
32. $2(-8)(-9)(10)$
33. $4(-7)(-4)(-12)$

34. **FLOODS** In 1993, the Mississippi River was so high that it caused the Illinois River to flow backward. If the Illinois River flowed at the rate of -1500 feet per hour, how far would the water travel in 24 hours?

35. **TEMPERATURE** During a 10-hour period, the temperature in Browning, Montana, changed at a rate of -10°F per hour, starting at 44°F. What was the ending temperature?

ALGEBRA Simplify each expression.

36. $-5 \cdot 7x$
37. $-8 \cdot 12y$
38. $6(-8a)$
39. $5(-11b)$
40. $-7s(-8t)$
41. $-12m(-9n)$
42. $2ab(3)(-7)$
43. $3x(5y)(-9)$
44. $-4(-p)(-q)$
45. $-8(-11b)(-c)$
46. $9(-2c)(3d)$
47. $-6j(3)(5k)$

ALGEBRA Evaluate each expression.

48. $-7n$, if $n = -4$
49. $9s$, if $s = -11$
50. ab, if $a = 9$ and $b = 8$
51. $-2xy$, if $x = -8$ and $y = 5$
52. $-16cd$, if $c = 4$ and $d = -5$
53. $18gh$, if $g = -3$ and $h = 4$

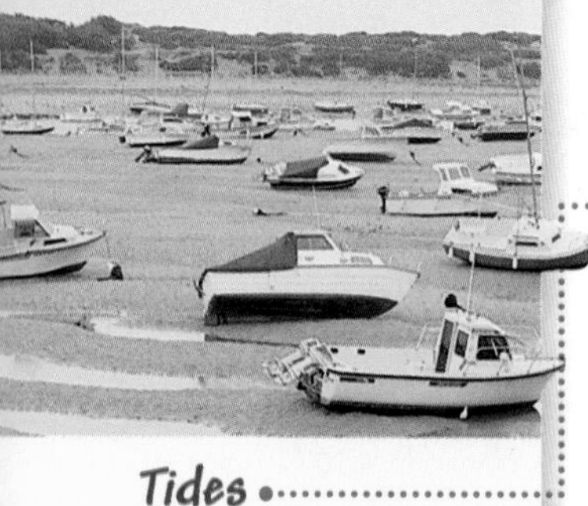

Tides

It takes about 6 hours for the ocean to move from low to high tide. High tide can change the width of the beach at a rate of -17 feet an hour.

TIDES For Exercises 54 and 55, use the information below and at the left. In Wrightsville, North Carolina, during low tide, the beachfront in some places is about 350 feet from the ocean to the homes. At high tide, the water is much closer to the homes.

54. What is the change in the width of the beachfront from low to high tide?

55. What is the distance from the ocean to the homes at high tide?

56. **CRITICAL THINKING** Write a rule that will help you determine the sign of the product if you are multiplying two or more integers.

57. **CRITICAL THINKING** Determine whether each statement is *true* or *false.* If false, give a counterexample. If true, give an example.
    a. Multiplication of integers is commutative.
    b. Multiplication of integers is associative.

58. **WRITING IN MATH** Answer the question that was posed at the beginning of the lesson.

How are the signs of factors and products related?

Include the following in your answer:

- a model of $2(-4)$,
- an explanation of why the product of a positive and a negative integer must be negative, and
- a pattern that explains why the product $-3(-3)$ is positive.

59. The product of two negative integers is—

(A) always negative. (B) always positive.

(C) sometimes negative. (D) never positive.

60. Which values complete the table at the right for $y = -3x$?

(A) $-6, -3, 0, 3$ (B) $-6, -2, 0, 2$

(C) $6, 2, 0, -2$ (D) $6, 3, 0, -3$

x	−2	−1	0	1
y				

Maintain Your Skills

Mixed Review

ALGEBRA **Evaluate each expression if $a = -2$, $b = -6$, and $c = 14$.** *(Lesson 2-3)*

61. $a - c$
62. $b - a$
63. $a - b$
64. $a + b + c$
65. $b - a + c$
66. $a - b - c$

67. **WEATHER RECORDS** The highest recorded temperature in Columbus, Ohio, is 104°F. The lowest recorded temperature is −22°F. What is the difference between the highest and lowest temperatures? *(Lesson 2-3)*

The cost of a trip to a popular amusement park can be determined with integers. Visit www.pre-alg.com/webquest to continue work on your WebQuest project.

Find each sum. *(Lesson 2-2)*

68. $-10 + 8 + 4$
69. $-4 + (-3) + (-7)$
70. $9 + (-14) + 2$

Refer to the coordinate system. Write the ordered pair that names each point. *(Lesson 1-6)*

71. E
72. C
73. B
74. F
75. D
76. A

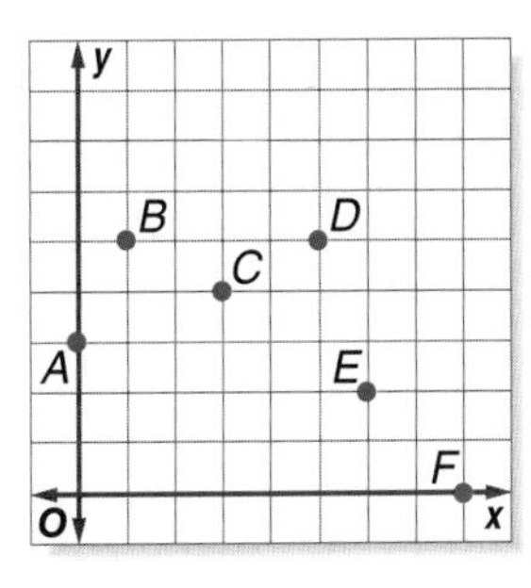

Find each sum or product mentally. *(Lesson 1-4)*

77. $3 \cdot 8 \cdot 20$
78. $8 + 98 + 102$
79. $5 \cdot 11 \cdot 10$

Getting Ready for the Next Lesson

BASIC SKILL **Find each quotient.**

80. $40 \div 8$
81. $90 \div 15$
82. $45 \div 3$
83. $105 \div 7$
84. $240 \div 6$
85. $96 \div 24$

2-5 Dividing Integers

What You'll Learn

- Divide integers.
- Find the average of a set of data.

Vocabulary

- average (mean)

Sunshine State Standards
MA.A.1.3.4-2, MA.A.3.3.1-1, MA.A.3.3.1-3, MA.A.3.3.2-1, MA.A.3.3.2-2, MA.A.3.3.3-1, MA.E.1.3.2-1

How is dividing integers related to multiplying integers?

You can find the quotient $-12 \div (-4)$ using a number line. To find how many groups of -4 there are in -12, show -12 on a number line. Then divide it into groups of -4.

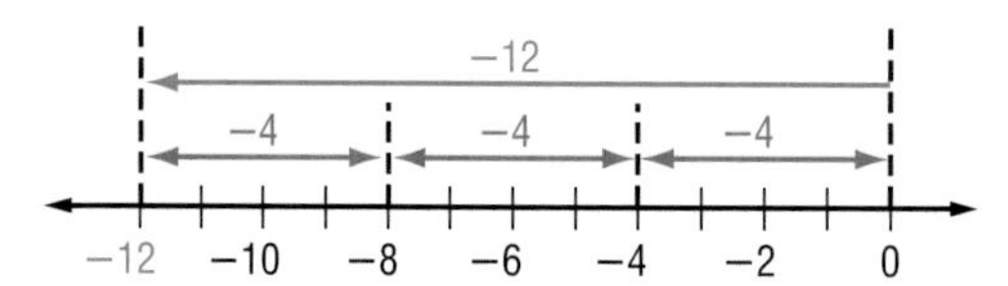

a. How many groups are there?

b. What is the quotient of $-12 \div (-4)$?

c. What multiplication sentence is also shown on the number line?

d. Draw a number line and find the quotient $-10 \div (-2)$.

DIVIDE INTEGERS You can find the quotient of two integers by using the related multiplication sentence.

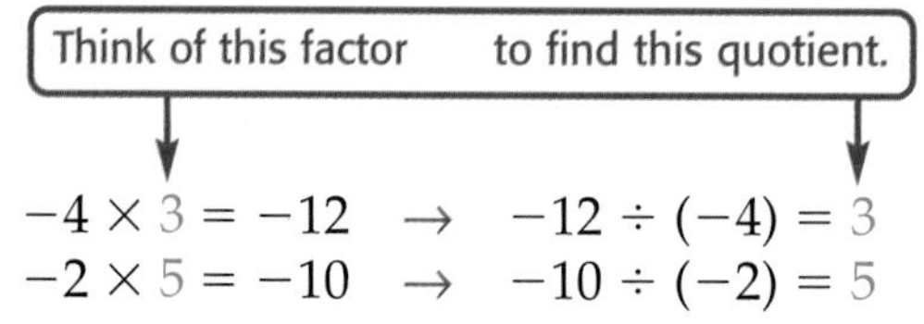

$$-4 \times 3 = -12 \rightarrow -12 \div (-4) = 3$$
$$-2 \times 5 = -10 \rightarrow -10 \div (-2) = 5$$

Reading Math

Parts of a Division Sentence In a division sentence, like $15 \div 5 = 3$, the number you are dividing, 15, is called the *dividend*. The number you are dividing by, 5, is called the *divisor*. The result, 3, is called the *quotient*.

In the division sentences $-12 \div (-4) = 3$ and $-10 \div (-2) = 5$, notice that the dividends and divisors are both negative. In both cases, the quotient is positive.

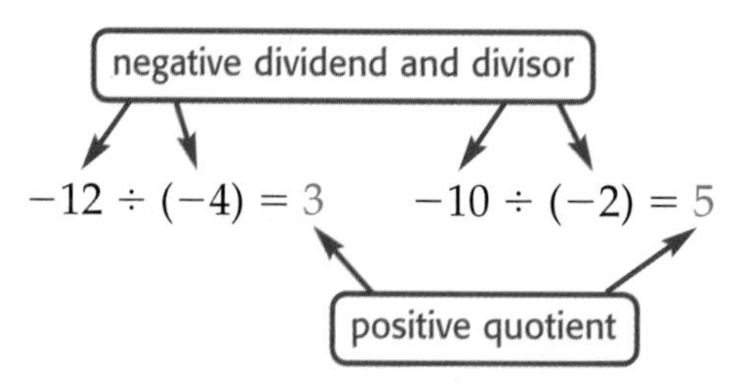

You already know that the quotient of two positive integers is positive.

$$12 \div 4 = 3 \qquad 10 \div 2 = 5$$

These and similar examples suggest the following rule for dividing integers with the same sign.

Key Concept — Dividing Integers with the Same Sign

- **Words** The quotient of two integers with the same sign is positive.
- **Examples** $-12 \div (-3) = 4$ $\qquad$ $12 \div 3 = 4$

Example 1 Divide Integers with the Same Sign

Find each quotient.

a. $\mathbf{-32 \div (-8)}$

$-32 \div (-8) = 4$ The dividend and the divisor have the same sign. The quotient is positive.

b. $\mathbf{\frac{75}{5}}$

$\frac{75}{5} = 75 \div 5$ The dividend and divisor have the same sign.

$= 15$ The quotient is positive.

What is the sign of the quotient of a positive and a negative integer? Look for a pattern in the following related sentences.

Think of this factor to find this quotient.

$-4 \times (-6) = 24 \rightarrow 24 \div (-4) = -6$

$2 \times (-9) = -18 \rightarrow -18 \div 2 = -9$

Notice that the signs of the dividend and divisor are different. In both cases, the quotient is negative.

different signs

$24 \div (-4) = -6$

$-18 \div 2 = -9$

negative quotient

different signs

These and other similar examples suggest the following rule.

Key Concept *Dividing Integers with Different Signs*

- **Words** The quotient of two integers with different signs is negative.
- **Examples** $-12 \div 4 = -3$ $12 \div (-4) = -3$

Concept Check How do you know the sign of the quotient of two integers?

Example 2 Divide Integers with Different Signs

Find each quotient.

a. $\mathbf{-42 \div 3}$

$-42 \div 3 = -14$ The signs are different. The quotient is negative.

b. $\mathbf{\frac{48}{-6}}$

$\frac{48}{-6} = 48 \div (-6)$ The signs are different. The quotient is negative.

$= -8$ Simplify.

ALGEBRA CONNECTION

You can use the rules for dividing integers to evaluate algebraic expressions.

Example 3 Evaluate Algebraic Expressions

Evaluate $ab \div (-4)$ if $a = -6$ and $b = -8$.

$ab \div (-4) = -6(-8) \div (-4)$ Replace a with -6 and b with -8.

$= 48 \div (-4)$ The product of -6 and -8 is positive.

$= -12$ The quotient of 48 and -4 is negative.

AVERAGE (MEAN) Division is used in statistics to find the **average**, or **mean**, of a set of data. To find the mean of a set of numbers, find the sum of the numbers and then divide by the number in the set.

Example 4 Find the Mean

a. Rachel had test scores of 84, 90, 89, and 93. Find the average (mean) of her test scores.

$\frac{84 + 90 + 89 + 93}{4} = \frac{356}{4}$ Find the sum of the test scores. Divide by the number of scores.

$= 89$ Simplify.

The average of her test scores is 89.

b. Find the average (mean) of -2, 8, 5, -9, -12, and -2.

$\frac{-2 + 8 + 5 + (-9) + (-12) + (-2)}{6} = \frac{-12}{6}$ Find the sum of the set of integers. Divide by the number in the set.

$= -2$ Simplify.

The average is -2.

Study Tip

Checking Reasonableness
The average must be between the greatest and least numbers in the set. Are the averages in Examples 4a and 4b reasonable?

You can refer to the following table to review operations with integers.

Concept Summary — *Operations with Integers*

Words	Examples
Adding Integers To add integers with the same sign, add their absolute values. Give the result the same sign as the integers. To add integers with different signs, subtract their absolute values. Give the result the same sign as the integer with the greater absolute value.	$-5 + (-4) = -9$ $5 + 4 = 9$ $-5 + 4 = -1$ $5 + (-4) = 1$
Subtracting Integers To subtract an integer, add its additive inverse.	$5 - 9 = 5 + (-9)$ or -4 $5 - (-9) = 5 + 9$ or 14
Multiplying Integers The product of two integers with the same sign is positive. The product of two integers with different signs is negative.	$5 \cdot 4 = 20$ $-5 \cdot (-4) = 20$ $-5 \cdot 4 = -20$ $5 \cdot (-4) = -20$
Dividing Integers The quotient of two integers with the same sign is positive. The quotient of two integers with different signs is negative.	$20 \div 5 = 4$ $-20 \div (-5) = 4$ $-20 \div 5 = -4$ $20 \div (-5) = -4$

Check for Understanding

Concept Check

1. **OPEN ENDED** Write an equation with three integers that illustrates dividing integers with different signs.
2. **Explain** how to find the average of a set of numbers.

Guided Practice **Find each quotient.**

3. $88 \div 8$
4. $-20 \div (-5)$
5. $-18 \div 6$
6. $\frac{-36}{-4}$
7. $\frac{70}{-7}$
8. $\frac{-81}{9}$

ALGEBRA **Evaluate each expression.**

9. $x \div 4$, if $x = -52$
10. $\frac{s}{t}$, if $s = -45$ and $t = 5$

Application

11. **WEATHER** The low temperatures for 7 days in January were −2, 0, 5, −1, −4, 2, and 0. Find the average for the 7-day period.

Practice and Apply

Homework Help

For Exercises	See Examples
12–17, 24, 25	1
18–23	2
26–31	3
32, 33	4

Extra Practice
See page 727.

Find each quotient.

12. $54 \div 9$
13. $45 \div 5$
14. $-27 \div (-3)$
15. $-64 \div (-8)$
16. $-72 \div (-9)$
17. $-60 \div (-6)$
18. $-77 \div 7$
19. $-300 \div 6$
20. $480 \div (-12)$
21. $\frac{132}{-12}$
22. $\frac{175}{-25}$
23. $\frac{143}{-13}$
24. What is −91 divided by −7?
25. Divide −76 by −4.

ALGEBRA **Evaluate each expression.**

26. $\frac{x}{-5}$, if $x = 85$
27. $\frac{108}{m}$, if $m = -9$
28. $\frac{c}{d}$, if $c = -63$ and $d = -7$
29. $\frac{s}{t}$, if $s = 52$ and $t = -4$
30. $xy \div (-3)$ if $x = 9$ and $y = -7$
31. $ab \div 6$ if $a = -12$ and $b = -8$
32. **STATISTICS** Find the average (mean) of 4, −8, 9, −3, −7, 10, and 2.
33. **BASKETBALL** In their first five games, the Jefferson Middle School basketball team scored 46, 52, 49, 53, and 45 points. What was their average number of points per game?

More About...

Energy
Energy providers use *degree days* to estimate the energy needed for heating and cooling.
Source: Michigan State University Extension

ENERGY **For Exercises 34–36, use the information below.**

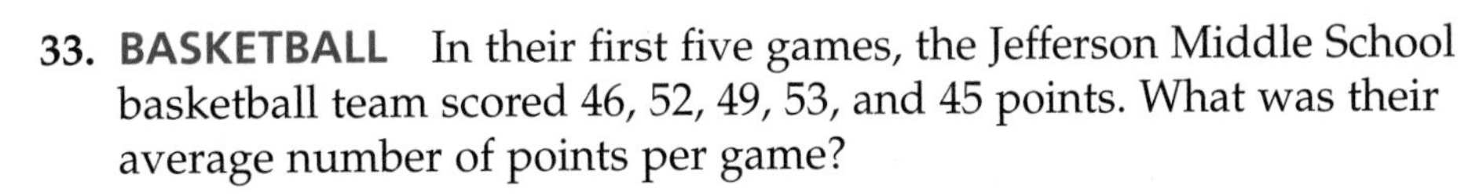

The formula $d = \left|65 - \frac{h + l}{2}\right|$ can be used to find degree days, where h is the high and l is the low temperature.

34. One day in July, Baltimore had a high of 81° and a low of 65°. Find the degree days.
35. One day in January, Milwaukee had a high of 8° and a low of 0°. Find the degree days.
36. **RESEARCH** Use the Internet or another resource to find the high and low temperature for your city for a day in January. Find the degree days.

37. CRITICAL THINKING Find values for x, y, and z, so that all of the following statements are true.

- $y > x$, $z < y$, and $x < 0$
- $z \div 2$ and $z \div 3$ are integers.
- $x \div z = -z$
- $x \div y = z$

38. CRITICAL THINKING Addition and multiplication are said to be closed for whole numbers, but subtraction and division are not. That is, when you add or multiply any two whole numbers, the result is a whole number. Which operations are closed for integers?

39. WRITING IN MATH Answer the question that was posed at the beginning of the lesson.

How is dividing integers related to multiplying integers?

Include the following in your answer:

- two related multiplication and division sentences, and
- an example of each case (same signs, different signs) of dividing integers.

40. On Saturday, the temperature fell 10° in 2 hours. Which expresses the temperature change per hour?

Ⓐ 5° Ⓑ −2° Ⓒ −5° Ⓓ −10°

41. Mark has quiz scores of 8, 7, 8, and 9. What is the lowest score he can get on the remaining quiz to have a final average (mean) score of at least 8?

Ⓐ 7 Ⓑ 8 Ⓒ 9 Ⓓ 10

Maintain Your Skills

Mixed Review

Find each difference or product. *(Lessons 2-3 and 2-4)*

42. $-8 - (-25)$ **43.** $75 - 114$ **44.** $2ab \cdot (-2)$ **45.** $(-10c)(5d)$

46. PATTERNS Find the next two numbers in the pattern 5, 4, 2, −1, … *(Lesson 1-1)*

Getting Ready for the Next Lesson

PREREQUISITE SKILL Use the grid to name the point for each ordered pair.
*(To review **ordered pairs**, see Lesson 1-6.)*

47. (1, 5) **48.** (6, 2)

49. (4, 5) **50.** (0, 3)

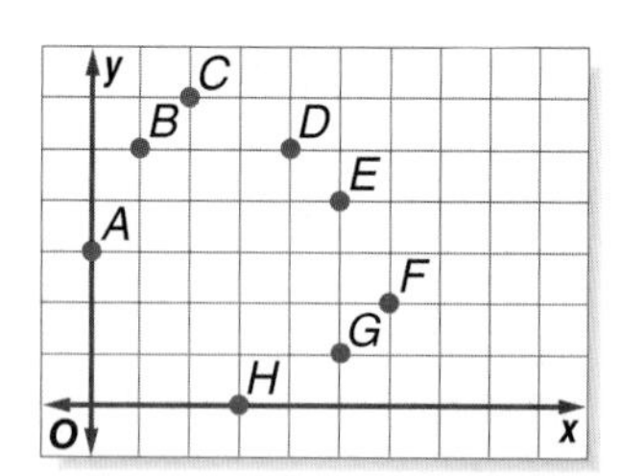

Practice Quiz 2

Lessons 2-4 and 2-5

Find each product. *(Lesson 2-4)*

1. $-12 \cdot 7$ **2.** $-6(-15)$ **3.** $-3(-7)(-6)$ **4.** $3(-8)(-5)$

Find each quotient. *(Lesson 2-5)*

5. $-124 \div 4$ **6.** $-90 \div (-6)$ **7.** $125 \div (-5)$ **8.** $-126 \div (-9)$

9. Simplify $4x(-5y)$. *(Lesson 2-4)*

10. Evaluate $-9a$ if $a = -6$. *(Lesson 2-4)*

2-6 The Coordinate System

Sunshine State Standards
MA.D.2.3.1-4, MA.C.3.3.2-1

What You'll Learn

- Graph points on a coordinate plane.
- Graph algebraic relationships.

Vocabulary

- quadrants

How is a coordinate system used to locate places on Earth?

A GPS, or Global Positioning System, can be used to find a location anywhere on Earth by identifying its latitude and longitude. Several cities are shown on the map below. For example, Sydney, Australia, is located at approximately 30°S, 150°E.

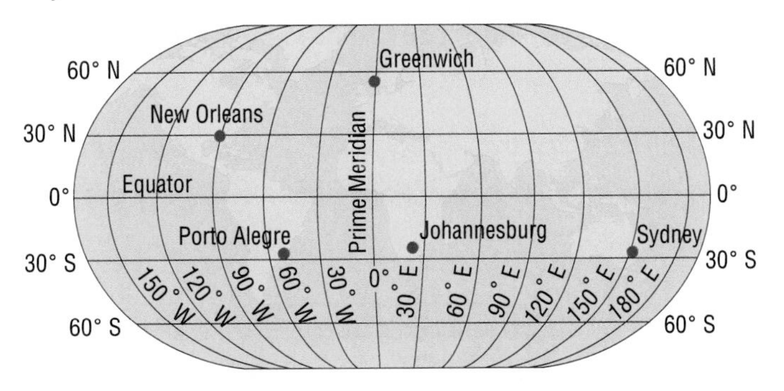

a. Latitude is measured north and south of the equator. What is the latitude of New Orleans?

b. Longitude is measured east and west of the prime meridian. What is the longitude of New Orleans?

c. What does the location 30°N, 90°W mean?

GRAPH POINTS Latitude and longitude are a kind of coordinate system. The coordinate system you used in Lesson 1-6 can be extended to include points below and to the left of the origin.

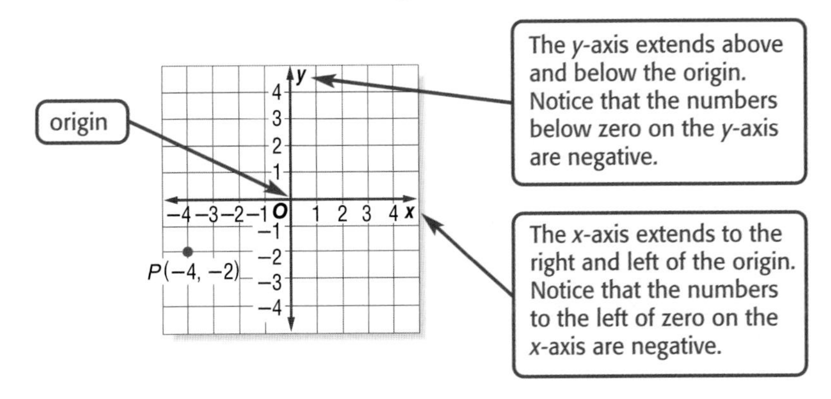

origin
y
x
O
P(−4, −2)
The y-axis extends above and below the origin. Notice that the numbers below zero on the y-axis are negative.
The x-axis extends to the right and left of the origin. Notice that the numbers to the left of zero on the x-axis are negative.

Recall that a point graphed on the coordinate system has an x-coordinate and a y-coordinate. The dot at the ordered pair $(-4, -2)$ is the graph of point P.

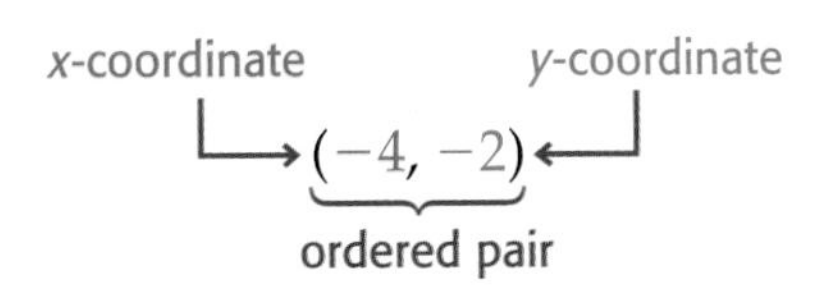

Example 1 Write Ordered Pairs

Write the ordered pair that names each point.

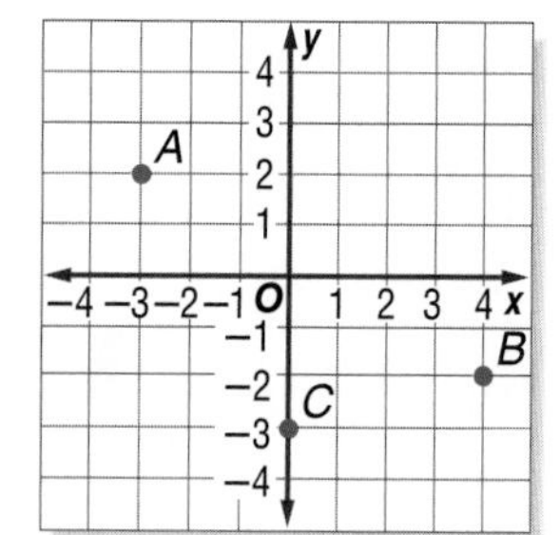

a. *A*

The x-coordinate is -3.
The y-coordinate is 2.
The ordered pair is $(-3, 2)$.

Study Tip

Ordered Pairs
Notice that the axes in an ordered pair (x, y) are listed in alphabetical order.

b. *B*

The x-coordinate is 4.
The y-coordinate is -2.
The ordered pair is $(4, -2)$.

c. *C*

The point lies on the y-axis, so its x-coordinate is 0.
The y-coordinate is -3. The ordered pair is $(0, -3)$.

The x-axis and the y-axis separate the coordinate plane into four regions, called **quadrants**. The axes and points on the axes are not located in any of the quadrants.

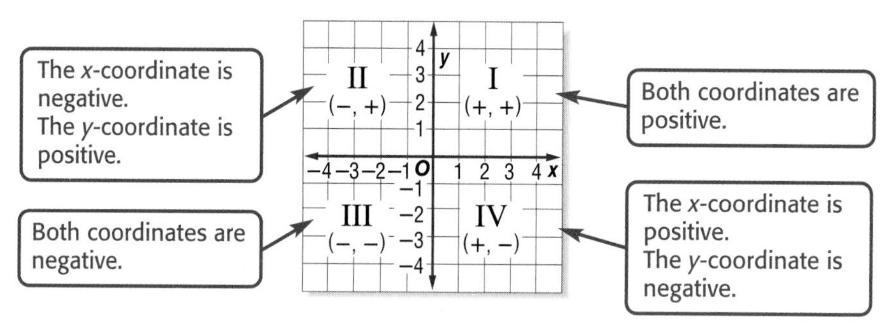

Example 2 Graph Points and Name Quadrant

Graph and label each point on a coordinate plane. Name the quadrant in which each point lies.

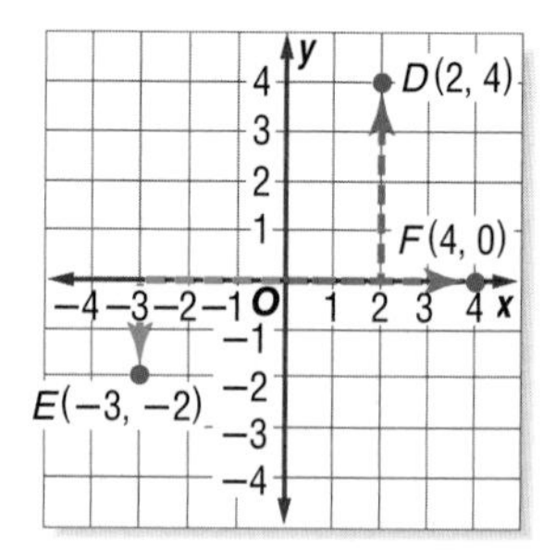

a. $D(2, 4)$

Start at the origin. Move 2 units right.
Then move 4 units up and draw a dot.
Point $D(2, 4)$ is in Quadrant I.

b. $E(-3, -2)$

Start at the origin. Move 3 units left.
Then move 2 units down and draw a dot.
Point $E(-3, -2)$ is in Quadrant III.

c. $F(4, 0)$

Start at the origin. Move 4 units right. Since the y-coordinate is 0, the point lies on the x-axis. Point $F(4, 0)$ is not in any quadrant.

Concept Check What parts of a coordinate graph do not lie in any quadrant?

GRAPH ALGEBRAIC RELATIONSHIPS You can use a coordinate graph to show relationships between two numbers.

Example 3 Graph an Algebraic Relationship

The sum of two numbers is 5. If x represents the first number and y represents the second number, make a table of possible values for x and y. Graph the ordered pairs and describe the graph.

First, make a table. Choose values for x and y that have a sum of 5.

$x + y = 5$		
x	y	(x, y)
2	3	(2, 3)
1	4	(1, 4)
0	5	(0, 5)
−1	6	(−1, 6)
−2	7	(−2, 7)

Then graph the ordered pairs on a coordinate plane.

The points on the graph are in a line that slants downward to the right. The line crosses the y-axis at $y = 5$.

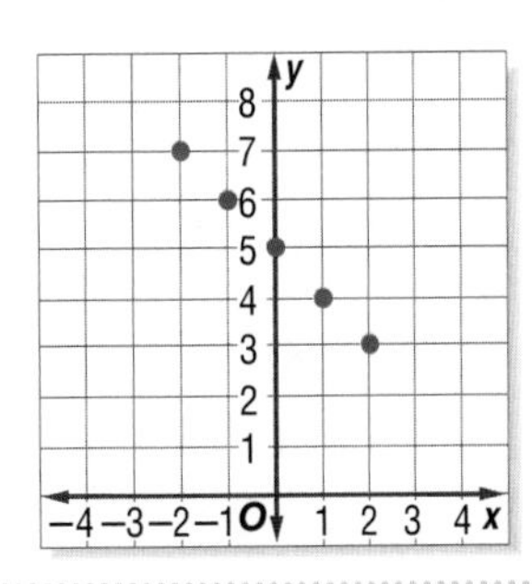

Check for Understanding

Concept Check

1. **Explain** why the point (3, 6) is different from the point (6, 3).
2. **OPEN ENDED** Name two ordered pairs whose graphs are *not* located in one of the four quadrants.
3. **FIND THE ERROR** Keisha says that if you interchange the coordinates of any point in Quadrant I, the new point would still be in Quadrant I. Jason says the new point would be in Quadrant 3. Who is correct? Explain your reasoning.

Guided Practice

Name the ordered pair for each point graphed at the right.

4. A
5. C
6. G
7. K

Graph and label each point on a coordinate plane. Name the quadrant in which each point is located.

8. $J(3, -4)$
9. $K(-2, 2)$
10. $L(0, 4)$
11. $M(-1, -2)$

Application

12. **ALGEBRA** Make a table of values and graph six ordered integer pairs where $x + y = 3$. Describe the graph.

Practice and Apply

Homework Help

For Exercises	See Examples
13–22	1
23–34, 41, 42	2
35–40, 43, 44	3

Extra Practice

See page 728.

Name the ordered pair for each point graphed at the right.

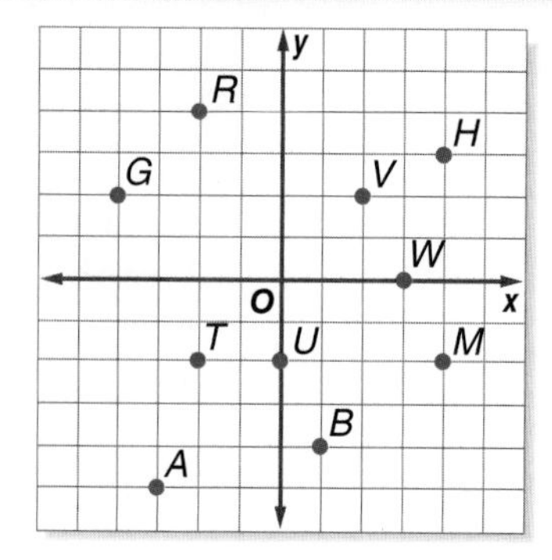

13. R **14.** G

15. M **16.** B

17. V **18.** H

19. U **20.** W

21. A **22.** T

Graph and label each point on a coordinate plane. Name the quadrant in which each point is located.

23. $A(4, 5)$ **24.** $K(-5, 1)$ **25.** $M(4, -2)$

26. $B(-5, -5)$ **27.** $S(2, -5)$ **28.** $R(-3, 5)$

29. $E(0, 3)$ **30.** $H(0, -3)$ **31.** $G(5, 0)$

32. $C(6, -1)$ **33.** $D(0, 0)$ **34.** $F(-4, 0)$

ALGEBRA **Make a table of values and graph six sets of ordered integer pairs for each equation. Describe the graph.**

35. $x + y = 5$ **36.** $x + y = -2$ **37.** $y = 2x$

38. $y = -2x$ **39.** $y = x + 2$ **40.** $y = x - 1$

Graph each point. Then connect the points in alphabetical order and identify the figure.

41. $A(0, 6)$, $B(4, -6)$, $C(-6, 2)$, $D(6, 2)$, $E(-4, -6)$, $F(0, 6)$

42. $A(5, 8)$, $B(1, 13)$, $C(5, 18)$, $D(9, 13)$, $E(5, 8)$, $F(5, 6)$, $G(3, 7)$, $H(3, 5)$, $I(7, 7)$, $J(7, 5)$, $K(5, 6)$, $L(5, 3)$, $M(3, 4)$, $N(3, 2)$, $P(7, 4)$, $Q(7, 2)$, $R(5, 3)$, $S(5, 1)$

43. Graph eight ordered integer pairs where $|x| > 3$. Describe the graph.

44. Graph all ordered integer pairs that satisfy the condition $|x| < 4$ and $|y| < 3$.

Reading Math

Vertex, Vertices

A *vertex* of a triangle is a point where two sides of a triangle meet. *Vertices* is the plural of *vertex*.

GEOMETRY **On a coordinate plane, draw a triangle *ABC* with vertices at $A(3, 1)$, $B(4, 2)$, and $C(2, 4)$. Then graph and describe each new triangle formed in Exercises 45–48.**

45. Multiply each coordinate of the vertices in triangle *ABC* by 2.

46. Multiply each coordinate of the vertices in triangle *ABC* by -1.

47. Add 2 to each coordinate of the vertices in triangle *ABC*.

48. Subtract 4 from each coordinate of the vertices in triangle *ABC*.

49. MAPS Find a map of your school and draw a coordinate grid on the map with the library as the center. Locate the cafeteria, principal's office, your math classroom, gym, counselor's office, and the main entrance on your grid. Write the coordinates of these places. How can you use these points to help visitors find their way around your school?

50. **CRITICAL THINKING** If the graph of A(x, y) satisfies the given condition, name the quadrant in which point A is located.

a. $x > 0, y > 0$ **b.** $x < 0, y < 0$ **c.** $x < 0, y > 0$

51. **CRITICAL THINKING** Graph eight sets of integer coordinates that satisfy $|x| + |y| > 3$. Describe the location of the points.

52. **WRITING IN MATH** Answer the question that was posed at the beginning of the lesson.

How is a coordinate system used to locate places on Earth?

Include the following in your answer:
- an explanation of how coordinates can describe a location, and
- a description of how latitude and longitude are related to the x- and y-axes on a coordinate plane. Include what corresponds to the origin on a coordinate plane.

53. On the coordinate plane at the right, what are the coordinates of the point that shows the location of the library?

Ⓐ (4, −2) Ⓑ (−2, −4)

Ⓒ (4, 2) Ⓓ (−4, −2)

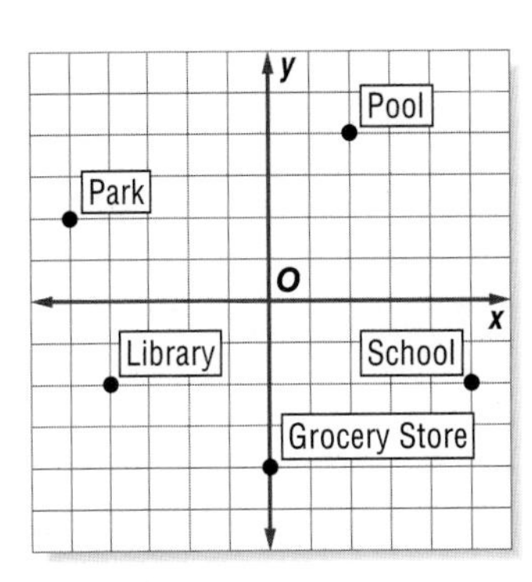

54. On the coordinate plane at the right, what location has coordinates (5, −2)?

Ⓐ Park Ⓑ School

Ⓒ Library Ⓓ Grocery Store

Maintain Your Skills

Mixed Review **Find each quotient.** *(Lesson 2-5)*

55. $-24 \div 8$ 56. $105 \div (-5)$ 57. $-400 \div (-50)$

ALGEBRA **Evaluate each expression if $f = -9$, $g = -6$, and $h = 8$.** *(Lesson 2-4)*

58. $-5fg$ 59. $2gh$ 60. $-10fh$

61. **WEATHER** In the newspaper, Amad read that the low temperature for the day was expected to be −5°F and the high temperature was expected to be 8°F. What was the difference in the expected high and low temperature? *(Lesson 2-3)*

ALGEBRA **Simplify each expression.** *(Lesson 1-4)*

62. $(a + 8) + 6$ 63. $4(6h)$ 64. $(n \cdot 7) \cdot 8$

65. $(b \cdot 9) \cdot 5$ 66. $(16 + 3y) + y$ 67. $0(4z)$

Chapter 2

Study Guide and Review

Vocabulary and Concept Check

absolute value (p. 58)
additive inverse (p. 66)
average (p. 82)
coordinate (p. 57)
inequality (p. 57)
integers (p. 56)
mean (p. 82)
negative number (p. 56)
opposites (p. 66)
quadrants (p. 86)

Complete each sentence with the correct term. Choose from the list above.

1. A(n) ________ is a number less than zero.
2. The four regions separated by the axes on a coordinate plane are called ________.
3. The number that corresponds to a point on the number line is called the ________ of that point.
4. An integer and its opposite are also called ________ of each other.
5. The set of ________ includes positive whole numbers, their opposites, and zero.
6. The ________ of a number is the distance the number is from zero on the number line.
7. A(n) ________ is a mathematical sentence containing $<$ or $>$.

Lesson-by-Lesson Review

2-1 Integers and Absolute Value

See pages 56–61.

Concept Summary

- Numbers on a number line increase as you move from left to right.
- The absolute value of a number is the distance the number is from zero on the number line.

Examples

1 **Replace the ● with $<$, $>$, or $=$ in -3 ● 2 to make a true sentence.**

Since -3 is to the left of 2, write $-3 < 2$.

2 **Evaluate $|-4|$.**

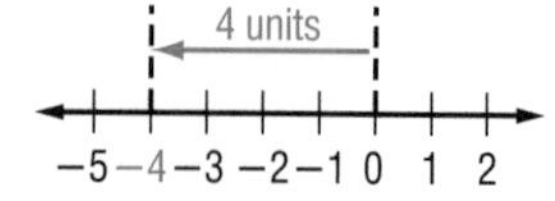

The graph of -4 is 4 units from 0. So, $|-4| = 4$.

Exercises **Replace each ● with $<$, $>$, or $=$ to make a true sentence.** *See Example 2 on page 57.*

8. 8 ● -8 **9.** -3 ● -3 **10.** -2 ● 0 **11.** -12 ● -21

Evaluate each expression. *See Example 4 on page 58.*

12. $|-32|$ **13.** $|25|$ **14.** $-|15|$ **15.** $|-8| + |-14|$

2-2 Adding Integers

See pages 64–68.

Concept Summary

- To add integers with the same sign, add their absolute values. Give the result the same sign as the integers.
- To add integers with different signs, subtract their absolute values. Give the result the same sign as the integer with the greater absolute value.

Examples **Find each sum.**

1 $-3 + (-4)$

$-3 + (-4) = -7$ The sum is negative.

2 $5 + (-2)$

$5 + (-2) = 3$ The sum is positive.

Exercises **Find each sum.** *See Examples 2, 4, and 6 on pages 64–66.*

16. $-6 + (-3)$ **17.** $-4 + (-1)$ **18.** $-2 + 7$
19. $4 + (-8)$ **20.** $6 + (-9) + (-8)$ **21.** $4 + (-7) + (-3) + (-4)$

2-3 Subtracting Integers

See pages 70–74.

Concept Summary

- To subtract an integer, add its additive inverse.

Examples **Find each difference.**

1 $-5 - 2$

$-5 - 2 = -5 + (-2)$ To subtract 2, add -2.
$= -7$

2 $8 - (-4)$

$8 - (-4) = 8 + 4$ To subtract -4, add 4.
$= 12$

Exercises **Find each difference.** *See Examples 1 and 2 on pages 70–71.*

22. $4 - 9$ **23.** $-3 - 5$ **24.** $7 - (-2)$ **25.** $-1 - (-6)$
26. $-7 - 8$ **27.** $6 - 10$ **28.** $-3 - (-7)$ **29.** $6 - (-3)$

2-4 Multiplying Integers

See pages 75–79.

Concept Summary

- The product of two integers with different signs is negative.
- The product of two integers with the same sign is positive.

Examples **Find each product.**

1 $6(-4)$

$6(-4) = -24$ The factors have different signs, so the product is negative.

2 $-8(-2)$

$-8(-2) = 16$ The factors have the same sign, so the product is positive.

Exercises **Find each product.** *See Examples 1 and 2 on pages 75–76.*

30. $-9(5)$ **31.** $11(-6)$ **32.** $-4(-7)$ **33.** $-3(-16)$

34. Simplify $-2a(4b)$. *See Example 5 on page 77.*

2-5 Dividing Integers

See pages 80–84.

Concept Summary

- The quotient of two integers with the same sign is positive.
- The quotient of two integers with different signs is negative.

Examples **Find each quotient.**

1 $-30 \div (-5)$.

$-30 \div (-5) = 6$ The signs are the same, so the quotient is positive.

2 $27 \div (-3)$

$27 \div (-3) = -9$ The signs are different, so the quotient is negative.

Exercises **Find each quotient.** *See Examples 1 and 2 on page 81.*

35. $-14 \div (-2)$ **36.** $-52 \div (-4)$ **37.** $-36 \div 9$ **38.** $88 \div (-4)$

39. Find the average (mean) of $-3, -6, 9, -3$, and 13. *See Example 4 on page 82*

2-6 The Coordinate System

See pages 85–89.

Concept Summary

- The x-axis and the y-axis separate the coordinate plane into four quadrants.
- The axes and points on the axes are not located in any of the quadrants.

Examples **Graph and label each point on a coordinate plane. Name the quadrant in which each point is located.**

1 $F(5, -3)$

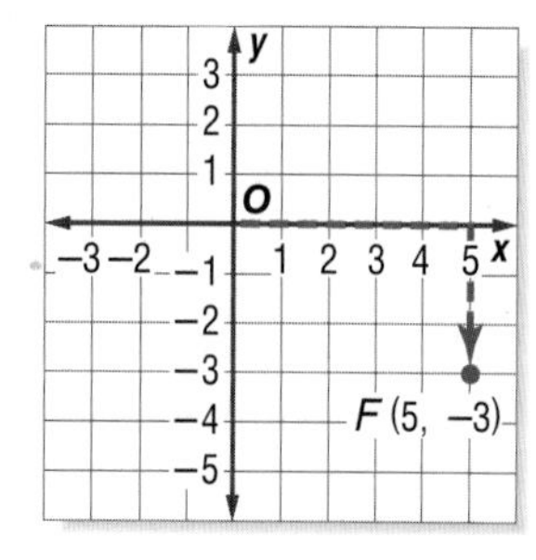

Point $F(5, -3)$ is in quadrant IV.

2 $G(0, 4)$

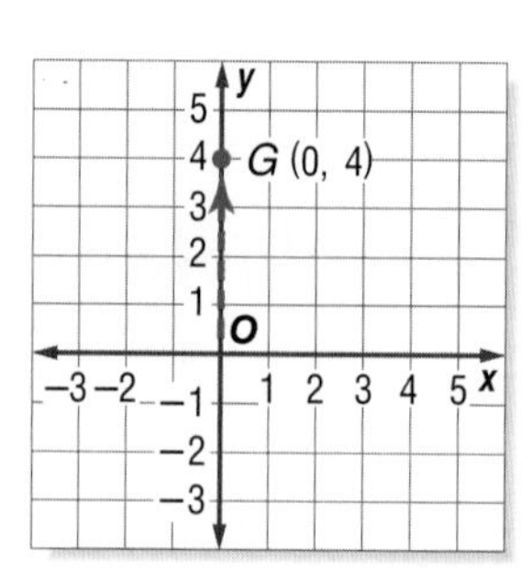

Point $G(0, 4)$ is not in any quadrant.

Exercises **Graph and label each point on a coordinate plane. Name the quadrant in which each point is located.** *See Example 2 on page 86.*

40. $A(4, 3)$ **41.** $J(-2, -4)$ **42** $K(-1, 3)$ **43.** $R(3, 0)$

Chapter 2

Practice Test

Vocabulary and Concepts

1. **Explain** how to add two integers with different signs.
2. **State** a rule used for subtracting integers.
3. **Graph** the set of integers $\{-6, 2, -1, 1\}$ on a number line.

Skills and Applications

Write two inequalities using the numbers in each sentence. Use the symbols < and >.

4. -5 is less than 2.
5. 12 is greater than -15.

Replace each ● with <, >, or = to make a true sentence.

6. -5 ● -3
7. -5 ● -14
8. 4 ● $|-7|$

Find each sum or difference.

9. $-4 + (-8)$
10. $-9 + 15$
11. $12 + (-15)$
12. $14 + (-7) + -11$
13. $4 - 13$
14. $8 - (-6)$
15. $-6 - (-10)$
16. $-14 - (-7)$

Find each product or quotient.

17. $6(-8)$
18. $-9(8)$
19. $-7(-5)$
20. $2(-4)(11)$
21. $54 \div (-9)$
22. $-64 \div (-4)$
23. $-250 \div 25$
24. $-144 \div (-6)$

ALGEBRA **Evaluate each expression if $a = -5$, $b = 3$, and $c = -10$.**

25. $ab - c$
26. $c \div a$
27. $4c + |a|$

Graph and label each point on a coordinate plane. Name the quadrant in which each point is located.

28. $D(-2, 4)$
29. $E(3, -4)$
30. $F(-1, -3)$

31. **WEATHER** The table shows the low temperatures during one week in Anchorage, Alaska. Find the average low temperature for the week.

Day	S	M	T	W	T	F	S
Temperature (°F)	−12	3	−7	0	−4	1	−2

32. **SPORTS** During the first play of the game, the Brownville Tigers football team lost seven yards. On each of the next three plays, an additional four yards were lost. Express the total yards lost at the end of the first four plays as an integer.

FCAT Practice

33. **STANDARDIZED TEST PRACTICE** Suppose Jason's home represents the origin on a coordinate plane. If Jason leaves his home and walks two miles west and then four miles north, what is the location of his destination as an ordered pair? In which quadrant is his destination?

Ⓐ $(-2, 4)$; II Ⓑ $(2, 4)$; I Ⓒ $(-2, -4)$; II Ⓓ $(4, -2)$; IV

Chapter 2 Standardized Test Practice

FCAT Practice

Part 1 Multiple Choice

Record your answers on the answer sheet provided by your teacher or on a sheet of paper.

1. The table below shows the number of cells present after a certain form of bacteria multiplies for a number of hours. How many cells will be present in five hours? (Lesson 1-1)

Number of Hours	Number of Cells
0	1
1	3
2	9
3	27

Ⓐ 81 Ⓑ 91
Ⓒ 243 Ⓓ 279

2. Suppose your sister has 3 more CDs than you do. Which equation represents the number of CDs that you have? Let y represent your CDs and s represent your sister's CDs. (Lesson 1-5)

Ⓐ $y = s + 3$ Ⓑ $y = s - 3$
Ⓒ $y = 3 - s$ Ⓓ $y = 3s$

3. Which expression represents the greatest integer? (Lesson 1-6)

Ⓐ $|4|$ Ⓑ $|-3|$
Ⓒ $|-8|$ Ⓓ -9

4. The water level of a local lake is normally 0 feet above sea level. In a flood, the water level rose 4 feet above normal. A month later, the water level had gone down 5 feet. Which integer best represents the water level at that time? (Lesson 2-1)

Ⓐ -3 Ⓑ -1
Ⓒ 4 Ⓓ 9

5. What is the sum of -5 and 2? (Lesson 2-2)

Ⓐ -7 Ⓑ -3
Ⓒ 3 Ⓓ 7

6. Find the value of x if $x = 7 - (-3)$. (Lesson 2-3)

Ⓐ -10 Ⓑ -4
Ⓒ 4 Ⓓ 10

7. If $t = -5$, what is the value of the expression $-3t + 7$? (Lesson 2-4)

Ⓐ -8 Ⓑ -6
Ⓒ 8 Ⓓ 22

8. If $a = -2$ and $b = 5$, what is the value of $\frac{b - 13}{a}$? (Lesson 2-5)

Ⓐ -4 Ⓑ -9
Ⓒ 9 Ⓓ 4

For Questions 9 and 10, use the following graph.

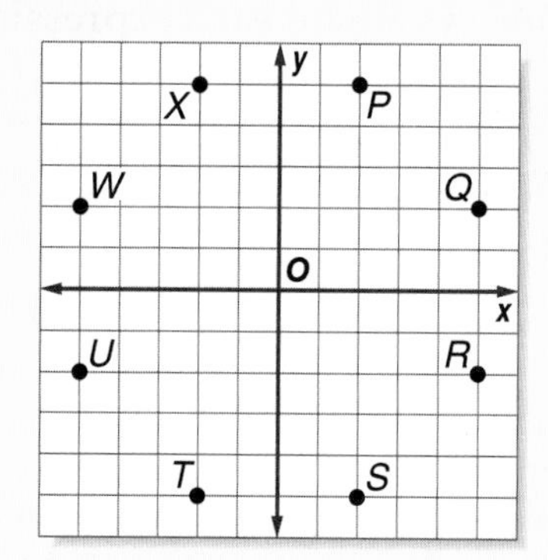

9. Which letter represents the ordered pair $(-2, 5)$? (Lesson 2-6)

Ⓐ R Ⓑ X
Ⓒ T Ⓓ W

10. Which ordered pair represents point U? (Lesson 2-6)

Ⓐ $(5, -2)$ Ⓑ $(-2, -5)$
Ⓒ $(-5, -2)$ Ⓓ $(-2, 5)$

Part 2 Short Response/Grid In

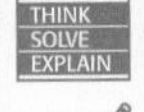

Record your answers on the answer sheet provided by your teacher or on a sheet of paper.

11. The bar graph shows the numbers of girls and boys in each grade at Muir Middle School. In which grade is the difference between the number of girls and the number of boys the greatest? (Prerequisite Skill, p. 722)

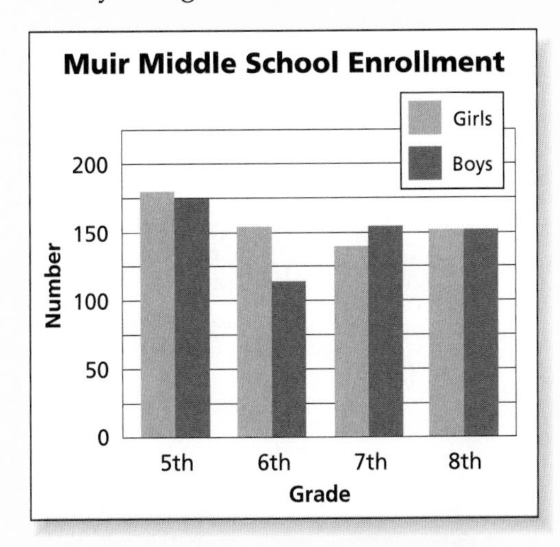

12. Nine less than a number is 15. Find the number. (Lesson 1-5)

13. The Springfield High School football team gained 7 yards on one play. On the next play, they lost 11 yards. Write an integer that represents the net result of these two plays. (Lesson 2-2)

14. The low temperature one winter night in Bismarck, North Dakota, was -15°F. The next day the high temperature was 3°F. How many degrees had the temperature risen? (Lesson 2-3)

15. The table below was used to change values of x into values of y.

x	$y = x - 7$
6	-1
7	0
8	1

What value of x can be used to obtain a y-value equal to 5? (Lesson 2-3)

16. The low temperatures in Minneapolis during four winter days were $+2$°F, -7°F, -12°F, and $+9$°F. What was the average low temperature during these four days? (Lesson 2-5)

FCAT Practice

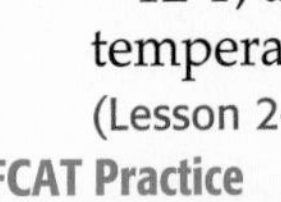

Part 3 Open Ended

Record your answers on a sheet of paper. Show your work.

17. On graph paper, graph the points $A(4, 2)$, $B(-3, 7)$, and $C(-3, 2)$. Connect the points to form a triangle. (Lesson 2-6)

a. Add 6 to the x-coordinate of each coordinate pair. Graph and connect the new points to form a new figure. Is the new figure the same size and shape as the original triangle? Describe how the size, shape, and position of the new triangle relate to the size, shape, and position of the original triangle.

b. If you add -6 to each original x-coordinate, and graph and connect the new points to create a new figure, how will the position of the new figure relate to that of the original one?

c. Multiply the y-coordinate of each original ordered pair by -1. Graph and connect the new points to form a new figure. Describe how the size, shape, and position of the new triangle relate to the size, shape, and position of the original triangle.

d. If you multiply each original x-coordinate by -1, and graph and connect the new points to create a new figure, how will the position of the new figure relate to that of the original one?

The Princeton Review **Test-Taking Tip**

Question 17 When answering open-ended items on standardized tests, follow these steps:
1. Read the item carefully.
2. Show all of your work. You may receive points for items that are only partially correct.
3. Check your work.

Chapter 3 Equations

What You'll Learn

- **Lessons 3-1 and 3-2** Use the Distributive Property to simplify expressions.
- **Lessons 3-3 and 3-4** Solve equations using the Properties of Equality.
- **Lessons 3-5 and 3-6** Write and solve two-step equations.
- **Lesson 3-7** Use formulas to solve real-world and geometry problems.

Key Vocabulary

- **equivalent expressions** (p. 98)
- **coefficient** (p. 103)
- **constant** (p. 103)
- **perimeter** (p. 132)
- **area** (p. 132)

Why It's Important

As you continue to study algebra, you will learn how to describe quantitative relationships using variables and equations. For example, the equation $d = rt$ shows the relationship between the variables d (distance), r (rate or speed), and t (time). *You will solve a problem about ballooning in Lesson 3-7.*

Getting Started

Prerequisite Skills To be successful in this chapter, you'll need to master these skills and be able to apply them in problem-solving situations. Review these skills before beginning Chapter 3.

For Lesson 3-1 — Multiply Integers

Find each product. *(For review, see Lesson 2-4.)*

1. $2(-3)$ **2.** $-4(3)$ **3.** $-5(-2)$ **4.** $-4(6)$

For Lesson 3-2 — Write Addition Expressions

Write each subtraction expression as an addition expression. *(For review, see Lesson 2-3.)*

5. $5 - 7$ **6.** $6 - 10$ **7.** $-5 - 9$ **8.** $11 - 10$

For Lessons 3-3 and 3-5 — Add Integers

Find each sum. *(For review, see Lesson 2-2.)*

9. $6 + (-9)$ **10.** $-8 + 4$ **11.** $4 + (-4)$ **12.** $7 + (-10)$

For Lesson 3-6 — Write Algebraic Expressions

Write an algebraic expression for each verbal expression. *(For review, see Lesson 1-3.)*

13. five more than twice a number

14. the difference of a number and 15

15. three less than a number

16. the quotient of a number and 10

FOLDABLES™
Study Organizer

Make this Foldable to help you organize information about expressions and equations. Begin with four sheets of $8\frac{1}{2}$" × 11" paper.

Step 1 Stack Pages

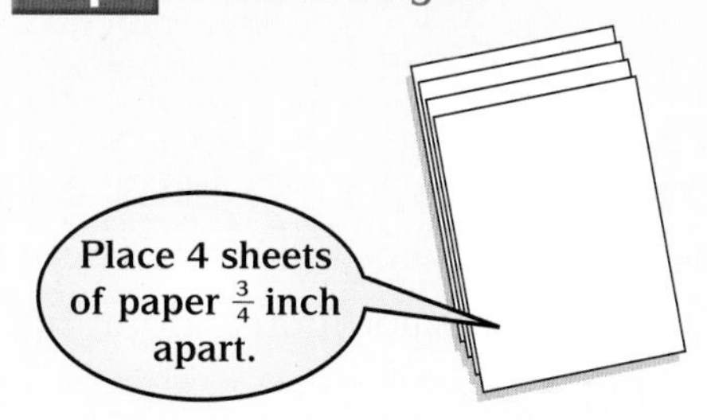
Place 4 sheets of paper $\frac{3}{4}$ inch apart.

Step 2 Roll Up Bottom Edges

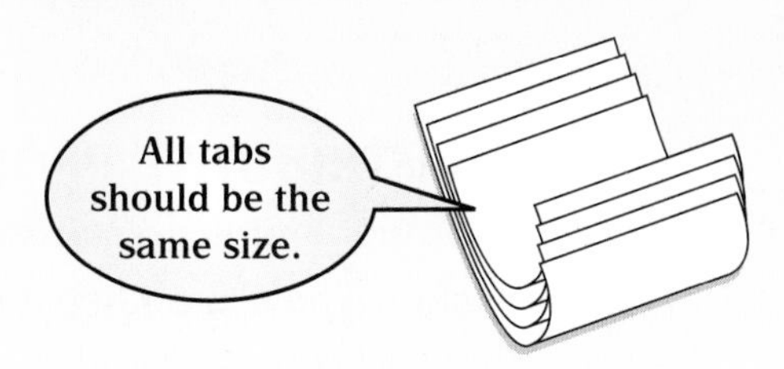
All tabs should be the same size.

Step 3 Crease and Staple

Staple along fold.

Step 4 Label

Label the tabs with topics from the chapter.

Chapter 3 Equations
1. Distributive Property
2. Simplifying Expressions
3. Equations: +, −
4. Equations: ×, ÷
5. Two-Step Equations
6. Writing Equations
7. Formulas

Reading and Writing As you read and study the chapter, record examples under each tab.

3-1 The Distributive Property

What You'll Learn

- Use the Distributive Property to write equivalent numerical expressions.
- Use the Distributive Property to write equivalent algebraic expressions.

Vocabulary

- equivalent expressions

How are rectangles related to the Distributive Property?

To find the area of a rectangle, multiply the length and width. You can find the total area of the blue and yellow rectangles in two ways.

Method 1

Put them together. Add the lengths, then multiply.

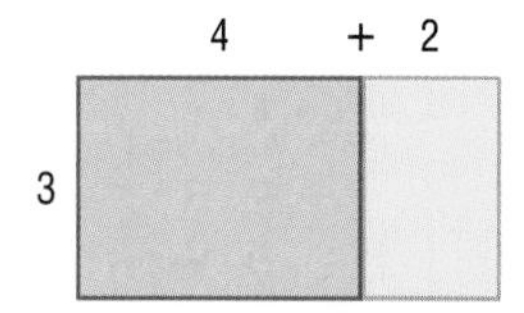

$3(4 + 2) = 3 \cdot 6$ Add.

$= 18$ Multiply.

Method 2

Separate them. Multiply to find each area, then add.

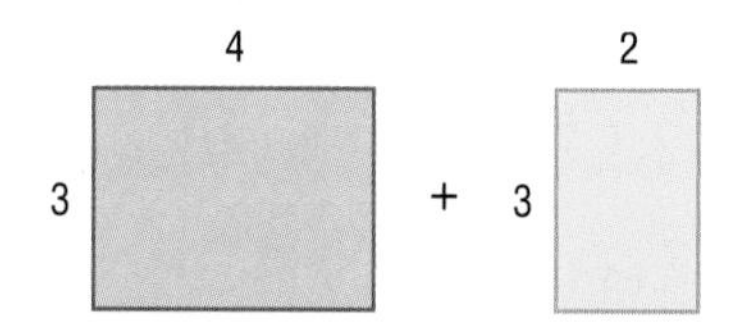

$3 \cdot 4 + 3 \cdot 2 = 12 + 6$ Multiply.

$= 18$ Add.

a. Draw a 2-by-5 and a 2-by-4 rectangle. Find the total area in two ways.

b. Draw a 4-by-4 and a 4-by-1 rectangle. Find the total area in two ways.

c. Draw any two rectangles that have the same width. Find the total area in two ways.

d. What did you notice about the total area in each case?

DISTRIBUTIVE PROPERTY The expressions $3(4 + 2)$ and $3 \cdot 4 + 3 \cdot 2$ are **equivalent expressions** because they have the same value, 18. This example shows how the **Distributive Property** combines addition and multiplication.

Reading Math

Distributive

Root Word: Distribute

To *distribute* means to deliver to each member of a group.

Key Concept — *Distributive Property*

- **Words** To multiply a number by a sum, multiply each number inside the parentheses by the number outside the parentheses.
- **Symbols** $a(b + c) = ab + ac$ $(b + c)a = ba + ca$
- **Examples** $3(4 + 2) = 3 \cdot 4 + 3 \cdot 2$ $(5 + 3)2 = 5 \cdot 2 + 3 \cdot 2$

Concept Check Name two operations that are combined by the Distributive Property.

Example 1 Use the Distributive Property

Use the Distributive Property to write each expression as an equivalent expression. Then evaluate the expression.

a. $2(6 + 4)$

$2(6 + 4) = 2 \cdot 6 + 2 \cdot 4$

$= 12 + 8$ Multiply.

$= 20$ Add.

b. $(8 + 3)5$

$(8 + 3)5 = 8 \cdot 5 + 3 \cdot 5$

$= 40 + 15$ Multiply.

$= 55$ Add.

Example 2 Use the Distributive Property to Solve a Problem

AMUSEMENT PARKS A one-day pass to an amusement park costs \$40. A round-trip bus ticket to the park costs \$5.

a. Write two equivalent expressions to find the total cost of a one-day pass and a bus ticket for 15 students.

Method 1 Find the cost for 1 person, then multiply by 15.

$15(\$40 + \$5)$ — cost for 1 person

Method 2 Find the cost of 15 passes and 15 tickets. Then add.

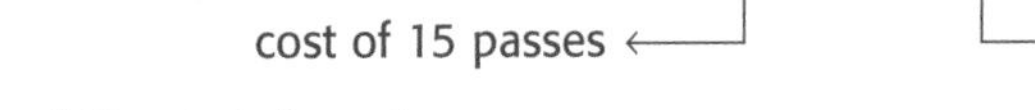

$15(\$40) + 15(\$5)$ — cost of 15 passes; cost of 15 tickets

b. Find the total cost.

Evaluate either expression to find the total cost.

$15(\$40 + 5) = 15(\$40) + 15(\$5)$ Distributive Property

$= \$600 + \$75$ Multiply.

$= \$675$ Add.

The total cost is \$675.

CHECK You can check your results by evaluating 15(\$45).

More About...

Amusement Parks

Attendance at U.S. amusement parks increased 22% in the 1990s. In 1999, over 300 million people attended these parks.

Source: *International Association of Amusement Parks and Attractions*

ALGEBRA CONNECTION

ALGEBRAIC EXPRESSIONS You can also model the Distributive Property by using algebra tiles.

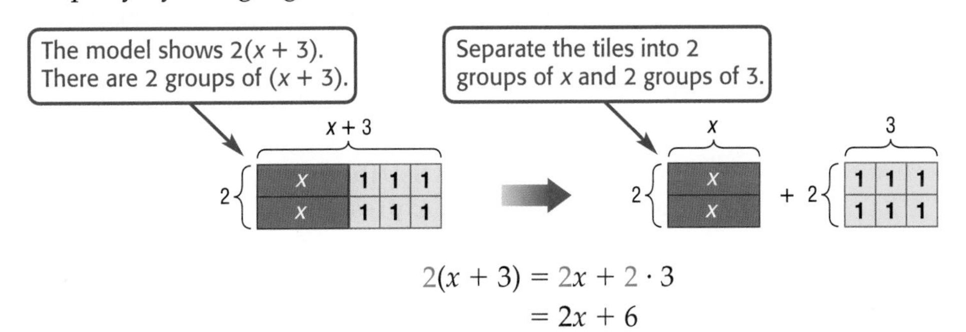

The model shows 2(x + 3). There are 2 groups of (x + 3).
Separate the tiles into 2 groups of x and 2 groups of 3.
x + 3
2
x
1 1 1
x
1 1 1
x
3
+ 2

$$2(x + 3) = 2x + 2 \cdot 3$$
$$= 2x + 6$$

The expressions $2(x + 3)$ and $2x + 6$ are equivalent expressions because no matter what x is, these expressions have the same value.

Concept Check Are $3(x + 1)$ and $3x + 1$ equivalent expressions? Explain.

Example 3 Simplify Algebraic Expressions

Use the Distributive Property to write each expression as an equivalent algebraic expression.

a. $3(x + 1)$

$3(x + 1) = 3x + 3 \cdot 1$

$= 3x + 3$ Simplify.

b. $(y + 4)5$

$(y + 4)5 = y \cdot 5 + 4 \cdot 5$

$= 5y + 20$ Simplify.

Example 4 Simplify Expressions with Subtraction

Use the Distributive Property to write each expression as an equivalent algebraic expression.

Study Tip

Look Back
To review **subtraction expressions**, see Lesson 2-3.

a. $2(x - 1)$

$2(x - 1) = 2[x + (-1)]$ Rewrite $x - 1$ as $x + (-1)$.

$= 2x + 2(-1)$ Distributive Property

$= 2x + (-2)$ Simplify.

$= 2x - 2$ Definition of subtraction

b. $-3(n - 5)$

$-3(n - 5) = -3[n + (-5)]$ Rewrite $n - 5$ as $n + (-5)$.

$= -3n + (-3)(-5)$ Distributive Property

$= -3n + 15$ Simplify.

Check for Understanding

Concept Check

1. **OPEN ENDED** Write an equation using three integers that is an example of the Distributive Property.
2. **FIND THE ERROR** Julia and Catelyn are using the Distributive Property to simplify $3(x + 2)$. Who is correct? Explain your reasoning.

Julia	Catelyn
$3(x + 2) = 3x + 2$	$3(x + 2) = 3x + 6$

Guided Practice

Use the Distributive Property to write each expression as an equivalent expression. Then evaluate it.

3. $5(7 + 8)$
4. $2(9 + 1)$
5. $(2 + 4)6$

ALGEBRA **Use the Distributive Property to write each expression as an equivalent algebraic expression.**

6. $4(x + 3)$
7. $(n + 2)3$
8. $8(y - 2)$
9. $-6(x - 5)$

Application

MONEY **For Exercises 10 and 11, use the following information.**
Suppose you work in a grocery store 4 hours on Friday and 5 hours on Saturday. You earn \$6.25 an hour.

10. Write two different expressions to find your wages.
11. Find the total wages for that weekend.

Practice and Apply

Homework Help

For Exercises	See Examples
12–23	1
24–25	2
26–33	3
34–47	4

Extra Practice

See page 728.

Use the Distributive Property to write each expression as an equivalent expression. Then evaluate it.

12. $2(6 + 1)$ **13.** $5(7 + 3)$ **14.** $(4 + 6)9$

15. $(4 + 3)3$ **16.** $(9 + 2)4$ **17.** $(8 + 8)2$

18. $7(3 - 2)$ **19.** $6(8 - 5)$ **20.** $-5(8 - 4)$

21. $-3(9 - 2)$ **22.** $(8 - 4)(-2)$ **23.** $(10 - 3)(-5)$

24. MOVIES One movie ticket costs \$7, and one small bag of popcorn costs \$3. Write two equivalent expressions for the total cost of four movie tickets and four bags of popcorn. Then find the cost.

25. SPORTS A volleyball uniform costs \$15 for the shirt, \$10 for the pants, and \$8 for the socks. Write two equivalent expressions for the total cost of 12 uniforms. Then find the cost.

ALGEBRA Use the Distributive Property to write each expression as an equivalent algebraic expression.

26. $2(x + 3)$ **27.** $5(y + 6)$ **28.** $3(n + 1)$

29. $7(y + 8)$ **30.** $(x + 3)4$ **31.** $(y + 2)10$

32. $(3 + y)6$ **33.** $(2 + x)5$ **34.** $3(x - 2)$

35. $9(m - 2)$ **36.** $8(z - 3)$ **37.** $15(s - 3)$

38. $(r - 5)6$ **39.** $(x - 3)12$ **40.** $(t - 4)5$

41. $(w - 10)2$ **42.** $-2(z + 4)$ **43.** $-5(a + 10)$

44. $-2(x - 7)$ **45.** $-5(w - 8)$ **46.** $(y - 4)(-2)$

47. $(a - 6)(-5)$ **48.** $2(x + y)$ **49.** $3(a + b)$

SHOPPING For Exercises 50 and 51, use the graphic.

50. Find the total amount spent by two teens and two adults during one average shopping trip.

51. Find the total amount spent by one teen and one adult during five average shopping trips.

USA TODAY Snapshots®

USA TODAY

Comparison shoppers

While teen-agers spend about 90 minutes on each trip to the mall, compared to 76 minutes for adults, they end up spending less money.

\$59.20
Adults

\$38.55
Teens

Source: International Council of Shopping Centers

By Marcy E. Mullins, USA TODAY

52. WRITING IN MATH Answer the question that was posed at the beginning of the lesson.

How are rectangles related to the Distributive Property?

Include the following in your answer:

- a drawing of two rectangles with the same width, and
- two different methods for finding the total area of the rectangles.

53. **CRITICAL THINKING** Is $3 + (x \cdot y) = (3 + x) \cdot (3 + y)$ a true statement? If so, explain your reasoning. If not, give a counterexample.

54. One ticket to a baseball game costs t dollars. A soft drink costs s dollars. Which expression represents the total cost of a ticket and soft drink for p people?

Ⓐ pst Ⓑ $p + (ts)$ Ⓒ $t(p + s)$ Ⓓ $p(t + s)$

55. Which equation is always true?

Ⓐ $5(a + b) = 5a + b$ Ⓑ $5(ab) = (5a)(5b)$

Ⓒ $5(a + b) = 5(b + a)$ Ⓓ $5(a + 0) = 5a + 5$

Extending the Lesson

MENTAL MATH The Distributive Property allows you to find certain products mentally. Replace one factor with the sum of a number and a multiple of ten. Then apply the Distributive Property.

Example Find $15 \cdot 12$ mentally.

$$15 \cdot 12 = 15(10 + 2) \quad \text{Think: 12 is 10 + 2.}$$
$$= 15 \cdot 10 + 15 \cdot 2 \quad \text{Distributive Property}$$
$$= 150 + 30 \quad \text{Multiply mentally.}$$
$$= 180 \quad \text{Add mentally.}$$

Rewrite each product so it is easy to compute mentally. Then find the product.

56. $7 \cdot 14$ 57. $8 \cdot 23$ 58. $9 \cdot 32$ 59. $16 \cdot 11$

60. $14 \cdot 12$ 61. $9 \cdot 103$ 62. $11 \cdot 102$ 63. $12 \cdot 1004$

Maintain Your Skills

Mixed Review

64. The table shows several solutions of the equation $x + y = 4$. *(Lesson 2-6)*

a. Graph the ordered pairs on a coordinate plane.

b. Describe the graph.

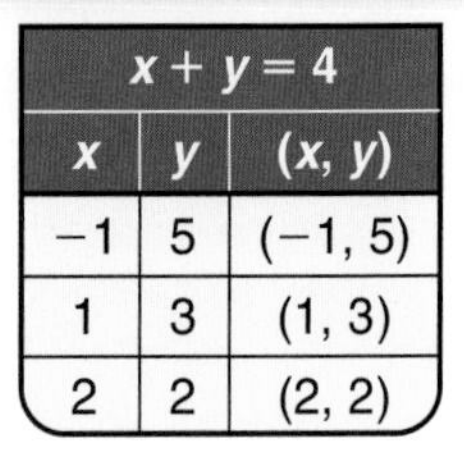

$x + y = 4$		
x	y	(x, y)
−1	5	(−1, 5)
1	3	(1, 3)
2	2	(2, 2)

65. **ALGEBRA** Evaluate $\frac{-4y}{x}$ if $x = 2$ and $y = -3$. *(Lesson 2-5)*

ALGEBRA Find the solution of each equation if the replacement set is {1, 2, 3, 4, 5}. *(Lesson 1-5)*

66. $2n + 3 = 9$ 67. $3n - 4 = 8$ 68. $4x - 9 = -5$

Find the next three terms in each pattern. *(Lesson 1-1)*

69. 5, 9, 13, 17, . . . 70. 20, 22, 26, 32, . . . 71. 5, 10, 20, 40, . . .

Getting Ready for the Next Lesson

PREREQUISITE SKILL Write each subtraction expression as an addition expression. *(To review **subtraction expressions**, see Lesson 2-3.)*

72. $5 - 3$ 73. $-8 - 4$ 74. $10 - 14$

75. $3 - 9$ 76. $-2 - (-5)$ 77. $-7 - 10$

3-2 Simplifying Algebraic Expressions

Sunshine State Standards
MA.D.2.3.1-1, MA.D.2.3.1-7, MA.D.2.3.2-1

Vocabulary

- term
- coefficient
- like terms
- constant
- simplest form
- simplifying an expression

What You'll Learn

- Use the Distributive Property to simplify algebraic expressions.

How can you use algebra tiles to simplify an algebraic expression?

You can use algebra tiles to represent expressions. You can also sort algebra tiles by their shapes and group them.

The drawing on the left represents the expression $2x + 3 + 3x + 1$. On the right, the algebra tiles have been sorted and combined.

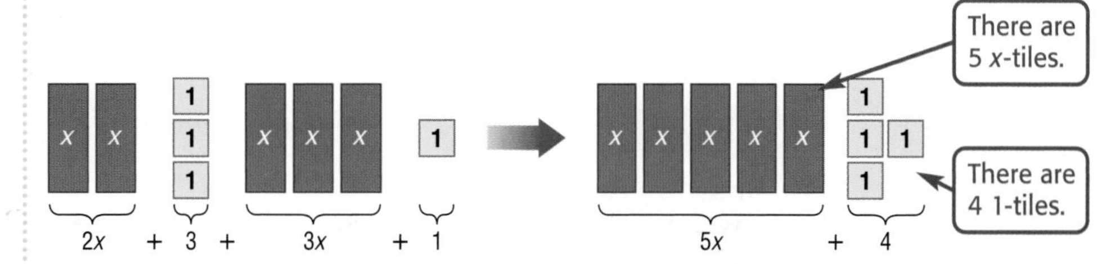

Therefore, $2x + 3 + 3x + 1 = 5x + 4$.

Model each expression with algebra tiles or a drawing. Then sort them by shape and write an expression represented by the tiles.

a. $3x + 2 + 4x + 3$

b. $2x + 5 + x$

c. $4x + 5 + 3$

d. $x + 2x + 4x$

SIMPLIFY EXPRESSIONS When plus or minus signs separate an algebraic expression into parts, each part is a **term**. The numerical part of a term that contains a variable is called the **coefficient** of the variable.

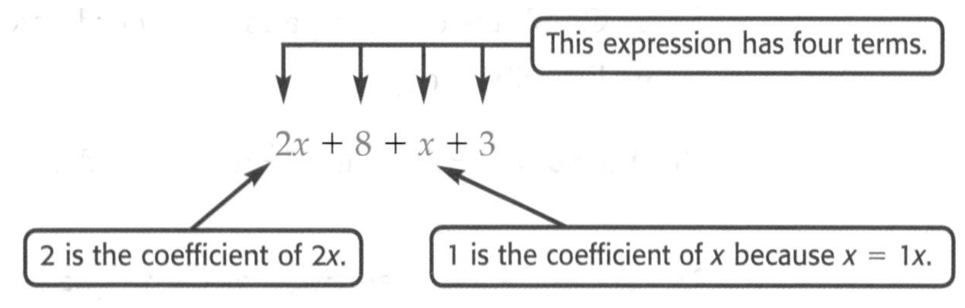

Reading Math

Constant
Everyday Meaning: unchanging
Math Meaning: fixed value in an expression

Like terms are terms that contain the same variables, such as $2n$ and $5n$ or $6xy$ and $4xy$. A term without a variable is called a **constant**. Constant terms are also like terms.

Like terms.

$$5y + 3 + 2y + 8y$$

Constant

Concept Check Are $5x$ and $5y$ like terms? Explain.

Rewriting a subtraction expression using addition will help you identify the terms of an expression.

Example 1 Identify Parts of Expressions

Identify the terms, like terms, coefficients, and constants in the expression $3x - 4x + y - 2$.

$3x - 4x + y - 2 = 3x + (-4x) + y + (-2)$ Definition of subtraction

$= 3x + (-4x) + 1y + (-2)$ Identity Property

The terms are $3x$, $-4x$, y, and -2. The like terms are $3x$ and $-4x$.

The coefficients are 3, -4, and 1. The constant is -2.

An algebraic expression is in **simplest form** if it has no like terms and no parentheses. When you use the Distributive Property to combine like terms, you are **simplifying the expression**.

Example 2 Simplify Algebraic Expressions

Study Tip

Equivalent Expressions
To check whether $2x + 8x$ and $10x$ are equivalent expressions, substitute any value for x and see whether the expressions have the same value.

Simplify each expression.

a. $2x + 8x$

$2x$ and $8x$ are like terms.

$2x + 8x = (2 + 8)x$ Distributive Property

$= 10x$ Simplify.

b. $6n + 3 + 2n$

$6n$ and $2n$ are like terms.

$6n + 3 + 2n = 6n + 2n + 3$ Commutative Property

$= (6 + 2)n + 3$ Distributive Property

$= 8n + 3$ Simplify.

c. $3x - 5 - 8x + 6$

$3x$ and $-8x$ are like terms. -5 and 6 are also like terms.

$3x - 5 - 8x + 6 = 3x + (-5) + (-8x) + 6$ Definition of subtraction

$= 3x + (-8x) + (-5) + 6$ Commutative Property

$= [3 + (-8)]x + (-5) + 6$ Distributive Property

$= -5x + 1$ Simplify.

d. $m + 3(n + 4m)$

$m + 3(n + 4m) = m + 3n + 3(4m)$ Distributive Property

$= m + 3n + 12m$ Associative Property

$= 1m + 3n + 12m$ Identity Property

$= 1m + 12m + 3n$ Commutative Property

$= (1 + 12)m + 3n$ Distributive Property

$= 13m + 3n$ Simplify.

Concept Check Expressions like $4(x - 3)$, $4x - 12$, and $x + 3x - 12$ are equivalent expressions. Which is in simplest form?

Example 3 Translate Verbal Phrases into Expressions

BASEBALL CARDS Suppose you and your brother collect baseball cards. He has 15 more cards in his collection than you have. Write an expression in simplest form that represents the total number of cards in both collections.

Words You have some cards. Your brother has 15 more.

Variables Let x = number of cards you have.

Let $x + 15$ = number of cards your brother has.

Expression To find the total, add the expressions.

$$\begin{aligned} x + (x + 15) &= (x + x) + 15 && \text{Associative Property} \\ &= (1x + 1x) + 15 && \text{Identity Property} \\ &= (1 + 1)x + 15 && \text{Distributive Property} \\ &= 2x + 15 && \text{Simplify.} \end{aligned}$$

The expression $2x + 15$ represents the total number of cards, where x is the number of cards you have.

More About. . .

Baseball Cards

Honus Wagner is considered by many to be baseball's greatest all-around player. In July, 2000, one of his baseball cards sold for $1.1 million.

Source: CMG Worldwide

Check for Understanding

Concept Check

1. **Define** *like terms.*
2. **OPEN ENDED** Write an expression containing three terms that is in simplest form. One of the terms should be a constant.
3. **FIND THE ERROR** Koko and John are simplifying the expression $5x - 4 + x + 2$.

Koko	John
5x - 4 + x + 2 =	5x - 4 + x + 2 =
6x - 2	5x - 2

Who is correct? Explain your reasoning.

Guided Practice

Identify the terms, like terms, coefficients, and constants in each expression.

4. $4x + 3 + 5x + y$
5. $2m - n + 6m$
6. $4y - 2x - 7$

Simplify each expression.

7. $6a + 2a$
8. $x + 9x + 3$
9. $6c + 4 + c + 8$
10. $7m - 2m$
11. $9y + 8 - 8$
12. $2x - 5 - 4x + 8$
13. $5 - 3(y + 7)$
14. $3x + 2y + 4y$
15. $x + 3(x + 4y)$

Application

16. **MONEY** You have saved some money. Your friend has saved \$20 more than you. Write an expression in simplest form that represents the total amount of money you and your friend have saved.

Practice and Apply

Homework Help

For Exercises	See Examples
17–22	1
23–49	2
50–53	3

Extra Practice

See page 728.

Identify the terms, like terms, coefficients, and constants in each expression.

17. $3 + 7x + 3x + x$
18. $y + 3y + 8y + 2$
19. $2a + 5c - a + 6a$
20. $5c - 2d + 3d - d$
21. $6m - 2n + 7$
22. $7x - 3y + 3z - 2$

Simplify each expression.

23. $2x + 5x$
24. $7b + 2b$
25. $y + 10y$
26. $5y + y$
27. $2a + 3 + 5a$
28. $4 + 2m + m$
29. $2y + 8 + 5y + 1$
30. $8x + 5 + 7 + 2x$
31. $5x - 3x$
32. $10b - 2b$
33. $4y - 5y$
34. $r - 3r$
35. $8 + x - 5x$
36. $6x + 4 - 7x$
37. $8y - 7 + 7$
38. $9x + 2 - 2$
39. $2x + 3 - 3x + 9$
40. $5t - 3 - t + 2$
41. $3(b + 2) + 2b$
42. $5(x + 3) + 8x$
43. $-3(a + 2) - a$
44. $-2(x + 3) + 2x$
45. $4x - 4(2 + x)$
46. $8a - 2(a - 7)$
47. $6m + 2n + 10m$
48. $-2y + x + 3y$
49. $c + 2(d - 5c)$

More About. . .

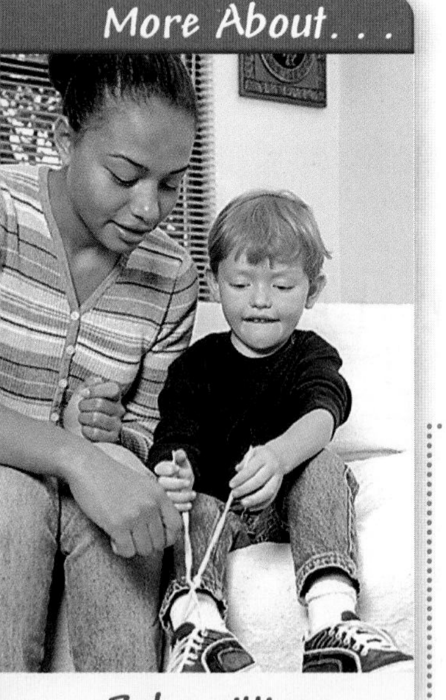

Baby-sitting

In a recent survey, 10% of students in grades 6–12 reported that most of their spending money came from baby-sitting.

Source: *USA WEEKEND*

For Exercises 50–53, write an expression in simplest form that represents the total amount in each situation.

50. **SCHOOL SUPPLIES** You bought 5 folders that each cost x dollars, a calculator for \$45, and a set of pens for \$3.

51. **SHOPPING** Suppose you buy 3 shirts that each cost s dollars, a pair of shoes for \$50, and jeans for \$30.

52. **BIRTHDAYS** Today is your friend's birthday. She is y years old. Her sister is 5 years younger.

53. **BABY-SITTING** Alicia earned d dollars baby-sitting. Her friend earned twice as much. You earned \$2 less than Alicia's friend earned.

GEOMETRY **You can find the perimeter of a geometric figure by adding the measures of its sides. Write an expression in simplest form for the perimeter of each figure.**

54.

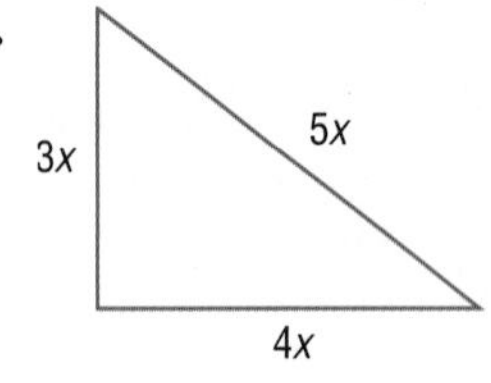

55.

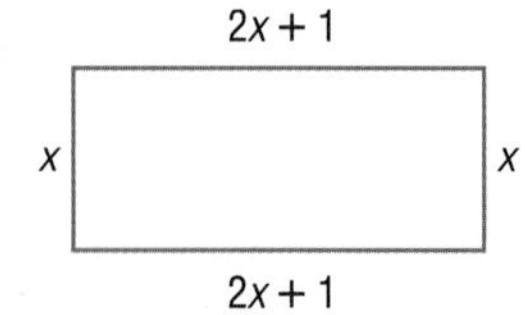

56. **WRITING IN MATH** Answer the question that was posed at the beginning of the lesson.

How can you use algebra tiles to simplify an algebraic expression?

Include the following in your answer:

- a drawing that shows how to simplify the expression $4x + 2 + 3x + 1$ using algebra tiles,
- a definition of *like terms,* and
- an explanation of how you use the Commutative and Distributive Properties to simplify $4x + 2 + 3x + 1$.

57. **CRITICAL THINKING** You use *deductive reasoning* when you base a conclusion on mathematical rules or properties. Indicate the property that justifies each step that was used to simplify $3(x + 4) + 5(x + 1)$.

a. $3(x + 4) + 5(x + 1) = 3x + 12 + 5x + 5$
b. $= 3x + 5x + 12 + 5$
c. $= 3x + 5x + 17$
d. $= 8x + 17$

FCAT Practice
Standardized Test Practice

58. Which expression is *not* equivalent to the other three?
Ⓐ $-6(x - 2)$ Ⓑ $x + 12 - 7x$
Ⓒ $-6x - 12$ Ⓓ $-x - 5x + 12$

59. Katie practiced the clarinet for m minutes. Her sister practiced 10 minutes less. Which expression represents the total time they spent practicing?
Ⓐ $m - 10$ Ⓑ $m + 10$ Ⓒ $2m - 10$ Ⓓ $2m + 10$

Maintain Your Skills

Mixed Review

ALGEBRA Use the Distributive Property to write each expression as an equivalent expression. *(Lesson 3-1)*

60. $3(a + 5)$ 61. $-2(y + 8)$ 62. $-3(x - 1)$

63. Name the quadrant in which $P(-5, -6)$ is located. *(Lesson 2-6)*

64. **CRUISES** The table shows the number of people who took a cruise in various years. Make a scatter plot of the data. *(Lesson 1-7)*

TICKET

People Taking Cruises

Year	1970	1980	1990	2000
Number (millions)	0.5	1.4	3.6	6.5

Source: Cruise Lines International Association

Evaluate each expression. *(Lesson 1-2)*

65. $2 + 3 \cdot 5$ 66. $8 \div 2 \cdot 4$ 67. $10 - 2 \cdot 4$

Getting Ready for the Next Lesson

PREREQUISITE SKILL Find each sum. *(To review **adding integers**, see Lesson 2-2.)*

68. $-5 + 4$ 69. $-8 + (-3)$ 70. $10 + (-1)$
71. $4 + (-9)$ 72. $11 + (-7)$ 73. $-4 + (-9)$

Practice Quiz 1

Lessons 3-1 and 3-2

Simplify each expression. *(Lessons 3-1 and 3-2)*

1. $6(x + 2)$ 2. $5(x - 7)$ 3. $6y - 4 + y$ 4. $2a + 4(a - 9)$

5. **SCHOOL** You spent m minutes studying on Monday. On Tuesday, you studied 15 more minutes than you did on Monday. Write an expression in simplest form that represents the total amount of time spent studying on Monday and Tuesday. *(Lesson 3-2)*

Algebra Activity

A Preview of Lessons 3-3 and 3-4

Sunshine State Standards
MA.D.2.3.1-3

Solving Equations Using Algebra Tiles

Activity 1

In a set of algebra tiles, 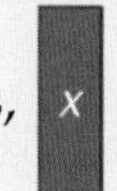 represents the variable x, [1] represents the integer 1,

and [−1] represents the integer −1. You can use algebra tiles and an equation mat to model equations.

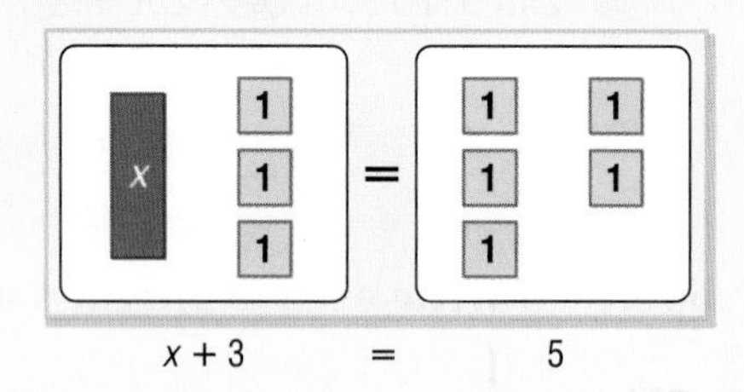

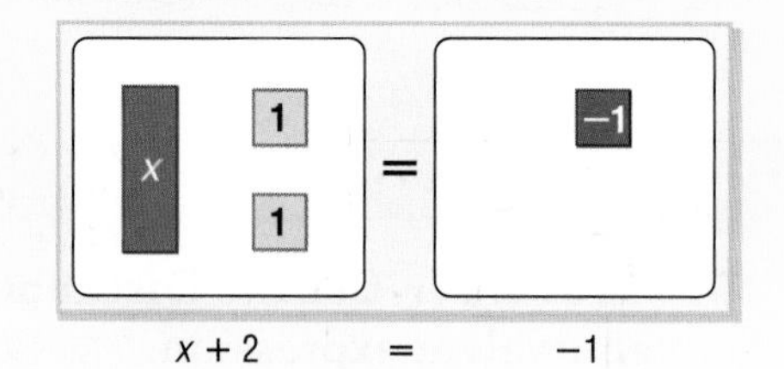

When you solve an equation, you are trying to find the value of x that makes the equation true. The following example shows how to solve $x + 3 = 5$ using algebra tiles.

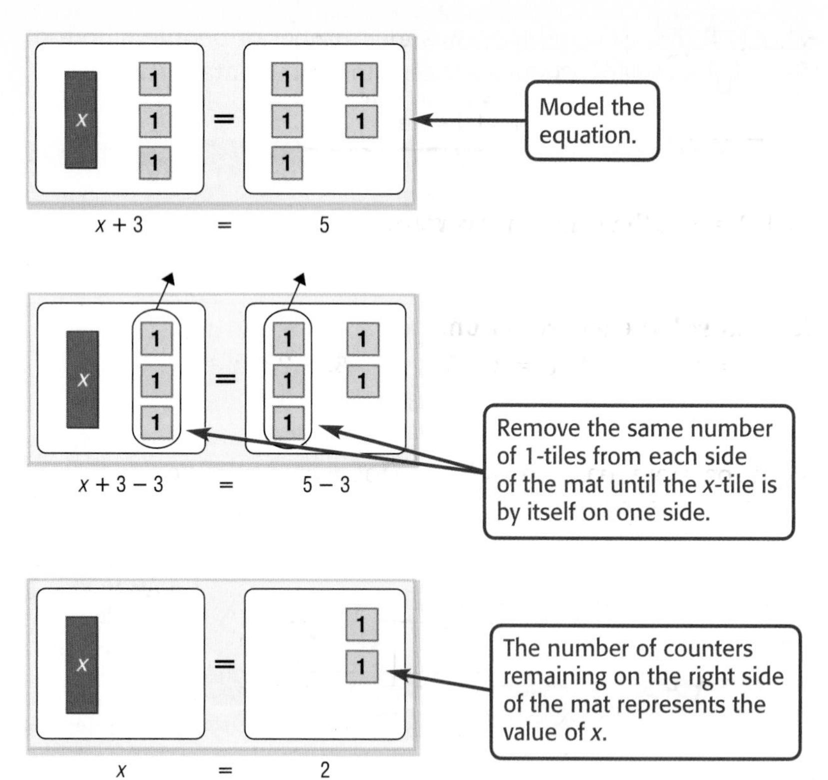

Therefore, $x = 2$. Since $2 + 3 = 5$, the solution is correct.

Model

Use algebra tiles to model and solve each equation.

1. $3 + x = 7$ **2.** $x + 4 = 5$ **3.** $6 = x + 4$ **4.** $5 = 1 + x$

Activity 2

Some equations are solved by using zero pairs. You may add or subtract a zero pair from either side of an equation mat without changing its value. The following example shows how to solve $x + 2 = -1$ by using zero pairs.

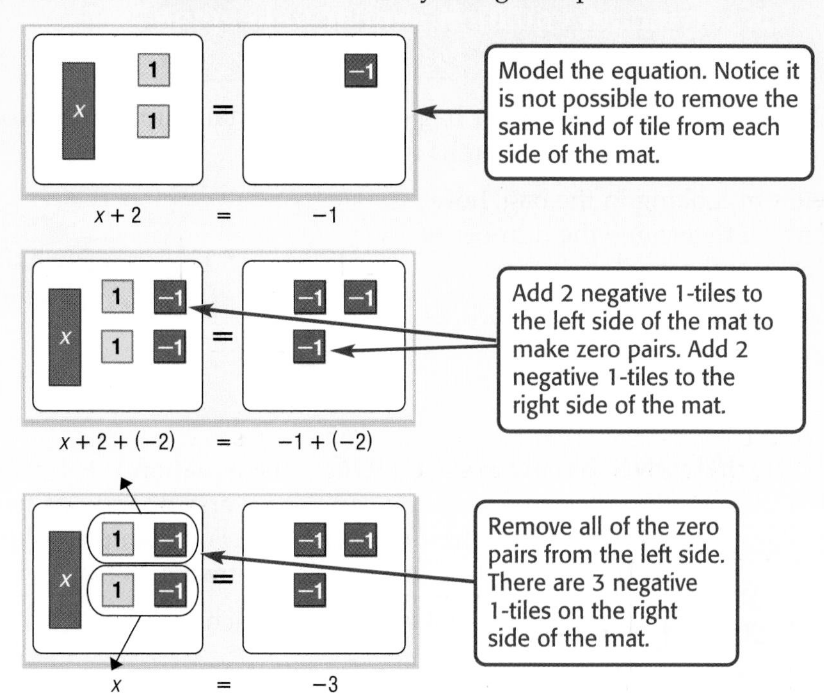

x
1
1
=
−1
Model the equation. Notice it is not possible to remove the same kind of tile from each side of the mat.
x + 2 = −1
Add 2 negative 1-tiles to the left side of the mat to make zero pairs. Add 2 negative 1-tiles to the right side of the mat.
x + 2 + (−2) = −1 + (−2)
Remove all of the zero pairs from the left side. There are 3 negative 1-tiles on the right side of the mat.
x = −3

Therefore, $x = -3$. Since $-3 + 2 = -1$, the solution is correct.

Model

Use algebra tiles to model and solve each equation.

5. $x + 2 = -2$ **6.** $x - 3 = 2$ **7.** $0 = x + 3$ **8.** $-2 = x + 1$

Activity 3

Some equations are modeled using more than one x-tile. The following example shows how to solve $2x = -6$ using algebra tiles.

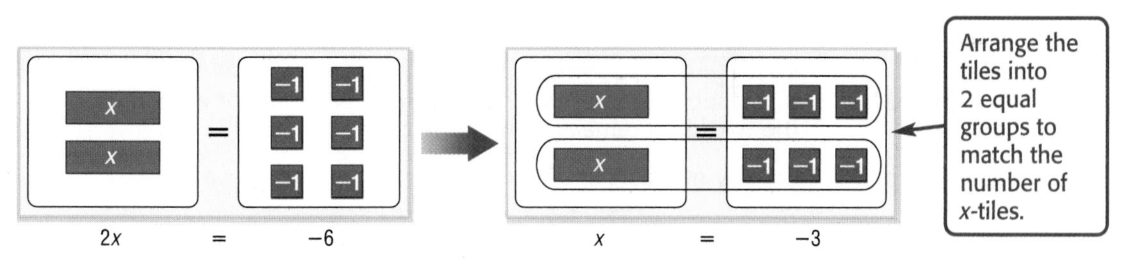

Arrange the tiles into 2 equal groups to match the number of x-tiles.

Therefore, $x = -3$. Since $2(-3) = -6$, the solution is correct.

Model

Use algebra tiles to model and solve each equation.

9. $3x = 3$ **10.** $2x = -8$ **11.** $6 = 3x$ **12.** $-4 = 2x$

3-3 Solving Equations by Adding or Subtracting

Sunshine State Standards
MA.A.3.3.1-2, MA.D.1.3.1-2, MA.D.2.3.1-3, M.A.D 2.3.2-1, MA.D.2.3.2-2

Vocabulary

- inverse operation
- equivalent equations

What You'll Learn

- Solve equations by using the Subtraction Property of Equality.
- Solve equations by using the Addition Property of Equality.

How is solving an equation similar to keeping a scale in balance?

On the balance below, the paper bag contains a certain number of blocks. (Assume that the paper bag weighs nothing.)

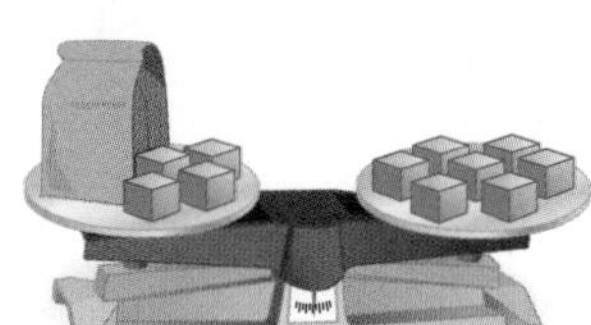

a. Without looking in the bag, how can you determine the number of blocks in the bag?

b. Explain why your method works.

SOLVE EQUATIONS BY SUBTRACTING The equation $x + 4 = 7$ is a model of the situation shown above. You can use inverse operations to solve the equation. **Inverse operations** "undo" each other. For example, to undo the addition of 4 in the expression $x + 4$, you would subtract 4.

To solve the equation $x + 4 = 7$, subtract 4 from each side.

$$x + 4 = 7$$
$$x + 4 - 4 = 7 - 4$$
$$x + 0 = 3$$
$$x = 3$$

Subtract 4 from the left side of the equation to isolate the variable.

Subtract 4 from the right side of the equation to keep it balanced.

The solution is 3.

You can use the **Subtraction Property of Equality** to solve any equation like $x + 4 = 7$.

Key Concept — *Subtraction Property of Equality*

- **Words** If you subtract the same number from each side of an equation, the two sides remain equal.
- **Symbols** For any numbers a, b, and c, if $a = b$, then $a - c = b - c$.
- **Examples**

$$5 = 5 \qquad\qquad x + 2 = 3$$
$$5 - 3 = 5 - 3 \qquad\qquad x + 2 - 2 = 3 - 2$$
$$2 = 2 \qquad\qquad x = 1$$

Concept Check Which integer would you subtract from each side of $x + 7 = 20$ to solve the equation?

The equations $x + 4 = 7$ and $x = 3$ are **equivalent equations** because they have the same solution, 3. When you solve an equation, you should always check to be sure that the first and last equations are equivalent.

Concept Check Are $x + 4 = 15$ and $x = 4$ equivalent equations? Explain.

Example 1 Solve Equations by Subtracting

Solve $x + 8 = -5$. Check your solution.

$x + 8 = -5$	Write the equation.
$x + 8 - 8 = -5 - 8$	Subtract 8 from each side.
$x + 0 = -13$	$8 - 8 = 0$, $-5 - 8 = -13$
$x = -13$	Identity Property; $x + 0 = x$

To check your solution, replace x with -13 in the original equation.

CHECK	$x + 8 = -5$	Write the equation.
	$-13 + 8 \stackrel{?}{=} -5$	Check to see whether this sentence is true.
	$-5 = -5$ ✓	The sentence is true.

The solution is –13.

Study Tip

Checking Equations It is always wise to check your solution. You can often use arithmetic facts to check the solutions of simple equations.

Example 2 Graph the Solutions of an Equation

Graph the solution of $16 + x = 14$ on a number line.

$16 + x = 14$	Write the equation.
$x + 16 = 14$	Commutative Property; $16 + x = x + 16$
$x + 16 - 16 = 14 - 16$	Subtract 16 from each side.
$x = -2$	Simplify.

The solution is -2. To graph the solution, draw a dot at -2 on a number line.

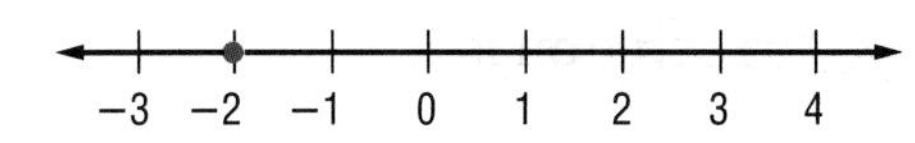

Study Tip

Look Back To review **graphing on a number line**, see Lesson 2-1.

SOLVE EQUATIONS BY ADDING Some equations can be solved by adding the same number to each side. This property is called the **Addition Property of Equality**.

Key Concept — *Addition Property of Equality*

- **Words** If you add the same number to each side of an equation, the two sides remain equal.
- **Symbols** For any numbers a, b, and c, if $a = b$, then $a + c = b + c$.
- **Examples**

$6 = 6$	$x - 2 = 5$
$6 + 3 = 6 + 3$	$x - 2 + 2 = 5 + 2$
$9 = 9$	$x = 7$

If an equation has a subtraction expression, first rewrite the expression as an addition expression. Then add the additive inverse to each side.

Example 3 Solve Equations by Adding

Solve $y - 7 = -25$.

$y - 7 = -25$	Write the equation.
$y + (-7) = -25$	Rewrite $y - 7$ as $y + (-7)$.
$y + (-7) + 7 = -25 + 7$	Add 7 to each side.
$y + 0 = -25 + 7$	Additive Inverse Property; $(-7) + 7 = 0$.
$y = -18$	Identity Property; $y + 0 = y$
The solution is -18.	Check your solution.

More About. . .

Aviation

On December 17, 1903, the Wright brothers made the first flights in a power-driven airplane. Orville's flight covered 120 feet, which was 732 feet shorter than Wilbur's.

Source: www.infoplease.com

Example 4 Use an Equation to Solve a Problem

AVIATION Use the information at the left. Write and solve an equation to find the length of Wilbur Wright's flight.

Words Orville's flight was 732 feet shorter than Wilbur's.

Variables Let x = the length of Wilbur's flight.

	Orville's flight	was	732 feet shorter than Wilbur's flight.
Equation	120	=	$x - 732$

Solve the equation.

$120 = x - 732$	Think of $x - 732$ as $x + (-732)$.
$120 + 732 = x - 732 + 732$	Add 732 to each side.
$852 = x$	Simplify.

Wilbur's flight was 852 feet.

Almost all standardized tests have items involving equations.

FCAT Practice

Standardized Test Practice

Example 5 Solve Equations

Multiple-Choice Test Item

What value of x makes $x - 4 = -2$ a true statement?

(A) 6 (B) 2 (C) -2 (D) -6

Read the Test Item To find the value of x, solve the equation.

Solve the Test Item

$x - 4 = -2$	Write the equation.
$x - 4 + 4 = -2 + 4$	Add 4 to each side.
$x = 2$	Simplify.

The answer is B.

The Princeton Review

Test-Taking Tip

Backsolving It may be easier to substitute each choice into the original equation until you get a true statement.

Check for Understanding

Concept Check

1. **Tell** what property you would use to solve $x - 15 = -3$.
2. **OPEN ENDED** Write two equations that are equivalent. Then write two equations that are *not* equivalent.

Guided Practice

ALGEBRA Solve each equation. Check your solution.

3. $x + 14 = 25$
4. $w + 4 = -10$
5. $16 = y + 20$
6. $n - 8 = 5$
7. $k - 25 = 30$
8. $r - 4 = -18$

ALGEBRA Graph the solution of each equation on a number line.

9. $-8 = x - 6$
10. $y - 3 = -1$

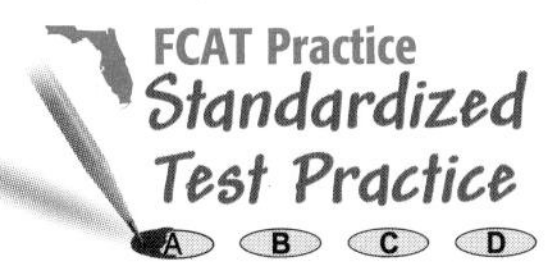

11. What value of x makes $x - 10 = -5$ a true statement?
   Ⓐ -5 Ⓑ 15 Ⓒ 5 Ⓓ -15

Practice and Apply

Homework Help

For Exercises	See Examples
12–32	1, 3
37–42	2
43–47	4

Extra Practice

See page 729.

ALGEBRA Solve each equation. Check your solution.

12. $y + 7 = 21$
13. $x + 5 = 18$
14. $m + 10 = -2$
15. $x + 5 = -3$
16. $a + 10 = -4$
17. $t + 6 = -9$
18. $y + 8 = 3$
19. $9 = 10 + b$
20. $k - 6 = 13$
21. $r - 5 = 10$
22. $8 = r - 5$
23. $19 = g - 5$
24. $x - 6 = -2$
25. $y - 49 = -13$
26. $-15 = x - 16$
27. $-8 = t - 4$
28. $23 + y = 14$
29. $59 = s + 90$
30. $x - 27 = -63$
31. $84 = r - 34$
32. $y - 95 = -18$

ALGEBRA Write and solve an equation to find each number.

33. The sum of a number and 9 is –2.
34. The sum of –5 and a number is –15.
35. The difference of a number and 3 is –6.
36. When 5 is subtracted from a number, the result is 16.

ALGEBRA Graph the solution of each equation on a number line.

37. $8 + w = 3$
38. $z - 5 = -8$
39. $8 = x - 2$
40. $9 + x = 12$
41. $-11 = y - 7$
42. $x - (-1) = 0$

43. **ELECTIONS** In the 2000 presidential election, Indiana had 12 electoral votes. That was 20 votes fewer than the number of electoral votes in Texas. Write and solve an equation to find the number of electoral votes in Texas.

44. **WEATHER** The difference between the record high and low temperatures in Charlotte, North Carolina, is 109°F. The record low temperature was −5°F. Write and solve an equation to find the record high temperature.

45. **RESEARCH** Use the Internet or another source to find record temperatures in your state. Use the data to write a problem.

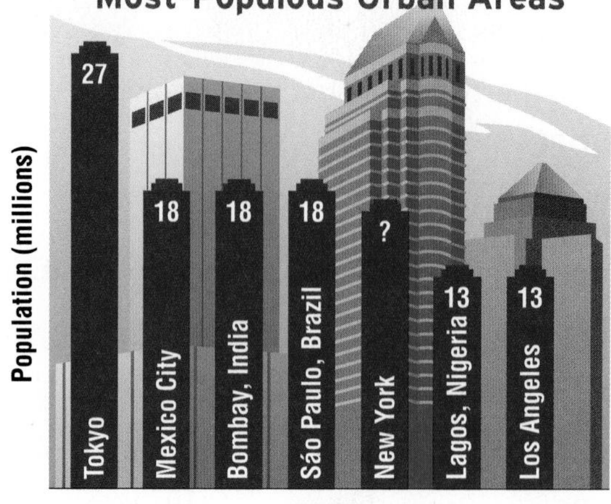

POPULATION For Exercises 46 and 47, use the graph and the following information. Tokyo's population is 10 million greater than New York's population. Los Angeles' population is 4 million less than New York's population.

46. Write two different equations to find New York's population.

47. Solve the equations.

Online Research **Data Update** How are the populations of these cities related today? Visit www.pre-alg.com/data_update to learn more.

48. CRITICAL THINKING Write two equations in which the solution is -5.

49. WRITING IN MATH Answer the question that was posed at the beginning of the lesson.

Why is solving an equation similar to keeping a scale in balance?

Include the following in your answer:

- a comparison of an equation and a balanced scale, and
- an explanation of the Addition and Subtraction Properties of Equality.

50. If $x + 4 = -2$, the numerical value of $-3x - 2$ is

Ⓐ -20. Ⓑ 16. Ⓒ 4. Ⓓ -8.

51. When 7 is subtracted from a number 5 times, the result is 3. What is the number?

Ⓐ 10 Ⓑ -2 Ⓒ 38 Ⓓ 35

Maintain Your Skills

Mixed Review

ALGEBRA Simplify each expression. *(Lessons 3-1 and 3-2)*

52. $-2(x + 5)$ **53.** $(t + 4)3$ **54.** $-4(x - 2)$

55. $6z - 3 - 10z + 7$ **56.** $2(x + 6) + 4x$ **57.** $3 - 4(m + 1)$

ALGEBRA What property is shown by each statement? *(Lessons 2-2 and 1-4)*

58. $9a + b = b + 9a$ **59.** $x[y + (-y)] = x(0)$ **60.** $6(3x) = (6 \cdot 3)x$

61. ALGEBRA Evaluate $9a + 4b$ if $a = 8$ and $b = 3$. *(Lesson 1-3)*

Getting Ready for the Next Lesson

PREREQUISITE SKILL Divide. *(To review **dividing integers**, see Lesson 2-5.)*

62. $-100 \div 10$ **63.** $50 \div (-2)$ **64.** $-49 \div (-7)$

65. $\frac{72}{-8}$ **66.** $\frac{-18}{-6}$ **67.** $\frac{-36}{9}$

3-4 Solving Equations by Multiplying or Dividing

What You'll Learn

- Solve equations by using the Division Property of Equality.
- Solve equations by using the Multiplication Property of Equality.

Sunshine State Standards
MA.B.2.3.2-2, MA.B.3.3.1-1, MA.D.1.3.1-2, MA.D.2.3.1-3, MA.D.2.3.2-2

How are equations used to find the U.S. value of foreign currency?

In Mexico, about 9 *pesos* can be exchanged for $1 of U.S. currency, as shown in the table.

U.S. Value ($)	Number of Pesos
1	$9(1) = 9$
2	$9(2) = 18$
3	$9(3) = 27$
4	$9(4) = 36$

In general, if we let d represent the number of U.S. dollars and p represent the number of pesos, then $9d = p$.

a. Suppose lunch in Mexico costs 72 pesos. Write an equation to find the cost in U.S. dollars.

b. How can you find the cost in U.S. dollars?

SOLVE EQUATIONS BY DIVIDING The equation $9x = 72$ is a model of the relationship described above. To undo the multiplication operation in $9x$, you would divide by 9.

To solve the equation $9x = 72$, divide each side by 9.

$$9x = 72$$
$$\frac{9x}{9} = \frac{72}{9}$$
$$1x = 8$$
$$x = 8$$

Divide the left side of the equation by 9 to undo the multiplication $9 \cdot x$.

Divide the right side of the equation by 9 to keep it balanced.

The solution is 8.

You can use the **Division Property of Equality** to solve any equation like $9x = 72$.

Key Concept — *Division Property of Equality*

- **Words** When you divide each side of an equation by the same nonzero number, the two sides remain equal.
- **Symbols** For any numbers a, b, and c, where $c \neq 0$, then $\frac{a}{c} = \frac{b}{c}$.
- **Examples**

$$14 = 14 \qquad 3x = -12$$
$$\frac{14}{7} = \frac{14}{7} \qquad \frac{3x}{3} = \frac{-12}{3}$$
$$2 = 2 \qquad x = -4$$

Example 1 Solve Equations by Dividing

Solve $5x = -30$. Check your solution and graph it on a number line.

$5x = -30$	Write the equation.
$\frac{5x}{5} = \frac{-30}{5}$	Divide each side by 5 to undo the multiplication in $5 \cdot x$.
$1x = -6$	$5 \div 5 = 1$, $-30 \div 5 = -6$
$x = -6$	Identity Property; $1x = x$

To check your solution, replace x with -6 in the original equation.

CHECK	$5x = -30$	Write the equation.
	$5(-6) \stackrel{?}{=} -30$	Check to see whether this statement is true.
	$-30 = -30$ ✓	The statement is true.

The solution is –6.

To graph the solution, draw a dot at –6 on a number line.

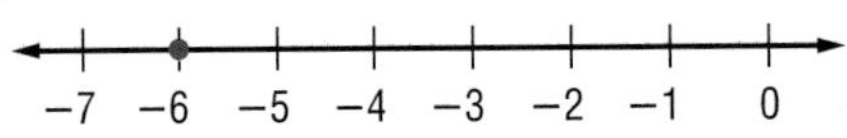

Concept Check Explain how you could find the value of x in $3x = 18$.

Example 2 Use an Equation to Solve a Problem

PARKS It costs \$3 per car to use the hiking trails along the Columbia River Highway. If income from the hiking trails totaled \$1275 in one day, how many cars entered the park?

Words \$3 times the number of cars equals the total.

Variables Let x represent the number of cars.

	The cost per car	times	the number of cars	equals	the total.
Equation	\$3	$\cdot$	x	=	\$1275

Solve the equation.

$3x = 1275$	Write the equation.
$\frac{3x}{3} = \frac{1275}{3}$	Divide each side by 3.
$x = 425$	Simplify.

CHECK	$3x = 1275$	Write the equation.
	$3(425) \stackrel{?}{=} 1275$	Check to see whether this statement is true.
	$1275 = 1275$ ✓	The statement is true.

Therefore, 425 cars entered the park.

More About. . .

Parks

The Columbia River Highway, built in 1913, is a historic route in Oregon that curves around twenty waterfalls through the Cascade Mountains.

Source: *USA TODAY*

Concept Check Suppose it cost \$5 per car to use the hiking trails and the total income was \$1275. What equation would you solve?

SOLVE EQUATIONS BY MULTIPLYING Some equations can be solved by multiplying each side by the same number. This property is called the **Multiplication Property of Equality**.

Key Concept — *Multiplication Property of Equality*

- **Words** When you multiply each side of an equation by the same number, the two sides remain equal.
- **Symbols** For any numbers a, b, and c, if $a = b$, then $ac = bc$.
- **Examples**

$8 = 8$	$\frac{x}{6} = 7$
$8(-2) = 8(-2)$	$\left(\frac{x}{6}\right)6 = (7)6$
$-16 = -16$	$x = 42$

Example 3 *Solve Equations by Multiplying*

Solve $\frac{y}{-4} = -9$. Check your solution.

$\frac{y}{-4} = -9$ Write the equation.

$\frac{y}{-4}(-4) = -9(-4)$ Multiply each side by -4 to undo the division in $\frac{y}{-4}$.

$y = 36$ Simplify.

CHECK $\frac{y}{-4} = -9$ Write the equation.

$\frac{36}{-4} \stackrel{?}{=} -9$ Check to see whether this statement is true.

$-9 = -9\ \checkmark$ The statement is true.

The solution is 36.

Study Tip

Division Expressions
Remember, $\frac{y}{-4}$ means *y divided by -4*.

Check for Understanding

Concept Check

1. **State** what property you would use to solve $\frac{x}{-9} = -36$.
2. **Explain** how to find the value of y in $-5y = -45$.
3. **OPEN ENDED** Write an equation of the form $ax = c$ where a and c are integers and the solution is 4.

Guided Practice **ALGEBRA Solve each equation. Check your solution.**

4. $4x = 24$
5. $-2a = 10$
6. $-42 = -7t$
7. $\frac{k}{3} = 9$
8. $\frac{y}{5} = -8$
9. $-11 = \frac{n}{-6}$

Application

10. **TOYS** A spiral toy that can bounce down a flight of stairs is made from 80 feet of wire. Write and solve an equation to find how many of these toys can be made from a spool of wire that contains 4000 feet.

Practice and Apply

Homework Help

For Exercises	See Examples
11–34, 39–44	1, 3
45–48	2

Extra Practice

See page 729.

ALGEBRA **Solve each equation. Check your solution.**

11. $3t = 21$
12. $8x = 72$
13. $-32 = 4y$
14. $5n = -95$
15. $-56 = -7p$
16. $-8j = -64$
17. $\frac{h}{4} = 6$
18. $\frac{c}{9} = 4$
19. $\frac{g}{-2} = -7$
20. $-42 = \frac{x}{-2}$
21. $11 = \frac{b}{-3}$
22. $\frac{h}{-7} = 20$
23. $45 = 5x$
24. $3u = 51$
25. $86 = -2v$
26. $-8a = 144$
27. $\frac{m}{45} = -3$
28. $\frac{d}{3} = -3$
29. $\frac{f}{-13} = -10$
30. $\frac{v}{-11} = -132$
31. $-116 = -4w$
32. $-68 = -4m$
33. $-21 = \frac{k}{8}$
34. $-56 = \frac{t}{9}$

ALGEBRA **Write and solve an equation for each sentence.**

35. The product of a number and 6 is −42.
36. The product of −7 and a number is −35.
37. The quotient of a number and −4 is 8.
38. When you divide a number by −5, the result is −2.

ALGEBRA **Graph the solution of each equation on a number line.**

39. $48 = -6x$
40. $-32t = 64$
41. $-6r = -18$
42. $-42 = -7x$
43. $\frac{n}{12} = 3$
44. $\frac{y}{-4} = -1$

More About. . .

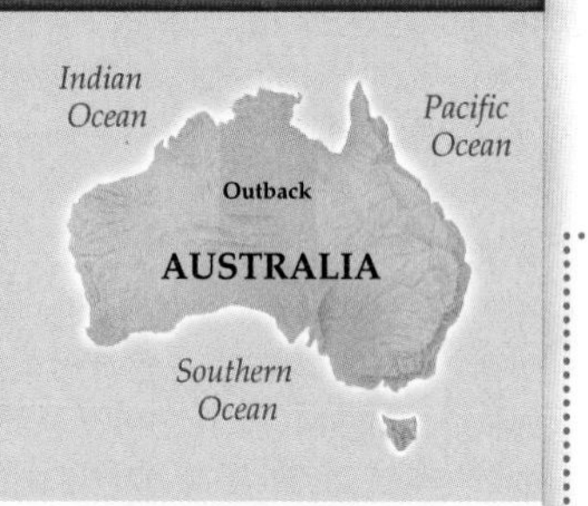

Ranching

Some students living in the Outback are so far from schools that they get their education by special radio programming. They mail in their homework and sometimes talk to teachers by two-way radio.

Source: *Kids Discover Australia*

45. **RANCHING** The largest ranch in the world is in the Australian Outback. It is about 12,000 square miles, which is five times the size of the largest United States ranch. Write and solve an equation to find the size of the largest United States ranch.

46. **RANCHING** In the driest part of an Outback ranch, each cow needs about 40 acres for grazing. Write and solve an equation to find how many cows can graze on 720 acres of land.

47. **PAINTING** A **person-day** is a unit of measure that represents one person working for one day. A painting contractor estimates that it will take 24 person-days to paint a house. Write and solve an equation to find how many painters the contractor will need to hire to paint the house in 6 days.

48. **MEASUREMENT** The chart shows several conversions in the customary system. Write and solve an equation to find each quantity.
    a. the number of feet in 132 inches
    b. the number of yards in 15 feet
    c. the number of miles in 10,560 feet

Customary System (length)
1 mile = 5280 feet
1 mile = 1760 yards
1 yard = 3 feet
1 foot = 12 inches
1 yard = 36 inches

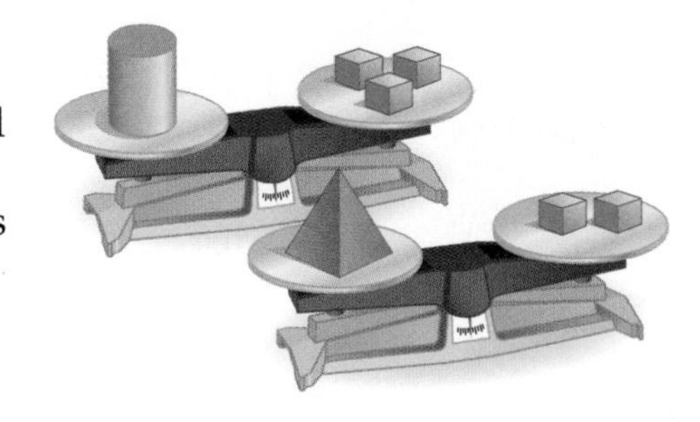

49. **CRITICAL THINKING** Suppose that one pyramid balances two cubes and one cylinder balances three cubes. Determine whether each statement is *true* or *false*. Justify your answers.

   a. One pyramid and one cube balance three cubes.

   b. One pyramid and one cube balance one cylinder.

   c. One cylinder and one pyramid balance four cubes.

50. **WRITING IN MATH** Answer the question that was posed at the beginning of the lesson.

   How are equations used to find the U.S. value of foreign currency?

   Include the following in your answer:
   - the cost in U.S. dollars of a 12-*pound* bus trip in Egypt, if 4 *pounds* can be exchanged for one U.S. dollar, and
   - the cost in U.S. dollars of a 3040-*schilling* hotel room in Austria, if 16 *schillings* can be exchanged for one U.S. dollar.

Online Research **Data Update** How many pounds and schillings can be exchanged for a U.S. dollar today? Visit www.pre-alg.com/data_update to learn more.

51. Solve $\frac{mx}{n} = p$ for x.

   (A) $x = \frac{m}{pn}$ (B) $x = \frac{pn}{m}$ (C) $x = \frac{p-n}{m}$ (D) $x = pn - m$

52. A number is divided by -6, and the result is 24. What is the original number?

   (A) -4 (B) 4 (C) -144 (D) 144

Maintain Your Skills

Mixed Review **ALGEBRA Solve each equation.** *(Lesson 3-3)*

53. $3 + y = 16$ 54. $29 = n + 4$ 55. $k - 12 = -40$

ALGEBRA Simplify each expression. *(Lesson 3-2)*

56. $4x + 7x$ 57. $2y + 6 + 5y$ 58. $3 - 2(y + 4)$

59. **ALGEBRA** Evaluate $3ab$ if $a = -6$ and $b = 2$. *(Lesson 2-4)*

60. Replace ● in -14 ● -4 with $<$, $>$, or $=$ to make a true sentence. *(Lesson 2-1)*

Getting Ready for the Next Lesson **PREREQUISITE SKILL Find each difference.**
*(To review **subtracting integers**, see Lesson 2-3.)*

61. $8 - (-2)$ 62. $-5 - 5$ 63. $-10 - (-8)$
64. $-18 - 4$ 65. $-3 - (-5)$ 66. $-45 - (-9)$
67. $-24 - (-5)$ 68. $-15 - (-15)$ 69. $-8 - 19$

3-5 Solving Two-Step Equations

Sunshine State Standards
MA.A.4.3.1-2, MA.D.1.3.1-2, MA.D.2.3.1-3, MA.D.2.3.2-2

Vocabulary

- two-step equation

What You'll Learn

- Solve two-step equations.

How can algebra tiles show the properties of equality?

The equation $2x + 1 = 9$, modeled below, can be solved with algebra tiles.

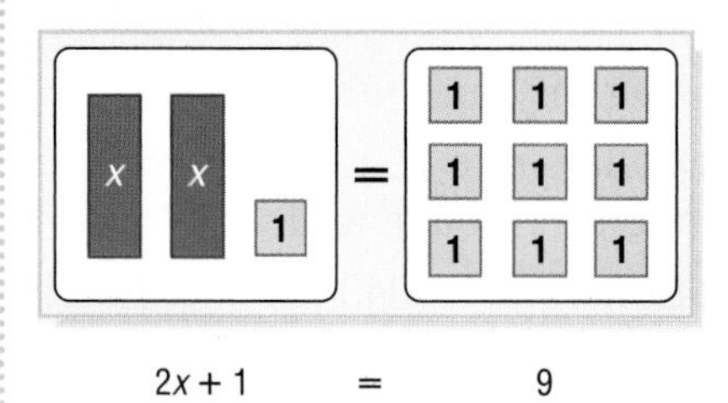

$2x + 1 \quad = \quad 9$

You can use the steps shown at the right to solve the equation.

Step 1 Remove 1 tile from each side of the mat.

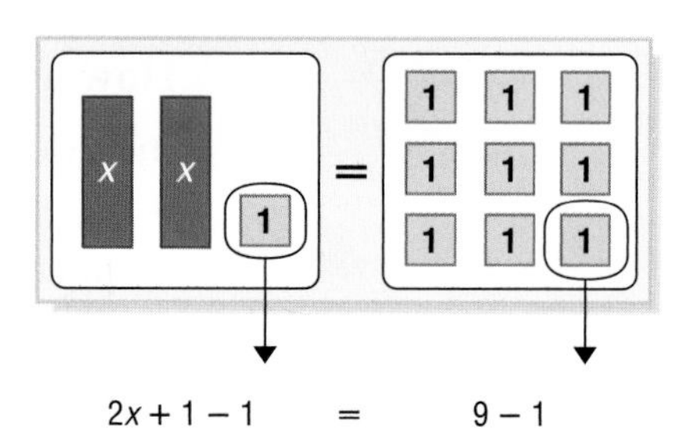

$2x + 1 - 1 \quad = \quad 9 - 1$

Step 2 Separate the remaining tiles into two equal groups.

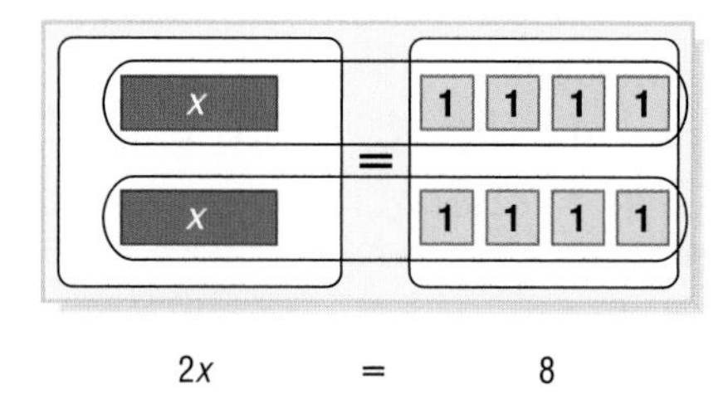

$2x \quad = \quad 8$

a. What property is shown by removing a tile from each side?

b. What property is shown by separating the tiles into two groups?

c. What is the solution of $2x + 1 = 9$?

SOLVE TWO-STEP EQUATIONS A **two-step equation** contains two operations. In the equation $2x + 1 = 9$, x is multiplied by 2 and then 1 is added. To solve two-step equations, use inverse operations to undo each operation in reverse order. You can solve $2x + 1 = 9$ in two steps.

Step 1 First, undo addition.

$$2x + 1 = 9$$
$$2x + 1 - 1 = 9 - 1 \quad \text{Subtract 1 from each side.}$$
$$2x = 8$$

Step 2 Then, undo multiplication.

$$2x = 8$$
$$\frac{2x}{2} = \frac{8}{2} \quad \text{Divide each side by 2.}$$
$$x = 4$$

The solution is 4.

Example 1 Solve Two-Step Equations

a. Solve $5x - 2 = 13$. Check your solution.

$5x - 2 = 13$	Write the equation.
$5x - 2 + 2 = 13 + 2$	Undo subtraction. Add to each side.
$5x = 15$	Simplify.
$\frac{5x}{5} = \frac{15}{5}$	Undo multiplication. Divide each side by 5.
$x = 3$	Simplify.

CHECK

$5x - 2 = 13$	Write the equation.
$5(3) - 2 \stackrel{?}{=} 13$	Check to see whether this statement is true.
$13 = 13$ ✓	The statement is true.

The solution is 3.

b. Solve $4 = \frac{n}{6} + 11$.

$4 = \frac{n}{6} + 11$	Write the equation.
$4 - 11 = \frac{n}{6} + 11 - 11$	Undo addition. Subtract 11 from each side.
$-7 = \frac{n}{6}$	Simplify.
$6(-7) = 6\left(\frac{n}{6}\right)$	Undo division. Multiply each side by 6.
$-42 = n$	Simplify.

The solution is -42. Check your solution.

Concept Check Explain how inverse operations can be used to solve a two-step equation.

Many real-world situations can be modeled with two-step equations.

Example 2 Use an Equation to Solve a Problem

SALES Mandy bought a DVD player. The sales clerk says that if she pays \$80 now, her monthly payments will be \$32. The total cost will be \$400. Solve $80 + 32x = 400$ to find how many months she will make payments.

$80 + 32x = 400$	Write the equation.
$80 - 80 + 32x = 400 - 80$	Subtract 80 from each side.
$32x = 320$	Simplify.
$\frac{32x}{32} = \frac{320}{32}$	Divide each side by 32.
$x = 10$	Simplify.

The solution is 10.

Therefore, Mandy will make payments for 10 months.

Study Tip

Checking Your Solution

Use estimation to determine whether your solution is reasonable: $80 + 30(10) = 380$. Since \$380 is close to \$400, the solution is reasonable.

Some two-step equations have terms with negative coefficients.

Example 3 Equations with Negative Coefficients

Solve $4 - x = 10$.

$4 - x = 10$	Write the equation.
$4 - 1x = 10$	Identity Property; $x = 1x$
$4 + (-1x) = 10$	Definition of subtraction
$-4 + 4 + (-1x) = -4 + 10$	Add -4 to each side.
$-1x = 6$	Simplify.
$\frac{-1x}{-1} = \frac{6}{-1}$	Divide each side by -1.
$x = -6$	Simplify.
The solution is -6.	Check your solution.

Sometimes it is necessary to combine like terms before solving.

Example 4 Combine Like Terms Before Solving

Study Tip

Mental Computation

You use the Distributive Property to simplify $1m - 5m$.

$1m - 5m = (1 - 5)m$
$= -4m$

You can also simplify the expression mentally.

Solve $m - 5m + 3 = 47$.

$m - 5m + 3 = 47$	Write the equation.
$1m - 5m + 3 = 47$	Identity Property; $m = 1m$
$-4m + 3 = 47$	Combine like terms, $1m$ and $-5m$.
$-4m + 3 - 3 = 47 - 3$	Subtract 3 from each side.
$-4m = 44$	Simplify.
$\frac{-4m}{-4} = \frac{44}{-4}$	Divide each side by -4.
$m = -11$	Simplify.
The solution is -11.	Check your solution.

Check for Understanding

Concept Check

1. **Explain** how you can work backward to solve a two-step equation.
2. **OPEN ENDED** Write a two-step equation that could be solved by using the Addition and Multiplication Properties of Equality.

Guided Practice

ALGEBRA Solve each equation. Check your solution.

3. $2x - 7 = 9$
4. $3t + 5 = 2$
5. $-16 = 6a - 4$
6. $\frac{y}{3} + 2 = 10$
7. $1 + \frac{k}{4} = -9$
8. $8 = \frac{n}{-7} - 5$
9. $3 - c = 7$
10. $2a - 8a = 24$
11. $8y - 9y + 6 = -4$

Application

12. **MEDICINE** For Jillian's cough, her doctor says that she should take eight tablets the first day and then four tablets each day until her prescription runs out. There are 36 tablets. Solve $8 + 4d = 36$ to find the number of days she will take four tablets.

Practice and Apply

Homework Help

For Exercises	See Examples
13–34	1
35–38	3
39–46	4
47–49	2

Extra Practice

See page 729.

ALGEBRA Solve each equation. Check your solution.

13. $3x + 1 = 7$
14. $5x - 4 = 11$
15. $4h + 6 = 22$
16. $8n + 3 = -5$
17. $37 = 4d + 5$
18. $9 = 15 + 2p$
19. $2n - 5 = 21$
20. $3j - 9 = 12$
21. $-1 = 2r - 7$
22. $12 = 5k - 8$
23. $10 = 6 + \frac{y}{7}$
24. $14 = 6 + \frac{n}{5}$
25. $3 + \frac{t}{2} = 35$
26. $13 + \frac{p}{3} = -4$
27. $\frac{k}{5} - 10 = 3$
28. $\frac{w}{8} - 4 = -7$
29. $8 = \frac{c}{-3} + 15$
30. $\frac{b}{-4} + 8 = -42$

ALGEBRA Find each number.

31. Five more than twice a number is 27. Solve [illegible].
32. Three less than four times a number is -7. Solve [illegible].
33. Ten less than the quotient of a number and 2 is 5. Solve [illegible].
34. Six more than the quotient of a number and 6 is -3. Solve [illegible]

ALGEBRA Solve each equation. Check your solution.

35. $8 - t = -25$
36. $3 - y = 13$
37. $-5 - b = 8$
38. $10 = -9 - x$
39. $2w - 4w = -10$
40. $3x - 5x = 22$
41. $x + 4x + 6 = 31$
42. $5r + 3r - 6 = 10$
43. $1 - 3y + y = 5$
44. $16 = w - 2w + 9$
45. $23 = 4t - 7 - t$
46. $-4 = -a + 8 - 2a$

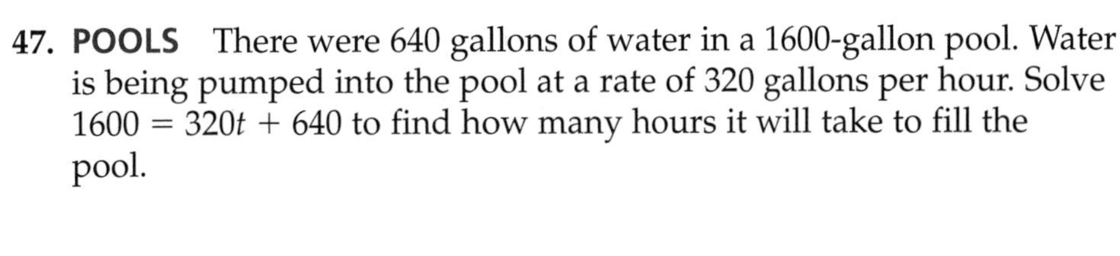

47. **POOLS** There were 640 gallons of water in a 1600-gallon pool. Water is being pumped into the pool at a rate of 320 gallons per hour. Solve $1600 = 320t + 640$ to find how many hours it will take to fill the pool.

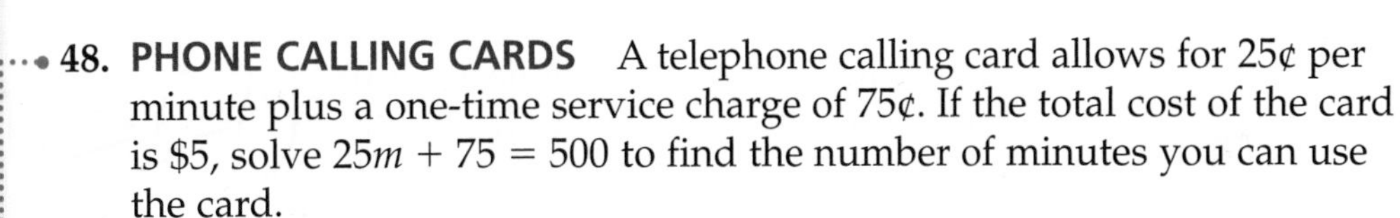

48. **PHONE CALLING CARDS** A telephone calling card allows for 25¢ per minute plus a one-time service charge of 75¢. If the total cost of the card is \$5, solve $25m + 75 = 500$ to find the number of minutes you can use the card.

More About. . .

Phone Calling Cards

Sales of prepaid phone cards skyrocketed from \$12 million in 1992 to \$1.9 billion in 1998.

Source: Atlantic ACM

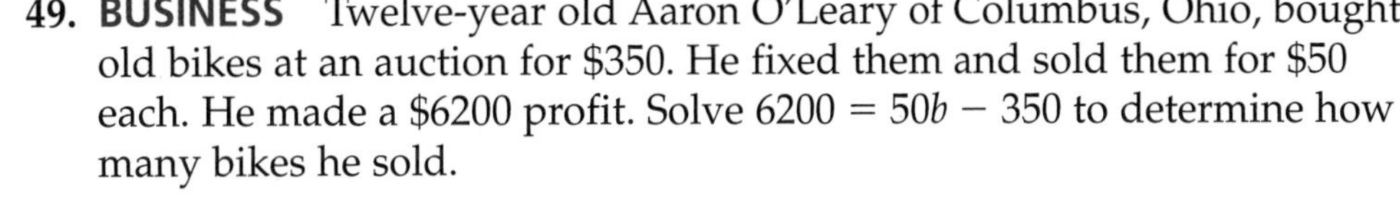

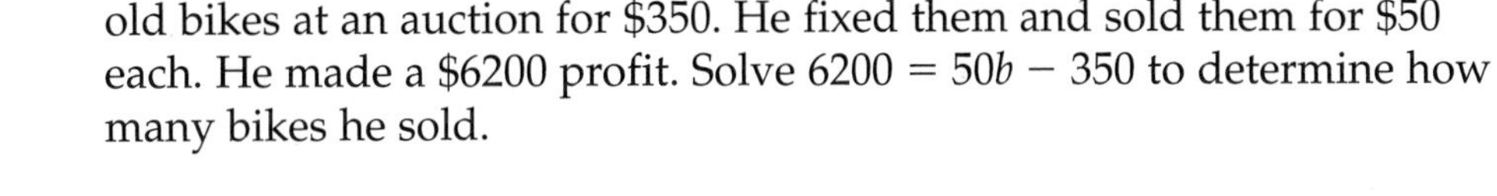

49. **BUSINESS** Twelve-year old Aaron O'Leary of Columbus, Ohio, bought old bikes at an auction for \$350. He fixed them and sold them for \$50 each. He made a \$6200 profit. Solve $6200 = 50b - 350$ to determine how many bikes he sold.

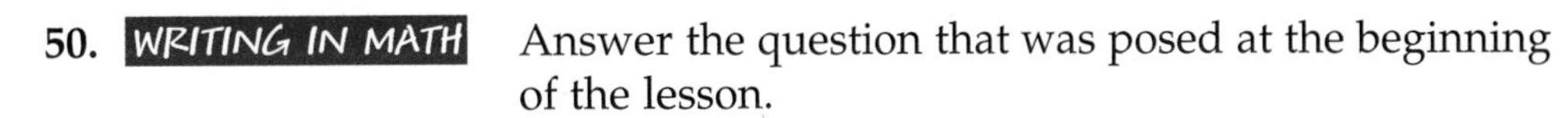

50. **WRITING IN MATH** Answer the question that was posed at the beginning of the lesson.

How can algebra tiles show the properties of equality?

Include the following in your answer:

- a drawing that shows how to solve $2x + 3 = 7$ using algebra tiles, and
- a list of the properties of equality that you used to solve $2x + 3 = 7$.

51. **CRITICAL THINKING** Write a two-step equation with a variable using the numbers 2, 5, and 8, in which the solution is 2.

52. **GRID IN** The charge to park at an art fair is a flat rate plus a per hour fee. The graph shows the charge for parking for up to 4 hours. If x represents the number of hours and y represents the total charge, what is the charge for parking for 7 hours?

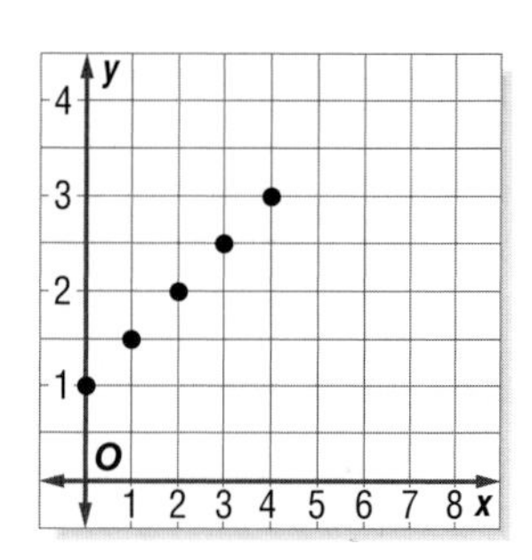

53. A local health club charges an initial fee of $45 for the first month and then a $32 membership fee each month after the first. The table shows the cost to join the health club for up to 6 months.

Months	1	2	3	4	5	6
Cost (dollars)	45	77	109	141	173	205

What is the cost to join the health club for 10 months?

(A) $215 (B) $320 (C) $333 (D) $450

Maintain Your Skills

Mixed Review

ALGEBRA Solve each equation. Check your solution. *(Lessons 3-3 and 3-4)*

54. $5y = 60$
55. $14 = -2n$
56. $\frac{x}{3} = -9$
57. $x - 4 = -6$
58. $-13 = y + 5$
59. $18 = 20 + x$

ALGEBRA Simplify each expression. *(Lesson 3-1)*

60. $4(x + 1)$
61. $-5(y + 3)$
62. $3(k - 10)$
63. $-9(y - 4)$
64. $7(a - 2)$
65. $-8(r - 5)$

Name the ordered pair for each point graphed on the coordinate plane at the right. *(Lesson 2-6)*

66. T
67. C
68. R
69. P

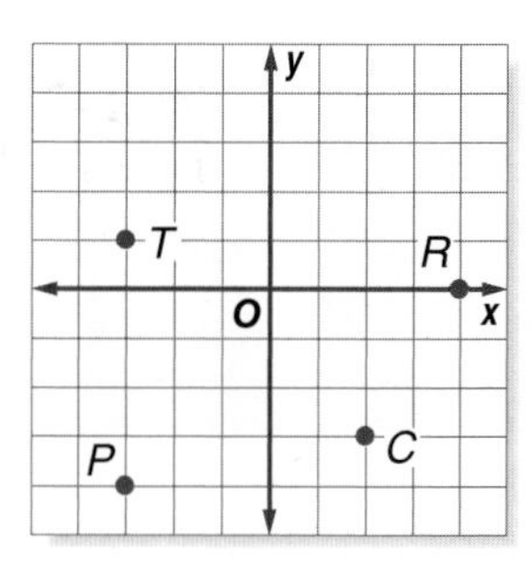

Getting Ready for the Next Lesson

PREREQUISITE SKILL Write an algebraic expression for each verbal expression. *(To review **algebraic expressions**, see Lesson 1-3.)*

70. two times a number less six
71. the quotient of a number and 15
72. the difference between twice a number and 8
73. twice a number increased by 10
74. the sum of $2x$, $7x$, and 4

Reading Mathematics

Translating Verbal Problems into Equations

An important skill in algebra is translating verbal problems into equations. To do this accurately, analyze the statements until you completely understand the relationships among the given information. Look for key words and phrases.

> Jennifer is 6 years older than Akira. The sum of their ages is 20.

You can explore a problem situation by asking and answering questions.

Questions	*Answers*
a. Who is older?	**a.** Jennifer
b. How many years older?	**b.** 6 years
c. If Akira is x years old, how old is Jennifer?	**c.** $x + 6$
d. What expression represents the phrase *the sum of their ages*?	**d.** $x + (x + 6)$
e. What equation represents the sentence *the sum of their ages is 20*?	**e.** $x + (x + 6) = 20$

Reading to Learn

For each verbal problem, answer the related questions.

1. Lucas is 5 inches taller than Tamika, and the sum of their heights is 137 inches.
  a. Who is taller?
  b. How many inches taller?
  c. If x represents Tamika's height, how tall is Lucas?
  d. What expression represents *the sum of their heights*?
  e. What equation represents the sentence *the sum of their heights is 137*?

2. There are five times as many students as teachers on the field trip, and the sum of students and teachers is 132.
  a. Are there more students or teachers?
  b. How many times more?
  c. If x represents the number of teachers, how many students are there?
  d. What expression represents *the sum of students and teachers*?
  e. What equation represents *the sum of students and teachers is 132*?

3-6 Writing Two-Step Equations

Sunshine State Standards
MA.A.3.3.3-1, MA.D.1.3.1-2, MA.D.1.3.2-2, MA.D.2.3.1-1, MA.D.2.3.1-3, MA.D.2.3.2-2

What You'll Learn

- Write verbal sentences as two-step equations.
- Solve verbal problems by writing and solving two-step equations.

How are equations used to solve real-world problems?

A phone company advertises that you can call anywhere in the United States for 4¢ per minute plus a monthly fee of 99¢ with their calling card. The table shows how to find your total monthly cost.

Time (minutes)	Monthly Cost (cents)
0	$4(0) + 99 = 99$
5	$4(5) + 99 = 119$
10	$4(10) + 99 = 139$
15	$4(15) + 99 = 159$
20	$4(20) + 99 = 179$

a. Let n represent the number of minutes. Write an expression that represents the cost when your call lasts n minutes.

b. Suppose your monthly cost was 299¢. Write and solve an equation to find the number of minutes you used the calling card.

c. Why is your equation considered to be a two-step equation?

WRITE TWO-STEP EQUATIONS In Chapter 1, you learned how to write verbal phrases as expressions.

Phrase the sum of 4 times some number and 99

Expression $4n + 99$

Study Tip

Expression
An expression is any combination of numbers and operations.

An equation is a statement that two expressions are equal. The expressions are joined with an equals sign. Look for the words *is, equals,* or *is equal to* when you translate sentences into equations.

Sentence The sum of 4 times some number and 99 is 299.

Equation $4n + 99 = 299$

Example 1 Translate Sentences into Equations

Translate each sentence into an equation.

Sentence	Equation
a. Six more than twice a number is -20.	$2n + 6 = -20$
b. Eighteen is 6 less than four times a number.	$18 = 4n - 6$
c. The quotient of a number and 5, increased by 8, is equal to 14.	$\frac{n}{5} + 8 = 14$

Concept Check What is the difference between an expression and an equation?

Example 2 Translate and Solve an Equation

Seven more than three times a number is 31. Find the number.

Words Seven more than three times a number is 31.

Variables Let n = the number.

Equation

$3n + 7 = 31$	Write the equation.
$3n + 7 - 7 = 31 - 7$	Subtract 7 from each side.
$3n = 24$	Simplify.
$n = 8$	Mentally divide each side by 3.

Therefore, the number is 8.

Study Tip

Solving Equations Mentally

When solving a simple equation like $3n = 24$, mentally divide each side by 3.

TWO-STEP VERBAL PROBLEMS There are many real-world situations in which you start with a given amount and then increase it at a certain rate. These situations can be represented by two-step equations.

Example 3 Write and Solve a Two-Step Equation

SCOOTERS Suppose you are saving money to buy a scooter that costs \$100. You have already saved \$60 and plan to save \$5 each week. How many weeks will you need to save?

Explore You have already saved \$60. You plan to save \$5 each week until you have \$100.

Plan Organize the data for the first few weeks in a table. Notice the pattern.

Week	Amount
0	$5(0) + 60 = 60$
1	$5(1) + 60 = 65$
2	$5(2) + 60 = 70$
3	$5(3) + 60 = 75$

Write an equation to represent the situation.

Let x = the number of weeks.

\$5 each week for x weeks	plus	amount already saved	equals	\$100.
$5x$	$+$	60	$=$	100

Solve

$5x + 60 = 100$	Write the equation.
$5x + 60 - 60 = 100 - 60$	Subtract 60 from each side.
$5x = 40$	Simplify.
$x = 8$	Mentally divide each side by 5.

You need to save \$5 each week for 8 weeks.

Examine If you save \$5 each week for 8 weeks, you'll have an additional \$40. The answer appears to be reasonable.

More About...

Scooters

Most scooters are made of aircraft aluminum. They can transport over 200 pounds easily, yet are light enough to carry.

Source: www.emazing.com

Example 4 Write and Solve a Two-Step Equation

OLYMPICS In the 2000 Summer Olympics, the United States won 9 more medals than Russia. Together they won 185 medals. How many medals did the United States win?

Words Together they won 185 medals.

Variables Let x = number of medals won by Russia.
Then $x + 9$ = number of medals won by United States.

Equation

$x + (x + 9) = 185$	Write the equation.
$(x + x) + 9 = 185$	Associative Property
$2x + 9 = 185$	Combine like terms.
$2x + 9 - 9 = 185 - 9$	Subtract 9 from each side.
$2x = 176$	Simplify.
$\frac{2x}{2} = \frac{176}{2}$	Divide each side by 2.
$x = 88$	Simplify.

Since x represents the number of medals won by Russia, Russia won 88 medals. The United States won 88 + 9 or 97 medals.

Study Tip

Alternative Method
Let x = number of U.S. medals. Then $x - 9$ = number of Russian medals.
$x + (x - 9) = 185$
$x = 97$
In this case, x is the number of U.S. medals, 97.

Check for Understanding

Concept Check

1. **List** three words or phrases in a verbal problem that can be translated into an equals sign in an equation.
2. **OPEN ENDED** Write a verbal sentence involving an unknown number and two operations.
3. **FIND THE ERROR** Alicia and Ben are translating the following sentence into an equation: *Three less than two times a number is 15.*

Alicia	Ben
$3 - 2x = 15$	$2x - 3 = 15$

Who is correct? Explain your reasoning.

Guided Practice **Translate each sentence into an equation. Then find each number.**

4. Three more than four times a number is 23.
5. Four less than twice a number is –2.

Applications **Solve each problem by writing and solving an equation.**

6. **METEOROLOGY** Suppose the current temperature is 17°F. It is expected to rise 3°F each hour for the next several hours. In how many hours will the temperature be 32°F?
7. **AGES** Lawana is five years older than her brother Cole. The sum of their ages is 37. How old is Lawana?

Practice and Apply

Homework Help

For Exercises	See Examples
8–19	1, 2
20, 21	3
22–24	4

Extra Practice

See page 730.

Translate each sentence into an equation. Then find each number.

8. Seven more than twice a number is 17.
9. Twenty more than three times a number is -4.
10. Four less than three times a number is 20.
11. Eight less than ten times a number is 82.
12. Ten more than the quotient of a number and -2 is three.
13. The quotient of a number and -4, less 8, is -42.
14. The difference between twice a number and 9 is 17.
15. The difference between three times a number and 8 is -2.
16. If 5 is decreased by 3 times a number, the result is –4.
17. If 17 is decreased by twice a number, the result is 5.
18. Three times a number plus twice the number plus 1 is –4.
19. Four times a number plus five more than three times the number is 47.

Career Choices

Meteorologist

Meteorologists are best known for forecasting the weather. However, they also work in the fields of air pollution, agriculture, air and sea transportation, and defense.

Online Research
For information about a career as a meteorologist, visit: www.pre-alg.com/careers

Solve each problem by writing and solving an equation.

20. **WILDLIFE** Your friend bought 3 bags of wild bird seed and an \$18 bird feeder. Each bag of birdseed costs the same amount. If your friend spent \$45, find the cost of one bag of birdseed.
21. **METEOROLOGY** The temperature is 8°F. It is expected to fall 5° each hour for the next several hours. In how many hours will the temperature be -7°F?
22. **FOOD SERVICE** You and your friend spent a total of \$15 for lunch. Your friend's lunch cost \$3 more than yours did. How much did you spend for lunch?
23. **POPULATION** By 2020, California is expected to have 2 million more senior citizens than Florida, and the sum of the number of senior citizens in the two states is expected to be 12 million. Find the expected senior citizen population of Florida in 2020.
24. **NATIVE AMERICANS** North Carolina's Native-American population is 22,000 greater than New York's. New York's Native-American population is 187,000 less than Oklahoma's. If the total population of all three is 437,000, find each state's Native-American population.

Native-American Populations (thousands)

State	Population
California	309
Arizona	256
New Mexico	163
Washington	103
Alaska	100
Texas	96
Michigan	60
Oklahoma	?
New York	?
North Carolina	?

Source: U.S. Census Bureau

25. **WRITE A PROBLEM** The table shows the expected population age 85 or older for certain states in 2020. Use the data to write a problem that can be solved by using a two-step equation.

Population (age 85 or older)	
State	Number (thousands)
CA	809
FL	735
TX	428
NY	418

26. **CRITICAL THINKING** If you begin with an even integer and count by two, you are counting *consecutive even integers*. Write and solve an equation to find two consecutive even integers whose sum is 50.

27. **WRITING IN MATH** Answer the question that was posed at the beginning of the lesson.

How are equations used to solve real-world problems?

Include the following in your answer:
- an example that starts with a given amount and increases, and
- an example that involves the sum of two quantities.

FCAT Practice
Standardized Test Practice
A B C D

28. Which verbal expression represents the phrase *three less than five times a number*?

(A) $3 - 5n$ (B) $n - 3$ (C) $5n - 3$ (D) $5 + n - 3$

29. The Bank of America building in San Francisco is 74 feet shorter than the Transamerica Pyramid. If their combined height is 1632 feet, how tall is the Transamerica Pyramid?

(A) 41 ft (B) 779 ft (C) 781 ft (D) 853 ft

Maintain Your Skills

Mixed Review

ALGEBRA Solve each equation. *(Lessons 3-3, 3-4, and 3-5)*

30. $6 - 2x = 10$
31. $-4x = -16$
32. $y - 7 = -3$

Evaluate each expression if $x = -12$, $y = 4$, and $z = -1$. *(Lesson 2-1)*

33. $|x| - 7$
34. $|x| + |y|$
35. $|z| - |x|$

36. Name the property shown by $(2 + 6) + 9 = 2 + (6 + 9)$. *(Lesson 1-4)*

Getting Ready for the Next Lesson

PREREQUISITE SKILL Solve each equation. Check your solution.
*(To review **solving equations**, see Lesson 3-4.)*

37. $2x = -8$
38. $24 = 6y$
39. $5w = -25$
40. $15s = 75$
41. $108 = 18x$
42. $25z = 175$

Practice Quiz 2

Lessons 3-3 through 3-6

ALGEBRA Solve each equation. *(Lessons 3-3, 3-4, and 3-5)*

1. $4h = -52$
2. $\frac{x}{-3} = 4$
3. $y - 5 = -23$
4. $2v - 11 = -5$

5. **ALGEBRA** Twenty more than three times a number is 32. Write and solve an equation to find the number. *(Lesson 3-6)*

3-7 Using Formulas

What You'll Learn

- Solve problems by using formulas.
- Solve problems involving the perimeters and areas of rectangles.

Sunshine State Standards
MA.A.3.3.3-1, MA.B.1.3.2-1, MA.D.2.3.1-3, MA.D.2.3.2-2

Vocabulary

- formula
- perimeter
- area

Why are formulas important in math and science?

The top recorded speed of a mallard duck in level flight is 65 miles per hour. You can make a table to record the distances that a mallard could fly at that rate.

Speed (mph)	Time (hr)	Distance (mi)
65	1	65
65	2	130
65	3	195
65	t	?

a. Write an expression for the distance traveled by a duck in t hours.

b. What disadvantage is there in showing the data in a table?

c. Describe an easier way to summarize the relationship between the speed, time, and distance.

FORMULAS A **formula** is an equation that shows a relationship among certain quantities. A formula usually contains two or more variables. One of the most commonly-used formulas shows the relationship between distance, rate (or speed), and time.

Words Distance equals the rate multiplied by the time.

Variables Let d = distance, r = rate, and t = time.

Equation $d = rt$

Example 1 Use the Distance Formula

SCIENCE What is the rate in miles per hour of a dolphin that travels 120 miles in 4 hours?

$d = rt$ Write the formula.

$120 = r \cdot 4$ Replace d with 120 and t with 4.

$\frac{120}{4} = \frac{r \cdot 4}{4}$ Divide each side by 4.

$30 = r$ Simplify.

The dolphin travels at a rate of 30 miles per hour.

Concept Check Name an advantage of using a formula to show a relationship among quantities.

PERIMETER AND AREA The distance around a geometric figure is called the **perimeter**. One method of finding the perimeter P of a rectangle is to add the measures of the four sides.

Key Concept — *Perimeter of a Rectangle*

- **Words** The perimeter of a rectangle is twice the sum of the length and width.
- **Symbols** $P = \ell + \ell + w + w$
  $P = 2\ell + 2w$ or $2(\ell + w)$
- **Model**

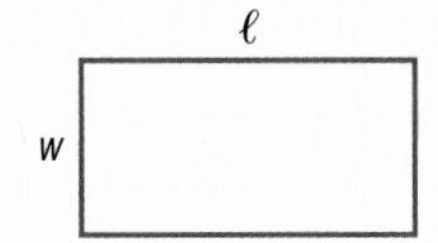

Example 2 Find the Perimeter of a Rectangle

Find the perimeter of the rectangle.

$P = 2(\ell + w)$	Write the formula.
$P = 2(11 + 5)$	Replace ℓ with 11 and w with 5.
$P = 2(16)$	Add 11 and 5.
$P = 32$	Simplify.

11 in.
5 in.

The perimeter is 32 inches.

Study Tip

Common Misconception
Although the length of a rectangle is usually greater than the width, it does not matter which side you choose to be the length.

Example 3 Find a Missing Length

The perimeter of a rectangle is 28 meters. Its width is 8 meters. Find the length.

$P = 2(\ell + w)$	Write the formula.
$P = 2\ell + 2w$	Distributive Property
$28 = 2\ell + 2(8)$	Replace P with 28 and w with 8.
$28 = 2\ell + 16$	Simplify.
$28 - 16 = 2\ell + 16 - 16$	Subtract 16 from each side.
$12 = 2\ell$	Simplify.
$6 = \ell$	Mentally divide each side by 2.

The length is 6 meters.

The measure of the surface enclosed by a figure is its **area**.

Key Concept — *Area of a Rectangle*

- **Words** The area of a rectangle is the product of the length and width.
- **Symbols** $A = \ell w$
- **Model**

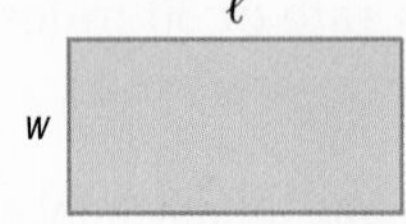

Example 4 Find the Area of a Rectangle

Find the area of a rectangle with length 15 meters and width 7 meters.

15 m

7 m

$A = \ell w$	Write the formula.
$A = 15 \cdot 7$	Replace ℓ with 15 and w with 7.
$A = 105$	Simplify.

The area is 105 square meters.

Example 5 Find a Missing Width

The area of a rectangle is 45 square feet. Its length is 9 feet. Find its width.

$A = \ell w$	Write the formula.
$45 = 9w$	Replace A with 45 and ℓ with 9.
$5 = w$	Mentally divide each side by 9.

The width is 5 feet.

Concept Check Which is measured in square units, area or perimeter?

Check for Understanding

Concept Check

1. **Write** the formula that shows the relationship among distance, rate, and time.
2. **Explain** the difference between the perimeter and area of a rectangle.
3. **OPEN ENDED** Draw and label a rectangle that has a perimeter of 18 inches.

Guided Practice

GEOMETRY Find the perimeter and area of each rectangle.

4. 8 ft, 3 ft

5. 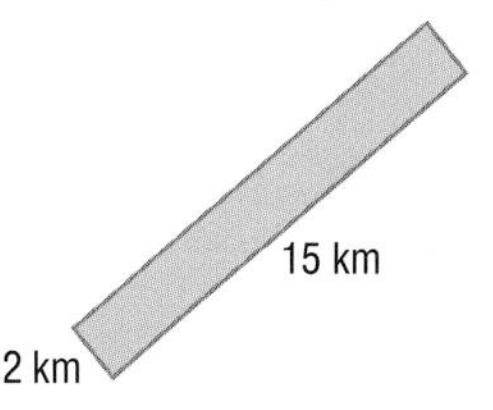

6. a rectangle with length 15 feet and width 6 feet

GEOMETRY Find the missing dimension in each rectangle.

7. 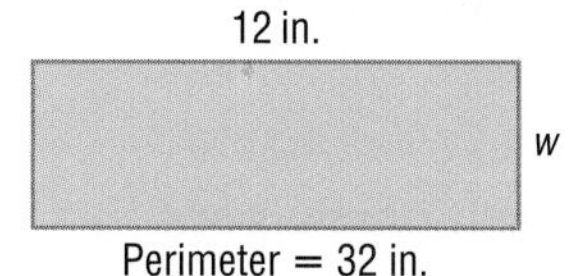

8. 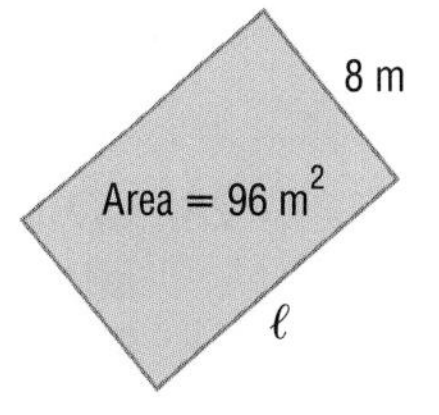

Application

9. **MILITARY** How long will it take an Air Force jet fighter to fly 5200 miles at 650 miles per hour?

Practice and Apply

Homework Help

For Exercises	See Examples
10–11	1
12–19	2, 4
20–27	3, 5

Extra Practice

See page 730.

10. **TRAVEL** Find the distance traveled by driving at 55 miles per hour for 3 hours.

11. **BALLOONING** What is the rate, in miles per hour, of a balloon that travels 60 miles in 4 hours?

GEOMETRY Find the perimeter and area of each rectangle.

12. 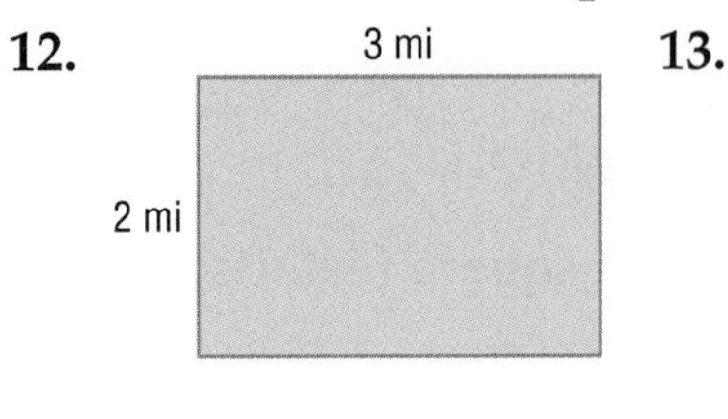

13. 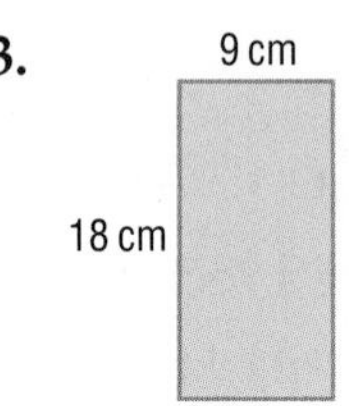

14. 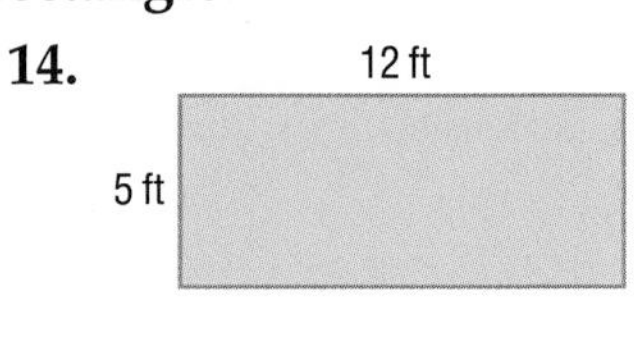

15. 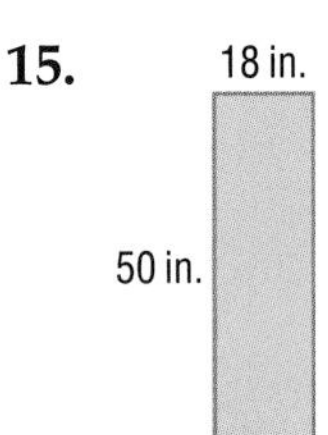

16. 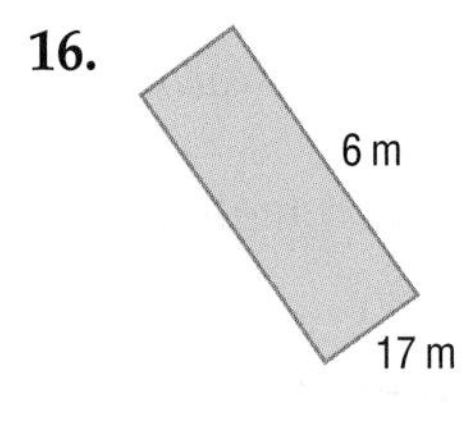

17. 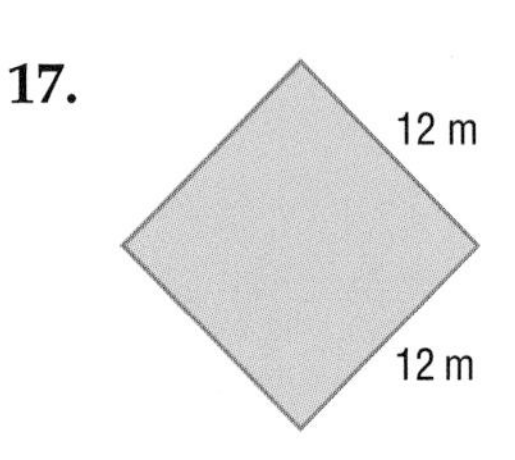

18. a rectangle that is 38 meters long and 10 meters wide

19. a square that is 5 meters on each side

GEOMETRY Find the missing dimension in each rectangle.

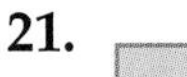

20. 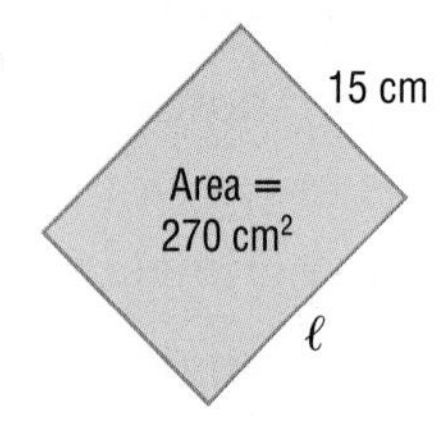

21. 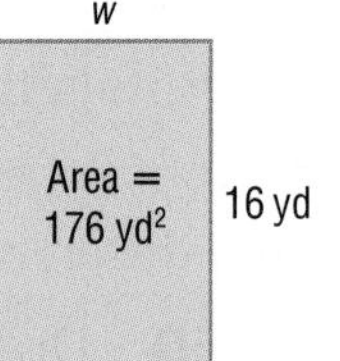

22. 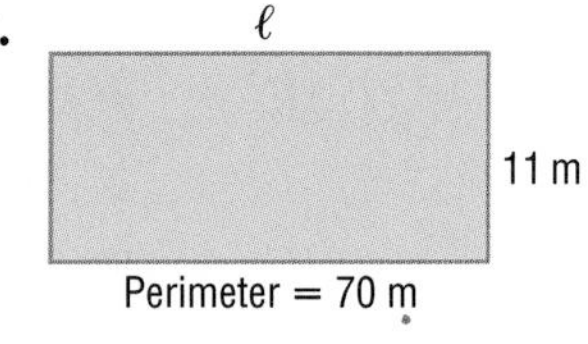

23. 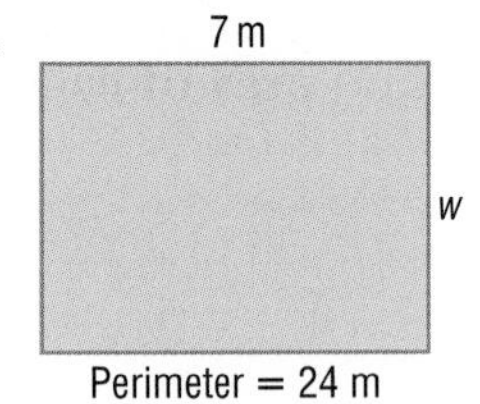

24. 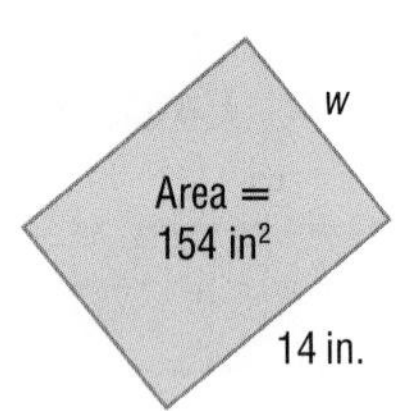

25. 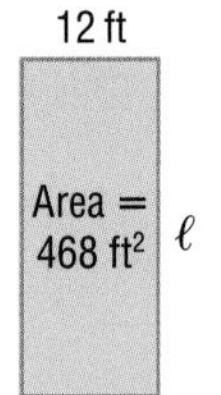

26. **GEOMETRY** The perimeter of a rectangle is 46 centimeters. Its width is 5 centimeters. Find the length.

27. **GEOMETRY** The area of a rectangle is 323 square yards. Its length is 17 yards. Find the width.

28. **COMMUNITY SERVICE** Each participant in a community garden is allotted a rectangular plot that measures 18 feet by 45 feet. How much fencing is needed to enclose each plot?

29. **SOCCER** Find the perimeter and area of the soccer field described at the left.

More About. . .

Soccer

Mia Hamm led her team to the Women's World Cup Championship in soccer in 1999. Soccer is played on a rectangular field that is usually 120 yards long and 75 yards wide.

Source: www.infoplease.com

Using a formula can help you find the cost of a vacation. Visit www.pre-alg.com/webquest to continue work on your WebQuest project.

For Exercises 30 and 31, translate each sentence into a formula.

30. **SALES** The sale price of an item s is equal to the list price ℓ minus the discount d.

31. **GEOMETRY** In a circle, the diameter d is twice the length of the radius r.

32. **RUNNING** The *stride rate* r of a runner is the number of strides n that he or she takes divided by the amount of time t, or $r = \frac{n}{t}$. The best runners usually have the greatest stride rate. Use the table to determine which runner has the greater stride rate.

Runner	Number of Strides	Time (s)
A	20	5
B	30	10

LANDSCAPING For Exercises 33 and 34, use the figure at the right.

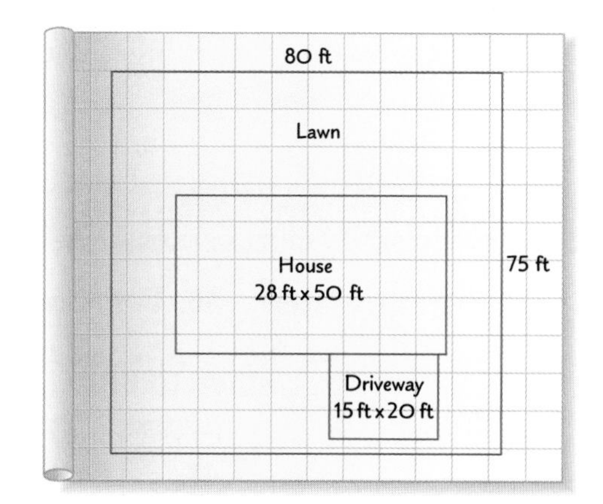

33. What is the area of the lawn?

34. Suppose your family wants to fertilize the lawn that is shown. If one bag of fertilizer covers 2500 square feet, how many bags of fertilizer should you buy?

BICYCLING For Exercises 35 and 36, use following information.
American Lance Armstrong won the 2000 Tour De France, completing the 2178-mile race in 92 hours 33 minutes 8 seconds.

35. Find Armstrong's average rate in miles per hour for the race.

36. Armstrong also won the 1999 Tour de France. He completed the 2213-mile race in 91 hours 32 minutes 16 seconds. Without calculating, determine which race was completed with a faster average speed. Explain.

GEOMETRY Draw and label the dimensions of each rectangle whose perimeter and area are given.

37. $P = 14$ ft, $A = 12$ ft^2

38. $P = 16$ m, $A = 12$ m^2

39. $P = 16$ cm, $A = 16$ cm^2

40. $P = 12$ in., $A = 8$ in^2

41. **CRITICAL THINKING** Is it *sometimes*, *always*, or *never* true that the perimeter of a rectangle is numerically greater than its area? Give an example.

42. **CRITICAL THINKING** An airplane flying at a rate of 500 miles per hour leaves Los Angeles. One-half hour later, a second airplane leaves Los Angeles in the same direction flying at a rate of 600 miles per hour. How long will it take the second airplane to overtake the first?

43. **WRITING IN MATH** Answer the question that was posed at the beginning of the lesson.

Why are formulas important in math and science?

Include the following in your answer:

- an example of a formula from math or science that you have used, and
- an explanation of how you used the formula.

FCAT Practice

Standardized Test Practice

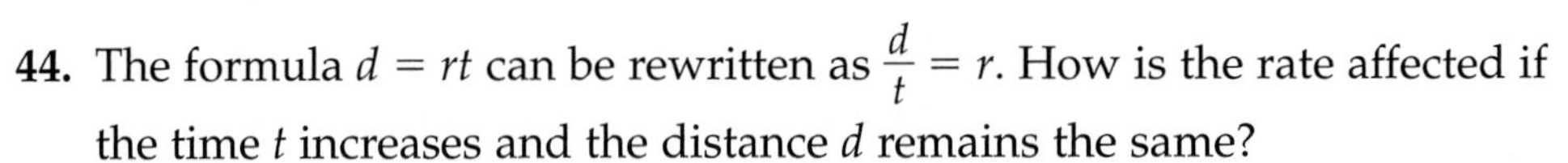

44. The formula $d = rt$ can be rewritten as $\frac{d}{t} = r$. How is the rate affected if the time t increases and the distance d remains the same?

(A) It increases. (B) It decreases.

(C) It remains the same. (D) There is not enough information.

45. The area of each square in the figure is 16 square units. Find the perimeter.

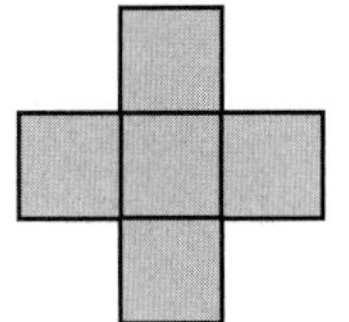

(A) 16 units (B) 32 units

(C) 48 units (D) 64 units

Maintain Your Skills

Mixed Review

46. Eight more than five times a number is 78. Find the number. *(Lesson 3-6)*

ALGEBRA Solve each equation. Check your solution. *(Lessons 3-3, 3-4, and 3-5)*

47. $-5x + 8 = 53$

48. $4y = -24$

49. $m + 5 = -3$

ALGEBRA Simplify each expression. *(Lesson 3-2)*

50. $3y + 5 - 2y$

51. $9 + x - 5x$

52. $3(r + 2) + 6r$

53. **LIGHT BULBS** The table shows the average life of an incandescent bulb for selected years. *(Lesson 1-6)*

a. Write a set of ordered pairs for the data.

b. State the domain and the range of the relation.

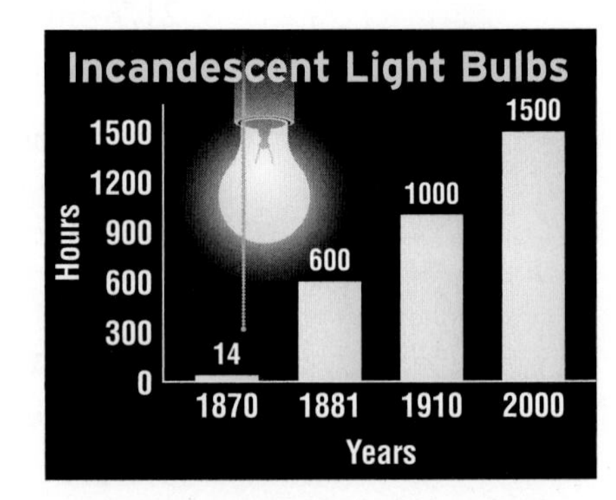

WebQuest Internet Project

Vacation Travelers Include More Families

It's time to complete your Internet project. Use the information and data you have gathered about the costs of lodging, transportation, and entertainment for each of the vacations. Prepare a brochure or Web page to present your project. Be sure to include graphs and/or tables in the presentation.

www.pre-alg.com/webquest

Spreadsheet Investigation

A Follow-Up of Lesson 3-7

Sunshine State Standards
MA.D.1.3.1-2, MA.D.1.3.1-4, MA.D.1.3.1-5, MA.D.1.3.2-2, MA.D.2.3.1-1

Perimeter and Area

A spreadsheet allows you to use formulas to investigate problems. When you change a numerical value in a cell, the spreadsheet recalculates the formula and automatically updates the results.

Example

Suppose a gardener wants to enclose a rectangular garden using part of a wall as one side and 20 feet of fencing for the other three sides. What are the dimensions of the largest garden she can enclose?

If w represents the length of each side attached to the wall, $20 - 2w$ represents the length of the side opposite the wall. These values are listed in column B. The areas are listed in column C.

Garden Dimensions

	A	B	C
1	Length of Fence	20	
2	Length of Side Attached to Wall	Length of Side Opposite Wall	Area
3	1	18	18
4	2	16	32
5	3	14	42
6	4	12	48
7	5	10	50
8	6	8	48
9	7	6	42
10	8	4	32
11	9	2	18
12			

Sheet1 Sheet2

The spreadsheet evaluates the formula $B1 - 2 \times A3$.

The spreadsheet evaluates the formula $A9 \times B9$.

The greatest possible area is 50 square feet. It occurs when the length of each side attached to the wall is 5 feet, and the length of the side opposite the wall is 10 feet.

Exercises

1. What is the area if the length of the side attached to the wall is 10 feet? 11 feet?
2. Are the answers to Exercise 1 reasonable? Explain.
3. Suppose you want to find the greatest area that you can enclose with 30 feet of fencing. Which cell should you modify to solve this problem?
4. Use a spreadsheet to find the dimensions of the greatest area you can enclose with 40 feet, 50 feet, and 60 feet of fencing.
5. **MAKE A CONJECTURE** Use any pattern you may have observed in your answers to Exercise 4 to find the dimensions of the greatest area you can enclose with 100 feet of fencing. Explain.

Chapter 3

Study Guide and Review

Vocabulary and Concept Check

area (p. 132)
coefficient (p. 103)
constant (p. 103)
Distributive Property (p. 98)
equivalent equations (p. 111)
equivalent expression (p. 98)
formula (p. 131)
inverse operations (p. 110)
like terms (p. 103)
perimeter (p. 132)
Properties of Equality
  Addition (p. 111)
  Division (p. 115)
Properties of Equality
  Multiplication (p. 117)
  Subtraction (p. 110)
simplest form (p. 104)
simplifying an expression (p. 104)
term (p. 103)
two-step equation (p. 120)

Complete each sentence with the correct term.

1. Terms that contain the same variables are called ________ .
2. The ________ of a geometric figure is the measure of the distance around it.
3. The ________ states that when you multiply each side of an equation by the same number, the two sides remain equal.
4. The equations $x + 3 = 8$ and $x = 5$ are ________ because they have the same solution.
5. You could use the ________ Property to rewrite $9(t - 2)$ as $9t - 18$.
6. In the term $4b$, 4 is the ________ of the expression.
7. The solution of $2y + 5 = 13$ is a ________ of a point on a number line.
8. The measure of the surface enclosed by a geometric figure is its ________ .
9. In the expression $10x + 6$, the number 6 is the ________ term.
10. Addition and subtraction are ________ because they "undo" each other.

Lesson-by-Lesson Review

3-1 The Distributive Property

See pages 98–102.

Concept Summary

- The Distributive Property combines addition and multiplication.
- For any numbers a, b, and c, $a(b + c) = ab + ac$ and $(b + c)a = ba + ca$.

Example Use the Distributive Property to rewrite $2(t - 3)$.

$2(t - 3) = 2[(t + (-3)]$	Rewrite $t - 3$ as $t + (-3)$.
$= 2t + 2(-3)$	Distributive Property
$= 2t + (-6)$	Simplify.
$= 2t - 6$	Definition of subtraction

Exercises Use the Distributive Property to rewrite each expression.
See Examples 3 and 4 on page 100.

11. $3(h + 6)$ **12.** $7(x + 2)$ **13.** $-5(k + 1)$ **14.** $-2(a + 8)$
15. $(t - 5)9$ **16.** $(x - 3)7$ **17.** $-2(b - 4)$ **18.** $-6(y - 3)$

3-2 Simplifying Algebraic Expressions

See pages 103–107.

Concept Summary

- Simplest form means no like terms and no parentheses.

Example **Simplify $9x + 3 - 7x$.**

$$\begin{aligned} 9x + 3 - 7x &= 9x + 3 + (-7x) && \text{Definition of subtraction} \\ &= 9x + (-7x) + 3 && \text{Commutative Property} \\ &= [9 + (-7)]x + 3 && \text{Distributive Property} \\ &= 2x + 3 && \text{Simplify.} \end{aligned}$$

Exercises **Simplify each expression.** *See Example 2 on page 104.*

19. $4a + 5a$ **20.** $3x + 7 + x$ **21.** $8(n - 1) - 10n$ **22.** $6w + 2(w + 9)$

3-3 Solving Equations by Adding or Subtracting

See pages 110–114.

Concept Summary

- When you add or subtract the same number from each side of an equation, the two sides remain equal.

Examples **1 Solve $x + 3 = 7$.**

$$\begin{aligned} x + 3 &= 7 \\ x + 3 - 3 &= 7 - 3 && \text{Subtract 3 from each side.} \\ x &= 4 \end{aligned}$$

2 Solve $y - 5 = -2$.

$$\begin{aligned} y - 5 &= -2 \\ y - 5 + 5 &= -2 + 5 && \text{Add 5 to each side.} \\ y &= 3 \end{aligned}$$

Exercises **Solve each equation.** *See Examples 1 and 3 on pages 111 and 112.*

23. $t + 5 = 8$ **24.** $12 = x + 4$ **25.** $k - 1 = 4$ **26.** $-7 = n - 6$

3-4 Solving Equations by Multiplying or Dividing

See pages 115–119.

Concept Summary

- When you multiply or divide each side of an equation by the same nonzero number, the two sides remain equal.

Examples **1 Solve $-5x = -30$.**

$$\begin{aligned} -5x &= -30 && \text{Write the equation.} \\ \frac{-5x}{-5} &= \frac{-30}{-5} && \text{Divide each side by } -5. \\ x &= 6 && \text{Simplify.} \end{aligned}$$

2 Solve $\frac{a}{-8} = 3$.

$$\begin{aligned} \frac{a}{-8} &= 3 && \text{Write the equation.} \\ -8\left(\frac{a}{-8}\right) &= -8(3) && \text{Multiply each side by } -8. \\ a &= -24 && \text{Simplify.} \end{aligned}$$

Exercises **Solve each equation.** *See Examples 1 and 3 on pages 116 and 117.*

27. $6n = 48$ **28.** $-3x = 30$ **29.** $\frac{t}{2} = -9$ **30.** $\frac{r}{-5} = -2$

• Extra Practice, see pages 728–730.
• Mixed Problem Solving, see page 760.

3-5 Solving Two-Step Equations

See pages 120–124.

Concept Summary

- To solve a two-step equation undo operations in reverse order.

Example **Solve $6k - 4 = 14$.**

$6k - 4 = 14$	Write the equation.
$6k - 4 + 4 = 14 + 4$	Undo subtraction. Add 4 to each side.
$6k = 18$	Simplify.
$k = 3$	Mentally divide each side by 6.

Exercises **Solve each equation.** *See Examples 1, 3, and 4 on pages 121 and 122.*

31. $6 + 2y = 8$ **32.** $3n - 5 = -17$ **33.** $\frac{t}{3} + 4 = 2$ **34.** $\frac{c}{9} - 3 = 2$

3-6 Writing Two-Step Equations

See pages 126–130.

Concept Summary

- The words *is*, *equals*, or *is equal to*, can be translated into an equals sign.

Example **Seven less than three times a number is −22. Find the number.**

$3n - 7 = -22$	Write the equation.
$3n - 7 + 7 = -22 + 7$	Add 7 to each side.
$3n = -15$	Simplify.
$n = -5$	Mentally divide each side by 3.

Exercises **Translate the sentence into an equation. Then find the number.** *See Examples 1 and 2 on pages 126 and 127.*

35. Three more than twice n is 53. **36.** Four times x minus 16 is 52.

3-7 Using Formulas

See pages 131–136.

Concept Summary

- Perimeter of a rectangle: $P = 2(\ell + w)$
- Area of a rectangle: $A = \ell w$

Example **Find the perimeter and area of a 14-meter by 6-meter rectangle.**

$P = 2(\ell + w)$	Formula for perimeter	$A = \ell w$	Formula for area
$P = 2(14 + 6)$	$\ell = 14$ and $w = 6$.	$A = 14 \cdot 6$	$\ell = 14$ and $w = 6$.
$P = 40$	Simplify.	$A = 84$	Simplify.

Exercises **Find the perimeter and area of each rectangle whose dimensions are given.** *See Examples 2 and 4 on pages 132 and 133.*

37. 8 feet by 9 feet **38.** 5 meters by 15 meters

Chapter 3 Practice Test

Vocabulary and Concepts

1. **OPEN ENDED** Give an example of two terms that are like terms and two terms that are *not* like terms.
2. **State** the Distributive Property in your own words.
3. **Explain** the difference between perimeter and area.

Skills and Applications

Simplify each expression.

4. $9x + 5 - x + 3$
5. $-3(a - 8)$
6. $10(y + 3) - 4y$

Solve each equation. Check your solution.

7. $19 = f + 5$
8. $-15 + z = 3$
9. $x - 7 = 16$
10. $g - 9 = -10$
11. $-8y = 72$
12. $\frac{n}{-30} = -6$
13. $25 = 2d - 9$
14. $4w - 18 = -34$
15. $6v + 10 = -62$
16. $-7 = \frac{d}{-5} + 1$
17. $7 - x = 18$
18. $b - 7b + 6 = -30$

Translate each sentence into an equation. Then find each number.

19. The quotient of a number and 8, decreased by 17 is -15.
20. Five less than 3 times a number is 25.

Find the perimeter and area of each rectangle.

21.
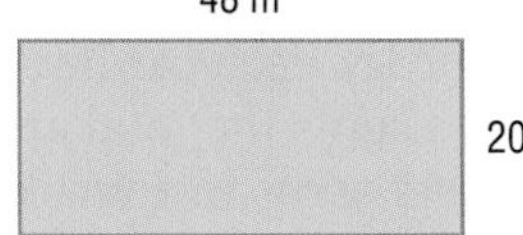

22.
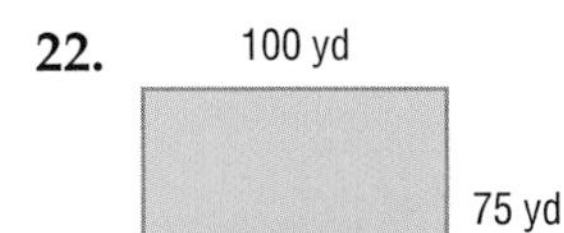

23. **ENTERTAINMENT** Suppose you pay \$15 per hour to go horseback riding. You ride 2 hours today and plan to ride 4 more hours this weekend.
 a. Write two different expressions to find the total cost of horseback riding.
 b. Find the total cost.

24. **HEIGHT** Todd is 5 inches taller than his brother. The sum of their heights is 139 inches. Find Todd's height.

25. **STANDARDIZED TEST PRACTICE** A carpet store advertises 16 square yards of carpeting for \$300, which includes the \$60 installation charge. Which equation could be used to determine the cost of one square yard of carpet x?
 Ⓐ $16x = 300$
 Ⓑ $x + 60 = 300$
 Ⓒ $60x + 16 = 300$
 Ⓓ $16x + 60 = 300$

www.pre-alg.com/chapter_test/fcat

Chapter 3

Standardized Test Practice

FCAT Practice

Part 1 Multiple Choice

Record your answers on the answer sheet provided by your teacher or on a sheet of paper.

1. Ms. Bauer notices that her car's gas tank is nearly empty. Gasoline costs $1.59 a gallon. *About* how many gallons can she buy with a $20 bill? (Prerequisite Skills, p. 714)

Ⓐ 10 Ⓑ 30

Ⓒ 12 Ⓓ 20

2. Which is equivalent to $3 \times 8 - 6 \div 2$? (Lesson 1-2)

Ⓐ 3 Ⓑ 9

Ⓒ 13 Ⓓ 21

3. Which statement illustrates the Commutative Property of Multiplication? (Lesson 1-4)

Ⓐ $5 + w + 8 = 5 + 8 + w$

Ⓑ $5 \cdot w \cdot 8 = 5 \cdot 8 \cdot w$

Ⓒ $(5 \cdot w) \cdot 8 = 5 \cdot (w \cdot 8)$

Ⓓ $(5 + w) + 8 = 5 + (w + 8)$

4. Which point on the graph below represents the ordered pair (4, 3)? (Lesson 1-6)

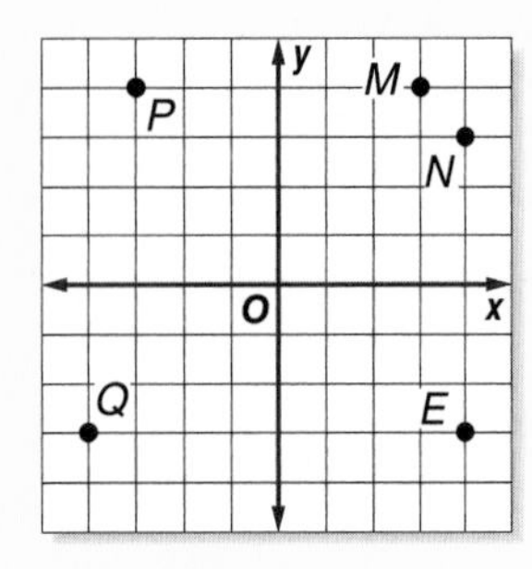

Ⓐ M Ⓑ N Ⓒ P Ⓓ Q

5. The temperature at 6:00 A.M. was −5°F. What was the temperature at 8:00 A.M. if it had risen 7 degrees? (Lesson 2-2)

Ⓐ 2°F Ⓑ −2°F

Ⓒ 12°F Ⓓ −12°F

6. Suppose points at (x, y) are graphed using the values in the table. Which statement is true about the graphs? (Lesson 2-6)

x	y
−1	5
−3	10
−5	6

Ⓐ The graphs are located in Quadrant I.

Ⓑ The graphs are located in Quadrant II.

Ⓒ The graphs are located in Quadrant III.

Ⓓ The graphs are located in Quadrant IV.

7. Which expression is equivalent to $5 \times 3 + 5 \times 12$? (Lesson 3-1)

Ⓐ $5 \times 8 \times 12$ Ⓑ $3 + (5 \times 12)$

Ⓒ $5 \times (3 + 12)$ Ⓓ $5 + (3 \times 12)$

8. Simplify $x - 4(x + 3)$. (Lesson 3-2)

Ⓐ $5x + 12$

Ⓑ $-3x - 12$

Ⓒ $-3x + 12$

Ⓓ $5x - 12$

9. Solve $y - (-4) = 6 - 8$. (Lesson 3-3)

Ⓐ −6 Ⓑ −2

Ⓒ 2 Ⓓ 6

10. Mr. Samuels is a car sales associate. He makes a salary of $400 per week. He also earns a bonus of $100 for each car he sells. Which equation represents the total amount of money Mr. Samuels earns in a week when he sells n cars? (Lesson 3-6)

Ⓐ $T = 400n + 100$

Ⓑ $T = n(100 + 400)$

Ⓒ $T = 100n + 400$

Ⓓ $T = 400 + 100 + n$

11. Tiffany's Gift Shop has fixed monthly expenses, E, of $1850. If the owner wants to make a profit, P, of $4000 next month, how many dollars in sales, S, does the shop need to earn? Use the formula $P = S - E$. (Lesson 3-7)

Ⓐ 2150 Ⓑ 4000

Ⓒ 5850 Ⓓ 7400

Test-Taking Tip

Question 15 This problem does not include a drawing. Make one. Your drawing will help you see how to solve the problem.

Part 2 Short Response/Grid In

Record your answers on the answer sheet provided by your teacher or on a sheet of paper.

12. The charge to enter a nature preserve is a flat amount per vehicle plus a fee for each person in the vehicle. The table shows the charge for vehicles holding up to 4 people.

Number of People	Charge (dollars)
1	1.50
2	2.00
3	2.50
4	3.00

What is the charge, in dollars, for a vehicle holding 8 people? (Lesson 1-1)

13. Evaluate $-2(-8 + 5)$. (Lesson 1-2)

14. Last week, the stock market rose 10 points in two days. What number expresses the average change in the stock market per day? (Lesson 2-1)

15. The ordered pairs $(-7, -2)$, $(-3, 5)$, and $(-3, -2)$ are coordinates of three of the vertices of a rectangle. What is the y-coordinate of the ordered pair that represents the fourth vertex? (Lesson 2-6)

16. What value of x makes $x - 4 = -2$ a true statement? (Lesson 3-3)

17. Cara filled her car's gas tank with 15 gallons of gas. Her car usually gets 24 miles per gallon. How many miles can she drive using 15 gallons of gas? (Lesson 3–7)

18. A mail-order greeting card company charges \$3 for each box of greeting cards plus a handling charge of \$2 per order. How many boxes of cards can you order from this company if you want to spend \$26? (Lesson 3-6)

19. Mr. Ruiz owns a health club and is planning to increase the floor area of the weight room. In the figure below, the rectangle with the solid border represents the floor area of the existing room, and the rectangle with the dashed border represents the floor area to be added. What will be the length, in feet, of the new weight room? (Lesson 3-7)

18 ft | 360 ft² | 180 ft²

FCAT Practice

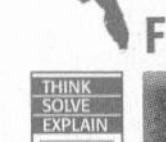

Part 3 Open Ended

Record your answers on a sheet of paper. Show your work.

20. The overnight low in Fargo, North Dakota, was -14°F. The high the next day was 6°F. (Lesson 3-3)

a. Draw a number line to represent the increase in temperature.

b. How many degrees did the temperature rise from the low to the high?

c. Explain how the concept of absolute value relates to this question.

21. In a school basketball game, each field goal is worth 2 points, and each free throw is worth 1 point. Josh heard the Springdale Stars scored a total of 63 points in their last game. Soledad says that they made a total of 12 free throws in that game. (Lesson 3-6)

a. Write an equation to represent the total points scored p. Use f for the number of free throws and g for the number of field goals.

b. Can both Josh and Soledad be correct? Explain.

UNIT 2

Algebra and Rational Numbers

Most of the numbers you encounter in the real world are *rational numbers*—fractions, decimals, and percents. In this unit, you will build on your foundation of algebra so that it includes rational numbers.

Chapter 4
Factors and Fractions

Chapter 5
Rational Numbers

Chapter 6
Ratio, Proportion, and Percent

WebQuest Internet Project

Kids Gobbling Empty Calories

"Teens are eating 150 more calories a day in snacks than they did two decades ago. And kids of all ages are munching on more of the richer goodies between meals than children did in the past."

Source: *USA TODAY*, April 30, 2001

In this project, you will be exploring how rational numbers are related to nutrition.

Log on to www.pre-alg.com/webquest. Begin your WebQuest by reading the Task.

Then continue working on your WebQuest as you study Unit 2.

Lesson	4-5	5-8	6-7
Page	173	242	301

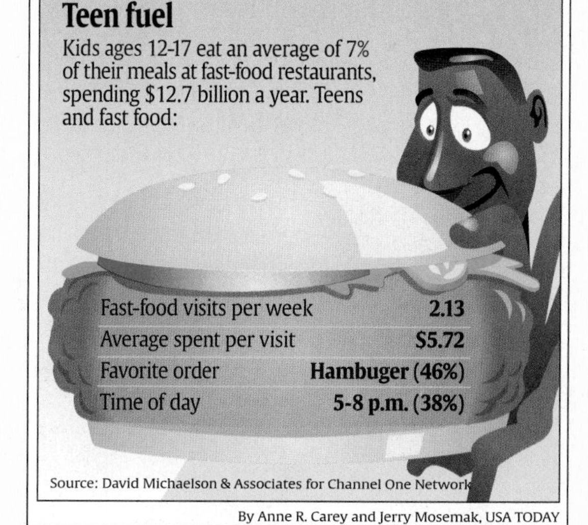

USA TODAY Snapshots®

Teen fuel

Kids ages 12-17 eat an average of 7% of their meals at fast-food restaurants, spending $12.7 billion a year. Teens and fast food:

Fast-food visits per week	2.13
Average spent per visit	$5.72
Favorite order	Hambuger (46%)
Time of day	5-8 p.m. (38%)

Source: David Michaelson & Associates for Channel One Network

By Anne R. Carey and Jerry Mosemak, USA TODAY

Chapter 4 Factors and Fractions

What You'll Learn

- **Lessons 4-1, 4-3, and 4-6** Identify, factor, multiply, and divide monomials.
- **Lessons 4-2 and 4-7** Evaluate expressions containing exponents.
- **Lesson 4-4** Factor algebraic expressions by finding the GCF.
- **Lesson 4-5** Simplify fractions using the GCF.
- **Lesson 4-8** Write numbers in scientific notation.

Key Vocabulary

- **factors** (p. 148)
- **monomial** (p. 150)
- **power** (p. 153)
- **prime factorization** (p. 160)
- **scientific notation** (p. 186)

Why It's Important

Fractions can be used to analyze and compare real-world data. For example, does a hummingbird or a tiger eat more, in relation to its size? You can use fractions to find the answer. *You will compare eating habits of these and other animals in Lesson 4-5.*

Getting Started

Prerequisite Skills To be successful in this chapter, you'll need to master these skills and be able to apply them in problem-solving situations. Review these skills before beginning Chapter 4.

For Lesson 4-1 — Distributive Property

Simplify. *(For review, see Lesson 3-1.)*

1. $2(x + 1)$ **2.** $3(n - 1)$ **3.** $-2(k + 8)$ **4.** $-4(x - 5)$

5. $6(2c + 4)$ **6.** $5(-3s + t)$ **7.** $7(a + b)$ **8.** $9(b - 2c)$

For Lesson 4-2 — Order of Operations

Evaluate each expression if $x = 2$, $y = 5$, and $z = -1$. *(For review, see Lesson 1-3.)*

9. $x + 12$ **10.** $z + (-5)$ **11.** $4y + 8$ **12.** $10 + 3z$

13. $(2 + y)9$ **14.** $6(x - 4)$ **15.** $3xy$ **16.** $2z + y$

For Lesson 4-8 — Product of Decimals

Find each product. *(For review, see page 715.)*

17. $4.5 \cdot 10$ **18.** $3.26 \cdot 100$ **19.** $0.1 \cdot 780$ **20.** $15 \cdot 0.01$

21. $3.9 \cdot 0.1$ **22.** $63.2 \cdot 0.1$ **23.** $0.01 \cdot 0.5$ **24.** $301.8 \cdot 0.001$

Make this Foldable to help you organize your notes about factors and fractions. Begin with four sheets of notebook paper.

Step 1 Fold

Fold four sheets of notebook paper in half from top to bottom.

Step 2 Cut and Staple

Cut along fold. Staple eight half-sheets together to form a booklet.

Step 3 Cut Tabs into Margin

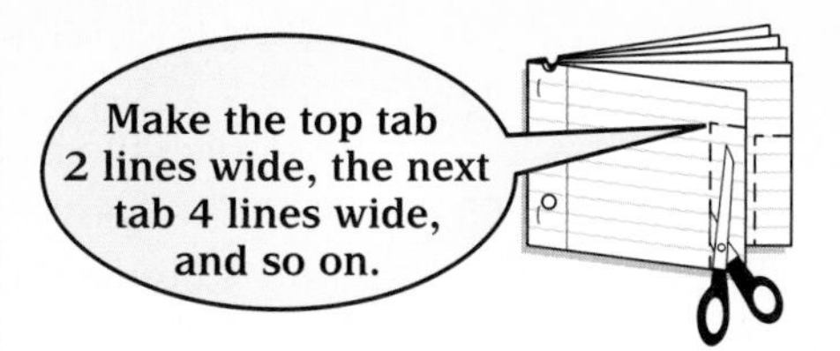

Make the top tab 2 lines wide, the next tab 4 lines wide, and so on.

Step 4 Label

Label each of the tabs with the lesson number and title.

Factors and Monomials

4-1

4-2

4-3

Reading and Writing As you read and study the chapter, write notes and examples on each page.

4-1 Factors and Monomials

Sunshine State Standards
MA.A.5.3.1-3

What You'll Learn

- Determine whether one number is a factor of another.
- Determine whether an expression is a monomial.

Vocabulary

- factors
- divisible
- monomial

How are side lengths of rectangles related to factors?

The rectangle at the right has an area of $9 \cdot 4$ or 36 square units.

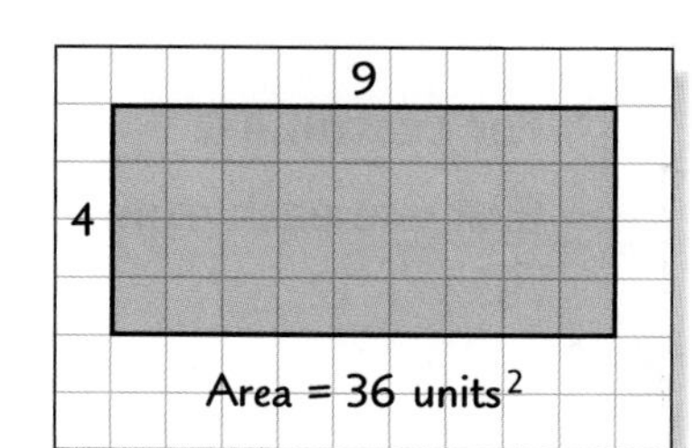

a. Use grid paper to draw as many other rectangles as possible with an area of 36 square units. Label the length and width of each rectangle.

b. Did you draw a rectangle with a length of 5 units? Why or why not?

c. List all of the pairs of whole numbers whose product is 36. Compare this list to the lengths and widths of all the rectangles that have an area of 36 square units. What do you observe?

d. Predict the number of rectangles that can be drawn with an area of 64 square units. Explain how you can predict without actually drawing them.

FIND FACTORS Two or more numbers that are multiplied to form a product are called **factors**.

$$4 \times 9 = 36$$

factors → 4, 9; product → 36

So, 4 and 9 are factors of 36 because they each divide 36 with a remainder of 0. We can say that 36 is **divisible** by 4 and 9. However, 5 is not a factor of 36 because $36 \div 5 = 7$ with a remainder of 1.

Sometimes you can test for divisibility mentally. The following rules can help you determine whether a number is divisible by 2, 3, 5, 6, or 10.

Reading Math

Even and Odd Numbers
A number that is divisible by 2 is called an *even number*. A number that is not divisible by 2 is called an *odd number*.

Concept Summary — Divisibility Rules

A number is divisible by:	Examples		Reasons
2 if the ones digit is divisible by 2.	54	→	4 is divisible by 2.
3 if the sum of its digits is divisible by 3.	72	→	$7 + 2 = 9$, and 9 is divisible by 3.
5 if the ones digit is 0 or 5.	65	→	The ones digit is 5.
6 if the number is divisible by 2 and 3.	48	→	48 is divisible by 2 and 3.
10 if the ones digit is 0.	120	→	The ones digit is 0.

Concept Check Is 51 divisible by 3? Why or why not?

Example 1 Use Divisibility Rules

Determine whether 138 is divisible by 2, 3, 5, 6, or 10.

Number	Divisible?	Reason
2	yes	The ones digit is 8, and 8 is divisible by 2.
3	yes	The sum of the digits is $1 + 3 + 8$ or 12, and 12 is divisible by 3.
5	no	The ones digit is 8, not 0 or 5.
6	yes	138 is divisible by 2 and 3.
10	no	The ones digit is not 0.

So, 138 is divisible by 2, 3, and 6.

More About. . .

Weddings

On average, it costs four times more to book reception sites in Los Angeles than in the Midwest.

Source: newschannel5.webpoint.com/wedding

Example 2 Use Divisibility Rules to Solve a Problem

WEDDINGS **A bride must choose whether to seat 5, 6, or 10 people per table at her reception. If there are 192 guests and she wants all the tables to be full, which should she choose?**

Seats Per Table	Yes/No	Reason
5	no	The ones digit of 192 does not end in 0 or 5, so 192 is not divisible by 5. There would be empty seats.
6	yes	192 is divisible by 2 and 3, so it is also divisible by 6. Therefore, all the tables would be full.
10	no	The ones digit of 192 does not end in 0, so 192 is not divisible by 10. There would be empty seats.

The bride should choose tables that seat 6 people.

You can also use the rules for divisibility to find the factors of a number.

Reading Math

Divisible/Factor

The following statements mean the same thing.

- 72 is divisible by 2.
- 2 is a factor of 72.

Example 3 Find Factors of a Number

List all the factors of 72.

Use the divisibility rules to determine whether 72 is divisible by 2, 3, 5, and so on. Then use division to find other factors of 72.

Number	72 Divisible by Number?	Factor Pairs
1	yes	$1 \cdot 72$
2	yes	$2 \cdot 36$
3	yes	$3 \cdot 24$
4	yes	$4 \cdot 18$
5	no	—
6	yes	$6 \cdot 12$
7	no	—
8	yes	$8 \cdot 9$
9	yes	$9 \cdot 8$

Use division to find the other factor in each factor pair. $72 \div 2 = 36$

You can stop finding factors when the numbers start repeating.

So, the factors of 72 are 1, 2, 3, 4, 6, 8, 9, 12, 18, 24, 36, and 72.

MONOMIALS A number such as 80 or an expression such as $8x$ is called a monomial. A **monomial** is a number, a variable, or a product of numbers and/or variables.

$$8 \cdot 10 = 80 \qquad 8 \cdot x = 8x$$

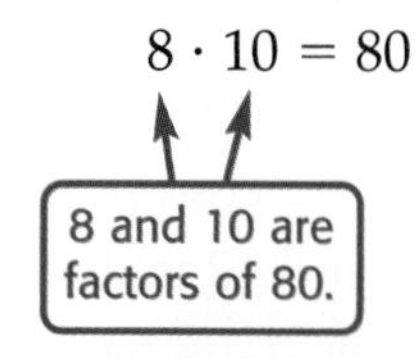

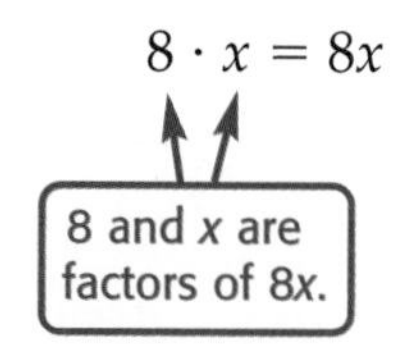

Reading Math

Monomial
The prefix *mono* means one. A monomial is an expression with one term.

Monomials		Not Monomials	
4	a number	$2 + x$	two terms are added
y	a variable	$5c - 6$	one term is subtracted from another term
$-2rs$	the product of a number and variables	$3(a + b)$	two terms are added $3(a + b) = 3a + 3b$

Concept Check Explain why $7q \cdot n$ is a monomial, but $7q + n$ is not.

Before you determine whether an expression is a monomial, be sure the expression is in simplest form.

Example 4 Identify Monomials

Determine whether each expression is a monomial.

a. $2(x - 3)$

$2(x - 3) = 2x + 2(-3)$ Distributive Property

$= 2x - 6$ Simplify.

This expression is not a monomial because it has two terms involving subtraction.

b. $-48xyz$

This expression is a monomial because it is the product of integers and variables.

Check for Understanding

Concept Check

1. **Explain** how you can mentally determine whether there is a remainder when 18,450 is divided by 6.
2. **Determine** whether 3 is a common factor of 125 and 132. Explain.
3. **OPEN ENDED** Use mental math, paper and pencil, or a calculator to find at least one number that satisfies each condition.
   a. a 3-digit number that is divisible by 2, 3, and 6
   b. a 4-digit number that is divisible by 3 and 5, but is not divisible by 10.
   c. a 3-digit number that is not divisible by 2, 3, 5, or 10

Guided Practice **Use divisibility rules to determine whether each number is divisible by 2, 3, 5, 6, or 10.**

4. 51 5. 146 6. 876 7. 3050

List all the factors of each number.

8. 203 9. 80 10. 115

ALGEBRA Determine whether each expression is a monomial. Explain why or why not.

11. 38 12. $2n - 2$ 13. $5(x + y)$ 14. $17(4)k$

Application

15. **CALENDARS** Years that are divisible by 4, called *leap years*, are 366 days long. Also, years ending in "00" that are divisible by 400 are leap years. Use the rule given below to determine whether 2000, 2004, 2015, 2018, 2022, and 2032 are leap years.

> If the last two digits form a number that is divisible by 4, then the number is divisible by 4.

Practice and Apply

Homework Help

For Exercises	See Examples
16–27	1
28–35	3
36–47	4
48–51	2

Extra Practice

See page 730.

Use divisibility rules to determine whether each number is divisible by 2, 3, 5, 6, or 10.

16. 39 17. 135 18. 82 19. 120
20. 250 21. 118 22. 378 23. 955
24. 5010 25. 684 26. 10,523 27. 24,640

List all the factors of each number.

28. 75 29. 114 30. 57 31. 65
32. 90 33. 124 34. 102 35. 135

ALGEBRA Determine whether each expression is a monomial. Explain why or why not.

36. m 37. 110 38. $s + t$ 39. $g - h$
40. $-12 + 12x$ 41. $3c + 6$ 42. $7(a + 1)$ 43. $4(2t - 1)$
44. $4b$ 45. $10(-t)$ 46. $-25abc$ 47. $8j(4k)$

More About. . .

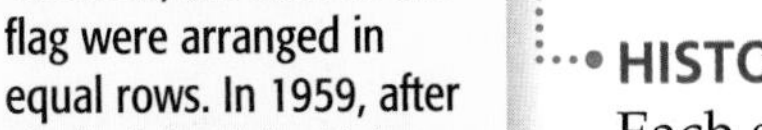

History

In 1912, when there were 48 states, the stars of the flag were arranged in equal rows. In 1959, after Alaska joined the Union, a new arrangement was proposed.

Source: www.usflag.org

MUSIC For Exercises 48 and 49, use the following information.
The band has 72 students who will march during halftime of the football game. For one drill, they need to march in rows with the same number of students in each row.

48. Can the whole band be arranged in rows of 7? Explain.

49. How many different ways could students be arranged? Describe the arrangements.

HISTORY For Exercises 50 and 51, use the following information.
Each star on the U.S. flag represents a state. As states joined the Union, the rectangular arrangement of the stars changed.

50. Use the information at the left to make a conjecture about how you think the stars of the flag were arranged in 1912 and in 1959.

51. **Research** What is the correct arrangement of stars on the U.S. flag? Explain why the arrangement is not rectangular.

Determine whether each statement is *sometimes, always,* or *never* true. Explain your reasoning.

52. A number that is divisible by 3 is also divisible by 6.

53. A number that has 10 as a factor is not divisible by 5.

54. A number that has a factor of 10 is an even number.

55. MONEY The homecoming committee can spend \$144 on refreshments for the dance. Soft drinks cost \$6 per case, and cookies cost \$4 per bag.

a. How many cases of soft drinks can they buy with \$144?

b. How many bags of cookies can they buy with \$144?

c. Suppose they want to buy approximately the same amounts of soft drinks and cookies. How many of each could they buy with \$144?

56. CRITICAL THINKING Write a number that satisfies each set of conditions.

a. the greatest three-digit number that is not divisible by 2, 3, or 10

b. the least three-digit number that is not divisible by 2, 3, 5, or 10

57. WRITING IN MATH Answer the question that was posed at the beginning of the lesson.

How are side lengths of rectangles related to factors?

Include the following in your answer:

- a drawing of a rectangle with its dimensions and area labeled, and
- a definition of *factors* and a description of the relationship between rectangle dimensions and factor pairs of a number.

58. Which number is divisible by 3?

Ⓐ 133 Ⓑ 444 Ⓒ 53 Ⓓ 250

59. Determine which expression is *not* a monomial.

Ⓐ $6d$ Ⓑ $6d \cdot 5$ Ⓒ $6d - 5$ Ⓓ 5

Maintain Your Skills

Mixed Review

GEOMETRY Find the perimeter and area of each rectangle. *(Lesson 3-7)*

60. Rectangle: 4.9 m by 3.5 m

61. Rectangle: 12 in. by 5 in.

ALGEBRA Translate each sentence into an equation. Then find each number. *(Lesson 3-6)*

62. Eight more than twice a number is −16.

63. Two less than 5 times a number equals 3.

ALGEBRA Solve each equation. Check your solution. *(Lesson 3-5)*

64. $2x - 1 = 9$ **65.** $14 = 8 + 3n$ **66.** $7 + \frac{k}{5} = -1$

Getting Ready for the Next Lesson

PREREQUISITE SKILL Find each product.

*(To review **multiplying integers**, see Lesson 2-4.)*

67. $4 \cdot 4 \cdot 4$ **68.** $10 \cdot 10 \cdot 10 \cdot 10$ **69.** $(-3)(-3)(-3)$

70. $(-2)(-2)(-2)(-2)$ **71.** $8 \cdot 8 \cdot 6 \cdot 6$ **72.** $(2)(2)(-5)(-5)(-5)$

4-2 Powers and Exponents

What You'll Learn

- Write expressions using exponents.
- Evaluate expressions containing exponents.

Why are exponents important in comparing computer data?

Vocabulary

- base
- exponent
- power
- standard form
- expanded form

Computer data are measured in small units called *bytes*. These units are based on factors of 2.

a. Write 16 as a product of factors of 2. How many factors are there?

b. How many factors of 2 form the product 128?

c. One megabyte is 1024 kilobytes. How many factors of 2 form the product 1024?

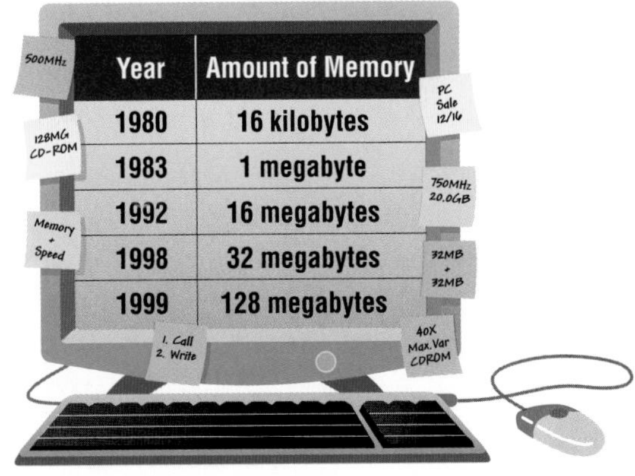

Personal Computers

Year	Amount of Memory
1980	16 kilobytes
1983	1 megabyte
1992	16 megabytes
1998	32 megabytes
1999	128 megabytes

Source: www.islandnet.com

Sunshine State Standards
MA.A.1.3.1-1, MA.A.1.3.2-2, MA.A.1.3.4-2, MA.A.2.3.1-1, MA.A.2.3.1-3, MA.A.3.3.2-4, MA.B.1.3.3-1, MA.B.1.3.3-3, MA.D.2.3.1-1, MA.D.2.3.1-6

EXPONENTS An expression like $2 \times 2 \times 2 \times 2$ can be written as a power. A power has two parts, a base and an exponent. An exponent is a shorter way of writing repeated multiplication. The expression $2 \times 2 \times 2 \times 2$ can be written as 2^4.

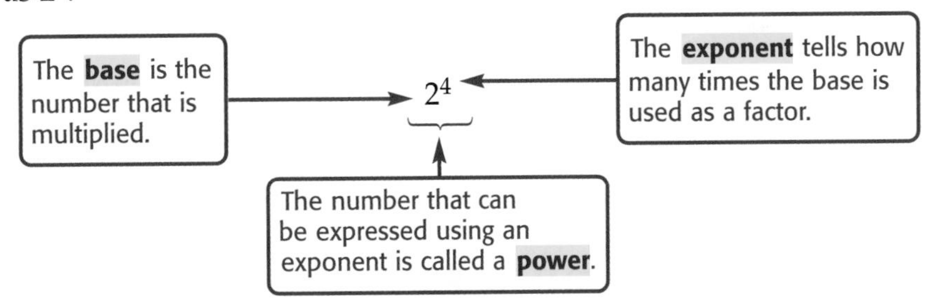

The **base** is the number that is multiplied.
2^4
The **exponent** tells how many times the base is used as a factor.
The number that can be expressed using an exponent is called a **power**.

The table below shows how to write and read powers with positive exponents.

Powers	Words	Repeated Factors
2^1	2 to the first power	2
2^2	2 to the second power or 2 squared	$2 \cdot 2$
2^3	2 to the third power or 2 cubed	$2 \cdot 2 \cdot 2$
2^4	2 to the fourth power or 2 to the fourth	$2 \cdot 2 \cdot 2 \cdot 2$
$\vdots$	$\vdots$	$\vdots$
2^n	2 to the *n*th power or 2 to the *n*th	$\underbrace{2 \cdot 2 \cdot 2 \cdot \ldots \cdot 2}_{n \text{ factors}}$

Study Tip

First Power
When a number is raised to the first power, the exponent is usually omitted. So 2^1 is written as 2.

Any number, except 0, raised to the zero power is defined to be 1.

$1^0 = 1$ $\quad 2^0 = 1$ $\quad 3^0 = 1$ $\quad 4^0 = 1$ $\quad 5^0 = 1$ $\quad x^0 = 1, x \neq 0$

Example 1 Write Expressions Using Exponents

Write each expression using exponents.

a. $3 \cdot 3 \cdot 3 \cdot 3 \cdot 3$

The base is 3. It is a factor 5 times, so the exponent is 5.
$3 \cdot 3 \cdot 3 \cdot 3 \cdot 3 = 3^5$

b. $t \cdot t \cdot t \cdot t$

The base is t. It is a factor 4 times, so the exponent is 4.
$t \cdot t \cdot t \cdot t = t^4$

c. $(-9)(-9)$

The base is -9. It is a factor 2 times, so the exponent is 2.
$(-9)(-9) = (-9)^2$

d. $(x + 1)(x + 1)(x + 1)$

The base is $x + 1$. It is a factor 3 times, so the exponent is 3.
$(x + 1)(x + 1)(x + 1) = (x + 1)^3$

e. $7 \cdot a \cdot a \cdot a \cdot b \cdot b$

First, group the factors with like bases. Then, write using exponents.

$7 \cdot a \cdot a \cdot a \cdot b \cdot b = 7 \cdot (a \cdot a \cdot a) \cdot (b \cdot b)$
$= 7a^3b^2$ — $a \cdot a \cdot a = a^3$ and $b \cdot b = b^2$

Study Tip

Common Misconception
$(-9)^2$ is not the same as -9^2.
$-9^2 = -1 \cdot 9^2$

Concept Check How would you write *ten to the fourth* using an exponent?

The number 13,548 is in **standard form** because it does not contain exponents. You can use place value and exponents to express a number in **expanded form**.

Example 2 Use Exponents in Expanded Form

Express 13,048 in expanded form.

Step 1 Use place value to write the value of each digit in the number.

$13{,}048 = 10{,}000 + 3000 + 0 + 40 + 8$
$= (1 \times 10{,}000) + (3 \times 1000) + (0 \times 100) + (4 \times 10) + (8 \times 1)$

Step 2 Write each place value as a power of 10 using exponents.

$13{,}048 = (1 \times 10^4) + (3 \times 10^3) + (0 \times 10^2) + (4 \times 10^1) + (8 \times 10^0)$

Recall that $10^0 = 1$.

EVALUATE EXPRESSIONS Since powers are forms of multiplication, they need to be included in the rules for order of operations.

Concept Summary — Order of Operations

	Words	Example
Step 1	Simplify the expressions inside grouping symbols. Start with the innermost grouping symbols.	$(3 + 4)^2 + 5 \cdot 2 = 7^2 + 5 \cdot 2$
Step 2	Evaluate all powers.	$= 49 + 5 \cdot 2$
Step 3	Do all multiplications or divisions in order from left to right.	$= 49 + 10$
Step 4	Do all additions or subtractions in order from left to right.	$= 59$

Follow the order of operations to evaluate algebraic expressions.

Example 3 Evaluate Expressions

Evaluate each expression.

a. 2^3

$2^3 = 2 \cdot 2 \cdot 2$ 2 is a factor 3 times.

$= 8$ Multiply.

b. $y^2 + 5$ **if** $y = -3$

$y^2 + 5 = (-3)^2 + 5$ Replace y with -3.

$= (-3)(-3) + 5$ -3 is a factor two times.

$= 9 + 5$ Multiply.

$= 14$ Add.

c. $3(x + y)^4$ **if** $x = -2$ **and** $y = 1$

$3(x + y)^4 = 3(-2 + 1)^4$ Replace x with -2 and y with 1.

$= 3(-1)^4$ Simplify the expression inside the parentheses.

$= 3(1)$ Evaluate $(-1)^4$.

$= 3$ Simplify.

Study Tip

Exponents
An exponent goes with the number, variable, or quantity in parentheses immediately preceding it.

- In $5 \cdot 3^2$, 3 is squared. $5 \cdot 3^2 = 5 \cdot 3 \cdot 3$
- In $(5 \cdot 3)^2$, $(5 \cdot 3)$ is squared. $(5 \cdot 3)^2 = (5 \cdot 3)(5 \cdot 3)$

Check for Understanding

Concept Check

1. **OPEN ENDED** Use exponents to write a numerical expression and an algebraic expression in which the base is a factor 5 times.
2. **Explain** how the expression *6 cubed* is repeated multiplication.
3. **Make a conjecture** about the value of 1^n and the value of $(-1)^n$ for any value of n. Explain.

Guided Practice

Write each expression using exponents.

4. $n \cdot n \cdot n$
5. $7 \cdot 7$
6. $3 \cdot 3 \cdot x \cdot x \cdot x \cdot x$
7. Express 2695 in expanded form.

ALGEBRA Evaluate each expression.

8. 2^4
9. $x^3 - 3$ if $x = -2$
10. $5(y - 1)^2$ if $y = 4$

Application

11. **SOUND** Fireworks can easily reach a sound of 169 decibels, which can be dangerous if prolonged. Write this number using exponents and a smaller base.

Practice and Apply

Homework Help

For Exercises	See Examples
12–23, 43, 47	1
24–27	2
28–42	3

Write each expression using exponents.

12. $4 \cdot 4 \cdot 4 \cdot 4 \cdot 4 \cdot 4$
13. 6
14. $(-5)(-5)(-5)$
15. $(-8)(-8)(-8)(-8)$
16. $k \cdot k$
17. $(-t)(-t)(-t)$
18. $(r \cdot r)(r \cdot r)$
19. $m \cdot m \cdot m \cdot m$
20. $a \cdot a \cdot b \cdot b \cdot b \cdot b$
21. $2 \cdot x \cdot x \cdot y \cdot y$
22. $7 \cdot 7 \cdot 7 \cdot n \cdot n \cdot n \cdot n$
23. $9 \cdot (p + 1) \cdot (p + 1)$

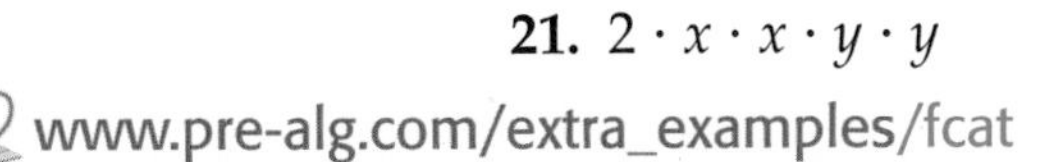

Extra Practice
See page 731.

Express each number in expanded form.

24. 452 **25.** 803 **26.** 6994 **27.** 23,781

ALGEBRA **Evaluate each expression if $a = 2$, $b = 4$, and $c = -3$.**

28. 7^2	**29.** 10^3	**30.** $(-9)^3$
31. $(-2)^5$	**32.** b^4	**33.** c^4
34. $5a^4$	**35.** ac^3	**36.** $b^0 - 10$
37. $c^2 + a^2$	**38.** $3a + b^3$	**39.** $a^2 + 3a - 1$
40. $b^2 - 2b + 6$	**41.** $3(b - 1)^4$	**42.** $2(3c + 7)^2$

43. TRAVEL Write each number in the graphic as a power.

44. Write *7 cubed times x squared* as repeated multiplication.

45. Write *negative eight, cubed* using exponents, as a product of repeated factors, and in standard form.

46. Without using a calculator, order 96, 96^2, 96^{10}, 96^5, and 96^0 from least to greatest. Explain your reasoning.

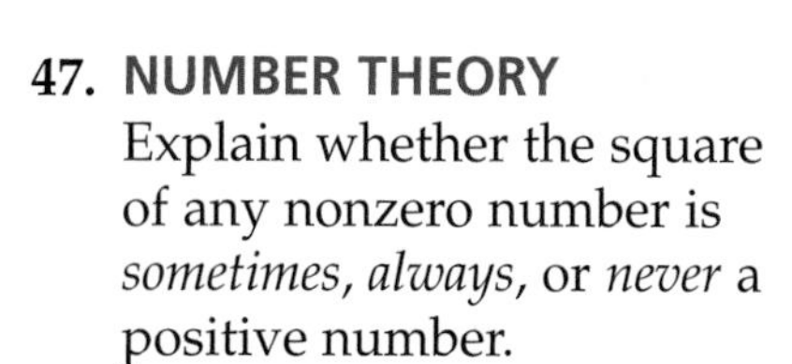

47. NUMBER THEORY Explain whether the square of any nonzero number is *sometimes*, *always*, or *never* a positive number.

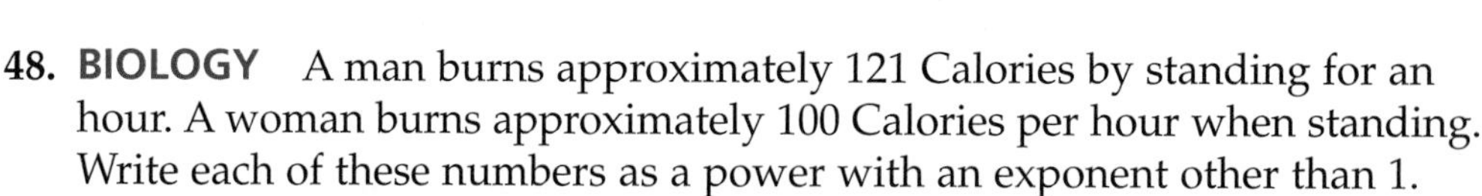

48. BIOLOGY A man burns approximately 121 Calories by standing for an hour. A woman burns approximately 100 Calories per hour when standing. Write each of these numbers as a power with an exponent other than 1.

More About. . .

History

The *so* noodles are about a yard long and as thin as a piece of yarn. Very few chefs still know how to make these noodles.

Source: *The Mathematics Teacher*

HISTORY **For Exercises 49–51, use the following information.**
In an ancient Chinese tradition, a chef stretches and folds dough to make long, thin noodles called *so*. After the first fold, he makes 2 noodles. He stretches and folds it a second time to make 4 noodles. Each time he repeats this process, the number of noodles doubles.

49. Use exponents to express the number of noodles after each of the first five folds.

50. Legendary chefs have completed as many as thirteen folds. How many noodles is this?

51. If the noodles are laid end to end and each noodle is 5 feet long, after how many of these folds will the length be more than a mile?

Replace each ● with <, >, or = to make a true statement.

52. 3^7 ● 7^3 **53.** 2^4 ● 4^2 **54.** 6^3 ● 4^4

GEOMETRY **For Exercises 55–57, use the cube below.**

55. The *surface area* of a cube is the sum of the areas of the faces. Use exponents to write an expression for the surface area of the cube.

56. The *volume* of a cube, or the amount of space that it occupies, is the product of the length, width, and height. Use exponents to write an expression for the volume of the cube.

57. If you double the length of each edge of the cube, are the surface area and volume also doubled? Explain.

58. **CRITICAL THINKING** Suppose the length of a side of a square is n units and the length of an edge of a cube is n units.
   a. If all the side lengths of a square are doubled, are the perimeter and the area of the square doubled? Explain.
   b. If all the side lengths of a square are tripled, show that the area of the new square is 9 times the area of the original square.
   c. If all the edge lengths of a cube are tripled, show that the volume of the new cube is 27 times the volume of the original cube.

59. **WRITING IN MATH** Answer the question that was posed at the beginning of the lesson.

Why are exponents important in comparing computer data?

Include the following in your answer:
- an explanation of how factors of 2 describe computer memory, and
- a sentence explaining the advantage of using exponents.

Online Research **Data Update** How many megabytes of memory are common today? Visit www.pre-alg.com/data_update to learn more.

FCAT Practice
Standardized Test Practice

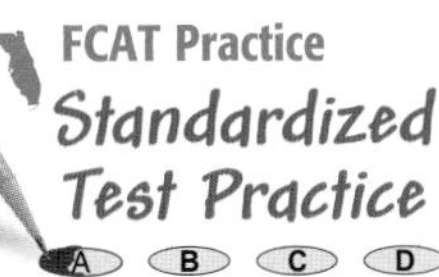

60. Write *ten million* as a power of ten.

(A) 10^5 (B) 10^6 (C) 10^7 (D) 10^8

61. What value of x will make $256 = 2^x$ true?

(A) 7 (B) 8 (C) 9 (D) 128

Maintain Your Skills

Mixed Review **State whether each number is divisible by 2, 3, 5, 6, or 10.** *(Lesson 4-1)*

62. 128 63. 370 64. 945

65. **METEOROLOGY** A tornado travels 300 miles in 2 hours. Use the formula $d = rt$ to find the tornado's speed in miles per hour. *(Lesson 3-7)*

ALGEBRA **Solve each equation. Check your solution.** *(Lesson 3-5)*

66. $2x + 1 = 7$ 67. $16 = 5k - 4$ 68. $\frac{n}{3} + 8 = 6$

69. **ALGEBRA** Simplify $4(y + 2) - y$. *(Lesson 3-2)*

Getting Ready for the Next Lesson **PREREQUISITE SKILL** **List all the factors for each number.**
*(To review **factoring**, see Lesson 4-1.)*

70. 11 71. 5 72. 9

73. 16 74. 19 75. 35

Algebra Activity

A Follow-Up of Lesson 4-2

Base 2

Activity

A computer contains a large number of tiny electronic switches that can be turned ON or OFF. The digits 0 and 1, also called *bits,* are the alphabet of computer language. This **binary** language uses a **base two** system of numbers.

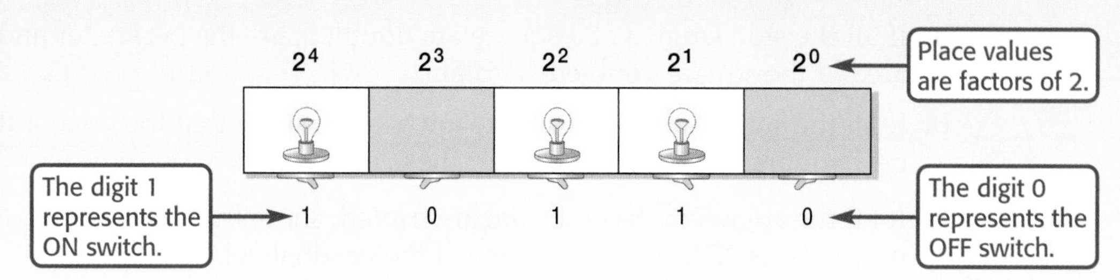

$$10110_2 = \underbrace{(1 \times 2^4)}_{16} + \underbrace{(0 \times 2^3)}_{0} + \underbrace{(1 \times 2^2)}_{4} + \underbrace{(1 \times 2^1)}_{2} + \underbrace{(0 \times 2^0)}_{0}$$

$$= 16 + 0 + 4 + 2 + 0$$

$$= 22$$

So, $10110_2 = 22_{10}$ or 22.

You can also reverse the process and express base ten numbers as equivalent numbers in base two.

Express the decimal number 13 as a number in base two.

Step 1 Make a base 2 place-value chart. Find the greatest factor of 2 that is less than 13. Place a 1 in that place value.

Step 2 Subtract 13 – 8 = 5. Now find the greatest factor of 2 that is less than 5. Place a 1 in that place value.

Step 3 Subtract 5 – 4 = 1. Place a 1 in that place value.

Step 4 There are no factors of 2 left, so place a 0 in any unfilled spaces.

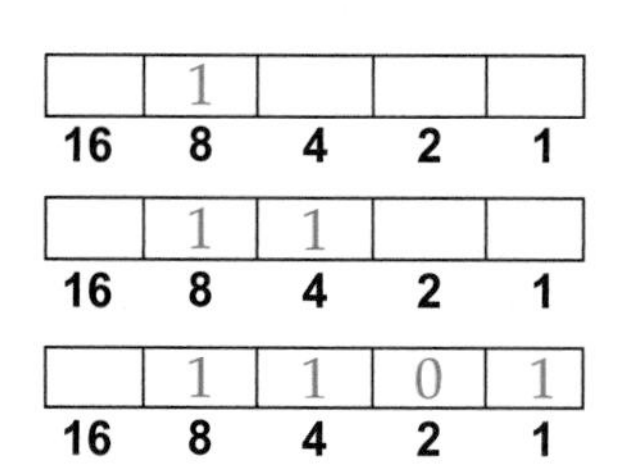

So, 13 in the base 10 system is equal to 1101 in the base 2 system. Or, $13 = 1101_2$.

Exercises

1. Express 1011_2 as an equivalent number in base 10.

Express each base 10 number as an equivalent number in base 2.

2. 6 **3.** 9 **4.** 15 **5.** 21

Extend the Activity

6. The first five place values for base 5 are shown. Any digit from 0 to 4 can be used to write a base 5 number. Write 179 in base 5.

625	125	25	5	1

7. OPEN ENDED Write 314 as an equivalent number in a base other than 2, 5, or 10. Include a place-value chart.

8. OPEN ENDED Choose a base 10 number and write it as an equivalent number in base 8. Include a place-value chart.

4-3 Prime Factorization

Sunshine State Standards
MA.A.1.3.3-3, MA.A.5.3.1-3

Vocabulary

- prime number
- composite number
- prime factorization
- factor tree
- factor

What You'll Learn

- Write the prime factorizations of composite numbers.
- Factor monomials.

How can models be used to determine whether numbers are prime?

There are two ways that 10 can be expressed as the product of whole numbers. This can be shown by using 10 squares to form rectangles.

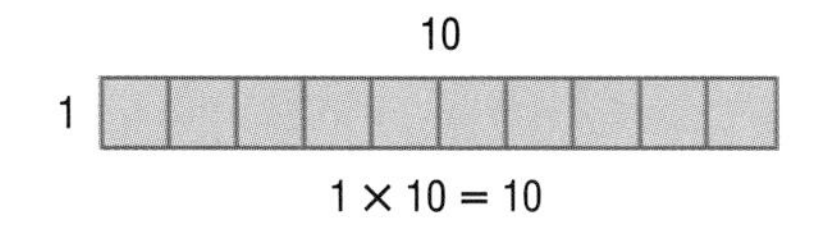

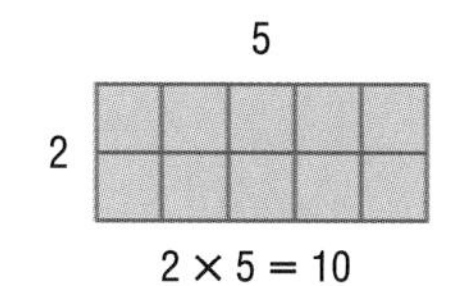

a. Use grid paper to draw as many different rectangular arrangements of 2, 3, 4, 5, 6, 7, 8, and 9 squares as possible.

b. Which numbers of squares can be arranged in more than one way?

c. Which numbers of squares can only be arranged one way?

d. What do the rectangles in part **c** have in common? Explain.

Reading Math

Composite
Everyday Meaning: materials that are made up of many substances
Math Meaning: numbers having many factors

PRIME NUMBERS AND COMPOSITE NUMBERS A **prime number** is a whole number that has exactly two factors, 1 and itself. A **composite number** is a whole number that has more than two factors. Zero and 1 are neither prime nor composite.

	Whole Numbers	Factors	Number of Factors
Prime Numbers	2	1, 2	2
	3	1, 3	2
	5	1, 5	2
	7	1, 7	2
Composite Numbers	4	1, 2, 4	3
	6	1, 2, 3, 6	4
	8	1, 2, 4, 8	4
	9	1, 3, 9	3
Neither Prime nor Composite	0	all numbers	infinite
	1	1	1

Example 1 Identify Numbers as Prime or Composite

a. Determine whether 19 is prime or composite.

Find factors of 19 by listing the whole number pairs whose product is 19.

$19 = 1 \times 19$

The number 19 has only two factors. Therefore, 19 is a prime number.

Study Tip

Mental Math
To determine whether a number is prime or composite, you can mentally use the rules for divisibility rather than listing factors.

b. Determine whether 28 is prime or composite.

Find factors of 28 by listing the whole number pairs whose product is 28.

$28 = 1 \times 28$
$28 = 2 \times 14$
$28 = 4 \times 7$

The factors of 28 are 1, 2, 4, 7, 14, and 28. Since the number has more than two factors, it is composite.

When a composite number is expressed as the product of prime factors, it is called the **prime factorization** of the number. One way to find the prime factorization of a number is to use a **factor tree**.

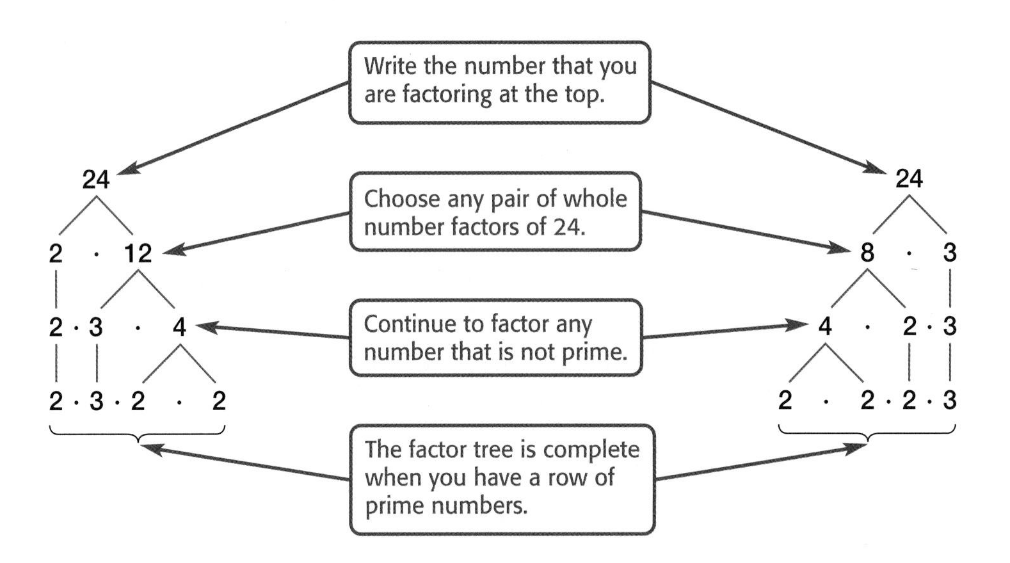

Study Tip

Commutative
The order of the factors does not matter because the operation of multiplication is commutative.

Both trees give the same prime factors, except in different orders. There is exactly one prime factorization of 24. The prime factorization of 24 is $2 \cdot 2 \cdot 2 \cdot 3$ or $2^3 \cdot 3$.

Concept Check Could a different factor tree have been used to write the prime factorization of 24? If so, would the result be the same?

Example 2 Write Prime Factorization

Write the prime factorization of 36.

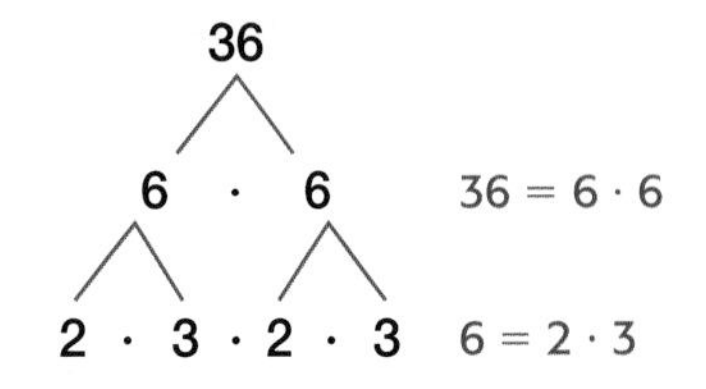

The factorization is complete because 2 and 3 are prime numbers.
The prime factorization of 36 is $2 \cdot 2 \cdot 3 \cdot 3$ or $2^2 \cdot 3^2$.

You can also use a strategy involving division called the *cake method* to find a prime factorization. The prime factorization of 210 is shown below using the cake method.

Study Tip

Remainders
If you get a remainder when using the cake method, choose a different prime number to divide the quotient.

Step 1
Begin with the smallest prime that is a factor of 210, in this case, 2. Divide 210 by 2.

$$2\overline{)210}\ \Rightarrow 105$$

→

Step 2
Divide the quotient 105 by the smallest possible prime factor, 3.

$$3\overline{)105}\ \Rightarrow 35, \quad 2\overline{)210}$$

→

Step 3
Repeat until the quotient is prime.

$$5\overline{)35}\ \Rightarrow 7, \quad 3\overline{)105}, \quad 2\overline{)210}$$

The prime factorization of 210 is $2 \cdot 3 \cdot 5 \cdot 7$. Multiply to check the result.

ALGEBRA CONNECTION

FACTOR MONOMIALS To **factor** a number means to write it as a product of its factors. A monomial can also be factored as a product of prime numbers and variables with no exponent greater than 1. Negative coefficients can be factored using −1 as a factor.

Example 3 Factor Monomials

Factor each monomial.

a. $8ab^2$

$8ab^2 = 2 \cdot 2 \cdot 2 \cdot a \cdot b^2$ $\quad 8 = 2 \cdot 2 \cdot 2$

$= 2 \cdot 2 \cdot 2 \cdot a \cdot b \cdot b$ $\quad a \cdot b^2 = a \cdot b \cdot b$

b. $-30x^3y$

$-30x^3y = -1 \cdot 2 \cdot 3 \cdot 5 \cdot x^3 \cdot y$ $\quad -30 = -1 \cdot 2 \cdot 3 \cdot 5$

$= -1 \cdot 2 \cdot 3 \cdot 5 \cdot x \cdot x \cdot x \cdot y$ $\quad x^3 \cdot y = x \cdot x \cdot x \cdot y$

Check for Understanding

Concept Check

1. **Explain** the difference between a prime and composite number.
2. **OPEN ENDED** Write a 2-digit number with prime factors that include 2 and 3.
3. **FIND THE ERROR** Cassidy and Francisca each factored 88.

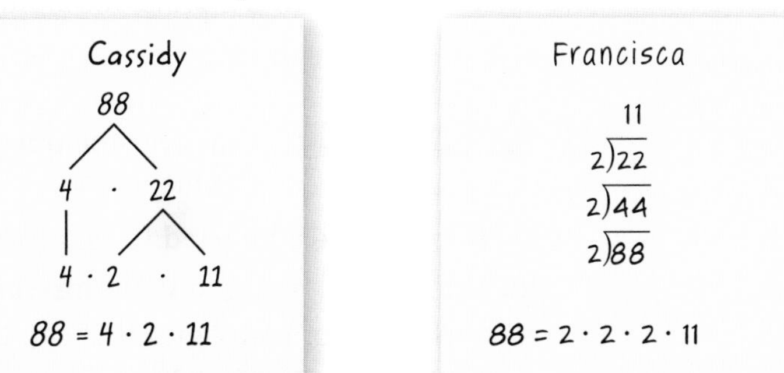

Who is correct? Explain your reasoning.

Guided Practice Determine whether each number is *prime* or *composite*.

4. 7 **5.** 23 **6.** 15

Write the prime factorization of each number. Use exponents for repeated factors.

7. 18 **8.** 39 **9.** 50

ALGEBRA Factor each monomial.

10. $4c^2$ **11.** $5a^2b$ **12.** $-70xyz$

Application

13. NUMBER THEORY One mathematical conjecture that is unproved states that there are infinitely many *twin primes*. Twin primes are prime numbers that differ by 2, such as 3 and 5. List all the twin primes that are less than 50.

Practice and Apply

Homework Help

For Exercises	See Examples
14–21	1
22–29	2
30–41	3

Extra Practice
See page 731.

Determine whether each number is *prime* or *composite*.

14. 21 **15.** 33 **16.** 23 **17.** 70
18. 17 **19.** 51 **20.** 43 **21.** 31

Write the prime factorization of each number. Use exponents for repeated factors.

22. 26 **23.** 81 **24.** 66 **25.** 63
26. 56 **27.** 100 **28.** 392 **29.** 110

ALGEBRA Factor each monomial.

30. $14w$ **31.** $9t^2$ **32.** $-7c^2$ **33.** $-25z^3$
34. $20st$ **35.** $-38mnp$ **36.** $28x^2y$ **37.** $21gh^3$
38. $13q^2r^2$ **39.** $64n^3$ **40.** $-75ab^2$ **41.** $-120r^2st^3$

42. Is the value of $n^2 - n + 41$ prime or composite if $n = 3$?

43. OPEN ENDED Write a monomial whose factors include -1, 5, and x.

44. TECHNOLOGY *Mersenne primes* are prime numbers in the form $2^n - 1$. In 1999, a computer programmer used special software to discover the largest prime number so far, $2^{6,972,593} - 1$. Write the prime factorization of each number, or write *prime* if the number is a Mersenne prime.

a. $2^5 - 1$ **b.** $2^6 - 1$ **c.** $2^7 - 1$ **d.** $2^8 - 1$

More About. . .

Technology

There are only 38 known Mersenne primes. Mathematicians continue to use computers to search for more of these numbers.

Source: www.mersenne.org

45. WRITING IN MATH Answer the question that was posed at the beginning of the lesson.

How can models be used to determine whether numbers are prime?

Include the following in your answer:

- the number of rectangles that can be drawn to represent prime and composite numbers, and
- an explanation of how one model can show that a number is *not* prime.

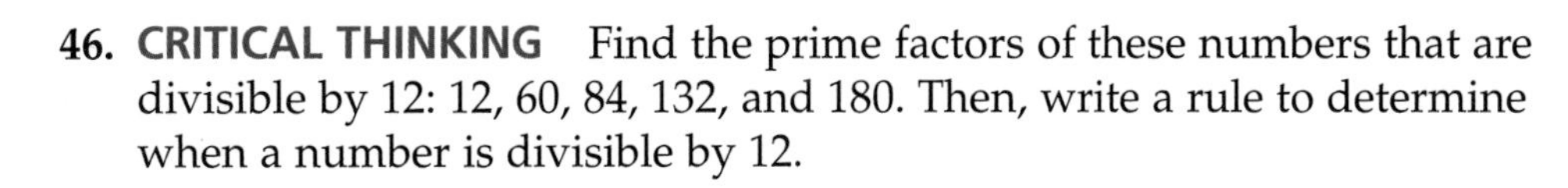

46. **CRITICAL THINKING** Find the prime factors of these numbers that are divisible by 12: 12, 60, 84, 132, and 180. Then, write a rule to determine when a number is divisible by 12.

47. Which table of values represents the following rule?
Add the input number to the square of the input number.

(A)

Input (x)	Output (y)
0	1
2	3
4	5

(B)

Input (x)	Output (y)
1	1
2	6
4	8

(C)

Input (x)	Output (y)
1	2
2	6
4	20

(D)

Input (x)	Output (y)
1	2
2	4
4	8

48. Determine which number is *not* a prime factor of 70.

(A) 2 (B) 5 (C) 7 (D) 10

Maintain Your Skills

Mixed Review

49. Write $(-5) \cdot (-5) \cdot (-5) \cdot h \cdot h \cdot k$ using exponents. *(Lesson 4-2)*

Determine whether each expression is a monomial. *(Lesson 4-1)*

50. $14cd$ 51. -5 52. $x - y$ 53. $3(1 + 3r)$

ALGEBRA Solve each equation. Check your solution. *(Lesson 3-4)*

54. $\frac{n}{8} = -4$ 55. $2x = -18$ 56. $30 = 6n$ 57. $-7 = \frac{y}{4}$

Getting Ready for the Next Lesson

PREREQUISITE SKILL Use the Distributive Property to rewrite each expression. *(To review the* ***Distributive Property****, see Lesson 3-1.)*

58. $2(n + 4)$ 59. $5(x - 7)$ 60. $-3(t + 4)$

61. $(a + 6)10$ 62. $(b - 3)(-2)$ 63. $8(9 - y)$

Practice Quiz 1

Lessons 4-1 through 4-3

Use divisibility rules to determine whether each number is divisible by 2, 3, 5, 6, or 10. *(Lesson 4-1)*

1. 105 2. 270 3. 511 4. 1368

5. **ALGEBRA** Evaluate $b^2 - 4ac$ if $a = -1$, $b = 5$, and $c = 3$. *(Lesson 4-2)*

6. **LITERATURE** In a story, a knight received a reward for slaying a dragon. He received 1 cent on the first day, 2 cents on the second day, 4 cents on the third day, and so on, continuing to double the amount for 30 days. *(Lesson 4-2)*
   a. Express his reward on each of the first three days as a power of 2.
   b. Express his reward on the 8th day as a power of 2. Then evaluate.

Factor each monomial. *(Lesson 4-3)*

7. $77x$ 8. $18st$ 9. $-23n^3$ 10. $30cd^2$

4-4 Greatest Common Factor (GCF)

What You'll Learn

- Find the greatest common factor of two or more numbers or monomials.
- Use the Distributive Property to factor algebraic expressions.

Sunshine State Standards
MA.A.5.3.1-1, MA.A.5.3.1-3

Vocabulary

- Venn diagram
- greatest common factor

How can a diagram be used to find the greatest common factor?

A **Venn diagram** shows the relationships among sets of numbers or objects by using overlapping circles in a rectangle.

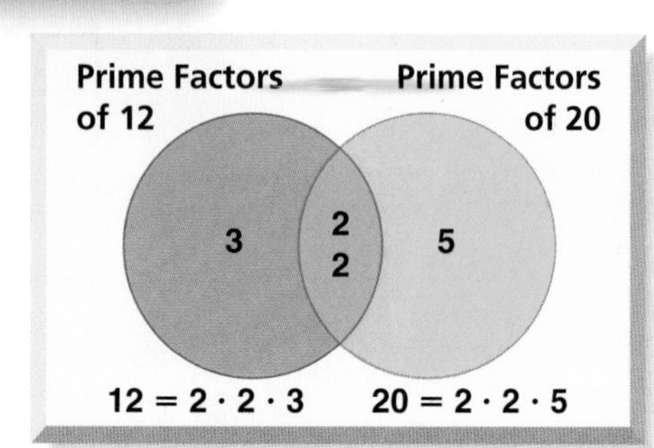

The Venn diagram at the right shows the prime factors of 12 and 20. The common prime factors are in both circles.

a. Which numbers are in both circles?

b. Find the product of the numbers that are in both circles.

c. Is the product also a factor of 12 and 20?

d. Make a Venn diagram showing the prime factors of 16 and 28. Then use it to find the common factors of the numbers.

GREATEST COMMON FACTOR Often, numbers have some of the same factors. The greatest number that is a factor of two or more numbers is called the **greatest common factor (GCF)**. Below are two ways to find the GCF of 12 and 20.

Example 1 Find the GCF

Find the GCF of 12 and 20.

Method 1 List the factors.

factors of 12: 1, 2, 3, (4), 6, 12
factors of 20: 1, 2, (4), 5, 10, 20

Common factors of 12 and 20: 1, 2, 4

The greatest common factor of 12 and 20 is 4.

Method 2 Use prime factorization.

$12 = \boxed{2} \cdot \boxed{2} \cdot 3$
$20 = \boxed{2} \cdot \boxed{2} \cdot 5$

Common prime factors of 12 and 20: 2, 2

The GCF is the product of the common prime factors.
$2 \cdot 2 = 4$

Again, the GCF of 12 and 20 is 4.

Study Tip

Choosing a Method
To find the GCF of two or more numbers, it is easier to
- list the factors if the numbers are small, or
- use prime factorization if the numbers are large.

Example 2 Find the GCF

Find the GCF of each set of numbers.

a. 30, 24

First, factor each number completely. Then circle the common factors.

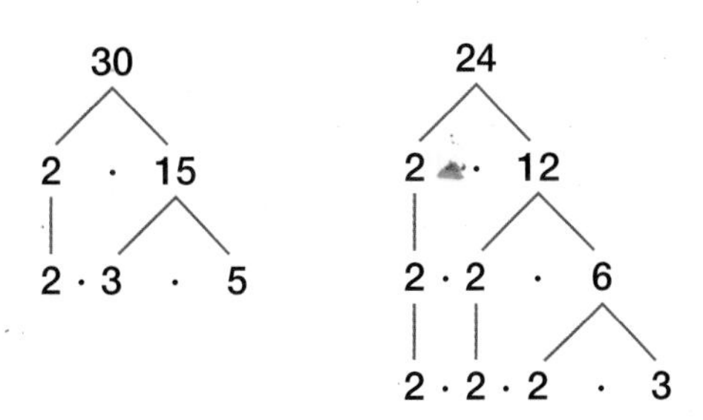

$30 = 2 \cdot 3 \cdot 5$
$24 = 2 \cdot 2 \cdot 2 \cdot 3$

The common prime factors are 2 and 3.

The GCF of 30 and 24 is $2 \cdot 3$ or 6.

Study Tip

Writing Prime Factors

Try to line up the common prime factors so that it is easier to circle them.

b. 54, 36, 45

$54 = 2 \cdot 3 \cdot 3 \cdot 3$
$36 = 2 \cdot 2 \cdot 3 \cdot 3$
$45 = 3 \cdot 3 \cdot 5$

The common prime factors are 3 and 3.

The GCF is $3 \cdot 3$ or 9.

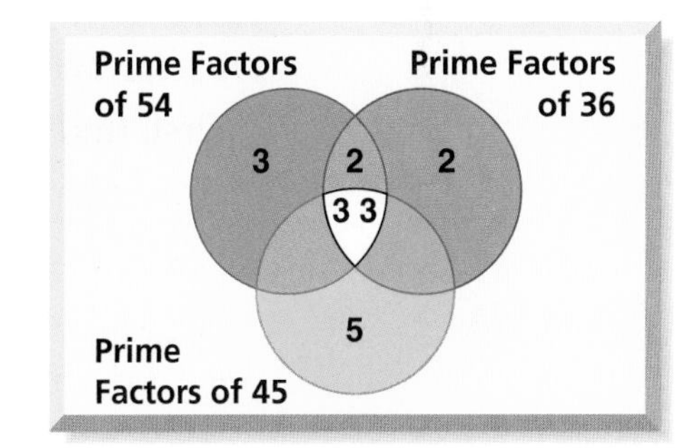

More About. . .

Track and Field

In some events such as sprints and the long jump, if the wind speed is greater than 2 meters per second, then the time or mark cannot be considered for record purposes.

Source: www.encarta.msn.com

Example 3 Use the GCF to Solve Problems

TRACK AND FIELD There are 208 boys and 240 girls participating in a field day competition.

a. What is the greatest number of teams that can be formed if each team has the same number of girls and each team has the same number of boys?

Find the GCF of 208 and 240.

$208 = 2 \cdot 2 \cdot 2 \cdot 2 \cdot 13$
$240 = 2 \cdot 2 \cdot 2 \cdot 2 \cdot 3 \cdot 5$

The common prime factors are 2, 2, 2, and 2.

The greatest common factor of 208 and 240 is $2 \cdot 2 \cdot 2 \cdot 2$ or 16. So, 16 teams can be formed.

b. How many boys and girls will be on each team?

$208 \div 16 = 13$
$240 \div 16 = 15$

So, each team will have 13 boys and 15 girls.

ALGEBRA CONNECTION

FACTOR ALGEBRAIC EXPRESSIONS You can also find the GCF of two or more monomials by finding the product of their common prime factors.

Example 4 Find the GCF of Monomials

Find the GCF of $16xy^2$ and $30xy$.

Completely factor each expression.

$16xy^2 = (2) \cdot 2 \cdot 2 \cdot 2 \cdot (x) \cdot (y) \cdot y$ Circle the common factors.
$30xy = (2) \cdot 3 \cdot 5 \cdot (x) \cdot (y)$

The GCF of $16xy^2$ and $30xy$ is $2 \cdot x \cdot y$ or $2xy$.

Study Tip

Look Back
To review the **Distributive Property**, see Lesson 3-1.

In Lesson 3-1, you used the Distributive Property to rewrite $2(x + 3)$ as $2x + 6$. You can also use this property to factor an algebraic expression such as $2x + 6$.

Example 5 Factor Expressions

Factor $2x + 6$.

First, find the GCF of $2x$ and 6.

$2x = (2) \cdot x$
$6 = (2) \cdot 3$ The GCF is 2.

Now write each term as a product of the GCF and its remaining factors.

$2x + 6 = 2(x) + 2(3)$
$= 2(x + 3)$ Distributive Property

So, $2x + 6 = 2(x + 3)$.

Concept Check Which property allows you to factor $3x + 9$?

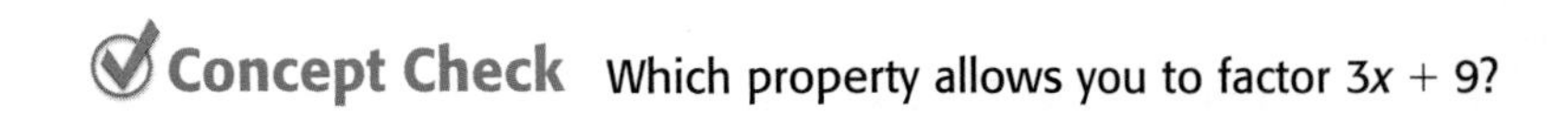

Check for Understanding

Concept Check

1. **Explain** how to find the greatest common factor of two or more numbers.
2. **OPEN ENDED** Name two different numbers whose GCF is 12.
3. **FIND THE ERROR** Christina and Jack both found the GCF of $2 \cdot 3^2 \cdot 11$ and $2^3 \cdot 5 \cdot 11$.

Christina

$2 \cdot 3^2 \cdot (11)$
$2^3 \cdot 5 \cdot (11)$

GCF = 11

Jack

$(2) \cdot 3^2 \cdot (11)$
$(2) \cdot 2 \cdot 2 \cdot 5 \cdot (11)$

GCF = 2 · 11 or 22

Who is correct? Explain your reasoning.

Guided Practice **Find the GCF of each set of numbers or monomials.**

4. 6, 8
5. 21, 45
6. 16, 56
7. 28, 42
8. 7, 30
9. 108, 144
10. 12, 24, 36
11. $14n$, $42n^2$
12. $36a^3b$, $56ab^2$

Factor each expression.

13. $3n + 9$
14. $t^2 + 4t$
15. $15 + 20x$

Application

16. **PARADES** In the parade, 36 members of the color guard are to march in front of 120 members of the high school marching band. Both groups are to have the same number of students in each row. Find the greatest number of students in each row.

Practice and Apply

Homework Help

For Exercises	See Examples
17–34	1, 2
35–42	4
44–52	5
53, 54	3

Extra Practice

See page 731.

Find the GCF of each set of numbers or monomials.

17. 12, 8
18. 3, 9
19. 24, 40
20. 21, 14
21. 20, 30
22. 12, 18
23. 18, 45
24. 22, 21
25. 16, 40
26. 42, 56
27. 30, 35
28. 12, 60
29. 116, 100
30. 135, 315
31. 9, 15, 24
32. 20, 21, 25
33. 20, 28, 36
34. 66, 90, 150
35. $12x$, $40x^2$
36. 18, $45mn$
37. $4st$, $10s$
38. $5ab$, $6b^2$
39. $14b$, $56b^2$
40. $30a^3b^2$, $24a^2b$
41. What is the greatest common factor of $32mn^2$, $16n$, and $12n^3$?
42. Name the GCF of $15v^2$, $70vw$, and $36w^2$.
43. Name two monomials whose GCF is $2x$.

Factor each expression.

44. $2x + 8$
45. $3r + 12$
46. $8 + 32a$
47. $6 + 3y$
48. $9 + 3t$
49. $14 + 21c$
50. $k^2 + 5k$
51. $4y - 16$
52. $5n - 10$

53. **PATTERNS** Consider the pattern 7, 14, 21, 28, 35, … .
    a. Find the GCF of the terms in the pattern. Explain how you know.
    b. Write the next two terms in the pattern.

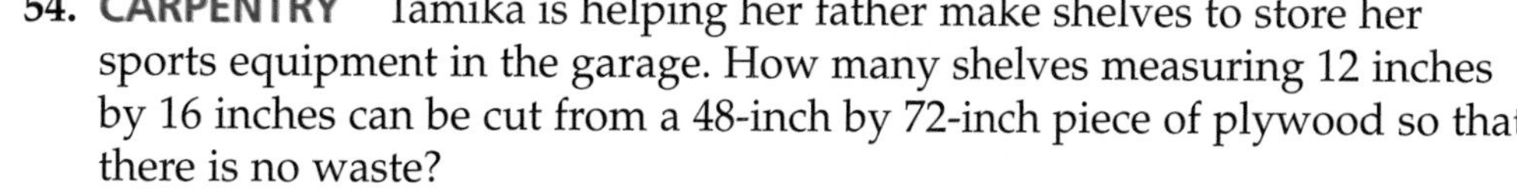

54. **CARPENTRY** Tamika is helping her father make shelves to store her sports equipment in the garage. How many shelves measuring 12 inches by 16 inches can be cut from a 48-inch by 72-inch piece of plywood so that there is no waste?

55. **DESIGN** Lauren is covering the surface of an end table with equal-sized ceramic tiles. The table is 30 inches long and 24 inches wide.
    a. What is the largest square tile that Lauren can use and not have to cut any tiles?
    b. How many tiles will Lauren need?

More About. . .

Design

A manufacturer in Istanbul, Turkey, uses 16th-century techniques to make ceramic tiles. It takes about two months to complete each tile.

Source: www.ceramics.about.com

56. **HISTORY** *The Nine Chapters on the Mathematical Art* is a Chinese math book written during the first century. It describes a procedure for finding the greatest common factors. Follow each step below to find the GCF of 86 and 110.

a. Subtract the lesser number, a, from the greater number, b.

b. If the result in part **a** is a factor of both numbers, it is the GCF. If the result is not a factor of both numbers, subtract the result from a or subtract a from the result so that the difference is a positive number.

c. Continue subtracting and checking the results until you find a number that is a factor of both numbers.

57. **CRITICAL THINKING** Can the GCF of a set of numbers be equal to one of the numbers? Give an example or a counterexample to support your answer.

58. **WRITING IN MATH** Answer the question that was posed at the beginning of the lesson.

How can a diagram be used to find the greatest common factor?

Include the following in your answer:

- a description of how a Venn diagram can be used to display the prime factorization of two or more numbers, and
- the part of a Venn diagram that is used to find the greatest common factor.

59. Write $6y + 21$ in factored form.

Ⓐ $6(y + 3)$ Ⓑ $2(3y + 7)$ Ⓒ $3(y + 7)$ Ⓓ $3(2y + 7)$

60. Find the GCF of $42x^2y$ and $38xy^2$.

Ⓐ $2x^2y$ Ⓑ $3xy$ Ⓒ $2xy$ Ⓓ $6x^2y^2$

Extending the Lesson

Two numbers are **relatively prime** if their only common factor is 1. Determine whether the numbers in each pair are relatively prime. Write *yes* or *no*.

61. 7 and 8 62. 13 and 11 63. 27 and 18

64. 20 and 25 65. 22 and 23 66. 8 and 12

Maintain Your Skills

Mixed Review

ALGEBRA **Factor each monomial.** *(Lesson 4-3)*

67. $9n$ 68. $15x^2$ 69. $-5jk$ 70. $22ab^3$

71. **ALGEBRA** Evaluate $7x^2 + y^3$ if $x = -2$ and $y = 4$. *(Lesson 4-2)*

Find each quotient. *(Lesson 2-5)*

72. $69 \div 23$ 73. $48 \div (-8)$ 74. $-24 \div (-12)$ 75. $-50 \div 5$

Getting Ready for the Next Lesson

PREREQUISITE SKILL **Find each equivalent measure.**

*(To review **converting measurements**, see pages 718–721.)*

76. 1 ft = __?__ in. 77. 1 yd = __?__ in. 78. 1 lb = __?__ oz

79. 1 day = __?__ h 80. 1 m = __?__ cm 81. 1 kg = __?__ g

4-5 Simplifying Algebraic Fractions

Sunshine State Standards
MA.A.5.3.1-3, MA.B.2.3.2-2, MA.B.2.3.2-3, MA.D.2.3.2-1

Vocabulary

- simplest form
- algebraic fraction

What You'll Learn

- Simplify fractions using the GCF.
- Simplify algebraic fractions.

How are simplified fractions useful in representing measurements?

You can use a fraction to compare a *part* of something to a *whole*. The figures below show what part 15 minutes is of 1 hour.

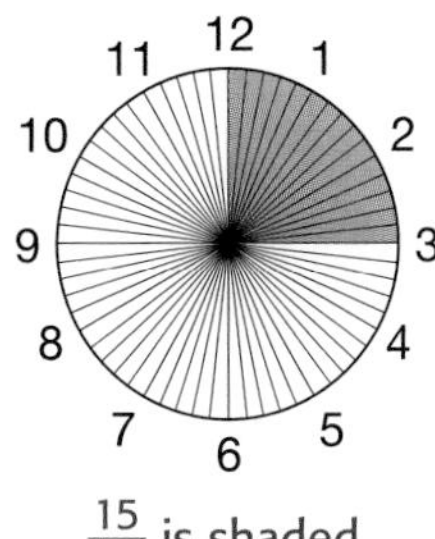

$\frac{15}{60}$ is shaded.

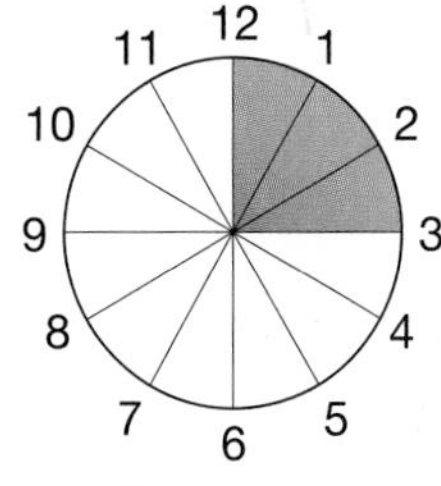

$\frac{3}{12}$ is shaded.

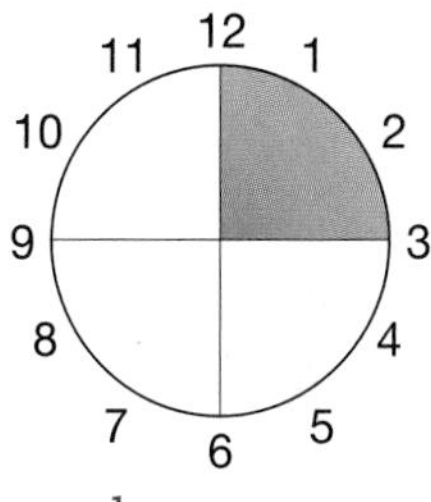

$\frac{1}{4}$ is shaded.

a. Are the three fractions equivalent? Explain your reasoning.

b. Which figure is divided into the least number of parts?

c. Which fraction would you say is written in simplest form? Why?

SIMPLIFY NUMERICAL FRACTIONS A fraction is in **simplest form** when the GCF of the numerator and the denominator is 1.

Fractions in Simplest Form	Fractions *not* in Simplest Form
$\frac{1}{4}, \frac{1}{3}, \frac{3}{4}, \frac{17}{50}$	$\frac{3}{12}, \frac{15}{60}, \frac{6}{8}, \frac{5}{20}$

One way to write a fraction in simplest form is to write the prime factorization of the numerator and the denominator. Then divide the numerator and denominator by the GCF.

Example 1 Simplify Fractions

Write $\frac{9}{12}$ in simplest form.

$9 = 3 \cdot 3$ Factor the numerator.
$12 = 2 \cdot 2 \cdot 3$ Factor the denominator.

The GCF of 9 and 12 is 3.

$\frac{9}{12} = \frac{9 \div 3}{12 \div 3}$ Divide the numerator and the denominator by the GCF.

$= \frac{3}{4}$ Simplest form

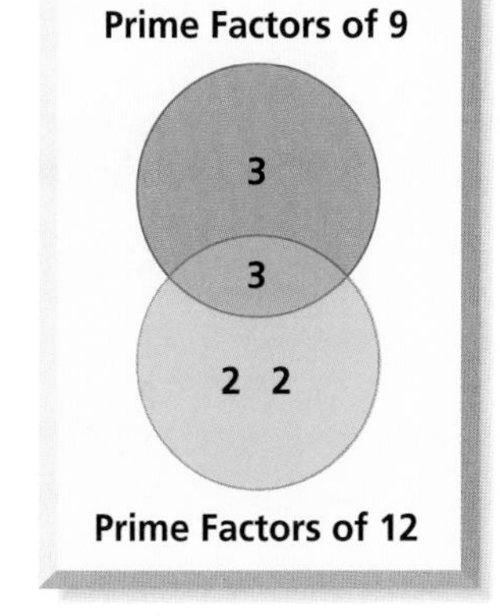

Study Tip

Use a Venn Diagram To simplify fractions, let one circle in a Venn diagram represent the numerator and let another circle represent the denominator. The number or product of numbers in the intersection is the GCF.

The division in Example 1 can be represented in another way.

$$\frac{9}{12} = \frac{\overset{1}{\cancel{3}} \cdot 3}{2 \cdot 2 \cdot \underset{1}{\cancel{3}}}$$ The slashes mean that the numerator and the denominator are both divided by the GCF, 3.

$$= \frac{3}{2 \cdot 2} \text{ or } \frac{3}{4}$$ Simplify.

Example 2 Simplify Fractions

Write $\frac{15}{60}$ in simplest form.

$$\frac{15}{60} = \frac{\overset{1}{\cancel{3}} \cdot \overset{1}{\cancel{5}}}{2 \cdot 2 \cdot \underset{1}{\cancel{3}} \cdot \underset{1}{\cancel{5}}}$$ Divide the numerator and denominator by the GCF, $3 \cdot 5$.

$$= \frac{1}{4}$$ Simplify.

Concept Check How do you know when a fraction is in simplest form?

Simplifying fractions is a useful tool in measurement.

Example 3 Simplify Fractions in Measurement

MEASUREMENT Eighty-eight feet is what part of 1 mile?

There are 5280 feet in 1 mile. Write the fraction $\frac{88}{5280}$ in simplest form.

$$\frac{88}{5280} = \frac{\overset{1}{\cancel{2}} \cdot \overset{1}{\cancel{2}} \cdot \overset{1}{\cancel{2}} \cdot \overset{1}{\cancel{11}}}{\underset{1}{\cancel{2}} \cdot \underset{1}{\cancel{2}} \cdot \underset{1}{\cancel{2}} \cdot 2 \cdot 2 \cdot 3 \cdot 5 \cdot \underset{1}{\cancel{11}}}$$ Divide the numerator and denominator by the GCF, $2 \cdot 2 \cdot 2 \cdot 11$.

$$= \frac{1}{60}$$ Simplify.

So, 88 feet is $\frac{1}{60}$ of a mile.

Study Tip

Alternative Method

You can also divide the numerator and denominator by common factors until the fraction is in simplest form.

$$\frac{88}{5280} = \frac{44}{2640} = \frac{22}{1320} = \frac{11}{660} \text{ or } \frac{1}{60}$$

ALGEBRA CONNECTION

SIMPLIFY ALGEBRAIC FRACTIONS A fraction with variables in the numerator or denominator is called an **algebraic fraction**. Algebraic fractions can also be written in simplest form.

Example 4 Simplify Algebraic Fractions

Simplify $\frac{21x^2y}{35xy}$.

$$\frac{21x^2y}{35xy} = \frac{3 \cdot \overset{1}{\cancel{7}} \cdot \overset{1}{\cancel{x}} \cdot x \cdot \overset{1}{\cancel{y}}}{5 \cdot \underset{1}{\cancel{7}} \cdot \underset{1}{\cancel{x}} \cdot \underset{1}{\cancel{y}}}$$ Divide the numerator and denominator by the GCF, $7 \cdot x \cdot y$.

$$= \frac{3x}{5}$$ Simplify.

Example 5 Simplify Algebraic Fractions

Multiple-Choice Test Item

Which fraction is $\frac{abc^3}{a^2b}$ written in simplest form?

(A) $\frac{bc^3}{a}$ (B) $\frac{bc^2}{a}$ (C) $\frac{c^3}{a}$ (D) $\frac{c^2}{a}$

The Princeton Review

Test-Taking Tip

Shortcuts You can solve some problems without much calculating if you understand the basic mathematical concepts. Look carefully at what is asked, and think of possible shortcuts for solving the problem.

Read the Test Item *In simplest form* means that the GCF of the numerator and the denominator is 1.

Solve the Test Item

$\frac{abc^3}{a^2b} = \frac{a\not{b}c^3}{a^2\not{b}}$

Without factoring, you can see that the variable b will not appear in the simplified fraction. That eliminates choices A and B.

$\frac{abc^3}{a^2b} = \frac{\overset{1}{\not{a}} \cdot \overset{1}{\not{b}} \cdot c \cdot c \cdot c}{\underset{1}{\not{a}} \cdot a \cdot \underset{1}{\not{b}}}$ Factor.

$= \frac{c^3}{a}$ Multiply.

The answer is C.

Check for Understanding

Concept Check

1. **Explain** what it means to express a fraction in simplest form.

2. **OPEN ENDED** Write examples of a numerical fraction and an algebraic fraction in simplest form and examples of a numerical fraction and an algebraic fraction not in simplest form.

Guided Practice

Write each fraction in simplest form. If the fraction is already in simplest form, write *simplified*.

3. $\frac{2}{14}$ 4. $\frac{9}{15}$ 5. $\frac{5}{11}$ 6. $\frac{25}{40}$ 7. $\frac{64}{68}$

ALGEBRA Simplify each fraction. If the fraction is already in simplest form, write *simplified*.

8. $\frac{x}{x^3}$ 9. $\frac{8a^2}{16a}$ 10. $\frac{12c}{15d}$ 11. $\frac{24}{5k}$

12. **MEASUREMENT** Nine inches is what part of 1 yard?

13. Which fraction is $\frac{25mn}{65n}$ written in simplest form?

(A) $\frac{2m}{6}$ (B) $\frac{5m}{13}$ (C) $\frac{5m}{13n}$ (D) $\frac{25mn}{65}$

Practice and Apply

Homework Help

For Exercises	See Examples
14–28	1, 2
30–41	4
42–43	3

Extra Practice

See page 732.

Write each fraction in simplest form. If the fraction is already in simplest form, write *simplified*.

14. $\frac{3}{18}$
15. $\frac{10}{12}$
16. $\frac{15}{21}$
17. $\frac{8}{36}$
18. $\frac{17}{20}$
19. $\frac{18}{44}$
20. $\frac{16}{64}$
21. $\frac{30}{37}$
22. $\frac{34}{38}$
23. $\frac{17}{51}$
24. $\frac{51}{60}$
25. $\frac{25}{60}$
26. $\frac{36}{96}$
27. $\frac{133}{140}$
28. $\frac{765}{2023}$

29. **AIRCRAFT** A model of Lindbergh's *Spirit of St. Louis* has a wingspan of 18 inches. The wingspan of the actual airplane is 46 feet. Write a fraction in simplest form comparing the wingspan of the model and the wingspan of the actual airplane. (*Hint:* convert 46 feet to inches.)

ALGEBRA Simplify each fraction. If the fraction is already in simplest form, write *simplified*.

30. $\frac{a}{a^4}$
31. $\frac{y^3}{y}$
32. $\frac{12m}{15m}$
33. $\frac{40d}{42d}$
34. $\frac{4k}{19m}$
35. $\frac{8t}{64t^2}$
36. $\frac{16n}{18n^2p}$
37. $\frac{28z^3}{16z}$
38. $\frac{6r}{15rs}$
39. $\frac{12cd}{19e}$
40. $\frac{30x^2}{51xy}$
41. $\frac{17g^2h}{51g}$

42. **MEASUREMENT** Fifteen hours is what part of one day?

43. **MEASUREMENT** Ninety-six centimeters is what part of a meter?

44. **MEASUREMENT** Twelve ounces is what part of a pound? (*Hint*: 1 lb = 16 oz)

Career Choices

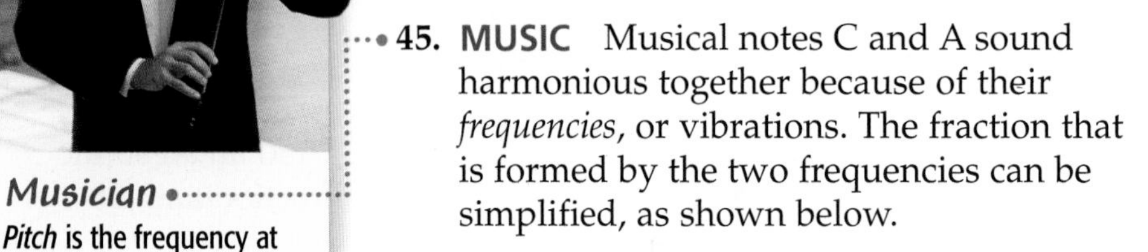

Musician

Pitch is the frequency at which an instrument's string vibrates when it is struck. To correct the pitch, a musician must increase or decrease tension in the strings.

Online Research For information about a career as a musician, visit: www.pre-alg.com/careers

45. **MUSIC** Musical notes C and A sound harmonious together because of their *frequencies*, or vibrations. The fraction that is formed by the two frequencies can be simplified, as shown below.

$$\frac{C}{A} = \frac{264}{440} \text{ or } \frac{3}{5}$$

When a fraction formed by two frequencies *cannot* be simplified, the notes sound like noise. Determine whether each pair of notes would sound harmonious together. Explain why or why not.

Note	Frequency (hz)
C	264
D	294
E	330
F	349
G	392
A	440
B	494
C	528

a. E and A
b. D and F
c. first C and last C

46. **ANIMALS** The table shows the average amount of food each animal can eat in a day and its average weight. What fraction of its weight can each animal eat per day?

Animal	Daily Amount of Food	Weight of Animal
elephant	450 lb	9000 lb
hummingbird	2 g	3 g
polar bear	25 lb	1500 lb
tiger	20 lb	500 lb

Source: *Animals as Our Companions, Wildlife Fact File*

Simplifying fractions will help you determine the portion of your daily allowance of fat grams. Visit www.pre-alg.com/webquest to continue work on your WebQuest project.

MONEY For Exercises 47–49, use the graph to write each fraction in simplest form.

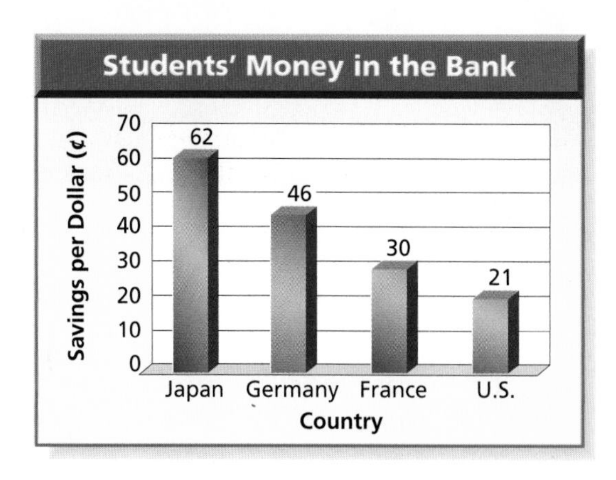

47. the fraction of a dollar that students in Japan save

48. the fraction of a dollar that students in the U.S. save

49. a fraction showing the amount of a dollar that students in the U.S. save compared to students in France

50. **CRITICAL THINKING** Is it true that $\frac{23}{53} = \frac{\overset{1}{\cancel{23}}}{\underset{1}{\cancel{53}}}$ or $\frac{2}{5}$? Explain.

51. **WRITING IN MATH** Answer the question that was posed at the beginning of the lesson.

How are simplified fractions useful in representing measurements?

Include the following in your answer:
- an explanation of how measurements represent parts of a whole, and
- examples of fractions that represent measurements.

52. Which fraction represents the shaded area written in simplest form?

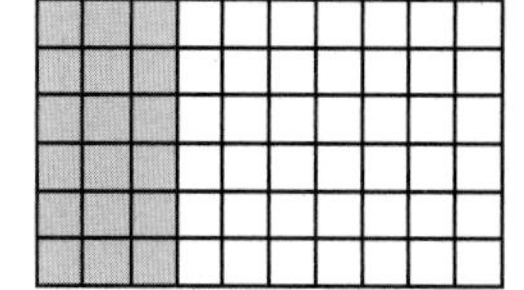

(A) $\frac{9}{30}$ (B) $\frac{3}{10}$ (C) $\frac{6}{20}$ (D) $\frac{30}{100}$

53. Write $\frac{15ab}{25b^2}$ in simplest form.

(A) $\frac{3a}{5b}$ (B) $\frac{15a}{25b}$ (C) $\frac{3ab}{5}$ (D) $\frac{15a}{25b^2}$

Maintain Your Skills

Mixed Review

Find the greatest common factor of each set of numbers or monomials. *(Lesson 4-4)*

54. 9, 15 55. 4, 12, 10 56. $40x^2$, $16x$ 57. $25a$, $30b$

Determine whether each number is *prime* or *composite*. *(Lesson 4-3)*

58. 13 59. 34 60. 99 61. 79

ALGEBRA **Solve each equation. Check your solution.** *(Lesson 3-3)*

62. $t - 18 = 24$ 63. $30 = 3 + y$ 64. $-7 = x + 11$

Getting Ready for the Next Lesson

PREREQUISITE SKILL **For each expression, use parentheses to group the numbers together and to group the powers with like bases together.** *(To review **properties of multiplication**, see Lesson 1-4.)*

Example: $a \cdot 4 \cdot a^3 \cdot 2 = (4 \cdot 2)(a \cdot a^3)$

65. $6 \cdot 7 \cdot k^3$ 66. $s \cdot t^2 \cdot s \cdot t$

67. $3 \cdot x^4 \cdot (-5) \cdot x^2$ 68. $5 \cdot n^3 \cdot p \cdot 2 \cdot n \cdot p$

Reading Mathematics

Powers

The phrase *the quantity* is used to indicate parentheses when reading expressions. Recall that an exponent indicates the number of times that the base is used as a factor. Suppose you are to write each of the following in symbols.

Words	Symbols	Examples (Let $x = 2$.)
three times x squared	$3x^2$	$3x^2 = 3 \cdot 2^2$ $= 3 \cdot 4$ Evaluate 2^2. $= 12$ Multiply $3 \cdot 4$.
three times x the quantity squared	$(3x)^2$	$(3x)^2 = (3 \cdot 2)^2$ $= 6^2$ Evaluate $3 \cdot 2$. $= 36$ Square 6.

In the expression $(3x)^2$, parentheses are used to show that $3x$ is used as a factor twice.

$$(3x)^2 = (3x)(3x)$$

The quantity can also be used to describe division of monomials.

Words	Symbols	Examples (Let $x = 2$.)
eight divided by x squared	$\frac{8}{x^2}$	$\frac{8}{x^2} = \frac{8}{2^2}$ $= \frac{8}{4}$ Evaluate 2^2. $= 2$ Divide $8 \div 4$.
eight divided by x the quantity squared	$\left(\frac{8}{x}\right)^2$	$\left(\frac{8}{x}\right)^2 = \left(\frac{8}{2}\right)^2$ $= 4^2$ Evaluate $8 \div 2$. $= 16$ Square 4.

Reading to Learn

State how you would read each expression.

1. $4a^2$ **2.** $(10x)^5$ **3.** $\frac{5}{n^3}$ **4.** $\left(\frac{4}{r}\right)^2$ **5.** $(m + n)^3$

6. $(a - b)^4$ **7.** $a - b^4$ **8.** $\frac{a}{b^4}$ **9.** $(4c^2)^3$ **10.** $\left(\frac{8}{c^2}\right)^3$

Determine whether each pair of expressions is equivalent. Write *yes* or *no*.

11. $4ab^5$ and $4(ab)^5$ **12.** $(2x)^3$ and $8x^3$ **13.** $(mn)^4$ and $m^4 \cdot n^4$

14. c^3d^3 and cd^3 **15.** $\frac{x}{y^2}$ and $\left(\frac{x}{y}\right)^2$ **16.** $\frac{n^2}{r^2}$ and $\left(\frac{n}{r}\right)^2$

4-6 Multiplying and Dividing Monomials

Sunshine State Standards
MA.A.1.3.4-2, MA.A.3.3.1-3

What You'll Learn

- Multiply monomials.
- Divide monomials.

How are powers of monomials useful in comparing earthquake magnitudes?

For each increase on the Richter scale, an earthquake's vibrations, or *seismic waves*, are 10 times greater. So, an earthquake of magnitude 4 has seismic waves that are 10 times greater than that of a magnitude 3.

Richter Scale	Times Greater than Magnitude 3 Earthquake	Written Using Powers
4	10	10^1
5	$10 \times 10 = 100$	$10^1 \times 10^1 = 10^2$
6	$10 \times 100 = 1000$	$10^1 \times 10^2 = 10^3$
7	$10 \times 1000 = 10{,}000$	$10^1 \times 10^3 = 10^4$
8	$10 \times 10{,}000 = 100{,}000$	$10^1 \times 10^4 = 10^5$

a. Examine the exponents of the factors and the exponents of the products in the last column. What do you observe?

b. **Make a conjecture** about a rule for determining the exponent of the product when you multiply powers with the same base. Test your rule by multiplying $2^2 \cdot 2^4$ using a calculator.

MULTIPLY MONOMIALS Recall that exponents are used to show repeated multiplication. You can use the definition of exponent to help find a rule for multiplying powers with the same base.

$$2^3 \cdot 2^4 = \underbrace{\overbrace{(2 \cdot 2 \cdot 2)}^{\text{3 factors}} \cdot \overbrace{(2 \cdot 2 \cdot 2 \cdot 2)}^{\text{4 factors}}}_{\text{7 factors}}$$

$$= 2^7$$

Notice the sum of the original exponents and the exponent in the final product. This relationship is stated in the following rule.

Study Tip

Common Misconception
When multiplying powers, do not multiply the bases.
$3^2 \cdot 3^4 = 3^6$, not 9^6

Key Concept — *Product of Powers*

- **Words** You can multiply powers with the same base by adding their exponents.
- **Symbols** $a^m \cdot a^n = a^{m+n}$
- **Example** $3^2 \cdot 3^4 = 3^{2+4}$ or 3^6

Example 1 Multiply Powers

Find $7^3 \cdot 7$.

$7^3 \cdot 7 = 7^3 \cdot 7^1$ — $7 = 7^1$

$= 7^{3+1}$ — The common base is 7.

$= 7^4$ — Add the exponents.

CHECK $7^3 \cdot 7 = (7 \cdot 7 \cdot 7)(7)$

$= 7 \cdot 7 \cdot 7 \cdot 7$ or 7^4 ✓

Concept Check Can you simplify $2^3 \cdot 3^3$ using the Product of Powers rule? Explain.

ALGEBRA CONNECTION

Monomials can also be multiplied using the rule for the product of powers.

Example 2 Multiply Monomials

Find each product.

a. $x^5 \cdot x^2$

$x^5 \cdot x^2 = x^{5+2}$ — The common base is x.

$= x^7$ — Add the exponents.

b. $(-4n^3)(2n^6)$

$(-4n^3)(2n^6) = (-4 \cdot 2)(n^3 \cdot n^6)$ — Use the Commutative and Associative Properties.

$= (-8)(n^{3+6})$ — The common base is n.

$= -8n^9$ — Add the exponents.

Study Tip

Look Back
To review the **Commutative and Associative Properties of Multiplication**, see Lesson 1-4.

DIVIDE MONOMIALS You can also write a rule for finding quotients of powers.

$$\frac{2^6}{2^1} = \frac{2 \cdot 2 \cdot 2 \cdot 2 \cdot 2 \cdot 2}{2}$$ ← 6 factors / ← 1 factor

$$= \frac{2 \cdot 2 \cdot 2 \cdot 2 \cdot 2 \cdot \overset{1}{\cancel{2}}}{\underset{1}{\cancel{2}}}$$ Divide the numerator and the denominator by the GCF, 2.

$$= 2^5$$ ← 5 factors — Simplify.

Compare the difference between the original exponents and the exponent in the final quotient. This relationship is stated in the following rule.

Key Concept — Quotient of Powers

- **Words** You can divide powers with the same base by subtracting their exponents.
- **Symbols** $\frac{a^m}{a^n} = a^{m-n}$, where $a \neq 0$
- **Example** $\frac{4^5}{4^2} = 4^{5-2}$ or 4^3

Example 3 Divide Powers

Find each quotient.

a. $\frac{5^7}{5^4}$

$\frac{5^7}{5^4} = 5^{7-4}$ The common base is 5.

$= 5^3$ Subtract the exponents.

b. $\frac{y^5}{y^3}$

$\frac{y^5}{y^3} = y^{5-3}$ The common base is y.

$= y^2$ Subtract the exponents.

Concept Check Can you simplify $\frac{x^7}{y^2}$ using the Quotient of Powers rule? Why or why not?

Example 4 Divide Powers to Solve a Problem

Reading Math

How Many/How Much

How many times faster indicates that division is to be used to solve the problem. If the question had said *how much faster*, then subtraction ($10^9 - 10^8$) would have been used to solve the problem.

COMPUTERS The table compares the processing speeds of a specific type of computer in 1993 and in 1999. Find how many times faster the computer was in 1999 than in 1993.

Year	Processing Speed (instructions per second)
1993	10^8
1999	10^9

Source: *The Intel Microprocessor Quick Reference Guide*

Write a division expression to compare the speeds.

$\frac{10^9}{10^8} = 10^{9-8}$ Subtract the exponents.

$= 10^1$ or 10 Simplify.

So, the computer was 10 times faster in 1999 than in 1993.

Check for Understanding

Concept Check

1. **State** whether you could use the Product of Powers rule, Quotient of Powers rule, or neither to find $m^5 \cdot n^4$. Explain.
2. **Explain** whether $4^8 \cdot 4^6$ and $4^4 \cdot 4^{10}$ are equivalent expressions.
3. **OPEN ENDED** Write a multiplication expression whose product is 5^3.

Guided Practice **Find each product or quotient. Express using exponents.**

4. $9^3 \cdot 9^2$
5. $a \cdot a^5$
6. $(n^4)(n^4)$
7. $-3x^2(4x^3)$
8. $\frac{3^8}{3^5}$
9. $\frac{10^5}{10^3}$
10. $\frac{x^3}{x}$
11. $\frac{a^{10}}{a^6}$

Application

12. **EARTHQUAKES** In 2000, an earthquake measuring 8 on the Richter scale struck Indonesia. Two months later, an earthquake of magnitude 5 struck northern California. How many times greater were the seismic waves in Indonesia than in California? (*Hint:* Let 10^8 and 10^5 represent the earthquakes, respectively.)

Practice and Apply

Homework Help

For Exercises	See Examples
13–24	1, 2
25–36	3
41–44	4

Extra Practice

See page 732.

Find each product or quotient. Express using exponents.

13. $3^3 \cdot 3^2$ **14.** $6 \cdot 6^7$ **15.** $d^4 \cdot d^6$ **16.** $10^4 \cdot 10^3$

17. $n^8 \cdot n$ **18.** $t^2 \cdot t^4$ **19.** $9^4 \cdot 9^5$ **20.** $a^6 \cdot a^6$

21. $2y \cdot 9y^4$ **22.** $(5r^3)(4r^4)$ **23.** $ab^5 \cdot 8a^2b^5$ **24.** $10x^3y \cdot (-2xy^2)$

25. $\frac{5^5}{5^2}$ **26.** $\frac{8^4}{8^3}$ **27.** $b^6 \div b^3$ **28.** $10^{10} \div 10^2$

29. $\frac{m^{20}}{m^8}$ **30.** $\frac{a^8}{a^8}$ **31.** $\frac{(-2)^6}{(-2)^5}$ **32.** $\frac{(-x)^5}{(-x)}$

33. $\frac{n^3(n^5)}{n^2}$ **34.** $\frac{s^7}{s \cdot s^2}$ **35.** $\left(\frac{k^3}{k}\right)\left(\frac{m^2}{m}\right)$ **36.** $\left(\frac{15}{5}\right)\left(\frac{n^9}{n}\right)$

37. the product of nine to the fourth power and nine cubed

38. the quotient of k to the fifth power and k squared

39. What is the product of 7^3, 7^5, and 7?

40. Find $a^4 \cdot a^6 \div a^2$.

CHEMISTRY **For Exercises 41–43, use the following information.**
The pH of a solution describes its acidity. Neutral water has a pH of 7. Each one-unit *decrease* in the pH means that the solution is 10 times more acidic. For example, a pH of 4 is 10 times more acidic than a pH of 5.

41. Suppose the pH of a lake is 5 due to acid rain. How much more acidic is the lake than neutral water?

42. Use the information at the left to find how much more acidic vinegar is than baking soda.

43. Cola is 10^4 times more acidic than neutral water. What is the pH value of cola?

More About. . .

Chemistry

The pH values of different kitchen items are shown below.

Item	pH
lemon juice	2
vinegar	3
tomatoes	4
baking soda	9

Source: *Biology*, Raven

44. LIFE SCIENCE When bacteria reproduce, they split so that one cell becomes two. The number of cells after t cycles of reproduction is 2^t.

a. *E. coli* reproduce very quickly, about every 15 minutes. If there are 100 *E. coli* in a dish now, how many will there be in 30 minutes?

b. How many times more *E. coli* are there in a population after 3 hours than there were after 1 hour?

GEOMETRY **For Exercises 45 and 46, use the information in the figures.**

45. How many times greater is the length of the edge of the larger cube than the smaller one?

46. How many times greater is the volume of the larger cube than the smaller one?

Volume = 2^3 cubic units

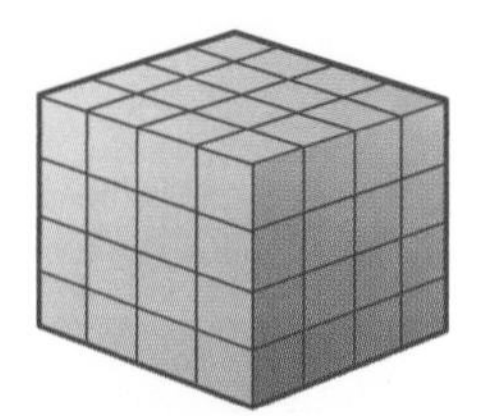
Volume = 2^6 cubic units

Find each missing exponent.

47. $(4^{\bullet})(4^3) = 4^{11}$ **48.** $\frac{t^{\bullet}}{t^2} = t^{14}$ **49.** $\frac{13^5}{13^{\bullet}} = 1$

50. **CRITICAL THINKING** Use the laws of exponents to show why the value of *any* nonzero number raised to the zero power equals 1.

51. **WRITING IN MATH** Answer the question that was posed at the beginning of the lesson.

How are powers of monomials useful in comparing earthquake magnitudes?

Include the following in your answer:
- a description of the Richter scale, and
- a comparison of two earthquakes of different magnitudes by using the Quotient of Powers rule.

52. Multiply $7xy$ and $x^{14}z$.

Ⓐ $7x^{15}yz$ Ⓑ $7x^{15}y$ Ⓒ $7x^{13}yz$ Ⓓ $x^{15}yz$

53. Find the quotient $a^5 \div a$.

Ⓐ a^5 Ⓑ a^4 Ⓒ a^6 Ⓓ a

Maintain Your Skills

Mixed Review

Write each fraction in simplest form. If the fraction is already in simplest form, write *simplified*. *(Lesson 4-5)*

54. $\frac{12}{40}$ 55. $\frac{20}{53}$ 56. $\frac{8n^2}{32n}$ 57. $\frac{6x^3}{4x^2y}$

Find the greatest common factor of each set of numbers or monomials. *(Lesson 4-4)*

58. 36, 4 59. 18, 28 60. 42, 54 61. $9a$, $10a^3$

62. Evaluate $|a| - |b| \cdot |c|$ if $a = -16$, $b = 2$, and $c = 3$. *(Lesson 2-1)*

63. **ENERGY** The graph shows the high temperature and the maximum amount of electricity that was used during each of fifteen summer days. Do the data show a *positive*, *negative*, or *no* relationship? Explain. *(Lesson 1-7)*

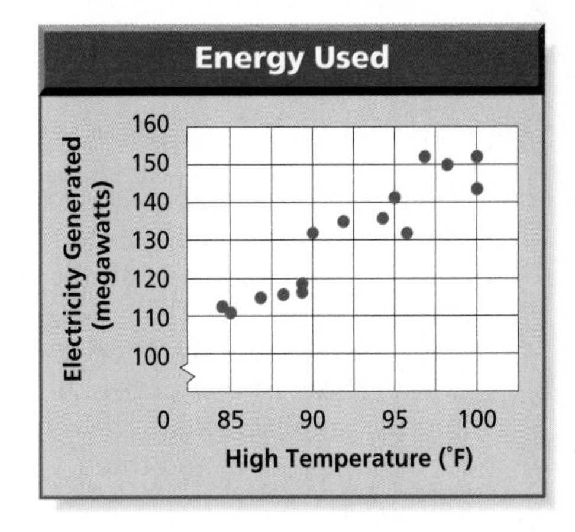

Getting Ready for the Next Lesson

PREREQUISITE SKILL **Evaluate each expression if $x = 10$, $y = -5$, and $z = 4$. Write as a fraction in simplest form.** *(To review **evaluating expressions**, see Lesson 1-3.)*

64. $\frac{1}{x}$ 65. $\frac{y}{50}$ 66. $\frac{z}{100}$

67. $\frac{1}{zy}$ 68. $\frac{1}{x \cdot x}$ 69. $\frac{1}{(z)(z)(z)}$

Algebra Activity

A Follow-Up of Lesson 4-6

Sunshine State Standards
MA.D.1.3.2-1, MA.D.2.3.1-6, MA.E.1.3.1-2

A Half-Life Simulation

A radioactive material such as uranium decomposes or decays in a regular manner best described as a *half-life*. A half-life is the time it takes for half of the atoms in the sample to decay.

Collect the Data

Step 1 Place 50 pennies heads up in a shoebox. Put the lid on the box and shake it up and down one time. This simulates one half-life.

Step 2 Open the lid of the box and remove all the pennies that are now tails up. In a table like the one at the right, record the number of pennies that remain.

Number of Half-Lives	Number of Pennies That Remain
1	
2	
3	
4	
5	

Step 3 Put the lid back on the box and again shake the box up and down one time. This represents another half-life.

Step 4 Open the lid. Remove all the tails up pennies. Count the pennies that remain.

Step 5 Repeat the half-life of decay simulation until less than five pennies remain in the shoebox.

Analyze the Data

1. On grid paper, draw a coordinate grid in which the x-axis represents the number of half-lives and the y-axis represents the number of pennies that remain. Plot the points (number of half-lives, number of remaining pennies) from your table.
2. Describe the graph of the data.

After each half-life, you expect to remove about one-half of the pennies. So, you expect about one-half to remain. The expressions at the right represent the average number of pennies that remain if you start with 50, after one, two, and three half-lives.

one half-life:	$50\left(\frac{1}{2}\right) = 50\left(\frac{1}{2}\right)^1$
two half-lives:	$50\left(\frac{1}{2}\right)\left(\frac{1}{2}\right) = 50\left(\frac{1}{2}\right)^2$
three half-lives:	$50\left(\frac{1}{2}\right)\left(\frac{1}{2}\right)\left(\frac{1}{2}\right) = 50\left(\frac{1}{2}\right)^3$

Make a Conjecture

3. Use the expressions to predict how many pennies remain after three half-lives. Compare this number to the number in the table above. Explain any differences.
4. Suppose you started with 1000 pennies. Predict how many pennies would remain after three half-lives.

4-7 Negative Exponents

What You'll Learn

Sunshine State Standards
MA.A.1.3.1-1, MA.A.1.3.2-2, MA.A.1.3.4-2, MA.A.2.3.1-1, MA.A.2.3.1-3, MA.D.2.3.1-6

- Write expressions using negative exponents.
- Evaluate numerical expressions containing negative exponents.

How do negative exponents represent repeated division?

Copy the table at the right.

a. Describe the pattern of the powers in the first column. Continue the pattern by writing the next two powers in the table.

b. Describe the pattern of values in the second column. Then complete the second column.

c. Verify that the powers you wrote in part **a** are equal to the values that you found in part **b**.

d. Determine how 3^{-1} should be defined.

Power	Value
2^6	64
2^5	32
2^4	16
2^3	8
2^2	4
2^1	2

ALGEBRA CONNECTION

NEGATIVE EXPONENTS Extending the pattern at the right shows that 2^{-1} can be defined as $\frac{1}{2}$.

$$2^2 = 4 \xrightarrow{\div 2} 2^1 = 2 \xrightarrow{\div 2} 2^0 = 1 \xrightarrow{\div 2} 2^{-1} = \frac{1}{2}$$

You can apply the Quotient of Powers rule and the definition of a power to $\frac{x^3}{x^5}$ and write a general rule about negative powers.

Method 1 Quotient of Powers

$$\frac{x^3}{x^5} = x^{3-5}$$
$$= x^{-2}$$

Method 2 Definition of Power

$$\frac{x^3}{x^5} = \frac{\overset{1}{\cancel{x}} \cdot \overset{1}{\cancel{x}} \cdot \overset{1}{\cancel{x}}}{\underset{1}{\cancel{x}} \cdot \underset{1}{\cancel{x}} \cdot \underset{1}{\cancel{x}} \cdot x \cdot x}$$
$$= \frac{1}{x \cdot x}$$
$$= \frac{1}{x^2}$$

Since $\frac{x^3}{x^5}$ cannot have two different values, you can conclude that $x^{-2} = \frac{1}{x^2}$. This suggests the following definition.

Key Concept — *Negative Exponents*

- **Symbols** $a^{-n} = \frac{1}{a^n}$, for $a \neq 0$ and any integer n
- **Example** $5^{-4} = \frac{1}{5^4}$

Example 1 Use Positive Exponents

Write each expression using a positive exponent.

a. 6^{-2}

$6^{-2} = \frac{1}{6^2}$ Definition of negative exponent

b. x^{-5}

$x^{-5} = \frac{1}{x^5}$ Definition of negative exponent

One way to write a fraction as an equivalent expression with negative exponents is to use prime factorization.

Example 2 Use Negative Exponents

Write $\frac{1}{9}$ as an expression using a negative exponent.

$\frac{1}{9} = \frac{1}{3 \cdot 3}$ Find the prime factorization of 9.

$= \frac{1}{3^2}$ Definition of exponent

$= 3^{-2}$ Definition of negative exponent

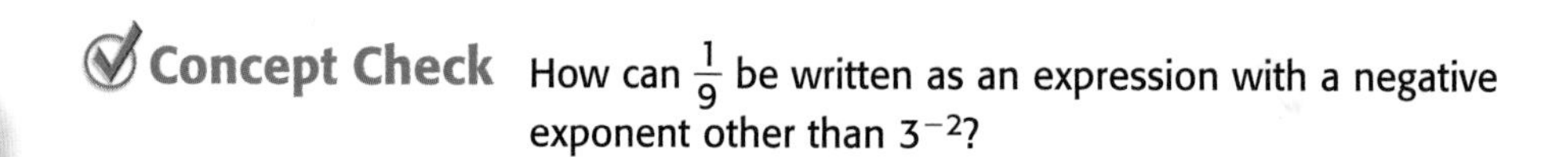

Concept Check How can $\frac{1}{9}$ be written as an expression with a negative exponent other than 3^{-2}?

Negative exponents are often used in science when dealing with very small numbers. Usually the number is a power of ten.

Example 3 Use Exponents to Solve a Problem

WATER A molecule of water contains two hydrogen atoms and one oxygen atom. A hydrogen atom is only 0.00000001 centimeter in diameter. Write the decimal as a fraction and as a power of ten.

The digit 1 is in the 100-millionths place.

$0.00000001 = \frac{1}{100{,}000{,}000}$ Write the decimal as a fraction.

$= \frac{1}{10^8}$ $100{,}000{,}000 = 10^8$

$= 10^{-8}$ Definition of negative exponent

More About. . .

Water

A single drop of water contains about 10^{20} molecules.

Source: www.composite.about.com

EVALUATE EXPRESSIONS Algebraic expressions containing negative exponents can be written using positive exponents and evaluated.

Example 4 Algebraic Expressions with Negative Exponents

Evaluate n^{-3} if $n = 2$.

$n^{-3} = 2^{-3}$ Replace n with 2.

$= \frac{1}{2^3}$ Definition of negative exponent

$= \frac{1}{8}$ Find 2^3.

Check for Understanding

Concept Check

1. **OPEN ENDED** Write a convincing argument that $3^0 = 1$ using the fact that $3^4 = 81$, $3^3 = 27$, $3^2 = 9$, and $3^1 = 3$.
2. **Order** 8^{-8}, 8^3 and 8^0 from greatest to least. Explain your reasoning.

Guided Practice

Write each expression using a positive exponent.

3. 5^{-2}
4. $(-7)^{-1}$
5. t^{-6}
6. n^{-2}

Write each fraction as an expression using a negative exponent other than −1.

7. $\frac{1}{3^4}$
8. $\frac{1}{9^2}$
9. $\frac{1}{49}$
10. $\frac{1}{8}$

ALGEBRA **Evaluate each expression if $a = 2$ and $b = -3$.**

11. a^{-5}
12. $(ab)^{-2}$

Application

13. **MEASUREMENT** A unit of measure called a *micron* equals 0.001 millimeter. Write this number using a negative exponent.

Practice and Apply

Homework Help

For Exercises	See Examples
14–27	1
28–35	2
36–39	3
40–43	4

Extra Practice

See page 732.

Write each expression using a positive exponent.

14. 4^{-1}
15. 5^{-3}
16. $(-6)^{-2}$
17. $(-3)^{-3}$
18. 3^{-5}
19. 10^{-4}
20. p^{-1}
21. a^{-10}
22. d^{-3}
23. q^{-4}
24. $2s^{-5}$
25. $\frac{1}{x^{-2}}$

For Exercises 26 and 27, write each expression using a positive exponent. Then write as a decimal.

26. A snowflake weighs 10^{-6} gram.
27. A small bird uses 5^{-4} Joules of energy to sing a song.

Write each fraction as an expression using a negative exponent other than −1.

28. $\frac{1}{9^4}$
29. $\frac{1}{5^5}$
30. $\frac{1}{8^3}$
31. $\frac{1}{13^2}$
32. $\frac{1}{100}$
33. $\frac{1}{81}$
34. $\frac{1}{27}$
35. $\frac{1}{16}$

Write each decimal using a negative exponent.

36. 0.1
37. 0.01
38. 0.0001
39. 0.00001

ALGEBRA **Evaluate each expression if $w = -2$, $x = 3$, and $y = -1$.**

40. x^{-4}
41. w^{-7}
42. 8^w
43. $(xy)^{-6}$

44. **PHYSICAL SCIENCE** A nanometer is equal to a billionth of a meter. The visible range of light waves ranges from 400 nanometers (violet) to 740 nanometers (red).

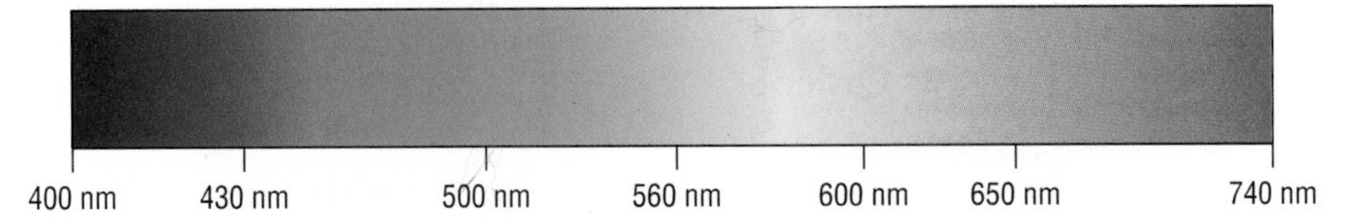

a. Write one billionth of a meter as a fraction and with a negative exponent.

b. Use the information at the left to express the greatest wavelength of an X ray in meters. Write the expression using a negative exponent.

More About. . .

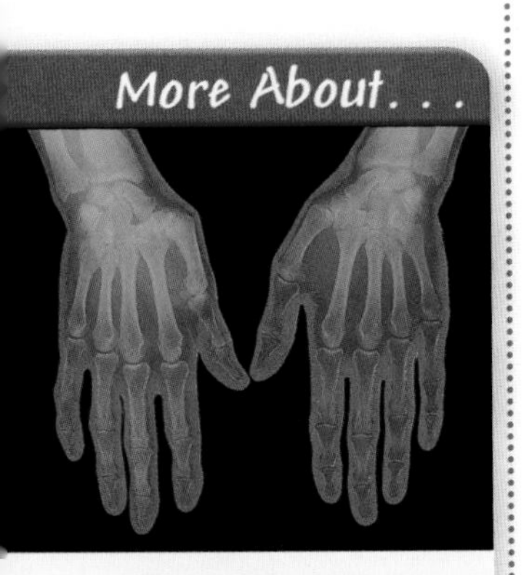

Physical Science

The wavelengths of X rays are between 1 and 10 nanometers.

Source: *Biology*, Raven

45. **ANIMALS** A common flea 2^{-4} inch long can jump about 2^3 inches high. How many times its body size can a flea jump?

46. **MEDICINE** Which type of molecule in the table has a greater mass? How many times greater is it than the other type?

Molecule	Mass (kg)
penicillin	10^{-18}
insulin	10^{-23}

Use the Product of Power and Quotient of Power rules to simplify each expression.

47. $x^{-2} \cdot x^{-3}$

48. $r^{-5} \cdot r^9$

49. $\frac{x^4}{x^7}$

50. $\frac{y^6}{y^{-10}}$

51. $\frac{a^4b^{-4}}{ab^{-2}}$

52. $\frac{36s^3t^5}{12s^6t^{-3}}$

53. **CRITICAL THINKING** Using what you learned about exponents, is $(x^3)^{-2} = (x^{-2})^3$? Why or why not?

54. **WRITING IN MATH** Answer the question that was posed at the beginning of the lesson.

How do negative exponents represent repeated division?

Include the following in your answer:

- an example of a power containing a negative exponent written in fraction form, and
- a discussion about whether the value of a fraction such as $\frac{1}{2^n}$ increases or decreases as the value of n increases.

FCAT Practice

Standardized Test Practice

55. Which is 15^{-5} written as a fraction?

(A) $\frac{1}{5^5}$ (B) $\frac{1}{15}$

(C) $\frac{1}{15^5}$ (D) $-\frac{1}{15^5}$

56. One square millimeter equals __?__ square centimeter(s). (*Hint*: 1 cm = 10 mm)

(A) 10^{-1} (B) 10^{-2}

(C) 10^{-3} (D) 10^3

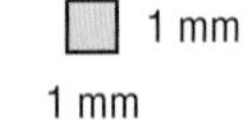

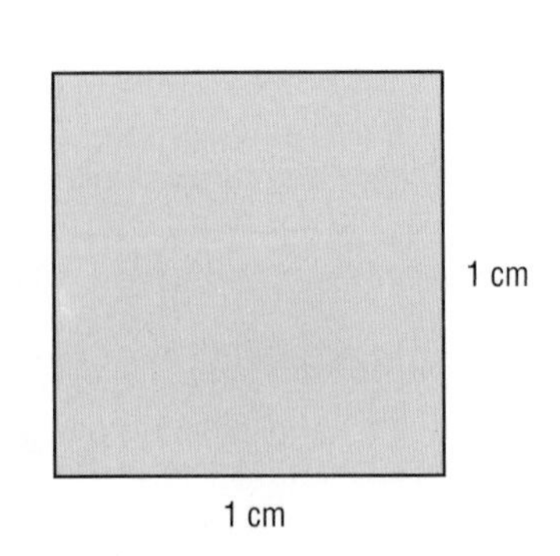

Extending the Lesson

Numbers less than 1 can also be expressed in expanded form.

Example: $0.568 = 0.5 + 0.06 + 0.008$
$= (5 \times 10^{-1}) + (6 \times 10^{-2}) + (8 \times 10^{-3})$

Express each number in expanded form.

57. 0.9 **58.** 0.24 **59.** 0.173 **60.** 0.5875

Maintain Your Skills

Mixed Review

Find each product or quotient. Express your answer using exponents. *(Lesson 4-6)*

61. $3^6 \cdot 3$ **62.** $x^2 \cdot x^4$ **63.** $\frac{5^5}{5^2}$

64. ALGEBRA Write $\frac{16n^3}{8n}$ in simplest form. *(Lesson 4-5)*

ALGEBRA Use the Distributive Property to rewrite each expression. *(Lesson 3-1)*

65. $8(y + 6)$ **66.** $(9 + k)(-2)$ **67.** $(n - 3)5$

68. Write the ordered pair that names point P. *(Lesson 2-6)*

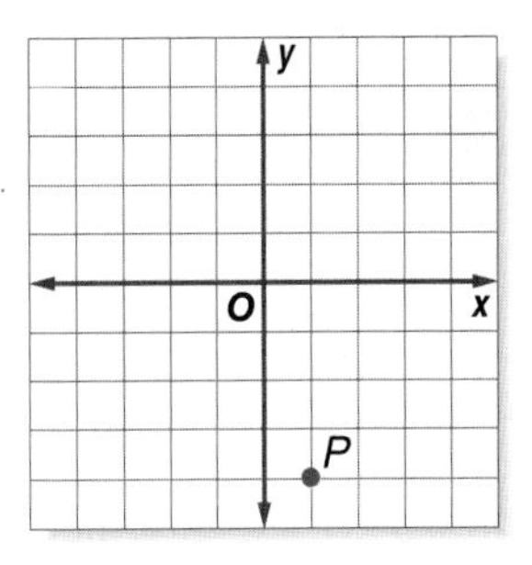

Getting Ready for the Next Lesson

PREREQUISITE SKILL Find each product.
*(To review **multiplying decimals**, see page 715.)*

69. 7.2×100 **70.** 1.6×1000 **71.** 4.05×10

72. 3.8×0.01 **73.** 5.0×0.0001 **74.** 9.24×0.1

Practice Quiz 2 — Lessons 4-4 through 4-7

Find the GCF of each set of numbers or monomials. *(Lesson 4-4)*

1. 15, 20 **2.** 24, 30 **3.** $2ab$, $6a^2$

Write each fraction in simplest form. *(Lesson 4-5)*

4. SCHOOL What fraction of days were you absent from school this nine-week period if you were absent twice out of 44 days?

5. COMMUNICATION What fraction of E-mail messages did you respond to, if you responded to 6 out of a total of 15 messages?

ALGEBRA Find each product or quotient. Express using exponents. *(Lesson 4-6)*

6. $4^2 \cdot 4^4$ **7.** $(n^4)(-2n^3)$ **8.** $\frac{q^9}{q^4}$

9. ALGEBRA Write b^{-6} as an expression using a positive exponent. *(Lesson 4-7)*

10. ALGEBRA Evaluate x^{-5} if $x = -2$. *(Lesson 4-7)*

4-8 Scientific Notation

What You'll Learn

- Express numbers in standard form and in scientific notation.
- Compare and order numbers written in scientific notation.

Sunshine State Standards
MA.A.1.3.1-1, MA.A.1.3.2-2, MA.A.1.3.4-3, MA.A.2.3.1-2

Vocabulary

- scientific notation

Why is scientific notation an important tool in comparing real-world data?

A compact disc or CD has a single spiral track that stores data. It circles from the inside of the disc to the outside. If the track were stretched out in a straight line, it would be 0.5 micron wide and over 5000 meters long.

Track Length	Track Width
5000 meters	0.5 micron

a. Write the track length in millimeters.

b. Write the track width in millimeters. (1 micron = 0.001 millimeter.)

SCIENTIFIC NOTATION When you deal with very large numbers like 5,000,000 or very small numbers like 0.0005, it is difficult to keep track of the place value. Numbers such as these can be written in **scientific notation**.

Key Concept — *Scientific Notation*

- **Words** A number is expressed in scientific notation when it is written as the product of a factor and a power of 10. The factor must be greater than or equal to 1 and less than 10.
- **Symbols** $a \times 10^n$, where $1 \leq a < 10$ and n is an integer
- **Examples** $5{,}000{,}000 = 5.0 \times 10^6$ $\quad 0.0005 = 5.0 \times 10^{-4}$

Concept Check Is 13.0×10^2 written in scientific notation? Why or why not?

Study Tip

Powers of Ten
To multiply by a power of 10,
- move the decimal point to the right if the exponent is positive, and
- move the decimal point to the left if the exponent is negative.

In each case, the exponent tells you how many places to move the decimal point.

You can express numbers that are in scientific notation in standard form.

Example 1 Express Numbers in Standard Form

Express each number in standard form.

a. 3.78×10^6

$3.78 \times 10^6 = 3.78 \times 1{,}000{,}000$ $\quad 10^6 = 1{,}000{,}000$

$= 3{,}780{,}000$ $\quad$ Move the decimal point 6 places to the right.

b. 5.1×10^{-5}

$5.1 \times 10^{-5} = 5.1 \times 0.00001$ $\quad 10^{-5} = 0.00001$

$= 0.000051$ $\quad$ Move the decimal point 5 places to the left.

To write a number in scientific notation, place the decimal point after the first nonzero digit. Then find the power of 10.

Example 2 Express Numbers in Scientific Notation

Express each number in scientific notation.

a. 60,000,000

$60,000,000 = 6.0 \times 10,000,000$ The decimal point moves 7 places.

$= 6.0 \times 10^7$ The exponent is positive.

b. 32,800

$32,800 = 3.28 \times 10,000$ The decimal point moves 4 places.

$= 3.28 \times 10^4$ The exponent is positive.

c. 0.0049

$0.0049 = 4.9 \times 0.001$ The decimal point moves 3 places.

$= 4.9 \times 10^{-3}$ The exponent is negative.

Study Tip

Positive and Negative Exponents

When the number is 1 or greater, the exponent is *positive*. When the number is between 0 and 1, the exponent is *negative*.

People who make comparisons or compute with extremely large or extremely small numbers use scientific notation.

Example 3 Use Scientific Notation to Solve a Problem

SPACE The table shows the planets and their distances from the Sun. Light travels 300,000 kilometers per second. Estimate how long it takes light to travel from the Sun to Pluto. (*Hint:* Recall that distance = rate × time.)

Planet	Distance from the Sun (km)
Mercury	5.80×10^7
Venus	1.03×10^8
Earth	1.55×10^8
Mars	2.28×10^8
Jupiter	7.78×10^8
Saturn	1.43×10^9
Uranus	2.87×10^9
Neptune	4.50×10^9
Pluto	5.90×10^9

Explore You know that the distance from the Sun to Pluto is 5.90×10^9 kilometers and that the speed of light is 300,000 kilometers per second.

Plan To find the time, solve the equation $d = rt$. Since you are estimating, round the distance 5.90×10^9 to 6.0×10^9. Write the rate 300,000 as 3.0×10^5.

Solve

$d = rt$ Write the formula.

$6.0 \times 10^9 = (3.0 \times 10^5)t$ Replace d with 6.0×10^9 and r with 3.0×10^5.

$\frac{6.0 \times 10^9}{3.0 \times 10^5} \approx \frac{(3.0 \times 10^5)t}{3.0 \times 10^5}$ Divide each side by 3.0×10^5.

$2.0 \times 10^4 \approx t$ Divide 6.0 by 3.0 and 10^9 by 10^5.

So, it would take about 2.0×10^4 seconds or about 6 hours for light to travel from the Sun to Pluto.

Examine Use estimation to check the reasonableness of these results.

Study Tip

Calculator

To enter a number in scientific notation on a calculator, enter the decimal portion, press [2nd] [EE], then enter the exponent. A calculator in Sci mode will display answers in scientific notation.

www.pre-alg.com/extra_examples/fcat

COMPARE AND ORDER NUMBERS To compare and order numbers in scientific notation, first compare the exponents. With positive numbers, any number with a greater exponent is greater. If the exponents are the same, compare the factors.

Example 4 *Compare Numbers in Scientific Notation*

SPACE Refer to the table in Example 3. Order Mars, Jupiter, Mercury, and Saturn from least to greatest distance from the Sun.

First, order the numbers according to their exponents. Then, order the numbers with the same exponent by comparing the factors.

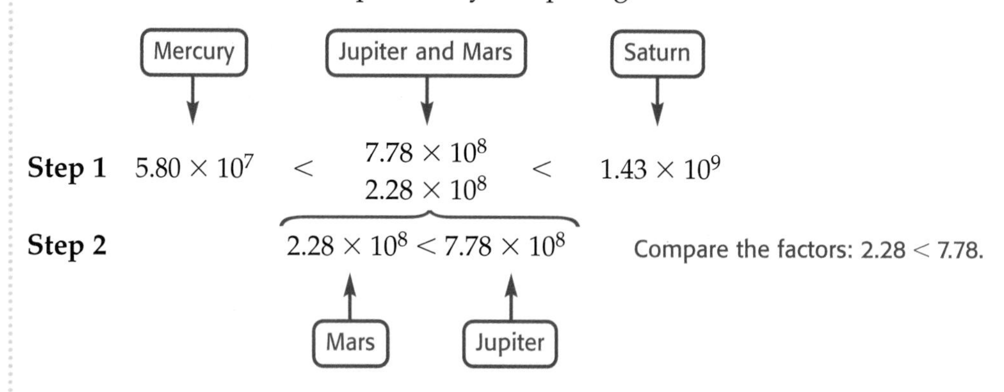

Mercury
Jupiter and Mars
Saturn
Step 1 $5.80 \times 10^7 < \begin{matrix} 7.78 \times 10^8 \\ 2.28 \times 10^8 \end{matrix} < 1.43 \times 10^9$
Step 2 $2.28 \times 10^8 < 7.78 \times 10^8$
Compare the factors: $2.28 < 7.78$.
Mars
Jupiter

So, the order is Mercury, Mars, Jupiter, and Saturn.

Check for Understanding

Concept Check

1. **Explain** the relationship between a number in standard form and the sign of the exponent when the number is written in scientific notation.
2. **OPEN ENDED** Write a number in standard form and then write the number in scientific notation, explaining each step that you used.

Guided Practice

Express each number in standard form.

3. 3.08×10^{-4}
4. 1.4×10^2
5. 8.495×10^5

Express each number in scientific notation.

6. 80,000,000
7. 697,000
8. 0.059
9. the diameter of a spider's thread, 0.001 inch

Applications

10. **SPACE** Refer to the table in Example 3 on page 187. To the nearest second, how long does it take light to travel from the Sun to Earth?
11. **SPACE** Rank the planets in the table at the right by diameter, from least to greatest.

Planet	Diameter (km)
Earth	1.276×10^4
Mars	6.790×10^3
Venus	1.208×10^4

Practice and Apply

Homework Help

For Exercises	See Examples
12–20	1
21–33	2
34	3
39–41	4

Extra Practice

See page 733.

Express each number in standard form.

12. 4.24×10^2
13. 5.72×10^4
14. 3.347×10^{-1}
15. 5.689×10^{-3}
16. 6.1×10^4
17. 9.01×10^{-2}
18. 1.399×10^5
19. 2.505×10^3
20. 1.5×10^{-4}

Express each number in scientific notation.

21. 2,000,000
22. 499,000
23. 0.006
24. 0.0125
25. 50,000,000
26. 39,560
27. 5,894,000
28. 0.000078
29. 0.000425
30. The flow rate of some Antarctic glaciers is 0.00031 mile per hour.
31. Humans blink about 6.25 million times a year.
32. The number of possible ways that a player can play the first four moves in a chess game is 3 billion.
33. A particle of dust floating in the air weighs 0.000000753 gram.

34. **SPACE** Refer to the table in Example 3 on page 187. To the nearest second, how long does it take light to travel from the Sun to Venus?

Choose the greater number in each pair.

35. $2.3 \times 10^5, 1.7 \times 10^5$
36. $1.8 \times 10^3, 1.9 \times 10^{-1}$
37. 5.2×10^2, 5000
38. $0.012, 1.6 \times 10^{-1}$

More About. . .

Oceans

In 2000, the International Hydrographic Organization named a fifth world ocean near Antarctica, called the *Southern Ocean*. It is larger than the Arctic Ocean and smaller than the Indian Ocean.

Source: www.geography.about.com

39. **OCEANS** Rank the oceans in the table at the right by area from least to greatest.

Ocean	Area (sq mi)
Arctic	5.44×10^6
Atlantic	3.18×10^7
Indian	2.89×10^7
Pacific	6.40×10^7

40. **MEASUREMENT** The table at the right shows the values of different prefixes that are used in the metric system. Write the units attometer, gigameter, kilometer, nanometer, petameter, and picometer in order from greatest to least measure.

Metric Measures	
Prefix	**Meaning**
atto	10^{-18}
giga	10^9
kilo	10^3
nano	10^{-9}
peta	10^{15}
pico	10^{-12}

41. Order 6.1×10^4, 6100, 6.1×10^{-5}, 0.0061, and 6.1×10^{-2} from least to greatest.

42. Write $(6 \times 10^0) + (4 \times 10^{-3}) + (3 \times 10^{-5})$ in standard form.

43. Write $(4 \times 10^4) + (8 \times 10^3) + (3 \times 10^2) + (9 \times 10^1) + (6 \times 10^0)$ in standard form.

Convert the numbers in each expression to scientific notation. Then evaluate the expression. Express in scientific notation and in decimal notation.

44. $\frac{20{,}000}{0.01}$
45. $\frac{(420{,}000)(0.015)}{0.025}$
46. $\frac{(0.078)(8.5)}{0.16(250{,}000)}$

More About. . .

Physical Science

The countries with the most volcanoes are:
United States, 157
Russia, 141
Indonesia, 127
Japan, 77

Source: *Kids Discover Volcanoes*

PHYSICAL SCIENCE **For Exercises 47 and 48, use the graph.** The graph shows the maximum amounts of lava in cubic meters per second that erupted from seven volcanoes in the last century.

Eruption Rates

Mount St. Helens, 1980	2.0×10^4
Ngauruhoe, 1975	2.0×10^3
Hekla, 1970	4.0×10^3
Agung, 1963	3.0×10^4
Bezymianny, 1956	2.0×10^5
Hekla, 1947	2.0×10^4
Santa Maria, 1902	4.0×10^4

Source: University of Alaska

47. Rank the volcanoes in order from greatest to least eruption rate.

48. How many times larger was the Santa Maria eruption than the Mount St. Helens eruptions?

Online Research **Data Update** How do the eruption rates of other volcanoes compare with those in the graph? Visit www.pre-alg.com/data_update to learn more.

49. **CRITICAL THINKING** In standard form, $3.14 \times 10^{-4} = 0.000314$, and $3.14 \times 10^4 = 31{,}400$. What is 3.14×10^0 in standard form?

50. WRITING IN MATH Answer the question that was posed at the beginning of the lesson.

Why is scientific notation an important tool in comparing real-world data?

Include the following in your answer:

- some real-world data that is written in scientific notation, and
- the advantages of using scientific notation to compare data.

FCAT Practice
Standardized Test Practice

51. If the bodies of water in the table are ordered from least to greatest area, which would be third in the list?

(A) Lake Huron (B) Lake Victoria
(C) Red Sea (D) Great Salt Lake

Body of Water	Area (km^2)
Lake Huron	5.7×10^4
Lake Victoria	6.9×10^4
Red Sea	4.4×10^5
Great Salt Lake	4.7×10^3

52. Which is 5.80×10^{-4} written in standard form?

(A) 58,000 (B) 5800 (C) 0.58 (D) 0.00058

Maintain Your Skills

Mixed Review **ALGEBRA** **Evaluate each expression if $s = -2$ and $t = 3$.** *(Lesson 4-7)*

53. t^{-4} **54.** s^{-5} **55.** 7^s

ALGEBRA **Find each product or quotient. Express using exponents.** *(Lesson 4-6)*

56. $4^4 \cdot 4^7$ **57.** $3a^2 \cdot 5a^2$ **58.** $c^5 \div c^2$

59. **BUSINESS** Online Book Distributors add a $2.50 shipping and handling charge to the total price of every order. If the cost of books in an order is c, write an expression for the total cost. *(Lesson 1-3)*

Study Guide and Review

Vocabulary and Concept Check

algebraic fraction (p. 170)
base (p. 153)
base two (p. 158)
binary (p. 158)
composite number (p. 159)
divisible (p. 148)
expanded form (p. 154)
exponent (p. 153)
factor (p. 161)
factors (p. 148)
factor tree (p. 160)
greatest common factor (GCF) (p. 164)
monomial (p. 150)
power (p. 153)
prime factorization (p. 160)
prime number (p. 159)
scientific notation (p. 186)
simplest form (p. 169)
standard form (p. 154)
Venn diagram (p. 164)

Determine whether each statement is *true* or *false*. If false, replace the underlined word or number to make a true statement.

1. A prime number is a whole number that has exactly two factors, 1 and itself.
2. Numbers expressed using exponents are called powers.
3. The number 7 is a factor of 49 because it divides 49 with a remainder of zero.
4. A monomial is a number, a variable, or a sum of numbers and/or variables.
5. The number 64 is a composite number.
6. The number 9,536 is written in standard form.
7. To write a fraction in simplest form, divide the numerator and the denominator by the GCF.
8. A fraction is in simplest form when the GCF of the numerator and the denominator is 2.

Lesson-by-Lesson Review

4-1 Factors and Monomials

See pages 148–152.

Concept Summary

- Numbers that are multiplied to form a product are called factors.
- The divisibility rules are useful in factoring numbers.

Determine whether 102 is divisible by 2, 3, 5, 6, or 10.

2: Yes, the ones digit is divisible by 2.
3: Yes, the sum of the digits is 3, and 3 is divisible by 3.
5: No, the ones digit is not 0 or 5.
6: Yes, the number is divisible by 2 and by 3.
10: No, the ones digit is not 0.

Exercises **Use divisibility rules to determine whether each number is divisible by 2, 3, 5, 6, or 10.** *See Example 1 on page 149.*

9. 111 **10.** 405 **11.** 635 **12.** 863
13. 582 **14.** 2124 **15.** 700 **16.** 4200

4-2 Powers and Exponents

See pages 153–157.

Concept Summary

- An exponent is a shorthand way of writing repeated multiplication.
- A number can be written in expanded form by using exponents.
- Follow the order of operations to evaluate algebraic expressions containing exponents.

Example **Evaluate $4(a + 2)^3$ if $a = -5$.**

$4(a + 2)^3 = 4(-5 + 2)^3$ Replace a with -5.

$= 4(-3)^3$ Simplify the expression inside the parentheses.

$= 4(-27)$ Evaluate $(-3)^3$.

$= -108$ Simplify.

Exercises **Evaluate each expression if $x = -3$, $y = 4$, and $z = -2$.**
See Example 3 on page 155.

17. 3^3 **18.** 10^4 **19.** $(-5)^2$ **20.** y^3
21. $10x^2$ **22.** xy^3 **23.** $7y^0z^4$ **24.** $2(3z + 4)^5$

4-3 Prime Factorization

See pages 159–163.

Concept Summary

- A prime number is a whole number that has exactly two factors, 1 and itself.
- A composite number is a whole number that has more than two factors.

Example **Write the prime factorization of 40.**

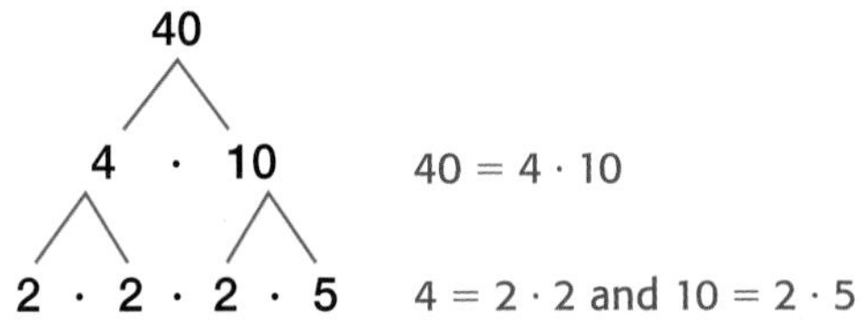

The prime factorization of 40 is $2 \cdot 2 \cdot 2 \cdot 5$ or $2^3 \cdot 5$.

Example **Factor $9s^3t^2$.**

$9s^3t^2 = 3 \cdot 3 \cdot s^3 \cdot t^2$ $9 = 3 \cdot 3$

$= 3 \cdot 3 \cdot s \cdot s \cdot s \cdot t \cdot t$ $s^3 \cdot t^2 = s \cdot s \cdot s \cdot t \cdot t$

Exercises **Write the prime factorization of each number. Use exponents for repeated factors.** *See Example 2 on page 160.*

25. 45 **26.** 55 **27.** 68 **28.** 200

Factor each monomial. *See Example 3 on page 161.*

29. $49k$ **30.** $-15n^2$ **31.** $26p^3$ **32.** $10a^2b$

4-4 Greatest Common Factor (GCF)

See pages 164–168.

Concept Summary

- The greatest number or monomial that is a factor of two or more numbers or monomials is the GCF.
- The Distributive Property can be used to factor algebraic expressions.

Examples

1 Find the GCF of $12a^2$ and $15ab$.

$12a^2 = 2 \cdot 2 \cdot \boxed{3} \cdot \boxed{a} \cdot a$

$15ab = \boxed{3} \cdot 5 \cdot \boxed{a} \cdot b$ The GCF of $12a^2$ and $15ab$ is $3 \cdot a$ or $3a$.

2 Factor $4n + 8$.

Step 1 Find the GCF of $4n$ and 8.

$4n = \boxed{2} \cdot \boxed{2} \cdot n$

$8 = \boxed{2} \cdot \boxed{2} \cdot 2$ The GCF is $2 \cdot 2$ or 4.

Step 2 Write the product of the GCF and its remaining factors.

$4n + 8 = 4(n) + 4(2)$ Rewrite each term using the GCF.

$= 4(n + 2)$ Distributive Property

Exercises Find the GCF of each set of numbers or monomials.
See Examples 2 and 4 on pages 165 and 166.

33. 6, 48 **34.** 16, 24 **35.** $4n, 5n^2$ **36.** $20c^3d, 12cd$

Factor each expression. *See Example 5 on page 166.*

37. $2t + 20$ **38.** $3x + 24$ **39.** $30 + 4n$

4-5 Simplifying Algebraic Fractions

See pages 169–173.

Concept Summary

- Algebraic fractions can be written in simplest form by dividing the numerator and the denominator by the GCF.

Example **Simplify $\frac{8np}{18n^2}$.**

$\frac{8np}{18n^2} = \frac{\overset{1}{\cancel{2}} \cdot 2 \cdot 2 \cdot \overset{1}{\cancel{n}} \cdot p}{\underset{1}{\cancel{2}} \cdot 3 \cdot 3 \cdot \underset{1}{\cancel{n}} \cdot n}$ Divide the numerator and the denominator by the GCF, $2 \cdot n$.

$= \frac{4p}{9n}$ Simplify.

Exercises Write each fraction in simplest form. If the fraction is already in simplest form, write *simplified*. *See Examples 2 and 4 on page 170.*

40. $\frac{6}{21}$ **41.** $\frac{24}{40}$ **42.** $\frac{15}{16}$ **43.** $\frac{30}{51}$

44. $\frac{st}{t^4}$ **45.** $\frac{23x}{32y}$ **46.** $\frac{9mn}{18n^2}$ **47.** $\frac{15ac^2}{24ab}$

• Extra Practice, see pages 730–733.
• Mixed Problem Solving, see page 761.

4-6 Multiplying and Dividing Monomials

See pages 175–179.

Concept Summary

- Powers with the same base can be multiplied by adding their exponents.
- Powers with the same base can be divided by subtracting their exponents.

Examples

1 Find $x^3 \cdot x^2$.

$x^3 \cdot x^2 = x^{3+2}$ The common base is x.

$= x^5$ Add the exponents.

2 Find $\frac{4^5}{4^3}$.

$\frac{4^5}{4^3} = 4^{5-3}$ The common base is 4.

$= 4^2$ Subtract the exponents.

Exercises **Find each product or quotient. Express using exponents.**
See Examples 1–3 on pages 176 and 177.

48. $8^4 \cdot 8^5$ **49.** $c \cdot c^3$ **50.** $\frac{3^7}{3^2}$ **51.** $\frac{r^{11}}{r^9}$ **52.** $7x \cdot 2x^6$

4-7 Negative Exponents

See pages 181–185.

Concept Summary

- For $a \neq 0$ and any integer n, $a^{-n} = \frac{1}{a^n}$.

Write 3^{-4} as an expression using a positive exponent.

$3^{-4} = \frac{1}{3^4}$ Definition of negative exponent

Exercises **Write each expression using a positive exponent.**
See Example 1 on page 182.

53. 7^{-2} **54.** 10^{-1} **55.** b^{-4} **56.** t^{-8} **57.** $(-4)^{-3}$

4-8 Scientific Notation

See pages 186–190.

Concept Summary

- A number in scientific notation contains a factor and a power of 10.

Examples

1 Express 3.5×10^{-2} in standard form.

$3.5 \times 10^{-2} = 3.5 \times 0.01$ $10^{-2} = 0.01$

$= 0.035$ Move the decimal point 2 places to the left.

2 Express 269,000 in scientific notation.

$269{,}000 = 2.69 \times 100{,}000$ The decimal point moves 5 places.

$= 2.69 \times 10^5$ The exponent is positive.

Exercises **Express each number in standard form.** *See Example 1 on page 186.*

58. 6.1×10^2 **59.** 2.9×10^{-3} **60.** 1.85×10^{-2} **61.** 7.045×10^4

Express each number in scientific notation. *See Example 2 on page 187.*

62. 1200 **63.** 0.008 **64.** 0.000319 **65.** 45,710,000

Chapter 4

Practice Test

Vocabulary and Concepts

1. **Explain** how to use the divisibility rules to determine whether a number is divisible by 2, 3, 5, 6, or 10.
2. **Explain** the difference between a prime number and a composite number.
3. **OPEN ENDED** Write an algebraic fraction that is in simplest form.

Skills and Applications

Determine whether each expression is a monomial. Explain why or why not.

4. $6xyz$
5. $-2m + 9$

Write each expression using exponents.

6. $3 \cdot 3 \cdot 3 \cdot 3$
7. $-2 \cdot -2 \cdot -2 \cdot a \cdot a \cdot a \cdot a$

Factor each expression.

8. $12r^2$
9. $50xy^2$
10. $7 + 21p$

Find the GCF of each set of numbers or monomials.

11. 70, 28
12. 36, 90, 180
13. $12a^3b$, $40ab^4$

Write each fraction in simplest form. If the fraction is already in simplest form, write *simplified*.

14. $\frac{57}{95}$
15. $\frac{240}{360}$
16. $\frac{56m^3n}{32mn}$

Find each product or quotient. Express using exponents.

17. $5^3 \cdot 5^6$
18. $(4x^7)(-6x^3)$
19. $w^9 \div w^5$

Write each expression using a positive exponent.

20. 4^{-2}
21. t^{-6}
22. $(yz)^{-3}$

Write each number in standard form.

23. 9.0×10^{-2}
24. 5.206×10^{-3}
25. 3.71×10^4

Write each number in scientific notation.

26. 345,000
27. 1,680,000
28. 0.00072

29. **BAKING** A recipe for butter cookies requires 12 tablespoons of sugar for every 16 tablespoons of flour. Write this as a fraction in simplest form.

FCAT Practice

30. **STANDARDIZED TEST PRACTICE** Earth is approximately 93 million miles away from the Sun. Express this distance in scientific notation.

Ⓐ 9.3×10^7 mi
Ⓑ 9.3×10^6 mi
Ⓒ 93×10^6 mi
Ⓓ 93,000,000 mi

www.pre-alg.com/chapter_test/fcat

Chapter 4 Standardized Test Practice

FCAT Practice

Part 1 Multiple Choice

Record your answers on the answer sheet provided by your teacher or on a sheet of paper.

1. Andy has 7 fewer computer games than Ling. Carlos has twice as many computer games as Andy. If Ling has x computer games, which of these represents the number of computer games that Carlos has? (Lesson 1-3)

Ⓐ $7 - 2x$ Ⓑ $x - 7$
Ⓒ $2x - 7$ Ⓓ $2(x - 7)$

2. Which of the following statements is *false,* when r, s, and t are different integers? (Lesson 1-4)

Ⓐ $(rs)t = r(st)$ Ⓑ $r + s = s + r$
Ⓒ $rs = sr$ Ⓓ $r - s = s - r$

3. The low temperatures during the past five days are given in the table. Find the average (mean) of the temperatures. (Lesson 2-5)

Day	1	2	3	4	5
Temperature (°F)	−2	0	4	5	4

Ⓐ 3°F Ⓑ 2.2°F Ⓒ 13°F Ⓓ 2.75°F

4. Which coordinates are most likely to be the coordinates of point P? (Lesson 2-6)

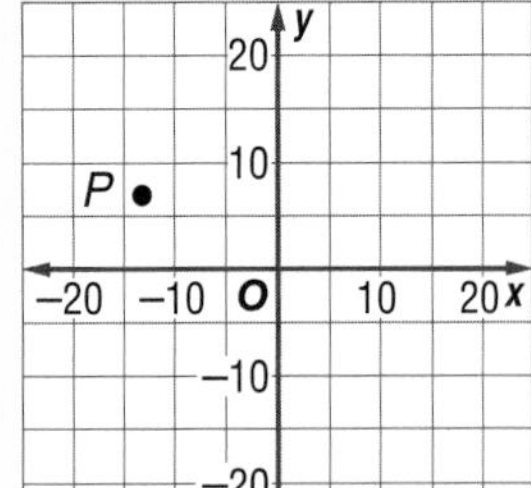
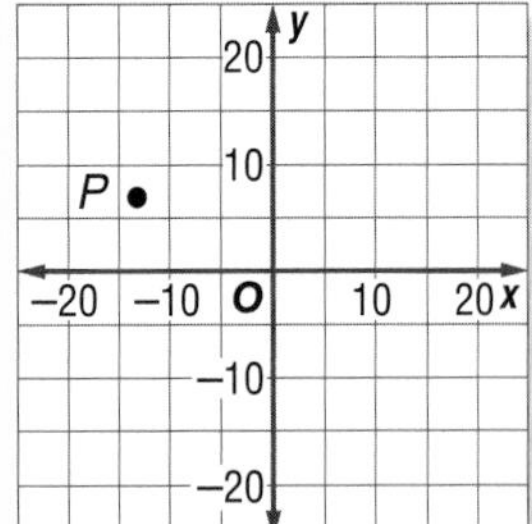

Ⓐ $(-13, 7)$
Ⓑ $(7, -13)$
Ⓒ $(13, 7)$
Ⓓ $(7, 13)$

The Princeton Review Test-Taking Tip

Question 2 When a multiple-choice question asks you to verify statements that include variables, you can substitute numbers into the variables to determine which statements are true and which are false.

5. Solve $3n - 6 = -39$ for n. (Lesson 3-5)

Ⓐ -15 Ⓑ -11
Ⓒ 11 Ⓓ 15

6. For every order purchased from an Internet bookstore, the shipping and handling charges include a base fee of \$5 plus a fee of \$3 per item purchased in the order. Which equation represents the shipping and handling charge for ordering n items? (Lesson 3-6)

Ⓐ $S = 5(3n)$ Ⓑ $S = 3n - 5$
Ⓒ $S = 5 + \frac{3}{n}$ Ⓓ $S = 5 + 3n$

7. The area of the rectangle below is 18 square units. Use the formula $A = \ell w$ to find its width. (Lesson 3-7)

Ⓐ $\frac{1}{3}$ unit
Ⓑ 2 units
Ⓒ 3 units
Ⓓ 12 units

8. Write the prime factorization of 84. (Lesson 4-3)

Ⓐ $2 \cdot 3 \cdot 7$
Ⓑ $4 \cdot 21$
Ⓒ $3 \cdot 4 \cdot 7$
Ⓓ $2 \cdot 2 \cdot 3 \cdot 7$

9. What is the greatest common factor of 28 and 42? (Lesson 4-4)

Ⓐ 2 Ⓑ 7
Ⓒ 14 Ⓓ 28

10. Write 3^{-3} as a fraction. (Lesson 4-7)

Ⓐ $-\frac{1}{9}$ Ⓑ $-\frac{1}{27}$
Ⓒ $\frac{1}{27}$ Ⓓ $\frac{1}{9}$

11. Asia is the largest continent. It has an area of 17,400,000 square miles. What is 17,400,000 expressed in scientific notation? (Lesson 4-8)

Ⓐ 174×10^5 Ⓑ 1.74×10^7
Ⓒ 174×10^7 Ⓓ 1.74×10^8

Aligned and verified by The Princeton Review

FCAT Practice

Part 2 Short Response/Grid In

Record your answers on the answer sheet provided by your teacher or on a sheet of paper.

12. A health club charges an initial fee of \$60 for the first month, and then a \$28 membership fee each month after the first month, as shown in the table. What is the total membership cost for 9 months? (Lesson 1-1)

Number of Months	Total Cost (\$)
1	60
2	88
3	116
4	144
5	172

13. The temperature in Concord at 5 P.M. was -3 degrees. By midnight, the temperature had dropped 9 degrees. What was the temperature at midnight? (Lesson 2-3)

14. A skating rink charges \$2.00 to rent a pair of skates and \$1.50 per hour of skating. Jeff wants to spend no more than \$8.00 and he needs to rent skates. How many hours can he skate? (Lesson 3-5)

15. At a birthday party, Maka gave 30 gel pens to her friends as prizes. Everyone got at least 1 gel pen. Six friends got just 1 gel pen each, 4 friends got 3 gel pens each for winning games, and the rest of the friends got 2 gel pens each. How many friends got 2 gel pens? (Lesson 3-6)

16. A table 8 feet long and 2 feet wide is to be covered for the school bake sale. If organizers want the covering to hang down 1 foot on each side, what is the area of the covering that they need? (Lesson 3-7)

17. What is the least 3-digit number that is divisible by 3 and 5? (Lesson 4-1)

18. Write $10 \cdot 10 \cdot 10 \cdot 10$ using an exponent. (Lesson 4-2)

19. Find the greatest common factor of 18, 44, and 12. (Lesson 4-4)

20. Write $\frac{1}{5 \cdot 5 \cdot 5}$ using a negative exponent. (Lesson 4-7)

21. The Milky Way galaxy is made up of about 200 billion stars, including the Sun. Write this number in scientific notation. (Lesson 4-8)

FCAT Practice

Part 3 Open Ended

Record your answers on a sheet of paper. Show your work.

22. Chandra plans to order CDs from an Internet shopping site. She finds that the CD prices are the same at three different sites, but that the shipping costs vary. The shipping costs include a fee per order, plus an additional fee for each item in the order, as shown in the table below. (Lesson 3-6)

Company	Shipping Cost	
	Per Order	Per Item
CDBargains	\$4.00	\$1.00
WebShopper	\$6.00	\$3.00
EverythingStore	\$2.50	\$1.50

a. For each company, write an equation that represents the shipping cost. In each of your three equations, use S to represent the shipping cost and n to represent the number of items purchased.

b. If Chandra orders 2 CDs, which company will charge the least for shipping? Use the equations you wrote and show your work.

c. If Chandra orders 10 CDs, which company will charge the least for shipping? Use the equations you wrote and show your work.

d. For what number of CDs do both CDBargains and EverythingStore charge the same amount for shipping and handling costs?

Chapter 5

Rational Numbers

What You'll Learn

- **Lessons 5-1 and 5-2** Write fractions as decimals and write decimals as fractions.
- **Lessons 5-3, 5-4, 5-5, and 5-7** Add, subtract, multiply, and divide rational numbers.
- **Lessons 5-6 and 5-9** Use the least common denominator to compare fractions and to solve equations.
- **Lesson 5-8** Use the mean, median, and mode to analyze data.
- **Lesson 5-10** Find the terms of arithmetic and geometric sequences.

Key Vocabulary

- **rational number** (p. 205)
- **algebraic fraction** (p. 211)
- **multiplicative inverse** (p. 215)
- **measures of central tendency** (p. 238)
- **sequence** (p. 249)

Why It's Important

Rational numbers are the numbers used most often in the real world. They include fractions, decimals, and integers. Understanding rational numbers is important in understanding and analyzing real-world occurrences, such as changes in barometric pressure during a storm. *You will compare the barometric pressure before and after a storm in Lesson 5-9.*

Getting Started

Prerequisite Skills To be successful in this chapter, you'll need to master these skills and be able to apply them in problem-solving situations. Review these skills before beginning Chapter 5.

For Lessons 5-1 through 5-4 — Multiply and Divide Integers

Find each product or quotient. If necessary, round to the nearest tenth. *(For review, see Lessons 2-4 and 2-5.)*

1. $3 \div 5$ **2.** $-1 \div 8$ **3.** $2 \cdot 17$

4. $-12 \cdot 3$ **5.** $-2 \div (-9)$ **6.** $-4(-6)$

7. $5(-15)$ **8.** $4 \div (-15)$ **9.** $-24 \div 14$

For Lesson 5-5 — Simplify Fractions Using the GCF

Write each fraction in simplest form. If the fraction is already in simplest form, write *simplified*. *(For review, see Lesson 4-5.)*

10. $\frac{5}{40}$ **11.** $\frac{12}{20}$ **12.** $\frac{14}{39}$ **13.** $\frac{36}{50}$

For Lessons 5-8 through 5-10 — Add and Subtract Integers

Find each sum or difference. *(For review, see Lessons 2-2 and 2-3.)*

14. $4 + (-9)$ **15.** $-10 + 16$ **16.** $20 - 12$ **17.** $19 - 32$

18. $7 + (-5)$ **19.** $26 - 11$ **20.** $(-3) + (-8)$ **21.** $-1 - (-10)$

Make this Foldable to record information about rational numbers. Begin with two sheets of $8\frac{1}{2}$" by 11" paper.

Step 1 Fold and Cut One Sheet

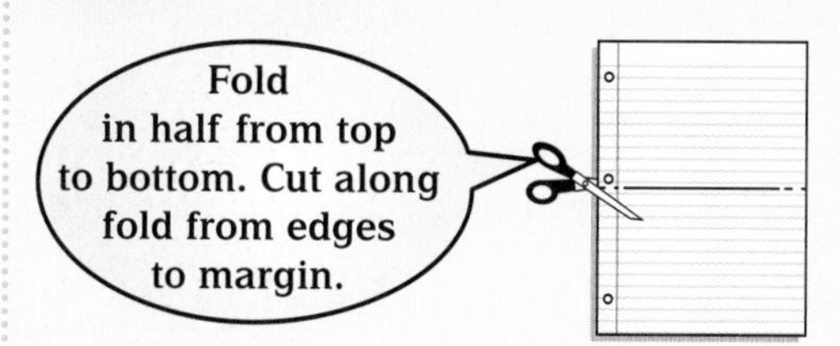

Fold in half from top to bottom. Cut along fold from edges to margin.

Step 2 Fold and Cut the Other Sheet

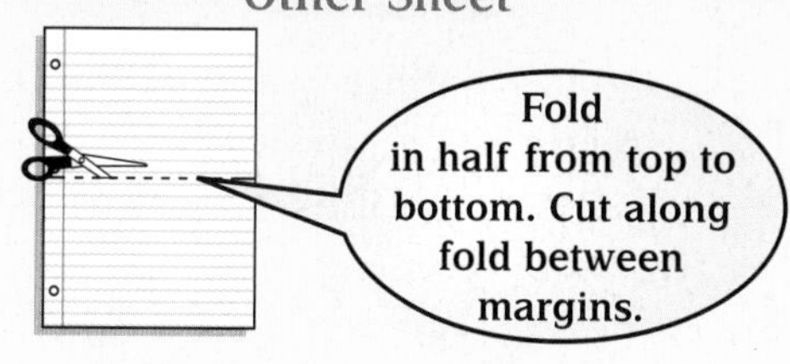

Fold in half from top to bottom. Cut along fold between margins.

Step 3 Fold

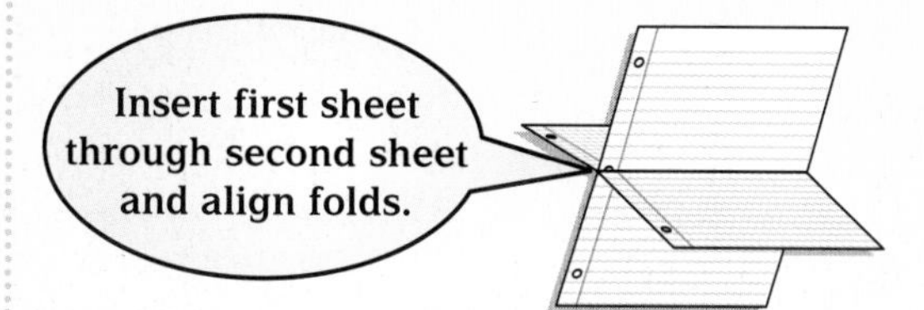

Insert first sheet through second sheet and align folds.

Step 4 Label

Label each page with a lesson number and title.
Chapter 5
Rational Numbers

Reading and Writing As you read and study the chapter, fill the journal with notes, diagrams, and examples for rational numbers.

5-1 Writing Fractions as Decimals

What You'll Learn

- Write fractions as terminating or repeating decimals.
- Compare fractions and decimals.

Sunshine State Standards
MA.A.1.3.2-1, MA.A.1.3.2-2, MA.A.1.3.3-2, MA.A.1.3.4-1

Vocabulary

- terminating decimal
- mixed number
- repeating decimal
- bar notation
- period

How were fractions used to determine the size of the first coins?

In the 18th century, a silver dollar contained $1 worth of silver. The sizes of all other coins were based on this coin.

a. A half dollar contained half the silver of a silver dollar. What was it worth?

b. Write the decimal value of each coin in the table.

c. Order the fractions in the table from least to greatest. (*Hint*: Use the values of the coins.)

Coin	Fraction of Silver of $1 Coin
quarter-dollar (quarter)	$\frac{1}{4}$
10-cent (dime)	$\frac{1}{10}$
half-dime* (nickel)	$\frac{1}{20}$

** In 1866, nickels were enlarged for convenience*

Reading Math

Terminating
Everyday Meaning: bringing to an end
Math Meaning: a decimal whose digits end

WRITE FRACTIONS AS DECIMALS Any fraction $\frac{a}{b}$, where $b \neq 0$, can be written as a decimal by dividing the numerator by the denominator. So, $\frac{a}{b} = a \div b$. If the division ends, or terminates, when the remainder is zero, the decimal is a **terminating decimal**.

Example 1 Write a Fraction as a Terminating Decimal

Write $\frac{3}{8}$ as a decimal.

Method 1 Use paper and pencil.

$$\begin{array}{r} 0.375 \\ 8\overline{)3.000} \\ \underline{-24} \\ 60 \\ \underline{-56} \\ 40 \\ \underline{-40} \\ 0 \end{array}$$

Division ends when the remainder is 0.

0.375 is a terminating decimal.

Method 2 Use a calculator.

3 [÷] 8 [ENTER] .375

$\frac{3}{8} = 0.375$

A **mixed number** such as $3\frac{1}{2}$ is the sum of a whole number and a fraction. Mixed numbers can also be written as decimals.

Example 2 Write a Mixed Number as a Decimal

Write $3\frac{1}{2}$ as a decimal.

$3\frac{1}{2} = 3 + \frac{1}{2}$ Write as the sum of an integer and a fraction.

$= 3 + 0.5$ $\frac{1}{2} = 0.5$

$= 3.5$ Add.

Not all fractions can be written as terminating decimals.

$\frac{2}{3} \rightarrow 3\overline{)2.000}$ = 0.666, with steps $-1\,8$, 20, -18, 20, -18, 2

The number 6 repeats.

The remainder after each step is 2.

CHECK 2 [÷] 3 [ENTER] .6666666667 ✓ The last digit is rounded.

So, $\frac{2}{3} = 0.6666666666\ldots$. This decimal is called a **repeating decimal**. You can use **bar notation** to indicate that the 6 repeats forever.

$0.6666666666\ldots = 0.\overline{6}$ The digit 6 repeats, so place a bar over the 6.

The **period** of a repeating decimal is the digit or digits that repeat. So, the period of $0.\overline{6}$ is 6.

Decimal	Bar Notation	Period
0.13131313...	$0.\overline{13}$	13
6.855555...	$6.8\overline{5}$	5
19.1724724...	$19.1\overline{724}$	724

Concept Check Is $0.\overline{75}$ a terminating or a repeating decimal? Explain.

Example 3 Write Fractions as Repeating Decimals

a. Write $-\frac{6}{11}$ as a decimal.

$-\frac{6}{11} \rightarrow 11\overline{)6.0000\ldots}$ = 0.5454... The digits 54 repeat.

So, $-\frac{6}{11} = -0.\overline{54}$.

b. Write $\frac{2}{15}$ as a decimal.

$\frac{2}{15} \rightarrow 15\overline{)2.0000\ldots}$ = 0.1333... The digit 3 repeats.

So, $\frac{2}{15} = 0.1\overline{3}$.

Study Tip

Mental Math
It will be helpful to memorize the following list of fraction-decimal equivalents.

$\frac{1}{2} = 0.5$ $\frac{3}{4} = 0.75$

$\frac{1}{3} = 0.\overline{3}$ $\frac{2}{5} = 0.4$

$\frac{1}{4} = 0.25$ $\frac{3}{5} = 0.6$

$\frac{1}{5} = 0.2$ $\frac{4}{5} = 0.8$

$\frac{2}{3} = 0.\overline{6}$

COMPARE FRACTIONS AND DECIMALS It may be easier to compare numbers when they are written as decimals.

Example 4 Compare Fractions and Decimals

Replace ● with <, >, or = to make $\frac{3}{5}$ ● 0.75 a true sentence.

$\frac{3}{5}$ ● 0.75 Write the sentence.

0.6 ● 0.75 Write $\frac{3}{5}$ as a decimal.

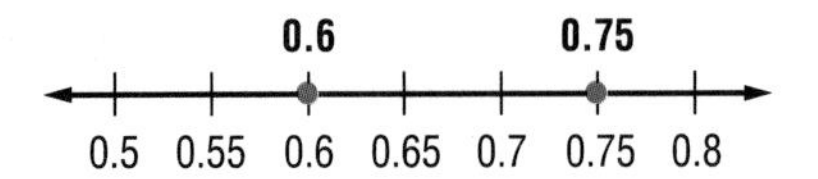

$0.6 < 0.75$ In the tenths place, $6 < 7$.

On a number line, 0.6 is to the left of 0.75, so $\frac{3}{5} < 0.75$.

Example 5 Compare Fractions to Solve a Problem

BREAKFAST In a survey of students, $\frac{13}{20}$ of the boys and $\frac{17}{25}$ of the girls make their own breakfast. Of those surveyed, do a greater fraction of boys or girls make their own breakfast?

Write the fractions as decimals and then compare the decimals.

boys: $\frac{13}{20} = 0.65$

girls: $\frac{17}{25} = 0.68$

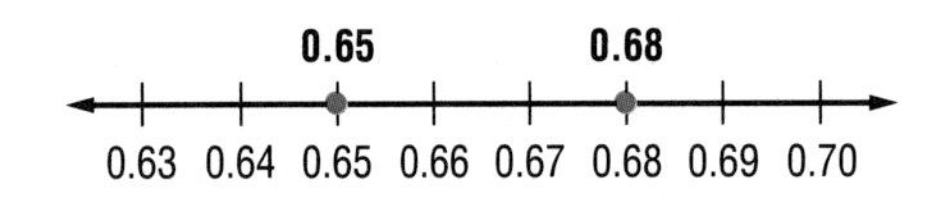

On a number line, 0.65 is to the left of 0.68. Since $0.65 < 0.68$, $\frac{13}{20} < \frac{17}{25}$. So, a greater fraction of girls make their own breakfast.

Check for Understanding

Concept Check

1. **Describe** the steps you should take to order $\frac{5}{8}$, 0.8, and $\frac{3}{5}$.
2. **Explain** how 0.5 and $0.\overline{5}$ are different. Which is greater?
3. **OPEN ENDED** Give an example of a repeating decimal whose period is 14.

Guided Practice

Write each fraction or mixed number as a decimal. Use a bar to show a repeating decimal.

4. $\frac{7}{8}$
5. $2\frac{2}{25}$
6. $-\frac{5}{9}$
7. $\frac{4}{15}$

Replace each ● with <, >, or = to make a true sentence.

8. $\frac{9}{10}$ ● 0.90
9. 0.3 ● $\frac{1}{3}$
10. $\frac{1}{4}$ ● $\frac{1}{3}$
11. $-\frac{3}{4}$ ● $-\frac{7}{8}$

Application

12. **TRUCKS** Of all the passenger trucks sold each year in the United States, $\frac{1}{5}$ are pickups and 0.17 are SUVs. Are more SUVs or pickups sold? Explain. **Source:** U.S. Department of Energy, EPA

Practice and Apply

Homework Help

For Exercises	See Examples
13–20	1, 2
21–28	3
32–43	4
46, 47	5

Extra Practice

See page 733.

Write each fraction or mixed number as a decimal. Use a bar to show a repeating decimal.

13. $\frac{1}{5}$
14. $\frac{3}{20}$
15. $\frac{8}{25}$
16. $-\frac{5}{8}$
17. $7\frac{3}{10}$
18. $1\frac{1}{2}$
19. $5\frac{1}{8}$
20. $-3\frac{3}{4}$
21. $\frac{1}{9}$
22. $-\frac{2}{9}$
23. $-\frac{5}{11}$
24. $\frac{4}{11}$
25. $\frac{1}{6}$
26. $\frac{7}{15}$
27. $\frac{5}{16}$
28. $\frac{7}{12}$

29. **ANIMALS** A marlin can swim $\frac{5}{6}$ mile in one minute. Write $\frac{5}{6}$ as a decimal rounded to the nearest hundredth.

30. **COMPUTERS** In a survey, 17 students out of 20 said they use a home PC as a reference source for school projects. Write 17 out of 20 as a decimal.
Source: NPD Online Research

31. Order $\frac{7}{8}$, 0.8, and $\frac{7}{9}$ from least to greatest.

Replace each ● with <, >, or = to make a true sentence.

32. $0.3 \bullet \frac{1}{4}$
33. $\frac{5}{8} \bullet 0.65$
34. $\frac{2}{5} \bullet 0.4$
35. $\frac{1}{3} \bullet \frac{1}{2}$
36. $\frac{1}{5} \bullet 0.\overline{5}$
37. $1\frac{1}{20} \bullet 1.01$
38. $\frac{7}{8} \bullet \frac{8}{9}$
39. $3\frac{4}{9} \bullet 3.\overline{4}$
40. $6.18 \bullet 6\frac{1}{5}$
41. $-0.75 \bullet -\frac{7}{9}$
42. $0.3\overline{4} \bullet \frac{34}{99}$
43. $-2\frac{1}{12} \bullet -2.09$

44. On a number line, would $\frac{11}{15}$ be graphed to the right or to the left of $\frac{3}{4}$? Explain.

45. Find a terminating and a repeating decimal between $\frac{1}{6}$ and $\frac{8}{9}$. Explain how you found them.

More About. . .

Animals

After being caught, a marlin can strip more than 300 feet of line from a fishing reel in less than five seconds.

Source: *Incredible Comparisons*

SCHOOL **For Exercises 46 and 47, use the graphic at the right and the information below.**

28% = 0.28	21% = 0.21
16% = 0.16	15% = 0.15
13% = 0.13	5% = 0.05

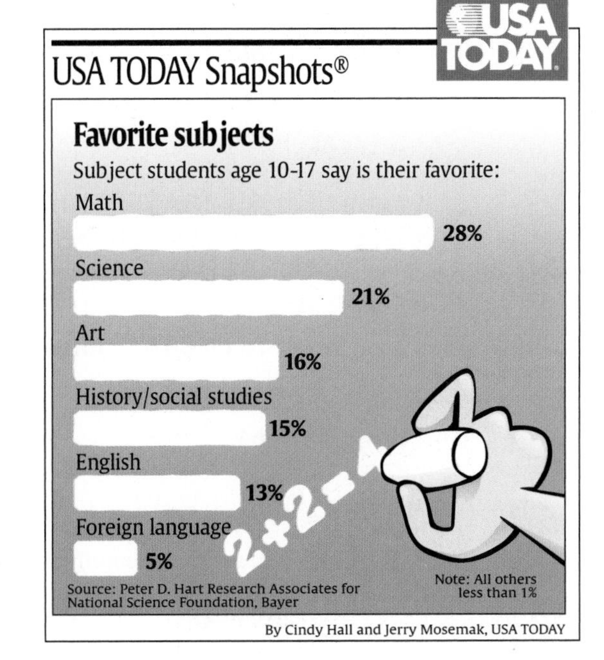

46. Did more or less than one-fourth of the students surveyed choose math as their favorite subject? Explain.

47. Suppose $\frac{1}{7}$ of the students in your class choose English as their favorite subject. How does this compare to the results of the survey? Explain.

48. **CRITICAL THINKING**

a. Write the prime factorization of each denominator in the fractions listed below.

$\frac{1}{2}, \frac{1}{3}, \frac{1}{4}, \frac{1}{5}, \frac{1}{6}, \frac{1}{8}, \frac{1}{9}, \frac{1}{10}, \frac{1}{12}, \frac{1}{15}, \frac{1}{20}$

b. Write the decimal equivalent of each fraction.

c. **Make a conjecture** relating prime factors of denominators and the decimal equivalents of fractions.

49. **WRITING IN MATH** Answer the question that was posed at the beginning of the lesson.

How were fractions used to determine the size of the first coins?

Include the following in your answer:

- a description of the first coins, and
- an explanation of why decimals rather than fractions are used in money exchange today.

50. Which decimal is equivalent to $\frac{1}{100}$?

(A) 0.001 (B) 0.01 (C) 0.1 (D) $0.\overline{1}$

51. Write the shaded portion of the figure at the right as a decimal.

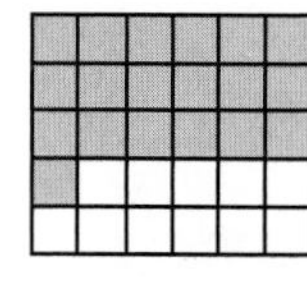

(A) 0.6 (B) $0.\overline{6}$

(C) 0.63 (D) $0.6\overline{3}$

Maintain Your Skills

Mixed Review

Write each number in scientific notation. *(Lesson 4-8)*

52. 854,000,000 53. 0.077 54. 0.00016 55. 925,000

Write each expression using a positive exponent. *(Lesson 4-7)*

56. 10^{-5} 57. $(-2)^{-7}$ 58. x^{-4} 59. y^{-3}

60. **ALGEBRA** Write $(a \cdot a \cdot a)(a \cdot a)$ using an exponent. *(Lesson 4-2)*

61. **TRANSPORTATION** A car can travel an average of 464 miles on one tank of gas. If the tank holds 16 gallons of gasoline, how many miles per gallon does it get? *(Lesson 3-7)*

ALGEBRA Solve each equation. Check your solution. *(Lesson 3-4)*

62. $4n = 32$ 63. $-64 = 2t$

64. $\frac{a}{5} = -9$ 65. $-8 = \frac{x}{-7}$

Getting Ready for the Next Lesson

PREREQUISITE SKILL Simplify each fraction.
*(To review **simplifying fractions**, see Lesson 4-5.)*

66. $\frac{4}{30}$ 67. $\frac{5}{65}$ 68. $\frac{36}{60}$ 69. $\frac{12}{18}$

70. $\frac{21}{24}$ 71. $\frac{16}{28}$ 72. $\frac{32}{48}$ 73. $\frac{125}{1000}$

5-2 Rational Numbers

Sunshine State Standards
MA.A.1.3.3-1, MA.A.1.3.3-2, MA.A.1.3.4-1

Vocabulary

- rational number

What You'll Learn

- Write rational numbers as fractions.
- Identify and classify rational numbers.

How are rational numbers related to other sets of numbers?

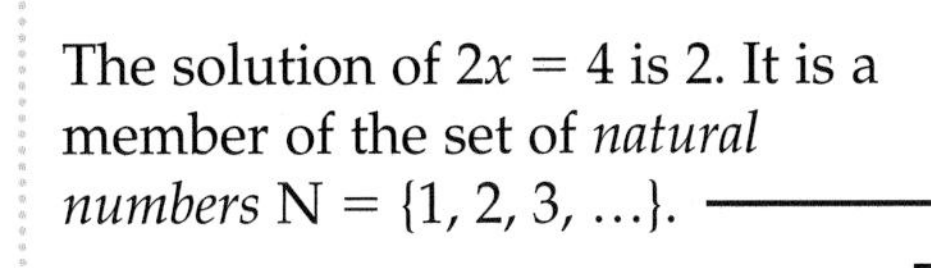

The solution of $2x = 4$ is 2. It is a member of the set of *natural numbers* N = {1, 2, 3, …}.

The solution of $x + 3 = 3$ is 0. It is a member of the set of *whole numbers* W = {0, 1, 2, 3, …}.

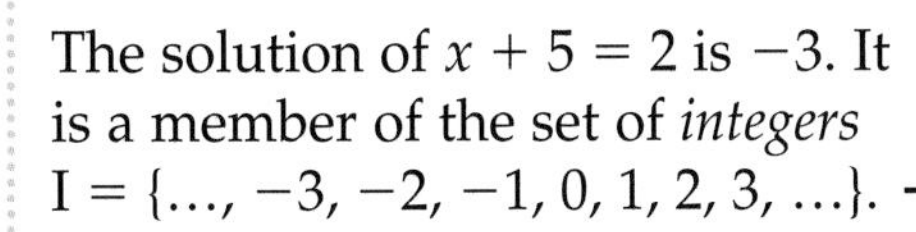

The solution of $x + 5 = 2$ is -3. It is a member of the set of *integers* I = {…, −3, −2, −1, 0, 1, 2, 3, …}.

The solution of $2x = 3$ is $\frac{3}{2}$, which is neither a natural number, a whole number, nor an integer. It is a member of the set of *rational numbers* Q.

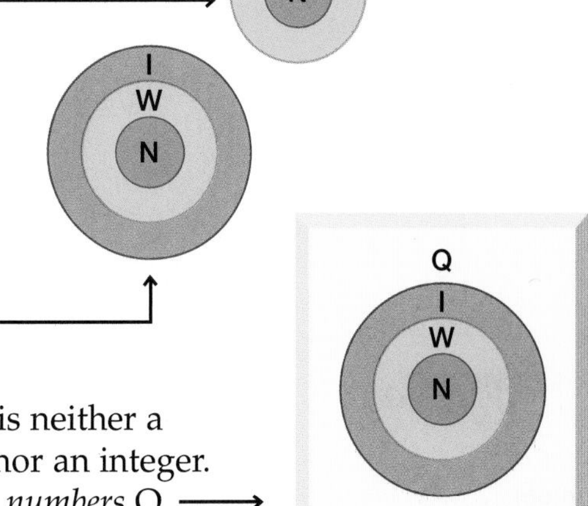

Rational numbers include fractions and decimals as well as natural numbers, whole numbers, and integers.

a. Is 7 a natural number? a whole number? an integer?

b. How do you know that 7 is also a rational number?

c. Is every natural number a rational number? Is every rational number a natural number? Give an explanation or a counterexample to support your answers.

Reading Math

Rational
Root Word: Ratio
A *ratio* is the comparison of two quantities by division. Recall that $\frac{a}{b} = a \div b$, where $b \neq 0$.

WRITE RATIONAL NUMBERS AS FRACTIONS A number that can be written as a fraction is called a **rational number**. Some examples of rational numbers are shown below.

$$0.75 = \frac{3}{4} \qquad -0.\overline{3} = -\frac{1}{3} \qquad 28 = \frac{28}{1} \qquad 1\frac{1}{4} = \frac{5}{4}$$

Example 1 Write Mixed Numbers and Integers as Fractions

a. Write $5\frac{2}{3}$ as a fraction.

$5\frac{2}{3} = \frac{17}{3}$ Write $5\frac{2}{3}$ as an improper fraction.

b. Write −3 as a fraction.

$-3 = \frac{-3}{1}$ or $-\frac{3}{1}$

Concept Check Explain why any integer n is a rational number.

Terminating decimals are rational numbers because they can be written as a fraction with a denominator of 10, 100, 1000, and so on.

Example 2 Write Terminating Decimals as Fractions

Write each decimal as a fraction or mixed number in simplest form.

a. 0.48

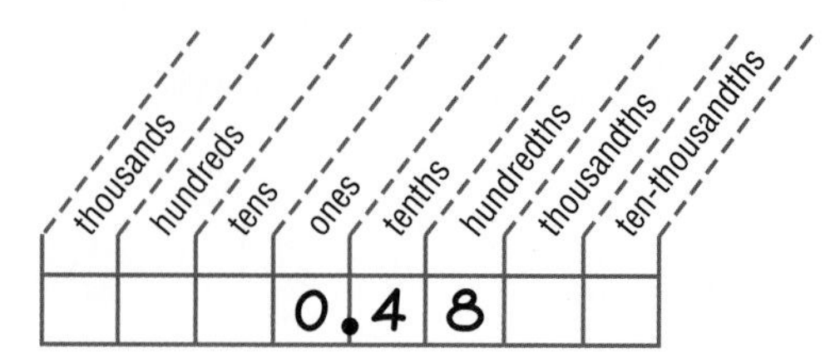

$0.48 = \frac{48}{100}$ 0.48 is 48 hundredths.

$= \frac{12}{25}$ Simplify. The GCF of 48 and 100 is 4.

b. 6.375

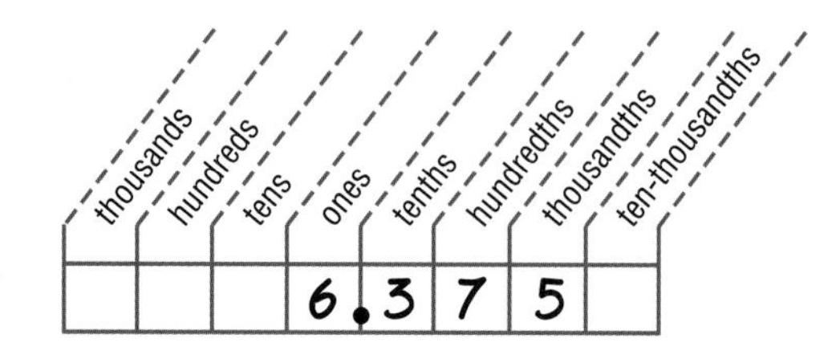

$6.375 = 6\frac{375}{1000}$ 6.375 is 6 and 375 thousandths.

$= 6\frac{3}{8}$ Simplify. The GCF of 375 and 1000 is 125.

Reading Math

Decimals

Use the word *and* to represent the decimal point.

- Read 0.375 as three hundred seventy-five thousandths.
- Read 300.075 as three hundred *and* seventy-five thousandths.

Any repeating decimal can be written as a fraction, so repeating decimals are also rational numbers.

Study Tip

Repeating Decimals

When *two* digits repeat, multiply each side by 100. Then subtract *N* from 100*N* to eliminate the repeating part.

Example 3 Write Repeating Decimals as Fractions

Write $0.\overline{8}$ as a fraction in simplest form.

$N = 0.888\ldots$ Let N represent the number.

$10N = 10(0.888\ldots)$ Multiply each side by 10 because one digit repeats.

$10N = 8.888\ldots$

Subtract N from $10N$ to eliminate the repeating part, 0.888... .

$10N = 8.888\ldots$

$-(N = 0.888\ldots)$

$9N = 8$ $10N - N = 10N - 1N$ or $9N$

$\frac{9N}{9} = \frac{8}{9}$ Divide each side by 9.

$N = \frac{8}{9}$ Simplify.

Therefore, $0.\overline{8} = \frac{8}{9}$.

CHECK 8 [÷] 9 [ENTER] .8888888889 ✓

IDENTIFY AND CLASSIFY RATIONAL NUMBERS All rational numbers can be written as terminating or repeating decimals. Decimals that are neither terminating nor repeating, such as the numbers below, are called *irrational* because they cannot be written as fractions. *You will learn more about irrational numbers in Chapter 9.*

$\pi = 3.141592654\ldots$ → The digits do not repeat.

$4.232232223\ldots$ → The same block of digits do not repeat.

The following model can help you classify rational numbers.

Concept Summary — *Rational Numbers*

- **Words** A rational number is any number that can be expressed as the quotient $\frac{a}{b}$ of two integers, a and b, where $b \neq 0$.
- **Model**

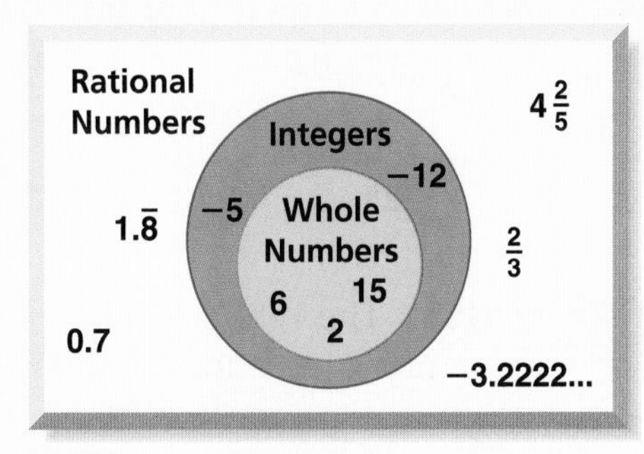

Example 4 *Classify Numbers*

Identify all sets to which each number belongs.

a. -6

-6 is an integer and a rational number.

b. $2\frac{4}{5}$

Because $2\frac{4}{5} = \frac{14}{5}$, it is a rational number. It is neither a whole number nor an integer.

c. **0.914114111...**

This is a nonterminating, nonrepeating decimal. So, it is not a rational number.

Check for Understanding

Concept Check

1. **Define** *rational number* in your own words.
2. **OPEN ENDED** Give an example of a number that is not rational. Explain why it is not rational.

Guided Practice

Write each number as a fraction.

3. $-2\frac{1}{3}$
4. 10

Write each decimal as a fraction or mixed number in simplest form.

5. 0.8
6. 6.35
7. $-0.\overline{7}$
8. $0.\overline{45}$

Identify all sets to which each number belongs.

9. -5
10. 6.05

Application

11. **MEASUREMENT** A *micron* is a unit of measure that is approximately 0.000039 inch. Express this as a fraction.

Practice and Apply

Homework Help

For Exercises	See Examples
12–15	1
16–21, 28–33	2
22–27	3
34–41	4

Extra Practice

See page 733.

Write each number as a fraction.

12. $5\frac{2}{3}$ **13.** $-1\frac{4}{7}$ **14.** -21 **15.** 60

Write each decimal as a fraction or mixed number in simplest form.

16. 0.4 **17.** 0.09 **18.** 5.22 **19.** 1.68

20. 0.625 **21.** 8.004 **22.** $0.\overline{2}$ **23.** $-0.333...$

24. $4.\overline{5}$ **25.** $5.\overline{6}$ **26.** $0.\overline{32}$ **27.** $2.\overline{25}$

28. **WHITE HOUSE** The White House covers an area of 0.028 square mile. What fraction of a square mile is this?

29. **RECYCLING** In 1999, 0.06 of all recycled newspapers were used to make tissues. What fraction is this?

More About. . .

Recycling

The portions of recycled newspapers used for other purposes are shown below.

Newsprint: 0.34
Exported for recycling: 0.22
Paperboard: 0.17
Other products: 0.18

Source: American Forest and Paper Association, Newspaper Association of America

GEOGRAPHY **Africa makes up $\frac{1}{5}$ of all the land on Earth. Use the table to find the fraction of Earth's land that is made up by other continents. Write each fraction in simplest form.**

30. Antarctica **31.** Asia

32. Europe **33.** North America

Continent	Decimal Portion of Earth's Land
Antarctica	0.095
Asia	0.295
Europe	0.07
North America	0.16

Source: *Incredible Comparisons*

Identify all sets to which each number belongs.

34. 4 **35.** -7 **36.** $-2\frac{5}{8}$ **37.** $\frac{6}{3}$

38. 15.8 **39.** 9.0202020... **40.** 1.2345... **41.** 30.151151115...

42. Write 125 thousandths as a fraction in simplest form.

43. Express *two hundred and nineteen hundredths* as a fraction or mixed number in simplest form.

Determine whether each statement is *sometimes, always,* or *never* true. Explain by giving an example or a counterexample.

44. An integer is a rational number.

45. A rational number is an integer.

46. A whole number is not a rational number.

47. **MANUFACTURING** A garbage bag has a thickness of 0.8 mil. This is 0.0008 inch. What fraction of an inch is this?

48. **GEOMETRY** Pi (π) to six decimal places has a value of 3.141592. Pi is often estimated as $\frac{22}{7}$. Is the estimate for π greater than or less than the actual value of π? Explain.

49. **MACHINERY** Will a steel peg 2.37 inches in diameter fit in a $2\frac{3}{8}$-inch diameter hole? How do you know?

50. **CRITICAL THINKING** Show that $0.999... = 1$.

51. **WRITING IN MATH** Answer the question that was posed at the beginning of the lesson.

How are rational numbers related to other sets of numbers?

Include the following in your answer:

- examples of numbers that belong to more than one set, and
- examples of numbers that are only rational.

FCAT Practice
Standardized Test Practice

52. There are infinitely many __?__ between S and T on the number line.

(A) rational numbers (B) integers

(C) whole numbers (D) natural numbers

53. Express 0.56 as a fraction in simplest form.

(A) $\frac{56}{100}$ (B) $\frac{28}{50}$ (C) $\frac{14}{25}$ (D) $\frac{7}{12}$

Maintain Your Skills

Mixed Review

Write each fraction or mixed number as a decimal. Use a bar to show a repeating decimal. *(Lesson 5-1)*

54. $\frac{2}{5}$ 55. $-7\frac{4}{5}$ 56. $-\frac{13}{20}$ 57. $2\frac{5}{9}$

Write each number in standard form. *(Lesson 4-8)*

58. 2×10^3 59. 3.05×10^6 60. 7.4×10^{-4} 61. 1.681×10^{-2}

62. **ALGEBRA** Write $\frac{12n^2}{3an}$ in simplest form. *(Lesson 4-5)*

63. Use place value and exponents to express 483 in expanded form. *(Lesson 4-2)*

Find the perimeter and area of each rectangle. *(Lesson 3-7)*

64.

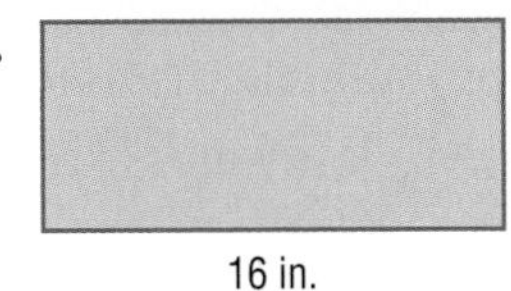

65.

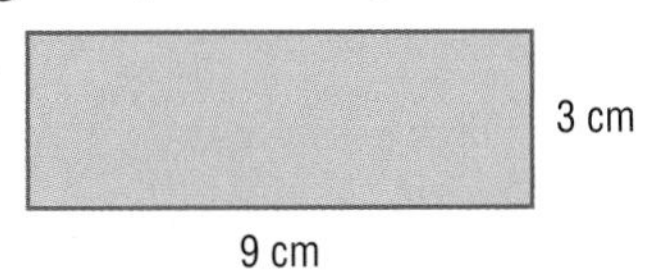

Use the Distributive Property to rewrite each expression. *(Lesson 3-1)*

66. $3(5 + 9)$ 67. $(8 + 1)2$ 68. $6(b - 5)$ 69. $(x + 4)7$

Getting Ready for the Next Lesson

PREREQUISITE SKILL **Estimate each product.**

*(To review **estimating products**, see page 717.)*

Example: $-3\frac{1}{4} \cdot 5\frac{7}{8} \approx -3 \cdot 6$ or -18

70. $1\frac{2}{3} \cdot 4\frac{1}{8}$ 71. $-5\frac{2}{5} \cdot 3\frac{4}{5}$ 72. $2\frac{1}{4} \cdot 2\frac{1}{9}$

73. $6\frac{7}{8} \cdot 1\frac{9}{10}$ 74. $9\frac{1}{8} \cdot \left(-4\frac{3}{4}\right)$ 75. $15\frac{5}{7} \cdot 2\frac{1}{3}$

5-3 Multiplying Rational Numbers

Sunshine State Standards
MA.A.1.3.3-2, MA.A.3.3.1-1, MA.B.1.3.2-1, MA.B.2.3.2-1, MA.D.2.3.1-1

Vocabulary

- algebraic fraction
- dimensional analysis

What You'll Learn

- Multiply fractions.
- Use dimensional analysis to solve problems.

How is multiplying fractions related to areas of rectangles?

To find $\frac{2}{3} \cdot \frac{3}{4}$, think of using an area model to find $\frac{2}{3}$ of $\frac{3}{4}$.

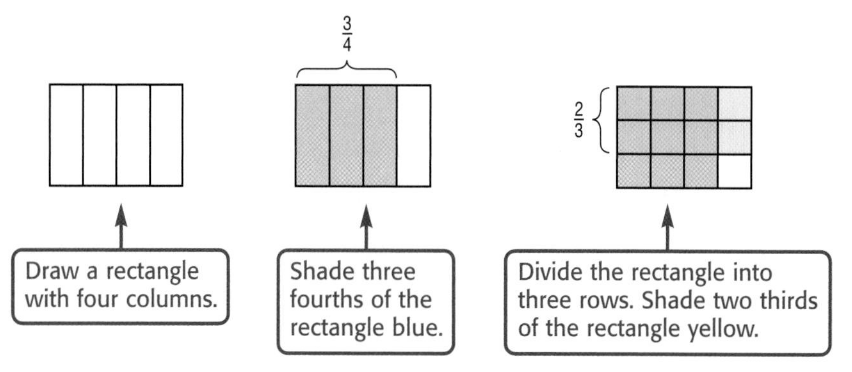

3/4
2/3
Draw a rectangle with four columns.
Shade three fourths of the rectangle blue.
Divide the rectangle into three rows. Shade two thirds of the rectangle yellow.

a. The overlapping green area represents the product of $\frac{2}{3}$ and $\frac{3}{4}$. What is the product?

Use an area model to find each product.

b. $\frac{1}{2} \cdot \frac{1}{3}$ **c.** $\frac{3}{5} \cdot \frac{1}{4}$ **d.** $\frac{3}{4} \cdot \frac{1}{3}$

e. What is the relationship between the numerators and denominators of the factors and the numerator and denominator of the product?

MULTIPLY FRACTIONS These and other similar models suggest the following rule for multiplying fractions.

Key Concept — *Multiplying Fractions*

- **Words** To multiply fractions, multiply the numerators and multiply the denominators.
- **Symbols** $\frac{a}{b} \cdot \frac{c}{d} = \frac{a \cdot c}{b \cdot d}$, where $b, d \neq 0$
- **Example** $\frac{1}{3} \cdot \frac{2}{5} = \frac{1 \cdot 2}{3 \cdot 5}$ or $\frac{2}{15}$

Example 1 *Multiply Fractions*

Find $\frac{2}{3} \cdot \frac{3}{4}$. Write the product in simplest form.

$\frac{2}{3} \cdot \frac{3}{4} = \frac{2 \cdot 3}{3 \cdot 4}$ ← Multiply the numerators. ← Multiply the denominators.

$= \frac{6}{12}$ or $\frac{1}{2}$ Simplify. The GCF of 6 and 12 is 6.

Study Tip

Look Back
To review **GCF**, see Lesson 4-4.

If the fractions have common factors in the numerators and denominators, you can simplify before you multiply.

Example 2 Simplify Before Multiplying

Find $\frac{4}{7} \cdot \frac{1}{6}$. Write the product in simplest form.

$\frac{4}{7} \cdot \frac{1}{6} = \frac{\overset{2}{\cancel{4}}}{7} \cdot \frac{1}{\underset{3}{\cancel{6}}}$ Divide 4 and 6 by their GCF, 2.

$= \frac{2 \cdot 1}{7 \cdot 3}$ Multiply the numerators and multiply the denominators.

$= \frac{2}{21}$ Simplify.

Example 3 Multiply Negative Fractions

Find $-\frac{5}{12} \cdot \frac{3}{8}$. Write the product in simplest form.

$-\frac{5}{12} \cdot \frac{3}{8} = \frac{-5}{\underset{4}{\cancel{12}}} \cdot \frac{\overset{1}{\cancel{3}}}{8}$ Divide 3 and 12 by their GCF, 3.

$= \frac{-5 \cdot 1}{4 \cdot 8}$ Multiply the numerators and multiply the denominators.

$= -\frac{5}{32}$ Simplify.

Study Tip

Negative Fractions

$-\frac{5}{12}$ can be written as $\frac{-5}{12}$ or as $\frac{5}{-12}$.

Example 4 Multiply Mixed Numbers

Find $1\frac{2}{5} \cdot 2\frac{1}{2}$. Write the product in simplest form.

$1\frac{2}{5} \cdot 2\frac{1}{2} = \frac{7}{5} \cdot \frac{5}{2}$ Rename $1\frac{2}{5}$ as $\frac{7}{5}$ and rename $2\frac{1}{2}$ as $\frac{5}{2}$.

$= \frac{7}{\underset{1}{\cancel{5}}} \cdot \frac{\overset{1}{\cancel{5}}}{2}$ Divide by the GCF, 5.

$= \frac{7 \cdot 1}{1 \cdot 2}$ Multiply.

$= \frac{7}{2}$ or $3\frac{1}{2}$ Simplify.

Study Tip

Estimation

You can estimate the product of mixed numbers.

- $1\frac{2}{5}$ is close to 1.
- $2\frac{1}{2}$ is close to 3.

So, $1\frac{2}{5} \cdot 2\frac{1}{2} \approx 1 \cdot 3$ or 3.

ALGEBRA CONNECTION

A fraction that contains one or more variables in the numerator or the denominator is called an **algebraic fraction**.

Example 5 Multiply Algebraic Fractions

Find $\frac{2a}{b} \cdot \frac{b^2}{d}$. Write the product in simplest form.

$\frac{2a}{b} \cdot \frac{b^2}{d} = \frac{2a}{\underset{1}{\cancel{b}}} \cdot \frac{\overset{1}{\cancel{b}} \cdot b}{d}$ The GCF of b and b^2 is b.

$= \frac{2ab}{d}$ Simplify.

DIMENSIONAL ANALYSIS **Dimensional analysis** is the process of including units of measurement when you compute. You can use dimensional analysis to check whether your answers are reasonable.

Example 6 Use Dimensional Analysis

SPACE TRAVEL The landing speed of the space shuttle is about 216 miles per hour. How far does the shuttle travel in $\frac{1}{3}$ hour during landing?

Words Distance equals the rate multiplied by the time.

Variables Let d = distance, r = rate, and t = time.

Formula $d = rt$ Write the formula.

$d = 216 \text{ miles per hour} \cdot \frac{1}{3} \text{ hour}$ Include the units.

$= \frac{\overset{72}{\cancel{216}} \text{ miles}}{1\ \cancel{\text{hour}}} \cdot \frac{1}{\underset{1}{\cancel{3}}} \cancel{\text{hour}}$ Divide by the common factors and units.

$= 72 \text{ miles}$ Simplify.

The space shuttle travels 72 miles in $\frac{1}{3}$ hour during landing.

CHECK The problem asks for the distance. When you divide the common units, the answer is expressed in miles. So, the answer is reasonable.

Concept Check What is the final unit when you multiply feet per second by seconds?

Check for Understanding

Concept Check

1. **OPEN ENDED** Choose two rational numbers whose product is a number between 0 and 1.
2. **FIND THE ERROR** Terrence and Marie are finding $\frac{5}{24} \cdot \frac{18}{25}$.

Terrence	Marie
$\frac{\overset{1}{\cancel{5}}}{\underset{4}{\cancel{24}}} \cdot \frac{\overset{3}{\cancel{18}}}{\underset{5}{\cancel{25}}} = \frac{3}{20}$	$\frac{\overset{1}{\cancel{5}}}{\underset{4}{\cancel{24}}} \cdot \frac{\overset{9}{\cancel{18}}}{\underset{5}{\cancel{25}}} = \frac{9}{20}$

Who is correct? Explain your reasoning.

Guided Practice **Find each product. Write in simplest form.**

3. $\frac{1}{4} \cdot \frac{3}{5}$
4. $\frac{1}{2}\left(-\frac{5}{6}\right)$
5. $-\frac{2}{3}\left(-\frac{5}{6}\right)$
6. $7\left(\frac{8}{21}\right)$
7. $3\frac{1}{4} \cdot \frac{2}{11}$
8. $-5\frac{1}{3} \cdot 3\frac{3}{8}$

ALGEBRA Find each product. Write in simplest form.

9. $\frac{2}{x} \cdot \frac{3x}{7}$
10. $\frac{a}{b} \cdot \frac{5b}{c}$
11. $\frac{4t}{9r} \cdot \frac{18r}{t^2}$

Application

12. **TRAVEL** A car travels 65 miles per hour for $3\frac{1}{2}$ hours. What is the distance traveled? Use the formula $d = rt$ and show how you can divide by the common units.

Practice and Apply

Homework Help

For Exercises	See Examples
13–24	1–3
25–33	4
37–42	5
47–50	6

Extra Practice

See page 734.

Find each product. Write in simplest form.

13. $\frac{6}{7} \cdot \frac{2}{7}$
14. $\frac{4}{9} \cdot \frac{2}{3}$
15. $\frac{1}{5}\left(-\frac{1}{8}\right)$
16. $-\frac{3}{4} \cdot \frac{3}{5}$
17. $\frac{5}{9} \cdot \frac{8}{25}$
18. $-\frac{1}{2}\left(-\frac{2}{7}\right)$
19. $\frac{2}{5} \cdot \frac{5}{6}$
20. $\frac{8}{9} \cdot \frac{27}{28}$
21. $\frac{3}{4}\left(-\frac{1}{3}\right)$
22. $-\frac{7}{8} \cdot \frac{2}{5}$
23. $\frac{3}{5} \cdot \frac{15}{24}$
24. $\frac{3}{32} \cdot \frac{24}{39}$
25. $2 \cdot \frac{7}{12}$
26. $\frac{6}{15}(-3)$
27. $6\frac{2}{3} \cdot \frac{1}{2}$
28. $\frac{5}{12} \cdot 3\frac{1}{9}$
29. $2\frac{2}{6} \cdot 6\frac{2}{7}$
30. $3\frac{1}{3} \cdot 2\frac{5}{8}$
31. $-6\frac{2}{3}\left(-1\frac{1}{2}\right)$
32. $1\frac{3}{7}\left(-9\frac{4}{5}\right)$
33. $-1\frac{1}{4} \cdot 3\frac{5}{9}$

MEASUREMENT **Complete.**

34. ___?___ feet = $\frac{5}{6}$ mile
(*Hint*: 1 mile = 5280 feet)

35. ___?___ ounces = $\frac{3}{8}$ pound
(*Hint*: 1 pound = 16 ounces)

36. $\frac{2}{3}$ hour = ___?___ minutes

37. $\frac{3}{4}$ yard = ___?___ inches

ALGEBRA **Find each product. Write in simplest form.**

38. $\frac{4a}{5} \cdot \frac{3}{a}$
39. $\frac{3x}{y} \cdot \frac{9y}{x}$
40. $\frac{12}{jk} \cdot \frac{3k}{4}$
41. $\frac{8}{c} \cdot \frac{c^2}{11}$
42. $\frac{n}{18} \cdot \frac{6}{n^4}$
43. $\frac{x}{2z} \cdot \frac{2z^3}{3}$

44. **ALGEBRA** Evaluate x^2 if $x = -\frac{1}{2}$.

45. **ALGEBRA** Evaluate $(xy)^2$ if $x = \frac{3}{4}$ and $y = -\frac{4}{5}$.

SCHOOL **For Exercises 46 and 47, use the graphic at the right.**

46. Five-eighths of an eighth grade class are boys. Predict approximately what fraction of the eighth graders are boys who talk about school at home. $\left(\textit{Hint}: 40\% = \frac{2}{5}\right)$

47. In a 12th grade class, five-ninths of the students are girls. Predict about what fraction of twelfth graders are girls who talk about school at home. $\left(\textit{Hint}: 33\% = \frac{33}{100}\right)$

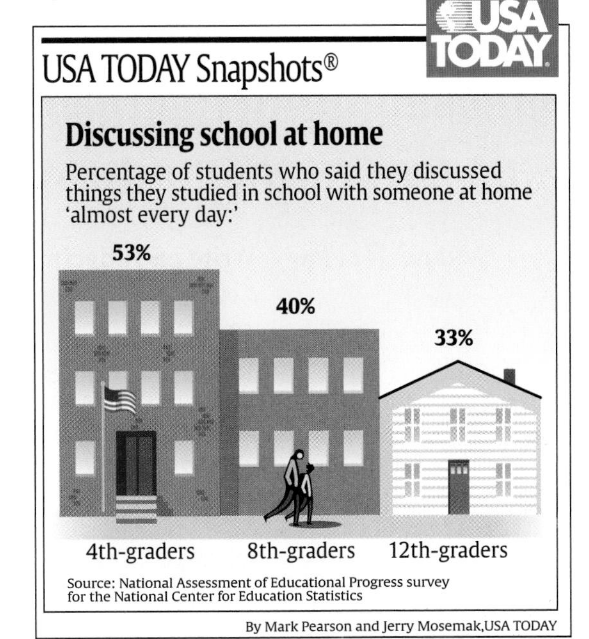

48. **GARDENING** Jamal's lawn is $\frac{2}{3}$ of an acre. If $7\frac{1}{2}$ bags of fertilizer are needed for 1 acre, how much will he need to fertilize his lawn?

More About. . .

Converting Measures

In 1998, a Mars probe was lost because scientists did not convert a customary measure of force to a metric measure.

Source: *Newsday*

CONVERTING MEASURES **Use dimensional analysis and the fractions in the table to find each missing measure.**

Conversion Factors	
Customary → Metric	Metric → Customary
$\frac{2.54 \text{ cm}}{1 \text{ in.}}$	$\frac{0.39 \text{ in.}}{1 \text{ cm}}$
$\frac{1.61 \text{ km}}{1 \text{ mi}}$	$\frac{0.62 \text{ mi}}{1 \text{ km}}$
$\frac{0.09 \text{ m}^2}{1 \text{ ft}^2}$	$\frac{10.76 \text{ ft}^2}{1 \text{ m}^2}$

49. 5 inches = __?__ centimeters

50. 10 kilometers = __?__ miles

51. 26.3 centimeters = __?__ inches

52. $8\frac{2}{3}$ square feet = __?__ square meters

53. CRITICAL THINKING Use the digits 3, 4, 5, 6, 8, and 9 to make true sentences.

a. $\frac{\square}{\square} \times \frac{\square}{\square} = \frac{6}{5}$ **b.** $\frac{\square}{\square} \times \frac{\square}{\square} = \frac{5}{8}$

54. WRITING IN MATH Answer the question that was posed at the beginning of the lesson.

How is multiplying fractions related to areas of rectangles?

Include the following in your answer:

- an area model of a multiplication problem involving fractions, and
- an explanation of how rectangles can be used to show multiplication of fractions.

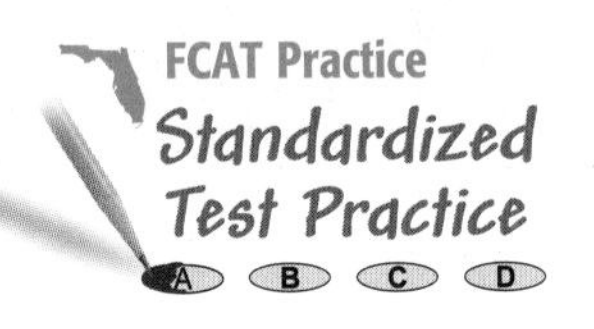

FCAT Practice

Standardized Test Practice

55. The product of $\frac{8}{15}$ and $\frac{3}{8}$ is a number

(A) between 0 and 1. (B) between 1 and 2.

(C) between 2 and 3. (D) greater than 3.

56. What is the equivalent length of a chain that is 52 feet long?

(A) 4 yards 5 feet (B) 4.5 yards

(C) 17 yards 1 foot (D) 17.1 yards

Extending the Lesson

Evaluate each expression if $n = 2$ and $p = -4$.

57. $5n^{-3}$ **58.** $7 \cdot 3^p$ **59.** $13n^{-1}p^{-2}$

Maintain Your Skills

Mixed Review

Write each decimal as a fraction or mixed number in simplest form. *(Lesson 5-2)*

60. 0.18 **61.** −0.2 **62.** 3.04 **63.** $0.\overline{7}$

Write each fraction or mixed number as a decimal. Use a bar to show a repeating decimal. *(Lesson 5-1)*

64. $\frac{17}{20}$ **65.** $\frac{1}{6}$ **66.** $2\frac{2}{11}$ **67.** $-4\frac{7}{8}$

68. ALGEBRA What is the product of x^2 and x^4? *(Lesson 4-6)*

Getting Ready for the Next Lesson

PREREQUISITE SKILL **Find the GCF of each pair of monomials.**

*(To review the **GCF of monomials**, see Lesson 4-4.)*

69. $8n, 16n$ **70.** $5ab, 8b$ **71.** $12t, 10t$

72. $2rs, 3rs$ **73.** $9k, 27$ **74.** $4p^2, 6p$

5-4 Dividing Rational Numbers

What You'll Learn

- Divide fractions using multiplicative inverses.
- Use dimensional analysis to solve problems.

How is dividing by a fraction related to multiplying?

The model shows $4 \div \frac{1}{3}$. Each of the 4 circles is divided into $\frac{1}{3}$-sections.

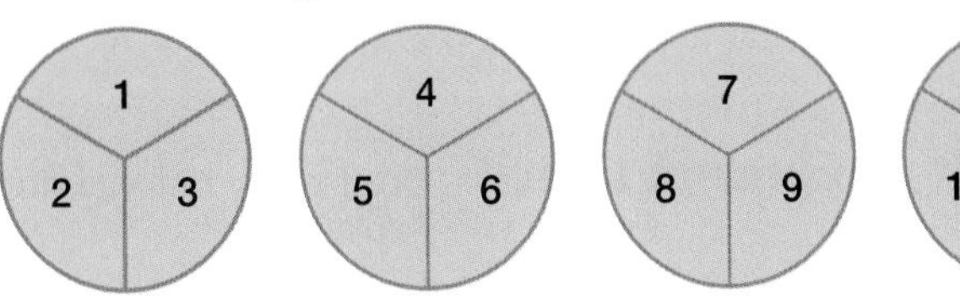

There are twelve $\frac{1}{3}$-sections, so $4 \div \frac{1}{3} = 12$. Another way to find the number of sections is $4 \times 3 = 12$.

Use a model to find each quotient. Then write a related multiplication problem.

a. $2 \div \frac{1}{3}$ **b.** $4 \div \frac{1}{2}$ **c.** $3 \div \frac{1}{4}$

d. Make a conjecture about how dividing by a fraction is related to multiplying.

Vocabulary

- multiplicative inverses
- reciprocals

Sunshine State Standards
MA.A.1.3.3-2, MA.A.3.3.1-1, MA.A.3.3.1-3, MA.A.4.3.1-2, MA.B.1.3.2-1, MA.B.2.3.2-1, MA.B.2.3.2-2

DIVIDE FRACTIONS Rational numbers have all of the properties of whole numbers and integers. Another property is shown by $\frac{1}{3} \cdot \frac{3}{1} = 1$. Two numbers whose product is 1 are called **multiplicative inverses** or **reciprocals**.

Key Concept — *Inverse Property of Multiplication*

- **Words** The product of a number and its multiplicative inverse is 1.
- **Symbols** For every number $\frac{a}{b}$, where $a, b \neq 0$, there is exactly one number $\frac{b}{a}$ such that $\frac{a}{b} \cdot \frac{b}{a} = 1$.
- **Example** $\frac{3}{4}$ and $\frac{4}{3}$ are multiplicative inverses because $\frac{3}{4} \cdot \frac{4}{3} = 1$.

Study Tip

Reading Math
Multiplicative inverse and *reciprocal* are different terms for the same concept. They may be used interchangeably.

Example 1 Find Multiplicative Inverses

Find the multiplicative inverse of each number.

a. $-\frac{3}{8}$

$-\frac{3}{8}\left(-\frac{8}{3}\right) = 1$ The product is 1.

The multiplicative inverse or reciprocal of $-\frac{3}{8}$ is $-\frac{8}{3}$.

b. $2\frac{1}{5}$

$2\frac{1}{5} = \frac{11}{5}$ Write as an improper fraction.

$\frac{11}{5} \cdot \frac{5}{11} = 1$ The product is 1.

The reciprocal of $2\frac{1}{5}$ is $\frac{5}{11}$.

Dividing by 2 is the same as multiplying by $\frac{1}{2}$, its multiplicative inverse. This is true for any rational number.

reciprocals

$6 \div 2 = 3$ $\quad 6 \cdot \frac{1}{2} = 3$

same result

Key Concept — Dividing Fractions

- **Words** To divide by a fraction, multiply by its multiplicative inverse.
- **Symbols** $\frac{a}{b} \div \frac{c}{d} = \frac{a}{b} \cdot \frac{d}{c}$, where $b, c, d \neq 0$
- **Example** $\frac{1}{4} \div \frac{5}{7} = \frac{1}{4} \cdot \frac{7}{5}$ or $\frac{7}{20}$

Concept Check Does every rational number have a multiplicative inverse? Explain.

Example 2 Divide by a Fraction

Find $\frac{1}{3} \div \frac{5}{9}$. Write the quotient in simplest form.

$\frac{1}{3} \div \frac{5}{9} = \frac{1}{3} \cdot \frac{9}{5}$ Multiply by the multiplicative inverse of $\frac{5}{9}$, $\frac{9}{5}$.

$= \frac{1}{\cancel{3}_1} \cdot \frac{\cancel{9}^3}{5}$ Divide 3 and 9 by their GCF, 3.

$= \frac{3}{5}$ Simplify.

Study Tip

Dividing By a Whole Number

When dividing by a whole number, always rename it as an improper fraction first. Then multiply by its reciprocal.

Example 3 Divide by a Whole Number

Find $\frac{5}{8} \div 6$. Write the quotient in simplest form.

$\frac{5}{8} \div 6 = \frac{5}{8} \div \frac{6}{1}$ Write 6 as $\frac{6}{1}$.

$= \frac{5}{8} \cdot \frac{1}{6}$ Multiply by the multiplicative inverse of $\frac{6}{1}$, $\frac{1}{6}$.

$= \frac{5}{48}$ Multiply the numerators and multiply the denominators.

Example 4 Divide by a Mixed Number

Find $-7\frac{1}{2} \div 2\frac{1}{10}$. Write the quotient in simplest form.

$-7\frac{1}{2} \div 2\frac{1}{10} = -\frac{15}{2} \div \frac{21}{10}$ Rename the mixed numbers as improper fractions.

$= -\frac{15}{2} \cdot \frac{10}{21}$ Multiply by the multiplicative inverse of $\frac{21}{10}$, $\frac{10}{21}$.

$= -\frac{\cancel{15}^5}{\cancel{2}_1} \cdot \frac{\cancel{10}^5}{\cancel{21}_7}$ Divide out common factors.

$= -\frac{25}{7}$ or $-3\frac{4}{7}$ Simplify.

You can divide algebraic fractions in the same way that you divide numerical fractions.

Example 5 Divide by an Algebraic Fraction

Find $\frac{3xy}{4} \div \frac{2x}{8}$. Write the quotient in simplest form.

$\frac{3xy}{4} \div \frac{2x}{8} = \frac{3xy}{4} \cdot \frac{8}{2x}$ Multiply by the multiplicative inverse of $\frac{2x}{8}$, $\frac{8}{2x}$.

$= \frac{\overset{1}{3\cancel{x}y}}{\underset{1}{\cancel{4}}} \cdot \frac{\overset{2}{\cancel{8}}}{\underset{1}{2\cancel{x}}}$ Divide out common factors.

$= \frac{6y}{2}$ or $3y$ Simplify.

More About. . .

Cheerleading

In the 1930s, both men and women cheerleaders began wearing cheerleading uniforms.

Source: www.umn.edu

DIMENSIONAL ANALYSIS Dimensional analysis is a useful way to examine the solution of division problems.

Example 6 Use Dimensional Analysis

CHEERLEADING How many cheerleading uniforms can be made with $22\frac{3}{4}$ yards of fabric if each uniform requires $\frac{7}{8}$-yard?

To find how many uniforms, divide $22\frac{3}{4}$ by $\frac{7}{8}$.

$22\frac{3}{4} \div \frac{7}{8} = 22\frac{3}{4} \cdot \frac{8}{7}$ Multiply by the reciprocal of $\frac{7}{8}$, $\frac{8}{7}$.

$= \frac{91}{4} \cdot \frac{8}{7}$ Write $22\frac{3}{4}$ as an improper fraction.

$= \frac{\overset{13}{\cancel{91}}}{\underset{1}{\cancel{4}}} \cdot \frac{\overset{2}{\cancel{8}}}{\underset{1}{\cancel{7}}}$ Divide out common factors.

$= 26$ Simplify.

So, 26 uniforms can be made.

CHECK Use dimensional analysis to examine the units.

$\text{yards} \div \frac{\text{yards}}{\text{uniform}} = \cancel{\text{yards}} \cdot \frac{\text{uniform}}{\cancel{\text{yards}}}$ Divide out the units.

$= \text{uniform}$ Simplify.

The result is expressed as uniforms. This agrees with your answer of 26 uniforms.

Check for Understanding

Concept Check

1. **Explain** how reciprocals are used in division of fractions.
2. **OPEN ENDED** Write a division expression that can be simplified by using the multiplicative inverse $\frac{7}{5}$.

Guided Practice **Find the multiplicative inverse of each number.**

3. $\frac{4}{5}$
4. -16
5. $3\frac{1}{8}$

Find each quotient. Write in simplest form.

6. $\frac{1}{2} \div \frac{6}{7}$
7. $-\frac{2}{3} \div \left(-\frac{5}{6}\right)$
8. $\frac{7}{9} \div \frac{2}{3}$
9. $7\frac{1}{3} \div 5$
10. $-\frac{8}{9} \div 3\frac{1}{5}$
11. $2\frac{1}{6} \div \left(-1\frac{1}{5}\right)$

ALGEBRA Find each quotient. Write in simplest form.

12. $\frac{14}{n} \div \frac{1}{n}$
13. $\frac{ab}{4} \div \frac{b}{6}$
14. $\frac{x^2}{5} \div \frac{ax}{2}$

Application

15. **CARPENTRY** How many boards, each 2 feet 8 inches long, can be cut from a board 16 feet long if there is no waste?

Practice and Apply

Homework Help

For Exercises	See Examples
16–21	1
22–39	2–4
40–45	5
46, 47	6

Extra Practice

See page 734.

Find the multiplicative inverse of each number.

16. $\frac{6}{11}$
17. $-\frac{1}{5}$
18. -7
19. 24
20. $5\frac{1}{4}$
21. $-3\frac{2}{9}$

Find each quotient. Write in simplest form.

22. $\frac{1}{4} \div \frac{3}{5}$
23. $\frac{2}{9} \div \frac{1}{4}$
24. $-\frac{1}{2} \div \frac{5}{6}$
25. $\frac{6}{11} \div \left(-\frac{4}{5}\right)$
26. $\frac{8}{9} \div \frac{4}{3}$
27. $\frac{7}{8} \div \frac{14}{15}$
28. $\frac{3}{4} \div \frac{3}{4}$
29. $\frac{2}{9} \div \left(-\frac{2}{9}\right)$
30. $\frac{3}{5} \div \frac{5}{9}$
31. $\frac{3}{10} \div \frac{1}{5}$
32. $12 \div \frac{4}{9}$
33. $-8 \div \frac{4}{5}$
34. $-\frac{5}{8} \div (-4)$
35. $6\frac{2}{3} \div 5$
36. $-1\frac{1}{9} \div \frac{2}{3}$
37. $-\frac{2}{3} \div \left(-\frac{1}{3}\right)$
38. $3\frac{3}{10} \div 1\frac{5}{6}$
39. $7\frac{1}{2} \div \left(-1\frac{1}{5}\right)$

ALGEBRA Find each quotient. Write in simplest form.

40. $\frac{a}{7} \div \frac{a}{42}$
41. $\frac{10}{3x} \div \frac{5}{2x}$
42. $\frac{c}{8} \div \frac{cd}{5}$
43. $\frac{5s}{t} \div \frac{6rs}{t}$
44. $\frac{k^3}{9} \div \frac{k}{24}$
45. $\frac{2s}{t^2} \div \frac{st^3}{8}$

46. **COOKING** How many $\frac{1}{4}$-pound hamburgers can be made from $2\frac{3}{4}$ pounds of ground beef?

47. **SEWING** How many 9-inch ribbons can be cut from $1\frac{1}{2}$ yards of ribbon?

ALGEBRA Evaluate each expression.

48. $m \div n$ if $m = -\frac{8}{9}$ and $n = \frac{7}{18}$
49. $r^2 \div s^2$ if $r = -\frac{3}{4}$ and $s = 1\frac{1}{3}$

More About. . .

Travel

Trains in Europe can travel faster than 125 miles per hour.

Source: www.geography.about.com

For Exercises 50 and 51, solve each problem. Then check your answer using dimensional analysis.

50. TRAVEL How long would it take a train traveling 80 miles per hour to go 280 miles?

51. FOOD The average young American woman drinks $1\frac{1}{2}$ cans of cola each day. At this rate, in how many days would it take to drink a total of 12 cans? **Source:** U.S. Department of Agriculture

52. CRITICAL THINKING

a. Divide $\frac{3}{4}$ by $\frac{1}{2}, \frac{1}{4}, \frac{1}{8}$, and $\frac{1}{12}$.

b. What happens to the quotient as the value of the divisor decreases?

c. **Make a conjecture** about the quotient when you divide $\frac{3}{4}$ by fractions that increase in value. Test your conjecture.

53. WRITING IN MATH Answer the question that was posed at the beginning of the lesson.

How is dividing by a fraction related to multiplying?

Include the following in your answer:

- a model of a whole number divided by a fraction, and
- an explanation of how division of fractions is related to multiplication.

FCAT Practice

Standardized Test Practice

54. Carla baby-sits for $2\frac{1}{4}$ hours and earns \$11.25. What is her rate?

Ⓐ \$4.00/h Ⓑ \$5.50/h Ⓒ \$4.50/h Ⓓ \$5.00/h

55. What is $\frac{3}{10}$ divided by $1\frac{4}{5}$?

Ⓐ $\frac{1}{2}$ Ⓑ $\frac{3}{8}$ Ⓒ $\frac{1}{6}$ Ⓓ $\frac{27}{50}$

Maintain Your Skills

Mixed Review **Find each product. Write in simplest form.** *(Lesson 5-3)*

56. $\frac{3}{5} \cdot \frac{1}{3}$ **57.** $\frac{2}{9} \cdot \frac{15}{16}$ **58.** $2\frac{4}{5} \cdot \frac{3}{8}$ **59.** $-\frac{5}{12} \cdot 1\frac{1}{7}$

Identify all sets to which each number belongs. *(Lesson 5-2)*

60. 16 **61.** −2.8888... **62.** $0.\overline{9}$ **63.** 5.121221222...

64. Write the prime factorization of 150. Use exponents for repeated factors. *(Lesson 4-3)*

65. ALGEBRA Solve $3x - 5 = 16$. *(Lesson 3-5)*

Getting Ready for the Next Lesson **PREREQUISITE SKILL** **Write each improper fraction as a mixed number in simplest form.** *(To review **simplifying fractions**, see Lesson 4-5.)*

66. $\frac{9}{4}$ **67.** $\frac{8}{7}$ **68.** $\frac{17}{2}$

69. $\frac{25}{4}$ **70.** $\frac{24}{5}$ **71.** $\frac{22}{6}$

72. $\frac{15}{6}$ **73.** $\frac{30}{18}$ **74.** $\frac{18}{15}$

5-5 Adding and Subtracting Like Fractions

Sunshine State Standards
MA.A.3.3.1-1, MA.A.3.3.1-3, MA.A.4.3.1-2, MA.B.2.3.2-1

What You'll Learn

- Add like fractions.
- Subtract like fractions.

Why are fractions important when taking measurements?

Measures of different parts of an insect are shown in the diagram. The sum of the parts is $\frac{6}{8}$ inch. Use a ruler to find each measure.

a. $\frac{1}{8}$ in. + $\frac{3}{8}$ in. **b.** $\frac{3}{8}$ in. + $\frac{4}{8}$ in.

c. $\frac{4}{8}$ in. + $\frac{4}{8}$ in. **d.** $\frac{6}{8}$ in. − $\frac{3}{8}$ in.

ADD LIKE FRACTIONS Fractions with the same denominator are called *like fractions*. The rule for adding like fractions is stated below.

Key Concept — Adding Like Fractions

- **Words** To add fractions with like denominators, add the numerators and write the sum over the denominator.
- **Symbols** $\frac{a}{c} + \frac{b}{c} = \frac{a+b}{c}$, where $c \neq 0$
- **Example** $\frac{1}{5} + \frac{2}{5} = \frac{1+2}{5}$ or $\frac{3}{5}$

Example 1 Add Fractions

Find $\frac{3}{7} + \frac{5}{7}$. Write the sum in simplest form. **Estimate:** $0 + 1 = 1$

$\frac{3}{7} + \frac{5}{7} = \frac{3+5}{7}$ The denominators are the same. Add the numerators.

$= \frac{8}{7}$ or $1\frac{1}{7}$ Simplify and rename as a mixed number.

Study Tip

Alternative Method
You can also stack the mixed numbers vertically to find the sum.

$$\begin{array}{r} 6\frac{5}{8} \\ +1\frac{1}{8} \\ \hline 7\frac{6}{8} \text{ or } 7\frac{3}{4} \end{array}$$

Example 2 Add Mixed Numbers

Find $6\frac{5}{8} + 1\frac{1}{8}$. Write the sum in simplest form. **Estimate:** $7 + 1 = 8$

$6\frac{5}{8} + 1\frac{1}{8} = (6 + 1) + \left(\frac{5}{8} + \frac{1}{8}\right)$ Add the whole numbers and fractions separately.

$= 7 + \frac{5+1}{8}$ Add the numerators.

$= 7\frac{6}{8}$ or $7\frac{3}{4}$ Simplify.

SUBTRACT LIKE FRACTIONS The rule for subtracting fractions with like denominators is similar to the rule for addition.

Key Concept *Subtracting Like Fractions*

- **Words** To subtract fractions with like denominators, subtract the numerators and write the difference over the denominator.
- **Symbols** $\frac{a}{c} - \frac{b}{c} = \frac{a-b}{c}$, where $c \neq 0$
- **Example** $\frac{5}{7} - \frac{1}{7} = \frac{5-1}{7}$ or $\frac{4}{7}$

Concept Check How is the rule for subtracting fractions with like denominators similar to the rule for adding fractions with like denominators?

Example 3 Subtract Fractions

Find $\frac{9}{20} - \frac{13}{20}$. Write the difference in simplest form. **Estimate:** $\frac{1}{2} - 1 = -\frac{1}{2}$

$\frac{9}{20} - \frac{13}{20} = \frac{9-13}{20}$ The denominators are the same. Subtract the numerators.

$= \frac{-4}{20}$ or $-\frac{1}{5}$ Simplify.

You can write the mixed numbers as improper fractions before adding or subtracting.

Example 4 Subtract Mixed Numbers

Evaluate $a - b$ if $a = 9\frac{1}{6}$ and $b = 5\frac{2}{6}$. **Estimate:** $9 - 5 = 4$

$a - b = 9\frac{1}{6} - 5\frac{2}{6}$ Replace a with $9\frac{1}{6}$ and b with $5\frac{2}{6}$.

$= \frac{55}{6} - \frac{32}{6}$ Write the mixed numbers as improper fractions.

$= \frac{23}{6}$ Subtract the numerators.

$= 3\frac{5}{6}$ Simplify.

ALGEBRA CONNECTION

You can use the same rules for adding or subtracting like algebraic fractions as you did for adding or subtracting like numerical fractions.

Example 5 Add Algebraic Fractions

Find $\frac{n}{8} + \frac{5n}{8}$. Write the sum in simplest form.

$\frac{n}{8} + \frac{5n}{8} = \frac{n+5n}{8}$ The denominators are the same. Add the numerators.

$= \frac{6n}{8}$ Add the numerators.

$= \frac{3n}{4}$ Simplify.

Check for Understanding

Concept Check

1. **Draw** a model to show the sum $\frac{2}{7} + \frac{4}{7}$.

2. **OPEN ENDED** Write a subtraction expression in which the difference of two fractions is $\frac{18}{25}$.

3. **FIND THE ERROR** Kayla and Ethan are adding $-2\frac{1}{8}$ and $-4\frac{3}{8}$.

Kayla

$-2\frac{1}{8} + \left(-4\frac{3}{8}\right) = -\frac{17}{8} + \left(-\frac{35}{8}\right)$
$= -\frac{52}{8}$ or $-6\frac{1}{2}$

Ethan

$-2\frac{1}{8} + \left(-4\frac{3}{8}\right) = \frac{17}{8} + \left(-\frac{35}{8}\right)$
$= -\frac{18}{8}$ or $-2\frac{1}{4}$

Who is correct? Explain your reasoning.

Guided Practice

Find each sum or difference. Write in simplest form.

4. $\frac{1}{7} + \frac{5}{7}$
5. $\frac{11}{14} - \frac{3}{14}$
6. $\frac{3}{10} + \frac{3}{10}$
7. $-\frac{1}{8} - \frac{5}{8}$
8. $-2\frac{4}{5} + \left(-\frac{2}{5}\right)$
9. $7\frac{1}{8} - \left(-1\frac{3}{8}\right)$

10. **ALGEBRA** Evaluate $x + y$ if $x = 2\frac{4}{9}$ and $y = 8\frac{7}{9}$.

ALGEBRA Find each sum or difference. Write in simplest form.

11. $\frac{6r}{11} + \frac{2r}{11}$
12. $\frac{19}{a} - \frac{12}{a}, a \neq 0$
13. $\frac{5}{3x} - \frac{6}{3x}, x \neq 0$

Application

14. **MEASUREMENT** Hoai was $62\frac{1}{8}$ inches tall at the end of school in June. He was $63\frac{7}{8}$ inches tall in September. How much did he grow during the summer?

Practice and Apply

Homework Help

For Exercises	See Examples
15–24	1–3
25–38, 45, 46	4
39–44	5

Extra Practice

See page 734.

Find each sum or difference. Write in simplest form.

15. $\frac{2}{5} + \frac{1}{5}$
16. $\frac{10}{11} - \frac{8}{11}$
17. $\frac{17}{18} - \frac{5}{18}$
18. $\frac{3}{10} + \frac{7}{10}$
19. $\frac{1}{12} + \left(-\frac{7}{12}\right)$
20. $\frac{9}{20} - \left(-\frac{7}{20}\right)$
21. $-\frac{3}{4} + \left(-\frac{3}{4}\right)$
22. $-\frac{13}{16} + \left(-\frac{9}{16}\right)$
23. $-\frac{7}{9} + \frac{5}{9}$
24. $-\frac{17}{20} + \frac{9}{20}$
25. $7\frac{2}{5} + 4\frac{2}{5}$
26. $-4\frac{5}{8} - \frac{3}{8}$
27. $5\frac{7}{9} - \left(-3\frac{5}{9}\right)$
28. $2\frac{5}{12} + \left(-2\frac{7}{12}\right)$
29. $2\frac{3}{8} - 1\frac{5}{8}$
30. $8\frac{9}{10} - 6\frac{1}{10}$
31. $7\frac{4}{7} - 2\frac{5}{7}$
32. $-8\frac{6}{11} - \left(-2\frac{5}{11}\right)$

33. Find $12\frac{7}{8} - 7\frac{3}{8} + 2\frac{5}{8}$.

34. Find $5\frac{5}{6} + 3\frac{5}{6} - 2\frac{1}{6}$.

ALGEBRA Evaluate each expression if $x = \frac{8}{15}$, $y = 2\frac{1}{15}$, and $z = \frac{11}{15}$. Write in simplest form.

35. $x + y$

36. $z + y$

37. $z - x$

38. $y - x$

ALGEBRA Find each sum or difference. Write in simplest form.

39. $\frac{x}{8} + \frac{4x}{8}$

40. $\frac{3r}{10} + \frac{3r}{10}$

41. $\frac{12}{m} - \frac{9}{m}$, $m \neq 0$

42. $\frac{10a}{3b} - \frac{7a}{3b}$, $b \neq 0$

43. $5\frac{4}{7}c - 3\frac{1}{7}c$

44. $-2\frac{1}{6}y + 8\frac{5}{6}y$

Career Choices

Carpenter

Carpenters must be able to make precise measurements and know how to add and subtract fractional measures.

Online Research For information about a career as a carpenter, visit: www.pre-alg.com/careers

45. **CARPENTRY** A 5-foot long kitchen countertop is to be installed between two walls that are $54\frac{5}{8}$ inches apart. How much of the countertop must be cut off so that it fits between the walls?

46. **SEWING** Chumani is making a linen suit. The portion of the pattern envelope that shows the yards of fabric needed for different sizes is shown at the right. If Chumani is making a size 6 jacket and skirt from 45-inch fabric, how much fabric should she buy?

Size	(6	8	10)
JACKET			
45"	$2\frac{5}{8}$	$2\frac{3}{4}$	$2\frac{3}{4}$
60"	2	2	2
SKIRT			
45"	$\frac{7}{8}$	$1\frac{1}{8}$	$1\frac{1}{8}$
60"	$\frac{7}{8}$	$\frac{7}{8}$	$\frac{7}{8}$

47. **GARDENING** Melanie's flower garden has a perimeter of 25 feet. She plans to add 2 feet 9 inches to the width and 3 feet 9 inches to the length. What is the new perimeter in feet?

48. **CRITICAL THINKING** The 7-piece puzzle at the right is called a *tangram*.

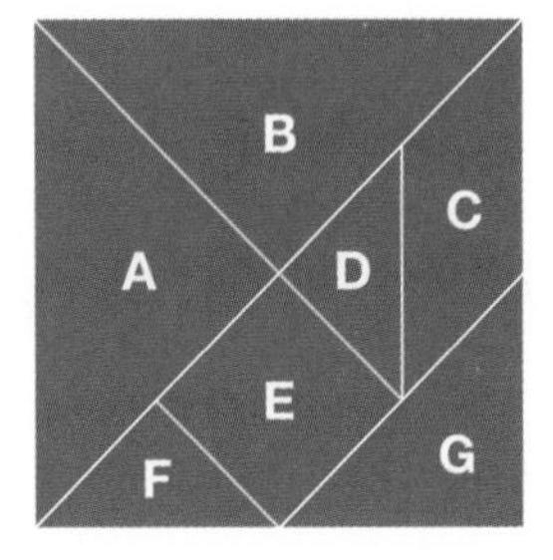

a. If the value of the entire puzzle is 1, what is the value of each piece?

b. How much is A + B?

c. How much is F + D?

d. How much is C + E?

e. Which pieces each equal the sum of E and G?

49. WRITING IN MATH Answer the question that was posed at the beginning of the lesson.

Why are fractions important when taking measurements?

Include the following in your answer:

- the fraction of an inch that each mark on a ruler or tape measure represents, and
- some real-world examples in which fractional measures are used.

50. Find $\frac{13}{20} - \frac{7}{20}$. Write in simplest form.

Ⓐ $\frac{6}{10}$　Ⓑ $\frac{3}{5}$　Ⓒ $\frac{6}{20}$　Ⓓ $\frac{3}{10}$

51. A piece of wood is $1\frac{9}{16}$ inches thick. A layer of padding $\frac{15}{16}$ inch thick is placed on top. What is the total thickness of the wood and the padding?

Ⓐ $2\frac{1}{2}$ in.　Ⓑ $1\frac{1}{2}$ in.　Ⓒ $1\frac{24}{16}$ in.　Ⓓ $1\frac{3}{8}$ in.

Maintain Your Skills

Mixed Review

Find each quotient. Write in simplest form. *(Lesson 5-4)*

52. $\frac{1}{6} \div \frac{3}{4}$　**53.** $-\frac{5}{8} \div \frac{1}{3}$　**54.** $\frac{2}{5} \div 1\frac{1}{2}$

Find each product. Write in simplest form. *(Lesson 5-3)*

55. $\frac{2}{5} \cdot \frac{3}{4}$　**56.** $\frac{1}{6} \cdot \left(-\frac{8}{9}\right)$　**57.** $\frac{4}{7} \cdot 2\frac{1}{3}$

58. Find the product of $4y^2$ and $8y^5$. *(Lesson 4-6)*

59. GEOMETRY Find the perimeter and area of the rectangle. *(Lesson 3-7)*

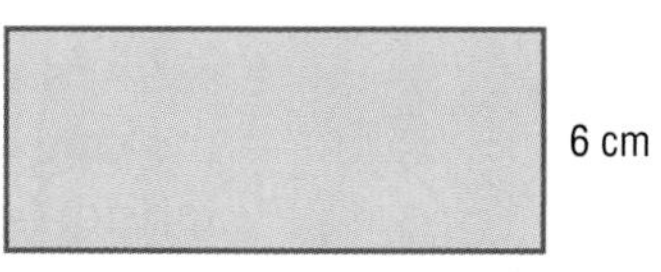

Getting Ready for the Next Lesson

PREREQUISITE SKILL Use exponents to write the prime factorization of each number or monomial. *(To review **prime factorization**, see Lesson 4-3.)*

60. 60　**61.** 175　**62.** 112

63. $12n$　**64.** $24s^2$　**65.** $42a^2b$

Practice Quiz 1

Lessons 5-1 through 5-5

Write each fraction or mixed number as a decimal. Use a bar to show a repeating decimal. *(Lesson 5-1)*

1. $\frac{4}{25}$　**2.** $-\frac{2}{9}$　**3.** $3\frac{1}{8}$

Write each decimal as a fraction or mixed number in simplest form. *(Lesson 5-2)*

4. -6.75　**5.** 0.12　**6.** 0.5555…

Simplify each expression. *(Lessons 5-3, 5-4, and 5-5)*

7. $\frac{5}{18} \cdot \frac{4}{15}$　**8.** $\frac{7}{8} \div \left(-\frac{1}{4}\right)$　**9.** $\frac{11}{12} - \frac{6}{12}$

10. ALGEBRA Find $6\frac{3}{5}a + 2\frac{4}{5}a$. Write in simplest form. *(Lesson 5-5)*

Reading Mathematics

Factors and Multiples

Many words used in mathematics are also used in everyday language. You can use the everyday meaning of these words to better understand their mathematical meaning. The table shows both meanings of the words *factor* and *multiple*.

Term	Everyday Meaning	Mathematical Meaning
factor	something that contributes to the production of a result • The weather was not a *factor* in the decision. • The type of wood is one *factor* that contributes to the cost of the table.	one of two or more numbers that are multiplied together to form a product
multiple	involving more than one or shared by many • *multiple* births • *multiple* ownership	the product of a quantity and a whole number

Source: *Merriam Webster's Collegiate Dictionary*

When you count by 2, you are listing the multiples of 2. When you count by 3, you are listing the multiples of 3, and so on, as shown in the table below.

Number	Factors	Multiples
2	1, 2	2, 4, 6, 8, …
3	1, 3	3, 6, 9, 12, …
4	1, 2, 4	4, 8, 12, 16, …

Notice that the mathematical meaning of each word is related to the everyday meaning. The word *multiple* means many, and in mathematics, a number has infinitely many multiples.

Reading to Learn

1. Write your own rule for remembering the difference between *factor* and *multiple*.
2. **RESEARCH** Use the Internet or a dictionary to find the everyday meaning of each word listed below. Compare them to the mathematical meanings of *factor* and *multiple*. Note the similarities and differences.
   a. factotum **b.** multicultural **c.** multimedia
3. Make lists of other words that have the prefixes fact- or multi-. Determine what the words in each list have in common.

5-6 Least Common Multiple

What You'll Learn

- Find the least common multiple of two or more numbers.
- Find the least common denominator of two or more fractions.

Sunshine State Standards
MA.A.1.3.2-1, MA.A.1.3.2-2, MA.A.1.3.3-2

Vocabulary

- multiple
- common multiples
- least common multiple (LCM)
- least common denominator (LCD)

How can you use prime factors to find the least common multiple?

A voter voted for both president and senator in the year 2000.

Candidate	Length of Term (years)
President	4
Senator	6

a. List the next three years in which the voter can vote for president.

b. List the next three years in which the voter can vote for senator.

c. What will be the next year in which the voter has a chance to vote for both president and senator?

LEAST COMMON MULTIPLE A **multiple** of a number is a product of that number and a whole number.

Multiples of 5
$5 \cdot 0 = 0$
$5 \cdot 1 = 5$
$5 \cdot 2 = 10$
$5 \cdot 3 = 15$
⋮ ⋮

Sometimes numbers have some of the same multiples. These are called **common multiples**.

Some common multiples of 4 and 6 are 0, 12, and 24.

multiples of 4: 0, 4, 8, 12, 16, 20, 24, 28, …
multiples of 6: 0, 6, 12, 18, 24, 30, 36, 42, …

The least of the *nonzero* common multiples is called the **least common multiple (LCM)**. So, the LCM of 4 and 6 is 12.

When numbers are large, an easier way of finding the least common multiple is to use prime factorization. The LCM is the smallest product that contains the prime factors of each number.

Study Tip

Prime Factors
If a prime factor appears in both numbers, use the factor with the greatest exponent.

Example 1 Find the LCM

Find the LCM of 108 and 240.

Number	Prime Factorization	Exponential Form
108	$2 \cdot 2 \cdot 3 \cdot 3 \cdot 3$	$2^2 \cdot 3^3$
240	$2 \cdot 2 \cdot 2 \cdot 2 \cdot 3 \cdot 5$	$2^4 \cdot 3 \cdot 5$

The prime factors of both numbers are 2, 3, and 5. Multiply the greatest power of 2, 3, and 5 appearing in either factorization.

$\text{LCM} = 2^4 \cdot 3^3 \cdot 5$

$= 2160$

So, the LCM of 108 and 240 is 2160.

Concept Check What is the LCM of 6 and 12?

The LCM of two or more monomials is found in the same way as the LCM of two or more numbers.

Example 2 The LCM of Monomials

Find the LCM of $18xy^2$ and $10y$.

$18xy^2 = 2 \cdot 3^2 \cdot x \cdot y^2$

$10y = 2 \cdot 5 \cdot y$

$\text{LCM} = 2 \cdot 3^2 \cdot 5 \cdot x \cdot y^2$ Multiply the greatest power of each prime factor.

$= 90xy^2$

The LCM of $18xy^2$ and $10y$ is $90xy^2$.

LEAST COMMON DENOMINATOR The **least common denominator (LCD)** of two or more fractions is the LCM of the denominators.

Example 3 Find the LCD

Find the LCD of $\frac{5}{9}$ and $\frac{11}{21}$.

$9 = 3^2$

$21 = 3 \cdot 7$

Write the prime factorization of 9 and 21. Highlight the greatest power of each prime factor.

$\text{LCM} = 3^2 \cdot 7$ Multiply.

$= 63$

The LCD of $\frac{5}{9}$ and $\frac{11}{21}$ is 63.

The LCD for algebraic fractions can also be found.

Example 4 Find the LCD of Algebraic Fractions

Find the LCD of $\frac{5}{12b^2}$ and $\frac{3}{8ab}$.

$12b^2 = 2^2 \cdot 3 \cdot b^2$

$8ab = 2^3 \cdot a \cdot b$

$\text{LCM} = 2^3 \cdot 3 \cdot a \cdot b^2$ or $24ab^2$

Thus, the LCD of $\frac{5}{12b^2}$ and $\frac{3}{8ab}$ is $24ab^2$.

Concept Check What is the LCD of $\frac{1}{x^2}$ and $\frac{1}{xy}$?

One way to compare fractions is to write them using the LCD. We can multiply the numerator and the denominator of a fraction by the same number, because it is the same as multiplying the fraction by 1.

Example 5 Compare Fractions

Replace ● with <, >, or = to make $\frac{1}{6}$ ● $\frac{7}{15}$ a true statement.

The LCD of the fractions is $2 \cdot 3 \cdot 5$ or 30. Rewrite the fractions using the LCD and then compare the numerators.

$\frac{1}{6} = \frac{1 \cdot 5}{2 \cdot 3 \cdot 5} = \frac{5}{30}$ Multiply the fraction by $\frac{5}{5}$ to make the denominator 30.

$\frac{7}{15} = \frac{7 \cdot 2}{3 \cdot 5 \cdot 2} = \frac{14}{30}$ Multiply the fraction by $\frac{2}{2}$ to make the denominator 30.

Since $\frac{5}{30} < \frac{14}{30}$, then $\frac{1}{6} < \frac{7}{15}$.

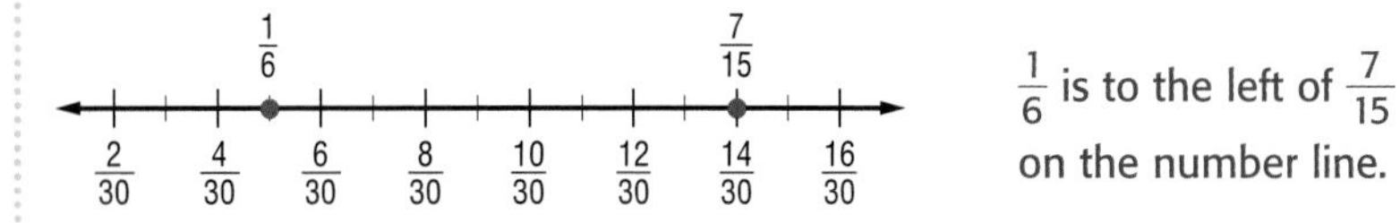

$\frac{1}{6}$ is to the left of $\frac{7}{15}$ on the number line.

Check for Understanding

Concept Check

1. **Compare and contrast** least common multiple (LCM) and least common denominator (LCD).
2. **OPEN ENDED** Write two fractions whose least common denominator (LCD) is 35.

Guided Practice

Find the least common multiple (LCM) of each pair of numbers or monomials.

3. 6, 8
4. 7, 9
5. 10, 14
6. 12, 30
7. 16, 24
8. $36ab$, $4b$

Find the least common denominator (LCD) of each pair of fractions.

9. $\frac{1}{2}, \frac{3}{8}$
10. $\frac{2}{3}, \frac{7}{10}$
11. $\frac{2}{25x}, \frac{13}{20x}$

Replace each ● with <, >, or = to make a true statement.

12. $\frac{1}{4}$ ● $\frac{3}{16}$
13. $\frac{10}{45}$ ● $\frac{2}{9}$
14. $\frac{5}{7}$ ● $\frac{7}{9}$

Application

15. **CYCLING** The front bicycle gear has 52 teeth and the back gear has 20 teeth. How many revolutions must each gear make for them to align again as shown? (*Hint:* First, find the number of teeth. Then divide to find the final answers.)

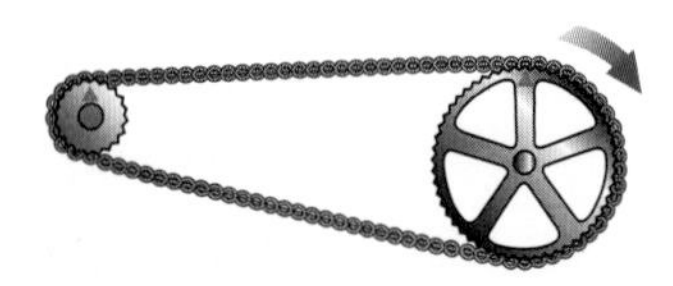

Practice and Apply

Homework Help

For Exercises	See Examples
16–29	1
30–33	2
34–39	3
40, 41	4
44–54	5

Extra Practice

See page 735.

Find the least common multiple (LCM) of each set of numbers or monomials.

16. 4, 10 **17.** 20, 12 **18.** 2, 9

19. 16, 3 **20.** 15, 75 **21.** 21, 28

22. 14, 28 **23.** 20, 50 **24.** 18, 32

25. 24, 32 **26.** 10, 20, 40 **27.** 7, 21, 84

28. 9, 12, 15 **29.** 45, 30, 35 **30.** $20c$, $12c$

31. $16a^2$, $14ab$ **32.** $7x$, $12x$ **33.** $75n^2$, $25n^4$

Find the least common denominator (LCD) of each pair of fractions.

34. $\frac{1}{4}, \frac{7}{8}$ **35.** $\frac{8}{15}, \frac{1}{3}$ **36.** $\frac{4}{5}, \frac{1}{2}$ **37.** $\frac{2}{5}, \frac{6}{7}$

38. $\frac{4}{9}, \frac{5}{12}$ **39.** $\frac{3}{8}, \frac{5}{6}$ **40.** $\frac{1}{3t}, \frac{4}{5t^2}$ **41.** $\frac{7}{8cd}, \frac{5}{16c^2}$

More About. . .

Planets

A spacecraft's launch date depends on planet alignment because the gravitational forces could affect the spacecraft's flight path.

Source: Memphis Space Center

42. PLANETS The table shows the number of Earth years it takes for some of the planets to revolve around the Sun. Find the least common multiple of the revolution times to determine approximately how often these planets align.

Planet	Revolution Time (Earth Years)
Jupiter	12
Saturn	30
Uranus	84

43. AUTO RACING One driver can circle a one-mile track in 30 seconds. Another driver takes 20 seconds. If they both start at the same time, in how many seconds will they be together again at the starting line?

Replace each ● with <, >, or = to make a true statement.

44. $\frac{1}{2}$ ● $\frac{5}{12}$ **45.** $\frac{7}{9}$ ● $\frac{5}{6}$ **46.** $\frac{3}{5}$ ● $\frac{4}{7}$

47. $\frac{21}{100}$ ● $\frac{1}{5}$ **48.** $\frac{17}{34}$ ● $\frac{1}{2}$ **49.** $\frac{12}{17}$ ● $\frac{36}{51}$

50. $\frac{8}{9}$ ● $\frac{19}{21}$ **51.** $-\frac{9}{11}$ ● $-\frac{5}{6}$ **52.** $-\frac{14}{15}$ ● $-\frac{9}{10}$

53. ANIMALS Of all the endangered species in the world, $\frac{7}{39}$ of the reptiles, and $\frac{5}{9}$ of the amphibians are in the United States. Is there a greater fraction of endangered reptiles or amphibians in the U.S.?

Online Research **Data Update** How many endangered species are in the United States today? Visit www.pre-alg.com/data_update to learn more.

54. TELEPHONES Eleven out of twenty people in Chicago, Illinois, have cellphones, and $\frac{14}{25}$ of the people in Anchorage, Alaska, have cellphones. In which city do a greater fraction of people have cellphones?

Source: Polk Research

55. Find two composite numbers between 10 and 20 whose least common multiple (LCM) is 36.

Determine whether each statement is *sometimes, always,* or *never* true. Give an example to support your answer.

56. The LCM of three numbers is one of the numbers.

57. If two numbers do not contain any factors in common, then the LCM of the two numbers is 1.

58. The LCM of two numbers is greater than the GCF of the numbers.

Study Tip

Relatively Prime
Recall that two numbers that are relatively prime have a GCF of 1.

59. CRITICAL THINKING

a. If two numbers are relatively prime, what is their LCM? Give two examples and explain your reasoning.

b. Determine whether the LCM of two whole numbers is *always, sometimes,* or *never* a multiple of the GCF of the same two numbers. Explain.

60. WRITING IN MATH Answer the question that was posed at the beginning of the lesson.

How can you use prime factors to find the least common multiple?

Include the following in your answer:

- a definition of least common multiple, and
- a description of the steps you take to find the LCM of two or more numbers.

FCAT Practice
Standardized Test Practice

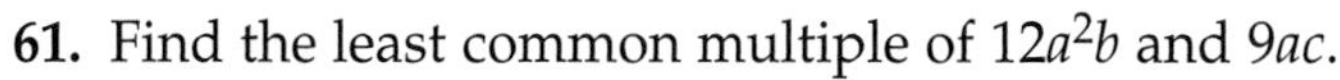

61. Find the least common multiple of $12a^2b$ and $9ac$.

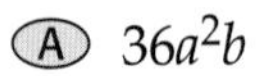

(A) $36a^2b$ (B) $36a^2bc$ (C) $3a^2bc$ (D) $3abc$

62. A $\frac{7}{8}$-inch wrench is too large to tighten a bolt. Of these, which is the next smaller size?

(A) $\frac{3}{4}$-inch (B) $\frac{5}{8}$-inch (C) $\frac{7}{16}$-inch (D) $\frac{13}{16}$-inch

Maintain Your Skills

Mixed Review

Find each sum or difference. Write in simplest form. *(Lesson 5-5)*

63. $\frac{7}{8} - \frac{3}{8}$ **64.** $3\frac{9}{11} - \frac{5}{11}$ **65.** $\frac{13}{14} + \frac{3}{14}$ **66.** $2\frac{5}{6} + 4\frac{1}{6}$

ALGEBRA Find each quotient. Write in simplest form. *(Lesson 5-4)*

67. $\frac{3}{n} \div \frac{1}{n}$ **68.** $\frac{x}{8} \div \frac{x}{6}$ **69.** $\frac{ac}{5} \div \frac{c}{d}$ **70.** $\frac{6k}{7m} \div \frac{3}{14m}$

71. ALGEBRA Translate *the sum of 7 and two times a number is 11* into an equation. Then find the number. *(Lesson 3-6)*

ALGEBRA Solve each equation. Check your solution. *(Lessons 3-3 and 3-4)*

72. $9 = x - 4$ **73.** $\frac{a}{-2} = 10$ **74.** $-5c = -105$

Getting Ready for the Next Lesson

PREREQUISITE SKILL Estimate each sum.
*(To review **estimating with fractions**, see page 716.)*

75. $\frac{3}{8} + \frac{3}{4}$ **76.** $\frac{9}{10} + \frac{14}{15}$ **77.** $\frac{4}{7} + 2\frac{1}{5}$

78. $5\frac{7}{8} + \frac{2}{3}$ **79.** $8\frac{3}{11} + 7\frac{2}{9}$ **80.** $20\frac{5}{16} + 6\frac{1}{9}$

Algebra Activity

A Follow-Up of Lesson 5-6

Juniper Green

Juniper Green is a game that was invented by a teacher in England.

Numbers 1-100

Getting Ready

This game is for two people, so students should divide into pairs.

Rules of the Game

- The first player selects an even number from the hundreds chart and circles it with a colored marker.
- The next player selects any remaining number that is a factor or multiple of this number and circles it.
- Players continue taking turns circling numbers, as shown below.
- When a player cannot select a number or circles a number incorrectly, then the game is over and the other player wins.

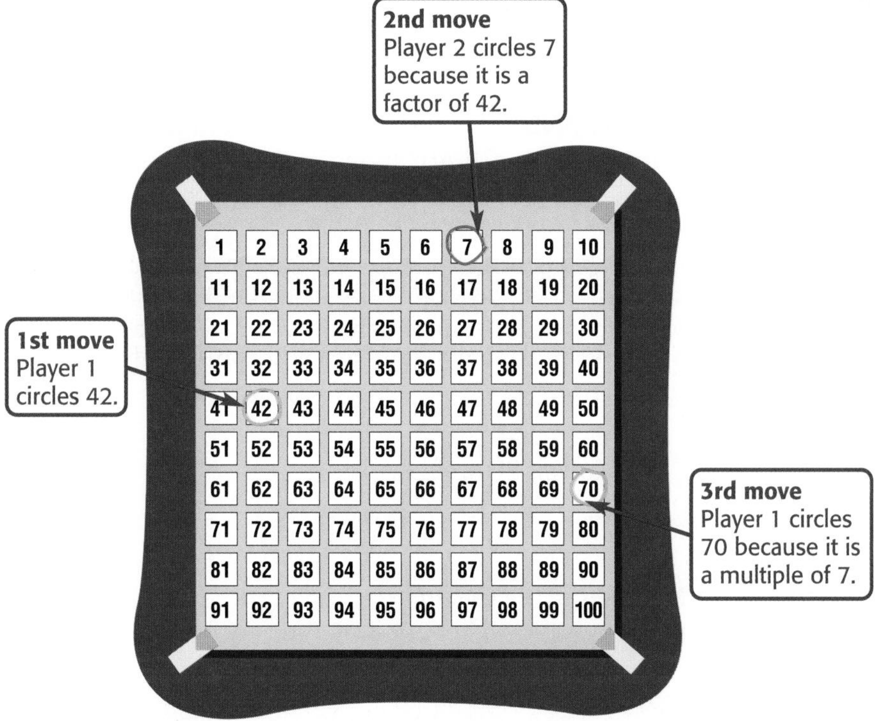
2nd move
Player 2 circles 7 because it is a factor of 42.
1st move
Player 1 circles 42.
3rd move
Player 1 circles 70 because it is a multiple of 7.

Analyze the Strategies

Play the game several times and then answer the following questions.

1. Why do you think the first player must select an even number? Explain.

2. Describe the kinds of moves that were made just before the game was over.

5-7 Adding and Subtracting Unlike Fractions

Sunshine State Standards
MA.A.1.3.3-2, MA.A.3.3.1-3, MA.A.4.3.1-2, MA.B.3.3.1-1

What You'll Learn

- Add unlike fractions.
- Subtract unlike fractions.

How can the LCM be used to add and subtract fractions with different denominators?

The sum $\frac{1}{2} + \frac{1}{3}$ is modeled at the right. We can use the LCM to find the sum.

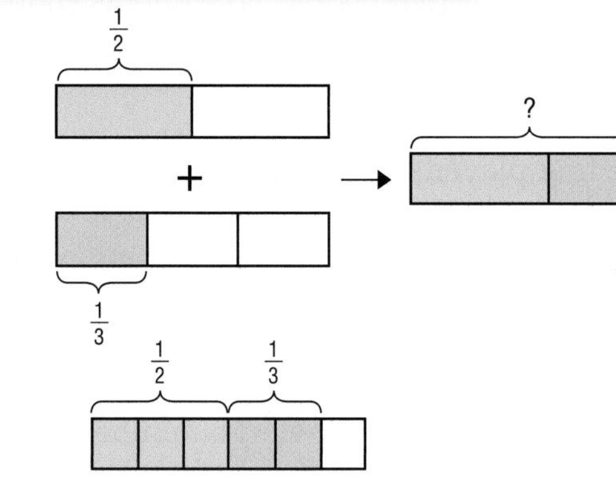

a. What is the LCM of the denominators?

b. If you partition the model into six parts, what fraction of the model is shaded?

c. How many parts are $\frac{1}{2}$? $\frac{1}{3}$?

d. Describe a model that you could use to add $\frac{1}{3}$ and $\frac{1}{4}$. Then use it to find the sum.

ADD UNLIKE FRACTIONS Fractions with different denominators are called *unlike fractions*. In the activity above, you used the LCM of the denominators to rename the fractions. You can use any common denominator.

Key Concept — Adding Unlike Fractions

- **Words** To add fractions with unlike denominators, rename the fractions with a common denominator. Then add and simplify.
- **Example** $\frac{1}{3} + \frac{2}{5} = \frac{1}{3} \cdot \frac{5}{5} + \frac{2}{5} \cdot \frac{3}{3}$

  $= \frac{5}{15} + \frac{6}{15}$ or $\frac{11}{15}$

Example 1 Add Unlike Fractions

Find $\frac{1}{4} + \frac{2}{3}$.

$\frac{1}{4} + \frac{2}{3} = \frac{1}{4} \cdot \frac{3}{3} + \frac{2}{3} \cdot \frac{4}{4}$ Use $4 \cdot 3$ or 12 as the common denominator.

$= \frac{3}{12} + \frac{8}{12}$ Rename each fraction with the common denominator.

$= \frac{11}{12}$ Add the numerators.

Concept Check Name a common denominator of $\frac{5}{9}$ and $\frac{4}{5}$.

You can rename unlike fractions using any common denominator. However, it is usually simpler to use the least common denominator.

Example 2 Add Fractions

Find $\frac{3}{8} + \frac{7}{12}$. **Estimate:** $\frac{1}{2} + \frac{1}{2} = 1$

$\frac{3}{8} + \frac{7}{12} = \frac{3}{8} \cdot \frac{3}{3} + \frac{7}{12} \cdot \frac{2}{2}$ The LCD is $2^3 \cdot 3$ or 24.

$= \frac{9}{24} + \frac{14}{24}$ Rename each fraction with the LCD.

$= \frac{23}{24}$ Add the numerators.

Example 3 Add Mixed Numbers

Find $1\frac{2}{9} + \left(-2\frac{1}{3}\right)$. Write in simplest form. **Estimate:** $1 + (-2) = -1$

$1\frac{2}{9} + \left(-2\frac{1}{3}\right) = \frac{11}{9} + \left(-\frac{7}{3}\right)$ Write the mixed numbers as improper fractions.

$= \frac{11}{9} + \left(-\frac{7}{3}\right) \cdot \frac{3}{3}$ Rename $-\frac{7}{3}$ using the LCD, 9.

$= \frac{11}{9} + \left(\frac{-21}{9}\right)$ Simplify.

$= \frac{-10}{9}$ Add the numerators.

$= -1\frac{1}{9}$ Simplify.

Study Tip

Negative Signs
When adding or subtracting a negative fraction, place the negative sign in the numerator.

SUBTRACT UNLIKE FRACTIONS The rule for subtracting fractions with unlike denominators is similar to the rule for addition.

Key Concept *Subtracting Unlike Fractions*

- **Words** To subtract fractions with unlike denominators, rename the fractions with a common denominator. Then subtract and simplify.
- **Example** $\frac{6}{7} - \frac{2}{3} = \frac{6}{7} \cdot \frac{3}{3} - \frac{2}{3} \cdot \frac{7}{7}$

  $= \frac{18}{21} - \frac{14}{21}$ or $\frac{4}{21}$

Example 4 Subtract Fractions

Find $\frac{5}{21} - \frac{6}{7}$.

$\frac{5}{21} - \frac{6}{7} = \frac{5}{21} - \frac{6}{7} \cdot \frac{3}{3}$ The LCD is 21.

$= \frac{5}{21} - \frac{18}{21}$ Rename $\frac{6}{7}$ using the LCD.

$= \frac{-13}{21}$ or $-\frac{13}{21}$ Subtract the numerators.

Example 5 Subtract Mixed Numbers

Find $6\frac{1}{2} - 4\frac{1}{5}$. Write in simplest form.

$6\frac{1}{2} - 4\frac{1}{5} = \frac{13}{2} - \frac{21}{5}$ Write the mixed numbers as improper fractions.

$= \frac{13}{2} \cdot \frac{5}{5} - \frac{21}{5} \cdot \frac{2}{2}$ Rename the fractions using the LCD.

$= \frac{65}{10} - \frac{42}{10}$ Simplify.

$= \frac{23}{10}$ or $2\frac{3}{10}$ Subtract.

Example 6 Use Fractions to Solve a Problem

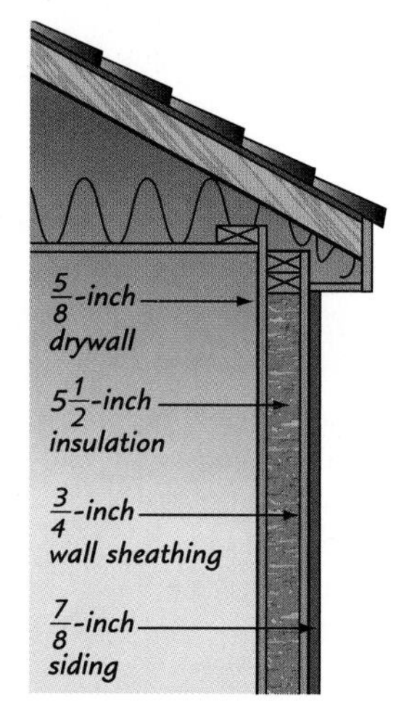

HOUSES **The diagram shows a cross-section of an outside wall. How thick is the wall?**

Explore You know the measure of each layer of the wall.

Plan Add the measures to find the total thickness of the wall. Estimate your answer.

$\frac{1}{2} + 5\frac{1}{2} + 1 + 1 = 8$

Solve $\frac{5}{8} + 5\frac{1}{2} + \frac{3}{4} + \frac{7}{8} = \frac{5}{8} + 5\frac{4}{8} + \frac{6}{8} + \frac{7}{8}$ Rename the fractions with the LCD, 8.

$= 5\frac{22}{8}$ Add the like fractions.

$= 7\frac{6}{8}$ or $7\frac{3}{4}$ in. Simplify.

The wall is $7\frac{3}{4}$ inches thick.

Examine Since $7\frac{3}{4}$ is close to 8, the answer is reasonable.

Check for Understanding

Concept Check

1. **Describe** the first step in adding or subtracting fractions with unlike denominators.
2. **OPEN ENDED** Write a real-world problem that you could solve by subtracting $2\frac{1}{8}$ from $15\frac{3}{4}$.
3. **FIND THE ERROR** José and Daniel are finding $\frac{9}{10} + \frac{7}{12}$.

José

$\frac{9}{10} + \frac{7}{12} = \frac{9}{10} \cdot \frac{12}{12} + \frac{7}{12} \cdot \frac{10}{10}$

Daniel

$\frac{9}{10} + \frac{7}{12} = \frac{9+7}{10+12}$

Who is correct? Explain your reasoning.

Guided Practice **Find each sum or difference. Write in simplest form.**

4. $\frac{1}{10} + \frac{1}{3}$
5. $-\frac{1}{6} + \frac{7}{18}$
6. $\frac{1}{4} - \frac{2}{3}$
7. $-\frac{7}{10} - \frac{2}{15}$
8. $6\frac{4}{5} + \left(-1\frac{3}{4}\right)$
9. $-9\frac{3}{4} - \left(-5\frac{1}{2}\right)$

Application 10. **SEWING** Jessica needs $1\frac{5}{8}$ yards of fabric to make a skirt and $3\frac{1}{2}$ yards to make a coat. How much fabric does she need in all?

Practice and Apply

Homework Help

For Exercises	See Examples
11–28	1–5
31–34	6

Extra Practice

See page 735.

Find each sum or difference. Write in simplest form.

11. $\frac{3}{5} + \frac{3}{10}$
12. $\frac{9}{26} + \frac{3}{13}$
13. $\frac{3}{7} + \left(-\frac{1}{4}\right)$
14. $-\frac{5}{8} + \left(-\frac{1}{3}\right)$
15. $\frac{7}{8} - \left(-\frac{3}{16}\right)$
16. $-\frac{2}{5} - \frac{7}{8}$
17. $\frac{3}{4} - \frac{5}{8}$
18. $\frac{5}{7} + \left(-\frac{10}{21}\right)$
19. $-\frac{1}{2} + \frac{3}{8}$
20. $-\frac{2}{3} + \frac{7}{12}$
21. $1\frac{2}{5} - \frac{1}{3}$
22. $\frac{7}{8} + 4\frac{1}{24}$
23. $-6\frac{2}{3} - \frac{8}{9}$
24. $2\frac{16}{30} - \frac{7}{15}$
25. $-4\frac{1}{6} + \left(-7\frac{11}{18}\right)$
26. $3\frac{1}{2} - \left(-7\frac{1}{3}\right)$
27. $-19\frac{3}{8} - \left(-4\frac{3}{4}\right)$
28. $-3\frac{2}{5} - \left(-2\frac{4}{7}\right)$

29. **ALGEBRA** Evaluate $x - y$ if $x = 4\frac{7}{18}$ and $y = 1\frac{1}{12}$.

30. **ALGEBRA** Solve $\frac{64}{143} - \frac{21}{208} = a$.

31. **EARTH SCIENCE** Did you know that water has a greater density than ice? Use the information in the table to find how much more water weighs per cubic foot.

1 Cubic Foot	Weight (lb)
water	$62\frac{1}{2}$
ice	$56\frac{9}{10}$

32. **GRILLING** Use the table to find the fraction of people who grill two, three, or four times per month.

33. **PUBLISHING** The length of a page in a yearbook is 10 inches. The top margin is $\frac{1}{2}$ inch, and the bottom margin is $\frac{3}{4}$ inch. What is the length of the page inside the margins?

34. **VOTING** In the class election, Murray received $\frac{1}{3}$ of the votes and Sara received $\frac{2}{5}$ of the votes. Makayla received the rest. What fraction of the votes did Makayla receive?

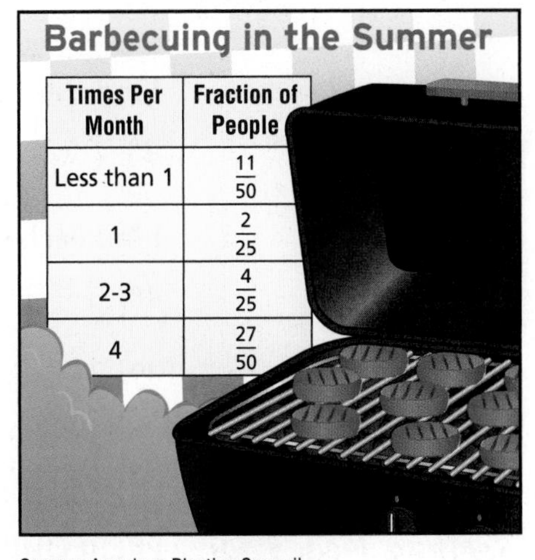

Barbecuing in the Summer

Times Per Month	Fraction of People
Less than 1	$\frac{11}{50}$
1	$\frac{2}{25}$
2-3	$\frac{4}{25}$
4	$\frac{27}{50}$

Source: American Plastics Council

35. **CRITICAL THINKING** A set of measuring cups has measures of 1 cup, $\frac{3}{4}$ cup, $\frac{1}{2}$ cup, $\frac{1}{3}$ cup, and $\frac{1}{4}$ cup. How could you get $\frac{1}{6}$ cup of milk by using these measures?

36. **CRITICAL THINKING** Do you think the rational numbers are closed under *addition, subtraction, multiplication,* or *division*? Explain.

37. **WRITING IN MATH** Answer the question that was posed at the beginning of the lesson.

How can the LCM be used to add and subtract fractions with different denominators?

Include the following in your answer:
- an example using the LCM, and
- an explanation of how prime factorization is a helpful way to add and subtract fractions that have different denominators.

38. For an art project, Halle needs $11\frac{3}{8}$ inches of red ribbon and $6\frac{7}{9}$ inches of white ribbon. Which is the best estimate for the total amount of ribbon that she needs?

Ⓐ 18 in. Ⓑ 26 in. Ⓒ 10 in. Ⓓ 8 in.

39. How much less is $\frac{6}{15}$ than $9\frac{1}{2}$?

Ⓐ $9\frac{27}{30}$ Ⓑ $9\frac{1}{10}$ Ⓒ $1\frac{11}{30}$ Ⓓ $9\frac{3}{10}$

Extending the Lesson

More than a thousand years ago, the Greeks wrote all fractions as the sum of *unit fractions.* A unit fraction is a fraction that has a numerator of 1, such as $\frac{1}{5}$, $\frac{1}{7}$, or $\frac{1}{4}$. Express each fraction below as the sum of two different unit fractions.

40. $\frac{7}{12}$ 41. $\frac{3}{5}$ 42. $\frac{2}{9}$

Maintain Your Skills

Mixed Review

Find the LCD of each pair of fractions. *(Lesson 5-6)*

43. $\frac{4}{9}, \frac{7}{12}$ 44. $\frac{3}{15t}, \frac{2}{5t}$ 45. $\frac{1}{3n}, \frac{7}{6n^3}$

Find each sum or difference. Write in simplest form. *(Lesson 5-5)*

46. $\frac{4}{7} + \frac{6}{7}$ 47. $2\frac{3}{4} + 6\frac{3}{4}$ 48. $\frac{7}{8} - \frac{5}{8}$

49. $3\frac{2}{5} - \frac{3}{5}$ 50. $4\frac{1}{6} + 5\frac{5}{6}$ 51. $8 - 6\frac{1}{5}$

52. Write the prime factorization of 124. *(Lesson 4-3)*

Getting Ready for the Next Lesson

PREREQUISITE SKILL Find each sum.
*(To review **adding integers**, see Lesson 2-2.)*

53. $24 + (-12) + 15$ 54. $(-2) + 5 + (-3)$

55. $4 + (-9) + (-9) + 5$ 56. $-10 + (-9) + (-11) + (-8)$

Algebra Activity

A Preview of Lesson 5-8

Sunshine State Standards
MA.E.1.3.2-1, MA.E.1.3.2-2, MA.E.1.3.2-3, MA.E.3.3.1-1, MA.E.3.3.1-2

Analyzing Data

Often, it is useful to describe or represent a set of data by using a single number. The table shows the daily maximum temperatures for twenty days during a recent April in Tampa, Florida.

One number to describe this data set might be 81. Some reasons for choosing this number are listed below.

- It occurs six times, more often than any other number.
- If the numbers are arranged in order from least to greatest, 81 falls in the center of the data set.

Tampa, Florida Maximum Temperatures (nearest °F)

81	79	82	82
81	82	81	73
84	72	84	73
82	81	72	66
75	81	81	82

Source: www.wunderground.com

There is an equal number of data above and below 81.

66 72 72 73 73 75 79 81 81 81 81 81 81 82 82 82 82 82 84 84

So, if you wanted to describe a typical high temperature for Tampa during April, you could say 81°F.

Collect the Data

Collect a group of data. Use one of the suggestions below, or use your own method.

- Research data about the weather in your city or in another city, such as temperatures, precipitation, or wind speeds.
- Find a graph or table of data in the newspaper or a magazine. Some examples include financial data, population data, and so on.
- Conduct a survey to gather some data about your classmates.
- Count the number of raisins in a number of small boxes.

Analyze the Data

1. Choose a number that best describes all of the data in the set.
2. Explain what your number means, and explain which method you used to choose your number.
3. Describe how your number might be useful in real life.

5-8 Measures of Central Tendency

What You'll Learn

Sunshine State Standards
MA.E.1.3.2-1, MA.E.1.3.2-2, MA.E.1.3.2-3, MA.E.3.3.1-1, MA.E.3.3.1-2

- Use the mean, median, and mode as measures of central tendency.
- Analyze data using mean, median, and mode.

Vocabulary

- measures of central tendency
- mean
- median
- mode

How are measures of central tendency used in the real world?

The *Iditarod* is a 1150-mile dogsled race across Alaska. The winning times for 1973–2000 are shown in the table.

Winning Times (days)						
20	21	15	19	17	15	15
14	12	16	13	13	18	12
11	11	11	11	13	11	11
11	9	9	9	9	10	9

Source: *Anchorage Daily News*

a. Which number appears most often?

b. If you list the data in order from least to greatest, which number is in the middle?

c. What is the sum of all the numbers divided by 28? If necessary, round to the nearest tenth.

d. If you had to give one number that best represents the winning times, which would you choose? Explain.

MEAN, MEDIAN, AND MODE When you have a list of numerical data, it is often helpful to use one or more numbers to represent the whole set. These numbers are called **measures of central tendency**. You will study three types.

Study Tip

Mean, Median, Mode

- The mean and median do not have to be part of the data set.
- If there is a mode, it is always a member of the data set.

Key Concept — *Measures of Central Tendency*

Statistic	Definition
mean	the sum of the data divided by the number of items in the data set
median	the middle number of the ordered data, or the mean of the middle two numbers
mode	the number or numbers that occur most often

Example 1 *Find the Mean, Median, and Mode*

SPORTS The heights of the players on the girls' basketball team are shown in the chart. Find the mean, median, and mode.

Height of Players (cm)		
130	154	148
155	172	153
160	162	140
149	151	150

$$\text{mean} = \frac{\text{sum of heights}}{\text{number of players}}$$

$$= \frac{130 + 154 + 148 + \ldots + 150}{12}$$

$$= \frac{1824}{12} \text{ or } 152$$

The mean height is 152 centimeters.

To find the median, order the numbers from least to greatest.

130, 140, 148, 149, 150, $\underbrace{151, 153}$, 154, 155, 160, 162, 172

$$\frac{151 + 153}{2} = 152$$

There is an even number of items. Find the mean of the two middle numbers.

The median height is 152.

There is no mode because each number in the set occurs once.

Example 2 Use a Line Plot

HURRICANES The line plot shows the number of Atlantic hurricanes that occurred each year from 1974–2000. Find the mean, median, and mode.

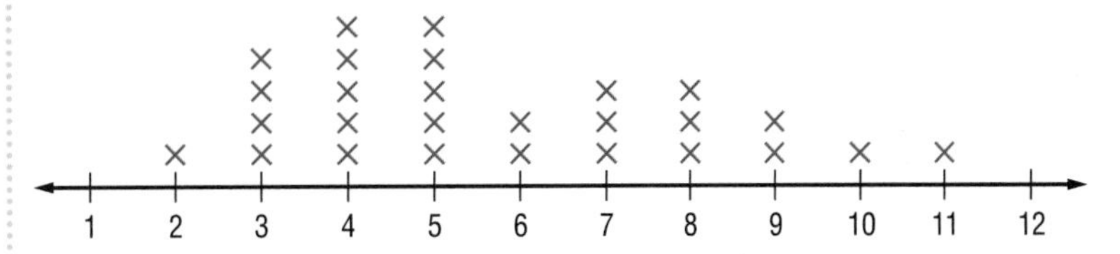

Source: Colorado State/Tropical Prediction Center

$$\text{mean} = \frac{2 + 3(4) + 4(5) + 5(5) + 6(2) + 7(3) + 8(3) + 9(2) + 10 + 11}{27} \approx 5.7$$

There are 27 numbers. So the middle number, which is the median, is the 14th number, or 5.

You can see from the graph that 4 and 5 both occur most often in the data set. So there are two modes, 4 and 5.

More About. . .

Hurricanes

The costliest hurricane to hit the U.S. mainland was Hurricane Andrew in 1992. It caused damages of $35 billion.

Source: www.explorezone.com

Concept Check If 4 were added to the data set, what would be the new mode?

A number in a set of data that is much greater or much less than the rest of the data is called an *extreme value*. An extreme value can affect the mean of the data.

Example 3 Find Extreme Values that Affect the Mean

NUTRITION The table shows the number of Calories per serving of each vegetable. Identify an extreme value and describe how it affects the mean.

Vegetable	Calories	Vegetable	Calories
asparagus	14	cauliflower	10
beans	30	celery	17
bell pepper	20	corn	66
broccoli	25	lettuce	9
cabbage	17	spinach	9
carrots	28	zucchini	17

The data value 66 appears to be an extreme value. Calculate the mean with and without the extreme value to find how it affects the mean.

mean with extreme value

$$\frac{\text{sum of values}}{\text{number of values}} = \frac{262}{12}$$

$$\approx 21.8$$

mean without extreme value

$$\frac{\text{sum of values}}{\text{number of values}} = \frac{196}{11}$$

$$\approx 17.8$$

The extreme value increases the mean by 21.8 − 17.8 or about 4.

ANALYZE DATA You can use measures of central tendency to analyze data.

Example 4 Use Mean, Median, and Mode to Analyze Data

HOURLY PAY Compare and contrast the central tendencies of the salaries for the two stores. Based on the averages, which store pays its employees better?

Hourly Salaries ($)	
Sports Superstore	Extreme Sports
7, 24, 7, 6, 8, 8, 8, 6	8, 9, 10, 10, 9, 8, 10, 10

Sports Superstore

mean:

$$= \frac{7 + 24 + 7 + 6 + 8 + 8 + 8 + 6}{8}$$

$= \$9.25$

median: 6, 6, 7, 7, 8, 8, 8, 24

$\frac{7 + 8}{2}$ or $7.50

mode: $8

Extreme Sports

mean:

$$= \frac{8 + 9 + 10 + 10 + 9 + 8 + 10 + 10}{8}$$

$= \$9.25$

median: 8, 8, 9, 9, 10, 10, 10, 10

$\frac{9 + 10}{2}$ or $9.50

mode: $10

The $24 per hour salary at Sports Superstore is an extreme value that increases the mean salary. However, the employees at Extreme Sports are generally better paid, as shown by the higher median and mode salaries.

If you know the value of the mean, you can work backward to find a missing value in the data set.

Example 5 Work Backward

Grid-In Test Item

Francisca needs an average score of 92 on five quizzes to earn an A. The mean of her first four scores was 91. What is the lowest score that she can receive on the fifth quiz to earn an A?

Read the Test Item To find the lowest score, write an equation to find the sum of the first four scores. Then write an equation to find the fifth score.

Solve the Test Item

Step 1 Find the sum of the first four scores x.

mean of first four scores → $91 = \frac{x}{4}$ ← sum of first four scores

$(91)4 = \left(\frac{x}{4}\right)4$ Multiply each side by 4.

$364 = x$ Simplify.

Step 2 Find the fifth score y.

$\text{mean} = \frac{\text{sum of the first four scores} + \text{fifth score}}{5}$ Write an equation.

$92 = \frac{364 + y}{5}$ Substitution

$460 = 364 + y$ Multiply each side by 5 and simplify.

$96 = y$ Subtract 364 from each side and simplify.

The Princeton Review

Test-Taking Tip

Substituting Check that your answer satisfies the conditions of the original problem.

Check for Understanding

Concept Check

1. **Explain** which measure of central tendency is most affected by an extreme value.
2. **OPEN ENDED** Write a set of data with at least four numbers that has a mean of 8 and a median that is *not* 8.

Guided Practice

Find the mean, median, and mode for each set of data. If necessary, round to the nearest tenth.

3. 4, 5, 7, 3, 9, 11, 23, 37
4. 7.2, 3.6, 9.0, 5.2, 7.2, 6.5, 3.6
5. 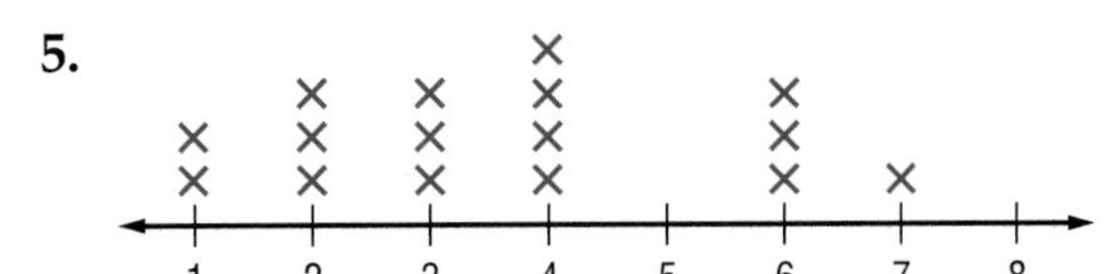

Application

VACATIONS For Exercises 6–8, use the table.

6. Find the mean, median, and mode.
7. Identify any extreme values and describe how they affect the mean.
8. Which statistic would you say best represents the data? Explain.

Annual Vacation Days	
Country	**Number of Days**
Brazil	34
Canada	26
France	37
Germany	35
Italy	42
Japan	25
Korea	25
United Kingdom	28
United States	13

Source: World Tourism Organization

FCAT Practice
Standardized Test Practice

9. Brad's average for five quizzes is 86. If he wants to have an average of 88 for six quizzes, what is the lowest score he can receive on his sixth quiz?

Practice and Apply

Homework Help

For Exercises	See Examples
10–13	1
14, 15	2
16	3
17, 18	4

Extra Practice

See page 735.

Find the mean, median, and mode for each set of data. If necessary, round to the nearest tenth.

10. 41, 37, 43, 43, 36
11. 2, 8, 16, 21, 3, 8, 9, 7, 6
12. 14, 6, 8, 10, 9, 5, 7, 13
13. 7.5, 7.1, 7.4, 7.6, 7.4, 9.0, 7.9, 7.1
14. 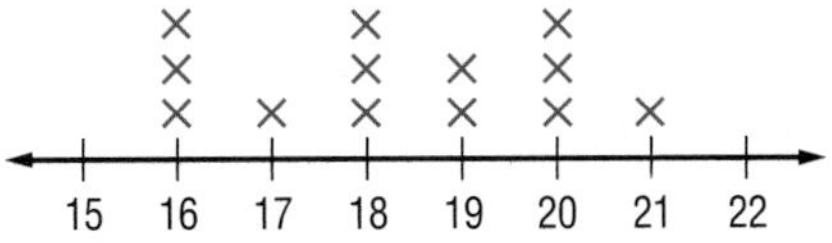 15 16 17 18 19 20 21 22
15. 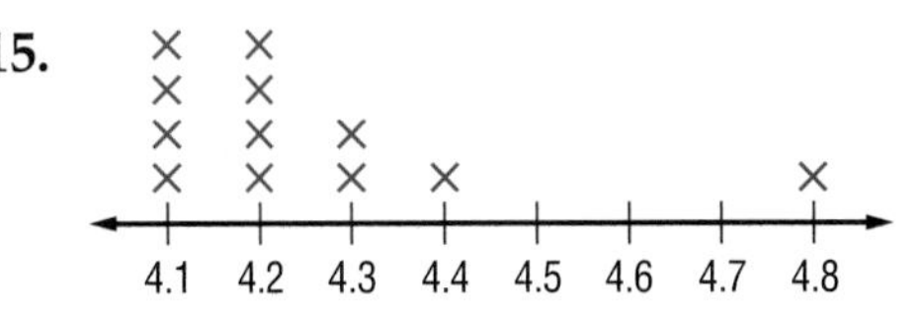

16. **BASKETBALL** Refer to the cartoon at the right. Which measure of central tendency would make opponents believe that the height of the team is much taller than it really is? Explain.

17. **TESTS** Which measure of central tendency best summarizes the test scores shown below? Explain.
97, 99, 95, 89, 99, 100, 87, 85, 89, 92, 96, 95, 60, 97, 85

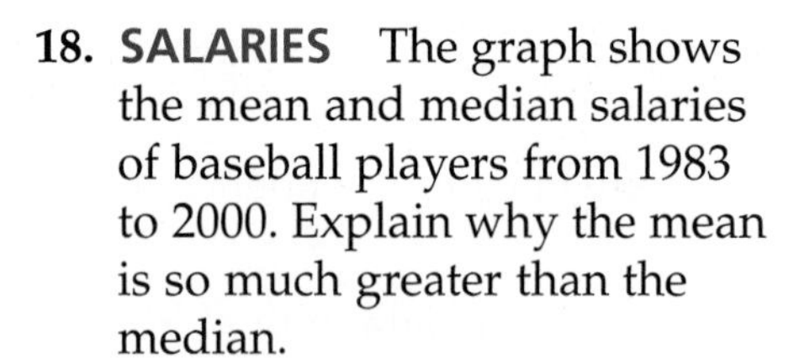

18. SALARIES The graph shows the mean and median salaries of baseball players from 1983 to 2000. Explain why the mean is so much greater than the median.

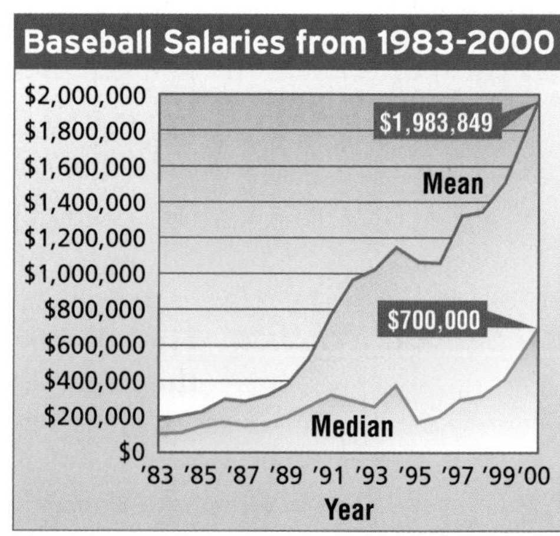

Source: USA TODAY research

Using measures of central tendency can help you analyze the data from fast-food restaurants. Visit www.pre-alg.com/webquest to continue work on your WebQuest project.

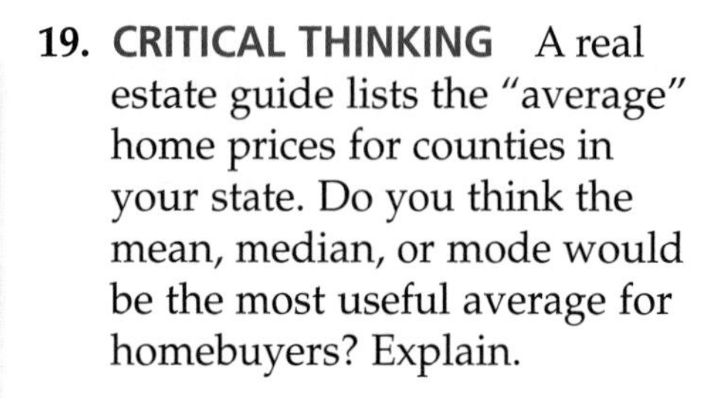

19. CRITICAL THINKING A real estate guide lists the "average" home prices for counties in your state. Do you think the mean, median, or mode would be the most useful average for homebuyers? Explain.

20. WRITING IN MATH Answer the question that was posed at the beginning of the lesson.

How are measures of central tendency used in the real world?

Include the following in your answer:

- examples of real-life data from home or school that can be described using the mean, median, or mode, and
- one or more newspaper articles in which averages are used.

21. If 18 were added to the data set below, which statement is true?

16, 14, 22, 16, 16, 18, 15, 25

(A) The mode increases. (B) The mean decreases.

(C) The mean increases. (D) The median increases.

22. Jonelle's tips as a waitress are shown in the table. On Friday, her tips were $74. Which measure of central tendency will change the *most* as a result?

(A) mean (B) median

(C) mode (D) no measure

Day	Tips
Monday	$36
Tuesday	$32
Wednesday	$40
Thursday	$36

Maintain Your Skills

Mixed Review **Find each sum or difference. Write in simplest form.** *(Lesson 5-7)*

23. $9\frac{2}{3} + \frac{1}{6}$ **24.** $\frac{7}{8} - \frac{3}{10}$ **25.** $-2\frac{3}{4} - 1\frac{1}{8}$

Replace each ● with <, >, or = to make a true statement. *(Lesson 5-6)*

26. $\frac{1}{2}$ ● $\frac{5}{12}$ **27.** $\frac{16}{50}$ ● $\frac{9}{30}$ **28.** $\frac{4}{5}$ ● $\frac{48}{60}$

ALGEBRA Solve each equation. Check your solution. *(Lessons 3-3 and 3-4)*

29. $y - 5 = 13$ **30.** $10 = 14 + n$ **31.** $-4w = 20$

Getting Ready for the Next Lesson **PREREQUISITE SKILL Find each quotient. If necessary, round to the nearest tenth.** *(To review **dividing decimals**, see page 715.)*

32. $25.6 \div 3$ **33.** $37 \div 4.7$ **34.** $30.5 \div 11.2$ **35.** $46.8 \div 15.6$

Graphing Calculator Investigation

A Follow-Up of Lesson 5-8

Sunshine State Standards
MA.E.1.3.2-1, MA.E.1.3.2-2, MA.E.1.3.2-3, MA.E.1.3.3-1 MA.E.1.3.3-2

Mean and Median

A graphing calculator is able to perform operations on large data sets efficiently. You can use a TI-83 Plus graphing calculator to find the mean and median of a set of data.

SURVEYS Fifteen seventh graders were surveyed and asked what was their weekly allowance (in dollars). The results of the survey are shown at the right.

20	10	5	5	10
15	5	10	5	10
5	5	5	5	5

Find the mean and median allowance.

Step 1 *Enter the data.*

- Clear any existing lists.

KEYSTROKES: STAT ENTER ▲ CLEAR ENTER

- Enter the allowances as L1.

KEYSTROKES: 20 ENTER 10 ENTER … 5 ENTER

Step 2 *Find the mean and median.*

- Display a list of statistics for the data.

KEYSTROKES: STAT ▶ ENTER ENTER

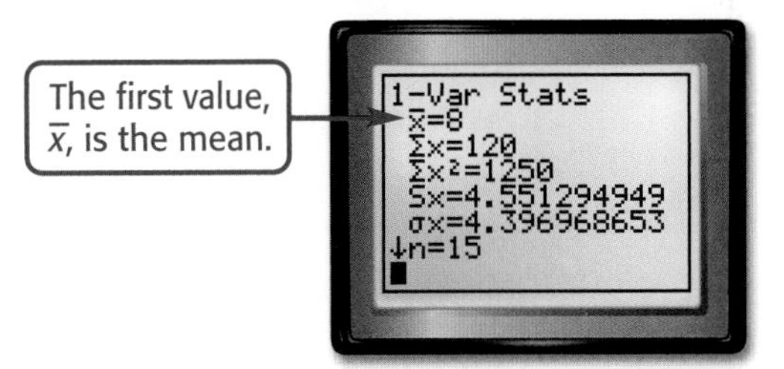

Use the down arrow key to locate "Med." The median allowance is \$5 and the mean allowance is \$8.

Exercises

Clear list L1 and find the mean and median of each data set. Round decimal answers to the nearest hundredth.

1. 6.4, 5.6, 7.3, 1.2, 5.7, 8.9
2. −23, −13, −16, −21, −15, −34, −22
3. 123, 423, 190, 289, 99, 178, 156, 217, 217
4. 8.4, 2.2, −7.3, −5.3, 6.7, −4.3, 5.1, 1.3, −1.1, −3.2, 2.2, 2.9, 1.4, 68
5. Look back at the medians found. When is the median a member of the data set?
6. Refer to Exercise 4.
   a. Which statistic best represents the data, the mean or median? Explain.
   b. Suppose the number 68 should have been 6.8. Recalculate the mean and median. Is there a significant difference between the first pair of values and the second pair?
   c. When there is an error in one of the data values, which statistic is least likely to be affected? Why?

www.pre-alg.com/other_calculator_keystrokes

5-9 Solving Equations with Rational Numbers

Sunshine State Standards
MA.A.3.3.2-1

What You'll Learn

- Solve equations containing rational numbers.

How are reciprocals used in solving problems involving music?

Musical sounds are made by vibrations. If n represents the number of vibrations for middle C, then the approximate vibrations for the other notes going up the scale are given below.

Notes	Middle C	D	E	F	G	A	B	C
Number of Vibrations	n	$\frac{9}{8}n$	$\frac{5}{4}n$	$\frac{4}{3}n$	$\frac{3}{2}n$	$\frac{5}{3}n$	$\frac{15}{8}n$	$\frac{2}{1}n$

a. A guitar string vibrates 440 times per second to produce the A above middle C. Write an equation to find the number of vibrations per second to produce middle C. If you multiply each side by 3, what is the result?

b. How would you solve the second equation you wrote in part **a**?

c. How can you combine the steps in parts **a** and **b** into one step?

d. How many vibrations per second are needed to produce middle C?

SOLVE ADDITION AND SUBTRACTION EQUATIONS You can solve rational number equations the same way you solved equations with integers.

Study Tip

Look Back
To review **solving equations**, see Lessons 3-3 and 3-4.

Example 1 Solve by Using Addition

Solve 2.1 = t − 8.5. Check your solution.

$2.1 = t - 8.5$ Write the equation.

$2.1 + 8.5 = t - 8.5 + 8.5$ Add 8.5 to each side.

$10.6 = t$ Simplify.

CHECK $2.1 = t - 8.5$ Write the original equation.

$2.1 \stackrel{?}{=} 10.6 - 8.5$ Replace t with 10.6.

$2.1 = 2.1 \checkmark$ Simplify.

Example 2 Solve by Using Subtraction

Solve $x + \frac{3}{5} = \frac{2}{3}$.

$x + \frac{3}{5} = \frac{2}{3}$ Write the equation.

$x + \frac{3}{5} - \frac{3}{5} = \frac{2}{3} - \frac{3}{5}$ Subtract $\frac{3}{5}$ from each side.

$x = \frac{2}{3} - \frac{3}{5}$ Simplify.

$x = \frac{10}{15} - \frac{9}{15}$ or $\frac{1}{15}$ Rename the fractions using the LCD and subtract.

Study Tip

Reciprocals
Recall that dividing by a fraction is the same as multiplying by its multiplicative inverse.

SOLVE MULTIPLICATION AND DIVISION EQUATIONS

To solve $\frac{1}{2}x = 3$, you can divide each side by $\frac{1}{2}$ or multiply each side by the multiplicative inverse of $\frac{1}{2}$, which is 2.

$\frac{1}{2}x = 3$	Write the equation.
$2 \cdot \frac{1}{2}x = 2 \cdot 3$	Multiply each side by 2.
$1x = 6$	Simplify.
$x = 6$	

The product of any number and its multiplicative inverse is 1.

Example 3 Solve by Using Division

Solve $-3y = 1.5$. Check your solution.

$-3y = 1.5$	Write the equation.
$\frac{-3y}{-3} = \frac{1.5}{-3}$	Divide each side by -3.
$y = -0.5$	Simplify.

CHECK

$-3y = 1.5$	Write the original equation.
$-3(-0.5) \stackrel{?}{=} 1.5$	Replace y with -0.5.
$1.5 = 1.5$ ✓	Simplify.

Example 4 Solve by Using Multiplication

a. Solve $5 = \frac{1}{4}y$. Check your solution.

$5 = \frac{1}{4}y$	Write the equation.
$4(5) = 4\left(\frac{1}{4}y\right)$	Multiply each side by 4.
$20 = y$	Simplify.

CHECK

$5 = \frac{1}{4}y$	Write the original equation.
$5 \stackrel{?}{=} \frac{1}{4}(20)$	Replace y with 20.
$5 = 5$ ✓	Simplify.

b. Solve $\frac{2}{3}x = 7$. Check your solution.

$\frac{2}{3}x = 7$	Write the equation.
$\frac{3}{2}\left(\frac{2}{3}x\right) = \frac{3}{2}(7)$	Multiply each side by $\frac{3}{2}$.
$x = \frac{21}{2}$	Simplify.
$x = 10\frac{1}{2}$	Simplify. Check the solution.

Concept Check What is the first step in solving $\frac{1}{8}r = \frac{1}{4}$?

Check for Understanding

Concept Check

1. **Name** the property of equality that you would use to solve $2 = \frac{3}{4} + x$.
2. **OPEN ENDED** Write an equation that can be solved by multiplying each side by 6.
3. **FIND THE ERROR** Grace and Ling are solving $0.3x = 4.5$.

Grace	Ling
$0.3x = 4.5$	$0.3x = 4.5$
$\frac{0.3x}{3} = \frac{4.5}{3}$	$\frac{0.3x}{0.3} = \frac{4.5}{0.3}$
$x = 1.5$	$x = 15$

Who is correct? Explain your reasoning.

Guided Practice

Solve each equation. Check your solution.

4. $y + 3.5 = 14.9$
5. $b - 5 = 13.7$
6. $\frac{3}{2} = w + \frac{3}{5}$
7. $c - \frac{3}{5} = \frac{5}{6}$
8. $x + \frac{5}{8} = 7\frac{1}{2}$
9. $4\frac{1}{6} = r + 6\frac{1}{4}$
10. $3.5a = 7$
11. $-\frac{1}{6}s = 15$
12. $9 = \frac{3}{4}g$

Application

13. **METEOROLOGY** When a storm struck, the barometric pressure was 28.79 inches. Meteorologists said that the storm caused a 0.36-inch drop in pressure. What was the barometric pressure before the storm?

Practice and Apply

Homework Help

For Exercises	See Examples
14–27, 37–40	1, 2
28–36	3, 4

Extra Practice

See page 736.

Solve each equation. Check your solution.

14. $y + 7.2 = 21.9$
15. $4.7 = a + 7.1$
16. $x - 5.3 = 8.1$
17. $n - 4.72 = 7.52$
18. $t + 3.17 = -3.17$
19. $a - 2.7 = 3.2$
20. $\frac{2}{3} = \frac{1}{8} + b$
21. $m + \frac{7}{12} = -\frac{5}{18}$
22. $g + \frac{2}{3} = 2$
23. $7 = \frac{2}{9} + k$
24. $n - \frac{3}{8} = \frac{1}{6}$
25. $x - \frac{2}{5} = -\frac{8}{15}$
26. $7\frac{1}{3} = c - \frac{4}{5}$
27. $-2 = \frac{3}{10} + f$
28. $\frac{7}{9}k = -\frac{5}{12}$
29. $4.1p = 16.4$
30. $8 = \frac{2}{3}d$
31. $0.4y = 2$
32. $\frac{1}{5}t = 9$
33. $4 = -\frac{1}{8}q$
34. $\frac{1}{3}n = \frac{2}{9}$
35. $\frac{5}{8} = \frac{1}{2}r$
36. $\frac{2}{3}a = 6$
37. $b - 1\frac{1}{2} = 4\frac{1}{4}$
38. $7\frac{1}{2} = r - 5\frac{2}{3}$
39. $3\frac{3}{4} + n = 6\frac{5}{8}$
40. $y + 1\frac{1}{3} = 3\frac{1}{18}$

41. **PUBLISHING** A newspaper is $12\frac{1}{4}$ inches wide and 22 inches long. This is $1\frac{1}{4}$ inches narrower and half an inch longer than the old edition. What were the previous dimensions of the newspaper?

42. AIRPORTS The graph shows the world's busiest cargo airports. What is the difference in cargo handling between Memphis and Tokyo?

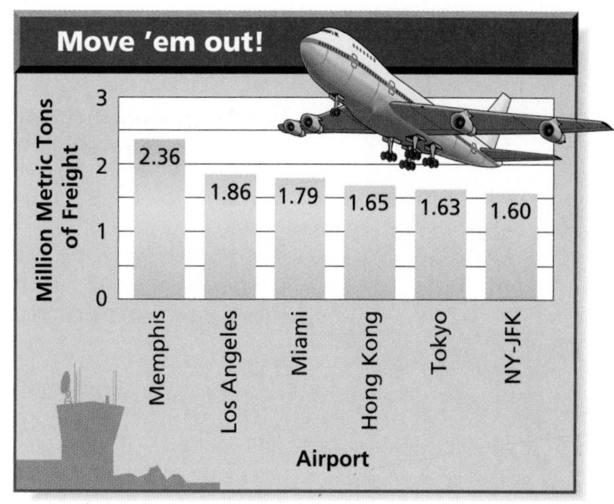

Source: Airports Council International

43. BUSINESS A store is going out of business. All of the items are marked $\frac{1}{3}$ off the ticketed price. How much would a shirt that was originally priced at \$24.99 cost now?

More About. . .

Airports

In 1995, there were 580 million airline passengers. In 2005, there will be an estimated 823 million airline passengers.

Source: Department of Transportation

44. COOKING Lucas made $2\frac{1}{2}$ batches of cookies for a bake sale and used $3\frac{3}{4}$ cups of sugar. How much sugar is needed for one batch of cookies?

45. TRAINS As a train begins to roll, the cars are "jerked" into motion. Slack is built into the couplings so that the engine does not have to move every car at once. If the slack built into each coupling is 3 inches, how many feet of slack is there between ten freight cars? (*Hint:* Do not include the coupling between the engine and the first car.)

46. CRITICAL THINKING The denominator of a fraction is 4 more than the numerator. If both the numerator and denominator are increased by 1, the resulting fraction equals $\frac{1}{2}$. Find the original fraction.

47. WRITING IN MATH Answer the question that was posed at the beginning of the lesson.

How are reciprocals used in solving problems involving music?

Include the following in your answer:

- an example of how fractions are used to compare musical notes, and
- an explanation of how reciprocals are useful in finding the number of vibrations per second needed to produce certain notes.

48. Find the value of z in $\frac{5}{6}z = \frac{3}{5}$.

Ⓐ $\frac{18}{25}$ Ⓑ $\frac{15}{30}$ Ⓒ $\frac{1}{2}$ Ⓓ $1\frac{7}{18}$

49. The area A of the triangle is $33\frac{3}{4}$ square centimeters. Use the formula $A = \frac{1}{2}bh$ to find the height h of the triangle.

h

9 cm

Ⓐ $3\frac{13}{18}$ cm Ⓑ $6\frac{1}{2}$ cm

Ⓒ $18\frac{13}{18}$ cm Ⓓ $7\frac{1}{2}$ cm

Maintain Your Skills

Mixed Review

Find the mean, median, and mode for each set of data. If necessary, round to the nearest tenth. *(Lesson 5-8)*

50. 2, 8, 5, 18, 3, 5, 6

51. 11, 12, 12, 14, 16, 11, 15

52. 0.9, 0.5, 0.7, 0.4, 0.3, 0.2

53. 56, 77, 60, 60, 72, 100

Find each sum or difference. Write in simplest form. *(Lesson 5-7)*

54. $\frac{3}{5} + \frac{1}{3}$

55. $\frac{5}{6} - \frac{1}{8}$

56. $-4\frac{1}{4} - \frac{1}{6}$

57. $\frac{5}{9} + \left(-\frac{1}{12}\right)$

58. $-3\frac{3}{4} + \left(-2\frac{1}{8}\right)$

59. $8\frac{9}{10} - 1\frac{1}{6}$

60. ALGEBRA Evaluate $a - b$ if $a = 9\frac{5}{6}$ and $b = 1\frac{1}{6}$. *(Lesson 5-5)*

61. HEALTH According to the National Sleep Foundation, teens should get approximately 9 hours of sleep each day. What fraction of the day is this? Write in simplest form. *(Lesson 4-5)*

ALGEBRA Solve each equation. Check your solution. *(Lesson 3-5)*

62. $3t - 6 = 15$

63. $8 = \frac{k}{-2} + 5$

64. $9n - 13n = 4$

Getting Ready for the Next Lesson

PREREQUISITE SKILL Divide. If necessary, write as a fraction in simplest form. *(To review **dividing integers**, see Lesson 2-5.)*

65. $-18 \div 3$

66. $24 \div (-2)$

67. $-20 \div (-4)$

68. $-55 \div (-5)$

69. $-12 \div 36$

70. $9 \div (-81)$

Practice Quiz 2

Lessons 5-6 through 5-9

Find the least common multiple of each set of numbers. *(Lesson 5-6)*

1. 8, 9

2. 12, 30

3. 2, 10

4. 6, 8

Find each sum or difference. Write in simplest form. *(Lesson 5-7)*

5. $\frac{3}{4} + \frac{1}{16}$

6. $-\frac{3}{7} + \frac{5}{14}$

7. $1\frac{1}{5} - \left(-\frac{7}{15}\right)$

8. $6\frac{3}{4} - 2\frac{1}{6}$

9. WEATHER The low temperatures on March 24 for twelve different cities are recorded at the right. Find the mean, median, and mode. If necessary, round to the nearest tenth. *(Lesson 5-8)*

Low Temperatures (°F)			
31	29	30	22
45	35	29	30
40	29	31	38

10. ALGEBRA Solve $a - 1\frac{1}{3} = 4\frac{1}{6}$. *(Lesson 5-9)*

5-10 Arithmetic and Geometric Sequences

Sunshine State Standards
MA.A.5.3.1-2

Vocabulary

- sequence
- arithmetic sequence
- term
- common difference
- geometric sequence
- common ratio

What You'll Learn

- Find the terms of arithmetic sequences.
- Find the terms of geometric sequences.

How can sequences be used to make predictions?

The table shows the distance a car moves during the time it takes to apply the brakes and while braking.

Speed (mph)	Reaction Distance (ft)	Braking Distance (ft)
20	20	20
30	30	45
40	40	80
50	50	125
60	60	180

a. What is the reaction distance for a car going 70 mph?

b. What is the braking distance for a car going 70 mph?

c. What is the difference in reaction distances for every 10-mph increase in speed?

d. Describe the braking distance as speed increases.

ARITHMETIC SEQUENCES A **sequence** is an ordered list of numbers. An **arithmetic sequence** is a sequence in which the difference between any two consecutive terms is the same. So, you can find the next term in the sequence by adding the same number to the previous term.

Each number is called a **term** of the sequence. → 20, 30, 40, 50, 60, …

+10 +10 +10 +10 ← The difference is called the **common difference**.

Example 1 Identify an Arithmetic Sequence

State whether the sequence 8, 5, 2, −1, −4, … is arithmetic. If it is, state the common difference and write the next three terms.

8, 5, 2, −1, −4 Notice that $5 - 8 = -3$, $2 - 5 = -3$, and so on.

−3 −3 −3 −3

The terms have a common difference of −3, so the sequence is arithmetic. Continue the pattern to find the next three terms.

−4, −7, −10, −13

−3 −3 −3

The next three terms of the sequence are −7, −10, and −13.

Example 2 Identify an Arithmetic Sequence

State whether the sequence 1, 2, 4, 7, 11, ... is arithmetic. If it is, write the next three terms of the sequence.

1, 2, 4, 7, 11

+1 +2 +3 +4 — The terms do not have a common difference.

The sequence is not arithmetic. However, if the pattern continues, the next three differences will be 5, 6, and 7.

11, 16, 22, 29

+5 +6 +7 — The next three terms are 16, 22, and 29.

GEOMETRIC SEQUENCES A **geometric sequence** is a sequence in which the quotient of any two consecutive terms is the same. So, you can find the next term in the sequence by multiplying the previous term by the same number.

1, 4, 16, 64, 256, ...

×4 ×4 ×4 ×4 ← The quotient is called the **common ratio**.

Concept Check Name the common ratio in the sequence 10, 20, 40, 80, 160, ...

Example 3 Identify Geometric Sequences

a. **State whether the sequence −2, 6, −18, 54, ... is geometric. If it is, state the common ratio and write the next three terms.**

−2, 6, −18, 54

×(−3) ×(−3) ×(−3)

Notice that $6 \div (-2) = -3$, $-18 \div 6 = -3$, and $54 \div (-18) = -3$.

The common ratio is −3, so the sequence is geometric. Continue the pattern to find the next three terms.

54, −162, 486, −1458

×(−3) ×(−3) ×(−3) — The next three terms are −162, 486, and −1458.

b. **State whether the sequence 20, 10, 5, $\frac{5}{2}$, $\frac{5}{4}$, ... is geometric. If it is, state the common ratio and write the next three terms.**

20, 10, 5, $\frac{5}{2}$, $\frac{5}{4}$

$\times\frac{1}{2}$ $\times\frac{1}{2}$ $\times\frac{1}{2}$ $\times\frac{1}{2}$

The common ratio is $\frac{1}{2}$ or 0.5, so the sequence is geometric. Continue the pattern to find the next three terms.

$\frac{5}{4}$, $\frac{5}{8}$, $\frac{5}{16}$, $\frac{5}{32}$

$\times\frac{1}{2}$ $\times\frac{1}{2}$ $\times\frac{1}{2}$ — The next three terms are $\frac{5}{8}$, $\frac{5}{16}$, and $\frac{5}{32}$.

Study Tip

Alternative Method

Multiplying by $\frac{1}{2}$ is the same as dividing by 2, so the following is also true.

20, 10, 5, $\frac{5}{2}$, $\frac{5}{4}$

÷2 ÷2 ÷2 ÷2

Check for Understanding

Concept Check

1. **Compare and contrast** arithmetic and geometric sequences.

2. **OPEN ENDED** Describe the terms of a geometric sequence whose common ratio is a fraction or decimal between 0 and 1. Then write four terms of such a sequence and name the common ratio.

Guided Practice

State whether each sequence is *arithmetic, geometric,* or *neither.* If it is arithmetic or geometric, state the common difference or common ratio and write the next three terms of the sequence.

3. 3, 7, 11, 15, …
4. 1, 3, 9, 27, …
5. 6, 8, 12, 18, …
6. 13, 8, 3, −2, …
7. $3, \frac{8}{3}, \frac{7}{3}, 2, \ldots$
8. $48, 12, 3, \frac{3}{4}, \ldots$

Application

9. **CARS** A new car is worth only about 0.82 of its value from the previous year during the first three years. Approximately how much will a $20,000 car be worth in 3 years? **Source:** www.caprice.com

Practice and Apply

Homework Help

For Exercises	See Examples
10–29	1–3

Extra Practice
See page 736.

State whether each sequence is *arithmetic, geometric,* or *neither.* If it is arithmetic or geometric, state the common difference or common ratio and write the next three terms of the sequence.

10. 2, 5, 8, 11, …
11. −6, 5, 16, 27, …
12. $\frac{1}{2}, 1, 2, 4, \ldots$
13. 2, 6, 18, 54, …
14. 18, 11, 4, −3, …
15. 25, 22, 19, 16, …
16. $4, 1, \frac{1}{4}, \frac{1}{16}, \ldots$
17. $-5, 1, -\frac{1}{5}, \frac{1}{25}, \ldots$
18. $\frac{1}{2}, 1, \frac{3}{2}, 2, \ldots$
19. $0, \frac{1}{6}, \frac{1}{3}, \frac{1}{2}, \ldots$
20. 0.75, 1.5, 2.25, …
21. 4.5, 4.0, 3.5, 3.0, …
22. 11, 14, 19, 26, …
23. 17, 16, 14, 11, …
24. 24, 12, 6, 3, …
25. $18, -6, 2, -\frac{2}{3}, \ldots$
26. 0.1, 0.3, 0.9, 2.7, …
27. $\frac{1}{2}, \frac{1}{4}, \frac{1}{8}, \frac{1}{16}, \ldots$

28. **PHYSICAL SCIENCE** A ball bounces back 0.75 of its height on every bounce. If a ball is dropped from 160 feet, how high does it bounce on the third bounce?

29. **TELEPHONE RATES** For an overseas call, WorldTel charges $6 for the first minute and then $3 for each additional minute.
    a. Is the cost an arithmetic or geometric sequence? Explain.
    b. How much would a 15-minute call cost?

30. **CRITICAL THINKING** The sum of four numbers in an arithmetic sequence is 42. What could the numbers be? Give two different examples.

31. **WRITING IN MATH** Answer the question that was posed at the beginning of the lesson.

    How can sequences be used to make predictions?

    Include the following in your answer:
    - a discussion of common difference and common ratio, and
    - examples of sequences occurring in nature.

More About. . .

Physical Science

If you put a baseball in a freezer for an hour, it will bounce only 0.8 as high as a baseball at room temperature.
Source: www.exploratorium.edu

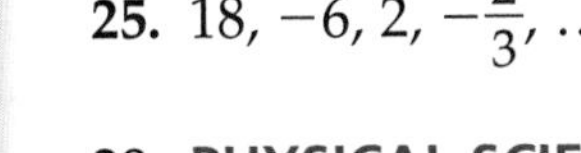

32. State the next term in the sequence 56, 48, 40, 32, ….

(A) 20 (B) 22 (C) 24 (D) 28

33. Which statement is true as the side length of a square increases?

Side Length	Perimeter	Area
1	4	1
2	8	4
3	12	9
4	16	16
5	20	25

(A) The perimeter values form an arithmetic sequence.

(B) The perimeter values form a geometric sequence.

(C) The area values form an arithmetic sequence.

(D) The area values form a geometric sequence.

Extending the Lesson

In an arithmetic sequence, d represents the common difference, a_1 represents the first term, a_2 represents the second term, and so on. For example, in the sequence 6, 9, 12, 15, 18, 21, …, $a_1 = 6$ and $d = 3$.

34. Use the information in the table to write an expression for finding the nth term of an arithmetic sequence.

Arithmetic Sequence	numbers	6	9	12	15	…	
	symbols	a_1	a_2	a_3	a_4	…	a_n
Expressed in Terms of d and the First Term	numbers	6 + 0(3)	6 + 1(3)	6 + 2(3)	6 + 3(3)	…	6 + (n − 1)(3)
	symbols	$a_1 + 0d$	$a_1 + 1d$	$a_1 + 2d$	$a_1 + 3d$	…	?

35. Use the expression you wrote in Exercise 34 to find the 9th term of an arithmetic sequence if $a_1 = 16$ and $d = 5$.

Maintain Your Skills

Mixed Review **ALGEBRA Solve each equation. Check your solution.** *(Lesson 5-9)*

36. $\frac{7}{6} = y + \frac{5}{12}$

37. $k - 4.1 = -9.38$

38. $40.3 = 6.2x$

39. $\frac{3}{4}b = 7\frac{1}{2}$

Find the mean, median, and mode for each set of data. If necessary, round to the nearest tenth. *(Lesson 5-8)*

40. 11, 45, 62, 12, 47, 8, 12, 35

41. 2.3, 3.6, 4.1, 3.6, 2.9, 3.0

42. **ALGEBRA** Find the LCD of $\frac{1}{6a^2}$ and $\frac{5}{9a^3}$. *(Lesson 5-6)*

43. **HOME REPAIR** Cole is installing shelves in his closet. Because of the shape of the closet, the three shelves measure $34\frac{3}{8}$ inches, $33\frac{7}{8}$ inches, and $34\frac{5}{8}$ inches. What length of lumber does he need to buy? *(Lesson 5-5)*

Find each product or quotient. Express using exponents. *(Lesson 4-6)*

44. $3^4 \cdot 3^2$

45. $\frac{b^5}{b^2}$

46. $2x^3(5x^2)$

Find the greatest common factor of each set of numbers. *(Lesson 4-4)*

47. 36, 42

48. 9, 24

49. 60, 45, 30

Algebra Activity

A Follow-Up of Lesson 5-10

Fibonacci Sequence

A special sequence that is neither arithmetic nor geometric is called the **Fibonacci sequence**. Each term in the sequence is the sum of the previous two terms, beginning with 1.

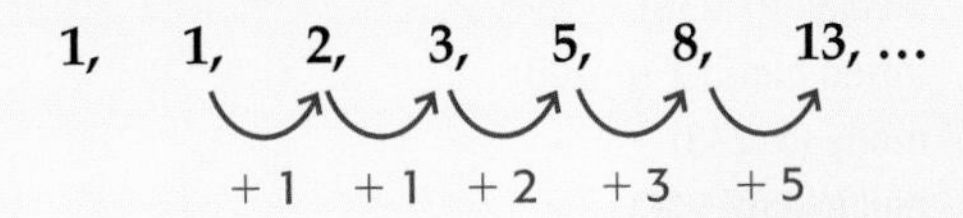

Collect the Data

- Examine an artichoke, a pineapple, a pinecone, or the seeds in the center of a sunflower.
- For each item, count the number of spiral rows and record your data in a table. If possible, count the rows that spiral up from left to right and count the rows that spiral up from right to left. ***Note:*** It may be helpful to use a marker to keep track of the rows as you are counting.

Analyze the Data

1. What do you notice about the number of rows in each item?

2. Compare your data with the data of the other students. How do they compare?

Make a Conjecture

3. Have a discussion with other students to determine the relationship between the number of rows in sunflowers, pinecones, pineapples, and artichokes and the Fibonacci sequence.

Extend the Activity

Numbers in an arithmetic sequence have a common difference, and numbers in a geometric sequence have a common ratio. The numbers in a Fibonacci sequence have a different kind of pattern.

4. Write the first fifteen terms in the Fibonacci sequence.

5. Use a calculator to divide each term by the previous term. Make a list of the quotients. If necessary, round to seven decimal places.

6. Describe the pattern in the quotients.

7. **RESEARCH** Find the definition of the **golden ratio**. What is the relationship between the numbers in the Fibonacci sequence and the golden ratio?

Chapter 5

Study Guide and Review

Vocabulary and Concept Check

algebraic fraction (p. 211)
arithmetic sequence (p. 249)
bar notation (p. 201)
common difference (p. 249)
common multiples (p. 226)
common ratio (p. 250)
dimensional analysis (p. 212)
Fibonacci sequence (p. 253)
geometric sequence (p. 250)
least common denominator (LCD) (p. 227)
least common multiple (LCM) (p. 226)
mean (p. 238)
measures of central tendency (p. 238)
median (p. 238)
mixed number (p. 200)
mode (p. 238)
multiple (p. 226)
multiplicative inverse (p. 215)
period (p. 201)
rational number (p. 205)
reciprocals (p. 215)
repeating decimal (p. 201)
sequence (p. 249)
term (p. 249)
terminating decimal (p. 200)

Choose the correct term to complete each sentence.

1. A mixed number is an example of a (whole, rational) number.
2. The decimal 0.900 is a (terminating, repeating) decimal.
3. $\frac{x}{15}$ is an example of an (algebraic fraction, integer).
4. The numbers 12, 15, and 18 are (factors, multiples) of 3.
5. To add unlike fractions, rename the fractions using the (LCD, GCF).
6. The product of a number and its multiplicative inverse is (0, 1).
7. To divide by a fraction, multiply the number by the (reciprocal, LCD) of the fraction.
8. The (median, mode) is the number that occurs most often in a set of data.
9. A common difference is found between terms in a(n) (arithmetic, geometric) sequence.

Lesson-by-Lesson Review

5-1 Writing Fractions as Decimals

See pages 200–204.

Concept Summary

- Any fraction or mixed number can be written as a terminating or repeating decimal.

Write $2\frac{7}{10}$ as a decimal.

$2\frac{7}{10} = 2 + \frac{7}{10}$ Write as the sum of an integer and a fraction.

$= 2 + 0.7$ or 2.7 Write $\frac{7}{10}$ as a decimal and add.

Exercises **Write each fraction or mixed number as a decimal. Use a bar to show a repeating decimal.** *See Examples 1–3 on pages 200 and 201.*

10. $\frac{7}{8}$ **11.** $\frac{9}{20}$ **12.** $\frac{2}{3}$ **13.** $-\frac{7}{15}$ **14.** $8\frac{3}{25}$ **15.** $6\frac{4}{11}$

www.pre-alg.com/vocabulary_review

5-2 Rational Numbers

See pages 205–209.

Concept Summary

- Any number that can be written as a fraction is a rational number.
- Decimals that are terminating or repeating are rational numbers.

Example **Write 0.16 as a fraction in simplest form.**

$0.16 = \frac{16}{100}$ 0.16 is 16 hundredths.

$= \frac{4}{25}$ Simplify. The GCF of 16 and 100 is 4.

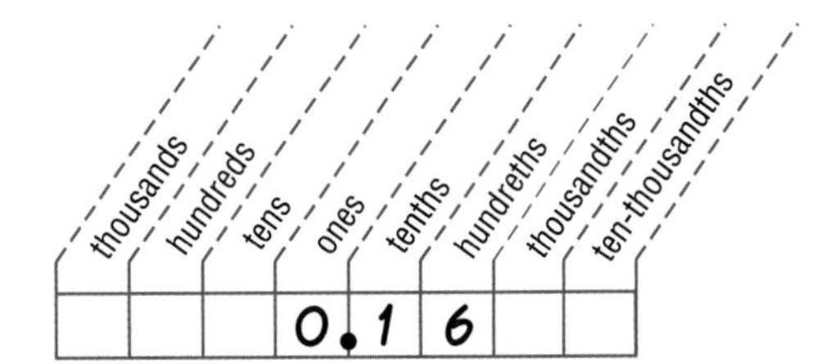

Exercises **Write each decimal as a fraction or mixed number in simplest form.** *See Examples 2 and 3 on page 206.*

16. 0.23
17. 0.6
18. -0.05
19. 0.125
20. 2.36
21. 4.44
22. -8.002
23. 0.555...
24. $0.\overline{3}$
25. $1.\overline{7}$
26. $0.\overline{72}$
27. $3.\overline{36}$

5-3 Multiplying Rational Numbers

See pages 210–214.

Concept Summary

- To multiply fractions, multiply the numerators and multiply the denominators.
- Dimensional analysis is a useful way to keep track of units while computing.

Example **Find $\frac{4}{5} \cdot 3\frac{1}{3}$. Write the product in simplest form.**

$\frac{4}{5} \cdot 3\frac{1}{3} = \frac{4}{5} \cdot \frac{10}{3}$ Rename $3\frac{1}{3}$ as an improper fraction.

$= \frac{4}{\not{5}_1} \cdot \frac{\not{10}^2}{3}$ Divide by the GCF, 5.

$= \frac{8}{3}$ or $2\frac{2}{3}$ Multiply and then simplify.

Exercises **Find each product. Write in simplest form.**
See Examples 1–5 on pages 210 and 211.

28. $\frac{2}{3} \cdot \frac{1}{8}$
29. $-\frac{7}{15} \cdot \frac{5}{9}$
30. $\frac{6}{11} \cdot \frac{2}{15}$
31. $8 \cdot \frac{4}{5}$
32. $-1\frac{5}{6} \cdot 9$
33. $\frac{6}{7} \cdot \frac{14}{9}$
34. $\frac{8}{9} \cdot \frac{5}{12}$
35. $2\frac{1}{4} \cdot \left(-\frac{4}{3}\right)$
36. $\frac{14}{15} \cdot 3\frac{2}{7}$
37. $2\frac{5}{6} \cdot 3\frac{1}{3}$
38. $\frac{ab}{4} \cdot \frac{2}{bc}$
39. $\frac{x^2}{r} \cdot \frac{r}{x}$

5-4 Dividing Rational Numbers

See pages 215–219.

Concept Summary

- The product of a number and its multiplicative inverse or reciprocal is 1.
- To divide by a fraction, multiply by its reciprocal.

Example **Find $\frac{6}{7} \div \frac{3}{4}$. Write the quotient in simplest form.**

$\frac{6}{7} \div \frac{3}{4} = \frac{\overset{2}{\cancel{6}}}{7} \cdot \frac{4}{\underset{1}{\cancel{3}}}$ Multiply by the reciprocal of $\frac{3}{4}$, $\frac{4}{3}$.

$= \frac{8}{7}$ or $1\frac{1}{7}$ Multiply and then simplify.

Exercises **Find each quotient. Write in simplest form.**
See Examples 2–5 on pages 216 and 217.

40. $\frac{4}{9} \div \frac{1}{3}$ **41.** $\frac{2}{5} \div \left(-\frac{1}{15}\right)$ **42.** $\frac{6}{13} \div \frac{8}{9}$ **43.** $5 \div \left(-1\frac{1}{3}\right)$

44. $\frac{11}{18} \div 4\frac{1}{2}$ **45.** $-3\frac{3}{5} \div \frac{6}{7}$ **46.** $\frac{n}{8} \div \frac{n}{32}$ **47.** $\frac{2}{7x} \div \frac{3}{2}$

5-5 Adding and Subtracting Like Fractions

See pages 220–224.

Concept Summary

- To add like fractions, add the numerators and write the sum over the denominator.
- To subtract like fractions, subtract the numerators and write the sum over the denominator.

Example **Find $3\frac{7}{8} + 9\frac{3}{8}$. Write the sum in simplest form.**

Estimate: $4 + 9 = 13$

$3\frac{7}{8} + 9\frac{3}{8} = (3 + 9) + \left(\frac{7}{8} + \frac{3}{8}\right)$ Add the whole numbers and fractions separately.

$= 12 + \frac{7 + 3}{8}$ The denominators are the same. Add the numerators.

$= 12\frac{10}{8}$ Simplify.

$= 13\frac{2}{8}$ or $13\frac{1}{4}$ Simplify.

Exercises **Find each sum or difference. Write in simplest form.**
See Examples 1–3 and 5 on pages 220 and 221.

48. $\frac{5}{18} + \frac{11}{18}$ **49.** $\frac{7}{9} + \left(-\frac{2}{9}\right)$ **50.** $\frac{19}{20} - \frac{17}{20}$ **51.** $1\frac{16}{21} - \frac{9}{21}$

52. $-\frac{12}{17} + \frac{10}{17}$ **53.** $8\frac{3}{10} - 5\frac{7}{10}$ **54.** $\frac{7t}{15} + \frac{t}{15}$ **55.** $\frac{5}{3x} - \frac{1}{3x}$

5-6 Least Common Multiple

See pages 226–230.

Concept Summary

- The LCM of two numbers is the least nonzero multiple common to both numbers.
- To compare fractions with unlike denominators, write the fractions using the LCD and compare the numerators.

Example **Replace ● with <, >, or = to make $\frac{7}{15}$ ● $\frac{5}{9}$ a true statement.**

The LCD is $3^2 \cdot 5$ or 45. Rewrite the fractions using the LCD.

$\frac{7}{15} \cdot \frac{3}{3} = \frac{21}{45}$ $\qquad$ $\frac{5}{9} \cdot \frac{5}{5} = \frac{25}{45}$

Since $21 < 25$, $\frac{21}{45} < \frac{25}{45}$. So, $\frac{7}{15} < \frac{5}{9}$.

Exercises **Find the least common multiple (LCM) of each pair of numbers or monomials.** *See Examples 1 and 2 on pages 226 and 227.*

56. 4, 18 **57.** 24, 20 **58.** $4a$, $6a$ **59.** $7c^2$, $21c$

Replace each ● with <, >, or = to make a true statement.
See Example 5 on page 228.

60. $\frac{3}{8}$ ● $\frac{5}{12}$ **61.** $\frac{2}{9}$ ● $\frac{4}{15}$ **62.** $\frac{5}{20}$ ● $\frac{1}{4}$ **63.** $\frac{3}{7}$ ● $\frac{8}{21}$

5-7 Adding and Subtracting Unlike Fractions

See pages 232–236.

Concept Summary

- To add or subtract fractions with unlike denominators, rename the fractions with the LCD. Then add or subtract.

Example **Find $\frac{7}{9} - \frac{5}{12}$.**

$\frac{7}{9} - \frac{5}{12} = \frac{7}{9} \cdot \frac{4}{4} - \frac{5}{12} \cdot \frac{3}{3}$ The LCD is $3^2 \cdot 2^2$ or 36.

$= \frac{28}{36} - \frac{15}{36}$ Rename the fractions using the LCD.

$= \frac{13}{36}$ Subtract the like fractions.

Exercises **Find each sum or difference. Write in simplest form.**
See Examples 2–5 on pages 233 and 234.

64. $\frac{1}{3} + \frac{5}{6}$ **65.** $\frac{11}{12} + \frac{3}{4}$ **66.** $\frac{7}{8} - \frac{5}{6}$ **67.** $3\frac{7}{12} - \frac{3}{4}$

68. $-\frac{3}{7} + \left(-\frac{11}{14}\right)$ **69.** $1\frac{2}{5} - \left(-\frac{1}{3}\right)$ **70.** $5\frac{1}{2} - 2\frac{2}{3}$ **71.** $-2\frac{1}{6} + 5\frac{1}{3}$

Chapter 5 **For More ...**

- Extra Practice, see pages 733–736.
- Mixed Problem Solving, see page 762.

5-8 Measures of Central Tendency

See pages 238–242.

Concept Summary

- The mean, median, and mode can be used to describe sets of data.

Example **Find the mean, median, and mode of 8, 4, 2, 2, and 10.**

mean: $\frac{8 + 4 + 2 + 2 + 10}{5}$ or 5.2 median: 4 mode: 2

Exercises **Find the mean, median, and mode for each set of data. If necessary, round to the nearest tenth.** *See Example 1 on page 238.*

72. 4, 5, 7, 3, 9, 11, 23, 37

73. 3.6, 7.2, 9.0, 5.2, 7.2, 6.5, 3.6

5-9 Solving Equations with Rational Numbers

See pages 244–248.

Concept Summary

- To solve an equation, use inverse operations to isolate the variable.

Example **Solve $1.6x = 8$. Check your solution.**

$1.6x = 8$ Write the equation.

$\frac{1.6x}{1.6} = \frac{8}{1.6}$ Divide each side by 1.6.

$x = 5$ Simplify.

CHECK $1.6x = 8$ Write the equation.

$1.6(5) \stackrel{?}{=} 8$ Replace x with 5.

$8 = 8$ ✓ Simplify.

Exercises **Solve each equation. Check your solution.**
See Examples 1–4 on pages 244 and 245.

74. $\frac{1}{2} = a + \frac{3}{8}$ **75.** $x - 1.5 = 1.75$ **76.** $0.2t = 6$ **77.** $2 = -\frac{4}{5}n$

5-10 Arithmetic and Geometric Sequences

See pages 249–252.

Concept Summary

- In an arithmetic sequence, the terms have a common difference.
- In a geometric sequence, the terms have a common ratio.

Example **State whether −8, −2, 4, 10, 16, ... is *arithmetic, geometric,* or *neither.* If it is arithmetic or geometric, write the next three terms.**

−8, −2, 4, 10, 16
+6 +6 +6 +6

The common difference is 6, so the sequence is arithmetic. The next three terms are 16 + 6 or 22, 22 + 6 or 28, and 28 + 6 or 34.

Exercises **State whether each sequence is *arithmetic, geometric,* or *neither.* If it is arithmetic or geometric, state the common difference or common ratio and write the next three terms of the sequence.** *See Examples 1–3 on pages 249 and 250.*

78. 4, 9, 14, 19, ... **79.** 1, 3, 9, 27, ... **80.** 32, 8, 2, $\frac{1}{2}$, ... **81.** −6, −5, −2, 3, ...

Chapter 5

Practice Test

Vocabulary and Concepts

1. **State** the difference between a terminating and a repeating decimal.
2. **Describe** how to add fractions with unlike denominators.
3. **Define** *geometric sequence.*

Skills and Applications

Write each fraction or mixed number as a decimal. Use a bar to show a repeating decimal.

4. $\frac{9}{20}$
5. $-\frac{7}{8}$
6. $4\frac{2}{9}$

Write each decimal as a fraction or mixed number in simplest form.

7. 0.24
8. 5.06
9. $-2.\overline{3}$

Replace each ● with <, >, or = to make a true sentence.

10. $0.6 \bullet \frac{2}{3}$
11. $-1\frac{5}{8} \bullet -1.6$

Find the least common denominator (LCD) of each pair of fractions.

12. $\frac{5}{6}, \frac{2}{9}$
13. $\frac{9}{4a^2}, \frac{2}{3ab}$

Find each product, quotient, sum, or difference. Write in simplest form.

14. $\frac{5}{8} \cdot \frac{6}{11}$
15. $\frac{5}{8} + \frac{1}{8}$
16. $\frac{7}{9} \div \frac{4}{15}$
17. $3\frac{5}{6} + 1\frac{2}{9}$
18. $\frac{ab}{9} \div \frac{b}{3}$
19. $\frac{11x}{3y} - \frac{8x}{3y}$

For Exercises 20 and 21, use the data set {20.5, 18.6, 16.3, 4.8, 19.1, 17.3, 20.5}.

20. Find the mean, median, and mode. If necessary, round to the nearest tenth.
21. Identify an extreme value and describe how it affects the mean.

Solve each equation. Check your solution.

22. $x + 4.3 = 9.8$
23. $12 = 0.75x$
24. $3.1m = 12.4$
25. $p - \frac{4}{5} = \frac{2}{3}$
26. $-\frac{3}{8} = \frac{5}{3}a$
27. $y - 2\frac{1}{4} = 1\frac{5}{6}$

State whether each sequence is *arithmetic, geometric,* or *neither.* If it is arithmetic or geometric, state the common difference or common ratio and write the next three terms of the sequence.

28. 2, 8, 32, 128, …
29. 5.5, 4.9, 4.3, 3.7, …
30. 1, 2, 4, 7, …

31. **TRAVEL** Max drives 6 hours at an average rate of 65 miles per hour. What is the distance Max travels? Use $d = rt$.

32. **SEWING** Allie needs $4\frac{2}{3}$ yards of lace to finish sewing the edges of a blanket. She only has $\frac{3}{4}$ of that amount. How much lace does Allie have?

FCAT Practice

33. **STANDARDIZED TEST PRACTICE** Write the sum of $-6\frac{1}{4}$ and $-9\frac{3}{20}$.

Ⓐ $-15\frac{2}{5}$ Ⓑ $-15\frac{1}{4}$ Ⓒ $-16\frac{2}{5}$ Ⓓ $-3\frac{1}{5}$

www.pre-alg.com/chapter_test/fcat

Chapter 5

Standardized Test Practice

FCAT Practice

Part 1 Multiple Choice

Record your answers on the answer sheet provided by your teacher or on a sheet of paper.

1. Kenzie paid for a CD with a $20 bill. She received 3 dollars, 3 dimes, and 2 pennies in change. How much did she pay for the CD? (Prerequisite Skill, p. 707)

Ⓐ $16.68 Ⓑ $16.88

Ⓒ $17.68 Ⓓ $17.88

2. A survey of 110 people asked which country they would most like to visit. The bar graph shows the data. How many people chose Canada, England, or Australia as the country they would most like to visit? (Prerequisite Skill, pp. 722–723)

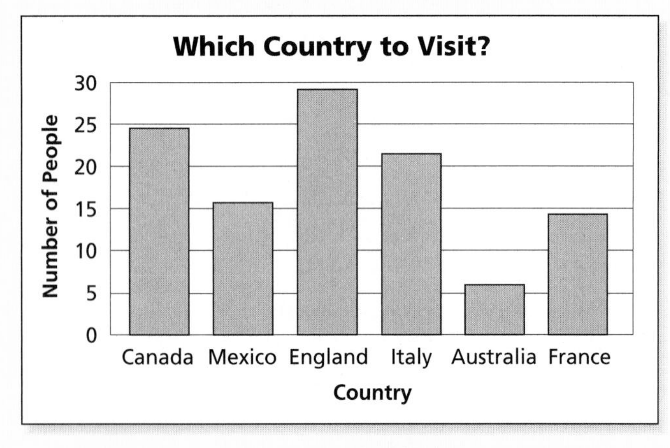

Ⓐ 52 Ⓑ 58

Ⓒ 68 Ⓓ 74

The Princeton Review Test-Taking Tip

Question 2 When an item includes a graph, scan the graph to see what kind of information it includes and how the information is organized. Don't try to memorize the information. Read each answer choice and compare it with the graph to see if the information in the answer choice is correct. Eliminate any wrong answer choices.

3. The low temperature overnight was -1°F. Each night for the next four nights, the low temperature was 7° lower than the previous night. What was the low temperature during the last night? (Lesson 2-4)

Ⓐ $-29°$ Ⓑ $-27°$ Ⓒ $-22°$ Ⓓ $-8°$

4. Write $\frac{6t^4}{18ts}$ in simplest form. (Lesson 4-5)

Ⓐ $\frac{1}{3}t^3s$ Ⓑ $3t^5s$

Ⓒ $\frac{t^3}{3s}$ Ⓓ $\frac{t^5}{3s}$

5. What is 8×10^{-2} in standard notation? (Lesson 4-8)

Ⓐ 0.008 Ⓑ 0.08

Ⓒ 0.8 Ⓓ 800

6. Add $\frac{2}{3} + \frac{1}{4} + \frac{5}{6}$. (Lesson 5-7)

Ⓐ $\frac{2}{3}$ Ⓑ $\frac{5}{4}$

Ⓒ $\frac{7}{4}$ Ⓓ $\frac{11}{6}$

7. Evaluate $\frac{1}{4}(2 - x) - x$ if $x = \frac{1}{2}$. (Lesson 5-7)

Ⓐ $-\frac{1}{2}$ Ⓑ $-\frac{1}{8}$ Ⓒ $\frac{1}{8}$ Ⓓ $\frac{1}{2}$

8. Antonia read four books that had the following number of pages: 324, 375, 420, 397. What is the mean number of pages in these books? (Lesson 5-8)

Ⓐ 375 Ⓑ 379

Ⓒ 380 Ⓓ 386

9. Which sequence is a geometric sequence with a common ratio of 2? (Lesson 5-10)

Ⓐ 2, −4, 8, −16, … Ⓑ 2, 4, 6, 8, 10, …

Ⓒ 3, 5, 7, 9, 11, … Ⓓ 3, 6, 12, 24, …

10. State the next three terms in the sequence 128, 32, 8, 2, …. (Lesson 5-10)

Ⓐ $\frac{1}{2}, \frac{1}{8}, \frac{1}{32}$ Ⓑ $\frac{1}{2}, \frac{1}{16}, \frac{1}{32}$

Ⓒ $\frac{1}{4}, \frac{1}{8}, \frac{1}{16}$ Ⓓ $\frac{1}{4}, \frac{1}{6}, \frac{1}{8}$

Part 2 Short Response/Grid In

Record your answers on the answer sheet provided by your teacher or on a sheet of paper.

11. Nate is cutting shelves from a board that is 15 feet long. Each shelf is 3 feet 4 inches long. What is the greatest number of shelves he can make from the board? (Prerequisite Skill, pp. 720–721)

12. Write the ordered pair that names point *L*. (Lesson 2-6)

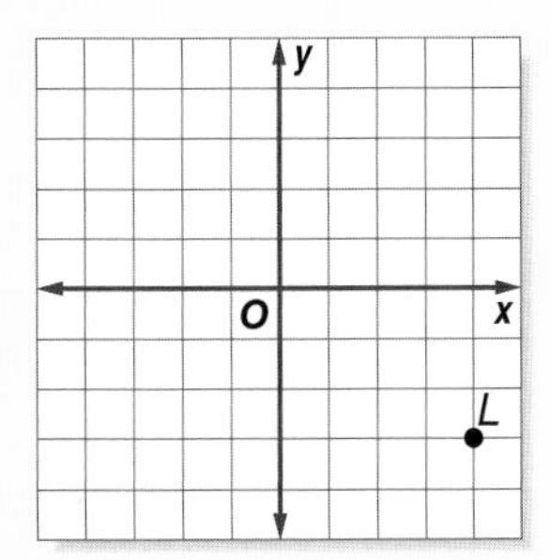

13. Find x if $3x + 4 = 28$. (Lesson 3-5)

14. Write an equation to represent the total number of Calories t in one box of snack crackers. The box of crackers contains 8 servings. Each serving has 125 Calories. (Lesson 3-6)

15. On Saturday, Juan plans to drive 275 miles at a rate of 55 miles per hour. How many hours will his trip take? Use the formula $d = rt$, where d represents the distance, r represents rate, and t represents the time. (Lesson 3-7)

16. Write $4^2 \times 5^3$ as a product of prime factors without using exponents. (Lesson 4-3)

17. The average American worker spends 44 minutes traveling to and from work each day. What fraction of the day is this? (Lesson 4-5)

18. The Saturn V rocket that took the Apollo astronauts to the moon weighed 6.526×10^6 pounds at lift-off. Write its weight in standard notation. (Lesson 4-8)

19. Write a decimal to represent the shaded portion of the figure below. (Lesson 5-1)

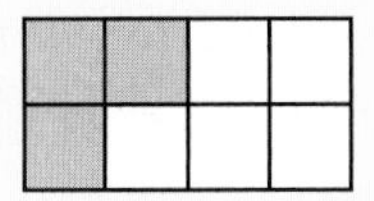

20. One elevator in a 40-story building is programmed to stop at every third floor. Another is programmed to stop at every fourth floor. Which floors in the building are served by both elevators? (Lesson 5-6)

21. Beth has $\frac{3}{4}$ cup of grated cheese. She needs $2\frac{1}{2}$ cups of grated cheese for making pizzas. How many more cups does she need? (Lesson 5-7)

22. The number of tickets sold for each performance of the Spring Music Fest are 352, 417, 307, 367, 433, and 419. What is the median number of tickets sold per performance? (Lesson 5-8)

FCAT Practice

Part 3 Open Ended

Record your answers on a sheet of paper. Show your work.

23. During seven regular season games, the Hawks basketball team scored the points shown in the table below. (Lesson 5-8)

Game	1	2	3	4	5	6	7
Points	68	60	73	74	64	78	73

a. Find the mean, median, and mode of these seven scores.

b. During the first playoff game after the regular season, the Hawks scored only 40 points. Find the mean, median, and mode of all eight scores.

c. Which of these three measures of central tendency—mean, median, or mode—changed the most as a result of the playoff score? Explain your answer.

d. Does the mean score or the median score best represent the team's scores for all eight games? Explain your answer.

www.pre-alg.com/standardized_test/fcat

Chapter 6 Ratio, Proportion, and Percent

What You'll Learn

- **Lesson 6-1** Write ratios as fractions and find unit rates.
- **Lessons 6-2 and 6-3** Use ratios and proportions to solve problems, including scale drawings.
- **Lesson 6-4** Write decimals and fractions as percents and vice versa.
- **Lessons 6-5, 6-6, 6-7, and 6-8** Estimate and compute with percents.
- **Lesson 6-9** Find simple probability.

Key Vocabulary

- **ratio** (p. 264)
- **rate** (p. 265)
- **proportion** (p. 270)
- **percent** (p. 281)
- **probability** (p. 310)

Why It's Important

The concept of proportionality is the foundation of many branches of mathematics, including geometry, statistics, and business math. Proportions can be used to solve real-world problems dealing with scale drawings, indirect measurement, predictions, and money. *You will solve a problem about currency exchange rates in Lesson 6-2.*

Getting Started

Prerequisite Skills To be successful in this chapter, you'll need to master these skills and be able to apply them in problem-solving situations. Review these skills before beginning Chapter 6.

For Lesson 6-1 — Convert Measurements

Complete each sentence. *(For review, see pages 718–721.)*

1. 2 ft = $\underline{\ ?\ }$ in. **2.** 4 yd = $\underline{\ ?\ }$ ft **3.** 2 mi = $\underline{\ ?\ }$ ft
4. 3 h = $\underline{\ ?\ }$ min **5.** 8 min = $\underline{\ ?\ }$ s **6.** 4 lb = $\underline{\ ?\ }$ oz
7. 2 T = $\underline{\ ?\ }$ lb **8.** 5 gal = $\underline{\ ?\ }$ qt **9.** 3 pt = $\underline{\ ?\ }$ c
10. 3 m = $\underline{\ ?\ }$ cm **11.** 5.8 m = $\underline{\ ?\ }$ cm **12.** 2 km = $\underline{\ ?\ }$ m
13. 5 cm = $\underline{\ ?\ }$ mm **14.** 2.3 L = $\underline{\ ?\ }$ mL **15.** 15 kg = $\underline{\ ?\ }$ g

For Lessons 6-2 and 6-3 — Multiply Decimals

Find each product. *(For review, see page 715.)*

16. 7(3.4) **17.** 6.1(8) **18.** 2.8×5.9 **19.** 1.6×8.4
20. 0.8×9.3 **21.** 0.6(0.3) **22.** 12.4(3.8) **23.** 15.2×0.2

For Lesson 6-9 — Write Fractions in Simplest Form

Simplify each fraction. If the fraction is already in simplest form, write *simplified*. *(For review, see Lesson 4-5.)*

24. $\frac{4}{8}$ **25.** $\frac{5}{15}$ **26.** $\frac{6}{10}$ **27.** $\frac{12}{25}$
28. $\frac{22}{20}$ **29.** $\frac{15}{16}$ **30.** $\frac{36}{42}$ **31.** $\frac{36}{48}$

FOLDABLES™
Study Organizer

Make this Foldable to help you organize information about fractions, decimals, and percents. Begin with a piece of lined paper.

Step 1 Fold in Thirds

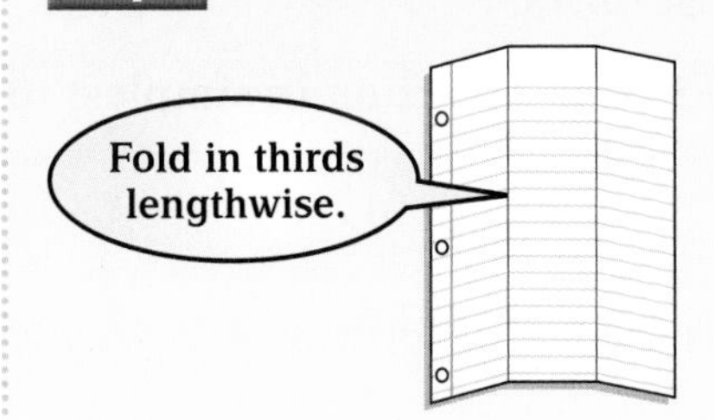
Fold in thirds lengthwise.

Step 2 Label

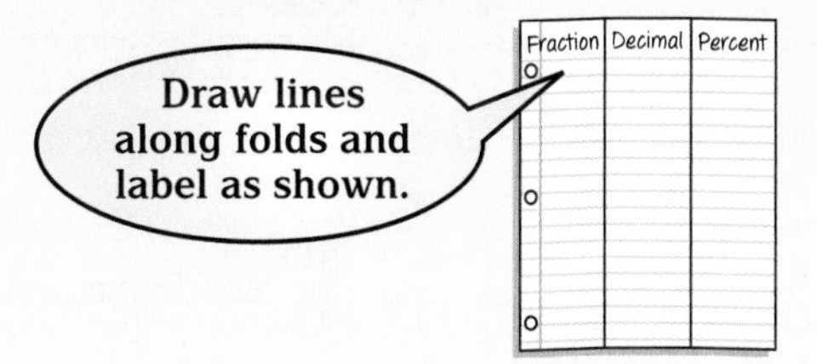
Draw lines along folds and label as shown.
Fraction
Decimal
Percent

Reading and Writing As you read and study the chapter, complete the table with the commonly-used fraction, decimal, and percent equivalents.

6-1 Ratios and Rates

What You'll Learn

- Write ratios as fractions in simplest form.
- Determine unit rates.

Vocabulary

- ratio
- rate
- unit rate

How are ratios used in paint mixtures?

The diagram shows a gallon of paint that is made using 2 parts blue paint and 4 parts yellow paint.

a. Which combination of paint would you use to make a smaller amount of the same shade of paint? Explain.

Combination A

Combination B

b. Suppose you want to make the same shade of paint as the original mixture? How many parts of yellow paint should you use for each part of blue paint?

Sunshine State Standards
MA.A.1.3.2-1, MA.A.1.3.2-2, MA.A.1.3.4-1, MA.A.3.3.2-1, MA.A.3.3.2-2, MA.B.2.3.2-1, MA.B.2.3.2-2, MA.B.2.3.2-3

WRITE RATIOS AS FRACTIONS IN SIMPLEST FORM A **ratio** is a comparison of two numbers by division. If a gallon of paint contains 2 parts blue paint and 4 parts yellow paint, then the ratio comparing the blue paint to the yellow paint can be written as follows.

2 to 4   2:4 $\frac{2}{4}$

Recall that a fraction bar represents division. When the first number being compared is less than the second, the ratio is usually written as a fraction in simplest form.

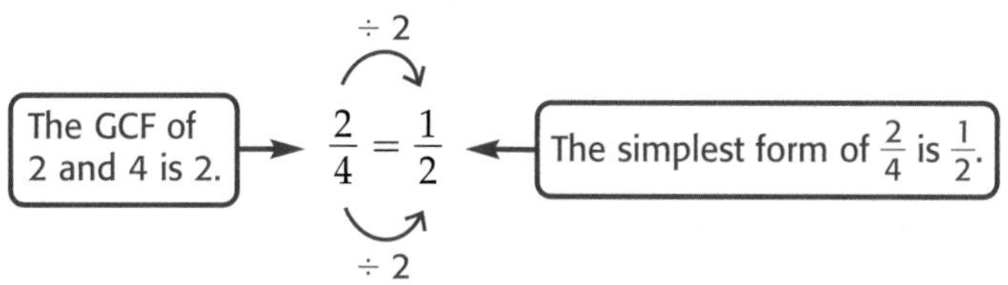

Study Tip

Look Back
To review how to write a fraction in **simplest form**, see Lesson 4-5.

Example 1 Write Ratios as Fractions

Express the ratio *9 goldfish out of 15 fish* as a fraction in simplest form.

$\frac{9}{15} = \frac{3}{5}$ (÷ 3)   Divide the numerator and denominator by the GCF, 3.

The ratio of goldfish to fish is 3 to 5. This means that for every 5 fish, 3 of them are goldfish.

When writing a ratio involving measurements, both quantities should have the same unit of measure.

Example 2 Write Ratios as Fractions

Express the ratio *3 feet to 16 inches* as a fraction in simplest form.

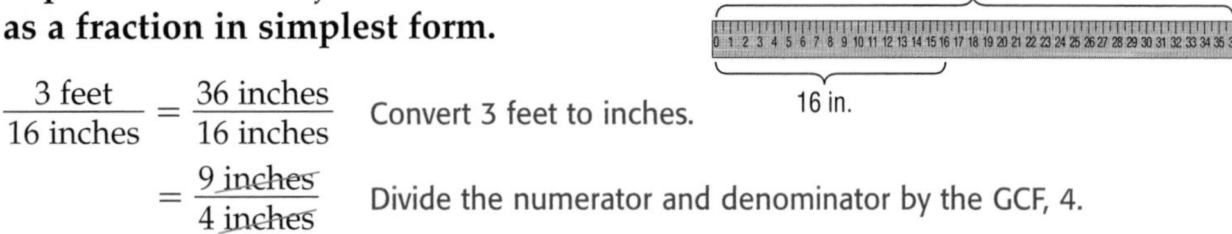

$\frac{3 \text{ feet}}{16 \text{ inches}} = \frac{36 \text{ inches}}{16 \text{ inches}}$ Convert 3 feet to inches.

$= \frac{9 \text{ inches}}{4 \text{ inches}}$ Divide the numerator and denominator by the GCF, 4.

Written in simplest form, the ratio is 9 to 4.

Concept Check Give an example of a ratio in simplest form.

FIND UNIT RATES A **rate** is a ratio of two measurements having different kinds of units. Here are two examples of rates.

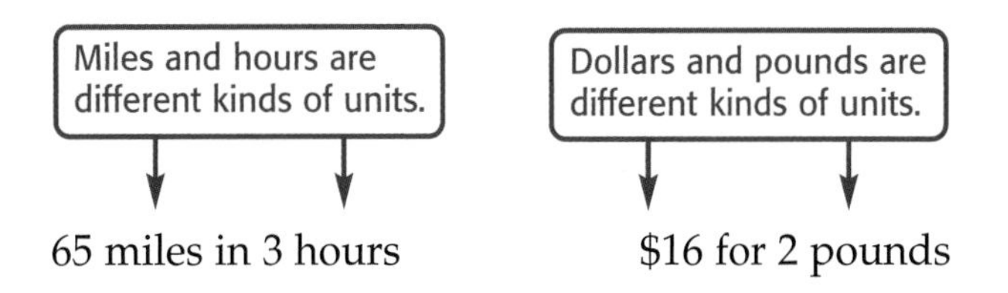

Miles and hours are different kinds of units.
Dollars and pounds are different kinds of units.

65 miles in 3 hours $16 for 2 pounds

When a rate is simplified so that it has a denominator of 1, it is called a **unit rate**. An example of a unit rate is $5 per pound, which means $5 per 1 pound.

Example 3 Find Unit Rate

SHOPPING A package of 20 recordable CDs costs $18, and a package of 30 recordable CDs costs $28. Which package has the lower cost per CD?

Find and compare the unit rates of the packages.

$\frac{18 \text{ dollars}}{20 \text{ CDs}} = \frac{0.9 \text{ dollars}}{1 \text{ CD}}$ (÷ 20) Divide the numerator and denominator by 20 to get a denominator of 1.

For the 20-pack, the unit rate is $0.90 per CD.

$\frac{28 \text{ dollars}}{30 \text{ CDs}} = \frac{0.9\overline{3} \text{ dollars}}{1 \text{ CD}}$ (÷ 30) Divide the numerator and denominator by 30 to get a denominator of 1.

For the 30-pack, the unit rate is $0.93 per CD.

So, the package that contains 20 CDs has the lower cost per CD.

Study Tip

Alternative Method
Another way to find the unit rate is to divide the cost of the package by the number of CDs in the package.

Concept Check Is $50 in 3 days a rate or a unit rate? Explain.

Study Tip

Look Back
To review **dimensional analysis,** see Lesson 5-3.

To convert a rate such as miles per hour to a rate such as feet per second, you can use dimensional analysis. Recall that this is the process of carrying units throughout a computation.

Example 4 Convert Rates

ANIMALS A grizzly bear can run 30 miles in 1 hour. How many feet is this per second?

You need to convert $\frac{30 \text{ mi}}{1 \text{ h}}$ to $\frac{\blacksquare \text{ ft}}{1 \text{ s}}$. There are 5280 feet in 1 mile and 3600 seconds in 1 hour. Write 30 miles per hour as $\frac{30 \text{ mi}}{1 \text{ h}}$.

$\frac{30 \text{ mi}}{1 \text{ h}} = \frac{30 \text{ mi}}{1 \text{ h}} \cdot \frac{5280 \text{ ft}}{1 \text{ mi}} \div \frac{3600 \text{ s}}{1 \text{ h}}$ — Convert miles to feet and hours to seconds.

$= \frac{30 \text{ mi}}{1 \text{ h}} \cdot \frac{5280 \text{ ft}}{1 \text{ mi}} \cdot \frac{1 \text{ h}}{3600 \text{ s}}$ — The reciprocal of $\frac{3600 \text{ s}}{1 \text{ h}}$ is $\frac{1 \text{ h}}{3600 \text{ s}}$.

$= \frac{\overset{1}{\cancel{30}} \cancel{\text{mi}}}{1 \cancel{\text{h}}} \cdot \frac{\overset{44}{\cancel{5280}} \text{ ft}}{1 \cancel{\text{mi}}} \cdot \frac{1 \cancel{\text{h}}}{\underset{\underset{1}{\cancel{120}}}{\cancel{3600}} \text{ s}}$ — Divide the common factors and units.

$= \frac{44 \text{ ft}}{\text{s}}$ — Simplify.

So, 30 miles per hour is equivalent to 44 feet per second.

Check for Understanding

Concept Check

1. **Draw** a diagram in which the ratio of circles to squares is 2:3.
2. **Explain** the difference between ratio and rate.
3. **OPEN ENDED** Give an example of a unit rate.

Guided Practice

Express each ratio as a fraction in simplest form.

4. 4 goals in 10 attempts
5. 15 dimes out of 24 coins
6. 10 inches to 3 feet
7. 5 feet to 5 yards

Express each ratio as a unit rate. Round to the nearest tenth, if necessary.

8. \$183 for 4 concert tickets
9. 9 inches of snow in 12 hours
10. 100 feet in 14.5 seconds
11. 254.1 miles on 10.5 gallons

Convert each rate using dimensional analysis.

12. 20 mi/h = ■ ft/min
13. 16 cm/s = ■ m/h

Application

GEOMETRY For Exercises 14 and 15, refer to the figure below.

14. Express the ratio of width to length as a fraction in simplest form.

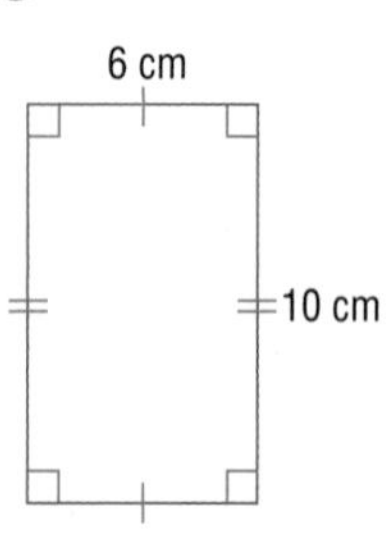

15. Suppose the width and length are each increased by 2 centimeters. Will the ratio of the width to length be the same as the ratio of the width to length of the original rectangle? Explain.

Practice and Apply

Homework Help

For Exercises	See Examples
16–27	1, 2
28–37	3
38–45	4
46, 47	3

Extra Practice
See page 736.

Express each ratio as a fraction in simplest form.

16. 6 ladybugs out of 27 insects
17. 14 girls to 35 boys
18. 18 cups to 45 cups
19. 12 roses out of 28 flowers
20. 7 cups to 9 pints
21. 9 pounds to 16 tons
22. 11 gallons to 11 quarts
23. 18 miles to 18 yards
24. 15 dollars out of 123 dollars
25. 17 rubies out of 118 gems
26. 155 apples to 75 oranges
27. 321 articles in 107 magazines

Express each ratio as a unit rate. Round to the nearest tenth, if necessary.

28. \$3 for 6 cans of tuna
29. \$0.99 for 10 pencils
30. 140 miles on 6 gallons
31. 68 meters in 15 seconds
32. 19 yards in 2.5 minutes
33. 25 feet in 3.2 hours
34. 236.7 miles in 4.5 days
35. 331.5 pages in 8.5 weeks

36. **MAGAZINES** Which costs more per issue, an 18-issue subscription for \$40.50 or a 12-issue subscription for \$33.60? Explain.

37. **SHOPPING** Determine which is less expensive per can, a 6-pack of soda for \$2.20 or a 12-pack of soda for \$4.25. Explain.

Convert each rate using dimensional analysis.

38. 45 mi/h = ■ ft/s
39. 18 mi/h = ■ ft/s
40. 26 cm/s = ■ m/min
41. 32 cm/s = ■ m/min
42. 2.5 qt/min = ■ gal/h
43. 4.8 qt/min = ■ gal/h
44. 4 c/min = ■ qt/h
45. 7 c/min = ■ qt/h

More About. . .

Population

In 2000, the population density of the United States was about 79.6 people per square mile.

Source: *The World Almanac*

46. **POPULATION** Population density is a unit rate that gives the number of people per square mile. Find the population density for each state listed in the table at the right. Round to the nearest whole number.

State	Population (2000)	Area (sq mi)
Alaska	626,932	570,374
New York	18,976,457	47,224
Rhode Island	1,048,319	1045
Texas	20,851,820	261,914
Wyoming	493,782	97,105

Source: U.S. Census Bureau

Online Research **Data Update** How has the population density of the states in the table changed since 2000? Visit www.pre-alg.com/data_update to learn more.

TRAVEL For Exercises 47 and 48, use the following information.
An airplane flew from Boston to Chicago to Denver. The distance from Boston to Chicago was 1015 miles and the distance from Chicago to Denver was 1011 miles. The plane traveled for 3.5 hours and carried 285 passengers.

47. About how fast did the airplane travel?

48. Suppose it costs \$5685 per hour to operate the airplane. Find the cost per person per hour for the flight.

49. CRITICAL THINKING Marty and Spencer each saved money earned from shoveling snow. The ratio of Marty's money to Spencer's money is 3:1. If Marty gives Spencer \$3, their ratio will be 1:1. How much money did Marty earn?

50. WRITING IN MATH Answer the question that was posed at the beginning of the lesson.

How are ratios used in paint mixtures?

Include the following in your answer:

- an example of a ratio of blue to yellow paint that would result in a darker shade of green, and
- an example of a ratio of blue to yellow paint that would result in a lighter shade of green.

51. Which ratio represents the same relationship as *for every 4 apples, 3 of them are green?*

Ⓐ 9:16 Ⓑ 3:4 Ⓒ 12:9 Ⓓ 6:8

52. Joe paid \$2.79 for a gallon of milk. Find the cost per quart of milk.

Ⓐ \$0.70 Ⓑ \$1.40 Ⓒ \$0.93 Ⓓ \$0.55

Extending the Lesson

53. Many objects such as credit cards or phone cards are shaped like golden rectangles.

> A *golden rectangle* is a rectangle in which the ratio of the length to the width is approximately 1.618 to 1. This ratio is called the **golden ratio**.

a. Find three different objects that are close to a golden rectangle. Make a table to display the dimensions and the ratio found in each object.

b. Describe how each ratio compares to the golden ratio.

c. RESEARCH Use the Internet or another source to find three places where the golden rectangle is used in architecture.

Maintain Your Skills

Mixed Review

State whether each sequence is *arithmetic, geometric,* or *neither.* Then state the common difference or common ratio and write the next three terms of the sequence. *(Lesson 5-10)*

54. –3, 6, –12, 24, … **55.** 12.1, 12.4, 12.7, 13, …

ALGEBRA Solve each equation. *(Lesson 5-9)*

56. $3.6 = x - 7.1$ **57.** $y + \frac{3}{4} = \frac{2}{3}$ **58.** $-4.8 = 6z$ **59.** $\frac{3}{8}w = 5$

60. Find the quotient of $1\frac{1}{7}$ and $-\frac{4}{7}$. *(Lesson 5-4)*

Write each number in scientific notation. *(Lesson 4-8)*

61. 52,000,000 **62.** 42,240 **63.** 0.038

64. Write $8 \cdot (k + 3) \cdot (k + 3)$ using exponents. *(Lesson 4-2)*

Getting Ready for the Next Lesson

PREREQUISITE SKILL Solve each equation.
*(To review **solving equations**, see Lesson 3-4.)*

65. $10x = 300$ **66.** $25m = 225$ **67.** $8k = 320$

68. $192 = 4t$ **69.** $195 = 15w$ **70.** $231 = 33n$

Reading Mathematics

Sunshine State Standards
MA.A.1.3.2-2

Making Comparisons

In mathematics, there are many different ways to compare numbers. Consider the information in the table.

Zoo	Size (acres)	Animals	Species
San Diego	100	4000	800
Houston	55	5000	700
Oakland	100	400	100
Columbus	400	11,000	700

The following types of comparison statements can be used to describe this information.

Difference Comparisons

- The Houston Zoo has 1000 more animals than the San Diego Zoo.
- The Columbus Zoo is 345 acres larger than the Houston Zoo.
- The Oakland Zoo has 700 less species of animals than the San Diego Zoo.

Ratio Comparisons

- The ratio of the size of the San Diego Zoo to the size of the Columbus Zoo is 1:4. So, the San Diego Zoo is one-fourth the size of the Columbus Zoo.
- The ratio of the number of animals at the San Diego Zoo to the number of animals at the Oakland Zoo is 4000:400 or 10:1. So, San Diego Zoo has ten times as many animals as the Oakland Zoo.

Reading to Learn

1. Refer to the zoo information above. Write a difference comparison and a ratio comparison statement that describes the information.

Refer to the information below. Identify each statement as a difference comparison or a ratio comparison.

Florida *The Sunshine State*
Total area: 59,928 sq mi
Land area: 53,937 sq mi
Land forested: 26,478.4 sq mi

Ohio *The Buckeye State*
Total area: 44,828 sq mi
Land area: 40,953 sq mi
Land forested: 12,580.8 sq mi

Source: *The World Almanac*

2. The area of Florida is about 15,000 square miles greater than the area of Ohio.

3. The ratio of the amount of land forested in Ohio to the amount forested in Florida is about 1 to 2.

4. More than one-fourth of the land in Ohio is forested.

6-2 Using Proportions

Sunshine State Standards
MA.A.3.3.2-1, MA.A.3.3.2-2, MA.A.3.3.2-3, MA.A.3.3.2-4, MA.B.2.3.1-1

Vocabulary

- proportion
- cross products

What You'll Learn

- Solve proportions.
- Use proportions to solve real-world problems.

How are proportions used in recipes?

For many years, Phyllis Norman was famous in her neighborhood for making her flavorful fruit punch. The recipe is shown at the right.

Fruit Punch
12 oz frozen lemonade concentrate
12 oz frozen grape juice concentrate
12 oz frozen orange juice concentrate
40 oz lemon-lime soda
84 oz water
Yields: 160 oz of punch

a. For each of the first four ingredients, write a ratio that compares the number of ounces of each ingredient to the number of ounces of water.

b. Double the recipe. (*Hint*: Multiply each number of ounces by 2.) Then write a ratio for the ounces of each of the first four ingredients to the ounces of water as a fraction in simplest form.

c. Are the ratios in part **a** and **b** the same? Why or why not?

PROPORTIONS To solve problems that relate to ratios, you can use a proportion. A **proportion** is a statement of equality of two ratios.

Key Concept — *Proportion*

- **Words** A proportion is an equation stating that two ratios are equal.
- **Symbols** $\frac{a}{b} = \frac{c}{d}$
- **Example** $\frac{2}{3} = \frac{6}{9}$

Consider the following proportion.

$$\frac{a}{b} = \frac{c}{d}$$

$$\frac{a}{\cancel{b}_1} \cdot \overset{1}{\cancel{b}}d = \frac{c}{\cancel{d}_1} \cdot b\overset{1}{\cancel{d}}$$ Multiply each side by bd to eliminate the fractions.

$$ad = cb$$ Simplify.

Study Tip

Properties
When you multiply each side of an equation by bd, you are using the Multiplication Property of Equality.

The products ad and cb are called the **cross products** of a proportion. Every proportion has two cross products.

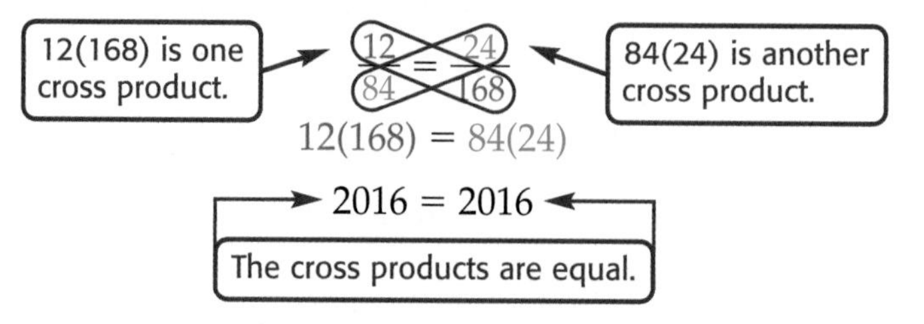

12(168) is one cross product.
$\frac{12}{84} = \frac{24}{168}$
84(24) is another cross product.
12(168) = 84(24)
2016 = 2016
The cross products are equal.

Concept Check Write a proportion whose cross products are equal to 18.

Cross products can be used to determine whether two ratios form a proportion.

Key Concept — *Property of Proportions*

- **Words** The cross products of a proportion are equal.
- **Symbols** If $\frac{a}{b} = \frac{c}{d}$, then $ad = bc$. If $ad = bc$, then $\frac{a}{b} = \frac{c}{d}$.

Example 1 Identify Proportions

Determine whether each pair of ratios forms a proportion.

a. $\frac{1}{3}, \frac{3}{9}$

$\frac{1}{3} \stackrel{?}{=} \frac{3}{9}$ Write a proportion.

$1 \cdot 9 \stackrel{?}{=} 3 \cdot 3$ Cross products

$9 = 9$ Simplify.

So, $\frac{1}{3} = \frac{3}{9}$.

b. $\frac{1.2}{4.0}, \frac{2}{5}$

$\frac{1.2}{4.0} \stackrel{?}{=} \frac{2}{5}$ Write a proportion.

$1.2 \cdot 5 \stackrel{?}{=} 4.0 \cdot 2$ Cross products

$6 \neq 8$ Simplify.

So, $\frac{1.2}{4.0} \neq \frac{2}{5}$.

Example 2 Solve Proportions

Solve each proportion.

a. $\frac{a}{25} = \frac{52}{100}$

$\frac{a}{25} = \frac{52}{100}$

$a \cdot 100 = 25 \cdot 52$ Cross products

$100a = 1300$ Multiply.

$\frac{100a}{100} = \frac{1300}{100}$ Divide.

$a = 13$

The solution is 13.

b. $\frac{12.5}{m} = \frac{15}{7.5}$

$\frac{12.5}{m} = \frac{15}{7.5}$

$12.5 \cdot 7.5 = m \cdot 15$ Cross products

$93.75 = 15m$ Multiply.

$\frac{93.75}{15} = \frac{15m}{15}$ Divide.

$6.25 = m$

The solution is 6.25.

Study Tip

Cross Products
When you find cross products, you are *cross multiplying*.

USE PROPORTIONS TO SOLVE REAL-WORLD PROBLEMS When you solve a problem using a proportion, be sure to compare the quantities in the same order.

Example 3 Use a Proportion to Solve a Problem

FOOD **Refer to the recipe at the beginning of the lesson. How much soda should be used if 16 ounces of each type of juice are used?**

Explore You know how much soda to use for 12 ounces of each type of juice. You need to find how much soda to use for 16 ounces of each type of juice.

Plan Write and solve a proportion using ratios that compare juice to soda. Let s represent the amount of soda to use in the new recipe.

(continued on the next page)

Solve $\frac{\text{juice in original recipe}}{\text{soda in original recipe}} = \frac{\text{juice in new recipe}}{\text{soda in new recipe}}$

$\frac{12}{40} = \frac{16}{s}$ Write a proportion.

$12 \cdot s = 40 \cdot 16$ Cross products

$12s = 640$ Multiply.

$\frac{12s}{12} = \frac{640}{12}$ Divide.

$s = 53\frac{1}{3}$ Simplify.

$53\frac{1}{3}$ ounces of soda should be used.

Explore Check the cross products. Since $12 \cdot 53\frac{1}{3} = 640$ and $40 \cdot 16 = 640$, the answer is correct.

More About...

Attractions

The world's largest baseball bat is located in Louisville, Kentucky. It is 120 feet long, has a diameter from 3.5 to 9 feet, and weighs 68,000 pounds.

Source: World's Largest Roadside Attractions

Proportions can also be used in measurement problems.

Example 4 Convert Measurements

ATTRACTIONS Louisville, Kentucky, is home to the world's largest baseball glove. The glove is 4 feet high, 10 feet long, 9 feet wide, and weighs 15 tons. Find the height of the glove in centimeters if 1 ft = 30.48 cm.

Let x represent the height in centimeters.

customary measurement → / metric measurement → $\frac{1 \text{ ft}}{30.48 \text{ cm}} = \frac{4 \text{ ft}}{x \text{ cm}}$ ← customary measurement / ← metric measurement

$1 \cdot x = 30.48 \cdot 4$ Cross products

$x = 121.92$ Simplify.

The height of the glove is 121.92 centimeters.

Check for Understanding

Concept Check

1. **Define** *proportion.*
2. **OPEN ENDED** Find two counterexamples for the statement *Two ratios always form a proportion.*

Guided Practice

Determine whether each pair of ratios forms a proportion.

3. $\frac{1}{4}, \frac{4}{16}$
4. $\frac{2.1}{3.5}, \frac{3}{7}$

ALGEBRA Solve each proportion.

5. $\frac{k}{35} = \frac{3}{7}$
6. $\frac{3}{t} = \frac{18}{24}$
7. $\frac{10}{8.4} = \frac{5}{m}$

Application

8. **PHOTOGRAPHY** A 3″ × 5″ photo is enlarged so that the length of the new photo is 7 inches. Find the width of the new photo.

Practice and Apply

Homework Help

For Exercises	See Examples
9–14	1
15–31	2
32–35, 36–42	3, 4

Extra Practice

See page 737.

Determine whether each pair of ratios forms a proportion.

9. $\frac{2}{3}, \frac{8}{12}$
10. $\frac{4}{2}, \frac{16}{5}$
11. $\frac{1.5}{5.0}, \frac{3}{9}$
12. $\frac{18}{2.4}, \frac{15}{2}$
13. $\frac{3.4}{1.6}, \frac{5.1}{2.4}$
14. $\frac{5.3}{15.9}, \frac{2.7}{8.1}$

ALGEBRA **Solve each proportion.**

15. $\frac{p}{6} = \frac{24}{36}$
16. $\frac{w}{11} = \frac{14}{22}$
17. $\frac{4}{10} = \frac{8}{a}$
18. $\frac{18}{12} = \frac{24}{q}$
19. $\frac{5}{h} = \frac{10}{30}$
20. $\frac{51}{z} = \frac{17}{7}$
21. $\frac{7}{45} = \frac{x}{9}$
22. $\frac{2}{15} = \frac{c}{72}$
23. $\frac{7}{5} = \frac{10.5}{b}$
24. $\frac{16}{7} = \frac{4.8}{h}$
25. $\frac{2}{9.4} = \frac{0.2}{v}$
26. $\frac{9}{7.2} = \frac{3.5}{k}$
27. $\frac{a}{0.28} = \frac{4}{1.4}$
28. $\frac{3}{14} = \frac{15}{m - 3}$
29. $\frac{16}{x + 5} = \frac{4}{5}$

30. Find the value of d that makes $\frac{5.1}{1.7} = \frac{7.5}{d}$ a proportion.

31. What value of m makes $\frac{6.5}{1.3} = \frac{m}{5.2}$ a proportion?

Write a proportion that could be used to solve for each variable. Then solve.

32. 8 pencils in 2 boxes
20 pencils in x boxes

33. 12 glasses in 3 crates
72 glasses in m crates

34. y dollars for 5.4 gallons
14 dollars for 3 gallons

35. 5 quarts for \$6.25
d quarts for \$8.75

OLYMPICS **For Exercises 36 and 37, use the following information.**
There are approximately 3.28 feet in 1 meter.

36. Write a proportion that could be used to find the distance in feet of the 110-meter dash.

37. What is the distance in feet of the 110-meter dash?

38. **PHOTOGRAPHY** Suppose an 8″ × 10″ photo is reduced so that the width of the new photo is 4.5 inches. What is the length of the new photo?

CURRENCY **For Exercises 39 and 40, use the following information and the table shown.**
The table shows the exchange rates for certain countries compared to the U.S. dollar on a given day.

Country	Rate
United Kingdom	0.667
Egypt	3.481
Australia	1.712
China	8.280

39. What is the cost of an item in U.S. dollars if it costs 14.99 in British pounds?

40. Find the cost of an item in U.S. dollars if it costs 12.50 in Egyptian pounds.

41. **SNACKS** The Skyway Snack Company makes a snack mix that contains raisins, peanuts, and chocolate pieces. The ingredients are shown at the right. Suppose the company wants to sell a larger-sized bag that contains 6 cups of raisins. How many cups of chocolate pieces and peanuts should be added?

42. **PAINT** If 1 pint of paint is needed to paint a square that is 5 feet on each side, how many pints must be purchased in order to paint a square that is 9 feet 6 inches on each side?

43. **CRITICAL THINKING** The Property of Proportions states that if $\frac{a}{b} = \frac{c}{d}$, then $ad = bc$. Write two proportions in which the cross products are ad and bc.

44. **WRITING IN MATH** Answer the question that was posed at the beginning of the lesson.

How are proportions used in recipes?

Include the following in your answer:

- an explanation telling how proportions can be used to increase or decrease the amount of ingredients needed, and
- an explanation of why adding 10 ounces to each ingredient in the punch recipe will not result in the same flavor of punch.

45. Jack is standing next to a flagpole as shown at the right. Jack is 6 feet tall. Which proportion could you use to find the height of the flagpole?

(A) $\frac{3}{6} = \frac{x}{12}$ (B) $\frac{x}{6} = \frac{3}{12}$

(C) $\frac{6}{3} = \frac{x}{12}$ (D) $\frac{3}{x} = \frac{12}{6}$

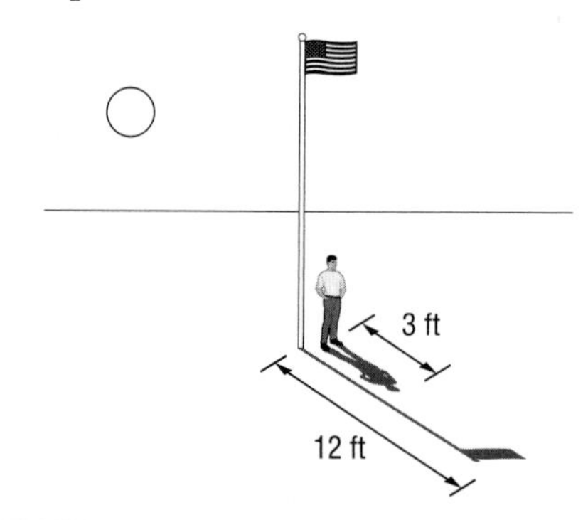

Maintain Your Skills

Mixed Review **Express each ratio as a unit rate. Round to the nearest tenth, if necessary.** *(Lesson 6-1)*

46. \$5 for 4 loaves of bread

47. 183.4 miles in 3.2 hours

48. Find the next three numbers in the sequence 2, 5, 8, 11, 14, ... *(Lesson 5-10)*

ALGEBRA **Find each quotient.** *(Lesson 5-4)*

49. $\frac{x}{5} \div \frac{x}{20}$

50. $\frac{3y}{4} \div \frac{5y}{8}$

51. $\frac{4z}{w} \div \frac{7yz}{w}$

Getting Ready for the Next Lesson **PREREQUISITE SKILL** **Complete each sentence.**
(To review ***converting measurements****, see pages 720 and 721.)*

52. 5 feet = ■ inches

53. 8.5 feet = ■ inches

54. 36 inches = ■ feet

55. 78 inches = ■ feet

Algebra Activity

A Follow-Up of Lesson 6-2

Sunshine State Standards
MA.A.3.3.2-1, MA.A.3.3.2-2, MA.A.4.3.1-1, MA.A.4.3.1-2

Capture-Recapture

Scientists often determine the number of fish in a pond, lake, or other body of water by using the *capture-recapture* method. A number of fish are captured, counted, carefully tagged, and returned to their habitat. The tagged fish are counted again and proportions are used to estimate the entire population. In this activity, you will model this estimation technique.

Collect the Data

Step 1 Copy the table below onto a sheet of paper.

Original Number Captured:		
Sample	**Recaptured**	**Tagged**
1		
2		
3		
4		
⋮		
10		
Total		

Step 2 Empty a bag of dried beans into a paper bag.

Step 3 Remove a handful of beans. Using a permanent marker, place an X on each side of each bean. These beans will represent the tagged fish. Record this number at the top of your table as the original number captured. Return the beans to the bag and mix.

Step 4 Remove a second handful of beans without looking. This represents the first sample of recaptured fish. Record the number of beans. Then count and record the number of beans that are tagged. Return the beans to the bag and mix.

Step 5 Repeat Step 4 for samples 2 through 10. Then use the results to find the total number of recaptured fish and the total number of tagged fish.

Analyze the Data

1. Use the following proportion to estimate the number of beans in the bag.

$$\frac{\text{original number captured}}{\text{total number in bag}} = \frac{\text{total number tagged}}{\text{total number recaptured}}$$

2. Count the number of beans in the bag. Compare the estimate to the actual number.

Make a Conjecture

3. Why is it a good idea to base a prediction on several samples instead of one sample?
4. Why does this method work?

6-3 Scale Drawings and Models

Sunshine State Standards
MA.B.1.3.4-1, MA.B.1.3.4-2, MA.B.2.3.1-1

Vocabulary

- scale drawing
- scale model
- scale
- scale factor

What You'll Learn

- Use scale drawings.
- Construct scale drawings.

How are scale drawings used in everyday life?

A set of landscape plans and a map are shown.

Designers use blueprints when planning landscapes.

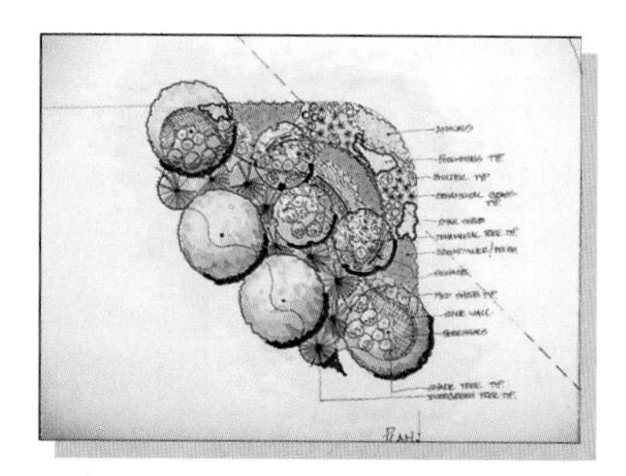

Maps are used to find actual distances between cities.

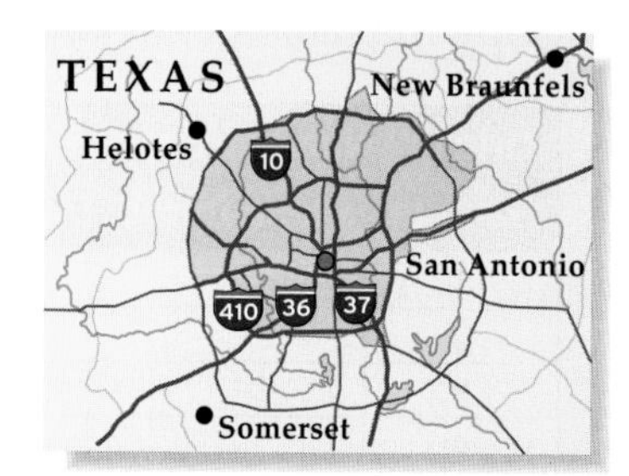

a. Suppose the landscape plans are drawn on graph paper and the side of each square on the paper represents 2 feet. What is the actual width of a rose garden if its width on the drawing is 4 squares long?

b. All maps have a scale. How can the scale help you estimate the distance between cities?

USE SCALE DRAWINGS AND MODELS A **scale drawing** or a **scale model** is used to represent an object that is too large or too small to be drawn or built at actual size. A few examples are maps, blueprints, model cars, and model airplanes.

Model cars are replicas of actual cars.

Concept Check Why are scale drawings or scale models used?

The **scale** gives the relationship between the measurements on the drawing or model and the measurements of the real object. Consider the following scales.

1 inch = 3 feet

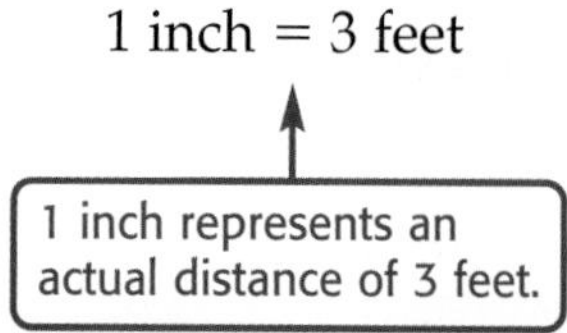

1 inch represents an actual distance of 3 feet.

1:24

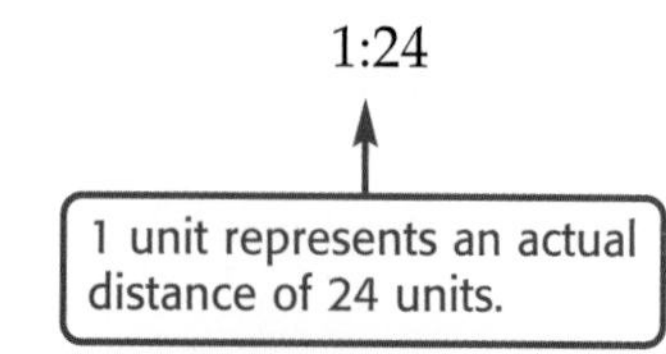

1 unit represents an actual distance of 24 units.

The ratio of a length on a scale drawing or model to the corresponding length on the real object is called the **scale factor**. Suppose a scale model has a scale of 2 inches = 16 inches. The scale factor is $\frac{2}{16}$ or $\frac{1}{8}$.

The lengths and widths of objects of a scale drawing or model are proportional to the lengths and widths of the actual object.

Study Tip

Scale Factor
When finding the scale factor, be sure to use the same units of measure.

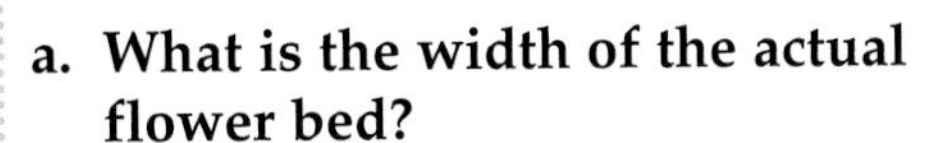

Example 1 Find Actual Measurements

DESIGN A set of landscape plans shows a flower bed that is 6.5 inches wide. The scale on the plans is 1 inch = 4 feet.

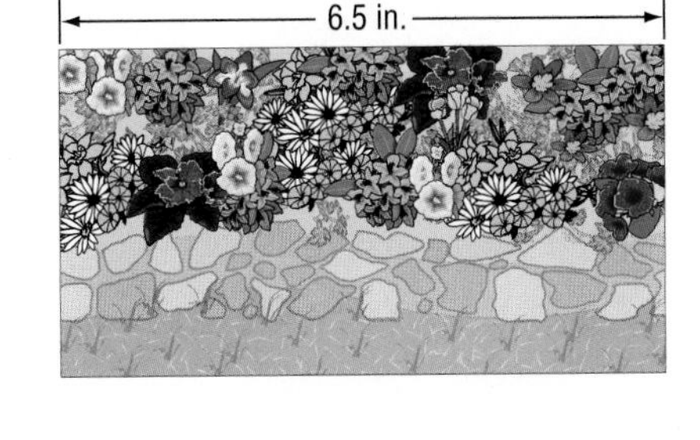

a. What is the width of the actual flower bed?

Let x represent the actual width of the flower bed. Write and solve a proportion.

plan width → / actual width → $\frac{1 \text{ inch}}{4 \text{ feet}} = \frac{6.5 \text{ inches}}{x \text{ feet}}$ ← plan width / ← actual width

$1 \cdot x = 4 \cdot 6.5$ Find the cross products.

$x = 26$ Simplify.

The actual width of the flower bed is 26 feet.

b. What is the scale factor?

To find the scale factor, write the ratio of 1 inch to 4 feet in simplest form.

$\frac{1 \text{ inch}}{4 \text{ feet}} = \frac{1 \text{ inch}}{48 \text{ inches}}$ Convert 4 feet to inches.

The scale factor is $\frac{1}{48}$. That is, each measurement on the plan is $\frac{1}{48}$ the actual measurement.

Example 2 Determine the Scale

ARCHITECTURE The inside of the Lincoln Memorial contains three chambers. The central chamber, which features a marble statue of Abraham Lincoln, has a height of 60 feet. Suppose a scale model of the chamber has a height of 4 inches. What is the scale of the model?

Write the ratio of the height of the model to the actual height of the statue. Then solve a proportion in which the height of the model is 1 inch and the actual height is x feet.

model height → / actual height → $\frac{4 \text{ inches}}{60 \text{ feet}} = \frac{1 \text{ inch}}{x \text{ feet}}$ ← model height / ← actual height

$4 \cdot x = 60 \cdot 1$ Find the cross products.

$4x = 60$ Simplify.

$\frac{4x}{4} = \frac{60}{4}$ Divide each side by 4.

$x = 15$ Simplify.

So, the scale is 1 inch = 15 feet.

More About. . .

Architecture

The exterior of the Lincoln Memorial features 36 columns that represent the states in the Union when Lincoln died in 1865. Each column is 44 feet high.

Source: www.infoplease.com

CONSTRUCT SCALE DRAWINGS To construct a scale drawing of an object, use the actual measurements of the object and the scale to which the object is to be drawn.

Example 3 Construct a Scale Drawing

INTERIOR DESIGN Antonio is designing a room that is 20 feet long and 12 feet wide. Make a scale drawing of the room. Use a scale of 0.25 inch = 4 feet.

Step 1 Find the measure of the room's length on the drawing. Let x represent the length.

$$\frac{0.25 \text{ inch}}{4 \text{ feet}} = \frac{x \text{ inches}}{20 \text{ feet}}$$ drawing length → / actual length → ... ← drawing length / ← actual length

$0.25 \cdot 20 = 4 \cdot x$ Find the cross products.

$5 = 4x$ Simplify.

$1.25 = x$ Divide each side by 4.

On the drawing, the length is 1.25 or $1\frac{1}{4}$ inches.

Step 2 Find the measure of the room's width on the drawing. Let w represent the width.

$$\frac{0.25 \text{ inch}}{4 \text{ feet}} = \frac{w \text{ inches}}{12 \text{ feet}}$$ drawing length → / actual length → ... ← drawing length / ← actual length

$0.25 \cdot 12 = 4 \cdot w$ Find the cross products.

$3 = 4w$ Simplify.

$\frac{3}{4} = \frac{4w}{4}$ Divide each side by 4.

$0.75 = w$ Simplify.

On the drawing, the width is 0.75 or $\frac{3}{4}$ inch.

Step 3 Make the scale drawing.
Use $\frac{1}{4}$-inch grid paper. Since $1\frac{1}{4}$ inches = 5 squares and $\frac{3}{4}$ inch = 3 squares, draw a rectangle that is 5 squares by 3 squares.

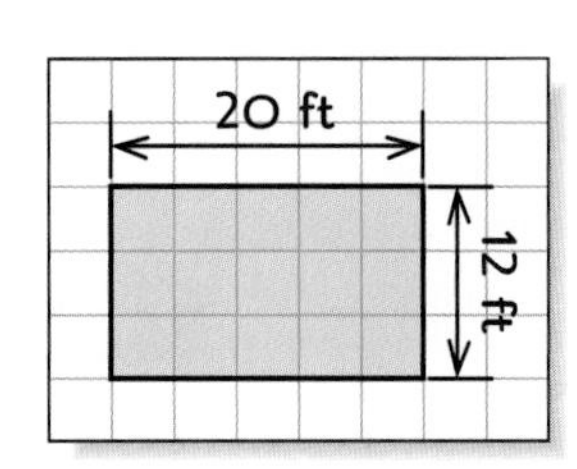

Career Choices

Interior Designer

Interior designers plan the space and furnish the interiors of places such as homes, offices, restaurants, hotels, hospitals, and even theaters. Creativity and knowledge of computer-aided design software is essential.

Online Research
For information about a career as an interior designer, visit: www.pre-alg.com/careers

Check for Understanding

Concept Check

1. **OPEN ENDED** Draw two squares in which the ratio of the sides of the first square to the sides of the second square is 1:3.

2. **FIND THE ERROR** Montega and Luisa are rewriting the scale 1 inch = 2 feet in $a{:}b$ form.

Who is correct? Explain your reasoning.

Guided Practice On a map of Pennsylvania, the scale is 1 inch = 20 miles. Find the actual distance for each map distance.

	From	To	Map Distance
3.	Pittsburgh	Perryopolis	2 inches
4.	Johnston	Homer City	$1\frac{3}{4}$ inches

Applications STATUES For Exercises 5 and 6, use the following information.
The Statue of Zeus at Olympia is one of the Seven Wonders of the World. On a scale model of the statue, the height of Zeus is 8 inches.

5. If the actual height of Zeus is 40 feet, what is the scale of the statue?
6. What is the scale factor?
7. **DESIGN** An architect is designing a room that is 15 feet long and 10 feet wide. Construct a scale drawing of the room. Use a scale of 0.5 in. = 10 ft.

Practice and Apply

Homework Help

For Exercises	See Examples
8–17	1
18, 19	1, 2
20	3

Extra Practice
See page 737.

On a set of architectural drawings for an office building, the scale is $\frac{1}{2}$ inch = 3 feet. Find the actual length of each room.

	Room	Drawing Distance
8.	Conference Room	7 inches
9.	Lobby	2 inches
10.	Mail Room	2.3 inches
11.	Library	4.1 inches
12.	Copy Room	2.2 inches
13.	Storage	1.9 inches
14.	Exercise Room	$3\frac{3}{4}$ inches
15.	Cafeteria	$8\frac{1}{4}$ inches

16. Refer to Exercises 8–15. What is the scale factor?
17. What is the scale factor if the scale is 8 inches = 1 foot?
18. **ROLLER COASTERS** In a scale model of a roller coaster, the highest hill has a height of 6 inches. If the actual height of the hill is 210 feet, what is the scale of the model?
19. **INSECTS** In an illustration of a honeybee, the length of the bee is 4.8 centimeters. The actual size of the honeybee is 1.2 centimeters. What is the scale of the drawing?

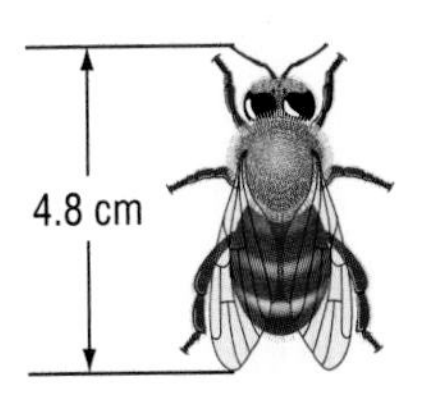

20. **GARDENS** A garden is 8 feet wide by 16 feet long. Make a scale drawing of the garden that has a scale of $\frac{1}{4}$ in. = 2 ft.

21. **CRITICAL THINKING** What does it mean if the scale factor of a scale drawing or model is less than 1? greater than 1? equal to 1?

22. **WRITING IN MATH** Answer the question that was posed at the beginning of the lesson.

How are scale drawings used in everyday life?

Include the following in your answer:

- an example of three kinds of scale drawings or models, and
- an explanation of how you use scale drawings in your life.

23. Which scale has a scale factor of $\frac{1}{18}$?

(A) 3 in. = 6 ft (B) 6 in. = 9 ft (C) 3 in. = 54 ft (D) 6 in. = 6 ft

24. A model airplane is built using a 1:16 scale. On the model, the length of the wing span is 5.8 feet. What is the actual length of the wing?

(A) 84.8 ft (B) 91.6 ft (C) 92.8 ft (D) 89.8 ft

Extending the Lesson

25. Two rectangles are shown. The ratio comparing their sides is 1:2.

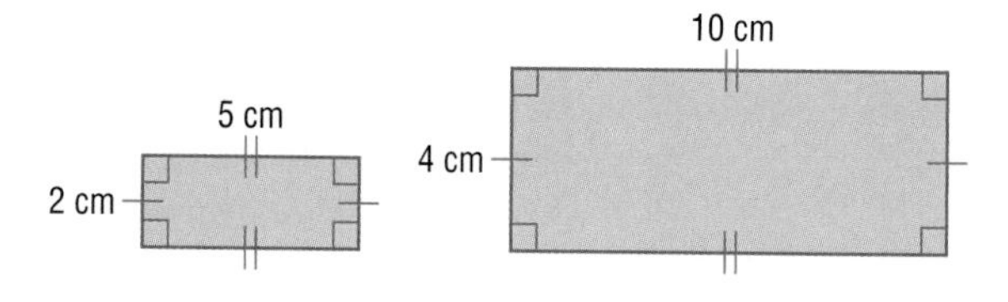

a. Write the ratio that compares their perimeters.

b. Write the ratio that compares their areas.

c. Find the perimeter and area of a 3-inch by 5-inch rectangle. Then make a conjecture about the perimeter and area of a 6-inch by 10-inch rectangle. Check by finding the actual perimeter and area.

Maintain Your Skills

Mixed Review

Solve each proportion. *(Lesson 6-2)*

26. $\frac{n}{20} = \frac{15}{50}$ 27. $\frac{14}{32} = \frac{x}{8}$ 28. $\frac{3}{2.2} = \frac{7.5}{y}$

Convert each rate using dimensional analysis. *(Lesson 6-1)*

29. 36 cm/s = ■ m/min 30. 66 gal/h = ■ qt/min

31. Find $1\frac{1}{4} + 4\frac{5}{6}$. Write the answer in simplest form. *(Lesson 5-7)*

ALGEBRA **Find each product or quotient. Express in exponential form.** *(Lesson 4-6)*

32. $4^3 \cdot 4^5$ 33. $3t^4 \cdot 6t$ 34. $7^{14} \div 7^8$ 35. $\frac{24m^5}{18m^2}$

36. **ALGEBRA** Find the greatest common factor of $14x^2y$ and $35xy^3$. *(Lesson 4-4)*

Getting Ready for the Next Lesson

PREREQUISITE SKILL **Simplify each fraction.**
*(To review **simplest form**, see Lesson 4-5.)*

37. $\frac{5}{100}$ 38. $\frac{25}{100}$ 39. $\frac{40}{100}$ 40. $\frac{52}{100}$

41. $\frac{78}{100}$ 42. $\frac{75}{100}$ 43. $\frac{82}{100}$ 44. $\frac{95}{100}$

6-4 Fractions, Decimals, and Percents

What You'll Learn

- Express percents as fractions and vice versa.
- Express percents as decimals and vice versa.

Vocabulary

- percent

Sunshine State Standards
MA.A.1.3.2-1, MA.A.1.3.2-2, MA.A.1.3.4-1, MA.A.3.3.2-1, MA.A.3.3.2-2, MA.A.3.3.2-3, MA.A.3.3.2-4

How are percents related to fractions and decimals?

A portion of each figure is shaded.

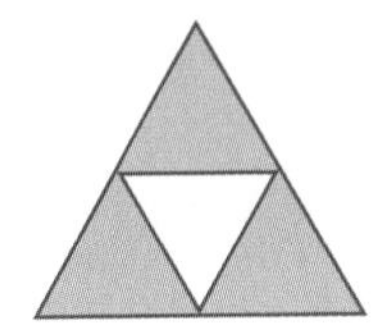
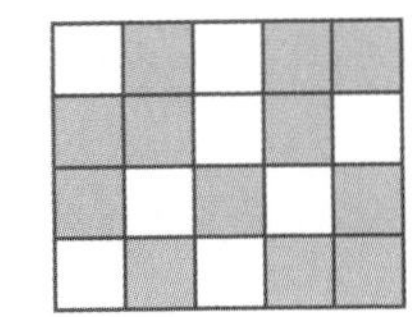
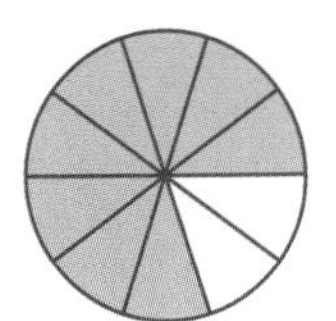

a. Write a ratio that compares the shaded region of each figure to its total region as a fraction in simplest form.

b. Rewrite each fraction using a denominator of 100.

c. Which figure has the greatest part of its area shaded?

d. Was it easier to compare the fractions in part **a** or part **b**? Explain.

Reading Math

Percent
Root Word: Cent
There are 100 *cents* in one dollar. *Percent* means *per hundred* or *hundredths.*

PERCENTS AND FRACTIONS A **percent** is a ratio that compares a number to 100. The meaning of 75% is shown at the right. In the figure, 75 out of 100 squares are shaded.

To write a percent as a fraction, express the ratio as a fraction with a denominator of 100. Then simplify if possible. Notice that a percent can be greater than 100% or less than 1%.

Example 1 Percents as Fractions

Express each percent as a fraction in simplest form.

a. 45%

$$45\% = \frac{45}{100}$$
$$= \frac{9}{20}$$

b. 120%

$$120\% = \frac{120}{100}$$
$$= \frac{6}{5} \text{ or } 1\frac{1}{5}$$

c. 0.5%

$$0.5\% = \frac{0.5}{100}$$
$$= \frac{0.5}{100} \cdot \frac{10}{10}$$
$$= \frac{5}{1000} \text{ or } \frac{1}{200}$$

Multiply by $\frac{10}{10}$ to eliminate the decimal in the numerator.

d. $83\frac{1}{3}\%$

$$83\frac{1}{3}\% = \frac{83\frac{1}{3}}{100}$$

The fraction bar indicates division.

$$= 83\frac{1}{3} \div 100$$
$$= \frac{\overset{5}{\cancel{250}}}{3} \cdot \frac{1}{\underset{2}{\cancel{100}}} \text{ or } \frac{5}{6}$$

To write a fraction as a percent, write an equivalent fraction with a denominator of 100.

Example 2 Fractions as Percents

Express each fraction as a percent.

a. $\frac{4}{5}$

$\frac{4}{5} = \frac{80}{100}$ or 80%

b. $\frac{9}{4}$

$\frac{9}{4} = \frac{225}{100}$ or 225%

PERCENTS AND DECIMALS Remember that *percent* means *per hundred.* In the previous examples, you wrote percents as fractions with 100 in the denominator. Similarly, you can write percents as decimals by dividing by 100.

Key Concept — *Percents and Decimals*

- To write a percent as a decimal, divide by 100 and remove the percent symbol.
- To write a decimal as a percent, multiply by 100 and add the percent symbol.

Example 3 Percents as Decimals

Express each percent as a decimal.

a. 28%

28% = 28%   Divide by 100 and remove the %.
= 0.28

b. 8%

8% = 08%   Divide by 100 and remove the %.
= 0.08

c. 375%

375% = 375%   Divide by 100 and remove the %.
= 3.75

d. 0.5%

0.5% = 00.5%   Divide by 100 and remove the %.
= 0.005

Study Tip

Mental Math
To divide a number by 100, move the decimal point two places to the left. To multiply a number by 100, move the decimal point two places to the right.

Example 4 Decimals as Percents

Express each decimal as a percent.

a. 0.35

0.35 = 0.35   Multiply by 100 and add the %.
= 35%

b. 0.09

0.09 = 0.09   Multiply by 100 and add the %.
= 9%

c. 0.007

0.007 = 0.007   Multiply by 100 and add the %.
= 0.7%

d. 1.49

1.49 = 1.49   Multiply by 100 and add the %.
= 149%

You have expressed fractions as decimals and decimals as percents. Fractions, decimals, and percents are all different names that represent the same number.

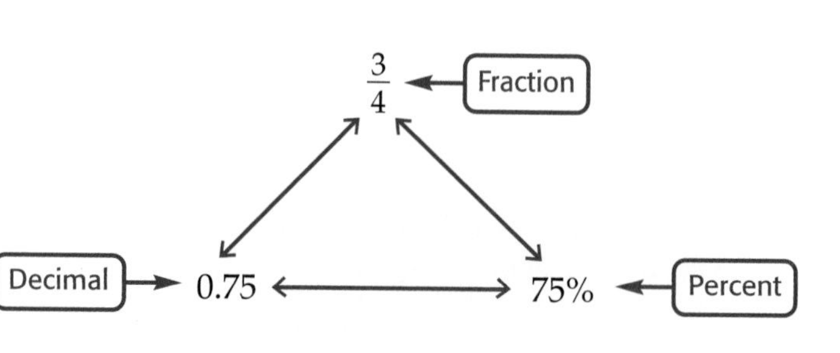

You can also express a fraction as a percent by first expressing the fraction as a decimal and then expressing the decimal as a percent.

Example 5 Fractions as Percents

Express each fraction as a percent. Round to the nearest tenth percent, if necessary.

a. $\frac{7}{8}$

$\frac{7}{8} = 0.875$

$= 87.5\%$

b. $\frac{2}{3}$

$\frac{2}{3} = 0.6666666\ldots$

$\approx 66.7\%$

c. $\frac{3}{500}$

$\frac{3}{500} = 0.006$

$= 0.6\%$

d. $\frac{15}{7}$

$\frac{15}{7} \approx 2.1428571$

$\approx 214.3\%$

Study Tip

Fractions

When the numerator of a fraction is less than the denominator, the fraction is less than 100%. When the numerator of a fraction is greater than the denominator, the fraction is greater than 100%.

Example 6 Compare Numbers

SHOES In a survey, one-fifth of parents said that they buy shoes for their children every 4–5 months while 27% of parents said that they buy shoes twice a year. Which of these groups is larger?

Write one-fifth as a percent. Then compare.

$\frac{1}{5} = 0.20$ or 20%

Since 27% is greater than 20%, the group that said they buy shoes twice a year is larger.

Check for Understanding

Concept Check

1. **Describe** two ways to express a fraction as a percent. Then tell how you know whether a fraction is greater than 100% or less than 1%.
2. **OPEN ENDED** Explain the method you would use to express $64\frac{1}{2}\%$ as a decimal.

Guided Practice

Express each percent as a fraction or mixed number in simplest form and as a decimal.

3. 30%
4. $12\frac{1}{2}\%$
5. 125%
6. 65%
7. 135%
8. 0.2%

Express each decimal or fraction as a percent. Round to the nearest tenth percent, if necessary.

9. 0.45
10. 1.3
11. 0.008
12. $\frac{1}{4}$
13. $\frac{12}{9}$
14. $\frac{3}{600}$

Application

15. **MEDIA** In a survey, 55% of those surveyed said that they get the news from their local television station while three-fifths said that they get the news from a daily newspaper. From which source do more people get their news?

Practice and Apply

Homework Help

For Exercises	See Examples
16–27	1, 3
28–39	2, 4, 5
40, 41	1
42	6

Extra Practice

See page 737.

Express each percent as a fraction or mixed number in simplest form and as a decimal.

16. 42% **17.** 88% **18.** $16\frac{2}{3}\%$ **19.** 87.5%

20. 150% **21.** 350% **22.** 18% **23.** 61%

24. 117% **25.** 223% **26.** 0.8% **27.** 0.53%

Express each decimal or fraction as a percent. Round to the nearest tenth percent, if necessary.

28. 0.51 **29.** 0.09 **30.** 3.21 **31.** 2.7

32. 0.0042 **33.** 0.0006 **34.** $\frac{7}{25}$ **35.** $\frac{9}{40}$

36. $\frac{10}{3}$ **37.** $\frac{14}{8}$ **38.** $\frac{15}{2500}$ **39.** $\frac{20}{1200}$

40. **GEOGRAPHY** Forty-six percent of the world's water is in the Pacific Ocean. What fraction is this?

41. **GEOGRAPHY** The Arctic Ocean contains 3.7% of the world's water. What fraction is this?

42. **FOOD** According to a survey, 22% of people said that mustard is their favorite condiment while two-fifths of people said that they prefer ketchup. Which group is larger? Explain.

More About. . .

Food

The three types of mustard commonly grown are white or yellow mustard, brown mustard, and Oriental mustard.

Source: Morehouse Foods, Inc.

Choose the greatest number in each set.

43. $\left\{\frac{2}{5}, 0.45, 35\%, 3 \text{ out of } 8\right\}$ **44.** $\left\{\frac{3}{4}, 0.70, 78\%, 4 \text{ out of } 5\right\}$

45. $\left\{19\%, \frac{3}{16}, 0.155, 2 \text{ to } 15\right\}$ **46.** $\left\{89\%, \frac{10}{11}, 0.884, 12 \text{ to } 14\right\}$

Write each list of numbers in order from least to greatest.

47. $\frac{2}{3}$, 61%, 0.69 **48.** $\frac{2}{7}$, 0.027, 27%

GEOMETRY **For Exercises 49 and 50, use the information and the figure shown.** Suppose that two fifths of the rectangle is shaded.

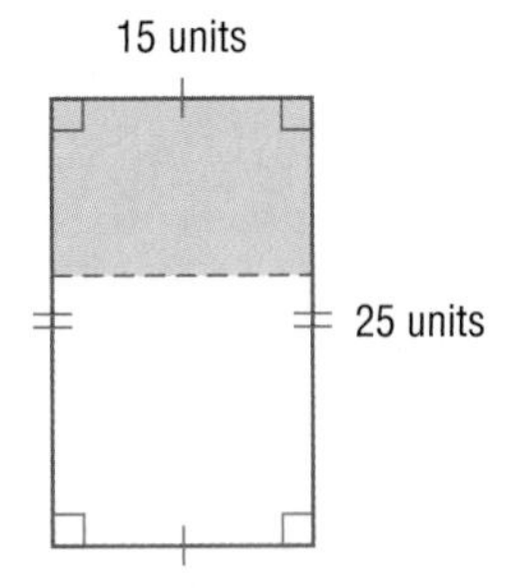

49. Write the decimal that represents the shaded region of the figure.

50. What is the area of the shaded region?

51. **CRITICAL THINKING** Find a fraction that satisfies the conditions below. Then write a sentence explaining why you think your fraction is or is not the only solution that satisfies the conditions.

- The fraction can be written as a percent greater than 1%.
- The fraction can be written as a percent less than 50%.
- The decimal equivalent of the fraction is a terminating decimal.
- The value of the denominator minus the value of the numerator is 3.

52. **CRITICAL THINKING** Explain why percents are rational numbers.

53. **WRITING IN MATH** Answer the question that was posed at the beginning of the lesson.

How are percents related to fractions and decimals?

Include the following in your answer:

- examples of figures in which 25%, 30%, 40%, and 65% of the area is shaded, and
- an explanation of why each percent represents the shaded area.

54. Assuming that the regions in each figure are equal, which figure has the greatest part of its area shaded?

(A) 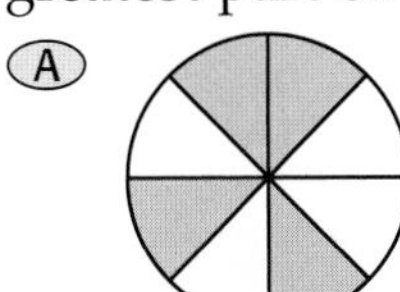

(B) 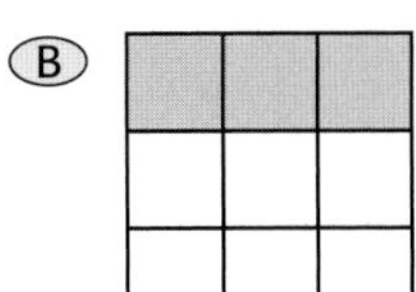

(C) 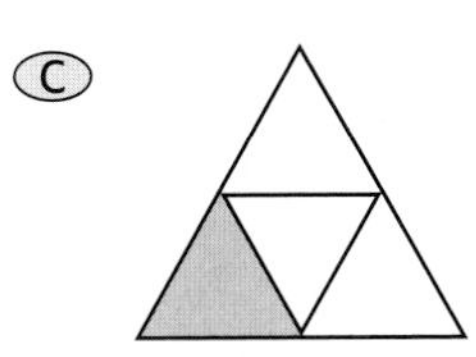

(D) 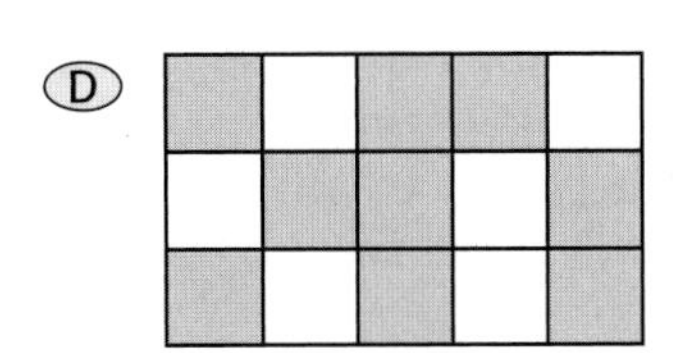

55. According to a survey, 85% of people eat a salad at least once a week. Which ratio represents this portion?

(A) 17 to 20 (B) 13 to 20 (C) 9 to 10 (D) 4 to 5

Maintain Your Skills

Mixed Review

Write the scale factor of each scale. *(Lesson 6-3)*

56. 3 inches = 18 inches

57. 2 inches = 2 feet

58. **ALGEBRA** Find the solution of $\frac{x}{54} = \frac{2}{3}$. *(Lesson 6-2)*

Find each product. Write in simplest form. *(Lesson 5-3)*

59. $\frac{4}{7} \cdot \frac{11}{12}$

60. $-\frac{3}{5} \cdot \frac{10}{18}$

61. $4 \cdot \frac{16}{52}$

62. Write 5.6×10^{-4} in standard form. *(Lesson 4-8)*

Determine whether each number is *prime* or *composite*. *(Lesson 4-3)*

63. 21

64. 47

65. 57

Getting Ready for the Next Lesson

PREREQUISITE SKILL **Solve each proportion.**

*(To review **proportions**, see Lesson 6-2.)*

66. $\frac{25}{4} = \frac{x}{100}$

67. $\frac{56}{7} = \frac{y}{100}$

68. $\frac{75}{8} = \frac{n}{100}$

69. $\frac{m}{10} = \frac{9.4}{100}$

70. $\frac{h}{350} = \frac{46}{100}$

71. $\frac{86.4}{k} = \frac{27}{100}$

Algebra Activity

A Preview of Lesson 6-5

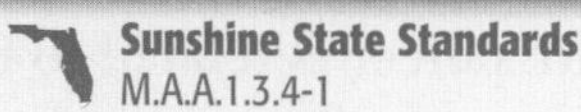

Sunshine State Standards
M.A.A.1.3.4-1

Using a Percent Model

Activity 1

When you see advertisements on television or in magazines, you are often bombarded with many claims. For example, you might hear that four out of five use a certain long-distance phone service. What percent does this represent?

You can find the percent by using a model.

Finding a Percent

Step 1	Step 2	Step 3
Draw a 10-unit by 1-unit rectangle on grid paper. Label the units on the right from 0 to 100, because percent is a ratio that compares a number to 100.	On the left side, mark equal units from 0 to 5, because 5 represents the whole quantity. Locate 4 on this scale.	Draw a horizontal line from 4 on the left side to the right side of the model. The number on the right side is the percent. Label the model as shown.

Using the model, you can see that the ratio *4 out of 5* is the same as 80%. So, according to this claim, 80% of people prefer the certain long-distance phone service.

Model

Draw a model and find the percent that is represented by each ratio. If it is not possible to find the exact percent using the model, estimate.

1. 6 out of 10

2. 9 out of 10

3. 2 out of 5

4. 3 out of 4

5. 9 out of 20

6. 8 out of 50

7. 2 out of 8

8. 3 out of 8

9. 2 out of 3

10. 5 out of 9

Activity 2

Suppose a store advertises a sale in which all merchandise is 20% off the original price. If the original price of a pair of shoes is $50, how much will you save?

In this case, you know the percent. You need to find what part of the original price you'll save.

You can find the part by using a similar model.

Finding a Part

Step 1	Step 2	Step 3
Draw a 10-unit by 1-unit rectangle on grid paper. Label the units on the right from 0 to 100 because percent is a ratio that compares a number to 100.	On the left side, mark equal units from 0 to 50, because 50 represents the whole quantity.	Draw a horizontal line from 20% on the right side to the left side of the model. The number on the left side is the part. Label the model as shown.

Using the model, you can see that 20% of 50 is 10. So, you will save $10 if you buy the shoes.

Model

Draw a model and find the part that is represented. If it is not possible to find an exact answer from the model, estimate.

11. 10% of 50

12. 60% of 20

13. 90% of 40

14. 30% of 10

15. 25% of 20

16. 75% of 40

17. 5% of 200

18. 85% of 500

19. $33\frac{1}{3}$% of 12

20. 37.5% of 16

6-5 Using the Percent Proportion

Sunshine State Standards
MA.A.1.3.4-1, MA.A.3.3.2-1, MA.A.3.3.2-2

Vocabulary

- percent proportion
- part
- base

What You'll Learn

- Use the percent proportion to solve problems.

Why are percents important in real-world situations?

Have you collected any of the new state quarters?

The quarters are made of a pure copper core and an outer layer that is an alloy of 3 parts copper and 1 part nickel.

a. Write a ratio that compares the amount of copper to the total amount of metal in the outer layer.

b. Write the ratio as a fraction and as a percent.

USE THE PERCENT PROPORTION In a **percent proportion**, one of the numbers, called the **part**, is being compared to the whole quantity, called the **base**. The other ratio is the percent, written as a fraction, whose base is 100.

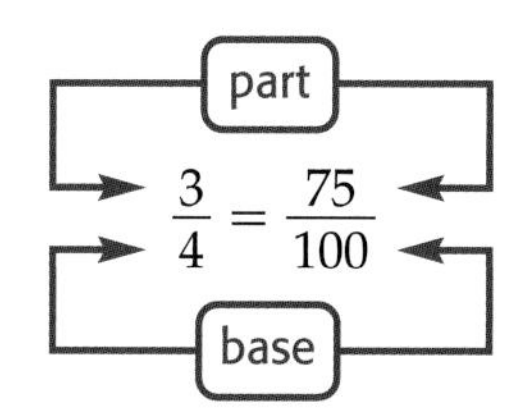

Key Concept — Percent Proportion

- **Words** $\frac{\text{part}}{\text{base}} = \frac{\text{percent}}{100}$
- **Symbols** $\frac{a}{b} = \frac{p}{100}$, where a is the part, b is the base, and p is the percent.

Example 1 Find the Percent

Five is what percent of 8?

Five is being compared to 8. So, 5 is the part and 8 is the base. Let p represent the percent.

$\frac{a}{b} = \frac{p}{100} \rightarrow \frac{5}{8} = \frac{p}{100}$ Replace a with 5 and b with 8.

$5 \cdot 100 = 8 \cdot p$ Find the cross products.

$500 = 8p$ Simplify.

$\frac{500}{8} = \frac{8p}{8}$ Divide each side by 8.

$62.5 = p$ So, 5 is 62.5% of 8.

Study Tip

Estimation
Five is a little more than one-half of eight. So, the answer should be a little more than 50%.

Concept Check In the percent proportion $\frac{15}{20} = \frac{75}{100}$, which number is the base?

Example 2 Find the Percent

What percent of 4 is 7?

Seven is being compared to 4. So, 7 is the part and 4 is the base. Let p represent the percent.

$\frac{a}{b} = \frac{p}{100} \rightarrow \frac{7}{4} = \frac{p}{100}$ Replace a with 7 and b with 4.

$7 \cdot 100 = 4 \cdot p$ Find the cross products.

$700 = 4p$ Simplify.

$\frac{700}{4} = \frac{4p}{4}$ Divide each side by 4.

$175 = p$ So, 175% of 4 is 7.

Study Tip

Base

In percent problems, the base usually follows the word *of*.

Log on for:

- Updated data
- More activities on finding a percent.

www.pre-alg.com/usa_today

Example 3 Apply the Percent Proportion

ENVIRONMENT The graphic shows the number of threatened species in the United States. What percent of the total number of threatened species are mammals?

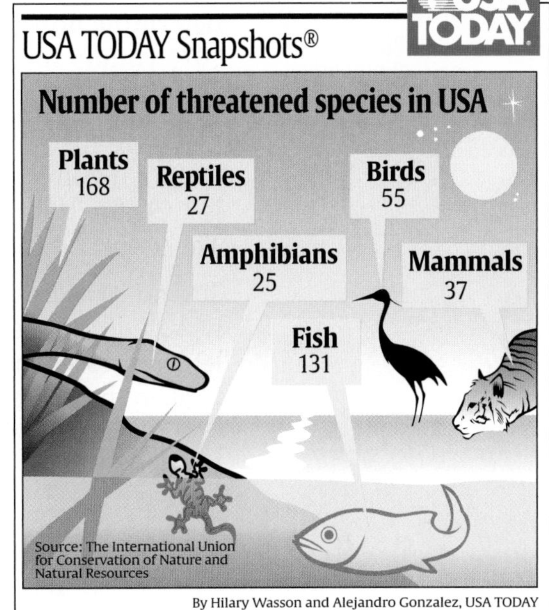

Compare the number of species of mammals, 37, to the total number of threatened species, 443. Let a represent the part, 37, and let b represent the base, 443, in the percent proportion. Let p represent the percent.

$\frac{a}{b} = \frac{p}{100} \rightarrow \frac{37}{443} = \frac{p}{100}$

$37 \cdot 100 = 443 \cdot p$

$3700 = 443p$ Simplify.

$\frac{3700}{443} = \frac{443p}{443}$ Divide each side by 443.

$8.4 \approx p$ Simplify.

So, about 8.4% of the total number of threatened species are mammals.

You can also use the percent proportion to find a missing part or base.

Concept Summary — Types of Percent Problems

Type	Example	Proportion
Find the Percent	3 is what percent of 4?	$\frac{3}{4} = \frac{p}{100}$
Find the Part	What number is 75% of 4?	$\frac{a}{4} = \frac{75}{100}$
Find the Base	3 is 75% of what number?	$\frac{3}{b} = \frac{75}{100}$

Example 4 Find the Part

What number is 5.5% of 650?

The percent is 5.5, and the base is 650. Let a represent the part.

$\frac{a}{b} = \frac{p}{100} \rightarrow \frac{a}{650} = \frac{5.5}{100}$ Replace b with 650 and p with 5.5.

$a \cdot 100 = 650 \cdot 5.5$ Find the cross products.

$100a = 3575$ Simplify.

$a = 35.75$ Mentally divide each side by 100.

So, 5.5% of 650 is 35.75.

Log on for:
- Updated data
- More activities on using the percent proportion.

www.pre-alg.com/usa_today

Example 5 Apply the Percent Proportion

CHORES **Use the graphic to determine how many of the 1074 youths surveyed do not clean their room because there is not enough time.**

USA TODAY
USA TODAY Snapshots®
Kids don't enjoy cleaning their rooms
When 1,074 youths 19 and under were asked why they don't clean their room more often, these were their responses[1]:
I don't like to
66%
Not enough time
29%
I'm too lazy
28%
The rest of the house is dirty
6%
No reply
2%
1 – More than one response allowed
Source: BSMG Worldwide
By Lori Joseph and Sam Ward, USA TODAY

The total number of youths is 1074. So, 1074 is the base. The percent is 29%.

To find 29% of 1074, let b represent the base, 1074, and let p represent the percent, 29%, in the percent proportion. Let a represent the part.

$\frac{a}{b} = \frac{p}{100} \rightarrow \frac{a}{1074} = \frac{29}{100}$

$a \cdot 100 = 1074 \cdot 29$

$100a = 31146$ Simplify.

$a = 311.46$ Mentally divide each side by 100.

So, about 311 youths do not clean their room because there is not enough time.

Example 6 Find the Base

Fifty-two is 40% of what number?

The percent is 40% and the part is 52. Let b represent the base.

$\frac{a}{b} = \frac{p}{100} \rightarrow \frac{52}{b} = \frac{40}{100}$ Replace a with 52 and p with 40.

$52 \cdot 100 = b \cdot 40$ Find the cross products.

$5200 = 40b$ Simplify.

$\frac{5200}{40} = \frac{40b}{40}$ Divide each side by 40.

$130 = b$ Simplify.

So, 52 is 40% of 130.

Check for Understanding

Concept Check

1. **OPEN ENDED** Write a proportion that can be used to find the percent scored on an exam that has 50 questions.

2. **FIND THE ERROR** Judie and Pennie are using a proportion to find what number is 35% of 21.

Judie

$$\frac{n}{21} = \frac{35}{100}$$

Pennie

$$\frac{21}{n} = \frac{35}{100}$$

Who is correct? Explain your reasoning.

Guided Practice

Use the percent proportion to solve each problem.

3. 16 is what percent of 40?
4. 21 is 30% of what number?
5. What is 80% of 130?
6. What percent of 5 is 14?

Applications

7. **BOOKS** Fifty-four of the 90 books on a shelf are history books. What percent of the books are history books?

8. **CHORES** Refer to Example 5 on page 290. How many of the 1074 youths surveyed do not clean their room because they do not like to clean?

Practice and Apply

Homework Help

For Exercises	See Examples
9–20	1, 2, 4, 6
21, 23, 24	3
22, 25	5

Extra Practice
See page 738.

Use the percent proportion to solve each problem. Round to the nearest tenth.

9. 72 is what percent of 160?
10. 17 is what percent of 85?
11. 36 is 72% of what number?
12. 27 is 90% of what number?
13. What is 44% of 175?
14. What is 84% of 150?
15. 52.2 is what percent of 145?
16. 19.8 is what percent of 36?
17. 14 is $12\frac{1}{2}$% of what number?
18. 36 is $8\frac{3}{4}$% of what number?
19. 7 is what percent of 3500?
20. What is 0.3% of 750?

21. **BIRDS** If 12 of the 75 animals in a pet store are parakeets, what percent are parakeets?

22. **FISH** Of the fish in an aquarium, 26% are angelfish. If the aquarium contains 50 fish, how many are angelfish?

SCIENCE **For Exercises 23 and 24, use the information in the table.**

23. What percent of the world's fresh water does the Antarctic Icecap contain?

24. **RESEARCH** Use the Internet or another source to find the total volume of the world's fresh and salt water. What percent of the world's total water supply does the Antarctic Icecap contain?

World's Fresh Water Supply

Source	Volume (mi^3)
Freshwater Lakes	30,000
All Rivers	300
Antarctic Icecap	6,300,000
Arctic Icecap and Glaciers	680,000
Water in the Atmosphere	3100
Ground Water	1,000,000
Deep-lying Ground Water	1,000,000
Total	**9,013,400**

Source: *Time Almanac*

www.pre-alg.com/self_check_quiz/fcat

25. **LIFE SCIENCE** Carbon constitutes 18.5% of the human body by weight. Determine the amount of carbon contained in a person who weighs 145 pounds.

26. **CRITICAL THINKING** A number n is 25% of some number a and 35% of a number b. Tell the relationship between a and b. Is $a < b$, $a > b$, or is it impossible to determine the relationship? Explain.

27. **WRITING IN MATH** Answer the question that was posed at the beginning of the lesson.

Why are percents important in real-world situations?

Include the following in your answer:

- an example of a real-world situation where percents are used, and
- an explanation of the meaning of the percent in the situation.

28. The table shows the number of people in each section of the school chorale. Which section makes up exactly 25% of the chorale?

School Chorale	
Section	**Number**
Soprano	16
Alto	15
Tenor	12
Bass	17

(A) Tenor (B) Alto

(C) Soprano (D) Bass

Maintain Your Skills

Mixed Review

Write each percent as a fraction in simplest form. *(Lesson 6-4)*

29. 42% 30. 56% 31. 120%

32. **MAPS** On a map of a state park, the scale is 0.5 inch = 1.5 miles. Find the actual distance from the ranger's station to the beach if the distance on the map is 1.75 inches. *(Lesson 6-3)*

Find each sum or difference. Write in simplest form. *(Lesson 5-5)*

33. $\frac{2}{9} + \frac{5}{9}$ 34. $\frac{11}{12} - \frac{3}{12}$ 35. $2\frac{5}{8} + \frac{7}{8}$

Getting Ready for the Next Lesson

PREREQUISITE SKILL **Find each product.**

*(To review **multiplying fractions**, see Lesson 5-3.)*

36. $\frac{1}{2} \times 14$ 37. $\frac{1}{4} \times 32$ 38. $\frac{1}{5} \times 15$

39. $\frac{2}{3} \times 9$ 40. $\frac{3}{4} \times 16$ 41. $\frac{5}{6} \times 30$

Practice Quiz 1

Lessons 6-1 through 6-5

1. Express $3.29 for 24 cans of soda as a unit rate. *(Lesson 6-1)*
2. What value of x makes $\frac{3}{4} = \frac{x}{68}$ a proportion? *(Lesson 6-2)*
3. **SCIENCE** A scale model of a volcano is 4 feet tall. If the actual height of the volcano is 12,276 feet, what is the scale of the model? *(Lesson 6-3)*
4. Express 352% as a decimal. *(Lesson 6-4)*
5. Use the percent proportion to find 32.5% of 60. *(Lesson 6-5)*

6-6 Finding Percents Mentally

What You'll Learn

Sunshine State Standards
MA.B.4.3.1-1, MA.B.4.3.1-2

- Compute mentally with percents.
- Estimate with percents.

How is estimation used when determining sale prices?

A sporting goods store is having a sale in which all merchandise is on sale at half off. A few regularly priced items are shown at the right.

a. What is the sale price of each item?

b. What percent represents half off?

c. Suppose the items are on sale for 25% off. Explain how you would determine the sale price.

FIND PERCENTS OF A NUMBER MENTALLY When working with common percents like 10%, 25%, 40%, and 50%, it may be helpful to use the fraction form of the percent. A few percent-fraction equivalents are shown.

0%	$12\frac{1}{2}\%$	25%	40%	50%	$66\frac{2}{3}\%$	75%	$87\frac{1}{2}\%$	100%
0	$\frac{1}{8}$	$\frac{1}{4}$	$\frac{2}{5}$	$\frac{1}{2}$	$\frac{2}{3}$	$\frac{3}{4}$	$\frac{7}{8}$	1

Some percents are used more frequently than others. So, it is a good idea to be familiar with these percents and their equivalent fractions.

Concept Summary — *Percent-Fraction Equivalents*

$20\% = \frac{1}{5}$	$10\% = \frac{1}{10}$	$25\% = \frac{1}{4}$	$12\frac{1}{2}\% = \frac{1}{8}$	$16\frac{2}{3}\% = \frac{1}{6}$
$40\% = \frac{2}{5}$	$30\% = \frac{3}{10}$	$50\% = \frac{1}{2}$	$37\frac{1}{2}\% = \frac{3}{8}$	$33\frac{1}{3}\% = \frac{1}{3}$
$60\% = \frac{3}{5}$	$70\% = \frac{7}{10}$	$75\% = \frac{3}{4}$	$62\frac{1}{2}\% = \frac{5}{8}$	$66\frac{2}{3}\% = \frac{2}{3}$
$80\% = \frac{4}{5}$	$90\% = \frac{9}{10}$		$87\frac{1}{2}\% = \frac{7}{8}$	$83\frac{1}{3}\% = \frac{5}{6}$

Example 1 *Find Percent of a Number Mentally*

Find the percent of each number mentally.

a. 50% of 32

$50\% \text{ of } 32 = \frac{1}{2} \text{ of } 32$ Think: $50\% = \frac{1}{2}$.

$= 16$ Think: $\frac{1}{2}$ of 32 is 16.

So, 50% of 32 is 16.

Study Tip

Look Back
To review **multiplying fractions**, see Lesson 5-3.

Find the percent of each number mentally.

b. 25% of 48

$25\% \text{ of } 48 = \frac{1}{4} \text{ of } 48$ Think: $25\% = \frac{1}{4}$.

$= 12$ Think: $\frac{1}{4}$ of 48 is 12.

So, 25% of 48 is 12.

c. 40% of 45

$40\% \text{ of } 45 = \frac{2}{5} \text{ of } 45$ Think: $40\% = \frac{2}{5}$.

$= 18$ Think: $\frac{1}{5}$ of 45 is 9. So, $\frac{2}{5}$ of 45 is 18.

So, 40% of 45 is 18.

ESTIMATE WITH PERCENTS Sometimes, an exact answer is not needed. In these cases, you can estimate. Consider the following model.

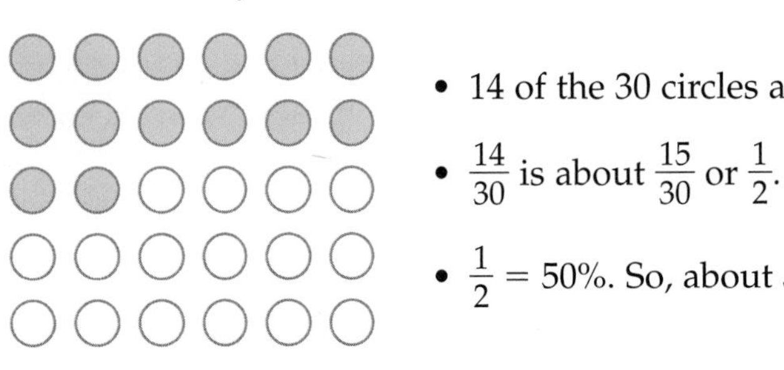

- 14 of the 30 circles are shaded.
- $\frac{14}{30}$ is about $\frac{15}{30}$ or $\frac{1}{2}$.
- $\frac{1}{2} = 50\%$. So, about 50% of the model is shaded.

The table below shows three methods you can use to estimate with percents. For example, let's estimate 22% of 237.

Method	Estimate 22% of 237.
Fraction	22% is a bit more than 20% or $\frac{1}{5}$. 237 is a bit less than 240. So, 22% of 237 is about $\frac{1}{5}$ of 240 or 48. Estimate: 48
1%	$22\% = 22 \times 1\%$ 1% of 237 = 2.37 or about 2. So, 22% of 237 is about 22×2 or 44. Estimate: 44
Meaning of Percent	22% means about 20 for every 100 or about 2 for every 10. 237 has 2 hundreds and about 4 tens. $(20 \times 2) + (2 \times 4) = 40 + 8$ or 48 Estimate: 48

Study Tip

Percents
To find 1% of any number, move the decimal point two places to the left.

You can use these methods to estimate the percent of a number.

Example 2 *Estimate Percents*

a. Estimate 13% of 120.

13% is about 12.5% or $\frac{1}{8}$.

$\frac{1}{8}$ of 120 is 15.

So, 13% of 120 is about 15.

b. Estimate 80% of 296.

80% is equal to $\frac{4}{5}$.

296 is about 300.

$\frac{4}{5}$ of 300 is 240.

So, 80% of 296 is about 240.

c. Estimate $\frac{1}{3}\%$ of 598.

$\frac{1}{3}\% = \frac{1}{3} \times 1\%$. 598 is almost 600.

1% of 600 is 6.

So, $\frac{1}{3}\%$ of 598 is about $\frac{1}{3} \times 6$ or 2.

d. Estimate 118% of 56.

118% means about 120 for every 100 or about 12 for every 10.

56 has about 6 tens.

$12 \times 6 = 72$

So, 118% of 56 is about 72.

Estimating percents is a useful skill in real-life situations.

Example 3 *Use Estimation to Solve a Problem*

MONEY Amelia takes a taxi from the airport to a hotel. The fare is $31.50. Suppose she wants to tip the driver 15%. What would be a reasonable amount of tip for the driver?

$31.50 is about $32.

$15\% = 10\% + 5\%$

10% of $32 is $3.20. Move the decimal point 1 place to the left.

5% of $32 is $1.60. 5% is one half of 10%.

So, 15% is about $3.20 + 1.60$ or $4.80.

A reasonable amount for the tip would be $5.

Study Tip

Finding Percents
To find 10% of a number, move the decimal point one place to the left.

Check for Understanding

Concept Check

1. **Explain** how to estimate 18% of 216 using the fraction method.
2. **Estimate** the percent of the figure that is shaded.

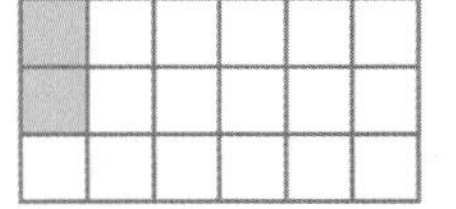

3. **OPEN ENDED** Tell which method of estimating a percent you prefer. Explain your decision.

Guided Practice **Find the percent of each number mentally.**

4. 75% of 64
5. 25% of 52
6. $33\frac{1}{3}\%$ of 27
7. 90% of 80

Estimate. Explain which method you used to estimate.

8. 20% of 61
9. 34% of 24
10. $\frac{1}{2}\%$ of 396
11. 152% of 14

Application

12. **MONEY** Lu Chan wants to leave a tip of 20% on a dinner check of $52.48. About how much should he leave?

Practice and Apply

Homework Help

For Exercises	See Examples
13–26	1
27–35	2
39, 40	3

Extra Practice

See page 738.

Find the percent of each number mentally.

13. 50% of 28
14. 75% of 16
15. 60% of 55
16. 20% of 105
17. $87\frac{1}{2}\%$ of 56
18. $16\frac{2}{3}\%$ of 42
19. $12\frac{1}{2}\%$ of 32
20. $66\frac{2}{3}\%$ of $24
21. 200% of 45
22. 150% of 54
23. 125% of 300
24. 175% of 200

MONEY For Exercises 25 and 26, use the following information.
In a recent year, the number of $1 bills in circulation in the United States was about 7 billion.

25. Suppose the number of $5 bills in circulation was 25% of the number of $1 bills. About how many $5 bills were in circulation?
26. If the number of $10 bills was 20% of the number of $1 bills, about how many $10 bills were in circulation?

Estimate. Explain which method you used to estimate.

27. 30% of 89
28. 25% of 162
29. 38% of 88
30. 81% of 25
31. $\frac{1}{4}\%$ of 806
32. $\frac{1}{5}\%$ of 40
33. 127% of 64
34. 140% of 95
35. 295% of 145

SPACE For Exercises 36–38, refer to the information in the table.

36. Which planet has a radius that measures about 50% of the radius of Mercury?
37. Name two planets such that the radius of one planet is about one-third the radius of the other planet.
38. Name two planets such that the mass of one planet is about 330% the mass of the other.

Radius and Mass of Each Planet

Planet	Radius (mi)	Mass
Mercury	1516	0.0553
Venus	3761	0.815
Earth	3960	1.000
Mars	2107	0.107
Jupiter	43,450	317.830
Saturn	36,191	95.160
Uranus	15,763	14.540
Neptune	15,304	17.150
Pluto	707	0.0021

Source: *The World Almanac*

39. **GEOGRAPHY** The United States has 88,633 miles of shoreline. Of the total amount, 35% is located in Alaska. About how many miles of shoreline are located in Alaska?
40. **GEOGRAPHY** About 8.5% of the total Pacific coastline is located in California. Use the information at the left to estimate the number of miles of coastline located in California.
41. **FOOD** A serving of shrimp contains 90 Calories and 7 of those Calories are from fat. About what percent of the Calories are from fat?
42. **FOOD** Fifty-six percent of the Calories in corn chips are from fat. Estimate the number of Calories from fat in a serving of corn chips if one serving contains 160 Calories.

More About. . .

Geography

There are four U.S. coastlines. They are the Atlantic, Gulf, Pacific, and Arctic coasts. Most of the coastline is located on the Pacific Ocean. It contains 40,298 miles.

Source: *The World Almanac*

43. **CRITICAL THINKING** In an election, 40% of the Democrats and 92.5% of the Republicans voted "yes". Of all of the Democrats and Republicans, 68% voted "yes". Find the ratio of Democrats to Republicans.

44. **WRITING IN MATH** Answer the question that was posed at the beginning of the lesson.

How is estimation used when determining sale prices?

Include the following in your answer:

- an example of a situation in which you used estimation to determine the sale price of an item, and
- an example of a real-life situation other than shopping in which you would use estimation with percents.

45. Which percent is greater than $\frac{3}{5}$ but less than $\frac{2}{3}$?

(A) 68% (B) 54% (C) 64% (D) 38%

46. Choose the best estimate for 26% of 362.

(A) 91 (B) 72 (C) 108 (D) 85

Maintain Your Skills

Mixed Review

Use the percent proportion to solve each problem. *(Lesson 6-5)*

47. What is 28% of 75?

48. 37.8 is what percent of 84?

49. **FORESTRY** The five states with the largest portion of land covered by forests are shown in the graphic. For each state, how many square miles of land are covered by forests?

State	Percent of land covered by forests	Area of state (square miles)
Maine	89.9%	35,387
New Hampshire	88.1%	9351
West Virginia	77.5%	24,231
Vermont	75.7%	9615
Alabama	66.9%	52,423

Source: The Learning Kingdom, Inc.

Express each decimal as a percent. *(Lesson 6-4)*

50. 0.27

51. 1.6

52. 0.008

Express each percent as a decimal. *(Lesson 6-4)*

53. 77%

54. 8%

55. 421%

56. 3.56%

ALGEBRA Solve each equation. Check your solution. *(Lesson 5-9)*

57. $n + 4.7 = 13.6$

58. $x + \frac{5}{6} = 2\frac{3}{8}$

59. $\frac{3}{7}r = -9$

60. **GEOMETRY** The perimeter of a rectangle is 22 feet. Its length is 7 feet. Find its width. *(Lesson 3-7)*

Getting Ready for the Next Lesson

PREREQUISITE SKILL Solve each equation. Check your solution.

*(To review **solving equations**, see Lesson 3-4.)*

61. $10a = 5$

62. $20m = 4$

63. $60h = 15$

64. $28g = 1.4$

65. $80w = 5.6$

66. $125n = 15$

6-7 Using Percent Equations

What You'll Learn

- Solve percent problems using percent equations.
- Solve real-life problems involving discount and interest.

Sunshine State Standards
MA.A.3.3.2-1, MA.A.3.3.2-2, MA.A.4.3.1-1, MA.A.4.3.1-2

Vocabulary

- percent equation
- discount
- simple interest

How is the percent proportion related to an equation?

As of July 1, 1999, 45 of the 50 U.S. states had a sales tax. The table shows the tax rate for four U.S. states.

State	Tax Rate (percent)
Alabama	4%
Connecticut	6%
New Mexico	5%
Texas	6.25%

Source: www.taxadmin.org

a. Use the percent proportion to find the amount of tax on a $35 purchase for each state.

b. Express each tax rate as a decimal.

c. Multiply the decimal form of the tax rate by $35 to find the amount of tax on the $35 purchase for each state.

d. How are the amounts of tax in parts **a** and **c** related?

PERCENT EQUATIONS The **percent equation** is an equivalent form of the percent proportion in which the percent is written as a decimal.

$$\frac{\text{Part}}{\text{Base}} = \text{Percent}$$ ← The percent is written as a decimal.

$$\frac{\text{Part}}{\cancel{\text{Base}}} \cdot \cancel{\text{Base}} = \text{Percent} \cdot \text{Base}$$ Multiply each side by the base.

$$\text{Part} = \text{Percent} \cdot \text{Base}$$ ← This form is called the percent equation.

Concept Summary — *The Percent Equation*

Type	Example	Equation
Missing Part	What number is 75% of 4?	$n = 0.75(4)$
Missing Percent	3 is what percent of 4?	$3 = n(4)$
Missing Base	3 is 75% of what number?	$3 = 0.75n$

Example 1 Find the Part

Find 52% of 85. **Estimate:** $\frac{1}{2}$ of 90 is 45.

You know that the base is 85 and the percent is 52%. Let n represent the part.

$n = 0.52(85)$ Write 52% as the decimal 0.52.

$n = 44.2$ Simplify.

So, 52% of 85 is 44.2.

Study Tip

Estimation
To determine whether your answer is reasonable, estimate before finding the exact answer.

Example 2 Find the Percent

28 is what percent of 70? **Estimate:** $\frac{28}{70} \approx \frac{25}{75}$ or $\frac{1}{3}$, which is $33\frac{1}{3}\%$.

You know that the base is 70 and the part is 28.
Let n represent the percent.

$28 = n(70)$

$\frac{28}{70} = n$ Divide each side by 70.

$0.4 = n$ Simplify.

So, 28 is 40% of 70. The answer makes sense compared to the estimate.

Example 3 Find the Base

18 is 45% of what number? **Estimate:** 18 is 50% of 36.

You know that the part is 18 and the percent is 45.
Let n represent the base.

$18 = 0.45n$ Write 45% as the decimal 0.45.

$\frac{18}{0.45} = \frac{0.45n}{0.45}$ Divide each side by 0.45.

$40 = n$ Simplify.

So, 18 is 45% of 40. The answer is reasonable since it is close to the estimate.

DISCOUNT AND INTEREST The percent equation can also be used to solve problems involving discount and interest. **Discount** is the amount by which the regular price of an item is reduced.

More About...

Skateboards

The popularity of the sport of skateboarding is increasing. An estimated 10,000,000 people worldwide participate in the sport.

Source: International Association of Skateboard Companies

Example 4 Find Discount

SKATEBOARDS Mateo wants to buy a skateboard. The regular price of the skateboard is $135. Suppose it is on sale at a 25% discount. Find the sale price of the skateboard.

Method 1

First, use the percent equation to find 25% of 135. **Estimate:** $\frac{1}{4}$ of 140 = 35

Let d represent the discount.

$d = 0.25(135)$ The base is 135 and the percent is 25%.

$d = 33.75$ Simplify.

Then, find the sale price.

$135 - 33.75 = 101.25$ Subtract the discount from the original price.

Method 2

A discount of 25% means the item will cost 100% − 25% or 75% of the original price. Use the percent equation to find 75% of 135.

Let s represent the sale price.

$s = 0.75(135)$ The base is 135 and the percent is 75%.

$s = 101.25$ Simplify.

The sale price of the skateboard will be $101.25.

Simple interest is the amount of money paid or earned for the use of money. For a savings account, interest is earned. For a credit card, interest is paid. To solve problems involving interest, use the following formula.

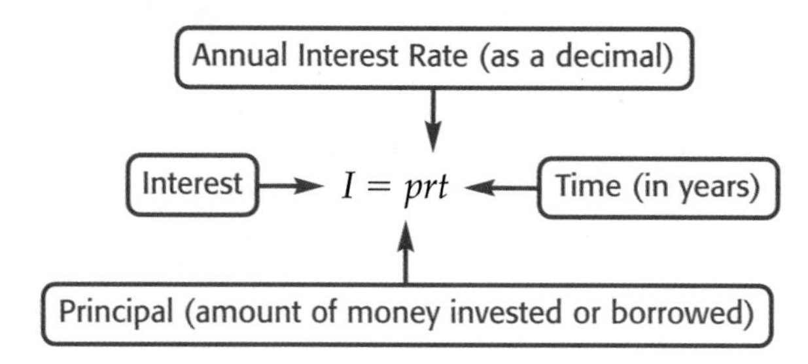

Reading Math

Formulas

The formula $I = prt$ is read *Interest is equal to principal times rate times time.*

Concept Check Name a situation where interest is earned and a situation where interest is paid.

Example 5 Apply Simple Interest Formula

BANKING Suppose Miguel invests $1200 at an annual rate of 6.5%. How long will it take until Miguel earns $195?

$I = prt$ Write the simple interest formula.

$195 = 1200(0.065)t$ Replace I with 195, p with 1200, and r with 0.065.

$195 = 78t$ Simplify.

$\frac{195}{78} = \frac{78t}{78}$ Divide each side by 78.

$2.5 = t$ Simplify.

Miguel will earn $195 in interest in 2.5 years.

Check for Understanding

Concept Check

1. **OPEN ENDED** Give an example of a situation in which using the percent equation would be easier than using the percent proportion.
2. **Define** *discount*.
3. **Explain** what I, p, r, and t represent in the simple interest formula.

Guided Practice

Solve each problem using the percent equation.

4. 15 is what percent of 60?
5. 30 is 60% of what number?
6. What is 20% of 110?
7. 12 is what percent of 400?
8. Find the discount for a $268 DVD player that is on sale at 20% off.
9. What is the interest on $8000 that is invested at 6% for $3\frac{1}{2}$ years? Round to the nearest cent.

Applications

10. **SHOPPING** A jacket that normally sells for $180 is on sale at a 35% discount. What is the sale price of the jacket?
11. **BANKING** How long will it take to earn $252 in interest if $2400 is invested at a 7% annual interest rate?

Practice and Apply

Homework Help

For Exercises	See Examples
12–27, 39	1–3
28–33	4
34–38	5

Extra Practice

See page 738.

Solve each problem using the percent equation.

12. 9 is what percent of 25?
13. 38 is what percent of 40?
14. 48 is 64% of what number?
15. 27 is 54% of what number?
16. Find 12% of 72.
17. Find 42% of 150.
18. 39.2 is what percent of 112?
19. 49.5 is what percent of 132?
20. What is 37.5% of 89?
21. What is 24.2% of 60?
22. 37.5 is what percent of 30?
23. 43.6 is what percent of 20?
24. 1.6 is what percent of 400?
25. 1.35 is what percent of 150?
26. 83.5 is 125% of what number?
27. 17.6 is $133\frac{1}{3}$% of what number?
28. **FOOD** A frozen pizza is on sale at a 25% discount. Find the sale price of the pizza if it normally sells for $4.85.
29. **CALCULATORS** Suppose a calculator is on sale at a 15% discount. If it normally sells for $29.99, what is the sale price?

Find the discount to the nearest cent.

30. 

31. $85 cordless phone, 20% off
32. $489 stereo, 15% off
33. 25% off a $74 baseball glove

Find the interest to the nearest cent.

34. 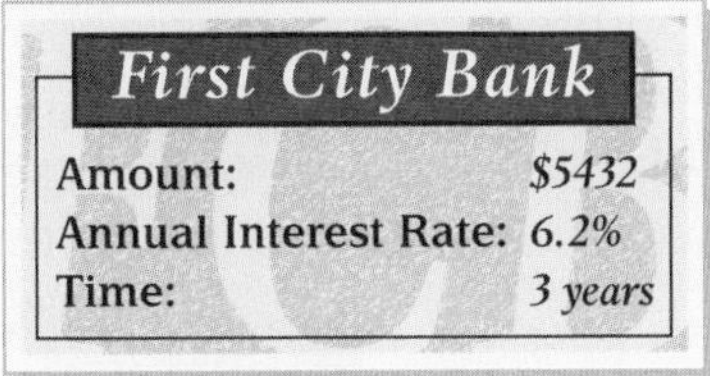

35. $4500 at 5.5% for $4\frac{1}{2}$ years
36. $3680 at 6.75% for $2\frac{1}{4}$ years
37. 5.5% for $1\frac{3}{4}$ years on $2543

WebQuest

The percent equation can help you analyze the nutritional value of food. Visit www.pre-alg.com/webquest to continue work on your WebQuest project.

38. **BANKING** What is the annual interest rate if $1600 is invested for 6 years and $456 in interest is earned?
39. **SPORTS** One season, a football team had 7 losses. This was 43.75% of the total games they played. How many games did they play?
40. **REAL ESTATE** A **commission** is a fee paid to a salesperson based on a percent of sales. Suppose a real estate agent earns a 3% commission. What commission would be earned for selling the house shown?

New Listing! Two-story home with 4 bedrooms and 2 bathrooms. Price: $130,000

41. **BUSINESS** To make a profit, stores try to sell an item for more than it paid for the item. The increase in price is called the **markup**. Suppose a store purchases paint brushes for $8 each. Find the markup if the brushes are sold for 15% over the price paid for them.

42. CRITICAL THINKING Determine whether $n\%$ of m is always equal to $m\%$ of n. Give examples to support your answer.

43. WRITING IN MATH Answer the question that was posed at the beginning of the lesson.

How is the percent proportion related to an equation?

Include the following in your answer:

- an explanation describing two methods for finding the amount of tax on an item, and
- an example of using both methods to find the amount of sales tax on an item.

44. What percent of 320 is 19.2?

Ⓐ 0.6% Ⓑ 60% Ⓒ 6% Ⓓ 0.06%

45. Ryan wants to buy a tent that costs $150 for his camping trip. The tent is on sale at a 30% discount. What will be the sale price of the tent?

Ⓐ 95 Ⓑ 105 Ⓒ 45 Ⓓ 110

Maintain Your Skills

Mixed Review

Estimate. Explain which method you used to estimate. *(Lesson 6-6)*

46. 47% of 84 **47.** 126% of 198 **48.** 9% of 514

Use the percent proportion to solve each problem. *(Lesson 6-5)*

49. What is 55% of 220? **50.** 50.88 is what percent of 96?

51. POPULATION The graphic shows the number of stories of certain buildings in Tulsa, Oklahoma. What is the mean of the data? *(Lesson 5-8)*

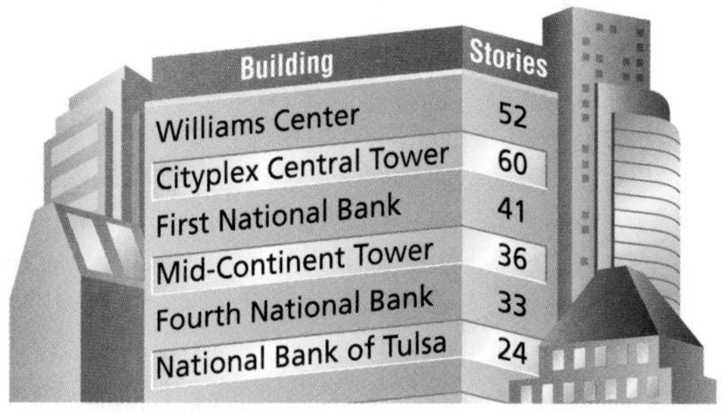

Building	Stories
Williams Center	52
Cityplex Central Tower	60
First National Bank	41
Mid-Continent Tower	36
Fourth National Bank	33
National Bank of Tulsa	24

Source: *The World Almanac*

52. List all the factors of 30. *(Lesson 4-1)*

GEOMETRY Find the perimeter of each rectangle. *(Lesson 3-7)*

53.

13 cm

6 cm

54.

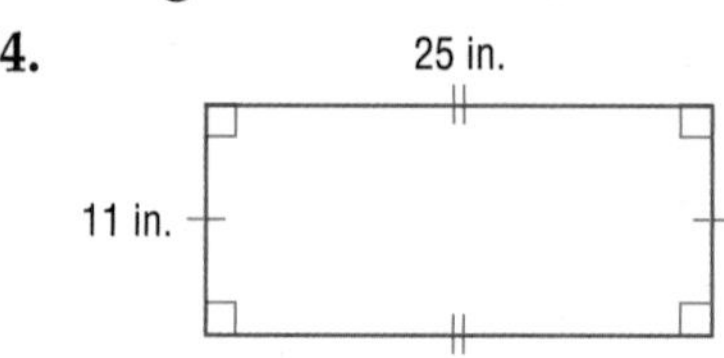

55. ALGEBRA Use the Distributive Property to rewrite $(w - 3)8$. *(Lesson 3-1)*

Getting Ready for the Next Lesson

PREREQUISITE SKILL Write each decimal as a percent.

*(To review **writing decimals as percents**, see Lesson 6-4.)*

56. 0.58 **57.** 0.89 **58.** 0.125

59. 1.56 **60.** 2.04 **61.** 0.224

Spreadsheet Investigation

A Follow-Up of Lesson 6-7

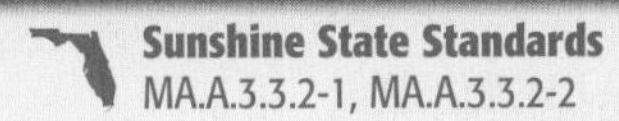

Sunshine State Standards
MA.A.3.3.2-1, MA.A.3.3.2-2

Compound Interest

Simple interest, which you studied in the previous lesson, is paid only on the initial principal of a savings account or a loan. **Compound interest** is paid on the initial principal and on interest earned in the past. You can use a spreadsheet to investigate the impact of compound interest.

SAVINGS Find the value of a $1000 savings account after five years if the account pays 6% interest compounded semiannually.

6% interest compounded semiannually means that the interest is paid twice a year, or every 6 months. The interest rate is 6% ÷ 2 or 3%.

Compound Interest

	A	B	C	D
1	RATE	0.03		
2				
3	PRINCIPAL	INTEREST	NEW PRINCIPAL	TIME (YR)
4	1000.00	30.00	1030.00	0.5
5	1030.00	30.90	1060.90	1.0
6	1060.90	31.83	1092.73	1.5
7	1092.73	32.78	1125.51	2.0
8	1125.51	33.77	1159.27	2.5
9	1159.27	34.78	1194.05	3.0
10	1194.05	35.82	1229.87	3.5
11	1229.87	36.90	1266.77	4.0
12	1266.77	38.00	1304.77	4.5
13	1304.77	39.14	1343.92	5.0

Sheet1 Sheet2 Sh

Edit

The rate is entered as a decimal.

The spreadsheet evaluates the formula A4 × B1.

The interest is added to the principal every 6 months. The spreadsheet evaluates the formula A4 + B4.

The value of the savings account after five years is $1343.92.

Model and Analyze

1. Suppose you invest $1000 for five years at 6% simple interest. How does the simple interest compare to the compound interest shown above?
2. Use a spreadsheet to find the amount of money in a savings account if $1000 is invested for five years at 6% interest compounded quarterly.
3. Suppose you leave $100 in each of three bank accounts paying 5% interest per year. One account pays simple interest, one pays interest compounded semiannually, and one pays interest compounded quarterly. Use a spreadsheet to find the amount of money in each account after three years.

Make a Conjecture

4. How does the amount of interest change if the compounding occurs more frequently?

6-8 Percent of Change

What You'll Learn

- Find percent of increase.
- Find percent of decrease.

Sunshine State Standards
MA.A.3.3.2-1, MA.A.3.3.2-2, MA.A.3.3.2-3, MA.A.3.3.2-4

Vocabulary

- percent of change
- percent of increase
- percent of decrease

How can percents help to describe a change in area?

Suppose the length of rectangle A is increased from 4 units to 5 units.

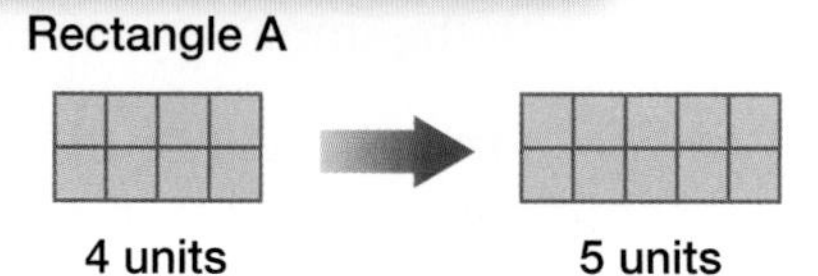

Rectangle A had an initial area of 8 square units. It increased to 10 square units. This is a change in area of 2 square units. The following ratio shows this relationship.

$$\frac{\text{change in area}}{\text{original area}} = \frac{2}{8} = \frac{1}{4} \text{ or } 25\%$$

This means that, compared to the original area, the new area increased by 25%.

Draw each pair of rectangles. Then compare the rectangles. Express the increase as a fraction and as a percent.

a. X: 2 units by 3 units
Y: 2 units by 4 units

b. G: 2 units by 5 units
H: 2 units by 6 units

c. J: 2 units by 4 units
K: 2 units by 5 units

d. P: 2 units by 6 units
Q: 2 units by 7 units

e. For each pair of rectangles, the change in area is 2 square units. Explain why the percent of change is different.

FIND PERCENT OF INCREASE A **percent of change** tells the percent an amount has increased or decreased in relation to the original amount.

Example 1 Find Percent of Change

Find the percent of change from 56 inches to 63 inches.

Step 1 Subtract to find the amount of change.

$63 - 56 = 7$ new measurement − original measurement

Step 2 Write a ratio that compares the amount of change to the original measurement. Express the ratio as a percent.

$$\text{percent of change} = \frac{\text{amount of change}}{\text{original measurement}}$$

$= \frac{7}{56}$ Substitution.

$= 0.125 \text{ or } 12.5\%$ Write the decimal as a percent.

The percent of change from 56 inches to 63 inches is 12.5%.

When an amount increases, as in Example 1, the percent of change is a **percent of increase**.

Example 2 Find Percent of Increase

FUEL In 1975, the average price per gallon of gasoline was \$0.57. In 2000, the average price per gallon was \$1.47. Find the percent of change.

Source: *The World Almanac*

Step 1 Subtract to find the amount of change.

$1.47 - 0.57 = 0.9$ new price − original price

Step 2 Write a ratio that compares the amount of change to the original price. Express the ratio as a percent.

$$\text{percent of change} = \frac{\text{amount of change}}{\text{original price}}$$

$$= \frac{0.9}{0.57} \quad \text{Substitution.}$$

$$\approx 1.58 \text{ or } 158\% \quad \text{Write the decimal as a percent.}$$

The percent of change is about 158%. In this case, the percent of change is a percent of increase.

Example 3 Find Percent of Increase

Multiple-Choice Test Item

Refer to the table shown. Which county had the greatest percent of increase in population from 1990 to 2000?

(A) Breckinridge (B) Bracken

(C) Calloway (D) Fulton

County	1990	2000
Breckinridge	16,312	18,648
Bracken	7766	8279
Calloway	30,735	34,177
Fulton	8271	7752

Read the Test Item

Percent of increase tells how much the population has increased in relation to 1990.

Solve the Test Item

Use a ratio to find each percent of increase. Then compare the percents.

- **Breckinridge**

$$\frac{18{,}648 - 16{,}312}{16{,}312} = \frac{2336}{16{,}312} \approx 0.1432 \text{ or } 14.3\%$$

- **Bracken**

$$\frac{8279 - 7766}{7766} = \frac{513}{7766} \approx 0.0661 \text{ or } 6.6\%$$

- **Calloway**

$$\frac{34{,}177 - 30{,}735}{30{,}735} = \frac{3442}{30{,}735} \approx 0.112 \text{ or } 11.2\%$$

- **Fulton**

Eliminate this choice because the population decreased.

Breckinridge County had the greatest percent of increase in population from 1990 to 2000. The answer is A.

The Princeton Review

Test-Taking Tip

If you are unsure of the correct answer, eliminate the choices you know are incorrect. Then consider the remaining choices.

PERCENT OF DECREASE When the amount decreases, the percent of change is negative. You can state a negative percent of change as a **percent of decrease**.

More About. . .

Stock Market

About 20 years ago, only 12.2% of Americans had money invested in the stock market. Today, more than 44% of Americans invest in the stock market.

Source: www.infoplease.com

Example 4 Find Percent of Decrease

STOCK MARKET One of the largest stock market drops on Wall Street occurred on October 19, 1987. On this day, the stock market opened at 2246.74 points and closed at 1738.42 points. What was the percent of change?

Step 1 Subtract to find the amount of change.

$1738.42 - 2246.74 = -508.32$ closing points – opening points

Step 2 Compare the amount of change to the opening points.

$$\text{percent of change} = \frac{\text{amount of change}}{\text{opening points}}$$

$$= \frac{-508.32}{2246.74} \quad \text{Substitution.}$$

$$\approx -0.226 \text{ or } -22.6\% \quad \text{Write the decimal as a percent.}$$

The percent of change is -22.6%. In this case, the percent of change is a percent of decrease.

Check for Understanding

Concept Check

1. **Explain** how you know whether a percent of change is a percent of increase or a percent of decrease.
2. **OPEN ENDED** Give an example of a percent of decrease.
3. **FIND THE ERROR** Scott and Mark are finding the percent of change when a shirt that costs $15 is on sale for $10.

Scott

$\frac{10 - 15}{10} = \frac{-5}{10}$ or -50%

Mark

$\frac{10 - 15}{15} = \frac{-5}{15}$ or $-33\frac{1}{3}\%$

Who is correct? Explain your reasoning.

Guided Practice

Find the percent of change. Round to the nearest tenth, if necessary. Then state whether the percent of change is a *percent of increase* or a *percent of decrease.*

4. from $50 to $67
5. from 45 in. to 18 in.
6. from 80 cm to 55 cm
7. from $228 to $251

8. **ANIMALS** In 2000, there were 356 endangered species in the U.S. One year later, 367 species were considered endangered. What was the percent of change?

9. Refer to Example 3 on page 305. Suppose in 10 years, the population of Calloway is 36,851. What will be the percent of change from 1990?

Ⓐ 19.9% Ⓑ 9.8% Ⓒ 10.7% Ⓓ 15.3%

Practice and Apply

Homework Help

For Exercises	See Examples
10–18, 20, 21	1, 2, 4
19	3

Extra Practice

See page 739.

Find the percent of change. Round to the nearest tenth, if necessary. Then state whether the percent of change is a *percent of increase* or a *percent of decrease*.

10. from 25 cm to 36 cm

11. from $10 to $27

12. from 68 min to 51 min

13. from 50 lb to 44 lb

14. from $135 to $120

15. from 257 m to 243 m

16. from 365 ft to 421 ft

17. from $289 to $762

18. **WEATHER** Seattle, Washington, receives an average of 6.0 inches of precipitation in December. In March, the average precipitation is 3.8 inches. What is the percent of change in precipitation from December to March?

19. **POPULATION** In 1990, the population of Alabama was 4,040,587. In 2000, the population was 4,447,100. Find the percent of change from 1990 to 2000.

20. Suppose 36 videos are added to a video collection that has 24 videos. What is the percent of change?

21. A biology class has 28 students. Four of the students transferred out of the class to take chemistry. Find the percent of change in the number of students in the biology class.

22. **BUSINESS** A restaurant manager wants to reduce spending on supplies 10% in January and an additional 15% in February. In January, the expenses were $2875. How much should the expenses be at the end of February?

23. **SCHOOL** Jiliana is using a copy machine to increase the size of a 2-inch by 3-inch picture of a spider. The enlarged picture needs to measure 3 inches by 4.5 inches.

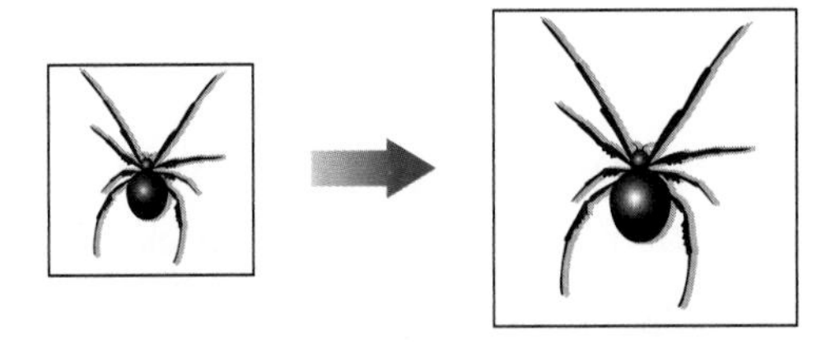

What enlargement setting on the copy machine should she use?

24. **CRITICAL THINKING** Explain why a 10% increase followed by a 10% decrease is less than the original amount if the original amount was positive.

25. WRITING IN MATH Answer the question that was posed at the beginning of the lesson.

How can percents help to describe a change in area?

Include the following in your answer:

- an explanation describing how you can tell whether the percent of increase will be greater than 100%, and
- an example of a model that shows an increase less than 100% and one that shows an increase greater than 100%.

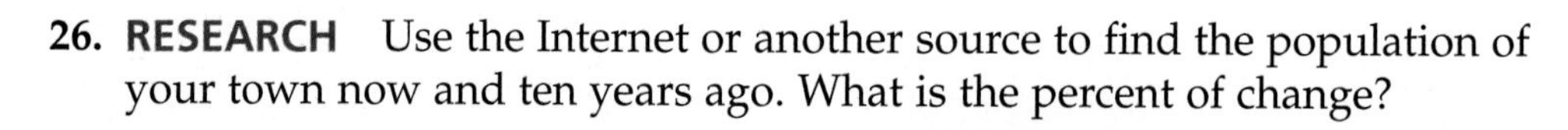

26. RESEARCH Use the Internet or another source to find the population of your town now and ten years ago. What is the percent of change?

For Exercises 27 and 28, refer to the information in the table.

Kennel Club Registrations

Breed	1998	1999
Labrador Retriever	157,936	157,897
Beagle	53,322	49,080
Maltese	18,013	16,358
Golden Retriever	65,681	62,652
Shih Tzu	38,468	34,576
Cocker Spaniel	34,632	29,958
Siberian Husky	21,078	18,106

27. What percent represents the percent of change in the number of beagles from 1998 to 1999?

Ⓐ −8.1% Ⓑ −7.5%

Ⓒ −9.7% Ⓓ −8.0%

28. Which breed had the largest percent of decrease?

Ⓐ Siberian Husky Ⓑ Cocker Spaniel

Ⓒ Golden Retriever Ⓓ Labrador Retriever

Maintain Your Skills

Mixed Review

29. Find the discount to the nearest cent for a television that costs $999 and is on sale at 15% off. *(Lesson 6-7)*

30. Find the interest on $1590 that is invested at 8% for 3 years. Round to the nearest cent. *(Lesson 6-7)*

31. A calendar is on sale at a 10% discount. What is the sale price if it normally sells for $14.95? *(Lesson 6-7)*

Estimate. Explain which method you used to estimate. *(Lesson 6-6)*

32. 60% of 134 **33.** 88% of 72 **34.** 123% of 32

Identify all of the sets to which each number belongs. *(Lesson 5-2)*

35. -8 **36.** $1\frac{1}{4}$ **37.** -5.63

Getting Ready for the Next Lesson

PREREQUISITE SKILL Write each fraction as a percent.
*(To review **writing fractions as percents**, see Lesson 6-4.)*

38. $\frac{3}{4}$ **39.** $\frac{1}{5}$ **40.** $\frac{2}{3}$ **41.** $\frac{5}{6}$ **42.** $\frac{3}{8}$

Practice Quiz 2

Lessons 6-6 through 6-8

Estimate. Explain which method you used to estimate. *(Lesson 6-6)*

1. 42% of 68 **2.** $66\frac{2}{3}$% of 34

3. Find the discount to the nearest cent on a backpack that costs $58 and is on sale at 25% off. *(Lesson 6-7)*

4. Find the interest to the nearest cent on $2500 that is invested at 4% for 2.5 years. *(Lesson 6-7)*

5. Find the percent of change from $0.95 to $2.45. *(Lesson 6-8)*

Algebra Activity

A Preview of Lesson 6-9

Sunshine State Standards
MA.E.3.3.1-3, MA.E.3.3.2-1, MA.E.3.3.2-2, MA.E.3.3.2-3, MA.E.3.3.2-4

Taking a Survey

The graph shows the results of a survey about what types of stores people in the United States shop at the most. Since it would be impossible to survey everyone in the country, a sample was used. A **sample** is a subgroup or subset of the population.

WHERE WE SHOP MOST

62.4% Discount stores
15.6% National chains
22.0% Conventional stores

Source: International Mass Retail Association

It is important to obtain a sample that is unbiased. An **unbiased** sample is a sample that is:

- representative of the larger population,
- selected at **random** or without preference, and
- large enough to provide accurate data.

To insure an unbiased sample, the following sampling methods may be used.

- **Random** The sample is selected at random.
- **Systematic** The sample is selected by using every nth member of the population.
- **Stratified** The sample is selected by dividing the population into groups.

Model and Analyze

Tell whether or not each of the following is a random sample. Then provide an explanation describing the strengths and weaknesses of each sample.

	Type of Survey	Location of Survey
1.	travel preference	mall
2.	time spent reading	library
3.	favorite football player	Miami Dolphins football game

4. Brad conducted a survey to find out which food people in his community prefer. He surveyed every second person that walked into a certain fast-food restaurant. Identify this type of sampling. Explain how the survey may be biased.

5. Suppose a study shows that teenagers who eat breakfast each day earn higher grades than teenagers who skip breakfast. Tell how you can use the stratified sampling technique to test this claim in your school.

6. Suppose you want to determine where students in your school shop the most.
   a. Formulate a hypothesis about where students shop the most.
   b. Design and conduct a survey using one of the sampling techniques described above.
   c. Organize and display the results of your survey in a chart or graph.
   d. Evaluate your hypothesis by drawing a conclusion based on the survey.

6-9 Probability and Predictions

What You'll Learn

- Find the probability of simple events.
- Use a sample to predict the actions of a larger group.

Sunshine State Standards
MA.E.2.3.1-1

Vocabulary

- outcomes
- simple event
- probability
- sample space
- theoretical probability
- experimental probability

How can probability help you make predictions?

A popular word game is played using 100 letter tiles. The object of the game is to use the tiles to spell words scoring as many points as possible. The table shows the distribution of the tiles.

Letter	Number of Tiles
E	12
A, I	9
O	8
N, R, T	6
D, L, S, U	4
G	3
B, C, F, H, M, P, V, W, Y, blank	2
J, K, Q, X, Z	1

a. Write the ratio that compares the number of tiles labeled E to the total number of tiles.

b. What percent of the tiles are labeled E?

c. What fraction of tiles is this?

d. Suppose a player chooses a tile. Is there a better chance of choosing a D or an N? Explain.

PROBABILITY OF SIMPLE EVENTS In the activity above, there are 27 possible tiles. These results are called **outcomes**. A **simple event** is one outcome or a collection of outcomes. For example, choosing a tile labeled E is a simple event.

You can measure the chances of an event happening with **probability**.

Study Tip

Probability
Each of the outcomes must be equally likely to happen.

Key Concept — *Probability*

- **Words** The probability of an event is a ratio that compares the number of favorable outcomes to the number of possible outcomes.
- **Symbols** $P(\text{event}) = \dfrac{\text{number of favorable outcomes}}{\text{number of possible outcomes}}$

The probability of an event is always between 0 and 1, inclusive. The closer a probability is to 1, the more likely it is to occur.

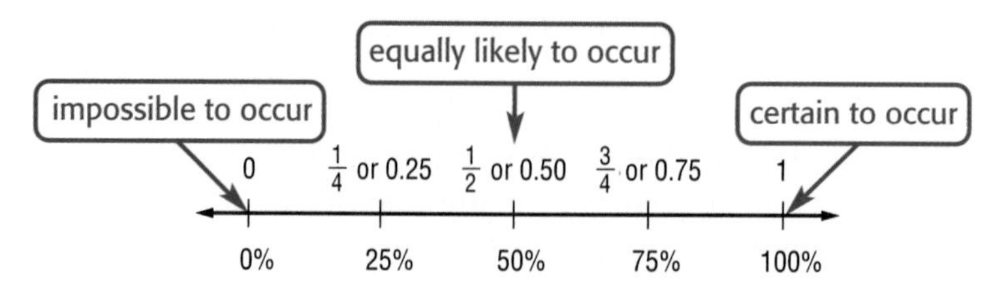

Concept Check Suppose there is a 45% chance that an event occurs. How likely is it that the event will occur?

Example 1 Find Probability

Suppose a number cube is rolled. What is the probability of rolling a prime number?

There are 3 prime numbers on a number cube: 2, 3, and 5.

There are 6 possible outcomes: 1, 2, 3, 4, 5, and 6.

$$P(\text{prime}) = \frac{\text{number of favorable outcomes}}{\text{number of possible outcomes}}$$

$$= \frac{3}{6} \text{ or } \frac{1}{2}$$

So, the probability of rolling a prime number is $\frac{1}{2}$ or 50%.

Reading Math

P(prime)

P(prime) is read as *the probability of rolling a prime number.*

The set of all possible outcomes is called the **sample space**. For Example 1, the sample space was {1, 2, 3, 4, 5, 6}. When you toss a coin, the sample space is {heads, tails}.

Example 2 Find Probability

Suppose two number cubes are rolled. Find the probability of rolling an even sum.

Make a table showing the sample space when rolling two number cubes.

	1	2	3	4	5	6
1	(1, 1)	(1, 2)	(1, 3)	(1, 4)	(1, 5)	(1, 6)
2	(2, 1)	(2, 2)	(2, 3)	(2, 4)	(2, 5)	(2, 6)
3	(3, 1)	(3, 2)	(3, 3)	(3, 4)	(3, 5)	(3, 6)
4	(4, 1)	(4, 2)	(4, 3)	(4, 4)	(4, 5)	(4, 6)
5	(5, 1)	(5, 2)	(5, 3)	(5, 4)	(5, 5)	(5, 6)
6	(6, 1)	(6, 2)	(6, 3)	(6, 4)	(6, 5)	(6, 6)

There are 18 outcomes in which the sum is even.

So, $P(\text{even sum}) = \frac{18}{36}$ or $\frac{1}{2}$.

This means there is a 50% chance of rolling an even sum.

The probabilities in Examples 1 and 2 are called theoretical probabilities. **Theoretical probability** is what *should* occur. **Experimental probability** is what *actually* occurs when conducting a probability experiment.

Example 3 Find Experimental Probability

The table shows the results of an experiment in which a coin was tossed. Find the experimental probability of tossing a coin and getting tails for this experiment.

Outcome	Tally	Frequency
Heads	𝍸 𝍸 IIII	14
Tails	𝍸 𝍸 I	11

$$\frac{\text{number of times tails occur}}{\text{number of possible outcomes}} = \frac{11}{14 + 11} \text{ or } \frac{11}{25}$$

The experimental probability of getting tails in this case is $\frac{11}{25}$ or 44%.

USE A SAMPLE TO MAKE PREDICTIONS Not all probability experiments are conducted using number cubes, coins, or spinners. For example, you can use an athlete's past performance to predict whether she will get a hit or make a basket. You can also use the results of a survey to predict the actions of a larger group.

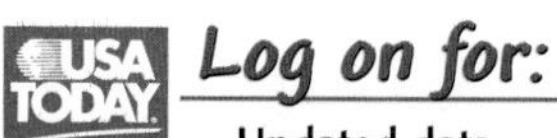

Log on for:
- Updated data
- More activities on Making Predictions

www.pre-alg.com/usa_today

Example 4 Make a Prediction

FOOD The graph shows the results of a survey. Out of a group of 450 people, how many would you expect to say that they prefer thin mint cookies?

USA TODAY Snapshots®

Monster cookies

Girl Scout cookie sales are an annual tradition from January to March in most of the USA. Last year's best-selling cookies in sales share:

26% Thin Mints

19% Samoas/Caramel deLites[1]

13% Tagalongs/Peanut Butter

12% Do-Si-Dos/Peanut Butter Sandwich[1]

11% Trefoils/Short-bread

1 – Same cookie with different name depending on baker

Source: Girl Scouts of the U.S.A.

By Anne R. Carey and Quin Tian, USA TODAY

The total number of people is 450. So, 450 is the base. The percent is 26%.

To find 26% of 450, let b represent the base, 450, and let p represent the percent, 26%, in the percent proportion. Let a represent the part.

$$\begin{array}{ll} \text{part} \rightarrow \\ \text{base} \rightarrow \end{array} \frac{a}{450} = \frac{26}{100} \leftarrow \text{percent}$$

$$100 \cdot a = 26 \cdot 450$$

$$100a = 11700 \quad \text{Simplify.}$$

$$a = 117 \quad \text{Mentally divide each side by 100.}$$

You can expect 117 people to say that they prefer thin mint cookies.

Check for Understanding

Concept Check

1. **Tell** what a probability of 0 means.
2. **Compare and contrast** theoretical and experimental probability.
3. **OPEN ENDED** Give an example of a situation in which the probability of the event is 25%.

Guided Practice

Ten cards are numbered 1 through 10, and one card is chosen at random. Determine the probability of each outcome. Express each probability as a fraction and as a percent.

4. $P(5)$
5. $P(\text{odd})$
6. $P(\text{less than } 3)$
7. $P(\text{greater than } 6)$

For Exercises 8 and 9, refer to the table in Example 2 on page 311. Determine each probability. Express each probability as a fraction and as a percent.

8. $P(\text{sum of 2 or 6})$
9. $P(\text{even or odd sum})$
10. Refer to Example 3 on page 311. Find the experimental probability of getting heads for the experiment.

Application

11. **FOOD** Maresha took a sample from a package of jellybeans and found that 30% of the beans were red. Suppose there are 250 jellybeans in the package. How many can she expect to be red?

Practice and Apply

Homework Help

For Exercises	See Examples
12–34,	1, 2
35, 36	3
37	4

Extra Practice

See page 739.

A spinner like the one shown is used in a game. Determine the probability of each outcome if the spinner is equally likely to land on each section. Express each probability as a fraction and as a percent.

12. $P(8)$
13. $P(\text{red})$
14. $P(\text{even})$
15. $P(\text{prime})$
16. $P(\text{greater than } 5)$
17. $P(\text{less than } 2)$
18. $P(\text{blue or } 11)$
19. $P(\text{not yellow})$
20. $P(\text{not red})$

There are 2 red marbles, 4 blue marbles, 7 green marbles, and 5 yellow marbles in a bag. Suppose one marble is selected at random. Find the probability of each outcome. Express each probability as a fraction and as a percent.

21. $P(\text{blue})$
22. $P(\text{yellow})$
23. $P(\text{not green})$
24. $P(\text{purple})$
25. $P(\text{red or blue})$
26. $P(\text{blue or yellow})$
27. $P(\text{not orange})$
28. $P(\text{not blue or not red})$

29. What is the probability that a calendar is randomly turned to the month of January or April?

30. Find the probability that today is November 31.

Suppose two spinners like the ones shown are spun. Find the probability of each outcome. (*Hint*: Make a table to show the sample space as in Example 2 on page 311.)

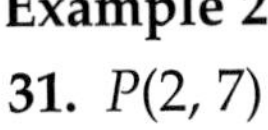

31. $P(2, 7)$
32. $P(\text{even, even})$

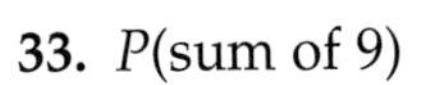

33. $P(\text{sum of } 9)$
34. $P(2, \text{greater than } 5)$

DRIVING **For Exercises 35 and 36, use the following information and the table shown.** The table shows the approximate number of licensed automobile drivers in the United States in a certain year. An automobile company is conducting a telephone survey using a list of licensed drivers.

Age	Drivers (millions)
19 and under	9
20–29	34
30–39	41
40–49	37
50–59	24
60–69	18
70 and over	17
Total	180

Source: U.S. Department of Transportation

35. Find the probability that a driver will be 19 years old or younger. Express the answer as a decimal rounded to the nearest hundredth and as a percent.

36. What is the probability that a randomly chosen driver will be 40–49 years old? Write the answer as a decimal rounded to the nearest hundredth and as a percent.

37. **FOOD** Refer to the graph. Out of 1200 people, how many would you expect to say they crave chocolate after dinner?

USA TODAY Snapshots®

Chocolate cravings

Time of day that adults say they crave chocolate:

Midafternoon	47%
Evening	42%
After dinner	37%
Lunch	21%
Before bed	18%
Midmorning	17%
Middle of the night	10%
Breakfast	9%

Source: Yankelovich Partners for the American Boxed Chocolate Manufacturers

By Cindy Hall and Alejandro Gonzalez, USA TODAY

38. **CRITICAL THINKING** In the English language, 13% of the letters used are E's. Suppose you are guessing the letters in a two-letter word of a puzzle. Would you guess an E? Explain.

39. WRITING IN MATH Answer the question that was posed at the beginning of the lesson.

How can probability help you make predictions?

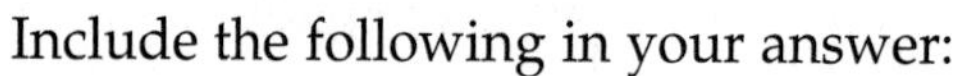

Include the following in your answer:

- an explanation telling the probability of choosing each letter tile, and
- an example of how you can use probability to make predictions.

FCAT Practice

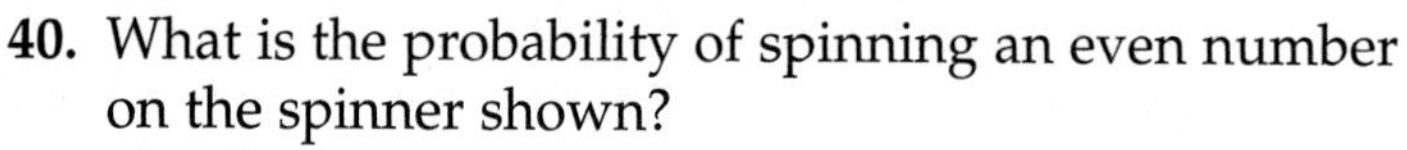

40. What is the probability of spinning an even number on the spinner shown?

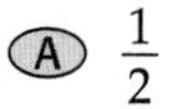 $\frac{1}{2}$ 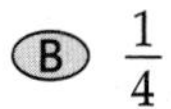 $\frac{1}{4}$ 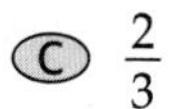 $\frac{2}{3}$ 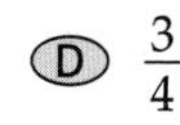 $\frac{3}{4}$

Maintain Your Skills

Mixed Review

41. Find the percent of change from 32 feet to 79 feet. Round to the nearest tenth, if necessary. Then state whether the percent of change is a *percent of increase* or a *percent of decrease*. *(Lesson 6-8)*

Solve each problem using an equation. Round to the nearest tenth. *(Lesson 6-7)*

42. 7 is what percent of 32?

43. What is 28.5% of 84?

ALGEBRA Find each product or quotient. Express your answer in exponential form. *(Lesson 4-6)*

44. $7^2 \cdot 7^3$

45. $x^4 \cdot 2x$

46. $\frac{8^{12}}{8^8}$

47. $\frac{36n^4}{14n^2}$

WebQuest Internet Project

Kids Gobbling Empty Calories

It is time to complete your project. Use the information and data you have gathered to prepare a brochure or Web page about the nutritional value of fast-food meals. Include the total Calories, grams of fat, and amount of sodium for five meals that a typical student would order from at least three fast-food restaurants.

www.pre-alg.com/webquest

Graphing Calculator Investigation

A Follow-Up of Lesson 6-9

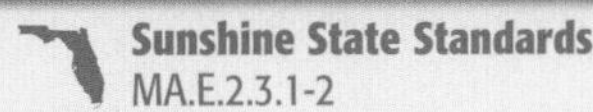

Probability Simulation

A random number generator can simulate a probability experiment. From the simulation, you can calculate experimental probabilities. Repeating a simulation may result in different probabilities since the numbers generated are different each time.

Example **Generate 30 random numbers from 1 to 6, simulating 30 rolls of a number cube.**

- Access the random number generator.
- Enter 1 as a lower bound and 6 as an upper bound for 30 trials.

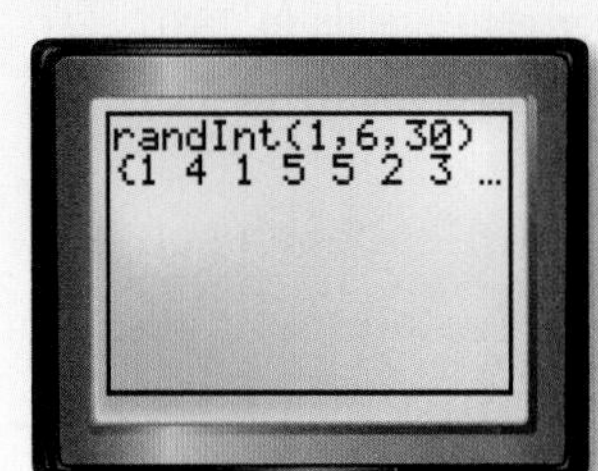

KEYSTROKES: MATH ◄ 5 1 , 6 , 30) ENTER

A set of 30 numbers ranging from 1 to 6 appears. Use the right arrow key to see the next number in the set. Record all 30 numbers, as a column, on a separate sheet of paper.

Exercises

1. Record how often each number on the number cube appeared.

a. Find the experimental probability of each number.

b. Compare the experimental probabilities with the theoretical probabilities.

2. Repeat the simulation of rolling a number cube 30 times. Record this second set of numbers in a column next to the first set of numbers. Each pair of 30 numbers represents a roll of two number cubes. Find the sum for each of the 30 pairs of rolls.

a. Find the experimental probability of each sum.

b. Compare the experimental probability with the theoretical probabilities.

3. Design an experiment to simulate 30 spins of a spinner that has equal sections colored red, white, and blue.

a. Find the experimental probability of each color.

b. Compare the experimental probabilities with the theoretical probabilities.

4. Suppose you play a game where there are three containers, each with ten balls numbered 0 to 9. Pick three numbers and then use the random number generator to simulate the game. Score 2 points if one number matches, 16 points if two numbers match, and 32 points if all three numbers match. Note: numbers can appear more than once.

a. Play the game if the order of your numbers *does not* matter. Total your score for 10 simulations.

b. Now play the game if the order of the numbers *does* matter. Total your score for 10 simulations.

c. With which game rules did you score more points?

www.pre-alg.com/other_calculator_keystrokes

Chapter 6

Study Guide and Review

Vocabulary and Concept Check

base (p. 288)
compound interest (p. 303)
cross products (p. 270)
discount (p. 299)
experimental probability (p. 311)
outcome (p. 310)
part (p. 288)
percent (p. 281)
percent equation (p. 298)
percent of change (p. 304)
percent of decrease (p. 306)
percent of increase (p. 305)
percent proportion (p. 288)
probability (p. 310)
proportion (p. 270)
random (p. 309)
rate (p. 265)
ratio (p. 264)
sample (p. 309)
sample space (p. 311)
scale (p. 276)
scale drawing (p. 276)
scale factor (p. 277)
scale model (p. 276)
simple event (p. 310)
simple interest (p. 300)
theoretical probability (p. 311)
unbiased (p. 309)
unit rate (p. 265)

Complete each sentence with the correct term.

1. A statement of equality of two ratios is called a(n) ____________.
2. A(n) ____________ is a ratio that compares a number to 100.
3. The ratio of a length on a scale drawing to the corresponding length on the real object is called the ____________.
4. The set of all possible outcomes is the ____________.
5. ____________ is what actually occurs when conducting a probability experiment.

Lesson-by-Lesson Review

6-1 Ratios and Rates

See pages 264–268.

Concept Summary

- A ratio is a comparison of two numbers by division.
- A rate is a ratio of two measurements having different units of measure.
- A rate that is simplified so that it has a denominator of 1 is called a unit rate.

Example **Express the ratio** ***2 meters to 35 centimeters*** **as a fraction in simplest form.**

$\frac{2 \text{ meters}}{35 \text{ centimeters}} = \frac{200 \text{ centimeters}}{35 \text{ centimeters}}$ Convert 2 meters to centimeters.

$= \frac{40 \text{ centimeters}}{7 \text{ centimeters}} \text{ or } \frac{40}{7}$ Divide the numerator and denominator by the GCF, 5.

Exercises **Express each ratio as a fraction in simplest form.**
See Examples 1 and 2 on pages 264 and 265.

6. 9 students out of 33 students
7. 12 hits out of 16 times at bat
8. 30 hours to 18 hours
9. 5 quarts to 5 gallons
10. 10 inches to 4 feet
11. 2 tons to 1800 pounds

www.pre-alg.com/vocabulary_review

6-2 Using Proportions

See pages 270–274.

Concept Summary

- A proportion is an equation stating two ratios are equal.
- If $\frac{a}{b} = \frac{c}{d}$, then $ad = bc$.

Example Solve $\frac{3}{7} = \frac{15}{x}$.

$\frac{3}{7} = \frac{15}{x}$	Write the proportion.
$3 \cdot x = 7 \cdot 15$	Cross products
$3x = 105$	Multiply.
$\frac{3x}{3} = \frac{105}{3}$	Divide each side by 3.
$x = 35$	The solution is 35.

Exercises **Solve each proportion.** *See Example 2 on page 271.*

12. $\frac{n}{12} = \frac{4}{3}$ **13.** $\frac{21}{x} = \frac{84}{120}$ **14.** $\frac{9}{7} = \frac{22.5}{y}$ **15.** $\frac{5}{7.5} = \frac{0.6}{k}$

6-3 Scale Drawings and Models

See pages 276–280.

Concept Summary

- A scale drawing or a scale model represents an object that is too large or too small to be drawn or built at actual size.
- The ratio of a length on a scale drawing or model to the corresponding length on the real object is called the scale factor.

Example **A scale drawing shows a pond that is 1.75 inches long. The scale on the drawing is 0.25 inch = 1 foot. What is the length of the actual pond?**

drawing length → / actual length → $\frac{0.25 \text{ in.}}{1 \text{ ft}} = \frac{1.75 \text{ in.}}{x \text{ ft}}$ ← drawing length / ← actual length

$0.25 \cdot x = 1 \cdot 1.75$	Find the cross products.
$0.25x = 1.75$	Simplify.
$x = 7$	Divide each side by 0.25.

The actual length of the pond is 7 feet.

Exercises **On the model of a ship, the scale is 1 inch = 12 feet. Find the actual length of each room.** *See Example 1 on page 277.*

	Room	Model Length
16.	Stateroom	0.9 in.
17.	Galley	3.8 in.
18.	Gym	6.0 in.

6-4 Fractions, Decimals, and Percents

See pages 281–285.

Concept Summary

- A percent is a ratio that compares a number to 100.
- Fractions, decimals, and percents are all different ways to represent the same number.

Examples

1 Express 60% as a fraction in simplest form and as a decimal.

$60\% = \frac{60}{100}$ or $\frac{3}{5}$ $\qquad 60\% = 60\%$ or 0.6

2 Express 0.38 as a percent.

$0.38 = 0.38$ or 38%

3 Express $\frac{5}{8}$ as a percent.

$\frac{5}{8} = 0.625$ or 62.5%

Exercises **Express each percent as a fraction or mixed number in simplest form and as a decimal.** *See Examples 1 and 3 on pages 281 and 282.*

19. 35% **20.** 42% **21.** 8% **22.** 19%
23. 120% **24.** 250% **25.** 62.5% **26.** 8.8%

Express each decimal or fraction as a percent. Round to the nearest tenth percent, if necessary. *See Examples 2, 4, and 5 on pages 282 and 283.*

27. 0.24 **28.** 0.03 **29.** 0.452 **30.** 1.9
31. $\frac{2}{5}$ **32.** $\frac{13}{22}$ **33.** $\frac{6}{80}$ **34.** $\frac{77}{225}$

6-5 Using the Percent Proportion

See pages 288–292.

Concept Summary

- If a is the part, b is the base, and p is the percent, then $\frac{a}{b} = \frac{p}{100}$.

Example

Forty-eight is 32% of what number?

$\frac{a}{b} = \frac{p}{100} \rightarrow \frac{48}{b} = \frac{32}{100}$ Replace a with 48 and p with 32.

$48 \cdot 100 = b \cdot 32$ Find the cross products.

$4800 = 32b$ Simplify.

$150 = b$ Divide each side by 32.

So, 48 is 32% of 150.

Exercises **Use the percent proportion to solve each problem.**
See Examples 1–6 on pages 288–290.

35. 18 is what percent of 45?
36. What percent of 60 is 39?
37. 23 is 92% of what number?
38. What is 74% of 110?
39. What is 80% of 62.5?
40. 36 is 15% of what number?

6-6 Finding Percents Mentally

See pages 293–297.

Concept Summary

- When working with common percents like 10%, 20%, 25%, and 50%, it is helpful to use the fraction form of the percent.

Examples

1 Find 20% of $45 mentally.

20% of $45 = \frac{1}{5}$ of 45 Think: $20\% = \frac{1}{5}$.

$= 9$ Think: $\frac{1}{5}$ of 45 is 9.

So, 20% of $45 is $9.

2 Estimate 32% of 150.

32% is about $33\frac{1}{3}\%$ or $\frac{1}{3}$.

$\frac{1}{3}$ of 150 is 50.

So, 32% of 150 is about 50.

Exercises Find the percent of each number mentally.

See Example 1 on pages 293 and 294.

41. 50% of 86
42. 20% of 55
43. 25% of 36
44. 40% of 75
45. $33\frac{1}{3}\%$ of 24
46. 90% of 60

Estimate. Explain which method you used to estimate.

See Example 2 on pages 294 and 295.

47. 48% of 32
48. 67% of 30
49. 20% of 51
50. 25% of 27
51. $\frac{1}{3}\%$ of 304
52. 147% of 200

6-7 Using Percent Equations

See pages 298–302.

Concept Summary

- The percent equation is an equivalent form of the percent proportion in which the percent is written as a decimal.
- Part = Percent · Base, where percent is in decimal form.

Example

119 is 85% of what number?

The part is 119, and the percent is 85%. Let n represent the base.

$119 = 0.85n$ Write 85% as the decimal 0.85.

$\frac{119}{0.85} = \frac{0.85n}{0.85}$ Divide each side by 0.85.

$140 = n$ So, 119 is 85% of 140.

Exercises Solve each problem using the percent equation.

See Examples 1–3 on pages 298 and 299.

53. 24 is what percent of 50?
54. 70 is 40% of what number?
55. What is 90% of 105?
56. What is 12.5% of 68?
57. 56 is 28% of what number?
58. 35.7 is what percent of 17?

• Extra Practice, see pages 736–739.
• Mixed Problem Solving, see page 763.

6-8 Percent of Change

See pages 304–308.

Concept Summary

- A percent of increase tells how much an amount has increased in relation to the original amount. (The percent will be positive.)
- A percent of decrease tells how much an amount has decreased in relation to the original amount. (The percent will be negative.)

Example **Find the percent of change from 36 pounds to 14 pounds.**

$$\text{percent of change} = \frac{\text{new weight} - \text{original weight}}{\text{original weight}} \quad \text{Write the ratio.}$$

$$= \frac{14 - 36}{36} \quad \text{Substitution}$$

$$= \frac{-22}{36} \quad \text{Subtraction}$$

$$\approx -0.611 \text{ or } -61.1\% \quad \text{Simplify.}$$

The percent of decrease is about 61.1%.

Exercises **Find the percent of change. Round to the nearest tenth, if necessary. Then state whether each change is a *percent of increase* or a *percent of decrease*.** *See Examples 1, 2, and 4 on pages 304–306.*

59. from 40 ft to 12 ft
60. from 80 cm to 96 cm
61. from 29 min to 54 min
62. from 80 lb to 77 lb

6-9 Probability and Predictions

See pages 310–314.

Concept Summary

- The probability of an event is a ratio that compares the number of favorable outcomes to the number of possible outcomes.

Example **Suppose a number cube is rolled. Find the probability of rolling a 5 or 6.**

Favorable outcomes: 5 and 6.
Possible outcomes: 1, 2, 3, 4, 5, and 6.

$$P(5 \text{ or } 6) = \frac{\text{number of favorable outcomes}}{\text{number of possible outcomes}}$$

$$= \frac{2}{6} \text{ or } \frac{1}{3}$$

So, the probability of rolling a 5 or 6 is $\frac{1}{3}$ or $33\frac{1}{3}\%$

Exercises **There are 2 blue marbles, 5 red marbles, and 8 green marbles in a bag. One marble is selected at random. Find the probability of each outcome.** *See Examples 1 and 2 on page 311.*

63. P(red)
64. P(green)
65. P(blue or green)
66. P(not blue)
67. P(yellow)
68. P(green, red, or blue)

Chapter 6 Practice Test

Vocabulary and Concepts

1. **Explain** the difference between a ratio and a rate.
2. **Describe** how to express a fraction as a percent.

Skills and Applications

Express each ratio as a fraction in simplest form.

3. 15 girls out of 40 students
4. 6 feet to 3 yards

Express each ratio as a unit rate. Round to the nearest tenth or cent.

5. 145 miles in 3 hours
6. $245 for 9 tickets
7. Convert 15 miles per hour to x feet per minute.
8. What value of y makes $\frac{8.4}{y} = \frac{1.2}{1.1}$ a proportion?

Express each percent as a fraction or mixed number in simplest form and as a decimal.

9. 36%
10. 52%
11. 225%
12. 315%
13. 0.6%
14. 0.4%

Express each decimal or fraction as a percent. Round to the nearest tenth percent, if necessary.

15. 0.47
16. 0.025
17. 5.38
18. $\frac{7}{20}$
19. $\frac{30}{22}$
20. $\frac{18}{4000}$

Use the percent proportion to solve each problem.

21. 36 is what percent of 80?
22. 35.28 is 63% of what number?

Estimate.

23. 25% of 82
24. 63% of 77
25. Find the interest on $2700 that is invested at 4% for $2\frac{1}{2}$ years.
26. Find the discount for a $135 coat that is on sale at 15% off.
27. Find the percent of change from 175 pounds to 140 pounds. Round to the nearest tenth.
28. There are 3 purple balls, 5 orange balls, and 8 yellow balls in a bowl. Suppose one ball is selected at random. Find P(orange).
29. **DESIGN** A builder is designing a swimming pool that is 8.5 inches in length on the scale drawing. The scale of the drawing is 1 inch = 6 feet. What is the length of the actual swimming pool?

FCAT Practice

30. **STANDARDIZED TEST PRACTICE** The table lists the reasons shoppers use online customer service. Out of 350 shoppers who own a computer, how many would you expect to say they use online customer service to track packages?

Reasons	Percent
Track Delivery	54
Product Information	24
Verify Shipping Charges	17
Transaction Help	16

Ⓐ 189　Ⓑ 84　Ⓒ 19　Ⓓ 154

www.pre-alg.com/chapter_test/fcat

Chapter 6 Standardized Test Practice

FCAT Practice

Part 1 Multiple Choice

Record your answers on the answer sheet provided by your teacher or on a sheet of paper.

1. Evaluate $x - y + z$ if $x = -6$, $y = 9$, and $z = -3$. (Lesson 2-3)

Ⓐ 0 Ⓑ −6

Ⓒ −18 Ⓓ 15

2. Which figure has an area of 192 cm^2? (Lesson 3-7)

Ⓐ
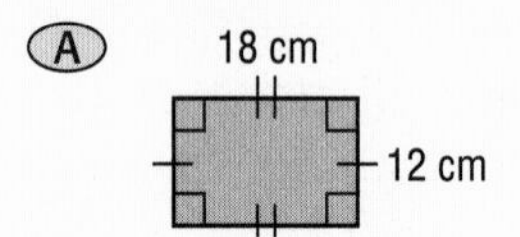

Ⓑ
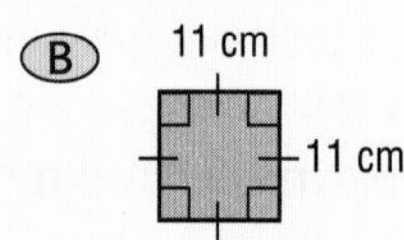

Ⓒ
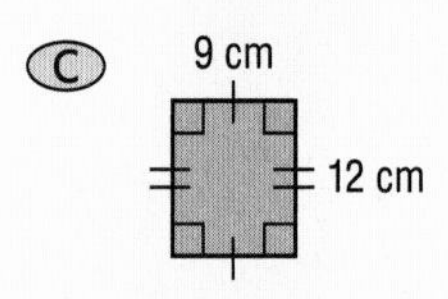

Ⓓ
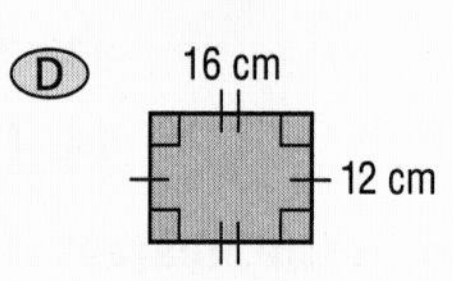

3. Which expression is *not* a monomial? (Lesson 4-1)

Ⓐ $5(-y)$ Ⓑ $8k$

Ⓒ $m - n$ Ⓓ $2x(-3y)$

4. Which fraction represents the ratio *8 apples to 36 pieces of fruit* in simplest form? (Lesson 6-1)

Ⓐ $\frac{1}{4}$ Ⓑ $\frac{4}{9}$

Ⓒ $\frac{2}{9}$ Ⓓ $\frac{1}{6}$

5. The ratio of girls to boys in a class is 5 to 4. Suppose there are 27 students in the class. How many of the students are girls? (Lesson 6-2)

Ⓐ 40 Ⓑ 15

Ⓒ 12 Ⓓ 9

6. A scale model of an airplane has a width of 13.5 inches. The scale of the model is 1 inch = 8 feet. What is the width of the actual airplane? (Lesson 6-3)

Ⓐ 110 ft Ⓑ 108 ft

Ⓒ 104 ft Ⓓ 115 ft

7. Randy, Eduardo, and Kelli took a quiz. For every 50 questions on the quiz, Randy answered 47 correctly. Eduardo answered 91% of the questions correctly. For every 10 questions on the quiz, Kelli answered 9 correctly. Who had the highest score?

Ⓐ Randy Ⓑ Eduardo

Ⓒ Kelli Ⓓ all the same

8. The graph shows the amount of canned food collected by the 9th grade classes at Hilltop High School.

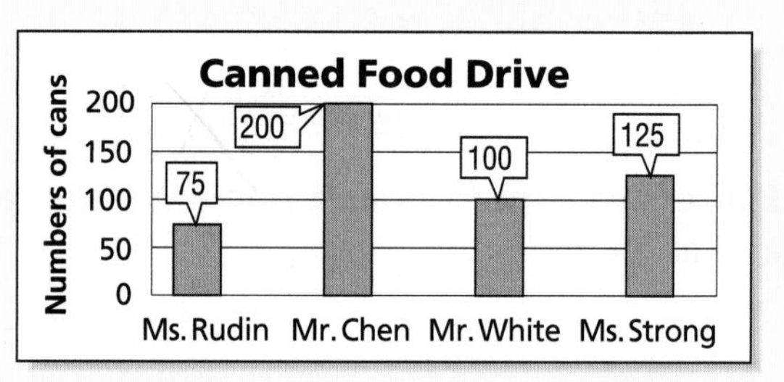

Of the total amount of cans collected, what percent did Mr. Chen's class collect? (Lesson 6-4)

Ⓐ 25% Ⓑ 33% Ⓒ 40% Ⓓ 50%

9. The table shows the increase in average salaries in each of the four major sports from the 1990–91 season to the 2000–01 season. (Lesson 6-8)

Sport	1990–91	2000–01
Hockey	$271,000	$1,400,000
Basketball	823,000	3,530,000
Football	430,000	1,200,000
Baseball	597,537	2,260,000

Source: *USA TODAY*

Which sport had a percent of increase in average salary of about 325%?

Ⓐ Hockey Ⓑ Basketball

Ⓒ Football Ⓓ Baseball

Test-Taking Tip

Question 8 To find what percent of the cans a class collected, you will first need to find the total number of cans collected by all of the 9th grade classes.

FCAT Practice

Part 2 Short Response/Grid In

Record your answers on the answer sheet provided by your teacher or on a sheet of paper.

10. Ana earns $6.80 per hour when she works on weekdays. She earns twice that amount per hour when she works on weekends. If Ana worked 4 hours on Tuesday, 4 hours on Thursday, and 5 hours on Saturday, then how much did she earn? (Prerequisite Skill, p. 713)

11. Juan and Julia decided to eat lunch at The Sub Shop. Juan ordered a veggie sub, lemonade, and a cookie. Julia ordered a ham sub, milk, and a cookie. What was the total cost of Juan and Julia's lunch? (Prerequisite Skill, p. 713)

Item	Cost
Veggie Sub	$3.89
Turkey Sub	$3.79
Ham Sub	$3.49
Soda	$1.25
Lemonade	$1.00
Milk	$0.79
Cookie	$0.99

12. What number should replace X in this pattern? (Lesson 4-2)

$4^0 = 1$
$4^1 = 4$
$4^2 = 16$
$4^3 = 64$
$4^4 = X$

13. Find the value of m in $\frac{3}{8}m = \frac{1}{4}$. (Lesson 5-9)

14. Nakayla purchased a package of 8 hamburger buns for $1.49. What is the ratio of the cost per hamburger bun? Round to the nearest penny. (Lesson 6-1)

15. What is 40% of 70? (Lesson 6-5)

16. Cameron purchased the portable stereo shown. About how much money did he save? (Lesson 6-7)

17. If you spin the spinner shown at the right, what is the probability that the arrow will stop at an even number? (Lesson 6-9)

FCAT Practice

Part 3 Open Ended

Record your answers on a sheet of paper. Show your work.

18. An electronics store is having a sale on certain models of televisions. Mr. Castillo would like to buy a television that is on sale. This television normally costs $679. (Lesson 6-7)

a. What price, not including tax, will Mr. Castillo pay if he buys the television on Saturday?

b. What price, not including tax, will Mr. Castillo pay if he buys the television on Wednesday?

c. How much money will Mr. Castillo save if he buys the television on a Saturday?

19. The graph shows the number of domain registrations for the years 1997–2000.

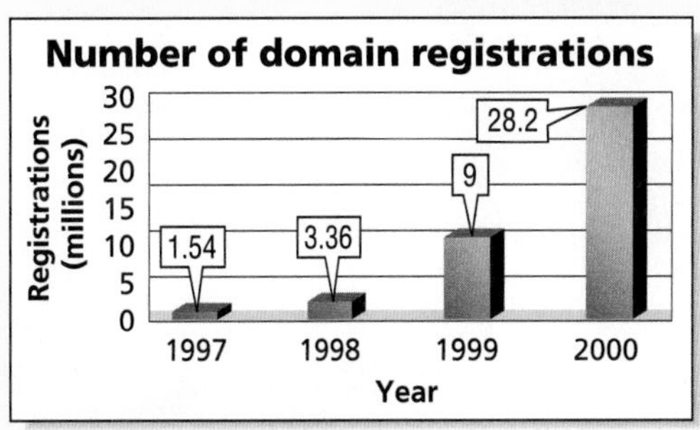

Source: Network Solutions (VeriSign)

Write a few sentences describing the percent of change in the number of domain registrations from one year to the next. (Lesson 6-8)

UNIT 3

Linear Equations, Inequalities, and Functions

Your study of algebra includes more than just solving equations. Many real-world situations can be modeled by equations and their graphs. In this unit, you will learn about functions and graphs.

Chapter 7
Equations and Inequalities

Chapter 8
Functions and Graphing

WebQuest Internet Project

Just for Fun

What do you like to do in your spare time—shop at the mall, attend a baseball or football game, go to the movies, ride the rides at an amusement park, or hike in the great outdoors?

In this project, you will be exploring how equations, functions, and graphs can help you examine how people spend their leisure time.

Log on to www.pre-alg.com/webquest. Begin your WebQuest by reading the Task.

Then continue working on your WebQuest as you study Unit 3.

Lesson	7-1	8-8
Page	333	411

USA TODAY Snapshots®
What fans pay after ticket
An average fan at a Major League Baseball game spends $15.40 on parking, food, drinks and souvenirs in addition to the $15 ticket price.
MLB
NBA
NHL
NFL
$15.40
$18.20
$18.25
$19.00
Source: American Demographics, Team Marketing Report
By Anne R. Carey and Marcy E. Mullins, USA TODAY

Chapter 7 Equations and Inequalities

What You'll Learn

- **Lessons 7-1 and 7-2** Solve equations with variables on each side and with grouping symbols.
- **Lesson 7-3** Write and graph inequalities.
- **Lessons 7-4 and 7-5** Solve inequalities using the Properties of Inequalities.
- **Lesson 7-6** Solve multi-step inequalities.

Key Vocabulary

- **null** or **empty set** (p. 336)
- **identity** (p. 336)
- **inequality** (p. 340)

Why It's Important

An equation is a statement that two expressions are equal. Sometimes, you want to know when one expression is greater or less than another. This kind of statement is an inequality. For example, you can solve an inequality to determine a healthy backpack weight. *You will solve problems involving backpacking in Lesson 7-6.*

Getting Started

Prerequisite Skills To be successful in this chapter, you'll need to master these skills and be able to apply them in problem-solving situations. Review these skills before beginning Chapter 7.

For Lesson 7-1 — Solve Two-Step Equations

Solve each equation. Check your solution. *(For review, see Lesson 3-5.)*

1. $2x + 5 = 13$ **2.** $4n - 3 = 5$ **3.** $16 = 8 + \frac{d}{3}$ **4.** $\frac{c}{-4} + 3 = -9$

For Lesson 7-4 — Add and Subtract Integers

Find each sum or difference. *(For review, see Lessons 2-2 and 2-3.)*

5. $-28 + (-16)$ **6.** $17 + (-25)$ **7.** $-13 + 24$

8. $36 + (-18)$ **9.** $31 - 48$ **10.** $-16 - 7$

11. $4 - (-12)$ **12.** $-23 - (-29)$ **13.** $-19 - (-5)$

For Lesson 7-5 — Multiply and Divide Integers

Find each product or quotient. *(For review, see Lessons 2-4 and 2-5.)*

14. $-6(8)$ **15.** $-3 \cdot 5$ **16.** $-6(-25)$

17. $2(-4)(-9)$ **18.** $64 \div (-32)$ **19.** $-15 \div 3$

20. $-12 \div (-3)$ **21.** $-6 \div (-6)$ **22.** $24 \div (-2)$

Make this Foldable to help you organize notes on equations and inequalities. Begin with a plain sheet of $8\frac{1}{2}$" by 11" paper.

Step 1 Fold in Half

Fold in half lengthwise.

Step 2 Fold in Sixths

Fold in thirds and then fold each third in half.

Step 3 Cut

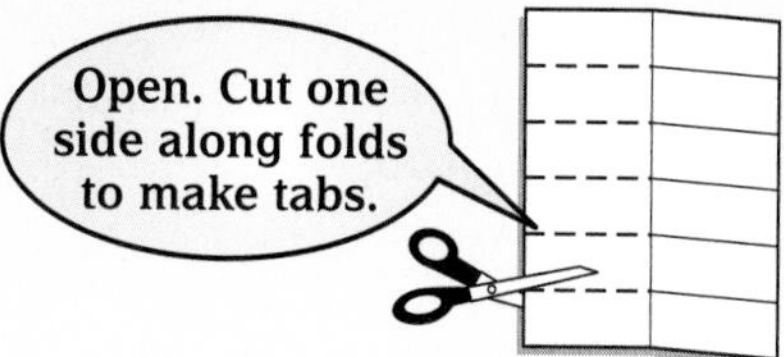

Open. Cut one side along folds to make tabs.

Step 4 Label

Label each tab with the lesson number as shown.

7-1
7-2
7-3
7-4
7-5
7-6

Reading and Writing As you read and study the chapter, write notes and examples under each tab.

Algebra Activity

A Preview of Lesson 7-1

Equations with Variables on Each Side

In Chapter 3, you used algebra tiles and an equation mat to solve equations in which the variable was on only one side of the equation. You can use algebra tiles and an equation mat to solve equations with variables on each side of the equation.

Activity 1

The following example shows how to solve $x + 3 = 2x + 1$ using algebra tiles.

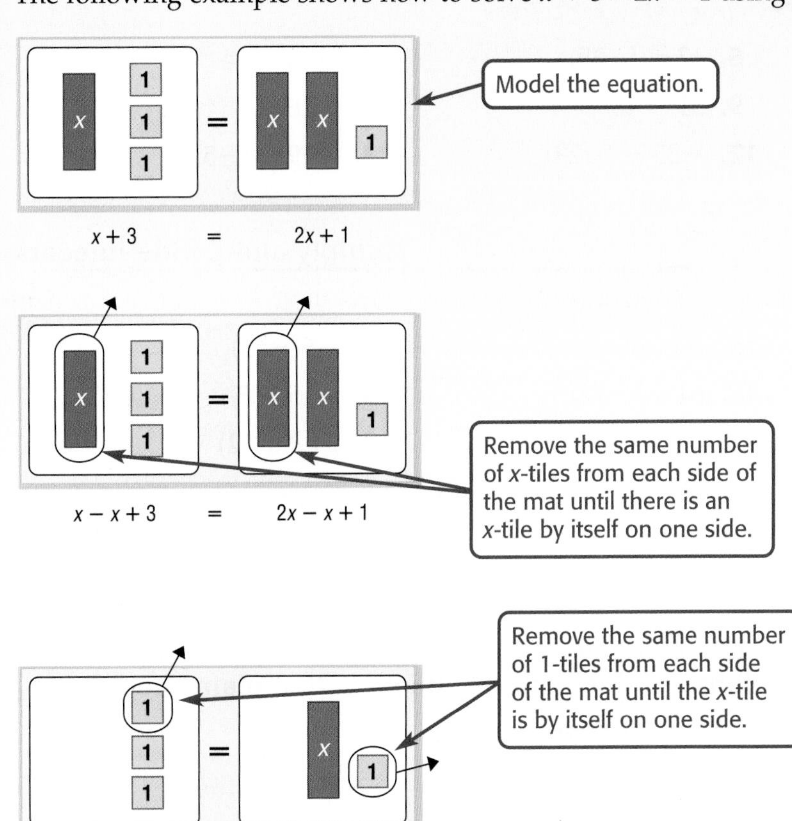

Model the equation.
x + 3 = 2x + 1
Remove the same number of x-tiles from each side of the mat until there is an x-tile by itself on one side.
x − x + 3 = 2x − x + 1
Remove the same number of 1-tiles from each side of the mat until the x-tile is by itself on one side.
2 = x

There are two 1-tiles on the left side of the mat and one x-tile on the right side. Therefore, $x = 2$. Since $2 + 3 = 2(2) + 1$, the solution is correct.

Model

Use algebra tiles to model and solve each equation.

1. $2x + 3 = x + 5$
2. $3x + 4 = 2x + 8$
3. $3x = x + 6$
4. $6 + x = 4x$
5. $2x - 4 = x - 6$
6. $5x - 1 = 4x - 5$

Analyze

7. Which property of equality allows you to remove a 1-tile from each side of the mat?

8. Explain why you can remove an x-tile from each side of the mat.

Activity 2

Some equations are solved by using zero pairs. Remember, you may add or subtract a zero pair from either side of an equation mat without changing its value. The following example shows how to solve $2x + 1 = x - 5$.

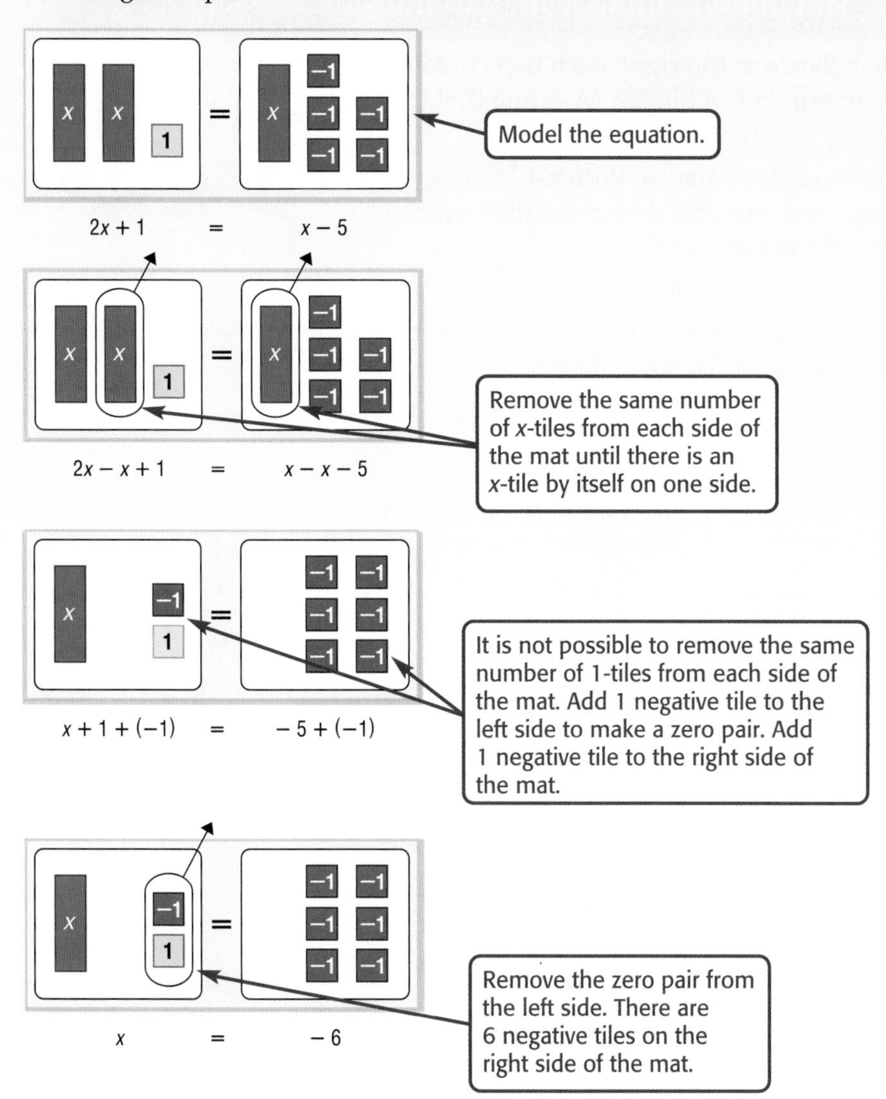

Model the equation.
2x + 1 = x − 5
Remove the same number of x-tiles from each side of the mat until there is an x-tile by itself on one side.
2x − x + 1 = x − x − 5
It is not possible to remove the same number of 1-tiles from each side of the mat. Add 1 negative tile to the left side to make a zero pair. Add 1 negative tile to the right side of the mat.
x + 1 + (−1) = − 5 + (−1)
Remove the zero pair from the left side. There are 6 negative tiles on the right side of the mat.
x = − 6

Therefore, $x = -6$. Since $2(-6) + 1 = -6 - 5$, the solution is correct.

Model

Use algebra tiles to model and solve each equation.

9. $2x + 3 = x - 5$ **10.** $3x - 2 = x + 6$ **11.** $x - 1 = 3x + 7$

12. $x + 6 = 2x - 3$ **13.** $2x + 4 = 3x - 2$ **14.** $4x - 1 = 2x + 5$

Analyze

15. Does it matter whether you remove x-tiles or 1-tiles first? Explain.

16. Explain how you could use models to solve $-2x + 5 = -x - 2$.

7-1 Solving Equations with Variables on Each Side

Sunshine State Standards
MA.A.3.3.2-2, MA.A.3.3.3-1, MA.D.1.3.2-2, MA.D.2.3.1-3, MA.D.2.3.2-1, MA.D.2.3.2-2

What You'll Learn

- Solve equations with variables on each side.

How is solving equations with variables on each side like solving equations with variables on one side?

On the balance at the right, each bag contains the same number of blocks. (Assume that the paper bag weighs nothing.)

a. The two sides balance. Without looking in a bag, how can you determine the number of blocks in each bag?

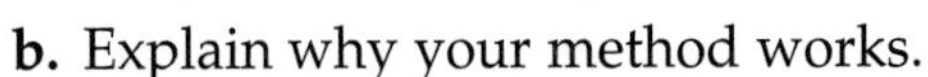

b. Explain why your method works.

c. Suppose x represents the number of blocks in the bag. Write an equation that is modeled by the balance.

d. Explain how you could solve the equation.

EQUATIONS WITH VARIABLES ON EACH SIDE To solve equations with variables on each side, use the Addition or Subtraction Property of Equality to write an equivalent equation with the variables on one side. Then solve the equation.

Study Tip

Look Back
To review **Addition and Subtraction Properties of Equality**, see Lesson 3-3.

Example 1 Equations with Variables on Each Side

Solve $2x + 3 = 3x$. Check your solution.

$2x + 3 = 3x$	Write the equation.
$2x - 2x + 3 = 3x - 2x$	Subtract $2x$ from each side.
$3 = x$	Simplify.

Subtract $2x$ from the left side of the equation to isolate the variable.

Subtract $2x$ from the right side of the equation to keep it balanced.

To check your solution, replace x with 3 in the original equation.

CHECK	$2x + 3 = 3x$	Write the equation.
	$2(3) + 3 \stackrel{?}{=} 3(3)$	Replace x with 3.
	$6 + 3 \stackrel{?}{=} 9$	Check to see whether this statement is true.
	$9 = 9$ ✓	The statement is true.

The solution is 3.

Concept Check What property allows you to add the same quantity to each side of an equation?

Example 2 Equations with Variables on Each Side

a. Solve $5x + 4 = 3x - 2$. Check your solution.

$5x + 4 = 3x - 2$	Write the equation.
$5x - 3x + 4 = 3x - 3x - 2$	Subtract $3x$ from each side.
$2x + 4 = -2$	Simplify.
$2x + 4 - 4 = -2 - 4$	Subtract 4 from each side.
$2x = -6$	Simplify.
$x = -3$	Mentally divide each side by 2.

CHECK

$5x + 4 = 3x - 2$	Write the equation.
$5(-3) + 4 \stackrel{?}{=} 3(-3) - 2$	Is this statement true?
$-11 = -11$ ✓	The solution checks.

The solution is -3.

b. Solve $2.4 + a = 2.5a - 4.5$.

$2.4 + a = 2.5a - 4.5$	Write the equation.
$2.4 + a - a = 2.5a - a - 4.5$	Subtract a from each side.
$2.4 = 1.5a - 4.5$	Simplify.
$2.4 + 4.5 = 1.5a - 4.5 + 4.5$	Add 4.5 to each side.
$6.9 = 1.5a$	Simplify.
$\frac{6.9}{1.5} = \frac{1.5a}{1.5}$	Divide each side by 1.5.
$4.6 = a$	Check your solution.

The solution is 4.6.

More About. . .

Videos

In 1980, only 1% of American households owned a VCR. Today, more than 80% do.

Source: Statistical Abstracts

You can use equations with variables on each side to solve problems.

Example 3 Use an Equation to Solve a Problem

VIDEOS A video store has two membership plans. Under plan A, a yearly membership costs \$30 plus \$1.50 for each rental. Under plan B, the yearly membership costs \$12 plus \$3 for each rental. What number of rentals results in the same yearly cost?

Let v represent the number of videos rented.

Words	\$30 plus \$1.50 for each video	\$12 plus \$3 for each video
Variables	$30 + 1.50v$	$12 + 3v$

Equation	$30 + 1.50v = 12 + 3v$	Write an equation.
	$30 + 1.5v - 1.5v = 12 + 3v - 1.5v$	Subtract $1.5v$ from each side.
	$30 = 12 + 1.5v$	Simplify.
	$30 - 12 = 12 - 12 + 1.5v$	Subtract 12 from each side.
	$18 = 1.5v$	Simplify.
	$\frac{18}{1.5} = \frac{1.5v}{1.5}$	Divide each side by 1.5.
	$12 = v$	Simplify.

The yearly cost is the same for 12 rentals.

Check for Understanding

Concept Check

1. **Name** the property of equality that allows you to subtract the same quantity from each side of an equation.

2. **OPEN ENDED** Write an example of an equation with variables on each side. State the steps you would use to isolate the variable.

Guided Practice

Solve each equation. Check your solution.

3. $4x - 8 = 5x$
4. $12x = 2x + 40$
5. $4x - 1 = 3x + 2$
6. $4k + 24 = 6k - 10$
7. $n + 0.4 = -n + 1$
8. $3.1w + 5 = 0.8 + w$

Application

9. **CAR RENTAL** Suppose you can rent a car from ABC Auto for either $25 a day plus $0.45 a mile or for $40 a day plus $0.25 a mile. What number of miles results in the same cost for one day?

Practice and Apply

Homework Help

For Exercises	See Examples
10–27, 30–33	1, 2
28, 29, 34–36	3

Extra Practice

See page 739.

Solve each equation. Check your solution.

10. $4x + 9 = 7x$
11. $6a = 26 + 4a$
12. $3y + 16 = 5y$
13. $n - 14 = 3n$
14. $8 - 3c = 2c - 2$
15. $3 - 4b = 10b + 10$
16. $7d - 13 = 3d + 7$
17. $2f - 6 = 7f + 24$
18. $-s + 4 = 7s - 3$
19. $4a - 2 = 7a - 6$
20. $12n - 24 = -14n + 28$
21. $13y - 18 = -5y + 36$
22. $12 + 1.5a = 3a$
23. $12.6 - x = 2x$
24. $2b + 6.2 = 13.2 - 8b$
25. $3c + 4.5 = 7.2 - 6c$
26. $12.4y + 14 = 6y - 2$
27. $4.3n - 1.6 = 2.3n + 5.2$

Define a variable and write an equation to find each number. Then solve.

28. Twice a number is 220 less than six times the number. What is the number?

29. Fourteen less than three times a number equals the number. What is the number?

Solve each equation. Check your solution.

30. $\frac{4}{5}y - 8 = \frac{2}{5}y + 16$
31. $\frac{3}{4}k + 16 = 2 - \frac{1}{8}k$
32. $\frac{x}{0.4} = 2x + 1.2$
33. $\frac{1}{3}b + 8 = \frac{1}{2}b - 4$

34. **GEOGRAPHY** The coastline of California is 46 miles longer than twice the length of Louisiana's coastline. It is also 443 miles longer than Louisiana's coastline. Find the lengths of the coastlines of California and Louisiana.

35. **CELLULAR PHONES** One cellular phone carrier charges $29.75 a month plus $0.15 a minute for local calls. Another carrier charges $19.95 a month and $0.29 a minute for local calls. For how many minutes is the cost of the plans the same?

More About. . .

Cellular Phones

The United States is divided into 734 cellular markets. There are more than 300 cellular and PCS phone companies in the United States.

Source: www.wirelessadvisor.com

The trends in attendance at various sporting events can be represented by equations. Visit www.pre-alg.com/webquest to continue work on your WebQuest project.

36. An empty bucket is put under two faucets. If one faucet is turned on alone, the bucket fills in 6 minutes. If the other faucet is turned on alone, the bucket fills in 4 minutes. If both are turned on, how many seconds will it take to fill the bucket?

37. **CRITICAL THINKING** Three times the quantity $y + 7$ equals four times the quantity $y - 2$. What value of y makes the sentence true?

38. **WRITING IN MATH** Answer the question that was posed at the beginning of the lesson.

How is solving equations with variables on each side like solving equations with variables on one side?

Include the following in your answer:
- examples of an equation with variables on each side and an equation with the variable on one side, and
- an explanation of how they are alike and how they are different.

39. Shoe World offers Olivia a temporary job during her spring break. The manager gives her a choice as to how she wants to be paid, but she must decide before she starts working. The choices are shown below.

	Pay per Hour	Pay for Each Dollar of Shoe Sales
Plan 1	\$3	15¢
Plan 2	\$4	10¢

Which equation shows what Olivia's sales would need to be in one hour to earn the same amount under either plan?

Ⓐ $3 + 0.15s = 4 + 0.10s$ Ⓑ $3s + 0.15 = 4s + 0.10$

Ⓒ $3 + 0.10s = 4 + 0.15s$ Ⓓ $3(s + 0.15) = 4(s + 0.10)$

40. What is the solution of $3x - 1 = x + 3$?

Ⓐ 1   Ⓑ 2   Ⓒ 3   Ⓓ 4

Extending the Lesson

41. **WEATHER** The formula $F = \frac{9}{5}C + 32$ is used for finding the Fahrenheit temperature when a Celsius temperature is known. Find the temperature where the Celsius and Fahrenheit scales are the same.

Maintain Your Skills

Mixed Review

42. **PROBABILITY** What is the probability of randomly choosing the letter T from the letters in PITTSBURGH? *(Lesson 6-9)*

43. Find the percent of increase from \$80 to \$90. *(Lesson 6-8)*

ALGEBRA Solve each problem using an equation. *(Lesson 6-7)*

44. 14 is what percent of 20?
45. Find 36% of 18.
46. 1.5 is 30% of what number?
47. Find 140% of 50.

Getting Ready for the Next Lesson

PREREQUISITE SKILL Use the Distributive Property to rewrite each expression as an equivalent algebraic expression.
(To review the ***Distributive Property****, see Lesson 3-1.)*

48. $4(x - 8)$
49. $3(2a + 9)$
50. $5(12 - x)$
51. $2(1.2c + 14)$
52. $8(-4k + 2.3)$
53. $\frac{1}{2}(n - 9)$

7-2 Solving Equations with Grouping Symbols

What You'll Learn

- Solve equations that involve grouping symbols.
- Identify equations that have no solution or an infinite number of solutions.

Sunshine State Standards
MA.A.3.3.2-2, MA.A.3.3.3-1, MA.D.1.3.2-2, MA.D.2.3.1-3, MA.D.2.3.2-1

Vocabulary
- null or empty set
- identity

Why is the Distributive Property important in solving equations?

Josh starts walking at a rate of 2 mph. One hour later, his sister Maria starts on the same path on her bike, riding at 10 mph.

The table shows expressions for the distance each has traveled after a given time.

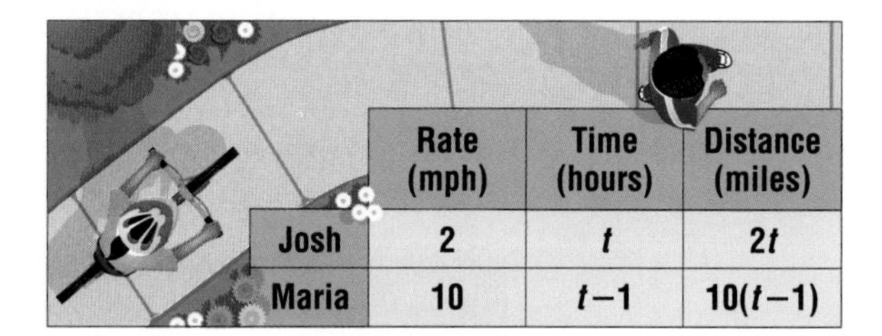

	Rate (mph)	Time (hours)	Distance (miles)
Josh	2	t	$2t$
Maria	10	$t-1$	$10(t-1)$

a. What does t represent?

b. Why is Maria's time shown as $t - 1$?

c. Write an equation that represents the time when Maria catches up to Josh. (*Hint*: They will have traveled the same distance.)

SOLVE EQUATIONS WITH GROUPING SYMBOLS To find how many hours it takes Maria to catch up to Josh, you can solve the equation $2t = 10(t - 1)$. First, use the Distributive Property to remove the grouping symbols.

Study Tip

Look Back
To review the **Distributive Property**, see Lesson 3-1.

Example 1 Solve Equations with Parentheses

a. Solve the equation $2t = 10(t - 1)$. Check your solution.

$2t = 10(t - 1)$ Write the equation.

$2t = 10(t) - 10(1)$ Use the Distributive Property.

$2t = 10t - 10$ Simplify.

$2t - 10t = 10t - 10t - 10$ Subtract $10t$ from each side.

$-8t = -10$ Simplify.

$\frac{-8t}{-8} = \frac{-10}{-8}$ Divide each side by -8.

$t = \frac{5}{4}$ or $1\frac{1}{4}$ Simplify.

CHECK Josh traveled $\frac{2 \text{ miles}}{\text{hour}} \cdot \frac{5 \text{ hour}}{4}$ or $2\frac{1}{2}$ miles.

Maria traveled one hour less than Josh. She traveled $\frac{10 \text{ miles}}{\text{hour}} \cdot \frac{1 \text{ hour}}{4}$ or $2\frac{1}{2}$ miles.

Therefore, Maria caught up to Josh in $\frac{1}{4}$ hour, or 15 minutes.

b. Solve $5(a - 4) = 3(a + 1.5)$.

$5(a - 4) = 3(a + 1.5)$	Write the equation.
$5a - 20 = 3a + 4.5$	Use the Distributive Property.
$5a - 20 + 20 = 3a + 4.5 + 20$	Add 20 to each side.
$5a = 3a + 24.5$	Simplify.
$5a - 3a = 3a - 3a + 24.5$	Subtract $3a$ from each side.
$2a = 24.5$	Simplify.
$\frac{2a}{2} = \frac{24.5}{2}$	Divide each side by 2.
$a = 12.25$	Simplify.

The solution is 12.25. Check your solution.

Study Tip

Alternative Method

You can also solve the equation by subtracting $3a$ from each side first, then adding 20 to each side.

Concept Check What property do you use to remove the grouping symbols from the equation $2(8 - a) = 4(a + 9)$?

Sometimes a geometric figure is described in terms of only one of its dimensions. To find the dimensions, you may have to solve an equation that contains grouping symbols.

Example 2 Use an Equation to Solve a Problem

GEOMETRY **The perimeter of a rectangle is 46 inches. Find the dimensions if the length is 5 inches greater than twice the width.**

Words The length is 5 inches greater than twice the width. The perimeter is 46 inches.

Variables Let w = the width.
Let $2w + 5$ = the length.

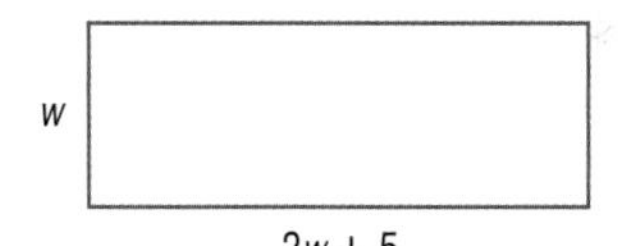

Study Tip

Look Back

To review **perimeter of a rectangle**, see Lesson 3-7.

2 times length + 2 times width = perimeter

Equation $2(2w + 5) + 2w = 46$

Solve $2(2w + 5) + 2w = 46$.

$2(2w + 5) + 2w = 46$	Write the equation.
$4w + 10 + 2w = 46$	Use the Distributive Property.
$6w + 10 = 46$	Simplify.
$6w + 10 - 10 = 46 - 10$	Subtract 10 from each side.
$6w = 36$	Simplify.
$w = 6$	Mentally divide each side by 6.

Evaluate $2w + 5$ to find the length.

$2(6) + 5 = 12 + 5$ or 17 Replace w with 6.

CHECK Add the lengths of the four sides.
$6 + 17 + 6 + 17 = 46$ ✓

The width is 6 inches. The length is 17 inches.

NO SOLUTION OR ALL NUMBERS AS SOLUTIONS Some equations have *no* solution. That is, no value of the variable results in a true sentence. The solution set is the **null** or **empty set**, shown by the symbol $\varnothing$ or $\{\}$.

Example 3 No Solution

Solve $3x + \frac{1}{3} = 3x - \frac{1}{2}$.

$3x + \frac{1}{3} = 3x - \frac{1}{2}$ Write the equation.

$3x - 3x + \frac{1}{3} = 3x - 3x - \frac{1}{2}$ Subtract $3x$ from each side.

$\frac{1}{3} = -\frac{1}{2}$ Simplify.

The sentence $\frac{1}{3} = -\frac{1}{2}$ is *never* true. So, the solution set is $\varnothing$.

Other equations may have every number as the solution. An equation that is true for every value of the variable is called an **identity**.

Example 4 All Numbers as Solutions

Solve $2(2x - 1) + 6 = 4x + 4$.

$2(2x - 1) + 6 = 4x + 4$ Write the equation.

$4x - 2 + 6 = 4x + 4$ Use the Distributive Property.

$4x + 4 = 4x + 4$ Simplify.

$4x + 4 - 4 = 4x + 4 - 4$ Subtract 4 from each side.

$4x = 4x$ Simplify.

$x = x$ Mentally divide each side by 4.

The sentence $x = x$ is *always* true. The solution set is all numbers.

Check for Understanding

Concept Check

1. **List** the steps you would take to solve the equation $2x + 3 = 4(x - 1)$.
2. **OPEN ENDED** Give an example of an equation that has no solution and an equation that is an identity.

Guided Practice

Solve each equation. Check your solution.

3. $3(a - 5) = 18$
4. $32 = 4(x + 9)$
5. $2(d + 6) = 3d - 1$
6. $6(n - 3) = 4(n + 2.1)$
7. $12 - h = -h + 3$
8. $3(2g + 4) = 6(g + 2)$

Application

9. **GEOMETRY** The perimeter of a rectangle is 20 feet. The width is 4 feet less than the length. Find the dimensions of the rectangle. Then find its area.

Practice and Apply

Homework Help

For Exercises	See Examples
10–19, 24, 25, 28, 29	1
20–23, 26, 27	3, 4
30–33	2

Extra Practice

See page 740.

Solve each equation. Check your solution.

10. $3(g - 3) = 6$
11. $3(x + 1) = 21$
12. $5(2c + 7) = 80$
13. $6(3d + 5) = 75$
14. $3(a - 3) = 2(a + 4)$
15. $3(s + 22) = 4(s + 12)$
16. $4(x - 2) = 3(1.5 + x)$
17. $3(a - 1) = 4(a - 1.5)$
18. $2(3.5n + 6) = 2.5n - 2$
19. $4.2x - 9 = 3(1.2x + 4)$
20. $4(f + 3) + 5 = 17 + 4f$
21. $3n + 4 = 5(n + 2) - 2n$
22. $8y - 3 = 5(y - 1) + 3y$
23. $2(x - 5) = 4x - 2(x + 5)$
24. $\frac{1}{2}(2n - 5) = 4n - 1$
25. $y - 2 = \frac{1}{3}(y + 6)$
26. $-3(4b - 10) = \frac{1}{2}(-24b + 60)$
27. $\frac{3}{4}a + 4 = \frac{1}{4}(3a + 16)$
28. $\frac{d}{0.4} = 2d + 1.24$
29. $\frac{a - 6}{12} = \frac{a - 2}{4}$

Find the dimensions of each rectangle. The perimeter is given.

30. $P = 460$ ft

w

w + 30

31. $P = 440$ yd

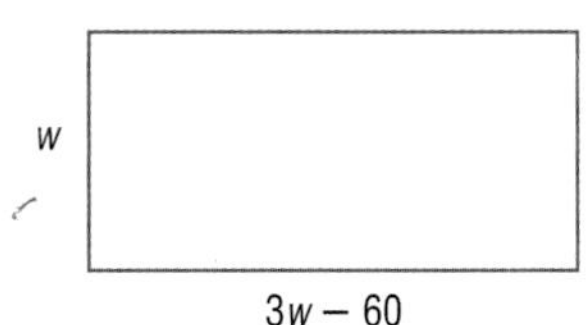

32. $P = 11$ m

w

2w – 2

33. **GEOMETRY** The perimeter of a rectangle is 32 feet. Find the dimensions if the length is 4 feet longer than three times the width. Then find the area of the rectangle.

34. **NUMBER THEORY** Three times the sum of three consecutive integers is 72. What are the integers?

35. **GEOMETRY** The triangle and the rectangle have the same perimeter. Find the dimensions of each figure. Then find the perimeter.

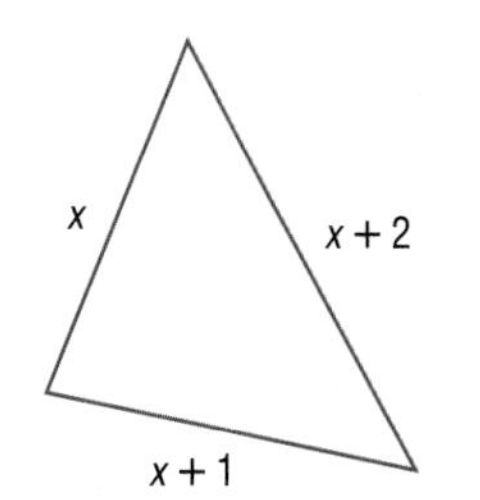

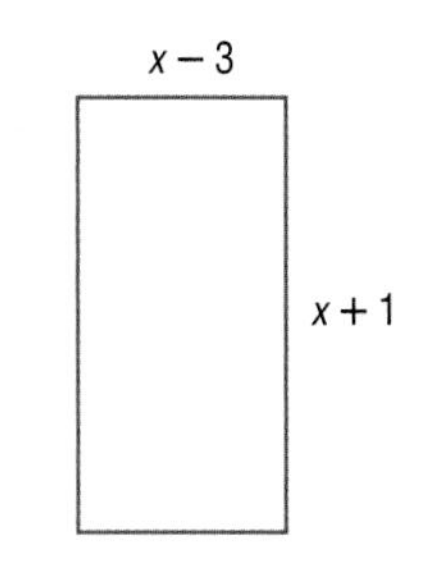

36. **BASKETBALL** Camilla has three times as many points as Lynn. Lynn has five more points than Kim. Camilla, Lynn, and Kim combined have twice as many points as Jasmine. If Jasmine has 25 points, how many points does each of the other three girls have?

37. **DECORATING** Suppose a rectangular room measures 15 feet long by 12 feet wide by 7 feet high and has two windows and two doors. Use the information at the left to find how many gallons of paint are needed to paint the room using two coats of paint.

More About. . .

Decorating

A gallon of paint covers about 350 square feet. Estimate the square footage by multiplying the combined wall lengths by wall height and subtracting 15 square feet for each window and door.

Source: *The Family Handyman Magazine Presents Handy Hints for Home, Yard, and Workshop*

38. **CRITICAL THINKING** An apple costs the same as 2 oranges. Together, an orange and a banana cost 10¢ more than an apple. Two oranges cost 15¢ more than a banana. What is the cost for one of each fruit?

39. **WRITING IN MATH** Answer the question that was posed at the beginning of the lesson.

Why is the Distributive Property important in solving equations?

Include the following in your answer:

- a definition of the Distributive Property, and
- a description of its use in solving equations.

FCAT Practice

Standardized Test Practice

40. Which equation is equivalent to $2(3x - 1) = 10 + 2x$?

Ⓐ $8x - 2 = 10$ Ⓑ $6x = 11 + 2x$

Ⓒ $4x - 2 = 10$ Ⓓ $6x - 1 = 10 + 2x$

41. Car X leaves Northtown traveling at a steady rate of 55 mph. Car Y leaves 1 hour later following Car X, traveling at a steady rate of 60 mph. Which equation can be used to determine how long after Car X leaves Car Y will catch up?

Ⓐ $55x = 60x - 1$ Ⓑ $55x = 60x$

Ⓒ $60x = 55(x - 1)$ Ⓓ $55x = 60(x - 1)$

Maintain Your Skills

Mixed Review

ALGEBRA Solve each equation. Check your solution. *(Lesson 7-1)*

42. $4x = 2x + 5$ 43. $3x + 5 = 7 - 2x$ 44. $1.5x + 9 = 3x - 3$

45. **PROBABILITY** Find the probability of choosing a girl's name at random from 20 girls' names and 30 boys' names. *(Lesson 6-9)*

Write each fraction as a decimal. Use a bar to show a repeating decimal. *(Lesson 5-1)*

46. $\frac{4}{10}$ 47. $\frac{3}{8}$ 48. $-\frac{1}{3}$ 49. $3\frac{6}{25}$ 50. $4\frac{5}{11}$

Getting Ready for the Next Lesson

PREREQUISITE SKILL Evaluate each expression.
(To review ***evaluating expressions****, see Lesson 1-3.)*

51. $x - 12, x = 5$ 52. $b + 11, b = -15$ 53. $4a, a = -6$

54. $2t + 8, t = -3$ 55. $\frac{24}{c}, c = -3$ 56. $\frac{3x}{4} + 2, x = 6$

Practice Quiz 1

Lessons 7-1 and 7-2

Define a variable and write an equation. Then solve. *(Lesson 7-1)*

1. Twice a number is 150 less than 5 times the number. What is the number?

Solve each equation. Check your solution. *(Lessons 7-1 and 7-2)*

2. $6y + 42 = 4y$ 3. $7m - 12 = 2.5m + 2$

4. $8(p - 4) = 2(2p + 1)$ 5. $b + 2(b + 5) = 3(b - 1) + 13$

Reading Mathematics

Meanings of At Most and At Least

The phrases *at most* and *at least* are used in mathematics. In order to use them correctly, you need to understand their meanings.

Phrase	Meaning	Mathematical Symbol
at most	• no more than • less than or equal to	$\leq$
at least	• no less than • greater than or equal to	$\geq$

Here is an example of one common use of each phrase, its meaning, and a mathematical expression for the situation.

Verbal Expression	You can spend *at most* \$20.
Meaning	You can spend \$20 or any amount less than \$20.
Mathematical Expression	$s \leq 20$, where s represents the amount you spend.

Verbal Expression	A person must be *at least* 18 to vote.
Meaning	A person who is 18 years old or any age older than 18 may vote.
Mathematical Expression	$a \geq 18$, where a represents age.

Notice that the word *or* is part of the meaning in each case.

Reading to Learn

1. Write your own rule for remembering the meanings of *at most* and *at least.*

For each expression, write the meaning. Then write a mathematical expression using $\leq$ or $\geq$.

2. You need to earn at least \$50 to help pay for a class trip.

3. The sum of two numbers is at most 6.

4. You want to drive at least 250 miles each day.

5. You want to hike 4 hours each day at most.

7-3 Inequalities

What You'll Learn

- Write inequalities.
- Graph inequalities.

Sunshine State Standards
MA.A.3.3-1, MA.D.1.3.2-2, MA.D.2.3.1-1, MA.D.2.3.1-3, MA.D.2.3.2-1

Vocabulary
- inequality

How can inequalities help you describe relationships?

If your age is *less than 6*, you eat free.

If your height is *more than 40 inches*, you can ride.

A speed of *35 or less* is legal.

a. Name three ages of children who can eat free at the restaurant. Does a child who is 6 years old eat free?

b. Name three heights of children who can ride the ride at the amusement park. Can a child who is 40 inches tall ride?

c. Name three speeds that are legal. Is a driver who is traveling at 35 mph driving at a legal speed?

WRITE INEQUALITIES A mathematical sentence that contains $<$ or $>$ is called an **inequality**.

Example 1 Write Inequalities with $<$ or $>$

Write an inequality for each sentence.

a. Your age is less than 6 years.

Variable Let a represent age.

Inequality $a < 6$

b. Your height is greater than 40 inches.

Variable Let h represent height.

Inequality $h > 40$

Some inequalities contain $\leq$ or $\geq$ symbols.

Example 2 Write Inequalities with $\leq$ or $\geq$

Write an inequality for each sentence.

a. Your speed is less than or equal to 35 miles per hour.

Variable Let s represent speed.

Inequality $s \leq 35$

b. Your speed is greater than or equal to 55 miles per hour.

Variable Let s represent speed.

Inequality $s \geq 55$

The table below shows some common verbal phrases and the corresponding mathematical inequalities.

Study Tip

Inequalities
Notice that $\leq$ and $\geq$ combine the symbol $<$ or $>$ with part of the symbol for equals, $=$.

Concept Summary — Inequalities

$<$	$>$	$\leq$	$\geq$
• is less than • is fewer than	• is greater than • is more than • exceeds	• is less than or equal to • is no more than • is at most	• is greater than or equal to • is no less than • is at least

Example 3 Use an Inequality

NUTRITION A food can be labeled low fat only if it has no more than 3 grams of fat per serving. Write an inequality to describe low fat foods.

Words Grams of fat per serving is no more than 3.

Variable Let f = number of grams of fat per serving.

Inequality $f \quad \leq \quad 3$

The inequality is $f \leq 3$.

Inequalities with variables are open sentences. When the variable in an open sentence is replaced with a number, the inequality may be true or false.

Reading Math

Symbols
Read the symbol $\not<$ as *is not less than*.

Example 4 Determine Truth of an Inequality

For the given value, state whether each inequality is *true* or *false*.

a. $s - 7 < 5$, $s = 14$

$s - 7 < 5$ Write the inequality.

$14 - 7 \overset{?}{<} 5$ Replace s with 14.

$7 \not< 5$ Simplify.

This sentence is false.

b. $12 \geq \frac{a}{2} + 2$, $a = 20$

$12 \geq \frac{a}{2} + 2$ Write the inequality.

$12 \overset{?}{\geq} \frac{20}{2} + 2$ Replace a with 20.

$12 \overset{?}{\geq} 10 + 2$ Simplify.

$12 \geq 12$ Simplify.

Although the inequality $12 > 12$ is false, the equation $12 = 12$ is true. Therefore, this sentence is true.

www.pre-alg.com/extra_examples/fcat

GRAPH INEQUALITIES Inequalities can be graphed on a number line. The graph helps you visualize the values that make the inequality true.

Example 5 Graph Inequalities

Graph each inequality on a number line.

a. $x > 4$

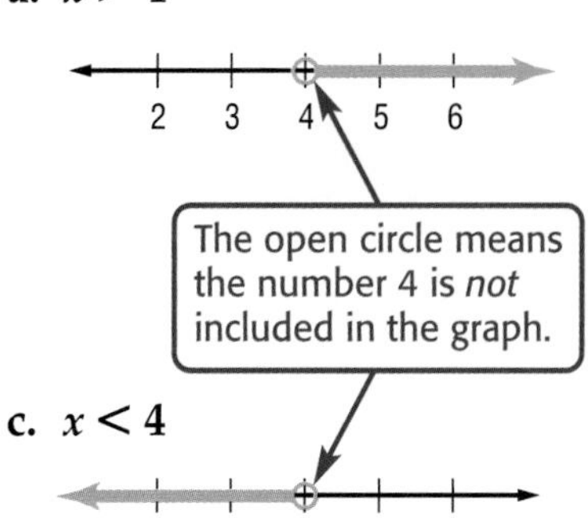

2 3 4 5 6
The open circle means the number 4 is *not* included in the graph.

b. $x \geq 4$

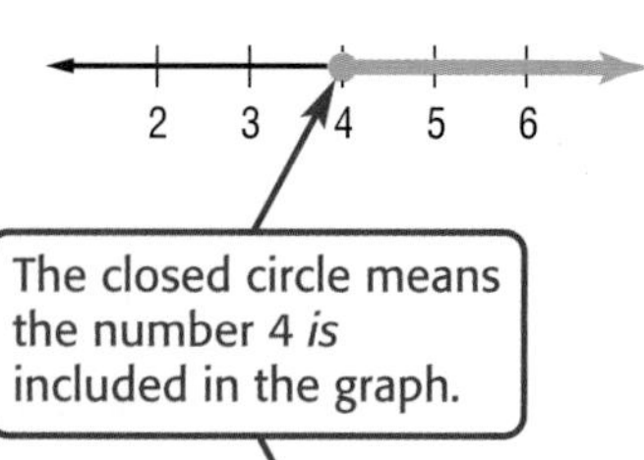

2 3 4 5 6
The closed circle means the number 4 *is* included in the graph.

c. $x < 4$

2 3 4 5 6

d. $x \leq 4$

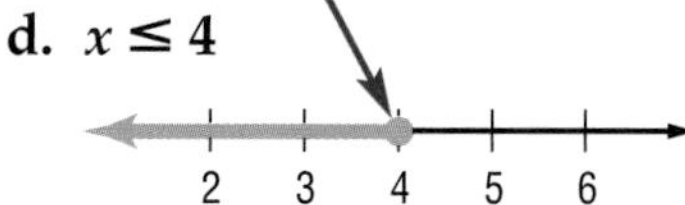

2 3 4 5 6

Example 6 Write an Inequality

Write the inequality for the graph.

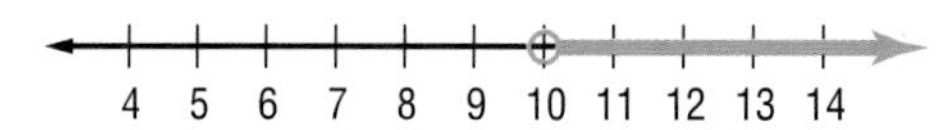

4 5 6 7 8 9 10 11 12 13 14

An open circle is on 10, so the point 10 is *not* included in the graph. The arrow points to the right, so the graph includes all numbers greater than 10. The inequality is $x > 10$.

Concept Check What symbols are used to write inequalities and what does each symbol mean?

Check for Understanding

Concept Check

1. **Explain** why a number line graph is a good way to represent an inequality.

2. **OPEN ENDED** Write four examples of inequalities using each of the symbols $<$, $>$, $\leq$, and $\geq$. Tell the meaning of each inequality.

Guided Practice

Write an inequality for each sentence.

3. A number increased by 14 is at least 25.
4. Five times some number is less than 65.

ALGEBRA For the given value, state whether the inequality is *true* or *false*.

5. $n + 4 > 6, n = 12$
6. $34 \leq 4r, r = 8$

Graph each inequality on a number line.

7. $n > 3$
8. $p \leq 5$
9. $x < 7$

Write the inequality for each graph.

10. 7 8 9 10 11 12 13 14 15

11. −22 −20 −18 −16 −14

Application **12. SAFETY** The elevators in an office building have been approved for a maximum load of 3600 pounds. Write an inequality to describe a safe load.

Practice and Apply

Homework Help

For Exercises	See Examples
13–16	1, 2
17–22	4
23–34	5
35–40	6
41–43	3

Extra Practice

See page 740.

Write an inequality for each sentence.

13. More than 18,000 fans attended the Kings' opening hockey game at the Staples Center in Los Angeles.

14. Kyle's earnings at $15 per hour were no more than $60.

15. The 10-km race time of 86 minutes was at least twice as long as the winner's time.

16. A savings account decreased by $75 is now less than $500.

ALGEBRA **For the given value, state whether each inequality is *true* or *false*.**

17. $18 - x > 4, x = 12$

18. $14 + n < 23, n = 8$

19. $5k > 35, k = 7$

20. $16 \leq 3c, c = 8$

21. $\frac{x}{3} \geq 2, x = 9$

22. $\frac{14}{c} < 7, c = 2$

Graph each inequality on a number line.

23. $a > 4$

24. $x > 6$

25. $n < 11$

26. $x < 5$

27. $t \geq 9$

28. $b \geq 8$

29. $d \leq 5$

30. $w \leq 8$

31. $x > -4$

32. $n \geq -3$

33. $x \leq -5$

34. $x < -2$

Write the inequality for each graph.

35. 7 8 9 10 11 12 13 14 15

36.

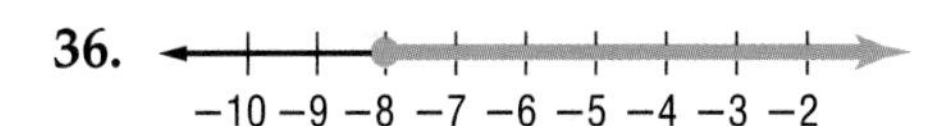

37. −7 −6 −5 −4 −3 −2 −1 0 1

38. −2 −1 0 1 2 3 4 5 6

39. −36 −34 −32 −30 −28

40. 4 5 6 7 8 9 10 11 12

HOMEWORK **For Exercises 41 and 42, use the graphic.**

41. Inali spends at least an hour more than the average time spent by boys on homework each week. Write an inequality for Inali's homework time.

42. Anna usually spends no more than the average time spent by girls on homework each week. Write an inequality to represent Anna's homework time.

USA TODAY Snapshots®

High school homework time

Students ages 14-18 say they spent a weekly average 6.1 hours on homework last school year, down from 6.6 in 1996-97. Weekly average hours reported by these groups:

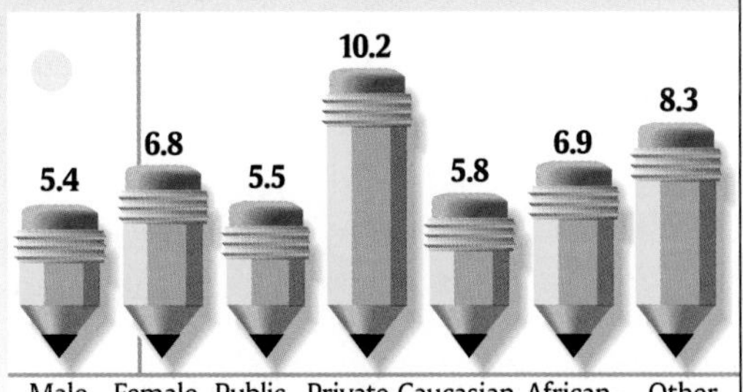

Male Female Public school student Private school student Caucasian African-American Other race/ ethnicity

Source: The State of Our Nation's Youth 1998-1999, Horatio Alger Association

By Cindy Hall and Sam Ward, USA TODAY

More About. . .

Sports

Almost twice as many boys play high school football as basketball, but basketball is offered at more high schools.

Source: National Association of State High School Associations

43. **SPORTS** There are more than 30,000 high school basketball and track programs in the United States. If there are 14,600 track programs, write and solve an inequality to determine the number of basketball programs.

44. Find a value for x that satisfies the inequality $0.6 < x < 0.75$.

45. **CRITICAL THINKING** In Chapter 1, you studied the Symmetric and Transitive Properties of Equality. Restate these properties using inequalities. Are the properties true for inequalities? If a property is not true, give a counterexample.

46. **WRITING IN MATH** Answer the question that was posed at the beginning of the lesson.

How can inequalities help you describe relationships?

Include the following in your answer:

- real-life examples using the four inequality symbols, and
- an explanation of the relationships described by each inequality.

FCAT Practice

Standardized Test Practice

47. Which inequality represents *a number decreased by 2 is at most 8*?

(A) $n - 2$ (B) $n - 2 \leq 8$

(C) $n - 2 \geq 8$ (D) $n - 2 > 8$

48. Which of the following is an inequality?

(A) $4 \leq x + 2$ (B) $x + 4 = 3$

(C) $x + 5 + y$ (D) $x - y$

Extending the Lesson

49. Graph the solutions for each compound inequality.

a. $y < -2$ or $y > 3$. (*Hint*: In a sentence, *or* means either part is true.)

b. $y \geq 0$ and $y \leq 5$ (*Hint*: In a sentence, *and* means both parts must be true.)

Maintain Your Skills

Mixed Review

ALGEBRA Solve each equation. Check your solution. *(Lesson 7-2)*

50. $2(3 + x) = 14$
51. $63 = 9(2y - 3)$
52. $3(n - 1) = 1.5(n + 2)$

53. **ALGEBRA** Four times a number minus 6 is equal to the sum of 3 times the number and 2. Define a variable and write an equation to find the number. *(Lesson 7-1)*

State whether each sequence is *arithmetic, geometric,* or *neither*. Then write the next three terms of each sequence. *(Lesson 5-10)*

54. $-4, -1, 2, 5, \ldots$
55. $-1, 2, -4, 8, \ldots$
56. $1, 2, 4, 7, \ldots$

Getting Ready for the Next Lesson

PREREQUISITE SKILL Solve each equation.
*(To review **solving equations**, see Lesson 3-3.)*

57. $x + 19 = 32$
58. $a + 7 = -3$
59. $26 + c = 19$
60. $44 - c = 26$
61. $y - 9.7 = 10.1$
62. $r - 1.6 = -0.6$

7-4

Solving Inequalities by Adding or Subtracting

What You'll Learn

Sunshine State Standards
MA.A.3.3.3-1, MA.D.1.3.2-2, MA.D.2.3.1-1, MA.D.2.3.1-3

- Solve inequalities by using the Addition and Subtraction Properties of Inequality.

How is solving an inequality similar to solving an equation?

On the balance at the right, the paper bag may contain some blocks.

The blocks and bag on the scale model an inequality because the two sides are not equal.

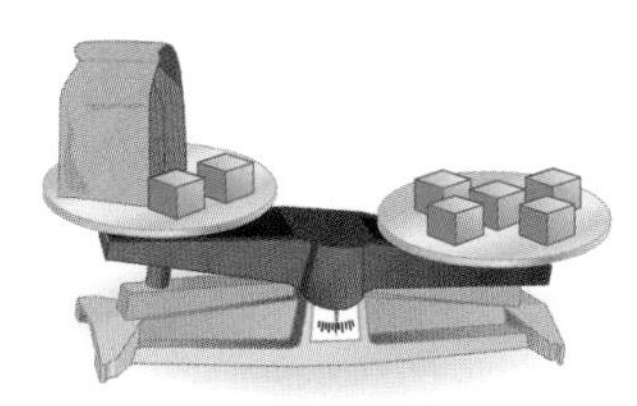

$x + 2 < 5$

The model shows the inequality $x + 2 < 5$. The side with the bag and 2 blocks weighs less than the side with 5 blocks.

a. How many blocks would be in the bag if the left side balanced the right side? (Assume that the paper bag weighs nothing.)

b. Explain how you determined your answer to part **a**.

c. What numbers of blocks can be in the bag to make the left side weigh *less than* the right side?

d. Write an inequality to represent your answer to part **c**.

SOLVE INEQUALITIES BY ADDING OR SUBTRACTING Solving an inequality means finding values for the variable that make the inequality true. In the example above, any number less than 3 is a solution. The solution is written as the inequality $x < 3$.

You can solve inequalities by using the Addition and Subtraction Properties of Inequalities.

Key Concept — *Addition and Subtraction Properties*

- **Words** When you add or subtract the same number from each side of an inequality, the inequality remains true.
- **Symbols** For all numbers a, b, and c,
  1. if $a > b$, then $a + c > b + c$ and $a - c > b - c$.
  2. if $a < b$, then $a + c < b + c$ and $a - c < b - c$.
- **Examples**

$$2 < 4 \qquad\qquad 6 > 3$$
$$2 + 3 < 4 + 3 \qquad\qquad 6 - 4 > 3 - 4$$
$$5 < 7 \qquad\qquad 2 > -1$$

These properties are also true for $a \geq b$ and $a \leq b$.

Study Tip

Inequalities
When you add or subtract any number from each side of an inequality, the inequality symbol remains the same.

Concept Check How are these properties similar to the Properties of Equality?

Example 1 Solve an Inequality Using Subtraction

Solve $x + 3 > 10$. Check your solution.

$x + 3 > 10$ Write the inequality.

$x + 3 - 3 > 10 - 3$ Subtract 3 from each side.

$x > 7$ Simplify.

To check your solution, try any number greater than 7.

CHECK $x + 3 > 10$ Write the inequality.

$8 + 3 \overset{?}{>} 10$ Replace x with 8.

$11 > 10$ ✓ This statement is true.

Any number greater than 7 will make the statement true. Therefore, the solution is $x > 7$.

Study Tip

Checking Solutions
Try a number less than 7 to show that it is *not* a solution.

Example 2 Solve an Inequality Using Addition

Solve $-6 \geq n - 5$. Check your solution.

$-6 \geq n - 5$ Write the inequality.

$-6 + 5 \geq n - 5 + 5$ Add 5 to each side.

$-1 \geq n$ Simplify.

CHECK You can check your result by replacing n in the original inequality with a number less than or equal to -1.

The solution is $-1 \geq n$ or $n \leq -1$.

Example 3 Graph Solutions of Inequalities

Solve $a + \frac{1}{2} < 2$. Graph the solution on a number line.

$a + \frac{1}{2} < 2$ Write the inequality.

$a + \frac{1}{2} - \frac{1}{2} < 2 - \frac{1}{2}$ Subtract $\frac{1}{2}$ from each side.

$a < \frac{4}{2} - \frac{1}{2}$ Rename 2 as a fraction with a denominator of 2.

$a < \frac{3}{2}$ or $1\frac{1}{2}$ Simplify.

The solution is $a < 1\frac{1}{2}$. Check your solution.

Graph the solution.

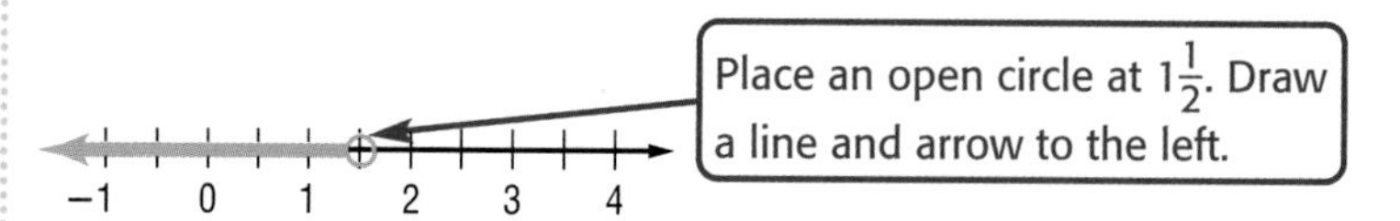
Place an open circle at $1\frac{1}{2}$. Draw a line and arrow to the left.
−1
0
1
2
3
4

Concept Check Does an inequality have only one solution? Explain.

Example 4 Use an Inequality to Solve a Problem

STATE FAIRS **Antonio has \$18 to ride go-carts and play games at the State Fair. If the go-carts cost \$5.50, what is the most he can spend on games?**

More About...

State Fairs

More than three million fairgoers are greeted each year by Big Tex, the symbol of the State Fair of Texas.

Source: www.bigtex.com

Explore We need to find the greatest amount of money Antonio can spend on games.

Plan Let x represent the amount Antonio can spend on games. Write an inequality to represent the problem. Recall that *at most* means *less than or equal to.*

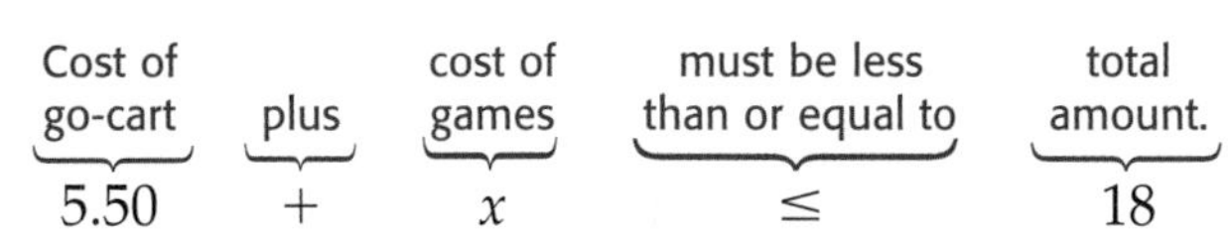

Cost of go-cart
plus
cost of games
must be less than or equal to
total amount.
5.50
+
x
≤
18

Solve

$5.5 + x \leq 18$	Write the inequality. ($5.50 = 5.5$)
$5.5 - 5.5 + x \leq 18 - 5.5$	Subtract 5.5 from each side.
$x \leq 12.5$	Simplify.

Examine Check by choosing an amount less than or equal to \$12.50, say, \$10. Then Antonio would spend \$5.50 + \$10 or \$15.50 in all. Since \$15.50 < \$18, the answer is reasonable.

So, the most Antonio can spend on games is \$12.50.

Check for Understanding

Concept Check

1. **Explain** when you would use addition and when you would use subtraction to solve an inequality.

2. **FIND THE ERROR** Dylan and Jada are using the statement *a minus three is greater than or equal to 15* to find values of *a*.

Dylan	Jada
$a - 3 \geq 15$	$a - 3 = 15$
$a - 3 + 3 \geq 15 + 3$	$a - 3 + 3 = 15 + 3$
$a \geq 18$	$a = 18$

Who is correct? Explain your reasoning.

3. **OPEN ENDED** Make up a problem whose solution is graphed below.

16 18 20 22 24

Guided Practice

Solve each inequality. Check your solution.

4. $x + 3 < 8$
5. $14 + y \geq 7$
6. $-13 \geq 9 + b$
7. $a - 5 > 6$
8. $c - (-2) \leq 3$
9. $-5 < t - 2$

Solve each inequality. Then graph the solution on a number line.

10. $h + 4 > 4$
11. $x - 6 \leq 4$

Application

12. **SAVINGS** Chris is saving money for a ski trip. He has \$62.50, but his goal is to save at least \$100. What is the least amount Chris needs to save to reach his goal?

Practice and Apply

Homework Help

For Exercises	See Examples
13–18, 25–28	1
19–24, 29, 30	2
31–42	3
43–46	4

Extra Practice

See page 740.

Solve each inequality. Check your solution.

13. $p + 7 < 9$
14. $t + 6 > -3$
15. $-14 \geq 8 + b$
16. $16 > -11 + k$
17. $3 \geq -2 + y$
18. $25 < n + (-12)$
19. $r - 5 \leq 2$
20. $a - 6 < 13$
21. $j - 8 \leq -12$
22. $-8 > h - 1$
23. $22 > w - (-16)$
24. $-30 \leq d + (-5)$
25. $1 + y \leq 2.4$
26. $2.9 < c + 7$
27. $f + (-4) \geq 1.4$
28. $z + (-2) > -3.8$
29. $b - \frac{3}{4} < 2\frac{1}{2}$
30. $g - 1\frac{2}{3} > 2\frac{1}{6}$

Solve each inequality. Then graph the solution on a number line.

31. $n + 4 < 9$
32. $t + 7 > 12$
33. $p + (-5) > -3$
34. $-3 + z > 2$
35. $-13 \geq x - 8$
36. $-32 \geq a + (-5)$
37. $33 \leq m - (-6)$
38. $k + 9 \geq -21$
39. $1\frac{1}{4} + b < 3$
40. $3 \leq \frac{1}{2} + a$
41. $4 \geq s - \frac{2}{3}$
42. $-\frac{3}{4} < w - 1$

43. **TRANSPORTATION** A certain minivan has a maximum carrying capacity of 1100 pounds. If the luggage weighs 120 pounds, what is the maximum weight allowable for passengers?

44. **BIOLOGY** Female killer whales usually weigh more than 3000 pounds and are up to 19 feet long. Suppose a female whale is 12 feet long. Write and solve an inequality to find how much longer the whale could grow.
Source: www.seaworld.org

Online Research **Data Update** Are there any whales in the world that are heavier or longer than the killer whale? Visit www.pre-alg.com/data_update to learn more.

Career Choices

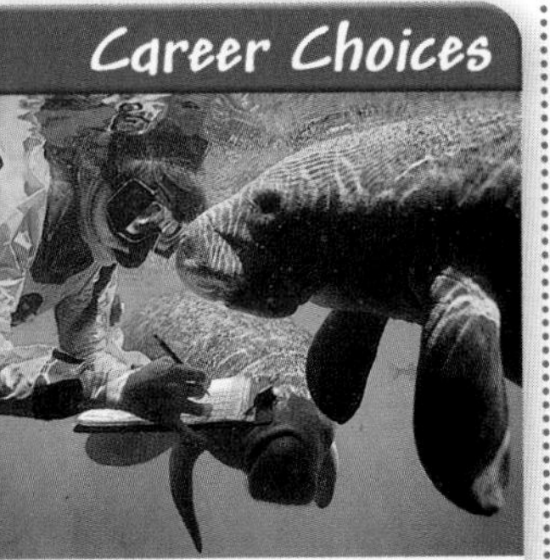

Marine Biologist

Marine biologists study the behavior, diseases, and life processes of salt-water organisms. Most marine biologists begin their studies with general biology.

Online Research For information about a career as a marine biologist, visit: www.pre-alg.com/careers

WEATHER For Exercises 45 and 46, use the diagram below.

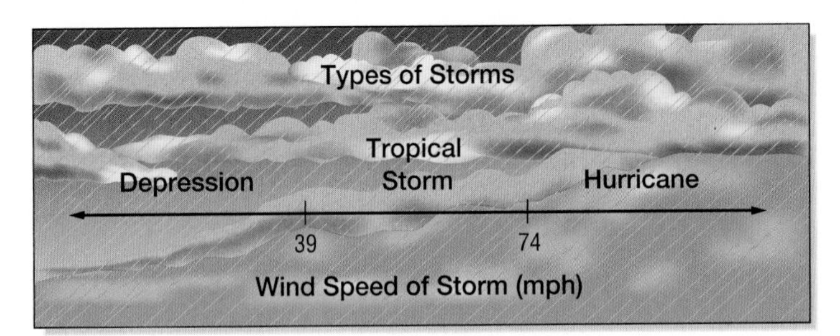

45. A hurricane has winds that are at least 74 miles per hour. Suppose a tropical storm has winds that are 42 miles per hour. Write and solve an inequality to find how much the winds must increase before the storm becomes a hurricane.

46. A *major storm* has wind speeds that are at least 110 miles per hour. Write and solve an inequality that describes how much greater these wind speeds are than the slowest hurricane.

47. **CRITICAL THINKING** Is it *always*, *sometimes*, or *never* true that $x - 1 < x$? Explain your answer.

48. **WRITING IN MATH** Answer the question that was posed at the beginning of the lesson.

How is solving an inequality similar to solving an equation?

Include the following in your answer:

- a description of what would happen if 3 blocks were removed from each side of the scale modeled at the right, and,
- a sentence that compares removing 3 blocks from each side of a scale and subtracting 3 from each side of an inequality.

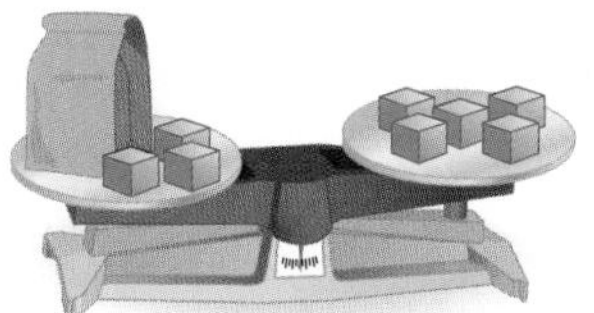

$x + 3 > 5$

49. Which inequality represents a temperature that is equal to or less than 42°?

(A) $t \geq 42$ (B) $t > 42$ (C) $t \leq 42$ (D) $t < 42$

50. Trevor has $25 to spend on a T-shirt and shorts for gym class. The shorts cost $14. Based on the inequality $14 + t \leq 25$, where t represents the cost of the T-shirt, what is the most Trevor can spend on the T-shirt?

(A) $9 (B) $10.99 (C) $11 (D) $11.50

Maintain Your Skills

Mixed Review

ALGEBRA For the given value, state whether each inequality is *true* or *false*. *(Lesson 7-3)*

51. $x - 5 > 4, x = 9$
52. $9 + a \leq 3, a = -7$
53. $\frac{x}{2} \geq 8, x = 4$
54. $6n < -4, n = -1$

55. **GEOMETRY** The perimeter of a rectangle is 24 centimeters. Find the dimensions if the length is 3 more than twice the width. *(Lesson 7-2)*

56. **GEOMETRY** Find the perimeter and area of the rectangle at the right. *(Lesson 3-7)*

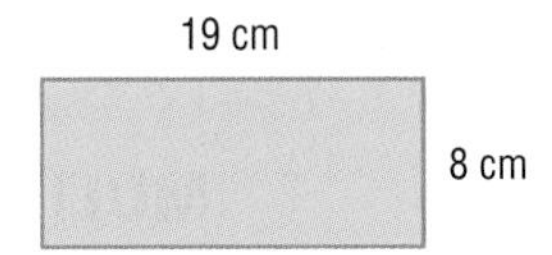

ALGEBRA Use the Distributive Property to rewrite each expression. *(Lesson 3-1)*

57. $4(2 + 8)$
58. $-2(n + 6)$
59. $5(x - 3.5)$
60. $(9 - d)(-3c)$

Find each difference. *(Lesson 2-3)*

61. $-15 - (-12)$
62. $8 - (-5)$
63. $-9 - 6$
64. $27 - 45$

Getting Ready for the Next Lesson

PREREQUISITE SKILL Solve each equation.
*(To review **solving equations**, see Lesson 3-4.)*

65. $-7x = 14$
66. $-3y = -27$
67. $5x = -20$
68. $\frac{d}{-3} = -6$
69. $\frac{c}{-4} = 12$
70. $\frac{a}{2} = -8$

7-5 Solving Inequalities by Multiplying or Dividing

What You'll Learn

Sunshine State Standards
MA.A.3.3.3-1, MA.D.1.3.2-2, MA.D.2.3.1-1, MA.D.2.3.1-3

- Solve inequalities by multiplying or dividing by a positive number.
- Solve inequalities by multiplying or dividing by a negative number.

How are inequalities used in studying space?

An astronaut in a space suit weighs about 300 pounds on Earth, but only 50 pounds on the moon because of weaker gravity.

$$\underbrace{300}_{\text{weight on Earth}} > \underbrace{50}_{\text{weight on moon}}$$

Location	Weight of Astronaut (lb)
Earth	300
Moon	50
Pluto	67
Mars	113
Neptune	407
Jupiter	796

If the astronaut and space suit each weighed half as much, would the inequality still be true? That is, would the astronaut's weight still be greater on Earth?

a. Divide each side of the inequality $300 > 50$ by 2. Is the inequality still true? Explain by using an inequality.

b. Would the weight of 5 astronauts be greater on Pluto or on Earth? Explain by using an inequality.

MULTIPLY OR DIVIDE BY A POSITIVE NUMBER The application above demonstrates how you can solve inequalities by using the Multiplication and Division Properties of Inequalities.

Key Concept — *Multiplication and Division Properties*

- **Words** When you multiply or divide each side of an inequality by the same positive number, the inequality remains true.
- **Symbols** For all numbers a, b, and c, where $c > 0$,
  1. if $a > b$, then $ac > bc$ and $\frac{a}{c} > \frac{b}{c}$.
  2. if $a < b$, then $ac < bc$ and $\frac{a}{c} < \frac{b}{c}$.
- **Examples**

$2 < 6$	$3 > -9$
$4(2) < 4(6)$	$\frac{3}{3} > \frac{-9}{3}$
$8 < 24$	$1 > -3$

These properties are also true for $a \geq b$ and $a \leq b$.

Study Tip

Positive Number
The inequality $c > 0$ means that c is a positive number.

Example 1 Multiply or Divide by a Positive Number

a. Solve $8x \leq 40$. Check your solution.

$8x \leq 40$ Write the inequality.

$\frac{8x}{8} \leq \frac{40}{8}$ Divide each side by 8.

$x \leq 5$ Simplify.

The solution is $x \leq 5$. You can check this solution by substituting 5 or a number less than 5 into the inequality.

b. Solve $\frac{d}{2} > 7$. Check your solution.

$\frac{d}{2} > 7$ Write the inequality.

$2\left(\frac{d}{2}\right) > 2(7)$ Multiply each side by 2.

$d > 14$ Simplify.

The solution is $d > 14$. You can check this solution by substituting a number greater than 14 into the inequality.

Example 2 Write an Inequality

Multiple-Choice Test Item

Ling earns \$8 per hour in the summer working at the zoo. Which inequality can be used to find how many hours he must work in a week to earn at least \$120?

Ⓐ $8x < 120$ Ⓑ $8x \leq 120$ Ⓒ $8x > 120$ Ⓓ $8x \geq 120$

The Princeton Review

Test-Taking Tip

Before taking a standardized test, review the meanings of phrases like *at least* and *at most*.

Read the Test Item

You are to write an inequality to represent a real-world problem.

Solve the Test Item

Let x represent the number of hours worked.

Amount earned per hour	times	number of hours	is at least	amount earned each week.
8	$\cdot$	x	$\geq$	120

The answer is D.

MULTIPLY OR DIVIDE BY A NEGATIVE NUMBER What happens when each side of an inequality is multiplied or divided by a negative number?

$-6 < 11$		$10 > 5$	
$-1(-6) \overset{?}{<} -1(11)$	Multiply each side by -1.	$\frac{10}{-5} \overset{?}{>} \frac{5}{-5}$	Divide each side by -5.
$6 \overset{?}{<} -11$	This inequality is false.	$-2 \overset{?}{>} -1$	This inequality is false.

The inequalities $6 < -11$ and $-2 > -1$ are both false. However, they would both be true if the inequality symbols were reversed. That is, change $<$ to $>$ and change $>$ to $<$.

$6 > -11$ → true $-2 < -1$ → true

This investigation suggests the following properties.

Key Concept — Multiplication and Division Properties

- **Words** When you multiply or divide each side of an inequality by the same negative number, the inequality symbol must be reversed for the inequality to remain true.
- **Symbols** For all numbers a, b, and c, where $c < 0$,
  1. if $a > b$, then $ac < bc$ and $\frac{a}{c} < \frac{b}{c}$.
  2. if $a < b$, then $ac > bc$ and $\frac{a}{c} > \frac{b}{c}$.
- **Examples**

$7 > 1$		$-4 < 16$
$-2(7) < -2(1)$	Reverse the symbols.	$\frac{-4}{-4} > \frac{16}{-4}$
$-14 < -2$		$1 > -4$

These properties are also true for $a \geq b$ and $a \leq b$.

Study Tip

Negative Number The inequality $c < 0$ means that c is a negative number.

Concept Check Explain why it is necessary to reverse the symbol when you multiply each side of an inequality by a negative number.

Example 3 Multiply or Divide by a Negative Number

Solve each inequality and check your solution. Then graph the solution on a number line.

a. $\frac{x}{-3} \leq 4$

$\frac{x}{-3} \leq 4$ Write the inequality.

$-3\left(\frac{x}{-3}\right) \geq -3(4)$ Multiply each side by -3 and reverse the symbol.

$x \geq -12$ Check this result.

CHECK You can check your result by replacing x in the original inequality with a number greater than -12.

Graph the solution, $x \geq -12$.

−16 −14 −12 −10 −8 −6

b. $-7x > -56$

$-7x > -56$ Write the inequality.

$\frac{-7x}{-7} < \frac{-56}{-7}$ Divide each side by -7 and reverse the symbol.

$x < 8$ Check this result.

Graph the solution, $x < 8$.

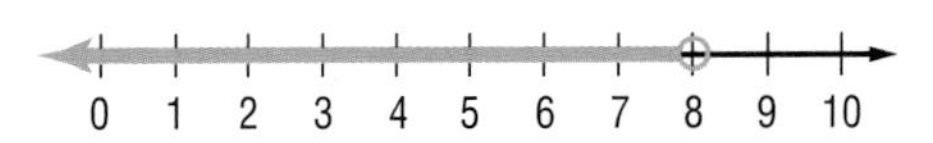

Check for Understanding

Concept Check

1. **List** the steps you would use to solve $\frac{y}{-12} < 6$.
2. **OPEN ENDED** Write an inequality that can be solved using the Division Property of Inequality, where the inequality symbol is *not* reversed.
3. **FIND THE ERROR** Brittany and Tamika each solved $-45 \geq 9k$.

Brittany	Tamika
$-45 \geq 9k$	$-45 \geq 9k$
$\frac{-45}{9} \leq \frac{9k}{9}$	$\frac{-45}{9} \geq \frac{9k}{9}$
$-5 \leq k$	$-5 \geq k$

Who is correct? Explain your reasoning.

Guided Practice

Solve each inequality and check your solution. Then graph the solution on a number line.

4. $2x < 8$
5. $3x \geq -6$
6. $-4t > -20$
7. $\frac{a}{5} > 10$
8. $-8 > \frac{k}{-0.4}$
9. $\frac{m}{-7} \leq 1.2$
10. $-\frac{s}{3} \leq -3.5$
11. $36 \geq -\frac{1}{2}y$
12. $-273 \geq -13z$

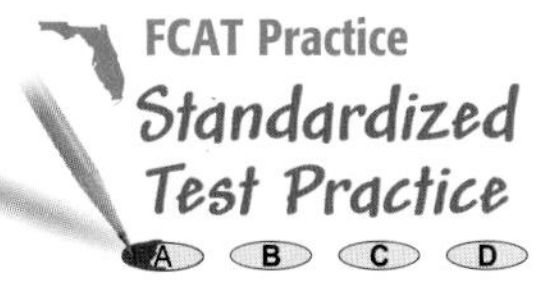
FCAT Practice
Standardized Test Practice

13. **EARNINGS** Julia delivers pizzas on weekends. Her average tip is $1.50 for each pizza that she delivers. How many pizzas must she deliver to earn at least $20 in tips?

Ⓐ 10 Ⓑ 13 Ⓒ 14 Ⓓ 20

Practice and Apply

Homework Help

For Exercises	See Examples
14–21, 32, 33	1
22–31, 34–37	3
42, 43	2

Extra Practice
See page 741.

Solve each inequality and check your solution. Then graph the solution on a number line.

14. $4x < 4$
15. $7y > 63$
16. $13a \geq -26$
17. $-15 \leq 5b$
18. $144 < 12d$
19. $15 \geq 3t$
20. $\frac{p}{6} > 5$
21. $7 \geq \frac{h}{14}$
22. $-3m > -33$
23. $-8z \leq -24$
24. $18 > -2g$
25. $-8 \leq -4w$
26. $6 > \frac{x}{-7}$
27. $\frac{r}{-2} < -2$
28. $\frac{y}{-3} < -7$
29. $\frac{k}{-2} < 9$
30. $-6a > -78$
31. $-25t \leq 400$
32. $\frac{y}{4} \geq 2.4$
33. $\frac{n}{5} \leq 0.8$
34. $-5 \leq \frac{c}{-4.5}$
35. $-19 > \frac{y}{-0.3}$
36. $-\frac{1}{3}x \geq -9$
37. $-36 < -\frac{1}{2}b$

38. **SOCCER** Tomás wants to spend less than $100 for a new soccer ball and shoes. The ball costs $24.
    a. Write an inequality to represent the amount left for shoes.
    b. What amount can he spend on shoes?

More About. . .

Swimming

Olympic athletes can swim 200 meters in less than a minute.

Source: www.infoplease.com

39. **SWIMMING** Nicole swims 40 meters per minute, and she wants to swim at least 2000 meters this morning.
    a. Write an inequality to represent how long she should swim.
    b. How many minutes should she swim?

40. **CRITICAL THINKING** The product of an integer and -7 is less than -84. Find the least integer that meets this condition.

41. **WRITING IN MATH** Answer the following question that was posed at the beginning of the lesson.

    How are inequalities used in studying space?

    Include the following in your answer:
    - inequalities comparing the weight of two astronauts on Mars and on the moon, and
    - an explanation of how the Multiplication and Division Properties of Inequality can be used to compare planets' gravities.

42. Which number is *not* a possible length of the rectangle if the area is less than 36 square inches?

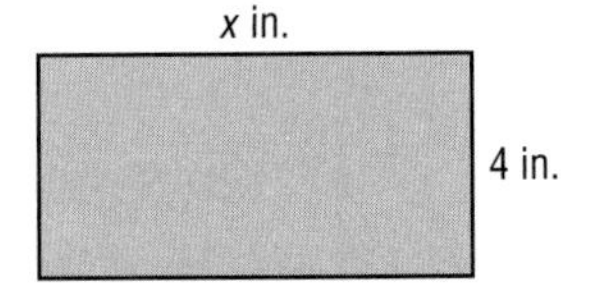

    (A) 6 (B) 7
    (C) 8 (D) 9

43. **GRID IN** Jessica is putting water into a 20-gallon fish tank using a 2-quart pitcher. How many pitchers of water will she need to fill the tank?

Maintain Your Skills

Mixed Review

ALGEBRA Solve each inequality. Check your solution. *(Lesson 7-4)*

44. $-4 + x > 23$
45. $c + 18 \leq -2$
46. $6 > n - 10$

47. Write an inequality for *2 times a number is at most 14.* *(Lesson 7-3)*

Find each product. Write in simplest form. *(Lesson 5-3)*

48. $\frac{1}{8} \cdot \frac{3}{4}$
49. $-\frac{3}{7} \cdot \frac{5}{9}$
50. $2\frac{1}{2} \cdot \left(-\frac{5}{6}\right)$
51. $\frac{ab}{2} \cdot \frac{4}{bc}$

Getting Ready for the Next Lesson

PREREQUISITE SKILL ALGEBRA Solve each equation.
*(To review **two-step equations**, see Lesson 3-5.)*

52. $2x + 3 = 9$
53. $5a - 6 = 14$
54. $3n - 8 = -26$
55. $\frac{t}{3} + 5 = 2$
56. $\frac{c}{4} - 1 = 4$
57. $\frac{d}{2} + 3 = 19$

Practice Quiz 2 — Lessons 7-3 through 7-5

Graph each inequality on a number line. *(Lesson 7-3)*

1. $x < -3$
2. $y \geq 5$

Solve each inequality. Check your solution. *(Lessons 7-4 and 7-5)*

3. $a - 26 \leq 14$
4. $46 + k > -8$
5. $115 \leq -9 + n$
6. $2.5 > 5r$
7. $\frac{r}{5} < -45$
8. $-\frac{s}{8} < -80$
9. $-12g \geq -84$
10. $5w \geq -2$

7-6 Solving Multi-Step Inequalities

Sunshine State Standards
MA.A.3.3.3-1, MA.D.1.3.2-2, MA.D.2.3.1-1, MA.D.2.3.1-3

What You'll Learn

- Solve inequalities that involve more than one operation.

How are multi-step inequalities used in backpacking?

Nearly 10 million Americans go backpacking each year. According to a fitness magazine, to avoid injury, three times the weight of your backpack and its contents should be less than your body weight.

a. Write an inequality that represents the relationship between body weight and a safe total backpack and contents weight.

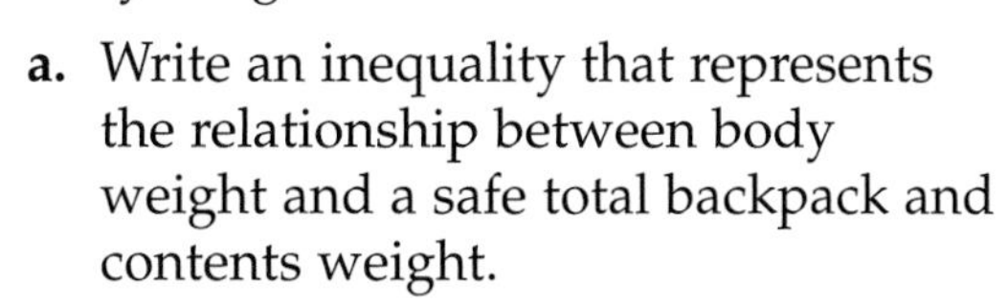

b. Suppose you weigh 120 pounds and your empty backpack weighs 5 pounds. Write an inequality that represents the maximum weight you can safely carry in the backpack.

INEQUALITIES WITH MORE THAN ONE OPERATION Some inequalities involve more than one operation. To solve the inequality, work backward to undo the operations, just as you did in solving multi-step equations.

Example 1 Solve a Two-Step Inequality

Solve $6x + 15 > 9$ and check your solution. Graph the solution on a number line.

$6x + 15 > 9$	Write the inequality.
$6x + 15 - 15 > 9 - 15$	Subtract 15 from each side.
$6x > -6$	Simplify.
$x > -1$	Mentally divide each side by 6.

To check your solution, try 0, a number greater than -1.

CHECK	$6x + 15 > 9$	Write the inequality.
	$6(0) + 15 > 9$	Replace x with 0.
	$0 + 15 > 9$	Simplify.
	$15 > 9$ ✓	The solution checks.

Graph the solution, $x > -1$.

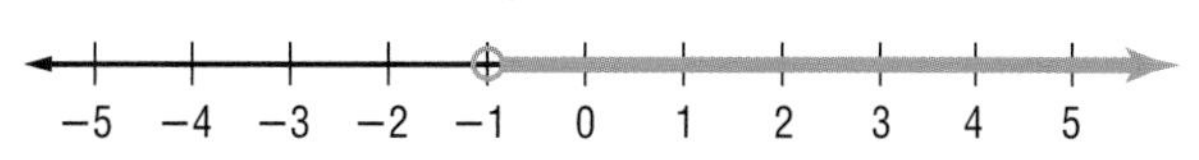

Study Tip

Common Misconception
Do not reverse the inequality sign just because there is a negative sign in the inequality. Only reverse the sign when you multiply or divide by a negative number.

Remember that you must reverse the inequality symbol if you multiply or divide each side of an inequality by a negative number.

Example 2 Reverse the Inequality Symbol

Solve $10 - 3a \leq 25 + 2a$ and check your solution. Graph the solution on a number line.

$10 - 3a \leq 25 + 2a$	Write the inequality.
$10 - 3a - 2a \leq 25 + 2a - 2a$	Subtract $2a$ from each side.
$10 - 5a \leq 25$	Simplify.
$10 - 10 - 5a \leq 25 - 10$	Subtract 10 from each side.
$-5a \leq 15$	Simplify.
$\frac{-5a}{-5} \geq \frac{15}{-5}$	Divide each side by −5 and change ≤ to ≥.
$a \geq -3$	Simplify.

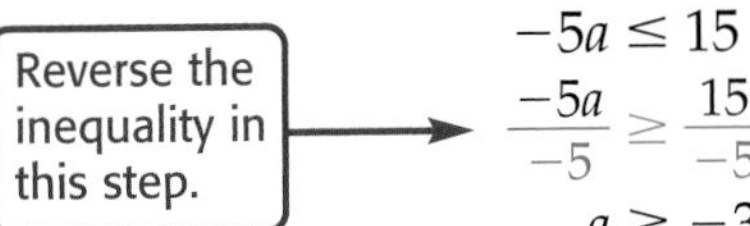
Reverse the inequality in this step.

CHECK

$10 - 3a \leq 25 + 2a$	Try −2, a number greater than −3.
$10 - 3(-2) \overset{?}{\leq} 25 + 2(-2)$	Replace a with −2.
$10 + 6 \overset{?}{\leq} 25 - 4$	Simplify.
$16 \leq 21$ ✓	The solution checks.

Graph the solution, $a \geq -3$.

−5 −4 −3 −2 −1 0 1 2 3 4 5

More About. . .

Backpacking

One of the most popular treks for backpackers is Yosemite National Park in California's High Sierra. It is possible to hike there for two weeks and never see a road!

Source: www.gorp.com

When inequalities contain grouping symbols, you can use the Distributive Property to begin simplifying the inequality.

Example 3 Inequalities with Grouping Symbols

BACKPACKING A person weighing 126 pounds has a 6-pound backpack. Refer to the application at the beginning of page 355. What is the maximum weight for the contents of the pack?

Let c represent the weight of the contents of the pack.

Words	3	times	weight of pack and contents	should be less than	body weight.
Inequality	3	·	$(6 + c)$	$<$	126

Solve the inequality.

$3(6 + c) < 126$	Write the inequality.
$18 + 3c < 126$	Use the Distributive Property.
$18 + 3c - 18 < 126 - 18$	Subtract 18 from each side.
$3c < 108$	Simplify.
$\frac{3c}{3} < \frac{108}{3}$	Divide each side by 3.
$c < 36$	Simplify.

The weight of the contents should be less than 36 pounds.

Concept Check How do you solve an inequality with more than one operation?

Check for Understanding

Concept Check

1. **OPEN ENDED** Explain how to check the solution of an inequality.
2. **Write** an inequality for the model at the right. Then solve the inequality.

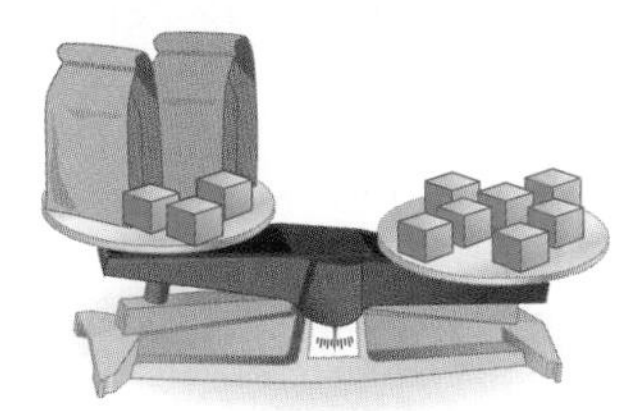

3. **FIND THE ERROR** Jerome and Ryan are beginning to solve $2(2y + 3) > y + 1$.

Jerome	Ryan
$2(2y+3) > y+1$	$2(2y+3) > y+1$
$4y+6 > y+1$	$4y+3 > y+1$

Who is correct? Explain your reasoning.

Guided Practice Solve each inequality and check your solution. Then graph the solution on a number line.

4. $3x + 4 \leq 31$
5. $2n + 5 > 11 - n$
6. $y + 1 \geq 4y + 4$
7. $16 - 2c < 14$
8. $-6.1n \geq 3.9n + 5$
9. $-4 \leq \frac{x}{4} - 6$
10. $-3(b - 1) > 18$
11. $\frac{1}{2}(2d + 3) < -8$

Application

12. **MONEY** Dante's telephone company charges \$10 a month plus \$0.05 for every minute or part of a minute. Dante wants his monthly bill to be under \$30. What is the greatest number of minutes he can talk?

Practice and Apply

Homework Help

For Exercises	See Examples
13–18, 25, 26	1
19, 20, 27, 28	2
21–24, 29, 30, 33–36	3

Extra Practice
See page 741.

Solve each inequality and check your solution. Then graph the solution on a number line.

13. $2x + 8 > 24$
14. $3y - 1 \leq 5$
15. $3 + 4c > -13$
16. $9 + 2p \leq 15$
17. $3x - 2 > 10 - x$
18. $c - 1 < 3c + 5$
19. $4 - 3k \leq 19$
20. $16 - 4n > 20$
21. $2(n + 3) < -4$
22. $2(d + 1) > 16$
23. $8 + 3b \leq 2(9 - b)$
24. $\frac{m}{2} + 9 \geq 5$
25. $2 + 0.3y \geq 11$
26. $0.5a - 1.4 \leq 2.1$
27. $\frac{1}{2}(6 - c) > 5$
28. $\frac{2}{3}(9 - x) < 3$

MORE PTS PLZ

29. Four times a number less 6 is greater than two times the same number plus 8. For what number or numbers is this true?
30. One-half of the sum of a number and 6 is less than 25. What is the number?

Solve each inequality and check your solution. Graph the solution on a number line.

31. $1.3n + 6.7 \geq 3.1n - 1.4$

32. $-5a + 3 > 3a + 23$

33. $-5(t + 4) \geq 3(t - 4)$

34. $8x - (x - 5) > x + 17$

35. $\frac{c + 8}{4} < \frac{5 - c}{9}$

36. $\frac{2(n + 1)}{7} \geq \frac{n + 4}{5}$

For Exercises 37–40, write and solve an inequality.

37. CANDY You buy some candy bars at \$0.55 each and one newspaper for \$0.35. How many candy bars can you buy with \$2?

38. SCHOOL Nate has scores of 85, 91, 89, and 93 on four tests. What is the least number of points he can get on the fifth test to have an average of at least 90?

39. SALES You earn \$2.00 for every magazine subscription you sell plus a salary of \$10 each week. How many subscriptions do you need to sell each week to earn at least \$40 each week?

40. REAL ESTATE A real estate agent receives a monthly salary of \$1500 plus a 4% commission on every home sold. For what amount of monthly sales will the agent earn at least \$5000?

Career Choices

Real Estate Agent

Real estate agents help people with one of the most important financial decisions of their lives—buying and selling a home. All states require prospective agents to pass a written test, which usually contains a section on mathematics.

Online Research
For information about a career as a real estate agent, visit: www.pre-alg.com/careers

41. CAR RENTAL The costs for renting a car from Able Car Rental and from Baker Car Rental are shown in the table. For what mileage does Baker have the better deal? Use the inequality $30 + 0.05x > 20 + 0.10x$. Explain why this inequality works.

Rental Car Costs

	Cost per Day	Cost per Mile
Able Car Rental	\$30	\$0.05
Baker Car Rental	\$20	\$0.10

42. HIKING You hike along the Appalachian Trail at 3 miles per hour. You stop for one hour for lunch. You want to walk at least 18 miles. How many hours should you expect to spend on the trail?

43. PHONE SERVICES Miko was asked by FoneCom to sign up for their service at \$15 per month plus \$0.10 per minute. Miko currently has BestPhone service at \$20 per month plus \$0.05 per minute. Miko figures that her monthly bill would be more with FoneCom. For how many minutes per month does she use the phone?

44. FUND-RAISERS The Booster Club sells football programs for \$1 each. The costs to make the programs are \$60 for page layout plus \$0.20 for printing each program. If they print 400 programs, how many programs must the Club sell to make at least \$200 profit?

45. CRITICAL THINKING Assume that k is an integer. Solve the inequality $10 - 2|k| > 4$.

46. **WRITING IN MATH** Answer the question that was posed at the beginning of the lesson.

How are multi-step inequalities used in backpacking?

Include the following in your answer:

- an explanation of what multi-step inequalities are, and
- a solution of the inequality you wrote for part **b** on page 355.

47. Which inequality represents *five more than twice a number is less than ten*?

Ⓐ $(5 + 2)n < 10$ Ⓑ $2n - 5 < 10$

Ⓒ $10 < 2n + 5$ Ⓓ $5 + 2n < 10$

48. Enola's scores on the first five science tests are shown in the table. Which inequality represents the score she must receive on the sixth test to have an average score of more than 88?

Test	Score
1	85
2	84
3	90
4	95
5	88

Ⓐ $s \geq 86$ Ⓑ $s \leq 88$

Ⓒ $s < 88$ Ⓓ $s > 86$

Extending the Lesson

49. The sum of three times a number and 5 lies between –10 and 8. Solve the *compound inequality* $-10 < 3x + 5 < 8$ to find the solution(s). (*Hint*: Any operation must be done to all three parts of the inequality.)

Maintain Your Skills

Mixed Review **ALGEBRA** **Solve each inequality. Check your solution.** *(Lessons 7-4 and 7-5)*

50. $20 < -9 + k$

51. $22 \leq -15 + y$

52. $6x < -27$

53. $-5n \geq -25$

54. $\frac{n}{-4} \leq -11$

55. $\frac{a}{-3} > 6.2$

56. If 12 of the 20 students in a class are boys, what percent are boys? *(Lesson 6-5)*

57. Write $\frac{1}{200}$ as a percent. *(Lesson 6-4)*

Express each ratio as a unit rate. *(Lesson 6-1)*

58. $5 for 2 loaves of bread

59. 200 miles on 12 gallons

60. 24 meters in 4 seconds

61. 9 monthly issues for $11.25

GEOMETRY **Find the missing dimension in each rectangle.** *(Lesson 3-7)*

62.

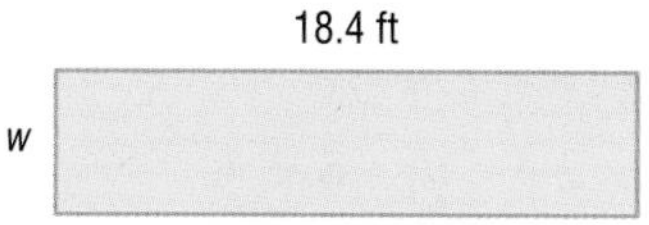

Perimeter = 49.6 ft

63.

ℓ

5.1 m

Area = 30.6 m^2

Chapter 7

Study Guide and Review

Vocabulary and Concept Check

identity (p. 336) inequality (p. 340) null or empty set (p. 336)

Determine whether each statement is *true* or *false*. If false, replace the underlined word or number to make a true statement.

1. When an equation has no solution, the solution set is the null set.
2. The inequality $n + 8 - 8 \geq 14 - 8$ demonstrates the Subtraction Property of Inequality.
3. An equation that is true for every value of the variable is called an inequality.
4. The inequality $\frac{x}{4}(4) < 7(4)$ demonstrates the Division Property of Inequality.
5. A mathematical sentence that contains < or > is called an empty set.
6. When the final result in solving an equation is $5 = -8$, the solution set is the null set.
7. When the final result in solving an equation is $x = x$, the solution set is all numbers.
8. To solve $3(x + 5) = 10$, use the Distributive Property to remove the parentheses.
9. The symbol $\geq$ means is less than or equal to.
10. A closed circle on a number line indicates that the point is included in the solution set for the inequality.

Lesson-by-Lesson Review

7-1 Solving Equations with Variables on Each Side

See pages 330–333.

Concept Summary

- Use the Addition or Subtraction Property of Equality to isolate the variables on one side of an equation.

Example Solve $7x = 3x - 12$.

$7x = 3x - 12$	Write the equation.
$7x - 3x = 3x - 3x - 12$	Subtract $3x$ from each side.
$4x = -12$	Simplify.
$x = -3$	Mentally divide each side by 4.

Exercises **Solve each equation. Check your solution.**

See Example 1 on page 330.

11. $2a + 9 = 5a$
12. $x - 4 = 3x$
13. $3y - 8 = y$
14. $19t = 26 + 6t$
15. $2 + 7n = 8 + n$
16. $5 + 6t = 10t - 7$
17. $-r + 4.2 = 8.8r + 14$
18. $12 + 1.5x = 9x$
19. $5b - 1 = 2.5b - 4$

www.pre-alg.com/vocabulary_review

7-2 Solving Equations with Grouping Symbols

See pages 334–338.

Concept Summary

- Use the Distributive Property to remove the grouping symbols.

Example **Solve $2(x + 3) = 15$.**

$2(x + 3) = 15$ Write the equation.
$2x + 6 = 15$ Use the Distributive Property.
$2x = 9$ Subtract 6 from each side and simplify.
$x = 4.5$ Divide each side by 2 and simplify.

Exercises **Solve each equation.** *See Examples 1–4 on pages 334–336.*

20. $4(k + 1) = 16$ **21.** $2(n - 5) = 8$ **22.** $11 + 2q = 2(q + 4)$

23. $\frac{1}{2}(t + 8) = \frac{3}{4}t$ **24.** $4(x + 2.5) = 3(7 + x)$ **25.** $3(x + 1) - 5 = 3x - 2$

7-3 Inequalities

See pages 340–344.

Concept Summary

- An inequality is a mathematical sentence that contains $<$, $>$, $\leq$, or $\geq$.

Example **State whether $n + 11 < 14$ is *true* or *false* for $n = 5$.**

$n + 11 < 14$ Write the inequality.
$5 + 11 \overset{?}{<} 14$ Replace n with 5.
$16 \not< 14$ Simplify. The sentence is false.

Exercises **For the given value, state whether each inequality is *true* or *false*.** *See Example 4 on page 341.*

26. $x + 4 > 9, x = 12$ **27.** $15 \leq 5n, n = 3$ **28.** $3n + 1 \geq 14, n = 4$

7-4 Solving Inequalities by Adding or Subtracting

See pages 345–349.

Concept Summary

- Solving an inequality means finding values for the variable that make the inequality true.

Example **Solve $x - 7 < 3$. Graph the solution on a number line.**

$x - 7 < 3$ Write the inequality.
$x - 7 + 7 < 3 + 7$ Add 7 to each side.
$x < 10$ Simplify.

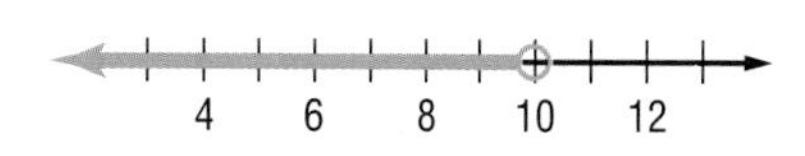

Exercises **Solve each inequality. Graph the solution on a number line.** *See Examples 1 and 2 on page 346.*

29. $b - 9 \geq 8$ **30.** $x + 4.8 \leq 2$ **31.** $t + \frac{1}{2} < 4$

Chapter 7 For More ...
- Extra Practice, see pages 739–741.
- Mixed Problem Solving, see page 764.

7-5 Solving Inequalities by Multiplying or Dividing

See pages 350–354.

Concept Summary

- When you multiply or divide each side of an inequality by a positive number, the inequality symbol remains the same.
- When you multiply or divide each side of an inequality by a negative number, the inequality symbol must be reversed.

Examples

1 Solve $\frac{a}{3} > 2$. Graph the solution on a number line.

$\frac{a}{3} > 2$ Write the inequality.

$3\left(\frac{a}{3}\right) > 3(2)$ Multiply each side by 3.

$a > 6$ Simplify.

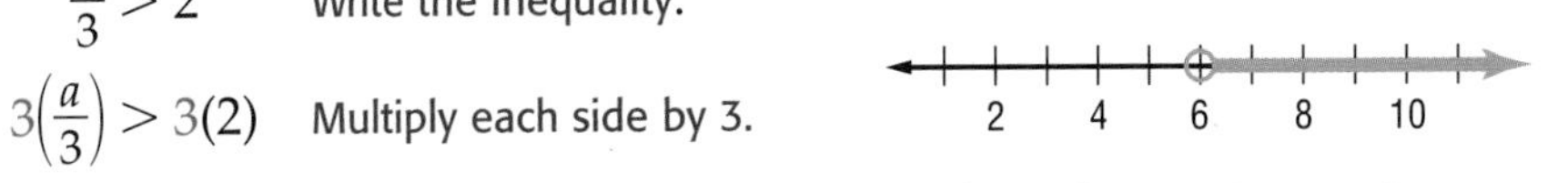

The solution is $a > 6$.

2 Solve $-2n \geq 26$. Graph the solution on a number line.

$-2n \geq 26$ Write the inequality.

$\frac{-2n}{-2} \leq \frac{26}{-2}$ Divide each side by -2 and reverse the symbol.

$n \leq -13$ Simplify.

−18 −16 −14 −12 −10

The solution is $n \leq -13$.

Exercises **Solve each inequality. Graph the solution on a number line.**
See Examples 1 and 3 on pages 351 and 352.

32. $\frac{n}{4} < 6$ **33.** $\frac{k}{1.7} \leq 3$ **34.** $0.5x > 3.2$

35. $-56 \geq 8y$ **36.** $9 > \frac{x}{-4}$ **37.** $-\frac{5}{6}a \leq 2$

7-6 Solving Multi-Step Inequalities

See pages 355–359.

Concept Summary

- To solve an inequality that involves more than one operation, work backward to undo the operations.

Example

Solve $4t + 7 < -5$.

$4t + 7 < -5$ Write the inequality.

$4t + 7 - 7 < -5 - 7$ Subtract 7 from each side.

$4t < -12$ Simplify.

$t < -3$ Mentally divide each side by 4.

The solution is $t < -3$.

Exercises **Solve each inequality.** *See Examples 1 and 2 on pages 355 and 356.*

38. $2x - 3 > 19$ **39.** $5n + 4 \leq 24$ **40.** $6 \geq \frac{r}{7} + 1$

41. $\frac{t}{-2} + 15 < 21$ **42.** $3(a + 8.4) > 30$ **43.** $\frac{1}{4} + 2b < 13 + 5b$

Chapter 7 Practice Test

Vocabulary and Concepts

1. **State** when to use an open circle and a closed circle in graphing an inequality.
2. **Describe** what happens to an inequality when each side is multiplied or divided by a negative number.

Skills and Applications

Solve each equation. Check your solution.

3. $7x - 3 = 10x$
4. $p - 9 = 4p$
5. $2.3n - 8 = 1.2n + 3$
6. $\frac{3}{8}y - 5 = \frac{5}{8}y - 3$
7. $6 + 2(x - 4) = 2(x - 1)$
8. $2(6 - 5d) = 8$
9. $8(2x - 9) = 4(5 + 4x)$
10. $4(a + 3) = 20$
11. $\frac{1}{3}(9b + 1) = b - 1$

Define a variable and write an equation to find each number. Then solve.

12. Eight more than three times a number equals four less than the number.
13. The product of a number and five is twelve more than the number.

14. **GEOMETRY** The perimeter of the rectangle is 22 feet. Find the dimensions of the rectangle.

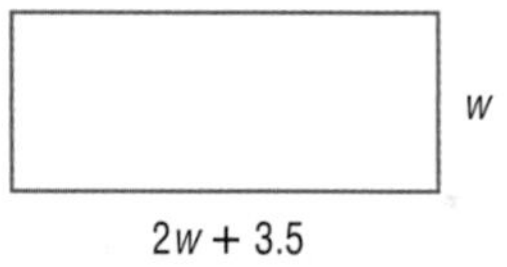

15. **SHOPPING** The cost of purchasing four shirts is at least $120. Write an inequality to describe this situation.

Write the inequality for each graph.

16. 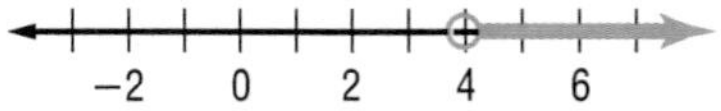

17. 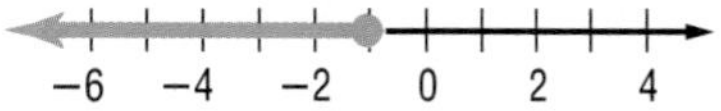

Solve each inequality and check your solution. Then graph the solution on a number line.

18. $-4 \geq p - 2$
19. $3x \geq 15$
20. $-42 < -0.6x$
21. $c - 3 \leq 4c + 9$
22. $7(3 - 2b) > 5b + 2$
23. $\frac{1}{2}(a + 4) > \frac{1}{4}(a - 8)$

24. **SALES** The Cookie Factory has a fixed cost of $300 per month plus $0.45 for each cookie sold. Each cookie sells for $0.95. How many cookies must be sold during one month for the profit to be at least $100?

FCAT Practice

25. **STANDARDIZED TEST PRACTICE** Danny earns $6.50 per hour working at a movie theater. Which inequality can be used to find how many hours he must work each week to earn at least $100 a week?

   (A) $6.50h < 100$
   (B) $6.50h > 100$
   (C) $6.50h \leq 100$
   (D) $6.50h \geq 100$

www.pre-alg.com/chapter_test/fcat

Chapter 7

Standardized Test Practice

FCAT Practice

Part 1 Multiple Choice

Record your answers on the answer sheet provided by your teacher or on a sheet of paper.

1. A delivery service calculates the cost c of shipping a package with the equation $c = 0.30w + 6$, where w is the weight of the package in pounds. Your package weighs at least 8 pounds. What is the lowest possible cost to ship your package? (Lesson 3-7)

Ⓐ \$6.30 Ⓑ \$8.40

Ⓒ \$14.30 Ⓓ \$30.00

2. The school band traveled on two buses with 36 students on each bus. At a lunch stop, two-thirds of the students on the first bus ate at Hamburger Haven, and the others ate at Taco Time. Three-fourths of the students on the second bus ate at Hamburger Haven. How many students in all ate at Hamburger Haven? (Lesson 5-3)

Ⓐ 24 Ⓑ 48

Ⓒ 51 Ⓓ 102

3. Shanté earned \$360 last summer. She spent $\frac{5}{9}$ of her earnings. How much money did she have left? (Lesson 5-3)

Ⓐ \$40 Ⓑ \$160

Ⓒ \$200 Ⓓ \$320

4. While exercising, Luke's heart is beating at 170 beats per minute. If he maintains this rate, about how many times will his heart beat in one hour? (Lesson 6-1)

Ⓐ 1000 Ⓑ 5000

Ⓒ 10,000 Ⓓ 100,000

The Princeton Review **Test-Taking Tip**

Questions 9 and 10 When an item requires you to solve an equation or inequality, plug in your solution to the original problem in order to check your answer.

5. Which of the circles has approximately the same fractional part shaded as that of the rectangle below? (Lesson 6-4)

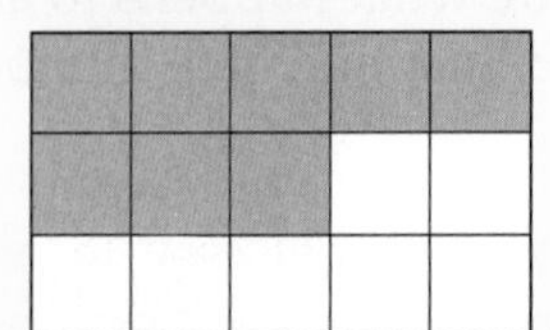

Ⓐ Ⓑ

Ⓒ Ⓓ

6. Which of the following statements is true? (Lesson 6-4)

Ⓐ $0.4 > 40\%$ Ⓑ $0.04 = 40\%$

Ⓒ $40\% \leq 0.04$ Ⓓ $40\% > 0.04$

7. A survey at the MegaMall showed that 15% of visitors attend a movie while at the mall. If 8700 people are at the mall, how many of these visitors are likely to attend a movie there? (Lesson 6-7)

Ⓐ 580 Ⓑ 870

Ⓒ 1305 Ⓓ 5800

8. Last year there were 1536 students at Cortéz Middle School. This year there are 5% more students. *About* how many students attend Cortéz this year? (Lesson 6-9)

Ⓐ 1550 Ⓑ 1600

Ⓒ 1650 Ⓓ 1700

9. If $5(x + 2) = 40$, what is the value of x? (Lesson 7-2)

Ⓐ 4 Ⓑ 6

Ⓒ 8 Ⓓ 10

10. Which of the following inequalities is equivalent to $\frac{x}{3} < 5$? (Lesson 7-5)

Ⓐ $x < \frac{5}{3}$ Ⓑ $x < 2$

Ⓒ $x > 2$ Ⓓ $x < 15$

Part 2 Short Response/Grid In

Record your answers on the answer sheet provided by your teacher or on a sheet of paper.

11. What is the value of $16 + 18 \div 2 \times 3$? (Lesson 1-2)

12. In 5 days, the stock market fell 25 points. What integer expresses the average change in the stock market per day? (Lesson 2-5)

13. Write $\frac{1}{5 \times 5 \times 5 \times 5}$ using a negative exponent. (Lesson 4-7)

14. Find $\frac{5}{12} - \frac{3}{8}$. (Lesson 5-4)

15. The high temperatures for five days in April are shown in the table below. What was the median high temperature? (Lesson 5-8)

High Temperatures	
Monday	45°
Tuesday	62°
Wednesday	57°
Thursday	41°
Friday	53°

16. To mix a certain color of paint, Alexis combines 5 liters of white paint, 2 liters of red paint, and 1 liter of blue paint. What is the ratio of white paint to the total amount of paint? (Lesson 6-1)

17. A box contains 42 pencils. Some are yellow, some are red, some are white, and some are black. If the probability of randomly selecting a red pencil is $\frac{3}{7}$, how many red pencils are in the box? (Lesson 6-2)

18. A city received a federal grant of \$350 million to build a light-rail system which actually cost \$625 million. What percent of the total cost was paid for with the Federal grant? Round to the nearest percent. (Lesson 6-7)

19. A leather jacket is on sale for 40% off the original price. The sale price is \$64 less than the original price. What was the original price of the jacket? (Lesson 6-8)

20. Find x if $8x - 12 = 5x + 6$. (Lesson 7-1)

21. Find the width w of the rectangle below if its perimeter is 88 meters. (Lesson 7-2)

$5w - 4$

w

22. Dakota earns \$8 per hour working at a landscaping company and wants to earn at least \$1200 this summer. What is the minimum number of hours he will have to work? (Lesson 7-5)

FCAT Practice

Part 3 Open Ended

Record your answers on a sheet of paper. Show your work.

23. At a post office, a customer bought an equal number of the following stamps: 1¢, 22¢, and 34¢. She also mailed a package that required \$2.80 in postage. The total bill was \$14. (Lesson 3-6)

a. Write an equation that describes this situation.

b. What does the variable in your equation represent?

c. Solve the equation. Show your work.

d. Write a sentence describing what the solution represents.

24. A magazine publisher collected data on subscription renewals and found that each year 3 out of 50 subscribers do *not* renew. The magazine currently has 24,000 subscribers. (Lesson 6-6)

a. What percent of subscribers do *not* renew their subscriptions?

b. What percent of subscribers per year do renew their subscriptions?

c. How many subscribers will likely renew their subscriptions this year?

Chapter 8

Functions and Graphing

What You'll Learn

- **Lesson 8-1** Use functions to describe relationships between two quantities.
- **Lessons 8-2, 8-3, 8-6, and 8-7** Graph and write linear equations using ordered pairs, the x- and y-intercepts, and slope and y-intercept.
- **Lessons 8-4 and 8-5** Find slopes of lines and use slope to describe rates of change.
- **Lesson 8-8** Draw and use best-fit lines to make predictions about data.
- **Lessons 8-9 and 8-10** Solve systems of linear equations and linear inequalities.

Key Vocabulary

- **function** (p. 369)
- **linear equation** (p. 375)
- **slope** (p. 387)
- **rate of change** (p. 393)
- **system of equations** (p. 414)

Why It's Important

You can often use functions to represent real-world data. For example, the winning times in Olympic swimming events can be shown in a scatter plot. You can then use the data points to write an equation representing the relationship between the year and the winning times. *You will use a function in Lesson 8-8 to predict the winning time in the women's 800-meter freestyle event for the 2008 Olympics.*

Getting Started

Prerequisite Skills To be successful in this chapter, you'll need to master these skills and be able to apply them in problem-solving situations. Review these skills before beginning Chapter 8.

For Lesson 8-1 **Relations**

Express each relation as a table. Then determine the domain and range.
(For review, see Lesson 1-6.)

1. $\{(0, 4), (-3, 3)\}$
2. $\{(-5, 11), (2, 1)\}$
3. $\{(6, 8), (7, 10), (8, 12)\}$
4. $\{(1, -9), (5, 12), (-3, -10)\}$
5. $\{(-8, 5), (7, -1), (6, 1), (1, -2)\}$

For Lesson 8-3 **The Coordinate System**

Use the coordinate grid to name the point for each ordered pair. *(For review, see Lesson 2-6.)*

6. $(-3, 0)$
7. $(3, -2)$
8. $(-4, -2)$
9. $(0, 4)$
10. $(4, 6)$
11. $(4, 0)$

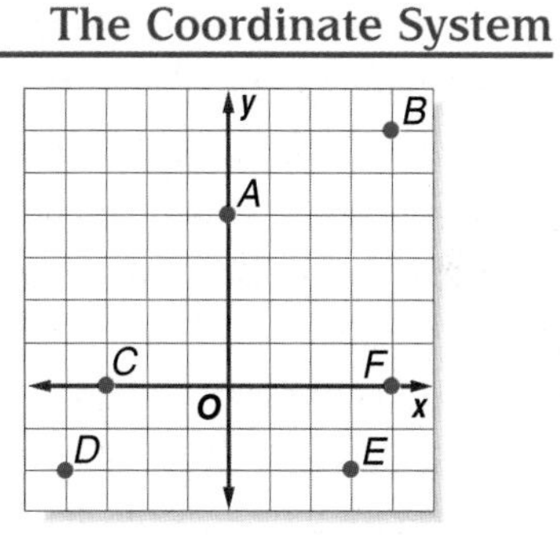
y
B
A
C
F
O
x
D
E

For Lesson 8-10 **Inequalities**

For the given value, state whether the inequality is *true* or *false*. *(For review, see Lesson 7-3.)*

12. $8y \geq 25, y = 4$
13. $18 < t + 12, t = 10$
14. $n - 15 > 7, n = 20$
15. $5 \geq 2x + 3, x = 1$
16. $12 \leq \frac{2}{3}n, n = 9$
17. $\frac{1}{2}x - 5 < 0, x = 8$

FOLDABLES™ Study Organizer

Make this Foldable to collect examples of functions and graphs. Begin with an 11" × 17" sheet of paper.

Step 1 Fold

Fold the short sides so they meet in the middle.

Step 2 Fold Again

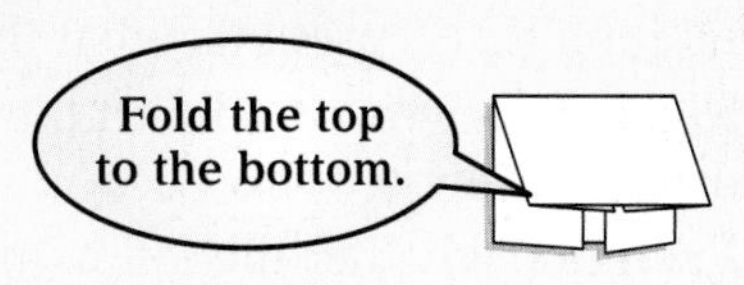
Fold the top to the bottom.

Step 3 Cut

Open. Cut along second fold to make four tabs. Staple a sheet of grid paper inside.

Step 4 Label

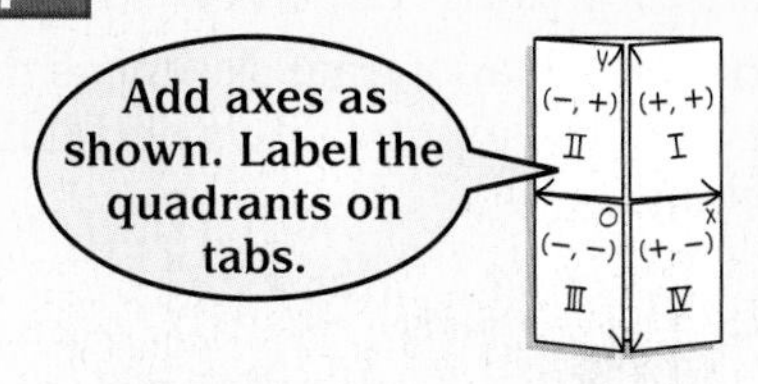
Add axes as shown. Label the quadrants on tabs.
(−, +)
II
(+, +)
I
(−, −)
III
(+, −)
IV

Reading and Writing As you read and study the chapter, draw examples of functions on the grid paper and write notes under the tabs.

Algebra Activity

A Preview of Lesson 8-1

Sunshine State Standards
MA.D.1.3.1-4, MA.D.1.3.1-5, MA.D.1.3.2-1

Input and Output

In a *function*, there is a relationship between two quantities or sets of numbers. You start with an input value, apply a function rule of one or more operations, and get an output value. In this way, each input is assigned exactly one output.

Collect the Data

Step 1 To make a *function machine*, draw three squares in the middle of a 3-by-5-inch index card, shown here in blue.

Step 2 Cut out the square on the left and the square on the right. Label the left "window" INPUT and the right "window" OUTPUT.

Step 3 Write a rule such as "$\times 2 + 3$" in the center square.

Step 4 On another index card, list the integers from -5 to 4 in a column close to the left edge.

−5
−4
−3
−2
−1
0
1
2
3
4

Input Rule Output
−5 ×2+3 −7
1
2
3
4

Step 5 Place the function machine over the number column so that -5 is in the left window.

Step 6 Apply the rule to the input number. The output is $-5 \times 2 + 3$, or -7. Write -7 in the right window.

Make a Conjecture

1. Slide the function machine down so that the input is -4. Find the output and write the number in the right window. Continue this process for the remaining inputs.
2. Suppose x represents the input and y represents the output. Write an algebraic equation that represents what the function machine does.
3. Explain how you could find the input if you are given a rule and the corresponding output.
4. Determine whether the following statement is *true* or *false*. Explain.
   The input values depend on the output values.
5. Write an equation that describes the relationship between the input value x and output value y in each table.

Input	Output
−1	−2
0	0
1	2
3	6

Input	Output
−2	2
−1	3
0	4
1	5

Extend the Activity

6. Write your own rule and use it to make a table of inputs and outputs. Exchange your table of values with another student. Use the table to determine each other's rule.

8-1 Functions

Sunshine State Standards
MA.D.1.3.1-1, MA.D.1.3.2-1

What You'll Learn

- Determine whether relations are functions.
- Use functions to describe relationships between two quantities.

Vocabulary

- function
- vertical line test

How can the relationship between actual temperatures and windchill temperatures be a function?

The table compares actual temperatures and windchill temperatures when the wind is blowing at 10 miles per hour.

Actual Temperature (°F)	Windchill Temperature (°F)
−10	−34
0	−22
10	−9
20	3

Source: *The World Almanac*

a. On grid paper, graph the temperatures as ordered pairs (actual, windchill).

b. Describe the relationship between the two temperature scales.

c. When the actual temperature is −20°F, which is the best estimate for the windchill temperature: −46°F, −28°F, or 0°F? Explain.

RELATIONS AND FUNCTIONS Recall that a relation is a set of ordered pairs. A **function** is a special relation in which each member of the domain is paired with *exactly* one member in the range.

Study Tip

Look Back
To review **relations**, **domain**, and **range**, see Lesson 1-6.

Relation	Diagram	Is the Relation a Function?
{(−10, −34), (0, −22), (10, −9), (20, 3)}	domain (x) range (y) −10 → −34 0 → −22 10 → −9 20 → 3	Yes, because each domain value is paired with exactly one range value.
{(−10, −34), (−10, −22), (10, −9), (20, 3)}	domain (x) range (y) −10 → −34 → −22 10 → −9 20 → 3	No, because −10 in the domain is paired with two range values, −34 and −22.

Since functions are relations, they can be represented using ordered pairs, tables, or graphs.

Example 1 Ordered Pairs and Tables as Functions

Determine whether each relation is a function. Explain.

a. {(−3, 1), (−2, 4), (−1, 7), (0, 10), (1, 13)}

This relation is a function because each element of the domain is paired with exactly one element of the range.

b.

x	5	3	2	0	−4	−6
y	1	3	1	3	−2	2

This is a function because for each element of the domain, there is only one corresponding element in the range.

Reading Math

Function

Everyday Meaning: a relationship in which one quality or trait depends on another. Height is a function of age.

Math Meaning: a relationship in which a range value depends on a domain value. y is a function of x.

Another way to determine whether a relation is a function is to use the **vertical line test**. Use a pencil or straightedge to represent a vertical line.

Place the pencil at the left of the graph. Move it to the right across the graph. If, for each value of x in the domain, it passes through no more than one point on the graph, then the graph represents a function.

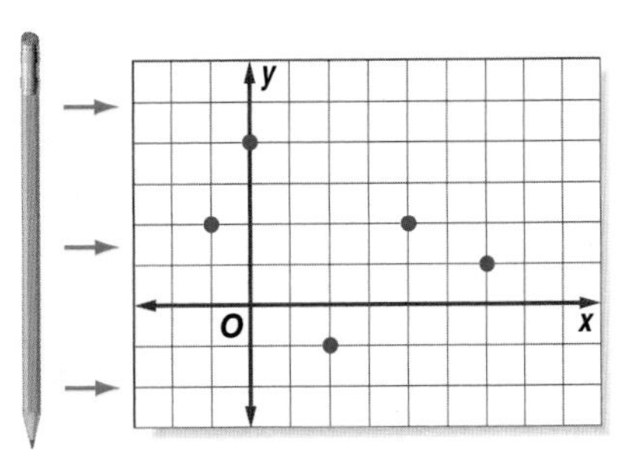

Example 2 Use a Graph to Identify Functions

Determine whether the graph at the right is a function. Explain your answer.

The graph represents a relation that is *not* a function because it does not pass the vertical line test. By examining the graph, you can see that when $x = 2$, there are three different y values.

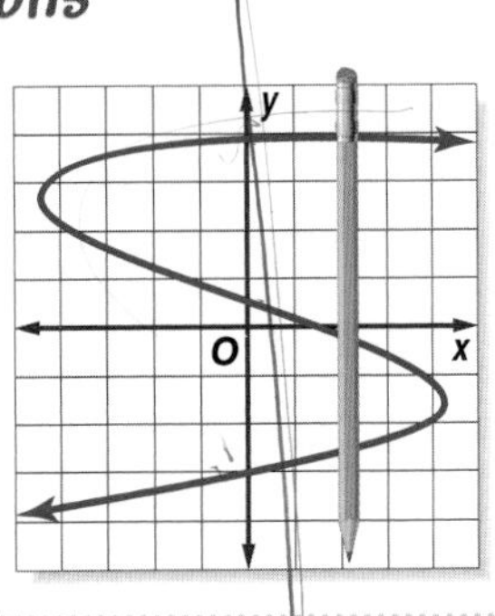

DESCRIBE RELATIONSHIPS A function describes the relationship between two quantities such as time and distance. For example, the distance you travel on a bike depends on how long you ride the bike. In other words, *distance is a function of time.*

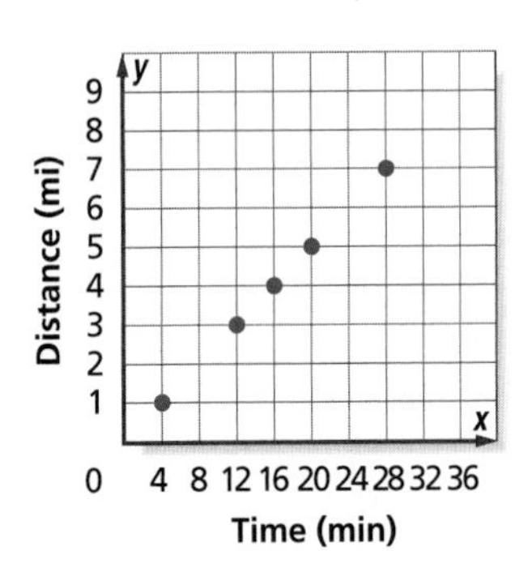

Example 3 Use a Function to Describe Data

SCUBA DIVING The table shows the water pressure as a scuba diver descends.

Depth (ft)	Water Pressure (lb/ft^3)
0	0
1	62.4
2	124.8
3	187.2
4	249.6
5	312.0

Source: www.infoplease.com

a. Do these data represent a function? Explain.

This relation is a function because at each depth, there is only one measure of pressure.

b. Describe how water pressure is related to depth.

Water pressure depends on the depth. As the depth increases, the pressure increases.

More About. . .

Scuba Diving

To prevent decompression sickness, or the "bends," it is recommended that divers ascend to the surface no faster than 30 feet per minute.

Source: www.mtsinai.org

Concept Check In Example 3, what is the domain and what is the range?

Check for Understanding

Concept Check

1. **Describe** three ways to represent a function. Show an example of each.
2. **Describe** two methods for determining whether a relation is a function.
3. **OPEN ENDED** Draw the graph of a relation that is not a function. Explain why it is not a function.

Guided Practice

Determine whether each relation is a function. Explain.

4. {(13, 5), (−4, 12), (6, 0), (13, 10)}
5. {(9.2, 7), (9.4, 11), (9.5, 9.5), (9.8, 8)}

6.

Domain	Range
−3	3
−1	−2
0	5
1	−4
2	3

7.

x	y
5	4
2	8
−7	9
2	12
5	14

8.

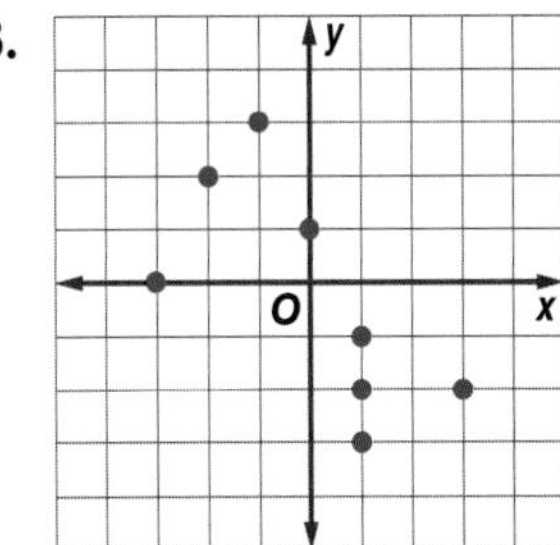

9.

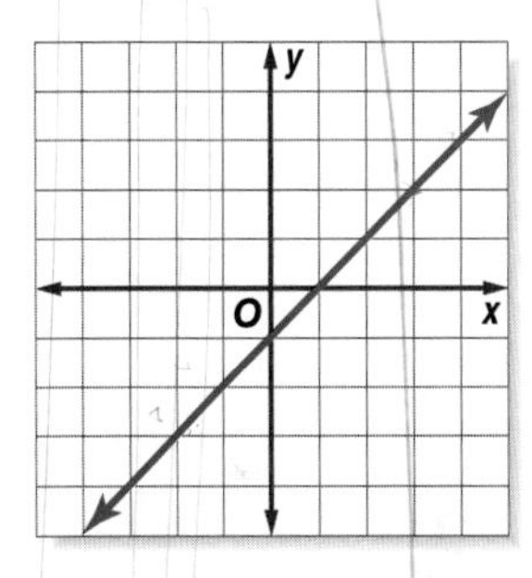

Application

WEATHER For Exercises 10 and 11, use the table that shows how various wind speeds affect the actual temperature of 30°F.

10. Do the data represent a function? Explain.
11. Describe how windchill temperatures are related to wind speed.

Wind Speed (mph)	Windchill Temperature (°F)
0	30
10	16
20	4
30	−2
40	−5

Source: *The World Almanac*

Practice and Apply

Homework Help

For Exercises	See Examples
12–19	1
20–23	2
24–27	3

Extra Practice

See page 741.

Determine whether each relation is a function. Explain.

12. {(−1, 6), (4, 2), (2, 36), (1, 6)}
13. {(−2, 3), (4, 7), (24, −6), (5, 4)}
14. {(9, 18), (0, 36), (6, 21), (6, 22)}
15. {(5, −4), (−2, 3), (5, −1), (2, 3)}

16.

Domain	Range
−4	−2
−2	1
0	2
3	1

17.

Domain	Range
−1	5
−2	5
−2	1
−6	1

Determine whether each relation is a function. Explain.

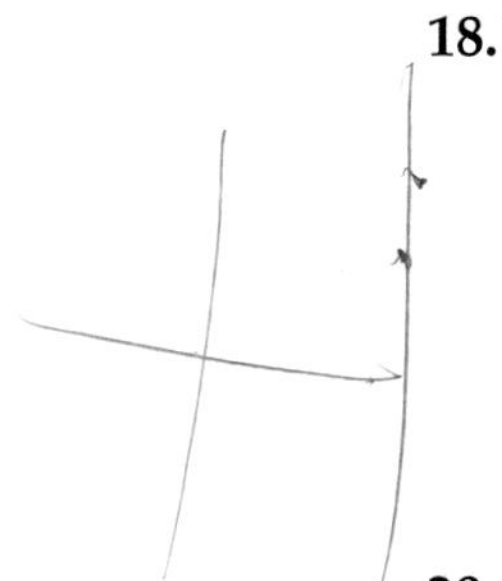

18.

x	y
−7	2
0	4
11	6
11	8
0	10

19.

x	y
14	5
15	10
16	15
17	20
18	25

20.

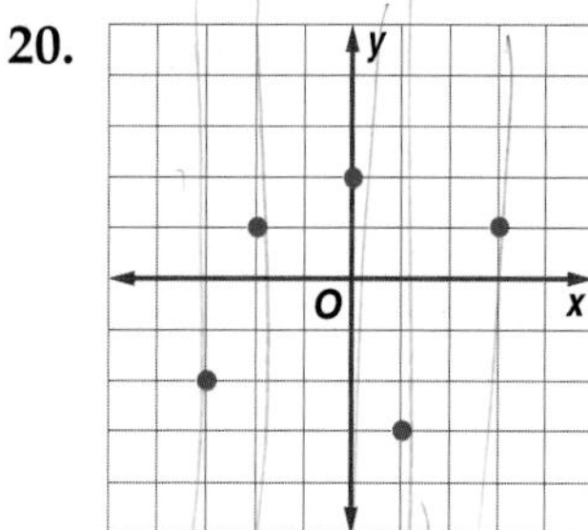

21.

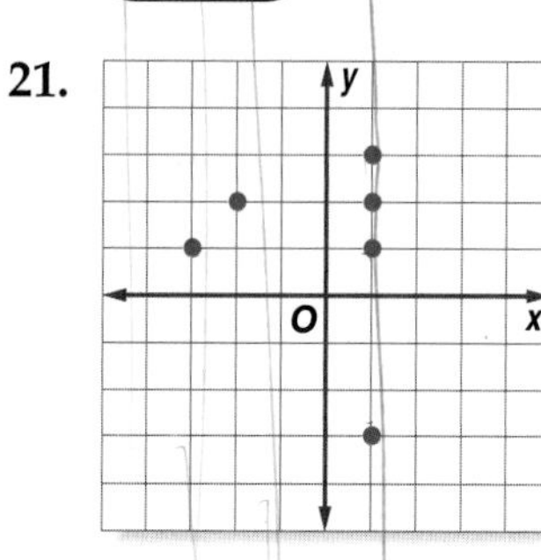

22.

23.

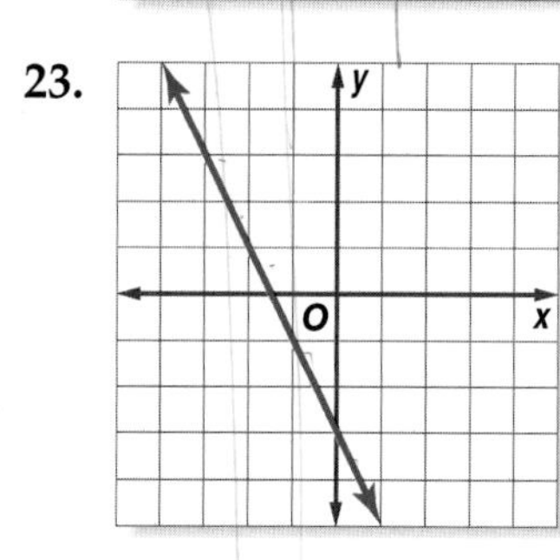

Study Tip

Trends

There may be general trends in sets of data. However, not every data point may follow the trend exactly.

FARMING **For Exercises 24–27, use the table that shows the number and size of farms in the United States every decade from 1950 to 2000.**

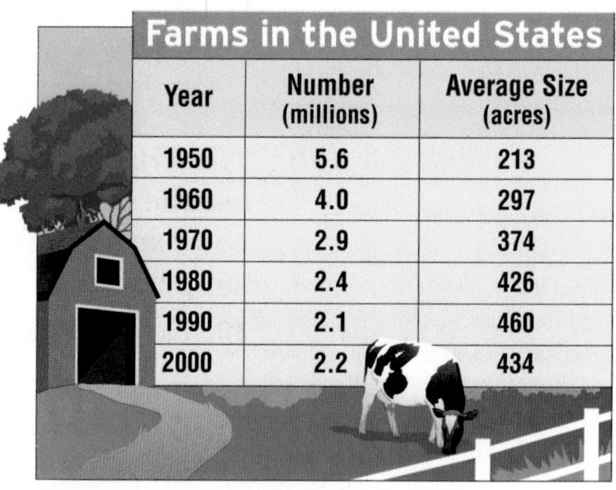

Farms in the United States

Year	Number (millions)	Average Size (acres)
1950	5.6	213
1960	4.0	297
1970	2.9	374
1980	2.4	426
1990	2.1	460
2000	2.2	434

Source: *The Wall Street Journal Almanac*

24. Is the relation (year, number of farms) a function? Explain.

25. Describe how the number of farms is related to the year.

26. Is the relation (year, average size of farms) a function? Explain.

27. Describe how the average size of farms is related to the year.

MEASUREMENTS **For Exercises 28 and 29, use the data in the table.**

28. Do the data represent a function? Explain.

29. Is there any relation between foot length and height?

Name	Foot Length (cm)	Height (cm)
Rosa	24	163
Tanner	28	182
Enrico	25	163
Jahad	24	168
Abbi	22	150
Cory	26	172

Tell whether each statement is *always*, *sometimes*, or *never* true. Explain.

30. A function is a relation.

31. A relation is a function.

32. **CRITICAL THINKING** The *inverse* of any relation is obtained by switching the coordinates in each ordered pair of the relation.

  a. Determine whether the inverse of the relation {(4, 0), (5, 1), (6, 2), (6, 3)} is a function.

  b. Is the inverse of a function *always*, *sometimes*, or *never* a function? Give an example to explain your reasoning.

33. **WRITING IN MATH** Answer the question that was posed at the beginning of the lesson.

  How can the relationship between actual temperatures and windchill temperatures be a function?

  Include the following in your answer:
  - an explanation of how actual temperatures and windchill temperatures are related for a given wind speed, and
  - a discussion about whether an actual temperature can ever have two corresponding windchill temperatures when the wind speed remains the same.

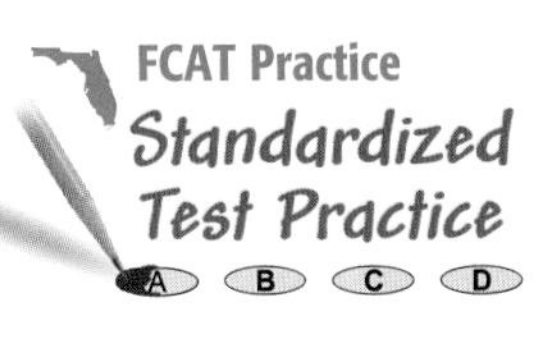

34. The relation {(2, 11), (−9, 8), (14, 1), (5, 5)} is *not* a function when which ordered pair is added to the set?

  Ⓐ (8, −9) Ⓑ (6, 11) Ⓒ (0, 0) Ⓓ (2, 18)

35. Which statement is true about the data in the table?

  Ⓐ The data represent a function.

  Ⓑ The data do not represent a function.

  Ⓒ As the value of x increases, the value of y increases.

  Ⓓ A graph of the data would not pass the vertical line test.

x	y
−4	−4
2	16
5	8
10	−4
12	15

Maintain Your Skills

Mixed Review

Solve each inequality. Check your solution. *(Lessons 7-5 and 7-6)*

36. $4y > 24$
37. $\frac{a}{3} < -7$
38. $18 \geq -2k$
39. $2x + 5 < 17$
40. $2t - 3 \geq 1.4t + 6$
41. $12r - 4 > 7 + 12r$

Solve each problem by using the percent equation. *(Lesson 6-7)*

42. 10 is what percent of 50?
43. What is 15% of 120?
44. Find 95% of 256.
45. 46.5 is 62% of what number?

46. State whether the sequence 120, 100, 80, 60, … is *arithmetic*, *geometric*, or *neither*. Then write the next three terms of the sequence. *(Lesson 5-10)*

Getting Ready for the Next Lesson

PREREQUISITE SKILL **Evaluate each expression if $x = 4$ and $y = -1$.**
*(To review **evaluating expressions**, see Lesson 1-3.)*

47. $3x + 1$
48. $2y$
49. $y + 6$
50. $-5x$
51. $2x - 8$
52. $3y - 4$

Graphing Calculator Investigation

A Preview of Lesson 8-2

Sunshine State Standards
MA.A.4.3.1-2, MA.B.1.3.2-1, MA.B.1.3.2-2, MA.D.1.3.1-5, MA.D.1.3.2-1, MA.D.2.3.1-5

Function Tables

You can use a TI-83 Plus graphing calculator to create function tables. By entering a function and the domain values, you can find the corresponding range values.

Use a function table to find the range of $y = 3n + 1$ if the domain is $\{-5, -2, 0, 0.5, 4\}$.

Step 1 *Enter the function.*

- The graphing calculator uses X for the domain values and Y for the range values. So, Y = 3X + 1 represents $y = 3n + 1$.
- Enter Y = 3X + 1 in the Y= list.

KEYSTROKES: [Y=] 3 [X,T,θ,n] [+] 1

Step 2 *Format the table.*

- Use TBLSET to select *Ask* for the independent variable and *Auto* for the dependent variable. Then you can enter any value for the domain.

KEYSTROKES: [2nd] [TBLSET] [▼] [▼] [▶] [ENTER] [▼] [ENTER]

Step 3 *Find the range by entering the domain values.*

- Access the table.

KEYSTROKES: [2nd] [TABLE]

- Enter the domain values.

KEYSTROKES: −5 [ENTER] −2 [ENTER] ... 4 [ENTER]

The range is $\{-14, -5, 1, 2.5, 13\}$.

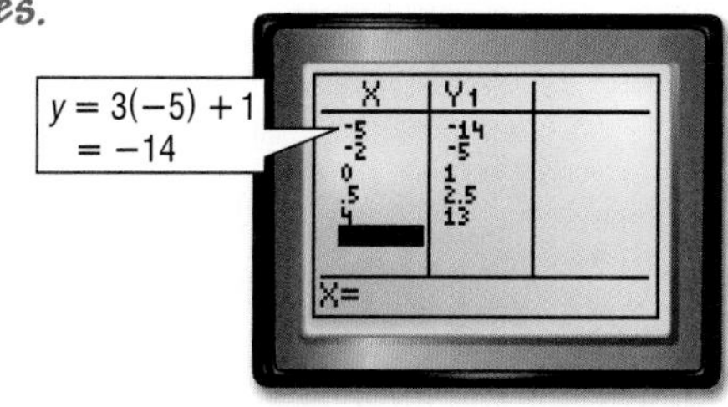

Exercises

Use the TABLE option on a graphing calculator to complete each exercise.

1. Consider the function $f(x) = -2x + 4$ and the domain values $\{-2, -1, 0, 1, 2\}$.
   a. Use a function table to find the range values.
   b. Describe the relationship between the X and Y values.
   c. If X is less than −2, would the value for Y be greater or less than 8? Explain.
2. Suppose you are using the formula $d = rt$ to find the distance d a car travels for the times t in hours given by $\{0, 1, 3.5, 10\}$.
   a. If the rate is 60 miles per hour, what function should be entered in the Y= list?
   b. Make a function table for the given domain.
   c. Between which two times in the domain does the car travel 150 miles?
   d. Describe how a function table can be used to better estimate the time it takes to drive 150 miles.
3. Serena is buying one packet of pencils for $1.50 and a number of fancy folders x for $0.40 each. The total cost y is given by $y = 1.50 + 0.40x$.
   a. Use a function table to find the total cost if Serena buys 1, 2, 3, 4, and 12 folders.
   b. Suppose plain folders cost $0.25 each. Enter $y = 1.50 + 0.25x$ in the Y= list as Y2. How much does Serena save if she buys pencils and 12 plain folders rather than pencils and 12 fancy folders?

www.pre-alg.com/other_calculator_keystrokes

8-2 Linear Equations in Two Variables

Sunshine State Standards
MA.C.3.3.2-1, MA.D.1.3.1-1, MA.D.1.3.1-2, MA.D.1.3.1-3, MA.D.1.3.2-1, MA.D.2.3.1-4

Vocabulary

- linear equation

What You'll Learn

- Solve linear equations with two variables.
- Graph linear equations using ordered pairs.

How can linear equations represent a function?

Peaches cost $1.50 per can.

a. Complete the table to find the cost of 2, 3, and 4 cans of peaches.

b. On grid paper, graph the ordered pairs (number, cost). Then draw a line through the points.

c. Write an equation representing the relationship between number of cans x and cost y.

Number of Cans (x)	$1.50x$	Cost (y)
1	1.50(1)	1.50
2		
3		
4		

SOLUTIONS OF EQUATIONS Functions can be represented in words, in a table, as ordered pairs, with a graph, and with an equation.

An equation such as $y = 1.50x$ is called a linear equation. A **linear equation** in two variables is an equation in which the variables appear in separate terms and neither variable contains an exponent other than 1.

Solutions of a linear equation are ordered pairs that make the equation true. One way to find solutions is to make a table. Consider the equation $y = -x + 8$.

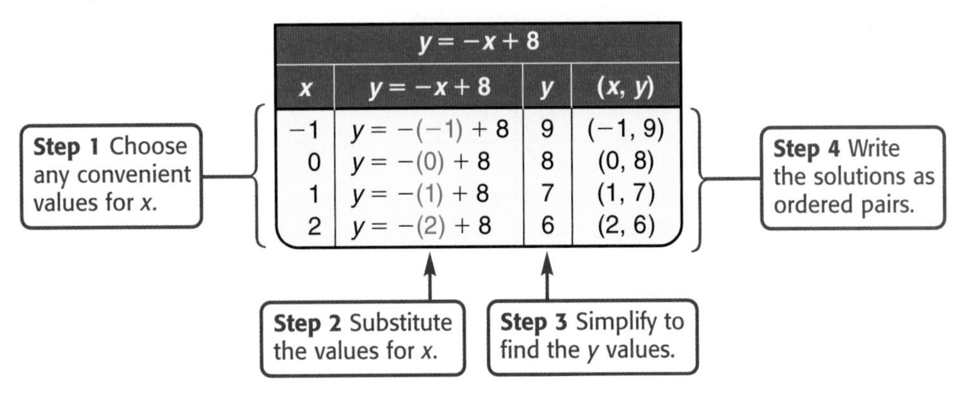

$y = -x + 8$			
x	$y = -x + 8$	y	(x, y)
-1	$y = -(-1) + 8$	9	$(-1, 9)$
0	$y = -(0) + 8$	8	$(0, 8)$
1	$y = -(1) + 8$	7	$(1, 7)$
2	$y = -(2) + 8$	6	$(2, 6)$

Step 1 Choose any convenient values for x.
Step 2 Substitute the values for x.
Step 3 Simplify to find the y values.
Step 4 Write the solutions as ordered pairs.

So, four solutions of $y = -x + 8$ are $(-1, 9)$, $(0, 8)$, $(1, 7)$, and $(2, 6)$.

Example 1 Find Solutions

Find four solutions of $y = 2x - 1$.

Choose four values for x. Then substitute each value into the equation and solve for y.

x	$y = 2x - 1$	y	(x, y)
0	$y = 2(0) - 1$	-1	$(0, -1)$
1	$y = 2(1) - 1$	1	$(1, 1)$
2	$y = 2(2) - 1$	3	$(2, 3)$
3	$y = 2(3) - 1$	5	$(3, 5)$

Four solutions are $(0, -1)$, $(1, 1)$, $(2, 3)$, and $(3, 5)$.

Study Tip

Choosing x Values It is often convenient to choose 0 as an x value to find a value for y.

Sometimes it is necessary to first rewrite an equation by solving for y.

Example 2 Solve an Equation for y

SHOPPING Fancy goldfish x cost \$3, and regular goldfish y cost \$1. Find four solutions of $3x + y = 8$ to determine how many of each type of fish Tyler can buy for \$8.

First, rewrite the equation by solving for y.

$3x + y = 8$ Write the equation.

$3x + y - 3x = 8 - 3x$ Subtract $3x$ from each side.

$y = 8 - 3x$ Simplify.

Choose four x values and substitute them into $y = 8 - 3x$. Four solutions are (0, 8), (1, 5), (2, 2), and (3, −1).

x	y = 8 − 3x	y	(x, y)
0	y = 8 − 3(0)	8	(0, 8)
1	y = 8 − 3(1)	5	(1, 5)
2	y = 8 − 3(2)	2	(2, 2)
3	y = 8 − 3(3)	−1	(3, −1)

(0, 8) → He can buy 0 fancy goldfish and 8 regular goldfish.

(1, 5) → He can buy 1 fancy goldfish and 5 regular goldfish.

(2, 2) → He can buy 2 fancy goldfish and 2 regular goldfish.

(3, −1) → This solution does not make sense, because there cannot be a negative number of goldfish.

Study Tip

Checking Solutions
Check solutions in the context of the original problem to be sure they make sense.

Concept Check Write the solution of $3x + y = 10$ if $x = 2$.

GRAPH LINEAR EQUATIONS A linear equation can also be represented by a graph. Study the graphs shown below.

Linear Equations

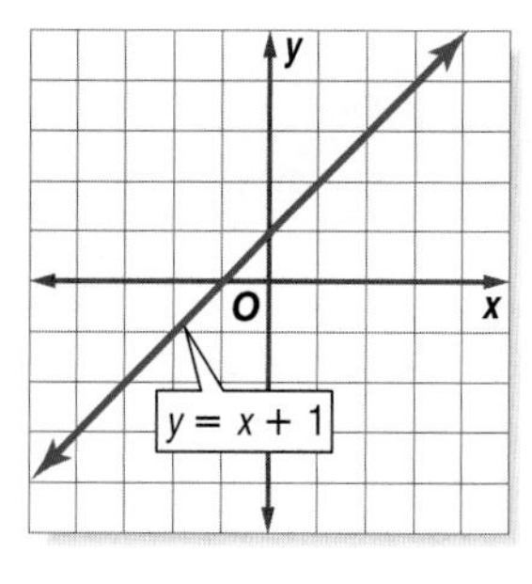

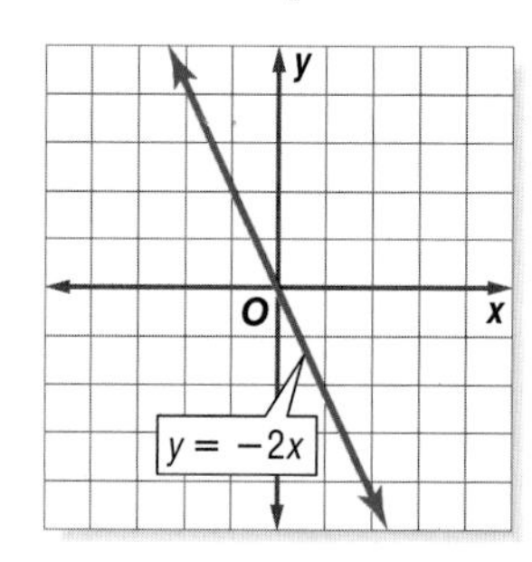

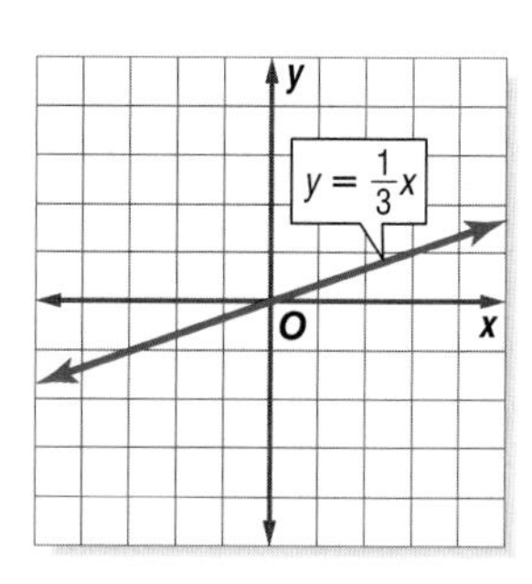

Nonlinear Equations

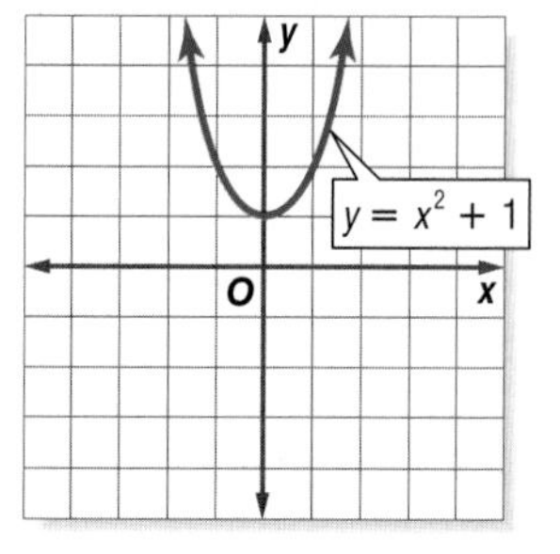

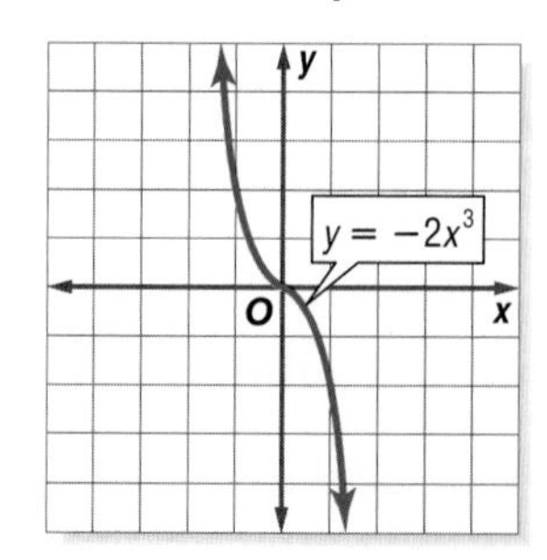

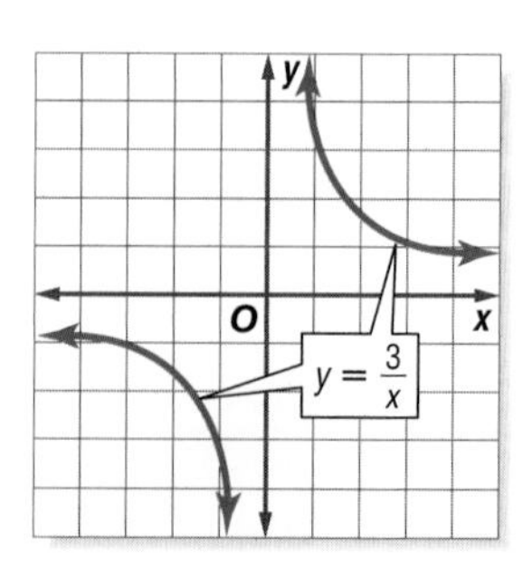

Notice that graphs of the linear equations are straight lines. This is true for all linear equations and is the reason they are called "linear." The coordinates of all points on a line are solutions to the equation.

Study Tip

Plotting Points
You can also graph just two points to draw the line and then graph one point to check.

To graph a linear equation, find ordered pair solutions, plot the corresponding points, and draw a line through them. It is best to find at least three points.

Example 3 Graph a Linear Equation

Graph $y = x + 1$ by plotting ordered pairs.

First, find ordered pair solutions. Four solutions are $(-1, 0)$, $(0, 1)$, $(1, 2)$, and $(2, 3)$.

x	y = x + 1	y	(x, y)
−1	y = −1 + 1	0	(−1, 0)
0	y = 0 + 1	1	(0, 1)
1	y = 1 + 1	2	(1, 2)
2	y = 2 + 1	3	(2, 3)

Plot these ordered pairs and draw a line through them. Note that the ordered pair for any point on this line is a solution of $y = x + 1$. The line is a complete graph of the function.

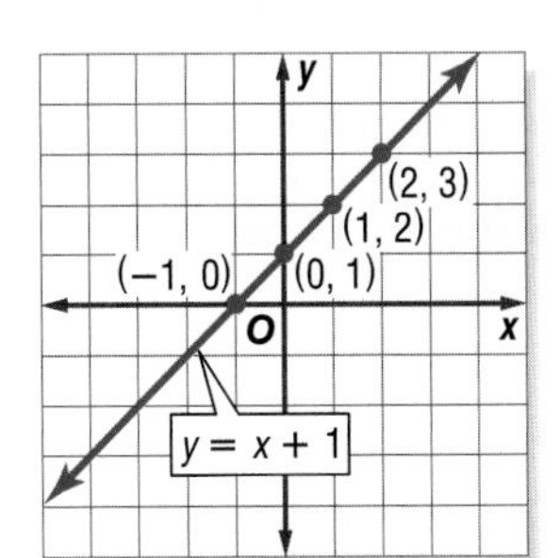

CHECK It appears from the graph that $(-2, -1)$ is also a solution. Check this by substitution.

$y = x + 1$ Write the equation.

$-1 \stackrel{?}{=} -2 + 1$ Replace x with -2 and y with -1.

$-1 = -1$ ✓ Simplify.

A linear equation is one of many ways to represent a function.

Concept Summary — *Representing Functions*

- **Words** The value of y is 3 less than the corresponding value of x.
- **Table**

x	y
0	−3
1	−2
2	−1
3	0

- **Graph**

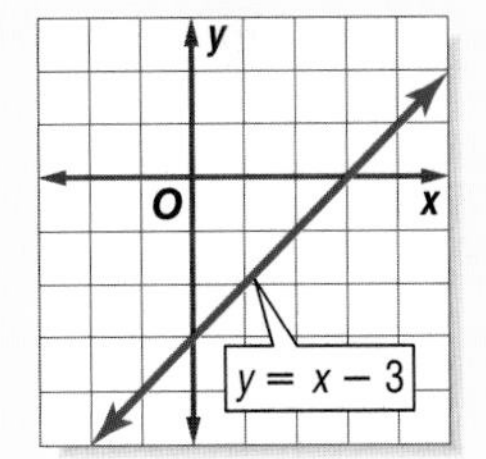

- **Ordered Pairs** (0, −3), (1, −2), (2, −1), (3, 0)
- **Equation** $y = x - 3$

Check for Understanding

Concept Check

1. **Explain** why a linear equation has infinitely many solutions.
2. **OPEN ENDED** Write a linear equation that has $(-2, 4)$ as a solution.

Guided Practice

3. Copy and complete the table. Use the results to write four solutions of $y = x + 5$. Write the solutions as ordered pairs.

x	x + 5	y
−3	−3 + 5	
−1		
0		
1		

Find four solutions of each equation. Write the solutions as ordered pairs.

4. $y = x + 8$
5. $y = 4x$
6. $y = 2x - 7$
7. $-5x + y = 6$

Graph each equation by plotting ordered pairs.

8. $y = x + 3$
9. $y = 2x - 1$
10. $x + y = 5$

Application

11. **SCIENCE** The distance y in miles that light travels in x seconds is given by $y = 186{,}000x$. Find two solutions of this equation and describe what they mean.

Practice and Apply

Homework Help

For Exercises	See Examples
12–25	1
26–29	2
30–41	3

Extra Practice
See page 742.

Copy and complete each table. Use the results to write four solutions of the given equation. Write the solutions as ordered pairs.

12. $y = x - 9$

x	$x - 9$	y
−1	−1 − 9	
0		
4		
7		

13. $y = 2x + 6$

x	$2x + 6$	y
−4	2(−4) + 6	
0		
2		
4		

Find four solutions of each equation. Write the solutions as ordered pairs.

14. $y = x + 2$
15. $y = x - 7$
16. $y = 3x$
17. $y = -5x$
18. $y = 2x - 3$
19. $y = 3x + 1$
20. $x + y = 9$
21. $x + y = -6$
22. $4x + y = 2$
23. $3x - y = 10$
24. $y = 8$
25. $x = -1$

MEASUREMENT **The equation $y = 0.62x$ describes the approximate number of miles y in x kilometers.**

26. Describe what the solution (8, 4.96) means.
27. About how many miles is a 10-kilometer race?

HEALTH **During a workout, a target heart rate y in beats per minute is represented by $y = 0.7(220 - x)$, where x is a person's age.**

28. Compare target heart rates of people 20 years old and 50 years old.
29. In which quadrant(s) would the graph of $y = 0.7(220 - x)$ make sense? Explain your reasoning.

Health

There are special music tapes that provide constant, non-stop beats during a workout. These tapes can be matched with a person's fitness level and workout pace.

Source: www.sportsmusic.com

Graph each equation by plotting ordered pairs.

30. $y = x + 2$
31. $y = x + 5$
32. $y = x - 4$
33. $y = -x - 6$
34. $y = -2x + 2$
35. $y = 3x - 4$
36. $x + y = 1$
37. $x - y = 6$
38. $2x + y = 5$
39. $3x - y = 7$
40. $x = 2$
41. $y = -3$

GEOMETRY **For Exercises 42–44, use the following information.**
The formula for finding the perimeter of a square with sides s units long is $P = 4s$.

42. Find three ordered pairs that satisfy this condition.
43. Draw the graph that contains these points.
44. Why do negative values of s make no sense?

Determine whether each relation or equation is linear. Explain.

45.

x	y
−1	−2
0	0
1	2
2	4

46.

x	y
−1	1
0	0
1	1
2	4

47.

x	y
−1	−1
0	−1
1	−1
2	−1

48. $3x + y = 20$

49. $y = x^2$

50. $y = 5$

51. CRITICAL THINKING Compare and contrast the functions shown in the tables. (*Hint:* Compare the change in values for each column.)

x	y
−1	−2
0	0
1	2
2	4

x	y
−1	1
0	0
1	1
2	4

52. WRITING IN MATH Answer the question that was posed at the beginning of the lesson.

How can linear equations represent a function?

Include the following in your answer:

- a description of four ways that you can represent a function, and
- an example of a linear equation that could be used to determine the cost of x pounds of bananas that are $0.49 per pound.

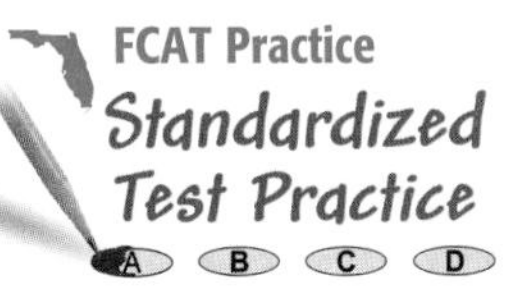

53. Identify the equation that represents the data in the table.

x	y
−2	11
0	5
1	2
3	−4

(A) $y = x + 5$ (B) $y = -3x + 5$

(C) $y = -5x + 1$ (D) $y = x + 13$

54. The graph of $2x - y = 4$ goes through which pair of points?

(A) $P(-2, -3), Q(0, 2)$ (B) $P(-2, -1), Q(2, -3)$

(C) $P(1, -2), Q(3, 2)$ (D) $P(-3, 2), Q(0, -4)$

Maintain Your Skills

Mixed Review

Determine whether each relation is a function. Explain. *(Lesson 8-1)*

55. $\{(2, 3), (3, 4), (4, 5), (5, 6)\}$

56. $\{(0, 6), (-3, 9), (4, 9), (-2, 1)\}$

57. $\{(11, 8), (13, -2), (11, 21)\}$

58. $\{(-0.1, 5), (0, 10), (-0.1, -5)\}$

Solve each inequality and check your solution. Graph the solution on a number line. *(Lesson 7-6)*

59. $3x + 4 < 16$

60. $9 - 2d \leq 23$

61. Evaluate $a \div b$ if $a = \frac{4}{7}$ and $b = \frac{2}{3}$. *(Lesson 5-4)*

Getting Ready for the Next Lesson

PREREQUISITE SKILL In each equation, find the value of y when $x = 0$.
*(To review **substitution**, see Lesson 1-5.)*

62. $y = 5x - 3$

63. $-x + y = 3$

64. $x + 2y = 12$

65. $4x - 5y = -20$

Reading Mathematics

Sunshine State Standards
MA.D.1.3.1-5

Language of Functions

Equations that are functions can be written in a form called *functional notation*, as shown below.

equation	**functional notation**
$y = 4x + 10$	$f(x) = 4x + 10$

Read $f(x)$ as f of x.

So, $f(x)$ is simply another name for y. Letters other than f are also used for names of functions. For example, $g(x) = 2x$ and $h(x) = -x + 6$ are also written in functional notation.

In a function, x represents the domain values, and $f(x)$ represents the range values.

range ↓ domain ↓
$f(x) = 4x + 10$

$f(3)$ represents the element in the range that corresponds to the element 3 in the domain. To find $f(3)$, substitute 3 for x in the function and simplify.

Read $f(3)$ as f of 3.

$f(x) = 4x + 10$ Write the function.
$f(3) = 4(3) + 10$ Replace x with 3.
$f(3) = 12 + 10$ or 22 Simplify.

So, the functional value of f for $x = 3$ is 22.

Reading to Learn

1. **RESEARCH** Use the Internet or a dictionary to find the everyday meaning of the word *function*. Write a sentence describing how the everyday meaning relates to the mathematical meaning.
2. Write your own rule for remembering how the domain and the range are represented using functional notation.
3. Copy and complete the table below.

x	$f(x) = 3x + 5$	$f(x)$
0	$f(0) = 3(0) + 5$	
1		
2		
3		

4. If $f(x) = 4x - 1$, find each value.
   a. $f(2)$ b. $f(-3)$ c. $f\left(\frac{1}{2}\right)$
5. Find the value of x if $f(x) = -2x + 5$ and the value of $f(x)$ is -7.

8-3 Graphing Linear Equations Using Intercepts

What You'll Learn

Sunshine State Standards
MA.C.3.3.2-3

- Find the x- and y-intercepts of graphs.
- Graph linear equations using the x- and y-intercepts.

Vocabulary

- x-intercept
- y-intercept

How can intercepts be used to represent real-life information?

The relationship between the temperature in degrees Fahrenheit F and the temperature in degrees Celsius C is given by the equation $F = \frac{9}{5}C + 32$. This equation is graphed at the right.

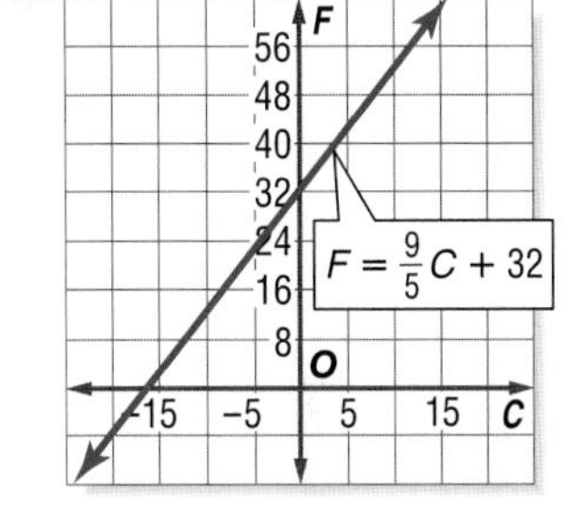

a. Write the ordered pair for the point where the graph intersects the y-axis. What does this point represent?

b. Write the ordered pair for the point where the graph intersects the x-axis. What does this point represent?

Reading Math

Intercept

Everyday Meaning: to interrupt or cut off

Math Meaning: the point where a coordinate axis crosses a line

FIND INTERCEPTS The **x-intercept** is the x-coordinate of a point where a graph crosses the x-axis. The y-coordinate of this point is 0.

The **y-intercept** is the y-coordinate of a point where a graph crosses the y-axis. The x-coordinate of this point is 0.

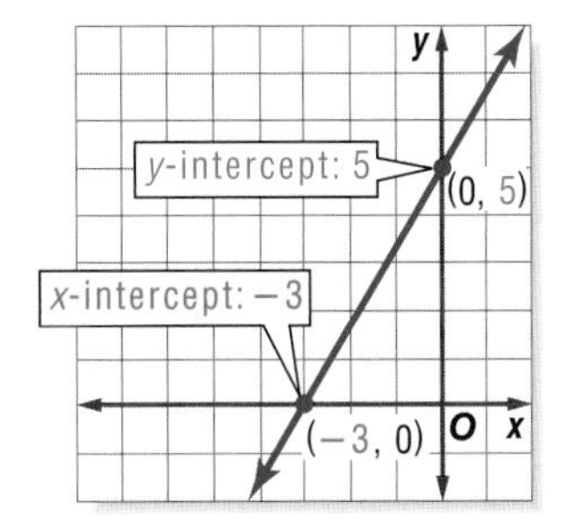

Example 1 Find Intercepts From Graphs

State the x-intercept and the y-intercept of each line.

a.

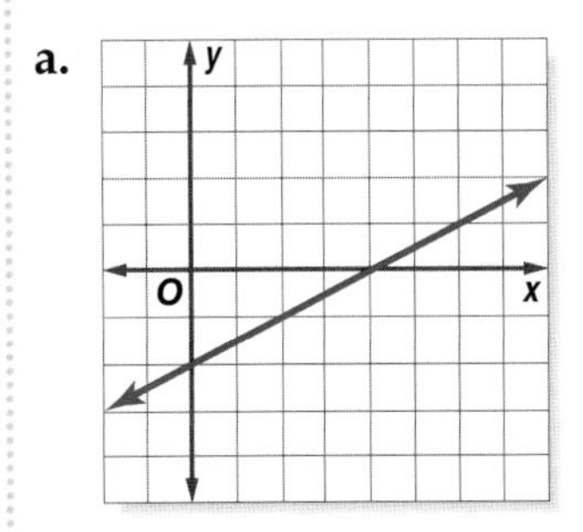

The graph crosses the x-axis at (4, 0). The x-intercept is 4. The graph crosses the y-axis at (0, −2). The y-intercept is −2.

b.

The graph crosses the x-axis at (3, 0). The x-intercept is 3. The graph does not cross the y-axis. There is no y-intercept.

Concept Check A graph passes through a point at (0, −10). Is −10 an x-intercept or a y-intercept?

You can also find the x-intercept and the y-intercept from an equation of a line.

Key Concept — *Intercepts of Lines*

- To find the x-intercept, let $y = 0$ in the equation and solve for x.
- To find the y-intercept, let $x = 0$ in the equation and solve for y.

Example 2 Find Intercepts from Equations

Find the x-intercept and the y-intercept for the graph of $y = x - 6$.

To find the x-intercept, let $y = 0$.

$y = x - 6$	Write the equation.
$0 = x - 6$	Replace y with 0.
$6 = x$	Simplify.

The x-intercept is 6. So, the graph crosses the x-axis at (6, 0).

To find the y-intercept, let $x = 0$.

$y = x - 6$	Write the equation.
$y = 0 - 6$	Replace x with 0.
$y = -6$	Simplify.

The y-intercept is -6. So, the graph crosses the y-axis at $(0, -6)$.

GRAPH EQUATIONS You can use the x- and y-intercepts to graph equations of lines.

In Example 2, we determined that the graph of $y = x - 6$ passes through (6, 0) and $(0, -6)$. To draw the graph, plot these points and draw a line through them.

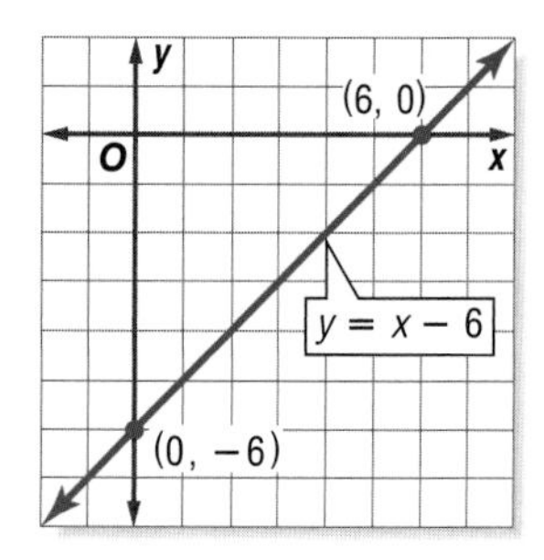

Example 3 Use Intercepts to Graph Equations

Graph $x + 2y = 4$ using the x- and y-intercepts.

Step 1
Find the x-intercept.

$x + 2y = 4$	Write the equation.
$x + 2(0) = 4$	Let $y = 0$.
$x = 4$	Simplify.

The x-intercept is 4, so the graph passes through (4, 0).

Step 2
Find the y-intercept.

$x + 2y = 4$	Write the equation.
$0 + 2y = 4$	Let $x = 0$.
$y = 2$	Divide each side by 2.

The y-intercept is 2, so the graph passes through (0, 2).

Step 3
Graph the points at (4, 0) and (0, 2) and draw a line through them.

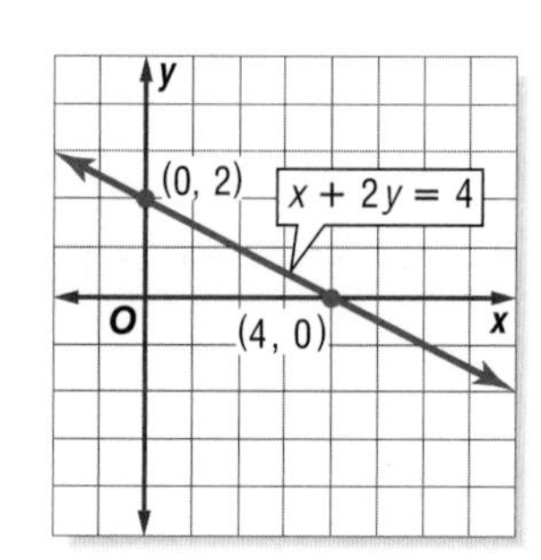

CHECK Choose some other point on the line and determine whether its ordered pair is a solution of $x + 2y = 4$.

Study Tip

Graphing Shortcuts
The ordered pairs of any two solutions can be used to graph a linear equation. However, it is often easiest to find the intercepts.

More About. . .

Earth Science

The temperature of the air is about 3°F cooler for every 1000 feet increase in altitude.

Source: www.hot-air-balloons.com

Example 4 Intercepts of Real-World Data

EARTH SCIENCE Suppose you take a hot-air balloon ride on a day when the temperature is 24°C at sea level. The equation $y = -6.6x + 24$ represents the temperature at x kilometers above sea level.

a. Use the intercepts to graph the equation.

Step 1 Find the x-intercept.

$y = -6.6x + 24$	Write the equation.
$0 = -6.6x + 24$	Replace y with 0.
$0 - 24 = -6.6x + 24 - 24$	Subtract 24 from each side.
$\frac{-24}{-6.6} = \frac{-6.6x}{-6.6}$	Divide each side by -6.6.
$3.6 \approx x$	The x-intercept is approximately 3.6.

Step 2 Find the y-intercept.

$y = -6.6x + 24$	Write the equation.
$y = -6.6(0) + 24$	Replace x with 0.
$y = 24$	The y-intercept is 24.

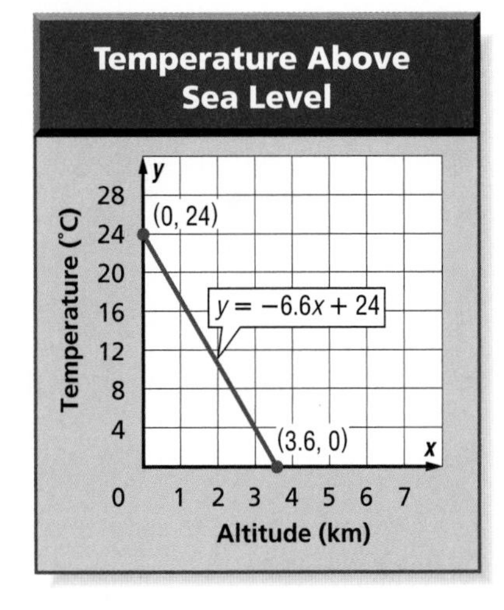

Step 3 Plot the points with coordinates (3.6, 0) and (0, 24). Then draw a line through the points.

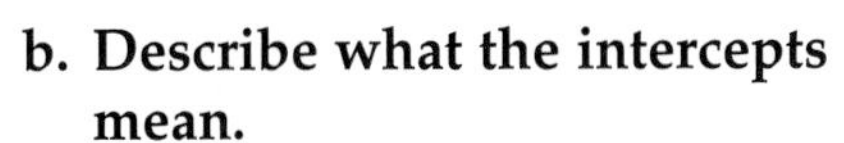

b. Describe what the intercepts mean.

The x-intercept 3.6 means that when the hot-air balloon is 3.6 kilometers above sea level, the temperature is 0°C. The y-intercept 24 means that the temperature at sea level is 24°C.

Some linear equations have just one variable. Their graphs are horizontal or vertical lines.

Example 5 Horizontal and Vertical Lines

Graph each equation using the x- and y-intercepts.

a. $y = 3$

Note that $y = 3$ is the same as $0x + y = 3$. The y-intercept is 3, and there is no x-intercept.

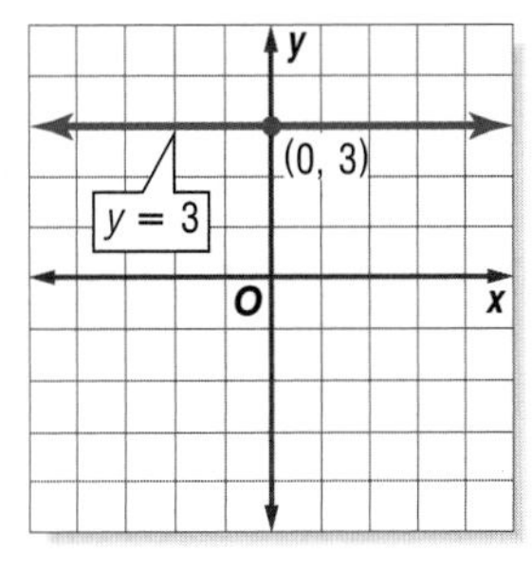

b. $x = -2$

Note that $x = -2$ is the same as $x + 0y = -2$. The x-intercept is -2, and there is no y-intercept.

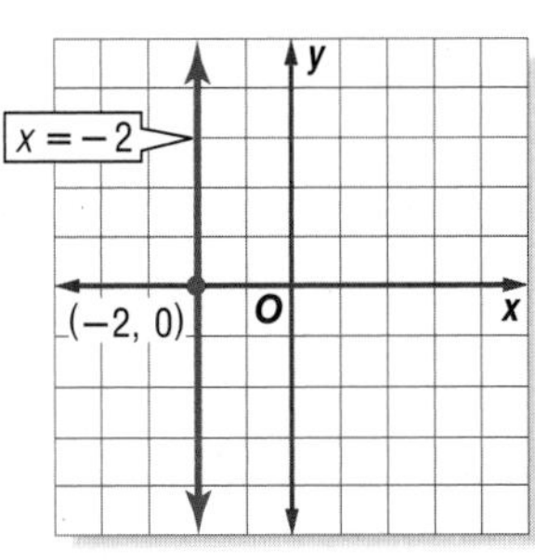

Reading Math

- $y = 3$ can be read *for all x, y = 3.*
- $x = -2$ can be read *for all y, x = −2.*

Check for Understanding

Concept Check

1. **Explain** how to find the x- and y-intercepts of a line given its equation.
2. **OPEN ENDED** Sketch the graph of a function whose x- and y-intercepts are both negative. Label the intercepts.

Guided Practice

State the x-intercept and the y-intercept of each line.

3. 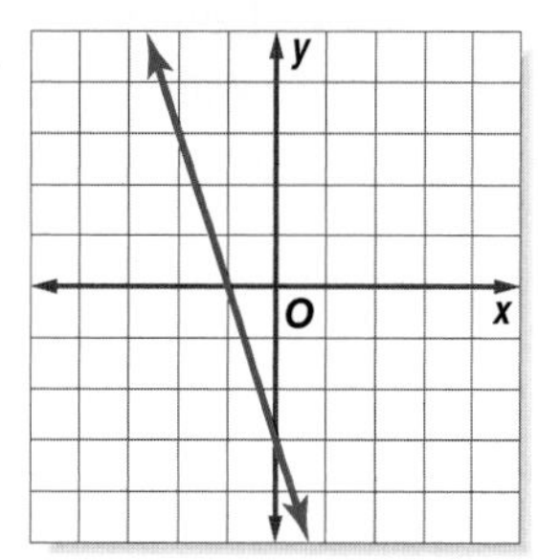

4. 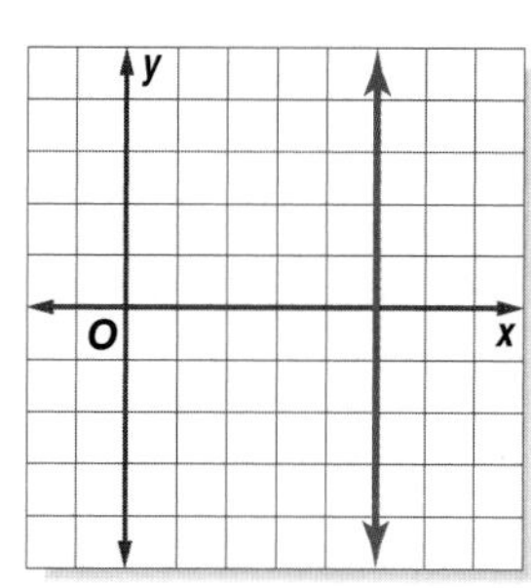

Find the x-intercept and the y-intercept for the graph of each equation.

5. $y = x + 4$
6. $y = 7$
7. $2x + 3y = 6$

Graph each equation using the x- and y-intercepts.

8. $y = x + 1$
9. $x - 2y = 6$
10. $x = -1$

Application

11. **BUSINESS** A lawn mowing service charges a base fee of \$3, plus \$6 per hour for labor. This can be represented by $y = 6x + 3$, where y is the total cost and x is the number of hours. Graph this equation and explain what the y-intercept represents.

Practice and Apply

Homework Help

For Exercises	See Examples
12–15	1
16–24	2
25–33	3, 5
34, 35	4

Extra Practice

See page 742.

State the x-intercept and the y-intercept of each line.

12. 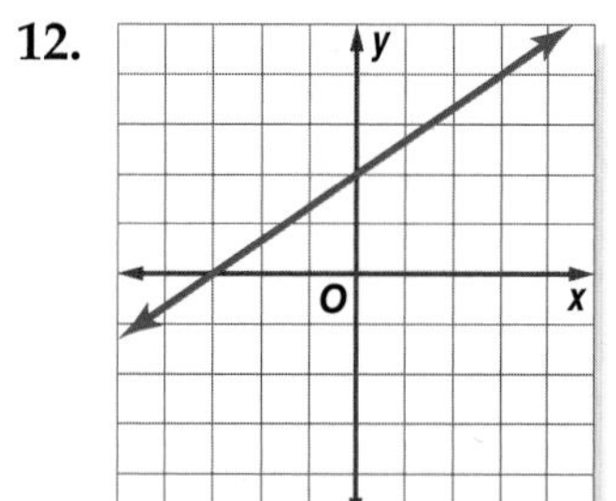

13. 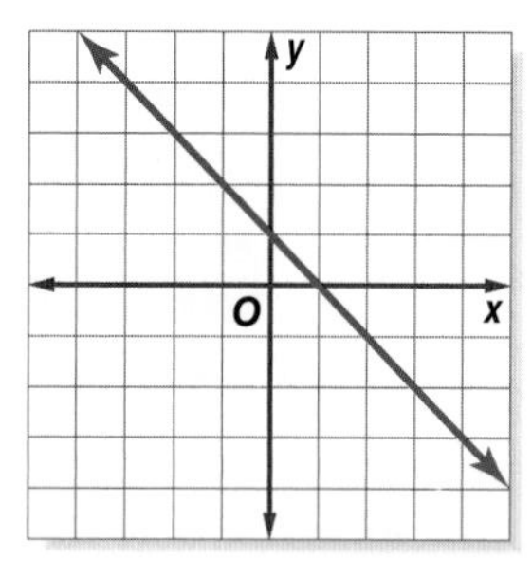

14. 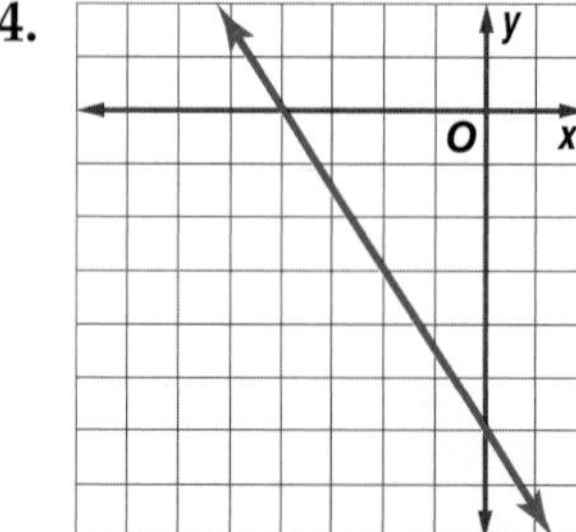

15. 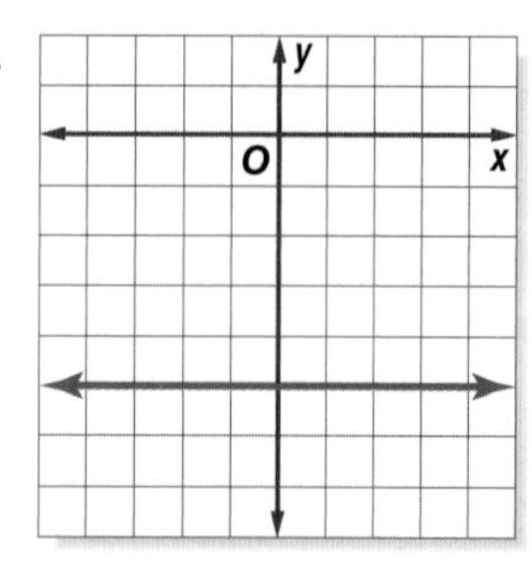

Find the x-intercept and the y-intercept for the graph of each equation.

16. $y = x - 1$
17. $y = x + 5$
18. $x = 9$
19. $y + 4 = 0$
20. $y = 2x + 10$
21. $x - 2y = 8$
22. $y = -3x - 12$
23. $4x + 5y = 20$
24. $6x + 7y = 12$

Graph each equation using the x- and y-intercepts.

25. $y = x + 2$ **26.** $y = x - 3$ **27.** $x + y = 4$

28. $y = 5x + 5$ **29.** $y = -2x + 4$ **30.** $x + 2y = -6$

31. $y = -2$ **32.** $x - 3 = 0$ **33.** $3x + 6y = 18$

34. **CATERING** For a luncheon, a caterer charges \$8 per person, plus a setup fee of \$24. The total cost of the luncheon y can be represented by $y = 8x + 24$, where x is the number of people. Graph the equation and explain what the y-intercept represents.

35. **MONEY** Jasmine has \$18 to buy books at the library used book sale. Paperback books cost \$3 each. The equation $y = 18 - 3x$ represents the amount of money she has left over if she buys x paperback books. Graph the equation and describe what the intercepts represent.

36. **GEOMETRY** The perimeter of a rectangle is 50 centimeters. This can be given by the equation $50 = 2\ell + 2w$, where ℓ is the length and w is the width. Name the x- and y-intercepts of the equation and explain what they mean.

37. **CRITICAL THINKING** Explain why you cannot graph $y = 2x$ by using intercepts only. Then draw the graph.

38. WRITING IN MATH Answer the question that was posed at the beginning of the lesson.

How can intercepts be used to represent real-life information?

Include the following in your answer:

- a graph showing a decrease in temperature, with the x-axis representing time and the y-axis representing temperature, and
- an explanation of what the intercepts mean.

39. What is the x-intercept of the graph of $y = 8x - 32$?

(A) -4 (B) 4 (C) -32 (D) 32

40. The graph of which equation does *not* have a y-intercept of 3?

(A) $2x + 3y = 9$ (B) $4x + y = 3$ (C) $x + 3y = 6$ (D) $x - 2y = -6$

Maintain Your Skills

Mixed Review **Find four solutions of each equation.** *(Lesson 8-2)*

41. $y = 2x + 7$ **42.** $y = -3x + 1$ **43.** $4x - y = -5$

Determine whether each relation is a function. *(Lesson 8-1)*

44. $\{(2, 12), (4, -5), (-3, -4), (11, 0)\}$

45. $\{(-4.2, 17), (-4.3, 16), (-4.3, 15), (-4.3, 14)\}$

Solve each inequality. *(Lesson 7-4)*

46. $y + 3 < 5$ **47.** $-2 + n > 10$ **48.** $7 \leq x + 8$

49. Express 0.028 as a percent. *(Lesson 6-4)*

Getting Ready for the Next Lesson **PREREQUISITE SKILL** **Subtract.** *(To review **subtracting integers**, see Lesson 2-3.)*

50. $-11 - 13$ **51.** $15 - 31$ **52.** $-26 - (-26)$ **53.** $9 - (-16)$

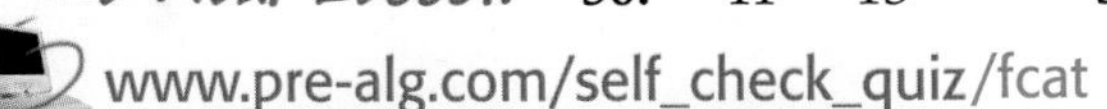

Algebra Activity

A Preview of Lesson 8-4

It's All Downhill

The steepness, or *slope*, of a hill can be described by a ratio.

$$\text{slope} = \frac{\text{vertical change}}{\text{horizontal change}} \begin{matrix} \leftarrow \text{height} \\ \leftarrow \text{length} \end{matrix}$$

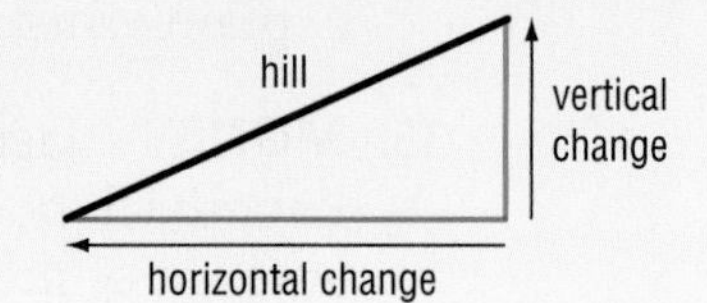

Collect the Data

Step 1 Use posterboard or a wooden board, tape, and three or more books to make a "hill."

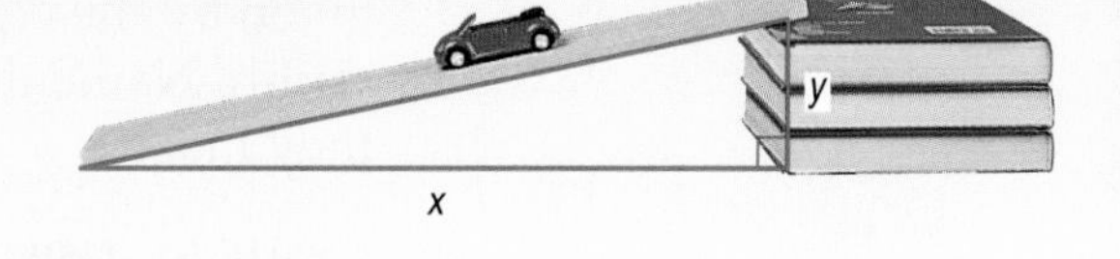

Step 2 Measure the height y and length x of the hill to the nearest $\frac{1}{2}$ inch or $\frac{1}{4}$ inch. Record the measurements in a table like the one below.

Hill	Height y (in.)	Length x (in.)	Car Distance (in.)	Slope $\frac{y}{x}$
1				
2				
3				

Step 3 Place a toy car at the top of the hill and let it roll down. Measure the distance from the bottom of the ramp to the back of the car when it stops. Record the distance in the table.

Step 4 For the second hill, increase the height by adding one or two more books. Roll the car down and measure the distance it rolls. Record the dimensions of the hill and the distance in the table.

Step 5 Take away two or three books so that hill 3 has the least height. Roll the car down and measure the distance it rolls. Record the dimensions of the hill and the distance in the table.

Step 6 Find the slopes of hills 1, 2, and 3 and record the values in the table.

Analyze the Data

1. How did the slope change when the height increased and the length decreased?
2. How did the slope change when the height decreased and the length increased?
3. **MAKE A CONJECTURE** On which hill would a toy car roll the farthest—a hill with slope $\frac{18}{25}$ or $\frac{25}{18}$? Explain by describing the relationship between slope and distance traveled.

Extend the Activity

4. Make a fourth hill. Find its slope and predict the distance a toy car will go when it rolls down the hill. Test your prediction by rolling a car down the hill.

8-4 Slope

What You'll Learn

- Find the slope of a line.

Sunshine State Standards
MA.C.3.3.2-2, MA.D.1.3.1-1, MA.D.1.3.2-1, MA.D.1.3.2-4

Vocabulary

- slope

How is slope used to describe roller coasters?

Some roller coasters can make you feel heavier than a shuttle astronaut feels on liftoff. This is because the speed and steepness of the hills increase the effects of gravity.

a. Use the roller coaster to write the ratio $\frac{\text{height}}{\text{length}}$ in simplest form.

b. Find the ratio of a hill that has the same length but is 14 feet higher than the hill above. Is this hill steeper or less steep than the original?

SLOPE **Slope** describes the steepness of a line. It is the ratio of the *rise*, or the vertical change, to the *run*, or the horizontal change.

$$\text{slope} = \frac{\text{rise}}{\text{run}} \quad \begin{matrix} \leftarrow \text{vertical change} \\ \leftarrow \text{horizontal change} \end{matrix}$$

$$= \frac{4}{3}$$

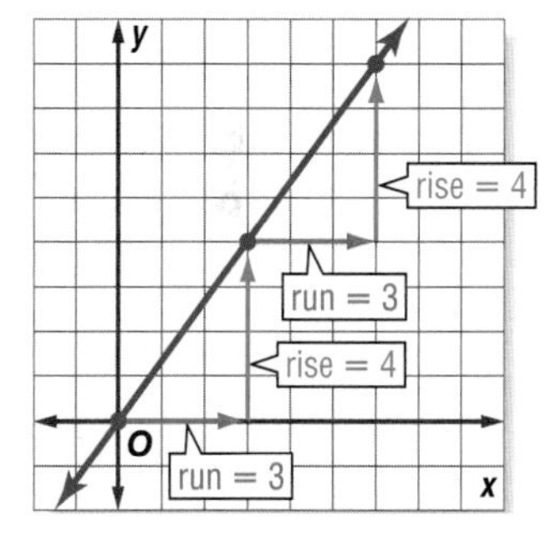

Note that the slope is the same for any two points on a straight line.

Example 1 Use Rise and Run to Find Slope

Find the slope of a road that rises 25 feet for every horizontal change of 80 feet.

$\text{slope} = \frac{\text{rise}}{\text{run}}$ Write the formula.

$= \frac{25 \text{ ft}}{80 \text{ ft}}$ rise = 25 ft, run = 80 ft

$= \frac{5}{16}$ Simplify.

The slope of the road is $\frac{5}{16}$ or 0.3125.

Concept Check What is the slope of a ramp that rises 2 inches for every horizontal change of 24 inches?

You can also find the slope by using the coordinates of any two points on a line.

Key Concept — **Slope**

- **Words** The slope m of a line passing through points at (x_1, y_1) and (x_2, y_2) is the ratio of the difference in y-coordinates to the corresponding difference in x-coordinates.
- **Symbols** $m = \frac{y_2 - y_1}{x_2 - x_1}$, where $x_2 \neq x_1$
- **Model**

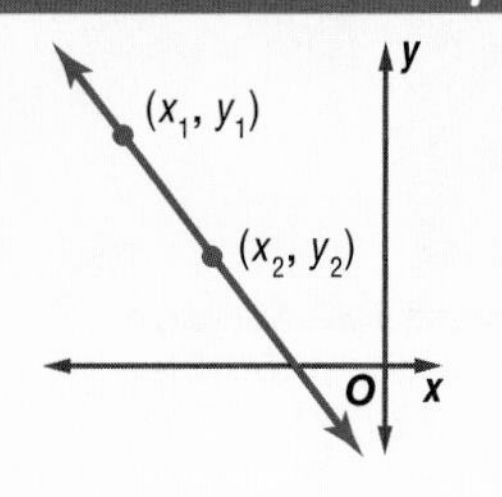

The slope of a line may be positive, negative, zero, or undefined.

Study Tip

Choosing Points

- Any two points on a line can be chosen as (x_1, y_1) and (x_2, y_2).
- The coordinates of both points must be used in the same order.

Check: In Example 2, let $(x_1, y_1) = (5, 3)$ and let $(x_2, y_2) = (2, 2)$, then find the slope.

Example 2 Positive Slope

Find the slope of the line.

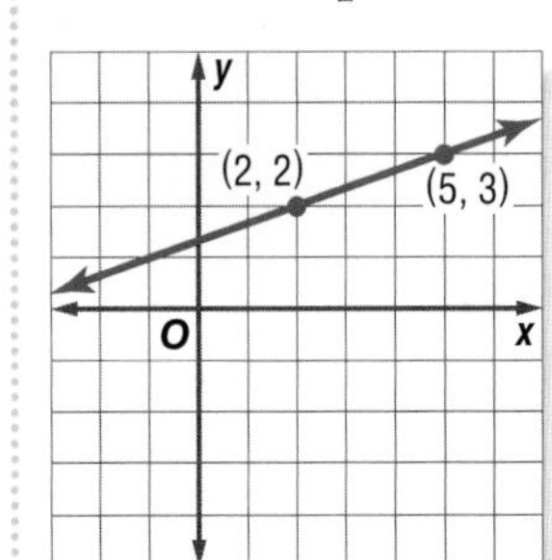

$m = \frac{y_2 - y_1}{x_2 - x_1}$ Definition of slope

$m = \frac{3 - 2}{5 - 2}$ $(x_1, y_1) = (2, 2)$, $(x_2, y_2) = (5, 3)$

$m = \frac{1}{3}$

The slope is $\frac{1}{3}$.

Example 3 Negative Slope

Find the slope of the line.

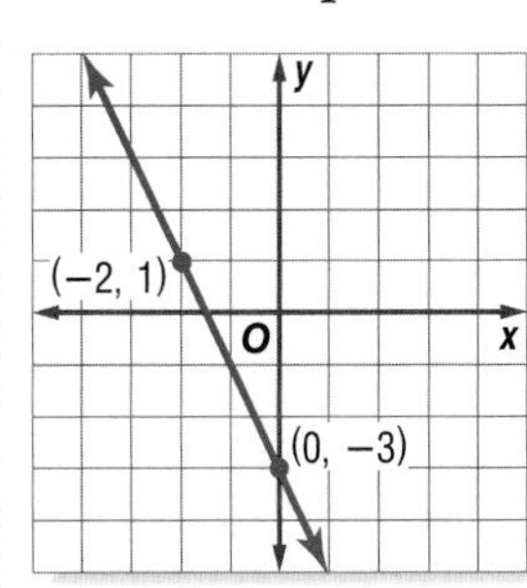

$m = \frac{y_2 - y_1}{x_2 - x_1}$ Definition of slope

$m = \frac{-3 - 1}{0 - (-2)}$ $(x_1, y_1) = (-2, 1)$, $(x_2, y_2) = (0, -3)$

$m = \frac{-4}{2}$ or -2

The slope is -2.

Example 4 Zero Slope

Find the slope of the line.

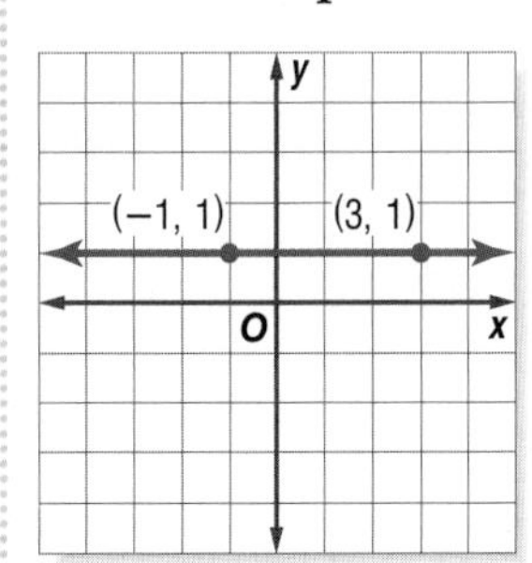

$m = \frac{y_2 - y_1}{x_2 - x_1}$ Definition of slope

$m = \frac{1 - 1}{3 - (-1)}$ $(x_1, y_1) = (-1, 1)$, $(x_2, y_2) = (3, 1)$

$m = \frac{0}{4}$ or 0

The slope is 0.

Example 5 Undefined Slope

Find the slope of the line.

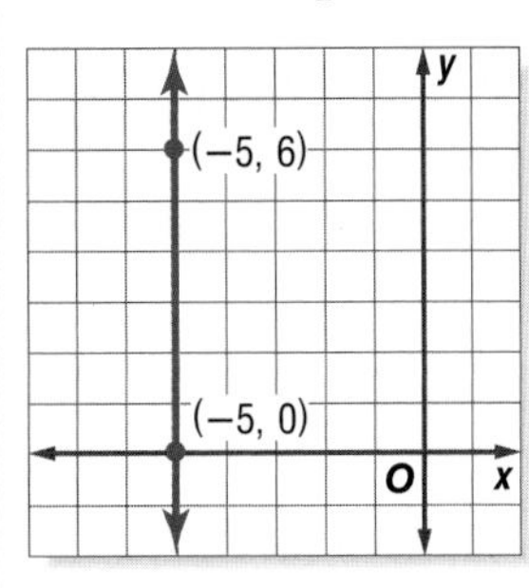

$m = \frac{y_2 - y_1}{x_2 - x_1}$ Definition of slope

$m = \frac{0 - 6}{-5 - (-5)}$ $(x_1, y_1) = (-5, 6)$, $(x_2, y_2) = (-5, 0)$

$m = \frac{-6}{0}$

Division by 0 is undefined. So, the slope is undefined.

The steepness of real-world inclines can be compared by using slope.

Example 6 Compare Slopes

Multiple-Choice Test Item

There are two major hills on a hiking trail. The first hill rises 6 feet vertically for every 42-foot run. The second hill rises 10 feet vertically for every 98-foot run. Which statement is true?

(A) The first hill is steeper than the second hill.

(B) The second hill is steeper than the first hill.

(C) Both hills have the same steepness.

(D) You cannot determine which hill is steeper.

The Princeton Review

Test-Taking Tip

Make a Drawing
Whenever possible, make a drawing that displays the given information. Then use the drawing to estimate the answer.

Read the Test Item To compare steepness of the hills, find the slopes.

Solve the Test Item

first hill

$\text{slope} = \frac{\text{rise}}{\text{run}}$

$= \frac{6 \text{ ft}}{42 \text{ ft}}$ rise = 6 ft, run = 42 ft

$= \frac{1}{7}$ or about 0.14

second hill

$\text{slope} = \frac{\text{rise}}{\text{run}}$

$= \frac{10 \text{ ft}}{98 \text{ ft}}$ rise = 10 ft, run = 98 ft

$= \frac{5}{49}$ or about 0.10

$0.14 > 0.10$, so the first hill is steeper than the second. The answer is A.

Check for Understanding

Concept Check

1. **Describe** your own method for remembering whether a horizontal line has 0 slope or an undefined slope.
2. **OPEN ENDED** Draw a line whose slope is $-\frac{1}{4}$.
3. **FIND THE ERROR** Mike and Chloe are finding the slope of the line that passes through $Q(-2, 8)$ and $R(11, 7)$.

Mike	Chloe
$m = \frac{8 - 7}{-2 - 11}$	$m = \frac{7 - 8}{11 - 2}$

Who is correct? Explain your reasoning.

Guided Practice

4. Find the slope of a line that decreases 24 centimeters vertically for every 30-centimeter horizontal increase.

Find the slope of each line.

5. 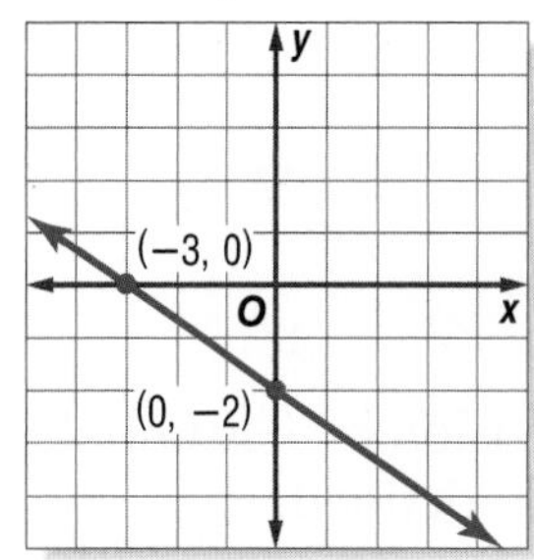

6. 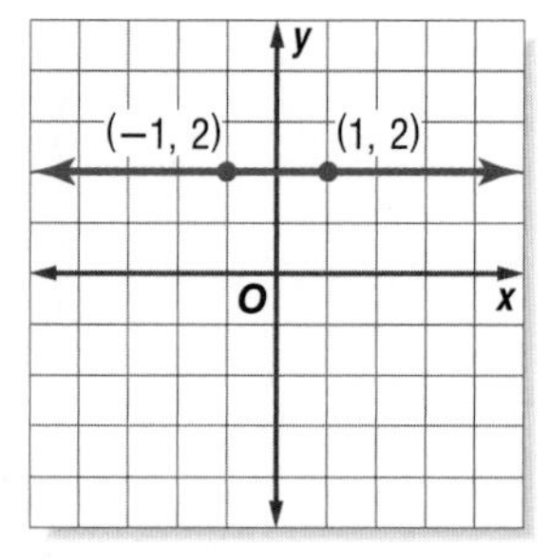

Find the slope of the line that passes through each pair of points.

7. $A(3, 4)$, $B(4, 6)$
8. $J(-8, 0)$, $K(-8, 10)$
9. $P(7, -1)$, $Q(9, -1)$
10. $C(-6, -4)$, $D(-8, -3)$

11. Which bike ramp is the steepest?

(A) 1

(B) 2

(C) 3

(D) 4

Bike Ramp	Height (ft)	Length (ft)
1	6	8
2	10	4
3	5	3
4	8	4

Practice and Apply

Homework Help

For Exercises	See Examples
12, 13	1
14–25	2–5
26, 27	6

Extra Practice

See page 742.

12. **CARPENTRY** In a stairway, the slope of the handrail is the ratio of the riser to the tread. If the tread is 12 inches long and the riser is 8 inches long, find the slope.

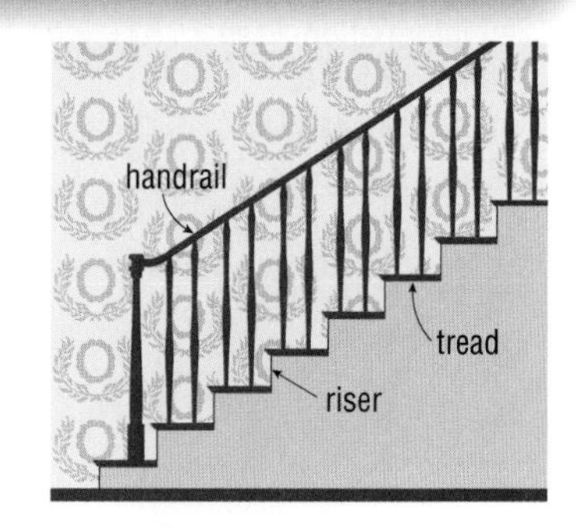

13. **HOME REPAIR** The bottom of a ladder is placed 4 feet away from a house and it reaches a height of 16 feet on the side of the house. What is the slope of the ladder?

Find the slope of each line.

14. 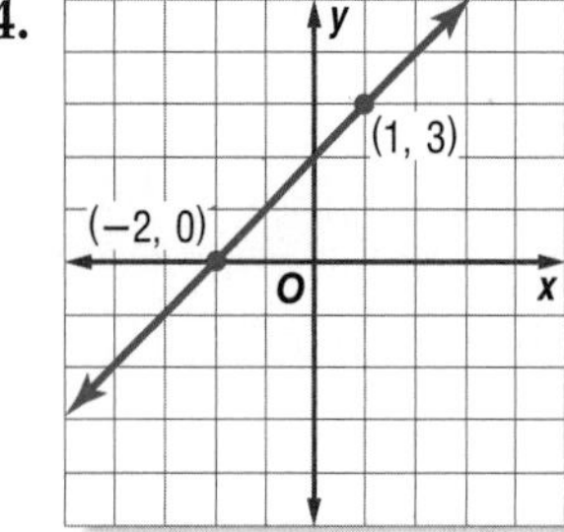

15. 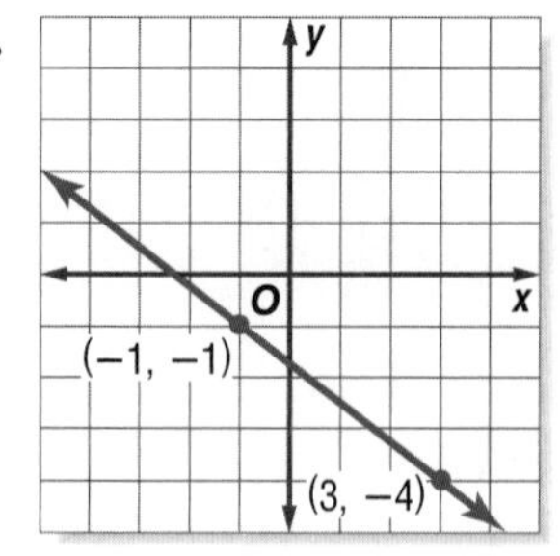

16. 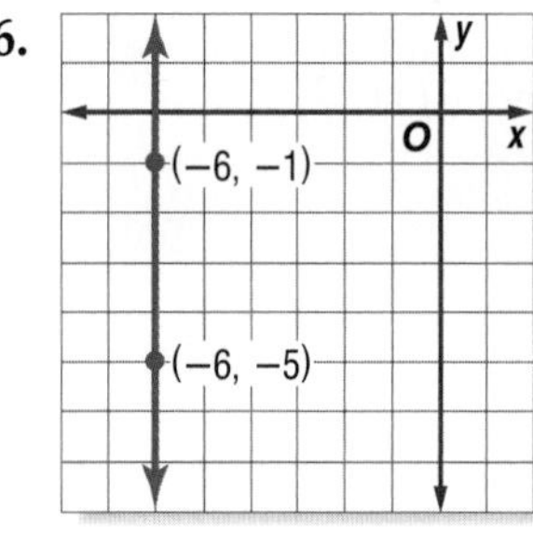

17. 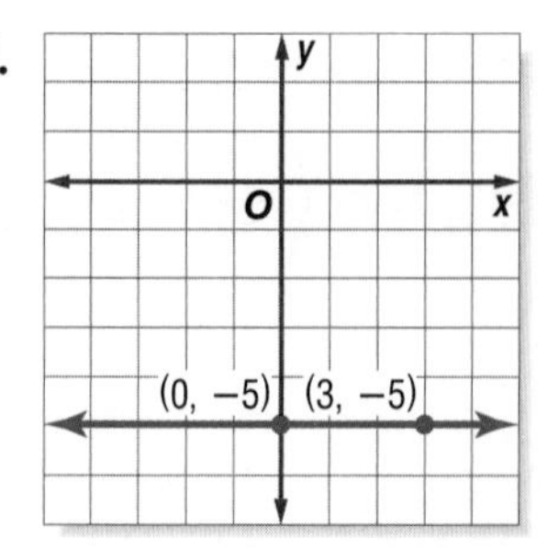

Find the slope of the line that passes through each pair of points.

18. $A(1, -3)$, $B(5, 4)$ **19.** $Y(4, -3)$, $Z(5, -2)$ **20.** $D(5, -1)$, $E(-3, 4)$

21. $J(-3, 6)$, $K(-5, 9)$ **22.** $N(2, 6)$, $P(-1, 6)$ **23.** $S(-9, -4)$, $T(-9, 8)$

24. $F(0, 1.6)$, $G(0.5, 2.1)$ **25.** $W\left(3\frac{1}{2}, 5\frac{1}{4}\right)$, $X\left(2\frac{1}{2}, 6\right)$

ENTERTAINMENT **For Exercises 26 and 27, use the graph.**

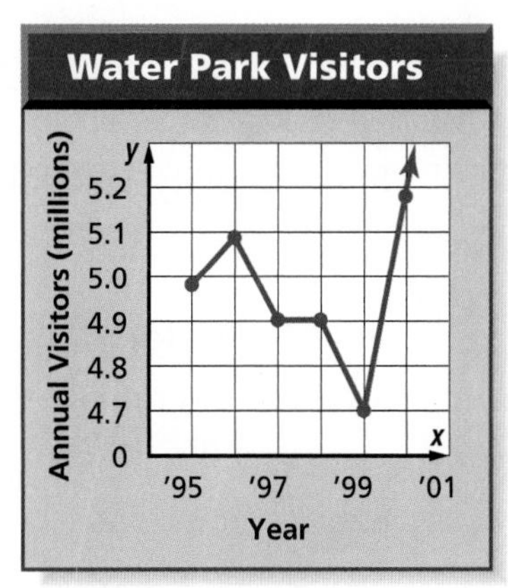

Source: *Amusement Business*

26. Which section of the graph shows the greatest increase in attendance? Describe the slope.

27. What happened to the attendance at the water park from 1996–1997? Describe the slope of this part of the graph.

28. **CRITICAL THINKING** The graph of a line goes through the origin (0, 0) and $C(a, b)$. State the slope of this line and explain how it relates to the coordinates of point C.

29. WRITING IN MATH Answer the question that was posed at the beginning of the lesson.

How is slope used to describe roller coasters?

Include the following in your answer:

- a description of slope, and
- an explanation of how changes in rise or run affect the steepness of a roller coaster.

30. Identify the graph that has a positive slope.

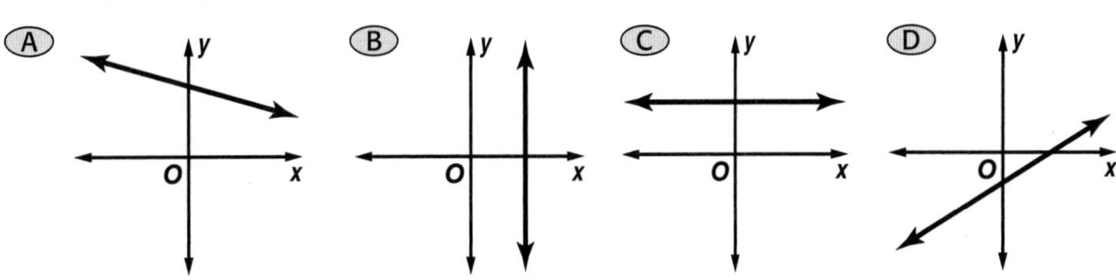

31. What is the slope of line LM given $L(9, -2)$ and $M(3, -5)$?

Ⓐ $\frac{1}{2}$ Ⓑ $-\frac{1}{2}$ Ⓒ 5 Ⓓ $-\frac{1}{3}$

Maintain Your Skills

Mixed Review **Find the x-intercept and the y-intercept for the graph of each equation.** *(Lesson 8-3)*

32. $y = x + 8$ **33.** $y = -3x + 6$ **34.** $4x - y = 12$

Find four solutions of each equation. Write the solutions as ordered pairs. *(Lesson 8-2)*

35. $y = 2x + 5$ **36.** $y = -3x$ **37.** $x + y = 7$

Getting Ready for the Next Lesson **PREREQUISITE SKILL** **Rewrite $y = kx$ by replacing k with each given value.** *(To review **substitution**, see Lesson 1-3.)*

38. $k = 5$ **39.** $k = -2$ **40.** $k = 0.25$ **41.** $k = \frac{1}{3}$

Algebra Activity

A Preview of Lesson 8-5

Sunshine State Standards
MA.C.3.3.2-2, MA.D.1.3.2-1, MA.D.1.3.2-4, MA.D.2.3.1-4

Slope and Rate of Change

In this activity, you will investigate the relationship between slope and rate of change.

Collect the Data

Step 1 On grid paper, make a coordinate grid of the first quadrant. Label the x-axis *Number of Measures* and label the y-axis *Height of Water (cm)*.

Step 2 Pour water into a drinking glass or a beaker so that it is more than half full.

Step 3 Use a ruler to find the initial height of the water and record the measurement in a table.

Step 4 Remove a tablespoon of water from the glass or beaker and record the new height in your table.

Step 5 Repeat Step 4 so that you have six measures.

Step 6 Fill the glass or beaker again so that it has the same initial height as in Step 3.

Step 7 Repeat Steps 4 and 5, using a $\frac{1}{8}$-cup measuring cup.

Analyze the Data

1. On the coordinate grid, graph the ordered pairs (number of measures, height of water) for each set of data. Draw a line through each set of points. Label the lines 1 and 2, respectively.
2. Compare the steepness of the two graphs. Which has a steeper slope?
3. What does the height of the water depend on?
4. What happens as the number of measures increases?
5. Did you empty the glass at a faster rate using a tablespoon or a $\frac{1}{8}$-cup? Explain.

Make a Conjecture

6. Describe the relationship between slope and the rate at which the glass was emptied.
7. What would a graph look like if you emptied a glass using a teaspoon? a $\frac{1}{4}$ cup? Explain.

Extend the Activity

8. Water is emptied at a constant rate from containers shaped like the ones shown below. Draw a graph of the water level in each of the containers as a function of time.

a. 

b. 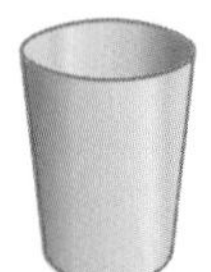

c. 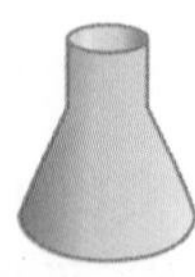

8-5 Rate of Change

What You'll Learn

- Find rates of change.
- Solve problems involving direct variation.

Sunshine State Standards
MA.B.1.3.2-2, MA.C.3.3.2-2, MA.D.1.3.1-1, MA.D.1.3.2-1, MA.D.1.3.2-4

Vocabulary
- rate of change
- direct variation
- constant of variation

How are slope and speed related?

A car traveling 55 miles per hour goes 110 miles in 2 hours, 165 miles in 3 hours, and 220 miles in 4 hours, as shown.

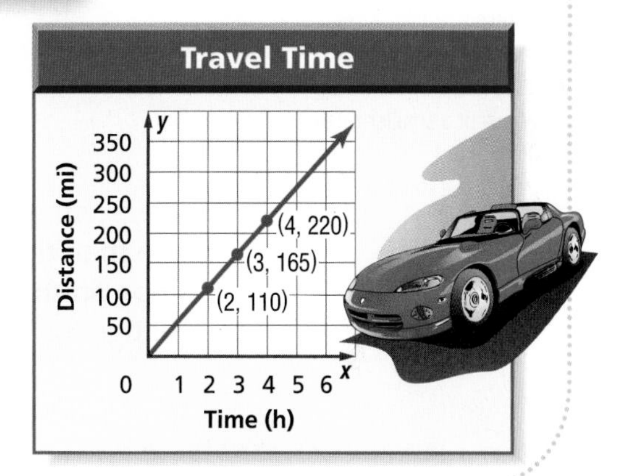

a. For every 1-hour increase in time, what is the change in distance?

b. Find the slope of the line.

c. **Make a conjecture** about the relationship between slope of the line and speed of the car.

RATE OF CHANGE A change in one quantity with respect to another quantity is called the **rate of change**. Rates of change can be described using slope.

$$\text{slope} = \frac{\text{change in } y}{\text{change in } x}$$

$$= \frac{55 \text{ mi}}{1 \text{ h}} \text{ or } 55 \text{ mi/h}$$

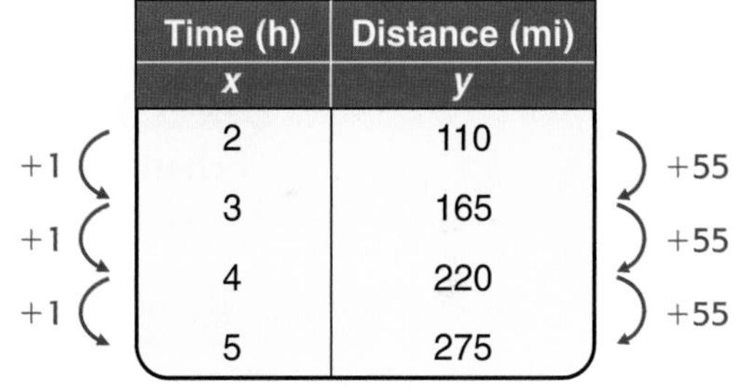

Time (h) x	Distance (mi) y
2	110
3	165
4	220
5	275

+1 +1 +1   +55 +55 +55

Each time x increases by 1, y increases by 55.

The slope of the line is the speed of the car.

You can find rates of change from an equation, a table of values, or a graph.

Example 1 Find a Rate of Change

TECHNOLOGY The graph shows the expected growth of subscribers to satellite radio for the first five years that it is introduced. Find the expected rate of change from Year 2 to Year 5.

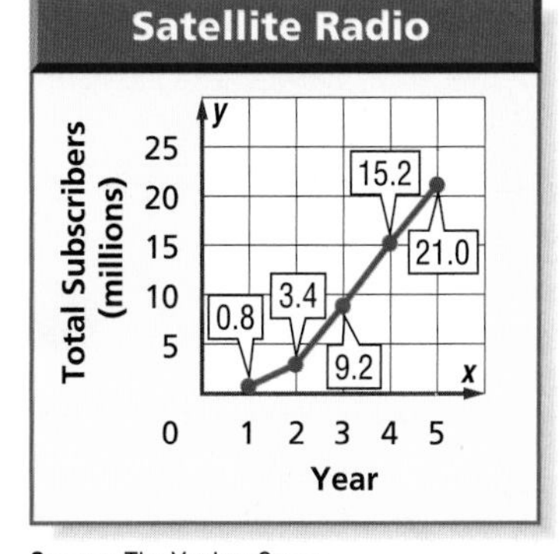

Source: The Yankee Group

rate of change = slope

$= \frac{y_2 - y_1}{x_2 - x_1}$ Definition of slope

$= \frac{21.0 - 3.4}{5 - 2}$ ← change in subscribers / ← change in time

≈ 5.9 Simplify.

So, the expected rate of change in satellite radio subscribers is an increase of about 5.9 million people per year.

The steepness of slopes is also important in describing rates of change.

Example 2 Compare Rates of Change

Study Tip

Slopes
- Positive slopes represent a rate of increase.
- Negative slopes represent a rate of decrease.
- Steeper slopes represent greater rates of change.
- Less steep slopes represent a smaller rate of change.

GEOMETRY The table shows how the perimeters of an equilateral triangle and a square change as side lengths increase. Compare the rates of change.

Side Length x	Perimeter y	
	Triangle	Square
0	0	0
2	6	8
4	12	16

triangle rate of change $= \dfrac{\text{change in } y}{\text{change in } x}$

$= \dfrac{6}{2}$ or 3 — For each side length increase of 2, the perimeter increases by 6.

square rate of change $= \dfrac{\text{change in } y}{\text{change in } x}$

$= \dfrac{8}{2}$ or 4 — For each side length increase of 2, the perimeter increases by 8.

The perimeter of a square increases at a faster rate than the perimeter of a triangle. A steeper slope on the graph indicates a greater rate of change for the square.

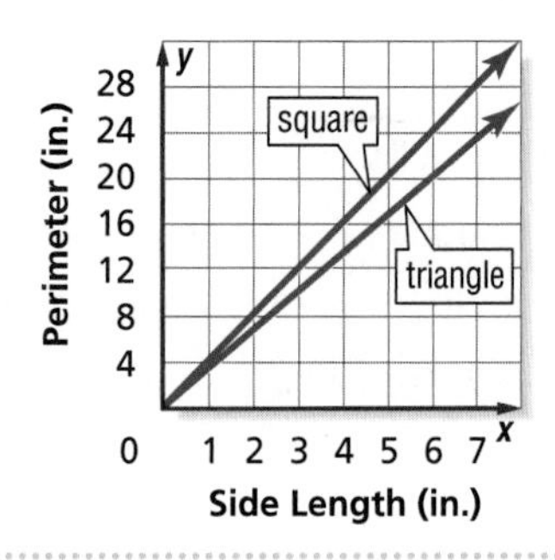

DIRECT VARIATION A special type of linear equation that describes rate of change is called a **direct variation**. The graph of a direct variation always passes through the origin and represents a proportional situation.

Key Concept — *Direct Variation*

- **Words** A direct variation is a relationship such that as x increases in value, y increases or decreases at a constant rate k.
- **Symbols** $y = kx$, where $k \neq 0$
- **Example** $y = 2x$
- **Model**

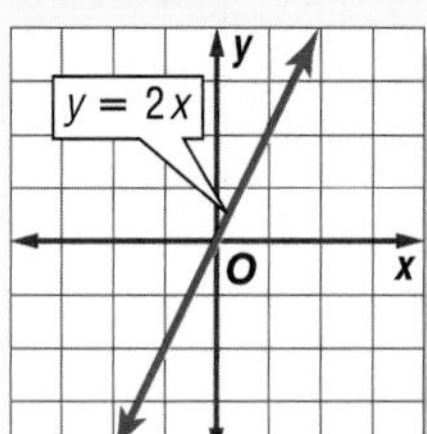

In the equation $y = kx$, k is called the **constant of variation**. It is the slope, or rate of change. We say that *y varies directly with x.*

Example 3 Write a Direct Variation Equation

Suppose y varies directly with x and $y = -6$ when $x = 2$. Write an equation relating x and y.

Step 1 Find the value of k.

$y = kx$ — Direct variation

$-6 = k(2)$ — Replace y with -6 and x with 2.

$-3 = k$ — Simplify.

Step 2 Use k to write an equation.

$y = kx$ — Direct variation

$y = -3x$ — Replace k with -3.

So, a direct variation equation that relates x and y is $y = -3x$.

The direct variation $y = kx$ can be written as $k = \frac{y}{x}$. In this form, you can see that the ratio of y to x is the same for any corresponding values of y and x.

Example 4 Use Direct Variation to Solve Problems

More About. . .

Pools

The Johnson Space Center in Houston, Texas, has a 6.2 million gallon pool used to train astronauts for space flight. It is 202 feet long, 102 feet wide, and 40 feet deep.

Source: www.jsc.nasa.gov

POOLS The height of the water as a pool is being filled is recorded in the table below.

a. Write an equation that relates time and height.

Step 1 Find the ratio of y to x for each recorded time. These are shown in the third column of the table. The ratios are approximately equal to 0.4.

Time (min) x	Height (in.) y	$k = \frac{y}{x}$
5	2.0	0.40
10	3.75	0.38
15	5.5	0.37
20	7.5	0.38

To the nearest tenth, $k \approx 0.4$.

Step 2 Write an equation.

$y = kx$ Direct variation

$y = 0.4x$ Replace k with 0.4.

So, a direct variation equation that relates the time x and the height of the water y is $y = 0.4x$.

b. Predict how long it will take to fill the pool to a height of 48 inches.

$y = 0.4x$ Write the direct variation equation.

$48 = 0.4x$ Replace y with 48.

$120 = x$ Divide each side by 0.4.

It will take about 120 minutes, or 2 hours to fill the pool.

Check for Understanding

Concept Check

1. **Describe** how slope, rate of change, and constant of variation are related by using $y = 60x$ as a model.
2. **OPEN ENDED** Draw a line that shows a 2-unit increase in y for every 1-unit increase in x. State the rate of change.
3. **FIND THE ERROR** Justin and Carlos are determining how to find rate of change from the equation $y = 4x + 5$.

Justin

The rate of change is the slope of its graph.

Carlos

There is no rate of change because the equation is not a direct variation.

Who is correct? Explain your reasoning.

Guided Practice **Find the rate of change for each linear function.**

4.

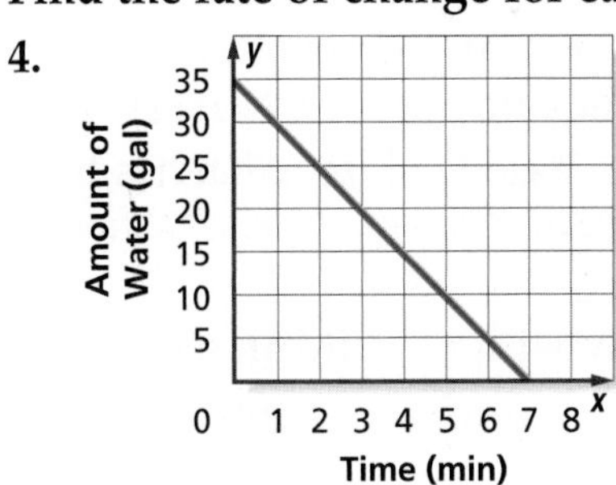

5.

Time (h) x	Wage ($) y
0	0
1	12
2	24
3	36

Suppose y varies directly with x. Write an equation relating x and y.

6. $y = 5$ when $x = -15$

7. $y = 24$ when $x = 4$

Application

8. PHYSICAL SCIENCE The length of a spring varies directly with the amount of weight attached to it. When a 25-gram weight is attached, a spring stretches to 8 centimeters.

a. Write a direct variation equation relating the weight x and the length y.

b. Estimate the length of a spring that has a 60-gram weight attached.

Practice and Apply

Homework Help

For Exercises	See Examples
9–12	1
13	2
14, 15	3
16, 17	4

Extra Practice
See page 743.

Find the rate of change for each linear function.

9.

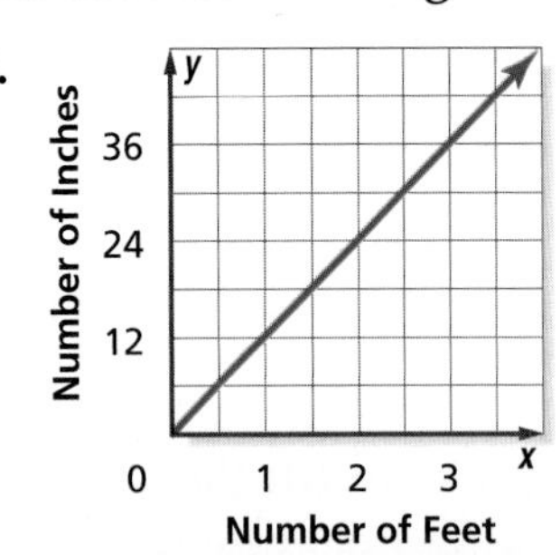

10.

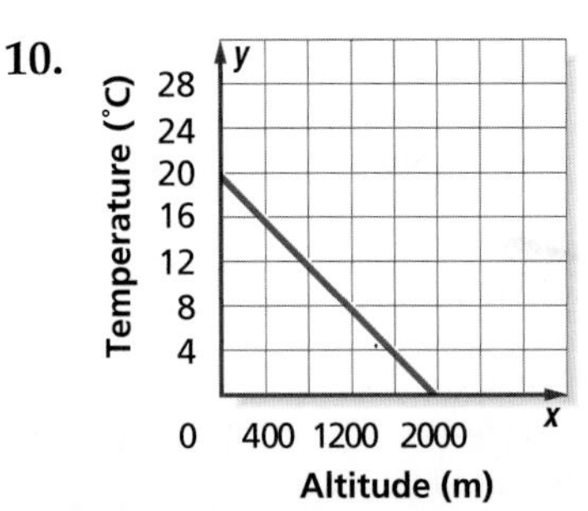

11.

Time (min) x	Temperature (°F) y
0	58
1	56
2	54
3	52

12.

Time (h) x	Distance (mi) y
0.0	0
0.5	25
1.5	75
3.0	150

13. ENDANGERED SPECIES The graph shows the populations of California condors in the wild and in captivity.

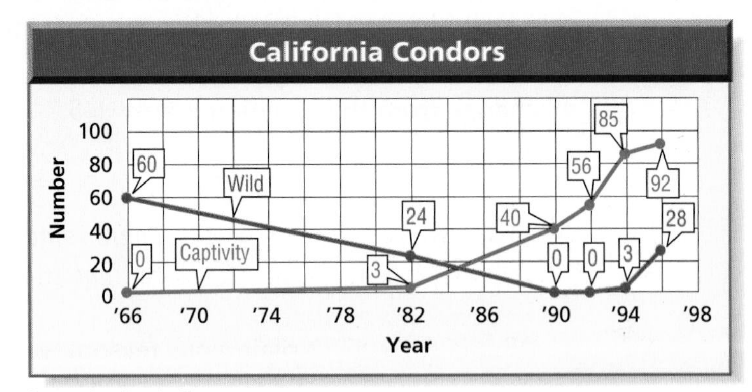

Source: Los Angeles Zoo

Write several sentences that describe how the populations have changed since 1966. Include the rate of change for several key intervals.

Online Research **Data Update** What has happened to the condor population since 1996? Visit www.pre-alg.com/data_update to learn more.

More About. . .

Endangered Species

California condors hold the record for being the largest flying land birds in North America.

Source: www.birding.about.com

Suppose y varies directly with x. Write an equation relating x and y.

14. $y = 8$ when $x = 4$

15. $y = -30$ when $x = 6$

16. $y = 9$ when $x = 24$

17. $y = 7.5$ when $x = 10$

18. **FOOD COSTS** The cost of cheese varies directly with the number of pounds bought. If 2 pounds cost $8.40, find the cost of 3.5 pounds.

19. **CONVERTING MEASUREMENTS** The number of centimeters in a measure varies directly as the number of inches. Write a direct variation equation that could be used to convert inches to centimeters.

Measure in Inches x	Measure in Centimeters y
1	2.54
2	5.08
3	7.62

20. **CRITICAL THINKING** Describe the rate of change for a graph that is a horizontal line and a graph that is a vertical line.

21. **WRITING IN MATH** Answer the question that was posed at the beginning of the lesson.

How are slope and speed related?

Include the following in your answer:

- a drawing of a graph showing distance versus time, and
- an explanation of how slope changes when speed changes.

22. A graph showing an increase in sales over time would have a(n)
    Ⓐ positive slope. Ⓑ negative slope.
    Ⓒ undefined slope. Ⓓ slope of 0.

23. Choose an equation that does *not* represent a direct variation.
    Ⓐ $y = x$ Ⓑ $y = 1$ Ⓒ $y = -5x$ Ⓓ $y = 0.9x$

Maintain Your Skills

Mixed Review

Find the slope of the line that passes through each pair of points. *(Lesson 8-4)*

24. $Q(-4, 4)$, $R(3, 5)$
25. $A(2, 6)$, $B(-1, 0)$

Graph each equation using the x- and y-intercepts. *(Lesson 8-3)*

26. $y = x + 5$
27. $y = -x + 1$
28. $2x + y = 4$

29. Estimate 20% of 72. *(Lesson 6-6)*

Getting Ready for the Next Lesson

PREREQUISITE SKILL Solve each equation for y.
*(To review **solving equations for a variable**, see Lesson 8-2.)*

30. $x + y = 6$
31. $3x + y = 1$
32. $-x + 5y = 10$

Practice Quiz 1

Lessons 8-1 through 8-5

Determine whether each relation is a function. Explain. *(Lesson 8-1)*

1. $\{(0, 5), (1, 2), (1, -3), (2, 4)\}$
2. $\{(-6, 3.5), (-3, 4.0), (0, 4.5), (3, 5.0)\}$

Graph each equation using ordered pairs. *(Lesson 8-2)*

3. $y = x - 4$
4. $y = 2x + 3$

Find the x-intercept and the y-intercept for the graph of each equation. *(Lesson 8-3)*

5. $y = x + 9$
6. $x + 2y = 12$
7. $4x - 5y = 20$

Find the slope of the line that passes through each pair of points. *(Lessons 8-4 and 8-5)*

8. $(1, 4), (0, 0)$
9. $(-2, 4), (3, -6)$
10. $(0, 2), (5, 2)$

8-6 Slope-Intercept Form

What You'll Learn

- Determine slopes and y-intercepts of lines.
- Graph linear equations using the slope and y-intercept.

Sunshine State Standards
MA.C.3.3.2-2, MA.C.3.3.2-3, MA.D.1.3.1-2, MA.D.1.3.2-2, MA.D.1.3.2-4, MA.D.2.3.1-5

Vocabulary

- slope-intercept form

How can knowing the slope and y-intercept help you graph an equation?

Copy the table.

Equation	Slope	y-intercept
$y = 2x + 1$		
$y = \frac{1}{3}x - 3$		
$y = -2x + 1$		

a. On the same coordinate plane, use ordered pairs or intercepts to graph each equation in a different color.

b. Find the slope and the y-intercept of each line. Complete the table.

c. Compare each equation with the value of its slope and y-intercept. What do you notice?

SLOPE AND y-INTERCEPT

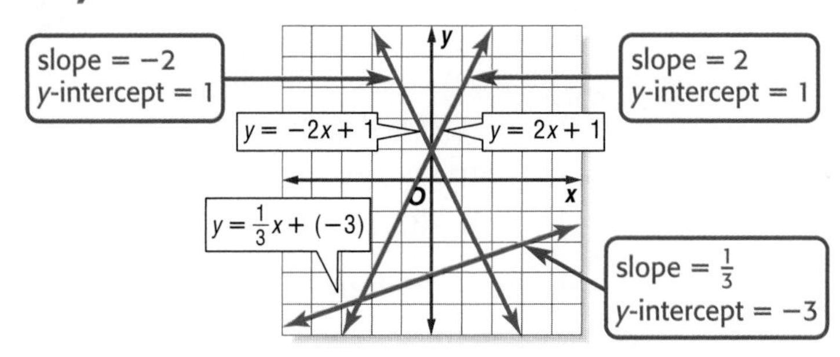

All the equations above are written in the form $y = mx + b$, where m is the slope and b is the y-intercept. This is called **slope-intercept form**.

$$y = mx + b$$

slope ↑ (m)   ↑ y-intercept (b)

Example 1 Find the Slope and y-Intercept

State the slope and the y-intercept of the graph of $y = \frac{3}{5}x - 7$.

$y = \frac{3}{5}x - 7$ — Write the original equation.

$y = \frac{3}{5}x + (-7)$ — Write the equation in the form $y = mx + b$.

↑ ↑

$y = mx + b$ — $m = \frac{3}{5}, b = -7$

The slope of the graph is $\frac{3}{5}$, and the y-intercept is -7.

Study Tip

Different Forms
Both equations below are written in slope-intercept form.
$y = x + (-2)$
$y = x - 2$

Concept Check What is the slope of $y = 8x + 6$?

Sometimes you must first write an equation in slope-intercept form before finding the slope and y-intercept.

Example 2 Write an Equation in Slope-Intercept Form

State the slope and the y-intercept of the graph of $5x + y = 3$.

$5x + y = 3$	Write the original equation.
$5x + y - 5x = 3 - 5x$	Subtract $5x$ from each side.
$y = -5x + 3$	Write the equation in slope-intercept form.
$\uparrow \quad \uparrow$	
$y = mx + b$	$m = -5, b = 3$

The slope of the graph is -5, and the y-intercept is 3.

GRAPH EQUATIONS You can use the slope-intercept form of an equation to easily graph a line.

Example 3 Graph an Equation

Graph $y = -\frac{1}{2}x - 4$ using the slope and y-intercept.

Step 1 Find the slope and y-intercept.

slope $= -\frac{1}{2}$ $\quad$ y-intercept $= -4$

Step 2 Graph the y-intercept point at $(0, -4)$.

Step 3 Write the slope $-\frac{1}{2}$ as $\frac{-1}{2}$. Use it to locate a second point on the line.

$m = \frac{-1}{2}$ $\quad$ ← change in y: down 1 unit; ← change in x: right 2 units

Another point on the line is at $(2, -5)$.

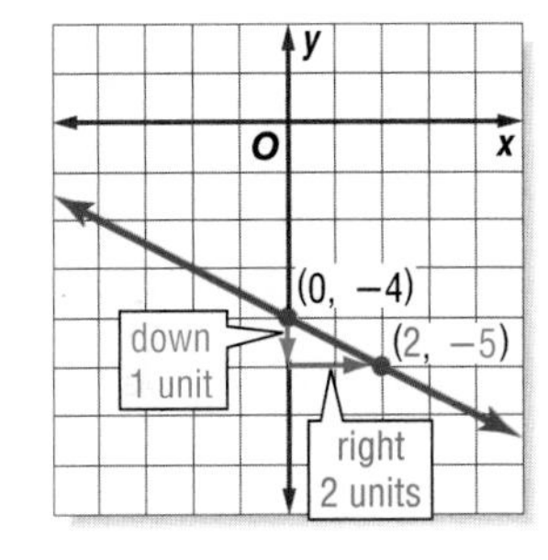

Step 4 Draw a line through the two points.

Career Choices

Business Owner

Business owners must understand the factors that affect cost and profit. Graphs are a useful way for them to display this information.

Online Research
For information about a career as a business owner, visit: www.pre-alg.com/careers

Example 4 Graph an Equation to Solve a Problem

BUSINESS A T-shirt company charges a design fee of \$24 for a pattern and then sells the shirts for \$12 each. The total cost y can be represented by the equation $y = 12x + 24$, where x represents the number of T-shirts.

a. Graph the equation.

First, find the slope and the y-intercept.

slope $= 12$

y-intercept $= 24$

Plot the point at $(0, 24)$. Then go up 12 and right 1. Connect these points.

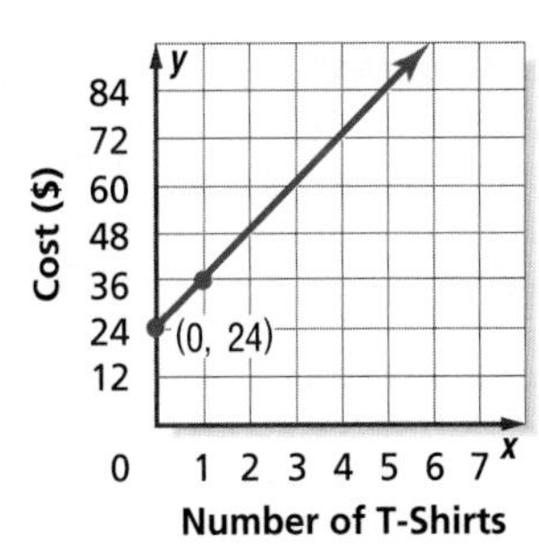

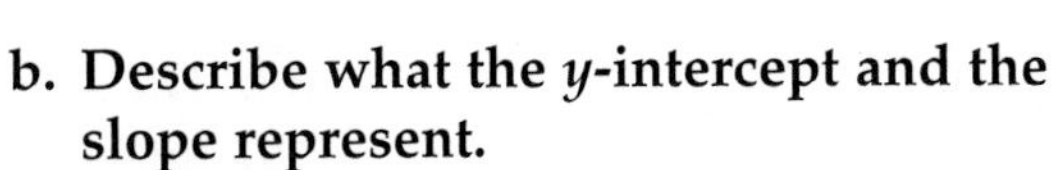

b. Describe what the y-intercept and the slope represent.

The y-intercept 24 represents the design fee. The slope 12 represents the cost per T-shirt, which is the rate of change.

Check for Understanding

Concept Check

1. **State** the value that tells you how many units to go up or down from the y-intercept if the slope of a line is $\frac{a}{b}$.
2. **OPEN ENDED** Draw the graph of a line that has a y-intercept but no x-intercept. What is the slope of the line?
3. **FIND THE ERROR** Carmen and Alex are finding the slope and y-intercept of $x + 2y = 8$.

Carmen	Alex
slope = 2	slope = $-\frac{1}{2}$
y-intercept = 8	y-intercept = 4

Who is correct? Explain your reasoning.

Guided Practice

State the slope and the y-intercept for the graph of each equation.

4. $y = x + 8$
5. $x + y = 0$
6. $x + 3y = 6$

Graph each equation using the slope and y-intercept.

7. $y = \frac{1}{4}x + 1$
8. $3x + y = 2$
9. $x - 2y = 4$

Application

BUSINESS Mrs. Allison charges $25 for a basic cake that serves 12 people. A larger cake costs an additional $1.50 per serving. The total cost can be given by $y = 1.5x + 25$, where x represents the number of additional slices.

10. Graph the equation.
11. Explain what the y-intercept and the slope represent.

Practice and Apply

Homework Help

For Exercises	See Examples
12–17	1, 2
18–31	3
32–34	4

Extra Practice

See page 743.

State the slope and the y-intercept for the graph of each equation.

12. $y = x + 2$
13. $y = 2x - 4$
14. $x + y = -3$
15. $2x + y = -3$
16. $5x + 4y = 20$
17. $y = 4$

Graph each line with the given slope and y-intercept.

18. slope = 3, y-intercept = 1
19. slope = $-\frac{3}{2}$, y-intercept = -1

Graph each equation using the slope and y-intercept.

20. $y = x + 5$
21. $y = -x + 6$
22. $y = 2x - 3$
23. $y = \frac{3}{4}x + 2$
24. $x + y = -3$
25. $x + y = 0$
26. $-2x + y = -1$
27. $5x + y = -3$
28. $x - 3y = -6$
29. $2x + 3y = 12$
30. $3x + 4y = 12$
31. $y = -3$

Hang Gliding
In the summer, hang gliders in the western part of the United States achieve altitudes of 5000 to 10,000 feet and fly for over 100 miles.
Source: www.ushga.org

HANG GLIDING **For Exercises 32–34, use the following information.**
The altitude in feet y of a hang glider who is slowly landing can be given by $y = 300 - 50x$, where x represents the time in minutes.

32. Graph the equation using the slope and y-intercept.
33. State the slope and y-intercept of the graph of the equation and describe what they represent.
34. Name the x-intercept and describe what it represents.
35. **CRITICAL THINKING** What is the x-intercept of the graph of $y = mx + b$? Explain how you know.
36. **WRITING IN MATH** Answer the question that was posed at the beginning of the lesson.

How can knowing the slope and y-intercept help you graph an equation?

Include the following in your answer:
- a description of how the slope-intercept form of an equation gives information, and
- an explanation of how you could write an equation for a line if you know the slope and y-intercept.

37. Which is $2x + 3y = 6$ written in slope-intercept form?
(A) $y = -\frac{2}{3}x - 2$ (B) $y = -\frac{2}{3}x + 2$
(C) $y = -\frac{3}{2}x + 2$ (D) $y = \frac{3}{2}x - 2$

38. What is the slope and y-intercept of the graph of $-x + 2y = 6$?
(A) $-\frac{1}{2}, 1$ (B) $1, 6$
(C) $1, 3$ (D) $\frac{1}{2}, 3$

Maintain Your Skills

Mixed Review **Suppose y varies directly with x. Write an equation relating x and y for each pair of values.** *(Lesson 8-5)*

39. $y = -36$ when $x = 9$
40. $y = 5$ when $x = 25$

Find the slope of the line that passes through each pair of points. *(Lesson 8-4)*

41. $A(3, 1), B(6, 7)$
42. $J(-2, 5), K(8, 5)$
43. $Q(2, 4), R(0, -4)$
44. Solve $4(r - 3) = 8$. *(Lesson 7-2)*
45. Six times a number is 28 more than twice the number. Write an equation and find the number. *(Lesson 7-1)*

Getting Ready for the Next Lesson **PREREQUISITE SKILL** **Simplify.** *(To review **order of operations**, see Lesson 1-2.)*

46. $2(18) - 1$
47. $(-2 - 4) \div 10$
48. $-1(6) + 8$
49. $5 - 8(-3)$
50. $(9 + 6) \div 3$
51. $3 - (-2)(4)$

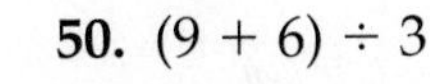

Graphing Calculator

A Follow-Up of Lesson 8-6

Families of Graphs

A graphing calculator is a valuable tool when investigating characteristics of linear functions. Before graphing, you must create a viewing window that shows both the x- and y-intercepts of the graph of a function.

You can use the standard viewing window $[-10, 10]$ scl: 1 by $[-10, 10]$ scl: 1 or set your own minimum and maximum values for the axes and the scale factor by using the WINDOW option.

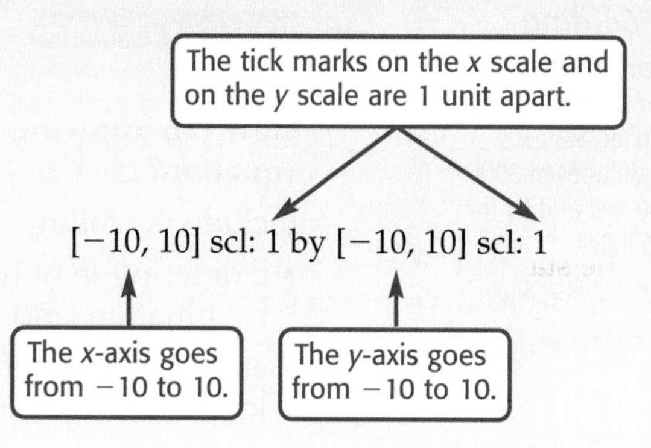

The tick marks on the x scale and on the y scale are 1 unit apart.
[−10, 10] scl: 1 by [−10, 10] scl: 1
The x-axis goes from −10 to 10.
The y-axis goes from −10 to 10.

You can use a TI-83 Plus graphing calculator to enter several functions and graph them at the same time on the same screen. This is useful when studying a **family of graphs**. A family of linear graphs is related by having the same slope or the same y-intercept.

Graph $y = 3x - 2$ and $y = 3x + 4$ in the standard viewing window and describe how the graphs are related.

Step 1 *Graph $y = 3x + 4$ in the standard viewing window.*

- Clear any existing equations from the Y= list.

  KEYSTROKES: [Y=] [CLEAR]

- Enter the equation and graph.

  KEYSTROKES: [Y=] 3 [X,T,θ,n] [+] 4 [ZOOM] 6

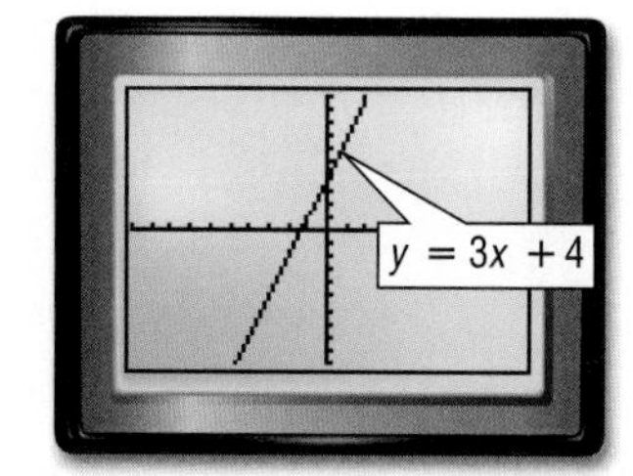

y = 3x + 4

Step 2 *Graph $y = 3x - 2$.*

- Enter the function $y = 3x - 2$ as Y2 with $y = 3x + 4$ already existing as Y1.

  KEYSTROKES: [Y=] 3 [X,T,θ,n] [−] 2

- Graph both functions in the standard viewing window.

  KEYSTROKES: [ZOOM] 6

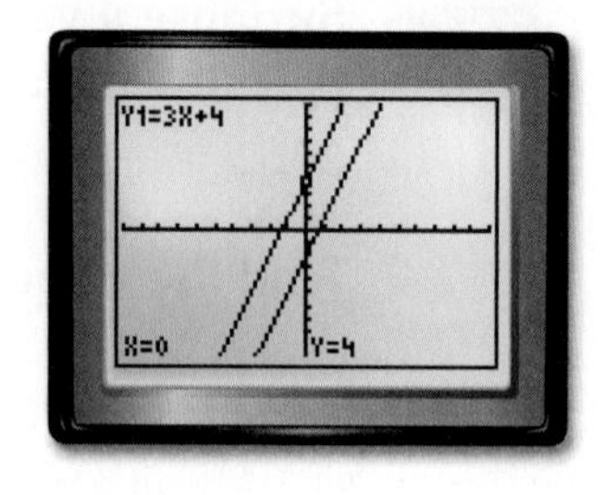

Y1=3X+4
X=0
Y=4

The first function graphed is Y1 or $y = 3x + 4$. The second function graphed is Y2 or $y = 3x - 2$. Press [TRACE]. Move along each function using the right and left arrow keys. Move from one function to another using the up and down arrow keys. The graphs have the same slope, 3, but different y-intercepts at 4 and -2.

 www.pre-alg.com/other_calculator_keystrokes

Investigation

Exercises

Graph $y = 2x - 5$, $y = 2x - 1$, and $y = 2x + 7$.

1. Compare and contrast the graphs.

2. How does adding or subtracting a constant c from a linear function affect its graph?

3. Write an equation of a line whose graph is parallel to $y = 3x - 5$, but is shifted up 7 units.

4. Write an equation of the line that is parallel to $y = 3x - 5$ and passes through the origin.

5. Four functions with a slope of 1 are graphed in the standard viewing window, as shown at the right. Write an equation for each, beginning with the left-most graph.

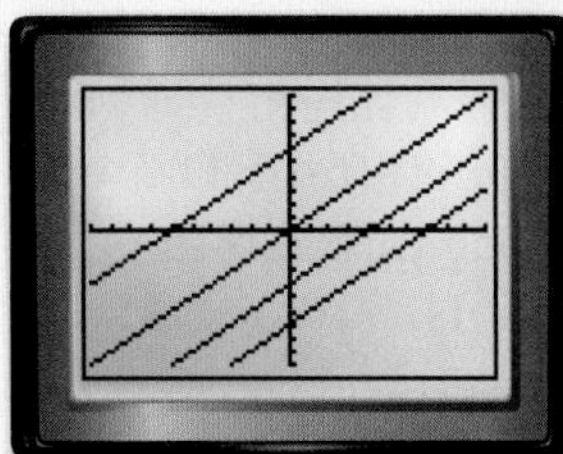

Clear all functions from the Y= menu and graph $y = \frac{1}{3}x$, $y = \frac{3}{4}x$, $y = x$, and $y = 4x$ in the standard viewing window.

6. How does the steepness of a line change as the coefficient for x increases?

7. Without graphing, determine whether the graph of $y = 0.4x$ or the graph of $y = 1.4x$ has a steeper slope. Explain.

Clear all functions from the Y= menu and graph $y = -4x$ and $y = 4x$.

8. How are these two graphs different?

9. How does the sign of the coefficient of x affect the slope of a line?

10. Clear Y2. Then with $y = -4x$ as Y1, enter $y = -x$ as Y2 and $y = -\frac{1}{2}x$ as Y3. Graph the functions and draw the three graphs on grid paper. How does the steepness of the line change as the absolute value of the coefficient of x increases?

11. The graphs of $y = 3x + 1$, $y = \frac{1}{2}x + 1$, and $y = -x + 1$ are shown at the right. Draw the graphs on the same coordinate grid and label each graph with its equation.

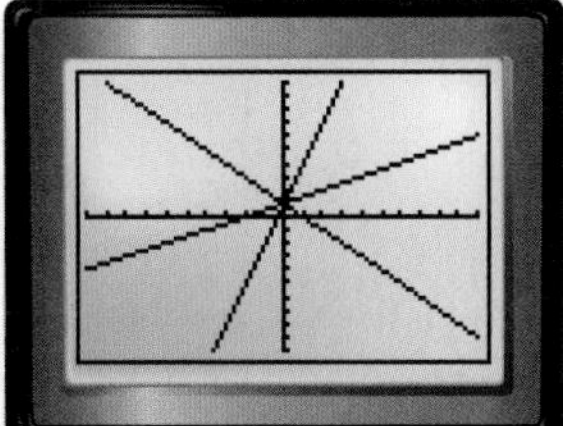

12. Describe the similarities and differences between the graph of $y = 2x - 3$ and the graph of each equation listed below.

a. $y = 2x + 3$

b. $y = -2x - 3$

c. $y = 0.5x + 3$

13. Write an equation of a line whose graph lies between the graphs of $y = -3x$ and $y = -6x$.

8-7 Writing Linear Equations

Sunshine State Standards
MA.C.3.3.2-2, MA.D.1.3.1-4, MA.D.1.3.2-2, MA.D.1.3.2-4

What You'll Learn

- Write equations given the slope and y-intercept, a graph, a table, or two points.

How can you model data with a linear equation?

You can determine the approximate outside temperature by counting the chirps of crickets, as shown in the table.

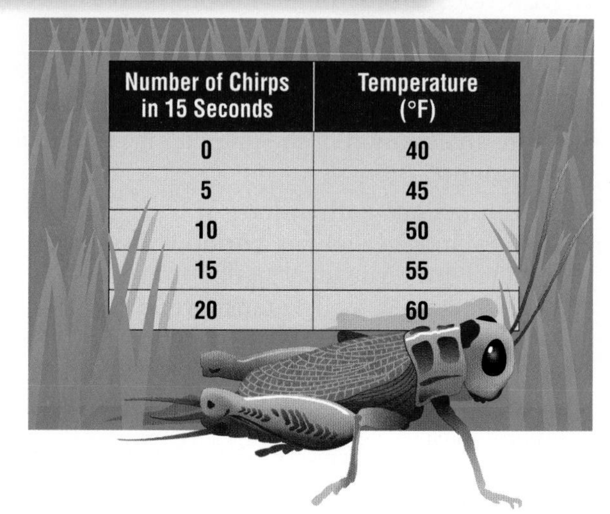

Number of Chirps in 15 Seconds	Temperature (°F)
0	40
5	45
10	50
15	55
20	60

a. Graph the ordered pairs (chirps, temperature). Draw a line through the points.

b. Find the slope and the y-intercept of the line. What do these values represent?

c. Write an equation in the form $y = mx + b$ for the line. Then translate the equation into a sentence.

WRITE EQUATIONS There are many different methods for writing linear equations. If you know the slope and y-intercept, you can write the equation of a line by substituting these values in $y = mx + b$.

Example 1 Write Equations From Slope and y-Intercept

Write an equation in slope-intercept form for each line.

a. slope = 4, y-intercept = −8

$y = mx + b$ Slope-intercept form

$y = 4x + (-8)$ Replace m with 4 and b with −8.

$y = 4x - 8$ Simplify.

b. slope = 0, y-intercept = 5

$y = mx + b$ Slope-intercept form

$y = 0x + 5$ Replace m with 0 and b with 5.

$y = 5$ Simplify.

c. slope = $-\frac{1}{2}$, y-intercept = 0

$y = mx + b$ Slope-intercept form

$y = -\frac{1}{2}x + 0$ Replace m with $-\frac{1}{2}$ and b with 0.

$y = -\frac{1}{2}x$ Simplify.

You can also write equations from a graph.

Study Tip

Check Equation
To check, choose another point on the line and substitute its coordinates for x and y in the equation.

Example 2 Write an Equation From a Graph

Write an equation in slope-intercept form for the line graphed.

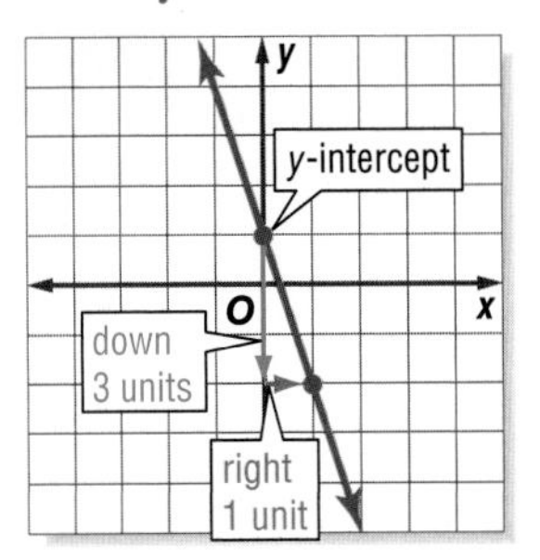

The y-intercept is 1. From (0, 1), you can go down 3 units and right 1 unit to another point on the line. So, the slope is $\frac{-3}{1}$, or -3.

$y = mx + b$ Slope-intercept form

$y = -3x + 1$ Replace m with -3 and b with 1.

In Lesson 8-3, you explored the relationship between altitude and temperature. You can write an equation for this relationship and use it to make predictions.

Example 3 Write an Equation to Solve a Problem

Study Tip

Use a Table
Translate the words into a table of values to help clarify the meaning of the slope.

	Alt. (m)	Temp. (°C)	
+305	0	30	−2
+305	305	28	−2
	610	26	

EARTH SCIENCE **On a summer day, the temperature at altitude 0, or sea level, is 30°C. The temperature decreases 2°C for every 305 meters increase in altitude.**

a. Write an equation to show the relationship between altitude x and temperature y.

Words Temperature decreases 2°C for every 305 meters increase in altitude.

Variables Let x = the altitude and let y = the temperature.

Equations Use $m = \frac{\text{change in } y}{\text{change in } x}$ and $y = mx + b$.

Step 1

Find the slope m.

$m = \frac{\text{change in } y}{\text{change in } x}$ ← $\frac{\text{change in temperature}}{\text{change in altitude}}$

$= \frac{-2}{305}$ ← decrease of −2°C / ← increase of 305 m

≈ -0.007 Simplify.

Step 2

Find the y-intercept b.

(x, y) = (altitude, temperature)
$= (0, b)$

When the altitude is 0, or sea level, the temperature is 30°C. So, the y-intercept is 30.

Step 3

Write the equation.

$y = mx + b$ Slope-intercept form

$y = -0.007x + 30$ Replace m with -0.007 and b with 30.

So, the equation that represents this situation is $y = -0.007x + 30$.

b. Predict the temperature for an altitude of 2000 meters.

$y = -0.007x + 30$ Write the equation.

$y = -0.007(2000) + 30$ Replace x with 2000.

$y \approx 16$ Simplify.

So, at an altitude of 2000 meters, the temperature is about 16°C.

You can also write an equation for a line if you know the coordinates of two points on a line.

Example 4 Write an Equation Given Two Points

Write an equation for the line that passes through (−2, 5) and (2, 1).

Step 1 Find the slope m.

$m = \frac{y_2 - y_1}{x_2 - x_1}$ Definition of slope

$m = \frac{5 - 1}{-2 - 2}$ or -1 $(x_1, y_1) = (-2, 5)$, $(x_2, y_2) = (2, 1)$

Step 2 Find the y-intercept b. Use the slope and the coordinates of either point.

$y = mx + b$ Slope-intercept form

$5 = -1(-2) + b$ Replace (x, y) with $(-2, 5)$ and m with -1.

$3 = b$ Simplify.

Step 3 Substitute the slope and y-intercept.

$y = mx + b$ Slope-intercept form

$y = -1x + 3$ Replace m with -1 and b with 3.

$y = -x + 3$ Simplify.

Example 5 Write an Equation From a Table

Use the table of values to write an equation in slope-intercept form.

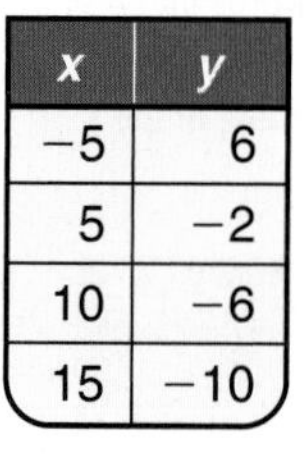

x	y
−5	6
5	−2
10	−6
15	−10

Step 1 Find the slope m. Use the coordinates of any two points.

$m = \frac{y_2 - y_1}{x_2 - x_1}$ Definition of slope

$m = \frac{-2 - 6}{5 - (-5)}$ or $-\frac{4}{5}$ $(x_1, y_1) = (-5, 6)$, $(x_2, y_2) = (5, -2)$

Step 2 Find the y-intercept b. Use the slope and the coordinates of any point.

$y = mx + b$ Slope-intercept form

$6 = -\frac{4}{5}(-5) + b$ Replace (x, y) with $(-5, 6)$ and m with $-\frac{4}{5}$.

$2 = b$ Simplify.

Step 3 Substitute the slope and y-intercept.

$y = mx + b$ Slope-intercept form

$y = -\frac{4}{5}x + 2$ Replace m with $-\frac{4}{5}$ and b with 2.

CHECK $y = -\frac{4}{5}x + 2$ Write the equation.

$-10 \stackrel{?}{=} -\frac{4}{5}(15) + 2$ Replace (x, y) with the coordinates of another point, $(15, -10)$.

$-10 = -10$ ✓ Simplify.

Study Tip

Alternate Strategy If a table includes the y-intercept, simply use this value and the slope to write an equation.

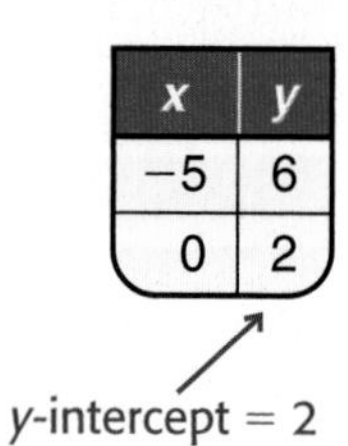

x	y
−5	6
0	2

Check for Understanding

Concept Check

1. **Explain** how to write the equation of a line if you are given a graph.
2. **OPEN ENDED** Choose a slope and y-intercept. Then graph the line.

Guided Practice

Write an equation in slope-intercept form for each line.

3. slope $= \frac{1}{2}$, y-intercept $= 1$
4. slope $= 0$, y-intercept $= -7$

5. 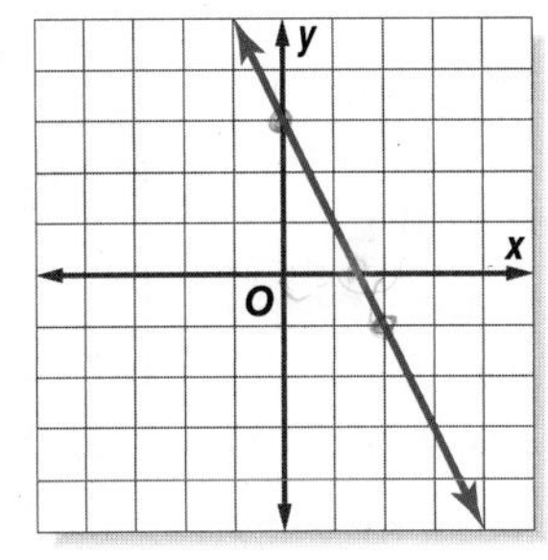

6. 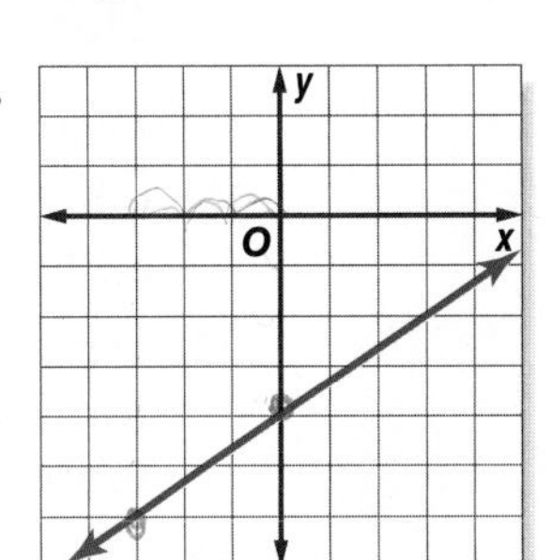

Write an equation in slope-intercept form for the line passing through each pair of points.

7. (2, 2) and (4, 3)
8. (3, −4) and (−1, 4)
9. Write an equation in slope-intercept form to represent the table of values.

x	−4	0	4	8
y	−4	−1	2	5

Application

10. **PICNICS** It costs \$50 plus \$10 per hour to rent a park pavilion.
    a. Write an equation in slope-intercept form that shows the cost y for renting the pavilion for x hours.
    b. Find the cost of renting the pavilion for 8 hours.

Practice and Apply

Homework Help

For Exercises	See Examples
11–16	1
17–22	2
23–28	4
29, 30	5
31–33	3

Extra Practice

See page 743.

Write an equation in slope-intercept form for each line.

11. slope $= 2$, y-intercept $= 6$
12. slope $= -4$, y-intercept $= 1$
13. slope $= 0$, y-intercept $= 5$
14. slope $= 1$, y-intercept $= -2$
15. slope $= -\frac{1}{3}$, y-intercept $= 8$
16. slope $= \frac{2}{5}$, y-intercept $= 0$

17. 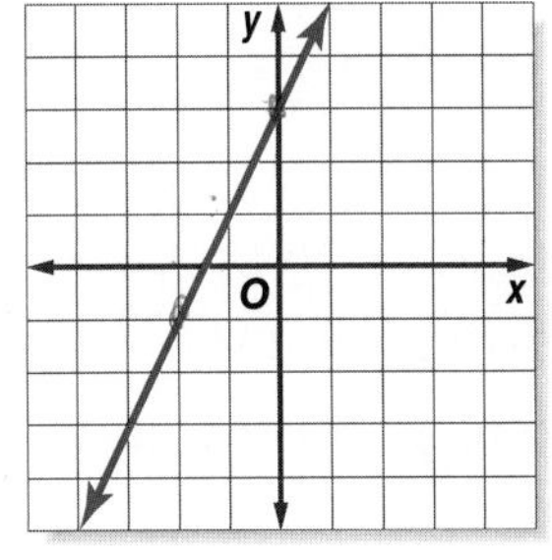

18.

19. 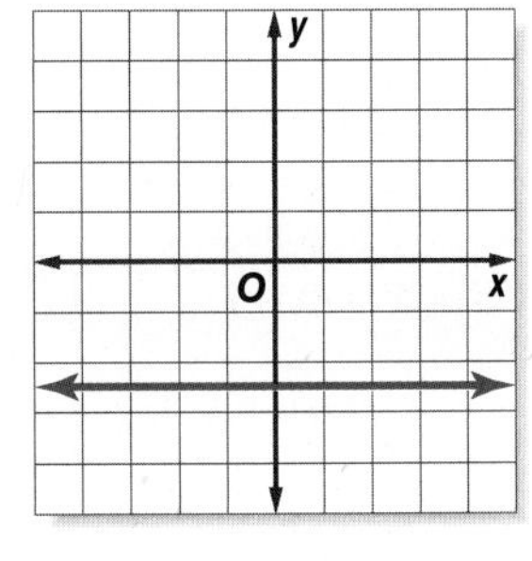

20.

21. 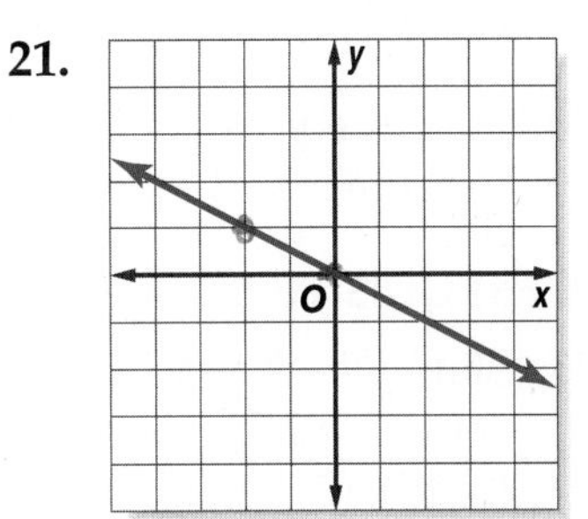

22. 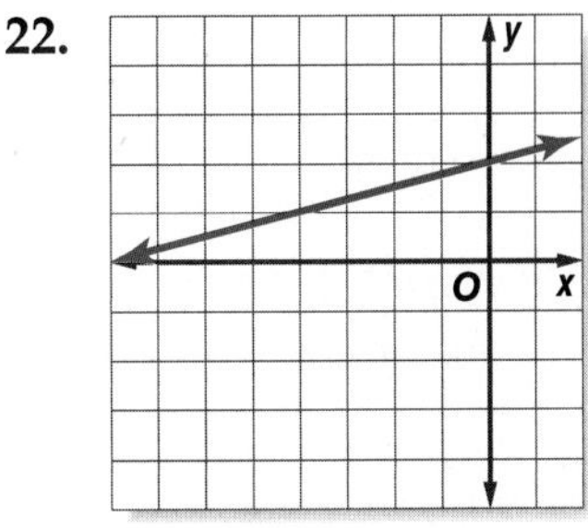

Write an equation in slope-intercept form for the line passing through each pair of points.

23. $(-2, -1)$ and $(1, 2)$ **24.** $(-4, 3)$ and $(4, -1)$ **25.** $(0, 0)$ and $(-1, 1)$

26. $(4, 2)$ and $(-8, -16)$ **27.** $(8, 7)$ and $(-9, 7)$ **28.** $(5, -6)$ and $(3, 2)$

Write an equation in slope-intercept form for each table of values.

29.

x	−1	0	1	2
y	−7	−3	1	5

30.

x	−3	−1	1	3
y	7	5	3	1

More About. . .

Sound

Coyotes communicate by using different barks and howls. Because of the way in which sound travels, a coyote is usually not in the area from which the sound seems to be coming.

Source: www.livingdesert.org

SOUND **For Exercises 31 and 32, use the table that shows the distance that sound travels through dry air at 0°C.**

Time(s)	Distance (ft)
x	y
0	0
1	1088
2	2176
3	3264

31. Write an equation in slope-intercept form to represent the data in the table. Describe what the slope means.

32. Estimate the number of miles that sound travels through dry air in one minute.

33. **CRITICAL THINKING** A CD player has a pre-sale price of $\$c$. Kim buys it at a 30% discount and pays 6% sales tax. After a few months, she sells it for $\$d$, which was 50% of what she paid originally.

a. Express d as a function of c.

b. How much did Kim sell it for if the pre-sale price was $50?

34. WRITING IN MATH Answer the question that was posed at the beginning of the lesson.

How can you model data with a linear equation?

Include the following in your answer:

- an explanation of how to find the y-intercept and slope by using a table.

FCAT Practice

Standardized Test Practice

35. Which equation is of a line that passes through $(2, -2)$ and $(0, 2)$?

Ⓐ $y = -3x$ Ⓑ $y = -2x + 3$ Ⓒ $y = -2x + 2$ Ⓓ $y = -3x + 2$

36. Which equation represents the table of values?

x	−4	−8	−12	−16
y	6	8	10	12

Ⓐ $y = -2x + 4$ Ⓑ $y = 2x + 6$

Ⓒ $y = -\frac{1}{2}x + 4$ Ⓓ $y = -x - 6$

Maintain Your Skills

Mixed Review **State the slope and the y-intercept for the graph of each equation.** *(Lesson 8-6)*

37. $y = 6x + 7$ **38.** $y = -x + 4$ **39.** $-3x + y = -2$

40. Suppose y varies directly as x and $y = 14$ when $x = 35$. Write an equation relating x and y. *(Lesson 8-5)*

Getting Ready for the Next Lesson **PREREQUISITE SKILL** *(To review **scatter plots**, see Lesson 1-7.)*

41. State whether a scatter plot containing the following set of points would show a *positive*, *negative*, or *no* relationship.
(0, 15), (2, 20), (4, 36), (5, 44), (4, 32), (3, 30), (6, 50)

8-8 Best-Fit Lines

Sunshine State Standards
MA.D.1.3.2-2

What You'll Learn

- Draw best-fit lines for sets of data.
- Use best-fit lines to make predictions about data.

Vocabulary

- best-fit line

How can a line be used to predict life expectancy for future generations?

The scatter plot shows the number of years people in the United States are expected to live, according to the year they were born.

a. Use the line drawn through the points to predict the life expectancy of a person born in 2010.

b. What are some limitations in using a line to predict life expectancy?

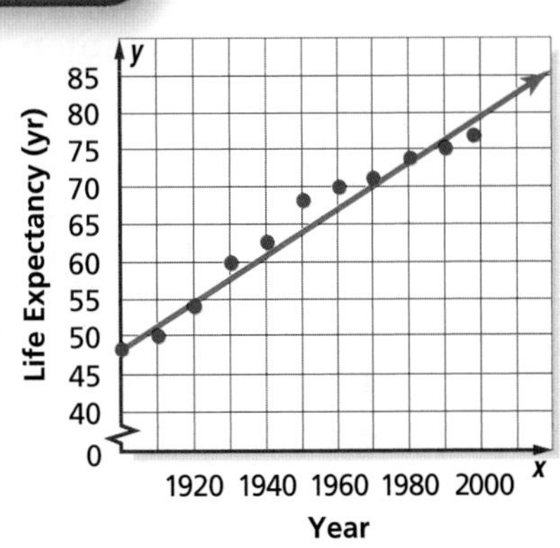

Source: *The World Almanac*

Study Tip

Estimation
Drawing a best-fit line using the method in this lesson is an estimation. Therefore, it is possible to draw different lines to approximate the same data.

BEST-FIT LINES When real-life data are collected, the points graphed usually do not form a straight line, but may approximate a linear relationship. A best-fit line can be used to show such a relationship. A **best-fit line** is a line that is very close to most of the data points.

Example 1 Make Predictions from a Best-Fit Line

MONEY The table shows the changes in the minimum wage since 1980.

a. **Make a scatter plot and draw a best-fit line for the data.**

Year	Wage ($/h)
1980	3.10
1981	3.35
1990	3.80
1996	4.25
1997	4.75
2000	5.15

Draw a line that best fits the data.

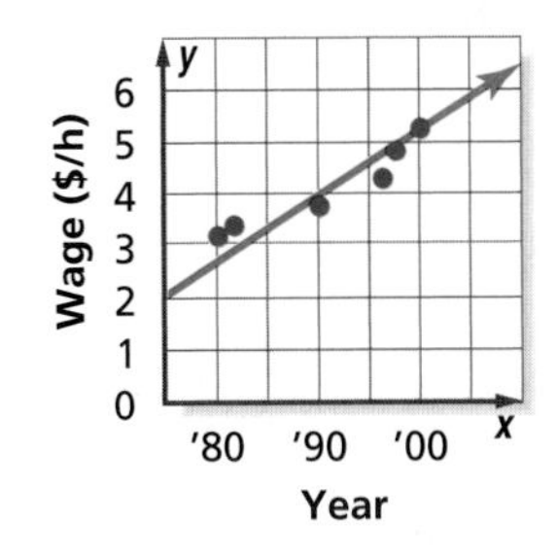

b. **Use the best-fit line to predict the minimum wage for the year 2010.**

Extend the line so that you can find the y value for an x value of 2010. The y value for 2010 is about 6.4. So, a prediction for the minimum wage in 2010 is approximately $6.40.

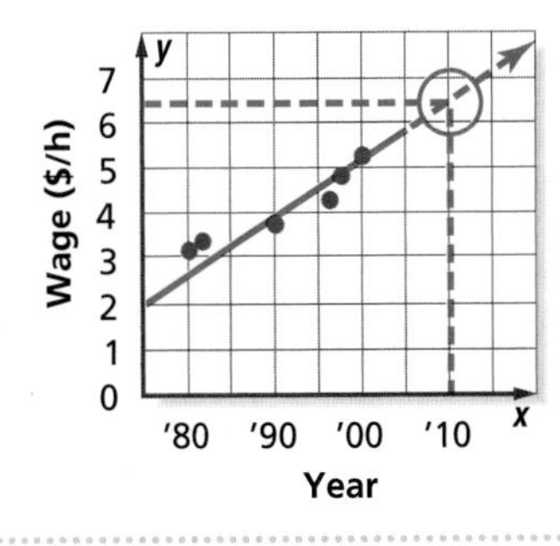

PREDICTION EQUATIONS You can also make predictions from the equation of a best-fit line.

Example 2 Make Predictions from an Equation

SWIMMING The scatter plot shows the winning Olympic times in the women's 800-meter freestyle event from 1968 through 2000.

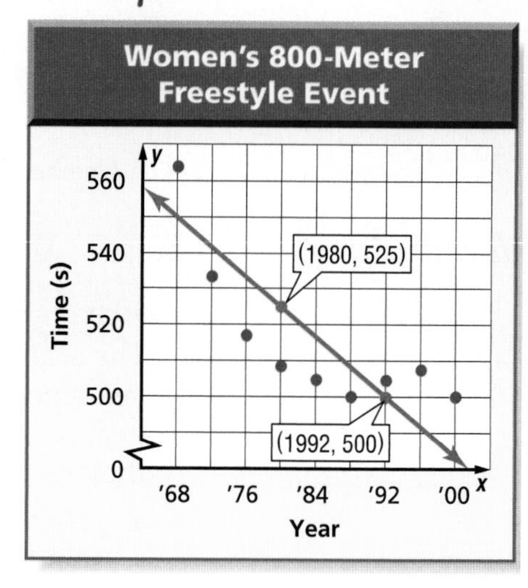

a. Write an equation in slope-intercept form for the best-fit line.

Step 1

First, select two points on the line and find the slope. Notice that the two points on the best-fit line are not original data points. We have chosen (1980, 525) and (1992, 500).

$m = \frac{y_2 - y_1}{x_2 - x_1}$	Definition of slope
$= \frac{525 - 500}{1980 - 1992}$	$(x_1, y_1) = (1992, 500)$, $(x_2, y_2) = (1980, 525)$
≈ -2.1	Simplify.

Step 2

Next, find the y-intercept.

$y = mx + b$	Slope-intercept form
$525 = -2.1(1980) + b$	Replace (x, y) with (1980, 525) and m with -2.1.
$4683 \approx b$	Simplify.

Step 3

Write the equation.

$y = mx + b$	Slope-intercept form
$y = -2.1x + 4683$	Replace m with -2.1 and b with 4683.

b. Predict the winning time in the women's 800-meter freestyle event in the year 2008.

$y = -2.1x + 4683$	Write the equation of the best-fit line.
$y = -2.1(2008) + 4683$	Replace x with 2008.
$y \approx 466.2$	Simplify.

A prediction for the winning time in the year 2008 is approximately 466.2 seconds or 7 minutes, 46.2 seconds.

Check for Understanding

Concept Check

1. **Explain** how to use a best-fit line to make a prediction.
2. **OPEN ENDED** Make a scatter plot with at least ten points that appear to be somewhat linear. Draw two different lines that could approximate the data.

Guided Practice

TECHNOLOGY For Exercises 3 and 4, use the table that shows the number of U.S. households with Internet access.

3. Make a scatter plot and draw a best-fit line.
4. Use the best-fit line to predict the number of U.S. households that will have Internet access in 2005.

Year	Number of Households (millions)
1995	9.4
1996	14.7
1997	21.3
1998	27.3
1999	32.7
2000	36.0

Source: *Wall Street Journal Almanac*

Application

SPENDING For Exercises 5 and 6, use the best-fit line that shows the billions of dollars spent by travelers in the United States.

5. Write an equation in slope-intercept form for the best-fit line.
6. Use the equation to predict how much money travelers will spend in 2008.

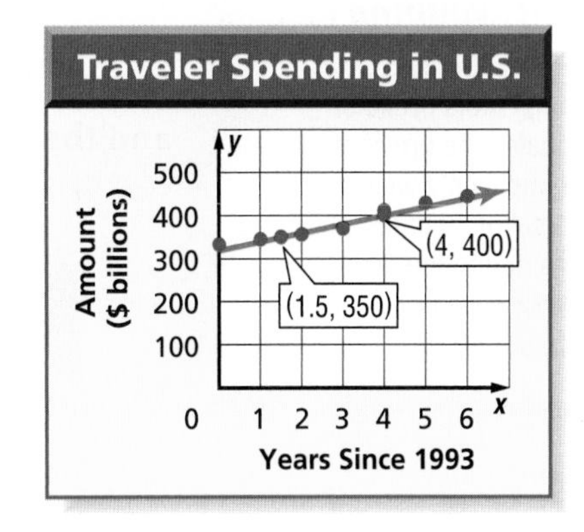

Practice and Apply

Homework Help

For Exercises	See Examples
7, 8, 12, 13	1
9–11, 14–17	2

Extra Practice

See page 744.

ENTERTAINMENT For Exercises 7 and 8, use the table that shows movie attendance in the United States.

7. Make a scatter plot and draw a best-fit line.
8. Use the best-fit line to predict movie attendance in 2005.

Year	Attendance (millions)
1993	1244
1994	1292
1995	1263
1996	1339
1997	1388
1998	1475
1999	1460

Source: *Wall Street Journal Almanac*

WebQuest

Best-fit lines can help make predictions about recreational activities. Visit www.pre-alg.com/webquest to continue work on your WebQuest project.

PRESSURE For Exercises 9–11, use the table that shows the approximate barometric pressure at various altitudes.

9. Make a scatter plot of the data and draw a best-fit line.
10. Write an equation for the best-fit line and use it to estimate the barometric pressure at 60,000 feet. Is the estimation reasonable? Explain.
11. Do you think that a line is the best model for this data? Explain.

Altitude (ft)	Barometric Pressure (in. mercury)
0	30
5000	25
10,000	21
20,000	14
30,000	9
40,000	6
50,000	3

Source: *New York Public Library Science Desk Reference*

More About. . .

Pole Vaulting

In 1964, thirteen competitors broke or equaled the previous Olympic pole vault record a total of 36 times. This was due to the new fiberglass pole.

Source: *Chance*

POLE VAULTING **For Exercises 12 and 13, use the table that shows the men's winning Olympic pole vault heights to the nearest inch.**

Year	Height (in.)
1976	217
1980	228
1984	226
1988	232
1992	228
1996	233
2000	232

Source: *The World Almanac*

12. Make a scatter plot and draw a best-fit line.

13. Use the best-fit line to predict the winning pole vault height in the 2008 Olympics.

EARTH SCIENCE **For Exercises 14–17, use the table that shows the latitude and the average temperature in July for five cities in the United States.**

City	Latitude (°N)	Average July High Temperature (°F)
Chicago, IL	41	73
Dallas, TX	32	85
Denver, CO	39	74
New York, NY	40	77
Duluth, MN	46	66

Source: *The World Almanac*

14. Make a scatter plot of the data and draw a best-fit line.

15. Describe the relationship between latitude and temperature shown by the graph.

16. Write an equation for the best-fit line you drew in Exercise 14.

17. Use your equation to estimate the average July temperature for a location with latitude 50° north.

18. **CRITICAL THINKING** The table at the right shows the percent of public schools in the United States with Internet access. Suppose you use (Year, Percent of Schools) to write a linear equation describing the data. Then you use (Years Since 1996, Percent of Schools) to write an equation. Is the slope or y-intercept of the graphs of the equations the same? Explain.

Year	Years Since 1996	Percent of Schools
1996	0	65
1997	1	78
1998	2	89
1999	3	95
2000	4	98

Source: National Center for Education Statistics

19. WRITING IN MATH Answer the question that was posed at the beginning of the lesson.

How can a line be used to predict life expectancy for future generations?

Include the following in your answer:

- a description of a best-fit line, and
- an explanation of how lines can represent sets of data that are not exactly linear.

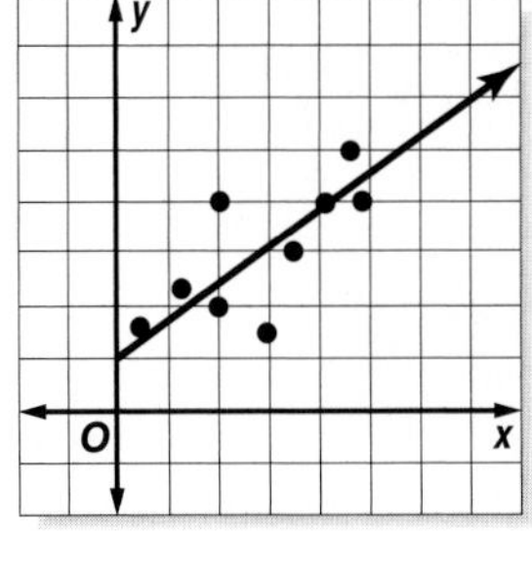

20. Use the best-fit line at the right to predict the value of y when $x = 7$.

Ⓐ 4 Ⓑ 6

Ⓒ 0 Ⓓ 7

21. Choose the correct statement about best-fit lines.

Ⓐ A best-fit line is close to most of the data points.

Ⓑ A best-fit line describes the exact coordinates of each point in the data set.

Ⓒ A best-fit line always has a positive slope.

Ⓓ A best-fit line must go through at least two of the data points.

Maintain Your Skills

Mixed Review

Write an equation in slope-intercept form for each line. *(Lesson 8-7)*

22. slope = 3, y-intercept = 5

23. slope = -2, y-intercept = 2

24.

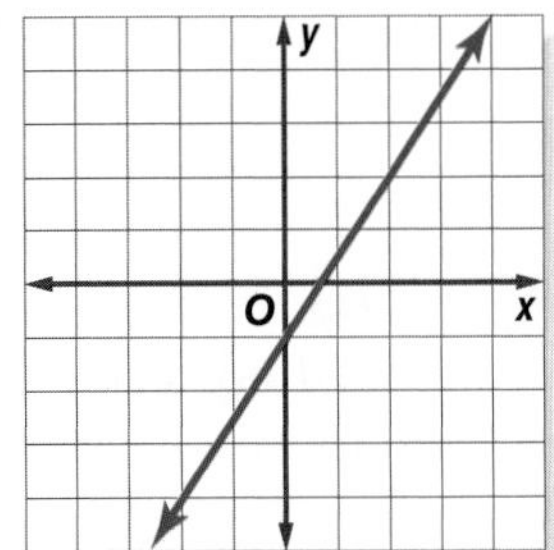

25.

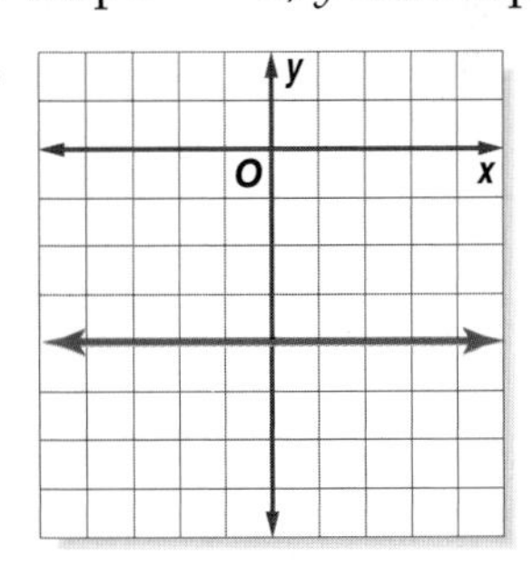

Graph each equation using the slope and y-intercept. *(Lesson 8-6)*

26. $y = x - 2$

27. $y = -x + 3$

28. $y = \frac{1}{2}x$

Solve each inequality and check your solution. *(Lesson 7-6)*

29. $3n - 11 \leq 10$

30. $8(x + 1) < 16$

31. $5d + 2 > d - 4$

32. SCHOOL Zach has no more than twelve days to complete his science project. Write an inequality to represent this sentence. *(Lesson 7-3)*

Solve each proportion. *(Lesson 6-2)*

33. $\frac{a}{3} = \frac{16}{24}$

34. $\frac{5}{10} = \frac{15}{x}$

35. $\frac{2}{16} = \frac{n}{36}$

36. Evaluate xy if $x = \frac{8}{9}$ and $y = \frac{12}{30}$. Write in simplest form. *(Lesson 5-3)*

Getting Ready for the Next Lesson

PREREQUISITE SKILL **Find the value of y in each equation by substituting the given value of x.** *(To review **substitution**, see Lesson 1-2.)*

37. $y = x + 1; x = 2$

38. $y = x + 5; x = -1$

39. $y = x - 4; x = 3$

40. $x + y = 2; x = 0$

41. $x + y = -1; x = 1$

42. $x + y = 0; x = 4$

8-9 Solving Systems of Equations

Sunshine State Standards
MA.C.3.3.2-3, MA.D.1.3.1-2, MA.D.1.3.2-1, MA.D.1.3.2-2, MA.D.2.3.1-1, MA.D.2.3.1-5

Vocabulary

- system of equations
- substitution

What You'll Learn

- Solve systems of linear equations by graphing.
- Solve systems of linear equations by substitution.

How can a system of equations be used to compare data?

Yolanda is offered two summer jobs, as shown in the table.

Job	Hourly Rate	Bonus
A	\$10	\$50
B	\$15	\$0

a. Write an equation to represent the income from each job. Let y equal the salary and let x equal the number of hours worked. (*Hint*: Income = hourly rate · number of hours worked + bonus.)

b. Graph both equations on the same coordinate plane.

c. What are the coordinates of the point where the two lines meet? What does this point represent?

SOLVE SYSTEMS BY GRAPHING The equations $y = 10x + 50$ and $y = 15x$ together are called a **system of equations**. The solution of this system is the ordered pair that is a solution of both equations, (10, 150).

$$y = 10x + 50 \qquad\qquad y = 15x$$
$$150 \stackrel{?}{=} 10(10) + 50 \qquad\qquad 150 \stackrel{?}{=} 15(10)$$
$$150 = 150 \checkmark \qquad\qquad 150 = 150 \checkmark$$

Replace (x, y) with (10, 150).

One method for solving a system of equations is to graph the equations on the same coordinate plane. The coordinates of the point where the graphs intersect is the solution of the system of equations.

Example 1 Solve by Graphing

Solve the system of equations by graphing.

$y = -x$

$y = x + 2$

The graphs appear to intersect at $(-1, 1)$. Check this estimate by substituting the coordinates into each equation.

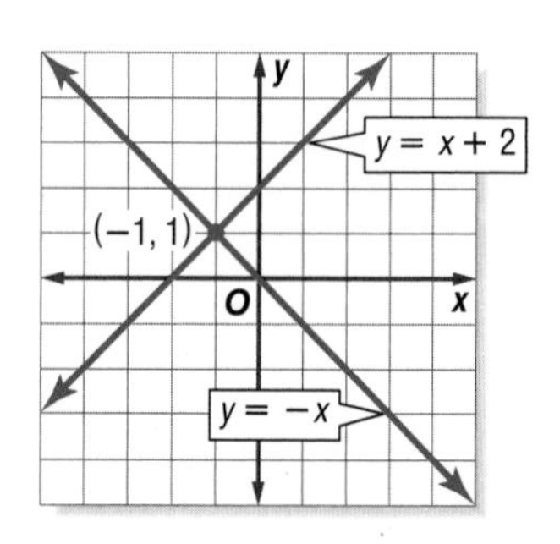

CHECK

$$y = -x \qquad\qquad y = x + 2$$
$$1 \stackrel{?}{=} -(-1) \qquad\qquad 1 \stackrel{?}{=} -1 + 2$$
$$1 = 1 \checkmark \qquad\qquad 1 = 1 \checkmark$$

The solution of the system of equations is $(-1, 1)$.

Concept Check Is (0, 0) a solution of the system of equations in Example 1? Explain.

Example 2 One Solution

ENTERTAINMENT **Video Planet offers two rental plans.**

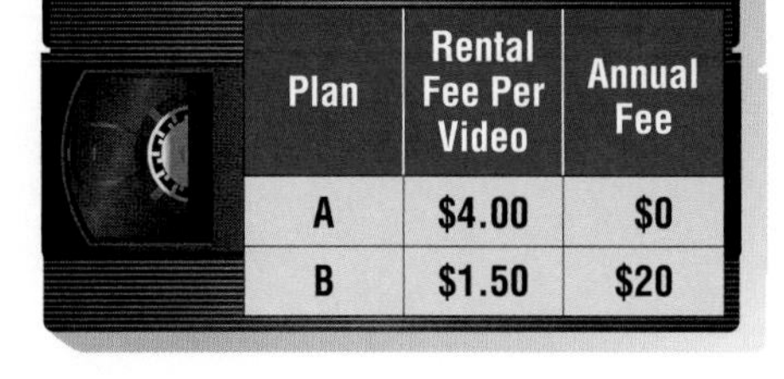

Plan	Rental Fee Per Video	Annual Fee
A	$4.00	$0
B	$1.50	$20

a. How many videos would Walt need to rent in a year for the plans to cost the same?

Explore You know the rental fee per video and the annual fee.

Plan Write an equation to represent each plan, and then graph the equations to find the solution.

Solve Let x = number of videos rented and let y = the total cost.

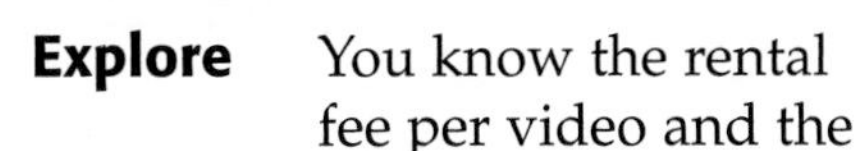

	total cost		rental fee times number of videos		annual fee
Plan A	y	=	$4x$	+	0
Plan B	y	=	$1.50x$	+	20

The graph of the system shows the solution is (8, 32). This means that if Walt rents 8 videos in a year, the plans cost the same, $32.

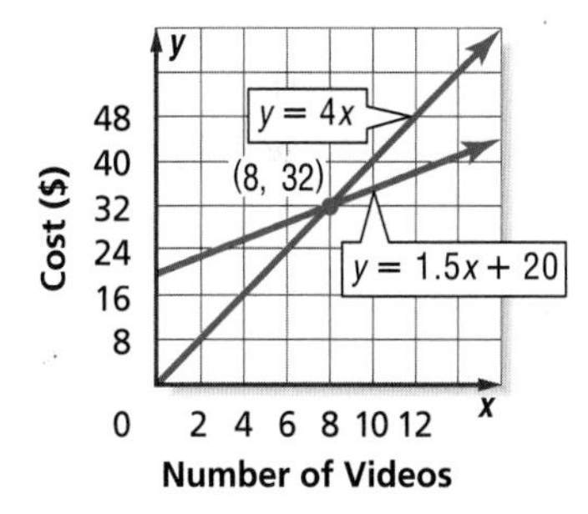

Examine Check by substituting (8, 32) into both equations in the system.

b. Which plan would cost less if Walt rents 12 videos in a year?

For $x = 12$, the line representing Plan B has a smaller y value. So, Plan B would cost less.

Example 3 No Solution

Solve the system of equations by graphing.

$y = 2x + 4$

$y = 2x - 1$

The graphs appear to be parallel lines. Since there is no coordinate pair that is a solution to both equations, there is no solution of this system of equations.

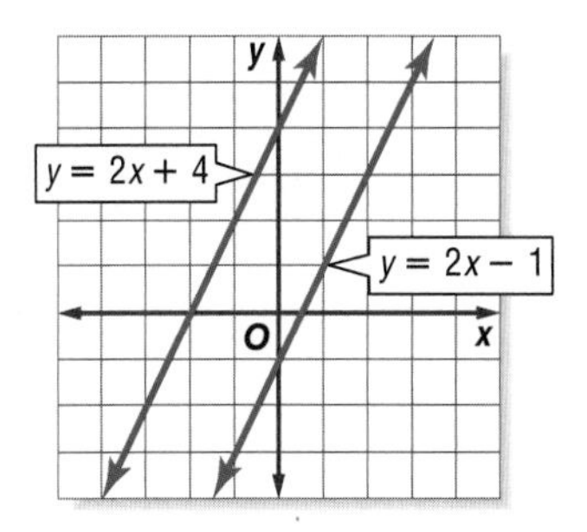

Study Tip

Slopes/Intercepts
When the graphs of a system of equations have:
- different slopes, there is exactly one solution,
- the same slope and different y-intercepts, there is no solution,
- the same slope and the same y-intercept, there are infinitely many solutions.

Example 4 Infinitely Many Solutions

Solve the system of equations by graphing.

$2y = x + 6$

$y = \frac{1}{2}x + 3$

Both equations have the same graph. Any ordered pair on the graph will satisfy both equations. Therefore, there are infinitely many solutions of this system of equations.

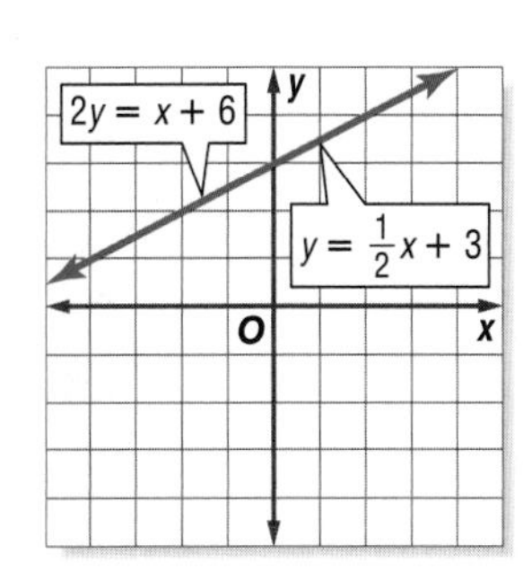

Concept Summary — Solutions to Systems of Equations

One Solution	No Solution	Infinitely Many Solutions
Intersecting Lines	Parallel Lines	Same Line

SOLVE SYSTEMS BY SUBSTITUTION A more accurate way to solve a system of equations is by using a method called **substitution**.

Example 5 Solve by Substitution

Solve the system of equations by substitution.

$y = x + 5$

$y = 3$

Since y must have the same value in both equations, you can replace y with 3 in the first equation.

$y = x + 5$	Write the first equation.
$3 = x + 5$	Replace y with 3.
$-2 = x$	Solve for x.

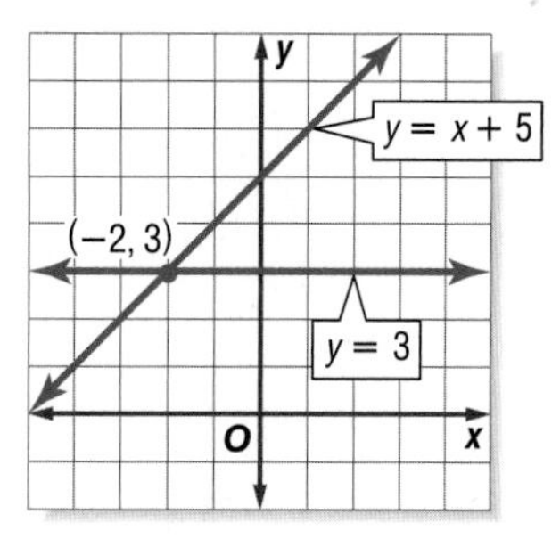

The solution of this system of equations is $(-2, 3)$. You can check the solution by graphing. The graphs appear to intersect at $(-2, 3)$, so the solution is correct.

Check for Understanding

Concept Check

1. **Explain** what is meant by a system of equations and describe its solution.
2. **OPEN ENDED** Draw a graph of a system of equations that has one solution, a system that has no solution, and a system that has infinitely many solutions.

Guided Practice

3. State the solution of the system of equations graphed at the right.

Solve each system of equations by graphing.

4. $y = 2x + 1$
   $y = -x + 1$
5. $x + y = 4$
   $x + y = 2$

Solve each system of equations by substitution.

6. $y = 3x - 4$
   $x = 0$
7. $x + y = 8$
   $y = 6$

Application

8. **GEOMETRY** The perimeter of a garden is 40 feet. If the width y equals 7 feet, write and solve a system of equations to find the length of the garden.

Practice and Apply

Homework Help

For Exercises	See Examples
9–17	1, 3, 4
18–23	5
24–26	2

Extra Practice

See page 744.

State the solution of each system of equations.

9. 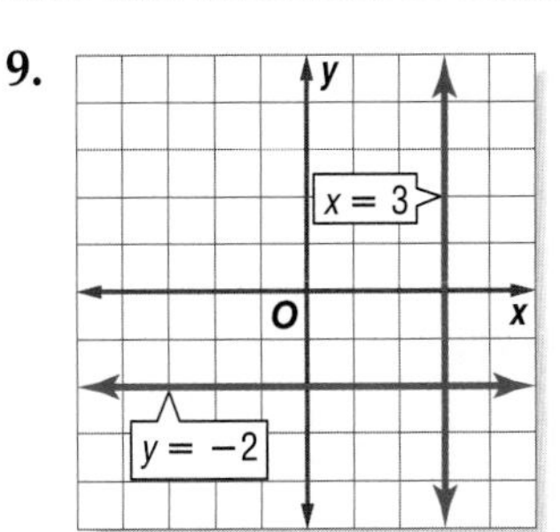

10. 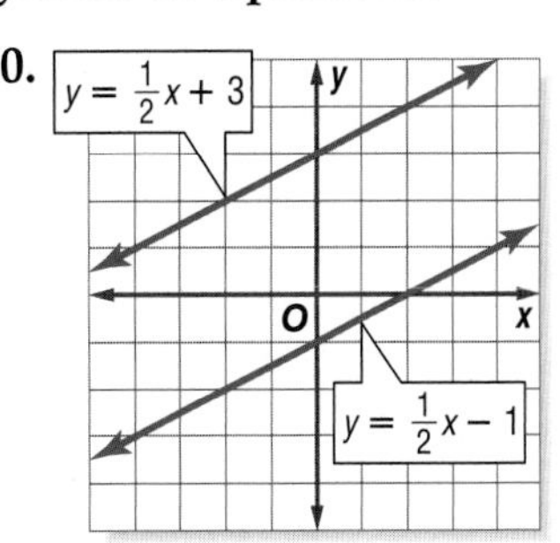

11. 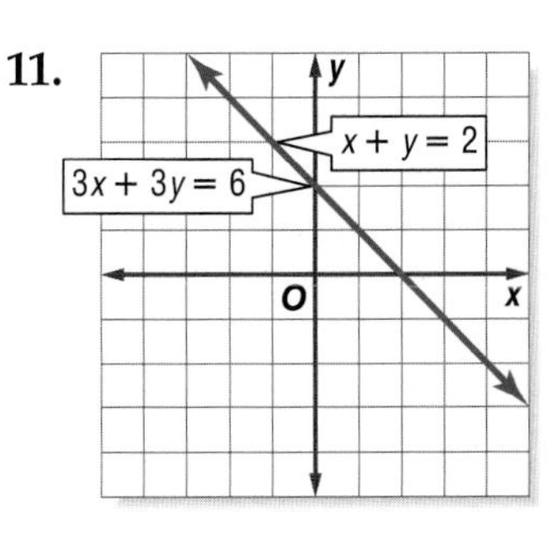

Solve each system of equations by graphing.

12. $y = -x$
    $y = x + 2$

13. $x + y = 6$
    $y = x$

14. $2x + y = 1$
    $y = -2x + 5$

15. $y = \frac{2}{3}x$
    $2x - 3y = 0$

16. $x + y = -4$
    $x - y = -4$

17. $y = -\frac{1}{2}x + 3$
    $y = -2x$

Solve each system of equations by substitution.

18. $y = x + 1$
    $x = 3$

19. $y = x + 2$
    $y = 0$

20. $y = 2x + 7$
    $y = -5$

21. $x + y = 6$
    $x = -4$

22. $2x + 3y = 5$
    $y = x$

23. $x + y = 9$
    $y = 2x$

More About. . .

Snowboards

In 1998, snowboarding became an Olympic event in Nagano, Japan, with a giant slalom and halfpipe competition.

Source: www.snowboarding.about.com

SNOWBOARDS **For Exercises 24–26, use the following information and the table.** Two Internet sites sell a snowboard for the same price, but have different shipping charges.

Shipping Charges		
Internet Site	Base Fee	Charge per Pound
A	\$5.00	\$1.00
B	\$2.00	\$1.50

24. Write a system of equations that represents the shipping charges y for x pounds. (*Hint*: Shipping charge = base fee + charge per pound · number of pounds.)

25. Solve the system of equations. Explain what the solution means.

26. If the snowboard weighs 8 pounds, which Internet site would be less expensive? Explain.

27. **CRITICAL THINKING** Two runners A and B are 50 meters apart and running at the same rate along the same path.
    a. If their rates continue, will the second runner ever catch up to the first? Draw a graph to explain why or why not.
    b. Draw a graph that represents the second runner catching up to the first. What is different about the two graphs that you drew?

28. **WRITING IN MATH** Answer the question that was posed at the beginning of the lesson.

How can a system of equations be used to compare data?

Include the following in your answer:

- a situation that can be modeled by a system of equations, and
- an explanation of what a solution to such a system of equations means.

29. Which system of equations represents the following verbal description? *The sum of two numbers is 6. The second number is three times greater than the first number.*

(A) $x + y = 6$, $x = 3 + y$ (B) $y = x + 6$, $y = x + 3$ (C) $x + y = 6$, $y = 3x$ (D) $x - y = 6$, $y = -3x$

30. Which equation, together with $x + y = 1$, forms a system that has a solution of $(-3, 4)$?

(A) $y = x$ (B) $x - y = 1$ (C) $y = x + 7$ (D) $-3x + 4y = 1$

Maintain Your Skills

Mixed Review

31. **SALES** Ice cream sales increase as the temperature outside increases. Describe the slope of a best-fit line that represents this situation. *(Lesson 8-8)*

Write an equation in slope-intercept form for the line passing through each pair of points. *(Lesson 8-7)*

32. (0, 1) and (3, 7)
33. (−2, 6) and (1, −3)
34. (8, 0) and (−8, −4)

Getting Ready for the Next Lesson

PREREQUISITE SKILL State whether each number is a solution of the given inequality. *(To review **inequalities**, see Lesson 7-3.)*

35. $1 < x + 3$; 0
36. $9 > t - 5$; 1
37. $2y \geq 2$; −1
38. $14 < 6 - n$; 0
39. $35 > 12 + k$; 12
40. $5n + 1 \leq 0$; −2

Practice Quiz 2

Lessons 8-6 through 8-9

State the slope and the y-intercept for the graph of each equation. *(Lesson 8-6)*

1. $y = -x + 8$
2. $y = 3x - 5$
3. $x + 2y = 6$

Write an equation in slope-intercept form for each line. *(Lesson 8-7)*

4. slope = 6, y-intercept = −7
5. slope = 0, y-intercept = 1
6. slope = 1, y-intercept = 0

7. **STATISTICS** The table shows the average age of the Women's U.S. Olympic track and field team. Make a scatter plot of the data and draw a best-fit line. *(Lesson 8-8)*

Year	1984	1988	1992	1996	2000
Average Age	24.6	25.0	27.3	28.7	29.2

Source: *Sports Illustrated*

Solve each system of equations by substitution. *(Lesson 8-9)*

8. $y = x - 1$, $y = 2$
9. $y = x + 5$, $x = 0$
10. $y = 2x + 4$, $y = -4$

8-10 Graphing Inequalities

What You'll Learn

- Graph linear inequalities.
- Describe solutions of linear inequalities.

Vocabulary

- boundary
- half plane

Sunshine State Standards
MA.D.1.3.1-1, MA.D.1.3.1-2, MA.D.1.3.2-1, MA.D.1.3.2-2, MA.D.1.3.2-3, MA.D.2.3.1-1, MA.D.2.3.1-5

How can shaded regions on a graph model inequalities?

Refer to the graph at the right.

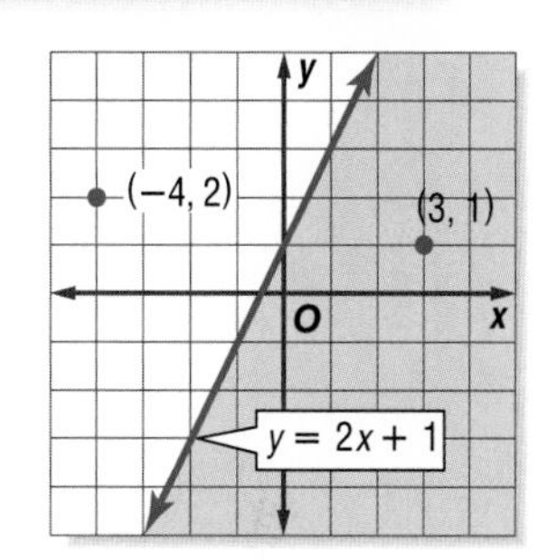

a. Substitute $(-4, 2)$ and $(3, 1)$ in $y > 2x + 1$. Which ordered pair makes the inequality true?

b. Substitute $(-4, 2)$ and $(3, 1)$ in $y < 2x + 1$. Which ordered pair makes the inequality true?

c. Which area represents the solution of $y < 2x + 1$?

GRAPH INEQUALITIES To graph an inequality such as $y > 2x - 3$, first graph the related equation $y = 2x - 3$. This is the **boundary**.

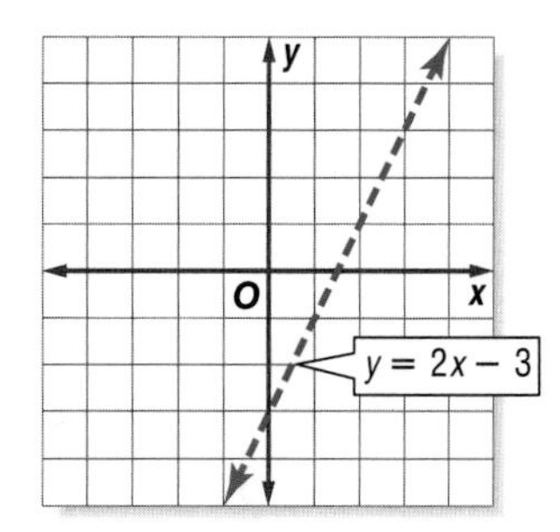

- If the inequality contains the symbol $\leq$ or $\geq$, then use a solid line to indicate that the boundary is included in the graph.
- If the inequality contains the symbol $<$ or $>$, then use a dashed line to indicate that the boundary is not included in the graph.

Next, test any point above or below the line to determine which region is the solution of $y > 2x - 3$. For example, it is easy to test (0, 0).

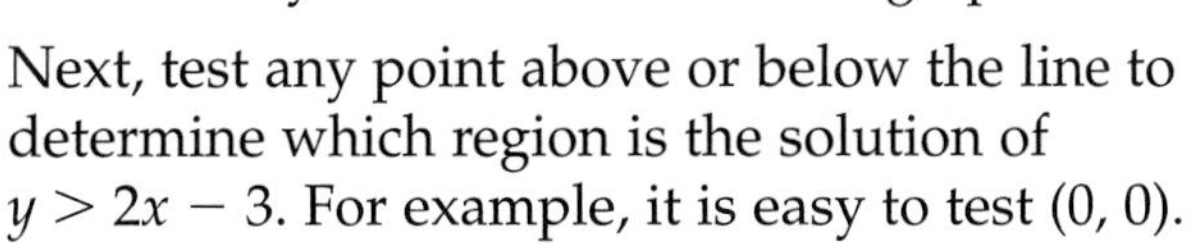

$y > 2x - 3$ Write the inequality.

$0 \overset{?}{>} 2(0) - 3$ Replace x with 0 and y with 0.

$0 > -3$ ✓ Simplify.

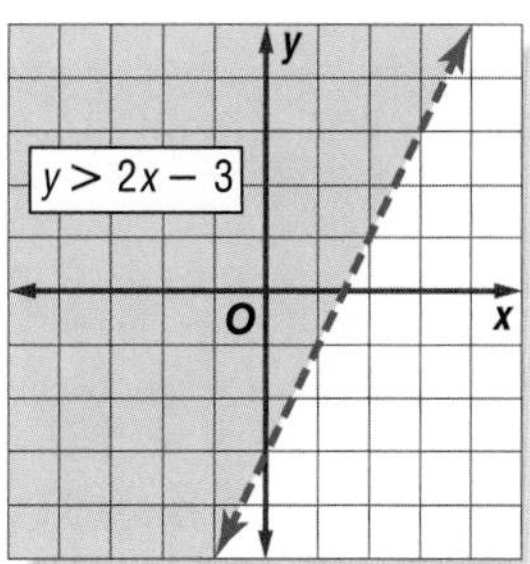

Since $0 > -3$ is true, (0, 0) is a solution of $y > 2x - 3$. Shade the region that contains the solution. This region is called a **half plane**. All points in this region are solutions of the inequality.

Example 1 Graph Inequalities

a. Graph $y < -x + 1$.

Graph $y = -x + 1$. Draw a dashed line since the boundary is not part of the graph.

Test (0, 0): $y < -x + 1$

$0 \overset{?}{<} -0 + 1$ Replace (x, y) with (0, 0).

$0 < 1$ ✓

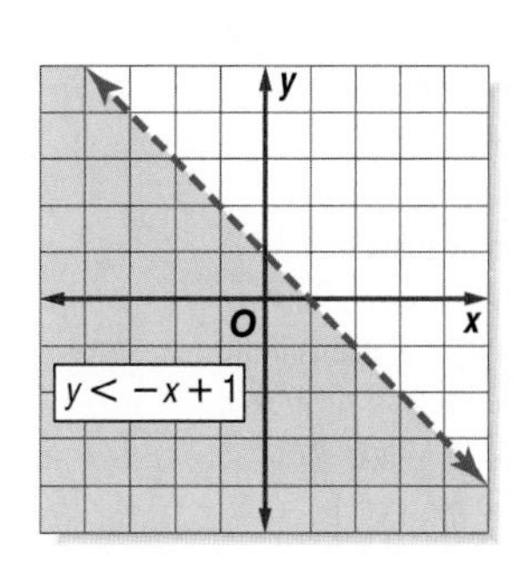

Thus, the graph is all points in the region below the boundary.

b. Graph $y \geq 2x + 2$.

Graph $y = 2x + 2$. Draw a solid line since the boundary is part of the graph.

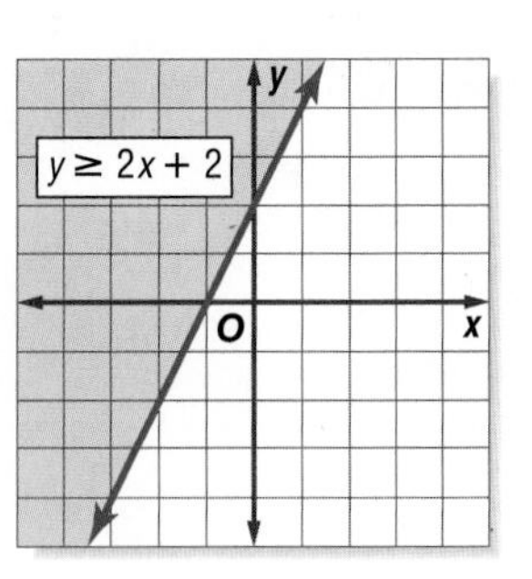

Test (0, 0): $y \geq 2x + 2$

$0 \overset{?}{\geq} 2(0) + 2$ Replace (x, y) with (0, 0).

$0 \geq 2$ not true

(0, 0) is not a solution, so shade the other half plane.

CHECK Test an ordered pair in the other half plane.

Concept Check Is (0, 2) a solution of $y \geq 2x + 2$? Explain.

FIND SOLUTIONS You can write and graph inequalities to solve real-world problems. In some cases, you may have to solve the inequality for y first and then graph the inequality.

Example 2 Write and Graph an Inequality to Solve a Problem

SCHOOL **Nathan has at most 30 minutes to complete his math and science homework. How much time can he spend on each?**

Step 1 Write an inequality.

Let x represent the time spent doing math homework and let y represent the time spent doing science homework.

Minutes doing math	plus	minutes doing science	is at most	30 minutes.
x	$+$	y	$\leq$	30

Step 2 Graph the inequality.

To graph the inequality, first solve for y.

$x + y \leq 30$ Write the inequality.

$y \leq -x + 30$ Subtract x from each side.

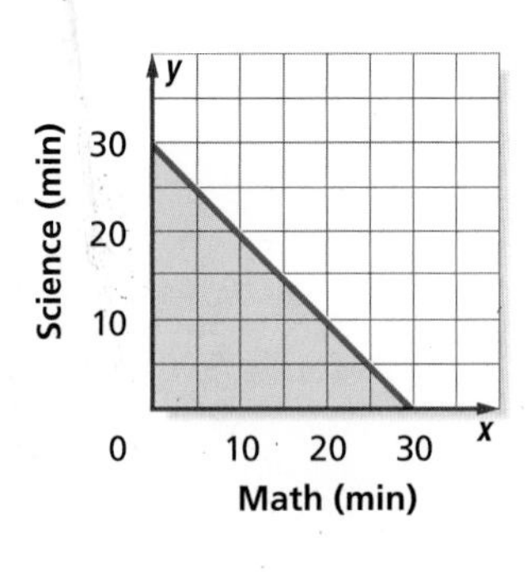

Graph $y \leq -x + 30$ as a solid line since the boundary is part of the graph. The origin is part of the graph since $0 \leq -0 + 30$. Thus, the coordinates of all points in the shaded region are possible solutions.

(10, 20) = 10 minutes on math, 20 minutes on science

(15, 15) = 15 minutes on math, 15 minutes on science

(10, 15) = 10 minutes on math, 15 minutes on science

(30, 0) = 30 minutes on math, 0 minutes on science

Note that the solutions are only in the first quadrant because negative values of time do not make sense.

Study Tip

Common Misconception
You may think that only whole number solutions are possible. Points in the shaded region such as (10.5, 18.5) are also solutions.

Check for Understanding

Concept Check

1. **Write** an inequality that describes the graph at the right.

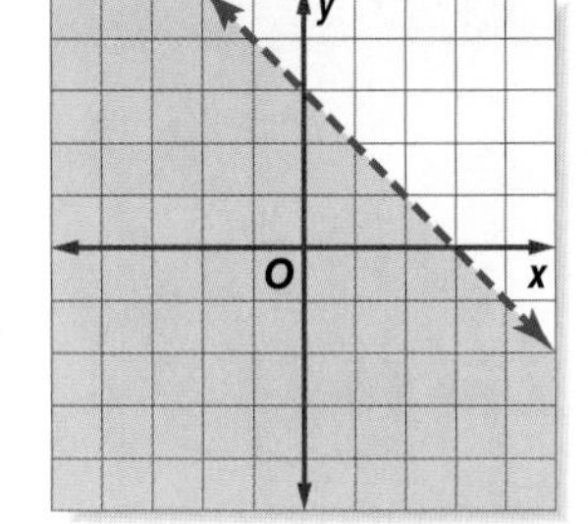

2. **Explain** how to determine which side of the boundary line to shade when graphing an inequality.
3. **List** three solutions of each inequality.
   a. $y < x$ b. $y \geq x - 3$
4. **OPEN ENDED** Write an inequality that has (1, 4) as a solution.

Guided Practice

Graph each inequality.

5. $y > x - 1$
6. $y \leq 2$
7. $y \geq -3x + 2$

Application

ENTERTAINMENT For Exercises 8 and 9, use the following information.
Adult passes for World Waterpark are \$25, and children's passes are \$15. A company is buying tickets for its employees and wants to spend no more than \$630.

8. Write an inequality to represent this situation.
9. Graph the inequality and use the graph to determine three possible combinations of tickets that the company could buy.

Practice and Apply

Homework Help

For Exercises	See Examples
10–21	1
22–32	2

Extra Practice
See page 744.

Graph each inequality.

10. $y \geq x$
11. $y < x$
12. $y < 1$
13. $y \geq -3$
14. $y > x - 1$
15. $y \leq x + 4$
16. $y \leq -x$
17. $y \geq 0$
18. $y > 2x + 3$
19. $y < 2x - 4$
20. $y < \frac{1}{2}x + 2$
21. $y \geq -\frac{1}{3}x - 1$

BUSINESS For Exercises 22–26, use the following information.
For a certain business to be successful, its monthly sales y must be at least \$3000 greater than its monthly costs x.

22. Write an inequality to represent this situation.
23. Graph the inequality.
24. Do points above or below the boundary line indicate a successful business? Explain how you know.
25. Would negative numbers make sense in this problem? Explain.
26. List two solutions.

CRAFTS For Exercises 27–29, use the following information.
Celia can make a small basket in 10 minutes and a large basket in 25 minutes. This month, she has no more than 24 hours to make these baskets for an upcoming craft fair.

27. Write an inequality to represent this situation.
28. Graph the inequality.
29. Use the graph to determine how many of each type of basket that she could make this month. List three possibilities.

Study Tip

Solutions
In Exercise 29, only whole-number solutions make sense since there cannot be parts of baskets.

RAFTING **For Exercises 30–32, use the following information.**
Collin's Rent-a-Raft rents Super Rafts for \$100 per day and Econo Rafts for \$40 per day. He wants to receive at least \$1500 per day renting out the rafts.

30. Write an inequality to represent this situation.

31. Graph the inequality.

32. Determine how many of each type of raft that Collin could rent out each day in order to receive at least \$1500. List three possibilities.

More About. . .

Rafting

More than 25 whitewater rafting companies are located on the American River in California. A typical fee is \$125 per person per day.

Source: www.rafting-whitewater.com

33. CRITICAL THINKING The solution of a *system of inequalities* is the set of all ordered pairs that satisfies *both* inequalities.

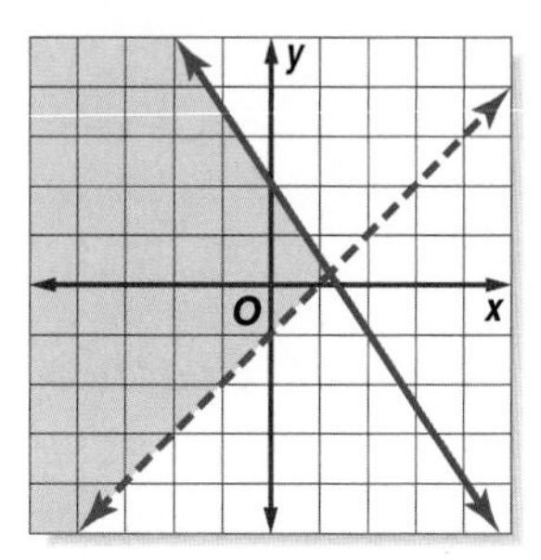

a. Write a system of inequalities for the graph at the right.

b. List three solutions of the system.

34. WRITING IN MATH Answer the question that was posed at the beginning of the lesson.

How can shaded regions on a graph model inequalities?

Include the following in your answer:

- a description of which points in the graph are solutions of the inequality.

35. Determine which ordered pair is a solution of $y - 8 > x$.

Ⓐ $(0, 2)$ Ⓑ $(2, 5)$ Ⓒ $(-9, 0)$ Ⓓ $(1, 4)$

36. Which inequality does *not* have the boundary included in its graph?

Ⓐ $y > x$ Ⓑ $y \leq 1$ Ⓒ $y \geq x + 5$ Ⓓ $y \geq 0$

Maintain Your Skills

Mixed Review **Solve each system of equations by graphing.** *(Lesson 8-9)*

37. $y = x + 3$
$x = 0$

38. $y = -x + 2$
$y = x + 4$

39. $y = -2x - 1$
$y = -x - 3$

40. Explain how you can make predictions from a set of ordered-pair data. *(Lesson 8-8)*

Write each fraction or mixed number as a decimal. Use a bar to show a repeating decimal. *(Lesson 5-1)*

41. $\frac{2}{5}$

42. $3\frac{7}{10}$

43. $-\frac{5}{9}$

Just for Fun
It is time to complete your project. Use the information and data you have gathered about recreational activities to prepare a Web page or poster. Be sure to include a scatter plot and a prediction for each activity.

www.pre-alg.com/webquest

Graphing Calculator Investigation

A Follow-Up of Lesson 8-10

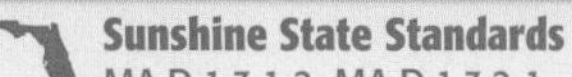

Sunshine State Standards
MA.D.1.3.1-2, MA.D.1.3.2-1, MA.D.1.3.2-2, MA.D.1.3.2-3, MA.D.2.3.1-5

Graphing Inequalities

You can use a TI-83 Plus graphing calculator to investigate the graphs of inequalities. Since the graphing calculator only shades between two functions, enter a lower boundary as well as an upper boundary for each inequality.

Graph two different inequalities on your graphing calculator.

Step 1 *Graph $y \leq -x + 4$.*

- Clear all functions from the Y= list.
  KEYSTROKES: [Y=] [CLEAR]
- Graph $y \leq -x + 4$ in the standard window.
  KEYSTROKES: [2nd] [DRAW] 7 [(–)] 10 [,] [(–)] [X,T,θ,n] [+] 4 [)] [ENTER]

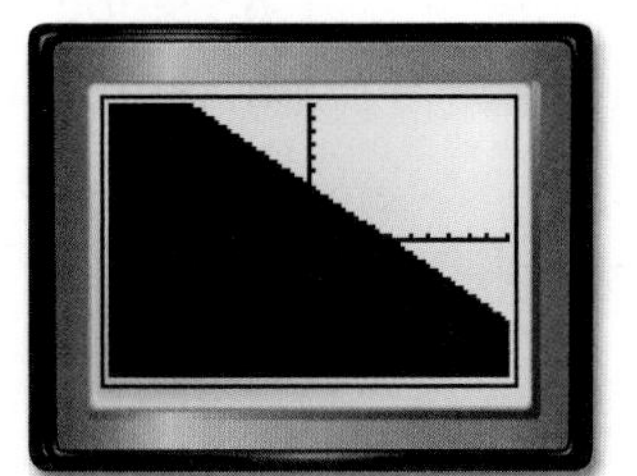

Ymin or -10 is used as the lower boundary and $y = -x + 4$ as the upper boundary. All ordered pairs in the shaded region satisfy the inequality $y \leq -x + 4$.

Step 2 *Graph $y \geq -x + 4$.*

- Clear the current drawing displayed.
  KEYSTROKES: [2nd] [DRAW] [ENTER]
- Graph $y \geq -x + 4$ in the standard window.
  KEYSTROKES: [2nd] [DRAW] 7 [(–)] [X,T,θ,n] [+] 4 [,] 10 [)] [ENTER]

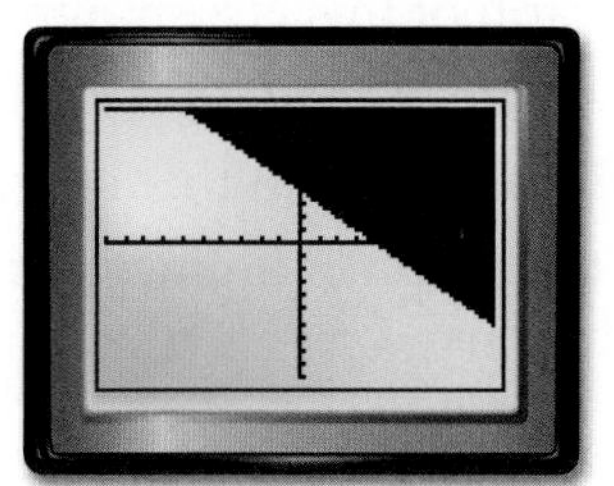

In this case, the lower boundary is $y = -x + 4$. The upper boundary is Ymax or 10. All ordered pairs in the shaded region satisfy the inequality $y \geq -x + 4$.

Exercises

1. Compare and contrast the two graphs shown above.
2. a. Graph $y \geq -2x - 6$ in the standard viewing window. Draw the graph on grid paper.
   b. What functions do you enter as the lower and upper boundaries?
   c. Use the graph to name four solutions of the inequality.

Use a graphing calculator to graph each inequality. Draw each graph on grid paper.

3. $y \leq x - 3$
4. $y \leq -1$
5. $x + y \geq 6$
6. $y \geq 3x$
7. $y \leq 0$
8. $y + 3 \leq -x$
9. $x + y \leq 5$
10. $2y - x \geq 2$

www.pre-alg.com/other_calculator_keystrokes

Chapter 8 Study Guide and Review

Vocabulary and Concept Check

best-fit line (p. 409)
boundary (p. 419)
constant of variation (p. 394)
direct variation (p. 394)
family of graphs (p. 402)
function (p. 369)
half plane (p. 419)
linear equation (p. 375)
rate of change (p. 393)
slope (p. 387)
slope-intercept form (p. 398)
substitution (p. 416)
system of equations (p. 414)
vertical line test (p. 370)
x-intercept (p. 381)
y-intercept (p. 381)

Choose the letter of the term that best matches each statement or phrase.

1. a relation in which each member of the domain is paired with exactly one member of the range
2. in a graph of an inequality, the line of the related equation
3. a value that describes the steepness of a line
4. a group of two or more equations
5. can be drawn through data points to approximate a linear relationship
6. a change in one quantity with respect to another quantity
7. a graph of this is a straight line
8. one type of method for solving systems of equations
9. a linear equation that describes rate of change
10. in the graph of an inequality, the region that contains all solutions

a. best-fit line
b. boundary
c. direct variation
d. function
e. half plane
f. linear equation
g. rate of change
h. slope
i. substitution
j. system of equations

Lesson-by-Lesson Review

8-1 Functions

See pages 369–373.

Concept Summary

- In a function, each member in the domain is paired with exactly one member in the range.

Determine whether {(−9, 2), (1, 5), (1, 10)} is a function. Explain.

domain (x)	range (y)
−9	2
1	5
	10

This relation is not a function because 1 in the domain is paired with two range values, 5 and 10.

Exercises **Determine whether each relation is a function. Explain.**
See Example 1 on page 369.

11. {(1, 12), (−4, 3), (6, 36), (10, 6)}
12. {(11.8, −9), (10.4, −2), (11.8, 3.8)}
13. {(0, 0), (2, 2), (3, 3), (4, 4)}
14. {(−0.5, 1.2), (3, 1.2), (2, 36)}

www.pre-alg.com/vocabulary_review

8-2 Linear Equations in Two Variables

See pages 375–379.

Concept Summary

- A solution of a linear equation is an ordered pair that makes the equation true.
- To graph a linear equation, plot points and draw a line through them.

Example **Graph $y = -x + 2$ by plotting ordered pairs.**

Find ordered pair solutions. Then plot and connect the points.

x	$-x + 2$	y	(x, y)
0	$-0 + 2$	2	(0, 2)
1	$-1 + 2$	1	(1, 1)
2	$-2 + 2$	0	(2, 0)
3	$-3 + 2$	-1	(3, -1)

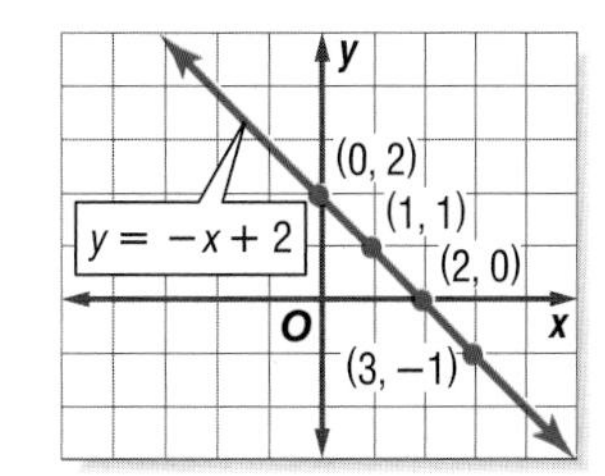

Exercises **Graph each equation by plotting ordered pairs.**

See Example 3 on page 377.

15. $y = x + 4$ **16.** $y = x - 2$ **17.** $y = -x$ **18.** $y = 2x$

19. $y = 3x + 2$ **20.** $y = -2x - 4$ **21.** $x + y = 4$ **22.** $x - y = -3$

8-3 Graphing Linear Equations Using Intercepts

See pages 381–385.

Concept Summary

- To graph a linear equation, you can find and plot the points where the graph crosses the x-axis and the y-axis. Then connect the points.

Example **Graph $-3x + y = 3$ using the x- and y-intercepts.**

$-3x + y = 3$

$-3x + 0 = 3$

$x = -1$ The x-intercept is -1.

$-3x + y = 3$

$-3(0) + y = 3$

$y = 3$ The y-intercept is 3.

Graph the points at $(-1, 0)$ and $(0, 3)$ and draw a line through them.

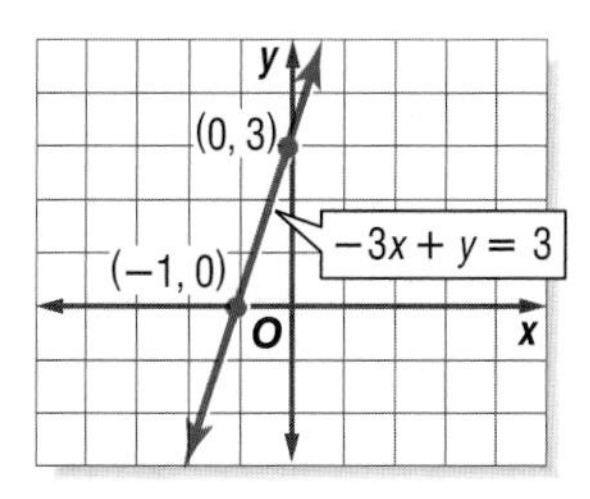

Exercises **Graph each equation using the x- and y-intercepts.**

See Example 3 on page 382.

23. $y = x + 2$ **24.** $y = x + 6$ **25.** $y = -x - 3$ **26.** $y = x - 1$

27. $x = -4$ **28.** $y = 5$ **29.** $x + y = -2$ **30.** $3x + y = 6$

8-4 Slope

See pages 387–391.

Concept Summary

- Slope is the ratio of the *rise*, or the vertical change, to the *run*, or the horizontal change.

Example **Find the slope of the line that passes through $A(0, 6)$, $B(4, -2)$.**

$m = \frac{y_2 - y_1}{x_2 - x_1}$ Definition of slope

$m = \frac{-2 - 6}{4 - 0}$ $(x_1, y_1) = (0, 6)$, $(x_2, y_2) = (4, -2)$

$m = \frac{-8}{4}$ or -2 The slope is -2.

Exercises **Find the slope of the line that passes through each pair of points.**
See Examples 2–5 on pages 388 and 389.

31. $J(3, 4)$, $K(4, 5)$
32. $C(2, 8)$, $D(6, 7)$
33. $R(7, 3)$, $B(-1, -4)$
34. $Q(2, 10)$, $B(4, 6)$
35. $X(-1, 5)$, $Y(-1, 9)$
36. $S(0, 8)$, $T(-3, 8)$

8-5 Rate of Change

See pages 393–397.

Concept Summary

- A change in one quantity with respect to another quantity is called the rate of change.
- Slope can be used to describe rates of change.

Example **Find the rate of change in population from 1990 to 2000 for Oakland, California, using the graph.**

Year	Population (1000s)
x	y
1990	372
2000	400

rate of change $= \frac{y_2 - y_1}{x_2 - x_1}$ Definition of slope

$= \frac{400 - 372}{2000 - 1990}$ ← change in population ← change in time

$= 2.8$ Simplify.

So, the rate of change in population was an increase of about 2.8 thousand, or 2800 people per year.

Exercises **Find the rate of change for the linear function represented in each table.** *See Example 1 on page 393.*

37.

Time (s)	Distance (m)
x	y
0	0
1	8
2	16

38.

Time (h)	Temperature (°F)
x	y
1	45
2	43
3	41

8-6 Slope-Intercept Form

See pages 398–401.

Concept Summary

- In the slope-intercept form $y = mx + b$, m is the slope and b is the y-intercept.

Example **State the slope and the y-intercept of the graph of $y = -2x + 3$.**

The slope of the graph is -2, and the y-intercept is 3.

Exercises **Graph each equation using the slope and y-intercept.**
See Example 3 on page 399.

39. $y = x + 4$ **40.** $y = -2x + 1$ **41.** $y = \frac{1}{3}x - 2$ **42.** $x + y = -5$

8-7 Writing Linear Equations

See pages 404–408.

Concept Summary

- You can write a linear equation by using the slope and y-intercept, two points on a line, a graph, a table, or a verbal description.

Example **Write an equation in slope-intercept form for the line having slope 4 and y-intercept -2.**

$y = mx + b$ Slope-intercept form

$y = 4x + (-2)$ Replace m with 4 and b with -2.

$y = 4x - 2$ Simplify.

Exercises **Write an equation in slope-intercept form for each line.**
See Example 1 on page 404.

43. slope $= -1$, y-intercept $= 3$ **44.** slope $= 6$, y-intercept $= -3$

8-8 Best-Fit Lines

See pages 409–413.

Concept Summary

- A best-fit line can be used to approximate data.

Example **Draw a best-fit line through the scatter plot.**

Draw a line that is close to as many data points as possible.

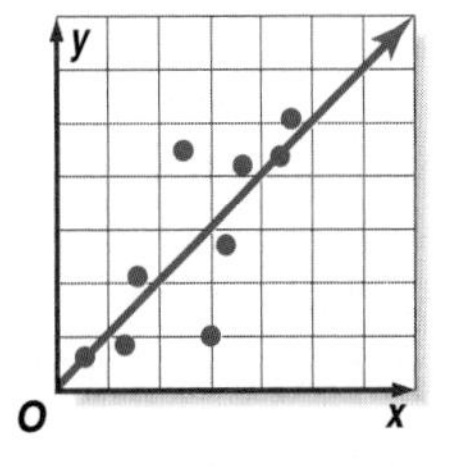

Exercises **The table shows the attendance for an annual art festival.** *See Example 1 on page 409.*

45. Make a scatter plot and draw a best-fit line.

46. Use the best-fit line to predict art festival attendance in 2008.

Year	Attendance
2000	2500
2001	2650
2002	2910
2003	3050

Chapter 8 **For More ...**

- Extra Practice, see pages 741–744.
- Mixed Problem Solving, see page 765.

8-9 Solving Systems of Equations

See pages 414–418.

Concept Summary

- The solution of a system of equations is the ordered pair that satisfies all equations in the system.

Examples

1 Solve the system of equations by graphing.

$y = x$

$y = -x + 2$

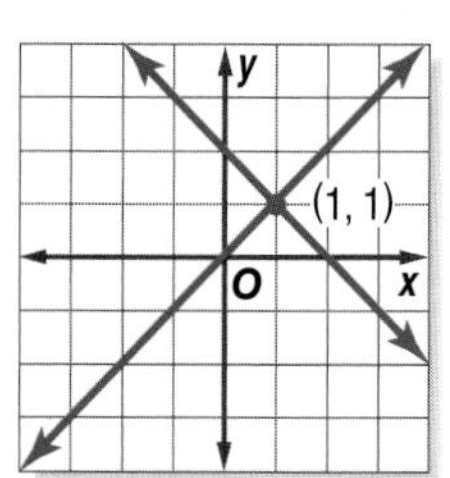

The graphs appear to intersect at (1, 1). The solution of the system of equations is (1, 1).

2 Solve the system $y = x - 1$ and $y = 3$ by substitution.

$y = x - 1$ Write the first equation.

$3 = x - 1$ Replace y with 3.

$4 = x$ Solve for x.

The solution of this system of equations is (4, 3). Check by graphing.

Exercises Solve each system of equations by graphing.
See Examples 1–4 on pages 414 and 415.

47. $y = x$, $y = 3$

48. $y = 2x + 4$, $x + y = -2$

49. $3x + y = 1$, $y = -3x + 5$

Solve each system of equations by substitution. *See Example 5 on page 416.*

50. $y = x + 6$, $y = -1$

51. $y = x$, $x = 4$

52. $y = 2x - 3$, $y = 0$

8-10 Graphing Inequalities

See pages 419–422.

Concept Summary

- To graph an inequality, first graph the related equation, which is the boundary.
- All points in the shaded region are solutions of the inequality.

Example

Graph $y < x + 2$.

Graph $y = x + 2$. Draw a dashed line since the boundary is not part of the graph. Test a point in the original inequality and shade the appropriate region.

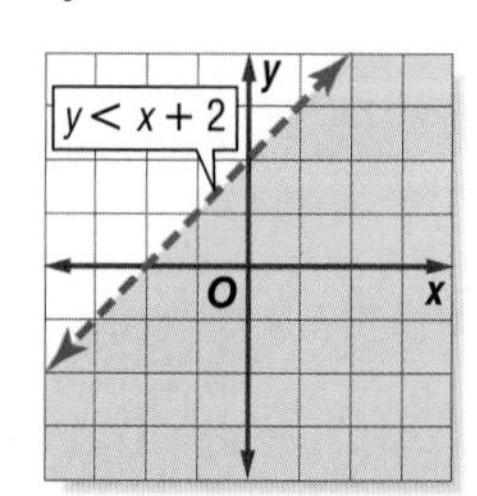

Exercises Graph each inequality. *See Example 1 on pages 419 and 420.*

53. $y \leq 3x - 1$ **54.** $y > \frac{1}{2}x - 3$ **55.** $y \geq -x + 4$ **56.** $y < -2x$

Chapter 8 Practice Test

Vocabulary and Concepts

1. **Describe** how to find the x-intercept and the y-intercept of a linear equation.
2. **State** how to choose which half plane to shade when graphing an inequality.
3. **OPEN ENDED** Write a system of equations and explain what the solution is.

Skills and Applications

Determine whether each relation is a function. Explain.

4. $\{(-3, 4), (2, 9), (4, -1), (-3, 6)\}$
5. $\{(1, 2), (4, -6), (-3, 5), (6, 2)\}$

Graph each equation by plotting ordered pairs.

6. $y = 2x + 1$
7. $3x + y = 4$

Find the x-intercept and y-intercept for the graph of each equation. Then, graph the equation using the x- and y-intercepts.

8. $y = x + 3$
9. $2x - y = 4$

Find the slope of the line that passes through each pair of points.

10. $A(2, 5), B(4, 11)$
11. $C(-4, 5), D(6, -3)$

12. Find the rate of change for the linear function represented in the table.

Hours Worked	1	2	3	4
Money Earned ($)	5.50	11.00	16.50	22.00

State the slope and y-intercept for the graph of each equation. Then, graph each equation using the slope and y-intercept.

13. $y = \frac{2}{3}x - 4$
14. $2x + 4y = 12$

15. Write an equation in slope-intercept form for the line with a slope of $\frac{3}{8}$ and y-intercept $= -2$.

16. Solve the system of equations $2x - y = 4$ and $4x + y = 2$.

17. Graph $y > 2x - 1$.

GARDENING **For Exercises 18 and 19, use the table at the right and the information below.**
The full-grown height of a tomato plant and the number of tomatoes it bears are recorded for five tomato plants.

Height (in.)	Number of Tomatoes
27	12
33	18
19	9
40	16
31	15

18. Make a scatter plot of the data and draw a best-fit line.
19. Use the best-fit line to predict the number of tomatoes a 43-inch tomato plant will bear.

FCAT Practice

20. **STANDARDIZED TEST PRACTICE** Which is a solution of $y \geq -2x + 5$?

Ⓐ $(1, -7)$ Ⓑ $(1, -3)$ Ⓒ $(-1, 3)$ Ⓓ $(-1, 7)$

Chapter 8

Standardized Test Practice

FCAT Practice

Part 1 Multiple Choice

Record your answers on the answer sheet provided by your teacher or on a sheet of paper.

1. A wooden rod is 5.5 feet long. If you cut the rod into 10 equal pieces, how many inches long will each piece be? (Prerequisite Skill, p. 715)

   (A) 0.55 in. (B) 5.5 in.
   (C) 6.6 in. (D) 66 in.

2. Which of the following is a true statement? (Lesson 2-1)

   (A) $\frac{9}{3} < \frac{3}{9}$ (B) $-\frac{3}{9} < -\frac{9}{3}$
   (C) $-\frac{9}{3} > \frac{3}{9}$ (D) $-\frac{3}{9} > -\frac{9}{3}$

3. You roll a cube that has 2 blue sides, 2 yellow sides, and 2 red sides. What is the probability that a yellow side will face upwards when the cube stops rolling? (Lesson 6-9)

   (A) $\frac{1}{6}$ (B) $\frac{1}{3}$
   (C) $\frac{1}{2}$ (D) $\frac{2}{3}$

4. Luis used 3 quarts of paint to cover 175 square feet of wall. He now wants to paint 700 square feet in another room. Which proportion could he use to calculate how many quarts of paint he should buy? (Lesson 6-3)

   (A) $\frac{175}{700} = \frac{x}{3}$ (B) $\frac{3}{700} = \frac{175}{x}$
   (C) $\frac{700}{3} = \frac{175}{x}$ (D) $\frac{3}{175} = \frac{x}{700}$

The Princeton Review Test-Taking Tip

Question 1 Pay attention to units of measurement, such as inches, feet, grams, and kilograms. A problem may require you to convert units; for example, you may be told the length of an object in feet and need to find the length of that object in inches.

5. A shampoo maker offered a special bottle with 30% more shampoo than the original bottle. If the original bottle held 12 ounces of shampoo, how many ounces did the special bottle hold? (Lesson 6-7)

   (A) 3.6 (B) 12.3 (C) 15.6 (D) 16

6. The graph shows the materials in a town's garbage collection. If the week's garbage collection totals 6000 pounds, how many pounds of garbage are *not* paper? (Lesson 6-7)

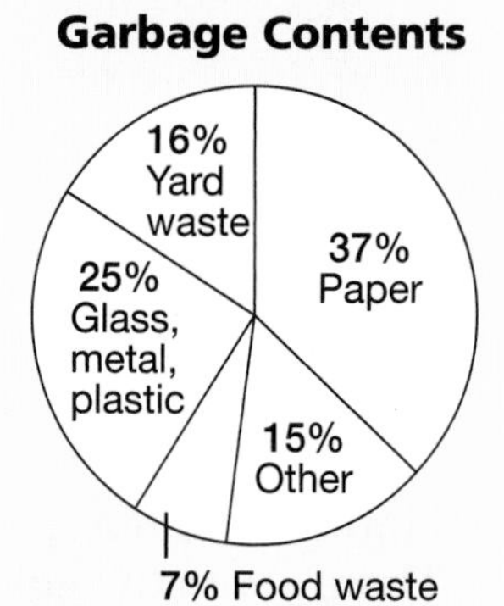

   (A) 2220 (B) 3780 (C) 5630 (D) 5963

7. The sum of an integer and the next greater integer is more than 51. Which of these could be the integer? (Lesson 7-6)

   (A) 23 (B) 24 (C) 25 (D) 26

8. The table represents a function between x and y. What is the missing number in the table? (Lesson 8-1)

x	y
1	3
2	■
4	9
6	13

   (A) 4 (B) 5
   (C) 6 (D) 7

9. Which equation is represented by the graph? (Lesson 8-7)

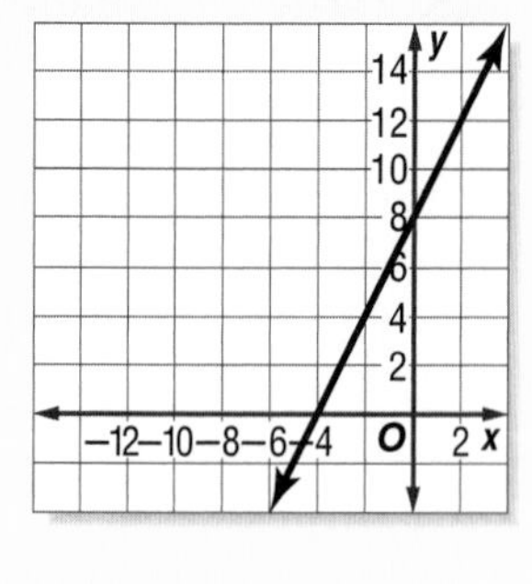

   (A) $y = 2x - 4$
   (B) $y = 2x + 8$
   (C) $y = \frac{1}{2}x - 8$
   (D) $y = \frac{1}{2}x + 4$

10. Which ordered pair is the solution of this system of equations? (Lesson 8-9)

$$2x + 3y = 7$$
$$3x - 3y = 18$$

   (A) (25, 1) (B) (2, 1)
   (C) (5, −1) (D) (7, 1)

Part 2 Short Response/Grid In

Record your answers on the answer sheet provided by your teacher or on a sheet of paper.

11. The graph shows the shipping charges per order, based on the number of items shipped in the order. What is the shipping charge for an order with 4 items? (Prerequisite Skill, pp. 722–723)

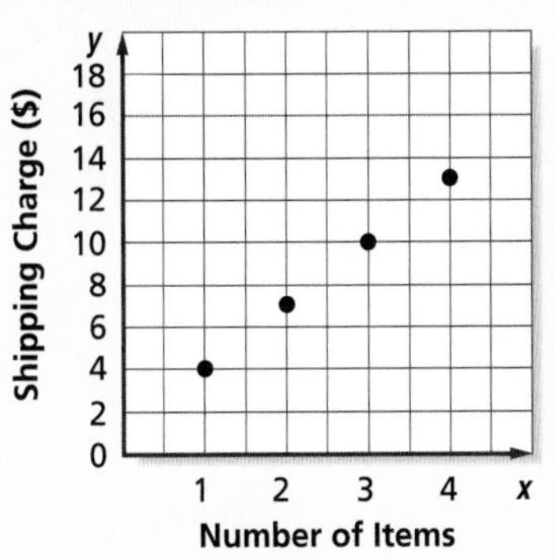

12. The length of a rectangle is 8 centimeters, and its perimeter is 24 centimeters. What is the area of the rectangle in square centimeters? (Lesson 3-5)

13. Write the statement *y is 5 more than one half the value of x* as an equation. (Lesson 3-6)

14. Write the number 0.09357 in scientific notation. Round your answer to two decimal places. (Lesson 4-8)

15. If $y = -\frac{3}{2}$, what is the value of x in $y = 4x - 3$? (Lesson 5-8)

16. Write $\frac{3}{5}$ as a percent. (Lesson 6-4)

17. In a diving competition, the diver in first place has a total score of 345.4. Ming has scored 68.2, 68.9, 67.5, and 71.7 for her first four dives and has one more dive remaining. Write an inequality to show the score x that Ming must receive on her fifth dive in order to overtake the diver in first place. (Lesson 7-4)

18. Ms. Vang drove at 30 mph for 30 minutes and at 56 mph for one hour and fifteen minutes. How far did she travel? (Lesson 8-5)

19. What is the slope of the line that contains (0, 4) and (2, 5)? (Lesson 8-6)

20. Find the solution of the system of equations graphed below. (Lesson 8-9)

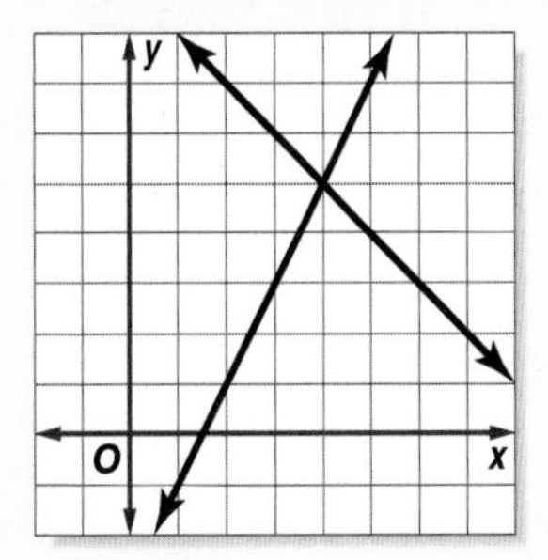

FCAT Practice

Part 3 Open Ended

Record your answers on a sheet of paper. Show your work.

21. Krishnan is considering three plans for cellular phone service. The plans each offer the same services for different monthly fees and different costs per minute. (Lesson 8-6)

Plan	Monthly Fee	Cost per Minute
X	$0	$0.24
Y	$15.95	$0.08
Z	$25.95	$0.04

a. For each plan, write an equation that shows the total monthly cost c for m minutes of calls.

b. What is the cost of each plan if Krishnan uses 100 minutes per month?

c. Which plan costs the least if Krishnan uses 100 minutes per month?

d. What is the cost of each plan if Krishnan uses 300 minutes per month?

e. Which plan costs the least if Krishnan uses 300 minutes per month?

UNIT 4

Applying Algebra to Geometry

Although they may seem like different subjects, algebra and geometry are closely related. In this unit, you will use algebra to solve geometry problems.

Chapter 9
Real Numbers and Right Triangles

Chapter 10
Two-Dimensional Figures

Chapter 11
Three-Dimensional Figures

WebQuest
Internet Project

Able to Leap Tall Buildings

The building with the tallest rooftop is the Sears Tower in Chicago, with a height of 1450 feet. However, the tallest building is the Petronas Twin Towers in Kuala Lumpur, Malaysia, whose architectural spires rise to 1483 feet.

In this project, you will be exploring how geometry and algebra can help you describe unusual or large structures of the world.

USA TODAY
Education

Log on to www.pre-alg.com/webquest. Begin your WebQuest by reading the Task.

Then continue working on your WebQuest as you study Unit 4.

Lesson	9-8	10-8	11-3
Page	481	542	571

USA TODAY Snapshots®
USA TODAY
World's tallest buildings
The ranking of the tallest skyscrapers is based on measuring the building from the sidewalk level of the main entrance to the structural top of the building. That includes spires, but not antennas or flagpoles.
1,483'
1,450'
1,380'
1
Petronas Twin Towers
Kuala Lumpur, Malaysia
2
Sears Tower
Chicago
3
Jin Mao Building
Shanghai
Source: Council on Tall Buildings and Urban Habitat
USA TODAY

Chapter

9 Real Numbers and Right Triangles

What You'll Learn

- **Lessons 9-1 and 9-2** Find and use squares and square roots and identify numbers in the real number system.
- **Lessons 9-3 and 9-4** Classify angles and triangles and find the missing angle measure of a triangle.
- **Lessons 9-5 and 9-6** Use the Pythagorean Theorem, the Distance Formula, and the Midpoint Formula.
- **Lesson 9-7** Identify and use properties of similar figures.
- **Lesson 9-8** Use trigonometric ratios to solve problems.

Key Vocabulary

- **angle** (p. 447)
- **triangle** (p. 453)
- **Pythagorean Theorem** (p. 460)
- **similar triangles** (p. 471)
- **trigonometric ratios** (p. 477)

Why It's Important

All of the numbers we use on a daily basis are real numbers. Formulas that contain real numbers can be used to solve real-world problems dealing with distance. For example, if you know the height of a lighthouse, you can use a formula to determine how far you can see from the top of the lighthouse. *You will solve a problem about lighthouses in Lesson 9-1.*

Getting Started

Prerequisite Skills To be successful in this chapter, you'll need to master these skills and be able to apply them in problem-solving situations. Review these skills before beginning Chapter 9.

For Lesson 9-2 — Compare Decimals

Replace each ● with <, >, or = to make a true statement. *(For review, see page 710.)*

1. 3.2 ● 3.5 $<$ **2.** 7.8 ● 7.7 $>$ **3.** 5.13 ● 5.16 $<$ **4.** 4.92 ● 4.89 $>$

5. 2.62 ● 2.6 $>$ **6.** 3.4 ● 3.41 $<$ **7.** 0.07 ● 0.7 $<$ **8.** 1.16 ● 1.06 $>$

For Lessons 9-4 and 9-7 — Solve Equations by Dividing

ALGEBRA Solve each equation. *(For review, see Lesson 3-4.)*

9. $3x = 24$ 8 **10.** $7y = 49$ 7 **11.** $120 = 2n$ 60 **12.** $54 = 6a$ 9

13. $90 = 10m$ 9 **14.** $144 = 12m$ 12 **15.** $15d = 165$ 11 **16.** $182 = 14w$ 13

For Lesson 9-6 — Exponents

Find the value of each expression. *(For review, see Lesson 4-2.)*

17. $(3 - 1)^2 + (4 - 2)^2$ 8 **18.** $(5 - 2)^2 + (6 - 3)^2$ 18 **19.** $(4 - 7)^2 + (3 - 8)^2$ 34

20. $(8 - 2)^2 + (3 - 9)^2$ 72 **21.** $(2 - 6)^2 + [(-8) - 1]^2$ 97 **22.** $(-7 - 2)^2 + [3 - (-4)]^2$ 130

Make this Foldable to help you organize information about real numbers and right triangles. Begin with three plain sheets of $8\frac{1}{2}$" by 11" paper.

Step 1 Fold

Fold to make a triangle. Cut off extra paper.

Step 2 Repeat

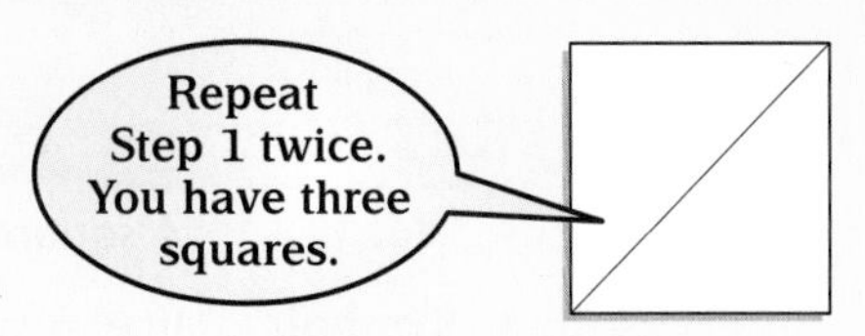

Repeat Step 1 twice. You have three squares.

Step 3 Stack and Staple

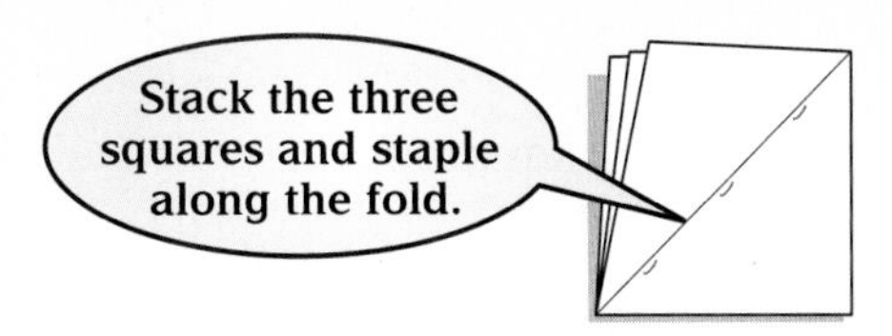

Stack the three squares and staple along the fold.

Step 4 Label

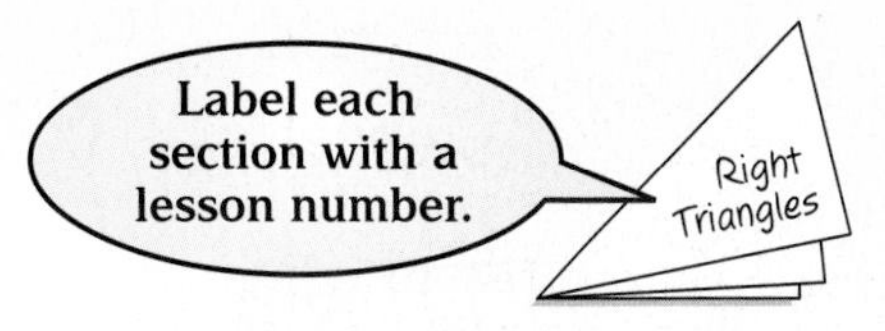

Label each section with a lesson number.
Right Triangles

Reading and Writing As you read and study the chapter, fill the pages with examples, diagrams, and formulas.

9-1 Squares and Square Roots

What You'll Learn

Sunshine State Standards
MA.A.1.3.2-2, MA.A.1.3.3-2, MA.A.1.3.4-2

- Find squares and square roots.
- Estimate square roots.

Vocabulary

- perfect square
- square root
- radical sign

How are square roots related to factors?

Values of x^2 are shown in the second column in the table. Guess and check to find the value of x that corresponds to x^2. If you cannot find an exact answer, estimate with decimals to the nearest tenth to find an approximate answer.

x	x^2
	25
	49
	169
	225
	8
	12
	65
	110

a. Describe the difference between the first four and the last four values of x.

b. Explain how you found an exact answer for the first four values of x.

c. How did you find an estimate for the last four values of x?

SQUARES AND SQUARE ROOTS Numbers like 25, 49, 169, and 225 are **perfect squares** because they are squares of whole numbers.

5×5 or 5^2 → 25

7×7 or 7^2 → 49

13×13 or 13^2 → 169

15×15 or 15^2 → 225

A **square root** of a number is one of two equal factors of the number. Every positive number has a positive square root and a negative square root. A negative number like -9 has no real square root because the square of a number is never negative.

Key Concept — *Square Root*

- **Words** A square root of a number is one of its two equal factors.
- **Symbols** If $x^2 = y$, then x is a square root of y.
- **Examples** Since $5 \cdot 5$ or $5^2 = 25$, 5 is a square root of 25.
  Since $(-5) \cdot (-5)$ or $(-5)^2 = 25$, -5 is a square root of 25.

A **radical sign**, $\sqrt{}$, is used to indicate the square root.

Example 1 Find Square Roots

Find each square root.

a. $\sqrt{36}$ $\sqrt{36}$ indicates the *positive* square root of 36.

Since $6^2 = 36$, $\sqrt{36} = 6$.

b. $-\sqrt{81}$ $\quad -\sqrt{81}$ indicates the *negative* square root of 81.

Since $9^2 = 81$, $-\sqrt{81} = -9$.

c. $\pm\sqrt{9}$ $\quad \pm\sqrt{9}$ indicates *both* square roots of 9.

Since $3^2 = 9$, $\sqrt{9} = 3$ and $-\sqrt{9} = -3$.

Reading Math

Plus or Minus Symbol The notation $\pm\sqrt{9}$ is read *plus or minus the square root of 9.*

Concept Check What does the radical sign indicate?

You can use a calculator to find an approximate square root of a number that is not a perfect square. The decimal portion of these square roots goes on forever.

Example 2 Calculate Square Roots

Use a calculator to find each square root to the nearest tenth.

a. $\sqrt{10}$

2nd [$\sqrt{\ }$] 10 ENTER 3.16227766 — Use a calculator.

$\sqrt{10} \approx 3.2$ — Round to the nearest tenth.

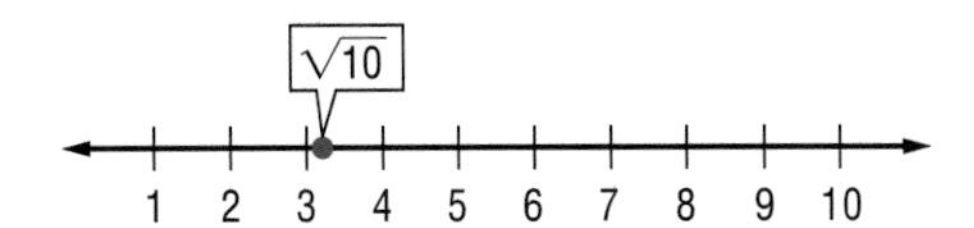

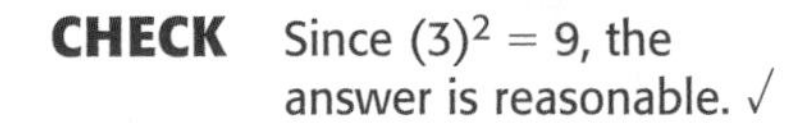
CHECK Since $(3)^2 = 9$, the answer is reasonable. ✓

Reading Math

Approximately Equal To Symbol The symbol $\approx$ is read *is approximately equal to.*

b. $-\sqrt{27}$

2nd [$\sqrt{\ }$] 27 ENTER 5.19615242 — Use a calculator.

$-\sqrt{27} \approx -5.2$ — Round to the nearest tenth.

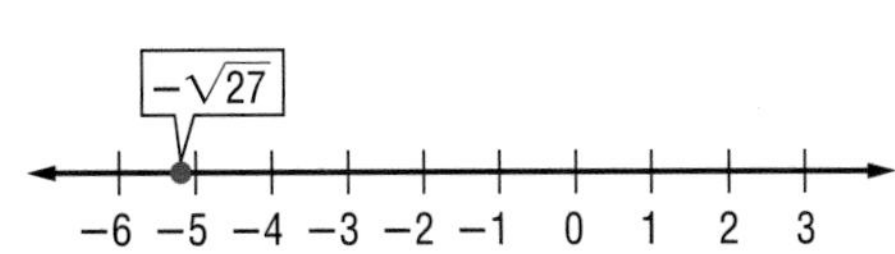

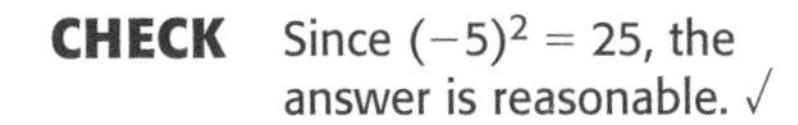
CHECK Since $(-5)^2 = 25$, the answer is reasonable. ✓

ESTIMATE SQUARE ROOTS You can also estimate square roots without using a calculator.

Example 3 Estimate Square Roots

Estimate each square root to the nearest whole number.

a. $\sqrt{38}$

Find the two perfect squares closest to 38. To do this, list some perfect squares.

1, 4, 9, 16, 25, 36, 49, …

36 and 49 are closest to 38.

$36 < 38 < 49$ — 38 is between 36 and 49.

$\sqrt{36} < \sqrt{38} < \sqrt{49}$ — $\sqrt{38}$ is between $\sqrt{36}$ and $\sqrt{49}$.

$6 < \sqrt{38} < 7$ — $\sqrt{36} = 6$ and $\sqrt{49} = 7$.

Since 38 is closer to 36 than 49, the best whole number estimate for $\sqrt{38}$ is 6.

b. $-\sqrt{175}$

Find the two perfect squares closest to 175. List some perfect squares.

$$\ldots, 100, 121, 144, 169, 196, \ldots$$

169 and 196 are closest to 175.

$-196 < -175 < -169$	-175 is between -196 and -169.
$-\sqrt{196} < -\sqrt{175} < -\sqrt{169}$	$-\sqrt{175}$ is between $-\sqrt{196}$ and $-\sqrt{169}$.
$-14 < -\sqrt{175} < -13$	$-\sqrt{196} = -14$ and $-\sqrt{169} = -13$.

Since -175 is closer to -169 than -196, the best whole number estimate for $-\sqrt{175}$ is -13. **CHECK** $-\sqrt{175} \approx -13.2$ ✓

Many formulas used in real-world applications involve square roots.

Example 4 Use Square Roots to Solve a Problem

More About...

Science

To estimate how far you can see from a point above the horizon, you can use the formula $D = 1.22 \times \sqrt{A}$ where D is the distance in miles and A is the altitude, or height, in feet.

SCIENCE Use the information at the left. The light on Cape Hatteras Lighthouse in North Carolina is 208 feet high. On a clear day, from about what distance on the ocean is the light visible? Round to the nearest tenth.

$D = 1.22 \times \sqrt{A}$	Write the formula.
$= 1.22 \times \sqrt{208}$	Replace A with 208.
$\approx 1.22 \times 14.42$	Evaluate the square root first.
≈ 17.5924	Multiply.

On a clear day, the light will be visible from about 17.6 miles.

Check for Understanding

Concept Check

1. **Explain** why every positive number has two square roots.
2. **List** three numbers between 200 and 325 that are perfect squares.
3. **OPEN ENDED** Write a problem in which the negative square root is not an integer. Then graph the square root.

Guided Practice

Find each square root, if possible.

4. $\sqrt{49}$
5. $-\sqrt{64}$
6. $\sqrt{-36}$

Use a calculator to find each square root to the nearest tenth.

7. $\sqrt{15}$
8. $-\sqrt{32}$

Estimate each square root to the nearest whole number. Do not use a calculator.

9. $\sqrt{66}$
10. $-\sqrt{103}$

Application

11. **SKYSCRAPERS** Refer to Example 4. Ryan is standing in the observation area of the Sears Tower in Chicago. About how far can he see on a clear day if the deck is 1353 feet above the ground?

Practice and Apply

Homework Help

For Exercises	See Examples
12–23	1
24–33	2
34–46	3
47, 48	4

Extra Practice

See page 745.

Find each square root, if possible.

12. $\sqrt{16}$
13. $\sqrt{36}$
14. $-\sqrt{1}$
15. $-\sqrt{25}$
16. $\sqrt{-4}$
17. $\sqrt{-49}$
18. $\sqrt{100}$
19. $\sqrt{196}$
20. $\pm\sqrt{256}$
21. $\pm\sqrt{324}$
22. $\sqrt{0.81}$
23. $\sqrt{2.25}$

Use a calculator to find each square root to the nearest tenth.

24. $\sqrt{15}$
25. $\sqrt{56}$
26. $-\sqrt{43}$
27. $-\sqrt{86}$
28. $\sqrt{180}$
29. $\sqrt{250}$
30. $-\sqrt{0.75}$
31. $-\sqrt{3.05}$

32. Find the negative square root of 1000 to the nearest tenth.

33. If $x = \sqrt{5000}$, what is the value of x to the nearest tenth?

34. The number $\sqrt{54}$ lies between which two consecutive whole numbers? Do not use a calculator.

Estimate each square root to the nearest whole number. Do not use a calculator.

35. $\sqrt{79}$
36. $\sqrt{95}$
37. $-\sqrt{54}$
38. $-\sqrt{125}$
39. $\sqrt{200}$
40. $\sqrt{396}$
41. $-\sqrt{280}$
42. $-\sqrt{490}$
43. $-\sqrt{5.25}$
44. $-\sqrt{17.3}$
45. $\sqrt{38.75}$
46. $\sqrt{140.57}$

ROLLER COASTERS **For Exercises 47–48, use the table shown and refer to Example 4 on page 438.**

47. On a clear day, how far can a person see from the top hill of the Vortex?

48. How far can a person see on a clear day from the top hill of the Titan?

Coaster	Maximum Height (ft)
Double Loop	95
The Villain	120
Mean Streak	161
Raptor	137
The Beast	110
Vortex	148
Titan	255
Shockwave	116

Source: Roller Coaster Database

Online Research **Data Update** How far can you see from the top of the tallest roller coaster in the United States? Visit www.pre-alg.com/data_update to learn more.

Complete Exercises 49 and 50 without the help of a calculator.

49. Which is greater, $\sqrt{65}$ or 9? Explain your reasoning.

50. Which is less, 11 or $\sqrt{120}$? Explain your reasoning.

51. **GEOMETRY** **The area of each square is given. Estimate the length of a side of each square to the nearest tenth.**

a. 109 in²

b. 203 cm²

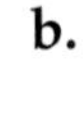

c. 70 m²

Simplify.

52. $2 + \sqrt{81} \times 3$
53. $\sqrt{256} \div 8 \times 5$
54. $(9 - \sqrt{36}) + (16 - \sqrt{100})$
55. $(\sqrt{1} + 7) \div (\sqrt{121} - 7)$

56. **GEOMETRY** Find the perimeter of a square that has an area of $\sqrt{2080}$ square meters. Round to the nearest tenth.

57. **CRITICAL THINKING** What are the possibilities for the ending digit of a number that has a whole number square root? Explain your reasoning.

58. **WRITING IN MATH** Answer the question that was posed at the beginning of the lesson.

How are square roots related to factors?

Include the following in your answer:

- an example of a number between 100 and 200 whose square root is a whole number, and
- an example of a number between 100 and 200 whose square root is a decimal that does not terminate.

59. Which statement is *not* true?

Ⓐ $6 < \sqrt{39} < 7$ Ⓑ $9 < \sqrt{89} < 10$

Ⓒ $-7 > -\sqrt{56} > -8$ Ⓓ $-4 < -\sqrt{17} < -5$

60. Choose the expression that is a rational number.

Ⓐ $-\sqrt{361}$ Ⓑ $\sqrt{125}$ Ⓒ $\sqrt{200}$ Ⓓ $\sqrt{325}$

Extending the Lesson

61. Squaring a number and finding the square root of a number are *inverse operations*. That is, one operation undoes the other operation. Name another pair of inverse operations.

62. Use inverse operations to evaluate each expression.

a. $(\sqrt{64})^2$ **b.** $(\sqrt{100})^2$ **c.** $(\sqrt{169})^2$

63. Use the pattern from Exercise 62 to find $(\sqrt{a})^2$.

Maintain Your Skills

Mixed Review

ALGEBRA Graph each inequality. *(Lesson 8-10)*

64. $y > x - 2$ 65. $y < -4$ 66. $y \geq 2x + 3$

ALGEBRA Solve each system of equations by substitution. *(Lesson 8-9)*

67. $y = x + 2$, $x = 5$ 68. $y = x + 3$, $y = 0$ 69. $y = 2x + 5$, $y = -3$

70. Determine whether the relation (4, −1), (3, 5), (−4, 1), (4, 2) is a function. Explain. *(Lesson 8-1)*

71. Suppose a number cube is rolled. What is the probability of rolling a 4 or a prime number? *(Lesson 6-9)*

Solve each proportion. *(Lesson 6-2)*

72. $\frac{n}{9} = \frac{15}{27}$ 73. $\frac{4}{b} = \frac{16}{36}$ 74. $\frac{7}{8.4} = \frac{0.5}{x}$

Getting Ready for the Next Lesson

PREREQUISITE SKILL Explain why each number is a rational number.
*(To review **rational numbers**, see Lesson 5-2.)*

75. $\frac{10}{2}$ 76. $1\frac{1}{2}$ 77. 0.75

78. $0.\overline{8}$ 79. 6 80. −7

9-2 The Real Number System

What You'll Learn

- Identify and compare numbers in the real number system.
- Solve equations by finding square roots.

Sunshine State Standards
MA.A.1.3.2-1, MA.A.1.3.2-2, MA.A.1.3.3-2, MA.A.1.3.4-1, MA.A.1.3.4-2

Vocabulary

- irrational numbers
- real numbers

How can squares have lengths that are not rational numbers?

In this activity, you will find the length of a side of a square that has an area of 2 square units.

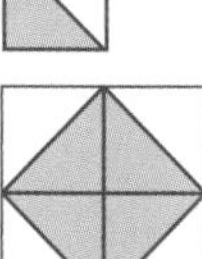

a. The small square at the right has an area of 1 square unit. Find the area of the shaded triangle.

b. Suppose eight triangles are arranged as shown. What shape is formed by the shaded triangles?

c. Find the total area of the four shaded triangles.

d. What number represents the length of the side of the shaded square?

IDENTIFY AND COMPARE REAL NUMBERS In Lesson 5-2, you learned that *rational numbers* can be written as fractions. A few examples of rational numbers are listed below.

$$-6 \quad 8\frac{2}{5} \quad 0.05 \quad -2.6 \quad 5.\overline{3} \quad -8.12121212\ldots \quad \sqrt{16}$$

Not all numbers are rational numbers. A few examples of numbers that are *not* rational are shown below. These numbers are not repeating or terminating decimals. They are called **irrational numbers**.

$$0.101001000100001\ldots \quad \sqrt{2} = 1.414213562\ldots \quad \pi = 3.14159\ldots$$

Key Concept — *Irrational Number*

An irrational number is a number that cannot be expressed as $\frac{a}{b}$, where a and b are integers and b does not equal 0.

Concept Check *True* or *False*? All square roots are irrational numbers.

The set of rational numbers and the set of irrational numbers together make up the set of **real numbers**. The Venn diagram at the right shows the relationship among the real numbers.

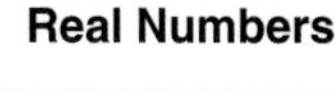

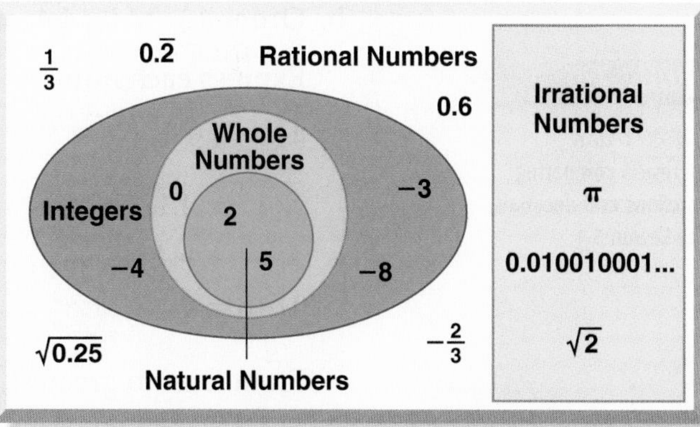

Example 1 Classify Real Numbers

Name all of the sets of numbers to which each real number belongs.

a. 9 — This number is a natural number, a whole number, an integer, and a rational number.

b. $0.\overline{3}$ — This repeating decimal is a rational number because it is equivalent to $\frac{1}{3}$. $1 \div 3 = 0.33333...$

c. $\sqrt{67}$ — $\sqrt{67} = 8.185352772...$ It is not the square root of a perfect square so it is irrational.

d. $-\frac{28}{4}$ — Since $-\frac{28}{4} = -7$, this number is an integer and a rational number.

e. $-\sqrt{121}$ — Since $-\sqrt{121} = -11$, this number is an integer and a rational number.

Concept Summary — Classification of Numbers

Example	Natural	Whole	Integer	Rational	Irrational	Real
0		√	√	√		√
4	√	√	√	√		√
−7			√	√		√
$\sqrt{25}$	√	√	√	√		√
$\sqrt{41}$					√	√
$\frac{3}{4}$				√		√
0.121212...				√		√
0.010110111...					√	√

Example 2 Compare Real Numbers on a Number Line

a. Replace ● with <, >, or = to make $\sqrt{34}$ ● $5\frac{3}{8}$ a true statement.

Express each number as a decimal. Then graph the numbers.

$\sqrt{34} = 5.830951895...$

$5\frac{3}{8} = 5.375$

$5\frac{3}{8}$ $\sqrt{34}$
5 5.1 5.2 5.3 5.4 5.5 5.6 5.7 5.8 5.9 6

Since $\sqrt{34}$ is to the right of $5\frac{3}{8}$, $\sqrt{34} > 5\frac{3}{8}$.

b. Order $4\frac{1}{2}$, $\sqrt{17}$, $4.\overline{4}$, and $\sqrt{16}$ from least to greatest.

Express each number as a decimal. Then compare the decimals.

$4\frac{1}{2} = 4.5$

$\sqrt{17} = 4.123105626...$

$4.\overline{4} = 4.444444444...$

$\sqrt{16} = 4$

$\sqrt{16}$ $\sqrt{17}$ $4.\overline{4}$ $4\frac{1}{2}$
4 4.1 4.2 4.3 4.4 4.5 4.6 4.7 4.8 4.9 5

From least to greatest, the order is $\sqrt{16}$, $\sqrt{17}$, $4.\overline{4}$, $4\frac{1}{2}$.

Study Tip

Look Back
To review **comparing fractions and decimals,** see Lesson 5-1.

SOLVE EQUATIONS BY FINDING SQUARE ROOTS

Some equations have irrational number solutions. You can solve some of these equations by taking the square root of each side.

Example 3 Solve Equations

Solve each equation. Round to the nearest tenth, if necessary.

a. $x^2 = 64$

$x^2 = 64$	Write the equation.
$\sqrt{x^2} = \sqrt{64}$	Take the square root of each side.
$x = \sqrt{64}$ or $x = -\sqrt{64}$	Find the positive and negative square root.
$x = 8$ or $x = -8$	

The solutions are 8 and -8.

Study Tip

Check Your Work
Check the reasonableness of the results by evaluating 9^2 and $(-9)^2$.
$9^2 = 81$
$(-9)^2 = 81$
Since 81 is close to 85, the solutions are reasonable.

b. $n^2 = 85$

$n^2 = 85$	Write the equation.
$\sqrt{n^2} = \sqrt{85}$	Take the square root of each side.
$n = \sqrt{85}$ or $n = -\sqrt{85}$	Find the positive and negative square root.
$n \approx 9.2$ or $n \approx -9.2$	Use a calculator.

The solutions are 9.2 and -9.2.

Check for Understanding

Concept Check

1. **Explain** the difference between a rational and irrational number.

2. **OPEN ENDED** Give an example of a number that is an integer and a rational number.

Guided Practice

Name all of the sets of numbers to which each real number belongs. Let N = natural numbers, W = whole numbers, Z = integers, Q = rational numbers, and I = irrational numbers.

3. 7
4. 0.5555...
5. $-\frac{3}{4}$
6. $\sqrt{12}$

Replace each ● with <, >, or = to make a true statement.

7. $6\frac{4}{5}$ ● $\sqrt{48}$
8. $-\sqrt{74}$ ● $-8.\overline{4}$

9. Order $3.\overline{7}$, $3\frac{3}{5}$, $\sqrt{13}$, $\frac{10}{3}$ from least to greatest.

ALGEBRA **Solve each equation. Round to the nearest tenth, if necessary.**

10. $y^2 = 25$
11. $m^2 = 74$

Application

12. **GEOMETRY** To find the radius r of a circle, you can use the formula $r = \sqrt{\frac{A}{\pi}}$ where A is the area of the circle. To the nearest tenth, find the radius of a circle that has an area of 40 square feet.

Practice and Apply

Homework Help

For Exercises	See Examples
13–28	1
33–42	2
45–61	3

Extra Practice

See page 745.

Name all of the sets of numbers to which each real number belongs. Let N = natural numbers, W = whole numbers, Z = integers, Q = rational numbers, and I = irrational numbers.

13. 8
14. 4
15. $\frac{2}{5}$
16. $\frac{1}{2}$
17. $0.\overline{2}$
18. 0.131313...
19. $\sqrt{10}$
20. $\sqrt{32}$
21. 1.247896...
22. 6.182576...
23. $-\frac{24}{8}$
24. $-\frac{56}{8}$
25. 2.8
26. 7.6
27. $-\sqrt{64}$
28. $-\sqrt{100}$

Determine whether each statement is *sometimes, always,* or *never* true.

29. A whole number is an integer.
30. An irrational number is a negative integer.
31. A repeating decimal is a real number.
32. An integer is a whole number.

Replace each ● with <, >, or = to make a true statement.

33. $5\frac{1}{4}$ ● $\sqrt{26}$
34. $\sqrt{80}$ ● 9.2
35. -3.3 ● $-\sqrt{10}$
36. $-\sqrt{18}$ ● $-4\frac{3}{8}$
37. $1\frac{1}{2}$ ● $\sqrt{2.25}$
38. $-\sqrt{6.25}$ ● $-\frac{5}{2}$

Order each set of numbers from least to greatest.

39. $5\frac{1}{4}, 2.\overline{1}, \sqrt{4}, \frac{6}{5}$
40. $4.\overline{23}, 4\frac{2}{3}, \sqrt{18}, \sqrt{16}$
41. $-10, -1.05, -\sqrt{105}, -10\frac{1}{2}$
42. $-\sqrt{14}, -4\frac{1}{10}, -3.8, -\frac{17}{4}$

Give a counterexample for each statement.

43. All square roots are irrational numbers.
44. All rational numbers are integers.

ALGEBRA Solve each equation. Round to the nearest tenth, if necessary.

45. $a^2 = 49$
46. $d^2 = 81$
47. $y^2 = 22$
48. $p^2 = 63$
49. $144 = w^2$
50. $289 = m^2$
51. $127 = b^2$
52. $300 = h^2$
53. $x^2 = 1.69$
54. $0.0016 = q^2$
55. $n^2 = 3.56$
56. $0.0058 = k^2$

57. If $(-a)^2 = 144$, what is the value of a?

58. What is the value of x to the nearest tenth if $x^2 - 4^2 = \sqrt{15^2}$?

59. Find the value of m if $\sqrt{256} = m^2$.

60. **WEATHER** Use the information at the left. Suppose a thunderstorm is 7 miles wide. How long will the storm last?

61. **FLOORING** A square room has an area of 324 square feet. The homeowners plan to cover the floor with 6-inch square tiles. How many tiles will be in each row on the floor?

More About. . .

Weather

To find the time t in hours that a thunderstorm will last, meteorologists can use the formula $t^2 = \frac{d^3}{216}$ where d is the distance across the storm in miles.

62. **CRITICAL THINKING** Tell whether the product of a rational number like 8 and an irrational number like 0.101001000… is rational or irrational. Explain your reasoning.

63. **WRITING IN MATH** Answer the question that was posed at the beginning of the lesson.

How can squares have lengths that are not rational numbers?

Include the following in your answer:

- an example of a square whose side length is irrational, and
- an example of a square that has a rational side length.

64. Which number can *only* be classified as a rational number?

(A) $\frac{1}{2}$ (B) $\sqrt{2}$ (C) -2 (D) 2

65. Which statement is *not* true?

(A) All integers are rational numbers.
(B) Every whole number is a natural number.
(C) All natural numbers are integers.
(D) Every real number is either a rational or irrational number.

Extending the Lesson

66. Heron's formula states that if the measures of the sides of a triangle are a, b, and c, the area $A = \sqrt{s(s-a)(s-b)(s-c)}$, where s is one-half the perimeter. Find the area of the triangle at the right. Round to the nearest tenth, if necessary.

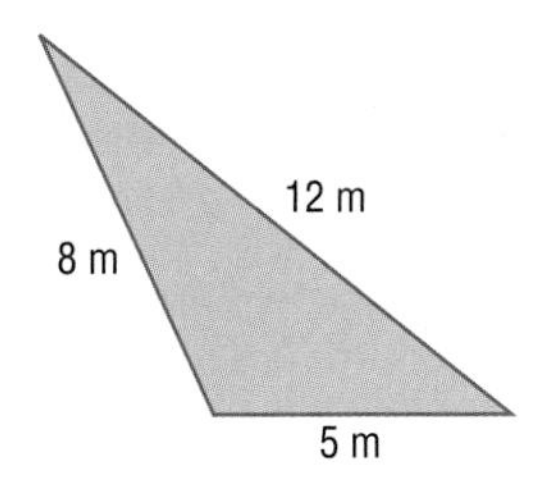

67. Find the area of a triangle whose sides measure 65 feet, 82 feet, and 95 feet.

68. **Make a Conjecture** Is the area of any triangle always a rational number? Support your answer with an example.

Maintain Your Skills

Mixed Review

Estimate each square root to the nearest whole number. *(Lesson 9-1)*

69. $\sqrt{54}$ 70. $-\sqrt{126}$ 71. $\sqrt{8.67}$

ALGEBRA List three solutions of each inequality. *(Lesson 8-10)*

72. $y < -x$ 73. $y \geq x + 5$ 74. $y \leq x - 3$

ALGEBRA Solve each inequality. *(Lesson 7-6)*

75. $3x + 2 > 17$ 76. $-2y + 9 \leq 3$

Express each ratio as a unit rate. Round to the nearest tenth, if necessary. *(Lesson 6-1)*

77. \$8 for 15 cupcakes 78. 120 miles on 4.3 gallons

Getting Ready for the Next Lesson

BASIC SKILL Use a ruler or straightedge to draw a diagram that shows how the hands on a clock appear at each time.

79. 3:00 80. 9:15 81. 10:00

82. 2:30 83. 7:45 84. 8:05

Reading Mathematics

Learning Geometry Vocabulary

Many of the words used in geometry are commonly used in everyday language. The everyday meanings of these words can be used to understand their mathematical meaning better.

The table below shows the meanings of some geometry terms you will use throughout this chapter.

Term	Everyday Meaning	Mathematical Meaning
ray	any of the thin lines, or beams, of light that appear to come from a bright source • a *ray* of light	a part of a line that extends from a point indefinitely in one direction
degree	extent, amount, or relative intensity • third *degree* burns	a common unit of measure for angles
acute	characterized by sharpness or severity • an *acute* pain	an angle with a measure that is greater than 0° and less than 90°
obtuse	not producing a sharp impression • an *obtuse* statement	an angle with a measure that is greater than 90° but less than 180°

Source: Merriam Webster's Collegiate Dictionary

Reading to Learn

1. Write a sentence using each term listed above. Be sure to use the everyday meaning of the term.

2. RESEARCH Use the Internet or a dictionary to find the everyday meaning of each term listed below. Compare them to their mathematical meaning. Note any similarities and/or differences.

a. midpoint **b.** converse **c.** indirect

3. RESEARCH Use the Internet or dictionary to determine which of the following words are used only in mathematics.

vertex equilateral similar scalene side isosceles

9-3 Angles

Sunshine State Standards
MA.B.1.3.2-1

What You'll Learn

- Measure and draw angles.
- Classify angles as acute, right, obtuse, or straight.

How are angles used in circle graphs?

The graph shows how voters prefer to vote.

a. Which method did half of the voters prefer?

b. Suppose 100 voters were surveyed. How many more voters preferred to vote in booths than on the Internet?

c. Each section of the graph shows an *angle*. A *straight angle* resembles a line. Which section shows a straight angle?

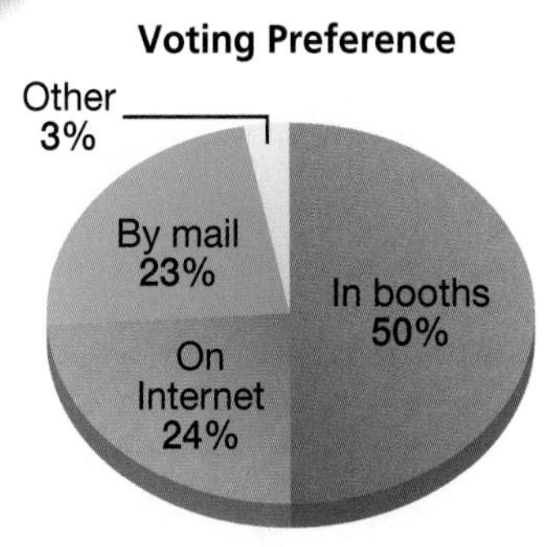
Voting Preference
Other 3%
By mail 23%
In booths 50%
On Internet 24%
Source: Princeton Survey Research

Vocabulary

- point
- ray
- line
- angle
- vertex
- side
- degree
- protractor
- acute angle
- right angle
- obtuse angle
- straight angle

MEASURE AND DRAW ANGLES The center of a circle represents a **point**. A point is a specific location in space with no size or shape.

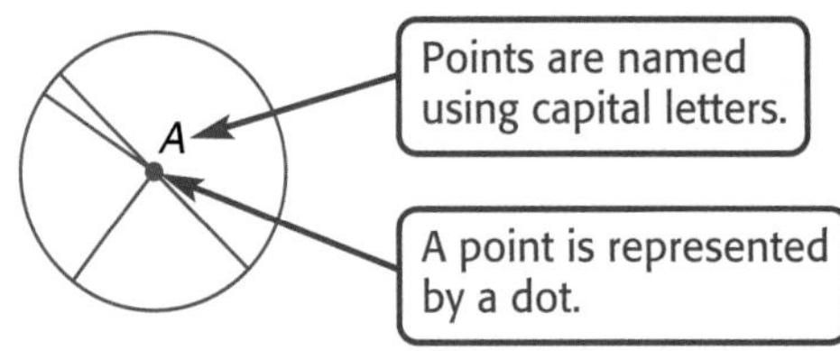
A
Points are named using capital letters.
A point is represented by a dot.

In the circle below, notice how the sides of each section begin at the center and extend in one direction. These are examples of **rays**. A **line** is a never-ending straight path extending in two directions.

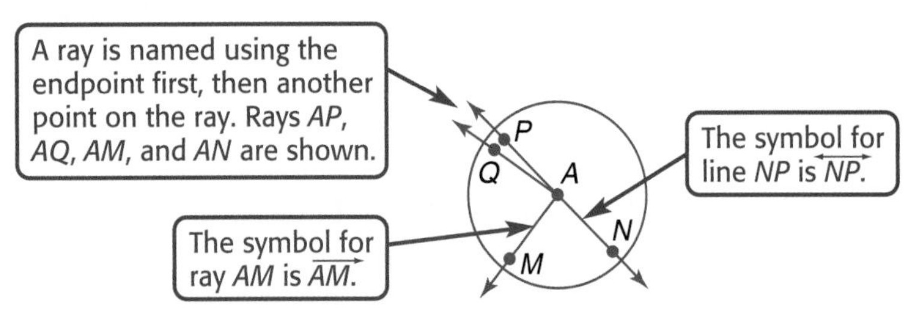
A ray is named using the endpoint first, then another point on the ray. Rays AP, AQ, AM, and AN are shown.
P
Q
A
N
M
The symbol for line NP is NP.
The symbol for ray AM is AM.

Two rays that have the same endpoint form an **angle**. The common endpoint is called the **vertex**, and the two rays that make up the angle are called the **sides** of the angle.

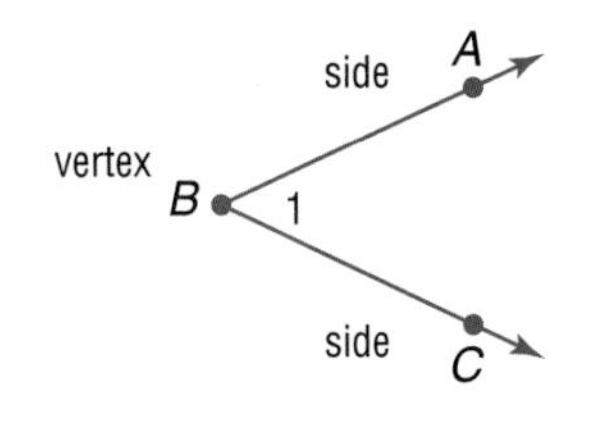
side
A
vertex
B
1
side
C

Study Tip

Angles
To name an angle using only the vertex or a number, no other angles can share the vertex of the angle being named.

The symbol $\angle$ represents angle. There are several ways to name the angle shown above.

- Use the vertex and a point from each side. $\angle ABC$ or $\angle CBA$
  The vertex is always the middle letter.
- Use the vertex only. $\angle B$
- Use a number. $\angle 1$

The most common unit of measure for angles is the **degree**. A circle can be separated into 360 arcs of the same length. An angle has a measurement of one degree if its vertex is at the center of the circle and the sides contain the endpoints of one of the 360 equal arcs.

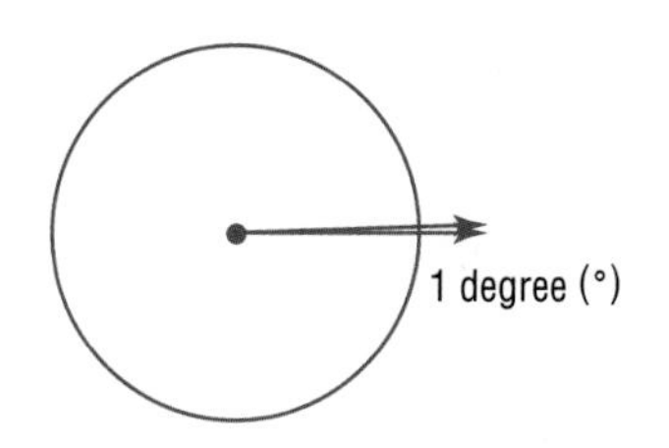

You can use a **protractor** to measure angles.

Example 1 Measure Angles

a. Use a protractor to measure $\angle CDE$.

Step 1 Place the center point of the protractor's base on vertex D. Align the straight side with side $\overrightarrow{DE}$ so that the marker for 0° is on the ray.

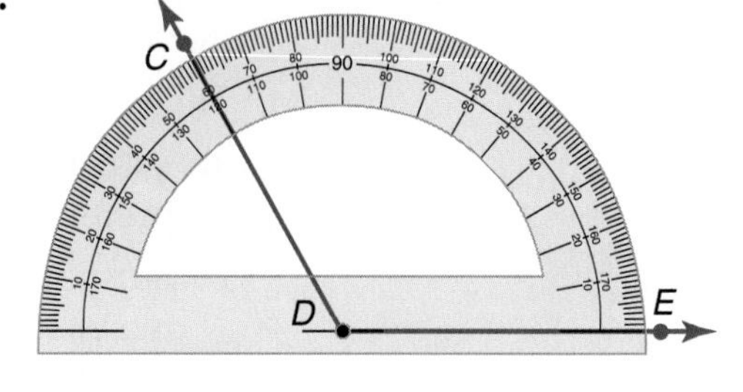

Step 2 Use the scale that begins with 0° at $\overrightarrow{DE}$. Read where the other side of the angle, $\overrightarrow{DC}$, crosses this scale.

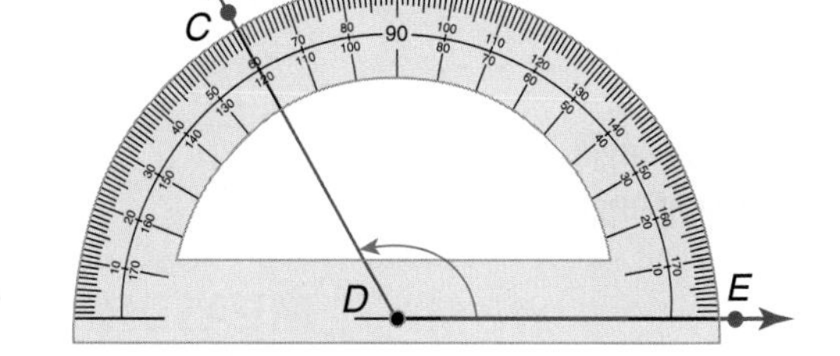

The measure of angle CDE is 120°. Using symbols, $m\angle CDE = 120°$.

Reading Math

Angle Measure
Read $m\angle CDE = 120°$ as *the measure of angle CDE is 120 degrees.*

b. Find the measures of $\angle KXN$, $\angle MXN$, and $\angle JXK$.

$m\angle KXN = 135°$
$\overrightarrow{XN}$ is at 0° on the right.

$m\angle MXN = 70°$
$\overrightarrow{XN}$ is at 0° on the right.

$m\angle JXK = 45°$
$\overrightarrow{XJ}$ is at 0° on the left.

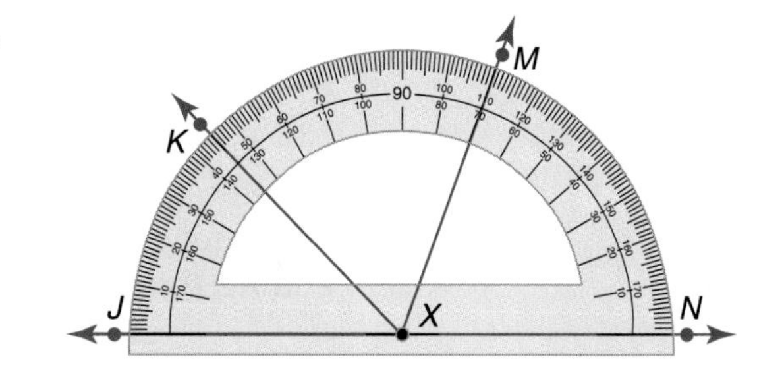

Protractors can also be used to draw an angle of a given measure.

Example 2 Draw Angles

Draw $\angle X$ having a measure of 85°.

Step 1 Draw a ray with endpoint X.

Step 2 Place the center point of the protractor on X. Align the mark labeled 0 with the ray.

Step 3 Use the scale that begins with 0. Locate the mark labeled 85. Then draw the other side of the angle.

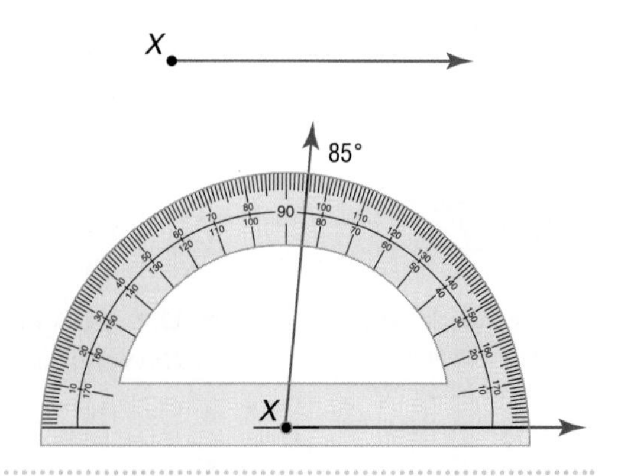

CLASSIFY ANGLES Angles can be classified by their degree measure.

Study Tip

Common Misconception
You may think that a right angle must always face to the right. However, a right angle can be in any position.

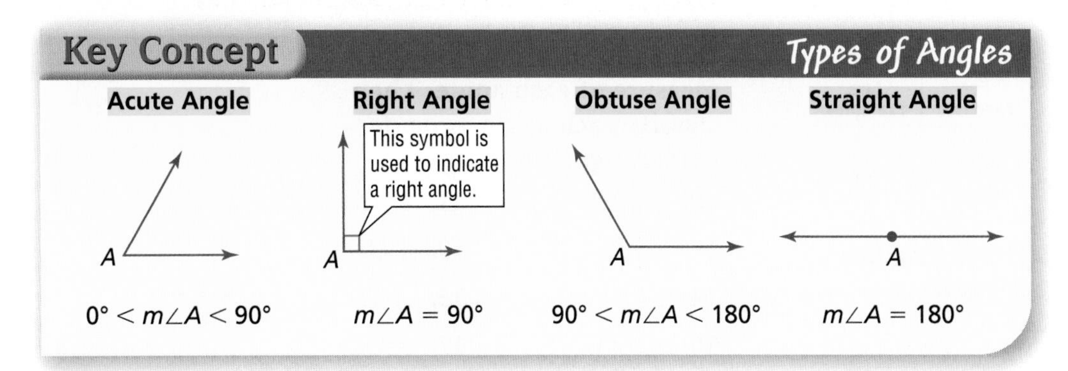

Key Concept
Types of Angles
Acute Angle
Right Angle
Obtuse Angle
Straight Angle
This symbol is used to indicate a right angle.
A
A
A
A
0° < m∠A < 90°
m∠A = 90°
90° < m∠A < 180°
m∠A = 180°

Example 3 Classify Angles

Classify each angle as *acute, obtuse, right,* or *straight.*

a.

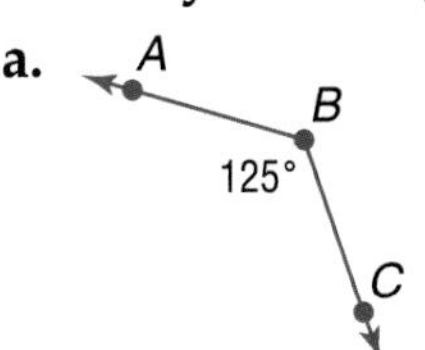

$m\angle ABC > 90°$.
So, $\angle ABC$ is obtuse.

b.

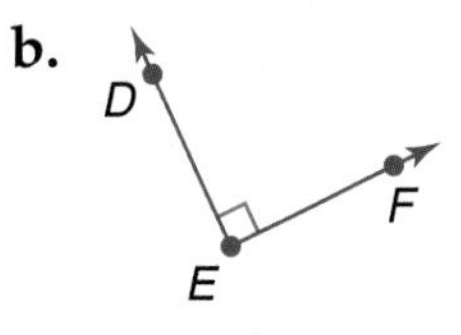

$m\angle DEF = 90°$.
So, $\angle DEF$ is right.

c.

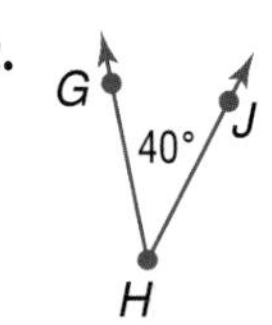

$m\angle GHJ < 90°$.
So, $\angle GHJ$ is acute.

Example 4 Use Angles to Solve a Problem

RACING The diagram shows the angle of the track at a corner of the Texas Motor Speedway in Fort Worth, Texas. Classify this angle.

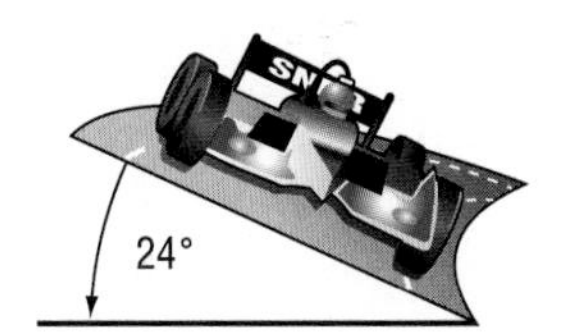

Since 24° is greater than 0° and less than 90°, the angle is acute.

Check for Understanding

Concept Check

1. **Name** the vertex and sides of the angle shown. Then name the angle in four ways.

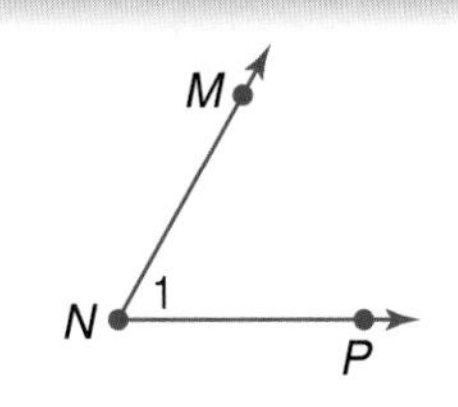

2. **OPEN ENDED** Draw an obtuse angle.

Guided Practice

Use a protractor to find the measure of each angle. Then classify each angle as *acute, obtuse, right,* or *straight.*

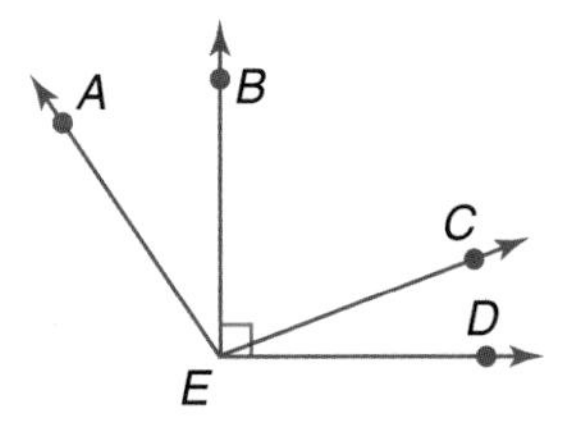

3. $\angle CED$
4. $\angle BED$
5. $\angle CEB$
6. $\angle AED$

Use a protractor to draw an angle having each measurement. Then classify each angle as *acute, obtuse, right,* or *straight.*

7. 55°
8. 140°

Application

9. **TIME** What type of angle is formed by the hands on a clock at 6:00?

Practice and Apply

Homework Help

For Exercises	See Examples
10–20	1, 3
21–30	2, 3
31, 32	4

Extra Practice
See page 745.

Use a protractor to find the measure of each angle. Then classify each angle as *acute*, *obtuse*, *right*, or *straight*.

10. $\angle XZY$ **11.** $\angle SZT$

12. $\angle SZY$ **13.** $\angle UZX$

14. $\angle TZW$ **15.** $\angle XZT$

16. $\angle UZV$ **17.** $\angle WZU$

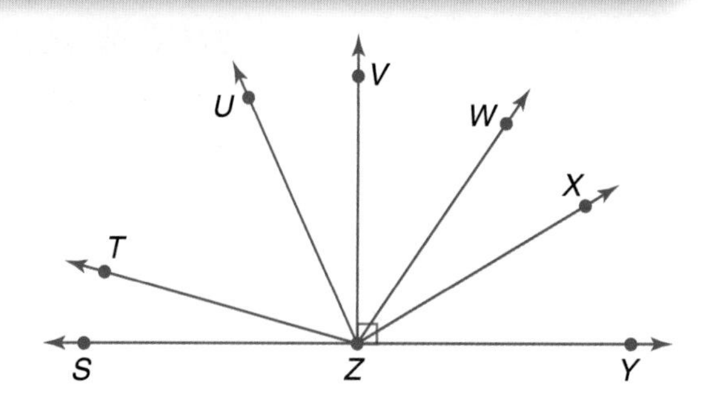

Use a protractor to find the measure of each angle.

18.

19.

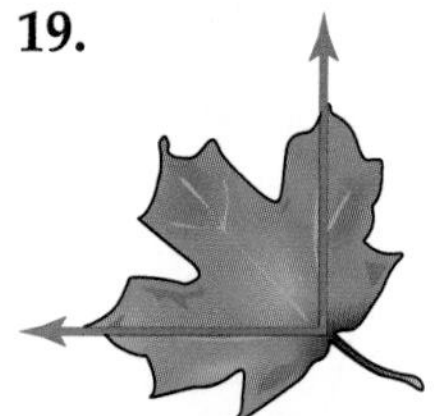

20.

Use a protractor to draw an angle having each measurement. Then classify each angle as *acute*, *obtuse*, *right*, or *straight*.

21. 40° **22.** 70° **23.** 65° **24.** 85° **25.** 95°

26. 110° **27.** 155° **28.** 140° **29.** 38° **30.** 127°

31. **BASEBALL** If you swing the bat too early or too late, the ball will probably go foul. The best way to hit the ball is at a right angle. Classify each angle shown.

a.

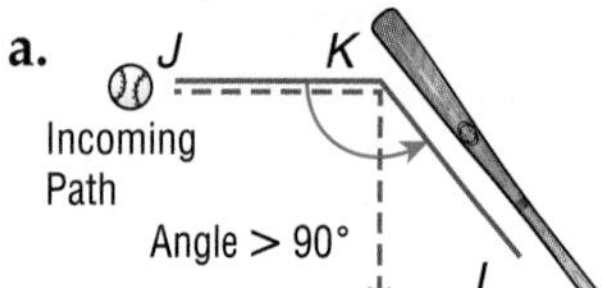

b.

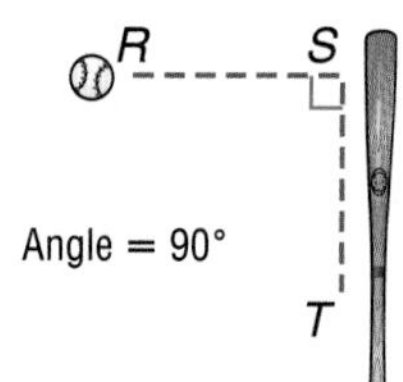

c.

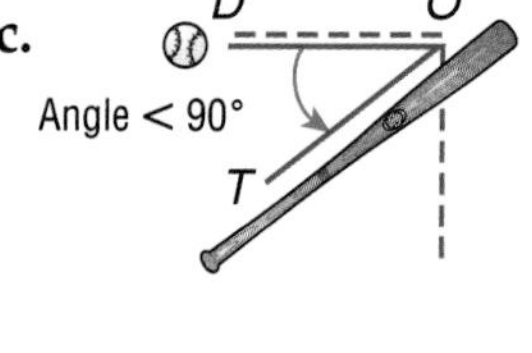

FITNESS **For Exercises 32–34, use the graphic.**

32. Classify each angle in the circle graph as *acute*, *obtuse*, *right*, or *straight*.

33. Find the measure of each angle of the circle graph.

34. Suppose 500 adults were surveyed. How many would you expect to exercise moderately?

35. **CRITICAL THINKING** In twelve hours, how many times do the hands of a clock form a right angle?

USA TODAY Snapshots®

Regular fitness activities

Nearly half of adults say they exercise regularly during an average week. Type of exercise they do:

Moderate (walking, yoga) 29%

Light (house-cleaning, gardening, golfing) 20%

No standard routine (take stairs instead of elevator) 13%

Don't exercise on a regular basis 19%

Intense (aerobics, running, swimming, biking) 19%

Source: Opinion Research Corp. for TOPS (Take Off Pounds Sensibly)

By Cindy Hall and Dave Merrill, USA TODAY

36. WRITING IN MATH Answer the question that was posed at the beginning of the lesson.

How are angles used in circle graphs?

Include the following in your answer:

- a classification of the angles in each section of the circle graph shown at the beginning of the lesson, and
- a range of percents that can be represented by an acute angle and an obtuse angle.

37. Which angle is an obtuse angle?

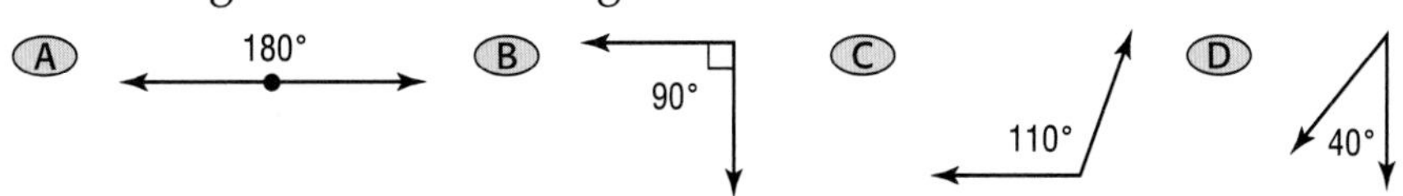

38. Which of the following is closest to the measure of $\angle XYZ$?

Ⓐ 45°

Ⓑ 55°

Ⓒ 135°

Ⓓ 145°

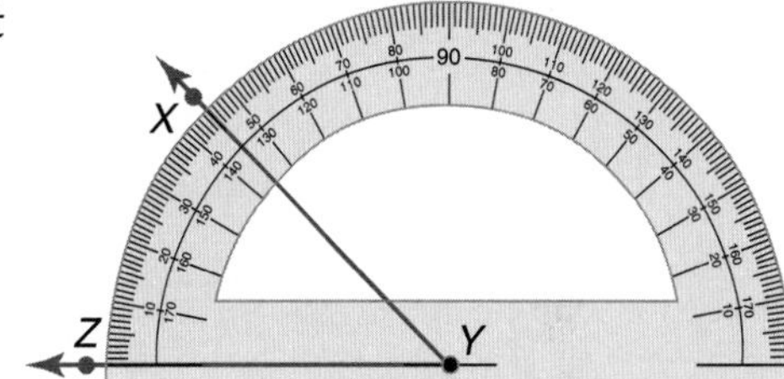

Maintain Your Skills

Mixed Review

Name all of the sets of numbers to which each real number belongs. Let N = natural numbers, W = whole numbers, Z = integers, Q = rational numbers, and I = irrational numbers. *(Lesson 9-2)*

39. -5 40. $0.\overline{4}$ 41. $\sqrt{63}$ 42. 7.4

Estimate each square root to the nearest whole number. *(Lesson 9-1)*

43. $\sqrt{18}$ 44. $\sqrt{79}$

45. Translate the sentence *A number decreased by seven is at least twenty-two.* *(Lesson 7-3)*

Getting Ready for the Next Lesson

PREREQUISITE SKILL Solve each equation.
(To review ***solving two-step equations****, see Lesson 3-5.)*

46. $18 + 57 + x = 180$ 47. $x + 27 + 54 = 180$ 48. $85 + x + 24 = 180$

49. $x + x + x = 180$ 50. $2x + 3x + 4x = 180$ 51. $2x + 3x + 5x = 180$

Practice Quiz 1 — *Lessons 9-1 through 9-3*

Find each square root, if possible. *(Lesson 9-1)*

1. $\sqrt{36}$ 2. $-\sqrt{169}$

3. **ALGEBRA** Solve $m^2 = 68$ to the nearest tenth. *(Lesson 9-2)*

Classify each angle measure as *acute, obtuse, right,* or *straight.* *(Lesson 9-3)*

4. 83° 5. 115°

Spreadsheet Investigation

A Follow-Up of Lesson 9-3

Circle Graphs and Spreadsheets

In the following example, you will learn how to use a computer spreadsheet program to graph the results of a probability experiment in a circle graph.

Example

The spinner like the one shown at the right was spun 20 times each for two trials. The data are shown below. Use a spreadsheet to make a circle graph of the results.

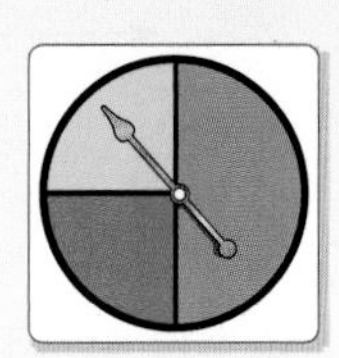

Step 1 Enter the data in a spreadsheet as shown.

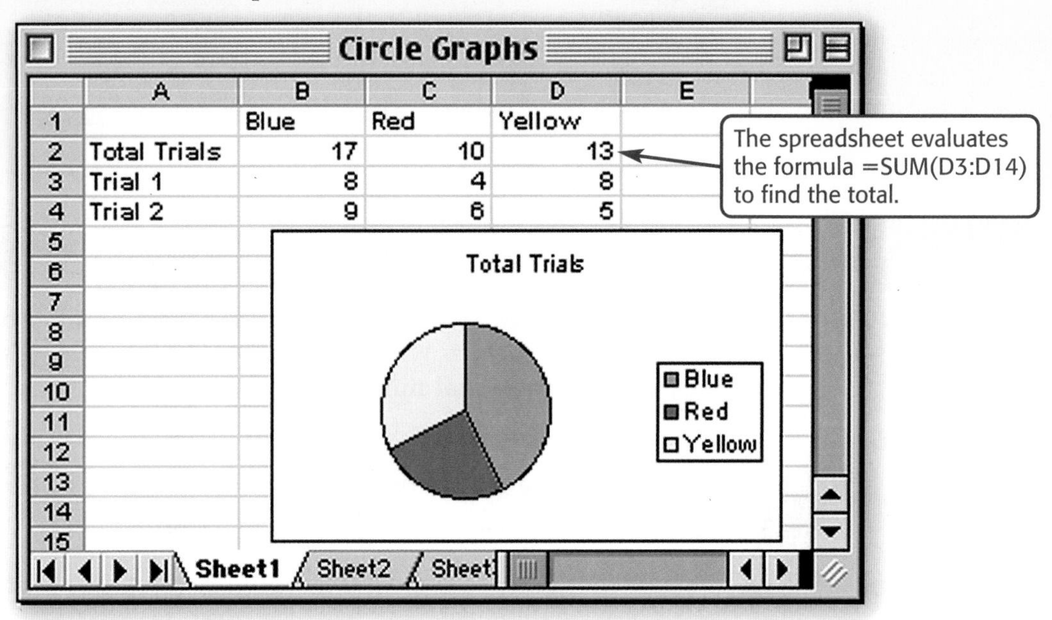

	A	B	C	D	E
1		Blue	Red	Yellow	
2	Total Trials	17	10	13	
3	Trial 1	8	4	8	
4	Trial 2	9	6	5	

Step 2 Select the data to be included in your graph. Then use the graph tool to create the graph. The spreadsheet will allow you to add titles, change colors, and so on.

Exercises

1. Describe the results you would theoretically expect for one trial of 20 spins. Explain your reasoning.
2. Make a spinner like the one shown above. Collect data for five trials of 20 spins each. Use a spreadsheet program to create a circle graph of the data.
3. A **central angle** is an angle whose vertex is the center of a circle and whose sides intersect the circle. After 100 spins, what kind of central angle would you theoretically expect for each section of the circle graph? Explain.
4. Predict how many trials of the experiment are required to match the theoretical results. Test your prediction.
5. When the theoretical results match the experimental results, what is true about the circle graph and the spinner?

9-4 Triangles

What You'll Learn

- Find the missing angle measure of a triangle.
- Classify triangles by angles and by sides.

Sunshine State Standards
MA.B.1.3.2-1, M.A.C.3.3.1-1

Vocabulary

- line segment
- triangle
- vertex
- acute triangle
- obtuse triangle
- right triangle
- congruent
- scalene triangle
- isosceles triangle
- equilateral triangle

How do the angles of a triangle relate to each other?

There is a relationship among the measures of the angles of a triangle.

a. Use a straightedge to draw a triangle on a piece of paper. Then cut out the triangle and label the vertices X, Y, and Z.

b. Fold the triangle as shown so that point Z lies on side XY as shown. Label $\angle Z$ as $\angle 2$.

c. Fold again so point X meets the vertex of $\angle 2$. Label $\angle X$ as $\angle 1$.

d. Fold so point Y meets the vertex of $\angle 2$. Label $\angle Y$ as $\angle 3$.

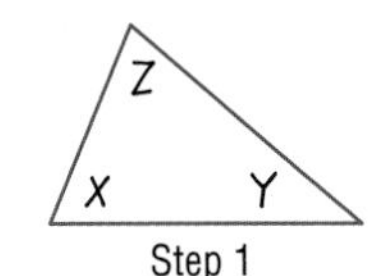

Step 1

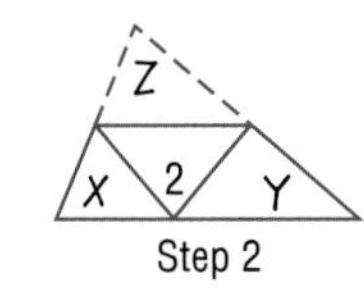

Step 2

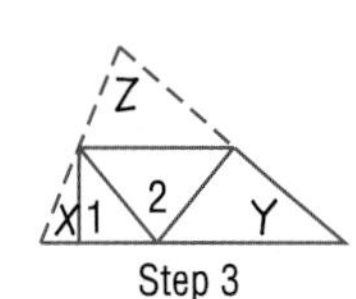

Step 3

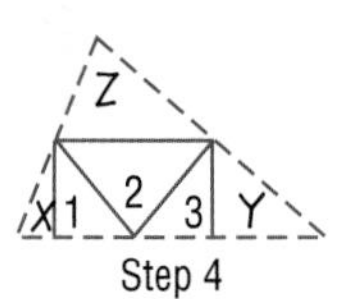

Step 4

e. Make a Conjecture about the sum of the measures of $\angle 1$, $\angle 2$, and $\angle 3$. Explain your reasoning.

ANGLE MEASURES OF A TRIANGLE A **line segment** is part of a line containing two endpoints and all of the points between them. A **triangle** is a figure formed by three line segments that intersect only at their endpoints.

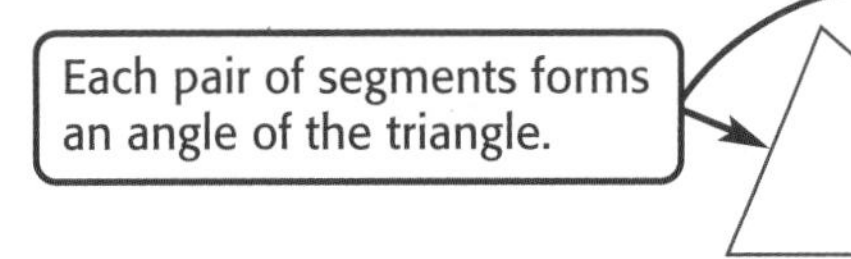

Each pair of segments forms an angle of the triangle.
The vertex of each angle is a vertex of the triangle. Vertices is the plural of vertex.

Reading Math

Line Segment
The symbol for line segment XY is $\overline{XY}$.

Triangles are named by the letters at their vertices. Triangle XYZ, written $\triangle XYZ$, is shown.

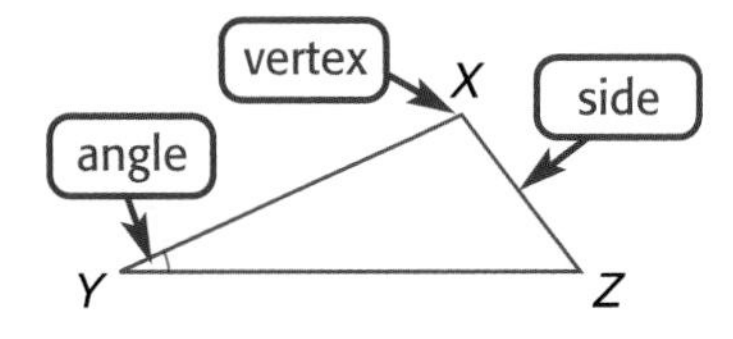

The sides are $\overline{XY}$, $\overline{YZ}$, and $\overline{XZ}$.
The vertices are X, Y, and Z.
The angles are $\angle X$, $\angle Y$, and $\angle Z$.

The activity above suggests the following relationship about the angles of any triangle.

Key Concept — Angles of a Triangle

- **Words** The sum of the measures of the angles of a triangle is 180°.
- **Model**

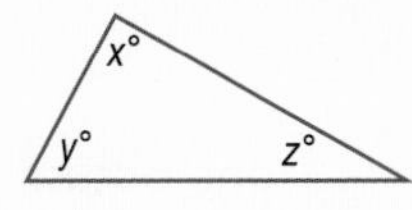

- **Symbols** $x + y + z = 180$

Example 1 Find Angle Measures

Find the value of x in $\triangle ABC$.

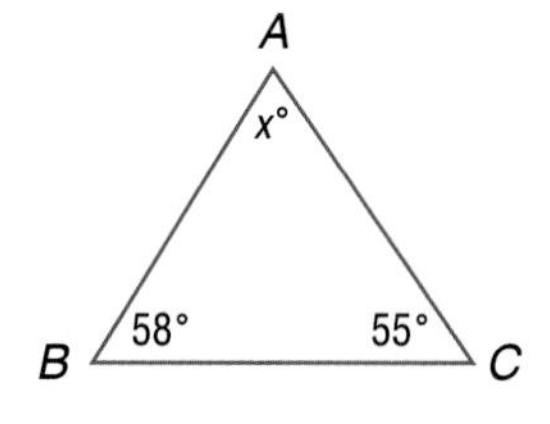

$m\angle A + m\angle B + m\angle C = 180$	The sum of the measures is 180.
$x + 58 + 55 = 180$	Replace $m\angle B$ with 58 and $m\angle C$ with 55.
$x + 113 = 180$	Simplify.
$x + 113 - 113 = 180 - 113$	Subtract 113 from each side.
$x = 67$	

The measure of $\angle A$ is 67°.

The relationships of the angles in a triangle can be represented using algebra.

Example 2 Use Ratios to Find Angle Measures

ALGEBRA **The measures of the angles of a certain triangle are in the ratio 1:4:7. What are the measures of the angles?**

Words The measures of the angles are in the ratio 1:4:7.

Variables Let x represent the measure of one angle, $4x$ the measure of a second angle, and $7x$ the measure of the third angle.

Equation

$x + 4x + 7x = 180$	The sum of the measures is 180.
$12x = 180$	Combine like terms.
$\frac{12x}{12} = \frac{180}{12}$	Divide each side by 12.
$x = 15$	Simplify.

Since $x = 15$, $4x = 4(15)$ or 60, and $7x = 7(15)$ or 105. The measures of the angles are 15°, 60°, and 105°.

CHECK $15 + 60 + 105 = 180$. So, the answer is correct. ✓

CLASSIFY TRIANGLES Triangles can be classified by their angles. All triangles have at least two acute angles. The third angle is either acute, obtuse, or right.

Study Tip

Equiangular
In an *equiangular* triangle, all angles have the same measure, 60°.

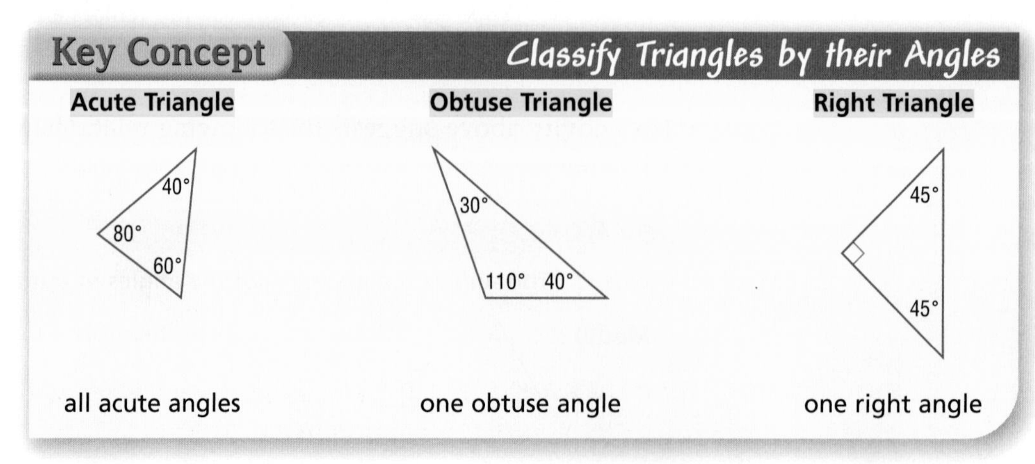

Key Concept
Classify Triangles by their Angles
Acute Triangle
Obtuse Triangle
Right Triangle
40°
80°
60°
30°
110°
40°
45°
45°
all acute angles
one obtuse angle
one right angle

Triangles can also be classified by their sides. **Congruent** sides have the same length.

Study Tip

Reading Math
Tick marks on the sides of a triangle indicate that those sides are congruent.

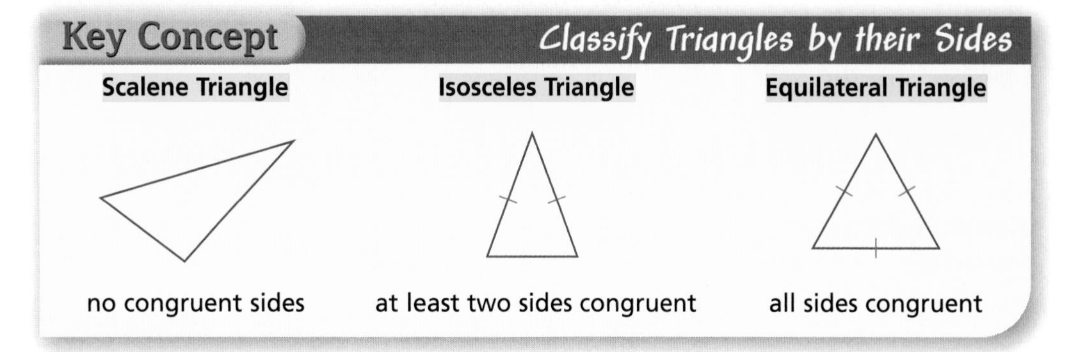

Key Concept — *Classify Triangles by their Sides*

Scalene Triangle	Isosceles Triangle	Equilateral Triangle
no congruent sides	at least two sides congruent	all sides congruent

Example 3 *Classify Triangles*

Classify each triangle by its angles and by its sides.

a.

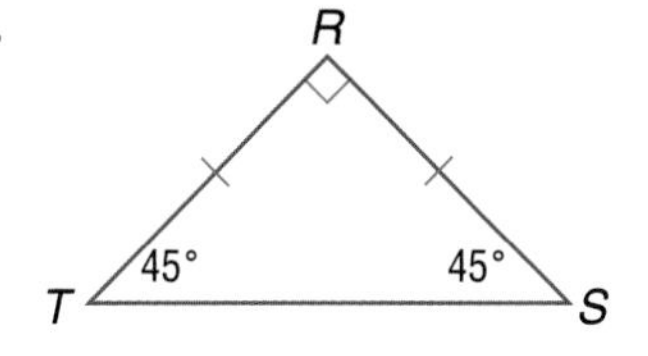

Angles
$\triangle RST$ has a right angle.

Sides
$\triangle RST$ has two congruent sides.
So, $\triangle RST$ is a right isosceles triangle.

b.

G
35°
120°
25°
I
H

Angles
$\triangle GHI$ has one obtuse angle.

Sides
$\triangle GHI$ has no two sides that are congruent.
So, $\triangle GHI$ is an obtuse scalene triangle.

Check for Understanding

Concept Check

1. **Compare** an isosceles triangle and an equilateral triangle.
2. **Name** a real-world object that is an equilateral triangle.
3. **OPEN ENDED** Draw an obtuse isosceles triangle.

Guided Practice **Find the value of x in each triangle. Then classify each triangle as *acute*, *right*, or *obtuse*.**

4.

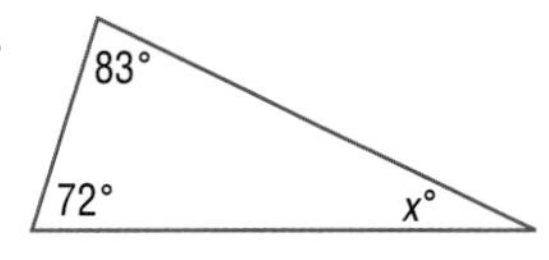

5.

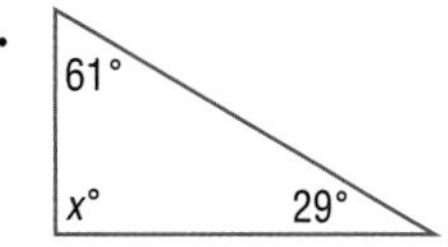

6.

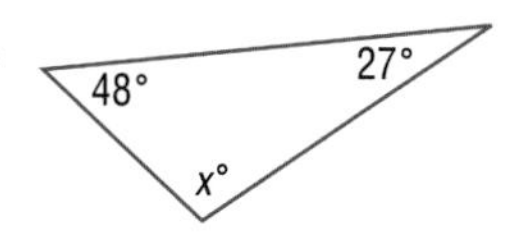

Classify each indicated triangle by its angles and by its sides.

7.

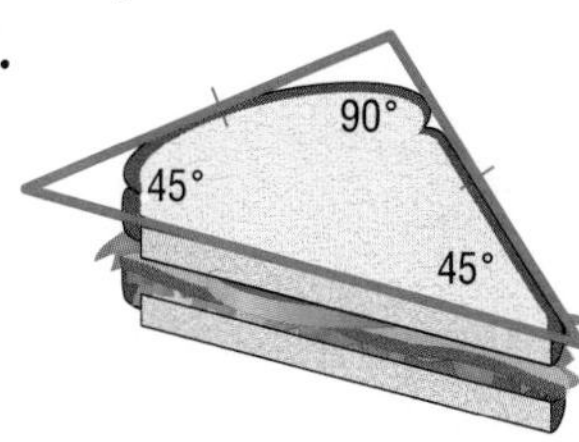

8.

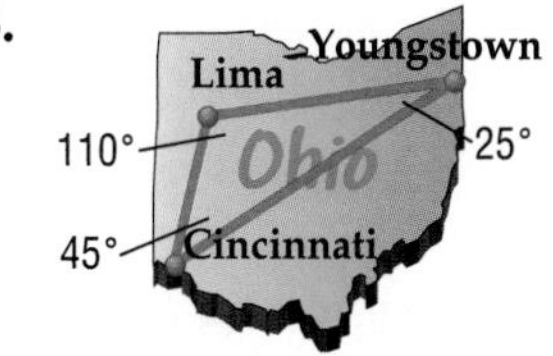

9.

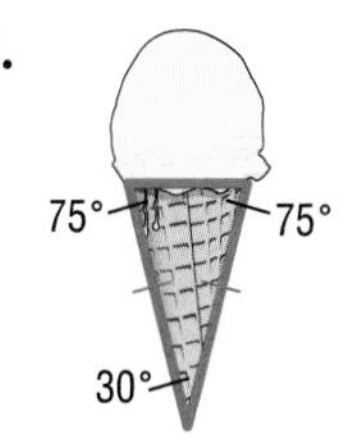

Application

10. **ALGEBRA** Triangle EFG has angles whose measures are in the ratio 2:4:9. What are the measures of the angles?

Practice and Apply

Homework Help

For Exercises	See Examples
11–16	1
17, 18	2
19–24	3

Extra Practice

See page 746.

Find the value of x in each triangle. Then classify each triangle as *acute*, *right*, or *obtuse*.

11. 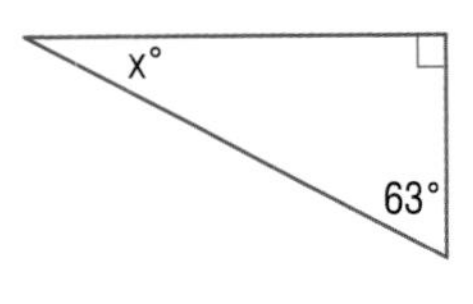

12. 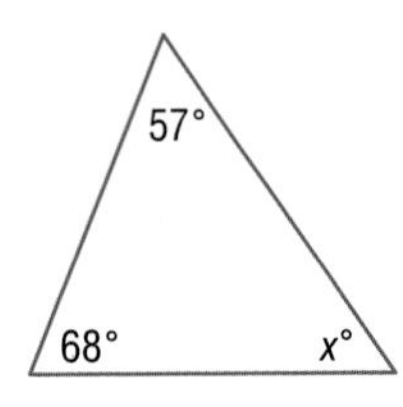

13. 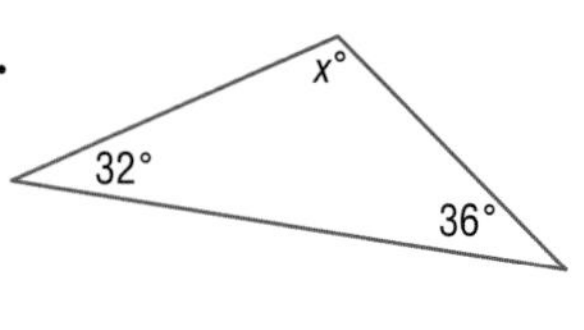

14. 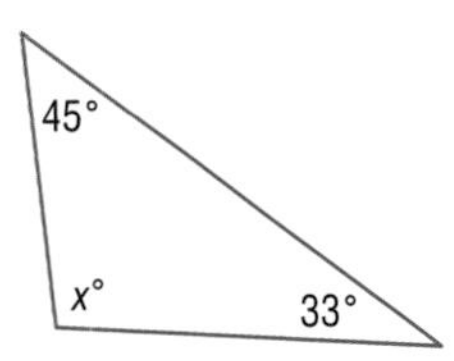

15. 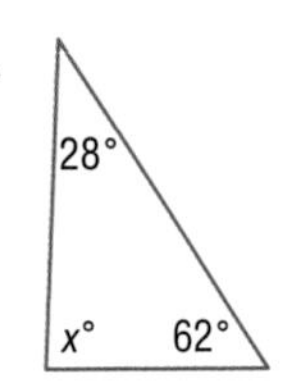

16. 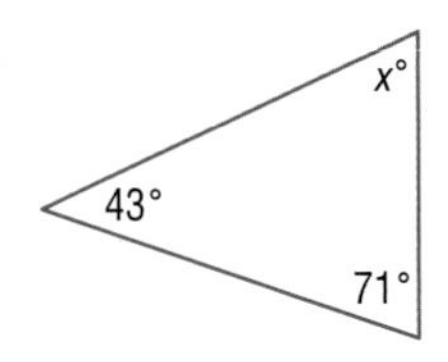

17. ALGEBRA The measures of the angles of a triangle are in the ratio 1:3:5. What is the measure of each angle?

18. ALGEBRA Determine the measures of the angles of $\triangle ABC$ if the measures of the angles are in the ratio 7:7:22.

Classify each indicated triangle by its angles and by its sides.

19. 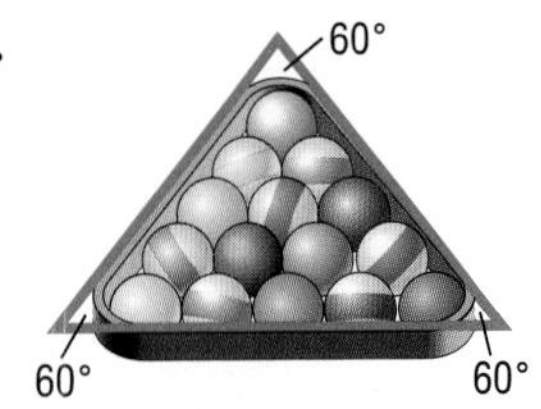

20. 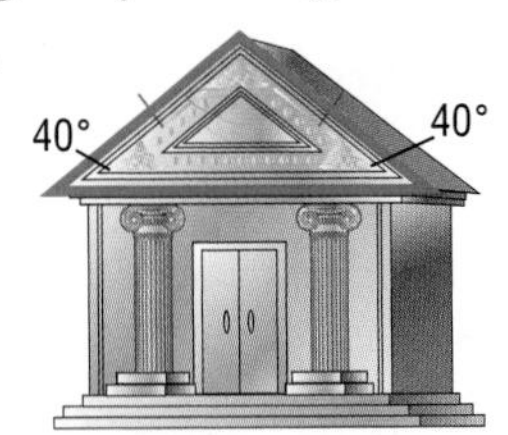

21. 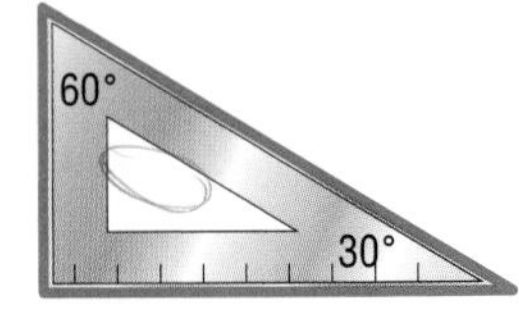

22. 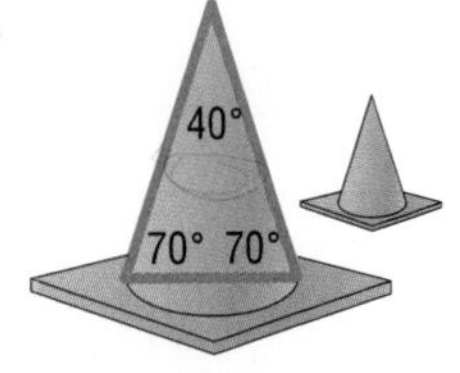

23. 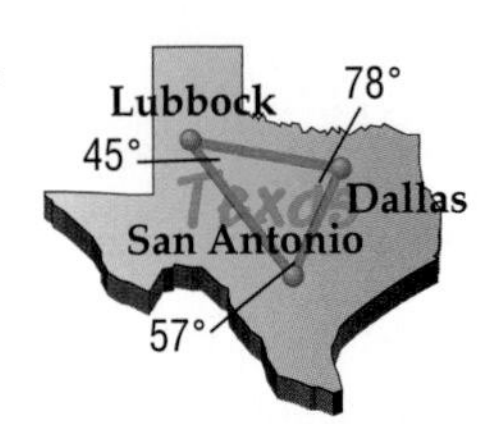

24. 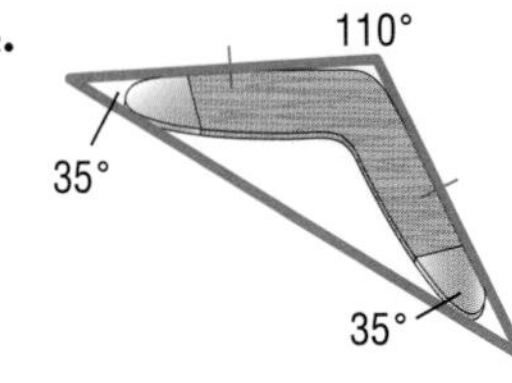

More About. . .

Billiards

According to the rules of billiards, each ball must weigh 5.5 to 6.0 ounces and have a diameter of $2\frac{1}{4}$ inches.

Source: www.bca-pool.com

Determine whether each statement is *sometimes, always,* or *never* true.

25. Equilateral triangles are isosceles triangles.

26. Isosceles triangles are equilateral triangles.

Estimate the measure of the angles in each triangle. Then classify each triangle as *acute, right,* or *obtuse*.

27. 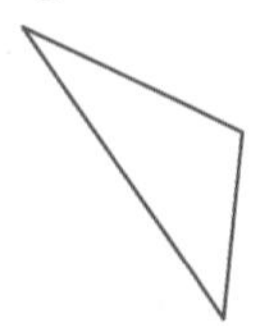

28. 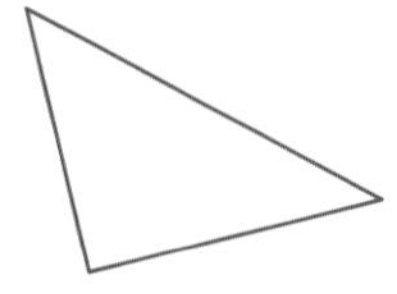

29. 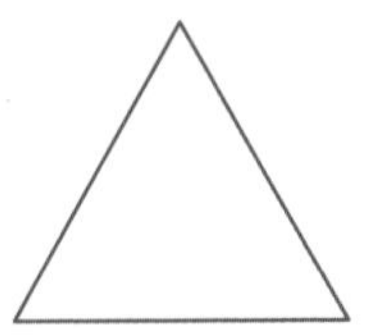

Sketch each triangle. If it is not possible to sketch the triangle, write *not possible*.

30. acute scalene

31. right equilateral

32. obtuse and *not* scalene

33. obtuse equilateral

ALGEBRA **Find the measures of the angles in each triangle.**

34. $5x°$ $3x°$ $x°$

35.

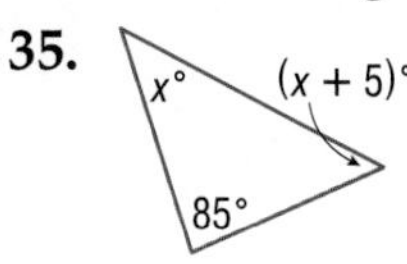

36.

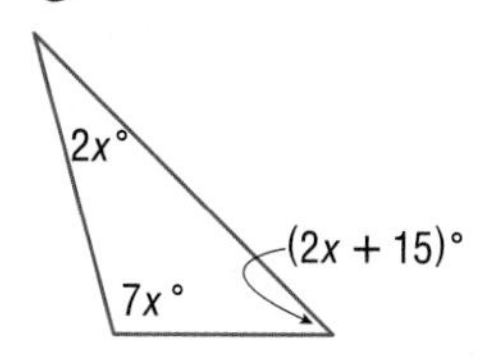

37. CRITICAL THINKING Numbers that can be represented by a triangular arrangement of dots are called *triangular numbers*. The first three triangular numbers are 1, 3, and 6. Find the next three triangular numbers.

38. WRITING IN MATH Answer the question that was posed at the beginning of the lesson.

How do the angles of a triangle relate to each other?

Include the following in your answer:

- an explanation telling why the sum of the angles is 180°, and
- drawings of two triangles with their angles labeled.

39. Refer to the figure shown. What is the measure of $\angle R$?

Ⓐ 28° Ⓑ 38°

Ⓒ 48° Ⓓ 180°

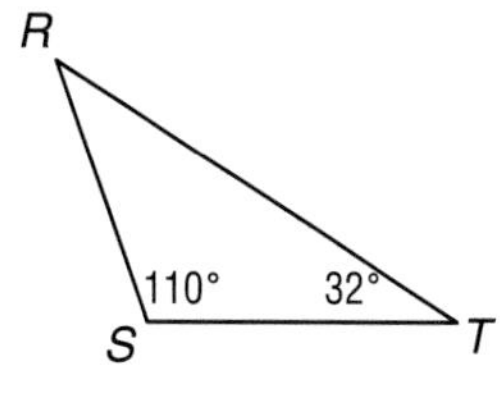

40. The measures of the angles of a triangle are 30°, 90°, and 60°. Which triangle most likely has these angle measures?

 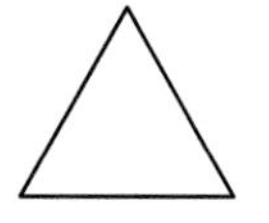  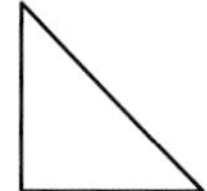  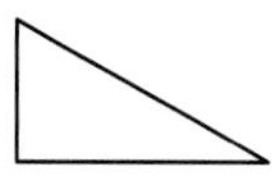 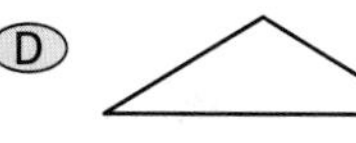

Maintain Your Skills

Mixed Review **Use a protractor to find the measure of each angle. Then classify each angle as *acute, obtuse, right,* or *straight*.** *(Lesson 9-3)*

41. $\angle ADC$ **42.** $\angle ADE$

43. $\angle CDE$ **44.** $\angle EDF$

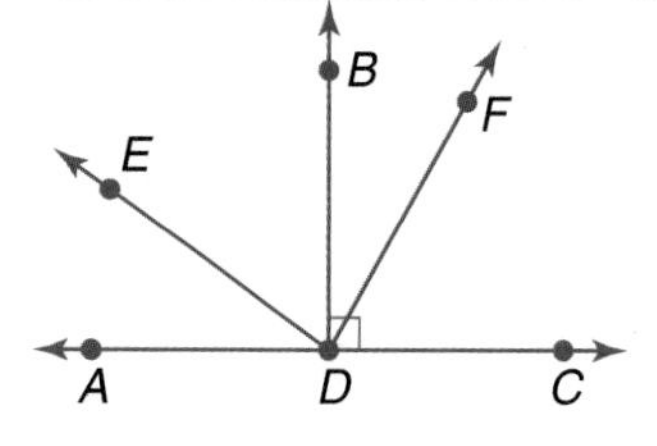

ALGEBRA **Solve each equation. Round to the nearest tenth, if necessary.** *(Lesson 9-2)*

45. $m^2 = 81$ **46.** $196 = y^2$ **47.** $84 = p^2$

48. Twenty-six is 25% of what number? *(Lesson 6-5)*

Getting Ready for the Next Lesson **PREREQUISITE SKILL** **Find the value of each expression.** *(To review **exponents**, see Lesson 4-2.)*

49. 12^2 **50.** 15^2 **51.** 18^2

52. 24^2 **53.** 27^2 **54.** 31^2

Algebra Activity

A Preview of Lesson 9-5

Sunshine State Standards
MA.C.3.3.1-2

The Pythagorean Theorem

Activity 1

To find the area of certain geometric figures, dot paper can be used. Consider the following examples.

Find the area of each shaded region if each square represents one square unit.

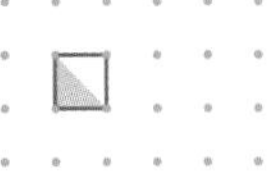

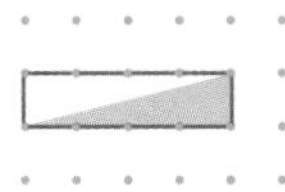

$A = \frac{1}{2}(1)$ or $\frac{1}{2}$ unit^2 $\quad$ $A = \frac{1}{2}(2)$ or 1 unit^2 $\quad$ $A = \frac{1}{2}(4)$ or 2 units^2

The area of other figures can be found by first separating the figure into smaller regions and then finding the sum of the areas of the smaller regions.

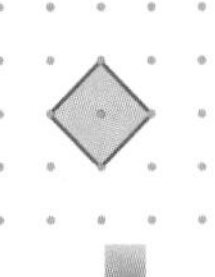

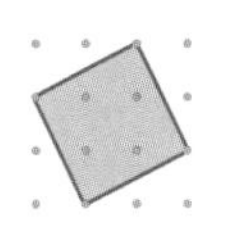

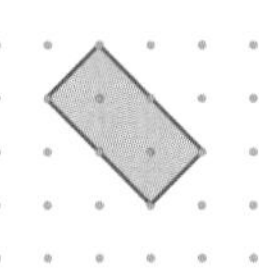

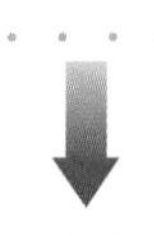

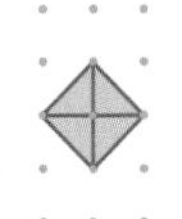

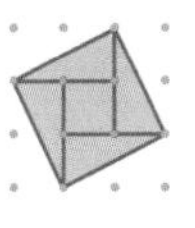

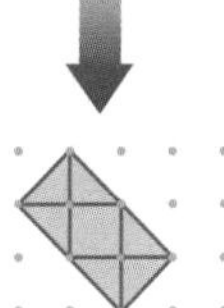

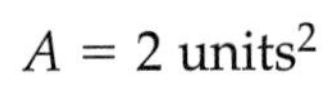

$A = 2$ units^2 $\quad$ $A = 5$ units^2 $\quad$ $A = 4$ units^2

Model

Find the area of each figure.

1. 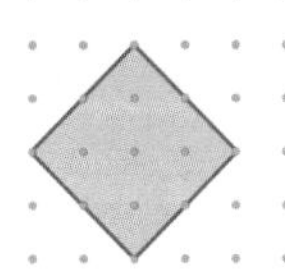

2. 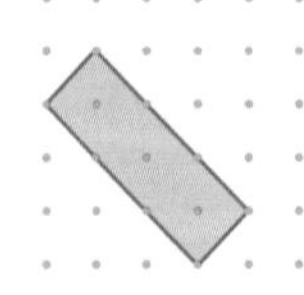

3. 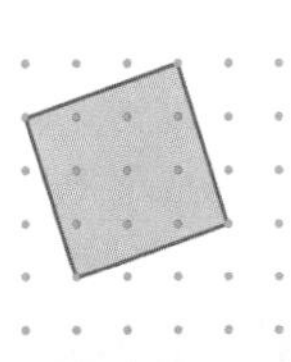

4. 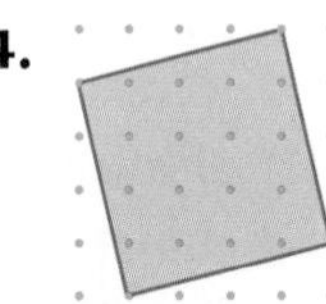

Activity 2

Let's investigate the relationship that exists among the sides of a right triangle. In each diagram shown, notice how a square is attached to each side of a right triangle.

Triangle 1

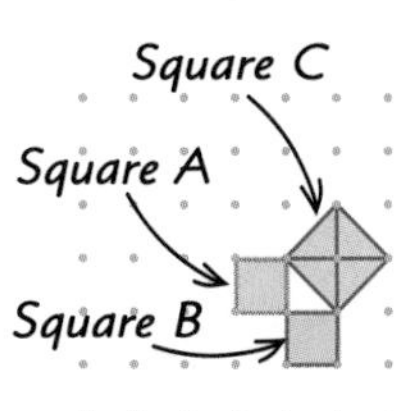

Triangle 2

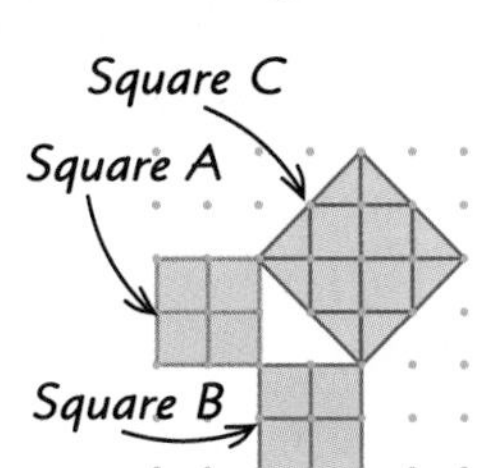

Triangle 3

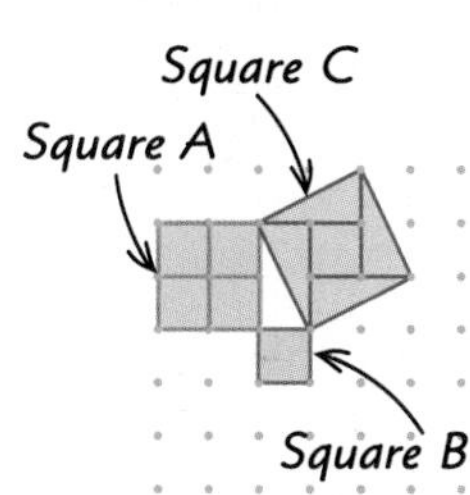

Triangle 4

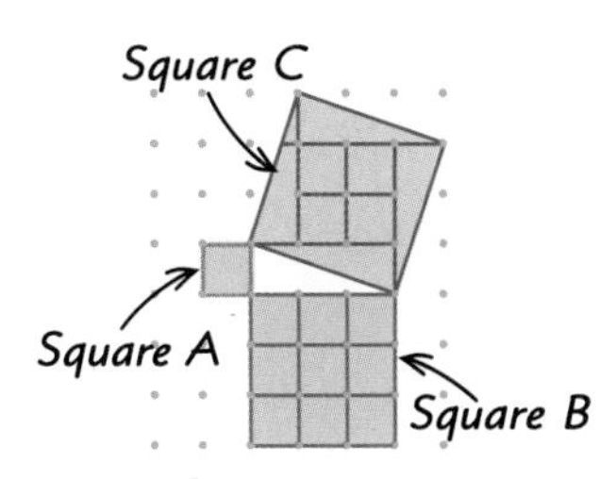

Triangle 5

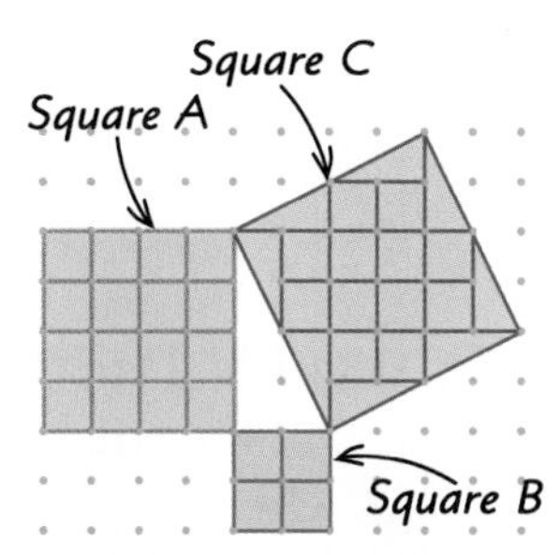

Copy the table. Then find the area of each square that is attached to the triangle. Record the results in your table.

Triangle	Area of Square A (units²)	Area of Square B (units²)	Area of Square C (units²)
1	1	1	2
2	4	4	8
3	4	1	5
4	1	9	10
5	16	4	20

Exercises

5. Refer to your table. How does the sum of the areas of square A and square B compare to the area of square C?

6. Refer to the diagram shown at the right. If the lengths of the sides of a right triangle are whole numbers such that $a^2 + b^2 = c^2$, the numbers a, b, and c are called a **Pythagorean Triple**. Tell whether each set of numbers is a Pythagorean Triple. Explain why or why not.

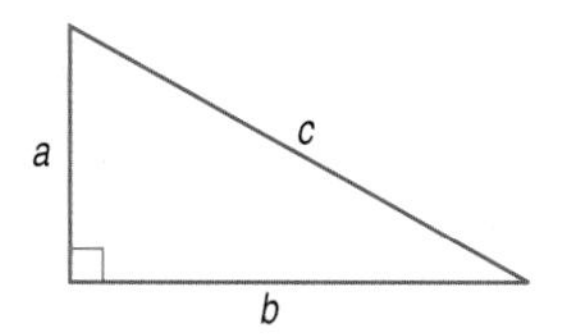

a. 3, 4, 5 **b.** 5, 7, 9 **c.** 6, 9, 12 **d.** 7, 24, 25

7. Write two different sets of numbers that are a Pythagorean Triple.

9-5 The Pythagorean Theorem

What You'll Learn

- Use the Pythagorean Theorem to find the length of a side of a right triangle.
- Use the converse of the Pythagorean Theorem to determine whether a triangle is a right triangle.

Vocabulary

- legs
- hypotenuse
- Pythagorean Theorem
- solving a right triangle
- converse

Sunshine State Standards
MA.C.3.3.1-2

How do the sides of a right triangle relate to each other?

In the diagram, three squares with sides 3, 4, and 5 units are used to form a right triangle.

a. Find the area of each square.

b. What relationship exists among the areas of the squares?

c. Draw three squares with sides 5, 12, and 13 units so that they form a right triangle. What relationship exists among the areas of these squares?

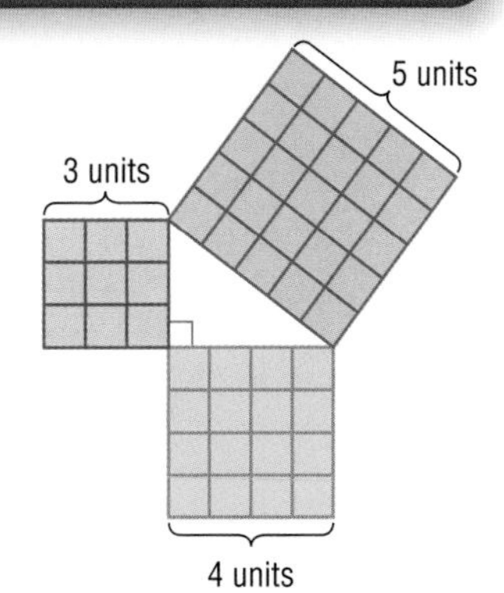

THE PYTHAGOREAN THEOREM In a right triangle, the sides that are adjacent to the right angle are called the **legs**. The side opposite the right angle is the **hypotenuse**.

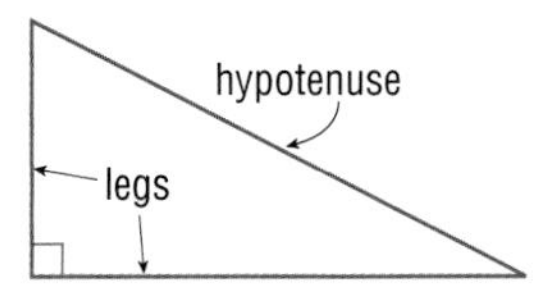

Study Tip

Hypotenuse
The hypotenuse is the longest side of a triangle.

The **Pythagorean Theorem** describes the relationship between the lengths of the legs and the hypotenuse. This theorem is true for *any* right triangle.

Key Concept — Pythagorean Theorem

- **Words** If a triangle is a right triangle, then the square of the length of the hypotenuse is equal to the sum of the squares of the lengths of the legs.
- **Model**

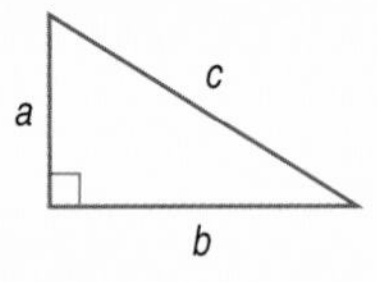

- **Symbols** $c^2 = a^2 + b^2$
- **Example** $5^2 = 3^2 + 4^2$
  $25 = 9 + 16$
  $25 = 25$

Example 1 Find the Length of the Hypotenuse

Find the length of the hypotenuse of the right triangle.

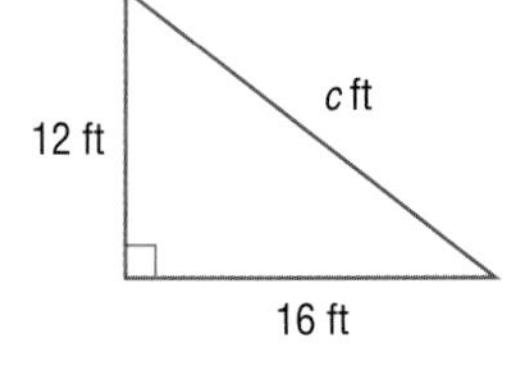

$c^2 = a^2 + b^2$	Pythagorean Theorem
$c^2 = 12^2 + 16^2$	Replace a with 12 and b with 16.
$c^2 = 144 + 256$	Evaluate 12^2 and 16^2.
$c^2 = 400$	Add 144 and 256.
$\sqrt{c^2} = \sqrt{400}$	Take the square root of each side.
$c = 20$	The length of the hypotenuse is 20 feet.

Study Tip

Look Back
To review **square roots,** see Lesson 9-1.

Concept Check Which side of a right triangle is the hypotenuse?

If you know the lengths of two sides of a right triangle, you can use the Pythagorean Theorem to find the length of the third side. This is called **solving a right triangle**.

Example 2 Solve a Right Triangle

Find the length of the leg of the right triangle.

$c^2 = a^2 + b^2$	Pythagorean Theorem
$14^2 = a^2 + 10^2$	Replace c with 14 and b with 10.
$196 = a^2 + 100$	Evaluate 14^2 and 10^2.
$196 - 100 = a^2 + 100 - 100$	Subtract 100 from each side.
$96 = a^2$	Simplify.
$\sqrt{96} = \sqrt{a^2}$	Take the square root of each side.

2nd [$\sqrt{\ }$] 96 ENTER 9.797958971

The length of the leg is about 9.8 centimeters.

Standardized tests often contain questions involving the Pythagorean Theorem.

Example 3 Use the Pythagorean Theorem

Multiple-Choice Test Item

A painter positions a 20-foot ladder against a house so that the base of the ladder is 4 feet from the house. About how high does the ladder reach on the side of the house?

(A) 17.9 ft (B) 18.0 ft (C) 19.6 ft (D) 20.4 ft

Test-Taking Tip

To help visualize the problem, it may be helpful to draw a diagram that represents the situation.

Read the Test Item

Make a drawing to illustrate the problem. The ladder, ground, and side of the house form a right triangle.

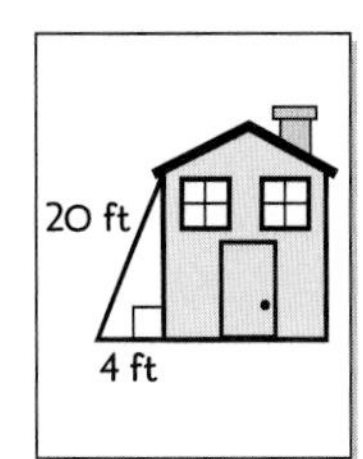

Solve the Test Item

Use the Pythagorean Theorem to find how high the ladder reaches on the side of the house.

$c^2 = a^2 + b^2$	Pythagorean Theorem
$20^2 = 4^2 + b^2$	Replace c with 20 and a with 4.
$400 = 16 + b^2$	Evaluate 20^2 and 4^2.
$400 - 16 = 16 + b^2 - 16$	Subtract 16 from each side.
$384 = b^2$	Simplify.
$\sqrt{384} = \sqrt{b^2}$	Take the square root of each side.
$19.6 \approx b$	Round to the nearest tenth.

The ladder reaches about 19.6 feet on the side of the house. The answer is C.

CONVERSE OF THE PYTHAGOREAN THEOREM The Pythagorean Theorem is written in *if-then* form. If you reverse the statements after *if* and *then*, you have formed the **converse** of the Pythagorean Theorem.

Pythagorean Theorem If a triangle is a right triangle, then $c^2 = a^2 + b^2$.

Converse If $c^2 = a^2 + b^2$, then a triangle is a right triangle.

You can use the converse to determine whether a triangle is a right triangle.

Example 4 Identify a Right Triangle

The measures of three sides of a triangle are given. Determine whether each triangle is a right triangle.

a. 9 m, 12 m, 15 m

$c^2 = a^2 + b^2$

$15^2 \stackrel{?}{=} 9^2 + 12^2$

$225 \stackrel{?}{=} 81 + 144$

$225 = 225$

The triangle is a right triangle.

b. 6 in., 7 in., 12 in.

$c^2 = a^2 + b^2$

$12^2 \stackrel{?}{=} 6^2 + 7^2$

$144 \stackrel{?}{=} 36 + 49$

$144 \neq 85$

The triangle is *not* a right triangle.

Check for Understanding

Concept Check

1. **OPEN ENDED** State the measures of three sides that could form a right triangle.
2. **FIND THE ERROR** Marcus and Allyson are finding the missing measure of the right triangle shown. Who is correct? Explain your reasoning.

Marcus

$c^2 = a^2 + b^2$

$15^2 = 9^2 + b^2$

$12 = b$

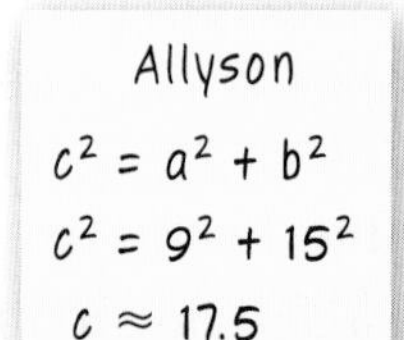

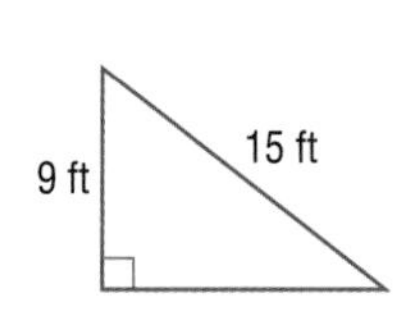

Guided Practice

Find the length of the hypotenuse in each right triangle. Round to the nearest tenth, if necessary.

3. 15 m, 20 m, c m

4. 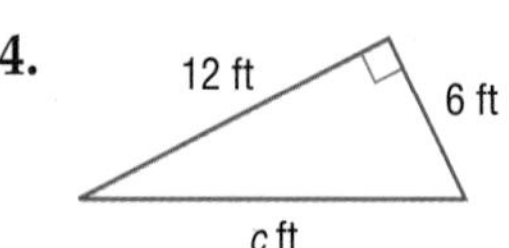

If c is the measure of the hypotenuse, find each missing measure. Round to the nearest tenth, if necessary.

5. $a = 8, b = ?, c = 17$
6. $a = ?, b = 24, c = 25$

The lengths of three sides of a triangle are given. Determine whether each triangle is a right triangle.

7. 5 cm, 7 cm, 8 cm
8. 10 ft, 24 ft, 26 ft

FCAT Practice
Standardized Test Practice

9. Kendra is flying a kite. The length of the kite string is 55 feet and she is positioned 40 feet away from beneath the kite. About how high is the kite?

Ⓐ 33.1 ft Ⓑ 37.7 ft Ⓒ 56.2 ft Ⓓ 68.0 ft

Practice and Apply

Homework Help

For Exercises	See Examples
10–15	1
16, 17	3
18–27	2
28–33	4

Extra Practice

See page 746.

Find the length of the hypotenuse in each right triangle. Round to the nearest tenth, if necessary.

10.

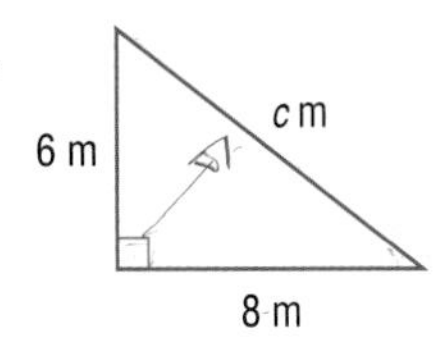

11.

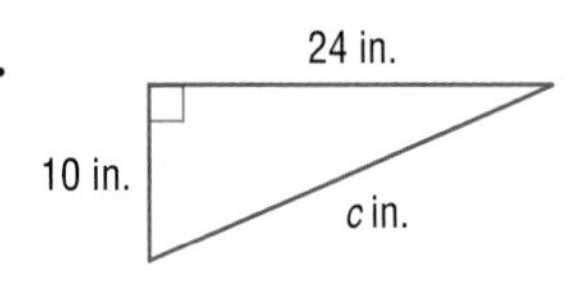

12.

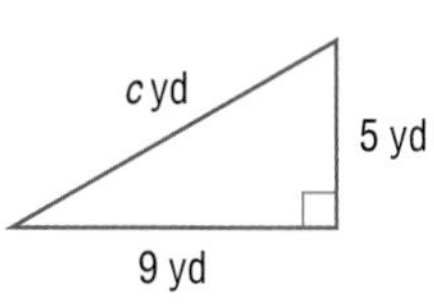

13.

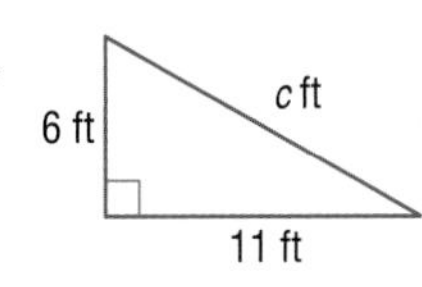

14.

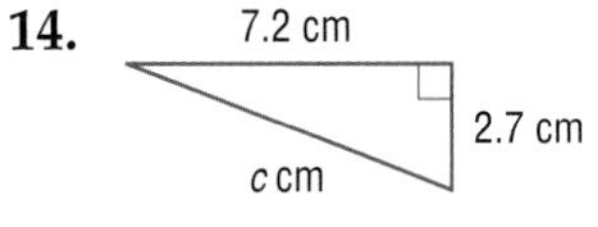

15.

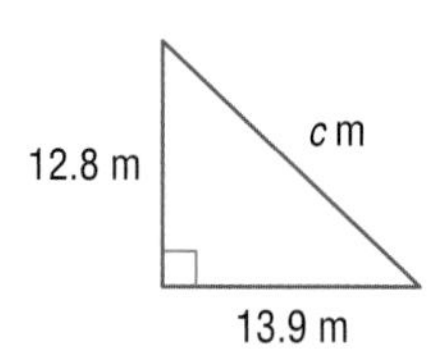

16. **GYMNASTICS** The floor exercise mat measures 40 feet by 40 feet. Find the measure of the diagonal.

17. **TELEVISION** The size of a television set is determined by the length of the diagonal of the screen. If a 35-inch television screen is 26 inches long, what is its height to the nearest inch?

More About. . .

Gymnastics

The floor exercise mat is a square of plywood covered by a 2-inch padding and mounted on 4" springs.

Source: The Gymnastics Place!

If c is the measure of the hypotenuse, find each missing measure. Round to the nearest tenth, if necessary.

18. $a = 9, b = ?, c = 41$

19. $a = ?, b = 35, c = 37$

20. $a = ?, b = 12, c = 19$

21. $a = 7, b = ?, c = 14$

22. $a = 27, b = ?, c = 61$

23. $a = ?, b = 73, c = 82$

24. $a = ?, b = \sqrt{123}, c = 22$

25. $a = \sqrt{177}, b = ?, c = 31$

Find each missing measure to the nearest tenth.

26.

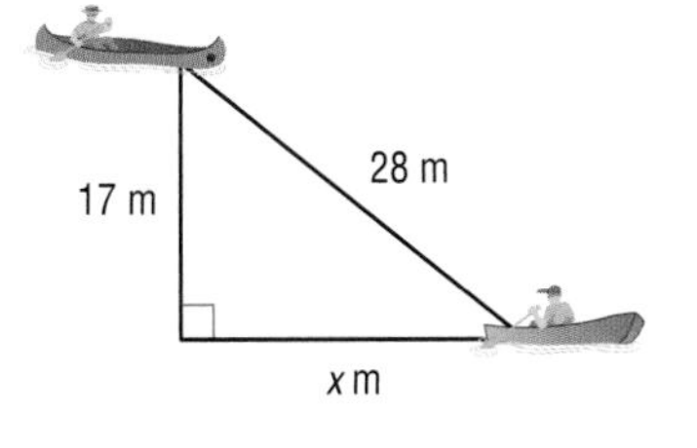

27.

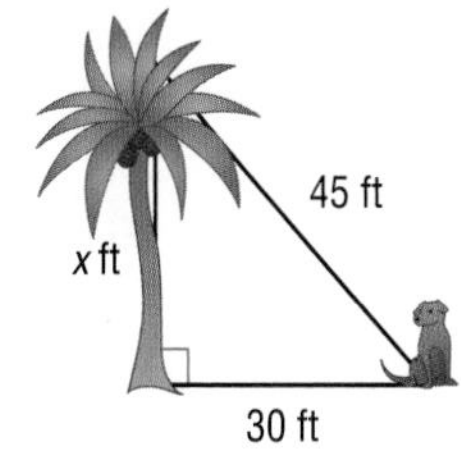

The lengths of three sides of a triangle are given. Determine whether each triangle is a right triangle.

28. $a = 5, b = 8, c = 9$

29. $a = 16, b = 30, c = 34$

30. $a = 18, b = 24, c = 30$

31. $a = 24, b = 28, c = 32$

32. $a = \sqrt{21}, b = 6, c = \sqrt{57}$

33. $a = 11, b = \sqrt{55}, c = \sqrt{177}$

ART **For Exercises 34 and 35, use the plasterwork design shown.**

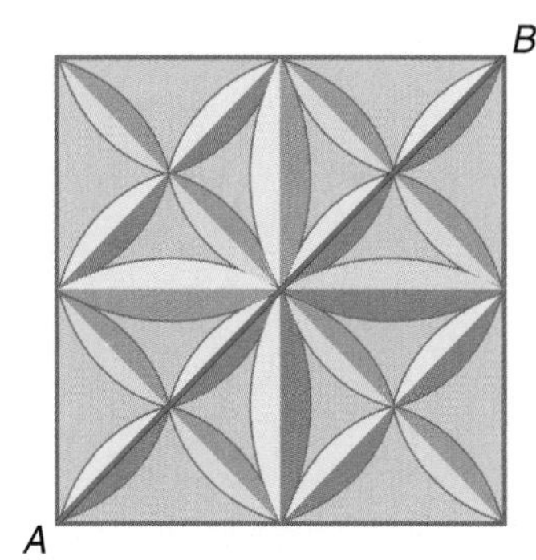

34. If the sides of the square measures 6 inches, what is the length of $\overline{AB}$?

35. What is the perimeter of the design if $\overline{AB}$ measures $\sqrt{128}$? (*Hint:* Use the guess-and-check strategy.)

36. **GEOMETRY** All angles *inscribed* in a semicircle are right angles. In the figure at the right, $\angle ACB$ is an inscribed right angle. If the length of $\overline{AB}$ is 17 and the length of $\overline{AC}$ is 8, find the length of $\overline{BC}$.

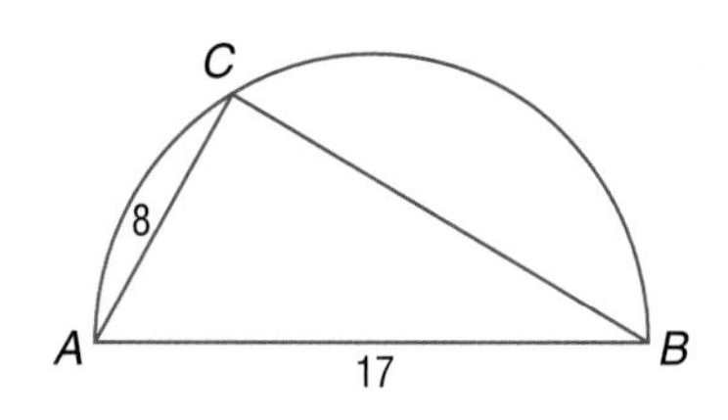

37. **CRITICAL THINKING** The hypotenuse of an isosceles right triangle is 8 inches. Is there enough information to find the length of the legs? If so, find the length of the legs. If not, explain why not.

38. **WRITING IN MATH** Answer the question that was posed at the beginning of the lesson.

How do the sides of a right triangle relate to each other?

Include the following in your answer:
- a right triangle with the legs and hypotenuse labeled, and
- an example of a set of numbers that represents the measures of the lengths of the legs and hypotenuse of a right triangle.

39. Which numbers represent the measures of the sides of a right triangle?

(A) 3, 3, 6 (B) 3, 4, 7 (C) 5, 7, 12 (D) 6, 8, 10

40. Which is the best estimate for the value of x?

(A) 5.0 (B) 7.9

(C) 8.3 (D) 16.0

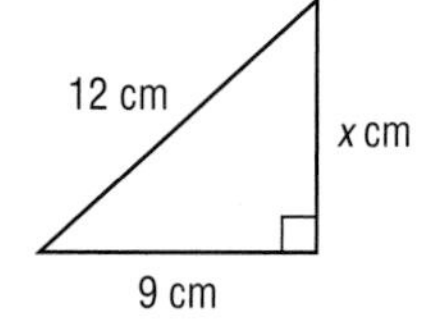

Extending the Lesson

GEOMETRY In the *rectangular prism* shown, $\overline{BD}$ is the diagonal of the base, and $\overline{FD}$ is the diagonal of the prism.

41. Find the measure of the diagonal of the base.

42. What is the measure of the diagonal of the prism to the nearest tenth?

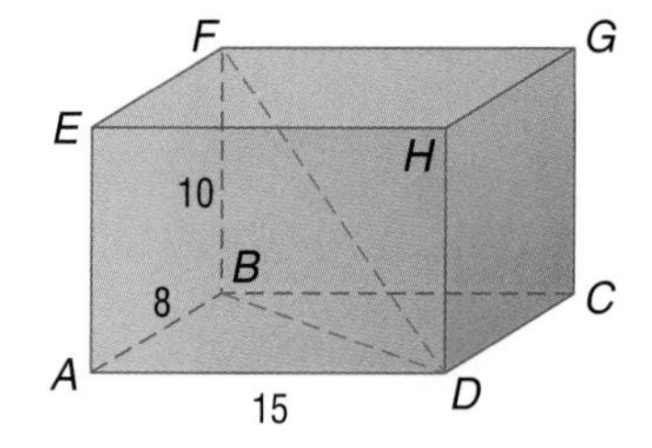

43. **MODELING** Measure the dimensions of a shoebox and use the dimensions to calculate the length of the diagonal of the box. Then use a piece of string and a ruler to check your calculation.

Maintain Your Skills

Mixed Review

Find the value of x in each triangle. Then classify each triangle as *acute*, *right*, or *obtuse*. *(Lesson 9-4)*

44.

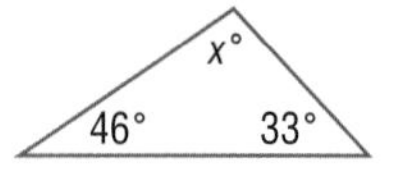

45.

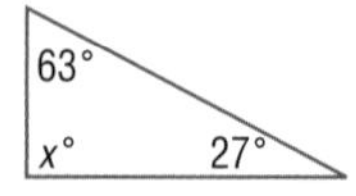

46.

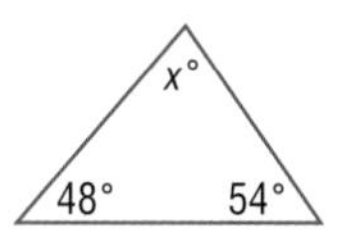

47. Use a protractor to draw an angle having a measure of 115°. *(Lesson 9-3)*

ALGEBRA Solve each inequality. *(Lesson 7-4)*

48. $x + 4 < 12$

49. $-15 \leq n - 6$

Getting Ready for the Next Lesson

PREREQUISITE SKILL Simplify each expression.
*(To review **order of operations** and **exponents**, see Lessons 1-2 and 4-2.)*

50. $(2 + 6)^2 + (-5 + 6)^2$

51. $(-4 + 3)^2 + (0 - 2)^2$

52. $[3 + (-1)]^2 + (8 - 4)^2$

Algebra Activity

A Follow-Up of Lesson 9-5

Graphing Irrational Numbers

In Lesson 2-1, you learned to graph integers on a number line. Irrational numbers can also be graphed on a number line. Consider the irrational number $\sqrt{53}$. To graph $\sqrt{53}$, construct a right triangle whose hypotenuse measures $\sqrt{53}$ units.

Step 1 Find two numbers whose squares have a sum of 53. Since $53 = 49 + 4$ or $7^2 + 2^2$, one pair that will work is 7 and 2. These numbers will be the lengths of the legs of the right triangle.

Step 2 Draw the right triangle

- First, draw a number line on grid paper.

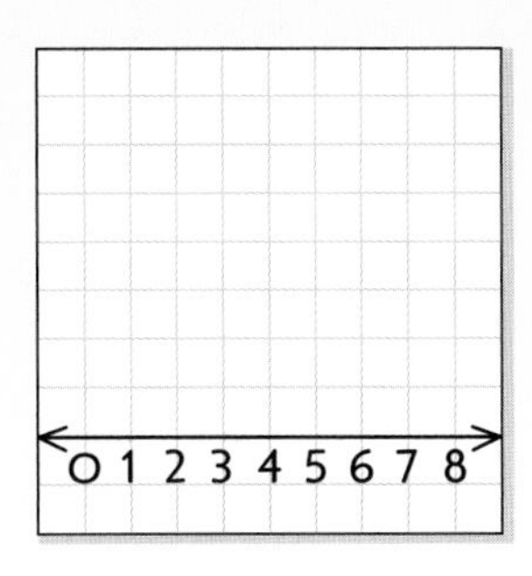

- Next, draw a right triangle whose legs measure 7 units and 2 units. Notice that this triangle can be drawn in two ways. Either way is correct.

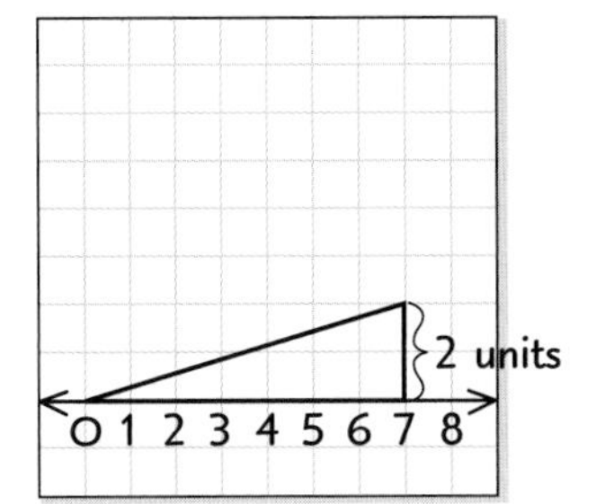

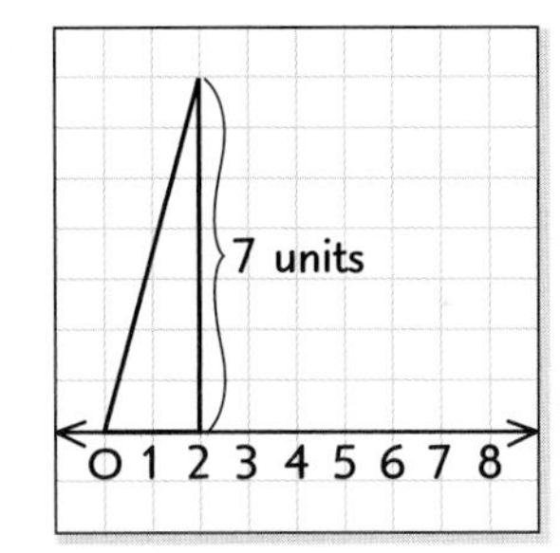

Step 3 Graph $\sqrt{53}$.

- Open your compass to the length of the hypotenuse.
- With the tip of the compass at 0, draw an arc that intersects the number line at point *B*.

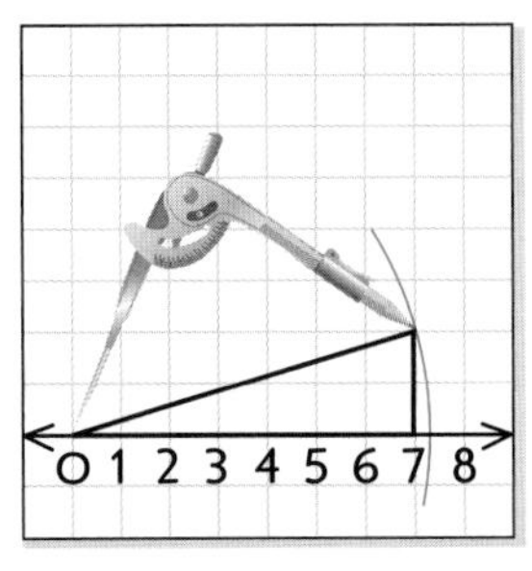

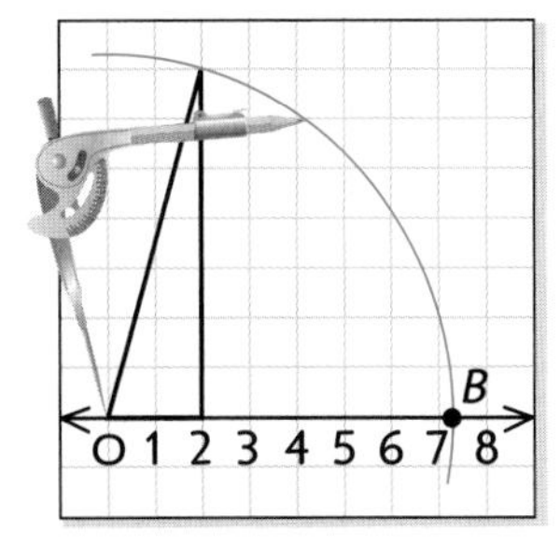

- The distance from 0 to *B* is $\sqrt{53}$ units. From the graph, $\sqrt{53} \approx 7.3$.

Model and Analyze

Use a compass and grid paper to graph each irrational number on a number line.

1. $\sqrt{5}$ **2.** $\sqrt{20}$ **3.** $\sqrt{45}$ **4.** $\sqrt{97}$

5. Describe two different ways to graph $\sqrt{34}$.

6. Explain how the graph of $\sqrt{2}$ can be used to locate the graph of $\sqrt{3}$.

9-6 The Distance and Midpoint Formulas

What You'll Learn

- Use the Distance Formula to determine lengths on a coordinate plane.
- Use the Midpoint Formula to find the midpoint of a line segment on the coordinate plane.

Vocabulary

- Distance Formula
- midpoint
- Midpoint Formula

How is the Distance Formula related to the Pythagorean Theorem?

The graph of points $N(3, 0)$ and $M(-4, 3)$ is shown. A horizontal segment is drawn from M, and a vertical segment is drawn from N. The intersection is labeled P.

a. Name the coordinates of P.

b. Find the distance between M and P.

c. Find the distance between N and P.

d. Classify $\triangle MNP$.

e. What theorem can be used to find the distance between M and N?

f. Find the distance between M and N.

THE DISTANCE FORMULA Recall that a line segment is a part of a line. It contains two endpoints and all of the points between the endpoints.

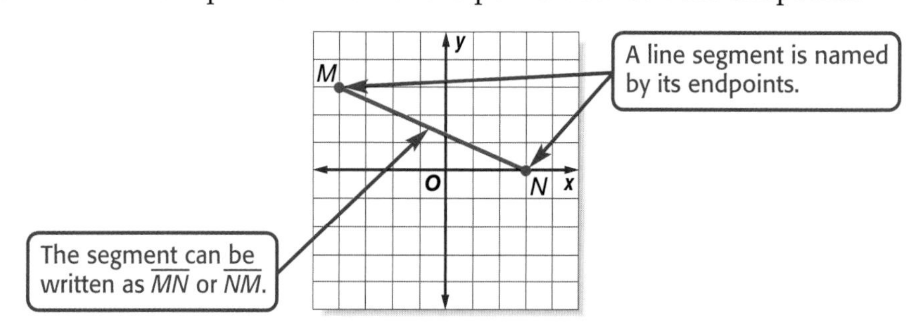
A line segment is named by its endpoints.
The segment can be written as $\overline{MN}$ or $\overline{NM}$.
M
N
O
x
y

To find the length of a segment on a coordinate plane, you can use the **Distance Formula**, which is based on the Pythagorean Theorem.

Key Concept — Distance Formula

- **Words** The distance d between two points with coordinates (x_1, y_1) and (x_2, y_2), is given by $d = \sqrt{(x_2 - x_1)^2 + (y_2 - y_1)^2}$.
- **Model**

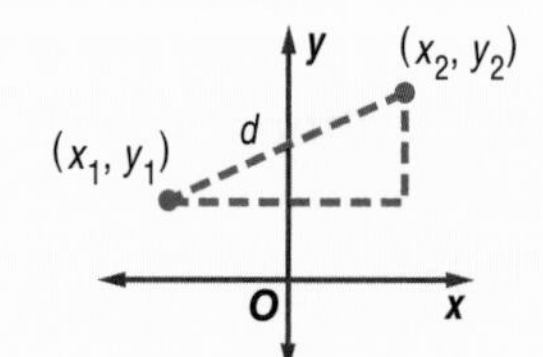
(x_1, y_1)
(x_2, y_2)
d
O
x
y

Study Tip

Look Back
To review the notation (x_1, y_1) and (x_2, y_2), see Lesson 8-4.

Concept Check How is a line segment named?

Example 1 Use the Distance Formula

Find the distance between $G(-3, 1)$ and $H(2, -4)$. Round to the nearest tenth, if necessary.

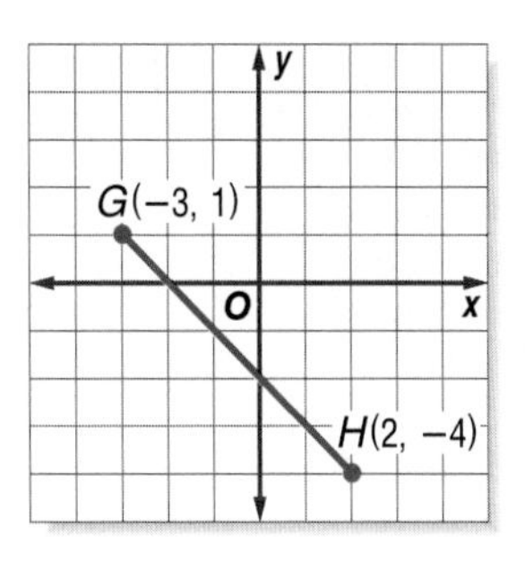

Use the Distance Formula.

$d = \sqrt{(x_2 - x_1)^2 + (y_2 - y_1)^2}$ Distance Formula

$GH = \sqrt{[2 - (-3)]^2 + (-4 - 1)^2}$ $(x_1, y_1) = (-3, 1), (x_2, y_2) = (2, -4)$

$GH = \sqrt{(5)^2 + (-5)^2}$ Simplify.

$GH = \sqrt{25 + 25}$ Evaluate 5^2 and $(-5)^2$.

$GH = \sqrt{50}$ Add 25 and 25.

$GH \approx 7.1$ Take the square root.

The distance between points G and H is about 7.1 units.

Study Tip

Substitution
You can use either point as (x_1, y_1). The distance will be the same.

The Distance Formula can be used to solve geometry problems.

Example 2 Use the Distance Formula to Solve a Problem

GEOMETRY Find the perimeter of $\triangle ABC$ to the nearest tenth.

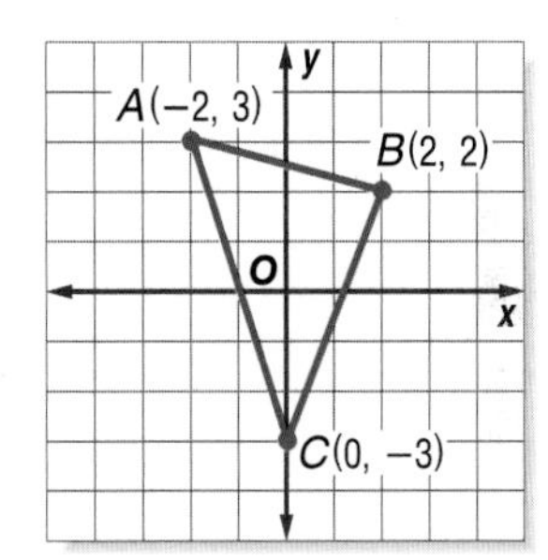

First, use the Distance Formula to find the length of each side of the triangle.

Side $\overline{AB}$: $A(-2, 3)$, $B(2, 2)$

$d = \sqrt{(x_2 - x_1)^2 + (y_2 - y_1)^2}$

$AB = \sqrt{[2 - (-2)]^2 + (2 - 3)^2}$

$AB = \sqrt{(4)^2 + (-1)^2}$

$AB = \sqrt{16 + 1}$

$AB = \sqrt{17}$

Side $\overline{BC}$: $B(2, 2)$, $C(0, -3)$

$d = \sqrt{(x_2 - x_1)^2 + (y_2 - y_1)^2}$

$BC = \sqrt{[0 - 2]^2 + (-3 - 2)^2}$

$BC = \sqrt{(-2)^2 + (-5)^2}$

$BC = \sqrt{4 + 25}$

$BC = \sqrt{29}$

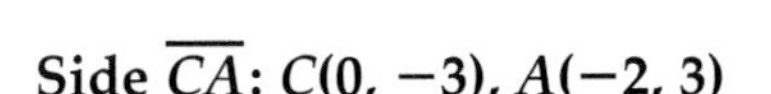

Side $\overline{CA}$: $C(0, -3)$, $A(-2, 3)$

$d = \sqrt{(x_2 - x_1)^2 + (y_2 - y_1)^2}$

$CA = \sqrt{[-2 - 0]^2 + (3 - (-3))^2}$

$CA = \sqrt{(-2)^2 + (6)^2}$

$CA = \sqrt{4 + 36}$

$CA = \sqrt{40}$

Then add the lengths of the sides to find the perimeter.

$\sqrt{17} + \sqrt{29} + \sqrt{40} \approx 4.123 + 5.385 + 6.325$

≈ 15.833

The perimeter is about 15.8 units.

Study Tip

Common Misconception
To find the sum of square roots, do *not* add the numbers inside the square root symbols.

$\sqrt{17} + \sqrt{29} + \sqrt{40} \neq \sqrt{86}$

Study Tip

Average
The x-coordinate of the midpoint is the *average* of the x-coordinates of the endpoints. The y-coordinate of the midpoint is the *average* of the y-coordinates of the endpoints.

THE MIDPOINT FORMULA On a line segment, the point that is halfway between the endpoints is called the **midpoint**.

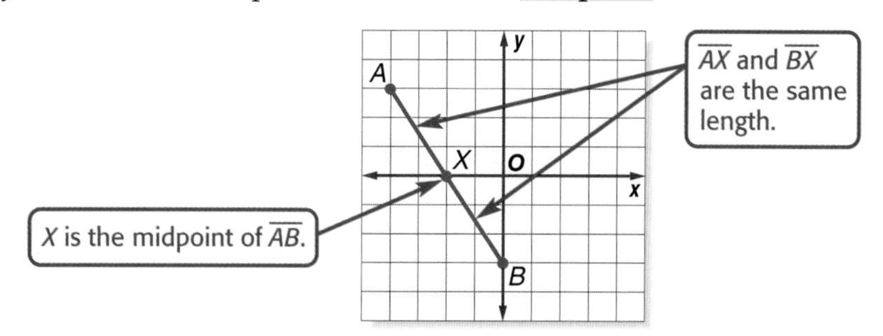

To find the midpoint of a segment on a coordinate plane, you can use the **Midpoint Formula**.

Key Concept — Midpoint Formula

- **Words** On a coordinate plane, the coordinates of the midpoint of a segment whose endpoints have coordinates at (x_1, y_1) and (x_2, y_2) are given by $\left(\frac{x_1 + x_2}{2}, \frac{y_1 + y_2}{2}\right)$.
- **Model**

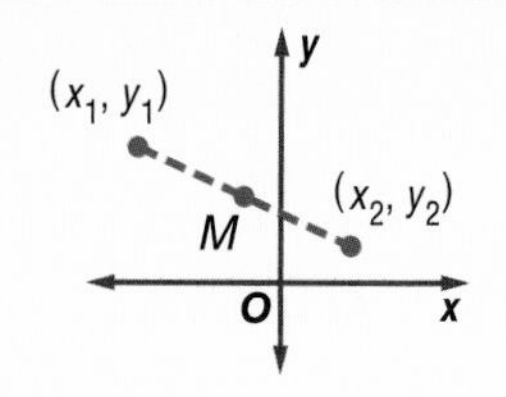

Example 3 Use the Midpoint Formula

Find the coordinates of the midpoint of $\overline{CD}$.

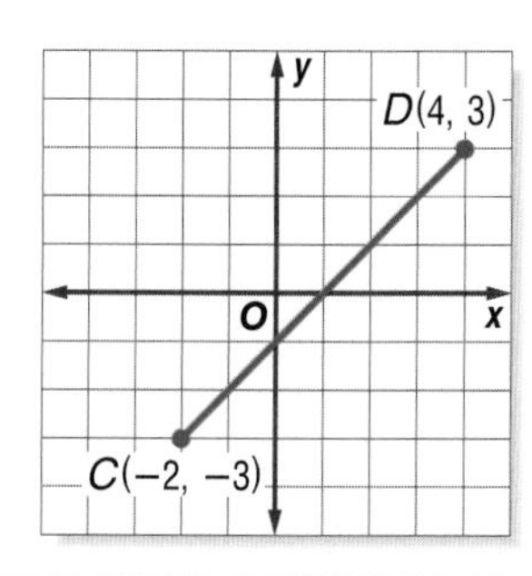

$$\text{midpoint} = \left(\frac{x_1 + x_2}{2}, \frac{y_1 + y_2}{2}\right) \quad \text{Midpoint Formula}$$

$$= \left(\frac{-2 + 4}{2}, \frac{-3 + 3}{2}\right) \quad \text{Substitution}$$

$$= (1, 0) \quad \text{Simplify.}$$

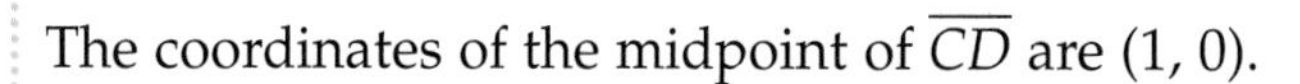

The coordinates of the midpoint of $\overline{CD}$ are (1, 0).

Check for Understanding

Concept Check

1. **Define** *midpoint*.
2. **Explain** the Distance Formula in your own words.
3. **OPEN ENDED** Draw any line segment on a coordinate system. Then find the midpoint of the segment.

Guided Practice

Find the distance between each pair of points. Round to the nearest tenth, if necessary.

4. $A(-1, 3)$, $B(8, -6)$
5. $M(4, -2)$, $N(-6, -7)$

The coordinates of the endpoints of a segment are given. Find the coordinates of the midpoint of each segment.

6. $B(4, 1)$, $C(-2, 5)$
7. $R(3, -6)$, $S(1, -4)$

Application

8. **GEOMETRY** Triangle EFG has vertices $E(1, 4)$, $F(-3, 0)$, and $G(4, -1)$. Find the perimeter of $\triangle EFG$ to the nearest tenth.

Practice and Apply

Homework Help

For Exercises	See Examples
9–16	1
17, 18	2
19–28	3

Extra Practice

See page 746.

Find the distance between each pair of points. Round to the nearest tenth, if necessary.

9. $J(5, -4), K(-1, 3)$
10. $C(-7, 2), D(6, -4)$
11. $E(-1, -2), F(9, -4)$
12. $V(8, -5), W(-3, -5)$
13. $S(-9, 0), T(6, -7)$
14. $M(0, 0), N(-7, -8)$
15. $Q\left(5\frac{1}{4}, 3\right), R\left(2, 6\frac{1}{2}\right)$
16. $A\left(-2\frac{1}{2}, 0\right), B\left(-8\frac{3}{4}, -6\frac{1}{4}\right)$

GEOMETRY **Find the perimeter of each figure.**

17.

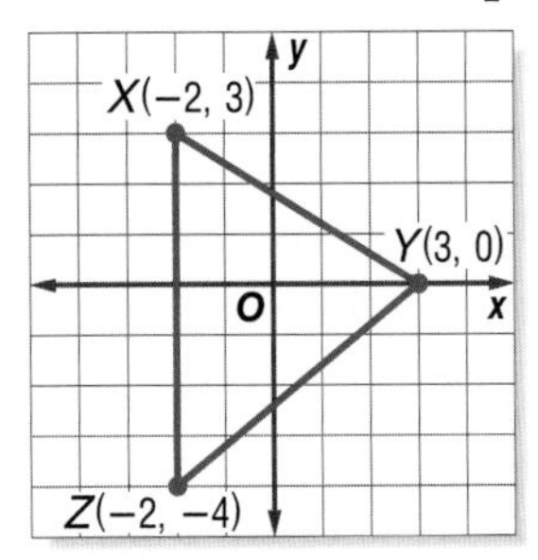

18.

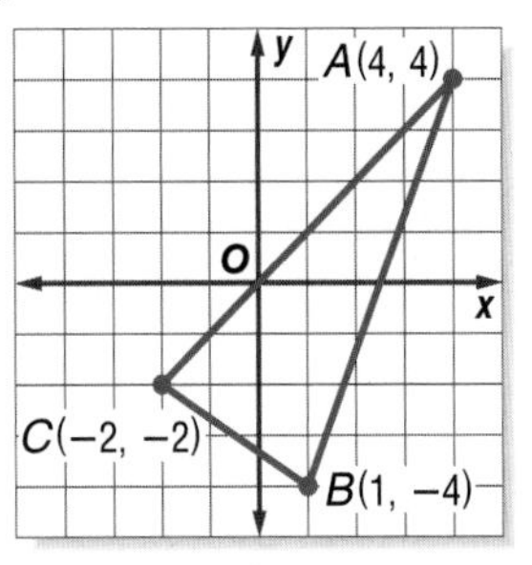

The coordinates of the endpoints of a segment are given. Find the coordinates of the midpoint of each segment.

19.

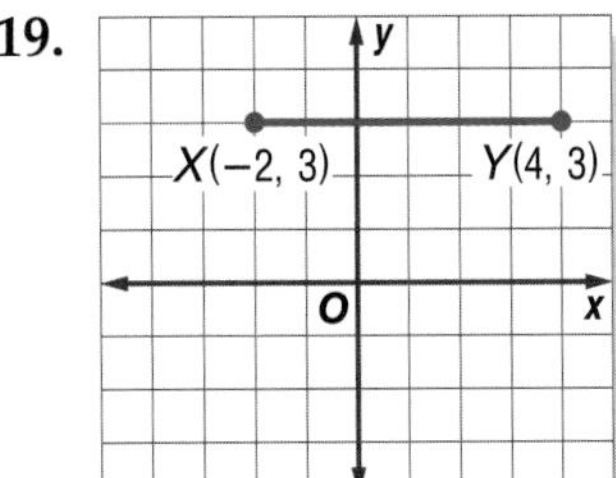

20.

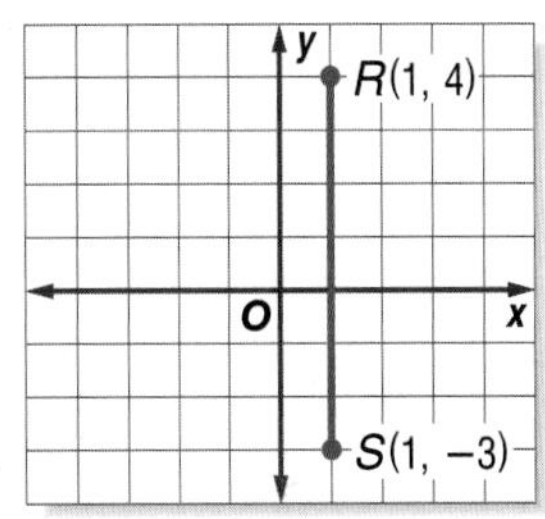

21. $A(6, 1), B(2, -5)$
22. $J(-3, 5), K(7, 9)$
23. $M(-1, -3), N(5, 7)$
24. $C(-4, 9), D(6, -5)$
25. $T(10, -3), U(-4, -5)$
26. $P(6, 11), Q(-4, -3)$
27. $F(15, -4), G(8, -6)$
28. $E(-12, -5), F(-3, -4)$

29. **GEOMETRY** Determine whether $\triangle MNP$ with vertices $M(3, -1)$, $N(-3, 2)$, and $P(6, 5)$ is isosceles. Explain your reasoning.

30. **GEOMETRY** Is $\triangle ABC$ with vertices $A(8, 4)$, $B(-2, 7)$, and $C(0, 9)$ a scalene triangle? Explain.

31. **CRITICAL THINKING** Suppose $C(8, -9)$ is the midpoint of $\overline{AB}$ and the coordinates of B are $(18, -21)$. What are the coordinates of A?

32. **WRITING IN MATH** Answer the question that was posed at the beginning of the lesson.

How is the Distance Formula related to the Pythagorean Theorem?

Include the following in your answer:

- a drawing showing how to use the Pythagorean Theorem to find the distance between two points on the coordinate system, and
- a comparison of the expressions $(x_2 - x_1)$ and $(y_2 - y_1)$ with the length of the legs of a right triangle.

FCAT Practice

Standardized Test Practice

A B C D

33. What are the coordinates of the midpoint of the line segment with endpoints $H(-2, 0)$ and $G(8, 6)$?

(A) $(-1, 7)$ (B) $(5, 3)$ (C) $(3, 3)$ (D) $(2, 2)$

34. Which expression shows how to find the distance between points $M(-5, -3)$ and $N(2, 3)$?

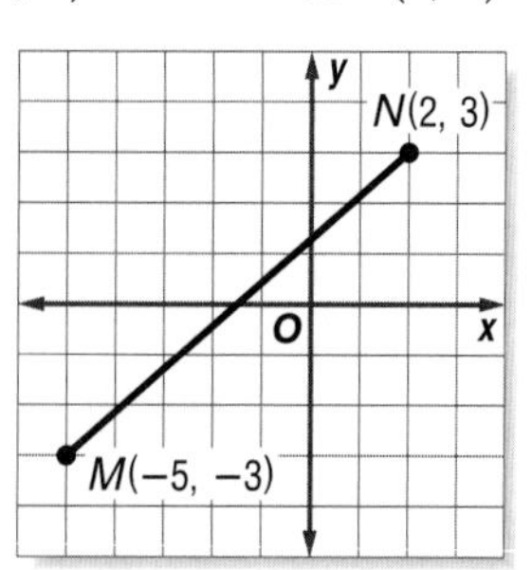

(A) $\sqrt{(2-5)^2 + (3-3)^2}$

(B) $\sqrt{[2-(-5)]^2 + [3-(-3)]^2}$

(C) $\sqrt{[2-(-5)]^2 + (3-3)^2}$

(D) $\sqrt{(3-2)^2 + [-3-(-5)]^2}$

Maintain Your Skills

Mixed Review **Find the length of the hypotenuse in each right triangle. Round to the nearest tenth, if necessary.** *(Lesson 9-5)*

35. 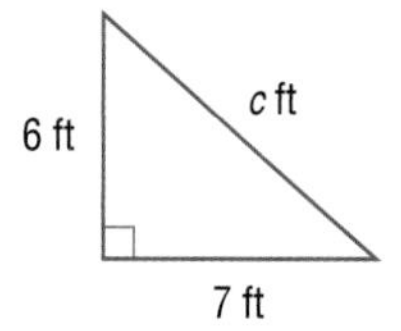

36. 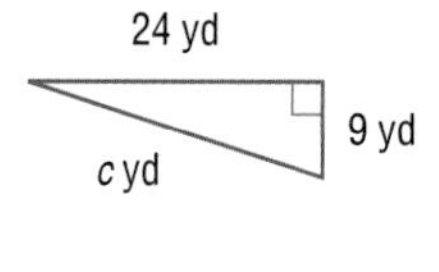

37. 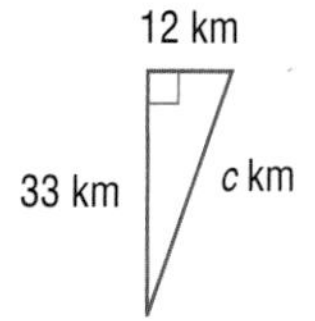

38. The measures of the angles of a triangle are in the ratio 1:4:5. Find the measure of each angle. *(Lesson 9-4)*

ALGEBRA State the slope and the y-intercept for the graph of each equation. *(Lesson 8-6)*

39. $y = 2x + 1$

40. $x + y = -4$

41. $4x + y = -6$

42. What number is 56% of 85? *(Lesson 6-7)*

Getting Ready for the Next Lesson **PREREQUISITE SKILL Solve each proportion.**

*(To review **proportions**, see Lesson 6-2.)*

43. $\frac{4}{16} = \frac{7}{x}$

44. $\frac{a}{15} = \frac{12}{60}$

45. $\frac{84}{m} = \frac{52}{13}$

46. $\frac{2.8}{h} = \frac{4.2}{12}$

47. $\frac{3.4}{85} = \frac{2.36}{n}$

48. $\frac{k}{5.5} = \frac{111}{15}$

Practice Quiz 2

Lessons 9-4 through 9-6

Classify each triangle by its angles and by its sides. *(Lesson 9-4)*

1. 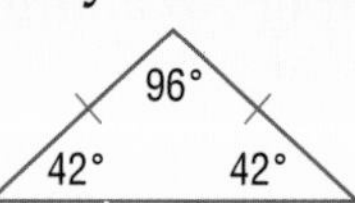

2. 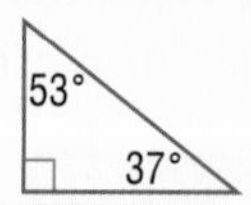

3. The lengths of three sides of a triangle are 34 meters, 30 meters, and 16 meters. Is the triangle a right triangle? Explain. *(Lesson 9-5)*

4. Find the distance between $A(-4, -1)$ and $B(7, -9)$ to the nearest tenth. *(Lesson 9-6)*

5. The coordinates of the endpoints of $\overline{MN}$ are $M(-2, 3)$ and $N(0, 7)$. What are the coordinates of the midpoint of $\overline{MN}$? *(Lesson 9-6)*

9-7 Similar Triangles and Indirect Measurement

What You'll Learn

- Identify corresponding parts and find missing measures of similar triangles.
- Solve problems involving indirect measurement using similar triangles.

Vocabulary

- similar triangles
- indirect measurement

Sunshine State Standards
MA.B.2.3.1-1, MA.C.1.3.1-1, MA.C.1.3.1-2, MA.C.1.3.1-4, MA.C.2.3.1-1, MA.C.2.3.1-2, MA.C.3.3.1-1

How can similar triangles be used to create patterns?

The triangle at the right is called *Sierpinski's triangle.* The triangle is made up of various equilateral triangles. The following activity investigates patterns similar to the ones in Sierpinski's triangle.

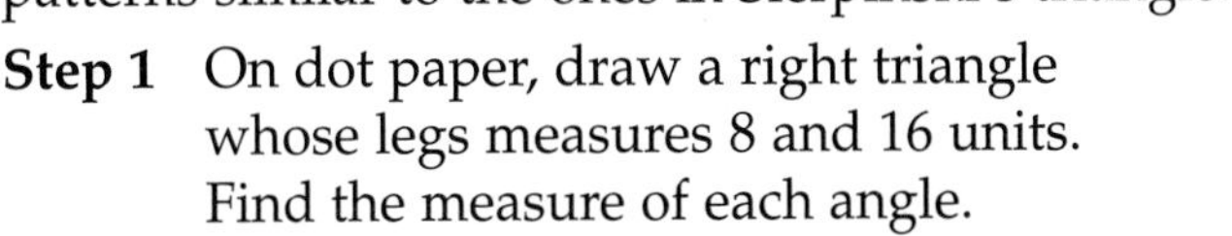

Step 1 On dot paper, draw a right triangle whose legs measures 8 and 16 units. Find the measure of each angle.

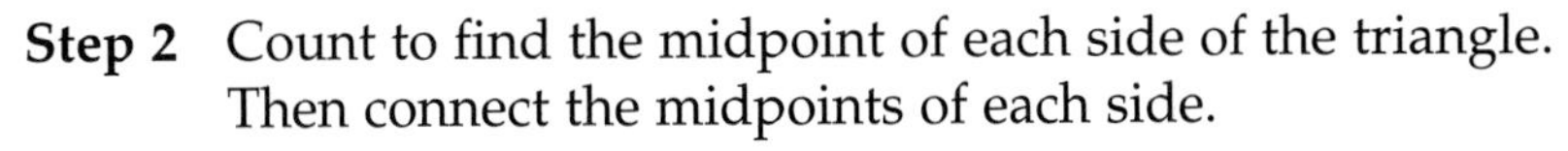

Step 2 Count to find the midpoint of each side of the triangle. Then connect the midpoints of each side.

Step 3 Shade the middle triangle.

Step 4 Repeat this process with each non-shaded triangle. Your triangles will resemble those shown below.

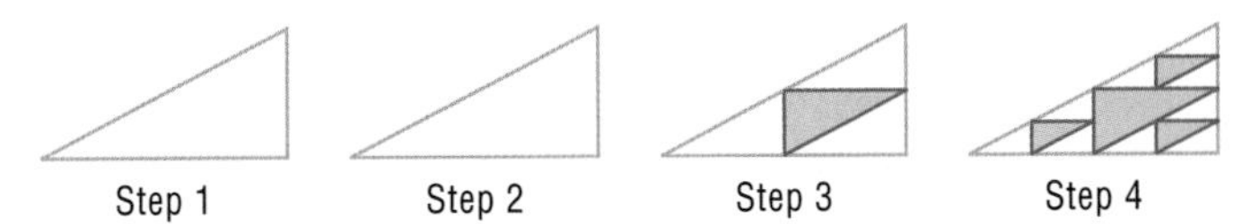

a. Compare the measures of the angles of each non-shaded triangle to the original triangle.

b. How do the lengths of the legs of the triangles compare?

CORRESPONDING PARTS Triangles that have the same shape but not necessarily the same size are called **similar triangles**. In the figure below, $\triangle ABC$ is similar to $\triangle XYZ$. This is written as $\triangle ABC \sim \triangle XYZ$.

Reading Math

Similar Symbol
The symbol $\sim$ is read *is similar to.*

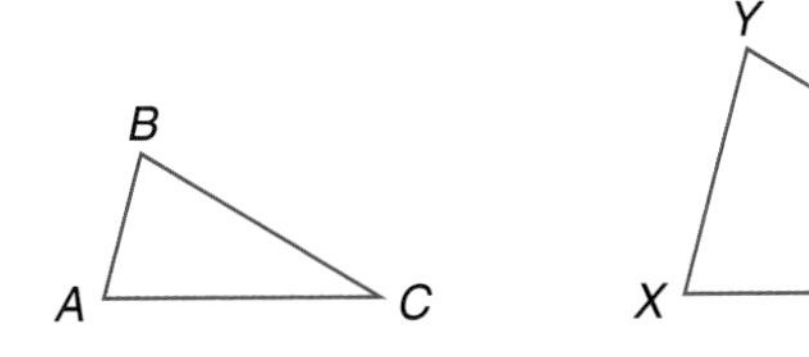

Similar triangles have corresponding angles and corresponding sides. Arcs are used to show congruent angles.

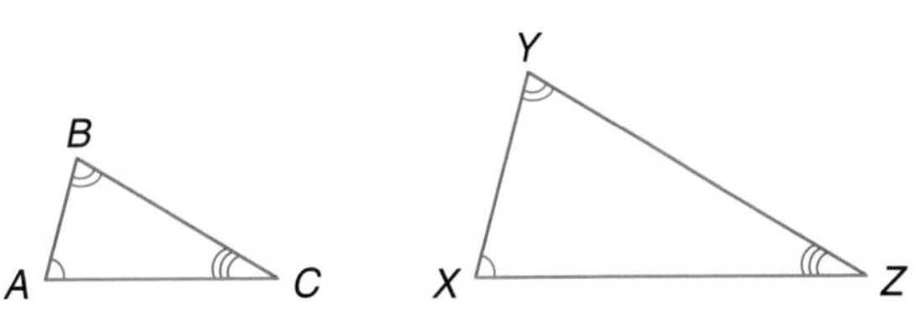

Corresponding Angles
$\angle A \leftrightarrow \angle X$ $\angle B \leftrightarrow \angle Y$ $\angle C \leftrightarrow \angle Z$

Corresponding Sides
$\overline{AB} \leftrightarrow \overline{XY}$ $\overline{BC} \leftrightarrow \overline{YZ}$ $\overline{AC} \leftrightarrow \overline{XZ}$

Concept Check Define similar triangles.

The following properties are true for similar triangles.

Key Concept — *Corresponding Parts of Similar Triangles*

- **Words** If two triangles are similar, then
  - the corresponding angles have the same measure, and
  - the corresponding sides are proportional.
- **Model**

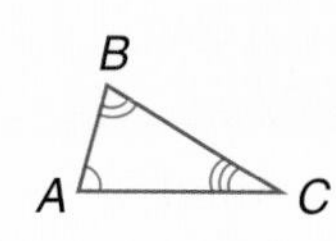

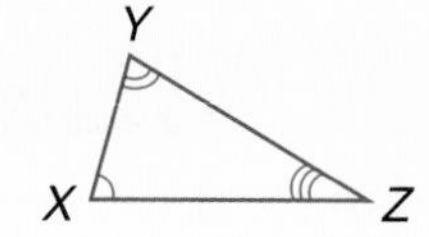

- **Symbols** $\angle A \cong \angle X$, $\angle B \cong \angle Y$, $\angle C \cong \angle Z$ and $\frac{AB}{XY} = \frac{BC}{YZ} = \frac{AC}{XZ}$

Reading Math

Segment Measure
The symbol AB means *the measure of segment AB.*

You can use proportions to determine the measures of the sides of similar triangles when some measures are known.

Example 1 Find Measures of Similar Triangles

If $\triangle JKM \sim \triangle RST$, what is the value of x?

The corresponding sides are proportional.

$\frac{JM}{RT} = \frac{KM}{ST}$ Write a proportion.

$\frac{3}{9} = \frac{5}{x}$ Replace JM with 3, RT with 9, KM with 5, and ST with x.

$3 \cdot x = 9 \cdot 5$ Find the cross products.

$3x = 45$ Simplify.

$x = 15$ Mentally divide each side by 3.

The value of x is 15.

Career Choices

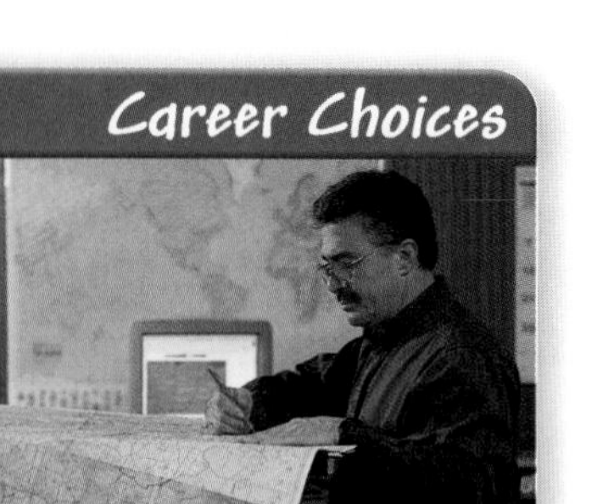

Cartographer
A cartographer gathers geographic, political, and cultural information and then uses this information to create graphic or digital maps of areas.

Online Research
For information about a career as a cartographer, visit: www.pre-alg.com/careers

INDIRECT MEASUREMENT The properties of similar triangles can be used to find measurements that are difficult to measure directly. This kind of measurement is called **indirect measurement**.

Example 2 Use Indirect Measurement

MAPS In the figure, $\triangle ABE \sim \triangle DCE$. Find the distance across the lake.

$\frac{EC}{EB} = \frac{CD}{BA}$ Write a proportion.

$\frac{96}{48} = \frac{x}{64}$ Replace EC with 96, EB with 48, CD with x, and BA with 64.

$96 \cdot 64 = 48 \cdot x$ Find the cross products.

$6144 = 48x$ Multiply.

$128 = x$ Divide each side by 48.

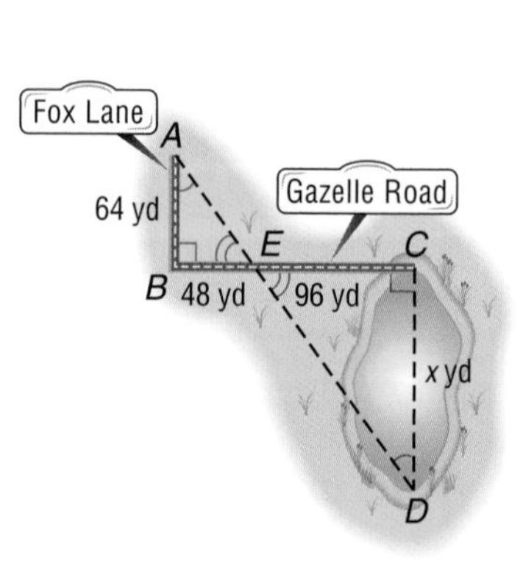

The distance across the lake is 128 yards.

The properties of similar triangles can be used to determine missing measures in shadow problems. This is called *shadow reckoning*.

Example 3 Use Shadow Reckoning

LANDMARKS Suppose the Space Needle in Seattle, Washington, casts a 220-foot shadow at the same time a nearby tourist casts a 2-foot shadow. If the tourist is 5.5 feet tall, how tall is the Space Needle?

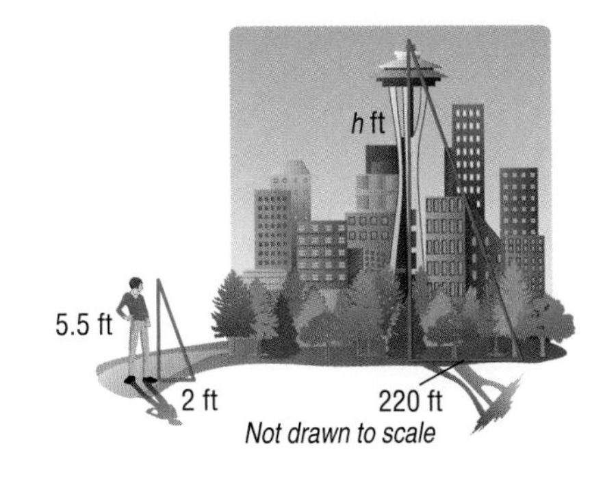

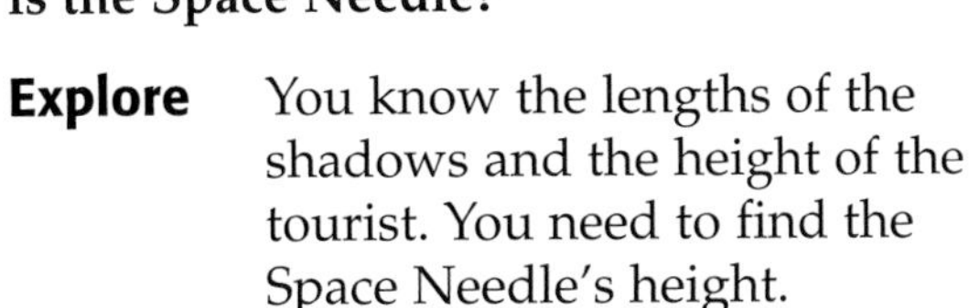

Explore You know the lengths of the shadows and the height of the tourist. You need to find the Space Needle's height.

Plan Write and solve a proportion.

Solve

$$\frac{5.5}{h} = \frac{2}{220}$$

tourist's height → 5.5; Space Needle's height → h; ← tourist's shadow (2); ← Space Needle's shadow (220)

$5.5 \cdot 220 = h \cdot 2$ Find the cross products.

$1210 = 2h$ Multiply.

$605 = h$ Mentally divide each side by 2.

The height of the Space Needle is 605 feet.

Examine The tourist's height is a little less than 3 times the length of his or her shadow. The space needle should be a little less than 3 times its shadow, or $3 \cdot 220$, which is 660 feet. So, 605 is reasonable.

Check for Understanding

Concept Check

1. **OPEN ENDED** Draw two similar triangles and label the vertices. Then write a proportion that compares the corresponding sides.
2. **Explain** indirect measurement in your own words.

Guided Practice

In Exercises 3 and 4, the triangles are similar. Write a proportion to find each missing measure. Then find the value of x.

3. 4.

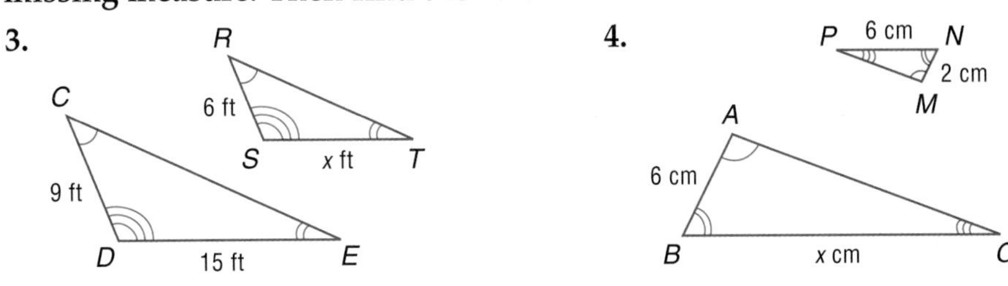

Applications

5. **MAPS** In the figure, $\triangle ABC \sim \triangle EDC$. Find the distance from Austintown to North Jackson.

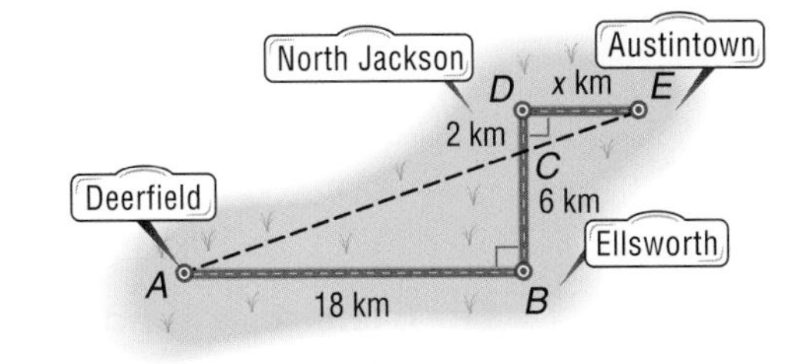

6. **SHADOWS** At the same time a 10-foot flagpole casts an 8-foot shadow, a nearby tree casts a 40-foot shadow. How tall is the tree?

Practice and Apply

Homework Help

For Exercises	See Examples
7–12	1
15, 16	2
17, 18	3

Extra Practice

See page 747.

In Exercises 7–12, the triangles are similar. Write a proportion to find each missing measure. Then find the value of x.

7.

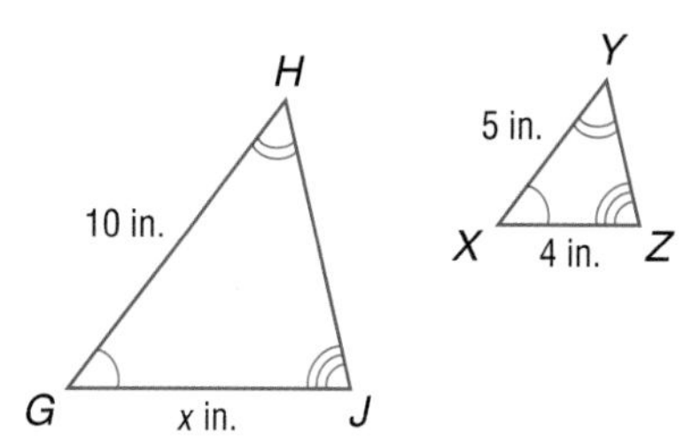

8.

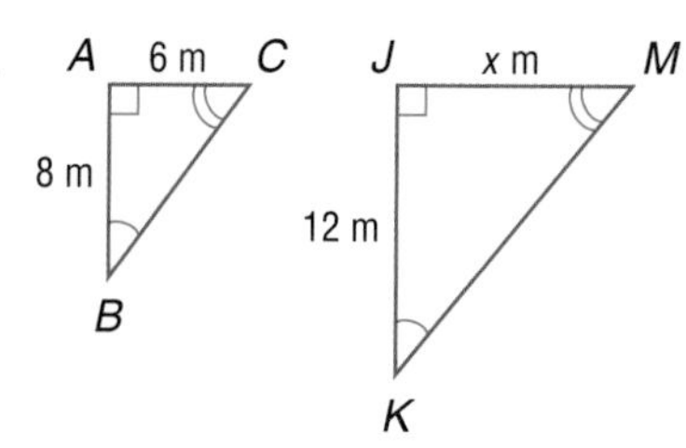

9.

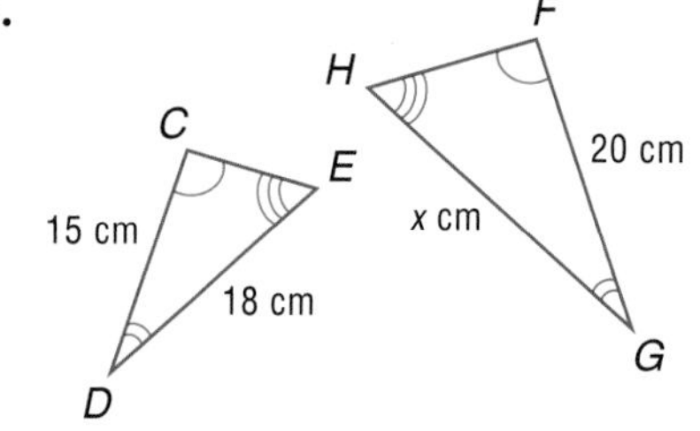

10.

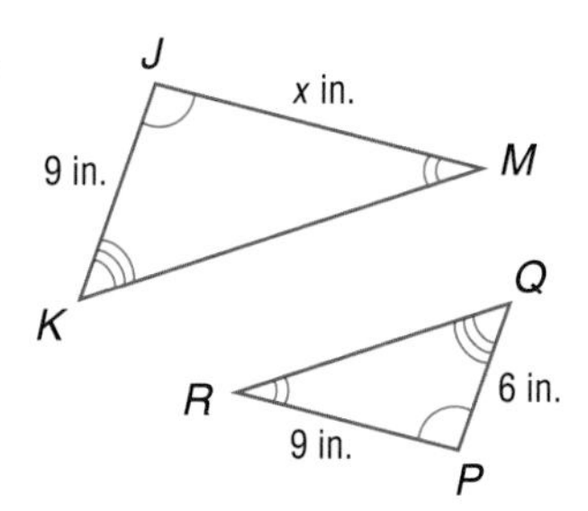

11.

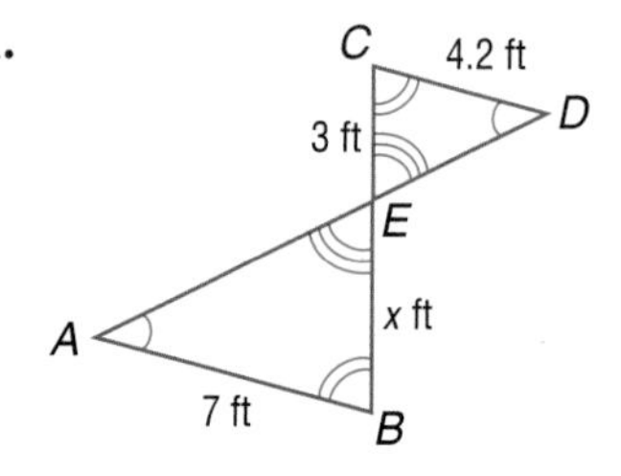

12.

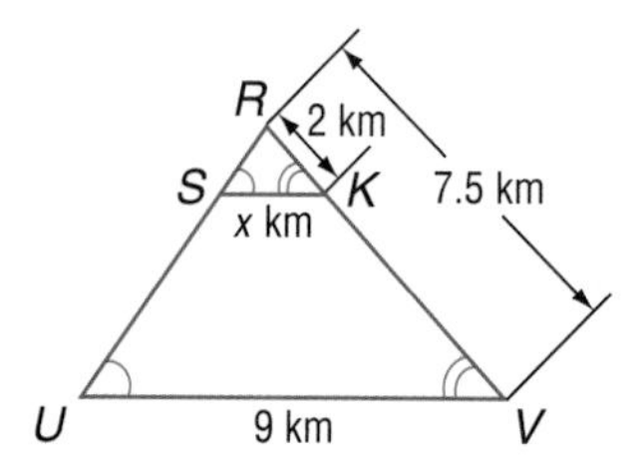

Determine whether each statement is *sometimes, always,* or *never* true.

13. The measures of corresponding angles in similar triangles are the same.

14. Similar triangles have the same shape and the same size.

In Exercises 15 and 16, the triangles are similar.

15. **PARKS** How far is the pavilion from the log cabin?

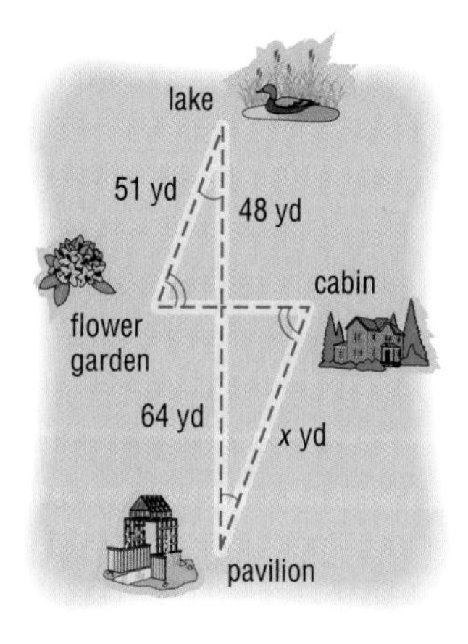

16. **ZOO** How far are the gorillas from the cheetahs?

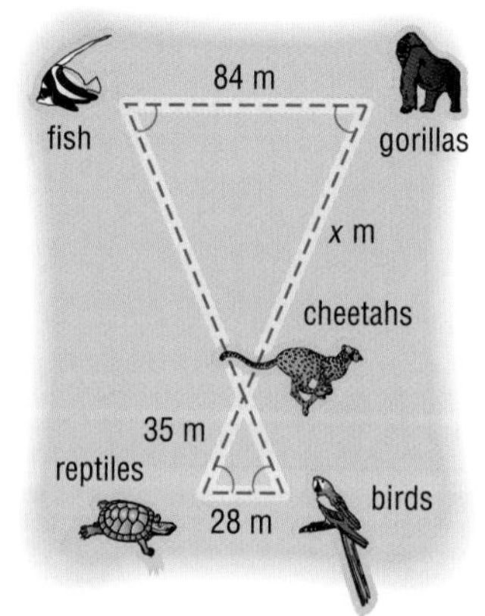

For Exercises 17 and 18, write a proportion. Then determine the missing measure.

17. **ANIMALS** At the same time a baby giraffe casts a 3.2-foot shadow, a 15-foot adult giraffe casts an 8-foot shadow. How tall is the baby giraffe?

More About. . .

Animals

The giraffe is the tallest of all living animals. At birth, the height of a giraffe is about 6 feet tall. An adult giraffe is about 18–19 feet tall.

Source: www.infoplease.com

18. **RIDES** Suppose a roller coaster casts a shadow of 31.5 feet. At the same time, a nearby Ferris wheel casts a 19-foot shadow. If the roller coaster is 126 feet tall, how tall is the Ferris wheel?

19. **CRITICAL THINKING** Michael is using three similar triangles to build a kite. One of the triangles measures 2 units by 3 units by 4 units. How large can the other two triangles be?

20. **WRITING IN MATH** Answer the question that was posed at the beginning of the lesson.

How can similar triangles be used to create patterns?

Include the following in your answer:
- an explanation telling how Sierpinski's triangle relates to similarity, and
- an example of the pattern formed when Steps 1–4 are performed on an acute scalene triangle.

21. The triangles shown are similar. Find the length of $\overline{JM}$ to the nearest tenth.

(A) 3.7 (B) 3.9
(C) 4.1 (D) 8.3

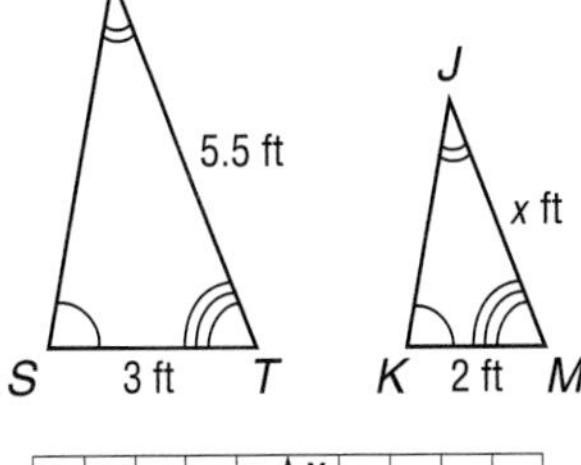

22. On the coordinate system, $\triangle MNP$ and two coordinates for $\triangle RST$ are shown. Which coordinate for point T will make $\triangle MNP$ and $\triangle RST$ similar triangles?

(A) $T(3, 1)$ (B) $T(1, -1)$
(C) $T(1, 2)$ (D) $T(1, 1)$

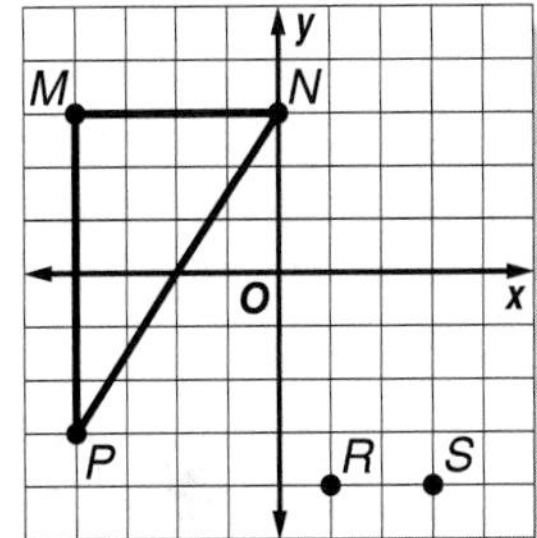

Maintain Your Skills

Mixed Review

Find the distance between each pair of points. Round to the nearest tenth, if necessary. *(Lesson 9-6)*

23. $S(2, 3), T(0, 6)$
24. $E(-1, 1), F(3, -2)$
25. $W(4, -6), V(-3, -5)$

If c is the measure of the hypotenuse, find each missing measure. Round to the nearest tenth, if necessary. *(Lesson 9-5)*

26. $a = 8, b = ?, c = 34$
27. $a = ?, b = 27, c = 82$

28. **ALGEBRA** Write an equation for the line that passes through points at $(1, -1)$ and $(-2, 11)$. *(Lesson 8-7)*

Write each percent as a fraction in simplest form. *(Lesson 6-4)*

29. 25%
30. $87\frac{1}{2}\%$
31. 150%

Getting Ready for the Next Lesson

PREREQUISITE SKILL Express each fraction as a decimal. Round to four decimal places, if necessary. *(To review **writing fractions as decimals**, see Lesson 5-1.)*

32. $\frac{12}{16}$
33. $\frac{10}{14}$
34. $\frac{20}{25}$
35. $\frac{9}{40}$

Algebra Activity

A Preview of Lesson 9-8

Ratios in Right Triangles

In the following activity, you will discover the special relationship among right triangles and their sides.

In any right triangle, the side **opposite** an angle is the side that is not part of the angle. In the triangle shown,

- side a is opposite $\angle A$,
- side b is opposite $\angle B$, and
- side c is opposite $\angle C$.

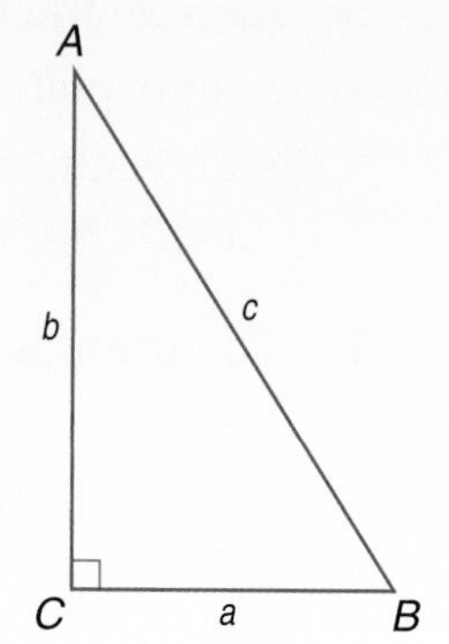

The side that is not opposite an angle and not the hypotenuse is called the **adjacent** side. In $\triangle ABC$,

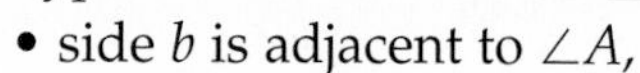

- side b is adjacent to $\angle A$,
- side a is adjacent to $\angle B$, and
- sides a and b are adjacent to $\angle C$.

Step 1 Copy the table shown.

Step 2 Draw right triangle XYZ in which $m\angle X = 30°$, $m\angle Y = 60°$, and $m\angle Z = 90°$.

Step 3 Find the length to the nearest millimeter of the leg opposite the angle that measures 30°. Record the length.

Step 4 Find the length of the leg adjacent to the 30° angle. Record the length.

	30° angle	60° angle
Length (mm) of opposite leg		
Length (mm) of adjacent leg		
Length (mm) of hypotenuse		
Ratio 1		
Ratio 2		
Ratio 3		

Step 5 Find the length of the hypotenuse. Record the length.

Step 6 Using the measurements and a calculator, find each of the following ratios to the nearest hundredth. Record the results.

$$\text{Ratio 1} = \frac{\text{opposite leg}}{\text{hypotenuse}} \quad \text{Ratio 2} = \frac{\text{adjacent leg}}{\text{hypotenuse}} \quad \text{Ratio 3} = \frac{\text{opposite leg}}{\text{adjacent leg}}$$

Step 7 Repeat the procedure for the 60° angle. Record the results.

Model and Analyze

1. Draw another 30°-60°-90° triangle with side lengths that are different than the one drawn in the activity. Then find the ratios for the 30°angle and the 60° angle.
2. **Make a conjecture** about the ratio of the sides of any 30°-60°-90° triangle.
3. Repeat the activity with a triangle whose angles measure 45°, 45°, and 90°.

9-8 Sine, Cosine, and Tangent Ratios

What You'll Learn

- Find sine, cosine, and tangent ratios.
- Solve problems by using the trigonometric ratios.

Vocabulary

- trigonometry
- trigonometric ratio
- sine
- cosine
- tangent

How are ratios in right triangles used in the real world?

In parasailing, a towrope is used to attach the parachute to the boat.

a. What type of triangle do the towrope, water, and height of the person above the water form?

b. Name the hypotenuse of the triangle.

c. What type of angle do the towrope and the water form?

d. Which side is opposite this angle?

e. Other than the hypotenuse, name the side adjacent to this angle.

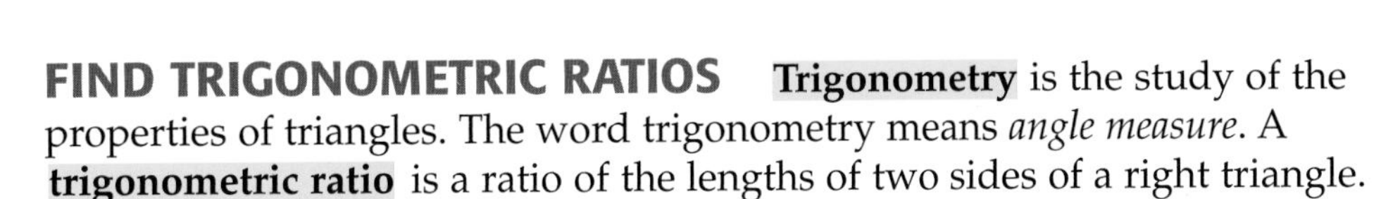

FIND TRIGONOMETRIC RATIOS **Trigonometry** is the study of the properties of triangles. The word trigonometry means *angle measure*. A **trigonometric ratio** is a ratio of the lengths of two sides of a right triangle.

The most common trigonometric ratios are the **sine**, **cosine**, and **tangent** ratios. These ratios are abbreviated as *sin*, *cos*, and *tan*, respectively.

Key Concept — Trigonometric Ratios

- **Words** If $\angle A$ is an acute angle of a right triangle,

$$\sin \angle A = \frac{\text{measure of leg opposite } \angle A}{\text{measure of hypotenuse}},$$

$$\cos \angle A = \frac{\text{measure of leg adjacent to } \angle A}{\text{measure of hypotenuse}}, \text{ and}$$

$$\tan \angle A = \frac{\text{measure of leg opposite } \angle A}{\text{measure of leg adjacent to } \angle A}.$$

- **Model**

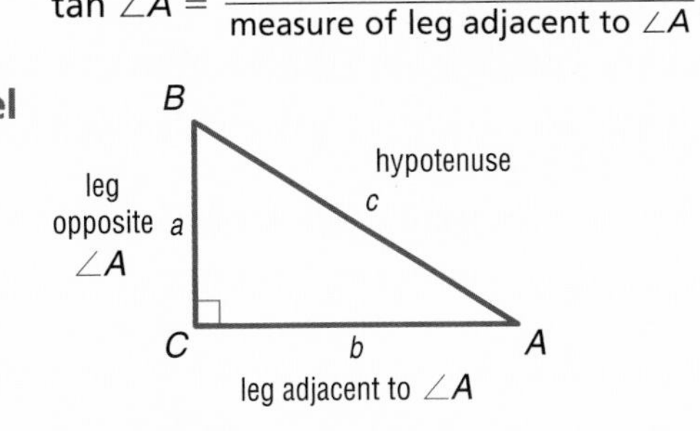

- **Symbols** $\sin A = \frac{a}{c}$ $\cos A = \frac{b}{c}$ $\tan A = \frac{a}{b}$

Reading Math

Trigonometry Terms

The notation sin *A* is read *the sine of angle A*. The notation cos *A* is read *the cosine of angle A*. The notation tan *A* is read *the tangent of angle A*.

All right triangles that have the same measure for $\angle A$ are similar. So, the value of the trigonometric ratio depends only on the measure of $\angle A$, not the size of the triangle. The trigonometric ratios are the same for angle A no matter what the size of the triangle.

Example 1 Find Trigonometric Ratios

Find sin P, cos P, and tan P.

$$\sin P = \frac{\text{measure of leg opposite } \angle P}{\text{measure of hypotenuse}}$$
$$= \frac{7}{25} \text{ or } 0.28$$

$$\cos P = \frac{\text{measure of leg adjacent to } \angle P}{\text{measure of hypotenuse}}$$
$$= \frac{24}{25} \text{ or } 0.96$$

$$\tan P = \frac{\text{measure of leg opposite } \angle P}{\text{measure of leg adjacent to } \angle P}$$
$$= \frac{7}{24} \text{ or } 0.2917$$

Study Tip

Rounding
When written as decimals, trigonometric ratios are often rounded to four decimal places.

You can use a calculator or a table of trigonometric ratios to find the sine, cosine, or tangent ratio for an angle with a given degree measure. Be sure that your calculator is in *degree* mode.

Example 2 Use a Calculator to Find Trigonometric Ratios

Find each value to the nearest ten thousandth.

a. sin 42°

SIN 42 ENTER 0.669130606

So, sin 42° is about 0.6691.

b. cos 65°

COS 65 ENTER 0.422618262

So, cos 65° is about 0.4226.

c. tan 78°

TAN 78 ENTER 4.704630109

So, tan 78° is about 4.7046.

APPLY TRIGONOMETRIC RATIOS Trigonometric ratios can be used to find missing measures in a right triangle if the measure of an acute angle and the length of one side of the triangle are known.

Example 3 Use Trigonometric Ratios

Find the missing measure. Round to the nearest tenth.

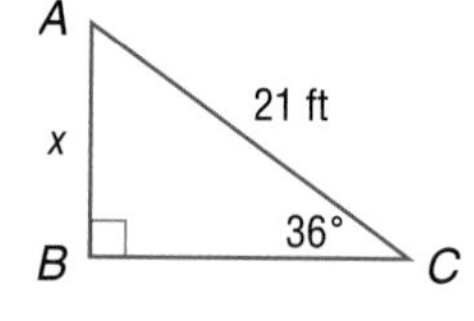

The measures of an acute angle and the hypotenuse are known. You need to find the measure of the side opposite the angle. Use the sine ratio.

$$\sin \angle C = \frac{\text{measure of leg opposite to } \angle C}{\text{measure of hypotenuse}}$$ Write the sine ratio.

$$\sin 36° = \frac{x}{21}$$ Substitution

$$21(\sin 36°) = 21 \cdot \frac{x}{21}$$ Multiply each side by 21.

21 × SIN 36 ENTER 12.3434903

$$12.3 \approx x$$ Simplify.

The measure of the side opposite the acute angle is about 12.3 feet.

Example 4 Use Trigonometric Ratios to Solve a Problem

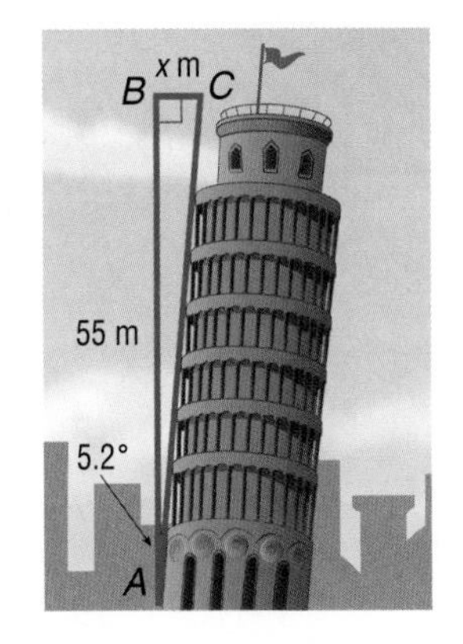

ARCHITECTURE The Leaning Tower of Pisa in Pisa, Italy, tilts about 5.2° from vertical. If the tower is 55 meters tall, how far has its top shifted from its original position?

Use the tangent ratio.

$\tan \angle A = \dfrac{\text{measure of leg opposite } \angle A}{\text{measure of leg adjacent to } \angle A}$ Write the tangent ratio.

$\tan 5.2° = \dfrac{x}{55}$ Substitution

$55(\tan 5.2°) = 55 \cdot \dfrac{x}{55}$ Multiply each side by 55.

55 [×] [TAN] 5.2 [ENTER] 5.00539211

$5.0 \approx x$ Simplify.

The top of the tower has shifted about 5.0 meters from its original position.

Check for Understanding

Concept Check

1. **OPEN ENDED** Compare and contrast the sine, cosine, and tangent ratios.
2. **FIND THE ERROR** Susan and Tadeo are finding the height of the hill.

Susan	Tadeo
$\sin 15° = \dfrac{x}{580}$	$\sin 15° = \dfrac{x}{560}$
$580(\sin 15°) = x$	$560(\sin 15°) = x$
$150 \approx x$	$145 \approx x$

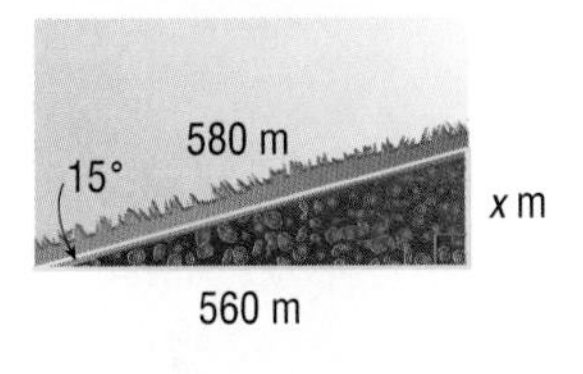

Who is correct? Explain your reasoning.

Guided Practice

Find each sine, cosine, or tangent. Round to four decimal places, if necessary.

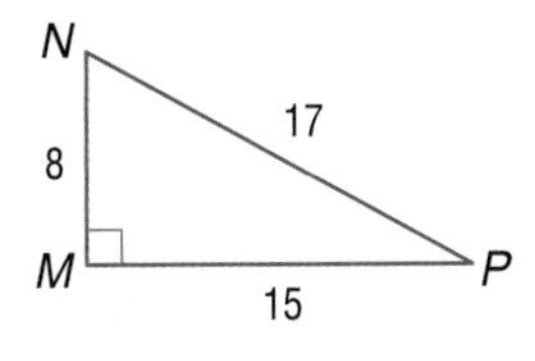

3. $\sin N$
4. $\cos N$
5. $\tan N$

Use a calculator to find each value to the nearest ten thousandth.

6. $\sin 52°$ 7. $\cos 19°$ 8. $\tan 76°$

For each triangle, find each missing measure to the nearest tenth.

9.
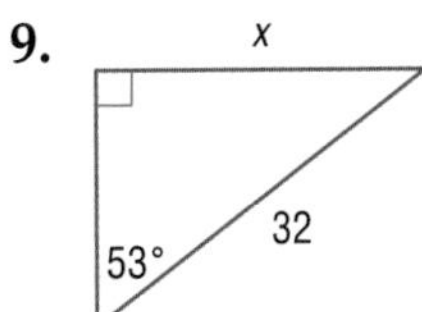

10. 21°, 73, x

11. x, 33°, 12

Application

12. **RECREATION** Miranda is flying a kite on a 50-yard string, which makes a 50° angle with the ground. How high above the ground is the kite?

Practice and Apply

Homework Help

For Exercises	See Examples
13–20	1
21–26	2
27–32	3
33–36	4

Extra Practice

See page 747.

Find each sine, cosine, or tangent. Round to four decimal places, if necessary.

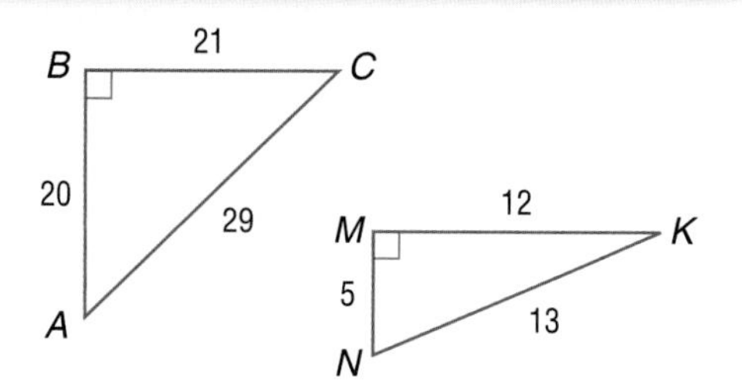

13. $\sin N$
14. $\sin K$
15. $\cos C$
16. $\cos A$
17. $\tan N$
18. $\tan C$

19. Triangle RST is shown. Find $\sin R$, $\cos R$, and $\tan R$.

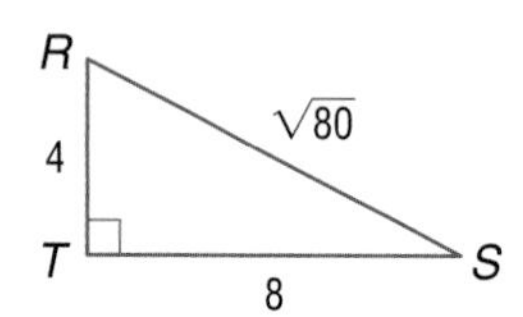

20. For $\triangle XYZ$, what is the value of $\sin X$, $\cos X$, and $\tan X$?

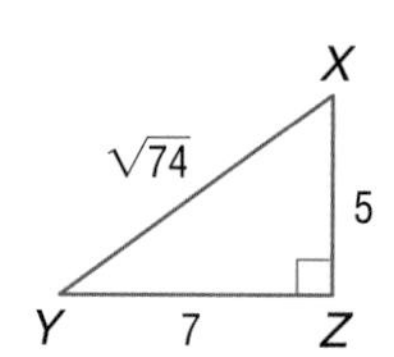

Use a calculator to find each value to the nearest ten thousandth.

21. $\sin 6°$
22. $\sin 51°$
23. $\cos 31°$
24. $\cos 87°$
25. $\tan 12°$
26. $\tan 66°$

For each triangle, find each missing measure to the nearest tenth.

27.

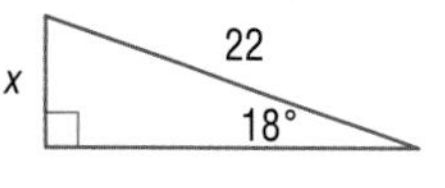

28.

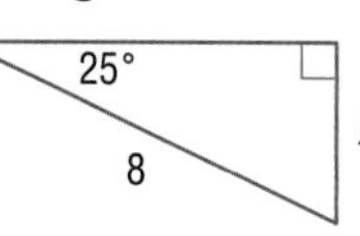

29.

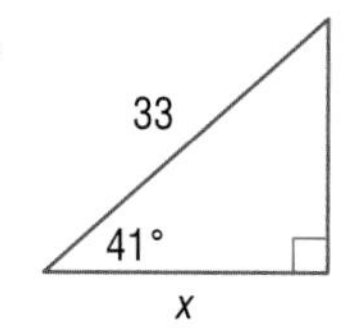

30.

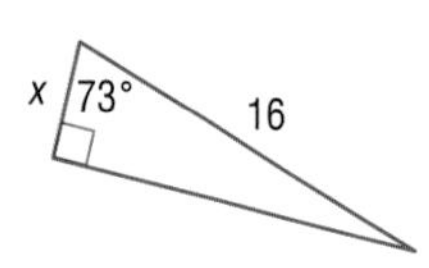

31.

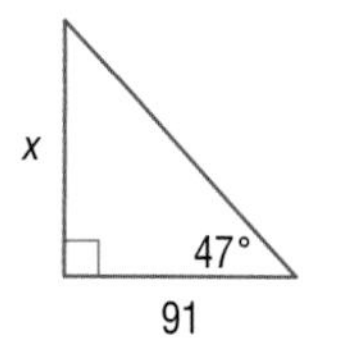

32.

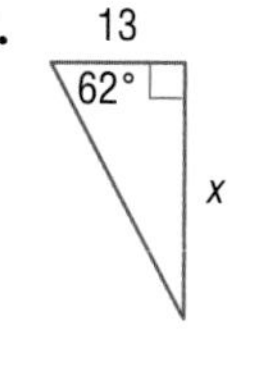

Career Choices

Civil Engineer

Civil engineers usually design and supervise the construction of roads, buildings, bridges, airports, dams, and water supply and sewage systems.

Online Research
For information about a career as a civil engineer, visit: www.pre-alg.com/careers

33. **SHADOWS** An *angle of elevation* is formed by a horizontal line and a line of sight above it. A flagpole casts a shadow 25 meters long when the angle of elevation of the Sun is 40°. How tall is the flagpole?

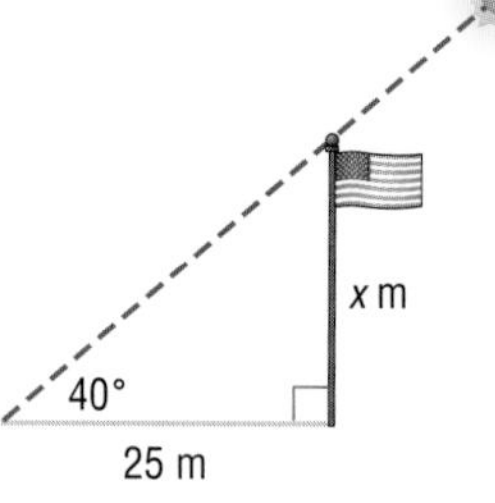

34. **SAFETY** The angle that a wheelchair ramp forms with the ground should not exceed 6°. What is the height of the ramp if it is 20 feet long? Round to the nearest tenth.

35. **BUILDINGS** Refer to the diagram shown. What is the height of the building?

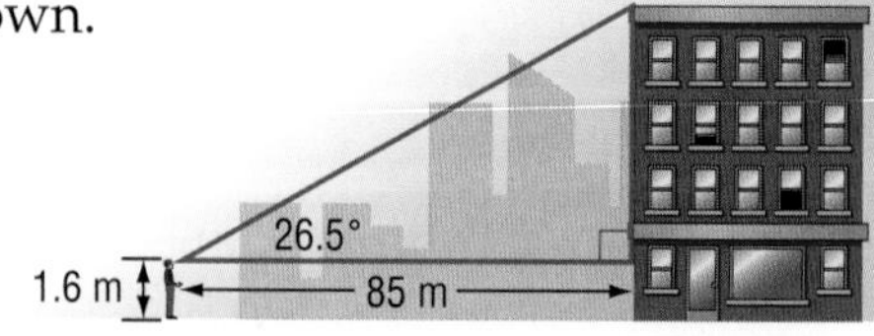

36. **CIVIL ENGINEERING** An exit ramp makes a 17° angle with the highway. Suppose the length of the ramp is 210 yards. What is the length of the base of the ramp? Round to the nearest tenth.

Find each measure. Round to the nearest tenth.

37. 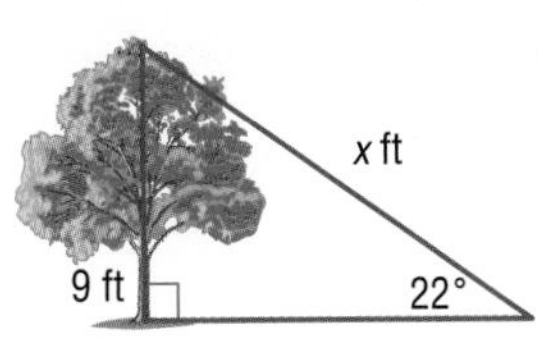

38. 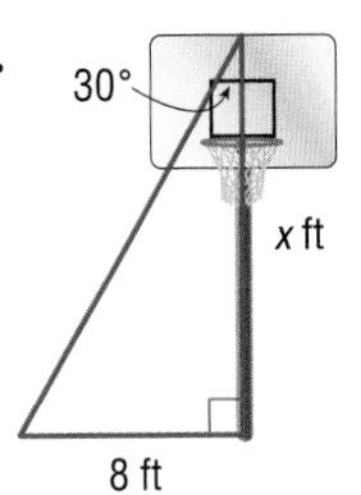

39. 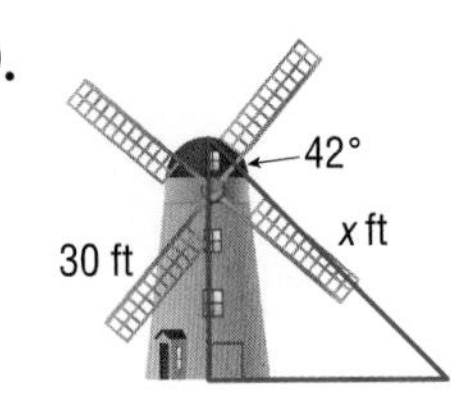

Trigonometric ratios are used to solve problems about the height of structures. Visit www.pre-alg.com/webquest to continue work on your WebQuest project.

CRITICAL THINKING **For Exercises 40–42, refer to the 45°-45°-90° and 30°-60°-90° triangles shown.**

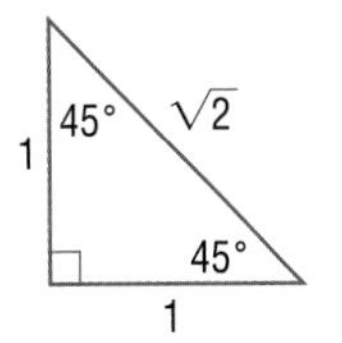

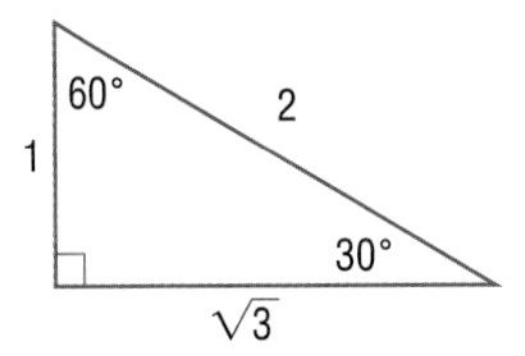

40. Complete the table shown. Round to four decimal places, if necessary.

x	30°	45°	60°
sin x			
cos x			
tan x			

41. Write a few sentences describing the similarities among the resulting sine and cosine values.

42. Which angle measure has the same sine and cosine ratio?

43. WRITING IN MATH Answer the question that was posed at the beginning of the lesson.

How are ratios in right triangles used in the real world?

Include the following in your answer:

- an explanation of two different methods for finding the measures of the lengths of the sides of a right triangle, and
- a drawing of a real-world situation that involves the sine ratio, with the calculations and solution included.

FCAT Practice
Standardized Test Practice

44. If the measure of the hypotenuse of a right triangle is 5 feet and $m\angle B = 58°$, what is the measure of the leg adjacent to $\angle B$?

(A) 4.2402 (B) 8.0017 (C) 0.1060 (D) 2.6496

45. Find the value of tan P to the nearest tenth.

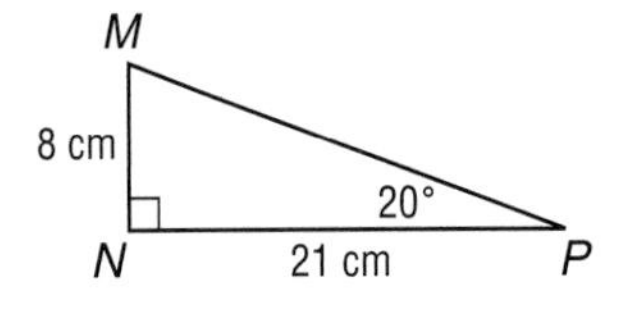

(A) 2.6 (B) 0.5

(C) 0.4 (D) 0.1

Maintain Your Skills

Mixed Review

46. At the same time a 40-foot silo casts a 22-foot shadow, a fence casts a 3.3-foot shadow. Find the height of the fence. *(Lesson 9-7)*

The coordinates of the endpoints of a segment are given. Find the coordinates of the midpoint of each segment. *(Lesson 9-6)*

47. $A(3, -4), C(0, 5)$ **48.** $M(-2, 1), N(6, -9)$ **49.** $R(-3, 0), S(-3, -8)$

ALGEBRA **Solve each inequality.** *(Lesson 7-5)*

50. $5m < 5$ **51.** $\frac{a}{-2} > 3$ **52.** $-4x \geq -16$

Graphing Calculator Investigation

A Follow-Up of Lesson 9-8

Finding Angles of a Right Triangle

A calculator can be used to find the measure of an acute angle of a right triangle if you know the measures of two sides of the triangle.

Example

The end of an exit ramp from an interstate highway is 22 feet higher than the highway. If the ramp is 630 feet long, what angle does it make with the highway?

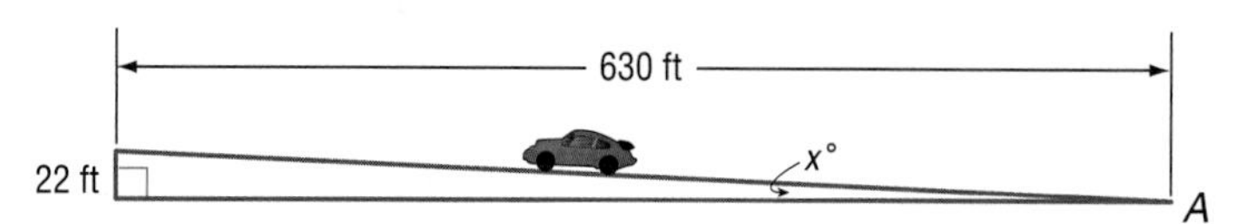

Step 1 Determine which trigonometric ratio is needed to solve the problem. Since you know the measure of the leg opposite $\angle A$ and the hypotenuse, use the sine ratio.

Step 2 Write the ratio.

$\sin \angle A = \frac{\text{opposite}}{\text{hypotenuse}}$ Sine Ratio

$\sin \angle A = \frac{22}{630}$ Substitution

Step 3 Use a calculator to find the measure of $\angle A$. The SIN^{-1} function will find the angle measure, given the value of its sine.

2nd [SIN^{-1}] 22 ÷ 630 ENTER 2.001211869

To the nearest degree, the measure of $\angle A$ is 2°.

Exercises

Use a calculator to find the measure of each acute angle. Round to the nearest degree.

1.

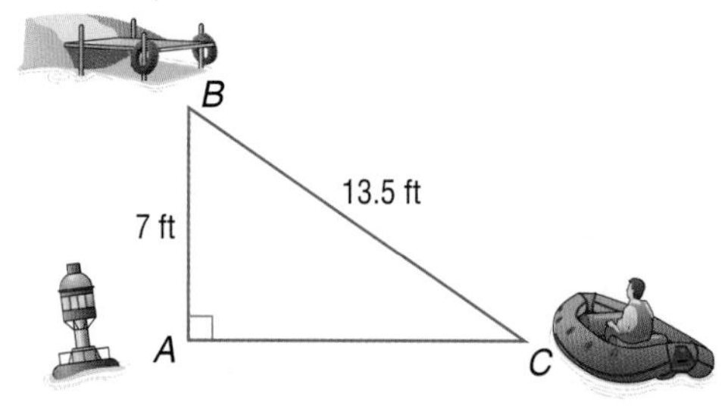

2.

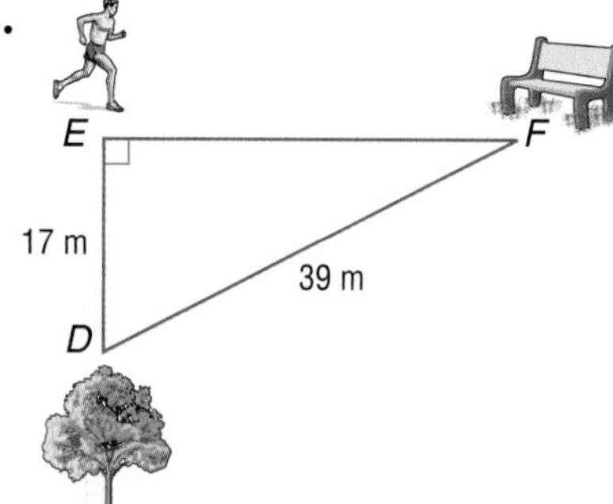

3. A flower garden is located 46 meters due west of an elm tree. A fountain is located 19 meters due south of the same elm tree. What are the measures of the angles formed by these three park features?

 www.pre-alg.com/other_calculator_keystrokes

Chapter 9

Study Guide and Review

Vocabulary and Concept Check

acute angle (p. 449)
acute triangle (p. 454)
adjacent (p. 476)
angle (p. 447)
congruent (p. 455)
converse (p. 462)
cosine (p. 477)
degree (p. 448)
Distance Formula (p. 466)
equilateral triangle (p. 455)
hypotenuse (p. 460)
indirect measurement (p. 472)
irrational numbers (p. 441)
isosceles triangle (p. 455)
legs (p. 460)
line (p. 447)
line segment (p. 453)
midpoint (p. 468)
Midpoint Formula (p. 468)
obtuse angle (p. 449)
obtuse triangle (p. 454)
opposite (p. 476)
perfect square (p. 436)
point (p. 447)
protractor (p. 448)
Pythagorean Theorem (p. 460)
radical sign (p. 436)
ray (p. 447)
real numbers (p. 441)
right angle (p. 449)
right triangle (p. 454)
scalene triangle (p. 455)
sides (p. 447)
similar triangles (p. 471)
sine (p. 477)
solving a right triangle (p. 461)
square root (p. 436)
straight angle (p. 449)
tangent (p. 477)
triangle (p. 453)
trigonometric ratio (p. 477)
trigonometry (p. 477)
vertex (pp. 447, 453)

Complete each sentence with the correct term. Choose from the list above.

1. The set of rational numbers and the set of irrational numbers make up the set of __?__.
2. A(n) __?__ measures between 0° and 90°.
3. A(n) __?__ triangle has one angle with a measurement greater than 90°.
4. A(n) __?__ has all sides congruent.
5. In a right triangle, the side opposite the right angle is the __?__.
6. Triangles that have the same shape but not necessarily the same size are called __?__.
7. A(n) __?__ is a ratio of the lengths of two sides of a right triangle.
8. A(n) __?__ is a part of a line that extends indefinitely in one direction.

Lesson-by-Lesson Review

9-1 Squares and Square Roots

See pages 436–440.

Concept Summary

- The square root of a number is one of two equal factors of the number.

Example **Find the square root of $-\sqrt{49}$.**

$-\sqrt{49}$ indicates the *negative* square root of 49.
Since $7^2 = 49$, $-\sqrt{49}, = -7$.

Exercises **Find each square root, if possible.** *See Example 1 on page 436.*

9. $\sqrt{36}$
10. $\sqrt{100}$
11. $-\sqrt{81}$
12. $\pm\sqrt{121}$
13. $\sqrt{-25}$
14. $-\sqrt{225}$

9-2 The Real Number System

See pages 441–445.

Concept Summary

- Numbers that cannot be written as terminating or repeating decimals are called irrational numbers.
- The set of rational numbers and the set of irrational numbers together make up the set of real numbers.

Example **Solve $x^2 = 72$. Round to the nearest tenth.**

$x^2 = 72$ Write the equation.

$\sqrt{x^2} = \sqrt{72}$ Take the square root of each side.

$x = \sqrt{72}$ or $x = -\sqrt{72}$ Find the positive and negative square root.

$x \approx 8.5$ or $x \approx -8.5$

Exercises **Solve each equation. Round to the nearest tenth, if necessary.**
See Example 3 on page 443.

15. $n^2 = 81$ **16.** $t^2 = 38$ **17.** $y^2 = 1.44$ **18.** $7.5 = r^2$

9-3 Angles

See pages 447–451.

Concept Summary

- An acute angle has a measure between 0° and 90°.
- A right angle measures 90°.
- An obtuse angle has a measure between 90° and 180°.
- A straight angle measures 180°.

Example **Use a protractor to find the measure of $\angle ABC$. Then classify the angle as *acute*, *obtuse*, *right*, or *straight*.**

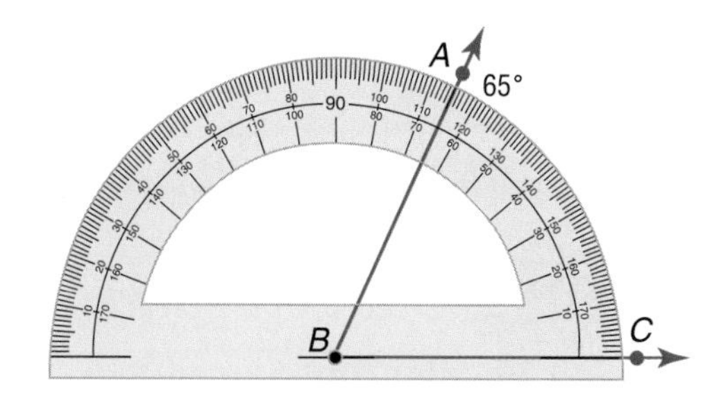

$\angle ABC$ appears to be acute. So, its measure should be between 0° and 90°.

$m\angle ABC = 65°$

Since $m\angle ABC < 90°$, $\angle ABC$ is acute.

Exercises **Use a protractor to find the measure of each angle. Then classify each angle as *acute, obtuse, right,* or *straight.***
See Examples 1 and 3 on pages 448 and 449.

19. **20.** **21.**

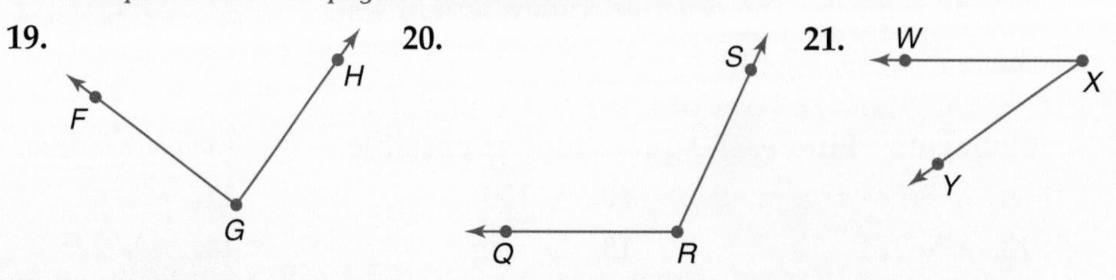

9-4 Triangles

See pages 453–457.

Concept Summary

- Triangles can be classified by their angles as acute, obtuse, or right and by their sides as scalene, isosceles, or equilateral.

Example **Classify the triangle by its angles and by its sides.**

$\triangle HJK$ has all acute angles and two congruent sides. So, $\triangle HJK$ is an acute isosceles triangle.

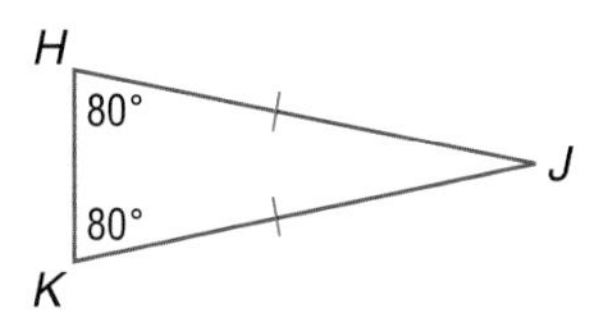

Exercises **Classify each triangle by its angles and by its sides.**
See Example 3 on page 455.

22.
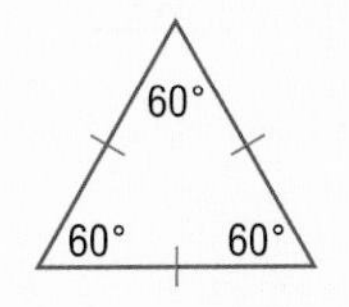

23.
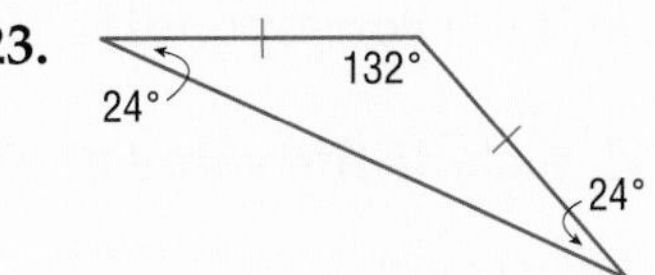

24.
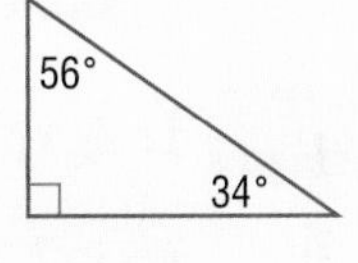

9-5 The Pythagorean Theorem

See pages 460–464.

Concept Summary

- Pythagorean Theorem: $c^2 = a^2 + b^2$

Example **Find the missing measure of the right triangle.**

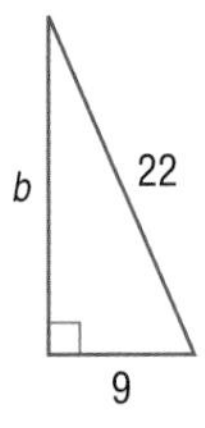

$c^2 = a^2 + b^2$	Pythagorean Theorem
$22^2 = 9^2 + b^2$	Replace c with 22 and a with 9.
$403 = b^2$	Simplify; subtract 81 from each side.
$20.1 \approx b$	Take the square root of each side.

Exercises **If c is the measure of the hypotenuse, find each missing measure. Round to the nearest tenth, if necessary.** *See Example 2 on page 461.*

25. $a = 6, b = ?, c = 15$ **26.** $a = ?, b = 2, c = 7$ **27.** $a = 18, b = ?, c = 24$

9-6 The Distance and Midpoint Formulas

See pages 466–470.

Concept Summary

- Distance Formula: $d = \sqrt{(x_2 - x_1)^2 + (y_2 - y_1)^2}$
- Midpoint Formula: $\left(\frac{x_1 + x_2}{2}, \frac{y_1 + y_2}{2}\right)$

Example **Find the distance between $A(-4, 0)$ and $B(2, 5)$.**

$d = \sqrt{(x_2 - x_1)^2 + (y_2 - y_1)^2}$	Distance Formula
$d = \sqrt{[2 - (-4)]^2 + (5 - 0)^2}$	$(x_1, y_1) = (-4, 0), (x_2, y_2) = (2, 5)$.

$d \approx 7.8$ The distance between points A and B is about 7.8 units.

• Extra Practice, see pages 745–747.
• Mixed Problem Solving, see page 766.

Exercises **Find the distance between each pair of points. Round to the nearest tenth, if necessary.** *See Example 1 on page 467.*

28. $J(0, 9)$, $K(2, 7)$ **29.** $A(-5, 1)$, $B(3, 6)$ **30.** $W(8, -4)$, $Y(3, 3)$

The coordinates of the endpoints of a segment are given. Find the coordinates of the midpoint of each segment. *See Example 3 on page 468.*

31. $M(8, 0)$, $N(-2, 10)$ **32.** $C(5, 9)$, $D(-7, 3)$ **33.** $Q(-6, 4)$, $R(6, -8)$

9-7 Similar Triangles and Indirect Measurement

See pages 471–475.

Concept Summary

- If two triangles are similar, then the corresponding angles have the same measure, and the corresponding sides are proportional.

Example **If $\triangle ABC \sim \triangle KLM$, what is the value of x?**

$\frac{AC}{KM} = \frac{BC}{LM}$ Write a proportion.

$\frac{x}{3} = \frac{2}{4}$ Substitution

$x = 1.5$ Find cross products and simplify.

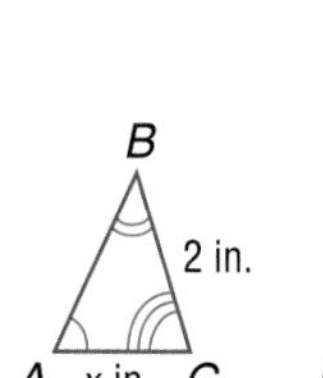

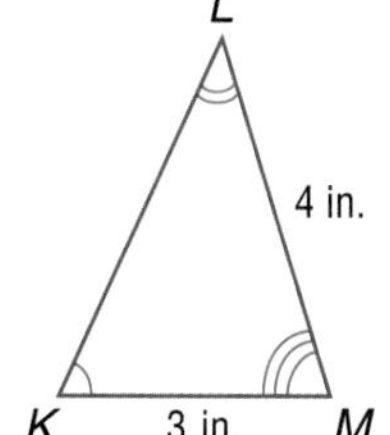

Exercises **In Exercises 34 and 35, the triangles are similar. Write a proportion to find each missing measure. Then find the value of x.** *See Example 1 on page 472.*

34.

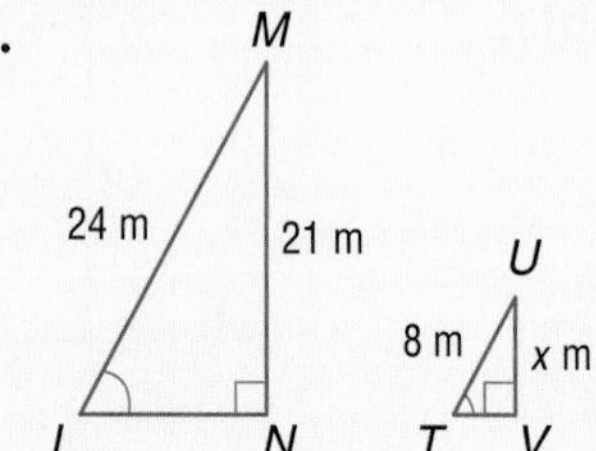

35.

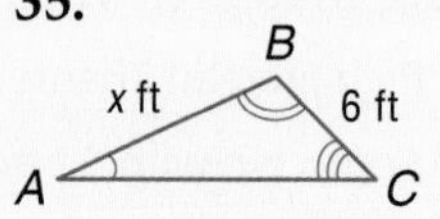

9-8 Sine, Cosine, and Tangent Ratios

See pages 477–481.

Concept Summary

- Trigonometric ratios compare the lengths of two sides of a right triangle.

Example **Find cos S.**

$\cos S = \frac{15}{25}$ or 0.6

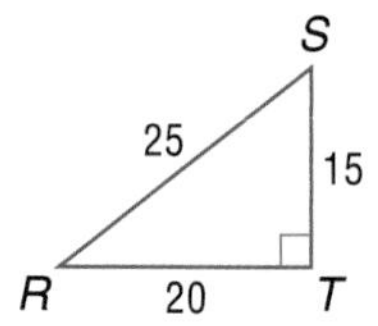

Exercises **Find each sine, cosine, or tangent in $\triangle RST$ above. Round to four decimal places, if necessary.** *See Example 1 on page 478.*

36. $\sin R$ **37.** $\tan S$ **38.** $\tan R$

Chapter 9 Practice Test

Vocabulary and Concepts

1. **OPEN ENDED** Give an example of a whole number, a natural number, an irrational number, a rational number, and an integer.
2. Explain how to classify a triangle by its angles and by its sides.

Skills and Applications

Find each square root, if possible.

3. $\sqrt{81}$
4. $-\sqrt{121}$
5. $\pm\sqrt{49}$
6. Without using a calculator, estimate $-\sqrt{42}$ to the nearest integer.

ALGEBRA Solve each equation. Round to the nearest tenth, if necessary.

7. $x^2 = 100$
8. $w^2 = 38$
9. Use a protractor to measure $\angle CAB$. Then classify the angle as *acute*, *obtuse*, *right*, or *straight*.

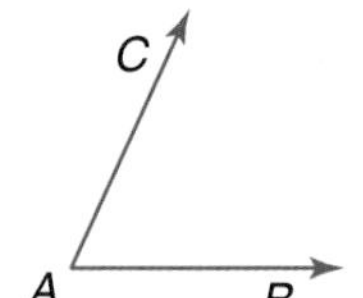

For Exercises 10 and 11, use the following information.
In $\triangle MNP$, $m\angle N = 87°$ and $m\angle P = 32°$.

10. Find the measure of $\angle M$.
11. Classify $\triangle MNP$ by its angles and by its sides.

If c is the measure of the hypotenuse, find each missing measure. Round to the nearest tenth, if necessary.

12. $a = 6, b = 8, c = ?$
13. $a = 15, b = ?, c = 32$
14. **HIKING** Brandon hikes 7 miles south and 4 miles west. How far is he from the starting point of his hike? Round to the nearest tenth.
15. Find the distance between $A(3, 8)$ and $B(-5, 2)$. Then find the coordinates of the midpoint of $\overline{AB}$.
16. **PARKS** In the map of the park, the triangles are similar. Find the distance to the nearest tenth from the playground to the swimming pool.

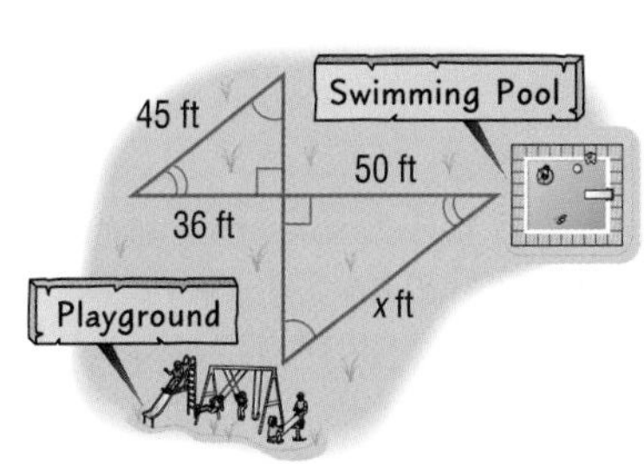

Find each sine, cosine, or tangent. Round to four decimal places, if necessary.

17. $\sin P$
18. $\tan M$
19. $\cos P$

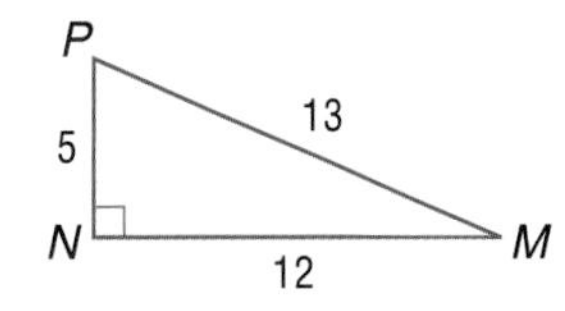

FCAT Practice

20. **STANDARDIZED TEST PRACTICE** Which statement is true?

Ⓐ $4 < \sqrt{10} < 3$
Ⓑ $-6 > -\sqrt{28} > -5$
Ⓒ $-7 > -\sqrt{59} > -8$
Ⓓ $7 < \sqrt{47} < 6$

Chapter 9 Standardized Test Practice

FCAT Practice

Part 1 Multiple Choice

Record your answers on the answer sheet provided by your teacher or on a sheet of paper.

1. When m and n are any two numbers, which of the following statements is true? (Lesson 1-4)

Ⓐ $m \cdot 0 = n$
Ⓑ $m \cdot n = n$
Ⓒ $n \cdot 1 = m$
Ⓓ $m + n = n + m$

2. How many units apart are the numbers -8 and 5 on a number line? (Lesson 2-1)

Ⓐ 13
Ⓑ 3
Ⓒ 14
Ⓓ 10

3. Find the length of a rectangle having a width of 9 feet and an area of 54 square feet. (Lesson 3-7)

Ⓐ 9 ft
Ⓑ 6 ft
Ⓒ 10 ft
Ⓓ 18 ft

4. Which of the following results in a negative number? (Lessons 4-2 and 4-7)

Ⓐ $(-2)^5$
Ⓑ $(5)^{-3}$
Ⓒ $-5 \cdot (-2)^5$
Ⓓ $(-3)^{-2} \cdot 5$

5. What is the value of x if $\frac{1}{2} + x = \frac{5}{6}$? (Lesson 5-7)

Ⓐ $\frac{1}{4}$
Ⓑ $\frac{2}{5}$
Ⓒ $\frac{2}{3}$
Ⓓ $\frac{1}{3}$

6. Elisa purchased two books that cost \$15.95 and \$6.95. The sales tax on the books was 6%. If she gave the sales clerk \$25, then how much change did she receive? (Lesson 6-9)

Ⓐ \$0.27
Ⓑ \$0.73
Ⓒ \$1.73
Ⓓ \$2.10

7. What is the value of t in $2s - t = s + 3t$ if $s = \frac{1}{2}$? (Lesson 7-1)

Ⓐ $\frac{1}{8}$
Ⓑ $\frac{1}{4}$
Ⓒ $\frac{1}{2}$
Ⓓ $\frac{3}{8}$

The Princeton Review Test-Taking Tip

Question 9 You can tell if a line has a negative or positive slope by looking at its graph. A line with a negative slope will slant down from left to right. A line with a positive slope will slant up from left to right.

8. The relation shown in the table is a function. Find the value of y when $x = 5$. (Lesson 8-2)

x	y
0	2
1	5
2	8
3	11

Ⓐ 13
Ⓑ 14
Ⓒ 17
Ⓓ 18

9. Which graph shows a line with a slope of –2? (Lesson 8-6)

Ⓐ
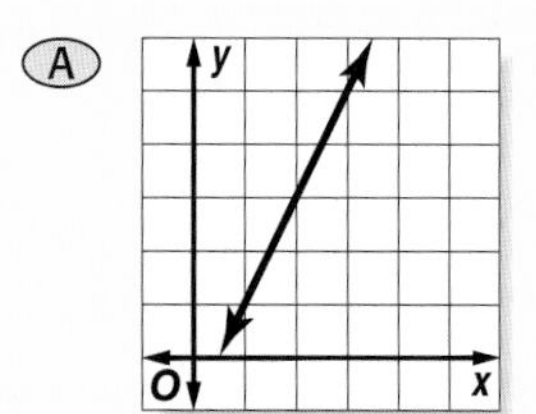

Ⓑ
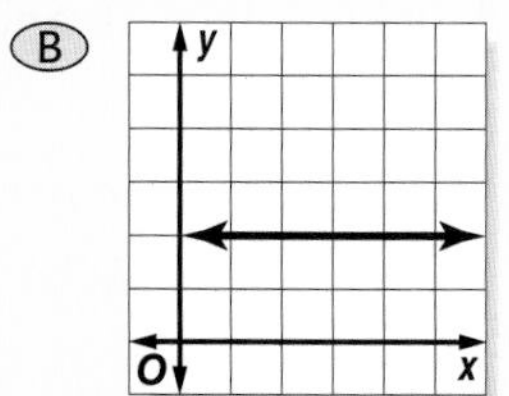

Ⓒ
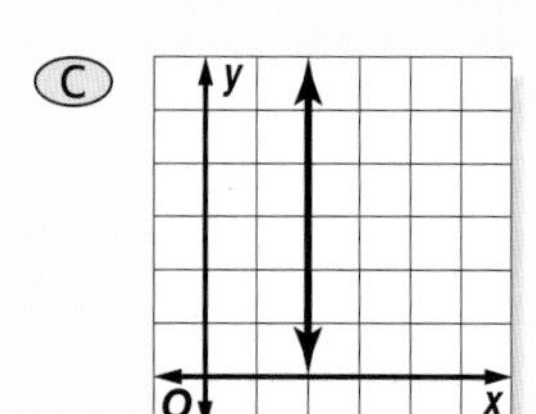

Ⓓ
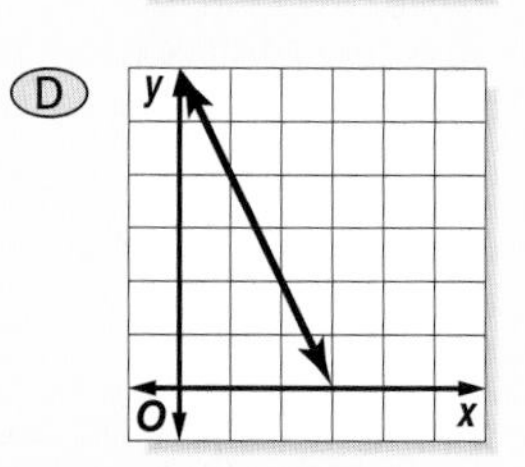

10. Which point on the number line shown is closest to $\sqrt{7}$? (Lesson 9-1)

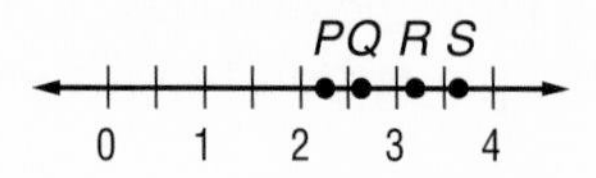

Ⓐ point P
Ⓑ point Q
Ⓒ point R
Ⓓ point S

11. Triangles ABC and DEF are similar triangles. What is the measure of side AB? (Lesson 9-5)

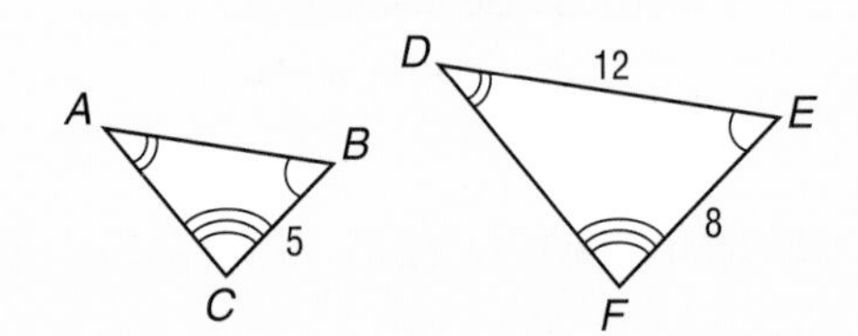

Ⓐ 3.33
Ⓑ 5.0
Ⓒ 6.5
Ⓓ 7.5

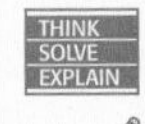

Part 2 Short Response/Grid In

Record your answers on the answer sheet provided by your teacher or on a sheet of paper.

12. When 8 is added to a number three times, the result is 27. Find the number. (Lesson 3-3)

13. Order the numbers 1.4×10^{-5}, 4.0×10^{2}, and 1.04×10^{-2} from least to greatest. (Lesson 4-8)

14. A conveyor belt moves at a rate of 6 miles in 4 hours. How many feet per minute does this conveyor belt move? (*Hint:* 1 mile = 5280 feet) (Lesson 6-1)

15. Write $\frac{11}{14}$ as a percent, and round to the nearest tenth. (Lesson 6-4)

16. If you spin the arrow on the spinner below, what is the probability that the arrow will land on an even number? (Lesson 6-9)

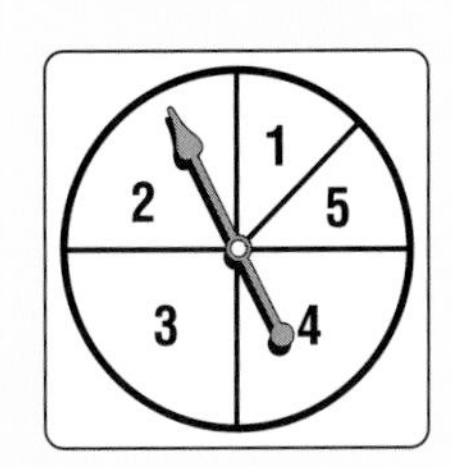

17. What is the least value of x in $y \leq 4x - 3$ if $y = 9$? (Lesson 7-6)

18. Write *y is less than one-half the value of x* as a mathematical statement. (Lesson 7-7)

19. What is the slope of the line shown? (Lesson 8-4)

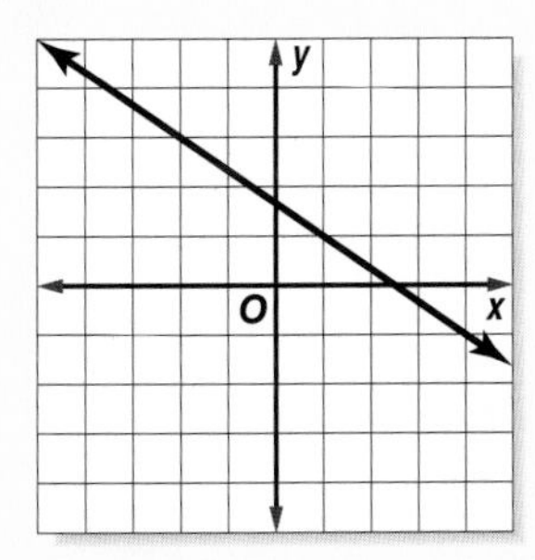

20. Find the positive square root of 196. (Lesson 9-1)

21. What type of angle is formed by the minute hand and hour hand when the time on the clock is 3:35? (Lesson 9-3)

22. What is the measure of $\angle X$? (Lesson 9-4)

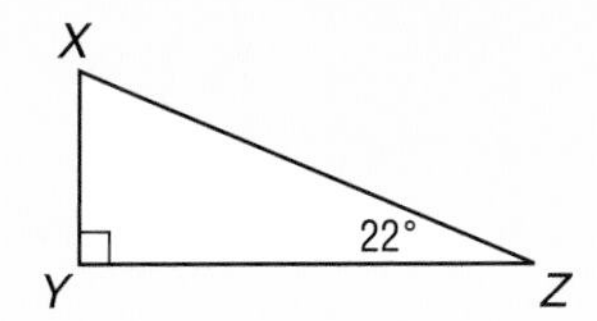

23. Find the distance between $A(-5, 4)$ and $B(6, -3)$ to the nearest tenth. (Lesson 9-6)

Part 3 Open Ended

Record your answers on a sheet of paper. Show your work.

24. The walls of a house usually meet to form a right angle. You can use string to determine whether two walls meet at a right angle.

a. Copy the diagram shown below. Then illustrate the following situation.

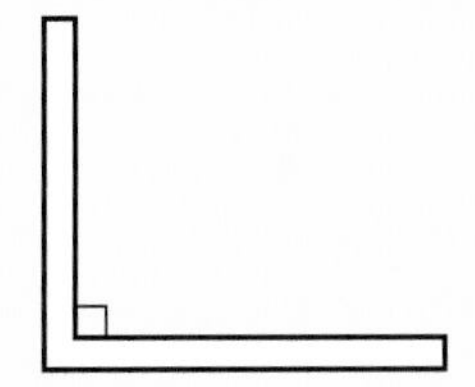

From a corner of the house, a 6-foot long piece of string is extended along one side of the wall, parallel to the floor. From the same corner, an 8-foot long piece of string is extended along the other wall, parallel to the floor.

b. If the walls of the house meet at a right angle, then what is the distance between the ends of the two pieces of string?

c. Draw an example of a situation where two walls of a house meet at an angle whose measure is greater than the measure of a right angle.

d. Suppose the length of the walls in part c are the same length as the walls in part a, and that the 6-foot and 8-foot pieces of string are extended from the same corner. Do you think the distance between the two ends will be the same as in part **b**? Explain your reasoning.

Chapter 10

Two-Dimensional Figures

What You'll Learn

- **Lesson 10-1** Identify the relationships of parallel and intersecting lines.
- **Lesson 10-2** Identify properties of congruent triangles.
- **Lesson 10-3** Identify and draw transformations.
- **Lessons 10-4 and 10-6** Classify and find angle measures of polygons.
- **Lessons 10-5, 10-7, and 10-8** Find the area of polygons and irregular figures, and find the area and circumference of circles.

Key Vocabulary

- **parallel lines** (p. 492)
- **transformation** (p. 506)
- **quadrilateral** (p. 513)
- **polygon** (p. 527)
- **circumference** (p. 533)

Why It's Important

Two-dimensional shapes are usually found in many real-world objects. The properties of two-dimensional figures can be used to solve real-world problems dealing with snowflakes. For example, if you know the exact shape of a snowflake, you can find the measure of an interior angle in a snowflake. *You will solve a problem about snowflakes in Lesson 10-6.*

Getting Started

Prerequisite Skills To be successful in this chapter, you'll need to master these skills and be able to apply them in problem-solving situations. Review these skills before beginning Chapter 10.

For Lessons 10-1 and 10-4 — Solve Equations

Solve each equation. *(For review, see Lessons 3-3 and 3-5.)*

1. $x + 46 = 90$ **2.** $x + 35 = 180$ **3.** $2x - 12 = 90$ **4.** $3x - 24 = 180$

5. $2x + 34 = 90$ **6.** $4x + 44 = 180$ **7.** $5x + 165 = 360$ **8.** $4x + 184 = 360$

For Lessons 10-5, 10-7, and 10-8 — Multiply Decimals

Find each product. Round to the nearest tenth, if necessary. *(For review, see page 715.)*

9. $(5.5)(8)$ **10.** $(7.5)(3.4)$ **11.** $(6.3)(11.4)$ **12.** $\frac{1}{2}(8)(2.5)$

13. $\frac{1}{2}(4.3)(5.8)$ **14.** $(3.14)(7)$ **15.** $(2)(3.14)(1.7)$ **16.** $2(3.1)(3.14)$

For Lesson 10-5 — Add Mixed Numbers

Find each sum. *(For review, see Lesson 5-7.)*

17. $5\frac{1}{2} + 4\frac{2}{3}$ **18.** $2\frac{1}{3} + 3\frac{3}{4}$ **19.** $1\frac{3}{8} + 2\frac{1}{2}$ **20.** $6\frac{1}{4} + 1\frac{5}{6}$

21. $3\frac{5}{8} + 1\frac{3}{4}$ **22.** $2\frac{3}{5} + 4\frac{7}{10}$ **23.** $2\frac{2}{3} + 3\frac{5}{9}$ **24.** $5\frac{2}{3} + 3\frac{4}{5}$

FOLDABLES™ Study Organizer

Make this Foldable to help you organize information about the characteristics of two-dimensional figures. Begin with four plain sheets of $8\frac{1}{2}$" by 11" paper, eight index cards, and glue.

Step 1 Fold

Step 2 Open and Fold Again

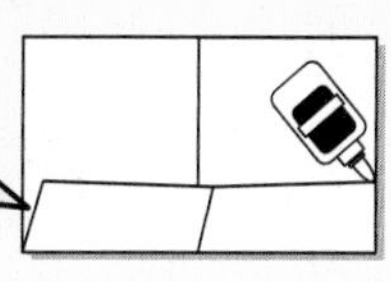

Fold the bottom to form a pocket. Glue edges.

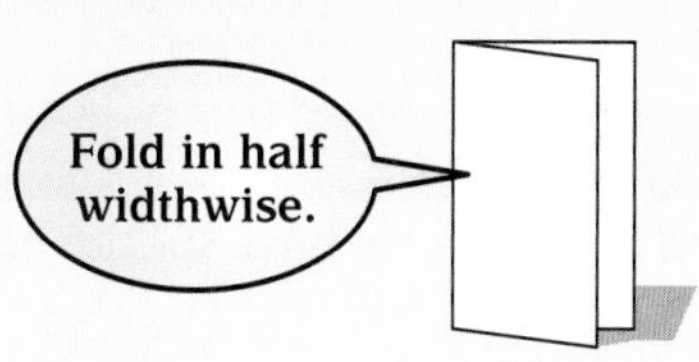

Step 3 Repeat Steps 1 and 2

Repeat three times. Then glue all four pieces together to form a booklet.

Step 4 Label

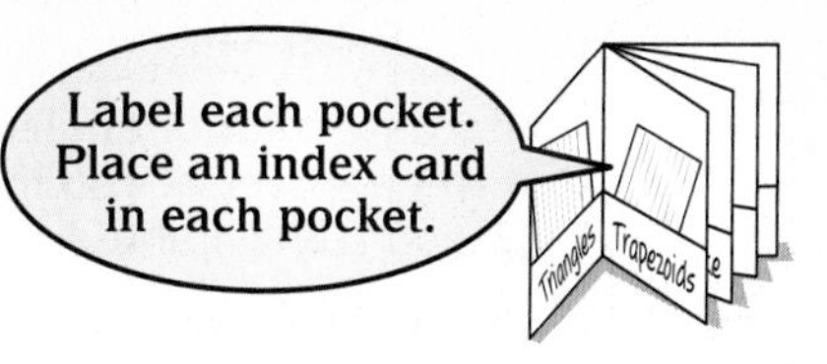

Label each pocket. Place an index card in each pocket.
Triangles
Trapezoids

Reading and Writing As you read and study the chapter, write the name of a two-dimensional figure on each index card, draw a diagram, and write a definition or describe the characteristics of each figure.

10-1 Line and Angle Relationships

What You'll Learn

- Identify the relationships of angles formed by two parallel lines and a transversal.
- Identify the relationships of vertical, adjacent, complementary, and supplementary angles.

Sunshine State Standards
MA.B.1.3.2-1, MA.C.2.3.1-1, MA.C.3.3.2-3

Vocabulary

- parallel lines
- transversal
- interior angles
- exterior angles
- alternate interior angles
- alternate exterior angles
- corresponding angles
- vertical angles
- adjacent angles
- complementary angles
- supplementary angles
- perpendicular lines

How are parallel lines and angles related?

Let's investigate what happens when two horizontal lines are intersected by a third line.

a. Trace two of the horizontal lines on a sheet of notebook paper. Then draw another line that intersects the horizontal lines.

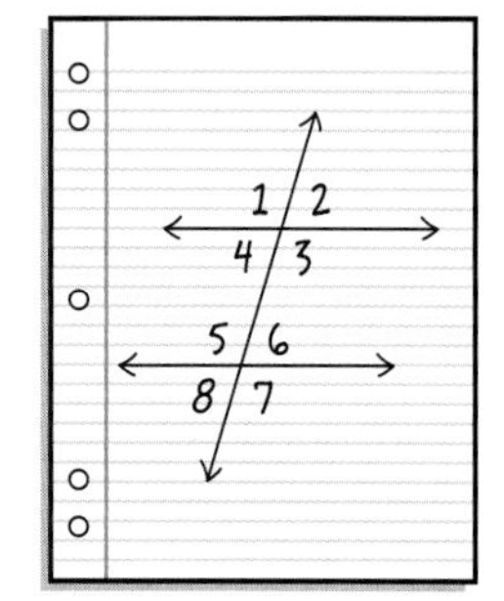

b. Label the angles as shown.

c. Find the measure of each angle.

d. What do you notice about the measures of the angles?

e. Which angles have the same measure?

f. What do you notice about the measures of the angles that share a side?

PARALLEL LINES AND A TRANSVERSAL In geometry, two lines in a plane that never intersect are **parallel lines**.

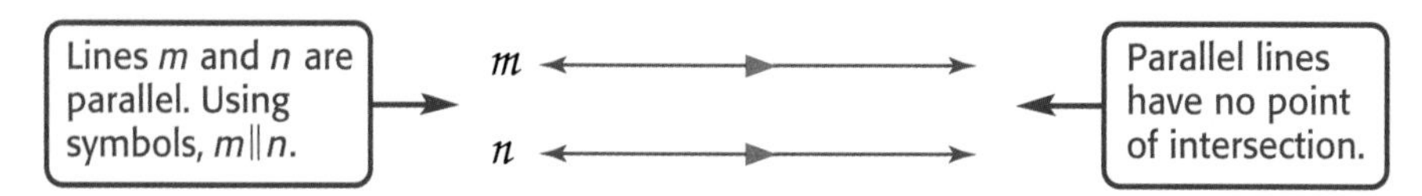

Study Tip

Parallel Lines
Arrowheads are often used in figures to indicate parallel lines.

When two parallel lines are intersected by a third line called a **transversal**, eight angles are formed.

Key Concept — *Names of Special Angles*

The eight angles formed by parallel lines and a transversal have special names.

- **Interior angles** lie inside the parallel lines.
  $\angle 3$, $\angle 4$, $\angle 5$, $\angle 6$
- **Exterior angles** lie outside the parallel lines.
  $\angle 1$, $\angle 2$, $\angle 7$, $\angle 8$
- **Alternate interior angles** are on opposite sides of the transversal and inside the parallel lines.
  $\angle 3$ and $\angle 5$, $\angle 4$ and $\angle 6$
- **Alternate exterior angles** are on opposite sides of the transversal and outside the parallel lines.
  $\angle 1$ and $\angle 7$, $\angle 2$ and $\angle 8$
- **Corresponding angles** are in the same position on the parallel lines in relation to the transversal.
  $\angle 1$ and $\angle 5$, $\angle 2$ and $\angle 6$, $\angle 3$ and $\angle 7$, $\angle 4$ and $\angle 8$

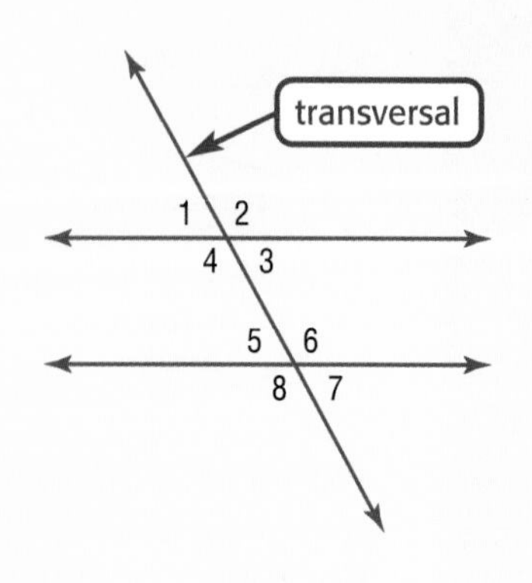

In Lesson 9-4, you learned that line segments are congruent if they have the same measure. Similarly, angles are congruent if they have the same measure.

Key Concept *Parallel Lines Cut by a Transversal*

If two parallel lines are cut by a transversal, then the following pairs of angles are congruent.

- Corresponding angles are congruent.
- Alternate interior angles are congruent.
- Alternate exterior angles are congruent.

Concept Check How many angles are formed when two parallel lines are intersected by a transversal?

Example 1 Find Measures of Angles

In the figure at the right, $m \parallel n$ and t is a transversal. If $m\angle 1 = 68°$, find $m\angle 5$ and $m\angle 6$.

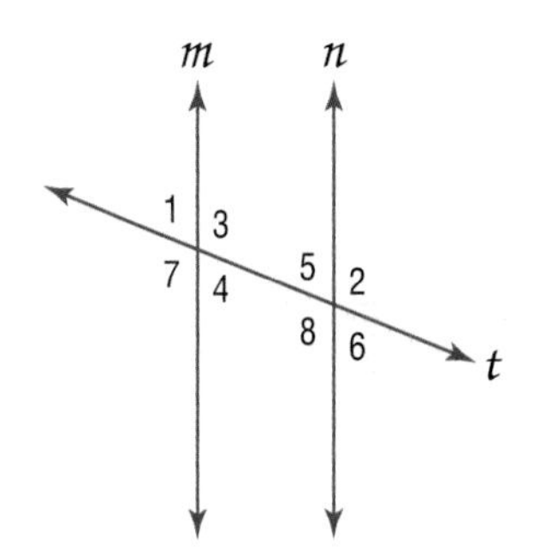

Since $\angle 1$ and $\angle 5$ are corresponding angles, they are congruent. So, $m\angle 5 = 68°$.

Since $\angle 1$ and $\angle 6$ are alternate exterior angles, they are congruent. So $m\angle 6 = 68°$.

INTERSECTING LINES AND ANGLES Other pairs of angles have special relationships. When two lines intersect, they form two pairs of opposite angles called **vertical angles**. Vertical angles are congruent. The symbol for *is congruent to* is $\cong$.

Reading Math

Congruent Angles Angle 1 *is congruent to* angle 2. This is written $\angle 1 \cong \angle 2$. The measure of $\angle 1$ *is equal to* the measure of $\angle 2$. This is written $m\angle 1 = m\angle 2$.

$\angle 1$ and $\angle 2$ are vertical angles. $\angle 1 \cong \angle 2$

$\angle 3$ and $\angle 4$ are vertical angles. $\angle 3 \cong \angle 4$

When two angles have the same vertex, share a common side, and do not overlap, they are **adjacent angles**.

$\angle 1$ and $\angle 2$ are adjacent angles.

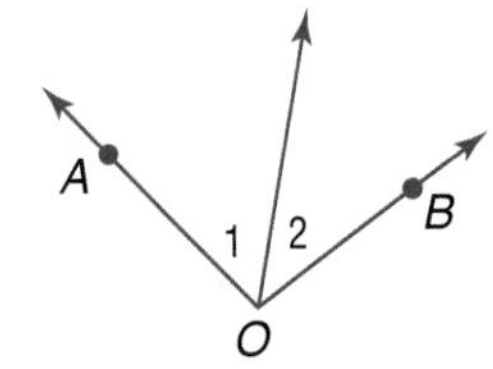

$m\angle AOB = m\angle 1 + m\angle 2$

If the sum of the measures of two angles is 90°, the angles are **complementary**.

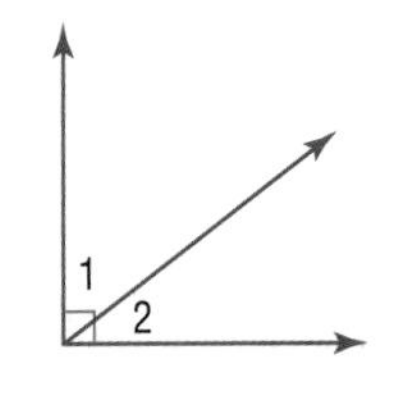

$m\angle 1 = 50°$, $m\angle 2 = 40°$
$m\angle 1 + m\angle 2 = 90°$

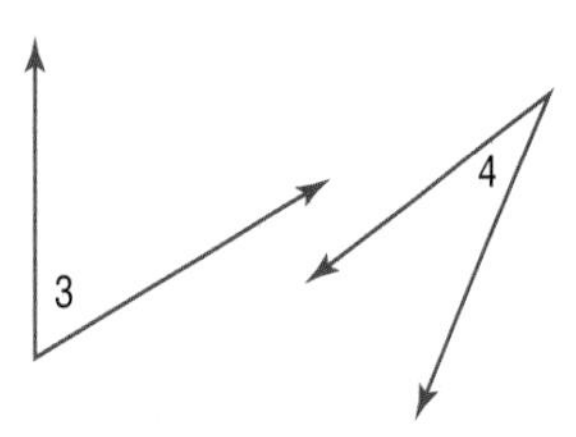

$m\angle 3 = 60°$, $m\angle 4 = 30°$
$m\angle 3 + m\angle 4 = 90°$

If the sum of the measures of two angles is 180°, the angles are **supplementary**.

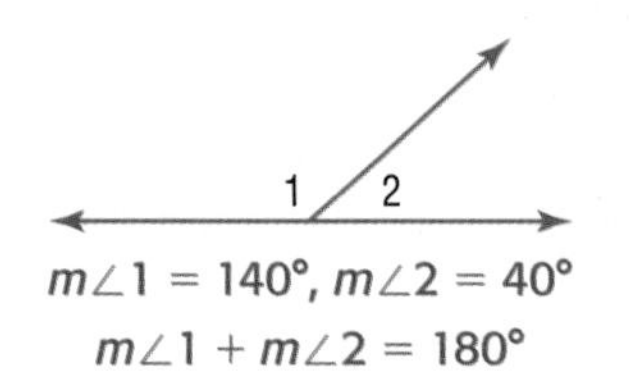

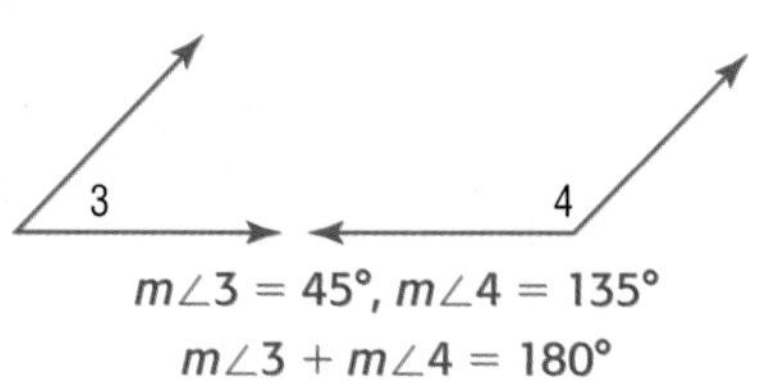

Lines that intersect to form a right angle are **perpendicular lines**.

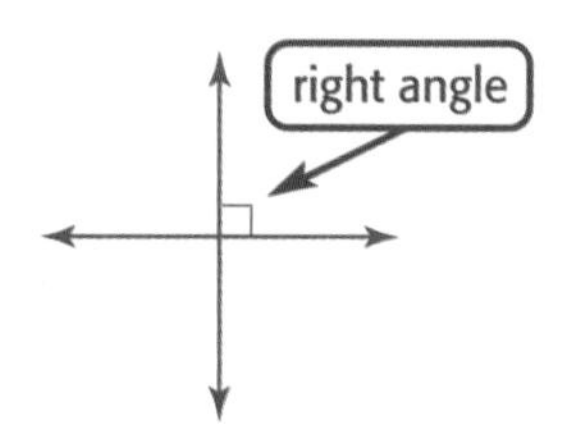

Example 2 Find a Missing Angle Measure

Multiple-Choice Test Item

If $m\angle B = 87°$ and $\angle A$ and $\angle B$ are supplementary, what is $m\angle A$?

(A) 3° (B) 87°

(C) 93° (D) 95°

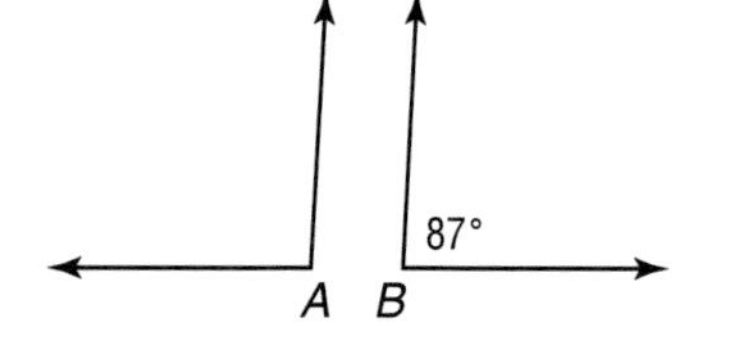

The Princeton Review

Test-Taking Tip

It is a good idea to review definitions of key terms, such as supplementary, before taking the test.

Read the Test Item

Since $\angle A$ and $\angle B$ are supplementary, $m\angle A + m\angle B = 180°$.

Solve the Test Item

$m\angle A + m\angle B = 180°$	Supplementary angles
$m\angle A + 87° = 180°$	Replace $m\angle B$ with 87°.
$m\angle A + 87° - 87° = 180° - 87°$	Subtract 87 from each side.
$m\angle A = 93°$	

The answer is C.

Example 3 Find Measures of Angles

ALGEBRA Angles *ABC* and *FGH* are complementary. If $m\angle ABC = x + 8$ and $m\angle FGH = x - 10$, find the measure of each angle.

Step 1 Find the value of x.

$m\angle ABC + m\angle FGH = 90°$	Complementary angles
$(x + 8) + (x - 10) = 90°$	Substitution
$2x - 2 = 90°$	Combine like terms.
$2x = 92°$	Add 2 to each side.
$x = 46°$	Divide each side by 2.

Step 2 Replace x with 46 to find the measure of each angle.

$m\angle ABC = x + 8$
$= 46 + 8$ or 54

$m\angle FGH = x - 10$
$= 46 - 10$ or 36

So, $m\angle ABC = 54°$ and $m\angle FGH = 36°$.

Study Tip

Checking Reasonableness of Results

To check your answer, add to see if the sum of the measures of the angles is 90. Since $54 + 36 = 90$, the answer is correct.

Example 4 Apply Angle Relationships

SAFETY A lifeguard chair is shown. If $m\angle 1 = 105°$, find $m\angle 4$ and $m\angle 6$.

Since $\angle 1$ and $\angle 4$ are vertical angles, they are congruent. So, $m\angle 4 = 105°$.

Since $\angle 6$ and $\angle 1$ are supplementary, the sum of their measures is 180°.

$180 - 105 = 75$. So, $m\angle 6 = 75°$.

Concept Summary — *Line and Angle Relationships*

Parallel Lines	Perpendicular Lines	Vertical Angles
$a \parallel b$	$m \perp n$	$\angle 1 \cong \angle 3$ $\angle 2 \cong \angle 4$
Adjacent Angles	**Complementary Angles**	**Supplementary Angles**
$m\angle ABC = m\angle 1 + m\angle 2$	$m\angle 1 + m\angle 2 = 90°$	$m\angle 1 + m\angle 2 = 180°$

Check for Understanding

Concept Check

1. **Explain** the difference between complementary and supplementary angles.
2. **OPEN ENDED** Draw a pair of adjacent, supplementary angles.

Guided Practice

In the figure at the right, $\ell \parallel m$ and k is a transversal. If $m\angle 1 = 56°$, find the measure of each angle.

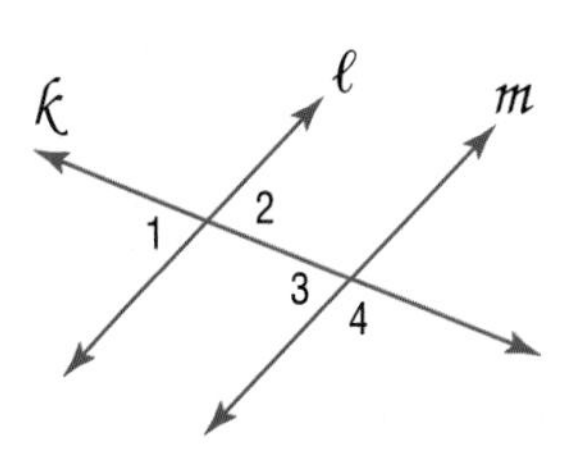

3. $\angle 2$
4. $\angle 3$
5. $\angle 4$

Find the value of x in each figure.

6. ($x°$, $140°$)

7.

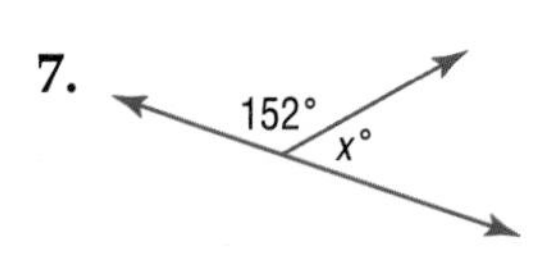

8. **ALGEBRA** If $m\angle N = 3x$ and $m\angle M = 2x$ and $\angle M$ and $\angle N$ are supplementary, what is the measure of each angle?

FCAT Practice *Standardized Test Practice*

9. If $m\angle B = 26°$ and $\angle A$ and $\angle B$ are complementary, what is $m\angle A$?

   Ⓐ 154° Ⓑ 90° Ⓒ 26° Ⓓ 64°

Practice and Apply

Homework Help

For Exercises	See Examples
10–15, 26–28	1, 4
16–21	2
22–25, 29–32	3

Extra Practice

See page 747.

In the figure at the right, $g \parallel h$ and t is a transversal. If $m\angle 4 = 53°$, find the measure of each angle.

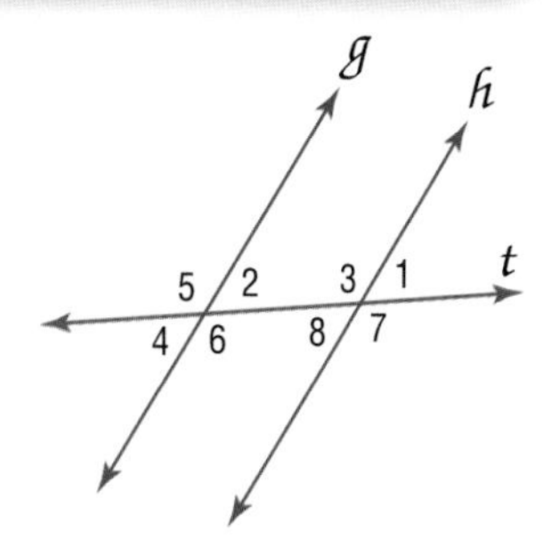

10. $\angle 1$
11. $\angle 5$
12. $\angle 7$
13. $\angle 8$
14. $\angle 2$
15. $\angle 3$

Find the value of x in each figure.

16. 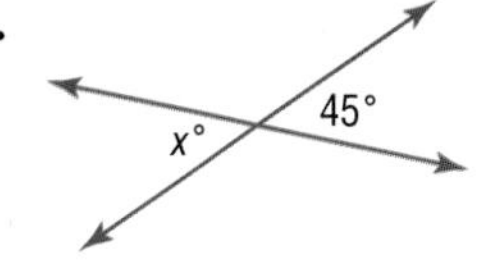

17. 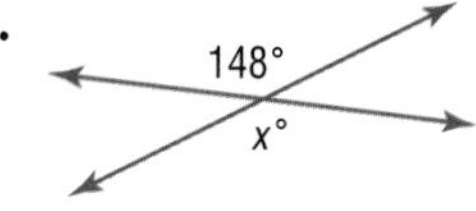

18. 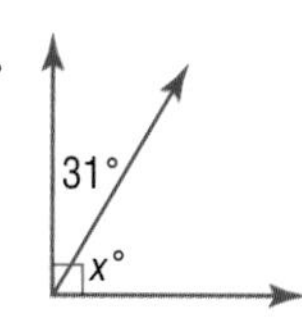

19. 

20. 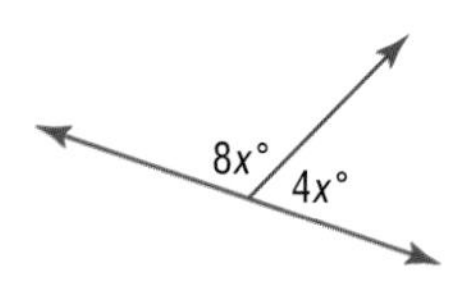

21. 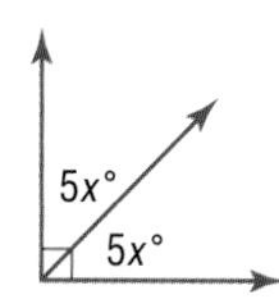

22. Find $m\angle A$ if $m\angle B = 17°$ and $\angle A$ and $\angle B$ are complementary.

23. Angles P and Q are supplementary. Find $m\angle P$ if $m\angle Q = 139°$.

24. **ALGEBRA** Angles J and K are complementary. If $m\angle J = x - 9$ and $m\angle K = x + 5$, what is the measure of each angle?

25. **ALGEBRA** Find $m\angle E$ if $\angle E$ and $\angle F$ are supplementary, $m\angle E = 2x + 15$, and $m\angle F = 5x - 38$.

26. **SAFETY** Refer to Example 4 on page 495. Find the measure of angles $\angle 2$, $\angle 3$, $\angle 5$, $\angle 7$, and $\angle 8$.

CONSTRUCTION For Exercises 27 and 28, use the following information and the diagram shown.

To measure the angle between a sloped ceiling and a wall, a carpenter uses a plumb line (a string with a weight attached).

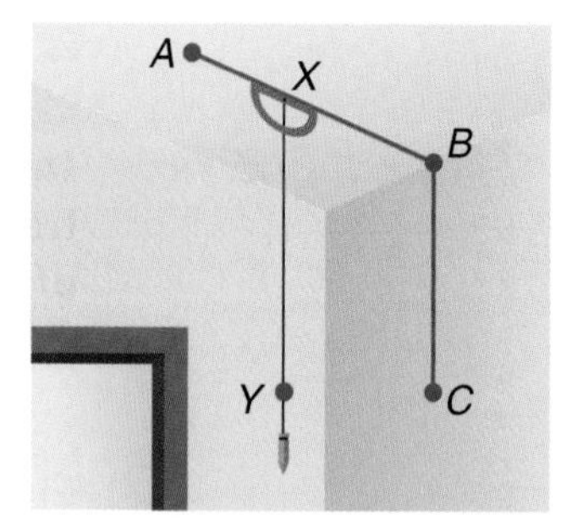

27. If $m\angle YXB = 68°$, what is $m\angle XBC$?

28. What type of angles are $\angle YXB$ and $\angle XBC$?

ALGEBRA In the figure at the right, $m \parallel \ell$ and t is a transversal. Find the value of x for each of the following.

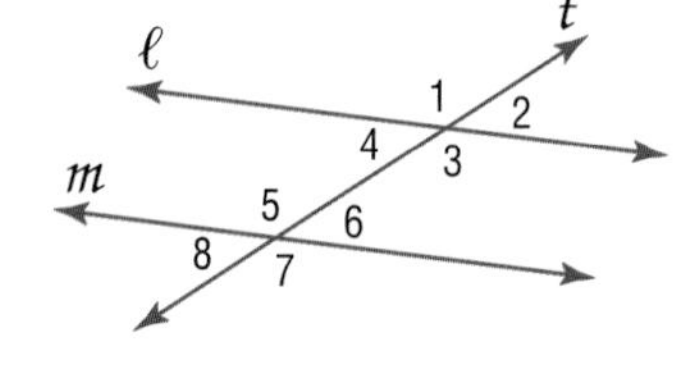

29. $m\angle 2 = 2x + 3$ and $m\angle 4 = 4x - 7$
30. $m\angle 8 = 4x - 32$ and $m\angle 5 = 5x + 50$
31. $m\angle 7 = 10x + 15$ and $m\angle 3 = 7x + 42$

32. **ALGEBRA** The measure of the supplement of an angle is 15° less than four times the measure of the complement. Find the measure of the angle.

33. **CRITICAL THINKING** Suppose two parallel lines are cut by a transversal. How are the interior angles on the same side of the transversal related?

34. **WRITING IN MATH** Answer the question that was posed at the beginning of the lesson.

How are parallel lines and angles related?

Include the following in your answer:

- a drawing of parallel lines intersected by a transversal, and
- a list of the congruent and supplementary angles.

For Exercises 35 and 36, use the diagram.

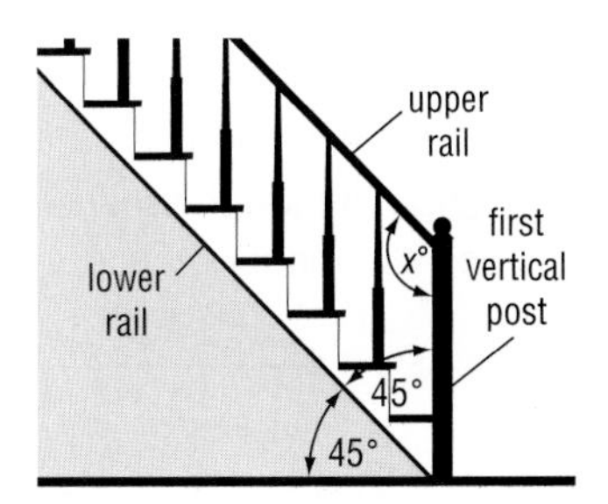

35. The upper rail is parallel to the lower rail. What is the measure of the angle formed by the upper rail and the first vertical post?

Ⓐ 135° Ⓑ 100°

Ⓒ 90° Ⓓ 45°

36. What is the measure of the angle formed by the second vertical post and the lower rail?

Ⓐ 100° Ⓑ 135° Ⓒ 45° Ⓓ 90°

Extending the Lesson

For Exercises 37–39, use the pairs of graphs shown at the right.

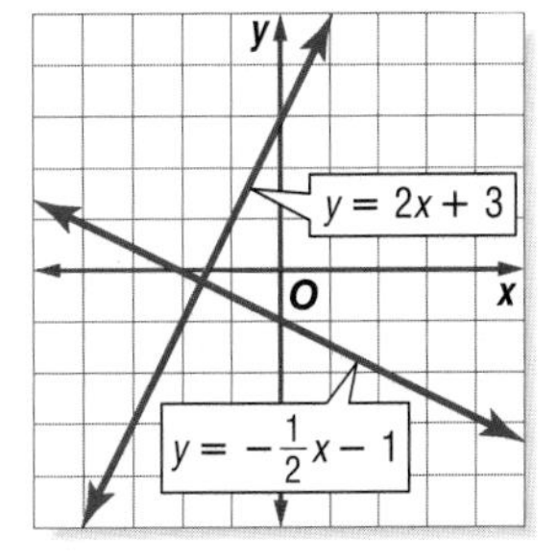

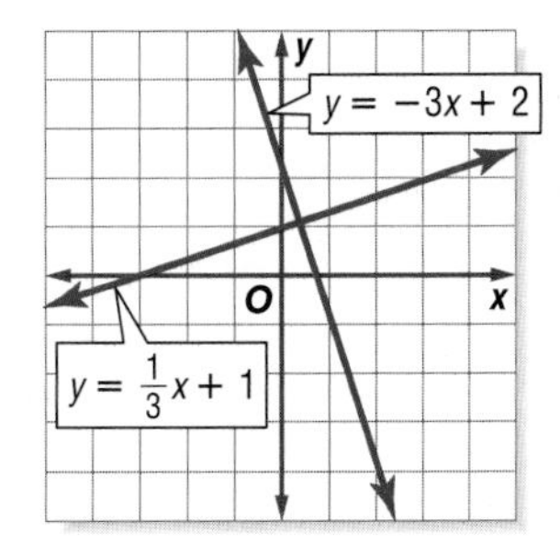

37. How are each pair of graphs related?

38. What seems to be true about the slopes of the graphs?

39. **Make a conjecture** about the slopes of the graphs of perpendicular lines.

Study Tip

Look Back
To review **slope**, see Lesson 8-4.

Maintain Your Skills

Mixed Review

Use a calculator to find each value to the nearest ten thousandth. *(Lesson 9-8)*

40. cos 21° 41. sin 63° 42. tan 38°

43. If $\triangle ABC \sim \triangle DEF$, what is the value of x? *(Lesson 9-7)*

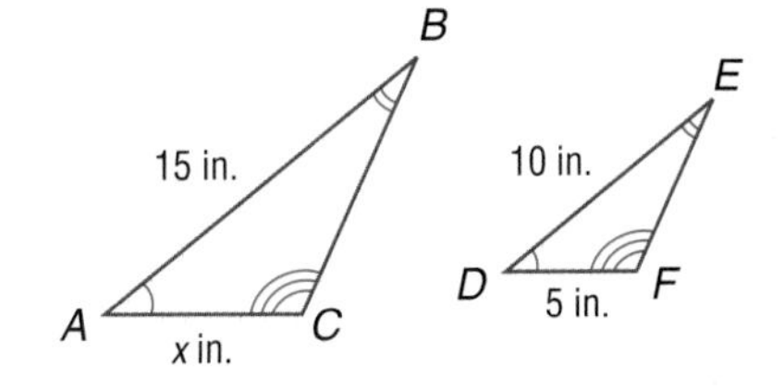

Simplify each expression. *(Lesson 3-2)*

44. $6a + (-18)a$

45. $-5m + (-4)m$

Getting Ready for the Next Lesson

PREREQUISITE SKILL Use a protractor to draw an angle having each measurement. *(To review **angles**, see Lesson 9-3.)*

46. 20° 47. 45° 48. 65° 49. 145° 50. 170°

Algebra Activity

A Follow-Up of Lesson 10-1

Sunshine State Standards
MA.C.2.3.1-1

Constructions

Activity 1 Construct a line segment congruent to a given line segment.

Step 1

Draw $\overline{AB}$. Then use a straightedge to draw $\overrightarrow{GH}$ so it is longer than $\overline{AB}$.

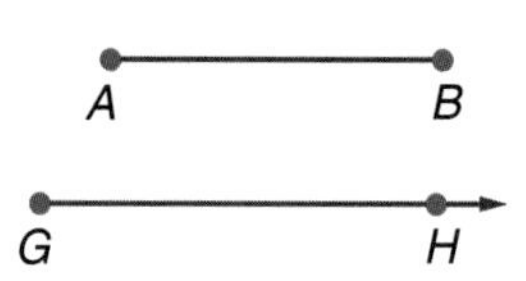

Step 2

Place the tip of the compass at A and the pencil tip at B.

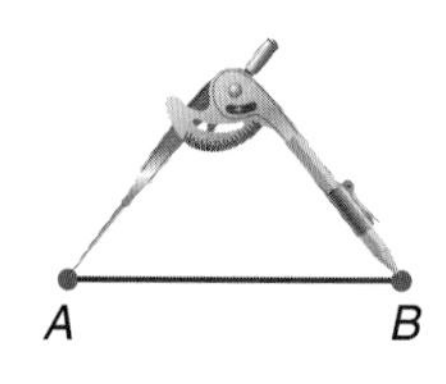

Step 3

Using this setting, place the tip at G. Draw an arc to intersect $\overrightarrow{GH}$. Label the intersection J. $\overline{GJ} \cong \overline{AB}$.

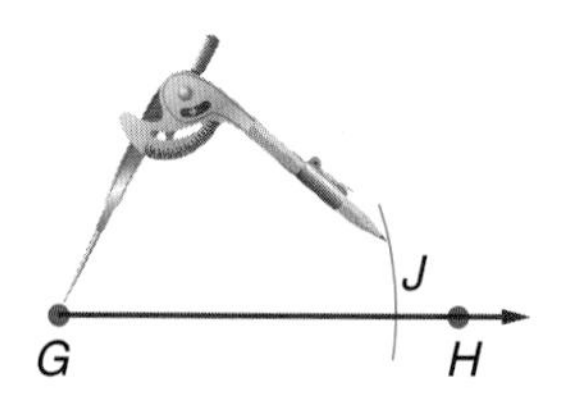

Activity 2 Construct an angle congruent to a given angle.

Step 1

Draw $\angle DEF$. Then use a straightedge to draw $\overrightarrow{JK}$.

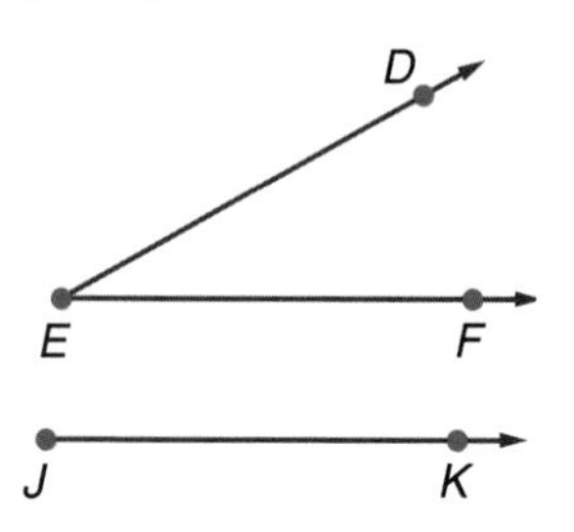

Step 2

Place the tip of the compass at E. Draw an arc to intersect both sides of $\angle DEF$ to locate points X and Y.

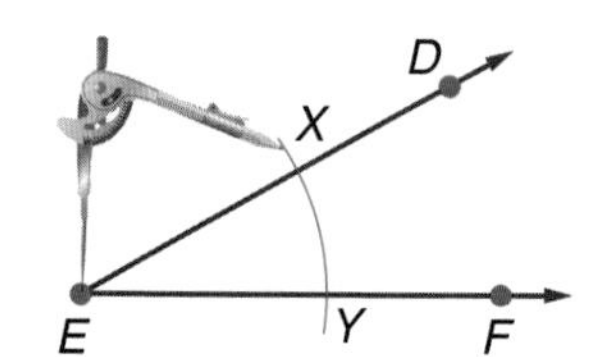

Step 3

Using this setting, place the compass at point J. Draw an arc to intersect $\overrightarrow{JK}$. Label the intersection A.

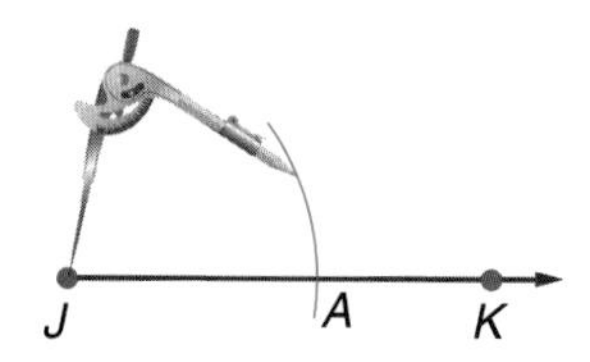

Step 4

Place the point of the compass on X and adjust so that the pencil tip is on Y.

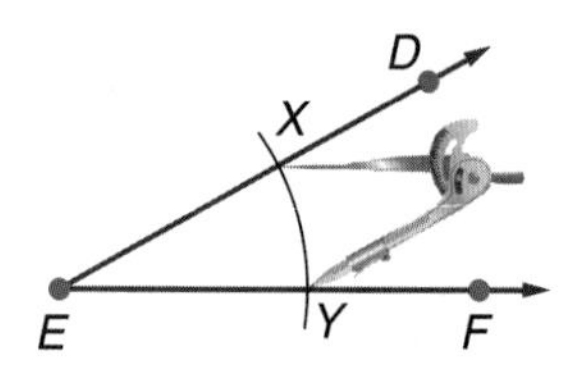

Step 5

Using this setting, place the compass at A and draw an arc to intersect the arc drawn in Step 3. Label the intersection M.

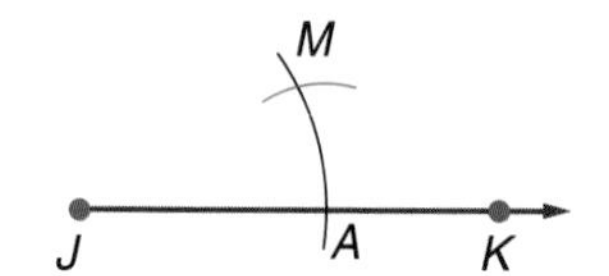

Step 6

Draw $\overrightarrow{JM}$. $\angle MJK \cong \angle DEF$.

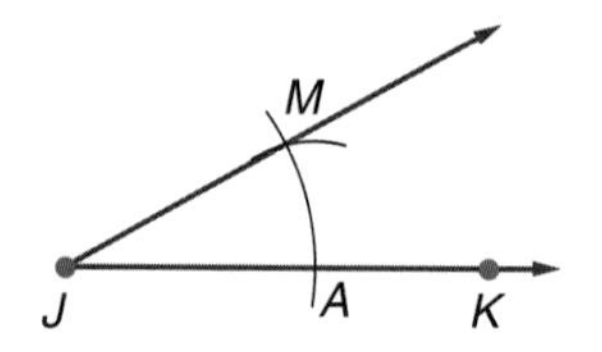

Model and Analyze

1. Draw a line segment. Construct a line segment congruent to the one drawn.

2. Draw an angle. Construct an angle congruent to the one drawn.

Activity 3 Construct the perpendicular bisector of a line segment.

Step 1

Draw $\overline{XY}$. Then place the compass at point X. Use a setting greater than one half of $\overline{XY}$. Draw an arc above and below $\overline{XY}$.

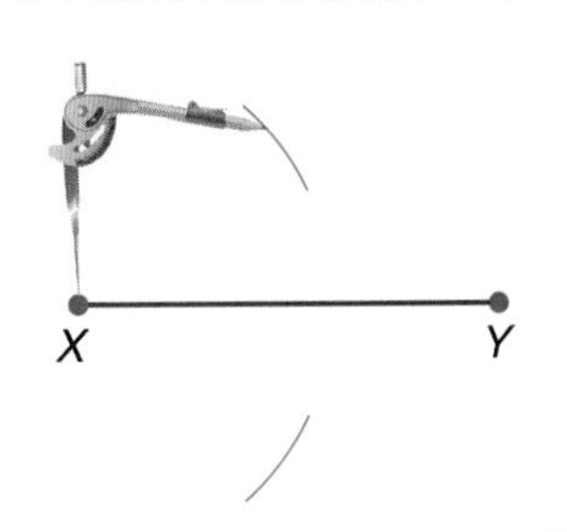

Step 2

Using this setting, place the compass at point Y. Draw an arc above and below $\overline{XY}$ as shown.

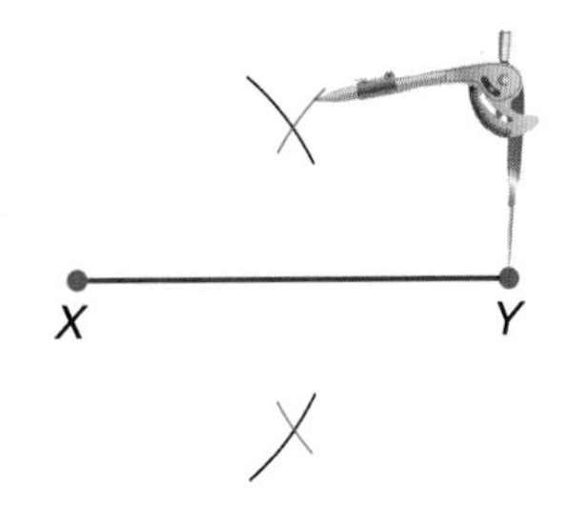

Step 3

Use a straightedge to align the two intersections. Draw a segment that intersects $\overline{XY}$. Label the intersection M.

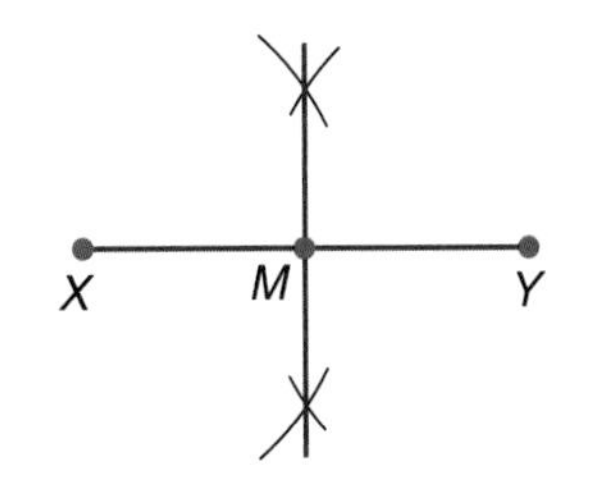

Activity 4 Construct the bisector of an angle.

Step 1

Draw $\angle MNP$. Then place the compass at point N and draw an arc that intersects both sides of the angle. Label the intersections X and Y.

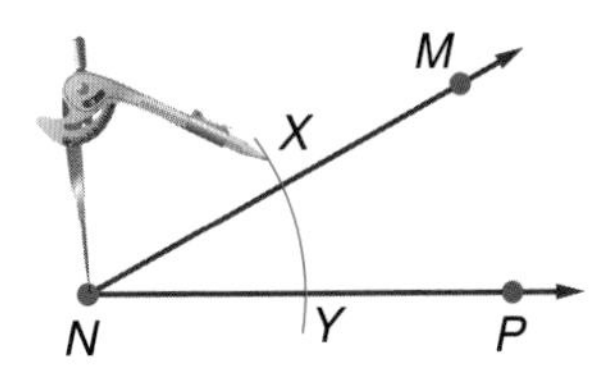

Step 2

With the compass at point X, draw an arc in the interior of $\angle N$. Using this setting, place the compass at point Y. Draw another arc. Label the intersection Q.

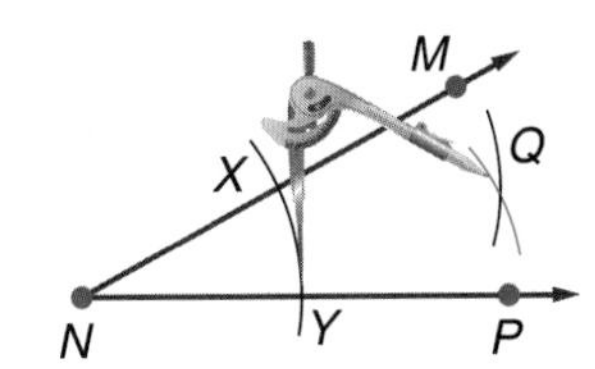

Step 3

Draw $\overrightarrow{NQ}$. $\overrightarrow{NQ}$ is the bisector of $\angle MNP$.

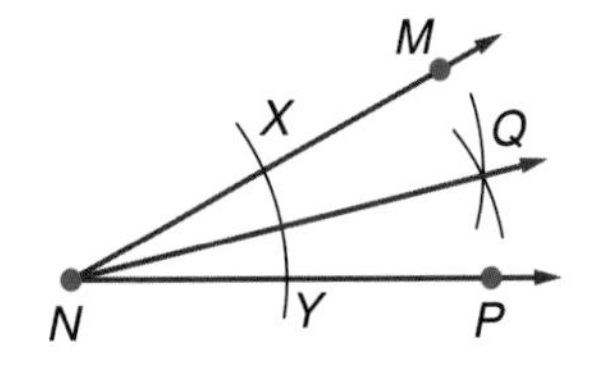

Model and Analyze

3. In Activity 3, use a ruler to measure $\overline{XM}$ and $\overline{MY}$. This construction *bisects* a segment. What do you think *bisects* means?
4. Draw a line segment. Construct the perpendicular bisector of the segment.
5. In Activity 4, what is true about the measures of $\angle MNQ$ and $\angle QNP$?
6. Why do we say $\overrightarrow{NQ}$ is the bisector of $\angle MNP$?

10-2 Congruent Triangles

Sunshine State Standards
MA.A.2.3.1-2, MA.C.1.3.1-1, MA.C.2.3.1-2

Vocabulary

- congruent
- corresponding parts

What You'll Learn

- Identify congruent triangles and corresponding parts of congruent triangles.

Where are congruent triangles present in nature?

Ivy is a type of climbing plant. Most ivy leaves have five major veins. In the photo shown, the veins outlined form two triangles.

a. Trace the triangles shown at the right onto a sheet of paper. Then label the triangles.

b. Measure and then compare the lengths of the sides of the triangles.

c. Measure the angles of each triangle. How do the angles compare?

d. Make a conjecture about the triangles.

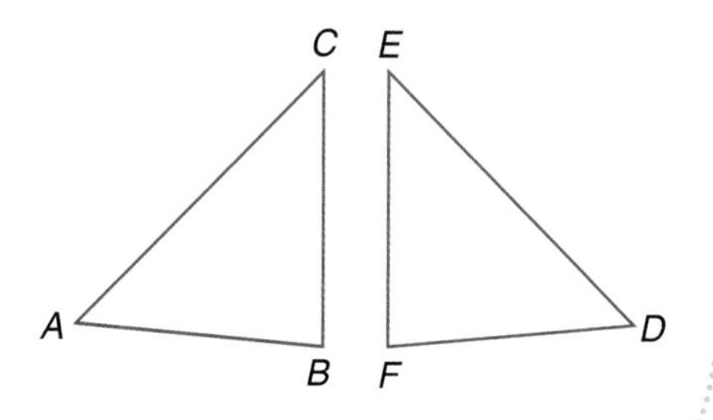
C
E
A
B
F
D

Reading Math

Corresponding
Everyday Meaning: *matching*
Math Meaning: *having the same position*

CONGRUENT TRIANGLES Figures that have the same size and shape are **congruent**. The parts of congruent triangles that "match" are **corresponding parts**.

Key Concept — *Corresponding Parts of Congruent Triangles*

- **Words** If two triangles are congruent, their corresponding sides are congruent and their corresponding angles are congruent.
- **Model**

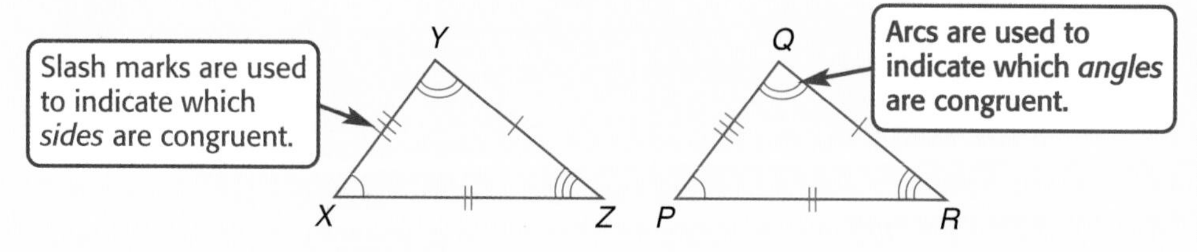
Slash marks are used to indicate which *sides* are congruent.
Y
X
Z
Q
P
R
Arcs are used to indicate which *angles* are congruent.

- **Symbols** Congruent Angles: $\angle X \cong \angle P$, $\angle Y \cong \angle Q$, $\angle Z \cong \angle R$
  Congruent Sides: $\overline{XY} \cong \overline{PQ}$, $\overline{YZ} \cong \overline{QR}$, $\overline{XZ} \cong \overline{PR}$

Concept Check What is true about the corresponding angles and sides of congruent triangles?

When writing a congruence statement, the letters must be written so that corresponding vertices appear in the same order. For example, for the diagram below, write $\triangle FGH \cong \triangle JKM$.

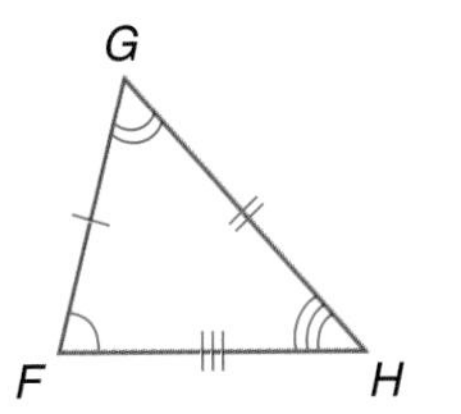

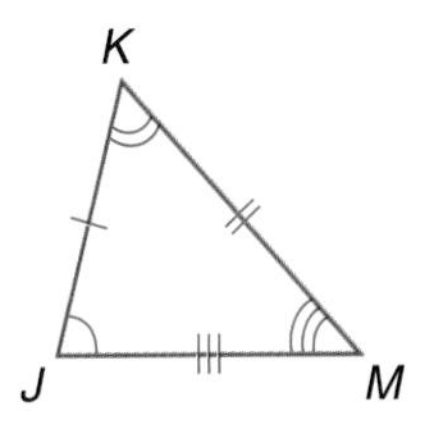

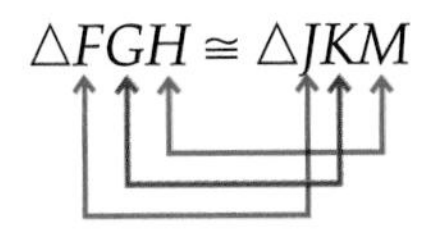

Vertex *F* corresponds to vertex *J*.
Vertex *G* corresponds to vertex *K*.
Vertex *H* corresponds to vertex *M*.

Study Tip

Congruence Statements

You can also write a congruence statement as $\triangle GHF \cong \triangle KMJ$, $\triangle HFG \cong \triangle MJK$, $\triangle FHG \cong \triangle JMK$, $\triangle GFH \cong \triangle KJM$, and $\triangle HGF \cong \triangle MKJ$.

Example 1 Name Corresponding Parts

Name the corresponding parts in the congruent triangles shown. Then complete the congruence statement.

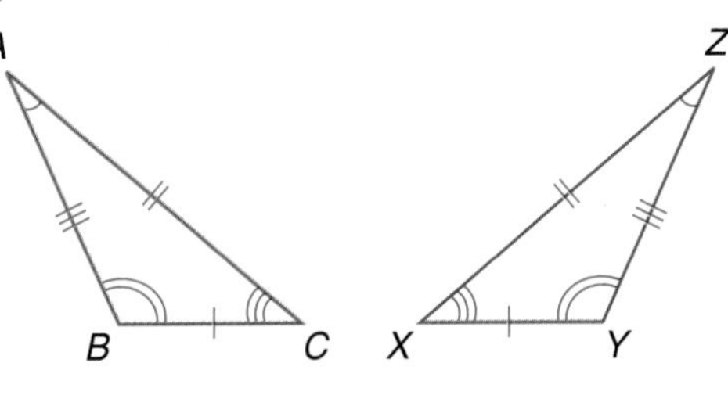

$\triangle ABC \cong$ __?__

Corresponding Angles
$\angle A \cong \angle Z$, $\angle B \cong \angle Y$, $\angle C \cong \angle X$

Corresponding Sides
$\overline{AB} \cong \overline{ZY}$, $\overline{BC} \cong \overline{YX}$, $\overline{CA} \cong \overline{XZ}$

One congruence statement is $\triangle ABC \cong \triangle ZYX$.

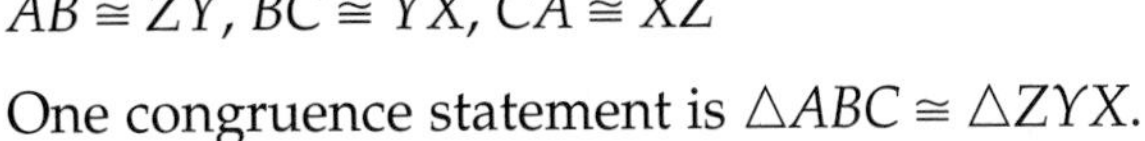

Congruence statements can be used to identify corresponding parts of congruent triangles.

Example 2 Use Congruence Statements

If $\triangle SRT \cong \triangle WVU$, complete each congruence statement.

$\angle S \cong$ __?__ $\angle U \cong$ __?__ $\angle R \cong$ __?__

$\overline{WV} \cong$ __?__ $\overline{RT} \cong$ __?__ $\overline{ST} \cong$ __?__

Explore You know the congruence statement. You need to find the corresponding parts.

Plan Use the order of the vertices in $\triangle SRT \cong \triangle WVU$ to identify the corresponding parts.

Solve $\triangle SRT \cong \triangle WVU$

$\angle S$ corresponds to $\angle W$ so $\angle S \cong \angle W$.
$\angle R$ corresponds to $\angle V$ so $\angle R \cong \angle V$.
$\angle T$ corresponds to $\angle U$ so $\angle T \cong \angle U$.

S corresponds to *W*, and *R* corresponds to *V* so $\overline{SR} \cong \overline{WV}$.
R corresponds to *V*, and *T* corresponds to *U* so $\overline{RT} \cong \overline{VU}$.
S corresponds to *W*, and *T* corresponds to *U* so $\overline{ST} \cong \overline{WU}$.

Examine Draw the triangles, using arcs and slash marks to show the congruent angles and sides.

You can use corresponding parts to find the measures of angles and sides in a figure that is congruent to a figure with known measures.

Example 3 Find Missing Measures

LANDSCAPING A brace is used to support a tree and help it to grow straight. In the figure, $\triangle TRS \cong \triangle ERS$.

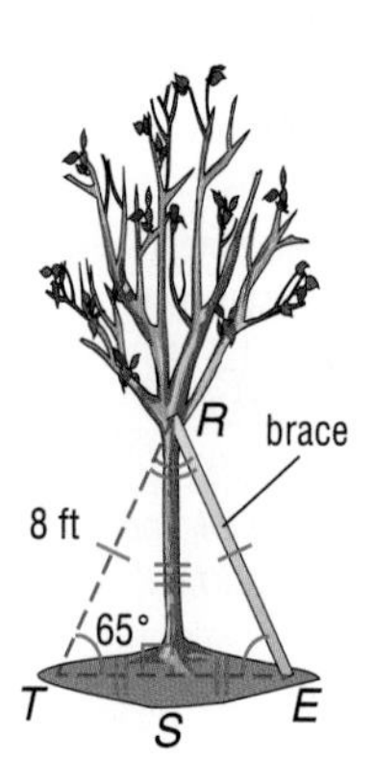

a. At what angle is the brace placed against the ground?

$\angle E$ and $\angle T$ are corresponding angles. So, they are congruent. Since $m\angle T = 65°$, $m\angle E = 65°$.

The brace is placed at a 65° angle with the ground.

b. What is the length of the brace?

$\overline{RE}$ corresponds to $\overline{RT}$. So, $\overline{RE}$ and $\overline{RT}$ are congruent. Since $RT = 8$ feet, $RE = 8$ feet.

The length of the brace is 8 feet.

Check for Understanding

Concept Check

1. **Explain** when two figures are congruent.
2. **OPEN ENDED** Draw and label a pair of congruent triangles. Be sure to mark the corresponding parts.

Guided Practice

For each pair of congruent triangles, name the corresponding parts. Then complete the congruence statement.

3.

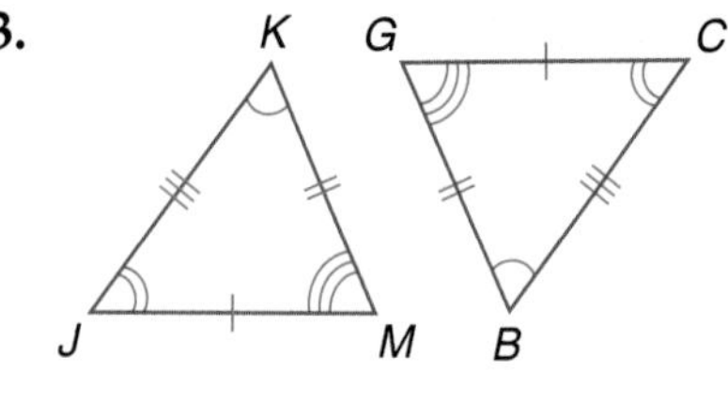

$\triangle KMJ \cong$ __?__

4.

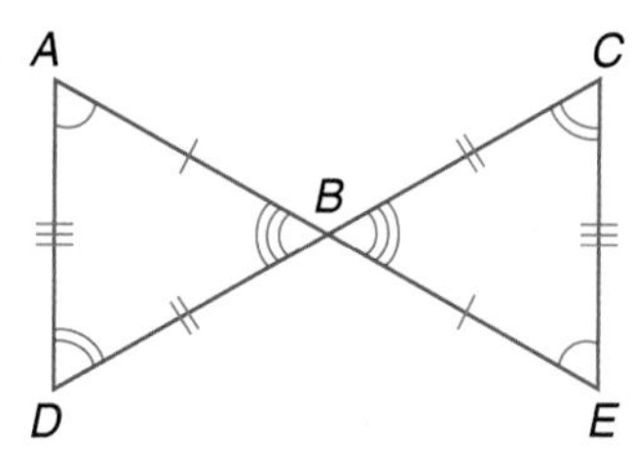

$\triangle CBE \cong$ __?__

Complete each congruence statement if $\triangle DKJ \cong \triangle NAM$.

5. $\angle J \cong$ __?__
6. $\angle A \cong$ __?__
7. $\overline{JD} \cong$ __?__
8. $\overline{AN} \cong$ __?__

Application

9. **TOWERS** A tower that supports high voltage power lines is shown at the right. In the tower, $\triangle ETC \cong \triangle YRC$. What is the length of $\overline{RC}$ if $EC = 10$ feet and $TC = 15$ feet?

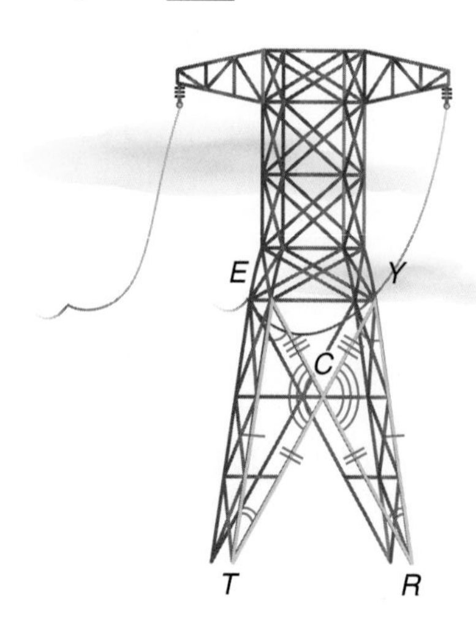

Practice and Apply

Homework Help

For Exercises	See Examples
10–13	1
16, 17, 27–29	3
18–26	2

Extra Practice
See page 748.

For each pair of congruent triangles, name the corresponding parts. Then complete the congruence statement.

10. 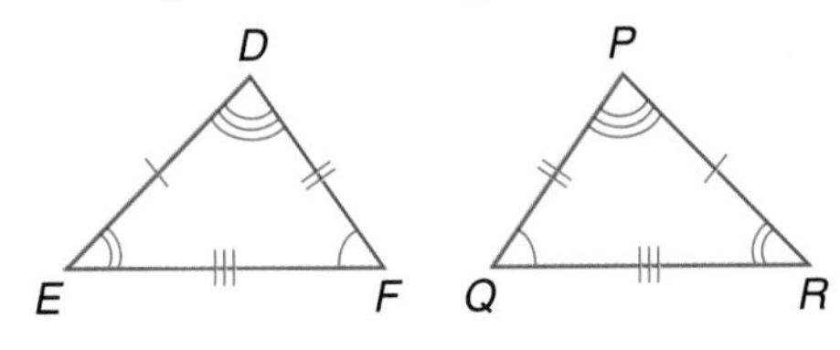

$\triangle DFE \cong$ __?__

11. 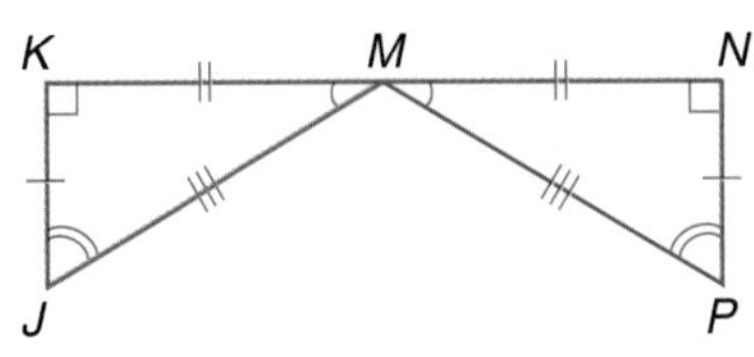

$\triangle KJM \cong$ __?__

12. 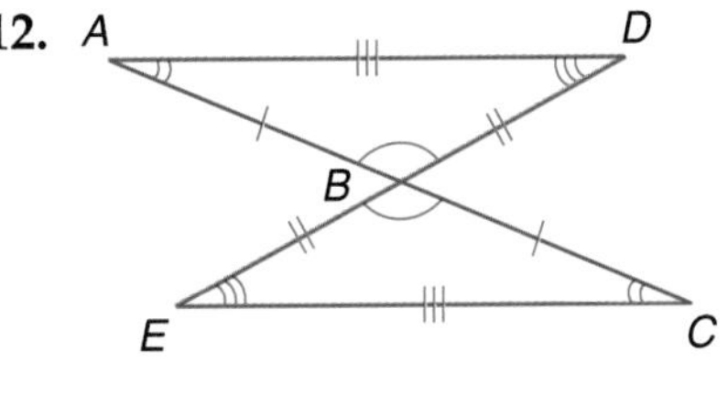

$\triangle DBA \cong$ __?__

13. 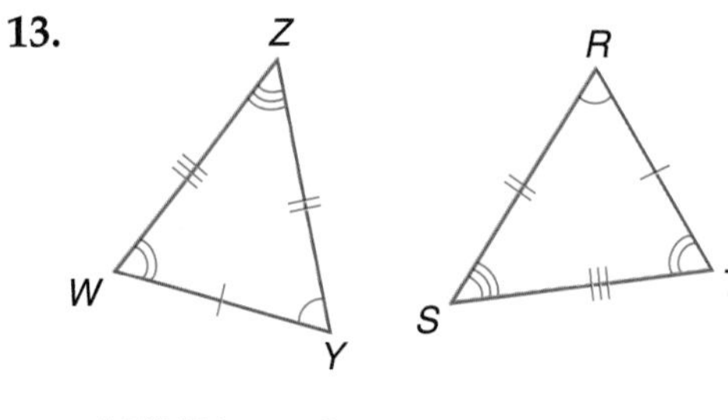

$\triangle ZWY \cong$ __?__

More About. . .

Architecture

Trusses were used in the construction of the Eiffel Tower in Paris, France. The tower contains more than 15,000 pieces of steel and 2.5 million rivets.

Source: www.paris.org

Tell whether each statement is *sometimes, always,* or *never* true.

14. If two triangles are congruent, then the perimeters are equal.

15. If the perimeters of two triangles are equal, then the triangles are congruent.

ARCHITECTURE **For Exercises 16 and 17, use the diagram of the roof truss at the right and the fact that $\triangle TRU \cong \triangle SRU$.**

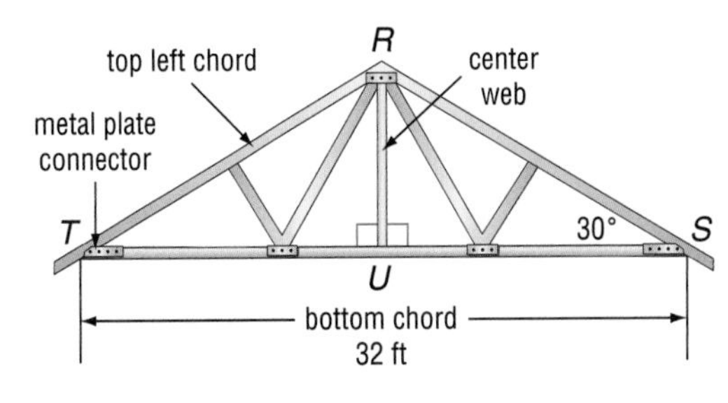

16. Find the distance from the left metal plate connector to the center web.

17. What is the measure of the angle formed by the top left chord and the bottom chord?

Complete each congruence statement if $\triangle FHG \cong \triangle CBD$ and $\triangle KMA \cong \triangle PRQ$.

18. $\angle G \cong$ __?__
19. $\angle C \cong$ __?__
20. $\angle M \cong \angle$ __?__
21. $\angle Q \cong$ __?__
22. $\overline{HG} \cong$ __?__
23. $\overline{BC} \cong$ __?__
24. $\overline{DC} \cong$ __?__
25. $\overline{GF} \cong$ __?__
26. $\angle GFH \cong \angle$ __?__

Find the value of x for each pair of congruent triangles.

27. 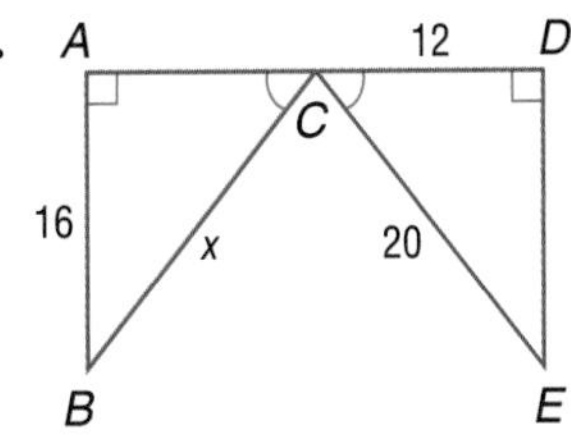

28. 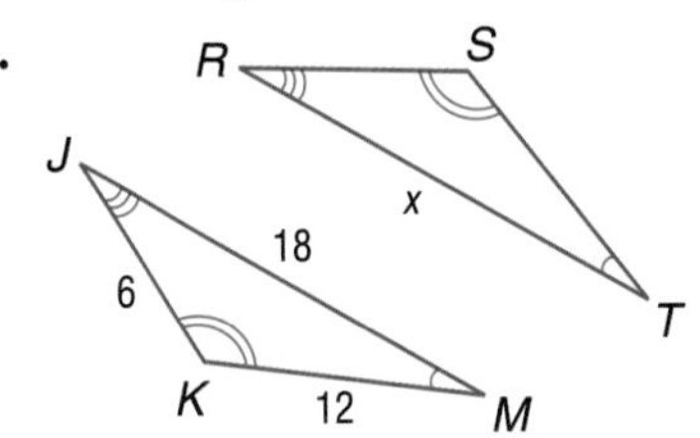

29. **ALGEBRA** If $\triangle ABC \cong \triangle XYZ$, what is the value of x?

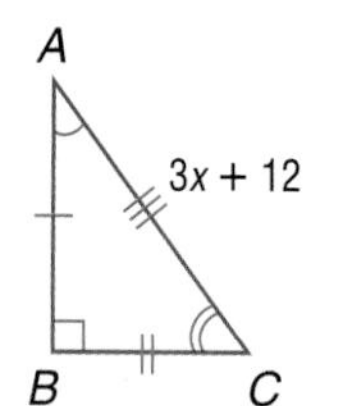

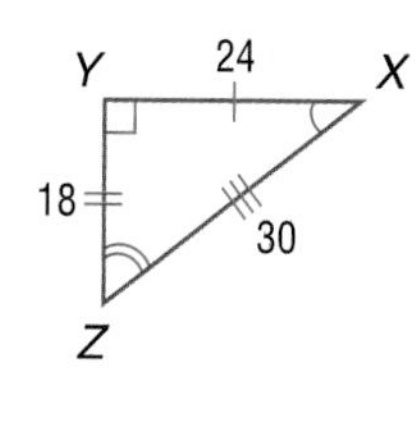

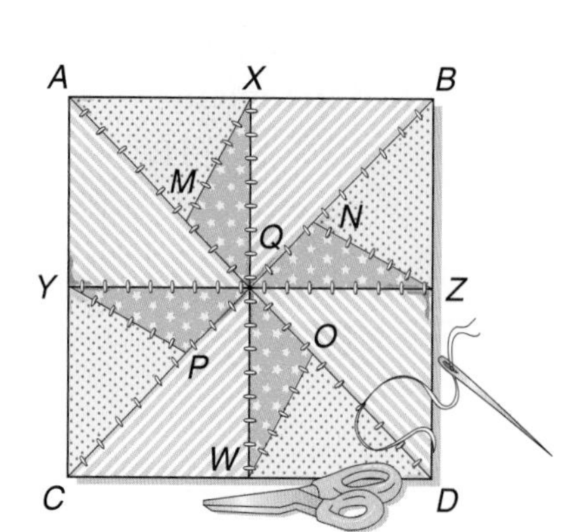

QUILTS **For Exercises 30–33, use the quilt pattern shown at the left. Name a triangle that appears to be congruent to each triangle listed.**

30. $\triangle XBQ$
31. $\triangle YQP$
32. $\triangle DWO$
33. $\triangle ZQN$

34. **CRITICAL THINKING** In the figure at the right, there are two pairs of congruent triangles. Write a congruence statement for each pair.

35. WRITING IN MATH Answer the question that was posed at the beginning of the lesson.

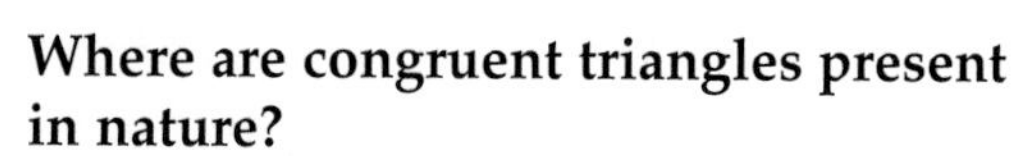

Where are congruent triangles present in nature?

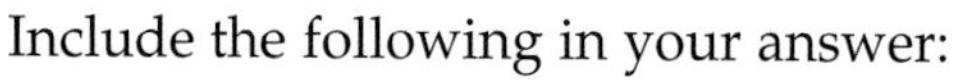

Include the following in your answer:
- a definition of congruent triangles, and
- two examples of objects in nature that contain congruent triangles.

36. Which of the following must be true if $\triangle ACD \cong \triangle EHF$?

Ⓐ $\angle C \cong \angle E$ Ⓑ $\overline{CA} \cong \overline{HF}$ Ⓒ $\overline{DC} \cong \overline{EF}$ Ⓓ $\angle CAD \cong \angle HEF$

37. Find the measure of $\overline{UY}$ if $\triangle XYZ \cong \triangle UYW$.

Ⓐ 10 Ⓑ 20
Ⓒ 24 Ⓓ 26

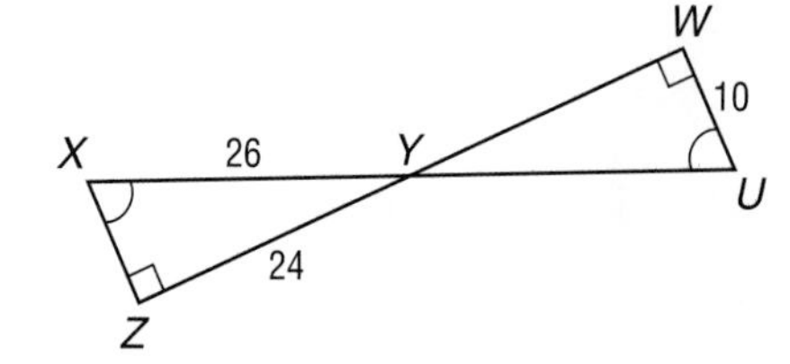

Maintain Your Skills

Mixed Review

38. Angles P and Q are supplementary. Find $m\angle P$ if $m\angle Q = 129°$. *(Lesson 10-1)*

Use a calculator to find each value to the nearest ten-thousandth. *(Lesson 9-8)*

39. cos 83°
40. sin 39.7°
41. tan 49.2°

42. Name the multiplicative inverse of $2\frac{1}{2}$. *(Lesson 5-4)*

43. **ALGEBRA** Solve $-5a - 6 = 24$. *(Lesson 3-5)*

Getting Ready for the Next Lesson

PREREQUISITE SKILL **Graph each point on a coordinate system.**
*(To review the **coordinate system**, see Lesson 2-6.)*

44. $A(2, 4)$
45. $J(-1, 3)$
46. $H(0, 5)$
47. $D(2, 0)$
48. $W(-2, -4)$
49. $T(-1, -3)$

Algebra Activity

A Preview of Lesson 10-3

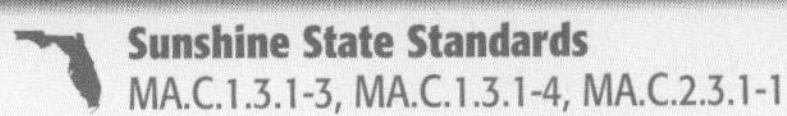

Sunshine State Standards
MA.C.1.3.1-3, MA.C.1.3.1-4, MA.C.2.3.1-1

Symmetry

Activity 1

Trace the outline of the butterfly shown. Then draw a line down the center of the butterfly. Notice how the two halves match. When this happens, a figure is said to have **line symmetry** and the line is called a *line of symmetry*. A figure that has line symmetry has **bilateral symmetry**.

Analyze

Determine whether each figure has line symmetry. If it does, trace the figure, and draw all lines of symmetry. If not, write *none*.

1. 

2. 

3. 

Activity 2

Copy the figure at the right. Then cut out the figure. Next, rotate the figure 90°, 180°, and 270°, about point C. What do you notice about the appearance of the figure in each rotation?

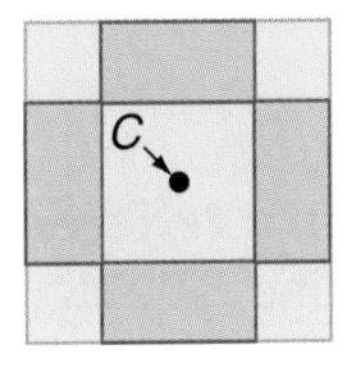

Any figure that can be turned or rotated less than 360° about a fixed point so that the figure looks exactly as it does in its original position has **rotational** or **turn symmetry**.

Analyze

Determine whether each figure has rotational symmetry. Write *yes* or *no*.

4. 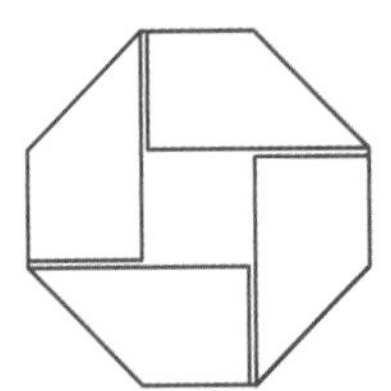

5. 

6. 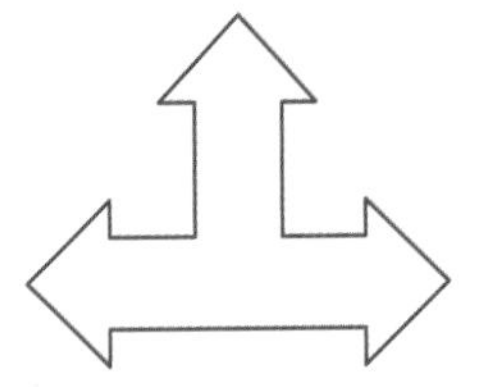

7. Name three objects that have both line symmetry and rotational symmetry.

10-3 Transformations on the Coordinate Plane

What You'll Learn

- Draw translations, rotations, and reflections on a coordinate plane.

Sunshine State Standards
MA.C.1.3.1-3, MA.C.2.3.1-3

How are transformations involved in recreational activities?

The physical motions used in recreational activities such as skateboarding, swinging, or riding a scooter, are related to mathematics.

a. Describe the motion involved in making a 180° turn on a skateboard.
b. Describe the motion that is used when swinging on a swing.
c. What type of motion does a scooter display when moving?

Vocabulary

- transformation
- translation
- reflection
- line of symmetry
- rotation

TRANSFORMATIONS A movement of a geometric figure is a **transformation**. Three types of transformations are shown below.

- In a **translation**, you slide a figure from one position to another without turning it. Translations are also called *slides*.

Translation

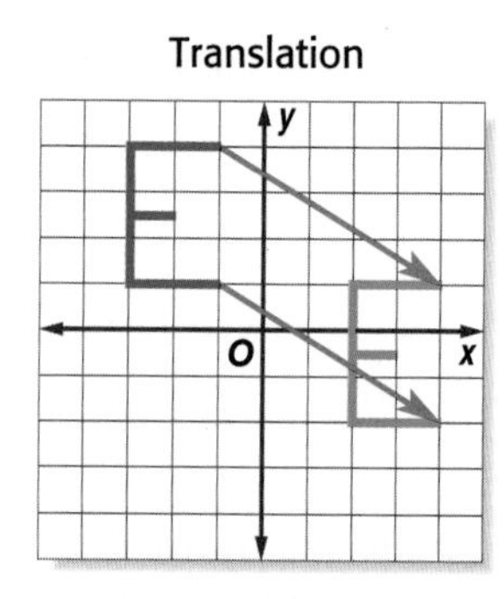

- In a **reflection**, you flip a figure over a line. The figures are mirror images of each other. Reflections are also called *flips*. The line is called a **line of symmetry**.

Reflection

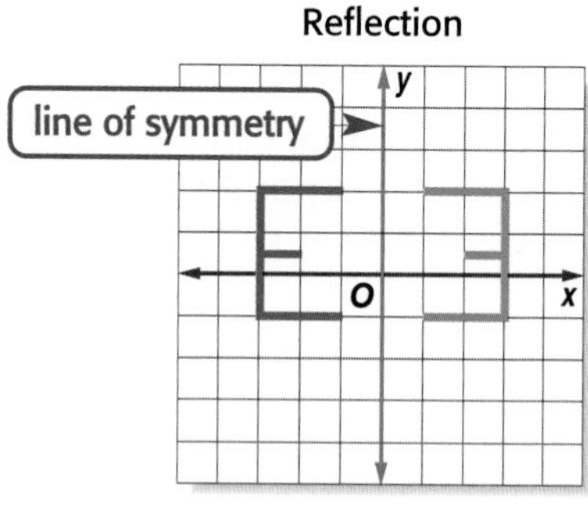

- In a **rotation**, you turn the figure around a fixed point. Rotations are also called *turns*.

Rotation

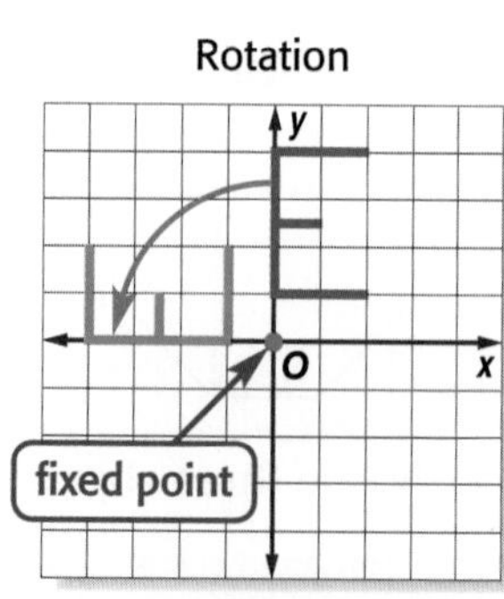

Study Tip

Common Misconception
In a translation, the order in which a figure is moved *does not* matter. For example, moving 3 units down and then 2 units across is the same as moving 2 units across and then 3 units down.

When translating a figure, every point of the original figure is moved the same distance and in the same direction.

Translation
4 units left

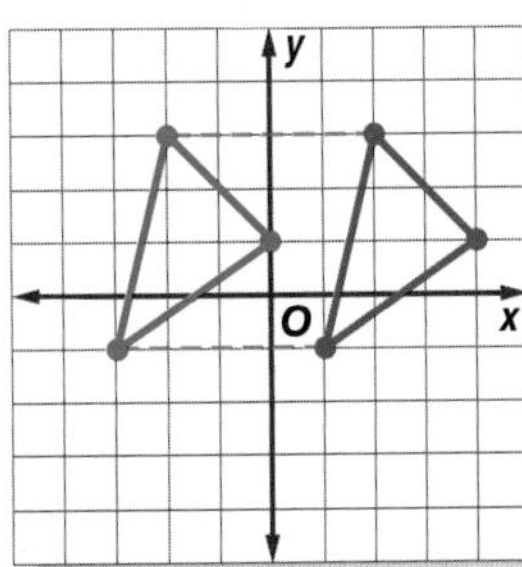

Translation
5 units down

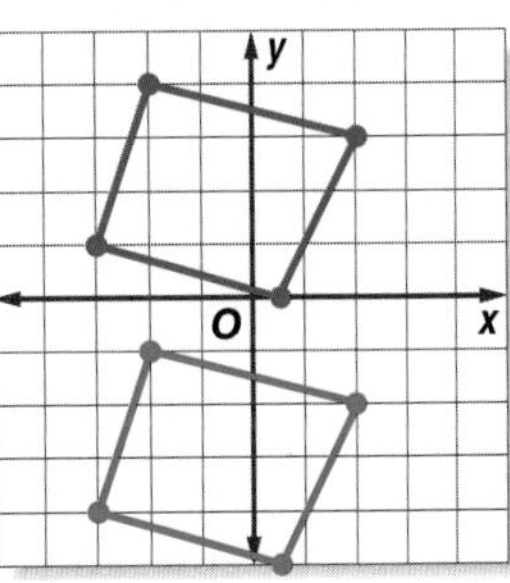

Translation
6 units right, 3 units up

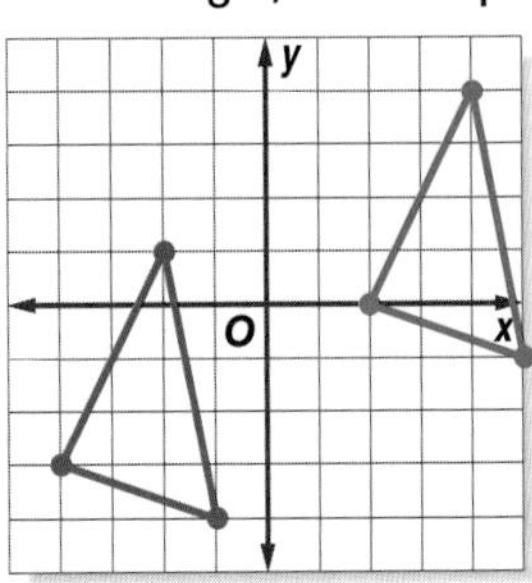

The following steps can be used to translate a point in the coordinate plane.

Key Concept — *Translation*

Step 1 Describe the translation using an ordered pair.

Step 2 Add the coordinates of the ordered pair to the coordinates of the original point.

Example 1 Translation in a Coordinate Plane

The vertices of $\triangle MNP$ are $M(4, -2)$, $N(0, 2)$, and $P(5, 2)$. Graph the triangle and the image of $\triangle MNP$ after a translation 5 units left and 3 units up.

This translation can be written as the ordered pair $(-5, 3)$. To find the coordinates of the translated image, add -5 to each x-coordinate and add 3 to each y-coordinate.

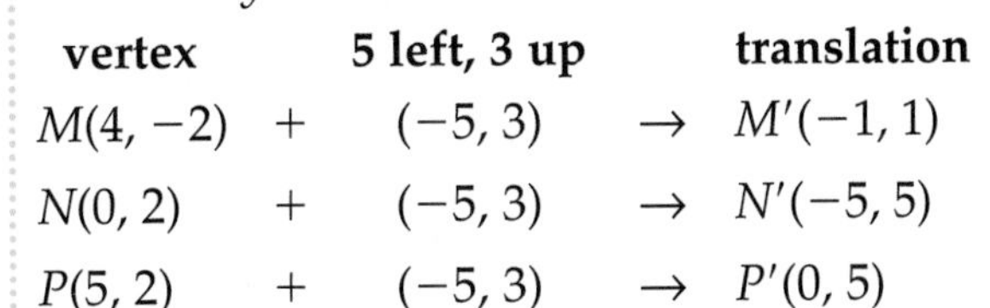

vertex		5 left, 3 up		translation
$M(4, -2)$	+	$(-5, 3)$	$\rightarrow$	$M'(-1, 1)$
$N(0, 2)$	+	$(-5, 3)$	$\rightarrow$	$N'(-5, 5)$
$P(5, 2)$	+	$(-5, 3)$	$\rightarrow$	$P'(0, 5)$

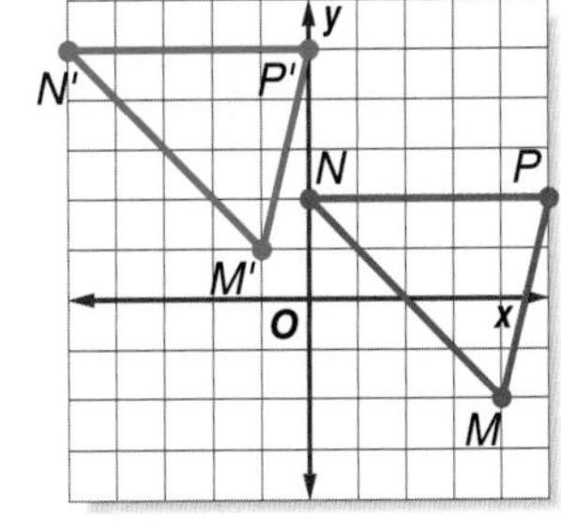

The coordinates of the vertices of $\triangle M'N'P'$ are $M'(-1, 1)$, $N'(-5, 5)$, and $P'(0, 5)$.

Study Tip

Notation
The notation M' is read *M prime.* It corresponds to point *M*.

When reflecting a figure, every point of the original figure has a corresponding point on the other side of the line of symmetry.

Reflection
over the **x**-axis

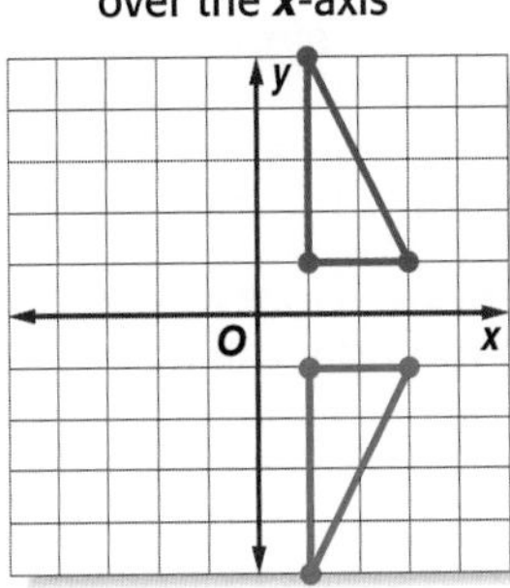

Reflection
over the **y**-axis

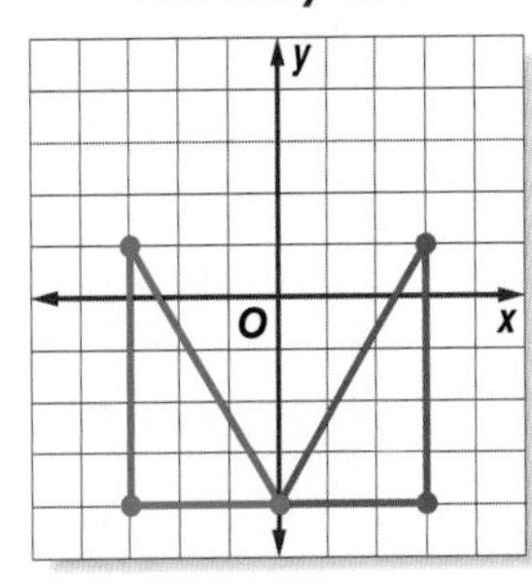

The following rules can be used to reflect a point over the x- or y-axis.

Key Concept — *Reflection*

- To reflect a point over the x-axis, use the same x-coordinate and multiply the y-coordinate by –1.
- To reflect a point over the y-axis, use the same y-coordinate and multiply the x-coordinate by –1.

Example 2 *Reflection in a Coordinate Plane*

The vertices of the figure below are $A(-2, 3)$, $B(0, 5)$, $C(3, 1)$, and $D(3, 3)$. Graph the figure and the image of the figure after a reflection over the x-axis.

To find the coordinates of the vertices of the image after a reflection over the x-axis, use the same x-coordinate and multiply the y-coordinate by –1.

vertex				reflection
$A(-2, 3)$	$\rightarrow$	$(-2, -1 \cdot 3)$	$\rightarrow$	$A'(-2, -3)$
$B(0, 5)$	$\rightarrow$	$(0, -1 \cdot 5)$	$\rightarrow$	$B'(0, -5)$
$C(3, 1)$	$\rightarrow$	$(3, -1 \cdot 1)$	$\rightarrow$	$C'(3, -1)$
$D(3, 3)$	$\rightarrow$	$(3, -1 \cdot 3)$	$\rightarrow$	$D'(3, -3)$

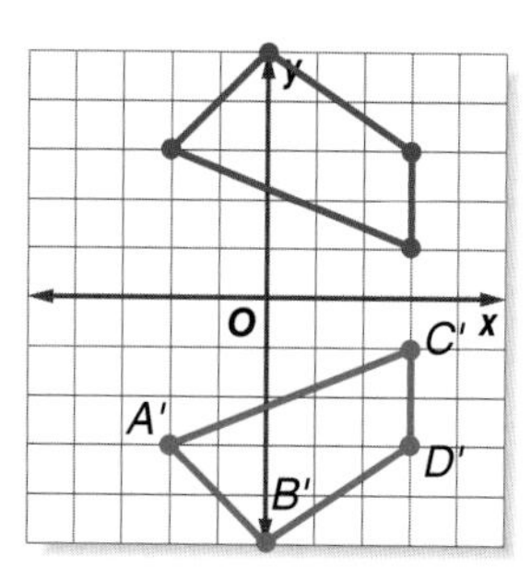

The coordinates of the vertices of the reflected figure are $A'(-2, -3)$, $B'(0, -5)$, $C'(3, -1)$, and $D'(3, -3)$.

The diagrams below show three of the ways a figure can be rotated.

Rotation of 90° clockwise

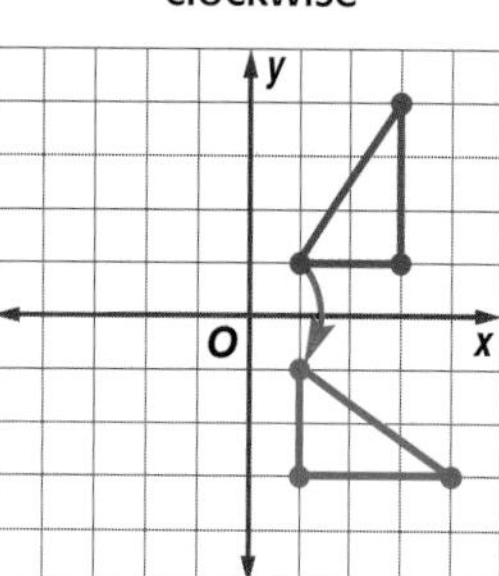

Rotation of 90° counterclockwise

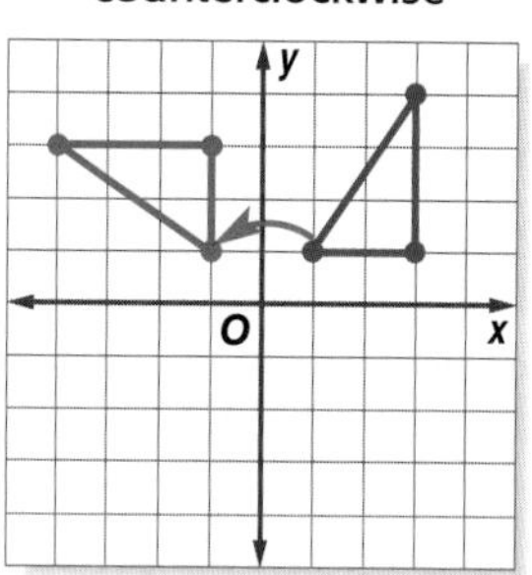

Rotation of 180°

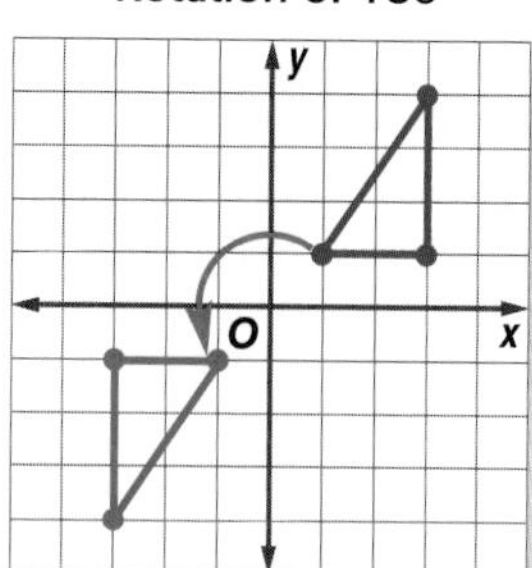

Reading Math

clockwise
The word clockwise refers to *the direction in which the hands of a clock rotate.*

counterclockwise
The word counterclockwise refers to *the direction opposite to that in which the hands of a clock rotate.*

You can use these rules to rotate a figure 90° clockwise, 90° counterclockwise, or 180° about the origin.

Key Concept — *Rotation*

- To rotate a figure 90° clockwise about the origin, switch the coordinates of each point and then multiply the new second coordinate by -1.
- To rotate a figure 90° counterclockwise about the origin, switch the coordinates of each point and then multiply the new first coordinate by -1.
- To rotate a figure 180° about the origin, multiply both coordinates of each point by -1.

Example 3 Rotations in a Coordinate Plane

A figure has vertices $J(1, 1)$, $K(4, 1)$, $M(1, 2)$, $N(4, 2)$, and $P(3, 4)$. Graph the figure and the image of the figure after a rotation of 90° counterclockwise.

To rotate the figure, switch the coordinates of each vertex and multiply the first by −1.

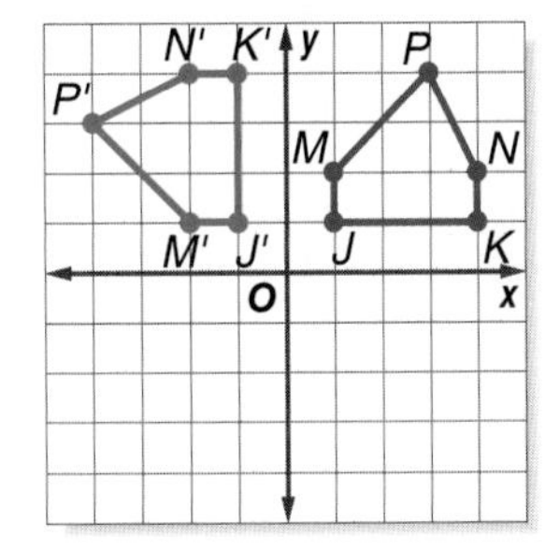

$J(1, 1) \rightarrow J'(-1, 1)$ $\quad N(4, 2) \rightarrow N'(-2, 4)$

$K(4, 1) \rightarrow K'(-1, 4)$ $\quad P(3, 4) \rightarrow P'(-4, 3)$

$M(1, 2) \rightarrow M'(-2, 1)$

The coordinates of the vertices of the rotated figure are $J'(-1, 1)$, $K'(-1, 4)$, $M'(-2, 1)$, $N'(-2, 4)$, and $P'(-4, 3)$.

Check for Understanding

Concept Check

1. **Write** a sentence to describe a figure that is translated by (5, −2).
2. **OPEN ENDED** Draw a triangle on grid paper. Then draw the image of the triangle after it is moved 5 units right and then rotated counterclockwise 90°.

Guided Practice

3. Rectangle *RSTU* is shown at the right. Graph the image of the rectangle after a translation 4 units right and 2 units down.

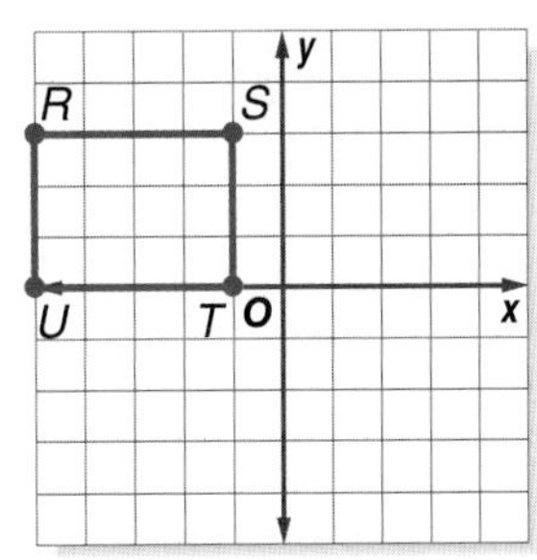

4. Suppose the figure graphed is reflected over the *y*-axis. Find the coordinates of the vertices after the reflection.

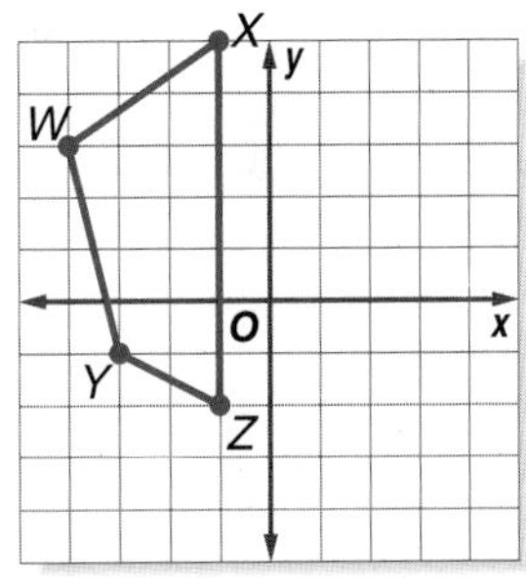

5. Triangle *ABC* is shown. Graph the image of $\triangle ABC$ after a rotation of 90° counterclockwise.

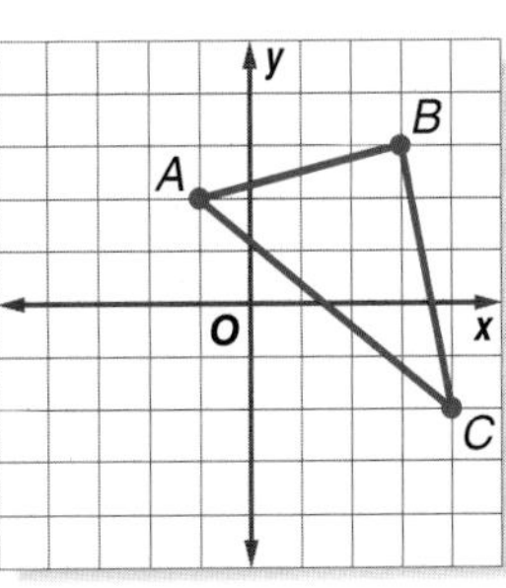

Application

6. **ART** Identify the type of transformation that is shown below.

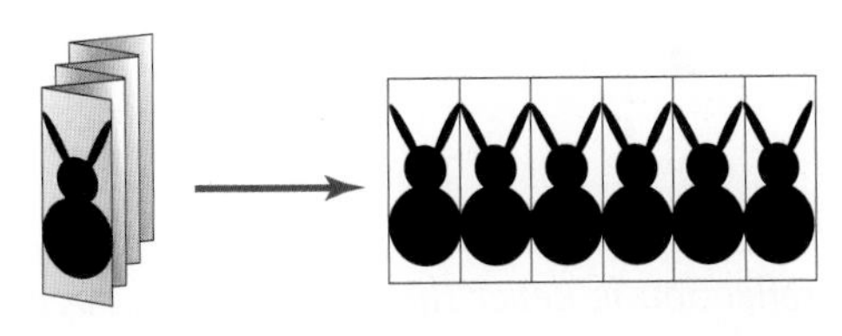

Practice and Apply

Homework Help

For Exercises	See Examples
7–10	1
11–14	2
15–17	3
18–20	1–3

Extra Practice

See page 748.

Find the coordinates of the vertices of each figure after the given translation. Then graph the translation image.

7. (2, 3)

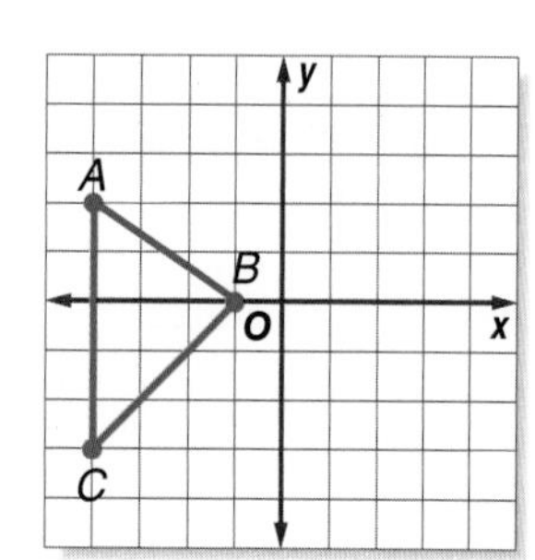

8. (−4, 3)

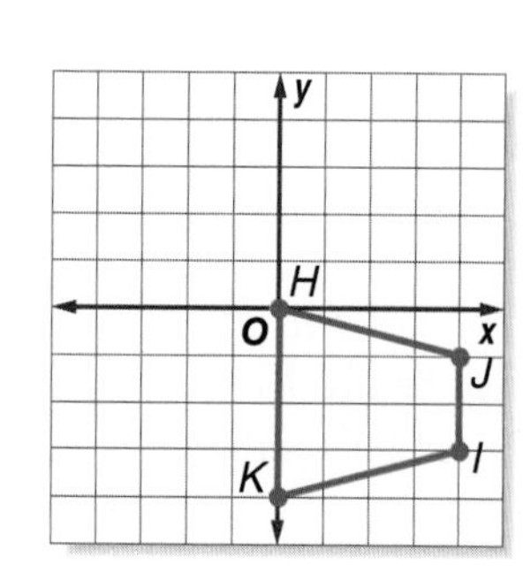

9. $\left(5, 2\frac{1}{2}\right)$

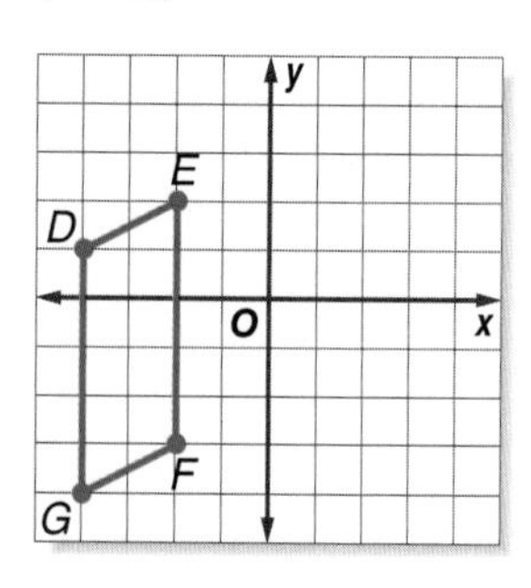

10. The vertices of a figure are *D*(1, 2), *E*(1, 4), *F*(−1, 2), *G*(−4, 4). Graph the image of the figure after a translation 4 units down.

Find the coordinates of the vertices of each figure after a reflection over the given axis. Then graph the reflection image.

11. *x*-axis

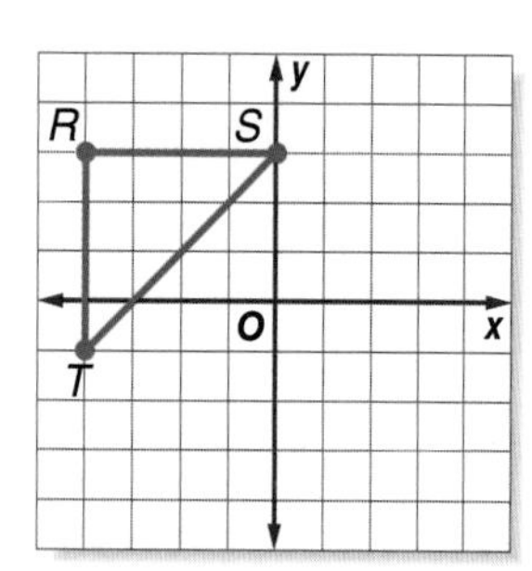

12. *y*-axis

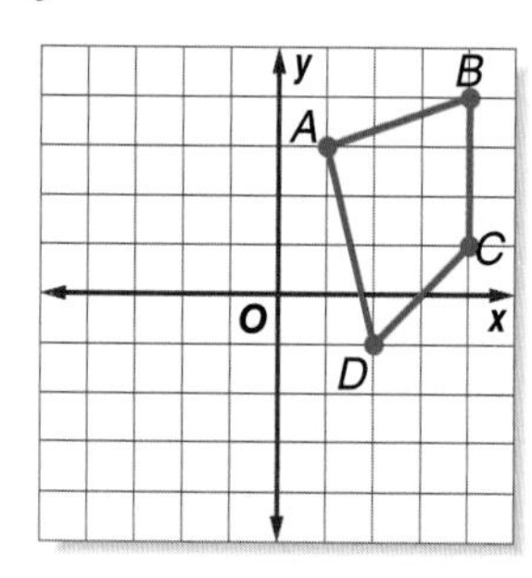

13. *x*-axis

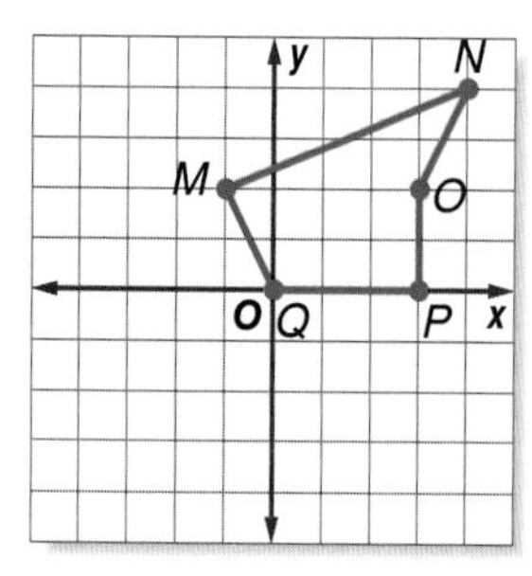

14. The vertices of a figure are *W*(−3, −3), *X*(0, −4), *Y*(4, −2), and *Z*(2, −1). Graph the image of its reflection over the *y*-axis.

For Exercises 15–17, use the graph shown.

15. Graph the image of the figure after a rotation of 90° counterclockwise.

16. Find the coordinates of the vertices of the figure after a 180° rotation.

17. Graph the image of the figure after a rotation of 90° clockwise.

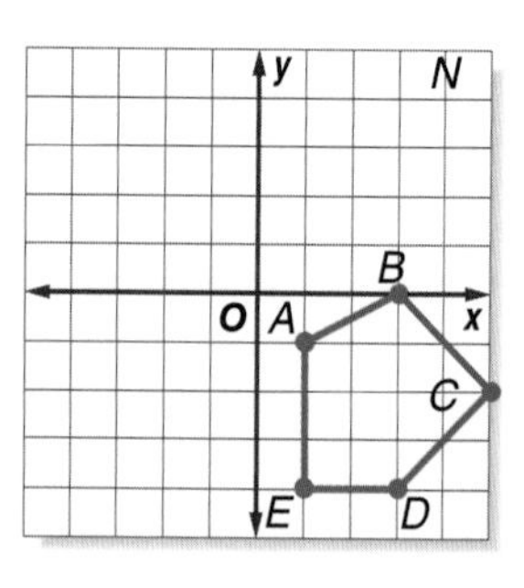

Identify each transformation as a *translation*, a *reflection*, or a *rotation*.

18.

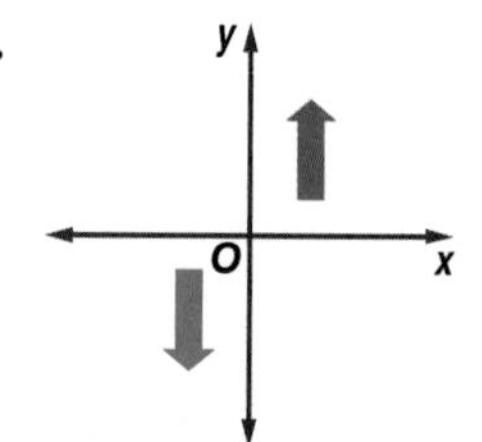

19.

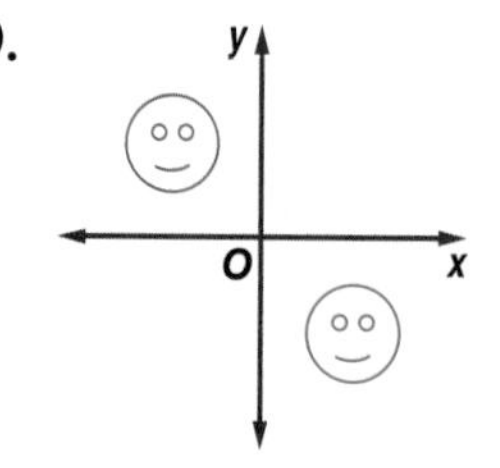

20.

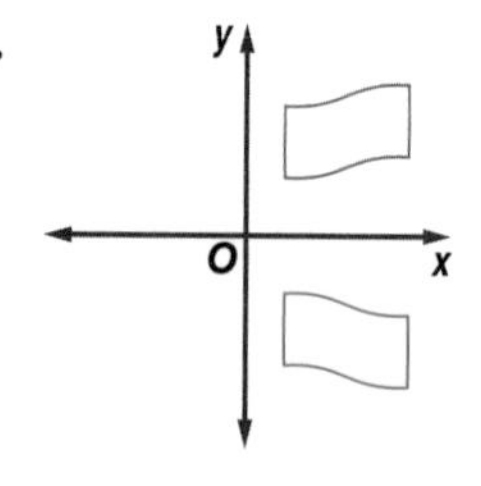

21. Give a counterexample for the following statement. *The image of a figure's reflection is never the same as the image of its translation.*

More About. . .

Games

In chess, each player has 16 game pieces, or chessmen. There are two rooks, two knights, two bishops, a queen, a king, and eight pawns.

Source: www.infoplease.com

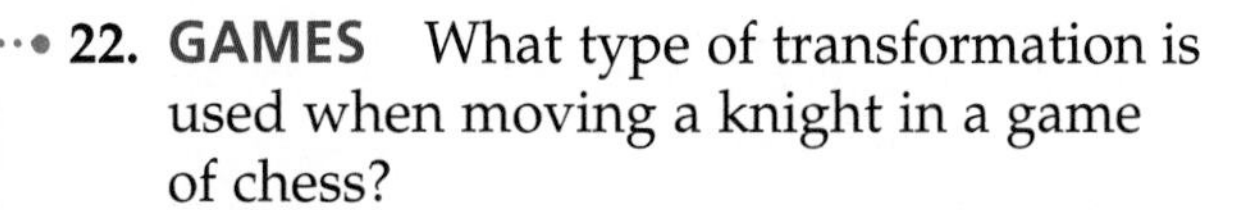

22. **GAMES** What type of transformation is used when moving a knight in a game of chess?

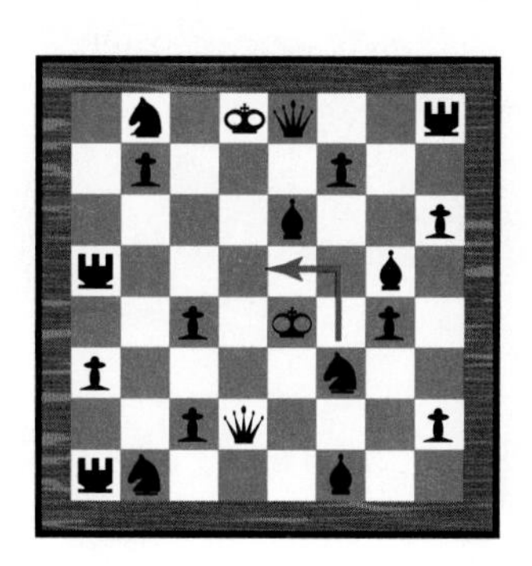

23. **MIRRORS** Which transformation exists when you look into a mirror?

24. **CRITICAL THINKING** After a rotation of 90° counterclockwise, the coordinates of the vertices of the image of $\triangle ABC$ are $A'(3, 2)$, $B'(0, 4)$, and $C'(5, 5)$. What were the coordinates of the vertices before the rotation?

25. WRITING IN MATH Answer the question that was posed at the beginning of the lesson.

How are the transformations involved in recreational activities?

Include the following in your answer:

- an explanation describing each type of transformation, and
- an explanation telling the type of transformation each recreational activity represents.

FCAT Practice

Standardized Test Practice

A B C D

For Exercises 26 and 27, suppose the figure shown is translated 4 units to the left and 3 units down.

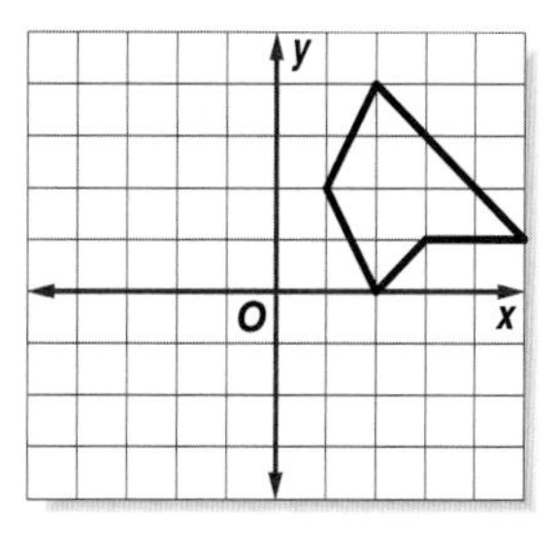

26. Which point is *not* a vertex of the translated image?

Ⓐ $(-2, -3)$ Ⓑ $(-4, -3)$

Ⓒ $(-3, -1)$ Ⓓ $(-2, 1)$

27. Which statement best describes this translation?

Ⓐ $(x, y) \rightarrow (x + 4, y + 3)$ Ⓑ $(x, y) \rightarrow (x - 4, y + 3)$

Ⓒ $(x, y) \rightarrow (x - 4, y - 3)$ Ⓓ $(x, y) \rightarrow (x + 4, y - 3)$

Maintain Your Skills

Mixed Review

Complete each congruence statement if $\triangle ABC \cong \triangle DEF$. *(Lesson 10-2)*

28. $\angle D \cong$ __?__ 29. $\overline{AC} \cong$ __?__ 30. $\overline{DE} \cong$ __?__

Find the value of x in each figure. *(Lesson 10-1)*

31.

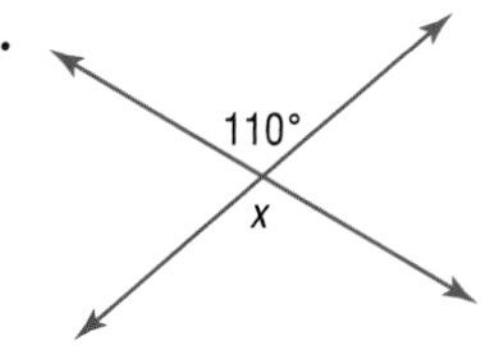

32.

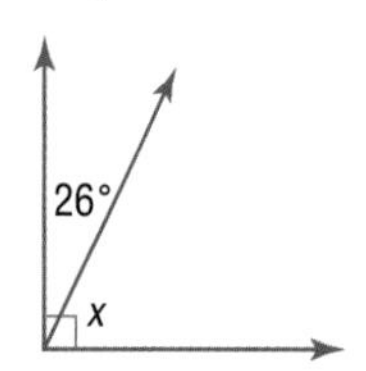

33. **ALGEBRA** Solve $x - 3.4 \geq 6.2$. Graph the solution on a number line. *(Lesson 7-4)*

34. Evaluate $|-4| - |3|$. *(Lesson 2-1)*

Getting Ready for the Next Lesson

PREREQUISITE SKILL Solve each equation.

*(To review **solving two-step equations**, see Lesson 3-5.)*

35. $2x + 134 = 360$ 36. $3x + 54 = 360$ 37. $5x + 125 = 360$

38. $4x + 92 = 360$ 39. $2x + 148 = 360$ 40. $6x + 102 = 360$

Algebra Activity

A Follow-Up of Lesson 10-3

Sunshine State Standards
MA.C.2.3.1-3

Dilations

In this activity, you will investigate **dilations**, which alter the size of a figure.

Collect the Data

Step 1 Draw and label a polygon on a coordinate plane. Trapezoid *ABCD* is shown.

Step 2 Suppose the scale factor is 2. Multiply the coordinates of each vertex by 2.

$A(-1, 2) \rightarrow A'(-2, 4)$ $B(0, 3) \rightarrow B'(0, 6)$
$C(4, 1) \rightarrow C'(8, 2)$ $D(2, -1) \rightarrow D'(4, -2)$

Step 3 Draw the new trapezoid.

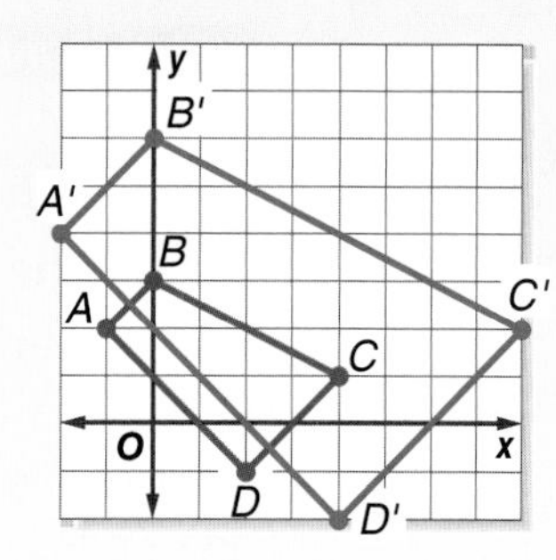

Analyze the Data

1. Use a protractor to measure the angles in each trapezoid. How do they compare?
2. Use a ruler to measure the sides of each trapezoid. How do they compare?
3. What ratio compares the measures of the corresponding sides?
4. Repeat the activity by multiplying the coordinates of trapezoid *ABCD* by $\frac{1}{2}$. Are the results the same? Explain.

Make a Conjecture

5. Explain how you know whether a dilation is a reduction or an enlargement.
6. Explain the difference between dilations and the other types of transformations.

Extend the Activity

Find the coordinates of the dilation image for the given scale factor, and graph the dilation image.

7. 3

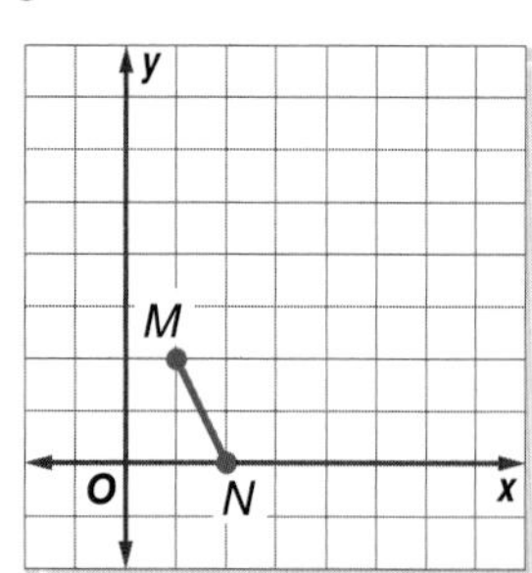

8. $\frac{1}{4}$

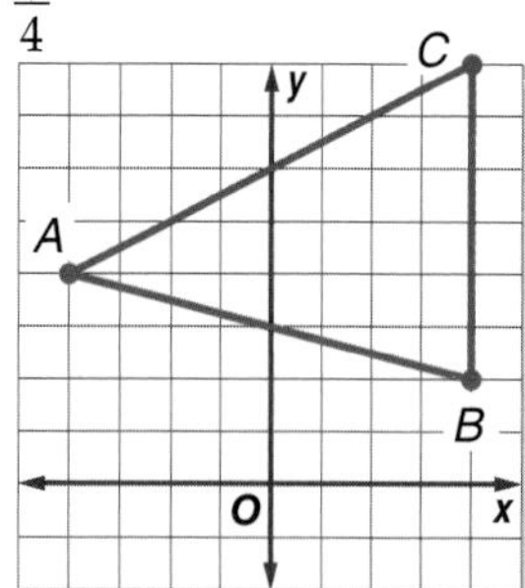

9. $1\frac{1}{2}$

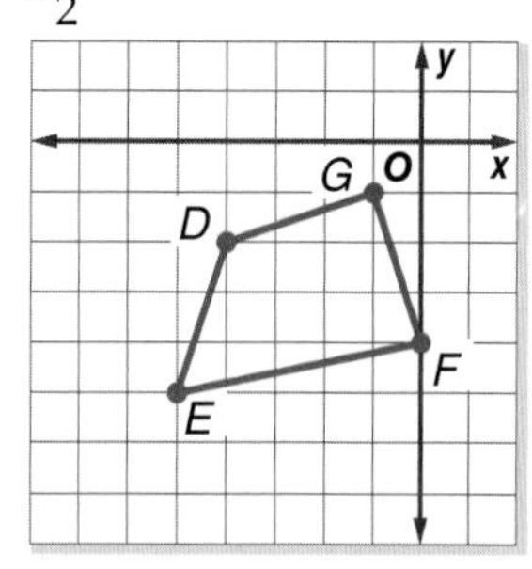

Identify each transformation as a *translation*, *rotation*, *reflection*, or *dilation*.

10.

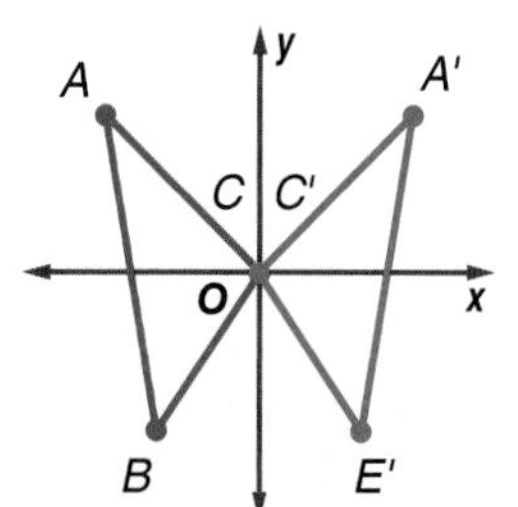

11.

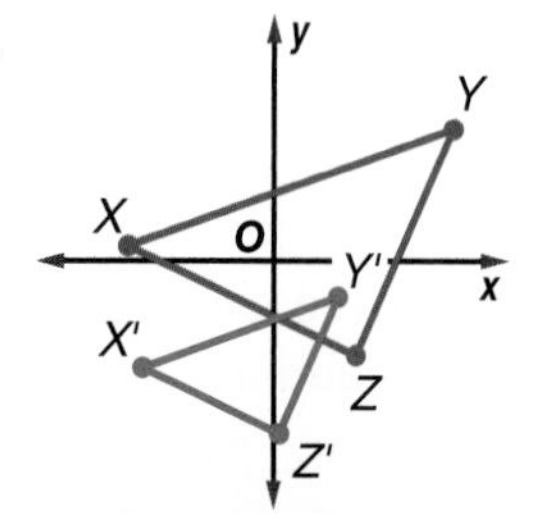

12.

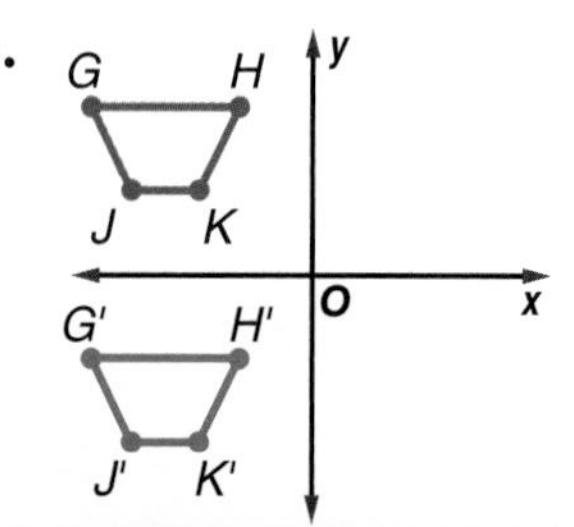

10-4 Quadrilaterals

Sunshine State Standards
MA.B.1.3.2-1, MA.C.1.3.1-1

What You'll Learn

- Find the missing angle measures of a quadrilateral.
- Classify quadrilaterals.

Vocabulary

- quadrilateral

How are quadrilaterals used in design?

Geometric figures are often used to create various designs. The design of a brick walkway is shown at the right. Notice how the walkway is formed using different-shaped bricks to create circles.

a. Describe the bricks that are used to create the smallest circles.

b. Describe how the shape of the bricks change as the circles get larger.

QUADRILATERALS Squares, rectangles, and trapezoids are examples of quadrilaterals. A **quadrilateral** is a closed figure with four sides and four vertices. The segments that make up a quadrilateral intersect only at their endpoints.

Quadrilaterals	*Not* Quadrilaterals

As with triangles, a quadrilateral can be named by its vertices. The quadrilateral below can be named quadrilateral $ABCD$.

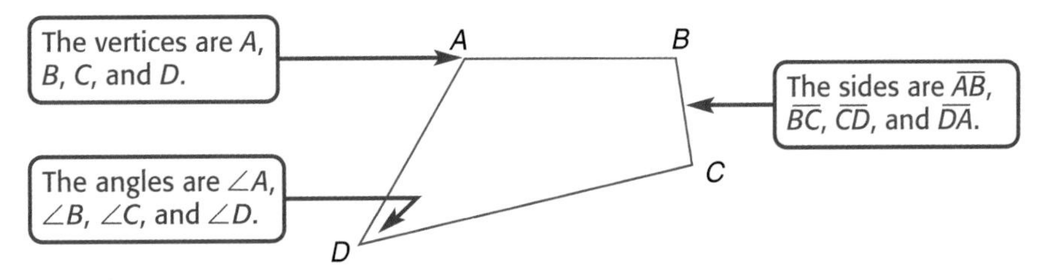

The quadrilateral shown above has many names. For example, it can also be named quadrilateral $BCDA$, quadrilateral $DABC$, or quadrilateral $CBAD$.

Study Tip

Naming Quadrilaterals When you name a quadrilateral, you can begin at any vertex. However, it is important to name vertices in order.

Concept Check How many vertices does a quadrilateral have?

A quadrilateral can be separated into two triangles. Since the sum of the measures of the angles of a triangle is 180°, the sum of the measures of the angles of a quadrilateral is 2(180°) or 360°.

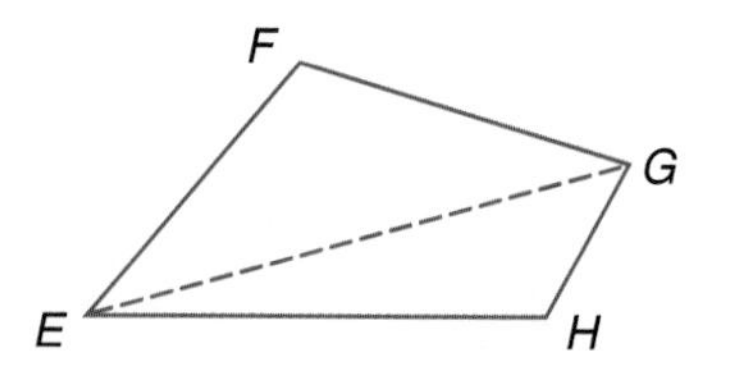

F
G
E
H

Key Concept — *Angles of a Quadrilateral*

The sum of the measures of the angles of a quadrilateral is 360°.

Example 1 Find Angle Measures

ALGEBRA Find the value of x. Then find each missing angle measure.

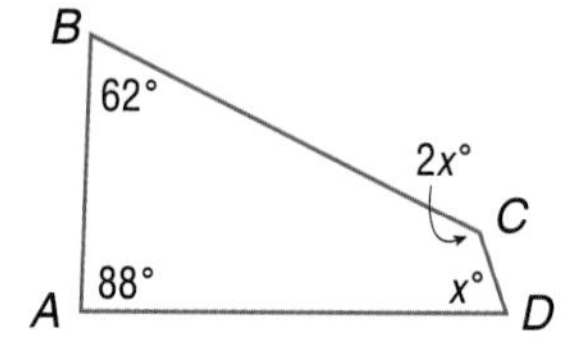

B
62°
2x°
C
A
88°
x°
D

Words The sum of the measures of the angles is 360°.

Variable Let $m\angle A$, $m\angle B$, $m\angle C$, and $m\angle D$ represent the measures of the angles.

Equation

$m\angle A + m\angle B + m\angle C + m\angle D = 360$	Angles of a quadrilateral
$88 + 62 + 2x + x = 360$	Substitution
$3x + 150 = 360$	Combine like terms.
$3x + 150 - 150 = 360 - 150$	Subtract 150 from each side.
$3x = 210$	Simplify.
$x = 70$	

The value of x is 70. So, $m\angle D = 70°$ and $m\angle C = 2(70)$ or 140°.

Study Tip

Check Your Work

To check the answer, find the sum of the measures of the angles. Since $88° + 62° + 140° + 70° = 360°$, the answer is correct.

CLASSIFY QUADRILATERALS The diagram below shows how quadrilaterals are related. Notice that it goes from the most general quadrilateral to the most specific.

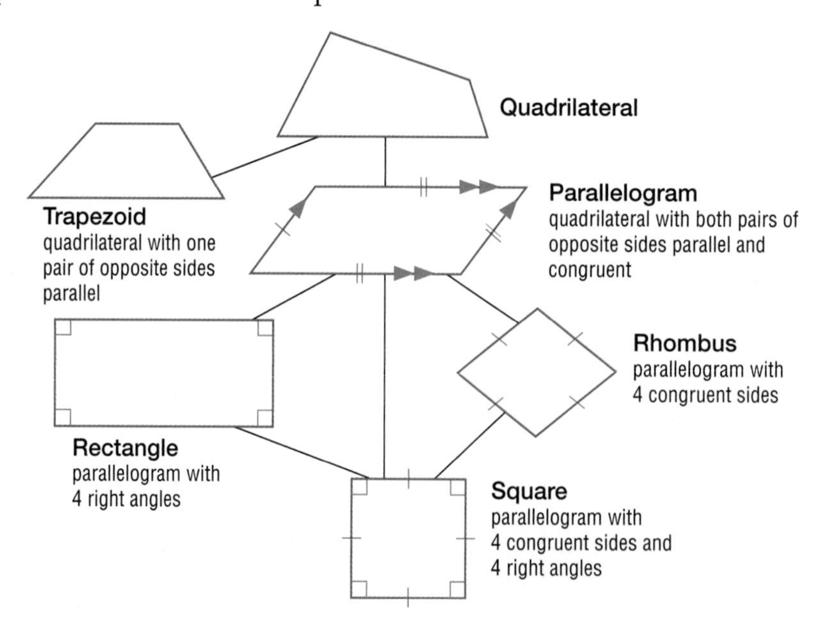

Quadrilateral
Trapezoid
quadrilateral with one pair of opposite sides parallel
Parallelogram
quadrilateral with both pairs of opposite sides parallel and congruent
Rhombus
parallelogram with 4 congruent sides
Rectangle
parallelogram with 4 right angles
Square
parallelogram with 4 congruent sides and 4 right angles

The best description of a quadrilateral is the one that is the most specific.

Example 2 Classify Quadrilaterals

Classify each quadrilateral using the name that best describes it.

a.

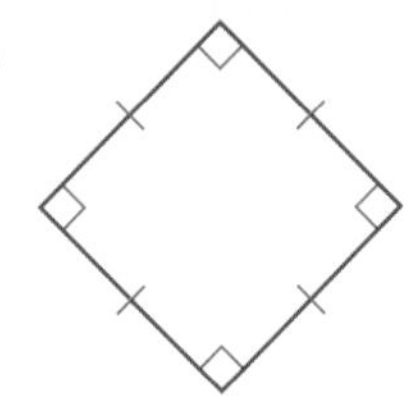

The quadrilateral has four congruent sides and four right angles. It is a square.

b.

The quadrilateral has opposite sides parallel and opposite sides congruent. It is a parallelogram.

c. ART Classify the quadrilaterals that are outlined in the painting at the right.

Each of the quadrilaterals has four right angles, but the four sides are not congruent. The quadrilaterals are rectangles.

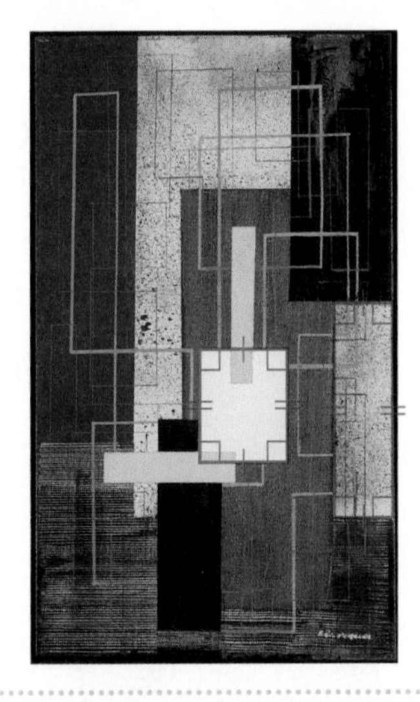

Irene Rice Periera. *Untitled.* 1951

Career Choices

Artist

Visual artists create paintings, sculptures, or illustrations to communicate ideas. Many aspects of geometry can be found in such artwork.

Online Research
For information about a career as a visual artist, visit: www.pre-alg.com/careers

Check for Understanding

Concept Check

1. **OPEN ENDED** Give a real-world example of a quadrilateral, a parallelogram, a rhombus, and a square.
2. **Describe** the characteristics of a rectangle and draw an example of one.

Guided Practice

ALGEBRA Find the value of x. Then find the missing angle measures.

3.

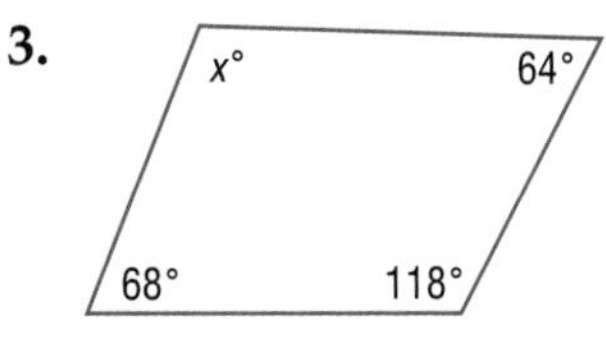

4.

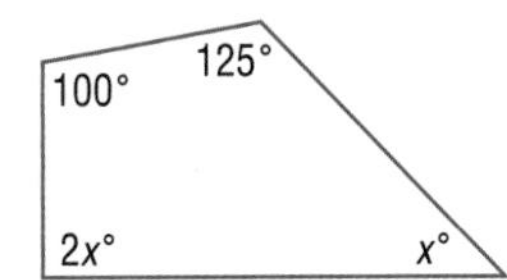

Classify each quadrilateral using the name that *best* describes it.

5.

6.

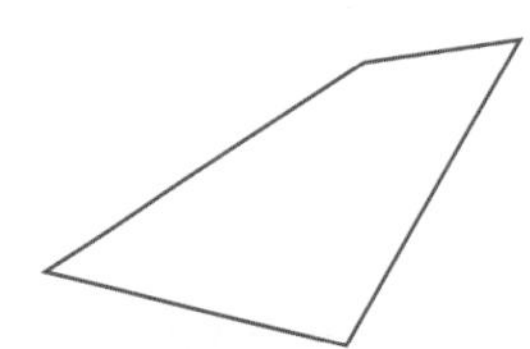

Application

7. **SPORTS** Classify the quadrilaterals that are found on the scoring region of a shuffleboard court.

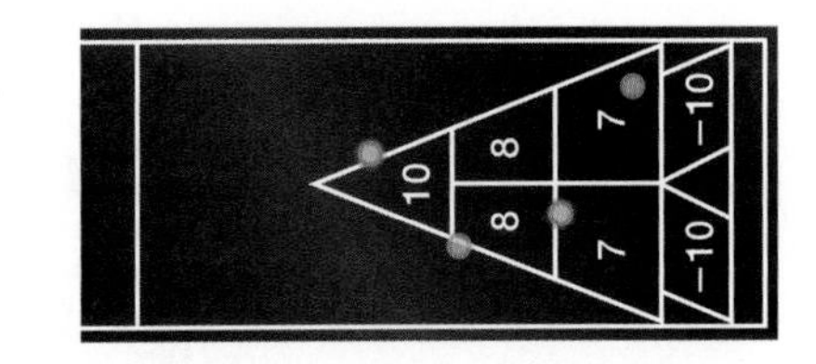

Practice and Apply

Homework Help

For Exercises	See Examples
8–13	1
14–22	2

Extra Practice
See page 748.

ALGEBRA Find the value of x. Then find the missing angle measures.

8.
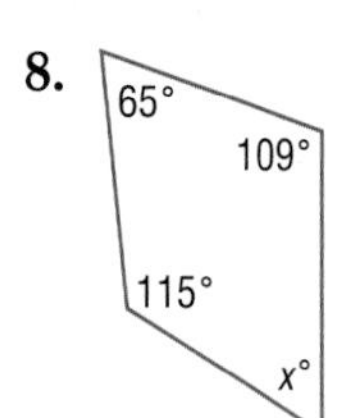

9.
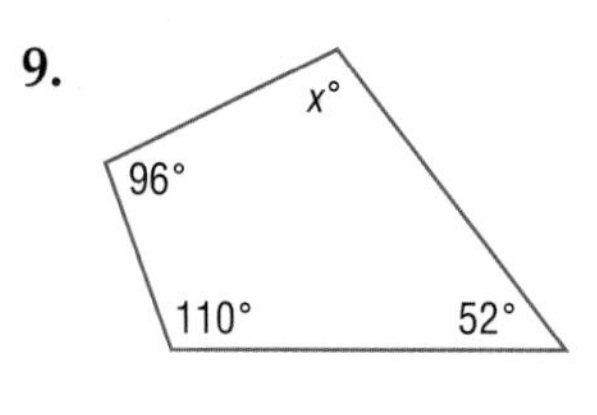

10.
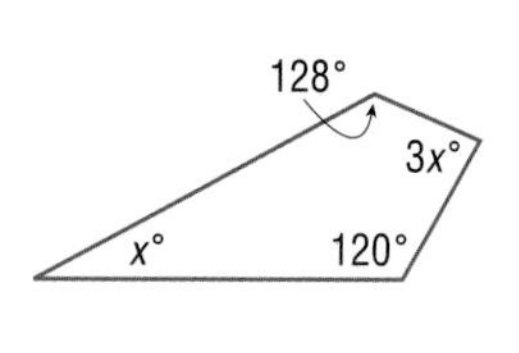

11.

12.
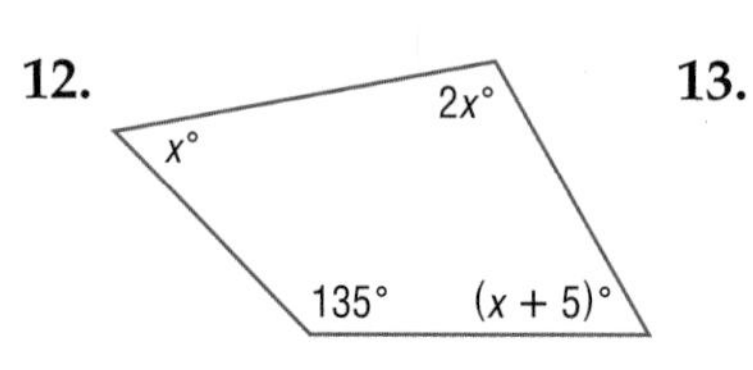

13.
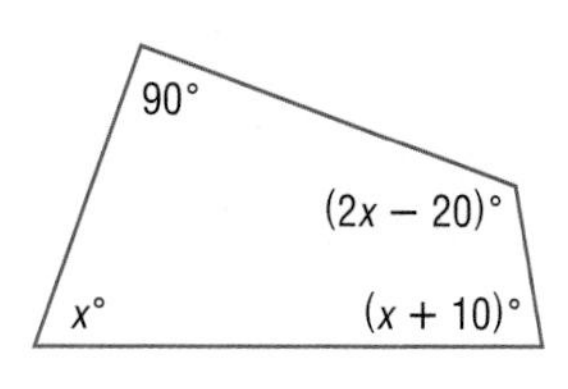

14. **COOKING** Name an item found in a kitchen that is rectangular in shape. Explain why the item is a rectangle.

15. **GAMES** Identify a board game that is played on a board that is shaped like a square. Explain why the board is a square.

Classify each quadrilateral using the name that *best* describes it.

16.
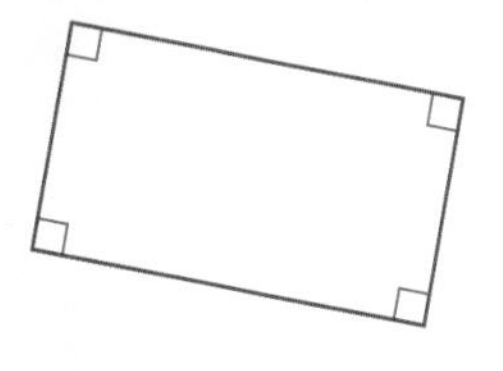

17.
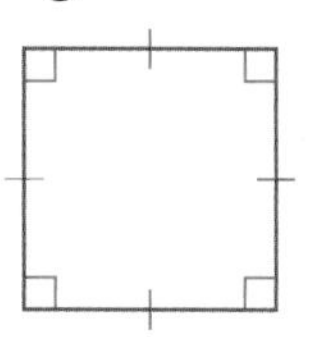

18.
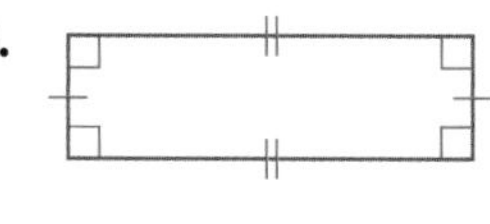

19.

20.
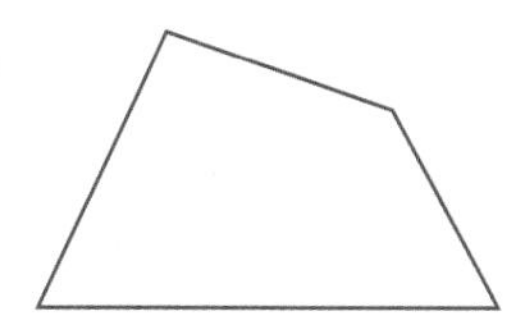

21.
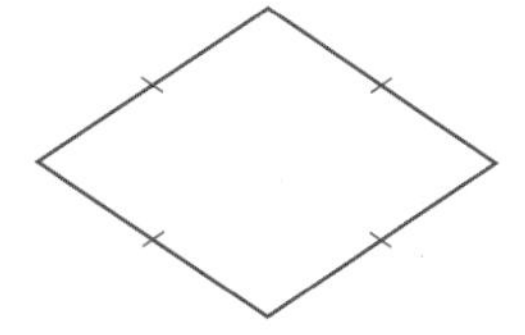

22. **ART** The abstract painting is an example of how shape and color are used in art. Write a few sentences describing the geometric shapes used by the artist.

Elizabeth Murray. *Painter's Progress,* 1981

Tell whether each statement is *sometimes, always,* or *never* true.

23. A square is a rhombus.

24. A parallelogram is a rectangle.

25. A rectangle is a square.

26. A parallelogram is a quadrilateral.

Make a drawing of each quadrilateral. Then classify each quadrilateral using the name that *best* describes it.

27. In quadrilateral $JKLM$, $m\angle J = 90°$, $m\angle K = 50°$, $m\angle L = 90°$, and $m\angle M = 130°$.

28. In quadrilateral $CDEF$, $\overline{CD}$ and $\overline{EF}$ are parallel, and $\overline{CF}$ and $\overline{DE}$ are parallel. Angle C is not congruent to $\angle D$.

29. **CRITICAL THINKING** An *equilateral* figure is one in which all sides have the same measure. An *equiangular* figure is one in which all angles have the same measure.

a. Is it possible for a quadrilateral to be equilateral without being equiangular? If so, explain with a drawing.

b. Is it possible for a quadrilateral to be equiangular without being equilateral? If so, explain with a drawing.

30. **WRITING IN MATH** Answer the question that was posed at the beginning of the lesson.

How are quadrilaterals used in design?

Include the following in your answer:

- an example of a real-world design that contains quadrilaterals, and
- an explanation of the figures used in the design.

31. Which figure is a rhombus?

Ⓐ 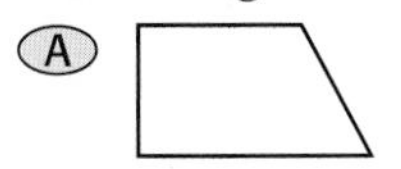 Ⓑ 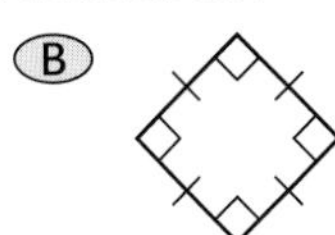 Ⓒ 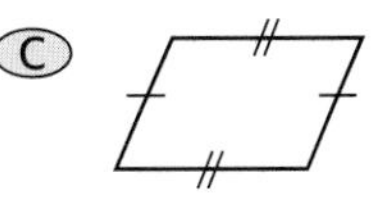 Ⓓ 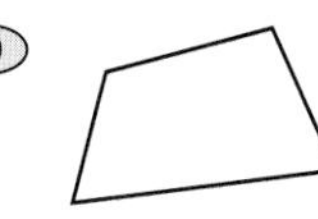

32. **GRID IN** Find the value of x in the figure at the right.

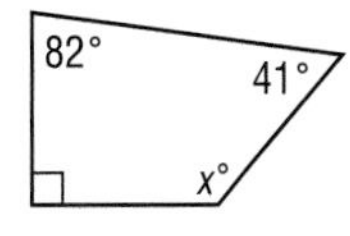

Maintain Your Skills

Mixed Review

33. A figure has vertices $D(1, 2)$, $E(1, 4)$, $F(-4, 4)$, and $G(-2, 2)$. Graph the figure and its image after a translation 4 units down. *(Lesson 10-3)*

Complete each congruence statement if $\triangle AKM \cong \triangle NDQ$. *(Lesson 10-2)*

34. $\angle K \cong$ __?__

35. $\overline{QN} \cong$ __?__

36. Find the discount for a \$45 shirt that is on sale for 20% off. *(Lesson 6-7)*

Getting Ready for the Next Lesson

PREREQUISITE SKILL Find each product.
(To review ***multiplying decimals****, see page 715.)*

37. (3)(4.8)

38. (5.4)(6)

39. (9.2)(3.1)

40. (10.5)(5.7)

Practice Quiz 1

Lessons 10-1 through 10-4

1. Angle A and $\angle B$ are complementary. Find $m\angle A$ if $m\angle B = 55°$. *(Lesson 10-1)*

2. Suppose $\triangle ABC \cong \triangle DEF$. Which angle is congruent to $\angle D$? *(Lesson 10-2)*

3. $\triangle QRS$ has vertices $Q(3, 3)$, $R(5, 6)$, and $S(7, 3)$. Find the coordinates for the vertices of the triangle after the figure is reflected over the x-axis. *(Lesson 10-3)*

ALGEBRA Find the value of x. Then find the missing angle measures. *(Lesson 10-4)*

4.

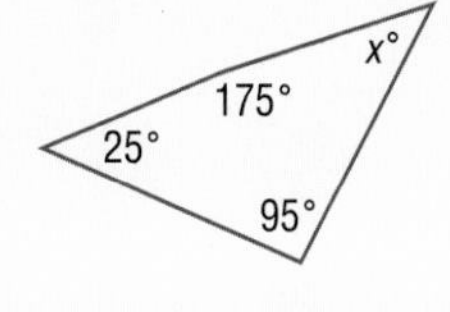

5.

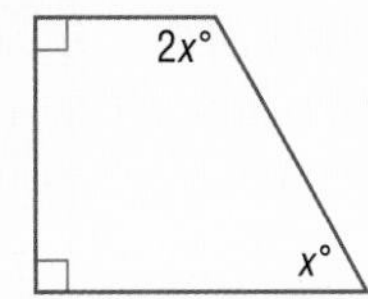

Algebra Activity

A Preview of Lesson 10-5

Sunshine State Standards
MA.B.1.3.1-1, MA.C.1.3.1-2

Area and Geoboards

Activity 1

One square on a geoboard has an area of one square unit.

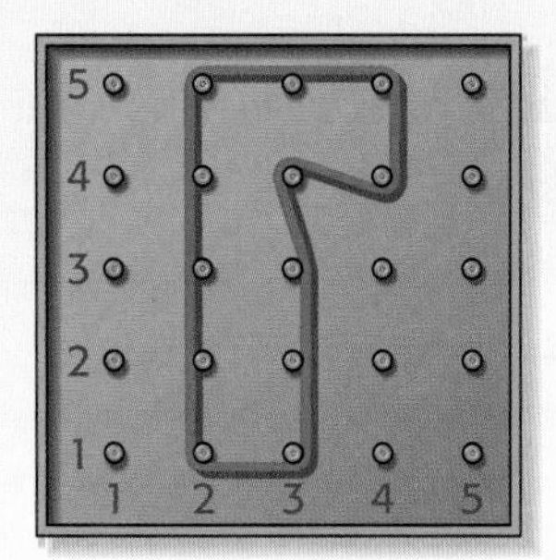

The area is about 5 square units.

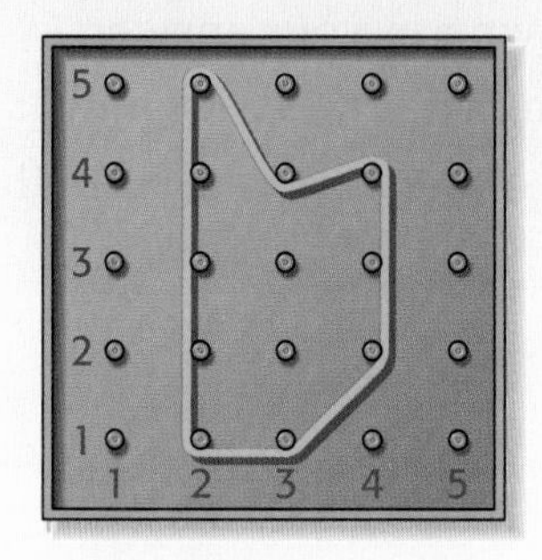

The area is about 6 square units.

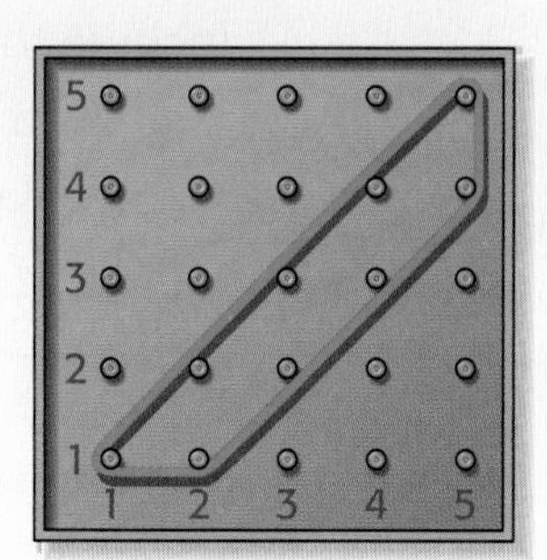

The area is 3.5 square units.

Model and Analyze

Find the area of each figure. Estimate, if necessary.

1.

2.

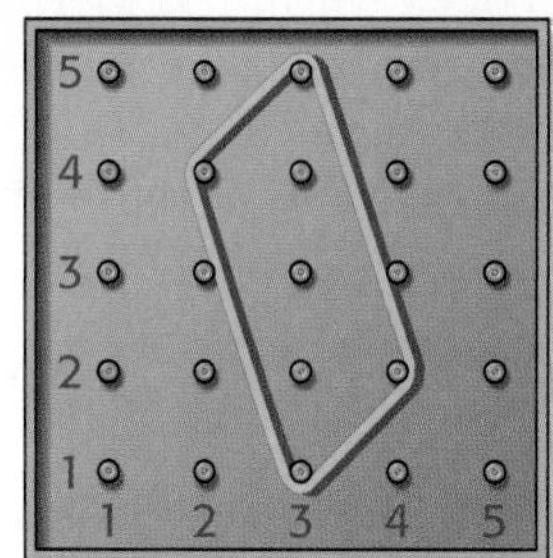

3.

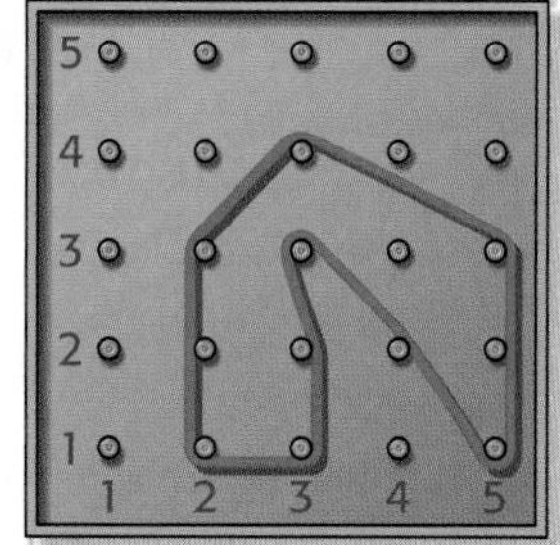

4. Explain how you found the area of each figure in Exercises 1–3.

5. Make a figure on the geoboard. Ask a classmate to find the area of the figure.

Activity 2

The following example shows how to find the area of a right triangle on a geoboard.

Step 1 First, make another triangle so that the two triangles form a rectangle.

Step 2 Then find the area of the rectangle.

Step 3 Next, divide by 2 to find the area of each triangle.

The area of the rectangle is 6 square units. So, the area of each triangle is 3 square units.

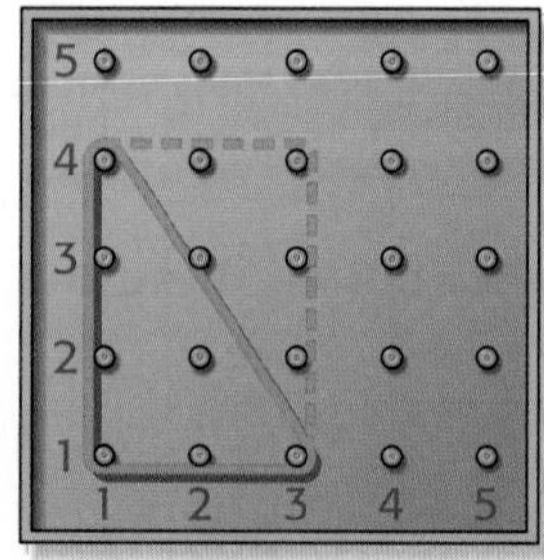

Model and Analyze

Find the area of each triangle.

6. 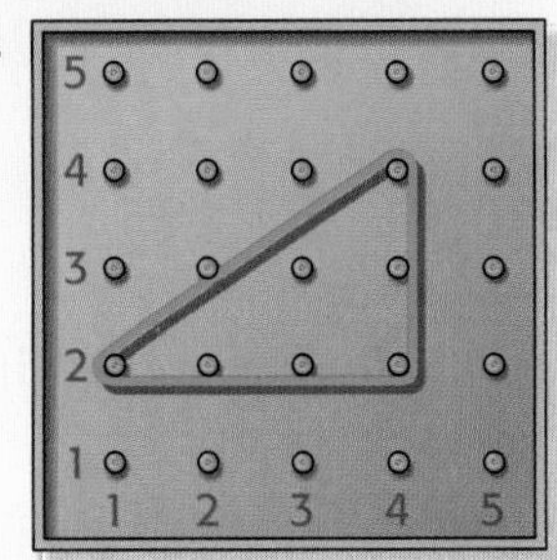

7. 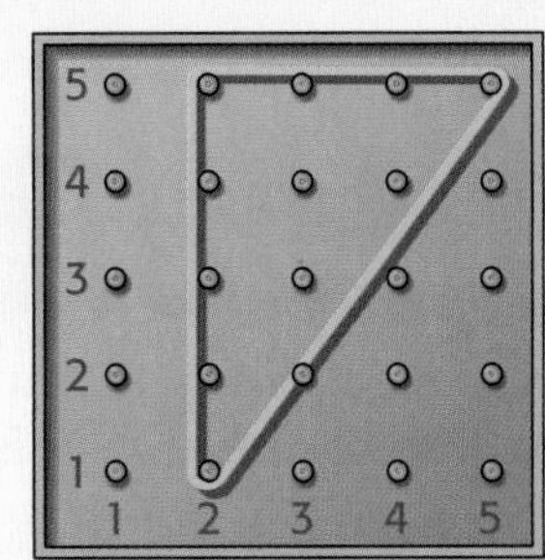

8. 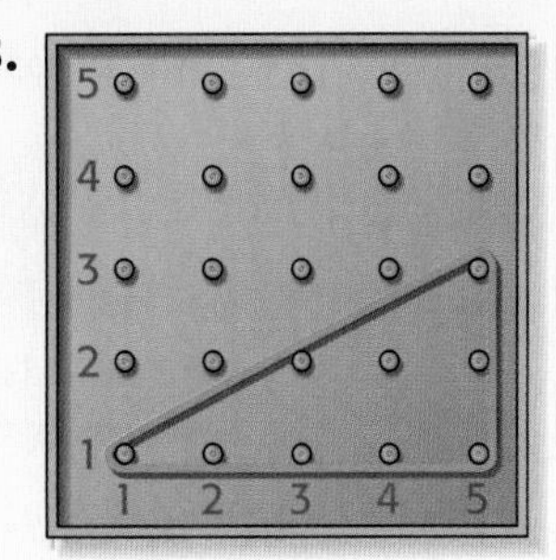

9. Make a right triangle on your geoboard. Find its area using this method.

10. Write a few sentences explaining how you found the area of the triangle.

Activity 3

Another way to find the area of a figure on a geoboard is to build a rectangle around the figure. Consider the following example.

Step 1 Make the triangle shown on a geoboard.

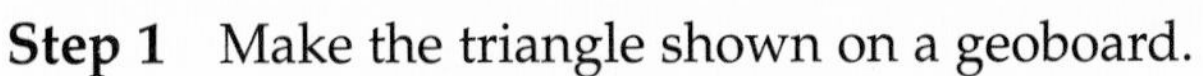

Step 2 Build a rectangle around the triangle.

Step 3 Subtract to find the area of the original triangle.

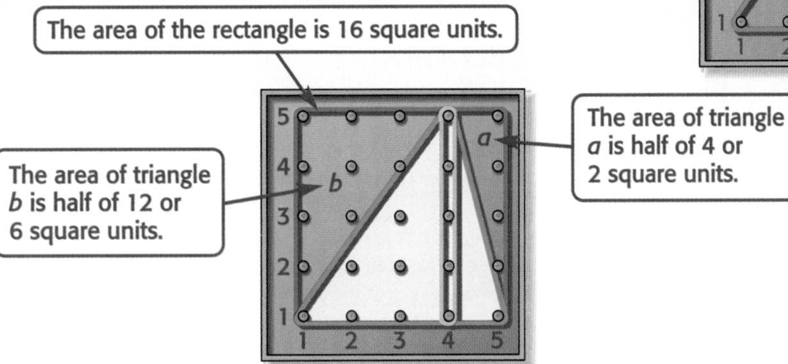
The area of the rectangle is 16 square units.
The area of triangle b is half of 12 or 6 square units.
The area of triangle a is half of 4 or 2 square units.
a
b

The area of the original triangle is $16 - 2 - 6$ or 8 square units.

Model and Analyze

Find the area of each figure by building a rectangle around the figure.

11. 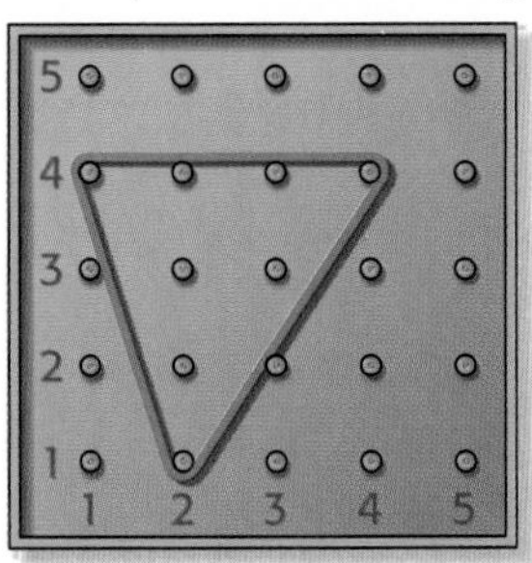

12. 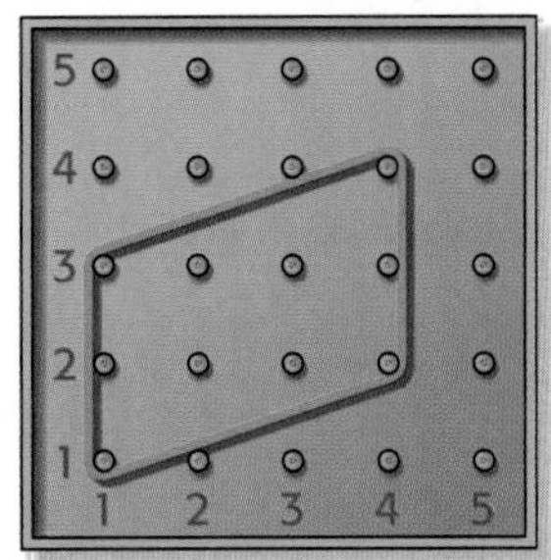

13. 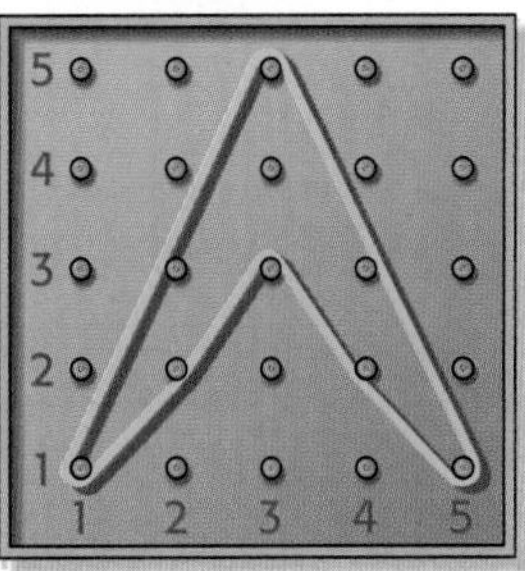

14. Make a figure on your geoboard. Find the area using this method.

10-5 Area: Parallelograms, Triangles, and Trapezoids

Sunshine State Standards
MA.B.1.3.1-1, MA.C.1.3.1-4

What You'll Learn

- Find area of parallelograms.
- Find the areas of triangles and trapezoids.

Vocabulary

- base
- altitude

How is the area of a parallelogram related to the area of a rectangle?

The area of a rectangle can be found by multiplying the length and width. The rectangle shown below has an area of 3×6 or 18 square units.

Suppose a triangle is cut from one side of the rectangle and moved to the other side.

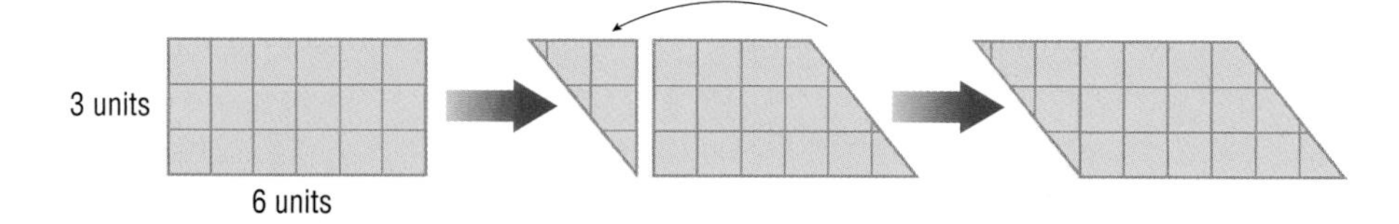

a. What figure is formed?

b. Compare the area of the rectangle to the area of the parallelogram.

c. What parts of a rectangle and parallelogram determine their area?

AREAS OF PARALLELOGRAMS The area of a parallelogram can be found by multiplying the measures of the base and the height.

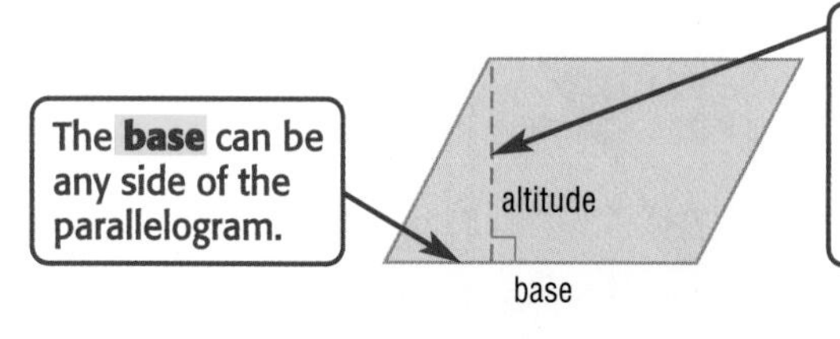

The **base** can be any side of the parallelogram.
altitude
base
The *height* is the length of an **altitude**, a line segment perpendicular to the bases with endpoints on the base and the side opposite the base.

Key Concept

- **Words** If a parallelogram has a base of b units and a height of h units, then the area A is bh square units.
- **Symbols** $A = bh$
- **Model**

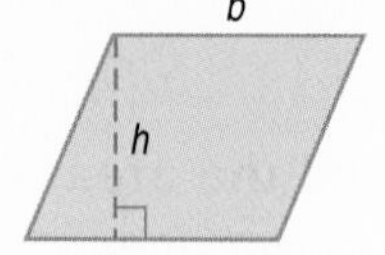

Example 1 Find Areas of Parallelograms

Find the area of each parallelogram.

a.

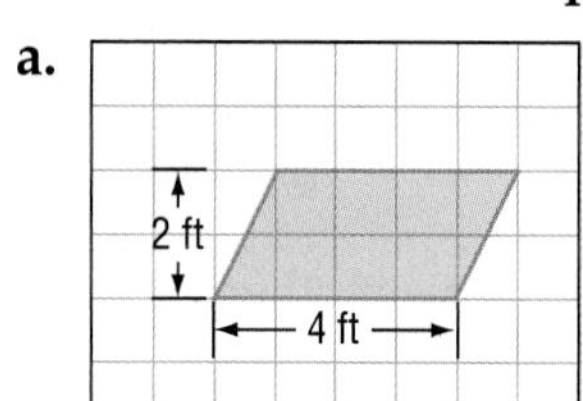

The base is 4 feet. The height is 2 feet.

$A = bh$	Area of a parallelogram
$A = 4 \cdot 2$	Replace b with 4 and h with 2.
$A = 8$	Multiply.

The area is 8 square feet.

b.

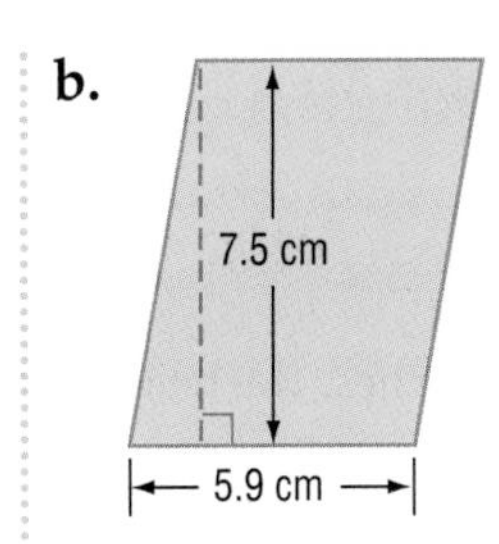

The base is 5.9 centimeters. The height is 7.5 centimeters.

$A = bh$	Area of a parallelogram
$A = (5.9)(7.5)$	Replace b with 5.9 and h with 7.5.
$A = 44.25$	Multiply.

The area is 44.25 square centimeters.

AREA OF TRIANGLES AND TRAPEZOIDS A diagonal of a parallelogram separates the parallelogram into two congruent triangles. The area of each triangle is one-half the area of the parallelogram.

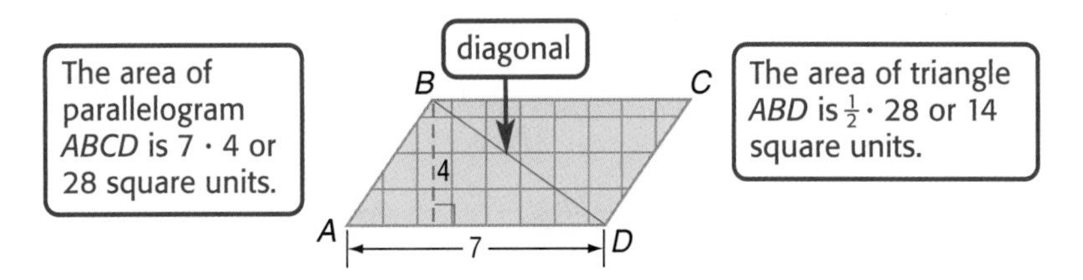

Using the formula for the area of a parallelogram, we can find the formula for the area of a triangle.

Key Concept — *Area of a Triangle*

- **Words** If a triangle has a base of b units and a height of h units, then the area A is $\frac{1}{2}bh$ square units.
- **Symbols** $A = \frac{1}{2}bh$
- **Model**

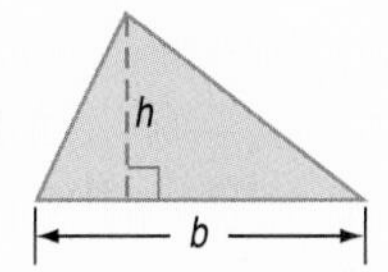

Concept Check The area of a triangle is one half of the area of what figure with the same height and base?

As with parallelograms, any side of a triangle can be used as a base. The height is the length of a corresponding altitude.

Study Tip

Altitudes
An altitude can be outside the triangle.

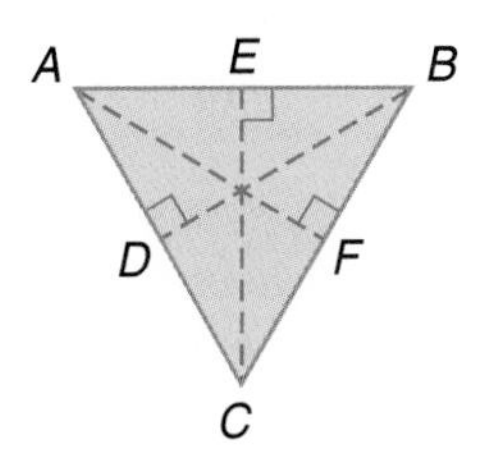

base	$\overline{AC}$	$\overline{AB}$	$\overline{BC}$
altitude	$\overline{BD}$	$\overline{CE}$	$\overline{AF}$

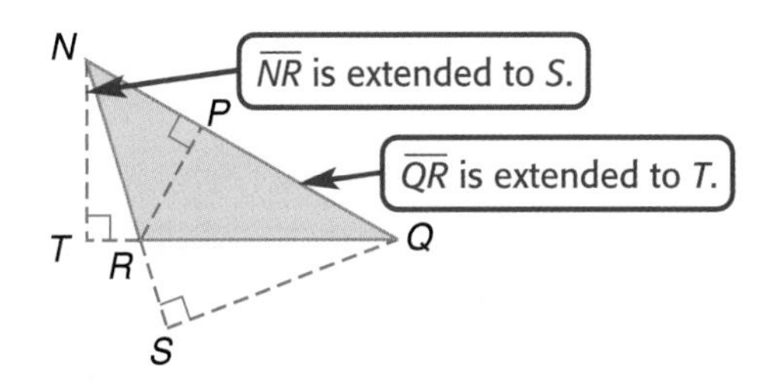

base	$\overline{NQ}$	$\overline{NR}$	$\overline{QR}$
altitude	$\overline{RP}$	$\overline{QS}$	$\overline{NT}$

Concept Check Which side of a triangle can be used as the base?

Example 2 Find Areas of Triangles

Find the area of each triangle.

a.

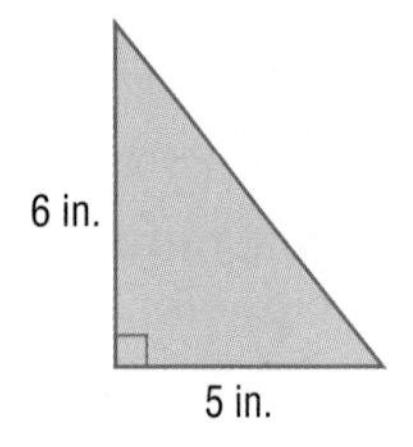

The base is 5 inches. The height is 6 inches.

$A = \frac{1}{2}bh$ — Area of a triangle

$A = \frac{1}{2}(5)(6)$ — Replace b with 5 and h with 6.

$A = \frac{1}{2}(30)$ — Multiply. $5 \times 6 = 30$

$A = 15$ — The area of the triangle is 15 square inches.

Study Tip

Alternative Method Multiplication is commutative and associative. So you can also find $\frac{1}{2}$ of 6 first and then multiply by 5.

b.

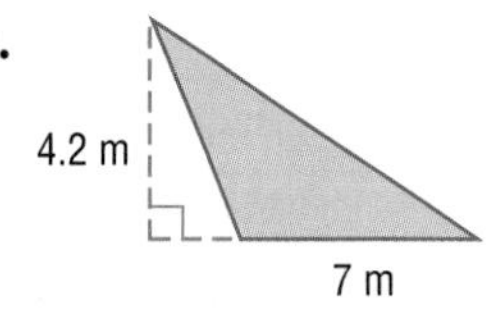

The base is 7 meters. The height is 4.2 meters.

$A = \frac{1}{2}bh$ — Area of a triangle

$A = \frac{1}{2}(7)(4.2)$ — Replace b with 7 and h with 4.2.

$A = \frac{1}{2}(29.4)$ — Multiply. $7 \times 4.2 = 29.4$

$A = 14.7$ — The area of the triangle is 14.7 square meters.

A trapezoid has two bases. The height of a trapezoid is the distance between the bases. A trapezoid can be separated into two triangles.

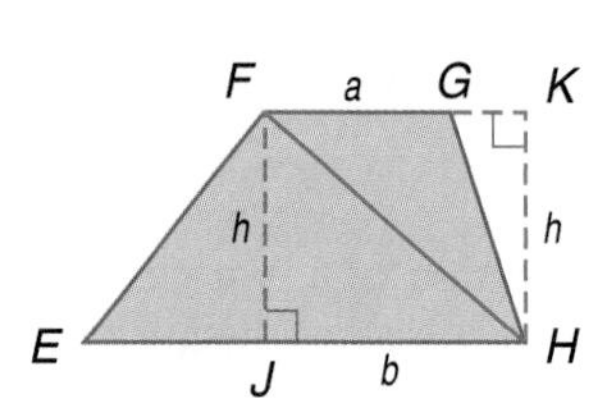

The triangles are $\triangle FGH$ and $\triangle EFH$.
The measure of a base of $\triangle FGH$ is a units.
The measure of a base of $\triangle EFH$ is b units.
The altitudes of the triangles, $\overline{FJ}$ and $\overline{HK}$, are congruent. Both are h units long.

$$\text{area of trapezoid } EFGH = \underbrace{\text{area of } \triangle FGH}_{\frac{1}{2}ah} + \underbrace{\text{area of } \triangle EFH}_{\frac{1}{2}bh}$$

$$= \frac{1}{2}ah + \frac{1}{2}bh$$

$$= \frac{1}{2}h(a + b) \quad \text{Distributive Property}$$

Key Concept — Area of a Trapezoid

- **Words** If a trapezoid has bases of a units and b units and a height of h units, then the area A of the trapezoid is $\frac{1}{2}h(a + b)$ square units.
- **Symbols** $A = \frac{1}{2}h(a + b)$
- **Model**

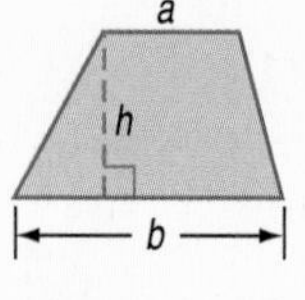

Example 3 Find Area of a Trapezoid

Find the area of the trapezoid.

The height is 4 inches.

The bases are $6\frac{1}{2}$ inches and $3\frac{1}{4}$ inches.

$A = \frac{1}{2}h(a + b)$ — Area of a trapezoid

$A = \frac{1}{2} \cdot 4\left(6\frac{1}{2} + 3\frac{1}{4}\right)$ — Replace h with 4, a with $6\frac{1}{2}$ and b with $3\frac{1}{4}$.

$A = \frac{1}{2} \cdot 4 \cdot 9\frac{3}{4}$ — $6\frac{1}{2} + 3\frac{1}{4} = 9\frac{3}{4}$

$A = \frac{1}{2} \cdot \frac{\cancel{4}}{1} \cdot \frac{39}{\cancel{4}}$ — Divide out the common factors.

$A = \frac{39}{2}$ or $19\frac{1}{2}$ — The area of the trapezoid is $19\frac{1}{2}$ square inches.

Study Tip

Look Back
To review **multiplying fractions**, see Lesson 5-3.

Example 4 Use Area to Solve a Problem

FLAGS The signal flag shown represents the number five. Find the area of the blue region.

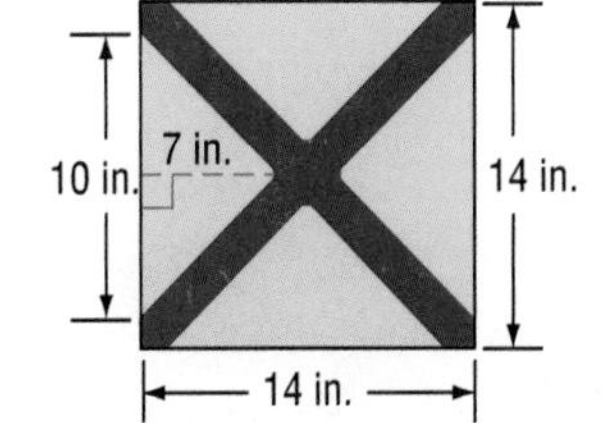

To find the area of the blue region, subtract the areas of the triangles from the area of the square.

Area of the square	**Area of each triangle**
$A = bh$	$A = \frac{1}{2}bh$
$A = 14 \cdot 14$	$A = \frac{1}{2} \cdot 10 \cdot 7$
$A = 196$	$A = 35$ $\quad \frac{1}{2} \cdot 10 = 5,\ 5 \cdot 7 = 35$

The total area of the triangles is 4(35) or 140 square inches. So, the area of the blue region is 196 − 140 or 56 square inches.

Check for Understanding

Concept Check

1. **OPEN ENDED** Draw and label a parallelogram that has an area of 24 square inches.
2. **Define** an *altitude* of a triangle and draw an example.

Guided Practice

Find the area of each figure.

3.

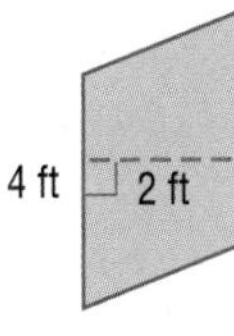

4.

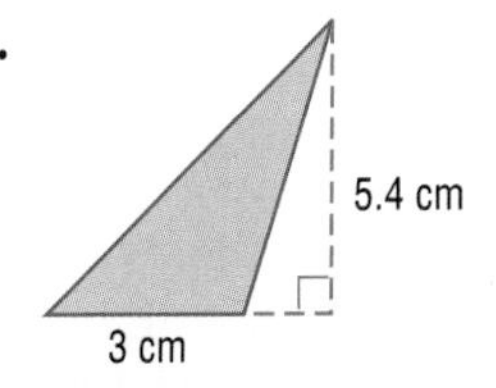

5.

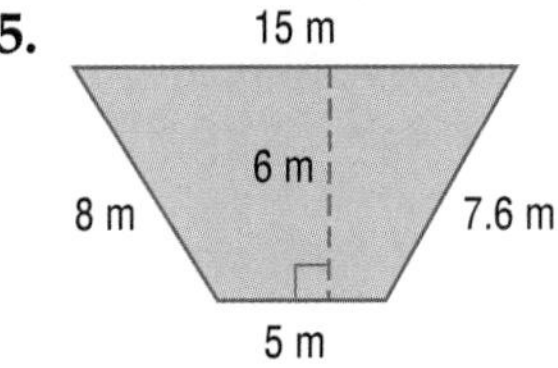

Application

6. **FLAGS** The flag shown at the right is the international signal for the number three. Find the area of the red region.

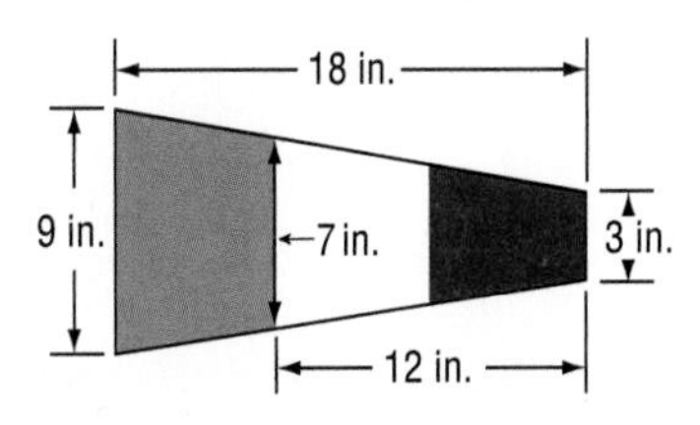

Practice and Apply

Homework Help

For Exercises	See Examples
7–20	1–3
21–25, 29	4

Extra Practice
See page 749.

Find the area of each figure.

7.
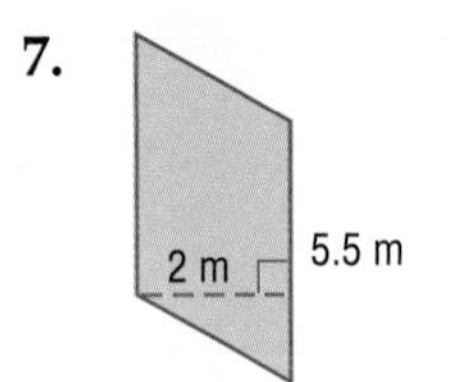

8.
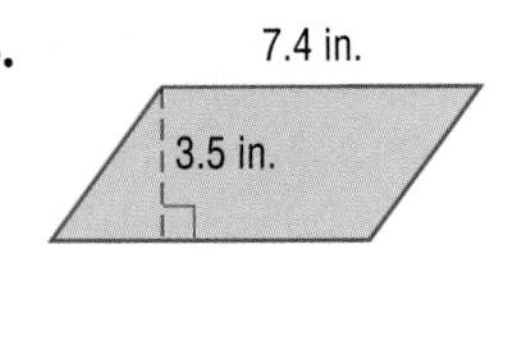

9.
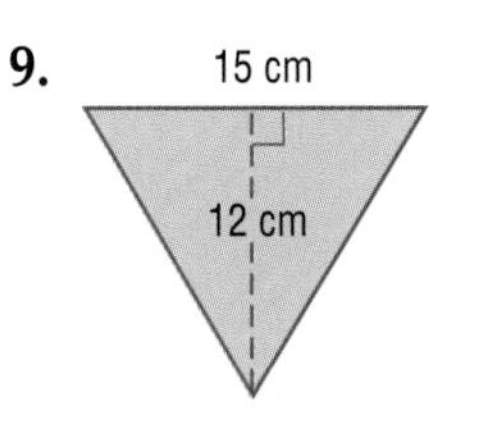

10.
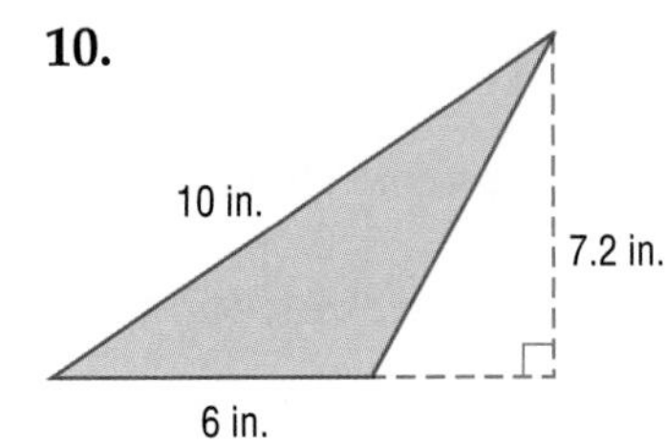

11.
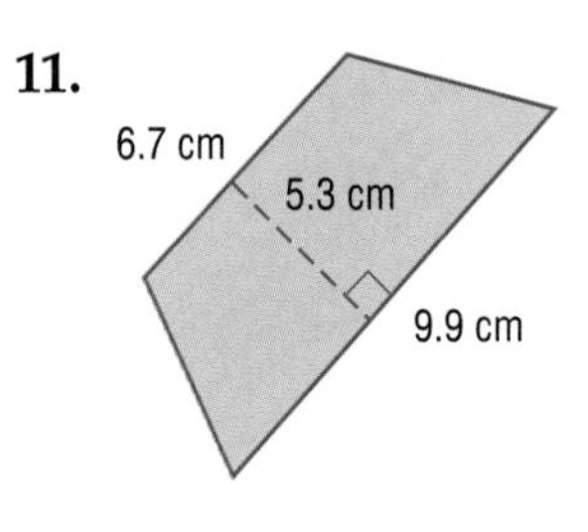

12.
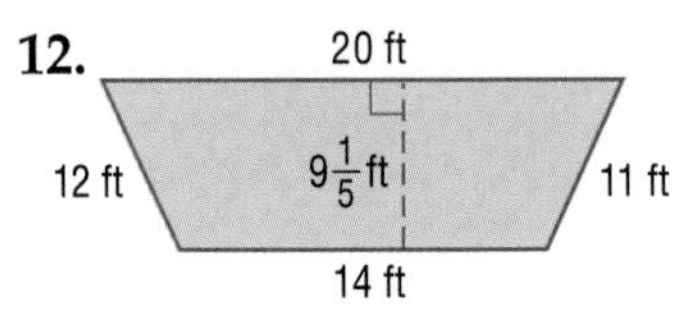

Find the area of each figure described.

13. triangle: base, 8 in.; height, 7 in.

14. trapezoid: height, 2 cm; bases, 3 cm, 6 cm

15. parallelogram: base, 3.8 yd; height, 6 yd

16. triangle: base, 9 ft; height, 3.2 ft

17. trapezoid: height, 3.5 m; bases, 10 m and 11 m

18. parallelogram: base, 5.6 km; height, 4.5 km

More About. . .

Geography

The largest U.S. state is Alaska. It has an area of 615,230 square miles. Rhode Island is the smallest state. It has an area of 1231 square miles.

Source: *The World Almanac*

GEOGRAPHY **For Exercises 19 and 20, use the approximate measurements to estimate the area of each state.**

19.

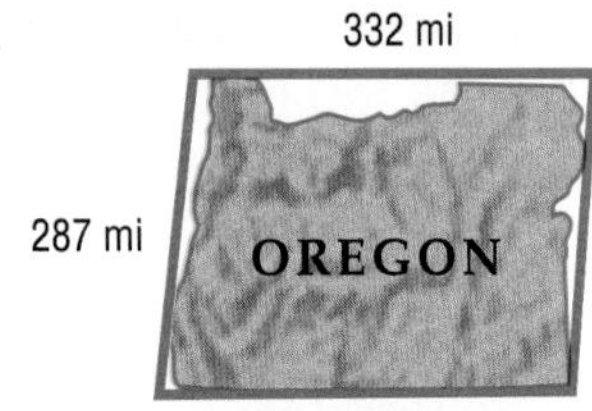

20.
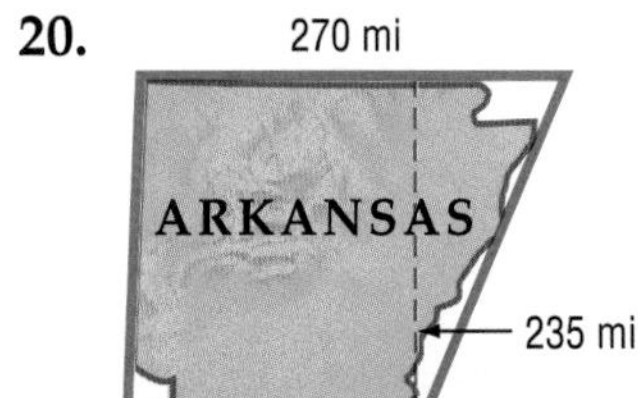

Find the area of each figure.

21.
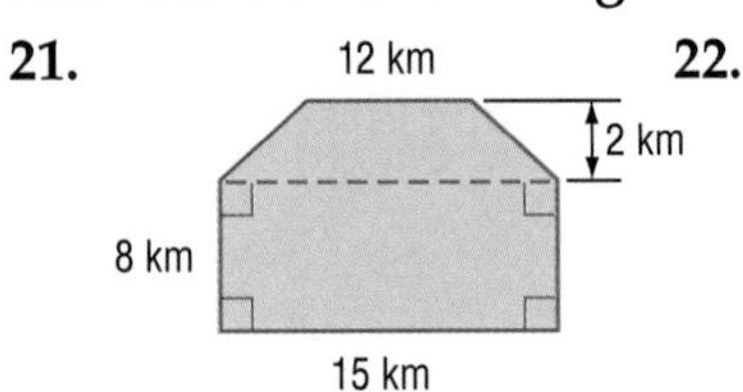

22.
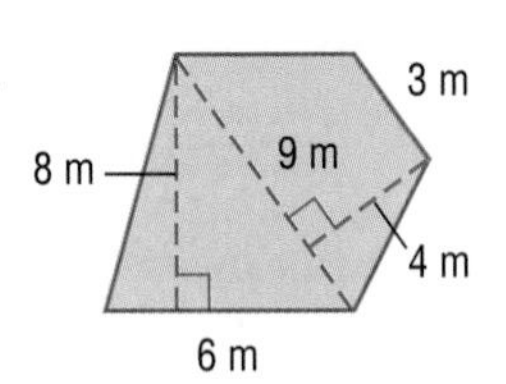

23.
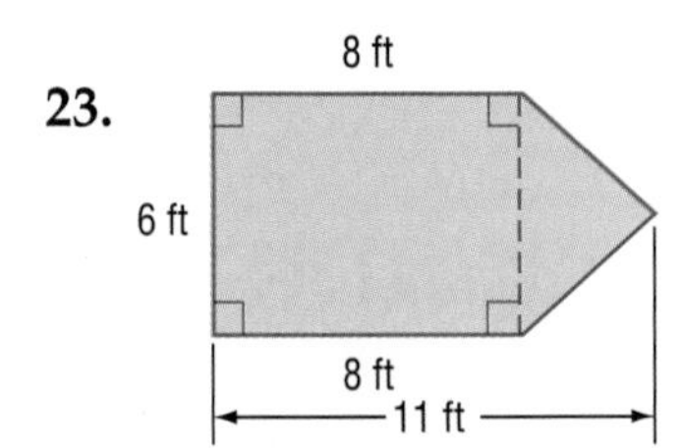

LAWNCARE **For Exercises 24 and 25, use the diagram shown and the following information.** Mrs. Malone plans to fertilize her lawn. The fertilizer she will be using indicates that one bag fertilizes 2000 square feet.

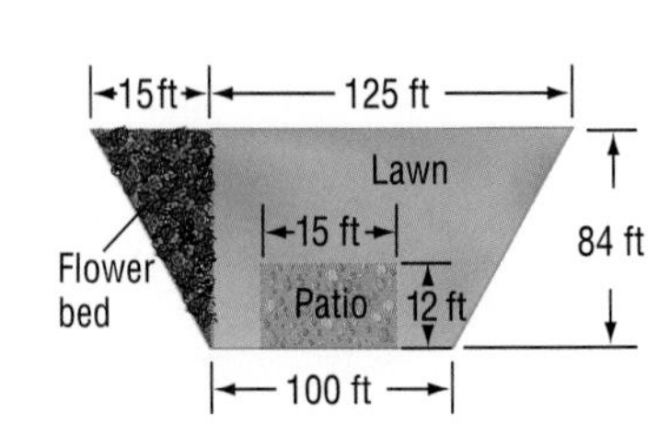

24. Find the area of the lawn.

25. How many bags of fertilizer should she buy?

26. Find the base of a parallelogram with a height of 9.2 meters and an area of 36.8 square meters.

27. Suppose a triangle has an area of 20 square inches and a base of 2.5 inches. What is the measure of the height?

28. A trapezoid has an area of 54 square feet. What is the measure of the height if the bases measure 16 feet and 8 feet?

29. **REAL ESTATE** The McLaughlins plan to build a house on the lot shown. If an acre is 43,560 square feet, what percent of an acre is the land?

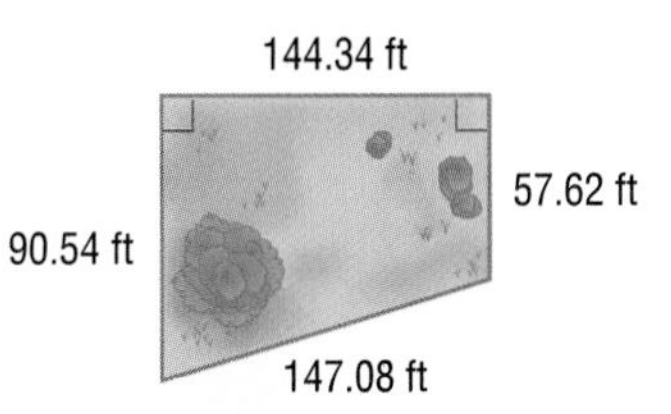

30. **CRITICAL THINKING** Explain how the formula for the area of a trapezoid can be used to find the formulas for the areas of parallelograms and triangles.

31. **WRITING IN MATH** Answer the question that was posed at the beginning of the lesson.

How is the area of a parallelogram related to the area of a rectangle?

Include the following in your answer:
- an explanation telling the similarities and differences between a rectangle and a parallelogram, and
- a diagram that shows how the area of a parallelogram is related to the area of a rectangle.

32. Which figure does *not* have an area of 48 square meters?

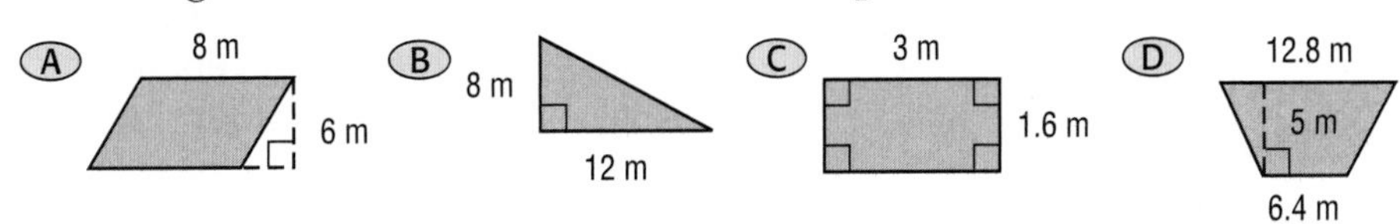

33. Square X has an area of 9 square feet. The sides of square Y are twice as long as the sides of square X. Find the area of square Y.

Ⓐ 18 ft^2 Ⓑ 36 ft^2 Ⓒ 9 ft^2 Ⓓ 6 ft^2

Maintain Your Skills

Mixed Review **Find the value of x. Then find the missing angle measures.** *(Lesson 10-4)*

34. $x°$ 60° 60° 120°

35.

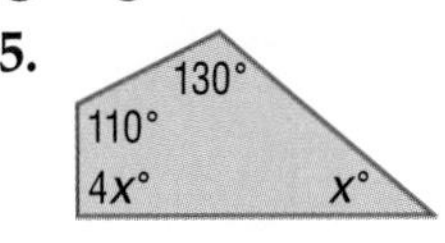

36. Triangle MNP has vertices $M(-1, 1)$, $N(5, 4)$, and $P(4, 1)$. Graph the image of $\triangle MNP$ after a translation 3 units left and 4 units down. *(Lesson 10-3)*

Find the percent of each number mentally. *(Lesson 6-6)*

37. 25% of 120 38. 75% of 160 39. 40% of 65

Getting Ready for the Next Lesson **PREREQUISITE SKILL Simplify each expression.**
*(To review **order of operations**, see Lesson 1-2).*

40. (5 − 2)180 41. (7 − 2)180 42. (10 − 2)180 43. (9 − 2)180

Reading Mathematics

Learning Mathematics Prefixes

The table shows some of the prefixes that are used in mathematics. These prefixes are also used in everyday language. In order to use each prefix correctly, you need to understand its meaning.

Prefix	Meaning	Everyday Words	Meaning
quad-	four	quadrennial quadruple quadruplet quadriceps	happening every four years a sum four times as great as another one of four offspring born at one birth a muscle with four points of origin
pent-	five	Pentagon pentagram pentathlon pentad	headquarters of the Department of Defense a five-pointed star a five-event athletic contest a group of five
hex-	six	hexapod hexagonal hexastich hexangular	having six feet having six sides a poem of six lines having six angles
hept-	seven	heptad heptagonal heptarchy heptastich	a group of seven having seven sides a government by seven rulers a poem of seven lines
oct-	eight	octopus octet octan octennial	a type of mollusk having eight arms a musical composition for eight instruments occurring every eight days lasting eight years
dec-	ten	decade decameter decathlon decare	a period of ten years ten meters a ten-event athletic contest a metric unit of area equal to 10 acres

Reading to Learn

1. Refer to the table above. For each prefix listed, choose one of the everyday words listed and write a sentence that contains the word.

2. **RESEARCH** Use the Internet, a dictionary, or another reference source to find a mathematical term that contains each of the prefixes listed. Write the definition of each term.

3. **RESEARCH** Use the Internet, a dictionary, or another reference source to find a different word that contains each prefix. Then define the term.

10-6 Polygons

Sunshine State Standards
MA.B.1.3.2-1, MA.C.1.3.1-1

Vocabulary

- polygon
- diagonal
- interior angles
- regular polygon

What You'll Learn

- Classify polygons.
- Determine the sum of the measures of the interior and exterior angles of a polygon.

How are polygons used in tessellations?

The tiled patterns below are called *regular tessellations.* Notice how the figures repeat to form patterns that contain no gaps or overlaps.

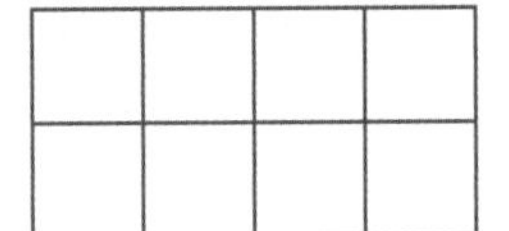

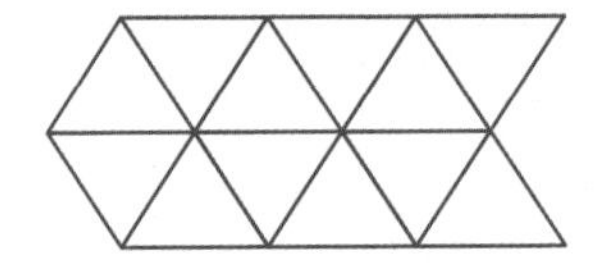

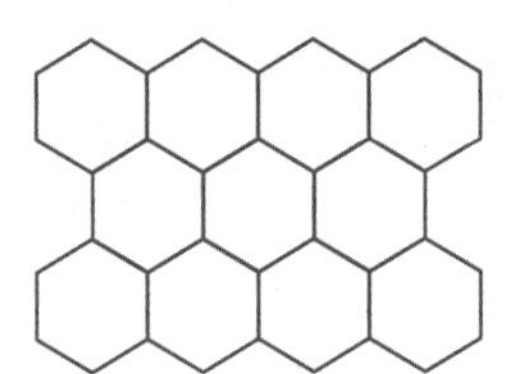

a. Which figure is used to create each tessellation?

b. Refer to the diagram at the right. What is the sum of the measures of the angles that surround the vertex?

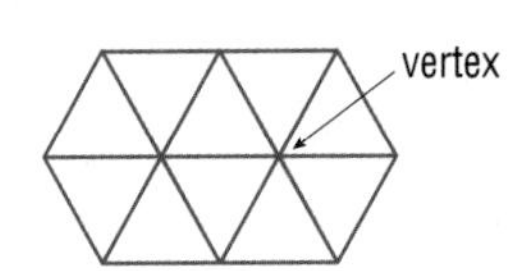

c. Does the sum in part **b** hold true for the square tessellation? Explain.

d. Make a conjecture about the sum of the measures of the angles that surround a vertex in the hexagon tessellation.

CLASSIFY POLYGONS A **polygon** is a simple, closed figure formed by three or more line segments. The figures below are examples of polygons.

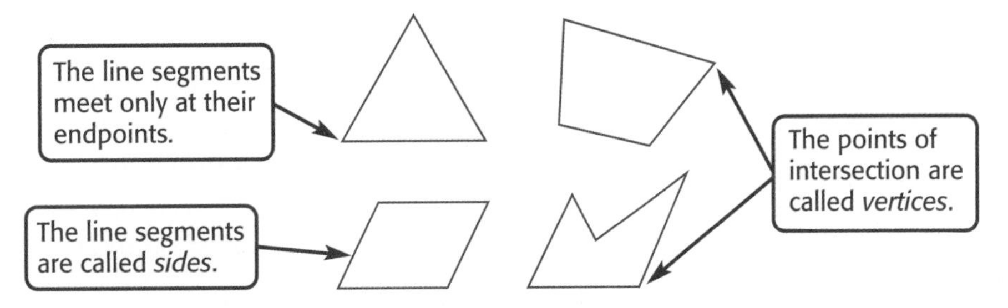

The following figures are *not* polygons.

This is not a polygon because it has a curved side.

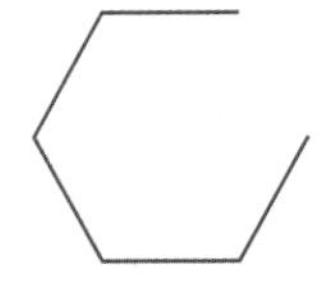

This is not a polygon because it is an open figure.

This is not a polygon because the sides overlap.

Concept Check Sketch a different figure that is *not* a polygon.

Study Tip

n-gon
A polygon with n sides is called an n-gon. For example, an octagon can also be called an 8-gon.

Polygons can be classified by the number of sides they have.

Number of Sides	Name of Polygon	Number of Sides	Name of Polygon
3	triangle	7	heptagon
4	quadrilateral	8	octagon
5	pentagon	9	nonagon
6	hexagon	10	decagon

Example 1 Classify Polygons

Classify each polygon.

a.

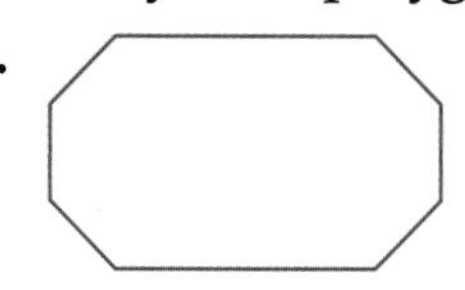

The polygon has 8 sides.
It is an octagon.

b.

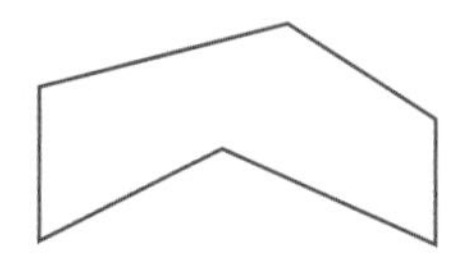

The polygon has 6 sides.
It is a hexagon.

MEASURES OF THE ANGLES OF A POLYGON A **diagonal** is a line segment in a polygon that joins two nonconsecutive vertices. In the diagram below, all possible diagonals from one vertex are shown.

pentagon hexagon heptagon octagon

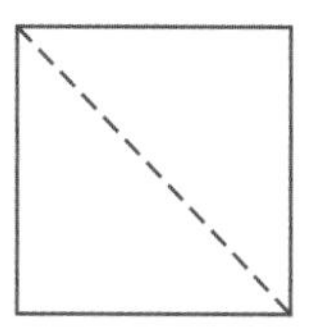

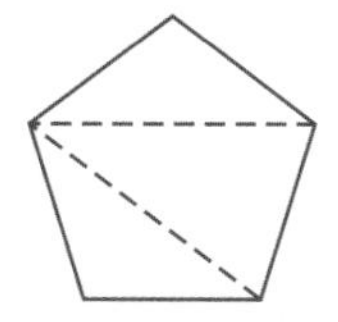

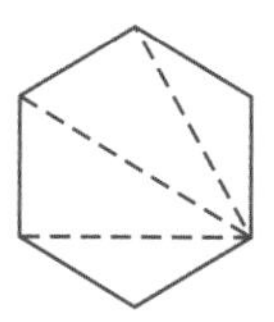

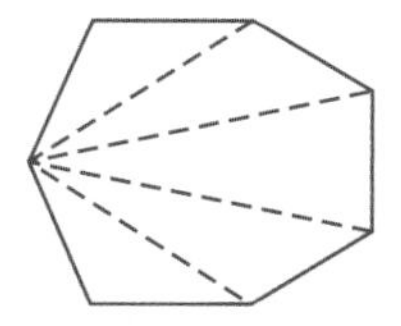

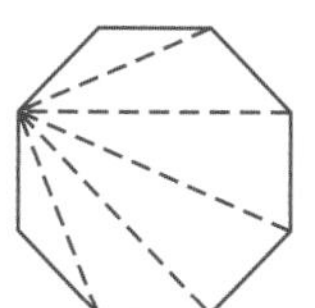

	Quadrilateral	Pentagon	Hexagon	Heptagon	Octagon
Sides	4	5	6	7	8
Diagonals	1	2	3	4	5
Triangles	2	3	4	5	6

Notice that the number of triangles is 2 less than the number of sides.

You can use the property of the sum of the measures of the angles of a triangle to find the sum of the measures of the interior angles of any polygon. An **interior angle** is an angle inside a polygon.

Key Concept — Interior Angles of a Polygon

If a polygon has n sides, then $n - 2$ triangles are formed. The sum of the degree measures of the interior angles of the polygon is $(n - 2)180$.

Example 2 Measures of Interior Angles

Find the sum of the measures of the interior angles of a heptagon.

A heptagon has 7 sides. Therefore, $n = 7$.

$(n - 2)180 = (7 - 2)180$ Replace n with 7.

$= 5(180)$ or 900 Simplify.

The sum of the measures of the interior angles of a heptagon is 900°.

A **regular polygon** is a polygon that is *equilateral* (all sides are congruent) and *equiangular* (all angles are congruent). Since the angles of a regular polygon are congruent, their measures are equal.

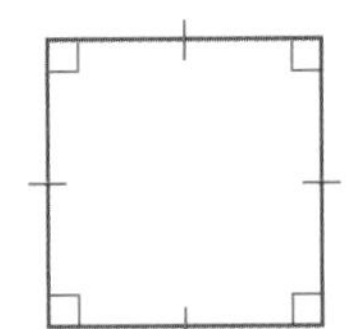

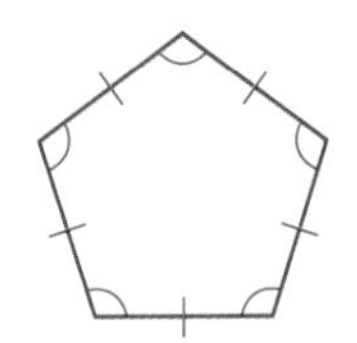

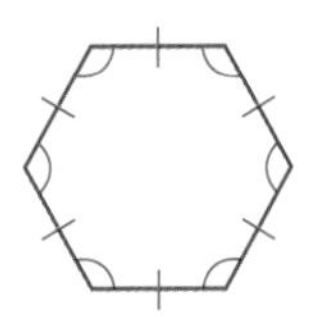

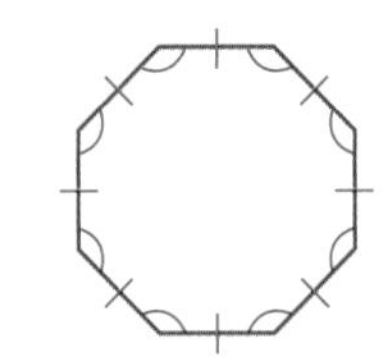

Example 3 Find Angle Measure of a Regular Polygon

SNOW Snowflakes are some of the most beautiful objects in nature. Notice how they are regular and hexagonal in shape. What is the measure of one interior angle in a snowflake?

Step 1 Find the sum of the measures of the angles.

A hexagon has 6 sides. Therefore, $n = 6$.

$(n - 2)180 = (6 - 2)180$ Replace n with 6.

$= 4(180)$ or 720 Simplify.

The sum of the measures of the interior angles is 720°.

Step 2 Divide the sum by 6 to find the measure of one angle.

$720 \div 6 = 120$

So, the measure of one interior angle in a snowflake is 120°.

More About. . .

Snow

Snowflakes are also called snow crystals. It is said that no two snowflakes are alike. They differ from each other in size, lacy structure, and surface markings.

Source: www.infoplease.com

Check for Understanding

Concept Check

1. **OPEN ENDED** Draw a polygon that is both equiangular and equilateral.
2. **Tell** why a square is a regular polygon.
3. **Explain** the relationship between the number of sides in a polygon and the number of triangles formed by each of the diagonals.

Guided Practice

Classify each polygon. Then determine whether it appears to be *regular* or *not regular*.

4. 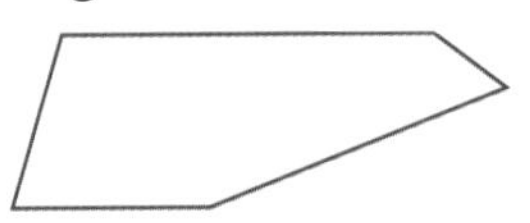

5. 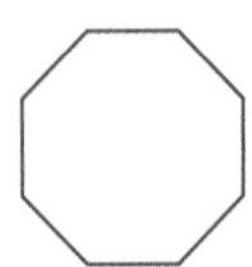

6. Find the sum of the measures of the interior angles of a nonagon.
7. What is the measure of each interior angle of a regular heptagon? Round to the nearest tenth.

Application

8. **TESSELLATIONS** Identify the polygons that are used to create the tessellation shown at the right.

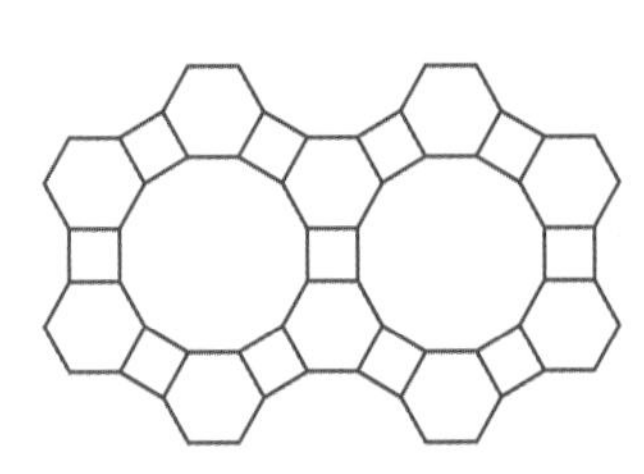

Practice and Apply

Homework Help

For Exercises	See Examples
9–14, 21	1
29, 30	
15–20	2
23–28	3

Extra Practice

See page 749.

Classify each polygon. Then determine whether it appears to be *regular* or *not regular*.

9. 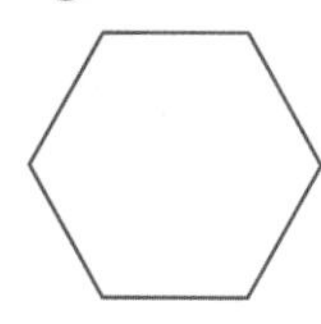

10. 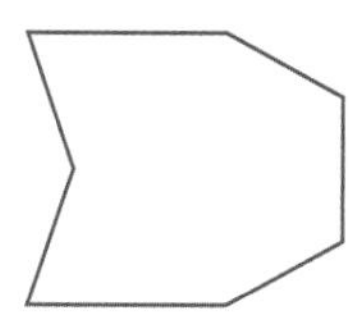

11. 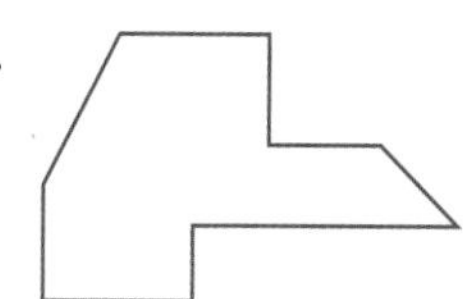

12. 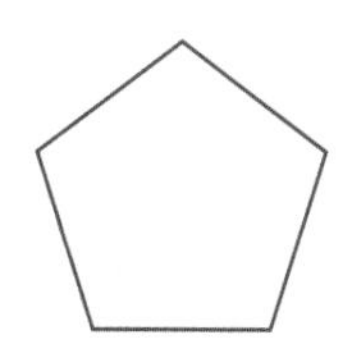

13. 

14. 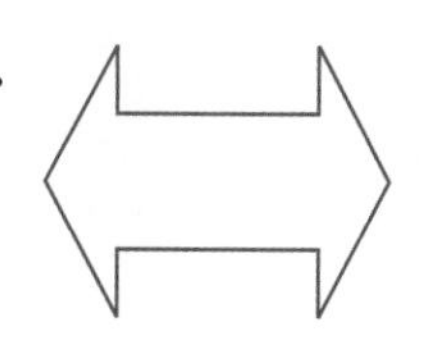

Find the sum of the measures of the interior angles of each polygon.

15. pentagon
16. octagon
17. decagon
18. hexagon
19. 18-gon
20. 23-gon

ART **For Exercises 21 and 22, use the painting shown.**

21. List five polygons used in the painting.

22. RESEARCH The title of the painting mentions the music symbol, *clef*. Use the Internet or another source to find a drawing of a clef. Is a clef a polygon? Explain.

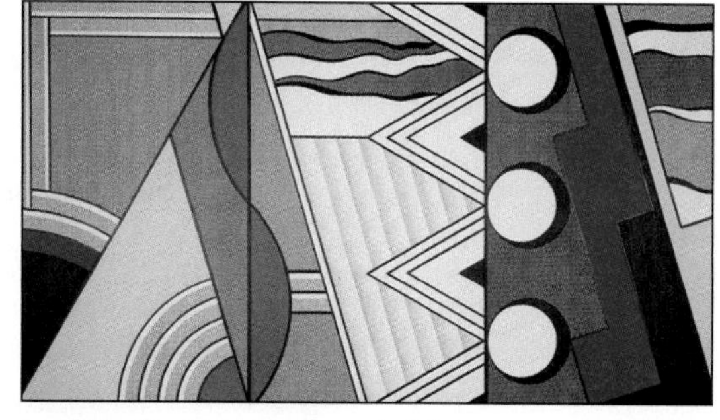

Roy Lichtenstein. *Modern Painting with Clef.* 1967

Find the measure of an interior angle of each polygon.

23. regular nonagon
24. regular pentagon
25. regular octagon
26. regular decagon
27. regular 12-gon
28. regular 25-gon

TESSELLATIONS **For Exercises 29 and 30, identify the polygons used to create each tessellation.**

29. 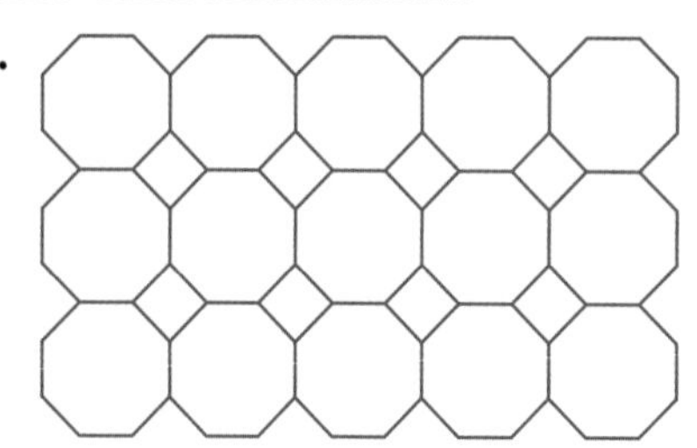

30. 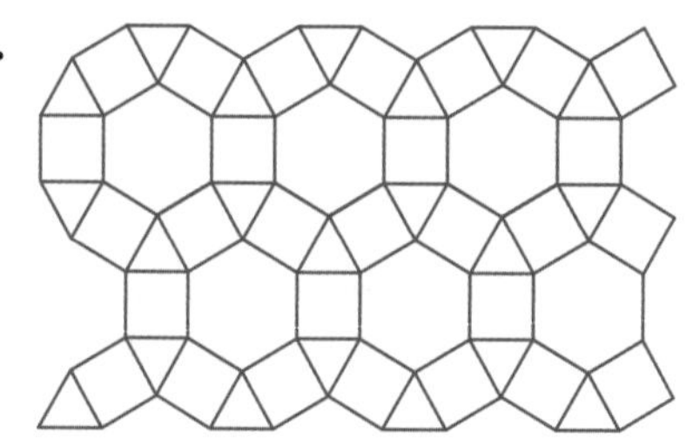

31. ART Refer to Exercise 8 on page 529. The tessellation design contains regular polygons. Find the perimeter of the design if the measure of the sides of the 12-gon is 5 centimeters.

32. What is the perimeter of a regular pentagon with sides 4.2 feet long?

33. Find the perimeter of a regular nonagon having sides $6\frac{1}{2}$ inches long.

34. **CRITICAL THINKING** Copy the dot pattern shown at the right. Then without lifting your pencil from the paper, draw four line segments that connect all of the points.

35. **WRITING IN MATH** Answer the question that was posed at the beginning of the lesson.

How are polygons used in tessellations?

Include the following in your answer:

- an example of a tessellation in which the pattern is formed using only one type of polygon, and
- an example of a tessellation in which the pattern is formed using more than one polygon.

FCAT Practice

Standardized Test Practice

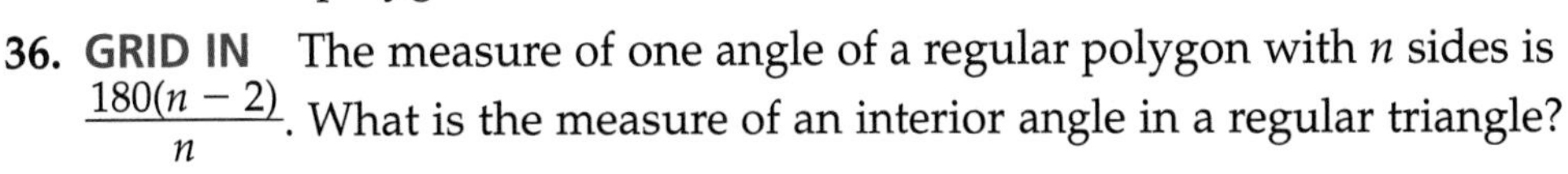

36. **GRID IN** The measure of one angle of a regular polygon with n sides is $\frac{180(n-2)}{n}$. What is the measure of an interior angle in a regular triangle?

37. Which figure best represents a regular polygon?

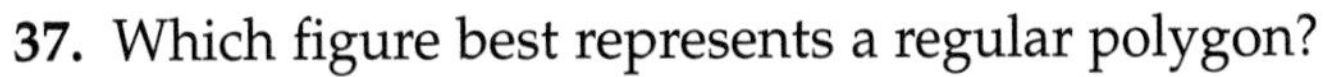

Ⓐ 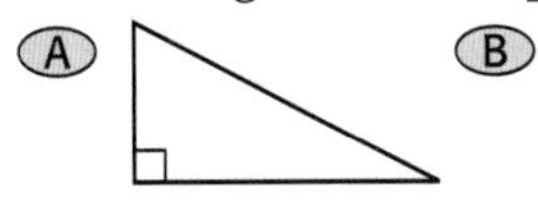 Ⓑ 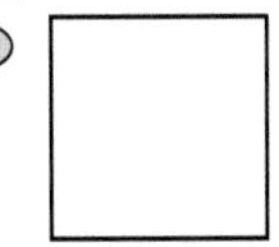 Ⓒ 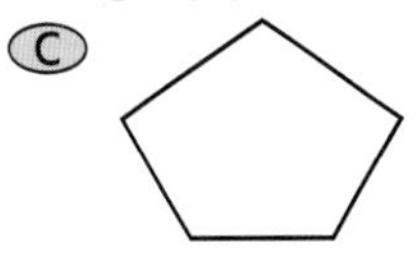 Ⓓ 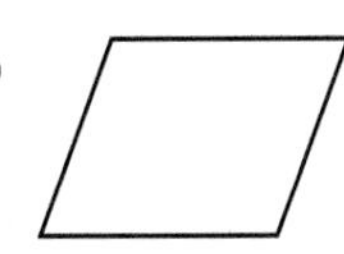

Extending the Lesson

EXTERIOR ANGLES When a side of a polygon is extended, an **exterior angle** is formed.

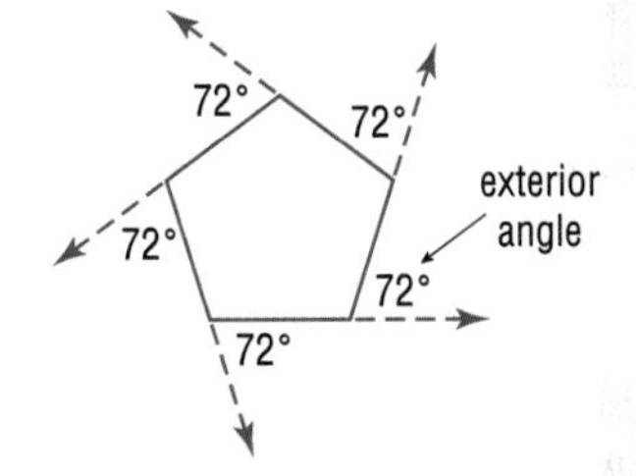

In any polygon, the sum of the measures of the exterior angles, one at each vertex, is 360.

Find the measure of each exterior angle of each regular polygon.

38. regular octagon
39. regular triangle
40. regular nonagon
41. regular hexagon
42. regular decagon
43. regular 12-gon

Maintain Your Skills

Mixed Review

Find the area of each figure described. *(Lesson 10-5)*

44. triangle: base, 9 in.; height, 6 in.
45. trapezoid: height, 3 cm; bases, 4 cm, 8 cm

Classify each quadrilateral using the name that *best* describes it. *(Lesson 10-4)*

46. 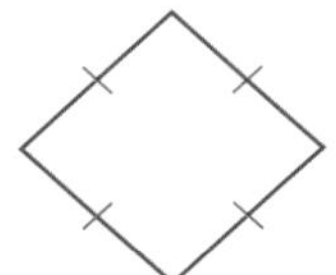

47. 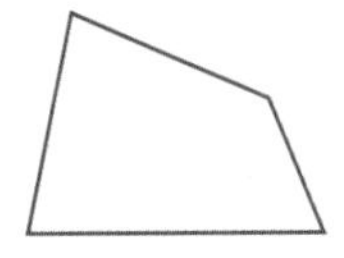

48. 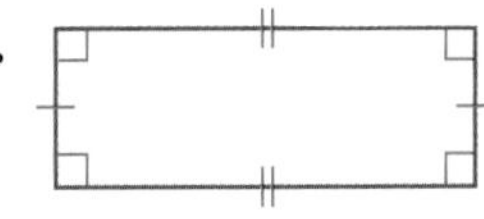

49. **ALGEBRA** Simplify $4.6x + 2.5x + 9.3x$. *(Lesson 3-2)*

Getting Ready for the Next Lesson

PREREQUISITE SKILL **Use a calculator to find each product. Round to the nearest tenth.** *(To review **rounding decimals**, see page 711.)*

50. $\pi \cdot 4.3$
51. $2 \cdot \pi \cdot 5.4$
52. $\pi \cdot 4^2$
53. $\pi(2.4)^2$

Algebra Activity

A Follow-Up of Lesson 10-6

Sunshine State Standards
MA.C.1.3.1-3, MA.C.2.3.2-1, MA.C.2.3.2-2

Tessellations

A tessellation is a pattern of repeating figures that fit together with no overlapping or empty spaces. Tessellations can be formed using transformations.

Activity 1 Create a tessellation using a translation.

Step 1 Draw a square. Then draw a triangle inside the top of the square as shown.

Step 2 Translate or slide the triangle from the top to the bottom of the square.

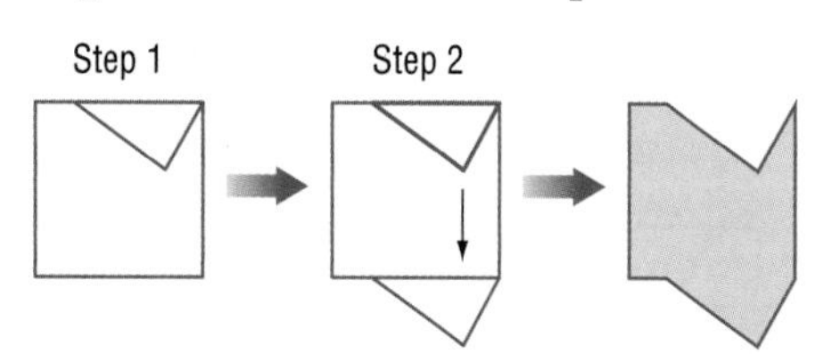

Step 3 Repeat this pattern unit to create a tessellation.

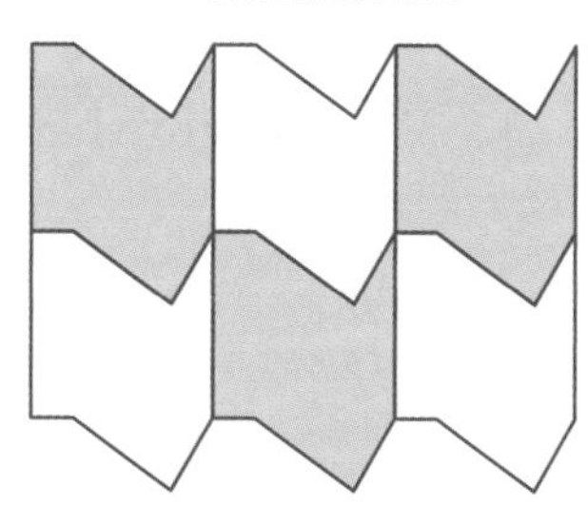

It is sometimes helpful to complete one pattern, cut it out, and trace it for the other pattern units.

Activity 2 Create a tessellation using a rotation.

Step 1 Draw an equilateral triangle. Then draw another triangle inside the left side of the triangle as shown below.

Step 2 Rotate the triangle so you can trace the change on the side as indicated.

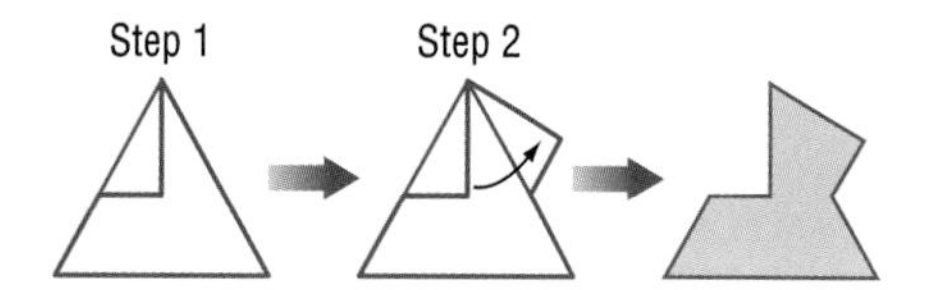

Step 3 Repeat this pattern unit to create a tessellation.

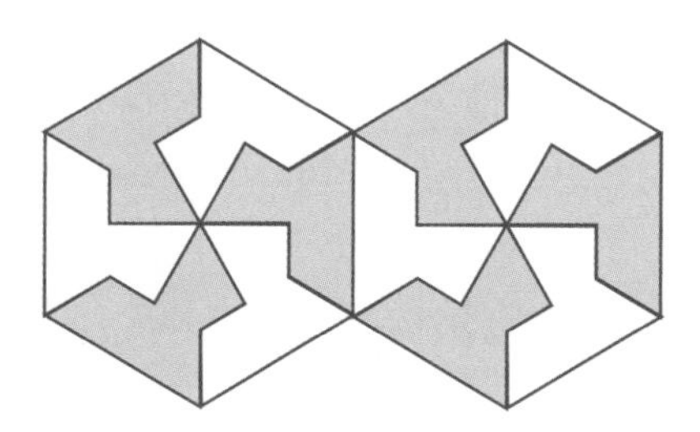

Model

Use a translation to create a tessellation for each pattern unit shown.

1. 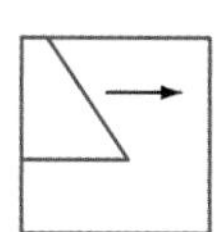

2. 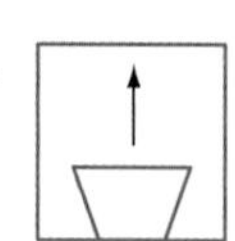

3. 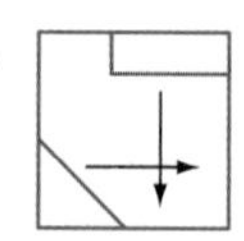

Use a rotation to create a tessellation for each pattern unit shown.

4. 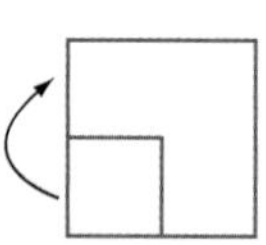

5. 

6. 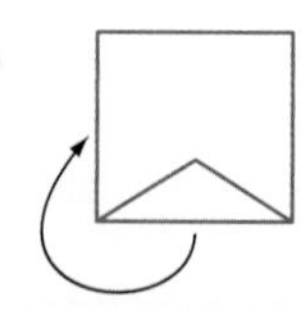

7. Make a tessellation that involves a translation, a rotation, or a combination of the two.

10-7 Circumference and Area: Circles

What You'll Learn

- Find circumference of circles.
- Find area of circles.

Sunshine State Standards
MA.B.1.3.1-1, MA.C.1.3.1-4

Vocabulary

- circle
- diameter
- center
- circumference
- radius
- π (pi)

How are circumference and diameter related?

Coins, paper plates, cookies, and CDs are all examples of objects that are circular in shape.

a. Collect three different-sized circular objects. Then copy the table shown.

Object	d	C	$\frac{C}{d}$
1			
2			
3			

b. Using a tape measure, measure each distance below to the nearest millimeter. Record your results.

- the distance across the circular object through its center (d)
- the distance around each circular object (C)

c. For each object, find the ratio $\frac{C}{d}$. Record the results in the table.

CIRCUMFERENCE OF CIRCLES A **circle** is the set of all points in a plane that are the same distance from a given point.

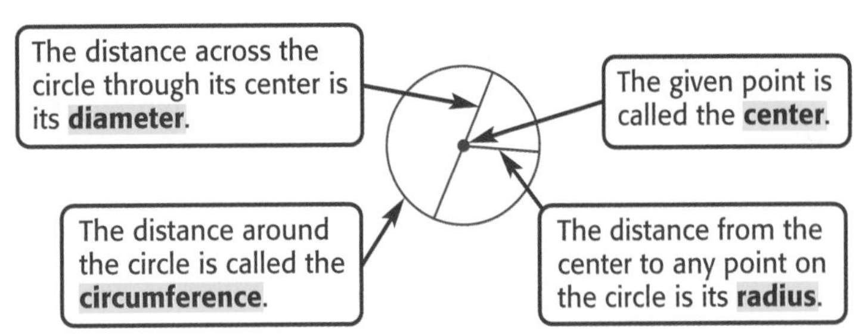

The distance across the circle through its center is its **diameter**.
The given point is called the **center**.
The distance around the circle is called the **circumference**.
The distance from the center to any point on the circle is its **radius**.

The relationship you discovered in the activity above is true for all circles. The ratio of the circumference of a circle to its diameter is always equal to 3.1415926... The Greek letter **π (pi)** stands for this number. Using this ratio, you can derive a formula for the circumference of a circle.

$\frac{C}{d} = \pi$ The ratio of the circumference to the diameter equals pi.

$\frac{C}{d} \cdot d = \pi \cdot d$ Multiply each side by d.

$C = \pi d$ Simplify.

Study Tip

Pi

Although π is an irrational number, 3.14 and $\frac{22}{7}$ are two generally accepted approximations for π.

Key Concept — Circumference of a Circle

- **Words** The circumference of a circle is equal to its diameter times π, or 2 times its radius times π.
- **Symbols** $C = \pi d$ or $C = 2\pi r$
- **Model**

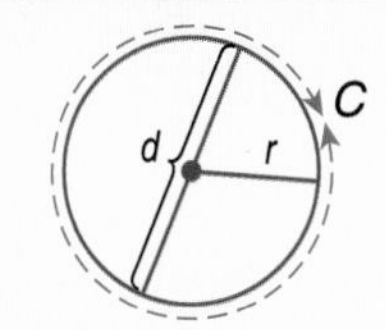

Concept Check Which term describes the distance from the center of a circle to any point on the circle?

Example 1 Find the Circumference of a Circle

Find the circumference of each circle to the nearest tenth.

a.

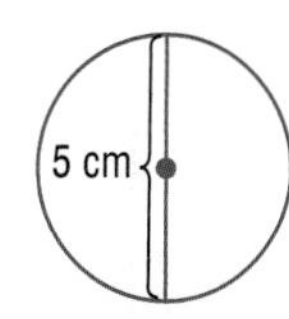

$C = \pi d$ Circumference of a circle

$C = \pi \cdot 5$ Replace d with 5.

$C = 5\pi$ Simplify. This is the *exact* circumference.

To estimate the circumference, use a calculator.

5 [×] [2nd] [π] [ENTER] 15.70796327

The circumference is about 15.7 centimeters.

b.

$C = 2\pi r$ Circumference of a circle

$C = 2 \cdot \pi \cdot 3.2$ Replace r with 3.2.

$C \approx 20.1$ Simplify. Use a calculator.

The circumference is about 20.1 feet.

Study Tip

Calculating with π
Unless otherwise specified, use a calculator to evaluate expressions involving π and then follow any instructions regarding rounding.

Many situations involve circumference and diameter of circles.

Example 2 Use Circumference to Solve a Problem

TREES A tree in Madison's yard was damaged in a storm. She wants to replace the tree with another whose trunk is the same size as the original tree. Suppose the circumference of the original tree was 14 inches. What should be the diameter of the replacement tree?

Explore You know the circumference of the original tree. You need to know the diameter of the new tree.

Plan Use the formula for the circumference of a circle to find the diameter.

Solve

$C = \pi d$ Circumference of a circle

$14 = \pi \cdot d$ Replace C with 14.

$\frac{14}{\pi} = d$ Divide each side by π.

$4.5 \approx d$ Simplify. Use a calculator.

The diameter of the tree should be about 4.5 inches.

Examine Check the reasonableness of the solution by replacing d with 4.5 in $C = \pi d$.

$C = \pi d$ Circumference of a circle

$C = \pi \cdot 4.5$ Replace d with 4.5.

$C \approx 14.1$ Simplify. Use a calculator.

The solution is reasonable.

AREAS OF CIRCLES A circle can be separated into parts as shown below. The parts can then be arranged to form a figure that resembles a parallelogram.

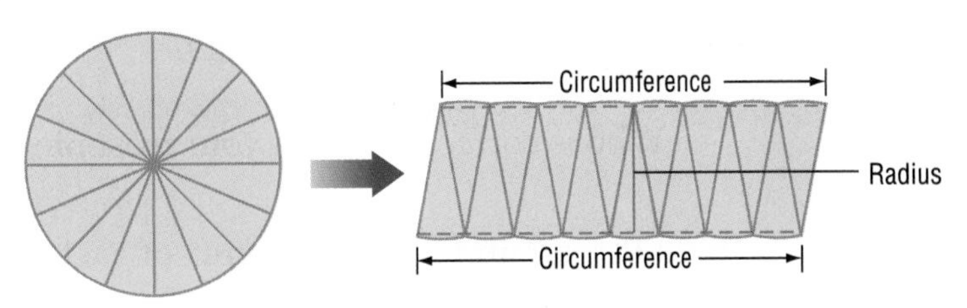

Since the circle has an area that is relatively close to the area of the figure, you can use the formula for the area of a parallelogram to find the area of a circle.

$A = bh$	Area of a parallelogram
$A = \left(\frac{1}{2} \times C\right) r$	The base of the parallelogram is one-half the circumference, and the radius is the height.
$A = \left(\frac{1}{2} \times 2\pi r\right) r$	Replace C with $2\pi r$.
$A = \pi \times r \times r$	Simplify.
$A = \pi r^2$	Replace $r \times r$ with r^2.

Study Tip

Look Back
To review **exponents**, see Lesson 4-2.

Key Concept — *Area of a Circle*

- **Words** The area of a circle is equal to π times the square of its radius.
- **Symbols** $A = \pi r^2$
- **Model**

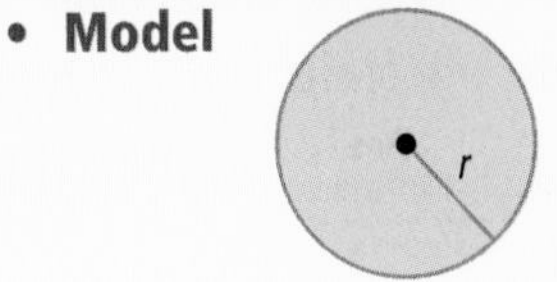

Study Tip

Estimation
To estimate the area of a circle, square the radius and then multiply by 3.

Example 3 Find Areas of Circles

Find the area of each circle. Round to the nearest tenth.

a. (circle with radius 6 in.)

$A = \pi r^2$	Area of a circle
$A = \pi \cdot 6^2$	Replace r with 6.
$A = \pi \cdot 36$	Evaluate 6^2.
$A \approx 113.1$	Use a calculator.

The area is about 113.1 square inches.

b. (circle with diameter 31 m)

$A = \pi r^2$	Area of a circle
$A = \pi \cdot (15.5)^2$	Replace r with 15.5.
$A = \pi \cdot 240.25$	Evaluate $(15.5)^2$.
$A \approx 754.8$	Use a calculator.

The area is about 754.8 square meters.

Check for Understanding

Concept Check

1. **Tell** how to find the circumference of a circle if you know the measure of the radius.
2. **OPEN ENDED** Draw and label a circle that has an area between 5 and 8 square units.

3. **FIND THE ERROR** Dario and Mark are finding the area of a circle with diameter 7.

Dario	Mark
$A \approx 153.9$ units^2	$A \approx 38.5$ units^2

Who is correct? Explain your reasoning.

Guided Practice **Find the circumference and area of each circle. Round to the nearest tenth.**

4. 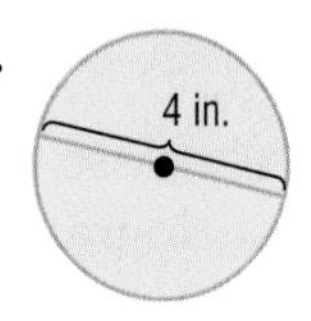

5. 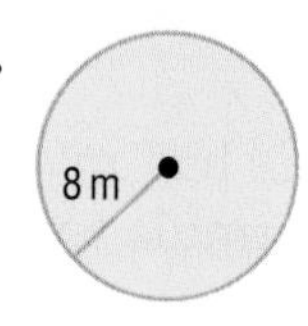

6. 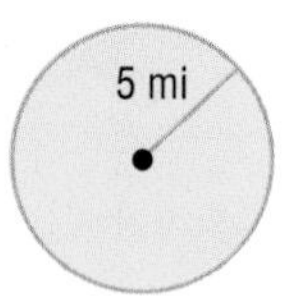

7. The radius is 1.3 kilometers.

8. The diameter is 6.1 centimeters.

Application 9. **MUSIC** During a football game, a marching band can be heard within a radius of 1.7 miles. What is the area of the neighborhood that can hear the band?

Practice and Apply

Homework Help

For Exercises	See Examples
10–19, 24, 26–35	1, 3
20–23, 25	2

Extra Practice
See page 749.

Find the circumference and area of each circle. Round to the nearest tenth.

10. 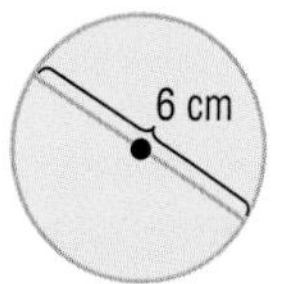

11. 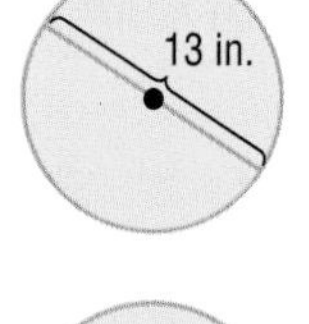

12. 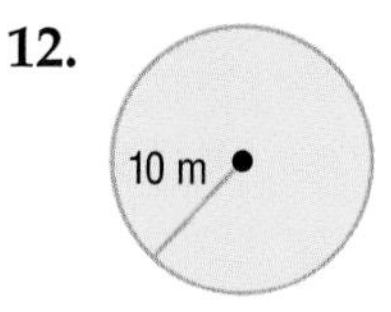

13. 21 km

14. $9\frac{1}{2}$ ft

15. 12.7 m

16. The radius is 4.5 meters.

17. The diameter is 7.3 centimeters.

18. The diameter is $7\frac{4}{5}$ feet.

19. The radius is $15\frac{3}{8}$ inches.

20. What is the diameter of a circle if its circumference is 25.8 inches? Round to the nearest tenth.

21. Find the radius of a circle if its circumference is 9.2 meters. Round to the nearest tenth.

22. Find the radius of a circle if its area is 254.5 square inches.

23. What is the diameter of a circle if its area is 132.7 square meters?

24. **BICYCLES** If a bicycle tire has a diameter of 27 inches, what is the distance the bicycle will travel in 10 rotations of the tire?

25. **SCIENCE** The circumference of Earth is about 25,000 miles. What is the distance to the center of Earth?

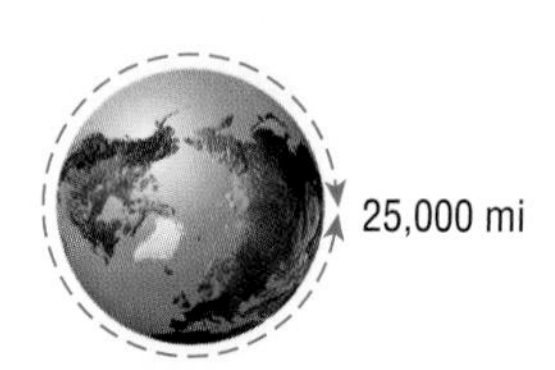

More About. . .

History

The Houston Astrodome opened on April 12, 1965. It is the first enclosed stadium built for baseball. The Houston Astrodome has a diameter of 216.4 meters.

Source: *Compton's Encyclopedia*

Match each circle described in the column on the left with its corresponding measurement in the column on the right.

26. radius: 4 units — **a.** circumference: 37.7 units

27. diameter: 7 units — **b.** area: 7.1 units^2

28. diameter: 3 units — **c.** area: 50.3 units^2

29. radius: 6 units — **d.** circumference: 22.0 units

30. MONUMENTS The Stonehenge monument in England is enclosed within a circular ditch that has a diameter of 300 feet. Find the area within the ditch to the nearest tenth.

31. HISTORY The dome of the Roman Pantheon has a diameter of 42.7 meters. Use the information at the left to find about how many times more area the Astrodome covers than the Pantheon.

FOOD For Exercises 32 and 33, use the following information and the graphic shown at the right.

Suppose the circle graph is redrawn onto a poster board so that the diameter of the graph is 9 inches.

32. How much space on the poster board will the circle graph cover?

33. How much of the total space will each section of the graph cover?

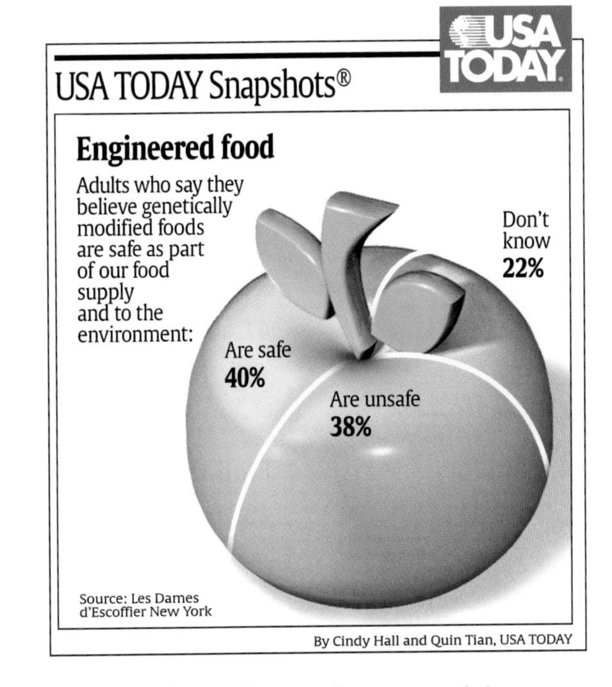

Study Tip

Look Back
To review **slope**, see Lesson 8-4.

34. FUNCTIONS Graph the circumference of a circle as a function of the diameter. Use values of d like 1, 2, 3, 4, and so on. What is the slope of this graph?

35. CRITICAL THINKING The numerical value of the area of a circle is twice the numerical value of the circumference. What is the radius of the circle? (*Hint*: Use a chart of values for radius, circumference, and area.)

36. WRITING IN MATH Answer the question that was posed at the beginning of the lesson.

How are circumference and diameter related?

Include the following in your answer:

- the ratio of the circumference to the diameter, and
- an explanation describing what happens to the circumference as the diameter increases or decreases.

37. The diameter of a circle is 8 units. What is the area of the circle if the diameter is doubled?

(A) 50.3 units^2 (B) 100.5 units^2 (C) 201.1 units^2 (D) 804.2 units^2

38. GRID IN The circumference of a circle is 18.8 meters. What is its area to the nearest tenth?

Extending the Lesson

CENTRAL ANGLES A **central angle** is an angle whose vertex is the center of the circle. It separates a circle into a *major arc* and a *minor arc*. An **inscribed angle** has its vertex on the circle and sides that are chords. A **chord** is a segment of a circle whose endpoints are on the circle.

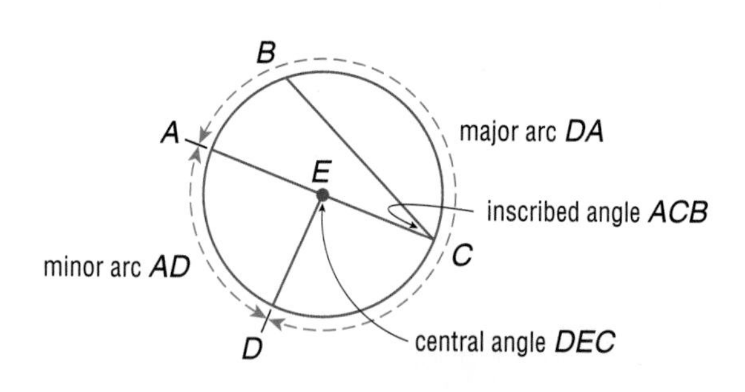

- The degree measure of a minor arc is the degree measure of the central angle.
- The measure of an inscribed angle equals one-half the measure of its intercepted arc.
- The degree measure of a major arc is 360 minus the degree measure of the central angle.

Refer to the diagram shown. Find the measures of the following angles and arcs.

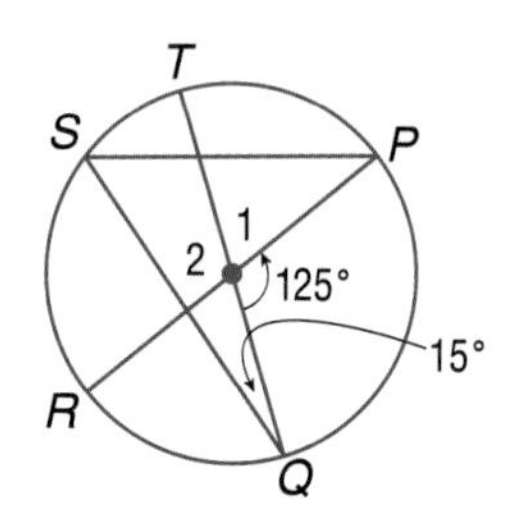

39. minor arc PQ

40. $\angle 1$

41. major arc QP

42. $\angle 2$

43. minor TR

44. minor RQ

45. $\angle PSQ$

46. minor SR

47. List three chords of the circle.

Maintain Your Skills

Mixed Review

Find the measure of an interior angle of each polygon. *(Lesson 10-6)*

48. regular hexagon

49. regular decagon

50. regular octagon

Find the area of each figure described. *(Lesson 10-5)*

51. trapezoid: height, 2 m; bases, 20 m and 18 m

52. parallelogram: base, 6 km; height, 8 km

53. **ALGEBRA** Solve $2x - 7 > 5x + 14$. *(Lesson 7-6)*

Getting Ready for the Next Lesson

PREREQUISITE SKILL **Find each sum.** *(To review **adding decimals**, see page 713.)*

54. $200 + 43.9$

55. $23.6 + 126.9$

56. $345.14 + 23.8$

Practice Quiz 2

Lessons 10-5 through 10-7

Find the area of each figure. *(Lesson 10-5)*

1. 13.2 cm; 17.1 cm

2.

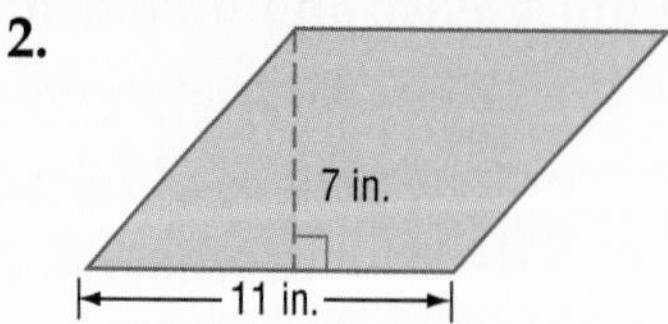

3.

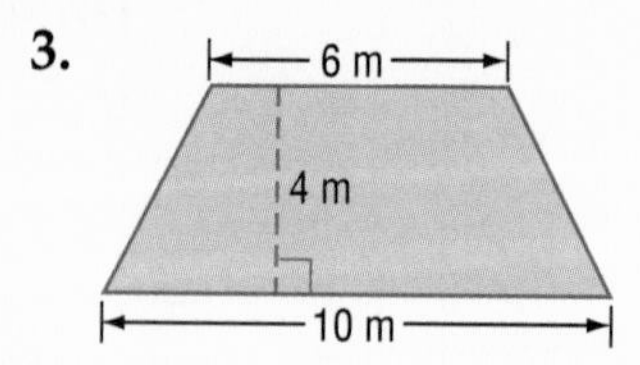

4. Find the sum of the measures of the interior angles of a 15-gon. *(Lesson 10-6)*

5. A circle has a radius of 4.7 inches. Find the circumference and area to the nearest tenth. *(Lesson 10-7)*

10-8 Area: Irregular Figures

Sunshine State Standards
MA.C.1.3.1-1, MA.C.1.3.1-2, MA.C.1.3.1-4

What You'll Learn

- Find area of irregular figures.

How can polygons help to find the area of an irregular figure?

California is the most populous state in the United States. It ranks third among the U.S. states in area.
Source: www.infoplease.com

In the diagram, the area of California is separated into polygons.

a. Identify the polygons.

b. Explain how polygons can be used to estimate the total land area.

c. What is the area of each region?

d. What is the total area?

AREA OF IRREGULAR FIGURES So far in this chapter, we have discussed the following area formulas.

Triangle	Trapezoid	Parallelogram	Circle
$A = \frac{1}{2}bh$	$A = \frac{1}{2}h(a + b)$	$A = bh$	$A = \pi r^2$

These formulas can be used to help you find the area of irregular figures. Some examples of irregular figures are shown.

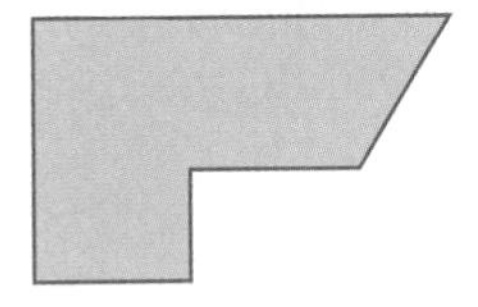
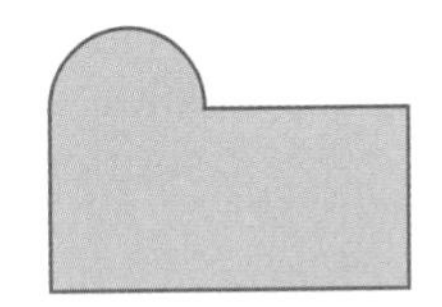

To find the area of an irregular figure, separate the irregular figure into figures whose areas you know how to find.

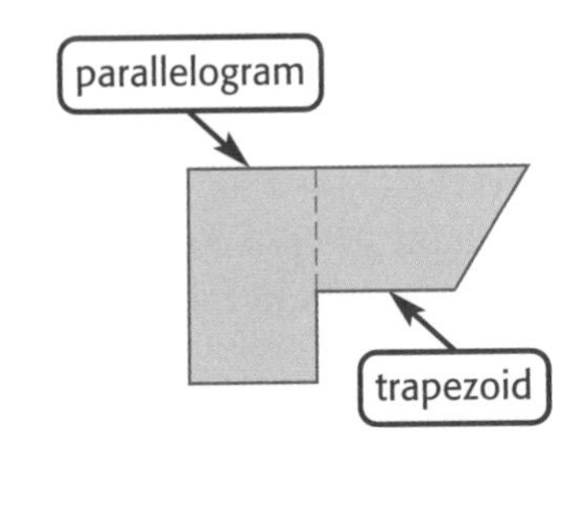

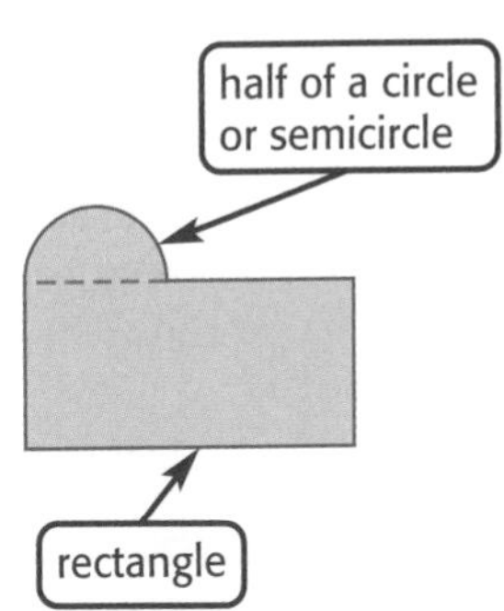

Study Tip

Irregular Figures
There can be more than one way to separate an irregular figure. For example, another way to separate the first figure at the right is shown below.

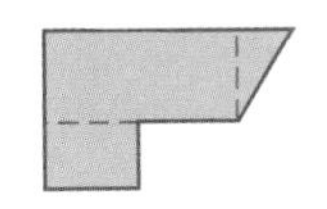

Concept Check Name three area formulas that can be used to find the area of an irregular figure.

Example 1 Find Area of Irregular Figures

Find the area of the figure to the nearest tenth.

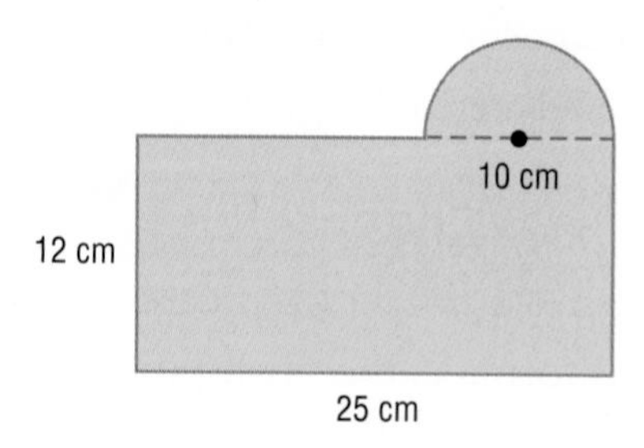

Explore You know the dimensions of the figure. You need to find its area.

Plan Solve a simpler problem. First, separate the figure into a parallelogram and a semicircle. Then find the sum of the areas of the figures.

Estimate: The area of the entire figure should be a little greater than the area of the rectangle. One estimate is 10×25 or 250.

Solve Area of Parallelogram

$A = bh$ Area of a parallelogram

$A = 25 \cdot 12$ Replace b with 25 and h with 12.

$A = 300$ Simplify.

Area of Semicircle

$A = \frac{1}{2}\pi r^2$ Area of a semicircle

$A = \frac{1}{2} \cdot \pi \cdot 5^2$ Replace r with 5.

$A \approx 39.3$ Simplify.

The area of the figure is $300 + 39.3$ or about 339.3 square centimeters.

Study Tip

Look Back

To review **diameter** and **radius**, see Lesson 10-7.

Examine Check the reasonableness of the solution by solving the problem another way. Separate the figure into two rectangles and a semicircle.

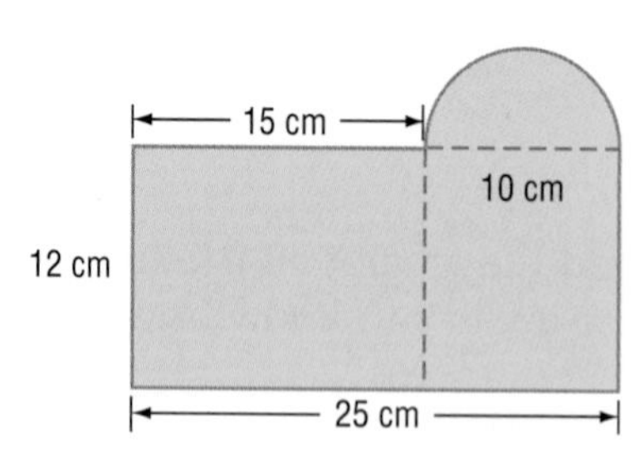

The area of one rectangle is $10 \cdot 12$ or 120 square centimeters, the area of the other rectangle is $12 \cdot 15$ or 180 square centimeters, and the area of the semicircle remains 39.3 square centimeters.

$120 + 180 + 39.3 = 339.3$

So, the answer is correct.

Many real-world situations involve finding the area of an irregular figure.

Example 2 Use Area of Irregular Figures

LANDSCAPE DESIGN Suppose one bag of mulch covers an area of about 9 square feet. How many bags of mulch will be needed to cover the flower garden?

Step 1 Find the area of the flower garden.

Area of rectangle

$A = bh$ Area of a rectangle

$A = 24 \cdot 38$ Replace b with 24 and h with 38.

$A = 912$ Simplify.

Area of parallelogram

$A = bh$ Area of a parallelogram

$A = 28 \cdot 8$ Replace b with 28 and h with 8.

$A = 224$ Simplify.

The area of the garden is $912 + 224$ or 1136 square feet.

Step 2 Find the number of bags of mulch needed.

$1136 \div 9 \approx 126.2$

So, 127 bags of mulch will be needed.

Check for Understanding

Concept Check

1. **OPEN ENDED** Draw two examples of irregular figures.
2. **Write** the steps you would use to find the area of the irregular figure shown at the right.

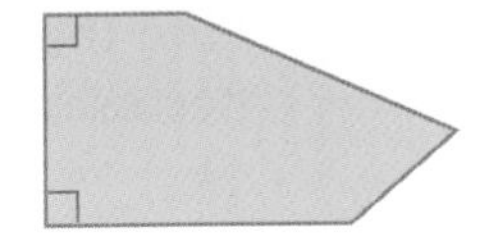

Guided Practice

Find the area of each figure. Round to the nearest tenth.

3.

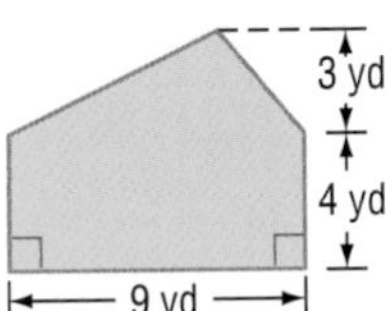

4.

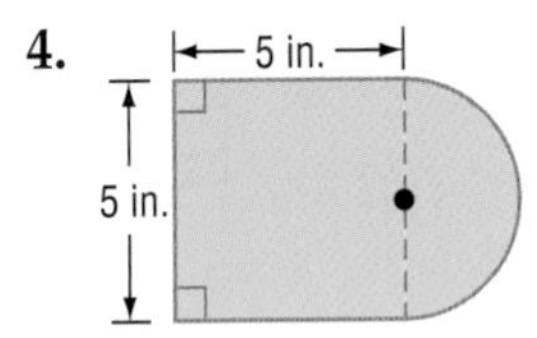

Application

HOME IMPROVEMENT For Exercises 5 and 6, use the diagram shown and the following information.

The Slavens are planning to stain their wood deck. One gallon of stain costs $19.95 and covers approximately 200 square feet.

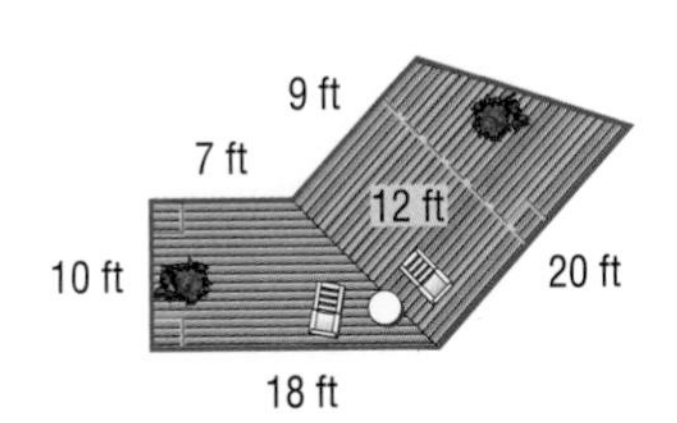

5. Suppose the Slavens need to apply only one coat of stain. How many gallons of stain will they need to buy?
6. Find the total cost of the stain, not including tax.

Practice and Apply

Homework Help

For Exercises	See Examples
7–15	1
19–22	2

Extra Practice

See page 750.

Find the area of each figure to the nearest tenth, if necessary.

7. 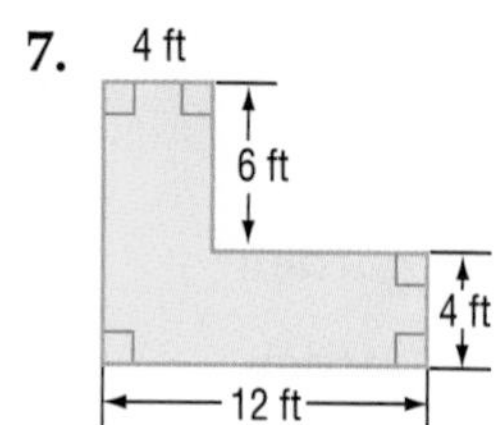

8. 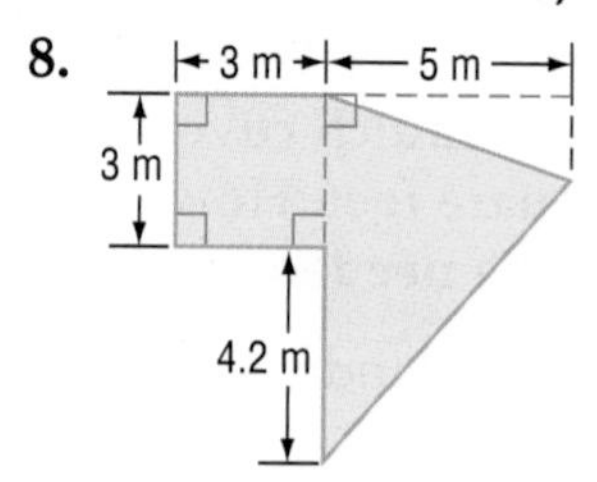

9. 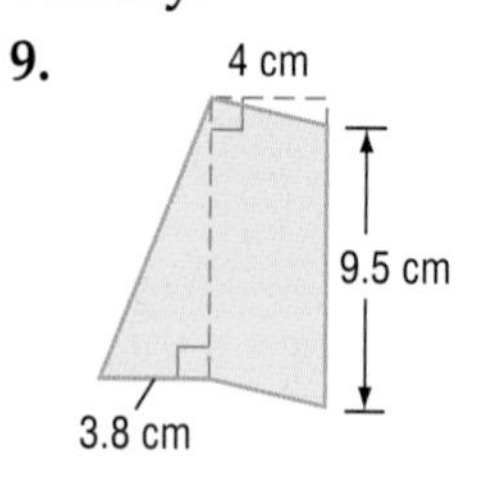

10. 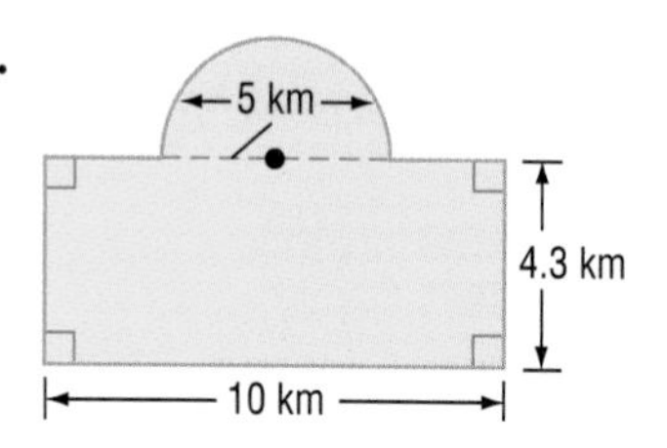

11. 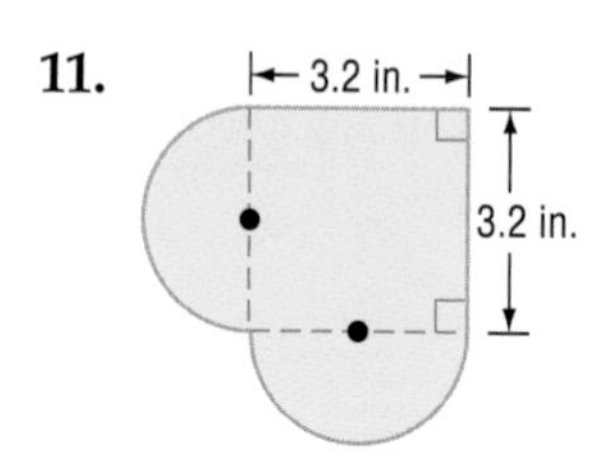

12.

13. 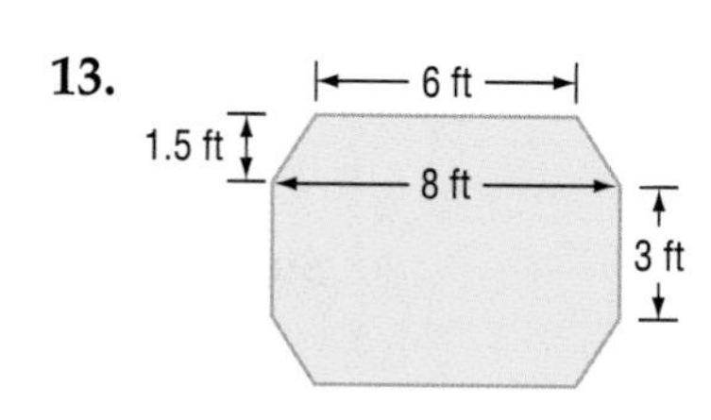

WebQuest

The formulas for area and circumference of circles are used in the design of various buildings. Visit www.pre-alg.com/webquest to continue work on your WebQuest project.

14. What is the area of a figure that is formed using a square with sides 8 meters and a semicircle with a diameter of 5.6 meters?

15. Find the area of a figure formed using a rectangle with base 3.5 yards and height 2.8 yards and a semicircle with radius 7 yards.

Find the area of each shaded region. Round to the nearest tenth, if necessary. (*Hint:* Find the total area and subtract the non-shaded area.)

16. 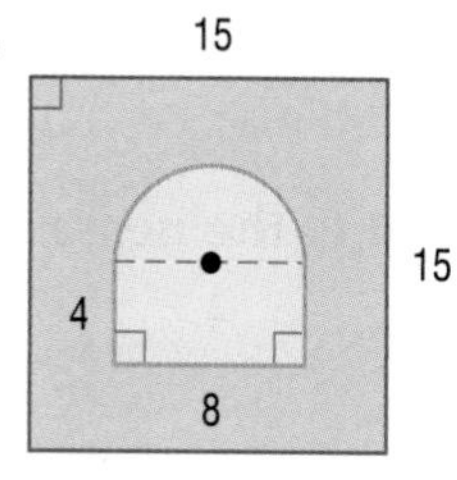

17. 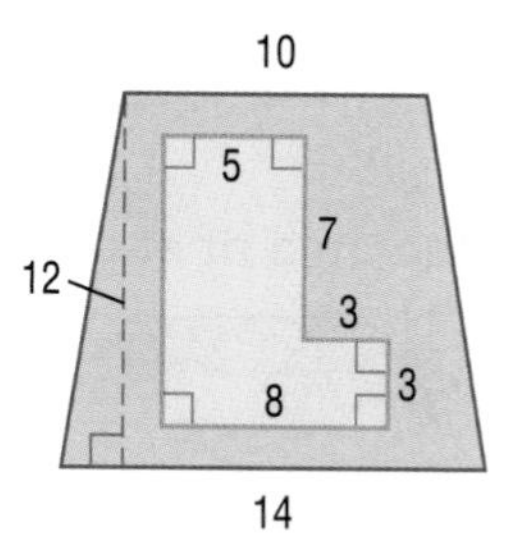

18. 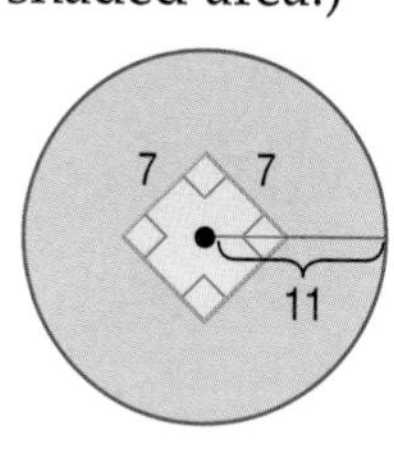

19. **SPORTS** In the diagram shown, a track surrounds a football field. To the nearest tenth, what is the area of the grass region inside the track?

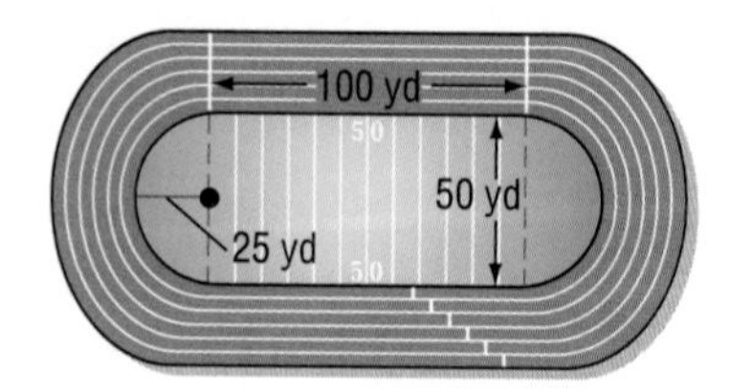

20. **WALKWAYS** A sidewalk forms a 3-foot wide border with the grass as shown. Find the area of the sidewalk.

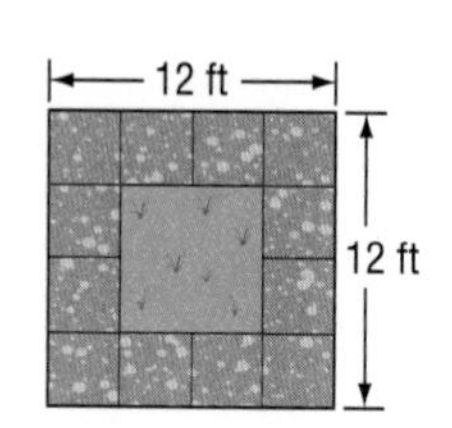

Geography

Oklahoma has the largest Native American population. The Oklahoma state flag represents the diversity of cultures in the state.

Source: www.infoplease.com

GEOGRAPHY **For Exercises 21–23, use the diagram shown at the right.**

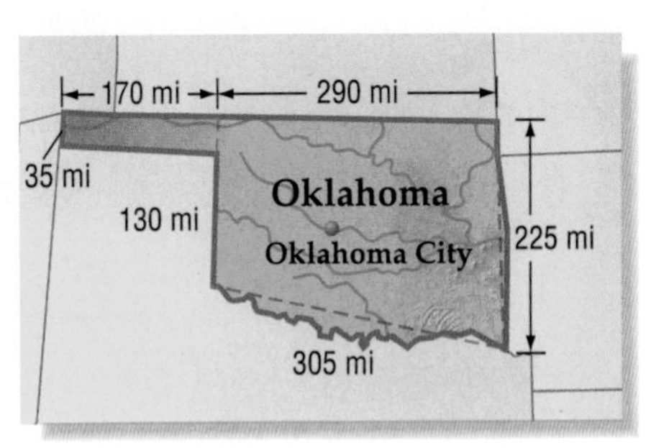

21. Tell how you would separate the irregular figure into polygons to find its area.

22. Use your method to find the total land area of Oklahoma.

23. **RESEARCH** Use the Internet or another source to find the actual total land area of Oklahoma.

Online Research **Data Update** How can you use irregular figures to estimate the total land area of your state? Visit www.pre-alg.com/data_update to learn more.

24. **CRITICAL THINKING** In the diagram, a patio that is 4 feet wide surrounds a swimming pool. What is the area of the patio?

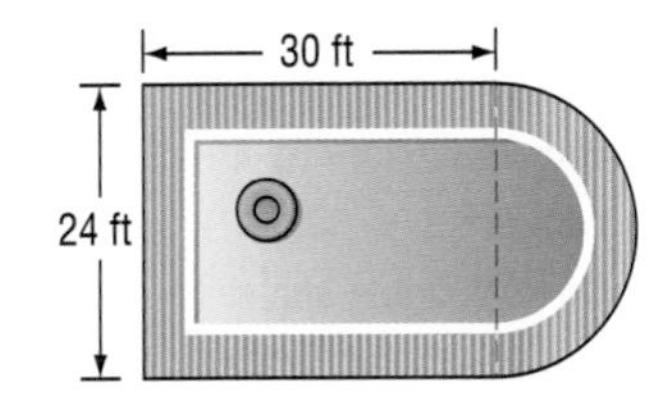

25. WRITING IN MATH Answer the question that was posed at the beginning of the lesson.

How can polygons help to find the area of an irregular figure?

Include the following in your answer:

- an example of an irregular figure, and
- an explanation as to how the figure can be separated to find its area.

For Exercises 26 and 27, refer to the diagram shown. Suppose 1 square unit equals 5 square feet.

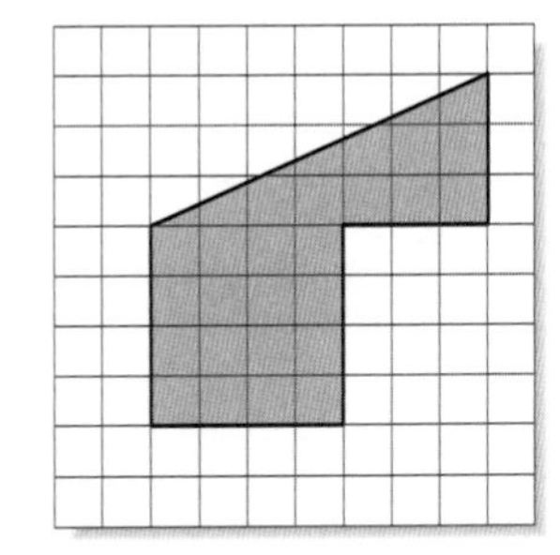

26. What is the area of the figure?

Ⓐ 26.5 ft^2 Ⓑ 34.5 ft^2

Ⓒ 132.5 ft^2 Ⓓ 185 ft^2

27. What is the area of the nonshaded region?

Ⓐ 73.5 ft^2 Ⓑ 140.5 ft^2

Ⓒ 367.5 ft^2 Ⓓ 473.5 ft^2

Maintain Your Skills

Mixed Review **Find the circumference and area of each circle. Round to the nearest tenth.** *(Lesson 10-7)*

28. The wheel on a game show has a diameter of 8.5 feet.

29. The radius is 7 centimeters. **30.** The diameter is 19 inches.

Find the sum of the measures of the interior angles of each polygon. *(Lesson 10-6)*

31. pentagon **32.** quadrilateral **33.** octagon

34. Draw an angle that measures 35°. *(Lesson 9-3)*

35. Simplify $(4x)(-6y)$. *(Lesson 3-2)*

Chapter 10

Study Guide and Review

Vocabulary and Concept Check

adjacent angles (p. 493)
alternate exterior angles (p. 492)
alternate interior angles (p. 492)
altitude (p. 520)
base (p. 520)
bilateral symmetry (p. 505)
center (p. 533)
circle (p. 533)
circumference (p. 533)
complementary (p. 493)
congruent (p. 500)
corresponding angles (p. 492)
corresponding parts (p. 500)
diagonal (p. 528)
diameter (p. 533)
dilation (p. 512)
exterior angles (p. 492, 531)
interior angles (pp. 492, 528)
line symmetry (p. 505)
line of symmetry (pp. 505, 506)
parallel lines (p. 492)
parallelogram (p. 514)
perpendicular lines (p. 494)
pi (p. 533)
polygon (p. 527)
quadrilateral (p. 513)
radius (p. 533)
reflection (p. 506)
regular polygon (p. 529)
rhombus (p. 514)
rotation (p. 506)
rotational symmetry (p. 505)
supplementary (p. 494)
tessellation (p. 532)
transformation (p. 506)
translation (p. 506)
transversal (p. 492)
trapezoid (p. 514)
turn symmetry (p. 505)
vertical angles (p. 493)

Choose the correct term to complete each sentence.

1. Two angles are (complementary, supplementary) if the sum of their measures is 180°.
2. A (rhombus, trapezoid) has four congruent sides.
3. In congruent triangles, the (corresponding angles, adjacent angles) are congruent.
4. In a (rotation, translation), a figure is turned around a fixed point.
5. A polygon in which all sides are congruent is called (equiangular, equilateral).

Lesson-by-Lesson Review

10-1 Angle Relationships

See pages 492–497.

Concept Summary

- When two parallel lines are cut by a transversal, the corresponding angles, the alternate interior angles, and the alternate exterior angles are congruent.
- Two angles are complementary if the sum of their measures is 90°.
- Two angles are supplementary if the sum of their measures is 180°.

Example **In the figure at the right, $\ell \parallel m$ and t is a transversal. If $m\angle 1 = 109°$, find $m\angle 7$.**

Since $\angle 1$ and $\angle 7$ are alternate exterior angles, they are congruent. So, $m\angle 7 = 109°$.

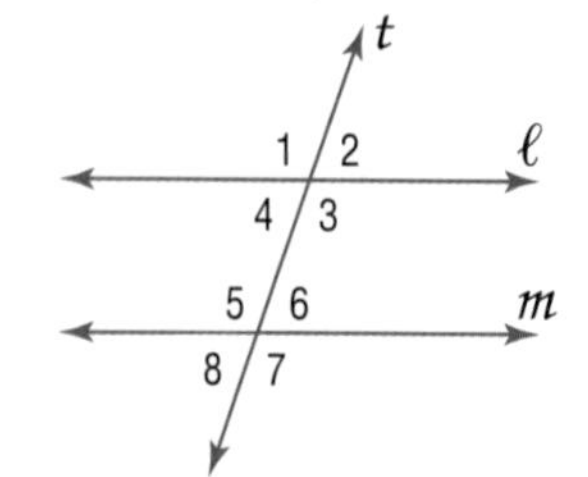

Exercises **Use the figure shown to find the measure of each angle.** *See Example 1 on page 493.*

6. $\angle 5$
7. $\angle 3$
8. $\angle 2$
9. $\angle 6$

10-2 Congruent Triangles

See pages 500–504.

Concept Summary

- Figures that have the same size and shape are congruent.
- The corresponding parts of congruent triangles are congruent.

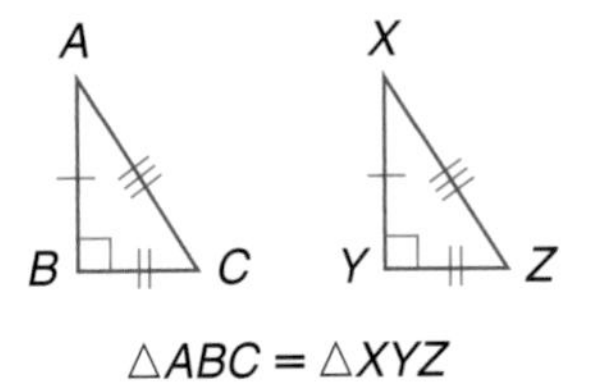

$\triangle ABC \cong \triangle XYZ$

Example **Use $\triangle ABC$ and $\triangle XYZ$ above to complete each congruence statement.**

$\angle B \cong$ ___?___ $\overline{YZ} \cong$ ___?___

B corresponds to Y, so $\angle B \cong \angle Y$.

$\overline{YZ}$ corresponds to $\overline{BC}$, so $\overline{YZ} \cong \overline{BC}$.

Exercises **Complete each congruence statement if $\triangle FGH \cong \triangle QRS$.**

See Example 2 on page 501.

10. $\angle F \cong$ ___?___ **11.** $\angle S \cong$ ___?___ **12.** $\angle R \cong$ ___?___

13. $\overline{GH} \cong$ ___?___ **14.** $\overline{HF} \cong$ ___?___ **15.** $\overline{RQ} \cong$ ___?___

10-3 Transformations on the Coordinate Plane

See pages 506–511.

Concept Summary

- Three types of transformations are translations, reflections, and rotations.

Examples

1 **The vertices of $\triangle JKL$ are $J(1, 2)$, $K(3, 2)$, and $L(1, -1)$. Graph the triangle and its image after a translation 3 units left and 2 units up.**

This translation can be written as $(-3, 2)$.

vertex		3 left, 2 up		translation
$J(1, 2)$	$+$	$(-3, 2)$	$\rightarrow$	$J'(-2, 4)$
$K(3, 2)$	$+$	$(-3, 2)$	$\rightarrow$	$K'(0, 4)$
$L(1, -1)$	$+$	$(-3, 2)$	$\rightarrow$	$L'(-2, 1)$

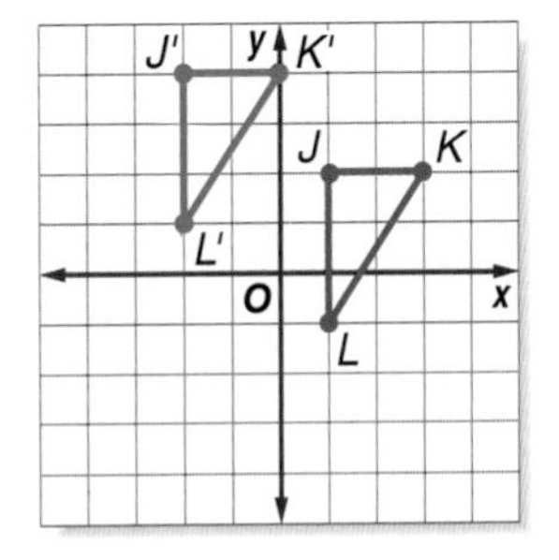

2 **The vertices of figure $ABCD$ are $A(1, -3)$, $B(4, -3)$, $C(1, -1)$, and $D(-2, -1)$. Graph the figure and its image after a reflection over the x-axis.**

Use the same x-coordinate and multiply the y-coordinate by -1.

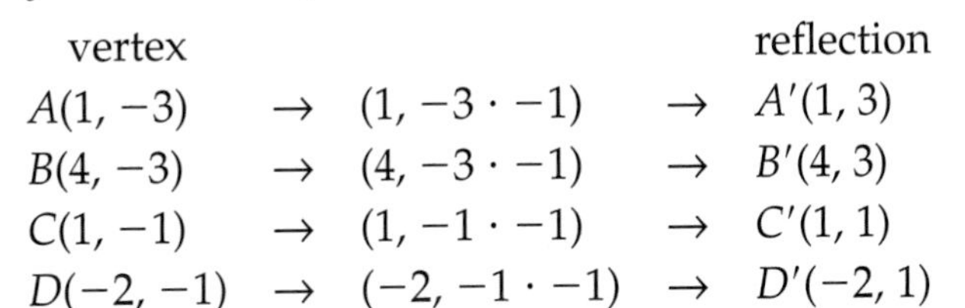

vertex				reflection
$A(1, -3)$	$\rightarrow$	$(1, -3 \cdot -1)$	$\rightarrow$	$A'(1, 3)$
$B(4, -3)$	$\rightarrow$	$(4, -3 \cdot -1)$	$\rightarrow$	$B'(4, 3)$
$C(1, -1)$	$\rightarrow$	$(1, -1 \cdot -1)$	$\rightarrow$	$C'(1, 1)$
$D(-2, -1)$	$\rightarrow$	$(-2, -1 \cdot -1)$	$\rightarrow$	$D'(-2, 1)$

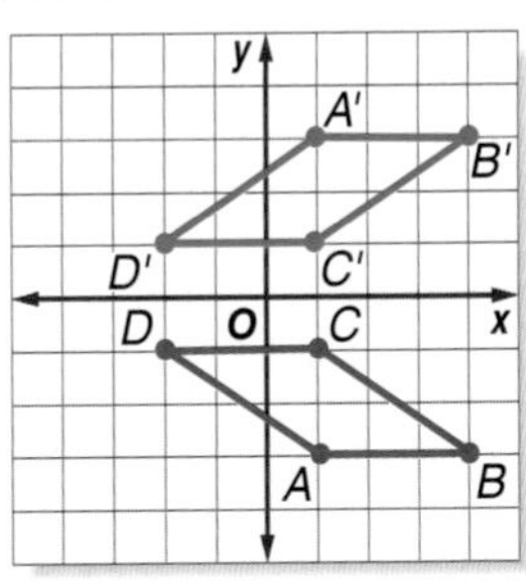

Example 3 **A triangle has vertices $T(-2, 0)$, $W(-4, -2)$, and $Z(-3, -4)$. Graph the triangle and its image after a rotation of 180° about the origin.**

Multiply both coordinates of each point by -1.

vertex		rotation
$T(-2, 0)$	$\rightarrow$	$T'(2, 0)$
$W(-4, -2)$	$\rightarrow$	$W'(4, 2)$
$Z(-3, -4)$	$\rightarrow$	$Z'(3, 4)$

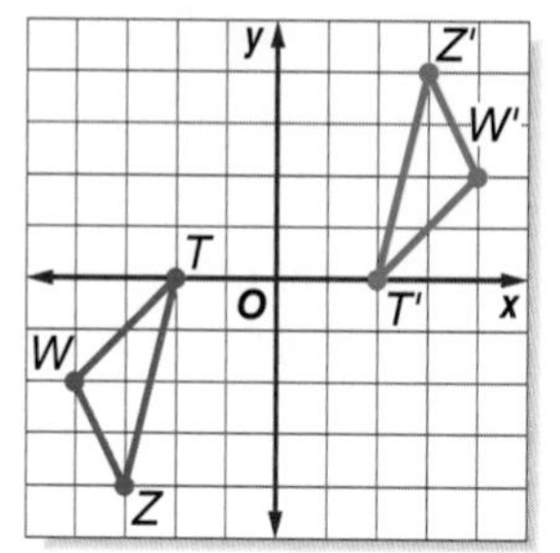

Exercises **Graph each figure and its image.**
See Examples 1–3 on pages 507–509.

16. The vertices of a rectangle are $C(0, 2)$, $D(2, 0)$, $F(-1, -3)$, and $G(-3, -1)$. The rectangle is translated 4 units right and two units down.

17. The vertices of a triangle are $H(-1, 4)$, $I(-4, -2)$, and $J(-2, -1)$. The triangle is reflected over the y-axis.

18. A triangle has vertices $N(1, 3)$, $P(3, 0)$, and $Q(1, -1)$. The triangle is rotated 90° clockwise.

10-4 Quadrilaterals

See pages 513–517.

Concept Summary

- The sum of the angle measures of a quadrilateral is 360°.
- A trapezoid, parallelogram, rhombus, square, and rectangle are examples of quadrilaterals.

Example **Find the value of x. Then find the missing angle measures.**

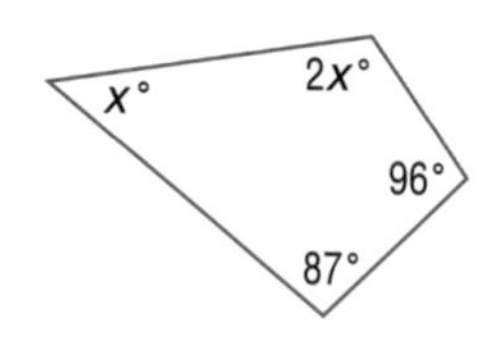

$x + 2x + 96 + 87 = 360$	Angles of a quadrilateral
$3x + 183 = 360$	Combine like terms.
$3x = 177$	Simplify.
$x = 59$	Divide each side by 3.

The value of x is 59. So, the missing angle measures are 59° and 2(59) or 118°

Exercises **Find the value of x. Then find the missing angle measures.**
See Example 1 on page 514.

19.
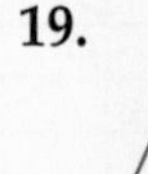
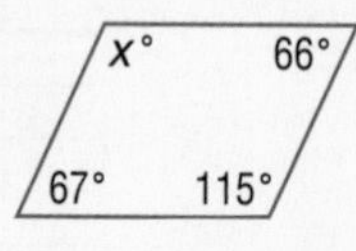

20.
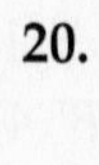
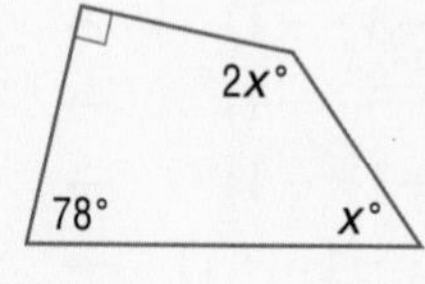

21.
110° 110°
x° x°

10-5 Area: Parallelograms, Triangles, and Trapezoids

See pages 520–525.

Concept Summary

- Area of a parallelogram: $A = bh$
- Area of a triangle: $A = \frac{1}{2}bh$
- Area of a trapezoid: $A = \frac{1}{2}h(a + b)$

Example **Find the area of the trapezoid.**

2 cm
1.8 cm
5 cm

$A = \frac{1}{2}h(a + b)$	Area of a trapezoid
$A = \frac{1}{2}(1.8)(2 + 5)$	Substitution
$A = \frac{1}{2} \cdot 1.8 \cdot 7$	Add 2 and 5.
$A = 6.3$	The area of the trapezoid is 6.3 square centimeters.

Exercises **Find the area of each figure.** *See Examples 1–3 on pages 520–523.*

22.

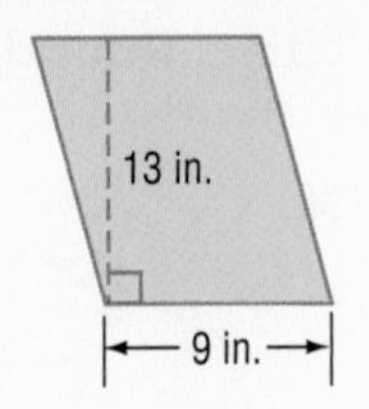

23.

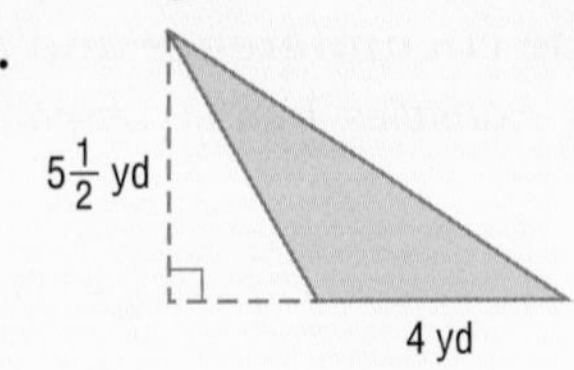

24.

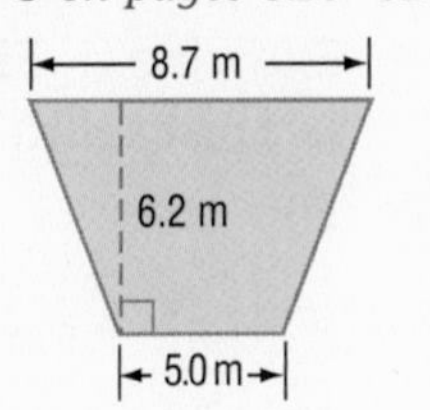

10-6 Polygons

See pages 527–531.

Concept Summary

- Polygons can be classified by the number of sides they have.
- If a polygon has n sides, then the sum of the degree measures of the interior angles of the polygon is $(n - 2)180$.

Example **Classify the polygon. Then find the sum of the measures of the interior angles.**

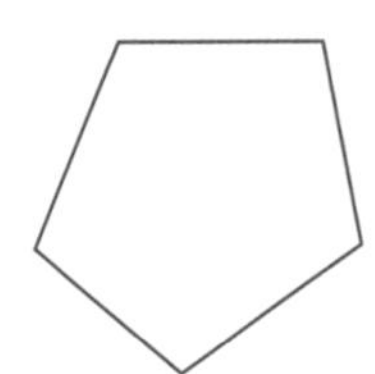

The polygon has 5 sides. It is a pentagon.

$(n - 2)180 = (5 - 2)180$	Replace n with 5.
$= 3(180)$ or 540	Simplify.

The sum of the measures of the angles is 540.

Exercises **Classify each polygon. Then find the sum of the measures of the interior angles.** *See Examples 1 and 2 on page 528.*

25.

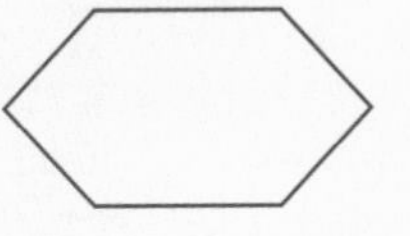

26.

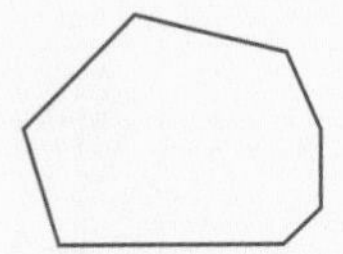

27.

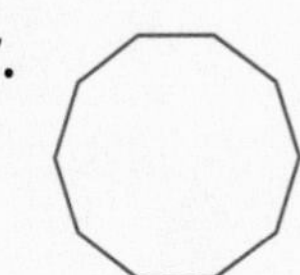

Chapter 10 **For More ...**

- Extra Practice, see pages 747–750.
- Mixed Problem Solving, see page 767.

10-7 Circumference and Area: Circles

See pages 533–538.

Concept Summary

- The circumference C of a circle with radius r is given by $C = 2\pi r$.
- The area A of a circle with radius r is given by $A = \pi r^2$.

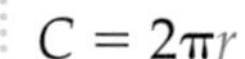

Example **Find the circumference and area of the circle. Round to the nearest tenth.**

15 m

$C = 2\pi r$ Circumference of a circle

$C = 2 \cdot \pi \cdot 7.5$ Replace r with 7.5.

$C \approx 47.1$ The circumference is about 47.1 inches.

$A = \pi r^2$ Area of a circle

$A = \pi \cdot 7.5^2$ Replace r with 7.5.

$A = \pi \cdot 56.25$ Evaluate 7.5^2.

$A \approx 176.7$ The area is about 176.7 square meters.

Exercises **Find the circumference and area of each circle. Round to the nearest tenth.** *See Examples 1 and 3 on pages 534 and 535.*

28.

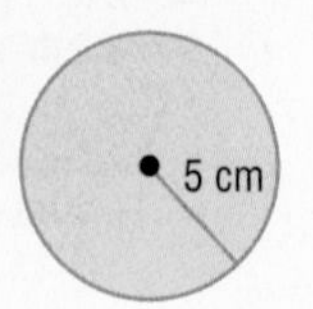

29.

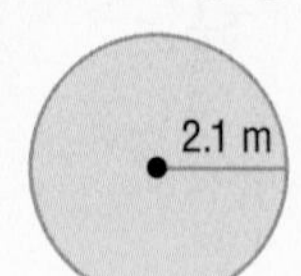

30.

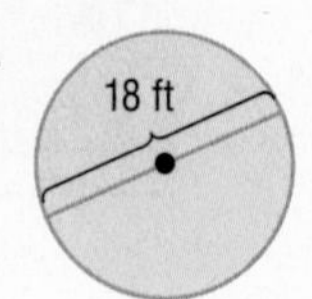

10-8 Area: Irregular Figures

See pages 539–543.

Concept Summary

- To find the area of an irregular figure, separate the irregular figure into figures whose areas you know how to find.

Example **Find the area of the figure.**

5 cm
3 cm
3 cm

Area of Parallelogram

$A = bh$

$A = 3(5)$ or 15

Area of Square

$A = s^2$

$A = 3^2$ or 9

The area of the figure is 15 + 9 or 24 square centimeters.

Exercises **Find the area of each figure. Round to the nearest tenth, if necessary.** *See Example 1 on page 540.*

31. **32.** **33.**

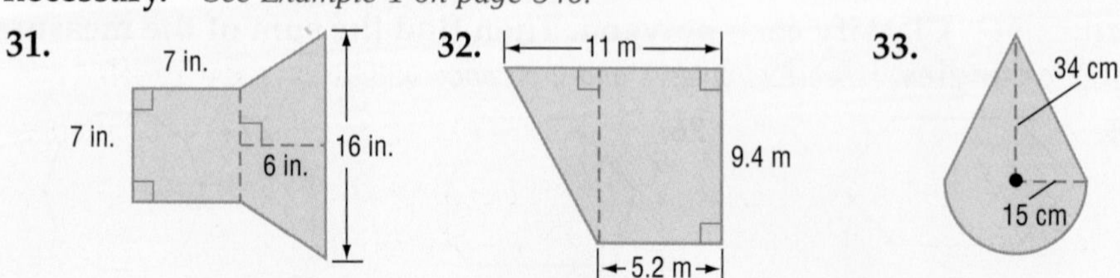

Chapter 10 Practice Test

Vocabulary and Concepts

1. **Draw** and label a diagram that represents each of the following.
   a. congruent triangles b. quadrilateral c. circumference
2. **Compare and contrast** complementary and supplementary angles.

Skills and Applications

In the figure at the right, $a \parallel b$, and c is a transversal. If $m\angle 5 = 58°$, find the measure of each angle.

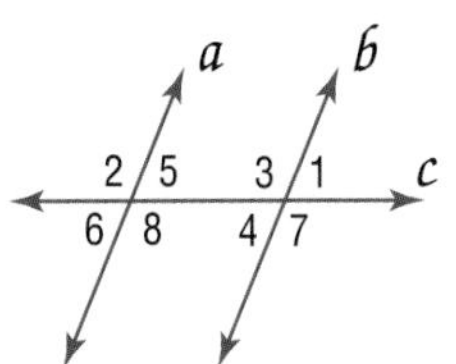

3. $\angle 6$ 4. $\angle 7$ 5. $\angle 4$ 6. $\angle 3$

Complete each congruence statement if $\triangle MNO \cong \triangle PRS$.

7. $\angle P \cong$ __?__ 8. $\overline{RS} \cong$ __?__ 9. $\angle MNO \cong$ __?__

Find the coordinates of the vertices of each figure after the given transformation. Then graph the transformation image.

10. reflection over the y-axis

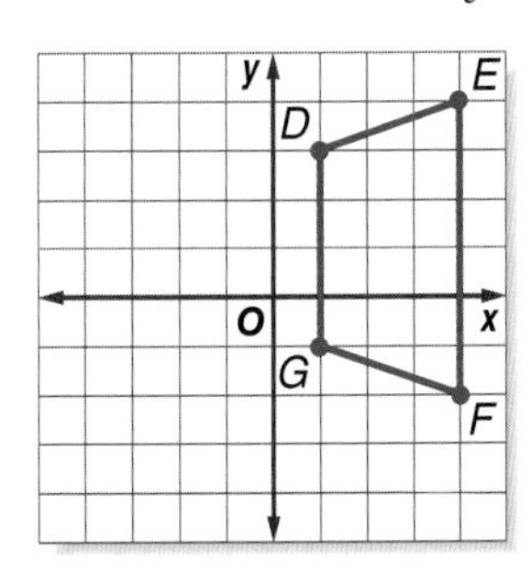

11. rotation of 90° clockwise

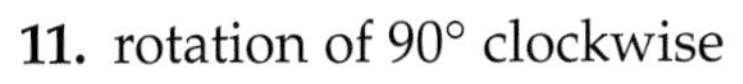

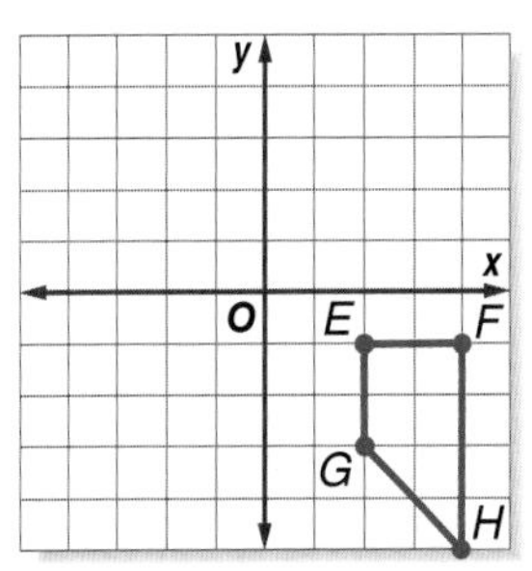

ALGEBRA Find the value of x. Then find the missing angle measure.

12.

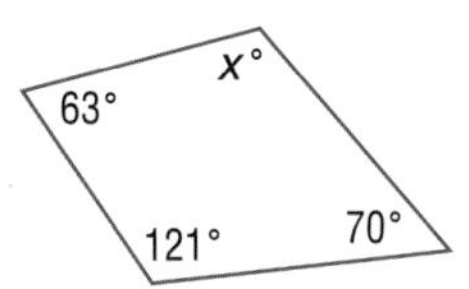

13.

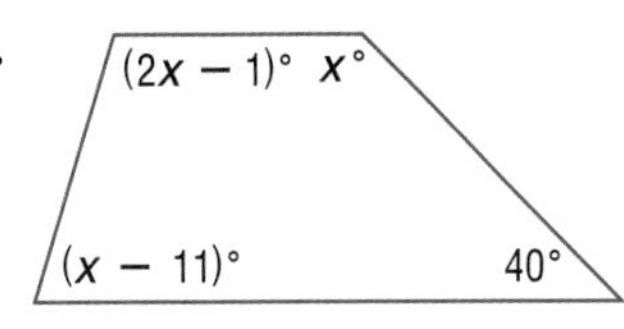

Find the area of each figure described.

14. triangle: base, 21 ft.; height, 16 ft
15. parallelogram: base, 7 ft; height, 2.5 ft

Classify each polygon. Find the sum of the measures of the interior angles.

16.

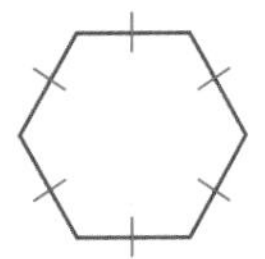

17.

Find the area and circumference of each circle. Round to the nearest tenth.

18. The radius is 3 miles.
19. The diameter is 10 inches.

FCAT Practice

20. **STANDARDIZED TEST PRACTICE** What is the diameter of a circle if its circumference is 54.8 meters? Round to the nearest tenth.
   (A) 8.7 m (B) 15.6 m (C) 17.4 m (D) 34.9 m

Chapter 10 Standardized Test Practice

FCAT Practice

Part 1 Multiple Choice

Record your answers on the answer sheet provided by your teacher or on a sheet of paper.

1. Which expression is equivalent to $\frac{5^5}{5^3}$? (Lessons 4-2 and 4-7)

Ⓐ 5^{-2} Ⓑ 5^2 Ⓒ 5^8 Ⓓ 5^{15}

2. The rectangles shown below are similar. Which proportion can be used to find x? (Lesson 6-2)

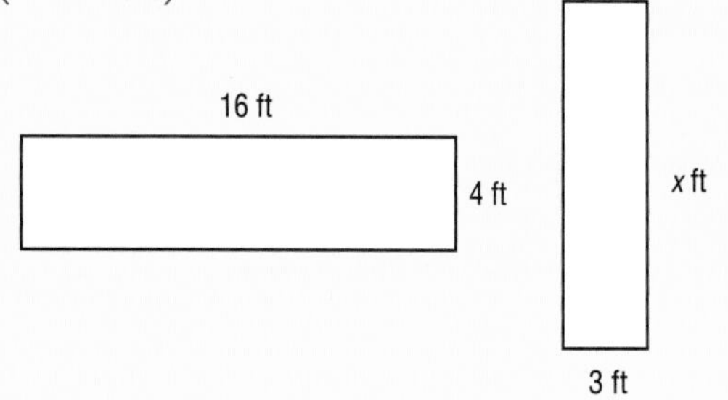

Ⓐ $\frac{x}{3} = \frac{4}{16}$ Ⓑ $\frac{3}{16} = \frac{4}{x}$

Ⓒ $\frac{16}{3} = \frac{x}{4}$ Ⓓ $\frac{16}{4} = \frac{x}{3}$

3. Suppose 2% of the containers made by a manufacturer are defective. If 750 of the containers are inspected, how many can be expected *not* to be defective? (Lesson 6-9)

Ⓐ 150 Ⓑ 730 Ⓒ 735 Ⓓ 748

4. Lawanda plans to paint each side of a cube either white or blue so that when the cube is tossed, the probability that the cube will land on a blue side is $\frac{1}{3}$. How many sides of the cube should she paint blue? (Lesson 6-9)

Ⓐ 1 Ⓑ 2 Ⓒ 3 Ⓓ 4

5. In the spreadsheet below, a formula applied to the values in columns A and B results in the values in column C. What is the formula? (Lesson 8-1)

Ⓐ $C = A - B$

Ⓑ $C = A - 2B$

Ⓒ $C = A + B$

Ⓓ $C = A + 2B$

	A	B	C
1	4	0	4
2	5	1	3
3	6	2	2
4	7	3	1

6. Triangle ABC is a right triangle. What is the length of the hypotenuse? (Lesson 9-5)

Ⓐ 12 units

Ⓑ 32 units

Ⓒ 60 units

Ⓓ 84 units

7. A graphic artist has designed the logo shown below. If rectangles A and B are parallel, then which statement is true? (Lesson 10-1)

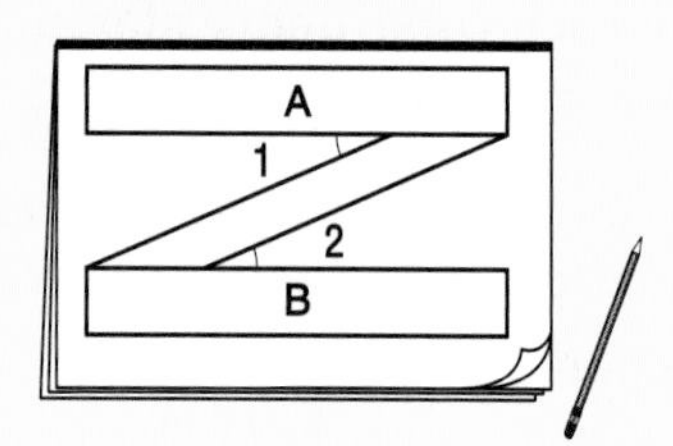

Ⓐ $m\angle 1 - m\angle 2 = 0°$

Ⓑ $m\angle 1 - m\angle 2 = 45°$

Ⓒ $m\angle 1 + m\angle 2 = 90°$

Ⓓ $m\angle 1 + m\angle 2 = 180°$

8. In quadrilateral $ABCD$, $m\angle A = 100°$, $m\angle B = 100°$, and $m\angle C = 90°$. Find $m\angle D$. (Lesson 10-4)

Ⓐ 70° Ⓑ 85° Ⓒ 90° Ⓓ 170°

9. Refer to the figure shown. What is the best estimate of the area of the circle inscribed in the square? (Lesson 10-7)

Ⓐ 7 in² Ⓑ 9 in²

Ⓒ 19 in² Ⓓ 28 in²

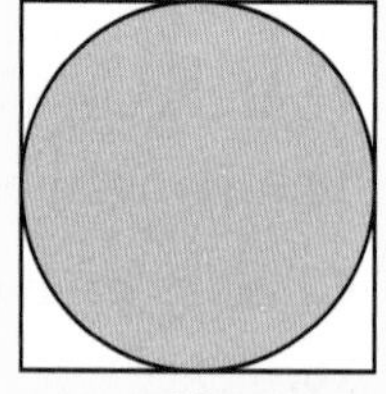

3 in.

Test-Taking Tip

Question 9 Most standardized tests include any necessary formulas in the test booklet. It helps to be familiar with formulas such as the area of a rectangle and the circumference of a circle, but use any formulas that are given to you.

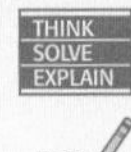

Part 2 Short Response/Grid In

Record your answers on the answer sheet provided by your teacher or on a sheet of paper.

10. If $x = 3$, what is the value of $\frac{14x + 6}{5x - 3}$? (Lesson 1-2)

11. Refer to the data set shown.

20 20 21 22 24 24 24 26

Which is greater, the median or the mode? (Lesson 5-8)

12. A magazine asked 512 students who use computers if they use e-mail. About 83% of the students said that they use e-mail. About how many of these students use e-mail? (Lesson 6-9)

13. Gracia is finding three consecutive whole numbers whose sum is 78. She uses the equation $n + (n + 1) + (n + 2) = 78$. What expression represents the greatest of the three numbers? (Lesson 7-2)

14. Solve $4x + 7 < 39$. (Lesson 7-6)

15. What is the slope of the line represented by the equation $3x + y = 5$? (Lesson 8-4)

16. Find the coordinates of the vertices for quadrilateral $WXYZ$ after a translation of $(-2, -1)$. (Lesson 10-3)

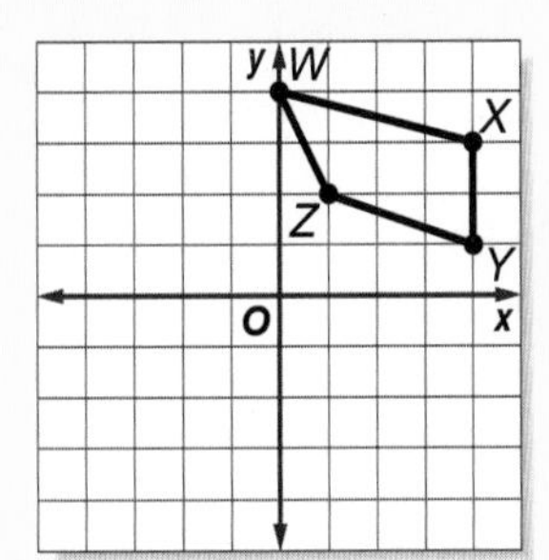

17. These two triangles are congruent. What is the value of x? (Lesson 9-4)

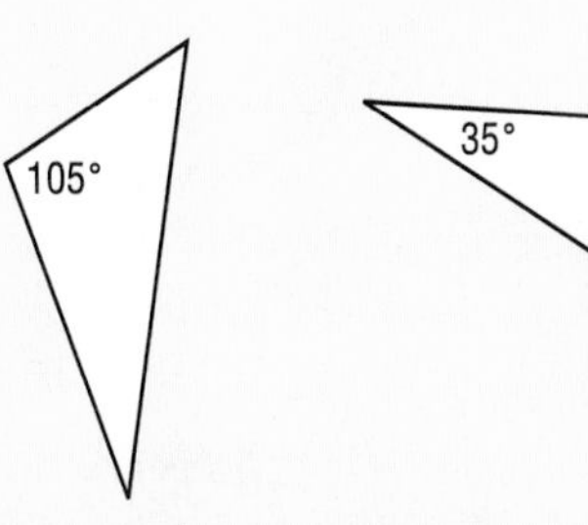

18. What name *best* classifies this quadrilateral? (Lesson 10-4)

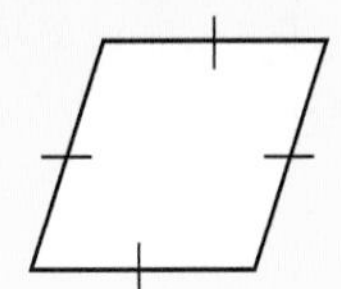

19. The diagram shows a kitchen counter with an area cut out for a sink. What is the area of the counter top? (Lesson 10-8)

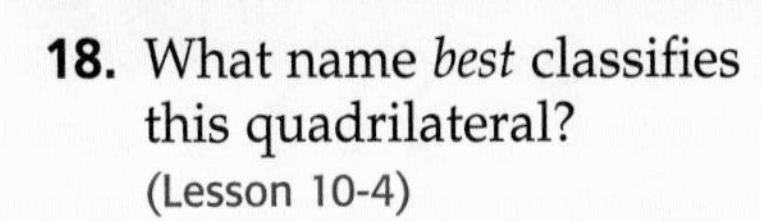

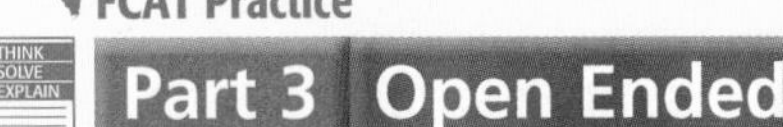

Part 3 Open Ended

Record your answers on a sheet of paper. Show your work.

20. Triangle JKM is shown. (Lesson 10-3)

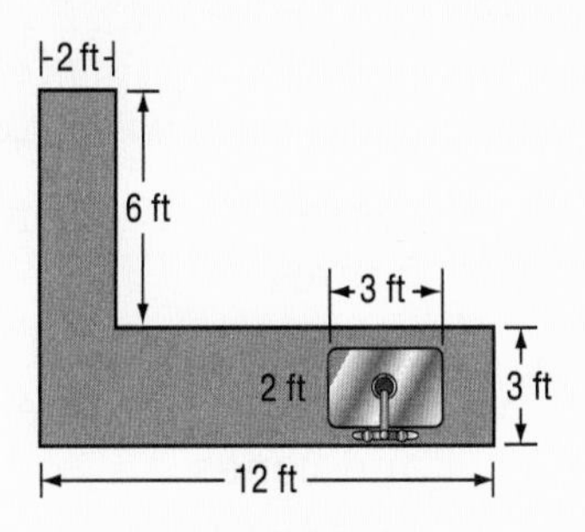

a. What are the coordinates of the vertices of $\triangle JKM$?

b. Copy $\triangle JKM$ onto a sheet of grid paper. Label the vertices.

c. Graph the image of $\triangle JKM$ after a translation 2 units left and 3 units down. Label the translated image $\triangle J'K'M'$.

d. Graph the image of $\triangle J'K'M'$ after a reflection over the y-axis. Label the reflection of $\triangle J'K'M'$ as $\triangle J''K''M''$.

e. On another sheet of grid paper, graph $\triangle JKM$. Then graph $\triangle JKM$ after a reflection over the y-axis. Label the reflection $\triangle J'K'M'$.

f. Graph the image of the reflection in part **e** after a translation 2 units right and 3 units down. Label the translated image $\triangle J''K''M''$.

g. Tell whether the image of $\triangle J''K''M''$ in part **d** is the same as the image of $\triangle J''K''M''$ in part **e**. Explain why or why not.

www.pre-alg.com/standardized_test/fcat

Chapter 11

Three-Dimensional Figures

What You'll Learn

- **Lesson 11-1** Identify three-dimensional figures.
- **Lessons 11-2 and 11-3** Find volumes of prisms, cylinders, pyramids, and cones.
- **Lessons 11-4 and 11-5** Find surface areas of prisms, cylinders, pyramids, and cones.
- **Lesson 11-6** Identify similar solids.
- **Lesson 11-7** Use precision and significant digits to describe measurements.

Key Vocabulary

- **polyhedron** (p. 556)
- **volume** (p. 563)
- **surface area** (p. 573)
- **similar solids** (p. 584)
- **precision** (p. 590)

Why It's Important

Three-dimensional figures have special characteristics. These characteristics are important when architects are designing buildings and other three-dimensional structures.
You will investigate the characteristics of architectural structures in Lesson 11-1.

Getting Started

Prerequisite Skills To be successful in this chapter, you'll need to master these skills and be able to apply them in problem-solving situations. Review these skills before beginning Chapter 11.

For Lesson 11-1 — Polygons

Determine whether each figure is a polygon. If it is, classify the polygon.
(For review, see Lesson 10-6.)

1.  
2. 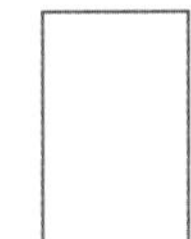
3. 
4. 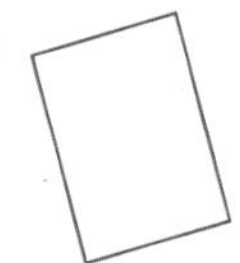

For Lessons 11-2 through 11-5 — Multiplying Rational Numbers

Find each product. *(For review, see Lesson 5-3.)*

5. $8.5 \cdot 2$

6. $3.2(3.2)10$

7. $\frac{1}{2} \cdot 14$

8. $\frac{1}{2}(6.4)(5)$

9. $\frac{1}{3}(50)(9.3)$

10. $\frac{1}{3}\left(\frac{1}{2} \cdot 3 \cdot 8\right)$

For Lesson 11-6 — Proportions

Determine whether each pair of ratios forms a proportion. *(For review, see Lesson 6-2.)*

11. $\frac{3}{8}, \frac{9}{24}$

12. $\frac{7}{2}, \frac{14}{6}$

13. $\frac{18}{32}, \frac{9}{16}$

14. $\frac{12}{15}, \frac{4}{5}$

15. $\frac{1.2}{5}, \frac{6}{25}$

16. $\frac{1.6}{2}, \frac{3.6}{6}$

Make this Foldable to help you organize information about surface area and volume of three-dimensional figures. Begin with a plain piece of 11" × 17" paper.

Step 1 Fold

Fold the paper in thirds lengthwise.

Step 2 Open and Fold

Fold a 2" tab along the short side. Then fold the rest in fourths.

Step 3 Label

Draw lines along folds and label as shown.

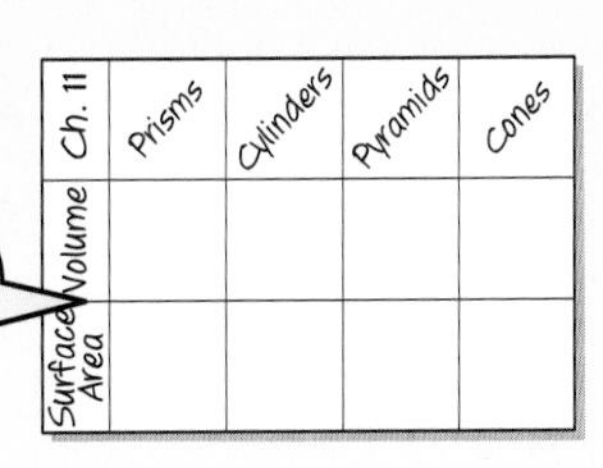

Reading and Writing As you read and study the chapter, write the formulas for surface area and volume and list the characteristics of each three-dimensional figure.

Geometry Activity

A Preview of Lesson 11-1

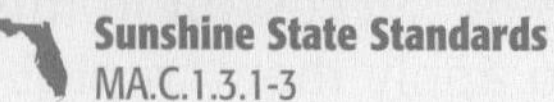

Sunshine State Standards
MA.C.1.3.1-3

Building Three-Dimensional Figures

Activity 1

Different views of a stack of cubes are shown at the right. A point of view is called a **perspective**. You can build or draw a three-dimensional figure using different perspectives.

When drawing figures, use isometric dot paper.

top side front

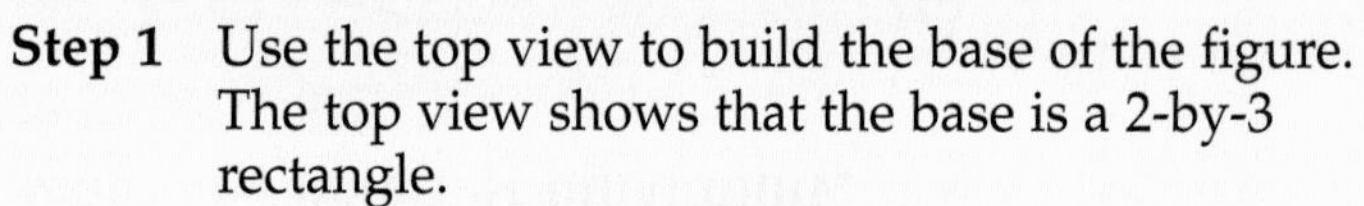

Step 1 Use the top view to build the base of the figure. The top view shows that the base is a 2-by-3 rectangle.

Step 2 Use the side view to complete the figure. The side view shows that the height of the first row is 1 unit, and the height of the second and third rows is 2 units.

Step 3 Use the front view to check the figure. The front view is a 2-by-2 square. This shows that the overall height and width of the figure is 2 units. So, the figure is correct.

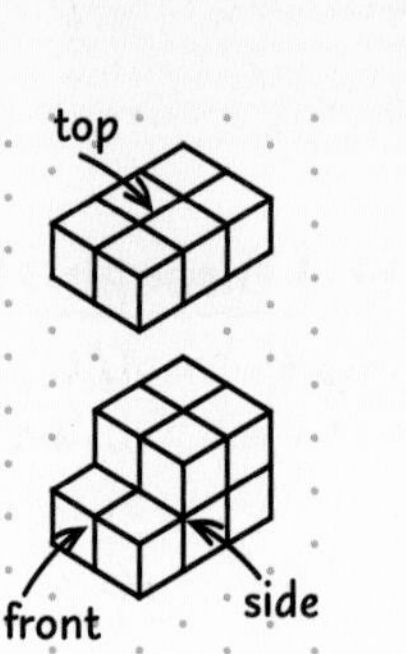

Model

The top view, a side view, and the front view of three-dimensional figures are shown. Use cubes to build each figure. Draw your model on isometric dot paper.

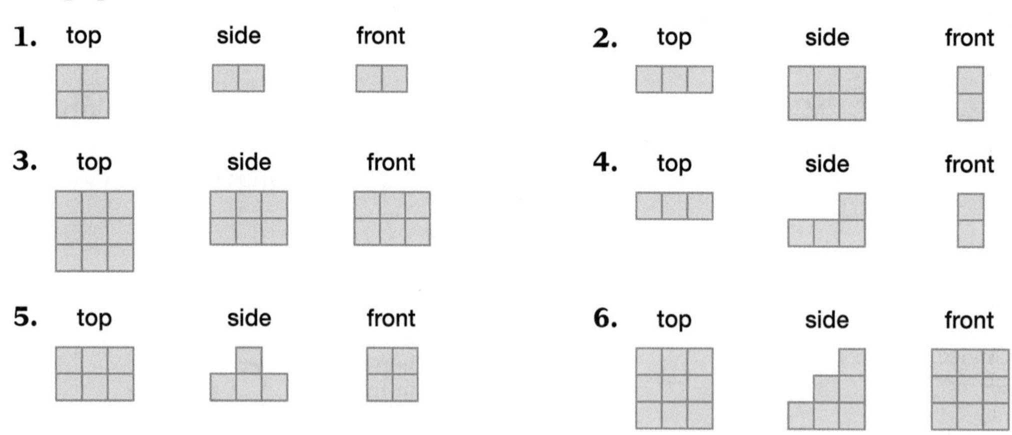

Draw and label the top view, a side view, and the front view for each figure.

7.

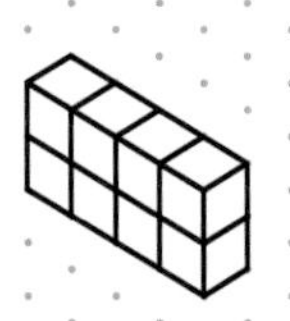

8.

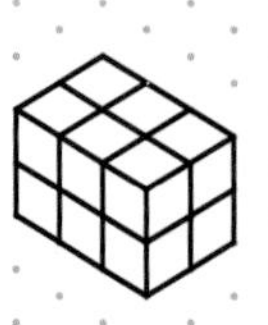

9.

Activity 2

Suppose you cut a cardboard box along its edges, open it up, and lay it flat. The result is a two-dimensional figure called a net. **Nets** are two-dimensional patterns for three-dimensional figures.

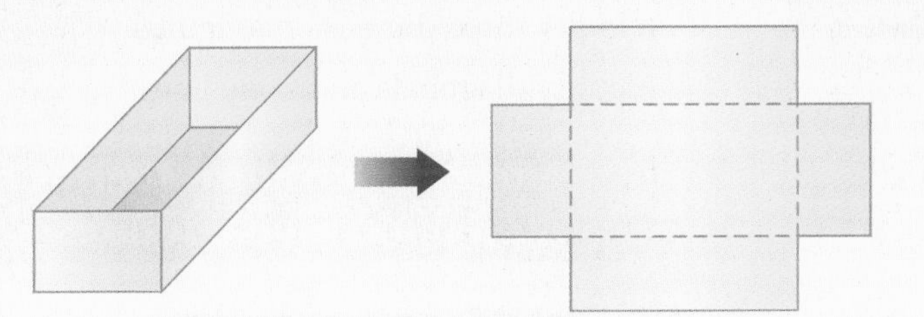

Nets can help you see the regions or *faces* that make up the surface of a figure. So, you can use a net to build a three-dimensional figure.

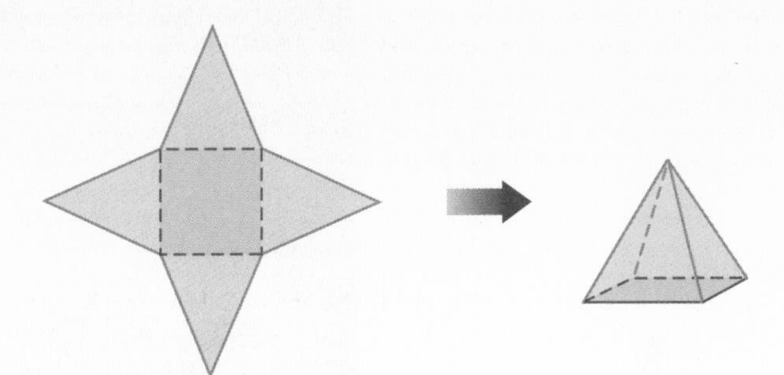

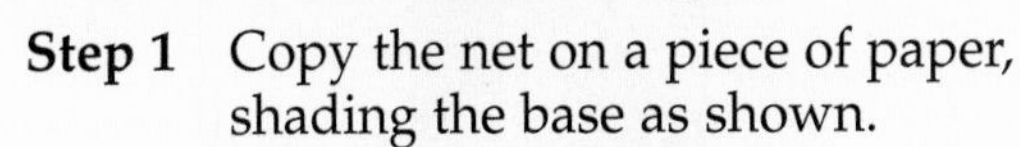

Step 1 Copy the net on a piece of paper, shading the base as shown.

Step 2 Use scissors to cut out the net.

Step 3 Fold on the dashed lines.

Step 4 Tape the sides together.

Different views of this figure are shown.

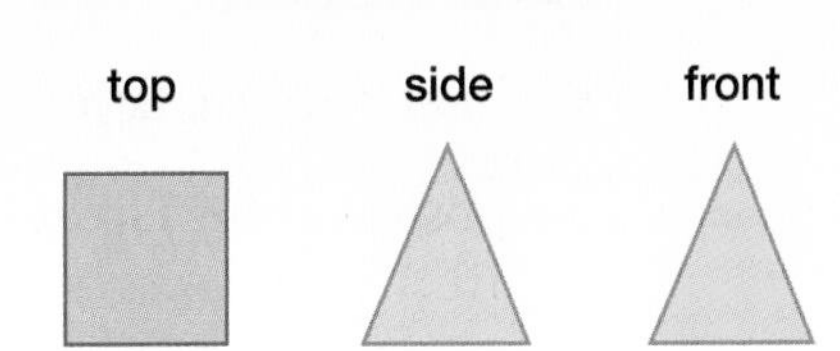

Model

Copy each net. Then cut out the net and fold on the dashed lines to make a 3-dimensional figure, using the purple areas as the bases. Sketch each figure, and draw and label the top view, a side view, and the front view.

10.

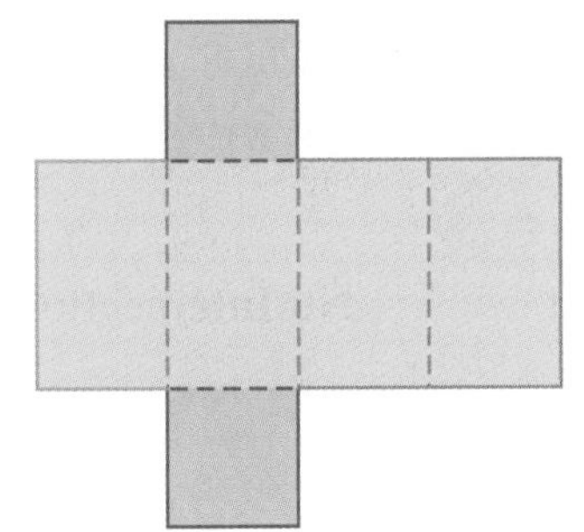

11.

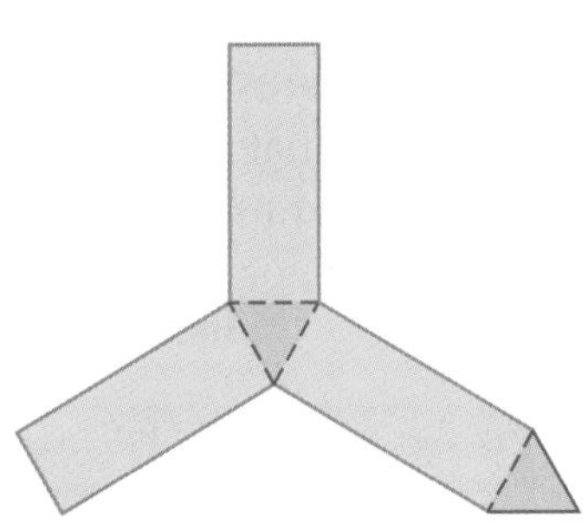

12.

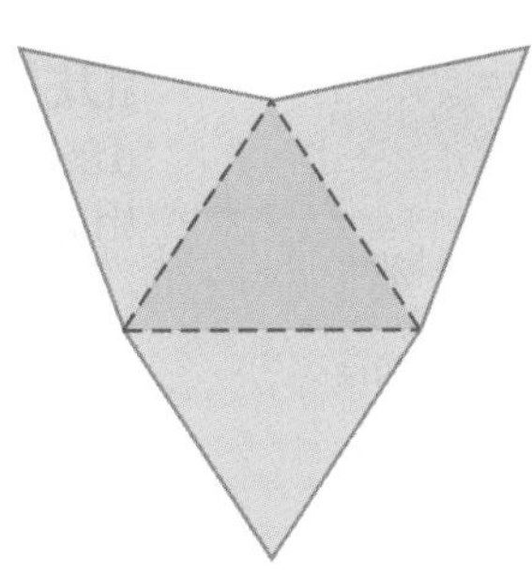

13.

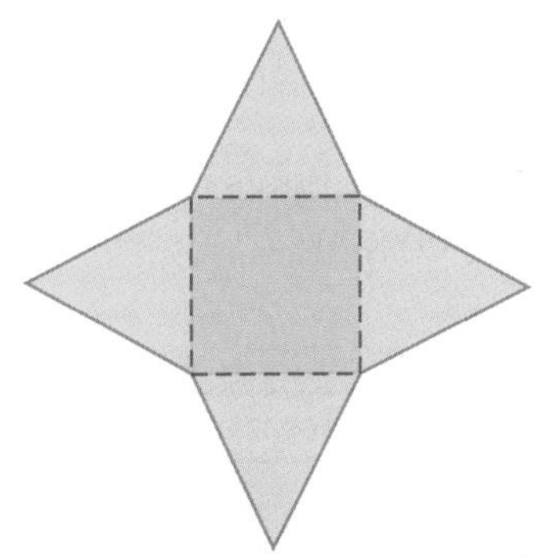

14.

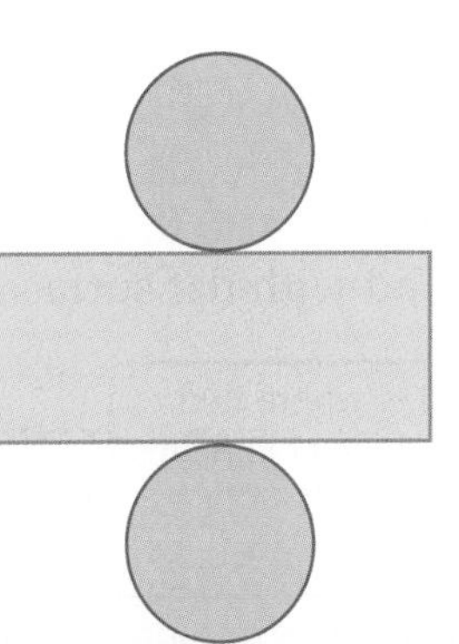

15.

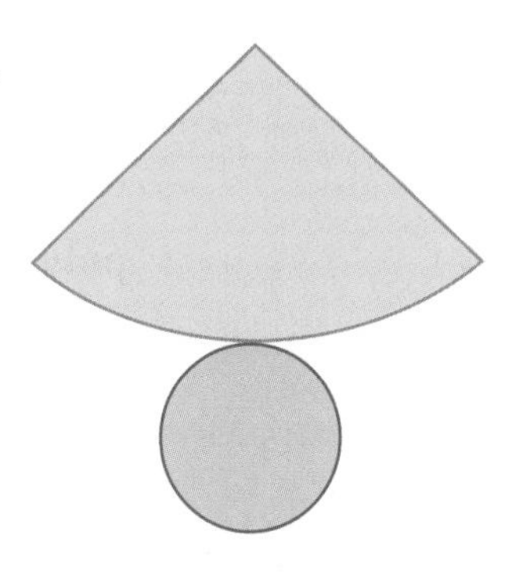

11-1 Three-Dimensional Figures

Sunshine State Standards
MA.C.1.3.1-2, MA.C.1.3.1-3, MA.C.1.3.1-4

Vocabulary

- plane
- solid
- polyhedron
- edge
- vertex
- face
- prism
- base
- pyramid
- skew lines

What You'll Learn

- Identify three-dimensional figures.
- Identify diagonals and skew lines.

How are 2-dimensional figures related to 3-dimensional figures?

Great Pyramid, Egypt

Inner Harbor & Trade Center, Baltimore

a. If you observed the Great Pyramid or the Inner Harbor & Trade Center from directly above, what geometric figure would you see?

b. If you stood directly in front of each structure, what geometric figure would you see?

c. Explain how you can see different polygons when looking at a 3-dimensional figure.

Study Tip

Dimensions
A two-dimensional figure has two dimensions, length and width. A three-dimensional figure has three dimensions, length, width, and depth (or height).

IDENTIFY THREE-DIMENSIONAL FIGURES A **plane** is a two-dimensional flat surface that extends in all directions. There are different ways that planes may be related in space.

Intersect in a Line

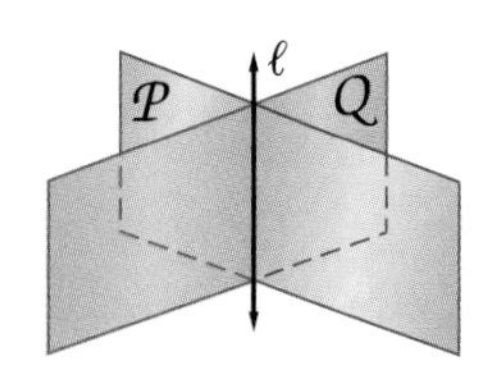

Intersect in a Point

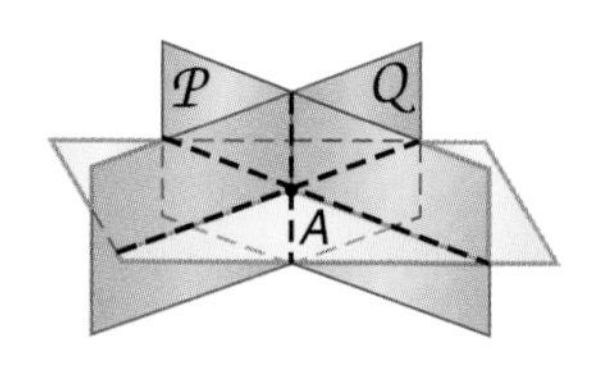

No Intersection

These are called *parallel planes*.

Intersecting planes can also form three-dimensional figures or **solids**. A **polyhedron** is a solid with flat surfaces that are polygons.

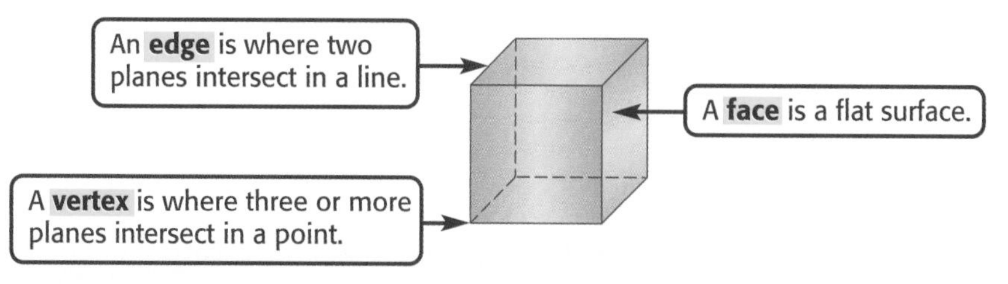

An **edge** is where two planes intersect in a line.
A **face** is a flat surface.
A **vertex** is where three or more planes intersect in a point.

A **prism** is a polyhedron with two parallel, congruent faces called **bases**. A **pyramid** is a polyhedron with one base that is any polygon. Its other faces are triangles.

Prisms and pyramids are named by the shape of their bases.

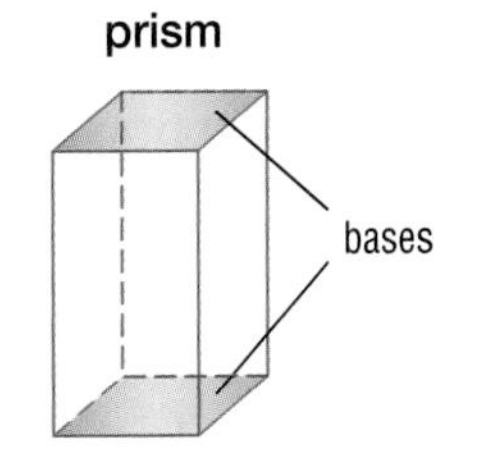

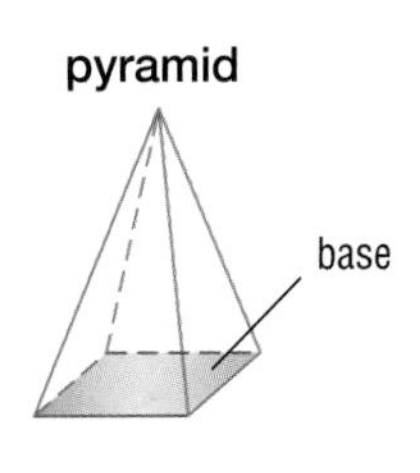

Study Tip

Common Misconception
In a rectangular prism, the bases do not have to be on the top and bottom. Any two parallel rectangles are bases. In a triangular pyramid, any face is a base.

Key Concept — *Polyhedrons*

Polyhedron	triangular prism	rectangular prism	triangular pyramid	rectangular pyramid
Number of Bases	2	2	1	1
Polygon Base	triangle	rectangle	triangle	rectangle
Figure				

Use the labels on the vertices to name a base or a face of a solid.

Example 1 Identify Prisms and Pyramids

Identify each solid. Name the bases, faces, edges, and vertices.

a.

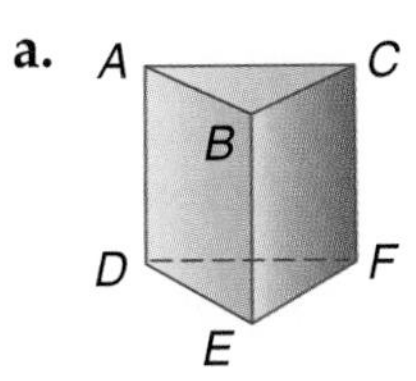

This figure has two parallel congruent bases that are triangles, *ABC* and *DEF*, so it is a triangular prism.

faces: *ABC*, *ADEB*, *BEFC*, *CFDA*, *DEF*

edges: $\overline{AB}$, $\overline{BC}$, $\overline{CA}$, $\overline{AD}$, $\overline{BE}$, $\overline{CF}$, $\overline{DE}$, $\overline{EF}$, $\overline{FD}$

vertices: *A*, *B*, *C*, *D*, *E*, *F*

b.

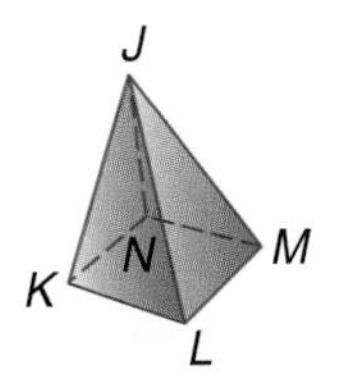

This figure has one rectangular base, *KLMN*, so it is a rectangular pyramid.

faces: *JKL*, *JLM*, *JMN*, *JNK*, *KLMN*

edges: $\overline{JK}$, $\overline{JL}$, $\overline{JM}$, $\overline{JN}$, $\overline{NK}$, $\overline{KL}$, $\overline{LM}$, $\overline{MN}$

vertices: *J*, *K*, *L*, *M*, *N*

Concept Check How many faces does a cube have?

DIAGONALS AND SKEW LINES

Skew lines are lines that are neither intersecting nor parallel. They lie in different planes. Line ℓ along the bridge below and line m along the river beneath it are skew.

> **Study Tip**
>
> *Diagonals*
> Three other diagonals could have been drawn in the prism at the right: $\overline{AG}$, $\overline{BH}$, and $\overline{DF}$.

We can use rectangular prisms to show skew lines. $\overline{EC}$ is a *diagonal* of the prism at the right because it joins two vertices that have no faces in common. $\overline{EC}$ is skew to $\overline{AD}$.

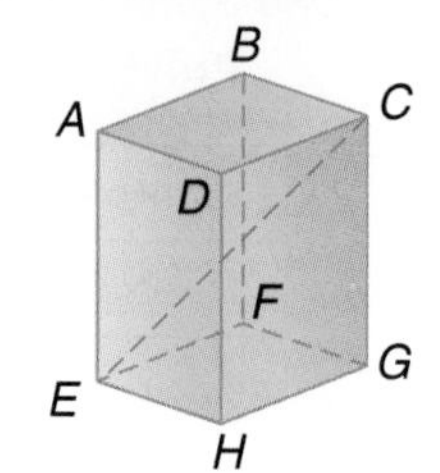

Example 2 Identify Diagonals and Skew Lines

Identify a diagonal and name all segments that are skew to it.

$\overline{MT}$ is a diagonal because vertex M and vertex T do not intersect any of the same faces.

$\overline{LP}$, $\overline{PN}$, $\overline{LQ}$, $\overline{NS}$, $\overline{QR}$, and $\overline{RS}$ are skew to $\overline{MT}$.

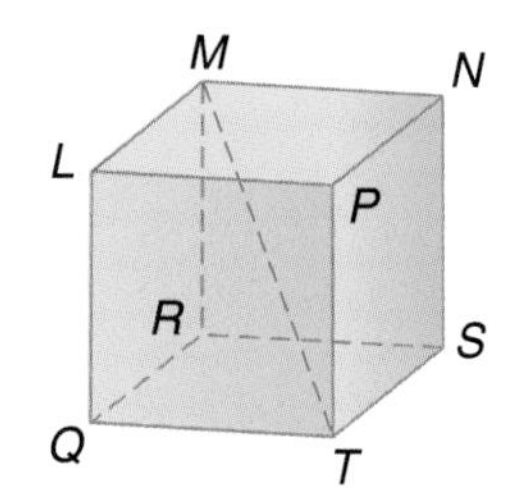

Example 3 Analyze Real-World Drawings

ARCHITECTURE An architect's sketch shows the plans for a new office building. Each unit on the drawing represents 40 feet.

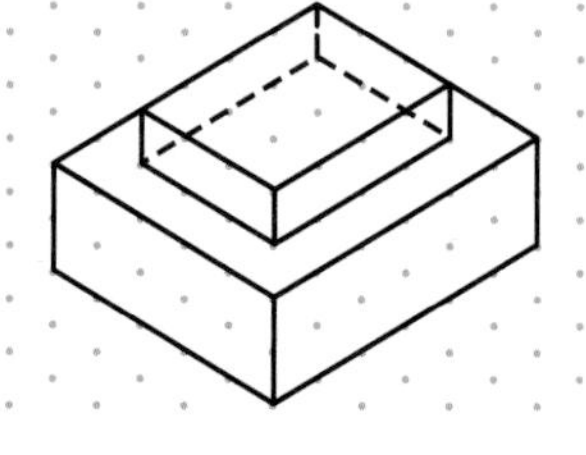

a. Draw a top view and find the area of the ground floor.

The drawing is 6 × 5, so the actual dimensions are 6(40) × 5(40) or 240 feet by 200 feet.

$A = \ell \cdot w$ Formula for area

$A = 240 \cdot 200$ or 48,000

The area of the ground floor is 48,000 square feet.

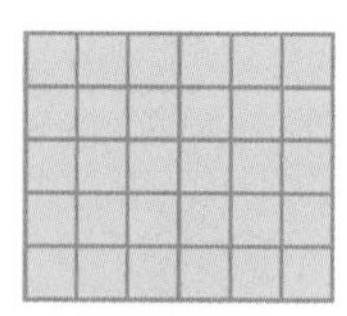

top view

b. How many floors are in the office building if each floor is 15 feet high?

You can see from the side view that the height of the building is 3 units.

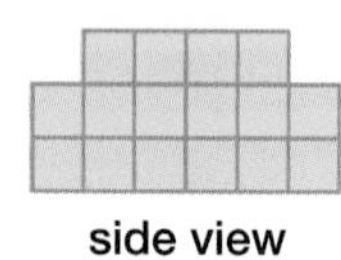

side view

total height: 3 units × 40 feet per unit = 120 feet

number of floors: 120 feet ÷ 15 feet per floor = 8 floors

There are 8 floors in the office building.

Check for Understanding

Concept Check

1. **Describe** the number of planes that form a square pyramid and discuss how the planes form edges and vertices of the pyramid.

2. **OPEN ENDED** Choose a solid object from your home and give an example and a nonexample of edges that form skew lines. Include drawings of your example and nonexample.

Guided Practice

Identify each solid. Name the bases, faces, edges, and vertices.

3. 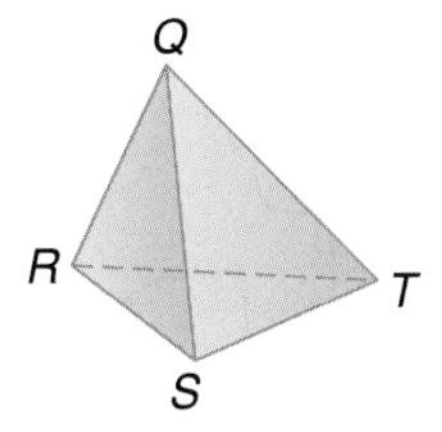

4. 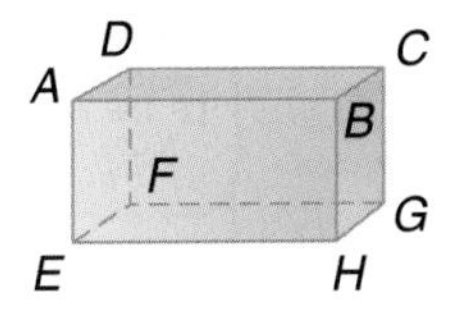

For Exercises 5 and 6, use the rectangular pyramid shown at the right.

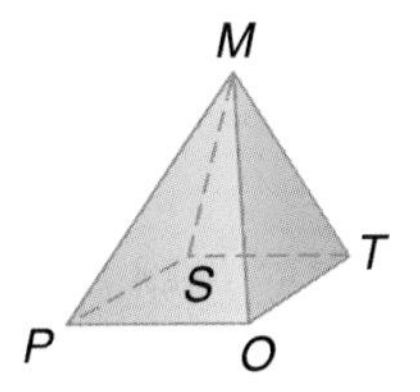

5. State whether $\overline{MP}$ and $\overline{TM}$ are *parallel*, *skew*, or *intersecting*.

6. Identify all lines skew to $\overline{PO}$.

Application

BUILDING **The sketch at the right shows the plans for porch steps.**

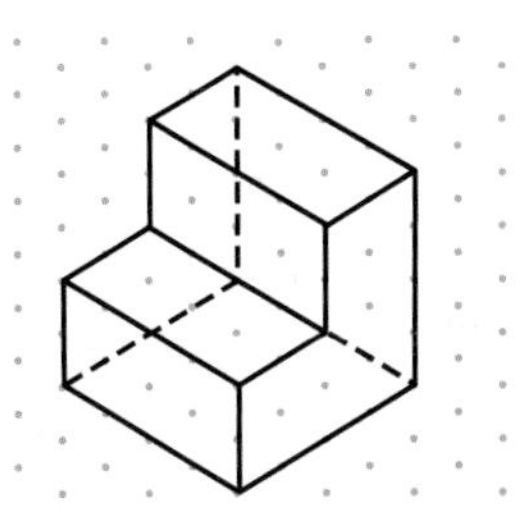

7. Draw and label the top, front, and side views.

8. If each unit on the drawing represents 4 inches, what is the height of the steps in feet?

Practice and Apply

Homework Help

For Exercises	See Examples
9–12, 19–21	1
13–16, 22–25	2
17, 18	3

Extra Practice

See page 750.

Identify each solid. Name the bases, faces, edges, and vertices.

9. 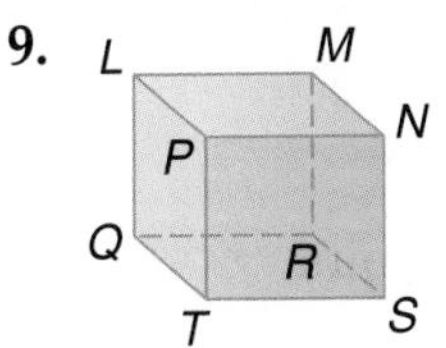

10. 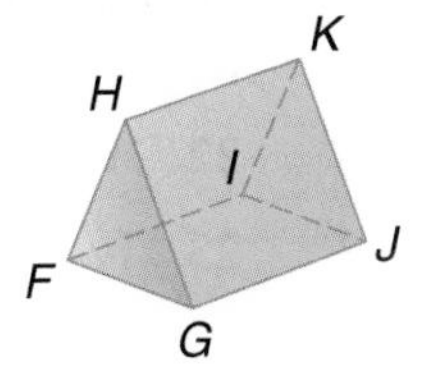

11. 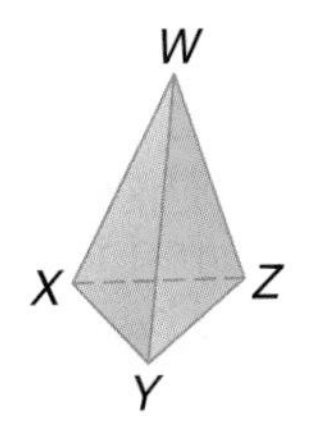

12.
A
B
E
D
C

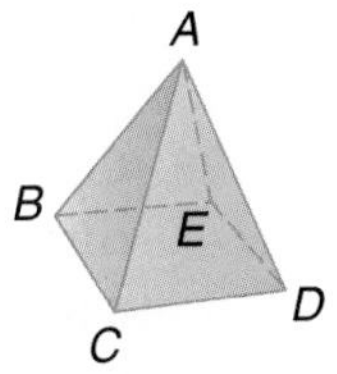

For Exercises 13–16, use the rectangular prism.

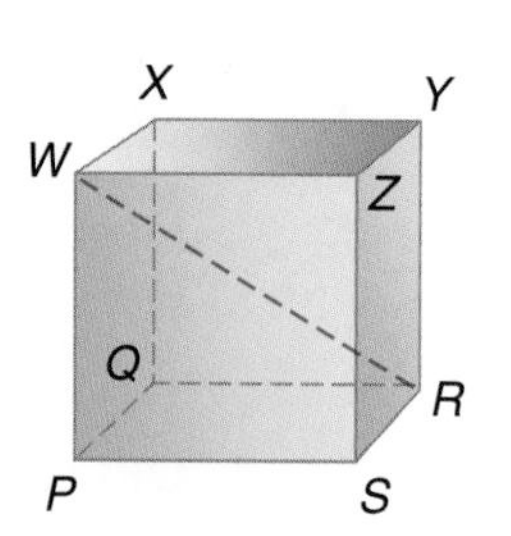

13. Identify a diagonal.
14. Name four segments skew to $\overline{QR}$.
15. State whether $\overline{WR}$ and $\overline{XY}$ are *parallel*, *skew*, or *intersecting*.
16. Name a segment that does *not* intersect the plane that contains $WXYZ$.

COMICS **For Exercises 17 and 18, use the comic below.**

SHOE

17. Which view of the Washington Monument is shown?
18. **RESEARCH** Use the Internet or another source to find a photograph of the Washington Monument. Draw and label the top, side, and front views.

ART **For Exercises 19 and 20, refer to Picasso's painting *The Factory, Horta de Ebro* shown at the right.**

19. Describe the polyhedrons shown in the painting.
20. Explain how the artist portrayed three-dimensional objects on a flat canvas.
21. **RESEARCH** Find other examples of art in which polyhedrons are shown. Describe the polyhedrons.

More About. . .

Art

Pablo Picasso (1881–1973) was one of the developers of a movement in art called *Cubism*. Cubist paintings are characterized by their angular shapes and sharp edges.

Source: World Wide Arts Resources

Determine whether each statement is *sometimes, always,* or *never* true. Explain.

22. Any two planes intersect in a line.
23. Two planes intersect in a single point.
24. A pyramid contains a diagonal.
25. Three planes do not intersect in a point.
26. **CRITICAL THINKING** Use isometric dot paper to draw a three-dimensional figure in which the front and top views have a line of symmetry but the side view does not. Then discuss whether the figure has bilateral symmetry or rotational symmetry.

27. **WRITING IN MATH** Answer the question that was posed at the beginning of the lesson.

How are 2-dimensional figures related to 3-dimensional figures?

Include the following in your answer:

- an explanation of the difference between 2-dimensional figures and 3-dimensional figures, and
- a description of how 2-dimensional figures can form a 3-dimensional figure.

28. Determine the intersection of the three planes at the right.

Ⓐ point

Ⓑ line

Ⓒ plane

Ⓓ no intersection

29. Which figure does *not* have the same dimensions as the other figures?

Ⓐ 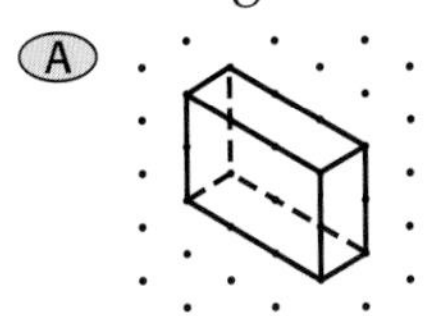

Ⓑ 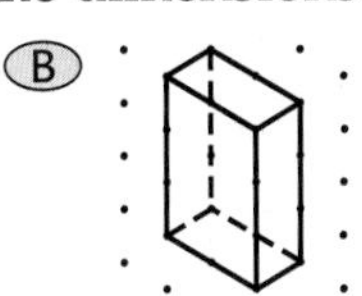

Ⓒ 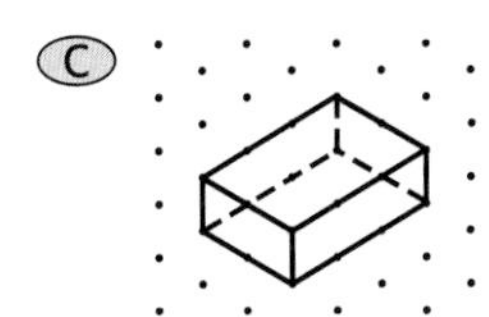

Ⓓ 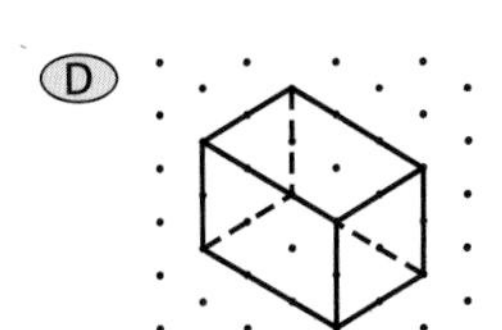

Maintain Your Skills

Mixed Review

30. Find the area of *ABCDEF* if each unit represents 1 square centimeter. *(Lesson 10-10)*

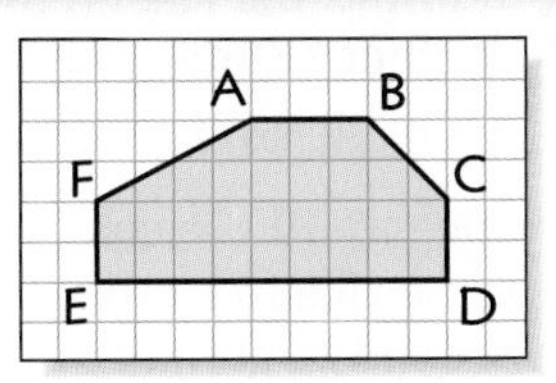

31. Find the circumference and the area of a circle whose radius is 6 centimeters. Round to the nearest tenth. *(Lesson 10-9)*

Use a calculator to find each ratio to the nearest ten thousandth. *(Lesson 9-7)*

32. $\sin 35°$ 33. $\sin 30°$ 34. $\cos 280°$

Solve each inequality. Check your solution. *(Lesson 7-4)*

35. $c + 4 < 12$ 36. $7 \geq t - 2$ 37. $-26 < n + (-15)$

38. $k + (-4) \geq 3.8$ 39. $y - \frac{1}{4} < 1\frac{1}{2}$ 40. $3\frac{1}{5} > a - \frac{3}{10}$

Getting Ready for the Next Lesson

PREREQUISITE SKILL Find the area of each triangle described.

*(To review **areas of triangles**, see Lesson 10-5.)*

41. base, 4 in.; height, 7 in.   42. base, 10 ft; height, 9 ft

43. base, 6.5 cm; height, 2 cm   44. base, 0.4 m; height, 1.3 m

Geometry Activity

A Preview of Lesson 11-2

Sunshine State Standards
MA.B.1.3.1-1, MA.B.1.3.3-1

Volume

In this activity, you will investigate volume by making containers of different shapes and comparing how much each container holds.

Activity

Collect the Data

Step 1 Use three 5-inch × 8-inch index cards to make three different containers, as shown below.

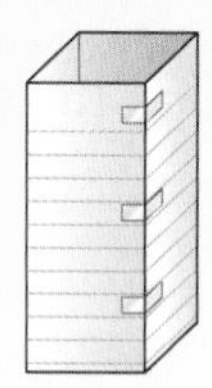

square base with 2-inch sides

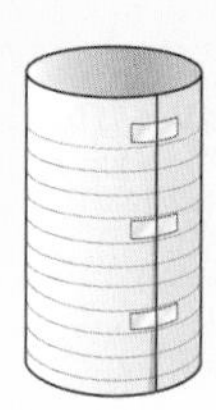

circular base with 8-inch circumference

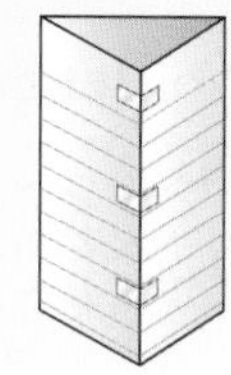

triangular base with sides 2 inches, 3 inches, and 3 inches

Step 2 Tape one end of each container to another card as a bottom, but leave the top open, as shown at the right.

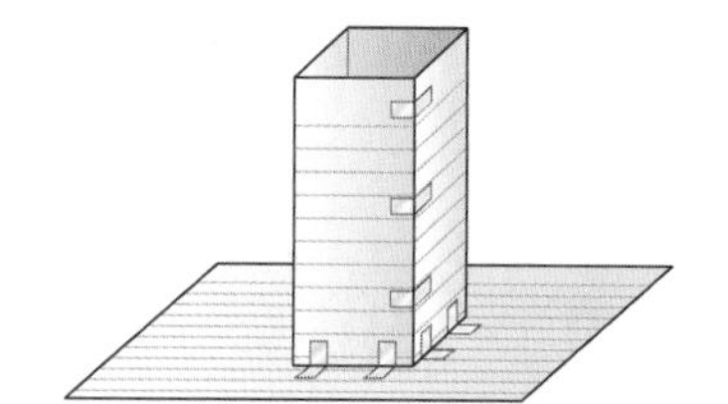

Step 3 Estimate which container would hold the most (have the greatest volume) and which would hold the least (have the least volume), or whether all the containers would hold the same amount.

Step 4 Use rice to fill the container that you believe holds the least amount. Then pour the rice from this container into another container. Does the rice fill the second container? Continue the process until you find out which container, if any, has the least volume and which has the greatest.

Analyze the Data

1. Which container holds the greatest amount of rice? Which holds the least amount?
2. How do the heights of the three containers compare? What is each height?
3. Compare the perimeters of the bases of each container. What is each base perimeter?
4. Trace the base of each container onto grid paper. Estimate the area of each base.
5. Which container has the greatest base area?
6. Does there appear to be a relationship between the area of the bases and the volume of the containers when the heights remain unchanged? Explain.

11-2 Volume: Prisms and Cylinders

What You'll Learn

- Find volumes of prisms.
- Find volumes of circular cylinders.

Sunshine State Standards
MA.B.1.3.1-1, MA.B.1.3.1-2, MA.B.1.3.3-1, MA.C.1.3.1-2, MA.C.1.3.1-4

Vocabulary
- volume
- cylinder

How is volume related to area?

The rectangular prism is built from 24 cubes.

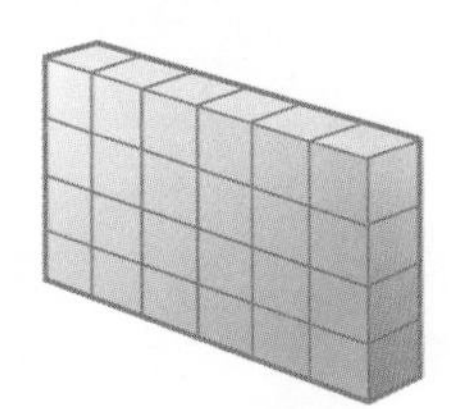

a. Build three more rectangular prisms using 24 cubes. Enter the dimensions and base areas in a table.

Prism	Length (units)	Width (units)	Height (units)	Area of Base (units^2)
1	6	1	4	6
2				
3				
4				

b. *Volume* equals the number of cubes that fill a prism. How is the volume of each prism related to the product of the length, width, and height?

c. **Make a conjecture** about how the area of the base B and the height h are related to the volume V of a prism.

Study Tip

Measures of Volume
A cubic centimeter (cm^3) is a cube whose edges measure 1 centimeter.

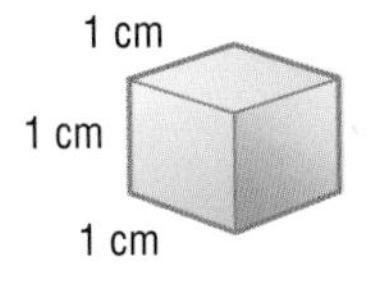

VOLUMES OF PRISMS The prism above has a volume of 24 cubic centimeters. **Volume** is the measure of space occupied by a solid region.

To find the volume of a prism, you can use the area of the base and the height, as given by the following formula.

Key Concept *Volume of a Prism*

- **Words** The volume V of a prism is the area of the base B times the height h.
- **Symbols** $V = Bh$
- **Models**

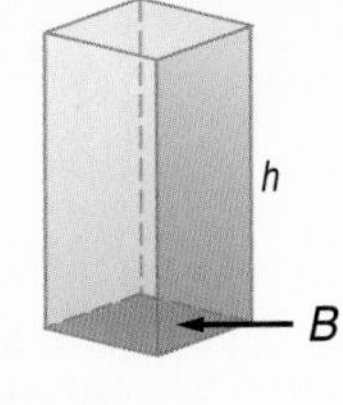

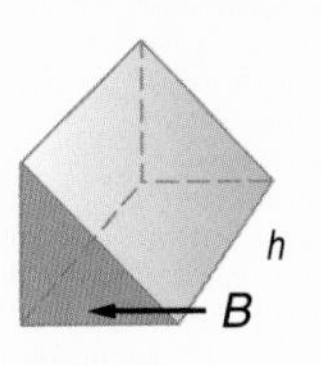

Example 1 Volume of a Rectangular Prism

Find the volume of the prism.

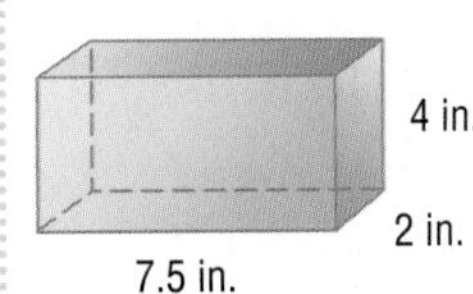

$V = Bh$ Formula for volume of a prism

$V = (\ell \cdot w)h$ The base is a rectangle, so $B = \ell \cdot w$.

$V = (7.5 \cdot 2)4$ $\ell = 7.5$, $w = 2$, $h = 4$

$V = 60$ Simplify.

The volume is 60 cubic inches.

Example 2 Volume of a Triangular Prism

Find the volume of the triangular prism.

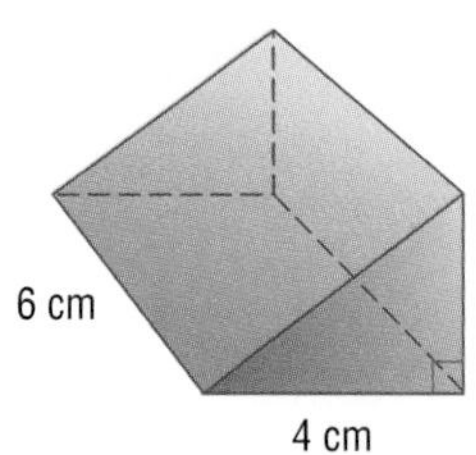

$V = Bh$	Formula for volume of a prism
$V = \left(\frac{1}{2} \cdot 4 \cdot 3\right)h$	B = area of base or $\frac{1}{2} \cdot 4 \cdot 3$
$V = \left(\frac{1}{2} \cdot 4 \cdot 3\right)6$	The height of the prism is 6 cm.
$V = 36$	Simplify.

The volume is 36 cubic centimeters.

Example 3 Height of a Prism

AQUARIUMS **A wall is being constructed to enclose three sides of an aquarium that is a rectangular prism 8 feet long and 5 feet wide. If the aquarium is to contain 220 cubic feet of water, what is its height?**

$V = Bh$	Formula for volume of a prism
$V = \ell \cdot w \cdot h$	Formula for volume of a rectangular prism
$220 = 8 \cdot 5 \cdot h$	Replace V with 220, ℓ with 8, and w with 5.
$220 = 40h$	Simplify.
$5.5 = h$	Divide each side by 40.

The height of the aquarium is 5.5 feet.

More About. . .

Aquariums
Shedd Aquarium in Chicago, Illinois, has a Caribbean reef exhibit that holds 90,000 gallons of water and contains more than 70 species of animals.
Source: www.shedd.org

To find the volume of a solid containing several prisms, break it down into simpler parts.

FCAT Practice
Standardized Test Practice
A B C D

Example 4 Volume of a Complex Solid

Multiple-Choice Test Item

Find the volume of the solid at the right.

(A) 180 ft^3
(B) 1320 ft^3
(C) 960 ft^3
(D) 1140 ft^3

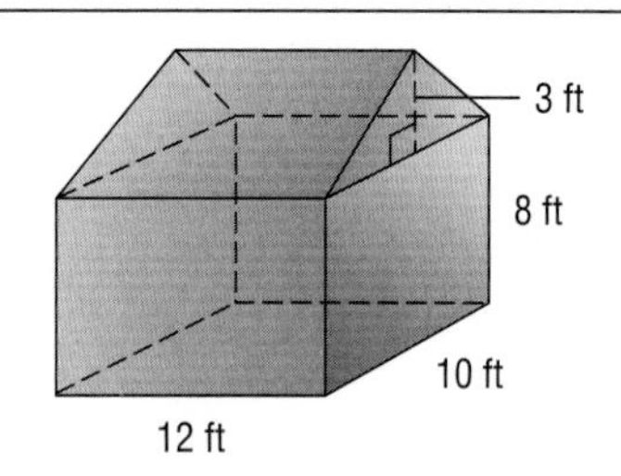

Read the Test Item

The solid is made up of a rectangular prism and a triangular prism. The *volume of the solid* is the sum of both volumes.

Solve the Test Item

Step 1 The volume of the rectangular prism is 12(10)(8) or 960 ft^3.

Step 2 In the triangular prism, the area of the base is $\frac{1}{2}(10)(3)$, and the height is 12. Therefore, the volume is $\frac{1}{2}(10)(3)(12)$ or 180 ft^3.

Step 3 Add the volumes.
$960 \text{ ft}^3 + 180 \text{ ft}^3 = 1140 \text{ ft}^3$

The answer is D.

The Princeton Review
Test-Taking Tip
Estimate You can eliminate A and C as answers because the volume of the rectangular prism is 960 ft^3, so the volume of the whole solid must be greater.

Reading Math

Cylinders

In this text, *cylinder* refers to a cylinder with a circular base.

VOLUMES OF CYLINDERS A **cylinder** is a solid whose bases are congruent, parallel circles, connected with a curved side. Like prisms, the volume of a cylinder is the product of the base area and the height.

Key Concept — *Volume of a Cylinder*

- **Words** The volume V of a cylinder with radius r is the area of the base B times the height h.
- **Symbols** $V = Bh$ or $V = \pi r^2 h$, where $B = \pi r^2$
- **Model**

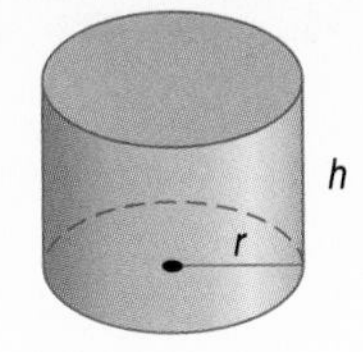

Example 5 Volume of a Cylinder

Study Tip

Unless specified otherwise, use 2nd [π] on a calculator to evaluate expressions involving π. Then follow any instructions regarding rounding.

Find the volume of each cylinder. Round to the nearest tenth.

a.

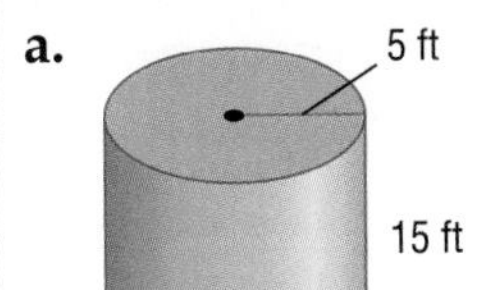

$V = \pi r^2 h$	Formula for volume of a cylinder
$V = \pi \cdot 5^2 \cdot 15$	Replace r with 5 and h with 15.
$V \approx 1178.1$	Simplify.

The volume is about 1178.1 cubic feet.

b. diameter of base 16.4 mm, height 20 mm

Since the diameter is 16.4 mm, the radius is 8.2 mm.

$V = \pi r^2 h$	Formula for volume of a cylinder
$V = \pi \cdot 8.2^2 \cdot 20$	Replace r with 8.2 and h with 20.
$V \approx 4224.8$	Simplify.

The volume is about 4224.8 cubic millimeters.

Check for Understanding

Concept Check

1. **OPEN ENDED** Describe a problem from an everyday situation in which you need to find the volume of a cylinder or a rectangular prism. Explain how to solve the problem.

2. **FIND THE ERROR** Eric says that doubling the length of each side of a cube doubles the volume. Marissa says the volume is eight times greater. Who is correct? Explain your reasoning.

Guided Practice

Find the volume of each solid. If necessary, round to the nearest tenth.

3.

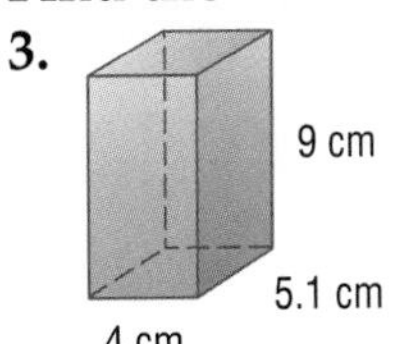

4.

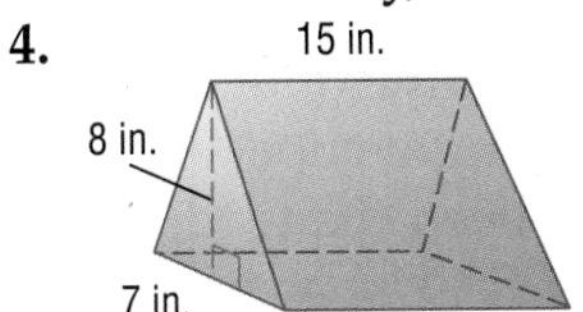

5.

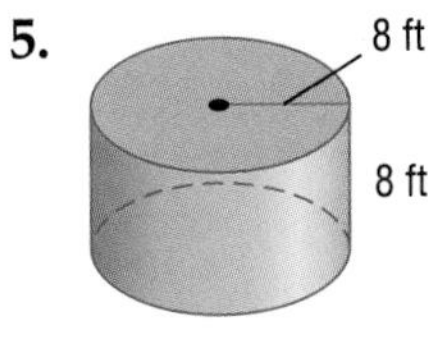

6. rectangular prism: length 6 in., width 6 in., height 9 in.

7. cylinder: radius 3 yd, height 10 yd

8. Find the height of a rectangular prism with a length of 3 meters, width of 1.5 meters, and a volume of 60.3 cubic meters.

9. **ENGINEERING** A cylindrical storage tank is being manufactured to hold at least 1,000,000 cubic feet of natural gas and have a diameter of no more than 80 feet. What height should the tank be to the nearest tenth of a foot?

10. Find the volume of the solid at the right.

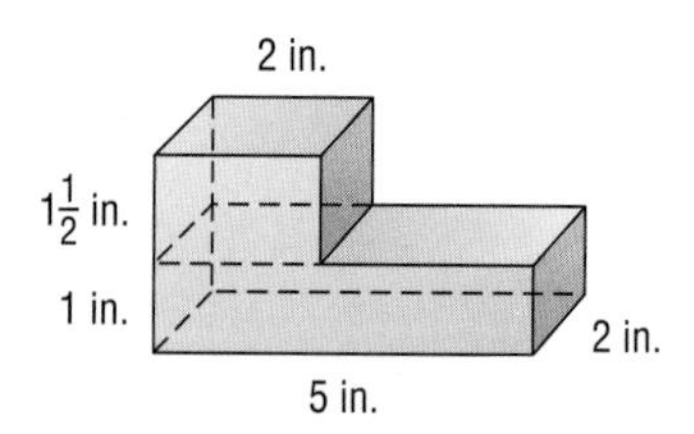

Ⓐ 6 in^3 Ⓑ 10 in^3

Ⓒ 13 in^3 Ⓓ 16 in^3

Practice and Apply

Homework Help

For Exercises	See Examples
11, 12, 17, 23–27	1
13, 14, 18	2
15, 16, 19, 28	5
21, 22	3

Extra Practice

See page 750.

Find the volume of each solid shown or described. If necessary, round to the nearest tenth.

11.

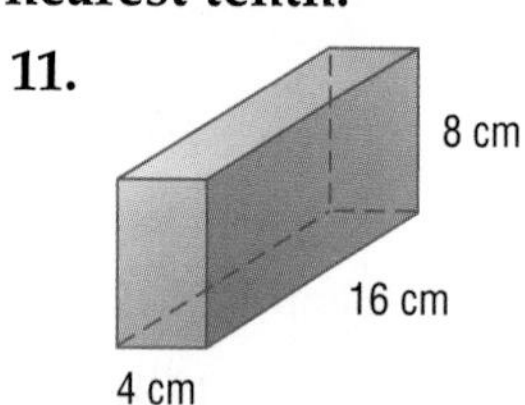

12.

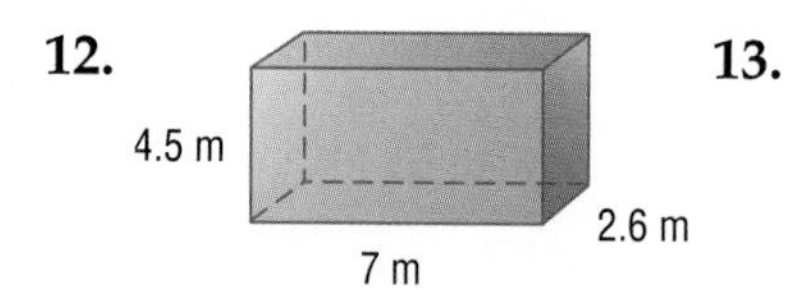

13.

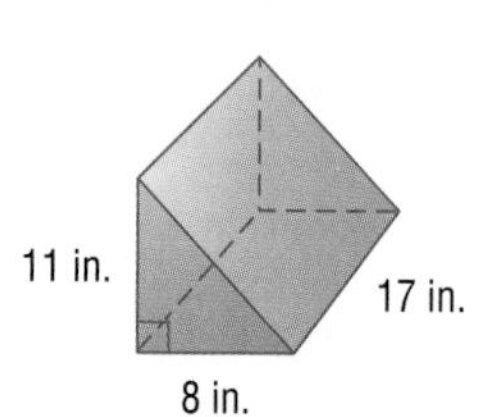

14.

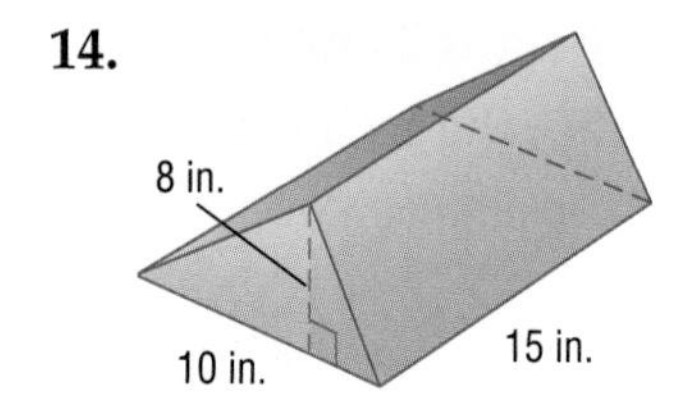

15.

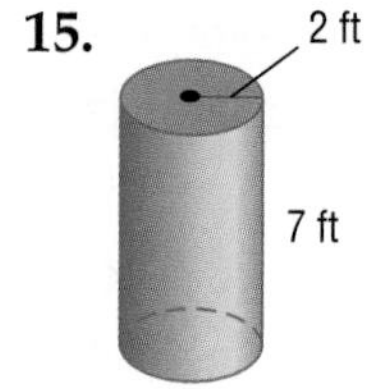

16.

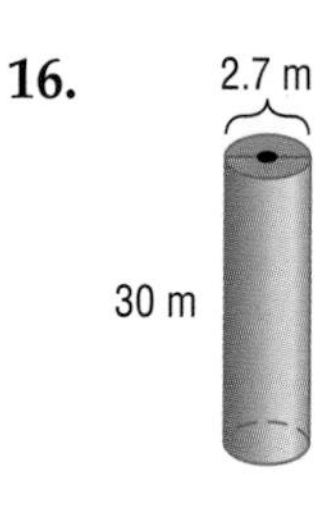

17. rectangular prism: length 3 mm, width 5 mm, height 15 mm

18. triangular prism: base of triangle 8 in., altitude of triangle 15 in., height of prism $6\frac{1}{2}$ in.

19. cylinder: $d = 2.6$ m, $h = 3.5$ m

20. octagonal prism: base area 25 m^2, height 1.5 m

21. Find the height of a rectangular prism with a length of 4.2 meters, width of 3.2 meters, and volume of 83.3 m^3.

22. Find the height of a cylinder with a radius of 2 feet and a volume of 28.3 ft^3.

CONVERTING UNITS OF MEASURE
For Exercises 23–25, use the cubes at the right.
The volume of the left cube is 1 yd^3. In the right cube, only the units have been changed. So, 1 yd^3 = 3(3)(3) or 27 ft^3. Use a similar process to convert each measurement.

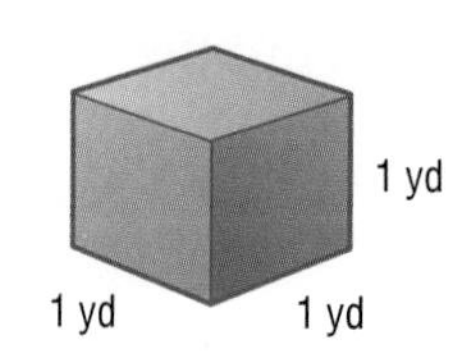

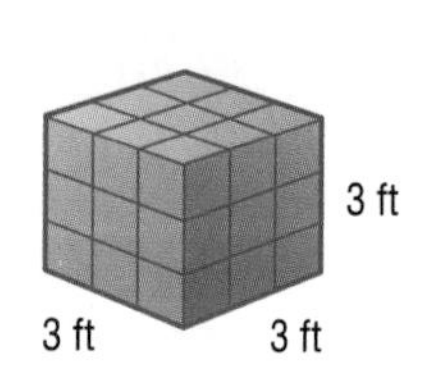

23. 1 ft^3 = ■ in^3

24. 1 cm^3 = ■ mm^3

25. 1 m^3 = ■ cm^3

26. **METALS** The *density* of gold is 19.29 grams per cubic centimeter. Find the mass in grams of a gold bar that is 2 centimeters by 3 centimeters by 2 centimeters.

27. **MICROWAVES** The inside of a microwave oven has a volume of 1.2 cubic feet and measures 18 inches wide and 10 inches long. To the nearest tenth, how deep is the inside of the microwave? (*Hint*: Convert 1.2 cubic feet to cubic inches.)

28. **BATTERIES** The current of an alkaline battery corresponds to its volume. Find the volume of each cylinder-shaped battery shown in the table. Write each volume in cm^3. (*Hint*: $1\ cm^3 = 1000\ mm^3$)

Battery Size	Diameter (mm)	Height (mm)
D	33.3	61.1
C	25.5	50.0
AA	14.5	50.5
AAA	10.5	44.5

29. **CRITICAL THINKING** An $8\frac{1}{2}$-by-11-inch piece of paper is rolled to form a cylinder. Will the volume be greater if the height is $8\frac{1}{2}$ inches or 11 inches, or will the volumes be the same? Explain your reasoning.

30. WRITING IN MATH Answer the question that was posed at the beginning of the lesson.

How is volume related to area?

Include the following in your answer:
- an explanation of why area is given in square units and volume is given in cubic units, and
- a description of why the formula for volume includes area.

FCAT Practice
Standardized Test Practice

31. Which is the best estimate for the volume of a cube whose sides measure 18.79 millimeters?

(A) $80\ mm^3$ (B) $800\ mm^3$ (C) $8000\ mm^3$ (D) $80{,}000\ mm^3$

32. Find the volume of the figure at the right.

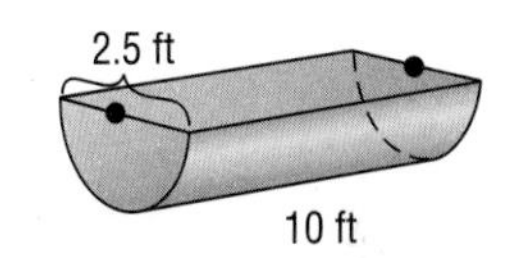

(A) $24.5\ ft^3$ (B) $20.5\ ft^3$
(C) $48\ ft^3$ (D) $49\ ft^3$

Maintain Your Skills

Mixed Review

33. Identify a pair of skew lines in the prism at the right. *(Lesson 11-1)*

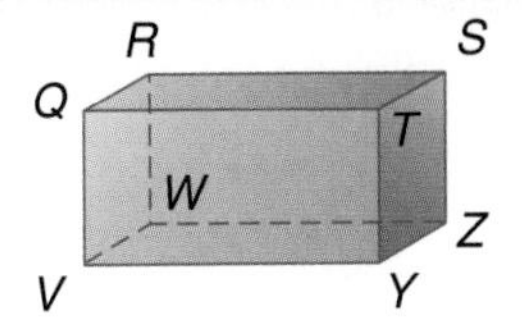

34. Estimate the area of the shaded figure to the nearest square unit. *(Lesson 10-8)*

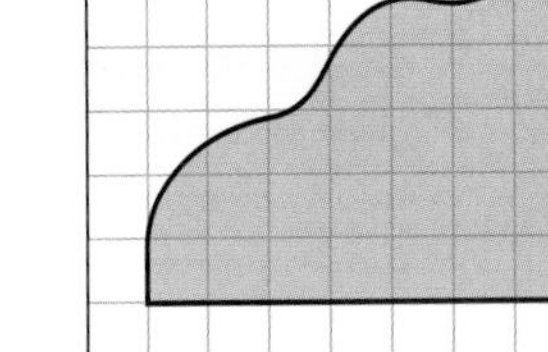

Solve each inequality. Check your solution. *(Lesson 7-4)*

35. $x + 5 > -3$ 36. $k + (-9) \geq 1.8$

Getting Ready for the Next Lesson

PREREQUISITE SKILL **Find each product.**
*(To review **multiplying fractions**, see Lesson 5-3.)*

37. $\frac{1}{3} \cdot 5 \cdot 15$ 38. $\frac{1}{3} \cdot 4 \cdot 9$ 39. $\frac{1}{3} \cdot 2 \cdot 2 \cdot 3$

40. $\frac{1}{3} \cdot 3 \cdot 4 \cdot 8$ 41. $\frac{1}{3} \cdot 2^2 \cdot 21$ 42. $\frac{1}{3} \cdot 3^2 \cdot 10$

11-3 Volume: Pyramids and Cones

Sunshine State Standards
MA.B.1.3.1-1, MA.B.1.3.1-2, MA.B.1.3.3-1, MA.C.1.3.1-4

Vocabulary

- cone

What You'll Learn

- Find volumes of pyramids.
- Find volumes of cones.

How is the volume of a pyramid related to the volume of a prism?

You can see that the volume of the pyramid shown at the right is less than the volume of the prism in which it sits.

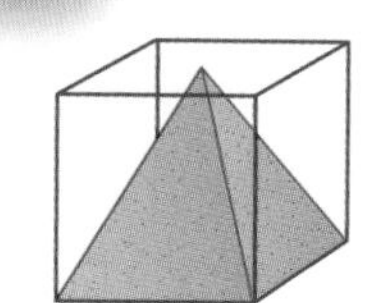

If the pyramid were made of sand, it would take three pyramids to fill a prism having the same base dimensions and height.

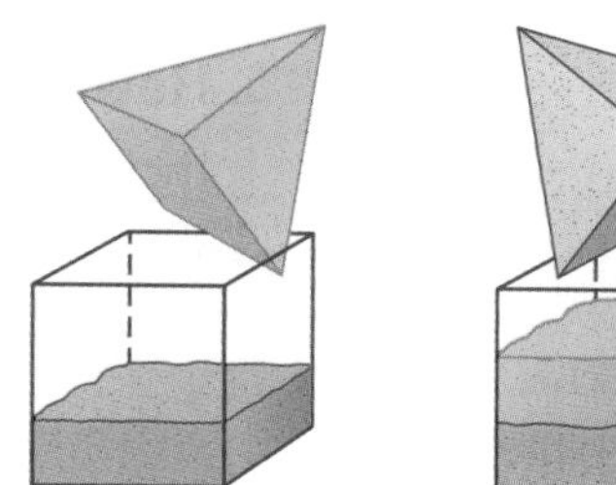

a. Compare the base areas and compare the heights of the prism and the pyramid.

b. How many times greater is the volume of the prism than the volume of one pyramid?

c. What fraction of the prism volume does one pyramid fill?

Study Tip

Height of Pyramid
The height of a pyramid is the distance from the vertex, perpendicular to the base.

VOLUMES OF PYRAMIDS A pyramid has one-third the volume of a prism with the same base area and height.

Key Concept — *Volume of a Pyramid*

- **Words** The volume V of a pyramid is one-third the area of the base B times the height h.
- **Symbols** $V = \frac{1}{3}Bh$
- **Model**

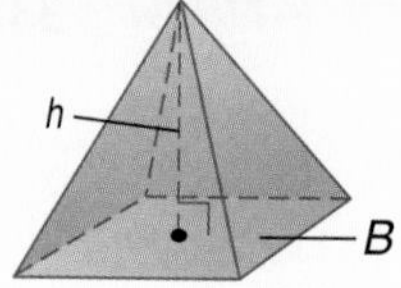

Example 1 *Volumes of Pyramids*

Find the volume of each pyramid. If necessary, round to the nearest tenth.

a.

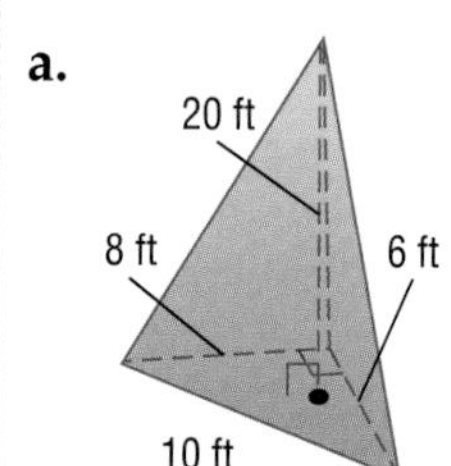

$V = \frac{1}{3}Bh$ — Formula for volume of a pyramid

$V = \frac{1}{3}\left(\frac{1}{2} \cdot 8 \cdot 6\right)h$ — The base is a triangle, so $B = \frac{1}{2} \cdot 8 \cdot 6$.

$V = \frac{1}{3}\left(\frac{1}{2} \cdot 8 \cdot 6\right)20$ — The height of the pyramid is 20 feet.

$V = 160$ — Simplify.

The volume is 160 cubic feet.

b. base area 125 cm², height 6.5 cm

$V = \frac{1}{3}Bh$ Formula for volume of a pyramid

$V = \frac{1}{3}(125)(6.5)$ Replace B with 125 and h with 6.5.

$V \approx 270.8$ Simplify.

The volume is about 270.8 cubic centimeters.

Reading Math

Cones In this text, *cone* refers to a circular cone.

VOLUMES OF CONES A **cone** is a three-dimensional figure with one circular base. A curved surface connects the base and the vertex.

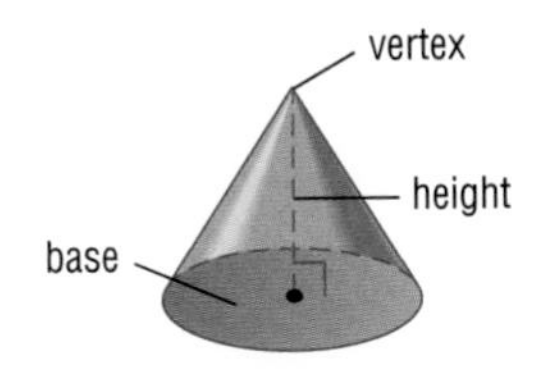

The volumes of a cone and a cylinder are related in the same way as the volumes of a pyramid and a prism are related.

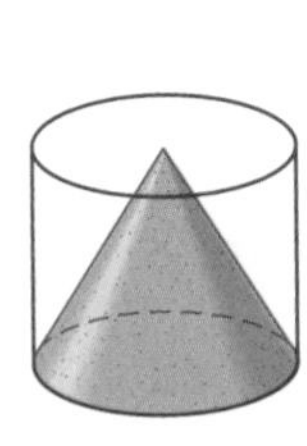

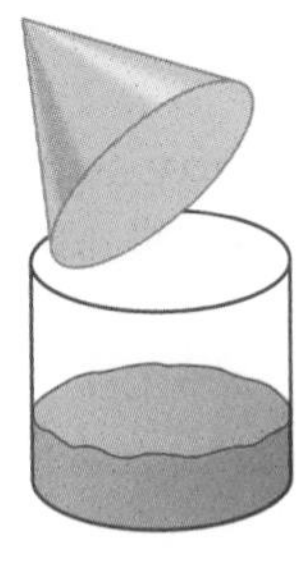

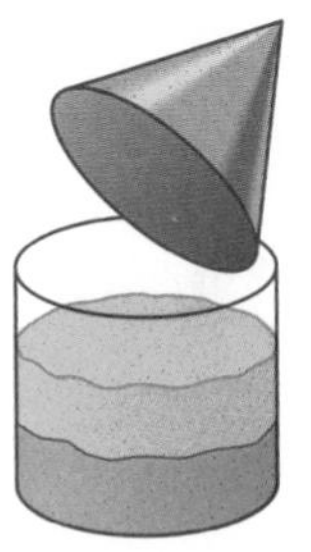

The volume of a cone is $\frac{1}{3}$ the volume of a cylinder with the same base area and height.

Key Concept — *Volume of a Cone*

- **Words** The volume V of a cone with radius r is one-third the area of the base B times the height h.
- **Symbols** $V = \frac{1}{3}Bh$ or $V = \frac{1}{3}\pi r^2 h$, where $B = \pi r^2$
- **Model**

Example 2 *Volume of a Cone*

Find the volume of the cone. Round to the nearest tenth.

$V = \frac{1}{3}\pi r^2 h$ Formula for volume of a cone

$V = \frac{1}{3} \cdot \pi \cdot 5^2 \cdot 12$ Replace r with 5 and h with 12.

$V \approx 314.2$ Simplify.

The volume is about 314.2 cubic centimeters.

Concept Check What is the volume of a cone whose base area is 86 meters squared and whose height is 3 meters?

Example 3 Use Volume to Solve Problems

HIGHWAY MAINTENANCE Salt and sand mixtures are often used on icy roads. When the mixture is dumped from a truck into the staging area, it forms a cone-shaped mound with a diameter of 10 feet and a height of 6 feet.

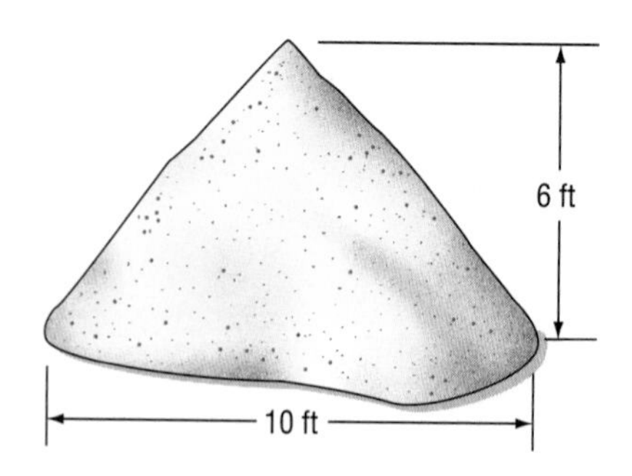

a. What is the volume of the salt-sand mixture?

Estimate: $\frac{1}{3} \cdot 3 \cdot 5^2 \cdot 6 = 150$

$V = \frac{1}{3}\pi r^2 h$ Formula for volume of a cone

$V = \frac{1}{3} \cdot \pi \cdot 5^2 \cdot 6$ Since $d = 10$, replace r with 5. Replace h with 6.

$V \approx 157$

The volume of the mixture is about 157 cubic feet.

Study Tip

Look Back
To review **dimensional analysis**, see Lesson 5-3.

b. How many square feet of roadway can be salted using the mixture in part a if 500 square feet can be covered by 1 cubic foot of salt?

$$\text{ft}^2 \text{ of roadway} = 157 \text{ ft}^3 \text{ mixture} \times \frac{500 \text{ ft}^2 \text{ of roadway}}{1 \text{ ft}^3 \text{ mixture}}$$

$$= 78{,}500 \text{ ft}^2 \text{ of roadway}$$

So, 78,500 square feet of roadway can be salted.

Check for Understanding

Concept Check

1. **Explain** why you can use πr^2 to find the area of the base of a cone.
2. **List** the formulas for volume that you have learned so far in this chapter. Tell what the variables represent and explain how you can remember which formula goes with which solid.
3. **OPEN ENDED** Draw and label a cone whose volume is between 100 cm^3 and 1000 cm^3.

Guided Practice **Find the volume of each solid. If necessary, round to the nearest tenth.**

4. 5. 6.

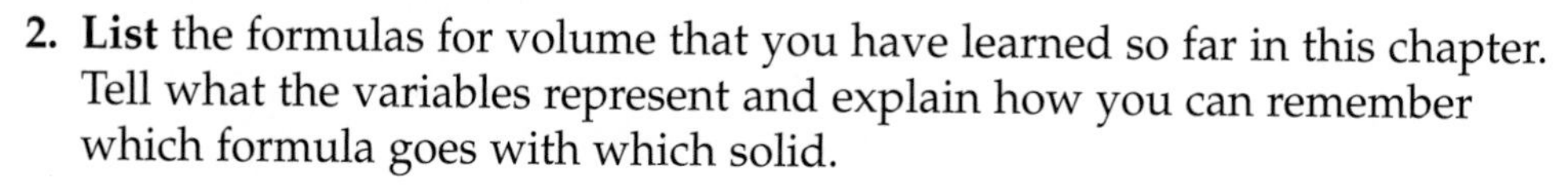

7. rectangular pyramid: length 9 ft, width 7 ft, height 18 ft
8. cone: radius 4 mm, height 6.5 mm

Application

9. **HISTORY** The Great Pyramid of Khufu in Egypt was originally 481 feet high and had a square base 756 feet on a side. What was its volume? Use an estimate to check your answer.

Practice and Apply

Homework Help

For Exercises	See Examples
10–22	1, 2
23–25	3

Extra Practice

See page 751.

Find the volume of each solid. If necessary, round to the nearest tenth.

10.

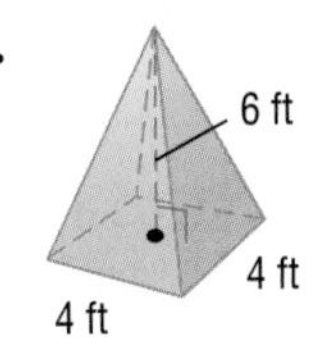

11.

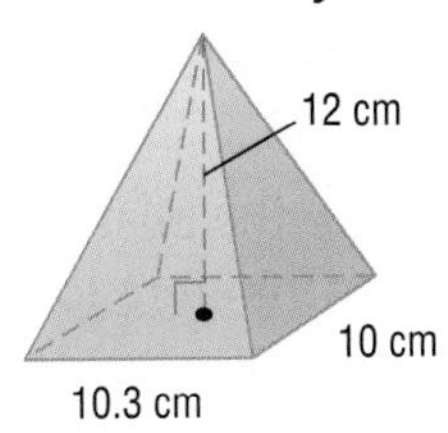

12.

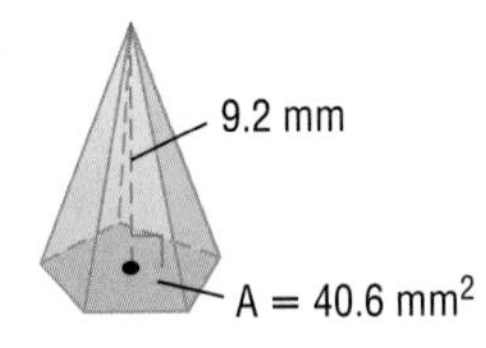

13.

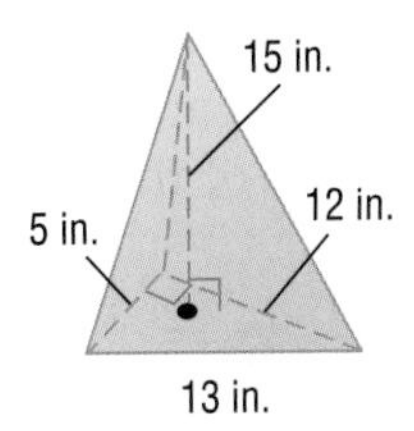

14.

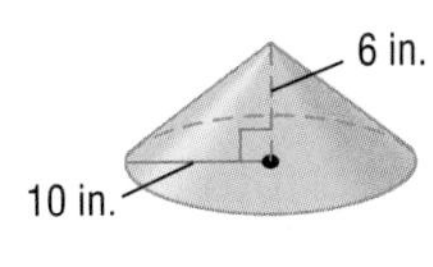

15.

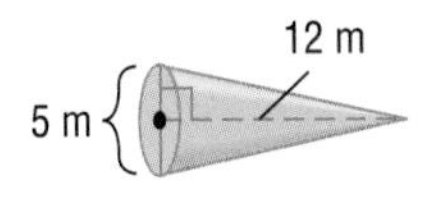

Finding the volumes of three-dimensional figures will help you analyze structures. Visit www.pre-alg.com/webquest to continue work on your WebQuest project.

16. square pyramid: length 5 in., height 6 in.

17. hexagonal pyramid: base area 125 cm^2, height 6.5 cm

18. cone: radius 3 yd, height 14 yd

19. cone: diameter 12 m, height 15 m

20.

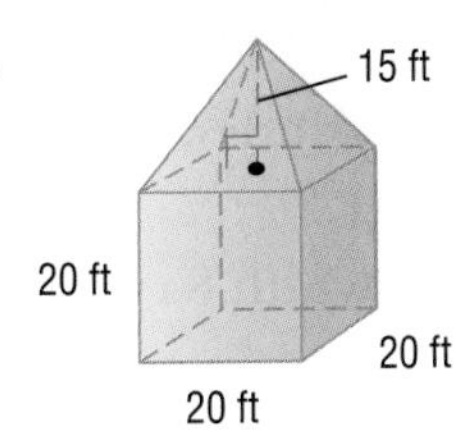

21.

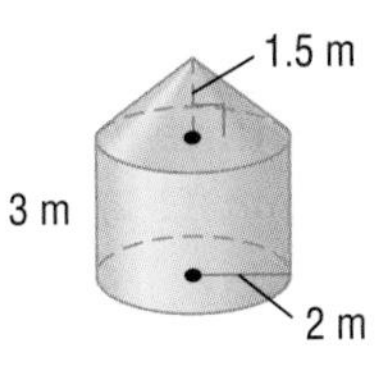

22.

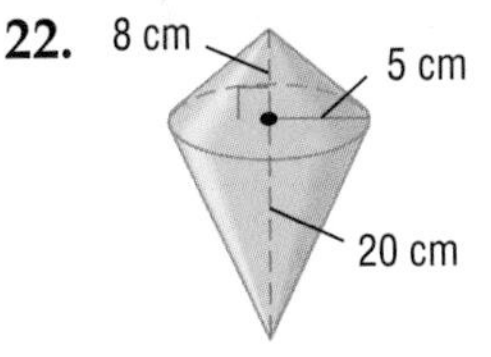

23. GEOLOGY A stalactite in the Endless Caverns in Virginia is cone-shaped. It is 4 feet long and has a diameter at its base of 1.5 feet.

a. Find the volume of the stalactite to the nearest tenth.

b. The stalactite is made of calcium carbonate, which weighs 131 pounds per cubic foot. What is the weight of the stalactite?

Geology

Scientists and explorers continue making expeditions to discover new passages in caverns.

Source: Endless Caverns, Inc.

24. SCIENCE In science, a standard funnel is shaped like a cone, and a buchner funnel is shaped like a cone with a cylinder attached to the base. Which funnel has the greatest volume?

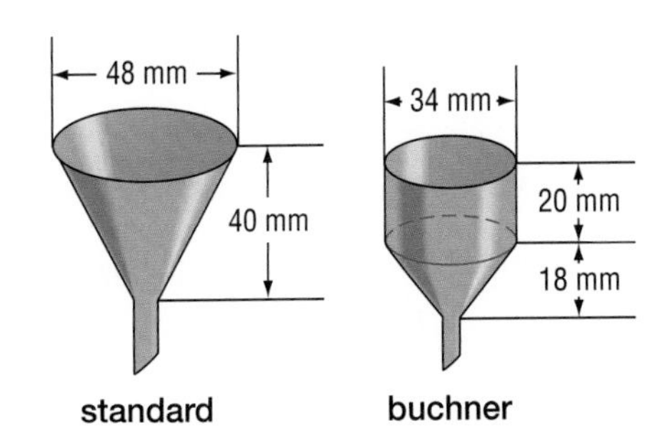

25. WRITING IN MATH Answer the question that was posed at the beginning of the lesson.

How is the volume of a pyramid related to the volume of a prism?

Include the following in your answer:

- a discussion of the similarities between the dimensions and base area of the pyramid and prism shown at the beginning of the lesson, and
- a description of how the formulas for the volume of a pyramid and the volume of a prism are similar.

26. **CRITICAL THINKING**

 a. If you double the height of a cone, how does the volume change?

 b. If you double the radius of the base of a cone, how does the volume change? Explain.

27. Choose the best estimate for the volume of a rectangular pyramid 4.9 centimeters long, 3 centimeters wide, and 7 centimeters high.

 Ⓐ 7 cm^3 Ⓑ 35 cm^3 Ⓒ 70 cm^3 Ⓓ 105 cm^3

28. The solids at the right have the same base area and height. If the cone is filled with water and poured into the cylinder, how much of the cylinder would be filled?

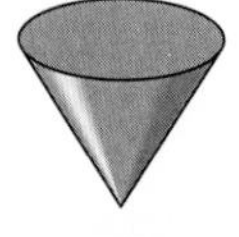

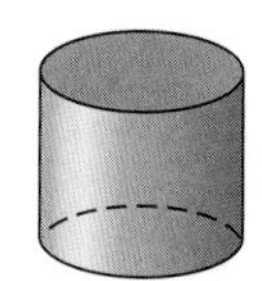

 Ⓐ $\frac{3}{4}$ Ⓑ $\frac{1}{2}$ Ⓒ $\frac{2}{3}$ Ⓓ $\frac{1}{3}$

Extending the Lesson **The volume V of a sphere with radius r is given by the formula $V = \frac{4}{3}\pi r^3$. Find the volume of each sphere to the nearest tenth.**

29. radius 6 cm 30. radius 1.4 in. 31. diameter 5.9 mm

Maintain Your Skills

Mixed Review **Find the volume of each prism or cylinder. If necessary, round to the nearest tenth.** *(Lesson 11-2)*

32. rectangular prism: length 4 cm, width 8 cm, height 2 cm

33. cylinder: diameter 1.6 in., height 5 in.

34. Identify the solid at the right. Name the bases, faces, edges, and vertices. *(Lesson 11-1)*

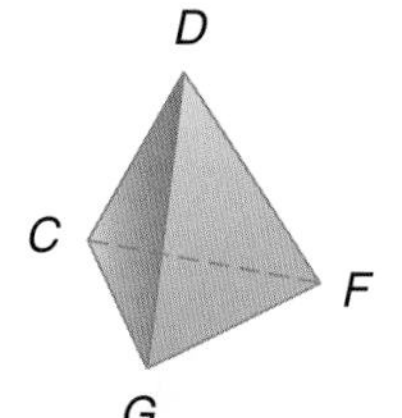

35. Find the distance between $A(3, 7)$ and $B(-2, 1)$. Round to the nearest tenth, if necessary. *(Lesson 9-6)*

Getting Ready for the Next Lesson **PREREQUISITE SKILL Estimate each product.**
*(To review **estimating products**, see page 714.)*

36. $4.9 \cdot 5.1 \cdot 3$ 37. $2 \cdot 1.7 \cdot 9$ 38. $2 \cdot \pi \cdot 6.8$

Practice Quiz 1 *Lessons 11-1 through 11-3*

Identify each solid. *(Lesson 11-1)*

1.

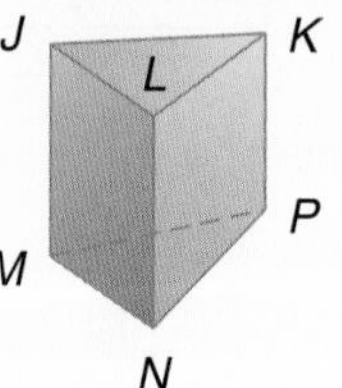

2.

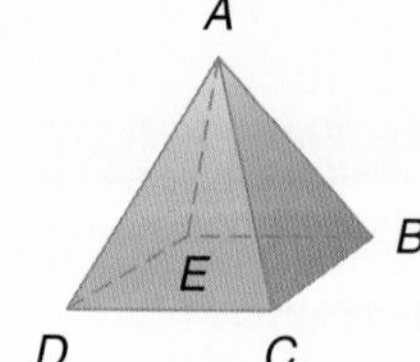

Find the volume of each solid. If necessary, round to the nearest tenth. *(Lesson 11-2)*

3. cylinder: radius 2 cm, height 1 cm

4. hexagonal prism: base area 42 ft^2, height 18 ft

5. **MINING** An open pit mine in the Elk mountain range is cone-shaped. The mine is 420 feet across and 250 feet deep. What volume of material was removed? *(Lesson 11-3)*

11-4 Surface Area: Prisms and Cylinders

What You'll Learn

- Find surface areas of prisms.
- Find surface areas of cylinders.

Sunshine State Standards
MA.B.1.3.1-1, MA.B.1.3.1-2, MA.B.1.3.3-1, MA.C.1.3.1-2, MA.C.1.3.1-4

Vocabulary

- surface area

How is the surface area of a solid different from its volume?

The sizes and prices of shipping boxes are shown in the table.

Box	Size (in.)	Price ($)
A	$8 \times 8 \times 8$	\$1.50
B	$15 \times 10 \times 12$	\$2.25
C	$20 \times 14 \times 10$	\$3.00

a. For each box, find the area of each face and the sum of the areas.

b. Find the volume of each box. Are these values the same as the values you found in part **a**? Explain.

SURFACE AREAS OF PRISMS If you open up a box or prism to form a net, you can see all the surfaces. The sum of the areas of these surfaces is called the **surface area** of the prism.

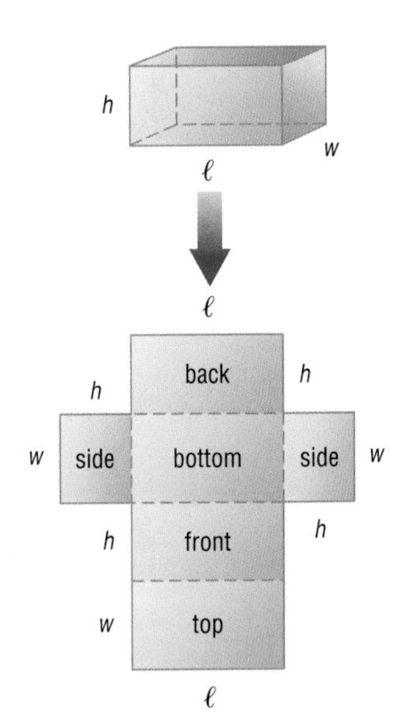

Faces	Area
top and bottom	$(\ell \cdot w) + (\ell \cdot w) = 2\ell w$
front and back	$(\ell \cdot h) + (\ell \cdot h) = 2\ell h$
two sides	$(w \cdot h) + (w \cdot h) = 2wh$
Sum of areas $\rightarrow$	$2\ell w + 2\ell h + 2wh$ or $2(\ell w + \ell h + wh)$

Key Concept — Surface Area of Rectangular Prisms

- **Words** The surface area S of a rectangular prism with length ℓ, width w, and height h is the sum of the areas of the faces.
- **Model**

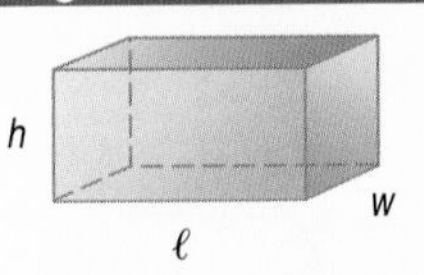

- **Symbols** $S = 2\ell w + 2\ell h + 2wh = 2(\ell w + \ell h + wh)$

Example 1 Surface Area of a Rectangular Prism

Find the surface area of the rectangular prism.

10 in., 14 in., 20 in.

$S = 2\ell w + 2\ell h + 2wh$ Write the formula.

$S = 2(20)(14) + 2(20)(10) + 2(14)(10)$ Substitution

$S = 1240$ Simplify.

The surface area of the prism is 1240 square inches.

Example 2 Surface Area of a Triangular Prism

Find the surface area of the triangular prism.

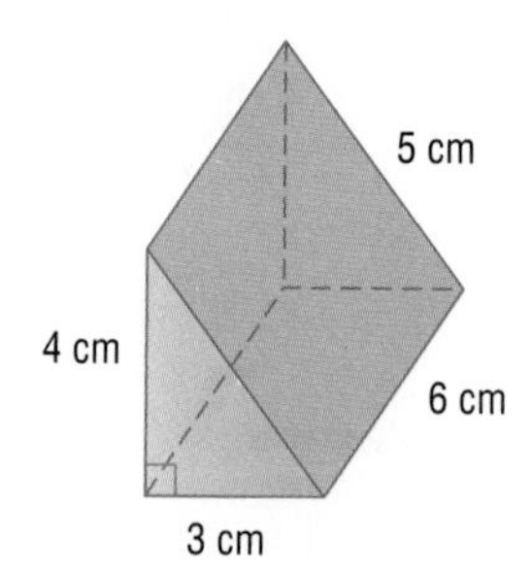

One way to easily see all of the surfaces of the prism is to draw a net on grid paper and label the dimensions of each face.

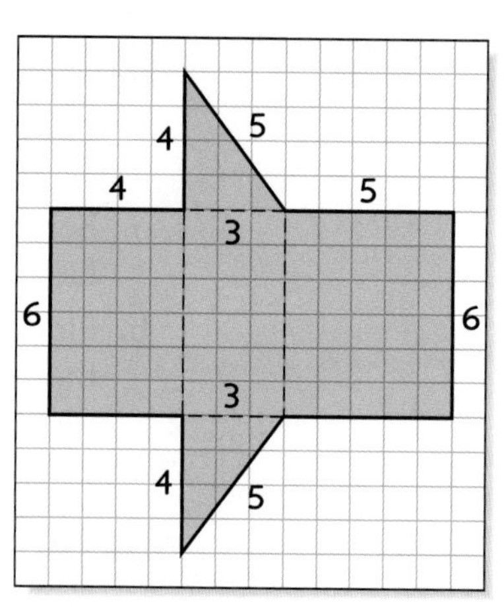

Find the area of each face.

bottom $3 \cdot 6 = 18$
left side $4 \cdot 6 = 24$
right side $5 \cdot 6 = 30$ } $A = \ell w$

two bases $2\left(\frac{1}{2} \cdot 3 \cdot 4\right) = 12$ $A = \frac{1}{2}bh$

Add to find the total surface area.
$18 + 24 + 30 + 12 = 84$

The surface area of the triangular prism is 84 square centimeters.

SURFACE AREAS OF CYLINDERS You can also find surface areas of cylinders. If you unroll a cylinder, its net is a rectangle and two circles.

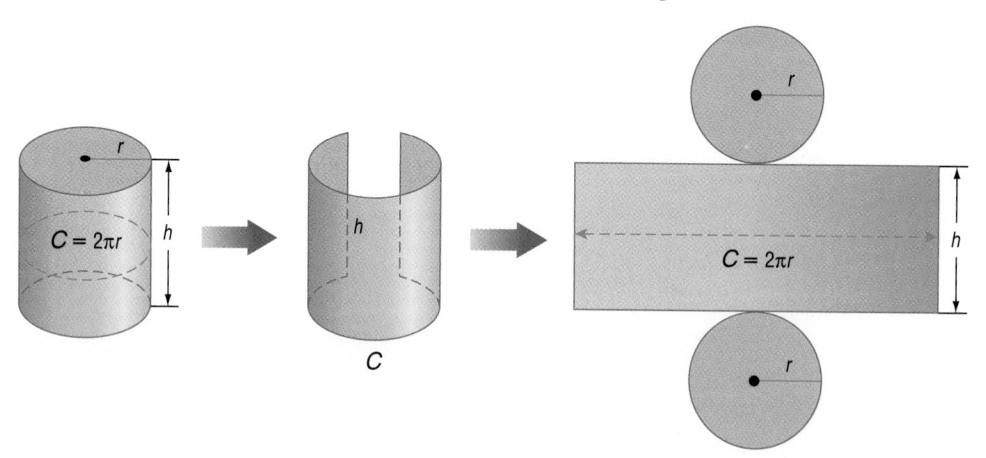

Study Tip

Circles and Rectangles
To see why $\ell = 2\pi r$, find the circumference of a soup can by using the formula $C = 2\pi r$. Then peel off its label and measure the length (minus the overlapping parts).

Model	Net
bases	congruent circles
curved surface	rectangle
height h	width of rectangle
circumference C	length of rectangle

The area of each circular base is πr^2. The area of the rectangular region is $\ell \cdot w$, or $2\pi r \cdot h$.

The surface area of a cylinder	equals	the area of two circular bases	plus	the area of the curved surface.
S	$=$	$2(\pi r^2)$	$+$	$2\pi rh$

Key Concept — Surface Area of Cylinders

- **Words** The surface area S of a cylinder with height h and radius r is the area of the two bases plus the area of the curved surface.
- **Model**

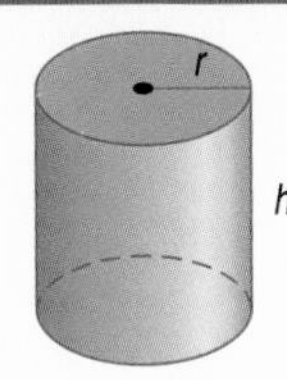

- **Symbols** $S = 2\pi r^2 + 2\pi rh$

Example 3 Surface Area of a Cylinder

Find the surface area of the cylinder. Round to the nearest tenth.

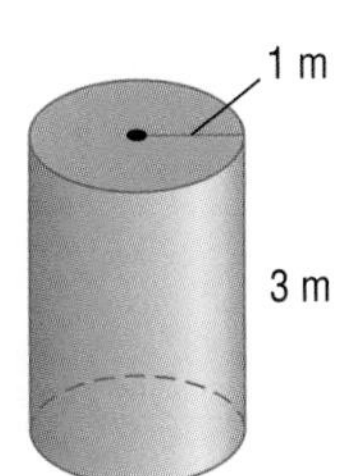

$S = 2\pi r^2 + 2\pi rh$	Formula for surface area of a cylinder
$S = 2\pi(1)^2 + 2\pi(1)(3)$	Replace r with 1 and h with 3.
$S \approx 25.1$	Simplify.

The surface area is about 25.1 square meters.

You can compare surface areas of prisms and cylinders.

Example 4 Compare Surface Areas

FRUIT DRINKS **Both containers hold about the same amount of pineapple juice. Does the box or the can have a greater surface area?**

Surface area of box

top/bottom + sides + front/back

$$S = 2\ell w + 2\ell h + 2wh$$
$$= 2(4 \cdot 7) + 2(4 \cdot 9) + 2(7 \cdot 9)$$
$$= 254$$

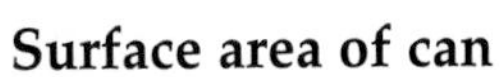

Surface area of can

top/bottom + curved surface

$$S = 2\pi r^2 + 2\pi rh$$
$$= 2\pi(3)^2 + 2\pi(3)(9)$$
$$\approx 226$$

Since $254 \text{ cm}^2 > 226 \text{ cm}^2$, the box has a greater surface area.

More About. . .

Fruit Drinks

A typical juice box is made of six layers of packaging. These layers include paper, foil, and special adhesives.

Source: www.dupont.com

Concept Check Why might a company prefer to sell juice in cans?

Check for Understanding

Concept Check

1. **Explain** why surface area is given in square units rather than cubic units.

2. **OPEN ENDED** Find the surface areas of a rectangular prism and a cylinder found in your home.

Guided Practice

Find the surface area of each solid shown or described. If necessary, round to the nearest tenth.

3. (cube: 5 ft, 5 ft, 5 ft)

4.

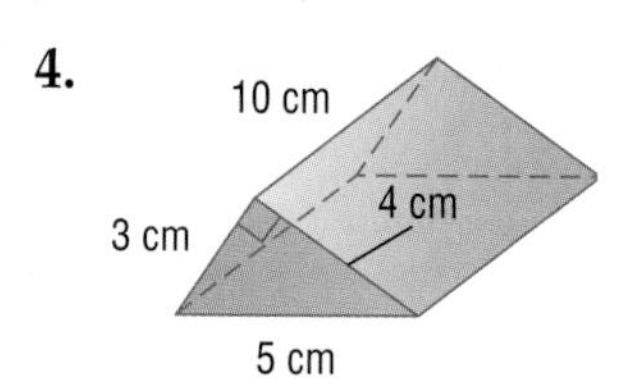

5.

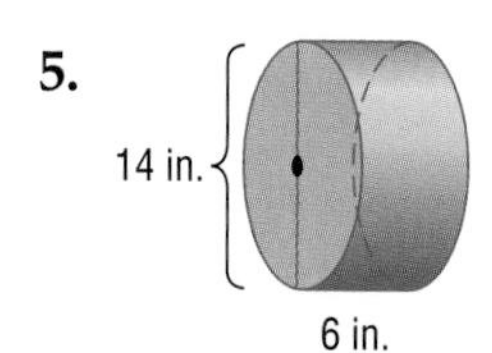

6. rectangular prism: length 3 cm, width 2 cm, height 1 cm

7. cylinder: radius 4 mm, height 1.6 mm

Application

8. **CRAFTS** Brianna sews together pieces of plastic canvas to make tissue box covers. For which tissue box will she use more plastic canvas to cover the sides and the top? Explain.

Box	Length (in.)	Width (in.)	Height (in.)
A	9	4	5
B	5	5	6

Practice and Apply

Homework Help

For Exercises	See Examples
9, 10, 15, 16	1
11, 12	2
13, 14, 17–19, 21, 22	3
20	4

Extra Practice

See page 751.

Find the surface area of each solid shown or described. If necessary, round to the nearest tenth.

9.

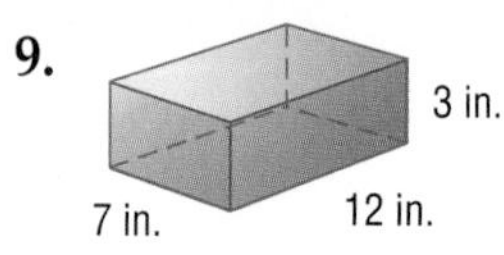

10.

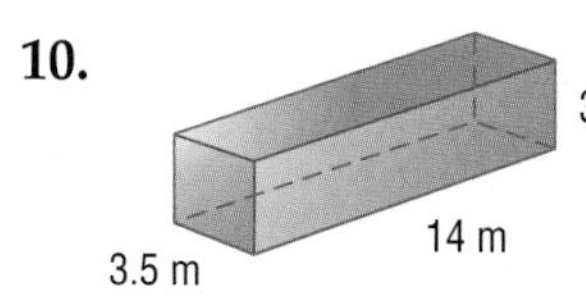

11.

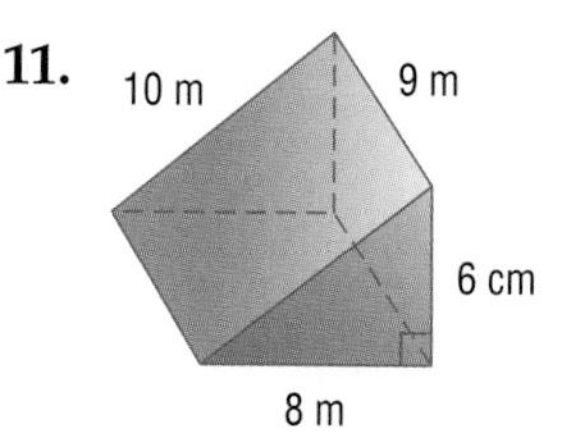

12.

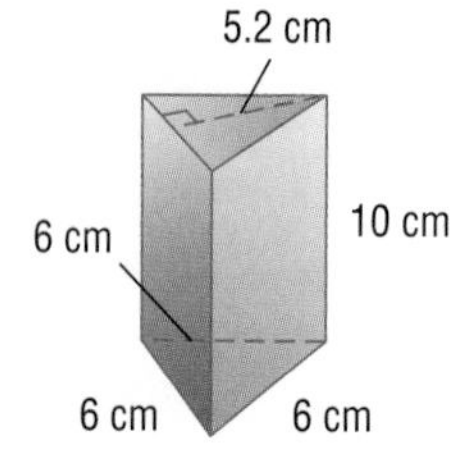

13.

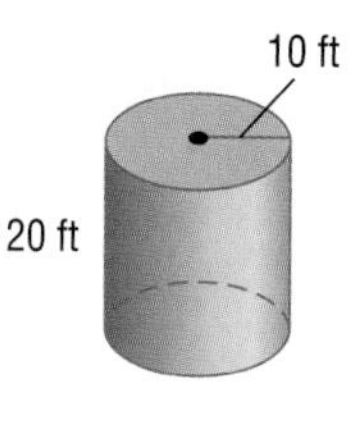

14.

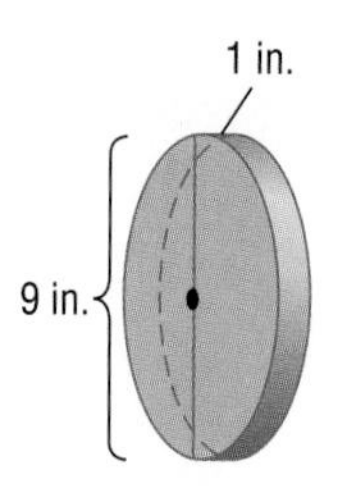

15. cube: side length 7 ft

16. rectangular prism: length 6.2 cm, width 4 cm, height 8.5 cm

17. cylinder: radius 5 in., height 15 in.

18. cylinder: diameter 4 m, height 20 m

19. Find the surface area of the complex solid at the right. Use estimation to check the reasonableness of your answer.

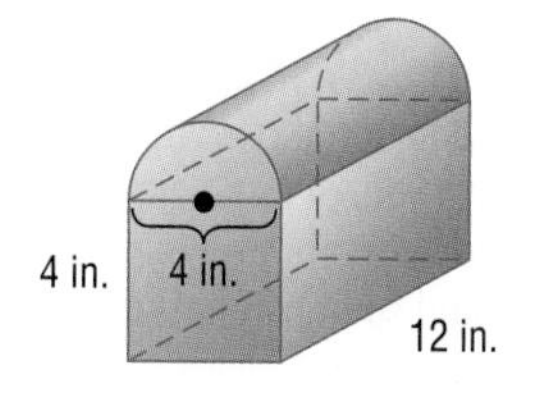

20. **AQUARIUMS** A standard 20-gallon aquarium tank is a rectangular prism that holds approximately 4600 cubic inches of water. The bottom glass needs to be 24 inches by 12 inches to fit on the stand.

a. Find the height of the aquarium to the nearest inch.

b. Find the total amount of glass needed in square feet for the five faces.

c. An aquarium with an octagonal-shaped base has sides that are 9 inches wide and 16 inches high. The area of the base is 392.4 square inches. Do the bottom and sides of this tank have a greater surface area than the rectangular tank? Explain.

POOLS **Vinyl liners cover the inside walls and bottom of the swimming pools whose top views are shown below. Find the area of the vinyl liner for each pool if they are 4 feet deep. Round to the nearest square foot.**

21.

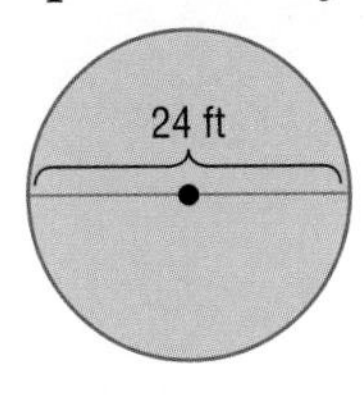

22.

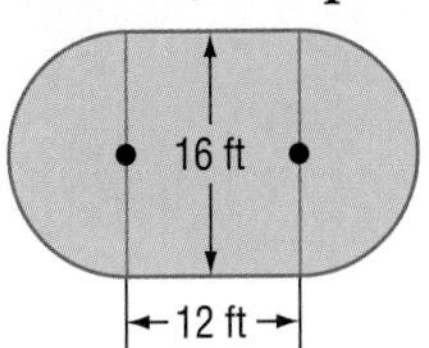

23. **CRITICAL THINKING** Suppose you double the length of the sides of a cube. How is the surface area affected?

24. WRITING IN MATH Answer the question that was posed at the beginning of the lesson.

How is the surface area of a solid different than its volume?

Include the following in your answer:

- a comparison of formulas for surface area and volume, and
- an explanation of the difference between surface area and volume.

FCAT Practice

Standardized Test Practice

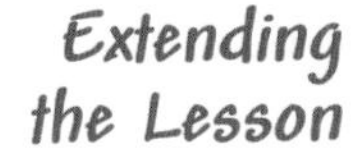

25. Find the surface area of a cylinder with a diameter of 15 centimeters and height of 2 centimeters.

Ⓐ 30 cm^2 Ⓑ 117.8 cm^2 Ⓒ 353.4 cm^2 Ⓓ 447.7 cm^2

26. How many 2-inch squares will completely cover a rectangular prism 10 inches long, 4 inches wide, and 6 inches high?

Ⓐ 40 Ⓑ 62 Ⓒ 240 Ⓓ 248

Extending the Lesson

If you make cuts in a solid, different 2-dimensional cross sections result, as shown at the right. Describe the cross section of each figure cut below.

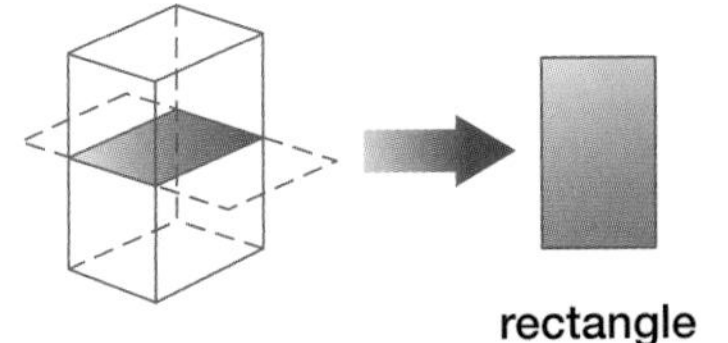

27.

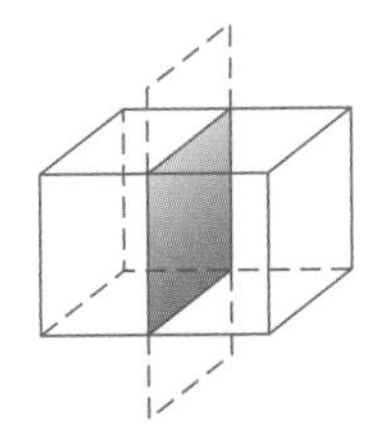

28.

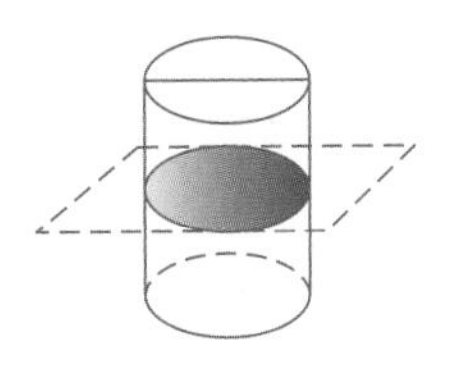

29.

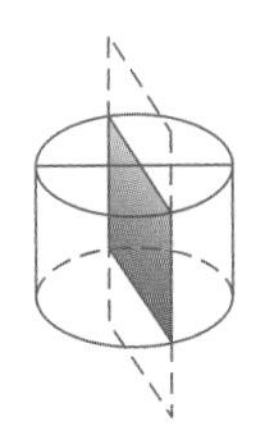

Maintain Your Skills

Mixed Review

Find the volume of each solid. If necessary, round to the nearest tenth.

(Lessons 11-2 and 11-3)

30. rectangular pyramid: length 6 ft, width 5 ft, height 7 ft

31. cylinder: diameter 6 in., height 20 in.

Getting Ready for the Next Lesson

PREREQUISITE SKILL **Find each product.**

(To review ***multiplying rational numbers****, see Lesson 5-3.)*

32. 10.3(8) **33.** 3.9(3.9) **34.** 12.3(9.2)(6)

35. $\frac{1}{2} \cdot 2.6$ **36.** $\frac{1}{2} \cdot 82 \cdot 90$ **37.** $\frac{1}{2}\left(6\frac{1}{2}\right)$

11-5 Surface Area: Pyramids and Cones

Sunshine State Standards
MA.B.1.3.1-1, MA.B.1.3.1-2, MA.C.1.3.1-4

Vocabulary

- lateral face
- slant height
- lateral area

What You'll Learn

- Find surface areas of pyramids.
- Find surface areas of cones.

How is surface area important in architecture?

The front of the Rock and Roll Hall of Fame in Cleveland, Ohio, is a glass pyramid.

a. The front triangle has a base of about 230 feet and height of about 120 feet. What is the area?

b. How could you find the total amount of glass used in the pyramid?

SURFACE AREAS OF PYRAMIDS The sides of a pyramid are called **lateral faces**. They are triangles that intersect at the vertex. The altitude or height of each lateral face is called the **slant height**.

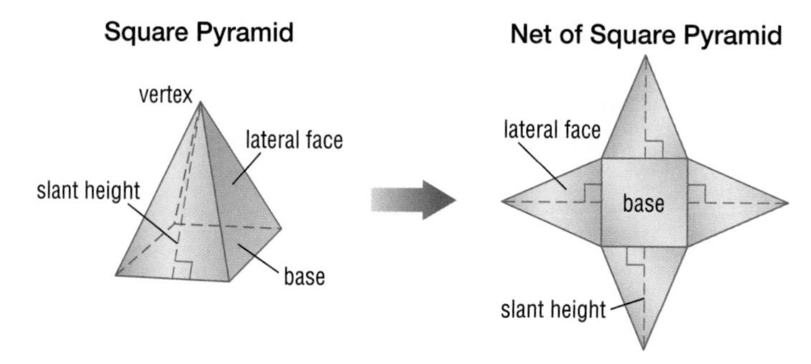

The sum of the areas of the lateral faces is the **lateral area** of a pyramid. The surface area of a pyramid is the lateral area plus the area of the base.

Example 1 Surface Area of a Pyramid

Find the surface area of the square pyramid.

Find the lateral area and the base area.

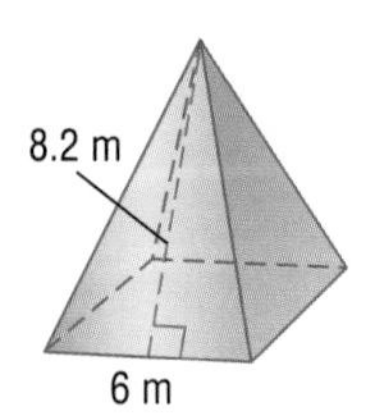

Area of each lateral face

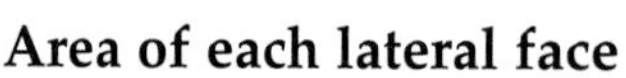

$A = \frac{1}{2}bh$ Area of a triangle

$A = \frac{1}{2}(6)(8.2)$ Replace b with 6 and h with 8.2.

$A = 24.6$ Simplify.

There are 4 faces, so the lateral area is 4(24.6) or 98.4 square meters.

Area of base

$A = s^2$ Area of a square

$A = 6^2$ or 36 Replace s with 6 and simplify.

The surface area of a pyramid	equals	the lateral area	plus	the area of the base.
S	$=$	98.4	$+$	36

The surface area of the pyramid is 134.4 square meters.

Example 2 Use Surface Area to Solve a Problem

ARCHITECTURE The Louvre museum in Paris has a huge square glass pyramid at the entrance with a slant height of about 92 feet. Its square base is 116 feet on each side. How much glass did it take to cover the pyramid?

Find the lateral area only, since the bottom of the pyramid is not covered in glass.

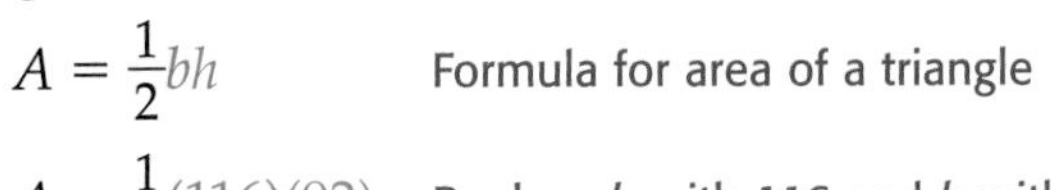

$A = \frac{1}{2}bh$ Formula for area of a triangle

$A = \frac{1}{2}(116)(92)$ Replace b with 116 and h with 92.

$A = 5336$ Simplify.

One lateral face has an area of 5336 square feet. There are 4 lateral faces, so the lateral area is $4 \cdot 5336$ or 21,344 square feet.

It took 21,344 square feet of glass to cover the pyramid.

Career Choices

Architect
Architects must be able to combine geometric solids and understand their properties to form unified structures.

Online Research
For information about a career as an architect, visit: www.pre-alg.com/careers

SURFACE AREAS OF CONES You can also find surface areas of cones. The net of a cone shows the regions that make up the cone.

Model of Cone

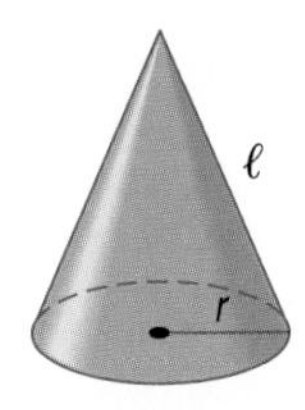

Net of Cone

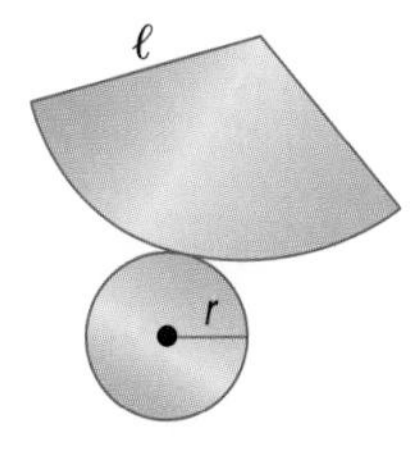

The lateral area of a cone with slant height ℓ is one-half the circumference of the base, $2\pi r$, times ℓ. So $A = \frac{1}{2} \cdot 2\pi r \cdot \ell$ or $A = \pi r\ell$. The base of the cone is a circle with area πr^2.

The surface area of a cone	equals	the lateral area	plus	the area of the base.
S	$=$	$\pi r\ell$	$+$	πr^2

Key Concept *Surface Area of a Cone*

- **Words** The surface area S of a cone with slant height ℓ and radius r is the lateral area plus the area of the base.
- **Symbols** $S = \pi r\ell + \pi r^2$
- **Model**

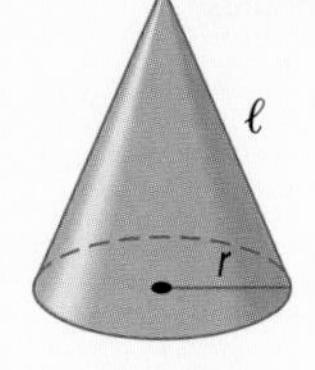

Example 3 Surface Area of a Cone

Find the surface area of the cone. Round to the nearest tenth.

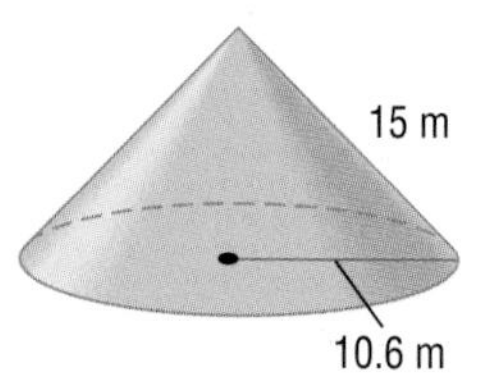

$S = \pi r\ell + \pi r^2$ — Formula for surface area of a cone

$S = \pi(10.6)(15) + \pi(10.6)^2$ — Replace r with 10.6 and ℓ with 15.

$S \approx 852.5$ — Simplify.

The surface area of the cone is about 852.5 square meters.

Concept Check What is the formula for the lateral area L of a cone?

Check for Understanding

Concept Check

1. **Describe** the difference between slant height and height of a pyramid and a cone.
2. **Explain** how to find the lateral area of a pyramid.
3. **OPEN ENDED** Describe a situation in everyday life when a person might use the formulas for the surface area of a cone or a pyramid.

Guided Practice

Find the surface area of each solid. If necessary, round to the nearest tenth.

4.

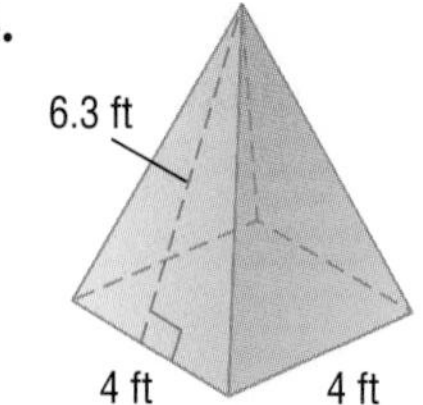

5.

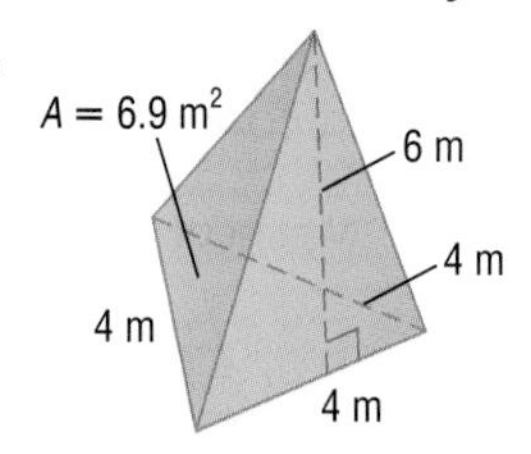

6.

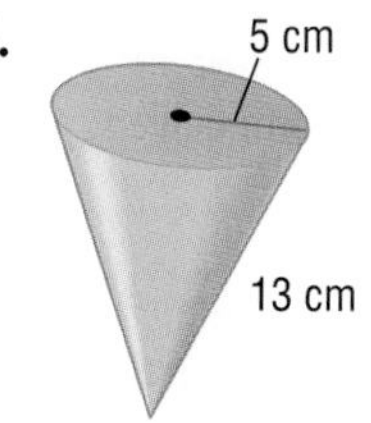

Application

7. **ARCHITECTURE** The small tower of a historic house is shaped like a regular hexagonal pyramid as shown at the right. How much roofing will be needed to cover this tower? (*Hint:* Do not include the base of the pyramid.)

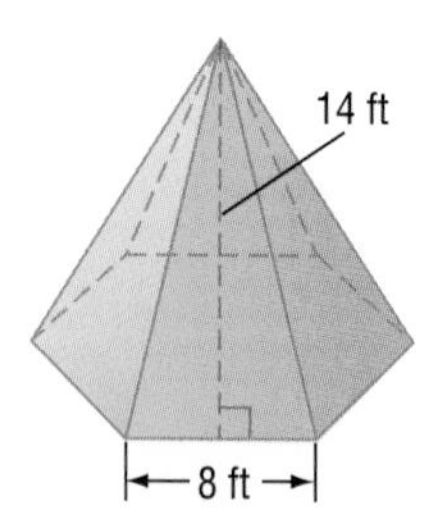

Practice and Apply

Homework Help

For Exercises	See Examples
8–11, 15, 16	1
12–14	3
17–19	2

Extra Practice

See page 751.

Find the surface area of each solid. If necessary, round to the nearest tenth.

8.

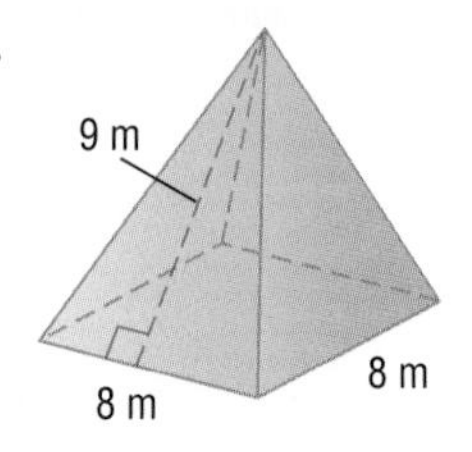

9.

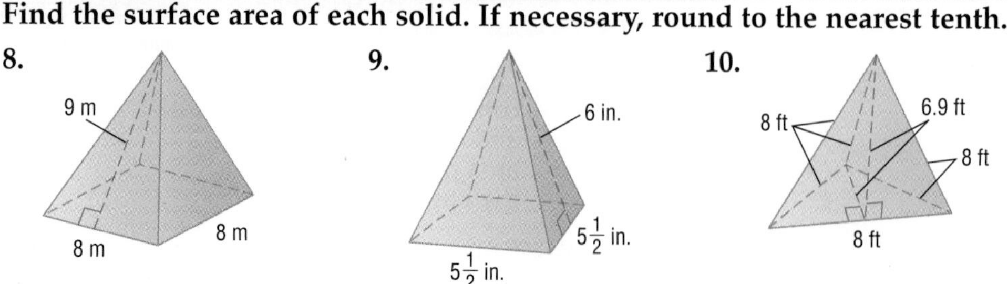

10.

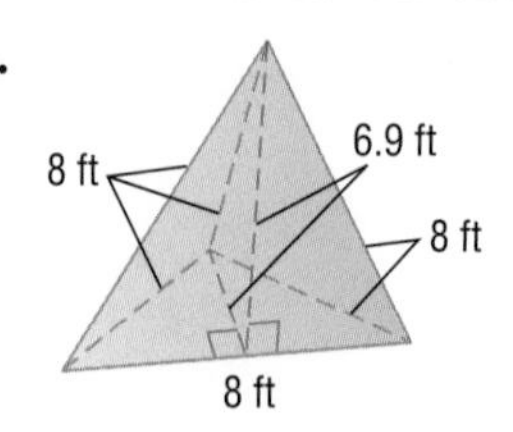

Study Tip

Triangular Pyramids
Assume that when the base of a pyramid is an equilateral triangle, all the faces are congruent.

Find the surface area of each solid. If necessary, round to the nearest tenth.

11.
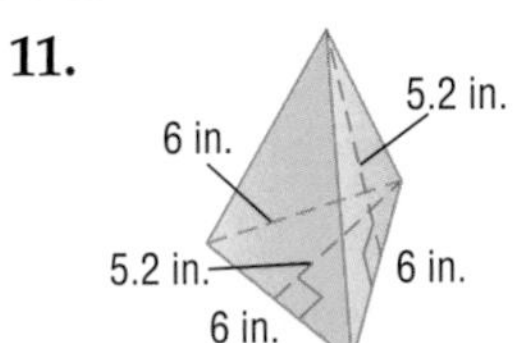

12.
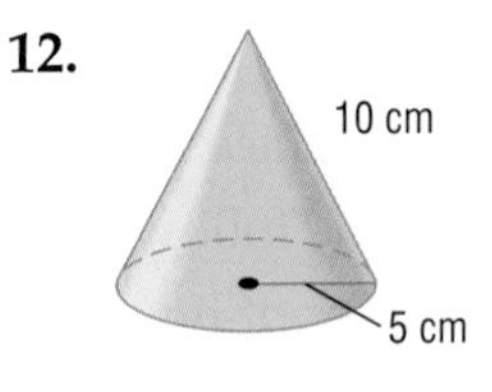

13.
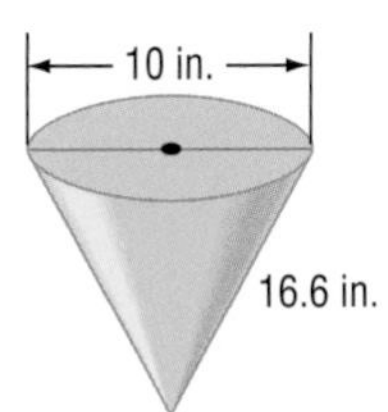

14.
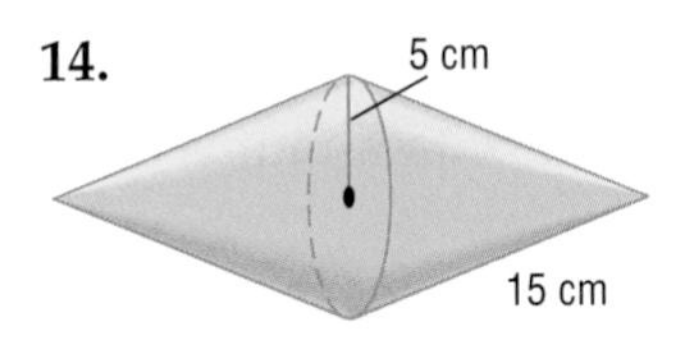

15.
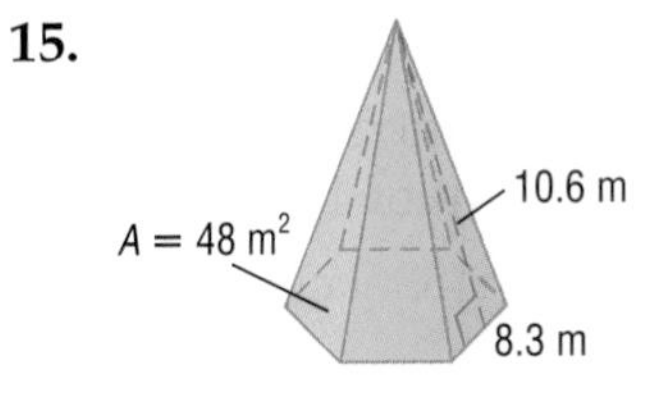

16.
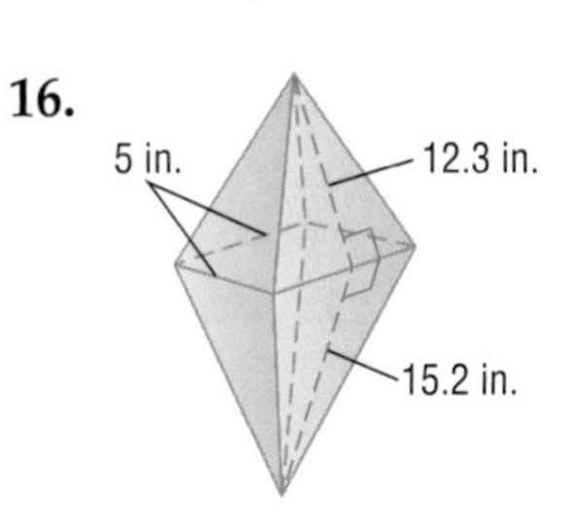

17. cone: radius 7.5 mm, slant height 14 mm

18. square pyramid: base side length 9 yd, slant height 8 yd

19. **FUND-RAISING** The cheerleaders are selling small megaphones decorated with the school mascot. There are two sizes, as shown at the right. What is the difference in the amount of plastic used in these two sizes? Round to the nearest square inch. (Note that a megaphone is open at the bottom.)

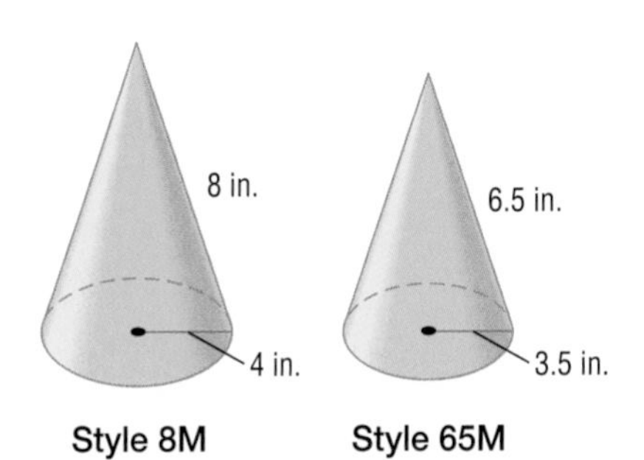

ARCHITECTURE **For Exercises 20 and 21, use the following information.**
A roofing company is preparing bids on two special jobs involving cone-shaped roofs. Roofing material is usually sold in 100-square-foot squares. For each roof, find the lateral surface area to the nearest square foot. Then determine the squares of roofing materials that would be needed to cover each surface.

20.

21.

22. **CRITICAL THINKING** A bar of lead in the shape of a rectangular prism 13 inches by 2 inches by 1 inch is melted and recast into 100 conical fishing sinkers. The sinkers have a diameter of 1 inch, a height of 1 inch, and a slant height of about 1.1 inches. Compare the total surface area of all the sinkers to the surface area of the original lead bar.

23. WRITING IN MATH Answer the question that was posed at the beginning of the lesson.

How is surface area important in architecture?

Include the following in your answer:

- examples of how surface area is used in architecture, and
- an explanation of why building contractors and architects need to know surface areas.

24. Find the surface area of a cone with a radius of 7 centimeters and slant height of 11.4 centimeters.

Ⓐ 153.9 cm^2 Ⓑ 250.7 cm^2 Ⓒ 272.7 cm^2 Ⓓ 404.6 cm^2

25. What is the lateral area of the square pyramid at the right if the slant height is 7 inches?

Ⓐ 17.5 in^2 Ⓑ 35 in^2

Ⓒ 70 in^2 Ⓓ 95 in^2

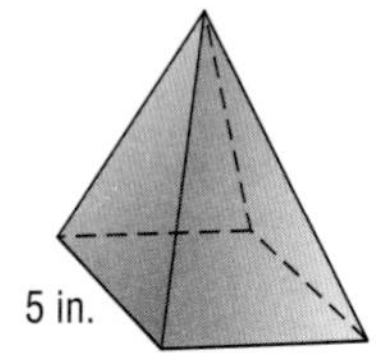

Extending the Lesson **The surface area S of a sphere with radius r is given by the formula $S = 4\pi r^2$. Find the surface area of each sphere to the nearest tenth.**

26. 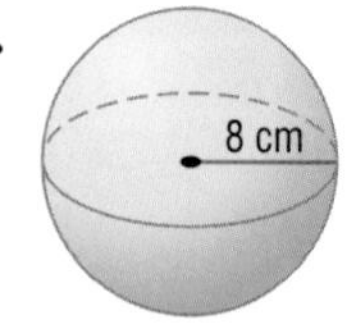

27. 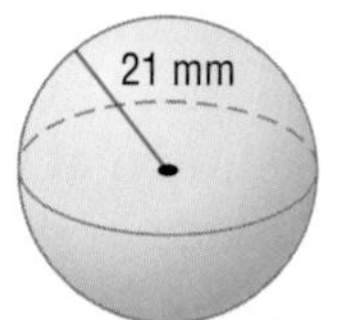

28. 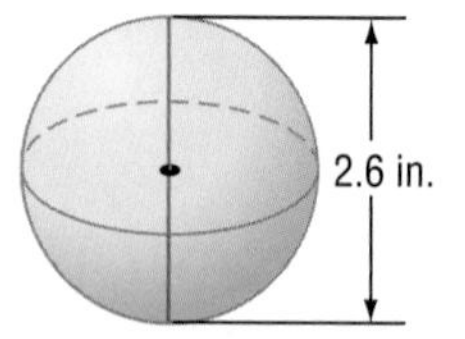

Maintain Your Skills

Mixed Review **Find the surface area of each solid. If necessary, round to the nearest tenth.** *(Lesson 11-4)*

29. rectangular prism: length 2 ft, width 1 ft, height 0.5 ft

30. cylinder: radius 4 cm, height 13.8 cm

31. Find the volume of a cone that has a height of 6 inches and radius of 2 inches. Round to the nearest tenth. *(Lesson 11-3)*

State the solution of each system of equations. *(Lesson 8-9)*

32. 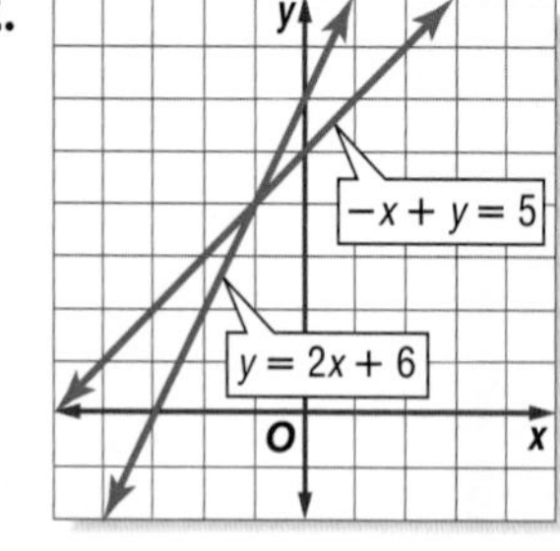

33. 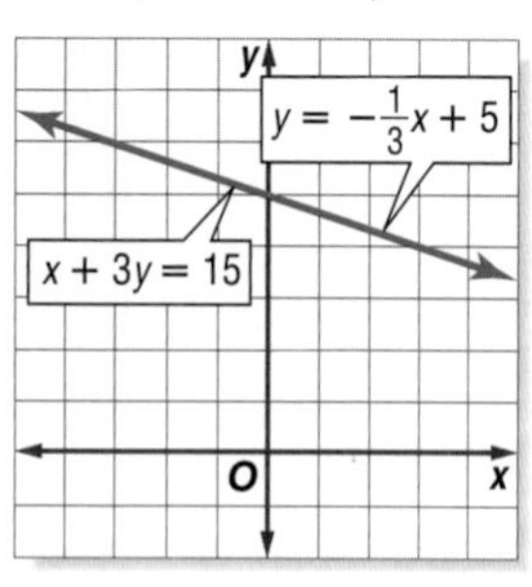

Getting Ready for the Next Lesson PREREQUISITE SKILL **Solve each proportion.**

*(To review **proportions**, see Lesson 6-2.)*

34. $\frac{1}{6} = \frac{x}{24}$

35. $\frac{9}{15} = \frac{n}{5}$

36. $\frac{t}{7} = \frac{40}{56}$

37. $\frac{1.6}{y} = \frac{9.6}{18}$

38. $\frac{4}{3.2} = \frac{w}{20}$

39. $\frac{18}{21} = \frac{2.7}{n}$

Geometry Activity

A Preview of Lesson 11-6

Sunshine State Standards
MA.B.1.3.1-1, MA.B.1.3.3-1, MA.B.1.3.3-2, MA.B.1.3.3-3, MA.C.1.3.1-2, MA.C.1.3.1-4, MA.C.3.3.1-1

Similar Solids

A model car is an exact replica of a real car, but much smaller. The dimensions of the model and the original are proportional. Therefore, these two objects are *similar solids*. The number of times that you increase or decrease the linear dimensions is called the scale factor.

You can use sugar cubes or centimeter blocks to investigate similar solids.

Activity 1

Collect the Data

- If each edge of a sugar cube is 1 unit long, then each face is 1 square unit and the volume of the cube is 1 cubic unit.
- Make a cube that has sides twice as long as the original cube.

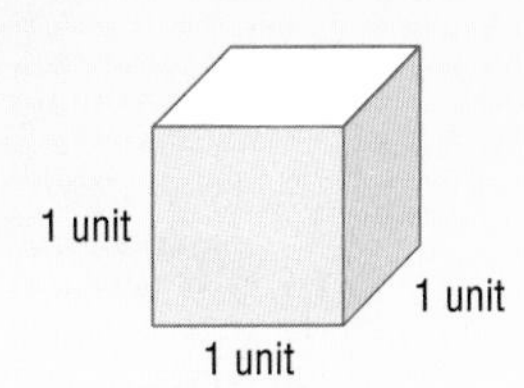

Analyze the Data

1. How many small cubes did you use?
2. What is the area of one face of the original cube?
3. What is the area of one face of the cube that you built?
4. What is the volume of the original cube?
5. What is the volume of the cube that you built?

Activity 2

Collect the Data

Build a cube that has sides three times longer than a sugar cube or centimeter block.

Analyze the Data

6. How many small cubes did you use?
7. What is the area of one face of the cube?
8. What is the volume of the cube?
9. Complete the table at the right.
10. What happens to the area of a face when the length of a side is doubled? tripled?
11. Considering the unit cube, if the scale factor is x, what is the area of one face? the surface area?
12. What happens to the volume of a cube when the length of a side is doubled? tripled?
13. Considering the unit cube, let the scale factor be x. Write an expression for the cube's volume.
14. **Make a conjecture** about the surface area and the volume of a cube if the sides are 4 times longer than the original cube.
15. **RESEARCH** the scale factor of a model car. Use the scale factor to estimate the surface area and volume of the actual car.

Scale Factor	Side Length	Area of a Face	Volume
1			
2			
3			

11-6 Similar Solids

What You'll Learn

- Identify similar solids.
- Solve problems involving similar solids.

Vocabulary

- similar solids

Sunshine State Standards
MA.B.1.3.3-1, MA.B.1.3.3-2, MA.B.1.3.3-3, MA.C.1.3.1-2, MA.C.1.3.1-4, MA.C.2.3.1-2, MA.C.3.3.1-1

How can linear dimensions be used to identify similar solids?

The model train below is $\frac{1}{87}$ the size of the original train.

a. The model boxcar is shaped like a rectangular prism. If it is 8.5 inches long and 1 inch wide, what are the length and width of the original train boxcar to the nearest hundredth of a foot?

b. A model tank car is 7 inches long and is shaped like a cylinder. What is the length of the original tank car?

c. **Make a conjecture** about the radius of the original tank car compared to the model.

IDENTIFY SIMILAR SOLIDS The cubes below have the same shape. The ratio of their corresponding edge lengths is $\frac{6}{2}$ or 3. We say that 3 is the scale factor.

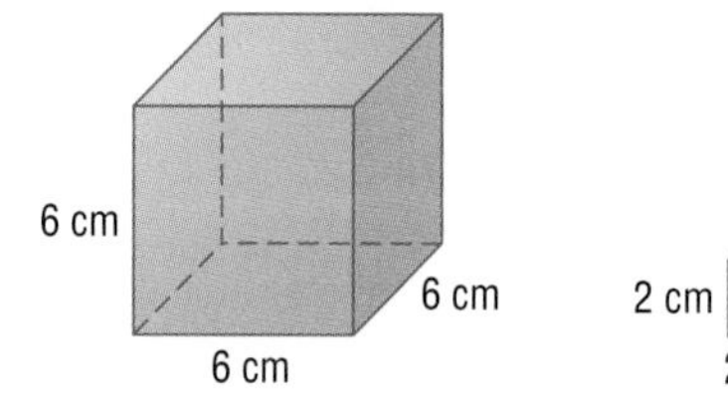

The cubes are **similar solids** because they have the same shape and their corresponding linear measures are proportional.

Study Tip

Look Back
To review **similar figures**, see Lesson 6-3.

Example 1 Identify Similar Solids

Determine whether each pair of solids is similar.

a.

32 cm
40 cm
1.25 cm
1.0 cm

$\frac{32}{1} \stackrel{?}{=} \frac{40}{1.25}$ Write a proportion comparing radii and heights.

$32(1.25) \stackrel{?}{=} 1(40)$ Find the cross products.

$40 = 40$ ✓ Simplify.

The radii and heights are proportional, so the cylinders are similar.

b. 14 in., 7 in.; 20 in., 12 in.

$\frac{14}{20} \stackrel{?}{=} \frac{7}{12}$ Write a proportion comparing corresponding edge lengths.

$14(12) \stackrel{?}{=} 20(7)$ Find the cross products.

$168 \neq 140$ Simplify.

The corresponding measures are not proportional, so the pyramids are not similar.

USE SIMILAR SOLIDS You can find missing measures if you know solids are similar.

Example 2 *Find Missing Measures*

The square pyramids at the right are similar. Find the height of pyramid B.

Pyramid A: 18 m, 14 m. Pyramid B: h, 28 m.

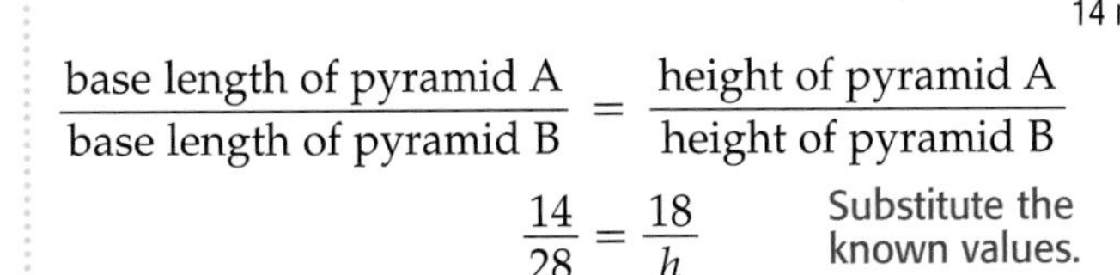

$$\frac{\text{base length of pyramid A}}{\text{base length of pyramid B}} = \frac{\text{height of pyramid A}}{\text{height of pyramid B}}$$

$\frac{14}{28} = \frac{18}{h}$ Substitute the known values.

$14h = 28(18)$ Find the cross products.

$h = 36$ Simplify.

The height of pyramid B is 36 meters.

The prisms at the right are similar with a scale factor of $\frac{3}{2}$.

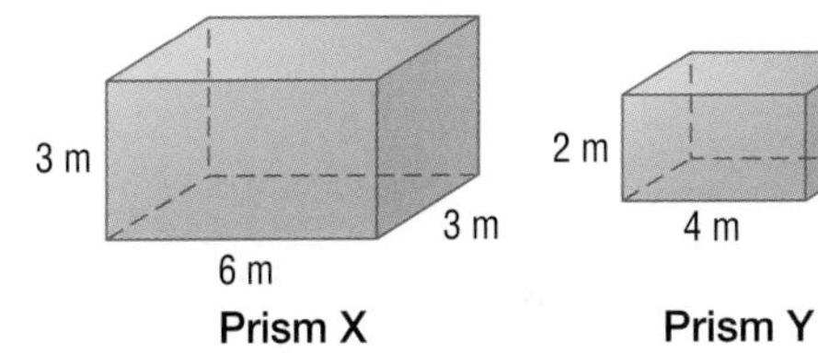

Prism	Surface Area	Volume
X	90 m^2	54 m^3
Y	40 m^2	16 m^3

Study Tip

Scale Factors
When the lengths of all dimensions of a solid are multiplied by a scale factor x, then the surface area is multiplied by x^2 and the volume is multiplied by x^3.

Notice the pattern in the following ratios.

$\frac{\text{surface area of prism X}}{\text{surface area of prism Y}} = \frac{90}{40}$ or $\frac{9}{4}$ → $\frac{9}{4} = \frac{3^2}{2^2}$

$\frac{\text{volume of prism X}}{\text{volume of prism Y}} = \frac{54}{16}$ or $\frac{27}{8}$ → $\frac{27}{8} = \frac{3^3}{2^3}$

This and other similar examples suggest that the following ratios are true for similar solids.

Key Concept *Ratios of Similar Solids*

- **Words** If two solids are similar with a scale factor of $\frac{a}{b}$, then the surface areas have a ratio of $\frac{a^2}{b^2}$ and the volumes have a ratio of $\frac{a^3}{b^3}$.
- **Model**

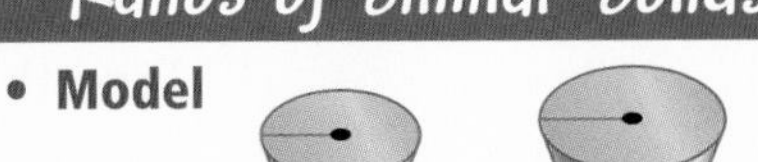

Solid A Solid B

Example 3 Use Similar Solids to Solve a Problem

SPACE TRAVEL **A scale model of the NASA space capsule is a combination of a truncated cone and cylinder. The small model built by engineers on a scale of 1 cm to 20 cm has a volume of 155 cm³. What is the volume of the actual space capsule?**

More About...

Space

Engineers need to know the volume of the actual space capsules in order to estimate the air pressure.

Source: www.space.about.com

Explore You know the scale factor $\frac{a}{b}$ is $\frac{1}{20}$ and the volume of the space capsule is 155 cm³.

Plan Since the volumes have a ratio of $\frac{a^3}{b^3}$ and $\frac{a}{b} = \frac{1}{20}$, replace a with 1 and b with 20 in $\frac{a^3}{b^3}$.

Solve

$$\frac{\text{volume of model}}{\text{volume of capsule}} = \frac{a^3}{b^3}$$ Write the ratio of volumes.

$$= \frac{1^3}{20^3}$$ Replace a with 1 and b with 20.

$$= \frac{1}{8000}$$ Simplify.

So, the volume of the capsule is 8000 times the volume of the model.

$8000 \cdot 155 \text{ cm}^3 = 1{,}240{,}000 \text{ cm}^3$

Examine Use estimation to check the reasonableness of this answer. $8000 \cdot 100 = 800{,}000$ and $8000 \cdot 200 = 1{,}600{,}000$, so the answer must be between 800,000 and 1,600,000. The answer 1,240,000 cm³ is reasonable.

Check for Understanding

Concept Check

1. **OPEN ENDED** Draw and label two cones that are similar. Explain why they are similar.
2. **Explain** how you can find the surface area of a larger cylinder if you know the surface area of a smaller cylinder that is similar to it and the scale factor.

Guided Practice

Determine whether each pair of solids is similar.

3.

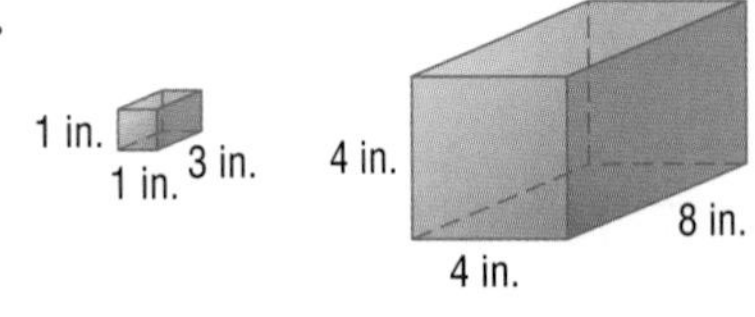

4.

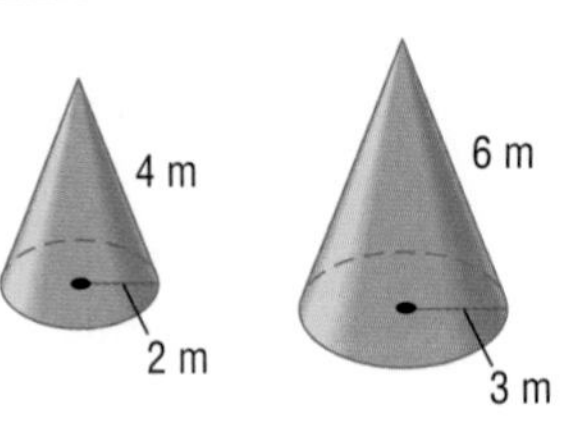

Find the missing measure for each pair of similar solids.

5.

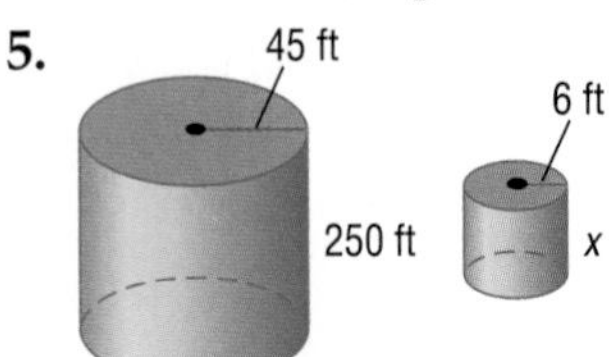

6.

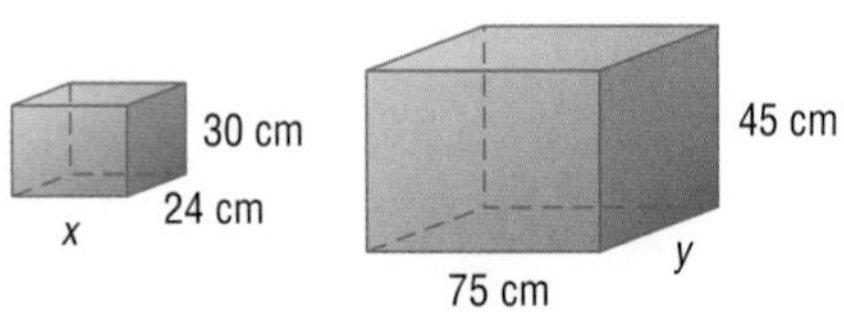

Application **ARCHITECTURE For Exercises 7–9, use the following information.**
A model for an office building is 60 centimeters long, 42 centimeters wide, and 350 centimeters high. On the model, 1 centimeter represents 1.5 meters.

7. How tall is the actual building in meters?

8. What is the scale factor between the model and the building?

9. Determine the volume of the building in cubic meters.

Practice and Apply

Homework Help

For Exercises	See Examples
10–13, 16–19	1
14, 15	2
20, 21	3

Extra Practice
See page 752.

Determine whether each pair of solids is similar.

10.

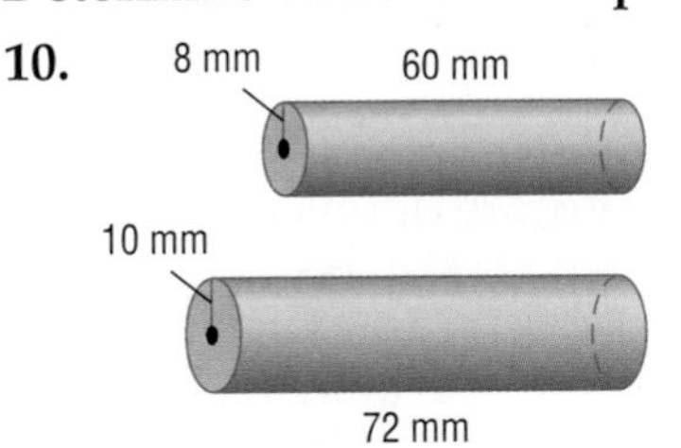

11.

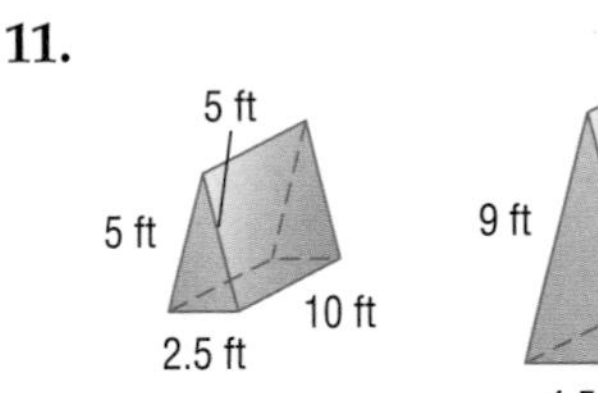

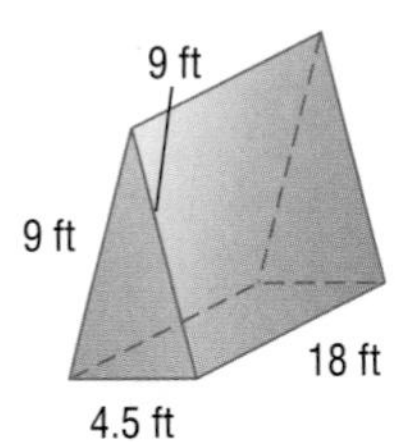

12.

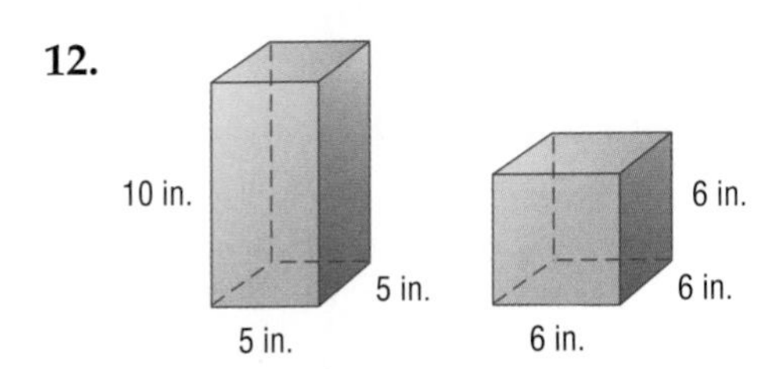

13.

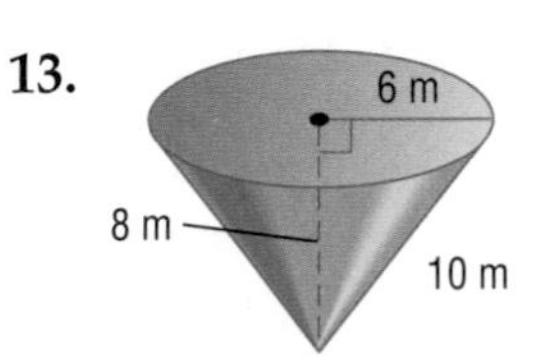

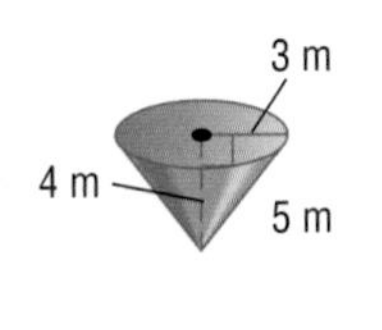

Find the missing measure for each pair of similar solids.

14.

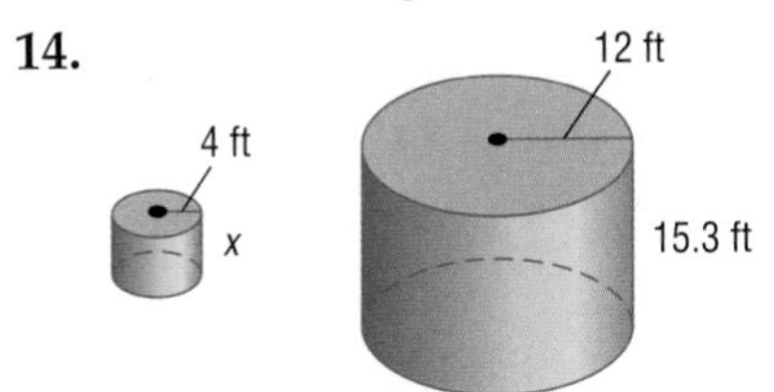

15.

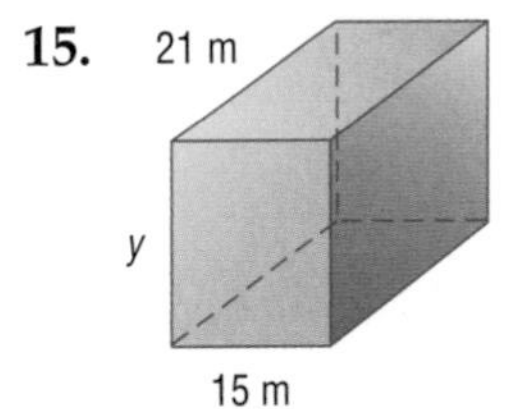

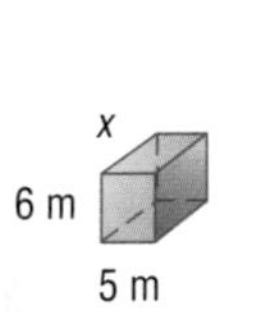

Determine whether each pair of solids is *sometimes, always,* or *never* similar. Explain.

16. two cubes

17. two prisms

18. a cone and a cylinder

19. two spheres

HISTORY The *Mankaure* pyramid in Egypt has a square base that is 110 meters on each side, a height of 68.8 meters, and a slant height of 88.5 meters. Suppose you want to construct a scale model of the pyramid using a scale of 4 meters to 2 centimeters.

20. How much material will you need to use?

21. How much greater is the volume of the actual pyramid to the volume of the model?

Online Research **Data Update** How have the dimensions of the Egyptian pyramids changed in thousands of years? Visit www.pre-alg.com/data_update to learn more.

22. CRITICAL THINKING The dimensions of a triangular prism are decreased so that the volume of the new prism is $\frac{1}{3}$ that of the original volume. Are the two prisms similar? Explain.

23. **WRITING IN MATH** Answer the question that was posed at the beginning of the lesson.

How can linear dimensions be used to identify similar solids?

Include the following in your answer:

- a description of the ratios needed for two solids to be similar, and
- an example of two solids that are *not* similar.

24. Which prism shown in the table is *not* similar to the other three?

Ⓐ prism A Ⓑ prism B

Ⓒ prism C Ⓓ prism D

Prism	Length	Width	Height
A	4	3	2
B	6	4.5	3
C	5	4	2
D	28	21	14

25. If the dimensions of a cone are doubled, the surface area

Ⓐ stays the same. Ⓑ is doubled.

Ⓒ is quadrupled. Ⓓ is 8 times greater.

Maintain Your Skills

Mixed Review **Find the surface area of each solid. If necessary, round to the nearest tenth.** *(Lessons 11-4 and 11-5)*

26.

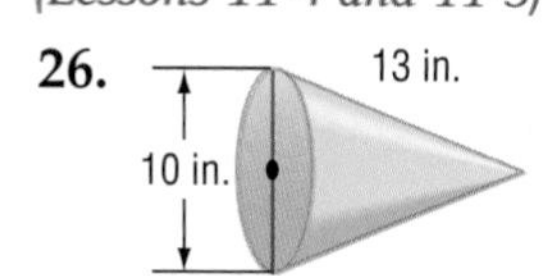

27.

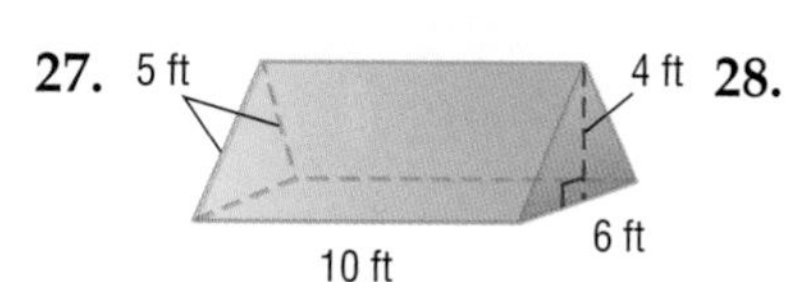

28.

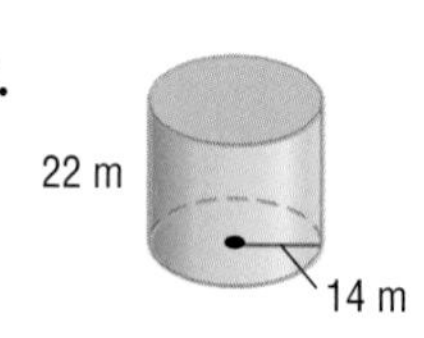

29. Angles J and K are complementary. Find $m\angle K$ if $m\angle J$ is 25°. *(Lesson 10-2)*

Solve each equation. Check your solution. *(Lesson 5-9)*

30. $r - 3.5 = 8$ 31. $\frac{2}{3} + y = \frac{1}{9}$ 32. $\frac{1}{4}a = 6$

Getting Ready for the Next Lesson **PREREQUISITE SKILL Find the value of each expression to the nearest tenth.** *(To review **rounding decimals**, see page 711.)*

33. $13.28 + 6.05$ 34. $8.99 - 1.2$ 35. $2.4 \cdot 2.5$

36. $55 \div 3.8$ 37. $6 + 1.9 + 1.45$ 38. $6.7(0.3)(1.8)$

Practice Quiz 2 *Lessons 11-4 through 11-6*

Find the surface area of each solid. If necessary, round to the nearest tenth. *(Lessons 11-4 and 11-5)*

1.

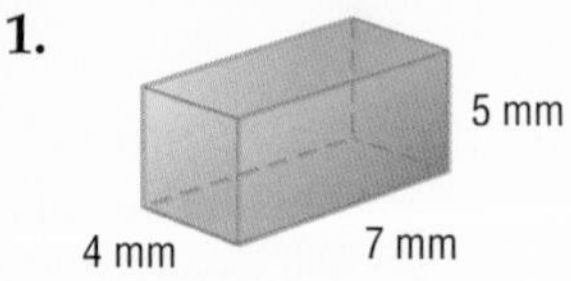

2.

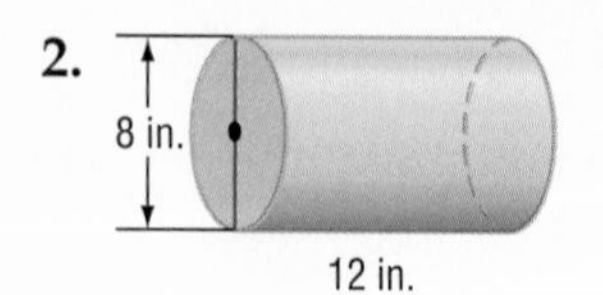

3.

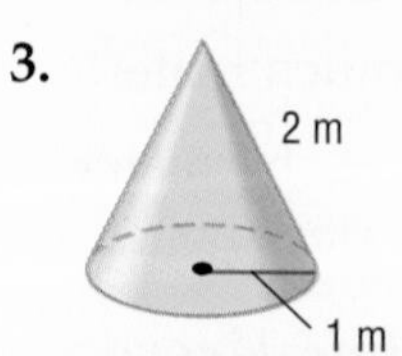

4.

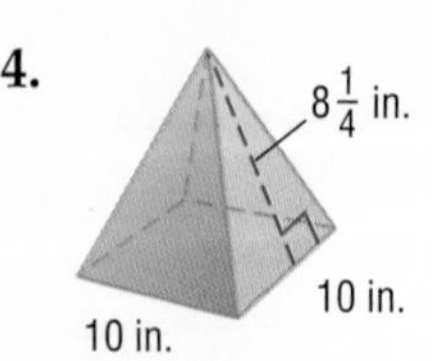

5. Are the cylinders described in the table similar? Explain your reasoning. *(Lesson 11-6)*

Cylinder	Diameter (mm)	Slant Height (mm)
A	24	21
B	16	14

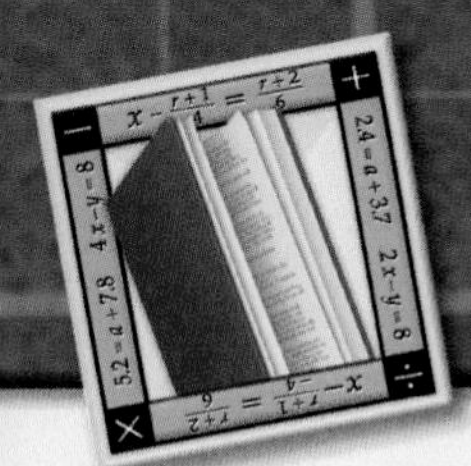

Reading Mathematics

Sunshine State Standards
MA.B.4.3.1-1, MA.B.4.3.1-3

Precision and Accuracy

In everyday language, *precision* and *accuracy* are used to mean the same thing. When measurement is involved, these two terms have different meanings.

Term	Definition	Example
precision	the degree of exactness in which a measurement is made	A measure of 12.355 grams is more precise than a measure of 12 grams.
accuracy	the degree of conformity of a measurement with the true value	Suppose the actual mass of an object is 12.355 grams. Then a measure of 12 grams is more accurate than a measure of 18 grams.

Reading to Learn

1. Describe in your own words the difference between accuracy and precision.
2. **RESEARCH** Use the Internet or other resources to find an instrument used in science that gives very precise measurements. Describe the precision of the instrument.
3. Use at least two different measuring instruments to measure the length, width, height, or weight of two objects in your home. Describe the measuring instruments that you used and explain which measurement was most precise.

Choose the correct term or terms to determine the degree of precision needed in each measurement situation.

4. In a travel brochure, the length of a cruise ship is described in (millimeters, meters).
5. The weight of a bag of apples in a grocery store is given to the nearest (tenth of a pound, tenth of an ounce).
6. In a science experiment, the mass of one drop of solution is found to the nearest 0.01 (gram, kilogram).
7. A person making a jacket measures the fabric to the nearest (inch, eighth of an inch).
8. **CONSTRUCTION** A construction company is ordering cement to complete all the sidewalks in a new neighborhood. Would the precision or accuracy be more important in the completion of their order? Explain.

11-7 Precision and Significant Digits

What You'll Learn

- Describe measurements using precision and significant digits.
- Apply precision and significant digits in problem-solving situations.

Vocabulary

- precision
- significant digits

Sunshine State Standards
MA.B.4.3.1-1, MA.B.4.3.1-2, MA.B.4.3.1-3, MA.B.4.3.1-4, MA.B.4.3.1-5, MA.B.4.3.2-1, MA.B.4.3.2-2

Why are all measurements really approximations?

Use cardboard to make three rulers 20 centimeters long, labeling the increments as shown at the right.

Ruler	Scale (cm)
1	0, 5, 10, 15, 20
2	0, 1, 2, 3, ..., 18, 19, 20
3	0, 0.1, 0.2, 0.3, ..., 19.8, 19.9, 20.0

a. Measure several objects (book widths, paper clips, pens...) using each ruler. Use a table to keep track of your measurements.

b. Analyze the measurements and determine which are most useful. Explain your reasoning.

PRECISION AND SIGNIFICANT DIGITS

The **precision** of a measurement is the exactness to which a measurement is made. Precision depends on the smallest unit of measure being used, or the *precision unit*. You can expect a measurement to be accurate to the nearest precision unit.

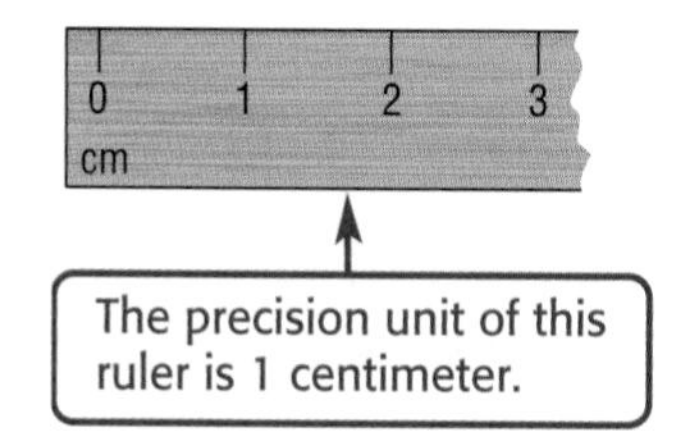

Example 1 Identify Precision Units

Identify the precision unit of the ruler at the right.

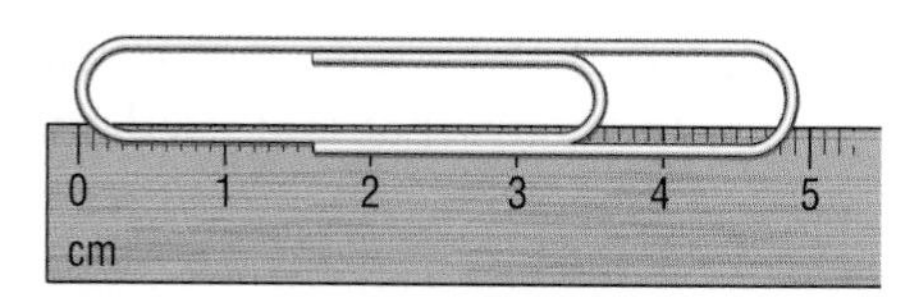

The precision unit is one tenth of a centimeter, or 1 millimeter.

One way to record a measure is to estimate to the nearest precision unit. A more precise method is to include all of the digits that are actually measured, plus one *estimated* digit. The digits you record when you measure this way are called significant digits. **Significant digits** indicate the precision of the measurement.

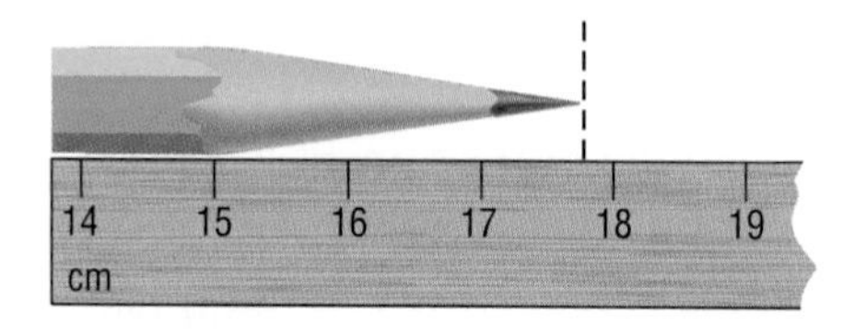

precision unit: 1 cm
actual measure: 17–18 cm
estimated measure: 17.7 cm

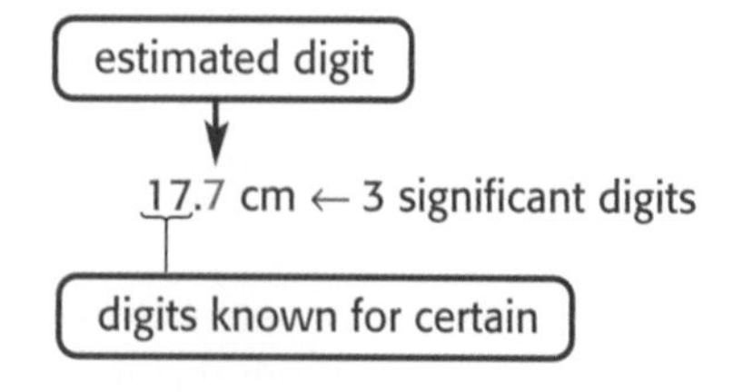

Study Tip

Precision
The precision unit of the measuring instrument determines the number of significant digits.

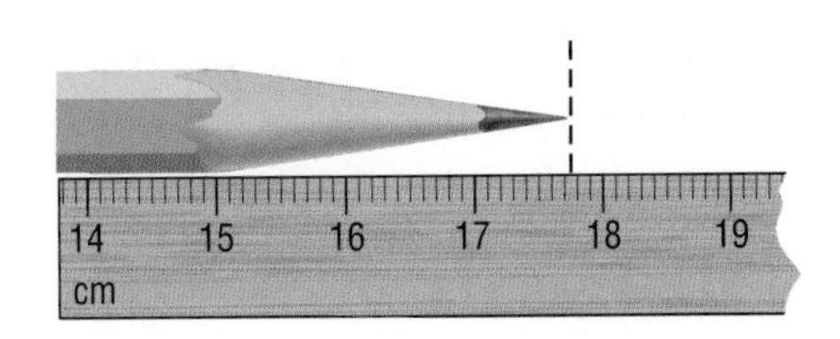

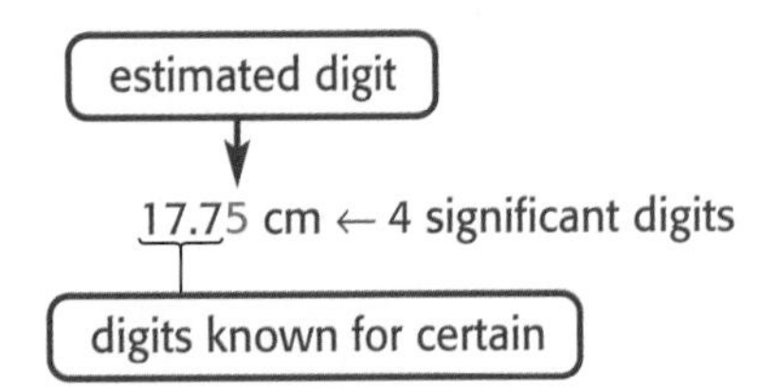

precision unit: 0.1 cm
actual measure: 17.7–17.8 cm
estimated measure: 17.75 cm

There are special rules for determining significant digits in a given measurement. If a number contains a decimal point, the number of significant digits is found by counting the digits from left to right, starting with the first *nonzero* digit and ending with the last digit.

Number		Number of Significant Digits	
1.23	→	3	All nonzero digits are significant.
10.05	→	4	Zeros between two significant digits are significant.
0.072	→	2	Zeros used to show place value of the decimal are not significant.
50.00	→	4	In a number with a decimal point, all zeros to the right of a nonzero digit are significant.

If a number does *not* contain a decimal point, the number of significant digits is found by counting the digits from left to right, starting with the first digit and ending with the last *nonzero* digit. For example, 8400 contains 2 significant digits, 8 and 4.

Study Tip

Precision Units
Since 150 miles has only two significant digits, the measure is precise to the nearest 10 miles. Therefore, the precision unit is 10 miles, not 1 mile.

Example 2 Identify Significant Digits

Determine the number of significant digits in each measure.

a. 20.98 centimeters
4 significant digits

b. 150 miles
2 significant digits

c. 0.007 gram
1 significant digit

d. 6.40 feet
3 significant digits

COMPUTE USING SIGNIFICANT DIGITS When adding or subtracting measurements, the sum or difference should have the *same precision* as the least precise measurement.

Study Tip

Common Misconception
Calculators give answers with as many digits as the display can show. Be sure to use the correct number of significant digits in your answer.

Example 3 Add Measurements

The sides of a triangle measure 14.35 meters, 8.6 meters, and 9.125 meters. Use the correct number of significant digits to find the perimeter.

$$\begin{array}{r l} 14.35 & \leftarrow \text{2 decimal places} \\ 8.6 & \leftarrow \text{1 decimal place} \\ +\ 9.125 & \leftarrow \text{3 decimal places} \\ \hline 32.075 & \end{array}$$

The least precise measurement, 8.6 meters, has one decimal place. So, round 32.075 to one decimal place, 32.1. The perimeter of the triangle is about 32.1 meters.

When multiplying or dividing measurements, the product or quotient should have the *same number of significant digits* as the measurement with the least number of significant digits.

Study Tip

Significant Digits
The least precise measure determines the number of significant digits in the sum or difference of measures. The measure with the fewest significant digits determines the number of significant digits in the product or quotient of measures.

Example 4 Multiply Measurements

What is the area of the bedroom shown at the right?

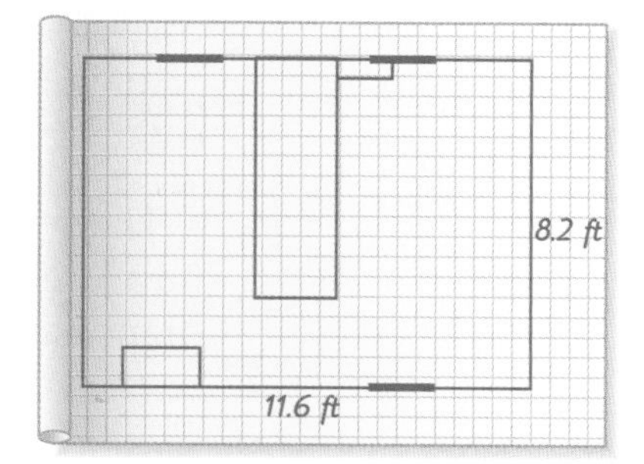

To find the area, multiply the length and the width.

$$\begin{array}{rl} 11.6 & \leftarrow \text{3 significant digits} \\ \times\ 8.2 & \leftarrow \text{2 significant digits} \\ \hline 95.12 & \leftarrow \text{4 significant digits} \end{array}$$

The answer cannot have more significant digits than the measurements of the length and width. So, round 95.12 cm^2 to 2 significant digits. The area of the bedroom is about 95 cm^2.

Check for Understanding

Concept Check

1. **FIND THE ERROR** A metal shelf is 0.0205 centimeter thick. Sierra says this measurement contains 2 significant digits. Josh says it contains 3 significant digits. Who is correct? Explain your reasoning.

2. **Choose** an instrument from the list at the right that is best for measuring each object.

yardstick
centimeter ruler
surveyor's tools
12-foot tape measure

   a. length of an envelope
   b. distance between two stoplights
   c. width of a kitchen
   d. height of a small child

3. **OPEN ENDED** Write a number that contains four digits, two of which are significant.

Guided Practice

4. Identify the precision unit of the scale at the right.

Determine the number of significant digits in each measure.

5. 2.30 cm
6. 50 yd
7. 0.801 mm

Calculate. Round to the correct number of significant digits.

8. 14.38 cm + 5.7 cm
9. 15.273 L − 8.2 L
10. 3.147 mm · 1.8 mm
11. 60.42 in. × 9.012 in.

Application

12. **MASONRY** A wall of bricks is 7.85 feet high and 13.0 feet wide. What is the area of the wall? Round to the correct number of significant digits.

Practice and Apply

Homework Help

For Exercises	See Examples
13, 14	1
15–22	2
23–38	3, 4

Extra Practice

See page 752.

Identify the precision unit of each measuring tool.

13. 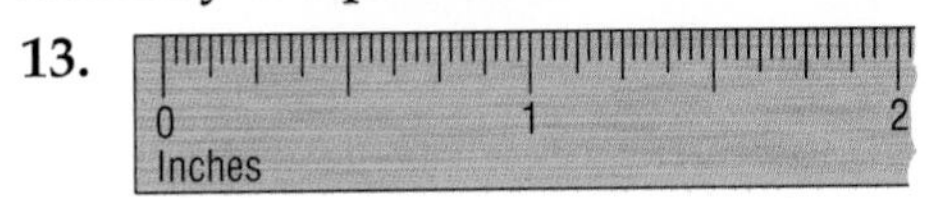

14. 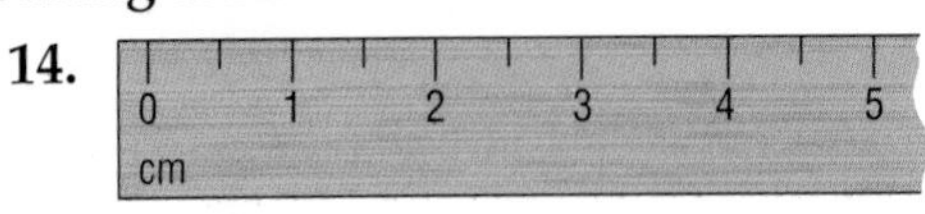

Determine the number of significant digits in each measure.

15. 925 g **16.** 40 km **17.** 2200 ft **18.** 53.6 in.

19. 0.01 mm **20.** 0.56 cm **21.** 18.50 m **22.** 4.0 L

Calculate. Round to the correct number of significant digits.

23. 27 in. + 18.2 in. **24.** 6.75 mm – 3.2 mm

25. 0.4 ft · 5.1 ft **26.** 7.30 yd × 1.61 yd

27. 29.307 m + 4.23 m + 50.93 m **28.** 127.2 g + 42.3 g – 5.7 g

29. 50.2 cm – 0.75 cm **30.** 18.160 L – 15 L

31. 5.327 m · 4.8 m **32.** 4.397 cm · 2.01 cm

33. MEASUREMENT Choose the best ruler for measuring an object to the nearest sixteenth of an inch. Explain your reasoning.

a. 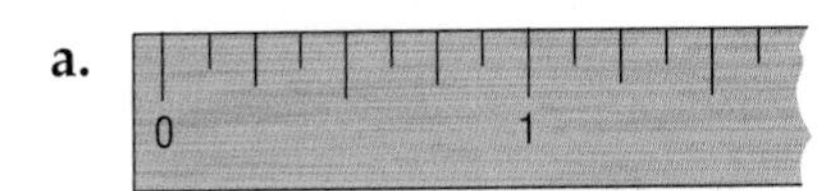

b. 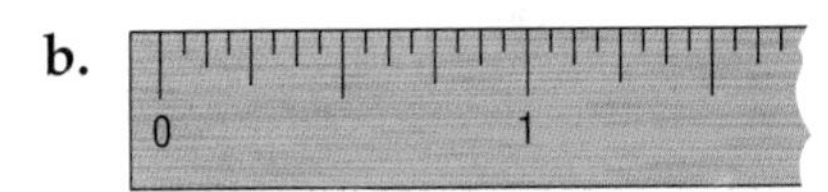

34. MEASUREMENT Order 0.40 mm, 40 mm, 0.4 mm, and 0.004 mm from most to least precise.

More About. . .

Oranges

The places that produce the most oranges are Brazil, the United States, Mexico, China, and Africa.

Source: The Brazilian Assoc. of Citrus Exporters

ORANGES For Exercises 35–37, refer to the graph at the right.

USA TODAY Snapshots®

Drought cuts orange crop

This year's U.S. orange production is expected to be down 5% from last year's 13-million-ton crop. Much of the drop is from reduced production in drought-stricken Florida.

Annual U.S. orange production:

15
10
5
0
8.9
12.4
(in millions of tons)
1992
2001

Source: National Agricultural Statistics Service, April 2001 forecast

By Quin Tian, USA TODAY

35. Are the numbers exact? Explain.

36. How many significant digits are used to describe orange production in 1992 and in 2001?

37. Write the number of tons of oranges produced in the United States in 2001 without using a decimal point. How many significant digits does this number have?

38. MEASUREMENT ERROR Mrs. Hernandez is covering the top of a kitchen shelf that is $26\frac{3}{8}$ inches long and $15\frac{1}{2}$ inches wide. She incorrectly measures the length to be 25 inches long. How will this error affect her calculations?

39. CRITICAL THINKING The sizes of Allen hex wrenches are 2.0 mm, 3.0 mm, 4.0 mm, and so on. Will they work with hexagonal bolts that are marked 2 mm, 3 mm, 4 mm, and so on? Explain.

40. **WRITING IN MATH** Answer the question that was posed at the beginning of the lesson.

Why are all measurements really approximations?

Include the following in your answer:

- an explanation of what determines the precision of a measurement, and
- an example of a real-life situation in which an exact solution is needed and a situation in which an approximate solution is sufficient.

41. Choose the measurement that is most precise.

Ⓐ 12 mm Ⓑ 12 cm Ⓒ 1.2 m Ⓓ 12 m

42. Which solution contains the correct number of significant digits for the product 2.80 mm · 0.1 mm?

Ⓐ 0.28 mm^2 Ⓑ 0.280 mm^2 Ⓒ 0.30 mm^2 Ⓓ 0.3 mm^2

Extending the Lesson

The **greatest possible error** is one-half the precision unit. It can be used to describe the actual measure. Refer to the paper clip in Example 1. It appears to be 4.9 centimeters long.

$$\text{greatest possible error} = \frac{1}{2} \cdot \text{precision unit}$$
$$= \frac{1}{2} \cdot 0.1 \text{ cm or } 0.05 \text{ cm}$$

The possible actual length of the paper clip is 0.05 centimeter less than or 0.05 centimeter greater than 4.9 centimeters, or between 4.85 and 4.95 centimeters.

43. **TRAVEL** The odometer on a car shows 132.8 miles traveled. Find the greatest possible error of the measurement and use it to determine between which two values is the actual distance traveled.

Maintain Your Skills

Mixed Review

44. Determine whether a cone with a height 14 centimeters and radius 8 centimeters is similar to a cone with a height of 12 centimeters and a radius of 6 centimeters. *(Lesson 11-6)*

Find the surface area of each solid. If necessary, round to the nearest tenth. *(Lesson 11-5)*

45.  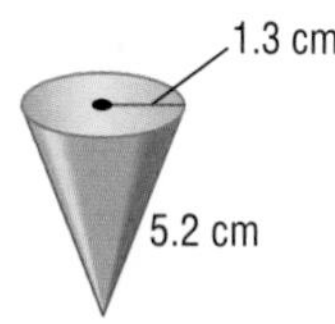

46. 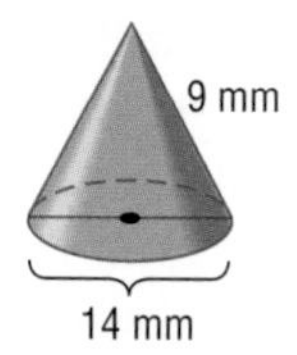

47. 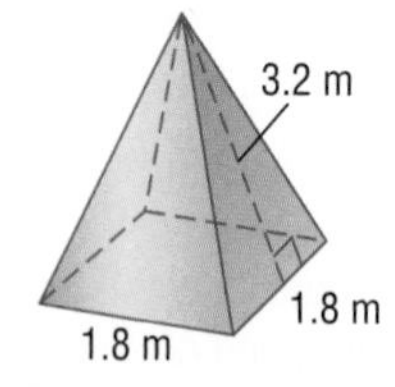

WebQuest Internet Project

Able to Leap Tall Buildings

It is time to complete your project. Use the information you have gathered about your building to prepare a report. Be sure to include information and facts about your building as well as a comparison of its size to some familiar item.

www.pre-alg.com/webquest

Chapter 11

Study Guide and Review

Vocabulary and Concept Check

base (p. 557)	plane (p. 556)	skew lines (p. 558)
cone (p. 569)	polyhedron (p. 556)	slant height (p. 578)
cylinder (p. 565)	precision (p. 590)	solid (p. 556)
edge (p. 556)	prism (p. 557)	surface area (p. 573)
face (p. 556)	pyramid (p. 557)	vertex (p. 556)
lateral area (p. 578)	significant digits (p. 590)	volume (p. 563)
lateral face (p. 578)	similar solids (p. 584)	

Determine whether each statement is *true* or *false*. If false, replace the underlined word or number to make a true statement.

1. The surface area of a pyramid is the sum of the areas of its lateral faces.
2. Volume is the amount of space that a solid contains.
3. The edge of a pyramid is the length of an altitude of one of its lateral faces.
4. A triangular prism has two bases.
5. A solid with two bases that are parallel circles is called a cone.
6. Prisms and pyramids are named by the shapes of their bases.
7. Figures that have the same shape and corresponding linear measures that are proportional are called similar solids.
8. Significant digits indicate the precision of a measurement.

Lesson-by-Lesson Review

11-1 Three-Dimensional Figures

See pages 556–561.

Concept Summary

- Prisms and pyramids are three-dimensional figures.

Identify the solid. Name the bases, faces, edges, and vertices.

There is one triangular base, so the solid is a triangular pyramid.

faces: JKL, JLM, JMK

edges: $\overline{JK}$, $\overline{JL}$, $\overline{JM}$, $\overline{KL}$, $\overline{LM}$, $\overline{MK}$

vertices: J, K, L, M

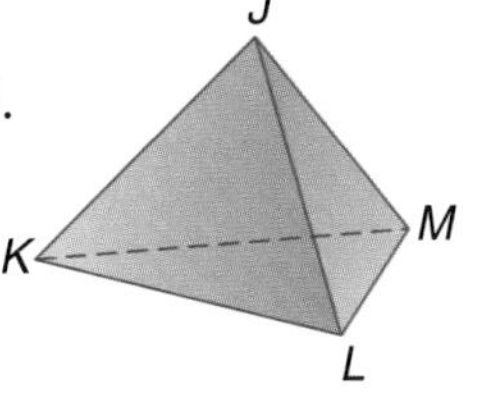

Exercises **Identify each solid. Name the bases, faces, edges, and vertices.**

See Example 1 on page 557.

9. 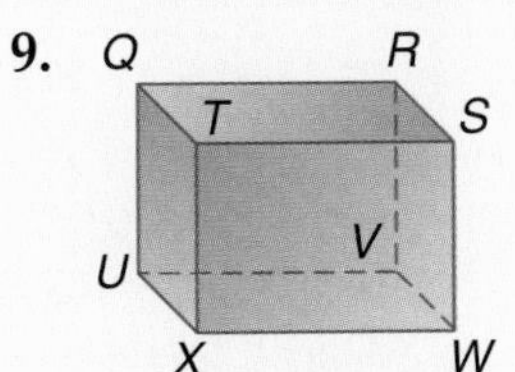

10. 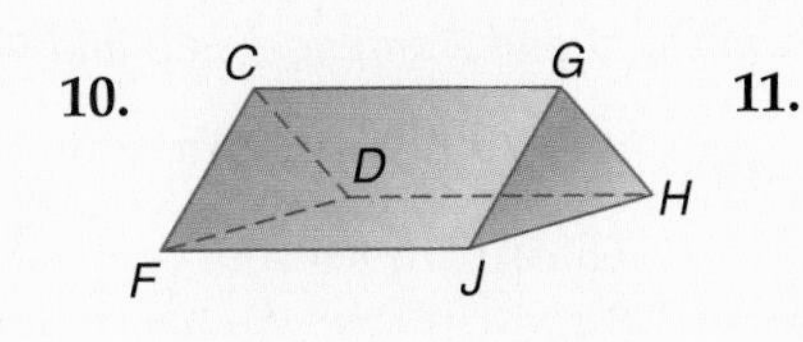

11. 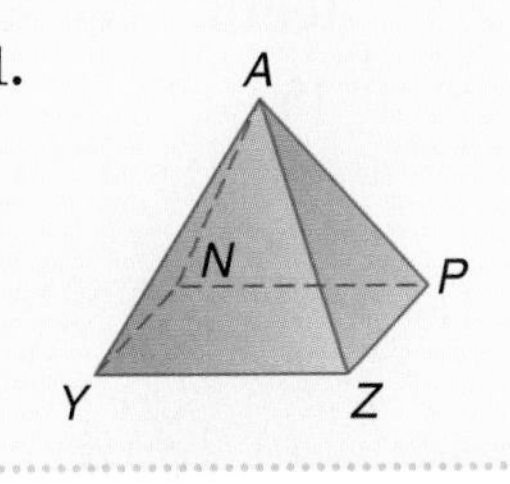

11-2 Volume: Prisms and Cylinders

See pages 563–567.

Concept Summary

- Volume is the measure of space occupied by a solid region.
- The volume of a prism or a cylinder is the area of the base times the height.

Example **Find the volume of the cylinder. Round to the nearest tenth.**

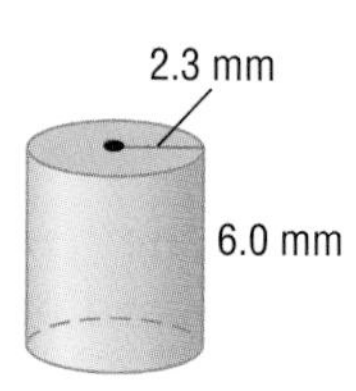

$V = \pi r^2 h$ Formula for volume of a cylinder

$V = \pi \cdot 2.3^2 \cdot 6.0$ Replace r with 2.3 and h with 6.0.

$V \approx 99.7$ Simplify.

The volume is about 99.7 cubic millimeters.

Exercises **Find the volume of each solid. If necessary, round to the nearest tenth.** *See Examples 1, 2, and 5 on pages 563–565.*

12.

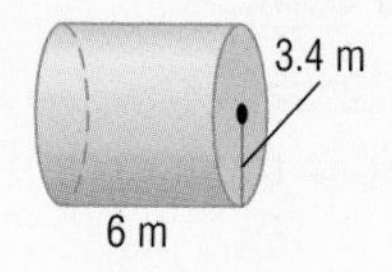

13.

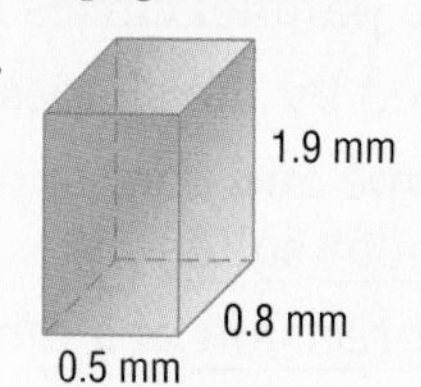

14.

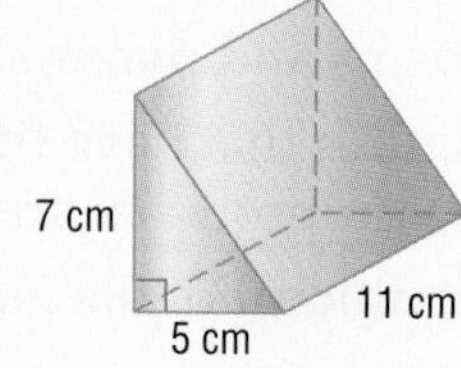

11-3 Volume: Pyramids and Cones

See pages 568–572.

Concept Summary

- The volume of a pyramid or a cone is one-third the area of the base times the height.

Example **Find the volume of the cone. Round to the nearest tenth.**

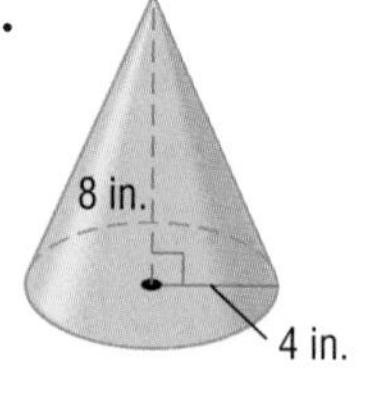

$V = \frac{1}{3}\pi r^2 h$ Formula for volume of a cone

$V = \frac{1}{3} \cdot \pi \cdot 4^2 \cdot 8$ Replace r with 4 and h with 8.

$V \approx 134.0$ Simplify.

The volume is about 134.0 cubic inches.

Exercises **Find the volume of each solid. If necessary, round to the nearest tenth.** *See Examples 1 and 2 on pages 568 and 569.*

15.

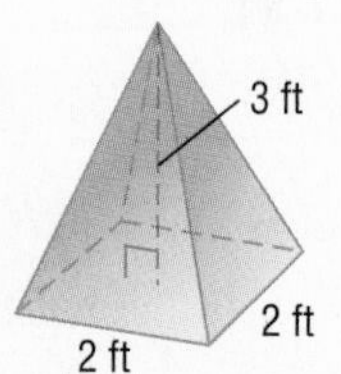

16.

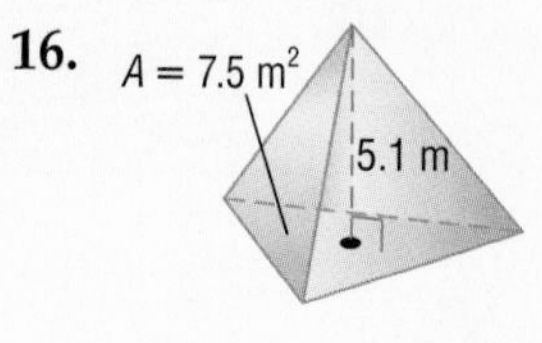

17.

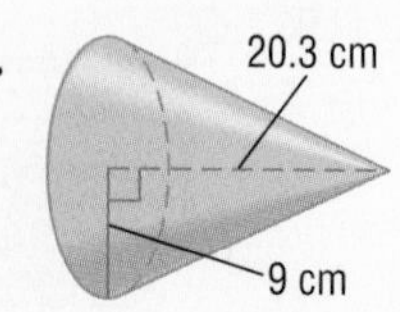

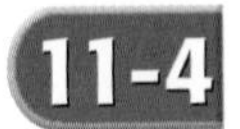

11-4 Surface Area: Prisms and Cylinders

See pages 573–577.

Concept Summary

- The surface area of a prism is the sum of the areas of the faces.
- The surface area of a cylinder is the area of the two bases plus the product of the circumference and the height.

Example **Find the surface area of the rectangular prism.**

$S = 2\ell w + 2\ell h + 2wh$ Write the formula.

$S = 2(5)(4) + 2(5)(2) + 2(4)(2)$ Substitution

$S = 76$ Simplify.

The surface area is 76 square inches.

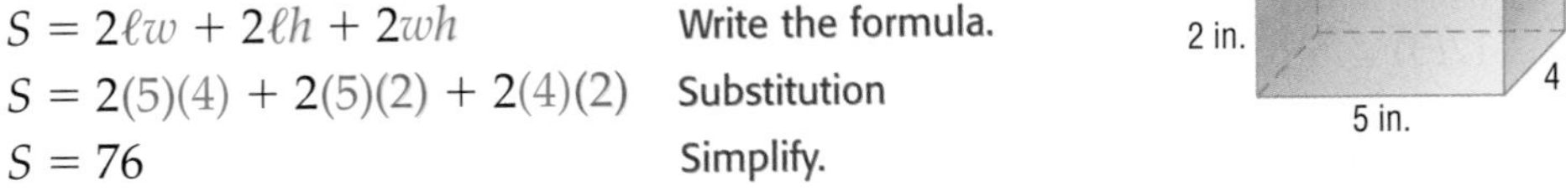

Exercises **Find the surface area of each solid. If necessary, round to the nearest tenth.** *See Examples 1–3 on pages 573–575.*

18.

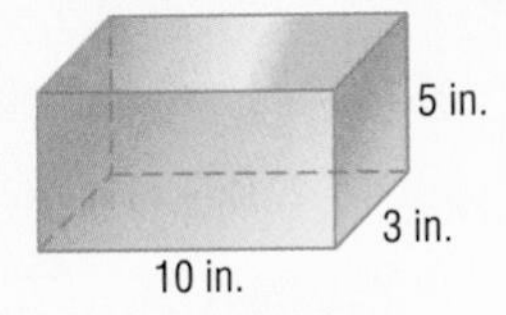

19.

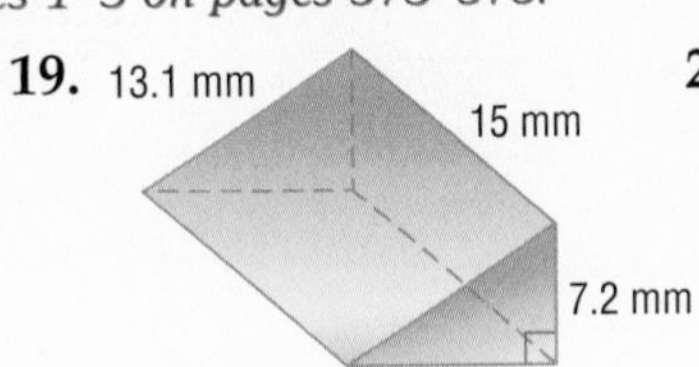

20.

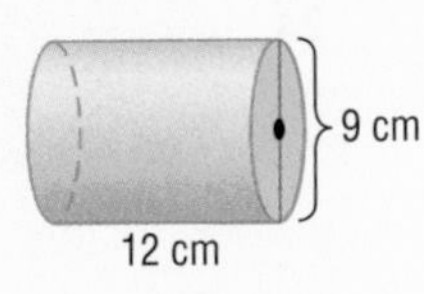

11-5 Surface Area: Pyramids and Cones

See pages 578–582.

Concept Summary

- The surface area of a pyramid or a cone is the sum of the lateral area and the base area.

Example **Find the surface area of the cone. Round to the nearest tenth.**

$S = \pi r\ell + \pi r^2$ Write the formula.

$S = \pi(1.5)(4) + \pi(1.5)^2$ Replace r with 1.5 and ℓ with 4.

$S \approx 25.9$ Simplify.

The surface area is about 25.9 square meters.

Exercises **Find the surface area of each solid. If necessary, round to the nearest tenth.** *See Examples 1 and 3 on pages 578 and 580.*

21.

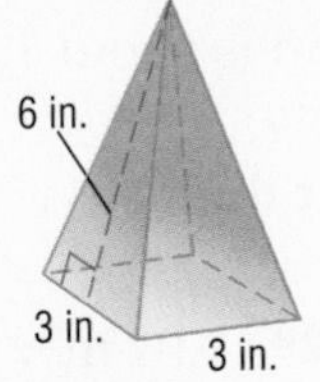

22.

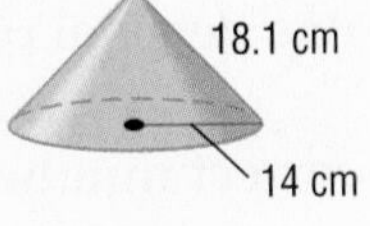

23.

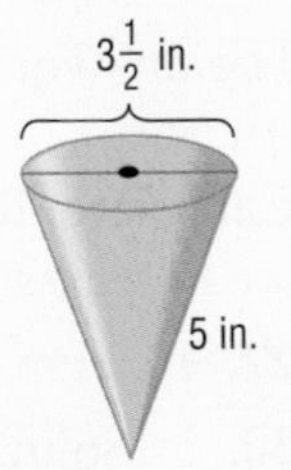

Chapter 11 For More ...

- Extra Practice, see pages 750–752.
- Mixed Problem Solving, see page 768.

11-6 Similar Solids

See pages 584–588.

Concept Summary

- Similar solids have the same shape and their corresponding linear measures are proportional.

Example **Determine whether the solids are similar.**

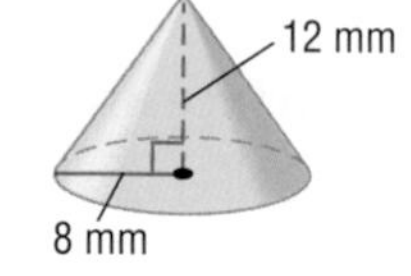

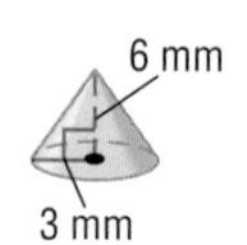

$\frac{12}{6} \stackrel{?}{=} \frac{8}{3}$ Write a proportion comparing radii and heights.

$12(3) \stackrel{?}{=} 6(8)$ Find the cross products.

$36 \neq 48$ Simplify.

The radii and heights are not proportional, so the cones are not similar.

Exercises **Determine whether each pair of solids is similar.**
See Example 1 on pages 584 and 585.

24.

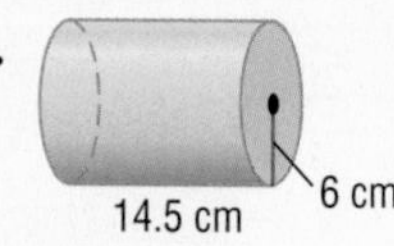

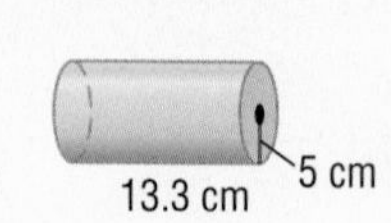

25.

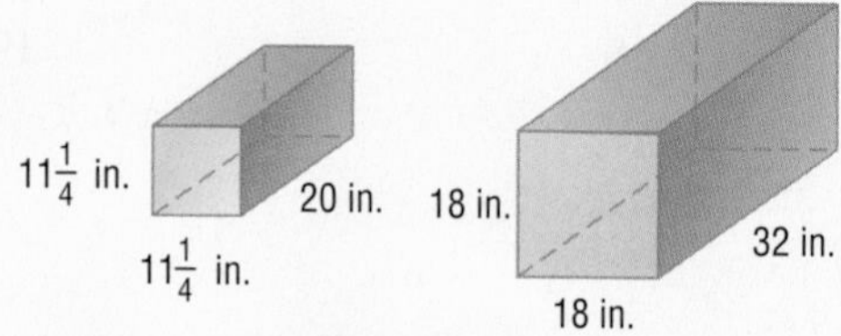

26. Find the missing measure for the pair of similar solids at the right.
See Example 2 on page 585.

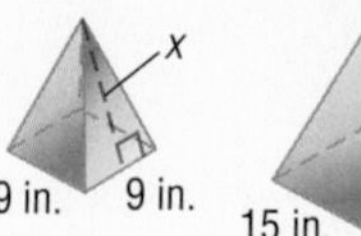

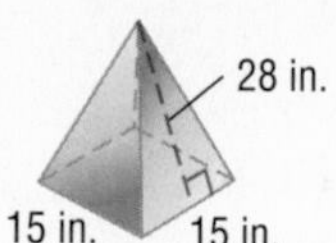

11-7 Precision and Significant Digits

See pages 590–594.

Concept Summary

- The precision of a measurement depends on the smallest unit of measure being used.
- Significant digits indicate the precision of a measurement.

Example **Find 0.5 m + 0.75 m. Round to the correct number of significant digits.**

$$\begin{array}{r} 0.5 \\ +\ 0.75 \\ \hline 1.25 \end{array}$$

0.5 ← 1 decimal place
0.75 ← 2 decimal places

The answer should have one decimal place. So, the sum is about 1.3 meters.

Calculate. Round to the correct number of significant digits.
See Examples 3 and 4 on pages 591 and 592.

27. 10.3 cm + 8.7 cm **28.** 25.71 kg − 11.2 kg **29.** 0.04 m + 0.9 m

30. 5.186 in. · 1.5 in. **31.** 32.0 ft · 30.4 ft **32.** 80.51 g − 6.01 g

Chapter 11

Practice Test

Vocabulary and Concepts

1. **Describe** the difference between a prism and a pyramid.
2. **Describe** the characteristics of similar solids.
3. **OPEN ENDED** Write a four-digit number that has three significant digits.

Skills and Applications

Identify each solid. Name the bases, faces, edges, and vertices.

4. 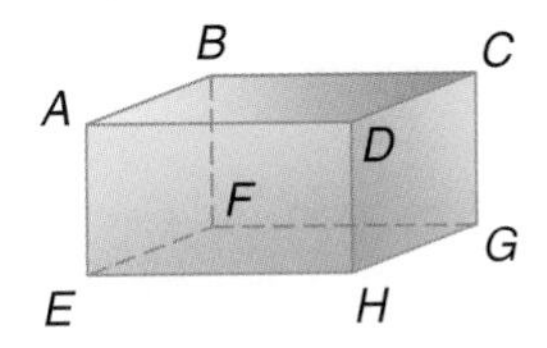

5. 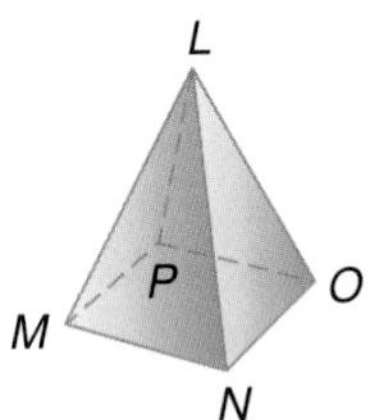

Find the volume of each solid. If necessary, round to the nearest tenth.

6. cylinder: radius 1.7 mm, height 8 mm
7. rectangular pyramid: length 14 in., width 8 in., height 5 in.
8. cube: length 9.2 cm
9. cone: diameter 26 ft, height 31 ft

Find the surface area of each solid. If necessary, round to the nearest tenth.

10. 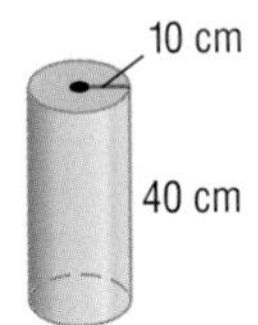

11. 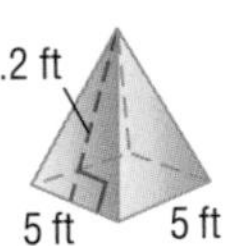

12. 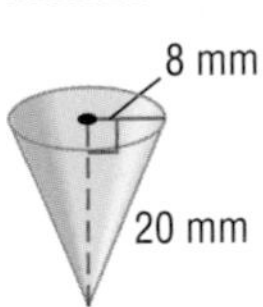

13. Determine whether the given pair of solids is similar. Explain.

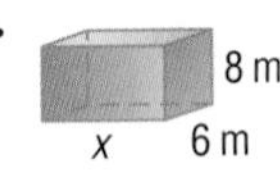

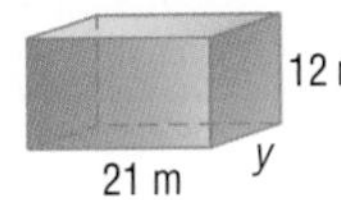

Find the missing measure for each pair of similar solids.

14. 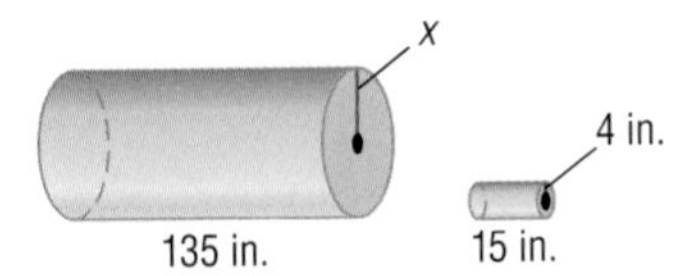

15.

Determine the number of significant digits in each measure.

16. 4500 mm
17. 0.036 in.

Calculate. Round to the correct number of significant digits.

18. 37.65 cm – 12.9 cm
19. 6.8 ft × 3.875 ft

FCAT Practice

20. **STANDARDIZED TEST PRACTICE** A model of a new grocery store is 15 inches long, 9 inches wide, and 7 inches high. The scale is 50 feet to 3 inches. Find the length of the actual store.

Ⓐ 45 ft Ⓑ 150 ft Ⓒ 250 ft Ⓓ 750 ft

Chapter 11 Standardized Test Practice

FCAT Practice

Part 1 Multiple Choice

Record your answers on the answer sheet provided by your teacher or on a sheet of paper.

1. If $-3x + 7 = -29$, then what is the value of x? (Lesson 3-5)

(A) -12 (B) -6

(C) 6 (D) 12

2. Austin wants to buy a fish tank so that his fish get as much oxygen as possible. The pet shop has four different fish tanks. The dimensions below represent the length and width of each fish tank. Which tank has the greatest surface area at the top? (Lesson 3-7)

(A) 22 in. × 18 in. (B) 24 in. × 16 in.

(C) 26 in. × 14 in. (D) 28 in. × 12 in.

3. The Hyde family's weekly food expenses for four consecutive weeks were $105.52, $98.26, $101.29, and $91.73. What is the mean of their weekly food expenses for those four weeks? (Lesson 5-7)

(A) $98.63 (B) $98.75

(C) $99.20 (D) $99.78

4. Students taste-tested three brands of instant hot cereal and chose their favorite brand. Which of these statements is *not* supported by the data in the table? (Lesson 6-1)

	Hot Cereal Brand		
	X	Y	Z
Girls	12	7	10
Boys	10	15	5

(A) Twice as many girls as boys chose Brand Z.

(B) The total number of students who chose Brand X is equal to the total number who chose Brand Y.

(C) Three times as many boys chose Brand Y as Brand Z.

(D) Half of the students who chose Brand Z were boys.

5. A sloppy-joe recipe for 12 servings calls for 2 pounds of ground beef. How many pounds of ground beef will be needed to make 30 servings? (Lesson 6-3)

(A) 2.5 lb (B) 4.5 lb (C) 5 lb (D) 6 lb

6. Three-fourths of a county's population is registered to vote. Only 50% of the registered voters in the county actually voted in the election. What fractional part of the county's population voted? (Lesson 6-4)

(A) $\frac{1}{2}$ (B) $\frac{3}{8}$ (C) $\frac{2}{3}$ (D) $\frac{3}{4}$

7. Rod has $10 to spend at an arcade. Each bag of popcorn at the arcade costs $3.25, and each video game costs $1.00. Which expression represents the amount of money Rod will have left after he buys one bag of popcorn and plays n video games? (Lesson 7-2)

(A) $10.00 + 3.25 + 1.00n$

(B) $10.00 - 3.25 - 1.00n$

(C) $3.25 - 1.00n - 10.00$

(D) $10.00 - 3.25n - 1.00n$

8. Cheyenne is 5 feet tall. She measures her shadow and a tree's shadow at the same time of day, as shown in the diagram below. How tall is the tree? (Lesson 9-7)

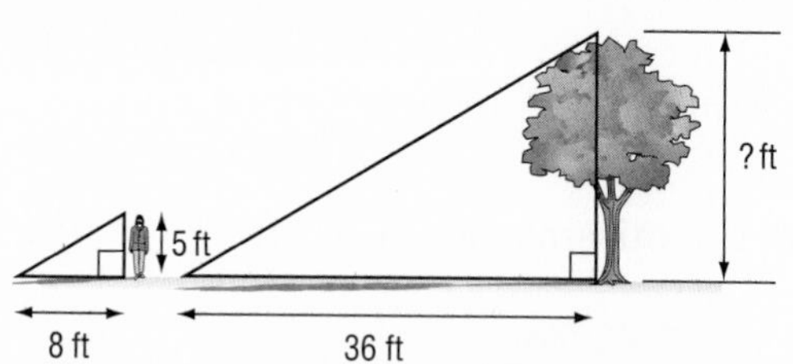

(A) 20 ft (B) 22.5 ft (C) 40 ft (D) 57.6 ft

Test-Taking Tip

Pace yourself. Do not spend too much time on any one question. If you're having difficulty answering a question, mark it in your test booklet and go on to the next question. Make sure that you also skip the question on your answer sheet. At the end of the test, go back and answer the questions that you skipped.

Part 2 Short Response/Grid In

Record your answers on the answer sheet provided by your teacher or on a sheet of paper.

9. Determine the range of the relation shown in the graph. (Lesson 1-6)

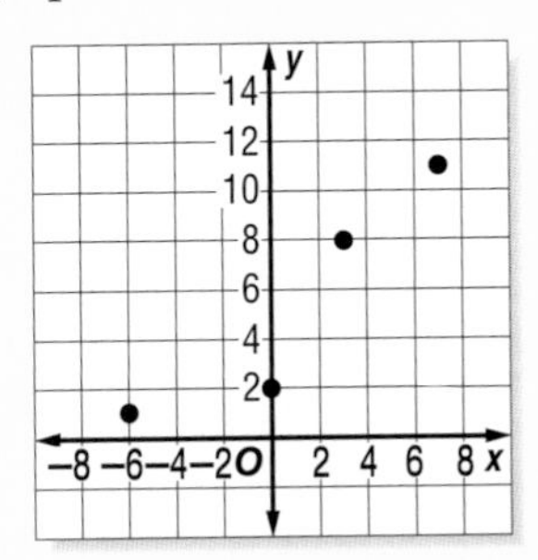

10. Simplify $-9s(4t)$. (Lesson 2-4)

11. Thang is enclosing a rectangular area for his dog. He bought enough wire fencing to enclose 308 square feet of space. If he makes the length of the rectangle 22 feet, what is the width in feet? (Lesson 3-7)

12. What is the value of x^{-2} for $x = 3$? (Lesson 4-7)

13. Which number is greater, 3.45×10^3 or 5.87×10^2? (Lesson 4-8)

14. Write the value of $\frac{5}{6} - \frac{1}{5} - \frac{1}{30}$ in simplest form. (Lesson 5-4)

15. Mr. Vazquez budgeted $300 for his home's January heating bill. The actual bill was $240. What percent of the $300 was left after Mr. Vazquez paid the heating bill? (Lesson 6-4)

16. What is the y-intercept of the graph of $y + 5 = 2x$? (Lesson 8-7)

17. Sarah is making a model house. The pitch of the roof is 35°. What is the measure, in degrees, of $\angle P$, the peak of the roof? (Lesson 10-3)

18. What is the circumference of the circle? Use $\pi = 3.14$ and round to the nearest tenth, if necessary. (Lesson 10-9)

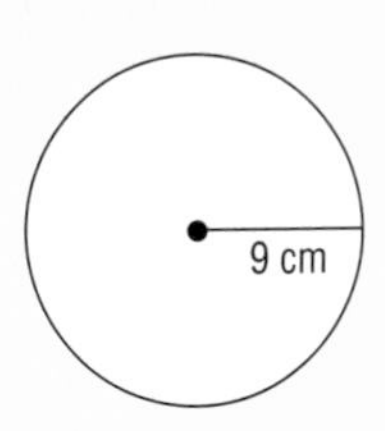

19. A concrete worker is making six cement steps. Each step is 4 inches high, 7 inches deep, and 20 inches wide. What volume of cement, in cubic inches, will be needed to make these steps? (Lesson 11-2)

20. The prisms below are similar. Find the height of the larger prism in centimeters. If necessary, round to the nearest tenth. (Lesson 11-6)

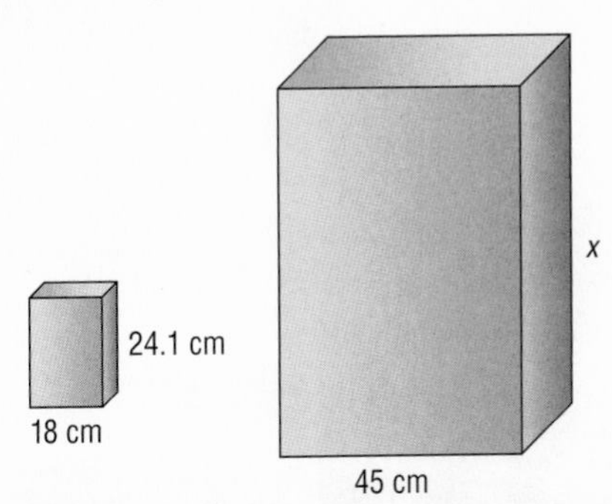

FCAT Practice

Part 3 Open Ended

Record your answers on a sheet of paper. Show your work.

21. A manufacturer ships its product in boxes that are 3 feet × 2 feet × 2 feet. The company needs to store some products in a warehouse space that is 32 feet long by 8 feet wide by 10 feet high. (Lesson 11-2)

a. What is the greatest number of boxes the company can store in this space? (All the boxes must be stored in the same position.)

b. What is the total volume of the stored boxes?

c. What is the volume of the storage space?

d. How much storage space is *not* filled with boxes?

UNIT 5

Extending Algebra to Statistics and Polynomials

In Unit 3, you learned about real-world data that can be represented by linear functions. In this unit, you will learn about real-world data that can be represented by nonlinear functions.

Chapter 12
More Statistics and Probability

Chapter 13
Polynomials and Nonlinear Functions

Will Family Farms Die Like Mom, Pop Stores?

"Once upon a time, in little towns … across the USA, every business was family owned. Mom and Pop ran the grocery store. Also the butcher shop. Drugstore, Movie house. Gas station. Most of their customers were area farm families."

Source: *USA TODAY*, November 1, 2000

In this project, you will be using statistics and functions to analyze farming or ranching in America.

Log on to www.pre-alg.com/webquest. Begin your WebQuest by reading the Task.

Then continue working on your WebQuest as you study Unit 5.

Lesson	12-4	13-5
Page	626	690

USA TODAY Snapshots®

Long days on the farm

More than one in five agriculture workers put in at least 60 hours a week on the job in August. Percentage of workers putting in 40 hours or more:

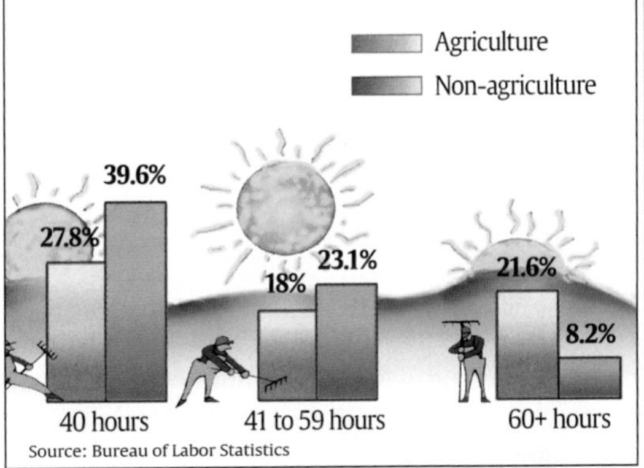

Source: Bureau of Labor Statistics

By Mark Pearson and Web Bryant, USA TODAY

Chapter 12 More Statistics and Probability

What You'll Learn

- **Lessons 12-1, 12-3, and 12-4** Display and interpret data in stem-and-leaf plots, box-and-whisker plots, and histograms.
- **Lesson 12-2** Find measures of variation.
- **Lesson 12-5** Recognize misleading statistics.
- **Lessons 12-6 and 12-7** Count outcomes using tree diagrams, the Fundamental Counting Principle, permutations, or combinations.
- **Lessons 12-8 and 12-9** Find probabilities and odds.

Key Vocabulary

- stem-and-leaf plot (p. 606)
- measures of variation (p. 612)
- box-and-whisker plot (p. 617)
- histogram (p. 623)
- odds (p. 646)

Why It's Important

Statistics is a branch of mathematics that involves the collection, presentation, and analysis of data. In statistics, graphs are usually used to present data. These graphs are important because they allow you to interpret the data easily. *You will display and interpret data about the United States government in Lesson 12-1.*

Getting Started

Prerequisite Skills To be successful in this chapter, you'll need to master these skills and be able to apply them in problem-solving situations. Review these skills before beginning Chapter 12.

For Lessons 12-1, 12-2, and 12-3 — Measures of Central Tendency

Find the mean, median, and mode for each set of data. Round to the nearest tenth, if necessary. *(For review, see Lesson 5-8.)*

1. 10, 15, 23

2. 21, 24, 24, 24, 42, 48

3. 3.2, 5.1, 6.5, 6.5

4. 2.2, 4.3, 5.4, 3.2, 4.8, 5.4, 6.2, 8.1

For Lesson 12-9 — Simple Probability

The spinner at the right is spun. Find each probability.
(For review, see Lesson 6-9.)

5. $P(\text{green})$ **6.** $P(4)$ **7.** $P(\text{even})$

8. $P(\text{not blue})$ **9.** $P(\text{prime})$ **10.** $P(\text{composite})$

For Lesson 12-9 — Compute with Fractions

Find each product, sum, or difference. *(For review, see Lessons 5-3 and 5-7.)*

11. $\frac{1}{6} + \frac{1}{3}$ **12.** $\frac{2}{3} - \frac{4}{9}$ **13.** $\frac{3}{4} \times \frac{1}{6}$ **14.** $\frac{5}{8} \times \frac{3}{4}$

15. $\frac{3}{8} \times \frac{4}{5} \times \frac{5}{9}$ **16.** $\frac{1}{2} \times \frac{5}{6} \times \frac{3}{4}$ **17.** $\frac{5}{12} + \frac{1}{12} - \frac{1}{3}$ **18.** $\frac{3}{8} + \frac{1}{2} - \frac{1}{4}$

Make this Foldable to help you study the topics of statistics and probability. Begin with a piece of notebook paper.

Step 1 Fold

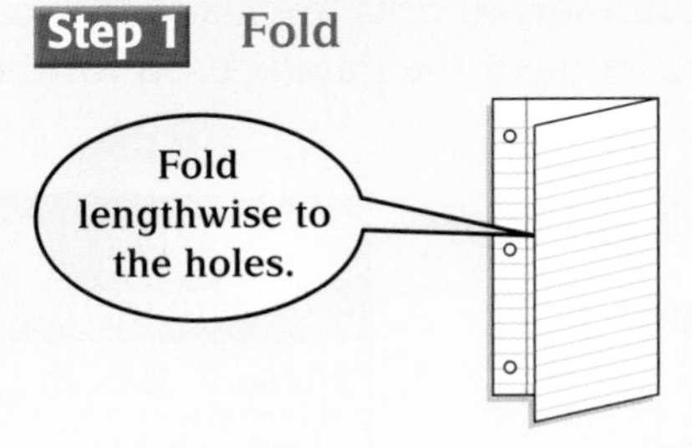

Fold lengthwise to the holes.

Step 2 Cut

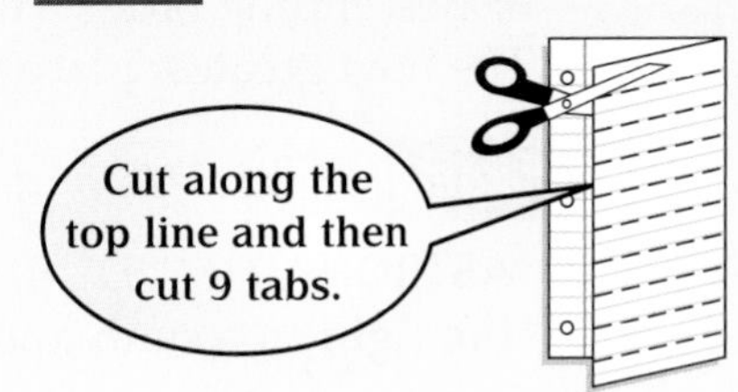

Cut along the top line and then cut 9 tabs.

Step 3 Label

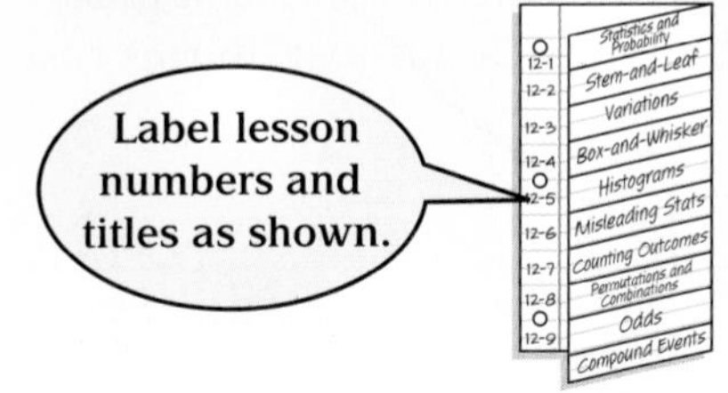

Reading and Writing As you read and study the chapter, you can write notes under the tabs.

12-1 Stem-and-Leaf Plots

Sunshine State Standards
MA.E.1.3.1-1, MA.E.1.3.1-2, MA.E.3.3.1-3

Vocabulary

- stem-and-leaf plot
- stems
- leaves
- back-to-back stem-and-leaf plot

What You'll Learn

- Display data in stem-and-leaf plots.
- Interpret data in stem-and-leaf plots.

How can stem-and-leaf plots help you understand an election?

The members of the Electoral College officially elect the President of the United States. These members are called electors. The number of electors for each state, including the District of Columbia, is shown.

Number of Electors							
AL: 9	DE: 3	IN: 12	MA: 12	NV: 4	CH: 21	TN: 11	WI: 11
AK: 3	DC: 3	IA: 7	MI: 18	NH: 4	OK: 8	TX: 32	WY: 3
AZ: 8	FL: 25	KS: 6	MN: 10	NJ: 15	OR: 7	UT: 5	
AR: 6	GA: 13	KY: 8	MS: 7	NM: 5	PA: 23	VT: 3	
CA: 54	HI: 4	LA: 9	MO: 11	NY: 33	RI: 4	VA: 13	
CO: 8	ID: 4	ME: 4	MT: 3	NC: 14	SC: 8	WA: 11	
CT: 8	IL: 22	MO: 10	NE: 5	ND: 3	SD: 3	WV: 5	

Source: *The World Almanac*

- Write each number on a self-stick note.
- Group the numbers: 0-9, 10-19, 20-29, 30-39, 40-49, 50-59.
- Organize the numbers in each group from least to greatest.

a. Is there an equal number of electors in each group? Explain.

b. Name an advantage of displaying the data in groups.

DISPLAY DATA One way to organize and display data is to use a stem-and-leaf plot. In a **stem-and-leaf plot**, numerical data are listed in ascending or descending order. The greatest place value of the data is used for the **stems**. The next greatest place value forms the **leaves**.

Example 1 Draw a Stem-and-Leaf Plot

ASTRONAUTS Display the data shown at the right in a stem-and-leaf plot.

Oldest U.S. Astronauts	
Astronaut	Age*
Roger K. Crouch	56
Don L. Lind	54
William G. Gregory	54
John H. Glenn	77
John E. Blaha	54
William E. Thornton	56
F. Story Musgrave	61
Karl G. Henize	58
Vance D. Brand	59
Henry W. Hartsfield	54

* At time of his last space shuttle flight

Source: *Top 10 of Everything*, 2001

Step 1 Find the least and the greatest number. Then identify the greatest place value digit in each number. In this case, tens.

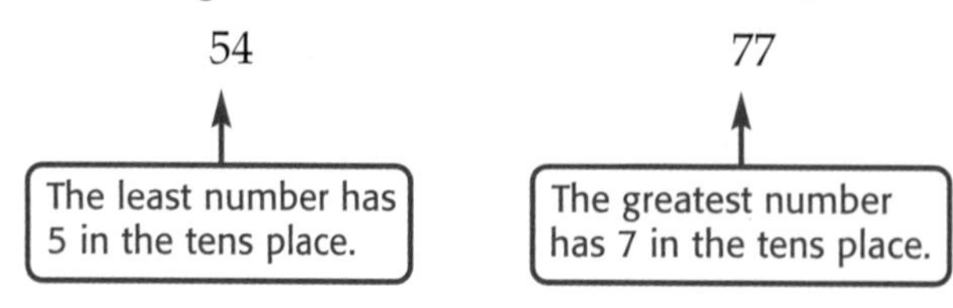
54
77
The least number has 5 in the tens place.
The greatest number has 7 in the tens place.

Step 2 Draw a vertical line and write the stems from 5 to 7 to the left of the line.

Stem	
5	
6	
7	

Step 3 Write the leaves to the right of the line, with the corresponding stem. For example, for 56, write 6 to the right of 5.

Stem	Leaf
5	6 4 4 4 6 8 9 4
6	1
7	7

Step 4 Rearrange the leaves so they are ordered from least to greatest. Then include a key or an explanation.

Stem	Leaf
5	4 4 4 4 6 6 8 9
6	1
7	7

The key tells what the stems and leaves represent. → *5|6 = 56 years*

More About. . .

Presidents

The first President was George Washington. He was 57 years old at the time of his inauguration. He served as President from 1789 to 1797 and earned $25,000 per year.

Source: www.infoplease.com

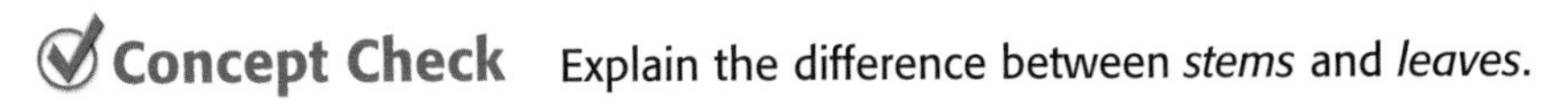

Concept Check Explain the difference between *stems* and *leaves*.

INTERPRET STEM-AND-LEAF PLOTS It is often easier to interpret data when they are displayed in a stem-and-leaf plot instead of a table. You can "see" how the data are distributed.

Example 2 Interpret Data

PRESIDENTS The stem-and-leaf plot lists the ages of the U.S. Presidents at the time of their inauguration. **Source:** *The World Almanac*

Stem	Leaf
4	2 3 6 6 7 8 9 9
5	0 0 1 1 1 1 2 2 4 4 4 4 4 5 5 5 5 6 6 6 7 7 7 7 8
6	0 1 1 1 2 4 4 6 8 9

5|0 = 50 years

a. In which interval do most of the ages occur?

Most of the data occurs in the 50–59 interval.

b. What is the age difference between the youngest and oldest President?

The youngest age is 42. The oldest age is 69. The difference between these ages is 69 – 42 or 27.

c. What is the median age of a President at inauguration?

The median, or the number in the middle, is 55.

Study Tip

Look Back

To review **mean**, **median**, and **mode**, see Lesson 5-8.

Two sets of data can be compared using a **back-to-back stem-and-leaf plot**. The back-to-back stem-and-leaf plot below shows the scores of two basketball teams for the games in one season.

Falcons		Cardinals
7 6 5 5 4 2 2 2	6	2 4
8 8 8 5 4	7	0 2 2 5 7 9
1 0 0	8	1 3 4 6 6 8 9 9

1|8 = 81 points *8|6 = 86 points*

The leaves for one set of data are on one side of the stem. The leaves for the other set of data are on the other side of the stem.

Example 3 Compare Two Sets of Data

WEATHER The average monthly temperatures for Helena, Montana, and Seattle, Washington, are shown.
Source: *The World Almanac*

Seattle, WA		Helena, MT
	2	0 1 6
	3	2 4
7 6 4 2 1	4	3 5
6 4 0	5	3 5
6 5 1 1	6	2 7 9

1|6 = 61° *4|5 = 45°*

a. Which city has lower monthly temperatures? Explain.

Helena; it experiences temperatures in the 20's and 30's.

b. Which city has more varied temperatures? Explain.

The data for Helena are spread out from the 20's to the 60's. The data for Seattle are clustered from the 40's to the 60's. So, Helena has the most varied temperatures.

Check for Understanding

Concept Check

1. **OPEN ENDED** Write a statement describing how the data in Example 2 on page 607 are distributed.
2. **Identify** the stems for the data set {48, 52, 46, 62, 51, 39, 41, 57, 68}.

Guided Practice

Display each set of data in a stem-and-leaf plot.

3.

Average Life Span					
Animal	Years	Animal	Years	Animal	Years
Asian Elephant	40	African Elephant	35	Lion	15
Horse	20	Red Fox	7	Chipmunk	6
Moose	12	Cow	15	Hippopotamus	41

Source: *The World Almanac*

4. **Summer Paralympic Games Participating Countries**

Year	'60	'64	'68	'72	'76	'80	'84	'88	'92	'96	'00
Countries	23	22	29	44	42	42	42	61	82	103	128

Source: www.paralympic.org

Applications

SCHOOL For Exercises 5–7, use the test score data shown at the right.

5. Find the lowest and highest scores.
6. What is the median score?
7. Write a statement that describes the data.

Pre-Algebra Test Scores

Stem	Leaf
5	0 9
6	4 5 7 8
7	0 4 4 5 5 6 7 8 8
8	2 3 3 5 7 8
9	0 1 5 5 9

5|9 = 59%

FOOD For Exercises 8 and 9, use the food data shown in the back-to-back stem-and-leaf plot.

8. What is the greatest number of fat grams in each sandwich?
9. In general, which type of sandwich has a lower amount of fat? Explain.

Fat (g) of Various Burgers and Chicken Sandwiches

Chicken		Burgers
8	0	
9 8 5 5 3 3	1	0 5 9
0	2	0 6
	3	0 3 6

8|0 = 8 g *2|6 = 26 g*

Practice and Apply

Homework Help

For Exercises	See Examples
10–15, 18	1
19–21	2
22–25	3

Extra Practice
See page 752.

Display each set of data in a stem-and-leaf plot.

10.

State Representatives Northeast Region	
State	**Number**
Connecticut	6
Maine	2
Massachusetts	10
New Hampshire	2
Rhode Island	2
Vermont	1
New Jersey	13
New York	31
Pennsylvania	21

Source: *The World Almanac*

11.

Detroit Tigers Statistics, 2001	
Player	**Runs**
D. Cruz	39
Higginson	84
Encarnacion	52
Inge	13
Magee	26
T. Clark	67
Simon	28
Halter	53
Easley	77
Palmer	34
Macias	62
Fick	62
Cedeno	79

Source: www.tigers.mlb.com

12. **Percent of Young Adults (18–24) in U.S. Living at Home**

Year	1960	1970	1980	1985	1990	1991	1992
Percent	52	54	54	60	58	60	60
Year	1993	1994	1995	1996	1997	1998	
Percent	59	60	58	59	60	59	

Source: Bureau of the Census

13. **Approximate Number of Students per Computer in U.S. Public Schools**

Year	'85–'86	'86–'87	'87–'88	'88–'89	'89–'90	'90–'91	'91–'92
Number	50	37	32	25	22	20	18
Year	'92–'93	'93–'94	'94–'95	'95–'96	'96–'97	'97–'98	'98–'99
Number	16	14	11	10	8	6	6

Source: *The World Almanac*

14. **Olympic Men's 400-m Hurdles Time(s), 1900–2000**

57.6	53.0	55.0	54.0	47.5	48.7
52.6	53.4	51.7	52.4	47.5	51.1
50.8	50.1	49.3	49.6	48.1	47.8
47.6	47.8	47.2	46.8		

Source: *The ESPN Sports Almanac*

15. **Heights (ft) of Tallest Buildings in Miami, Florida**

789	400	625	400	487	405
510	480	425	484	450	456
520	764				

Source: *The World Almanac*

Tell whether each statement is *sometimes*, *always*, or *never* true.

16. A back-to-back stem-and-leaf plot has two sets of data.

17. A basic stem-and-leaf plot has two keys.

HEALTH **For Exercises 18–21, use the graphic shown.**

18. Display the data in a stem-and-leaf plot.
19. What is the greatest percent of people who exercise daily?
20. In how many of the cities do fewer than 30% of the people exercise daily?
21. Write a sentence that describes the data.

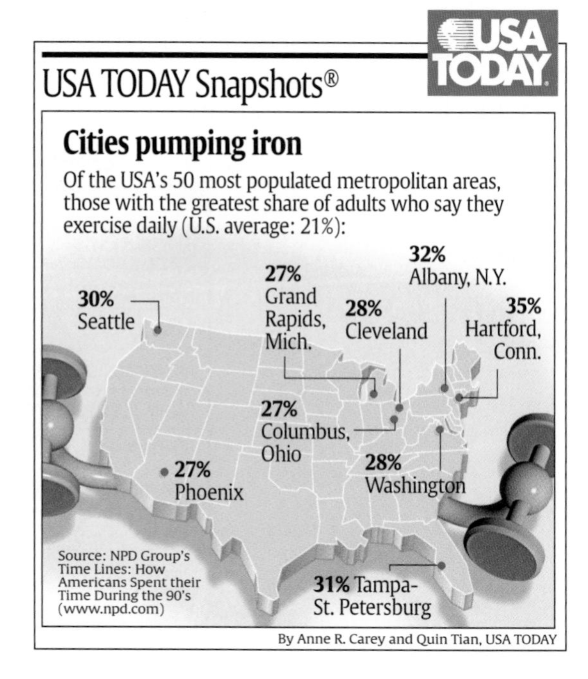

BASKETBALL **For Exercises 22–25, use the information shown in the back-to-back stem-and-leaf plot.** **Source:** USA TODAY

NCAA Woman's Basketball Statistics
Overall Games Won, 2000–2001

Big Ten Conference		Big East Conference
8 4	0	5 8 9 9
9 8 8 7 7 6 4 0	1	2 2 3 5 6 9 9
4	2	0 4 5
8∣1 = 18 wins		*1∣5 = 15 wins*

22. What is the greatest number of games won by a Big Ten Conference team?
23. What is the least number of games won by a Big East Conference team?
24. How many teams are in the Big East Conference?
25. Compare the average number of games won by each conference.

More About...

Basketball

The Louisiana Tech women's basketball team has the best-winning percentage in Division I. Over a 27-year period, the team has 768 wins and 128 losses.

Source: www.infoplease.com

RESEARCH **For Exercises 26 and 27, use the Internet, a newspaper, or another reference source to gather data about a topic that interests you.**

26. Make a stem-and-leaf plot of the data.
27. Write a sentence that describes the data.

28. WRITING IN MATH Answer the question that was posed at the beginning of the lesson.

How can stem-and-leaf plots help you understand an election?

Include the following in your answer:

- a stem-and-leaf plot that displays the number of electors for each state and the District of Columbia,
- a statement describing how the data in the stem-and-leaf plot are distributed, and
- an explanation telling how a presidential candidate might use the display.

29. **CRITICAL THINKING** Suppose you have a table and a stem-and-leaf plot that display the same data.

a. For which display is it easier to find the median? Explain.

b. For which display is it easier to find the mean? Explain.

c. For which display is it easier to find the mode? Explain.

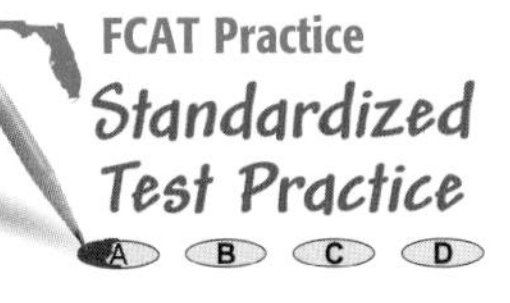

30. What are the stems for the data {13, 34, 37, 25, 25, 35, 52, 28}?

(A) {2, 3, 4, 5, 7, 8}　　(B) {1, 2, 3, 4, 5}

(C) {1, 2, 3, 5, 6}　　(D) {0, 1, 2, 3, 5}

31. The back-to-back stem-and-leaf plot shows the amount of protein in certain foods. Which of the following is a true statement?

Amount of Protein (g)

Dairy Products		Legumes, Nuts, Seeds
9 8 8 7 7 5 2 2	0	5 6 9
0	1	4 5 8
6	2	
	3	9

6 | 2 = 26 grams　　*3 | 9 = 39 grams*

Source: *The World Almanac*

(A) The median amount of protein in dairy products is 9 grams.

(B) The difference between the greatest and least amount of protein in dairy products is 28 grams.

(C) The average amount of protein in legumes, nuts, and seeds is more than the average amount in dairy products.

(D) The greatest amount of protein in legumes, nuts, and seeds is 93 grams.

Maintain Your Skills

Mixed Review

32. A triangle has sides that measure 12.38 inches, 7.5 inches, and 6.185 inches. Find the perimeter of the triangle using the correct number of significant digits. *(Lesson 11-7)*

Determine whether each pair of solids is similar. *(Lesson 11-6)*

33.

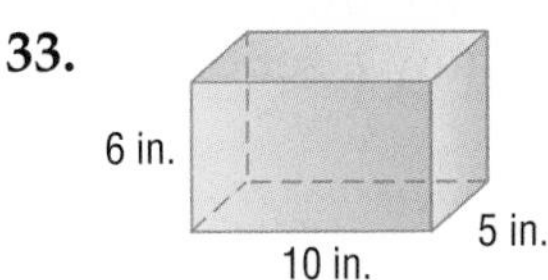

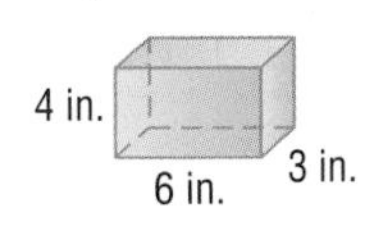

34.

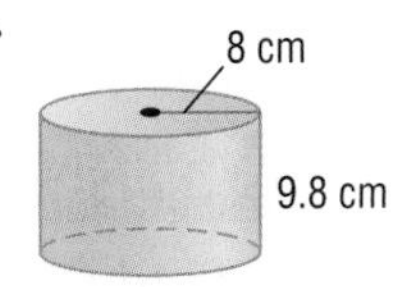

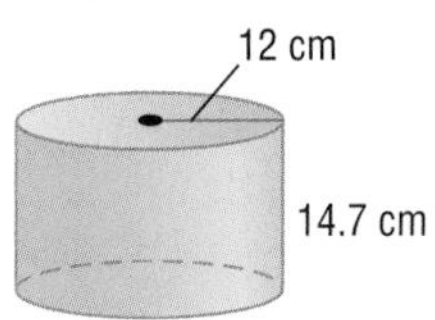

35. Find the circumference and area of a circle with a radius of 10 feet. Round to the nearest tenth. *(Lesson 10-7)*

Express each decimal or fraction as a percent. Round to the nearest tenth percent, if necessary. *(Lesson 6-4)*

36. 0.36　　37. 2.47　　38. 0.019　　39. 0.0065

40. $\frac{6}{25}$　　41. $\frac{4}{7}$　　42. $\frac{15}{8}$　　43. $\frac{24}{1500}$

Getting Ready for the Next Lesson

PREREQUISITE SKILL Find the median for each set of data. If necessary, round to the nearest tenth. *(To review **median**, see Lesson 5-8.)*

44. 23, 45, 21, 35, 28　　45. 18, 9, 2, 4, 6, 15, 13, 6, 1

46. 78, 54, 50, 64, 39, 45　　47. 0.4, 1.3, 0.8, 2.6, 0.3, 1.8, 0.2, 2.1

12-2 Measures of Variation

What You'll Learn

- Find measures of variation.
- Use measures of variation to interpret and compare data.

Sunshine State Standards
MA.E.1.3.2-2, MA.E.1.3.2-3, MA.E.3.3.1-3

Vocabulary

- measures of variation
- range
- quartiles
- lower quartile
- upper quartile
- interquartile range

Why are measures of variation important in interpreting data?

The race that attracts the largest audience in auto racing is the Daytona 500. The average speed of each winning car from 1990 to 2001 is shown.

Car Driver	Speed (mph)	Car Driver	Speed (mph)
Derrike Cope	166	Dale Jarrett	154
Ernie Irvan	148	Jeff Gordon	148
Davey Allison	160	Dale Earnhardt	173
Dale Jarrett	155	Jeff Gordon	162
Sterling Marlin	157	Dale Jarrett	156
Sterling Marlin	142	Michael Waltrip	162

Source: *The World Almanac*

a. What is the fastest speed?
b. What is the slowest speed?
c. Find the difference between these two speeds.
d. Write a sentence comparing the fastest winning average speed and the slowest winning average speed.

MEASURES OF VARIATION In statistics, **measures of variation** are used to describe the distribution of the data. One measure of variation is the range. The **range** of a set of data is the difference between the greatest and the least values of the set. It describes how a set of data varies.

Example 1 Range

Find the range of each set of data.

a. {5, 11, 16, 8, 4, 7, 15, 6}

The greatest value is 16, and the least value is 4.
So, the range is $16 - 4$ or 12.

b.

Stem	Leaf
5	4 4 4 4 6 6 8 9
6	1
7	7

6|1 = 61

The greatest value is 77, and the least value is 54.
So, the range is $77 - 54$ or 23.

In a set of data, the **quartiles** are the values that divide the data into four equal parts. Recall that the median of a set of data separates the set in half.

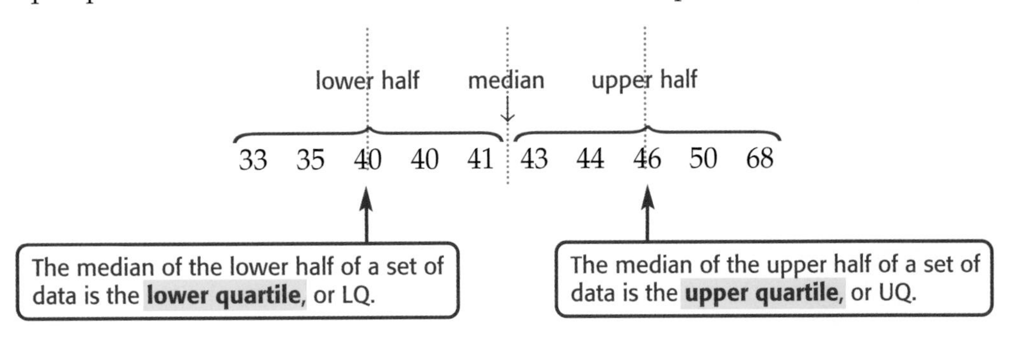

lower half
median
upper half
33 35 40 40 41 43 44 46 50 68
The median of the lower half of a set of data is the **lower quartile**, or LQ.
The median of the upper half of a set of data is the **upper quartile**, or UQ.

Concept Check Into how many parts do the quartiles divide a set of data?

The upper and lower quartiles can be used to find another measure of variation called the **interquartile range**.

Key Concept — *Interquartile Range*

- **Words** The interquartile range is the range of the middle half of a set of data. It is the difference between the upper quartile and the lower quartile.
- **Symbols** Interquartile range = UQ − LQ

Study Tip

Statistics

A small interquartile range means that the data in the middle of the set are close in value. A large interquartile range means that the data in the middle are spread out, or vary.

Example 2 *Interquartile Range*

Find the interquartile range for each set of data.

a. {27, 37, 21, 54, 47, 35}

Step 1 List the data from least to greatest. Then find the median.

21 27 35 37 47 54

median = $\frac{35 + 37}{2}$ or 36

Step 2 Find the upper and lower quartiles.

lower half: 21 27 35 | upper half: 37 47 54

LQ = 27, median between 35 and 37, UQ = 47

The interquartile range is 47 − 27 or 20.

b. {7, 12, 3, 2, 11, 9, 6, 4, 8}

Step 1 List the data from least to greatest. Then find the median.

2 3 4 6 7 8 9 11 12

median = 7

Step 2 Find the upper and lower quartiles.

lower half: 2 3 4 6 | 7 (median) | upper half: 8 9 11 12

The interquartile range is 10 − 3.5 or 6.5.

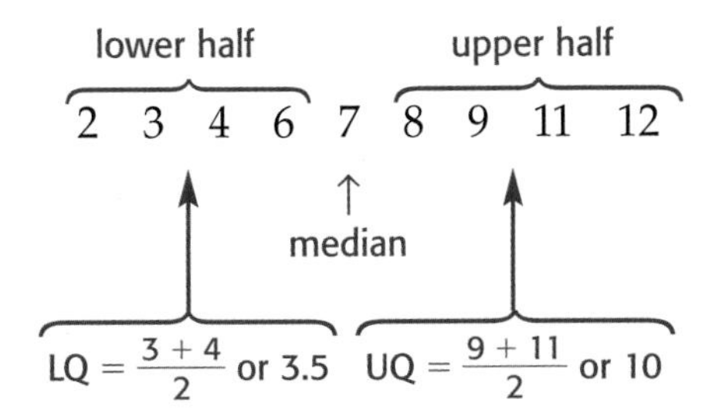

LQ = $\frac{3+4}{2}$ or 3.5
UQ = $\frac{9+11}{2}$ or 10

More About. . .

Traffic Laws

In 1974, the national speed limit was 55 mph. Today, the speed limits for the 50 states range from 55 to 75 mph.

Source: www.infoplease.com

USE MEASURES OF VARIATION You can use measures of variation to interpret and compare data.

Example 3 Interpret and Compare Data

TRAFFIC LAWS The maximum allowable speed limits for certain western and eastern states are listed in the stem-and-leaf plot.

Western States		Eastern States
5	5	
5 5	6	5 5 5 5 5 5 5 5 5
5 5 5 5 5 5 5 5 5 0 0	7	0 0 0
0 | 7 = 70 mph		*6 | 5 = 65 mph*

Source: www.infoplease.com

a. What is the median speed limit for each region?

The median speed limit for the western states is 75.

The median speed limit for the eastern states is 65.

b. Compare the western states' range with the eastern states' range.

The range for the east is 70 − 65 or 5 mph, and the range for the west is 75 − 55 or 20 mph. So, the speed limits in the west vary more.

Check for Understanding

Concept Check

1. **Explain** how the range of a set of data differs from the interquartile range of a set of data.
2. **Define** *upper quartile* and *lower quartile*.
3. **OPEN ENDED** Write a list of at least 12 numbers that has an interquartile range of ten.

Guided Practice

Find the range and interquartile range for each set of data.

4. {26, 48, 12, 32, 41, 35}

5.

Stem	Leaf
7	2 3 6 6 9
8	0 0 1
9	9

Application

SCIENCE For Exercises 6–8, use the information in the table.

6. Which planet's day length divides the data in half?
7. What is the median length of day for the planets?
8. Write a sentence describing how the lengths of days vary.

Planet	Length of Day* (Earth hours)
Mercury	1416
Venus	5832
Earth	24
Mars	25
Jupiter	10
Saturn	11
Uranus	17
Neptune	16
Pluto	154

*The lengths are approximate.

Source: *The World Almanac*

Practice and Apply

Homework Help

For Exercises	See Examples
9–16	1, 2
17–20	3

Extra Practice
See page 753.

Find the range and interquartile range for each set of data.

9. {65, 64, 73, 34, 15, 43, 92}

10. {9, 13, 25, 9, 1, 5, 6, 8}

11. {68°, 74°, 65°, 55°, 75°, 82°, 32°, 69°, 70°, 77°}

12. {$25, $21, $55, $43, $10, $89, $39, $91, $44, $76, $58}

13.

Stem	Leaf
0	1 2 2 5
1	3 4 7 8 9 9 9
2	6 6

2|6 = 26

14.

Stem	Leaf
4	0
5	0 1 1 5 7 7 7 8
6	7 7 9

5|7 = 57

15. Find the interquartile range for {213, 226, 204, 215, 210, 362, 119}.

16. Determine the range of the middle half of the data set {30.2, 29.3, 35.3, 30.1, 28.5, 31.6, 27.5, 21.2}.

WEATHER **For Exercises 17 and 18, use the data in the table.**

Average Temperature (°F)

City	Feb.	July	City	Feb.	July
Asheville, NC	39	73	Louisville, KY	36	77
Atlanta, GA	45	79	Oklahoma City, OK	41	82
Birmingham, AL	46	80	Portland, OR	44	68
Fresno, CA	51	82	Syracuse, NY	24	70
Houston, TX	54	83	Tampa, FL	62	82
Indianapolis, IN	30	75	Washington, DC	34	76
Little Rock, AR	44	82			

Source: *The World Almanac*

17. Find the interquartile range for each month's set of data.

18. Which month has more consistent temperatures? Justify your answer.

BASEBALL **For Exercises 19 and 20, use the data in the stem-and-leaf plot.**

Home Runs Hit by League Leaders, 1960–2001

National League		American League
	2	2
9 9 8 8 8 7 7 7 6 6 6 5 1	3	2 2 2 2 3 6 6 7 9 9 9
4 4 4 3 1 0 0 0 0 0	4	0 0 0 0 1 2 3 3 4 4 4 4
9 9 9 8 8 8 7 7 7 6 6 5 5	4	5 5 6 6 7 8 8 9 9 9 9
2 2 0	5	0 1 2 2 6 6
5	6	1
3 0	7	

7|4 = 47 home runs *5|6 = 56 home runs*

Source: *The World Almanac*

19. Find the range, median, upper quartile, lower quartile, and the interquartile range for each set of data.

20. Write a few sentences that compare the data.

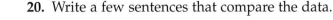

21. **CRITICAL THINKING** Write a set of data that satisfies each condition.

a. 12 pieces of data, a median of 60, an interquartile range of 20

b. 12 pieces of data, a median of 60, an interquartile range of 50

c. Compare the measures of variation for each set of data in parts **a** and **b**. What conclusions can be drawn about the sets of data?

22. **WRITING IN MATH** Answer the question that was posed at the beginning of the lesson.

Why are measures of variation important in interpreting data?

Include the following in your answer:

- the median, range, and interquartile range for the set of data, and
- an explanation telling what the median, range, and interquartile range convey about the speeds of the winning cars.

FCAT Practice

Standardized Test Practice

23. Find the median for the set {43, 49, 91, 42, 94, 73, 93, 67, 55, 54, 78, 82}.

(A) 68 (B) 52 (C) 70 (D) 83

24. Which sentence best describes the data shown in the table?

Height (ft) of Mountains in Alaska and Colorado					
Colorado			Alaska		
14,238	14,433	14,309	14,163	14,410	20,320
14,083	14,264	14,197	16,550	14,530	14,831
14,269	14,196	14,150	17,400	16,237	14,070
14,165	14,420	14,246	15,885	14,573	16,390
14,286	14,265	14,361	15,638	14,730	16,286

Source: *The World Almanac*

(A) The heights of the mountains in Alaska vary by 6200 feet.

(B) The heights of the mountains in Colorado are clustered around the median height.

(C) The median height of a mountain in Alaska is 16,000 feet.

(D) The heights of the mountains in Colorado tend to be less consistent than the heights of the mountains in Alaska.

Maintain Your Skills

Mixed Review

25. Display the data set {\$12, \$15, \$18, \$21, \$14, \$37, \$27, \$9} in a stem-and-leaf plot. *(Lesson 12-1)*

26. Calculate 27.08 mm + 6.5 mm. Round to the correct number of significant digits. *(Lesson 11-7)*

Find the volume of each cone described. Round to the nearest tenth. *(Lesson 11-3)*

27. radius 7 cm, height 9 cm

28. diameter 8.4 yd, height 6.5 yd

29. The circumference of a circle is 9.82 feet. Find the radius of the circle to the nearest tenth. *(Lesson 10-7)*

Getting Ready for the Next Lesson

PREREQUISITE SKILL Order each set of decimals from least to greatest.

*(To review **ordering decimals**, see page 710.)*

30. 5.6, 5.3, 4.8, 4.3, 5.0, 4.9

31. 0.3, 1.4, 0.6, 1.5, 0.2, 0.8, 1.2

32. 45.2, 50.7, 46.0, 45.4, 40.6

33. 10.9, 11.4, 9.8, 10.5, 11.2, 9.9

12-3 Box-and-Whisker Plots

Sunshine State Standards
MA.E.1.3.1-1, MA.E.1.3.1-2, MA.E.3.3.1-3

Vocabulary

- box-and-whisker plot

What You'll Learn

- Display data in a box-and-whisker plot.
- Interpret data in a box-and-whisker plot.

How can box-and-whisker plots help you interpret data?

The table shows the average monthly temperatures for two cities.

Average Monthly Temperatures (°F)

	J	F	M	A	M	J	J	A	S	O	N	D
Tampa, FL	60	62	67	71	77	81	82	82	81	75	68	62
Caribou, ME	9	12	25	38	51	61	66	63	54	43	31	15

a. Find the low, high, and the median temperature, and the upper and lower quartile for each city.

b. Draw a number line extending from 0 to 85. Label every 5 units.

c. About one-half inch above the number line, plot the data found in part **a** for Tampa. About three-fourths inch above the number line, plot the data for Caribou.

d. Write a few sentences comparing the average monthly temperatures.

DISPLAY DATA A **box-and-whisker plot** divides a set of data into four parts using the median and quartiles. A *box* is drawn around the quartile values, and *whiskers* extend from each quartile to the extreme data points.

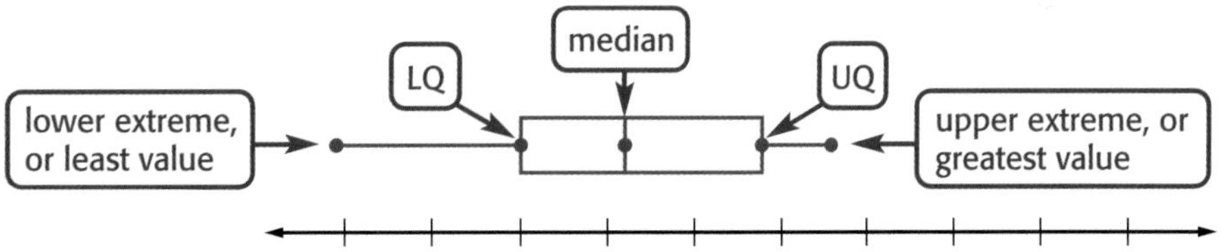

median
LQ
UQ
lower extreme, or least value
upper extreme, or greatest value

Study Tip

Common Misconception
You may think that the median always divides the box in half. However, the median may not divide the box in half because the data may be clustered toward one quartile.

Example 1 Draw a Box-and-Whisker Plot

GEOGRAPHY The amount of coastline for states along the Atlantic Coast is shown. Display the data in a box-and-whisker plot.

Atlantic Coast Coastline

State	Amount (mi)	State	Amount (mi)
Delaware	28	New Jersey	130
Florida	580	New York	127
Georgia	100	North Carolina	301
Maine	228	Rhode Island	40
Maryland	31	South Carolina	187
Massachusetts	192	Virginia	112
New Hampshire	13		

Source: www.infoplease.com

(continued on the next page)

Step 1 Find the least and greatest number. Then draw a number line that covers the range of the data.

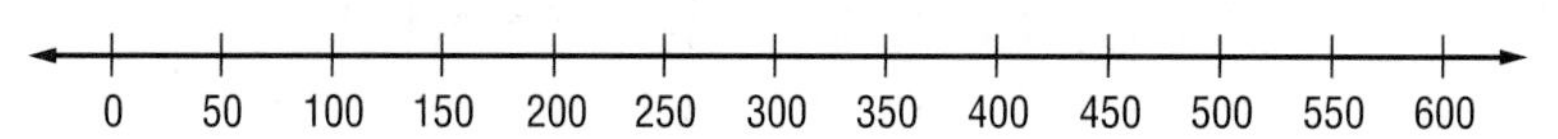

Step 2 Find the median, the extremes, and the upper and lower quartiles. Mark these points above the number line.

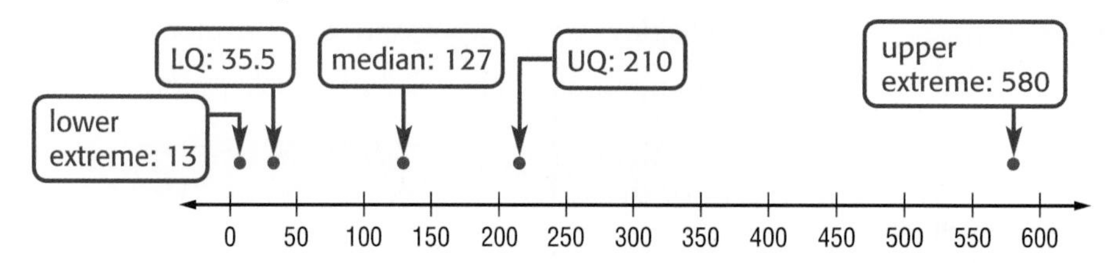

Step 3 Draw a box and the whiskers.

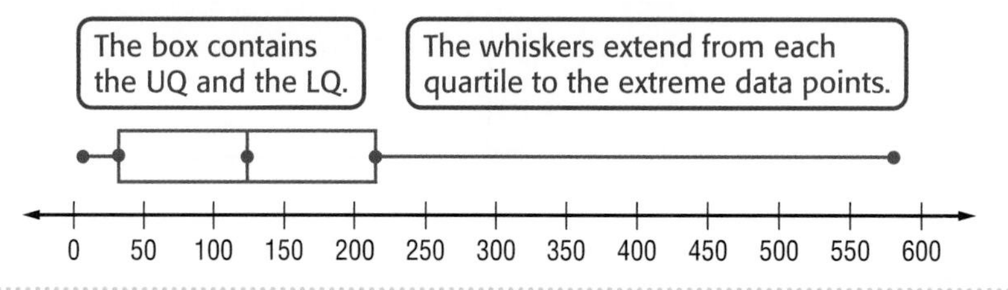

Concept Check What are the extreme values of a set of data?

Study Tip

Box-and-Whisker Plots

If the length of the whisker or box is short, the values of the data in that part are concentrated. If the length of the whisker or box is long, the values of the data in that part are spread out.

INTERPRET BOX-AND-WHISKER PLOTS Box-and-whisker plots separate data into four parts. Even though the parts may differ in length, each part contains 25% of the data.

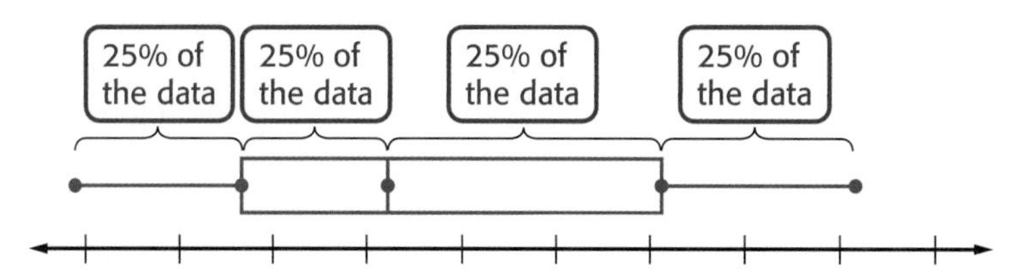

Data displayed in a box-and-whisker plot can be easily interpreted.

Example 2 Interpret Data

EDUCATION **Refer to the information shown below.**

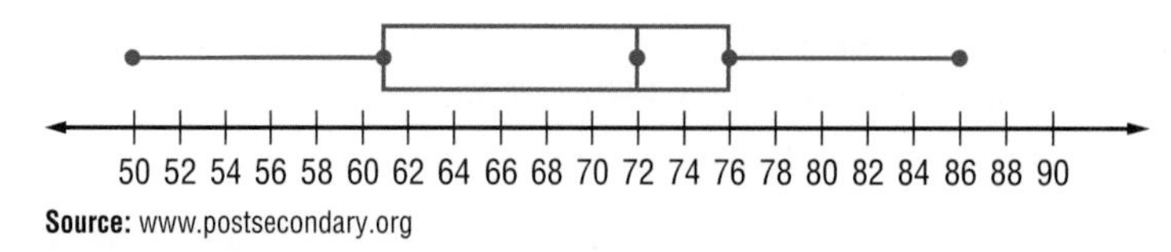

Source: www.postsecondary.org

a. What is the smallest percent of students graduating in any state?

The smallest percent of students graduating in any state is 50%.

b. Half of the states have a graduation rate under what percent?

Half of the states have graduation rates under 72%.

c. What does the length of the box-and-whisker plot tell about the data?

The length of the box-and-whisker plot is long. This tells us that the values of the data are spread out.

Double box-and-whisker plots can be used to compare two sets of data. Notice that one number line is used to display both plots.

Example 3 Compare Two Sets of Data

ANIMALS The weight, in pounds, for Asiatic black bears and pandas is displayed below. How do the weights of Asiatic black bears compare to pandas?

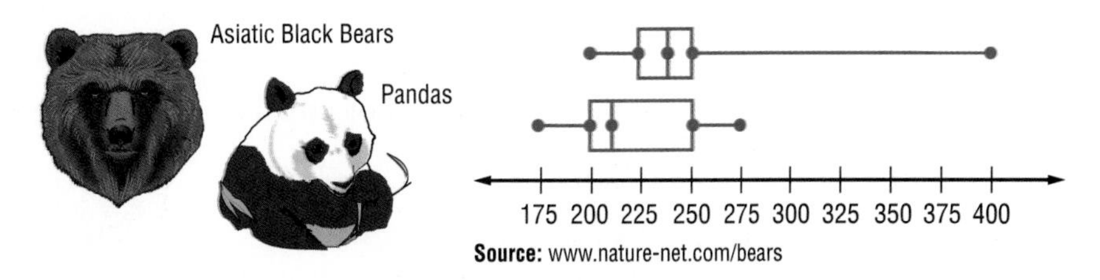

Most Asiatic black bears weigh between 225 and 250 pounds. However, some weigh as much as 400 pounds. Most pandas weigh between 200 and 250 pounds. However, some weigh up to 275 pounds. Thus, the weights of the Asiatic black bears vary more than pandas.

Check for Understanding

Concept Check

1. **Tell** which points the two whiskers of a box plot connect.
2. **Explain** how a box-and-whisker plot separates a set of data.
3. **OPEN ENDED** Write a set of data that, when displayed in a box-and-whisker plot, will result in a long box and short whiskers.

Guided Practice

Draw a box-and-whisker plot for each set of data.

4. 25, 30, 27, 35, 19, 23, 25, 22, 40, 34, 20
5. $15, $22, $29, $30, $32, $50, $26, $22, $36, $31

Applications

OLYMPICS For Exercises 6 and 7, use the data shown in the table.

Summer Olympic Games 1924–2000
Winning Times for Men's Marathon

Year	1924	1928	1932	1936	1948	1952	1956	1960	1964
Time (min)	161	153	152	149	155	143	145	135	132
Year	1968	1972	1976	1980	1984	1988	1992	1996	2000
Time (min)	140	132	130	131	129	131	133	133	130

Source: *The ESPN Sports Almanac*

6. Make a box-and-whisker plot for the data.
7. Write a sentence describing what the length of the box-and-whisker plot tells about the winning times for the men's marathon.

TRAVEL **For Exercises 8 and 9, use the box-and-whisker plots shown.**

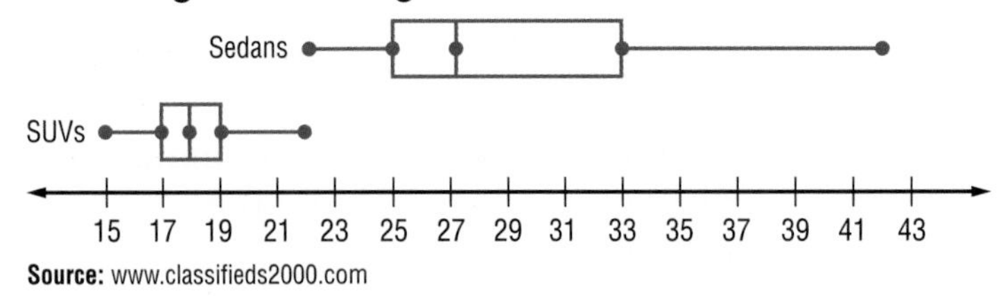

8. Which types of vehicles tend to be less fuel-efficient?

9. Compare the most fuel-efficient SUV to the least fuel-efficient sedan.

Practice and Apply

Homework Help

For Exercises	See Examples
10–15	1
16–18	2
19	3

Extra Practice

See page 753.

Draw a box-and-whisker plot for each set of data.

10. 65, 92, 74, 61, 55, 35, 88, 99, 97, 100, 96

11. 60, 60, 120, 80, 68, 90, 100, 69, 104, 99, 130

12. 80, 72, 42, 40, 63, 51, 55, 78, 81, 73, 77, 65, 67, 68, 59

13. $95, $105, $85, $122, $165, $55, $100, $158, $174, $162

14.

Magnitudes of Recent Major Earthquakes		
6.8	6.2	6.2
5.9	6.1	6.8
6.9	6.5	6.7
6.1	6.9	7.4
7.5	6.3	5.9
6.1	5.8	7.6
7.3	7.8	7.9

Source: *The World Almanac*

15.

Average Points Scored per Game for NBA Scoring Leaders		
33.1	37.1	29.8
30.7	35.0	29.3
32.3	32.5	30.4
28.4	33.6	29.6
30.6	31.5	28.7
32.9	30.1	26.8
30.3	32.6	29.7

Source: *The World Almanac*

More About. . .

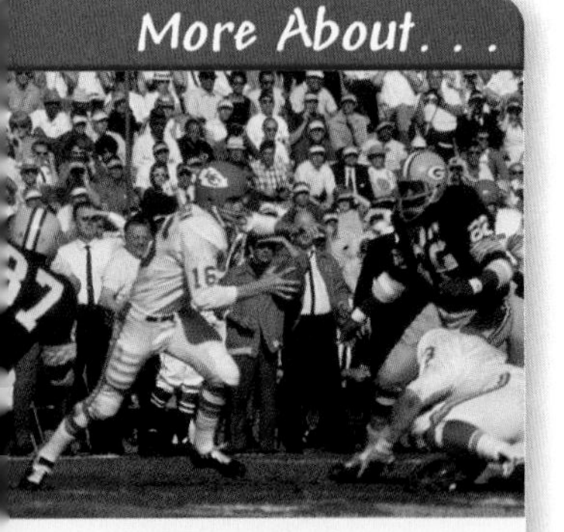

Sports

The first Super Bowl was held in 1967 in the Los Angeles Coliseum. The Green Bay Packers played the Kansas City Chiefs and won 35-10.

Source: *The World Almanac*

SCHOOL **For Exercises 16–18, use the box-and-whisker plot shown.**

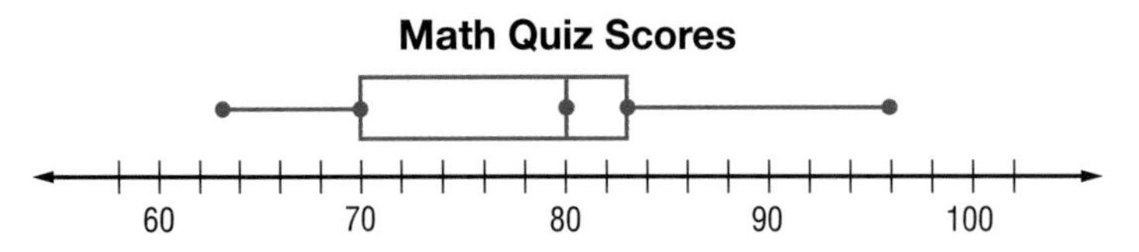

16. What was the highest quiz score?

17. What percent of the students scored between 80 and 96?

18. Based on the plot, how did the students' scores vary?

19. **SPORTS** The number of games won by the teams in each conference of the National Football League is displayed below. Write a few sentences that compare the data.

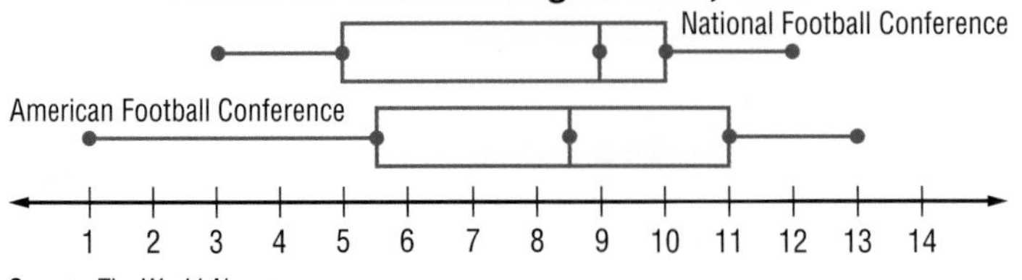

20. **CRITICAL THINKING** Write a set of data that contains 12 values for which the box-and-whisker plot has no whiskers.

21. **WRITING IN MATH** Answer the question that was posed at the beginning of the lesson.

How can box-and-whisker plots help you interpret data?

Include the following in your answer:

- box-and-whisker plots that display the temperature data for each city,
- a description of the temperatures in Tampa and Caribou, and
- an advantage of displaying data in a box-and-whisker plot instead of in a table.

For Exercises 22 and 23, use the box-and-whisker plot shown.

Highest Recorded Wind Speeds (mph) in the U.S.

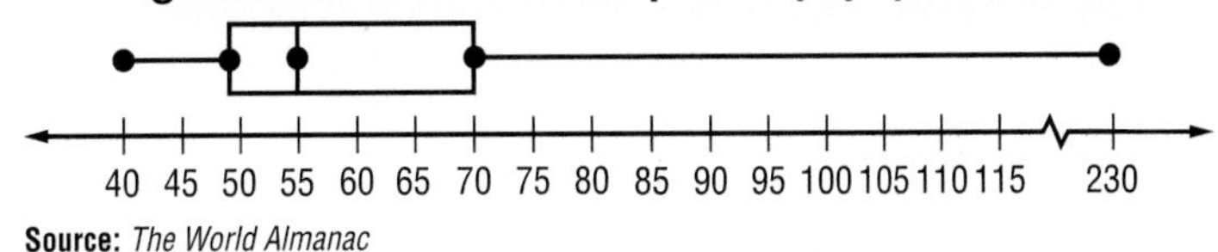

Source: *The World Almanac*

22. What is the least highest recorded wind speed?

Ⓐ 40 mph Ⓑ 55 mph Ⓒ 70 mph Ⓓ 230 mph

23. What percent of the wind speeds range from 40 to 70 mph?

Ⓐ 25% Ⓑ 50% Ⓒ 75% Ⓓ 100%

Extending the Lesson

Data that are more than 1.5 times the interquartile range from the quartiles are called **outliers**. Consider the data set shown.

15 22 [22] 26 27 29 30 31 [32] 36 50

The interquartile range is $32 - 22$ or 10. The outliers are the values more than 1.5(10) or 15 from the quartiles.

$$22 - 15 = 7 \qquad 32 + 15 = 47$$

The limits for the outliers are 7 and 47. So, there is one outlier, 50.

Determine whether any outliers exist for each data set.

24. 67, 75, 89, 72, 56, 65, 70

25. 27, 30, 36, 35, 37, 46, 31, 4, 29, 38, 30

Maintain Your Skills

Mixed Review

For Exercises 26 and 27, use the set of data {2.4, 2.1, 4.8, 2.7, 1.4, 3.9}.

26. What is the range and interquartile range for the data? *(Lesson 12-2)*

27. Display the data in a stem-and-leaf plot. *(Lesson 12-1)*

Getting Ready for the Next Lesson

PREREQUISITE SKILL **For Exercises 28 and 29, refer to the table shown.**

*(To review **analyzing data**, see pages 722 and 723.)*

28. How many people were surveyed?

29. How many people spend more than 7 hours a week on recreational activities?

Weekly Recreation Time

Time (h)	Tally	Frequency
0–3	III	3
4–7	𝍸 III	8
8–11	𝍸 IIII	9
12–15	𝍸	5

Graphing Calculator Investigation

A Follow-Up of Lesson 12-3

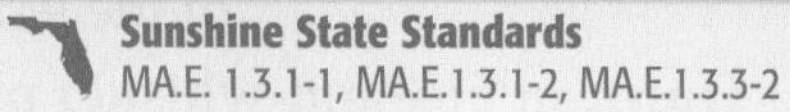

Sunshine State Standards
MA.E. 1.3.1-1, MA.E.1.3.1-2, MA.E.1.3.3-2

Box-and-Whisker Plots

You can use a TI-83 Plus graphing calculator to create box-and-whisker plots.

Example

The table shows the ages of the students in two karate classes.

Class	Age (years)														
A	39	33	37	26	39	25	39	40	27	25	35	31	29	28	35
B	19	26	40	19	20	32	16	24	24	16	27	23	22	25	16

Make box-and-whisker plots for the ages in Class A and in Class B.

Step 1 *Enter the data.*

- Clear any existing data.

  KEYSTROKES: STAT ENTER ▲ CLEAR ENTER

- Enter the Class A ages in L1 and the Class B ages in L2.

KEYSTROKES: *Review entering a list on page 45.*

Step 2 *Format the graph.*

- Turn on two statistical plots.

  KEYSTROKES: *Review statistical plots on page 45.*

- For Plot 1, select the box-and-whisker plot and L1 as the Xlist.

  KEYSTROKES: ▼ ▶ ▶ ▶ ENTER ▼ 2nd L1 ENTER

- Repeat for Plot 2, using L2 as the Xlist, to make a box-and-whisker plot for Class B.

Step 3 *Graph the box-and-whisker plots.*

- Display the graph.

  KEYSTROKES: ZOOM 9

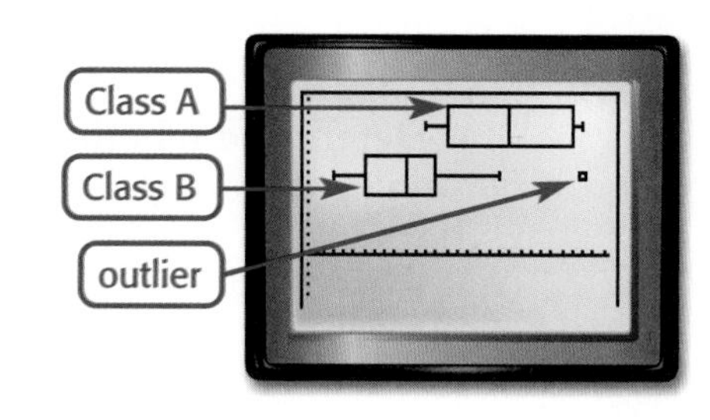

Press TRACE. Move from one plot to the other using the up and down arrow keys. The right and left arrow keys allow you to find the least value, greatest value, and quartiles.

Exercises

1. What are the least, greatest, quartile, and median values for Classes A and B?
2. What is the interquartile range for Class A? Class B?
3. Are there any outliers? How does the graphing calculator show them?
4. a. Estimate the percent of Class A members who are high school students.

   b. Estimate the percent of Class B members who are high school students.
5. If you were a high school student, which class would you join? Explain.

 www.pre-alg.com/other_calculator_keystrokes

12-4 Histograms

What You'll Learn

- Display data in a histogram.
- Interpret data in a histogram.

Sunshine State Standards
MA.E. 1.3.1-1, MA.E.1.3.1-2

Vocabulary

- histogram

How are histograms similar to frequency tables?

The number of counties for each state in the United States is displayed in the table shown. This table is a *frequency table.*

a. What does each tally mark represent?

b. What does the last column represent?

c. What do you notice about the intervals that represent the counties?

Number of Counties in Each State		
Counties	Tally	Frequency
1–25	𝍸𝍸 III	13
26–50	𝍸 II	7
51–75	𝍸𝍸 II	12
76–100	𝍸𝍸 II	12
101–125	IIII	4
126–150		0
151–175	I	1
176–200		0
201–225		0
226–250		0
251–275	I	1

Source: *The World Almanac*

DISPLAY DATA Another type of graph that can be used to display data is a histogram. A **histogram** uses bars to display numerical data that have been organized into equal intervals.

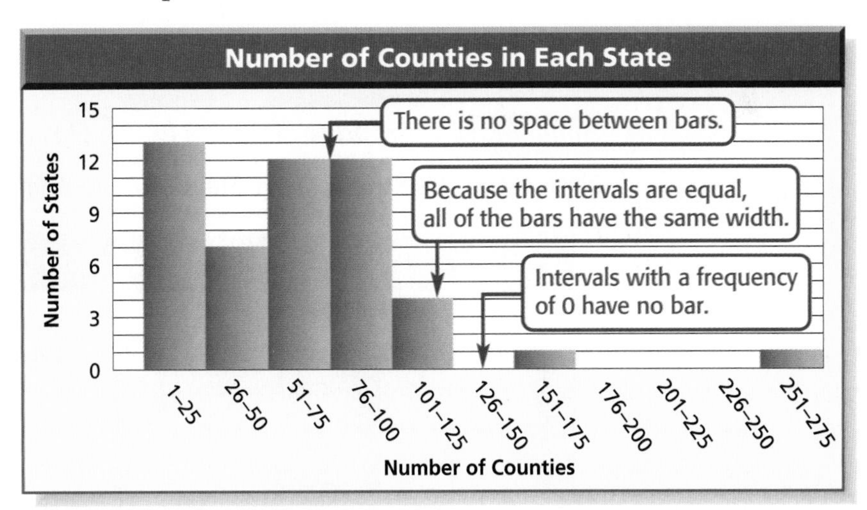

Example 1 Draw a Histogram

WATER PARKS The frequency table shows certain water park admission costs. Display the data in a histogram.

Water Park Admission		
Cost ($)	Tally	Frequency
8–15	𝍸	5
16–23	𝍸 II	7
24–31	IIII	4
32–39		0
40–47	II	2

Step 1 Draw and label a horizontal and vertical axis as shown. Include the title.

(continued on the next page)

Study Tip

Break in Scale

The symbol -√- means there is a break in the scale. The scale from 0 to 7 has been omitted.

Step 2 Show the intervals from the frequency table on the horizontal axis and an interval of 1 on the vertical axis.

Step 3 For each cost interval, draw a bar whose height is given by the frequency.

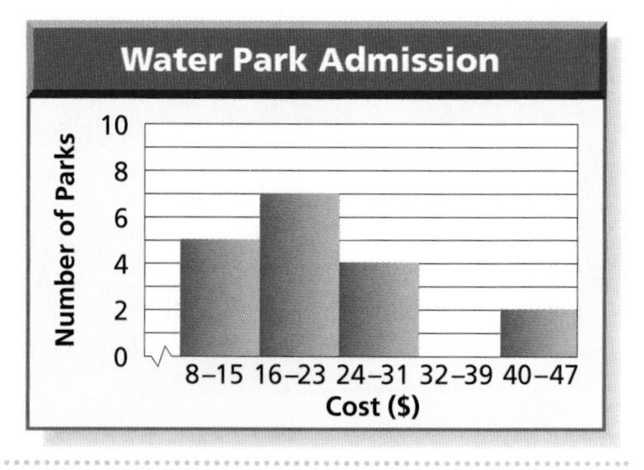

INTERPRET HISTOGRAMS A histogram gives a better visual display of data than a frequency table. Thus, it is easier to interpret data displayed in a histogram.

Example 2 Interpret Data

SCHOOL **Refer to the histogram at the right.**

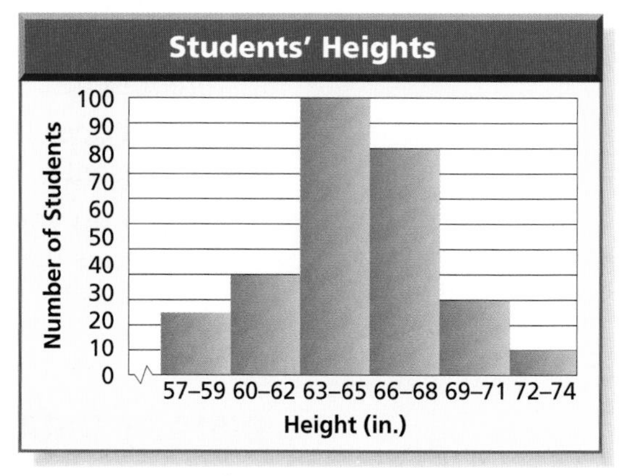

a. How many students are at least 69 inches tall?

Since 30 students are 69–71 inches tall, and 10 students are 72–74 inches tall, 30 + 10 or 40 students are at least 69 inches tall.

b. Is it possible to tell the height of the tallest student?

No, you can only tell that the tallest student is between 72 and 74 inches.

Reading Math

At Least

Recall that *at least* means *is greater than or equal to*.

Histograms can also be used to compare data.

Example 3 Compare Two Sets of Data

OLYMPICS **Use the histograms below to answer the question.**

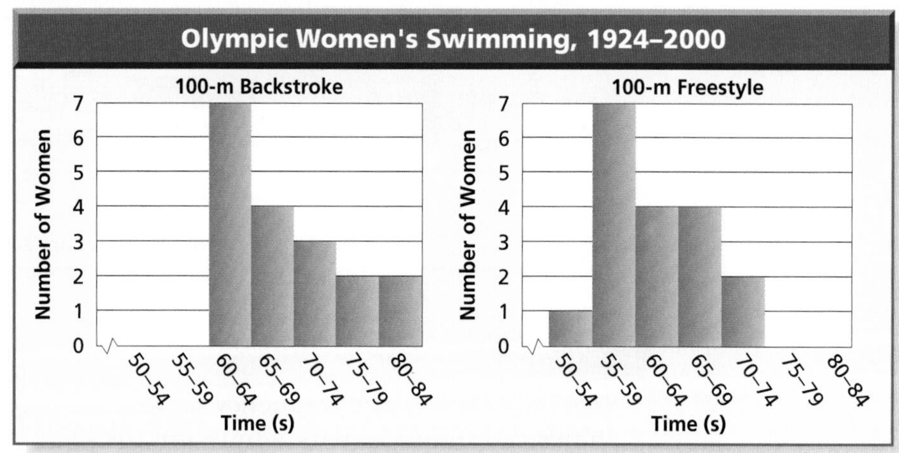

Source: *The World Almanac*

Which event has more winning times less than 1 minute?

The 100-meter freestyle has 1 + 7 or 8 athletes with a winning time less than 1 minute while the 100-meter backstroke has none.

Check for Understanding

Concept Check

1. **Explain** why there are no spaces between the bars of a histogram.
2. **OPEN ENDED** Tell how a histogram gives a better visual display than a frequency table.

Guided Practice

Display each set of data in a histogram.

3.

Pet Survey		
Pets	Tally	Frequency
1–3	卌 卌 卌 卌 I	21
4–6	卌 II	7
7–9	II	2
10–12		0
13–15	I	1

4.

Test Scores		
Score	Tally	Frequency
95–100	卌	5
89–94	卌 卌 II	12
83–88	卌 IIII	9
77–82	卌 I	6
71–76	IIII	4

Applications

ROLLER COASTERS **For Exercises 5–8, use the histogram shown.**

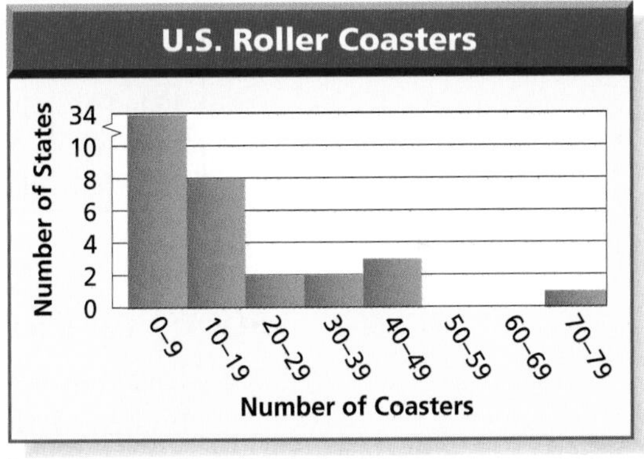

Source: The Roller Coaster Database

5. Describe the data.
6. Which interval has the most roller coasters?
7. Why is there a jagged line in the vertical axis?
8. How many states have no roller coasters? Explain.

Online Research **Data Update** How has the number of roller coasters in the United States changed since 2000? Visit www.pre-alg.com/data_update to learn more.

VACATIONS **For Exercises 9 and 10, use the histograms below.**

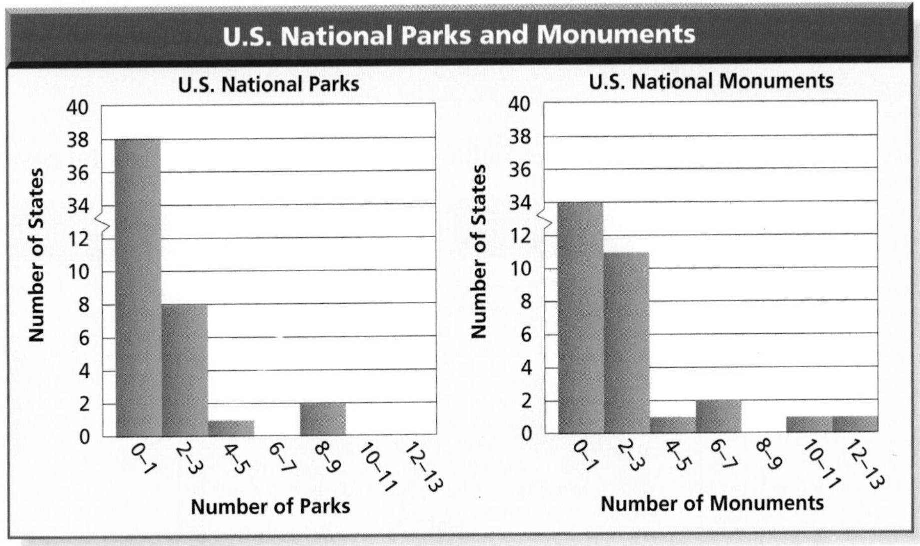

Source: www.infoplease.com

9. Are there more states with two or more national parks or two or more national monuments?
10. How many more states have either one or no national parks than either one or no national monuments?

Practice and Apply

Homework Help

For Exercises	See Examples
11–14	1
15–17	2
18–20	3

Extra Practice

See page 753.

Display each set of data in a histogram.

11.

Weekly Study Time		
Time (hr)	Tally	Frequency
0–3	||	2
4–6	|||	3
7–9	𝍸 |||	8
10–12	𝍸 𝍸 ||	12
13–15	𝍸 𝍸	10

12.

Weekly Allowance		
Amount	Tally	Frequency
\$0–\$5	𝍸 𝍸 |	11
\$6–\$11	𝍸 ||||	9
\$12–\$17	𝍸 |||	8
\$18–\$23	|||	3
\$24–\$29	𝍸	5

13.

Touchdowns in a Season		
Amount	Tally	Frequency
80–96	𝍸 𝍸	10
97–113	𝍸	5
114–130	||||	4
131–147	||	2
148–164		0
165–181	|	1

14.

Goals in a Season		
Amount	Tally	Frequency
65–69	𝍸 |	6
70–74	𝍸 ||	7
75–79	|||	3
80–84		0
85–89	|||	3
90–94	|	1

FOOD **For Exercises 15–17, use the data in the histogram.**

15. How many restaurants sell chicken sandwiches that cost under \$3?

16. How many restaurants were surveyed?

17. What percent of the restaurants surveyed sell chicken sandwiches that cost between \$2.00 and \$2.49?

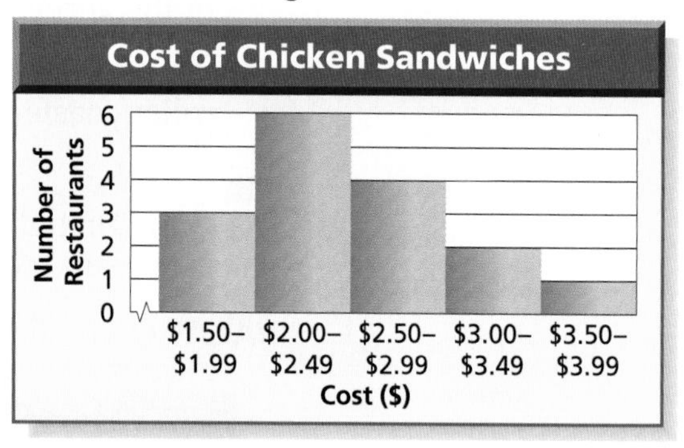

Frequency tables and histograms can help you analyze data. Visit www.pre-alg.com/webquest to continue work on your WebQuest project.

ARCHITECTURE **For Exercises 18–20, use the histograms below.**

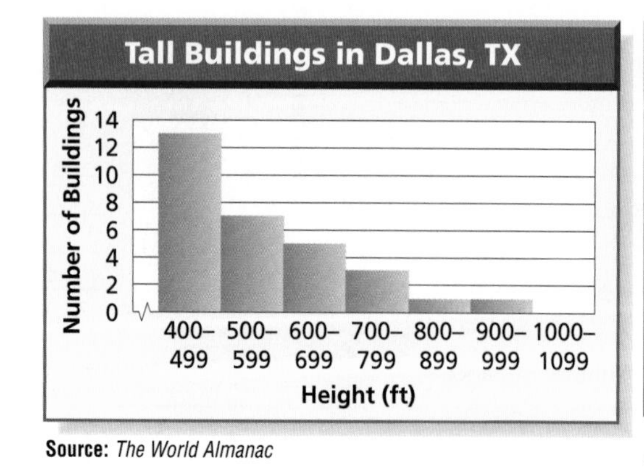

Source: *The World Almanac*

Tall Buildings in Los Angeles, CA

Number of Buildings: 0, 2, 4, 6, 8, 10, 12, 14

400–499, 500–599, 600–699, 700–799, 800–899, 900–999, 1000–1099

Height (ft)

Source: *The World Almanac*

18. Which city has more tall buildings?

19. Which city has the fewer number of buildings over 600 feet?

20. Compare the heights of the tall buildings in the two cities.

CRITICAL THINKING **For Exercises 21–25, determine whether each statement is *true* or *false*.**

21. You can determine the median from a box-and-whisker plot.

22. You can determine the range from a stem-and-leaf plot.

23. You can reconstruct the original data from a histogram.

24. You can reconstruct the original data from a box-and-whisker plot.

25. You can determine the interval in which the median falls in a histogram.

26. WRITING IN MATH Answer the question that was posed at the beginning of the lesson.

How are histograms similar to frequency tables?

Include the following in your answer:

- examples of a histogram and a frequency table, and
- an explanation describing how data are displayed in each.

The histogram shows the ages of the students in a drama club.

Drama Club Students

Number of Students: 0, 2, 4, 6, 8

Age: 12–13, 14–15, 16–17, 18–19

27. How old are the oldest students?

Ⓐ 18–19 Ⓑ 16–19

Ⓒ 17–18 Ⓓ 16–18

28. What is the total number of students in the drama club?

Ⓐ 20 Ⓑ 22

Ⓒ 24 Ⓓ 26

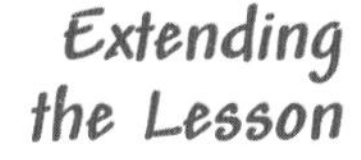

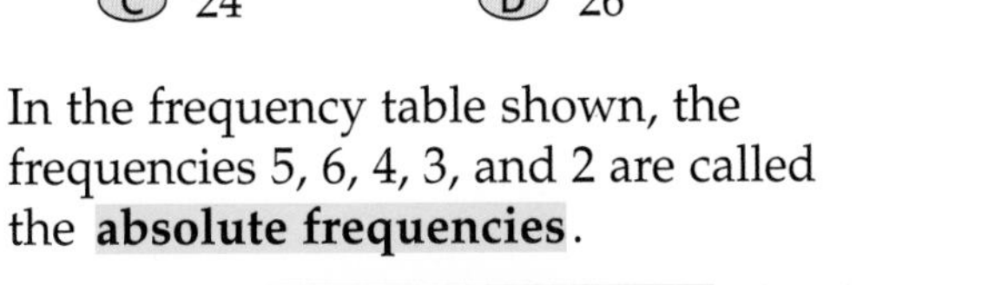

In the frequency table shown, the frequencies 5, 6, 4, 3, and 2 are called the **absolute frequencies**.

To find the **relative frequencies**, divide each absolute frequency by the total number of items. For the data shown, the relative frequencies are $\frac{5}{20}$ or $\frac{1}{4}$, $\frac{6}{20}$ or $\frac{3}{10}$, $\frac{4}{20}$ or $\frac{1}{5}$, $\frac{3}{20}$, and $\frac{2}{20}$ or $\frac{1}{10}$.

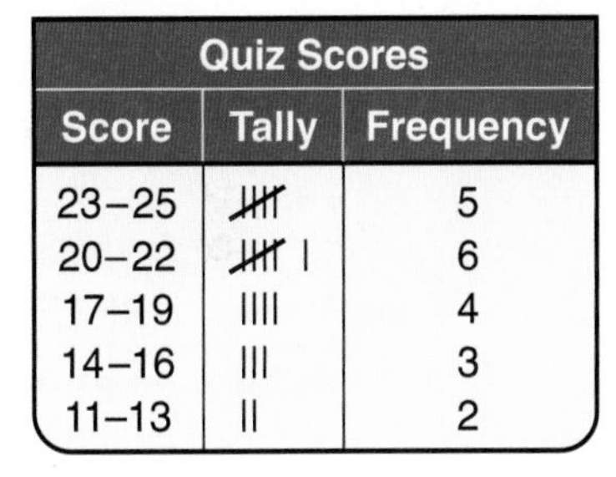

Quiz Scores		
Score	Tally	Frequency
23–25	𝍸	5
20–22	𝍸 I	6
17–19	IIII	4
14–16	III	3
11–13	II	2

The **cumulative frequencies** are the sums of all preceding frequencies. For the data shown, the cumulative frequencies are 5, 5 + 6 or 11, 5 + 6 + 4 or 15, 5 + 6 + 4 + 3 or 18, and 5 + 6 + 4 + 3 + 2 or 20.

Find the absolute, relative, and cumulative frequency for each data set.

29.

Record High Temperature (°F)		
Temp.	Tally	Frequency
100–104	II	2
105–109	𝍸 IIII	9
110–114	𝍸 𝍸 𝍸 II	17
115–119	𝍸 𝍸 II	12
120–124	𝍸 II	7
125–129	II	2

30.

Record Wind Speeds (mph)		
Speed	Tally	Frequency
30–36	𝍸 𝍸 I	11
37–43	𝍸 𝍸 𝍸 𝍸	20
44–50	𝍸 𝍸 𝍸 𝍸 I	21
51–57	𝍸 𝍸 II	12
58–64	I	1
65–71	I	1

When the middles of the intervals on a histogram are connected with line segments, a **frequency polygon** is formed. A frequency polygon is shown.

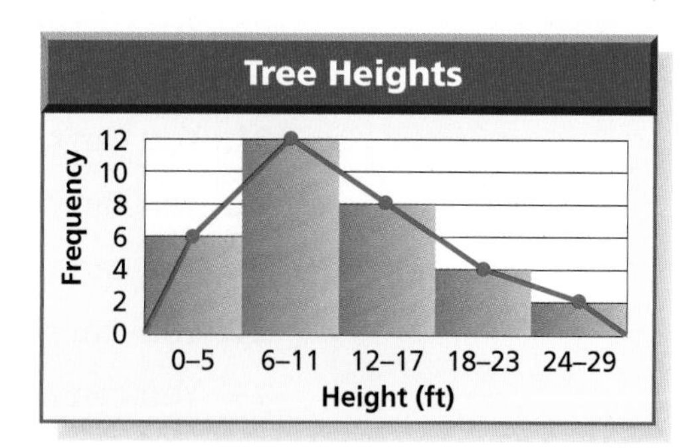

31. Make a frequency polygon of the data shown below.

Average Length (mi) of U.S. States*									
330	1480	400	260	770	380	110	100	500	300
570	390	270	310	400	380	380	320	250	190
490	400	340	300	630	430	490	190	150	370
330	500	340	220	400	360	283	40	260	380
440	790	350	160	430	360	240	310	360	

Source: *The World Almanac*

* Hawaii is not listed.

Maintain Your Skills

Mixed Review **PRESIDENTS** **For Exercises 32 and 33, use the data shown below.**

Ages of Past Presidents at Time of Death

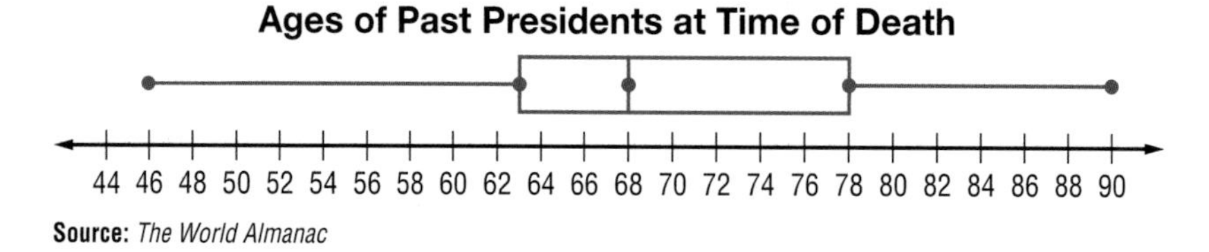

Source: *The World Almanac*

32. What percent of presidents died by the time they were 78 years old? *(Lesson 12-3)*

33. Find the range and interquartile range for the data. *(Lesson 12-2)*

Getting Ready for the Next Lesson **PREREQUISITE SKILL** **For Exercises 34 and 35, use the graph shown.**

*(To review **bar graphs**, see pages 722 and 723.)*

34. Which jeans cost the most?

35. How does the cost of Brand B compare to the cost of Brand C?

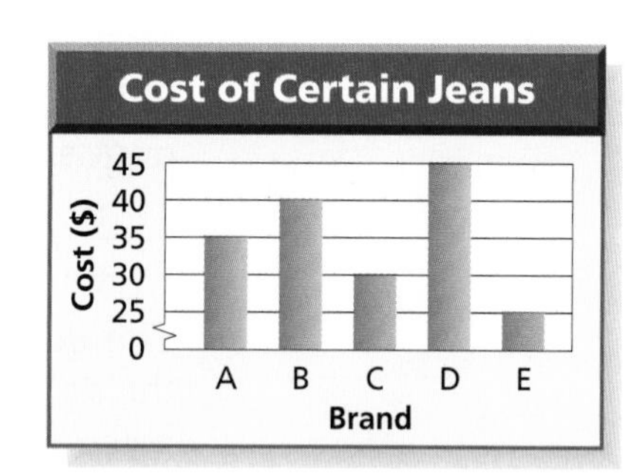

Practice Quiz 1

Lessons 12-1 through 12-4

For Exercises 1–5, use the data in the table.

$20	$28	$18	$89	$55	$28
$30	$86	$19	$42	$19	$16
$26	$43	$28	$22	$32	$40

1. Display the data in a stem-and-leaf plot. *(Lesson 12-1)*

2. Find the range and interquartile range. *(Lesson 12-2)*

3. Display the data in a box-and-whisker plot. *(Lesson 12-3)*

4. What percent of the costs were less than $20? *(Lesson 12-3)*

5. Display the data in a histogram. *(Lesson 12-4)*

Graphing Calculator Investigation

A Follow-Up of Lesson 12-4

Sunshine State Standards
MA.E.1.3.1-1, MA.E.1.3.1-2, MA.E.1.3.3-2

Histograms

You can use a TI-83 Plus graphing calculator to make a histogram.

PRESIDENTS **The list below shows the ages of the first 43 presidents at the time of inauguration.**

57	61	57	57	58	57	61	54	68	51	49
64	50	48	65	52	56	46	54	49	50	47
55	55	54	42	51	56	55	51	54	51	60
62	43	55	56	61	52	69	64	46	54	

Make a histogram to show the age distribution.

Step 1 *Enter the data.*

- Clear any existing data in list L1.

  KEYSTROKES: STAT ENTER ▲ CLEAR ENTER

- Enter the ages in L1.

  KEYSTROKES: *Review entering a list on page 45.*

Step 2 *Format the graph.*

- Turn on the statistical plot.

  KEYSTROKES: 2nd [STAT PLOT] ENTER ENTER

- Select the histogram and L1 as the Xlist.

  KEYSTROKES: ▼ ▶ ▶ ENTER ▼ 2nd L1 ENTER

Step 3 *Graph the histogram.*

Set the viewing window so the x-axis goes from 40 to 75 in increments of 5, and the y-axis goes from -5 to 15 in increments of 1. So, $[40, 75]$ scl: 5 by $[-5, 15]$ scl: 1. Then graph.

KEYSTROKES: WINDOW 40 ENTER 75 ENTER 5 ENTER -5 ENTER 15 ENTER 1 ENTER GRAPH

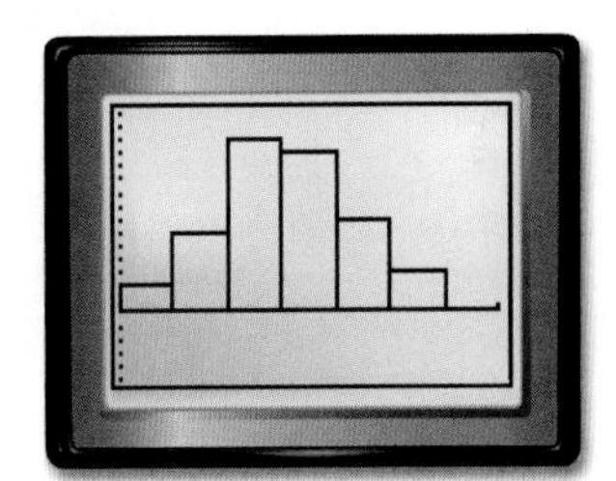

Exercises

1. Press Trace. Find the frequency of each interval using the right and left arrow keys.
2. Discuss why the domain is from 40 to 75 for this data set.
3. How does the graphing calculator determine the size of the intervals?
4. At inauguration, how many presidents have been at least 45, but less than 65?
5. What percent of presidents falls in the interval of Exercise 4?
6. Can you tell from the histogram how many presidents were inaugurated at age 52? Explain.
7. Refer to Example 2 on page 607. How does the stem-and-leaf plot compare to the histogram you have graphed here? Which graph is easier to read?

www.pre-alg.com/other_calculator_keystrokes

12-5 Misleading Statistics

Sunshine State Standards
MA.E.1.3.1-1, MA.E.1.3.1-2, MA.E.3.3.2-1, MA.E.3.3.2-2, MA.E.3.3.2-3, MA.E.3.3.2-4

What You'll Learn

- Recognize when statistics are misleading.

How can graphs be misleading?

The graphs below show the monthly sales for one year for a company.

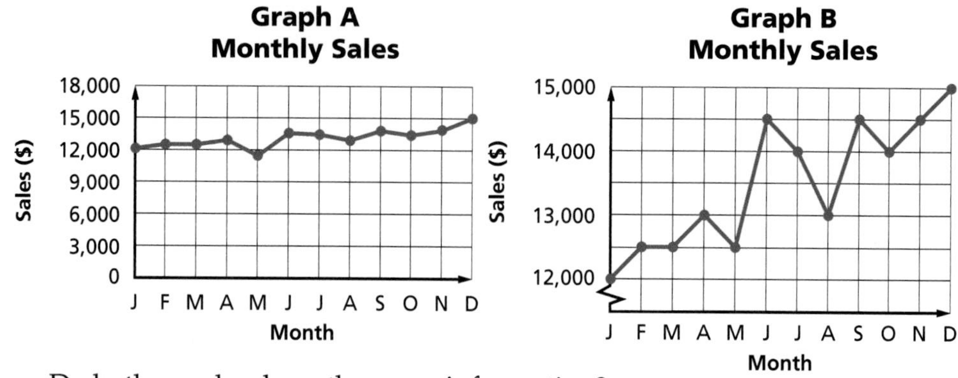

a. Do both graphs show the same information?

b. Which graph suggests a dramatic increase in sales from May to June?

c. Which graph suggests steady sales?

d. How are the graphs similar? How are they different?

Study Tip

Statistics
A graph is also misleading if there is no title, there are no labels on either scale, and the vertical axis does not include zero.

MISLEADING GRAPHS Two line graphs that represent the same data may look quite different. Consider the graphs above. Different vertical scales are used. So, each graph gives a different visual impression.

Example 1 Misleading Graphs

TRAVEL The graphs show the growth of the cruise industry.

a. Why do the graphs look different?

The vertical scales differ.

b. Which graph appears to show a greater increase in the growth of the cruise industry? Explain.

Graph B; the size of the ship makes the increase appear more dramatic because both the height and width of the ship are increasing.

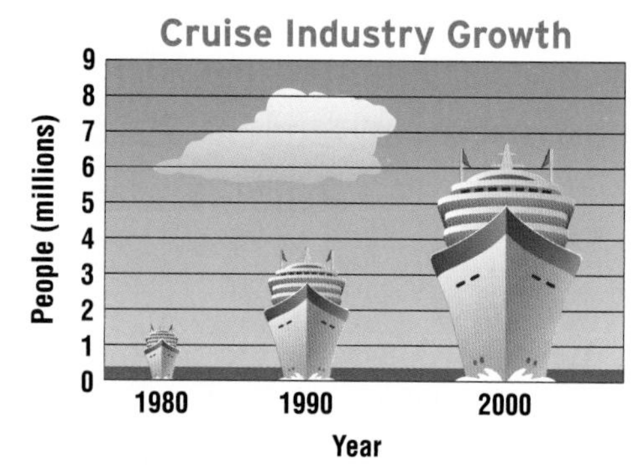

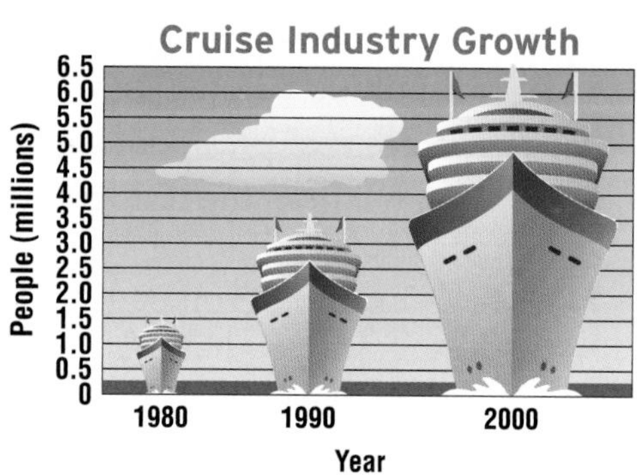

Bar graphs can also be misleading.

Example 2 Misleading Bar Graphs

Study Tip

Labels and Scales When interpreting any graph, be sure to read the labels and the scales in addition to looking at the graph.

READING The graph shows the amount of time people spend reading the newspaper each day. Explain why the graph is misleading.

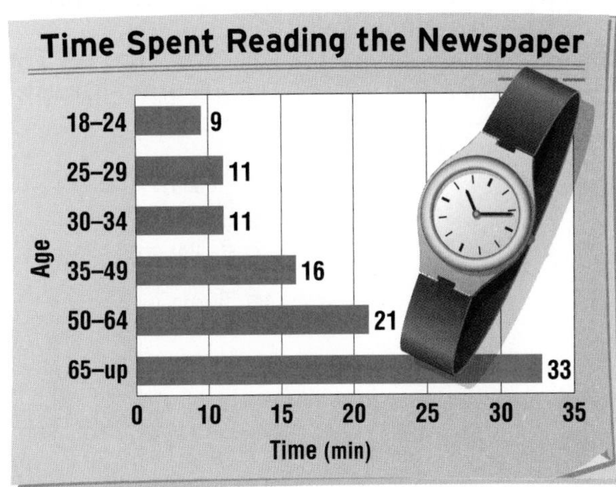

The inconsistent vertical scale and horizontal scale cause the data to be misleading.

The graph gives the impression that people aged 65 and up read the newspaper six times longer than those aged 18–24. By using the horizontal scale, you can see that it is only about 3.5 times longer.

Check for Understanding

Concept Check

1. **Name** two ways a graph can be misleading.

2. **OPEN ENDED** Find an example of a misleading graph in a newspaper or magazine. Explain why it is misleading.

Guided Practice **SCHOOL For Exercises 3 and 4, refer to the graphs below.**

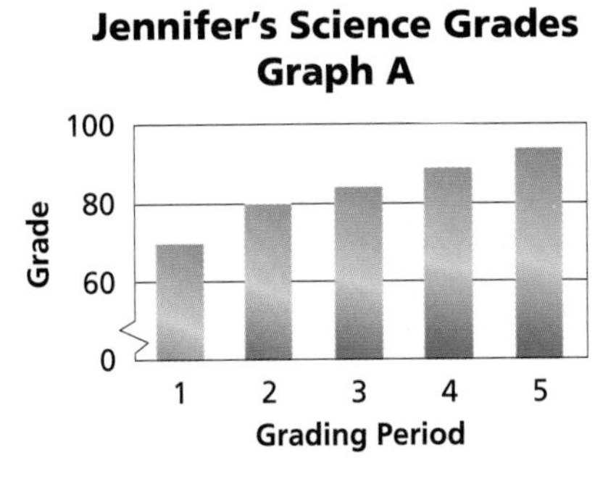

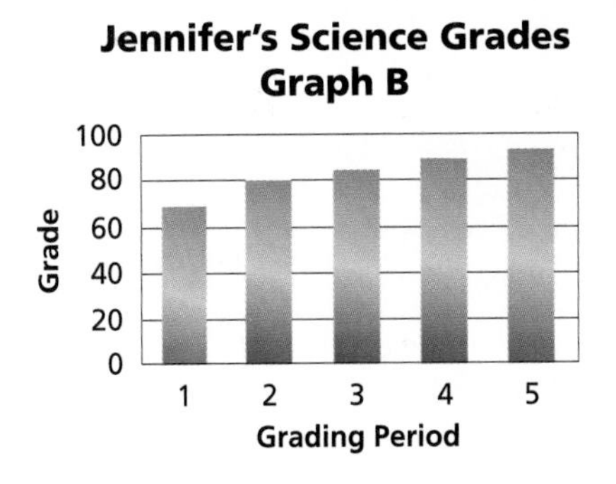

3. Explain why the graphs look different.

4. Which graph appears to show Jennifer's grades improving more? Explain.

Application

5. **COMMUNICATION** The graph shows how the number of area codes in the U.S. have increased over the years. Tell why the graph is misleading.

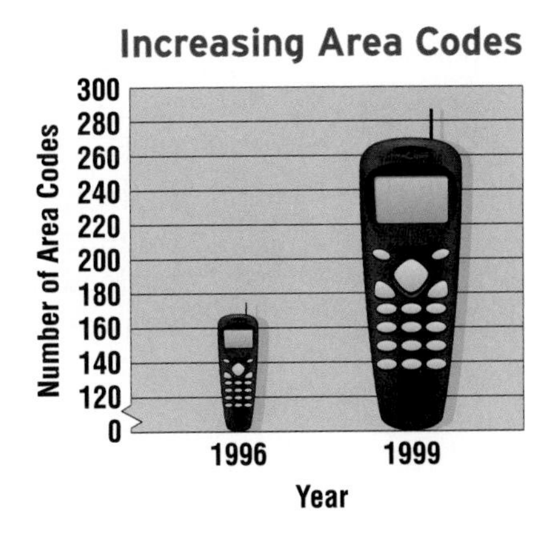

Practice and Apply

Homework Help

For Exercises	See Examples
6–10	1, 2

Extra Practice

See page 754.

MOVIES For Exercises 6 and 7, refer to the graphs below.

Top Five All Time Movies
Graph A

Amount Made (millions): 0, 300, 400, 500, 600, 700

A B C D E

Movie

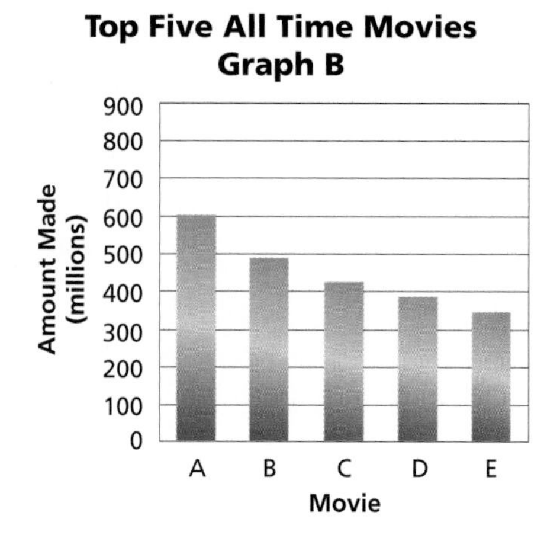

6. Which graph gives the impression that the top all-time movie made far more money than any other top all-time movie?
7. Which graph shows that movie C made nearly as much money as the other top movies?

More About. . .

Jobs

The highest unemployment rate in the United States was in 1933 during the great depression. The unemployment rate averaged 25.2%. About 15 million people were jobless.

Source: *The World Almanac*

JOBS For Exercises 8 and 9, use the graphs below.

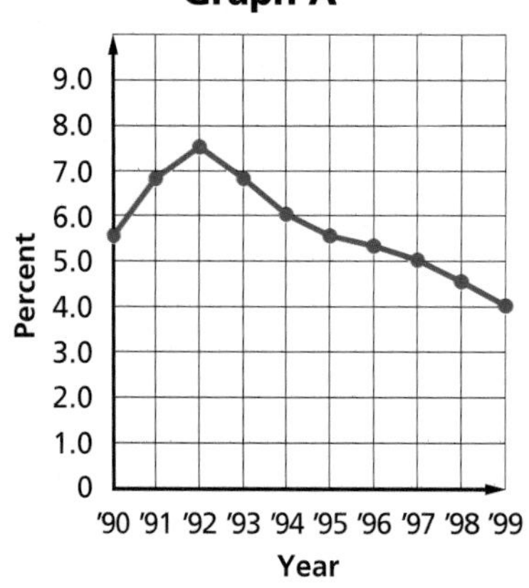

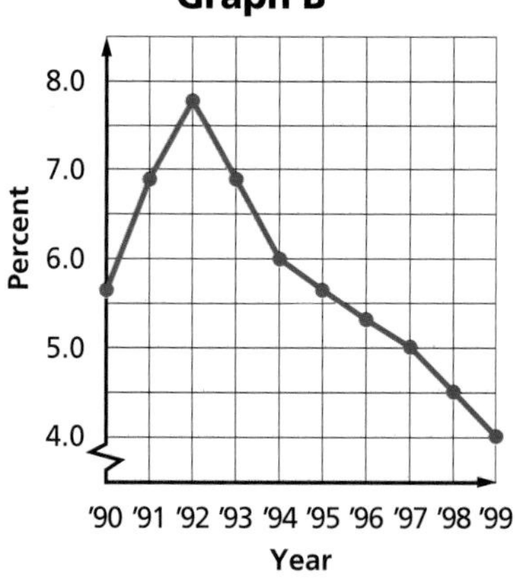

8. What causes the graphs to differ in their appearance?
9. Which graph appears to show that unemployment rates have decreased rapidly since 1992? Explain your reasoning.
10. **TRAVEL** The distance adults drive each week is shown in the graph at the right. Is the graph misleading? Explain your reasoning.

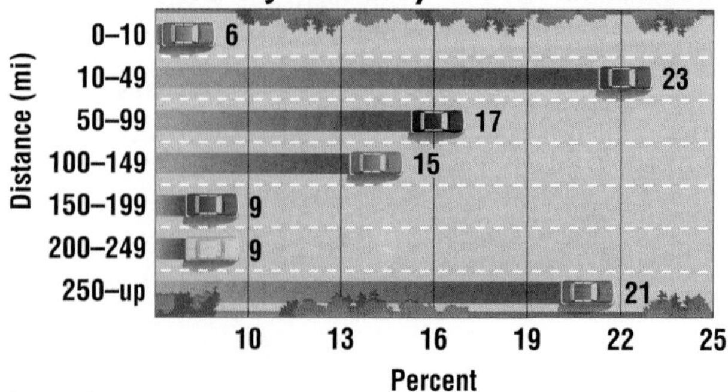

11. **CRITICAL THINKING** The table shows the number of yearly passengers on certain commuter trains.

a. Draw a graph that shows a slow increase in the number of passengers.

b. Redraw the graph so that it shows a rapid increase in the number of passengers.

Commuter Train Passengers

Year	Amount (millions)
1996	45.9
1997	48.5
1998	54.0
1999	58.3

Source: *Time Almanac*

12. WRITING IN MATH Answer the question that was posed at the beginning of the lesson.

How can graphs be misleading?

Include the following in your answer:

- an example of a graph that is misleading, and
- a discussion of how to redraw the graph so it is not misleading.

13. Which sentence is a true statement about the data in the graph?

Ⓐ From 1995 to 1996, the amount spent on dining out doubled.

Ⓑ The amount spent in 1998 was about 1.2 times the amount spent in 1994.

Ⓒ The amount spent on dining out from 1995 to 1997 increased by three times.

Ⓓ From 1997 to 1998 the amount spent on dining out increased by about 20%.

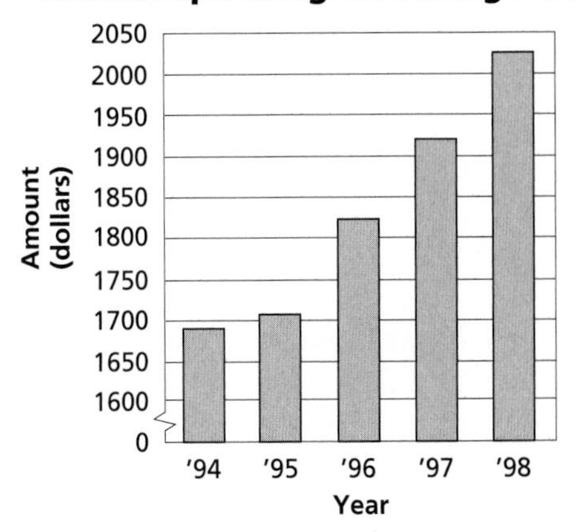

Maintain Your Skills

Mixed Review

14. Display the data shown in a histogram. *(Lesson 12-4)*

15. Draw a box-and-whisker plot for {56°, 43°, 38°, 42°, 50°, 47°, 41°, 55°}. *(Lesson 12-3)*

Book Survey

Books Read	Tally	Frequency
0–2	~~IIII~~ III	8
3–5	IIII	4
6–8	~~IIII~~ ~~IIII~~	10
9–11	~~IIII~~	5
12–14	III	3

Find the area of each figure described. *(Lesson 10-5)*

16. triangle: base, 6 feet; height, 4.2 feet

17. trapezoid: height, 5.8 meters; bases, 4 meters, 3 meters

Getting Ready for the Next Lesson

A bag contains 3 yellow marbles, 2 blue marbles, and 7 purple marbles. Suppose one marble is selected at random. Find the probability of each outcome. Express each probability as a fraction and as a percent.

*(To review **probability**, see Lesson 6-9)*

18. P(yellow)
19. P(blue)
20. P(not purple)
21. P(not blue)
22. P(blue or purple)
23. P(yellow or not blue)

Reading Mathematics

Sunshine State Standards
MA.E. 3.3.2-4

Dealing with Bias

In statistics, a sample is biased if it favors certain outcomes or parts of the population over others. While taking a random survey is the best way to eliminate bias or favoritism, there are still many ways in which a survey and its responses can be biased.

Voluntary Response
Consider a survey where people call or write in. Those who take the time to voluntarily respond usually have strong opinions on an issue. This may result in bias.

Response Bias
Some surveys are biased because either the people participating in the survey are influenced by the interviewer or the people do not give accurate responses.

Nonresponse
Consider a survey where the selected individuals cannot be contacted or they refuse to cooperate. Since the survey does not include a portion of the population, bias results.

Poorly Worded Questions
A survey is biased if it contains questions that are worded to influence people's responses.

Reading to Learn

Tell whether each situation may result in bias. Explain your reasoning.

1. Suppose a bakery wants to know what percent of households makes baked goods from scratch. A sample is taken of 300 households. An interviewer goes from door to door between 9 A.M. and 4 P.M.

2. A telephone survey of 500 urban households is taken. The interviewer asks, "Does anyone in your household use public transportation?"

3. A radio station is conducting a survey as to whether people want a law that prohibits the use of computers for downloading music files. The radio announcer gives a number to call to answer *yes* or *no*. Of the responses, 85% said they do not want this law.

4. An interviewer states, "Due to heavy traffic, should another lane be added to Main Street?"

12-6 Counting Outcomes

Sunshine State Standards
MA.E.2.3.1-2

Vocabulary

- tree diagram
- Fundamental Counting Principle

What You'll Learn

- Use tree diagrams or the Fundamental Counting Principle to count outcomes.
- Use the Fundamental Counting Principle to find the probability of an event.

How can you count the number of skateboard designs that are available from a catalog?

The basic model of a skateboard has 5 choices for decks and 3 choices for wheel sets, as shown at the right. How many different skateboards are possible?

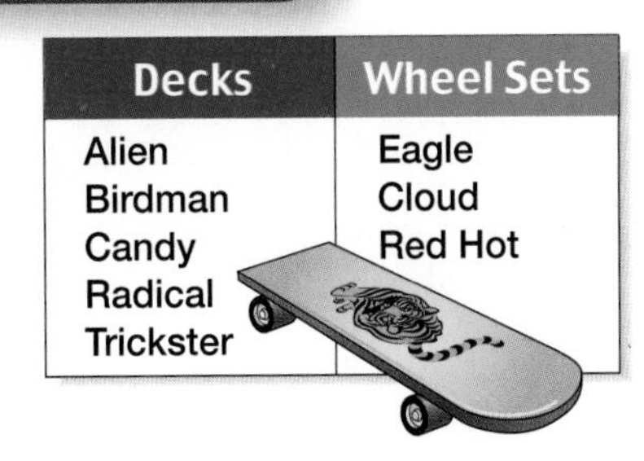

Decks	Wheel Sets
Alien	Eagle
Birdman	Cloud
Candy	Red Hot
Radical	
Trickster	

a. Write the names of each deck choice on 5 sticky notes of one color. Write the names of each type of wheel on 3 notes of another color.

b. Choose one deck note and one wheel note. One possible skateboard is Alien, Eagle.

c. Make a list of all the possible skateboards.

d. How many different skateboard designs are possible?

COUNTING OUTCOMES To solve the skateboard problem above, you can look at a simpler problem. Suppose there are only three deck choices, Birdman, Alien, or Candy, and only two wheel choices, Eagle or Cloud. You can draw a **tree diagram** to represent the possible outcomes.

Study Tip

Look Back To review **outcomes**, see Lesson 6-9.

Example 1 Use a Tree Diagram to Count Outcomes

How many different skateboards can be made from three deck choices and two wheel choices?

You can draw a diagram to find the number of possible skateboards.

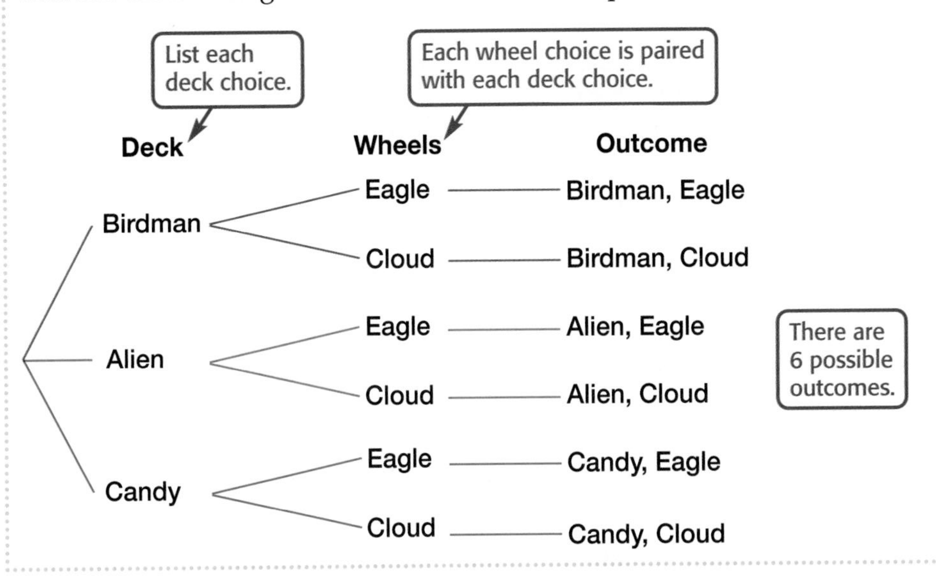

In Example 1, notice that the product of the number of decks and the number of types of wheels, $3 \cdot 2$, is the same as the number of outcomes, 6. The **Fundamental Counting Principle** relates the number of outcomes to the number of choices.

Key Concept — Fundamental Counting Principle

- **Words** If event M can occur in m ways and is followed by event N that can occur in n ways, then the event M followed by N can occur in $m \cdot n$ ways.
- **Example** If there are 5 possible decks and 3 possible sets of wheels, then there are $5 \cdot 3$ or 15 possible skateboards.

Concept Check How many outcomes are possible if you toss a coin and roll a 6-sided number cube?

You can also use the Fundamental Counting Principle when there are more than two events.

Example 2 Use the Fundamental Counting Principle

SKIING **When you rent ski equipment at Bridger Peaks Ski Resort, you choose from 4 different types of ski boots, 5 lengths of skis, and 2 types of poles. How many different outfits are possible?**

Use the Fundamental Counting Principle.

The number of types of boots	times	the number of lengths of skis	times	the number of types of poles	equals	the number of possible outcomes.
4	$\times$	5	$\times$	2	$=$	40

There are 40 possible different sets of boots, skis, and poles.

Study Tip

Multiplying More than Two Factors
Remember, when you multiply, you can change the order of the factors. For example, in $4 \times 5 \times 2$ you can multiply 5×2 first, then multiply the product, 10, by 4 to get 40.

FIND THE PROBABILITY OF AN EVENT When you know the number of outcomes, you can find the probability that an event will occur.

Example 3 Find Probabilities

a. Jasmine is going to toss two coins. What is the probability that she will toss one head and one tail?

First find the number of outcomes.

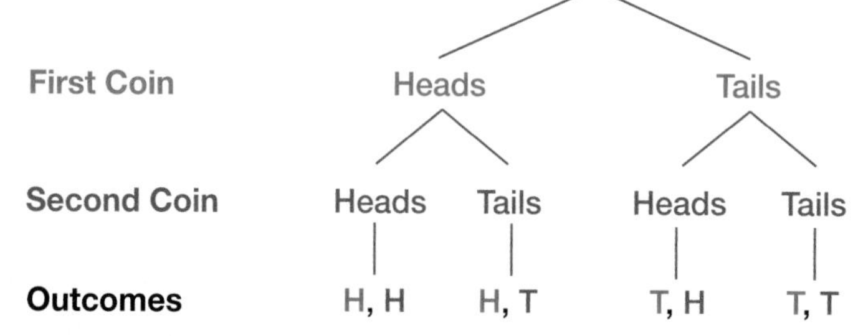

There are four possible outcomes.

Study Tip

Look Back
You can review **probability** in Lesson 6-9.

Look at the tree diagram. There are two outcomes that have one head and one tail.

$$P(\text{one head, one tail}) = \frac{\text{number of favorable outcomes}}{\text{number of possible outcomes}}$$
$$= \frac{2}{4} \text{ or } \frac{1}{2}$$

The probability that Jasmine will toss one head and one tail is $\frac{1}{2}$.

b. What is the probability of winning a state lottery game where the winning number is made up of four digits from 0 to 9 chosen at random?

First, find the number of possible outcomes. Use the Fundamental Counting Principle.

choices for the 1st digit	times	choices for the 2nd digit	times	choices for the 3rd digit	times	choices for the 4th digit	equals	total number of outcomes
10	×	10	×	10	×	10	=	10,000

There are 10,000 possible outcomes. There is 1 winning number. So, the probability of winning with one ticket is $\frac{1}{10{,}000}$. This probability can also be written as a decimal, 0.0001, or a percent, 0.01%.

Check for Understanding

Concept Check

1. **Compare and contrast** using a tree diagram and using the Fundamental Counting Principle to find numbers of outcomes.

2. **OPEN ENDED** Give an example of a situation that would have 12 outcomes.

3. **Explain** how to find the probability of an order containing chicken filling from a choice of burrito or taco with chicken, beef, or bean filling.

Guided Practice

The spinner at the right is spun twice.

4. Draw a tree diagram to represent the situation. How many outcomes are possible?

5. What is the probability of spinning two blues?

A coin is tossed, and a six-sided number cube is rolled.

6. How many outcomes are possible?

7. What is the probability of tails and an odd number?

8. Four coins are tossed. How many outcomes are possible?

Application

9. **FOOD SERVICE** Hastings Cafeteria serves toast, a muffin, or a bagel with coffee, milk, or orange juice. How many different breakfasts of one bread and one beverage are possible?

Practice and Apply

Homework Help

For Exercises	See Examples
10–13	1
14–19, 24	2
20–23	3

Extra Practice

See page 754.

Draw a tree diagram to find the number of outcomes for each situation.

10. Each spinner shown at the right is spun once.

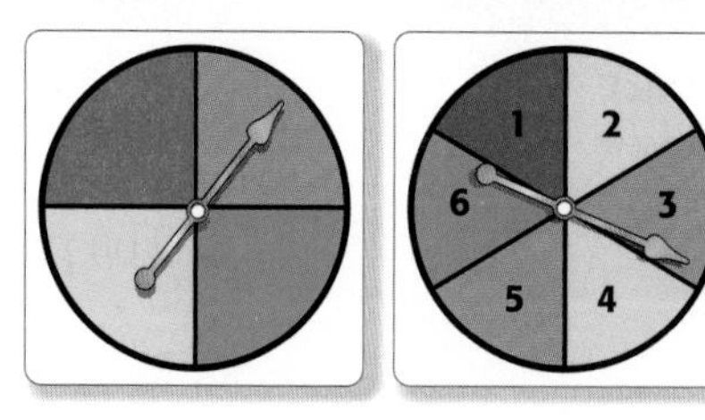

11. Three coins are tossed.

12. A restaurant offers three types of pasta with two types of sauce and a choice of meatball or sausage.

13. Andrew has a choice of a blue, yellow, white, or striped shirt with a choice of black, navy, or tan pants.

Find the number of possible outcomes for each situation.

14. School sweatshirts come in four sizes and four colors.

15. A number cube is rolled twice.

16. Two coins are tossed and a number cube is rolled.

17. A car comes with two or four doors, a four- or six-cylinder engine, and a choice of six exterior colors.

18. A quiz has five true-false questions.

19. There are four answer choices for each of five multiple-choice questions on a quiz.

Find the probability of each event.

20. Three coins are tossed. What is the probability of two heads and one tail?

21. Two six-sided number cubes are rolled. What is the probability of getting a 3 on exactly one of the number cubes?

22. An 8-sided die is rolled three times. What is the probability of getting three 7s?

23. What is the probability of winning a lottery game where the winning number is made up of five digits from 0 to 9 chosen at random?

More About...

Skateboards

Skateboarding events have been part of the X (or Extreme) Games since they were first held in June of 1995.

Source: www.infoplease.com

24. **SKATEBOARDS** How many different deluxe skateboards are possible from 10 choices of decks, 8 choices of trucks (the axles that hold the wheels on), and 12 choices of wheels?

25. **GAMES** Suppose you play a game where each player rolls two number cubes and records the sum. The first player chooses whether to win with an even or an odd sum. Should the player choose even or odd? Explain your reasoning.

26. **COMPUTERS** The table shows the features you can choose to customize a computer.

Processor	RAM	External Drive	Printer	Color
regular	64 MB	regular	basic	red
high-speed	128 MB	extra-capacity	standard	gray
	256 MB		deluxe	green
			fax edition	white
				lime

a. How many customized computers include the deluxe printer?

b. How many customized computers include 256 MB of RAM and a high-speed processor?

27. CRITICAL THINKING A certain state uses a system to design motor vehicle license plates that allows for 6,760,000 different plates using a total of six digits and letters. Find a way to use the digits 0–9 and letters A–Z to produce this exact number of arrangements.

28. WRITING IN MATH Answer the question that was posed at the beginning of the lesson.

How can you count the number of skateboard designs that are available from a catalog?

Include the following in your answer:

- the strategy you used to find all possible skateboard designs,
- the relationship of the number of designs to the number of wheels and decks, and
- how the number of skateboards would change if the number of types of decks was doubled.

29. A 4-character password uses the letters of the alphabet. Each letter can be used more than once, but the letter I is not used at all. How many different passwords are possible?

(A) 100 (B) 13,800

(C) 303,600 (D) 390,625

30. Elena has 6 sweaters, 4 pairs of pants, and 3 pairs of shoes. How many different outfits of one sweater, one pair of pants, and one pair of shoes can she make?

(A) 13 (B) 24

(C) 27 (D) 72

Maintain Your Skills

Mixed Review

31. STATISTICS Describe a situation that might cause a line graph to be misleading. *(Lesson 12-5)*

ANIMALS For Exercises 32–34, use the histogram. *(Lesson 12-4)*

32. How many years are there in each interval?

33. Which interval has the greatest number of animals?

34. How many of the animals have a life span more than 20 years?

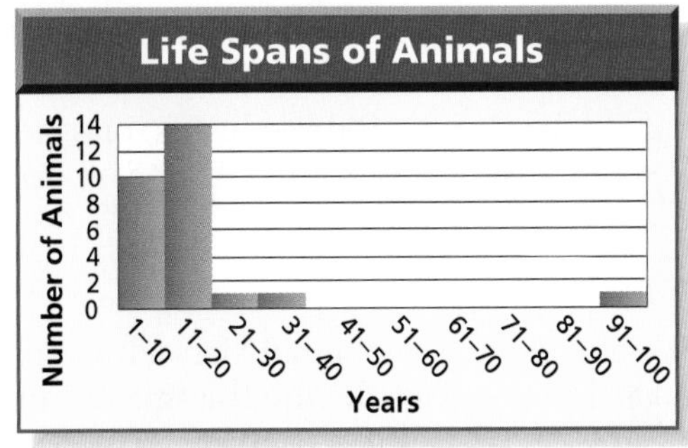

Source: *The World Almanac*

ALGEBRA Use the slope and the y-intercept to graph each equation. *(Lesson 8-3)*

35. $y = 3x + 1$ **36.** $y = \frac{1}{2}x - 2$ **37.** $y = 5$ **38.** $2x - y = 4$

Getting Ready for the Next Lesson

PREREQUISITE SKILL Simplify. *(To review* ***simplifying fractions****, see Lesson 4-5.)*

39. $\frac{6 \cdot 5}{2 \cdot 1}$ **40.** $\frac{5 \cdot 4 \cdot 3}{3 \cdot 2 \cdot 1}$ **41.** $\frac{8 \cdot 7}{2 \cdot 1}$ **42.** $\frac{6 \cdot 5 \cdot 4 \cdot 3}{4 \cdot 3 \cdot 2 \cdot 1}$

Algebra Activity

A Follow-Up of Lesson 12-6

Probability and Pascal's Triangle

Collect Data

Step 1 Copy and complete the tree diagram shown below listing all possible outcomes if you toss a penny and a dime.

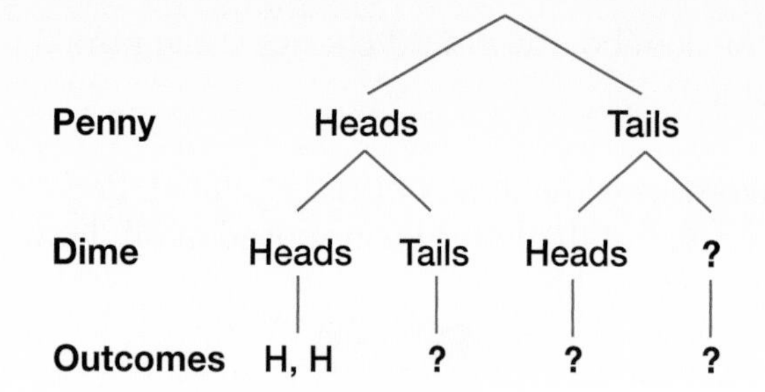

Step 2 Make another tree diagram showing the possible outcomes if you toss a penny, a nickel, and a dime.

Step 3 Make a third tree diagram to show the outcomes for tossing a penny, a nickel, a dime, and a quarter.

Analyze the Data

1. For tossing two coins, how many outcomes are there? How many have one head and one tail?
2. Find P(two heads), P(one head, one tail) and P(two tails). Do not simplify.
3. For tossing three coins, how many outcomes are there? How many have two heads and one tail? one head and two tails?
4. Find P(three heads), P(two heads, one tail), P(one head, two tails), and P(three tails). Do not simplify.
5. For tossing four coins, how many outcomes are there? How many have three heads and one tail? two heads and two tails? one head and three tails?
6. Find P(four heads), P(three heads, one tail), P(two heads, two tails), P(one head, three tails), and P(four tails). Do not simplify.

Make a Conjecture

Pascal was a French mathematician who lived in the 1600s. He is known for the triangle of numbers at the right, called Pascal's triangle.

1	Row 0
1 1	Row 1
1 2 1	Row 2
1 3 3 1	Row 3
1 4 6 4 1	Row 4

7. Examine the rows of Pascal's triangle. Explain how the numbers in each row are related to tossing coins. (*Hint*: Row 2 relates to tossing two coins.)

Extend the Activity

8. Use Pascal's triangle to find the probabilities for tossing five coins.
9. Find other patterns in Pascal's triangle.

12-7 Permutations and Combinations

What You'll Learn

- Use permutations.
- Use combinations.

Vocabulary

- permutation
- factorial
- combination

Why is order sometimes important when determining outcomes?

Lenora, Michael, Ned, Olivia, and Patrick are running for president and treasurer of the class. How many pairs are possible for the two offices?

$$\underbrace{5}_{\text{choices for president}} \underbrace{\times}_{\text{times}} \underbrace{4}_{\text{choices for treasurer}} \underbrace{=}_{\text{equals}} \underbrace{20}_{\text{possible pairs}}$$

Use the Fundamental Counting Principle.

There are 20 possible pairs. How is the number of pairs different for five students running for two student council seats, where order is not important?

a. Make a list of all possible pairs for class offices. (*Note:* Lenora-Michael is different than Michael-Lenora.)

b. How does the Fundamental Counting Principle relate to the number of pairs you found?

c. Make another list for student council seats. (*Note:* For this list, Lenora-Michael is the same as Michael-Lenora.)

d. How does the answer in part **a** compare to the answer in part **c**?

Reading Math

Permutation

Root Word: Permute

Permute means to change the order or arrangement of, especially to arrange the order in all possible ways.

USE PERMUTATIONS An arrangement or listing in which order is important is called a **permutation**. The symbol $P(5, 2)$ represents the number of permutations of 5 things taken 2 at a time, as in 5 students running for 2 offices.

5 choices for president
4 choices left for treasurer

$$P(5, 2) = 5 \cdot 4$$

5 students — Choose 2.

Example 1 Use a Permutation

a. SWIMMING How many ways can six swimmers be arranged on a four-person relay team?

On a relay team, the order of the swimmers is important. This arrangement is a permutation.

6 swimmers — Choose 4.

$$P(6, 4) = 6 \cdot 5 \cdot 4 \cdot 3$$
$$= 360$$

6 choices for 1st person
5 choices for 2nd person
4 choices for 3rd person
3 choices for 4th person

There are 360 possible arrangements.

b. How many four-digit numbers can be made from the digits 1, 3, 5, and 7 if each digit is used only once?

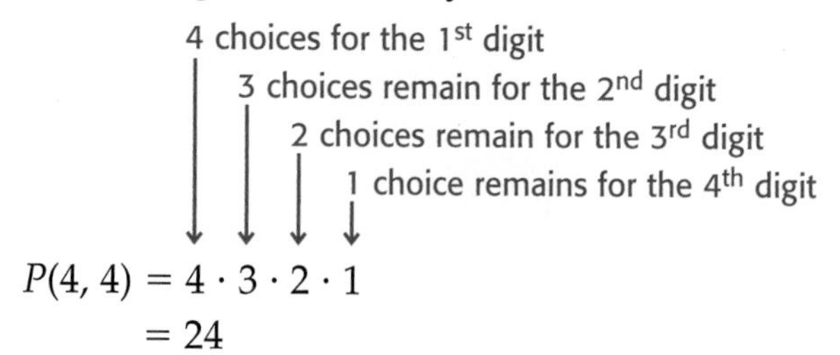

4 choices for the 1st digit
3 choices remain for the 2nd digit
2 choices remain for the 3rd digit
1 choice remains for the 4th digit

$P(4, 4) = 4 \cdot 3 \cdot 2 \cdot 1$
$= 24$

Reading Math

The Symbol !
Read 4! as *four factorial.*

The number of permutations in Example 1b can be written as 4!. It means $4 \cdot 3 \cdot 2 \cdot 1$. The notation n **factorial** means the product of all counting numbers beginning with n and counting backward to 1.

Example 2 Factorial Notation

Find the value of 5!.

$5! = 5 \cdot 4 \cdot 3 \cdot 2 \cdot 1$ Multiply 5 and all of the counting numbers less than 5.
$= 120$

USE COMBINATIONS Sometimes order is not important. For example, *pepperoni, mushrooms,* and *onions* is the same as *onions, pepperoni,* and *mushrooms* when you order a pizza. An arrangement or listing where order is *not* important is called a **combination**.

Example 3 Use a Combination

SCHOOL COLORS How many ways can students choose two school colors from red, blue, white and gold?

Since order is not important, this arrangement is a combination.

First, list all of the permutations of red, blue, white, and gold taken two at a time. Then cross off arrangements that are the same as another one.

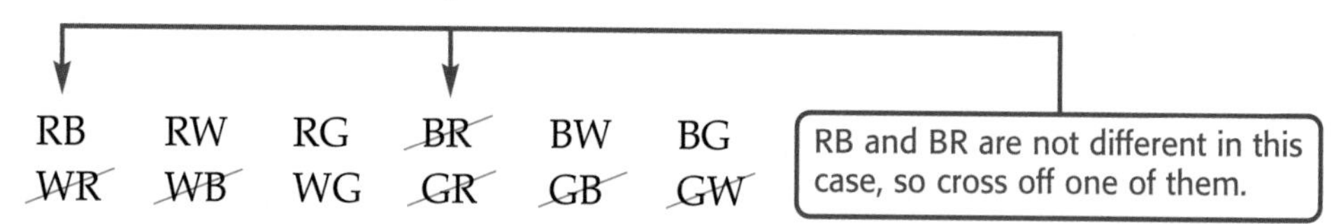

RB RW RG BR BW BG
WR WB WG GR GB GW
RB and BR are not different in this case, so cross off one of them.

There are only six *different* arrangements. So, there are six ways to choose two colors from a list of four colors.

Example 4

FLOWERS How many ways can three flowers be chosen from tulips, daffodils, lilies, and roses?

The arrangement is a combination because order is not important.

TDL	TDR	TLR	~~TLD~~	~~TRD~~	~~TRL~~
DLR	~~DLT~~	~~DRT~~	~~DRL~~	~~DTL~~	~~DTR~~
~~LRT~~	~~LRD~~	~~LTD~~	~~LTR~~	~~LDR~~	~~LDT~~
~~RTD~~	~~RTL~~	~~RDL~~	~~RDT~~	~~RLT~~	~~RLD~~

First, list all of the permutations. Then cross off the arrangements that are the same.

There are 4 ways to choose three flowers from a list of four flowers.

You can find the number of combinations of items by dividing the number of permutations of the set of items by the number of ways each smaller set can be arranged.

Reading Math

Combination Notation

$C(4, 2)$ is read *the number of combinations of 4 things taken 2 at a time.*

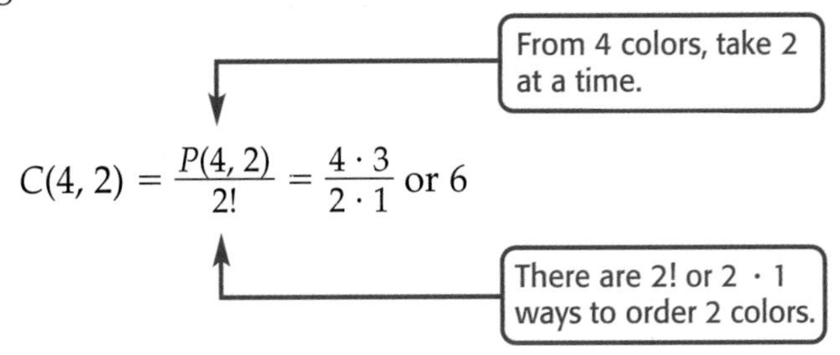

From 4 colors, take 2 at a time.
$C(4, 2) = \frac{P(4, 2)}{2!} = \frac{4 \cdot 3}{2 \cdot 1}$ or 6
There are 2! or 2 · 1 ways to order 2 colors.

Example 5 Use a Combination to Solve a Problem

GEOMETRY Find the number of line segments that can be drawn between any two vertices of an octagon.

Explore An octagon has 8 vertices.

Plan The segment connecting vertex A to vertex C is the same as the segment connecting C to A, so this is a *combination*. Find the combination of 8 vertices taken 2 at a time.

Solve

$$C(8, 2) = \frac{P(8, 2)}{2!}$$

$$= \frac{8 \cdot 7}{2 \cdot 1} \text{ or } 28$$

Examine Draw an octagon and all the segments connecting any two vertices. Check to see that there are 28 segments.

Be sure to count the sides of the octagon.

Check for Understanding

Concept Check

1. **OPEN ENDED** Write a problem that can be solved by finding the value of $P(4, 3)$.
2. **Compare and contrast** $5 \cdot 4 \cdot 3$ and 5!.
3. **FIND THE ERROR** Sindu thinks choosing five CDs from a collection of 30 to take to a party is a permutation. Sarah thinks it is a combination. Who is correct? Explain your reasoning.

Guided Practice

4. Find the value of 6!.

Tell whether each situation is a *permutation* or *combination*. Then solve.

5. How many ways can 5 people be arranged in a line?
6. How many programs of 4 musical pieces can be made from 8 possible pieces?
7. How many ways can a 3-player team be chosen from 9 students?
8. How many ways can 6 different flowers be chosen from 12 different flowers?

Application

9. **FOOD** A pizza shop has 12 toppings to choose from. How many different 3-topping pizzas can be ordered?

Practice and Apply

Homework Help

For Exercises	See Examples
10–17	1, 3, 4
18–21	2
22, 23	5

Extra Practice

See page 754.

Tell whether each situation is a *permutation* or *combination*. Then solve.

10. How many ways can 6 cars line up for a race?

11. How many different flags can be made from the colors red, blue, green, and white if each flag has three vertical stripes?

12. How many ways can 4 shirts be chosen from 10 shirts to take on a trip?

13. How many ways can you buy 2 DVDs from a display of 15?

14. How many 3-digit numbers can you write using the digits 6, 7, and 8 exactly once in each number?

15. There are 12 paintings in a show. How many ways can the paintings take first, second, and third place?

16. How many 5-card hands can be dealt from a standard deck of 52 cards?

17. How many ways can you choose 3 flavors of ice cream from a choice of 14 flavors?

Find each value.

18. 7! **19.** 8! **20.** 10! **21.** 11!

22. **GEOMETRY** Twelve points are marked on a circle. How many different line segments can be drawn between any two of the points?

23. **HANDSHAKES** Nine people gather for a meeting. Each person shakes hands with every other person exactly once. How many handshakes will take place?

More About. . .

Amusement Parks

An amusement park in Ohio is known for its thrilling roller coasters. There are fourteen roller coasters at the park.

Source: www.cedarpoint.com

AMUSEMENT PARKS **For Exercises 24 and 25, use the information at the left.**

24. Suppose you only have time to ride eight of the coasters. How many ways are there to ride eight coasters if order is important?

25. How many ways are there to ride eight of the coasters if order is not important?

26. **LICENSE PLATES** North Carolina issues general license plates with three letters followed by four numbers. (The first number cannot be zero.) Numbers can repeat, but letters cannot. How many license plates can North Carolina generate with this format?

FLOWERS **For Exercises 27–29, use the following information.**

Three roses are to be placed in a vase. The color choices are red, pink, white, yellow, and orange.

27. How many different 3-rose combinations can be made from the 5 roses?

28. What is the probability that 3 roses selected at random will include pink, white, and yellow?

29. What is the probability that 3 roses selected at random will *not* include red?

30. **CRITICAL THINKING** Is the value of $P(x, y)$ *sometimes*, *always*, or *never* greater than the value of $C(x, y)$? (Assume neither x nor y equals 1 and $x \neq y$.)

31. **WRITING IN MATH** Answer the question that was posed at the beginning of the lesson.

Why is order sometimes important when determining outcomes?

Include the following in your answer:

- how the number of pairs differed when order was not important, and
- an example of a situation where order is important and one where order is not important.

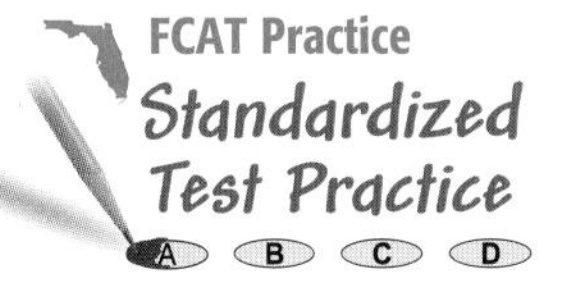

32. Refer to the table. How many different slates of officers could be made if a slate consists of one candidate for each office?

President	Secretary	Treasurer
Tracy	Glenn	Ariel
Marta	Ling	Sherita
José		Wesley

(A) 3 (B) 6

(C) 9 (D) 18

33. Caitlyn knows a phone number begins with 444 and the last four digits are 3, 2, 1, and 0, but she does not remember in what order. What is the greatest number of calls she would have to make to get the right number?

(A) 4 (B) 12 (C) 24 (D) 72

Maintain Your Skills

Mixed Review

34. Draw a tree diagram to find the number of outcomes for rolling two number cubes. *(Lesson 12-6)*

35. Is the graph at the right misleading? Explain your reasoning. *(Lesson 12-5)*

Ticket Sales

Number of Tickets: 300, 250, 200, 150, 100

Class: A, B, C

Name all of the sets of numbers to which each real number belongs. Let N = natural numbers, W = whole numbers, Z = integers, Q = rational numbers, and I = irrational numbers. *(Lesson 9-2)*

36. 5 37. 0.262626...

38. -81 39. $\sqrt{20}$

PREREQUISITE SKILL **Write each ratio in simplest form.**

(To review ***simplifying ratios****, see Lesson 6-1.)*

40. 16:14 41. 4:32 42. 4:48 43. 44:8 44. 45:55

Practice Quiz 2

Lessons 12-5 through 12-7

1. **STATISTICS** Describe a situation that might cause a bar graph to be misleading. *(Lesson 12-5)*
2. Draw a tree diagram to represent the possible combinations of blue or tan shorts with a red, white, or yellow shirt. *(Lesson 12-6)*
3. How many outcomes are possible for a quiz with 8 true-false questions? *(Lesson 12-6)*
4. How many ways can the letters of the word STUDY be arranged? *(Lesson 12-7)*
5. **GEOMETRY** Find the number of line segments that can be drawn between any two vertices of a hexagon. *(Lesson 12-7)*

12-8 Odds

Sunshine State Standards
MA.E.2.3.2-1

Vocabulary

- odds

What You'll Learn

- Find the odds of a simple event.

How are odds related to probability?

Suppose you play the following game.

- Roll two number cubes.
- If the sum of the numbers you roll is 6 or less, you win. If the sum is *not* 6 or less, you lose.

Do you think you will win or lose more often? Play the game to find out.

a. Roll the number cubes 50 times. Record whether you win or lose each time.

b. What is the experimental probability of winning the game based on your results in part **a**?

c. Write the ratio of wins to losses using your results in part **a**.

FIND ODDS You can find theoretical probabilities for rolling a certain sum from two 6-sided number cubes.

$$P(6 \text{ or less}) = \frac{15}{36} \text{ or } \frac{5}{12} \qquad P(\textit{not } 6 \text{ or less}) = \frac{21}{36} \text{ or } \frac{7}{12}$$

Study Tip

Rolling Number Cubes

There are 36 ways to roll two 6-sided number cubes. 15 of the results have a sum of 6 or less. 21 results have a sum greater than 6. See Lesson 6-9 for a table of possible outcomes.

Another way to describe the chance of an event occurring is with odds. The **odds** in favor of an event is the ratio that compares the number of ways the event can occur to the ways that the event *cannot* occur.

ways to occur → 15; ways to not occur → 21

odds of sum of 6 or less → 15:21 or 5:7

Read 15:21 as 15 to 21.

Key Concept — Definition of Odds

The odds in favor of an outcome is the ratio of the number of ways the outcome can occur to the number of ways the outcome cannot occur.

Odds in favor = number of successes : number of failures

The odds against an outcome is the ratio of the number of ways the outcome cannot occur to the number of ways the outcome can occur.

Odds against = number of failures : number of successes

Concept Check If the odds in favor of an outcome are 2:3, what are the odds *against* the same outcome?

Example 1 Find Odds

a. Find the odds of a sum less than 4 if a pair of number cubes are rolled.

There are $6 \cdot 6$ or 36 sums possible for rolling a pair of number cubes.
There are 3 sums less than 4. They are (1, 1), (1, 2), and (2, 1).
There are 36 – 3 or 33 sums that are not less than 4.

Odds of rolling a sum less than 4

= number of ways to roll a sum less than 4 to number of ways to roll any other sum

= 3 : 33 or 1:11

The odds of rolling a sum less than 4 are 1:11.

Odds Notation
Read 1:11 as *1 to 11*.

b. A bag contains 6 red marbles, 4 blue marbles, and 2 gold marbles. What are the odds against drawing a blue marble from the bag?

There are 12 − 4 or 8 marbles that are not blue.

Odds against drawing a blue marble

= number of ways to draw a marble that is not blue to number of ways to draw a blue marble

= 8 : 4 or 2:1

The odds *against* drawing a blue marble are 2:1.

FCAT Practice

Standardized Test Practice

Example 2 Use Odds

Multiple-Choice Test Item

Enrico got positive results 6 out of the 18 times he conducted a science experiment. Based on these results, what are the odds that he will get a positive result the next time he conducts the experiment?

(A) 3 to 9

(B) 9 to 3

(C) 1 to 2

(D) 2 to 1

Read the Test Item

To find the odds, compare the number of success to the number of failures.

Solve the Test Item

The experiment was conducted 18 times.

6 results were positive.

18 − 6 or 12 results were not positive.

successes : failures = 6 to 12 or 1 to 2

The answer is C.

The Princeton Review

Test-Taking Tip

Preparing for Tests
As part of your preparation for a standardized test, review basic definitions such as *odds* and *probability*.

How do the number of successes and failures relate to the number of total possible outcomes?

Check for Understanding

Concept Check

1. **Explain** how to find the odds of an event occurring.
2. **OPEN ENDED** Describe a situation in real life that uses odds.
3. **FIND THE ERROR** Hoshi says that the probability of getting a 2 on one roll of a number cube is 1 out of 6. Nashoba says that the odds of getting a 2 on one roll of a number cube are 1:6. Who is correct? Explain your reasoning.

Guided Practice

Find the odds of each outcome if a number cube is rolled.

4. a number less than 2
5. a multiple of 3
6. a number greater than 3
7. not a 5

8. Ramon found that 2 out of 8 promotional cards at a fast-food restaurant were instant winners. Based on these results, what are the odds against the next card being an instant winner?

   (A) 1 to 4 (B) 4 to 1 (C) 1 to 3 (D) 3 to 1

Practice and Apply

Homework Help

For Exercises	See Examples
9–25	1
26–29	2

Extra Practice
See page 755.

Find the odds of each outcome if the spinner at the right is spun.

9. yellow
10. orange
11. not green
12. not red or green

Find the odds of each outcome if a pair of number cubes are rolled.

13. an odd sum
14. an even sum
15. a sum that is a multiple of 4
16. a sum less than 2
17. a sum that is a prime number
18. a sum that is a composite number
19. *not* a sum of 8
20. not a sum of 9 or 10
21. a sum of 6 with a 4 on one number cube
22. an even sum or a sum greater than 6
23. a sum that is *not* 6, 7, or 8

A card is selected from a standard deck of 52 cards.

24. What are the odds of selecting a red queen?
25. What are the odds of *not* selecting a diamond *or* an ace?
26. The probability of having all girls in an eight-child family is $\frac{1}{256}$. What are the odds in favor of an eight-girl family? the odds against?
27. A family had 13 children, and they were all boys. The odds in favor of this type of family are 1:8191. What is the probability of a 13-child family having all boys?

Career Choices

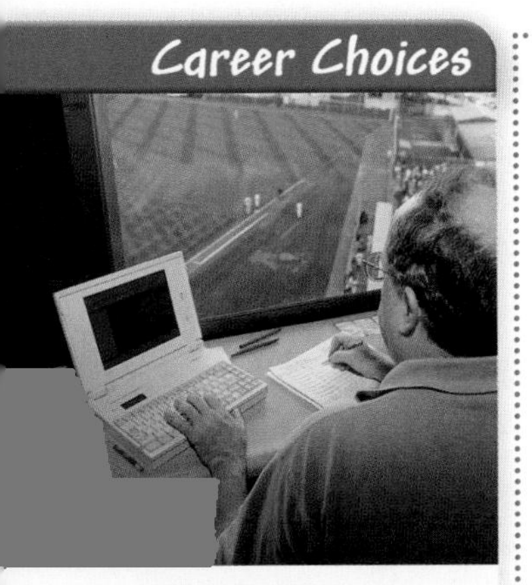

Statistician

Statisticians use their knowledge of statistical methods in many subject areas, including biology, business, engineering, medicine, psychology, and education.

Online Research
For information about a career as a statistician, visit:
www.pre-alg.com/careers

STATISTICS For Exercises 28–30, use the graphic shown below.

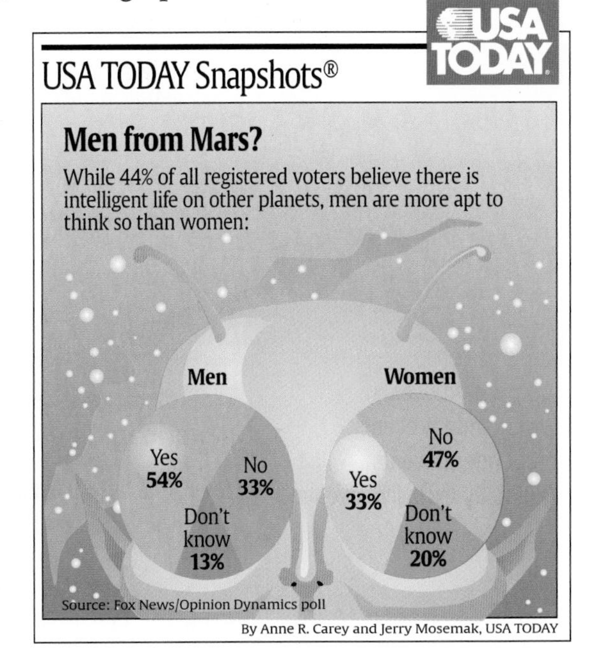

USA TODAY Snapshots®

Men from Mars?

While 44% of all registered voters believe there is intelligent life on other planets, men are more apt to think so than women:

Source: Fox News/Opinion Dynamics poll

By Anne R. Carey and Jerry Mosemak, USA TODAY

28. If you select a man at random from a group, what are the odds that he believes in aliens?

29. If you select a woman at random from a group, what are the odds that she does *not* believe in aliens?

30. In a group of 500 men, how many can be expected to believe in aliens?

31. **CRITICAL THINKING** A carnival game consists of rolling three 8-sided dice. The odds for winning are listed as 8:512. Do you think the odds given are correct? Explain.

32. WRITING IN MATH Answer the question that was posed at the beginning of the lesson.

How are odds related to probability?

Include the following in your answer:

- a comparison of your results from parts **b** and **c** on page 646, and
- an explanation of how you can find the odds of an event if you know its probability.

FCAT Practice
Standardized Test Practice

33. A set of cards is numbered 1 through 40. A card is drawn at random. What are the odds that the card drawn is greater than 25?

Ⓐ 3 to 5 Ⓑ 3 to 8 Ⓒ 5 to 3 Ⓓ 5 to 8

34. Judie made a basket 15 of the 25 times she shot the basketball. Based on this record, what would be the odds *against* Judie making a basket the next time she shoots the ball?

Ⓐ 3 to 5 Ⓑ 5 to 3 Ⓒ 3 to 2 Ⓓ 2 to 3

Maintain Your Skills

Mixed Review

35. How many ways can a family of four be seated in a row of four chairs at the theater if the father sits in the aisle seat? *(Lesson 12-7)*

36. How many outcomes are possible for rolling three number cubes? *(Lesson 12-6)*

Solve each inequality. Then graph the solution on a number line. *(Lessons 7-5 and 7-6)*

37. $2a - 3 \geq 9$ **38.** $4c + 4 > 32$ **39.** $-2y + 3 < 9$

Getting Ready for the Next Lesson

PREREQUISITE SKILL **Find each product.**
*(To review **multiplying fractions**, see Lesson 5-3.)*

40. $\frac{1}{6} \cdot \frac{1}{3}$ **41.** $\frac{2}{3} \cdot \frac{3}{6}$ **42.** $\frac{1}{3} \cdot \frac{1}{3} \cdot \frac{1}{3}$ **43.** $\frac{3}{8} \cdot \frac{2}{7} \cdot \frac{1}{6}$

12-9 Probability of Compound Events

Sunshine State Standards
MA.E.2.3.1-2

What You'll Learn

- Find the probability of independent and dependent events.
- Find the probability of mutually exclusive events.

Vocabulary

- compound events
- independent events
- dependent events
- mutually exclusive events

How are compound events related to simple events?

Place two red counters and two white counters in a paper bag. Then complete the following activity.

Step 1 Without looking, remove a counter from the bag and record its color. Place the counter back in the bag.

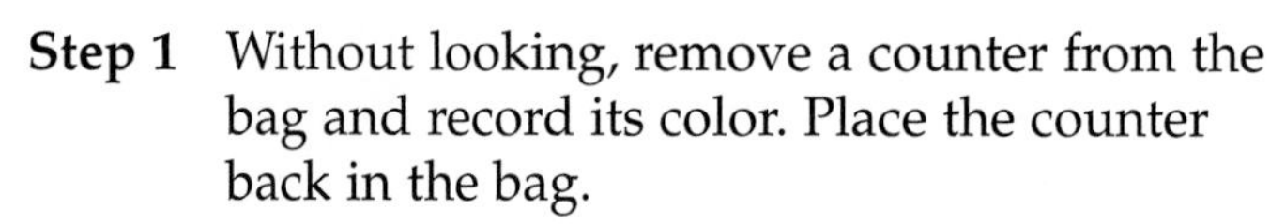

Step 2 Without looking, remove a second counter and record its color. The two colors are one trial. Place the counter back in the bag.

Step 3 Repeat until you have 50 trials. Count and record the number of times you chose a red counter, followed by a white counter.

a. What was your experimental probability for the red then white outcome?

b. Would you expect the probability to be different if you did not place the first counter back in the bag? Explain your reasoning.

PROBABILITIES OF INDEPENDENT AND DEPENDENT EVENTS

A **compound event** consists of two or more simple events. The activity above finds P(red *and* white), the probability of choosing a red counter, followed by a white counter.

The results from the activity above are independent events. In **independent events**, the outcome of one event does *not* influence the outcome of a second event.

$P(\text{red on 1st draw}) = \frac{2}{4} \text{ or } \frac{1}{2}$ ← There are 4 counters and 2 of them are red.

$P(\text{white on 2nd draw}) = \frac{2}{4} \text{ or } \frac{1}{2}$ ← You replaced the first counter. There are still 4 counters and 2 are white.

The probability of two independent events can be found by using multiplication.

Reading Math

Probability Notation
Read $P(A \text{ and } B)$ as *the probability of A followed by B.*

Key Concept — Probability of Two Independent Events

- **Words** The probability of two independent events is found by multiplying the probability of the first event by the probability of the second event.
- **Symbols** $P(A \text{ and } B) = P(A) \cdot P(B)$
- **Example** $P(\text{red and white}) = \frac{1}{2} \cdot \frac{1}{2} \text{ or } \frac{1}{4}$

Example 1 Probability of Independent Events

GAMES In some versions of the board game Parchisi, your piece returns to Start if you roll three doubles in a row. What is the probability of rolling three doubles in a row?

The events are independent since each roll of the number cubes does not affect the outcome of the next roll.

There are six ways to roll doubles, (1, 1), (2, 2), and so on, and there are 36 ways to roll two number cubes. So, the probability of rolling doubles on a toss of the number cubes is $\frac{6}{36}$ or $\frac{1}{6}$.

$$P(\text{three doubles}) = P(\text{doubles on 1st roll}) \cdot P(\text{doubles on 2nd roll}) \cdot P(\text{doubles on 3rd roll})$$

$$= \frac{1}{6} \cdot \frac{1}{6} \cdot \frac{1}{6}$$

$$= \frac{1}{216}$$

The probability of rolling three doubles in a row is $\frac{1}{216}$.

More About. . .

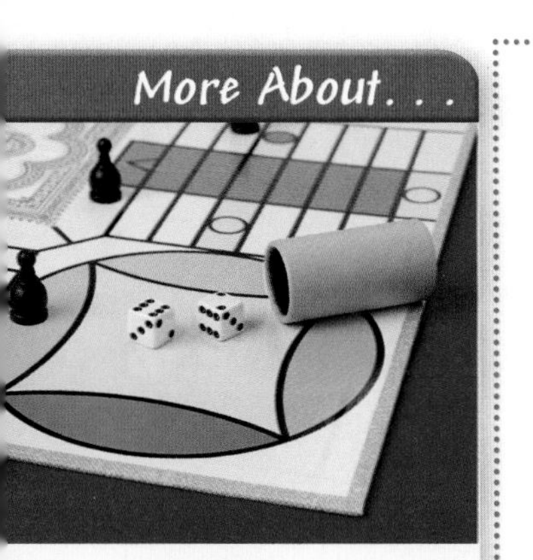

Games

Forms of the game Parchisi (also known as Parchesi or Parcheesi) have been in existence since the 4th century A.D.

Source: www.boardgames.about.com

If the outcome of one event affects the outcome of a second event, the events are called **dependent events**. In the opening activity, if you do not replace the first counter, the events are dependent events.

$P(\text{red on } 1^{st} \text{ draw}) = \frac{1}{2}$

$P(\text{white on } 2^{nd} \text{ draw}) = \frac{2}{3}$ ← If you do not replace the counter, there are three counters left and two of them are white.

Key Concept — Probability of Two Dependent Events

- **Words** If two events, *A* and *B*, are dependent, then the probability of both events occurring is the product of the probability of *A* and the probability of *B* after *A* occurs.
- **Symbols** $P(A \text{ and } B) = P(A) \cdot P(B \text{ following } A)$
- **Example** $P(\text{red and white, without replacement}) = \frac{1}{2} \cdot \frac{2}{3}$ or $\frac{1}{3}$

Concept Check How can you tell whether events are independent or dependent?

Example 2 Probability of Dependent Events

Reiko takes two coins at random from the 3 quarters, 5 dimes, and 2 nickels in her pocket. What is the probability that she chooses a quarter followed by a dime?

$$P(\text{quarter and dime}) = \frac{3}{10} \cdot \frac{5}{9}$$

3 of 10 coins are quarters.

5 of 9 remaining coins are dimes.

$$= \frac{15}{90} \text{ or } \frac{1}{6}$$

MUTUALLY EXCLUSIVE EVENTS

If two events cannot happen at the same time, they are said to be **mutually exclusive**. For example, when you roll two number cubes, you cannot roll a sum that is both 5 and even.

Second Number Cube (columns); **First Number Cube** (rows)

+	1	2	3	4	5	6
1	2	3	4	5	6	7
2	3	4	5	6	7	8
3	4	5	6	7	8	9
4	5	6	7	8	9	10
5	6	7	8	9	10	11
6	7	8	9	10	11	12

The probability of two mutually exclusive events is found by adding.

$$P(5 \text{ or even}) = P(5) + P(\text{even})$$

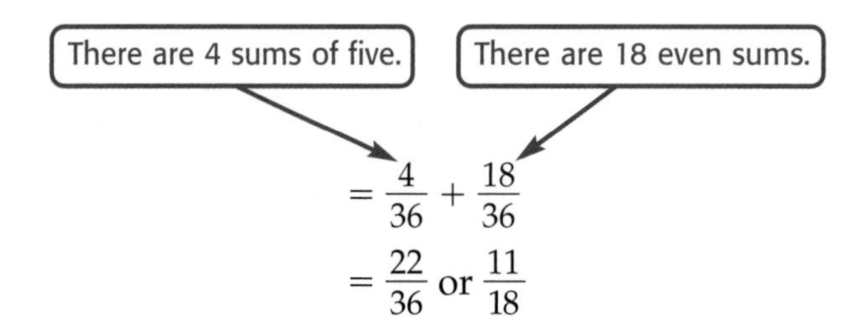
There are 4 sums of five.
There are 18 even sums.

$$= \frac{4}{36} + \frac{18}{36}$$

$$= \frac{22}{36} \text{ or } \frac{11}{18}$$

Study Tip

Look Back
To review **adding fractions with like denominators**, see Lesson 5-5.

Key Concept — *Probability of Mutually Exclusive Events*

- **Words** The probability of one or the other of two mutually exclusive events can be found by adding the probability of the first event to the probability of the second event.
- **Symbols** $P(A \text{ or } B) = P(A) + P(B)$
- **Example** $P(5 \text{ or even}) = \frac{4}{36} + \frac{18}{36} \text{ or } \frac{11}{18}$

Example 3 *Probability of Mutually Exclusive Events*

The spinner at the right is spun. What is the probability that the spinner will stop on blue or an even number?

6
1
5
2
4
3

The events are mutually exclusive because the spinner cannot stop on both blue and an even number at the same time.

$$P(\text{blue or even}) = P(\text{blue}) + P(\text{even})$$

$$= \frac{1}{6} + \frac{1}{2}$$

$$= \frac{4}{6} \text{ or } \frac{2}{3}$$

The probability that the spinner will stop on blue or an even number is $\frac{2}{3}$.

Concept Check Using the spinner in Example 3, are the events (green and even) mutually exclusive? Explain.

Check for Understanding

Concept Check

1. **Compare and contrast** independent and dependent events.
2. **OPEN ENDED** Write an example of two mutually exclusive events.
3. **Describe** how to find the probability of the second of two dependent events.

Guided Practice

A number cube is rolled and the spinner is spun. Find each probability.

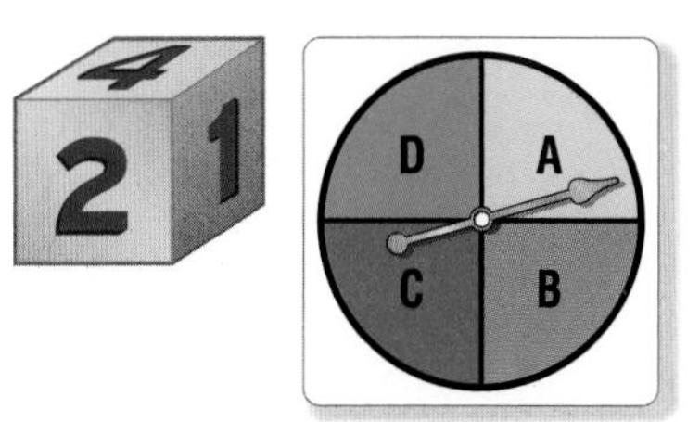

4. *P*(an odd number and a B)
5. *P*(a composite number and a vowel)

A card is drawn from a deck of eight cards numbered from 1 to 8. The card is not replaced and a second card is drawn. Find each probability.

6. *P*(5 and 2)
7. *P*(two odd numbers)
8. A card is drawn from a standard deck of 52 cards. What is the probability that it is a diamond or a club?
9. There are 3 books of poetry, 5 history books, and 4 books about animals on a shelf. If a book is chosen at random, what is the probability of choosing a book about history or animals?

Application

10. **GAMES** Jack is playing a board game that involves rolling two number cubes. He needs to roll a sum of 5 or 8 to land on an open space. What is the probability that he will land on an open space?

Practice and Apply

Homework Help

For Exercises	See Examples
11–14	1
15–20	2
21–26	3

Extra Practice

See page 755.

A number cube is rolled and the spinner is spun. Find each probability.

11. *P*(3 and E)
12. *P*(an even number and A)
13. *P*(a prime number and a vowel)
14. *P*(an odd number and a consonant)

There are 3 red marbles, 4 green marbles, 2 yellow marbles, and 5 blue marbles in a bag. Once a marble is drawn, it is not replaced. Find the probability of each outcome.

15. two yellow marbles in a row
16. two blue marbles in a row
17. a blue then a green marble
18. a yellow then a red marble
19. a blue marble, a yellow marble, and then a red marble
20. three green marbles in a row

An eight-sided die is rolled. Find the probability of each outcome.

21. P(3 or even)

22. P(6 or prime)

A card is drawn from the cards shown. Find the probability of each outcome.

23. P(3 or multiple of 2)

24. P(4 or greater than 5)

25. P(odd or even)

26. P(2 or 6)

27. A bag contains six blue marbles and three red marbles. A marble is drawn, it is replaced, and another marble is drawn. What is the probability of drawing a red marble and a blue marble in either order?

28. FAMILIES Each time a baby is born, the chance for either a boy or a girl is one-half. Find the probability that a family of four children has four girls.

INTERNET USE **For Exercises 29 and 30, use the graphic.**

29. What is the probability that a teen chosen at random has used the Internet for both games and studying in the last 30 days? Write the probability as a decimal to the nearest hundredth.

30. What is the probability that a teen chosen at random has used the Internet to both browse and use e-mail in the last 30 days? Write the probability as a percent to the nearest percent.

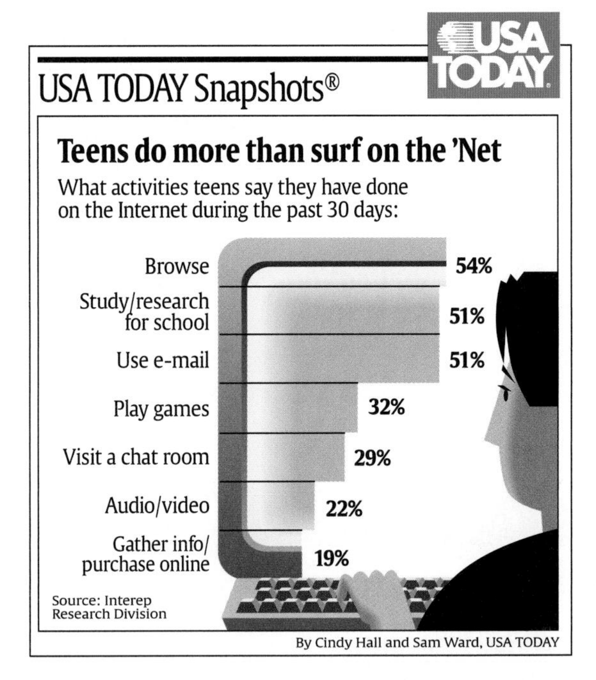

More About. . .

Internet

It is estimated that by the year 2005, one billion people worldwide will be connected to the Internet.

Source: *The World Almanac*

31. CRITICAL THINKING There are 9 marbles in a bag. Some are red, some are white, and some are blue. The probability of selecting a red marble, a white marble, and then a blue marble is $\frac{1}{21}$.

a. How many of each color are in the bag?

b. Explain why there is more than one correct answer to part **a**.

32. WRITING IN MATH Answer the question that was posed at the beginning of the lesson.

How are compound events related to simple events?

Include the following in your answer:

- an explanation of how the results of drawing two counters relate to the probability for drawing one counter, and
- the difference between independent and dependent events.

33. A bag contains three green balls, two blue balls, four pink balls, and one yellow ball, all the same size. Marcus chooses a ball at random, then without replacing it chooses a second ball. What is the probability that Marcus chooses a blue ball followed by a yellow ball?

(A) $\frac{1}{50}$ (B) $\frac{1}{45}$ (C) $\frac{2}{45}$ (D) $\frac{3}{20}$

34. If a card is drawn at random from a standard deck of 52 cards, what is the probability that the card drawn is a jack, queen, or king?

(A) $\frac{1}{13}$ (B) $\frac{1}{12}$ (C) $\frac{3}{52}$ (D) $\frac{3}{13}$

Extending the Lesson

35. When two events are *inclusive*, they can happen at the same time. To find the probability of inclusive events, add the probabilities of the events and subtract the probability of both events happening. Find P(green or even) for the spinner shown at the right.

Maintain Your Skills

Mixed Review **A card is drawn from a standard deck of 52 cards.** *(Lesson 12-8)*

36. Find the odds of selecting a red card.

37. Find the odds of selecting a jack, queen, or king.

38. How many different teams of 3 players can be chosen from 8 players? *(Lesson 12-7)*

39. How many license plates can be made from 3 letters (A–Z) and 3 numbers (0–9)? *(Lesson 12-7)*

In the figure at the right, $a \parallel b$. Find the measure of each angle. *(Lesson 10-1)*

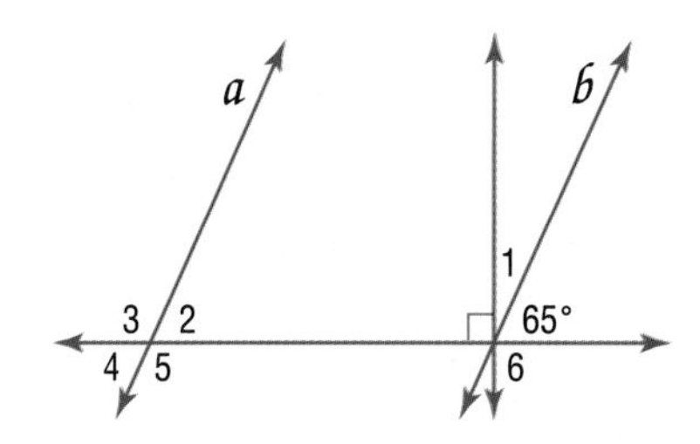

40. $\angle 1$
41. $\angle 2$
42. $\angle 3$
43. $\angle 4$
44. $\angle 5$
45. $\angle 6$

In a right triangle, if a and b are the measures of the legs and c is the measure of the hypotenuse, find each missing measure. Round to the nearest tenth. *(Lesson 9-5)*

46. $a = 6, b = 8$
47. $a = 7, b = 40$
48. $a = 8, c = 15$
49. $b = 63, c = 65$

50. **BLOOD TYPES** The distribution of blood types in a random survey is shown in the table. If there are 625 students at Ford Middle School, how many would you expect to have Type A blood? *(Lesson 6-7)*

Distribution of Blood Types

Type	Percent
A	40
B	9
AB	4
O	47

Algebra Activity

A Follow-Up of Lesson 12-9

Sunshine State Standards
MA.E.2.3.1-1, MA.E.3.3.1-1, MA.E.3.3.1-2, MA.E.3.3.1-3, MA.E.3.3.2-1

Simulations

You can use a **simulation** to act out a situation so that you can see outcomes. For many problems, you can conduct a simulation of the outcomes by using items such as a number cube, a coin, or a spinner. The items or combination of items used should have the same number of outcomes as the number of possible outcomes of the situation.

Activity 1

A quiz has 10 true-false questions. The correct answers are T, F, F, T, T, T, F, F, T, F. You need to correctly answer 7 or more questions to pass the quiz. Is tossing a coin to decide your answers a good strategy for taking the quiz?

Since two choices are available for each answer, tossing a coin is a reasonable activity to simulate guessing the answers.

Collect the Data

Step 1 Toss a coin and record the answer for each question. Write T(true) for tails and F(false) for heads.

Step 2 Repeat the simulation three times.

Step 3 Shade the cells with the correct answers. A sample for one simulation is shown in the table below.

Answers	T	F	F	T	T	T	F	F	T	F	Number Correct
Simulation 1	F	T	T	F	F	F	T	T	T	F	2
Simulation 2											
Simulation 3											

Analyze the Data

1. Based on the simulations, is tossing a coin a good way to take the quiz? Explain.

Extend the Activity

Use a simulation to act out the problem.

2. A restaurant includes prizes with children's meals. Six different prizes are available. There is an equally likely chance of getting each prize each time.
   a. Use a number cube to simulate this problem. Let each number represent one of the prizes. Conduct a simulation until you have one of each number.
   b. Based on your simulation, how many meals must be purchased in order to get all six different prizes?

Activity 2

Logan usually makes three out of every four free throws he attempts during a basketball game. Conduct the following experiment to simulate the probability of Logan's making two free throws in a row.

Collect the Data

Step 1 Put 20 red and white counters in a bag. Use red to represent a basket and white to represent a miss. The probability Logan will make a free throw is $\frac{3}{4}$, or $\frac{15}{20}$. So, use 15 red counters and 5 white counters.

Step 2 Conduct a simulation for 25 free throws.

Step 3 Without looking, draw a counter from the bag and record its color. Replace the counter and draw a second counter.

Step 4 Repeat 25 times and record the results of the simulation in a chart like the one shown below.

Misses the first shot	Makes the first shot, misses the second	Makes both shots

Analyze the Data

3. Calculate the experimental probability that Logan makes two free throws in a row.
4. How do the results in Activity 2 compare to the theoretical probability that Logan will make two free throws in a row? (*Hint*: These are independent events.)

Model the Data

5. Trevor usually makes four out of every five free throws he attempts. Calculate the theoretical probability that Trevor will make two free throws in a row.
6. To simulate the probability of Trevor making two free throws in a row, Drew puts 50 red and blue marbles in a bag. How many red and how many blue marbles should Drew use? Explain your reasoning.
7. Conduct a simulation for this situation. Compare the theoretical probability with the experimental probability.

Extend the Activity

8. There are three gumball machines numbered 1, 2, and 3 in a video arcade. Each machine contains an equal number of gumballs, some orange, some red, and some green. Ali thinks that her chance of getting a green gumball is the same from each machine. She conducted an experiment in which she bought 20 gumballs from each of the three machines. She got 4 green gumballs from machine 1, 8 from machine 2, and 12 from machine 3. Does this data support Ali's hypothesis? Explain why or why not.

Study Guide and Review

Vocabulary and Concept Check

back-to-back stem-and-leaf plot (p. 607)
box-and-whisker plot (p. 617)
combination (p. 642)
compound events (p. 650)
dependent events (p. 651)
factorial (p. 642)
Fundamental Counting Principle (p. 636)
histogram (p. 623)
independent events (p. 650)
interquartile range (p. 613)
leaves (p. 606)
lower quartile (p. 613)
measures of variation (p. 612)
mutually exclusive events (p. 652)
odds (p. 646)
permutation (p. 641)
quartiles (p. 613)
range (p. 612)
simulation (p. 656)
stem-and-leaf plot (p. 606)
stems (p. 606)
tree diagram (p. 635)
upper quartile (p. 613)

Choose the letter of the term that best matches each statement or phrase.

1. an arrangement or listing in which order is important
2. the ratio of the number of ways an event can occur to the ways that the event cannot occur
3. two or more events that cannot occur at the same time
4. an arrangement or listing in which order is not important
5. the median of the lower half of a set of data

a. combination
b. permutation
c. mutually exclusive events
d. lower quartile
e. odds

Lesson-by-Lesson Review

12-1 Stem-and-Leaf Plots

See pages 606–611.

Concept Summary

- A stem-and-leaf plot can be used to organize and display data.

Example **Display the data below in a stem-and-leaf plot.**

Lobster Length (mm)			
75	76	80	77
77	77	79	84
80	76	78	69
79	66	84	85

Stem	Leaf
6	6 9
7	5 6 6 7 7 7 8 9 9
8	0 0 4 4 5

$7|6 = 76$ mm

Exercises **Display each set of data in a stem-and-leaf plot.**
(See Example 1 on pages 606 and 607.)

6.

Height of Girls on Soccer Team (in.)			
58	62	59	60
61	65	57	56
55	59	62	61

7.

Price of Juice (¢)			
85	45	75	60
60	50	55	75
45	50	60	60
55	75	85	60

8.

Theater Attendance			
110	112	140	124
128	145	119	129
118	124	140	123
146	142	120	114

www.pre-alg.com/vocabulary_review

12-2 Measures of Variation

See pages 612–616.

Concept Summary

- The range is the difference between the greatest and the least values of a data set.
- The interquartile range is the range of the middle half of a set of data.

Example **Find the range and interquartile range for the set of data {18, 11, 26, 28, 15, 21, 20, 20, 15, 23, 19}.**

lower half: 11, 15, 15, 18, 19; 20; upper half: 20, 21, 23, 26, 28 — List the data from least to greatest.

LQ = 15, median = 20, UQ = 23

The range is 28 − 11 or 17. The interquartile range is 23 − 15 or 8.

Exercises **Find the range and interquartile range for each set of data.**
(See Examples 1 and 2 on pages 612 and 613.)

9. {42, 45, 38, 27, 41, 39, 50}
10. {7, 6, 1, 3, 4, 4, 5, 8, 11, 8, 5}
11. {58°, 64°, 72°, 62°, 74°, 80°, 65°, 70°}
12.

Stem	Leaf
2	4 4 9
3	0 2 3 3 6 7
4	5 7 8 9 9 9

3|2 = 32

12-3 Box-and-Whisker Plots

See pages 617–621.

Concept Summary

- A box-and-whisker plot separates data into four parts.

Example **Use the box-and-whisker plot shown to find the percent of New York City marathons that were held on days that had a high temperature greater than 72.5°F.**

Each of the four parts represents 25% of the data, so 25% of the marathons had a high temperature greater than 72.5°F.

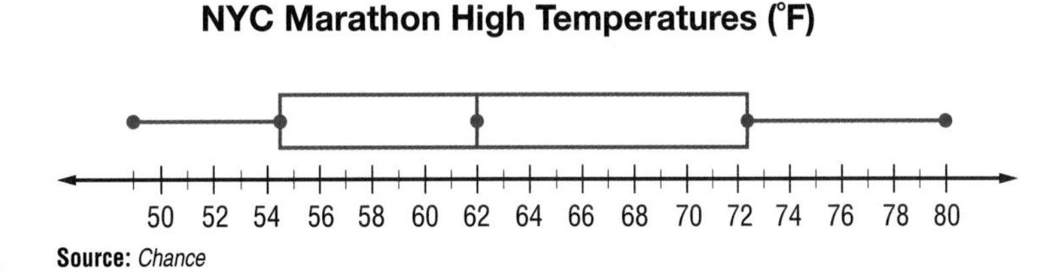

Exercises **For Exercises 13–15, use the box-and-whisker plot shown above.**
(See Example 2 on page 618.)

13. What was the highest temperature?
14. Half the marathons were held on days having at least what high temperature?
15. What percent of the marathons were held on days that had a high temperature between 54.5°F and 72.5°F?

12-4 Histograms

See pages 623–628.

Concept Summary

- A histogram displays data that have been organized into equal intervals.

Example **Display the set of data in a histogram.**

Boys' 50-Yard Dash		
Time (s)	**Tally**	**Frequency**
6.0–6.4	|||	3
6.5–6.9	𝍸		6
7.0–7.4	𝍸 |||	8
7.5–7.9	𝍸 |||	8
8.0–8.4	𝍸	5

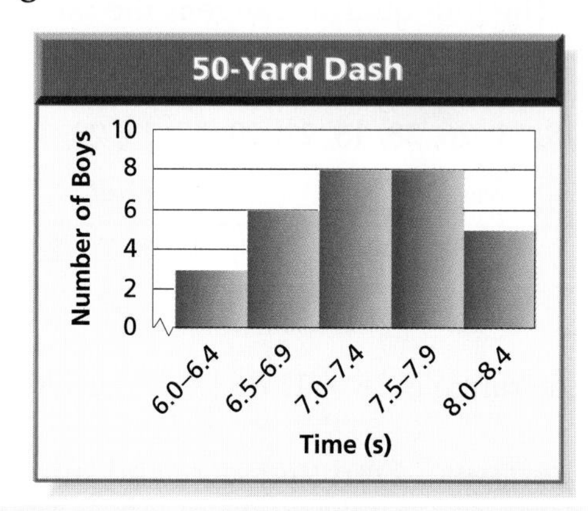

Exercises

16. The frequency table shows the results of a reading survey. Display the data in a histogram. *(See Example 1 on page 623.)*

Books Read in a Month		
Books	**Tally**	**Frequency**
0–1	𝍸 𝍸 𝍸	15
2–3	𝍸 𝍸 𝍸 |||	18
4–5	𝍸 𝍸 ||	12

12-5 Misleading Statistics

See pages 630–633.

Concept Summary

- Graphs that do not have a title or labels on the scales may be misleading.
- Graphs that use different vertical scales may be misleading.

Example **Explain why the graphs look different.**

The vertical scales are different.

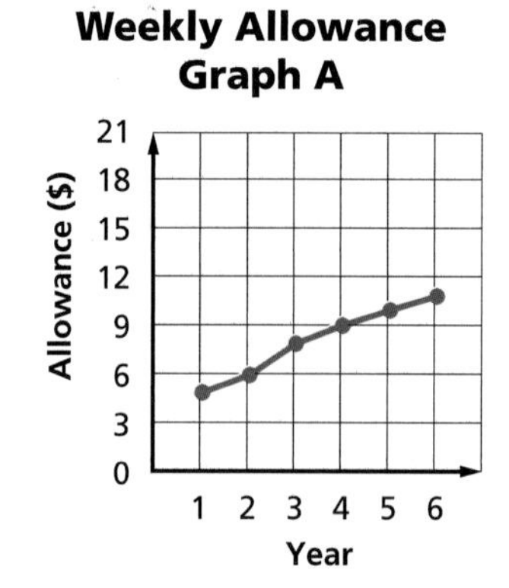

Weekly Allowance
Graph B

Allowance ($): 11, 10, 9, 8, 7, 6, 5, 0
Year: 1 2 3 4 5 6

Exercises **Refer to the graphs shown.** *(See Example 1 on page 630.)*

17. Which graph seems to show a slight increase in weekly allowances?

18. Which graph seems to show a dramatic increase in weekly allowance?

12-6 Counting

See pages 635–639.

Concept Summary

- The Fundamental Counting Principle relates the number of outcomes to the number of choices.

Example **A number cube is rolled three times. Find the number of possible outcomes.**

Outcomes on the first roll	times	outcomes on the second roll	times	outcomes on the third roll	equals	possible outcomes.
6	×	6	×	6	=	216

There are 216 possible outcomes.

Exercises **Find the number of possible outcomes for each situation.**
(See Example 2 on page 636.)

19. Four coins are tossed.

20. A tennis shoe comes in men's and women's sizes; cross training, walking, and running styles; blue, black, or white colors.

12-7 Permutations and Combinations

See pages 641–645.

Concept Summary

- Permutation: order is important.
- Combination: order is *not* important.

Examples

1 How many ways can 8 horses place first, second, and third in a race?

The order is important, so this is a permutation.

8 horses → ← Choose 3.

$P(8, 3) = 8 \cdot 7 \cdot 6$ ← 8 choices for first place; 7 choices for second place; 6 choices for third place

$= 336$

There are 336 ways for 8 horses to place first, second, and third.

2 An ice cream shop has 5 toppings from which to choose. How many different 2-topping sundaes are possible?

The order is not important, so this is a combination.

$C(5, 2) = \frac{P(5, 2)}{2!} = \frac{5 \cdot 4}{2 \cdot 1}$ or 10 different sundaes

Exercises **Tell whether each situation is a *permutation* or *combination*. Then solve.** *(See Examples 1 and 3 on pages 641 and 642.)*

21. How many ways can 3-person teams be chosen from 14 students?

22. How many 5-digit security codes are possible if each digit is a number from 0 to 9?

23. How many ways can you choose 2 team colors from a total of 7 colors?

• Extra Practice, see pages 752–755.
• Mixed Problem Solving, see page 769.

12-8 Odds

See pages 646–649.

Concept Summary

- The odds in favor of an event is the ratio that compares the number of ways the event can occur to the ways that the event *cannot* occur.

Example **Find the odds of spinning a red if the spinner at the right is spun.**

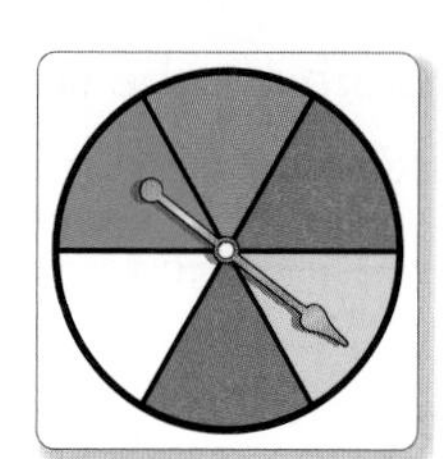

odds of spinning red

= number of ways to spin red to number of ways to spin any color other than red

= 2 : 4

= 1:2

The odds of spinning a red are 1:2.

Exercises **Find the odds of each outcome if the spinner at the right is spun.**
(See Example 1 on page 647.)

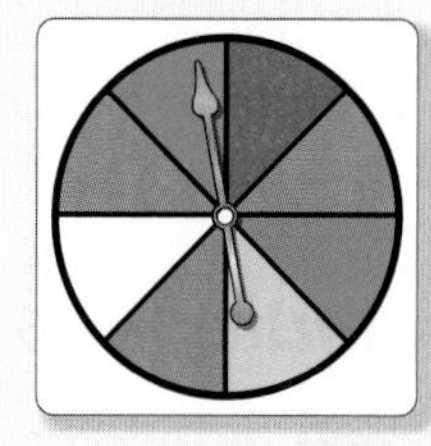

24. yellow
25. blue
26. not green
27. white or red

12-9 Probability of Compound Events

See pages 650–655.

Concept Summary

- A compound event consists of two or more simple events.
- When the outcome of one event does *not* affect the outcome of a second event, these are called independent events.
- When the outcome of one event *does* affect the outcome of a second event, these are called dependent events.

Example **There are 3 red, 4 purple, and 2 green marbles in a bag. Find the probability of randomly drawing a purple marble and then a green marble without replacement.**

P(purple, then green) = P(purple on 1st draw) · P(green on 2nd draw)

$$= \frac{4}{9} \cdot \frac{2}{8} \text{ or } \frac{1}{9}$$

The probability of drawing a purple marble and then a green marble is $\frac{1}{9}$.

Exercises **A card is drawn from a deck of six cards numbered from 1 to 6. Find each probability.** *(See Examples 1 and 3 on pages 651 and 652.)*

28. P(odd number or 2)

29. The card is not replaced, and a second card is drawn. Find P(3 and 6).

30. The card is replaced, and a second card is drawn. Find P(4 and 2).

Chapter 12 Practice Test

Vocabulary and Concepts

1. **OPEN ENDED** Describe a situation that involves a permutation and a situation that involves a combination.
2. **Explain** what it means when two events are mutually exclusive.

Skills and Applications

For Exercises 3–5, use the table shown.

Students' Heights (in.)							
70	81	59	69	78	68	75	73
76	62	67	74	75	64	60	58

3. Display the data in a stem-and-leaf plot.
4. What is the median height of the students?
5. In which interval do most of the heights occur?

For Exercises 6–8, use the stem-and-leaf plot shown.

Stem	Leaf
7	1 2 2 3 7
8	0 0 4 4 4 9
9	3 5 6 6 6 8 8 9

8|4 = $84

6. Find the range for the data.
7. Display the data in a box-and-whisker plot.
8. Find the interquartile range.

9. Display the data shown at the right in a histogram.

Length of Bus Ride to School		
Time (min)	Tally	Frequency
0–9	IIII	4
10–19	𝍸 II	7
20–29	𝍸 I	6

10. Find the number of possible outcomes for a choice of fish, chicken, pork, or beef and a choice of green beans, asparagus, or mixed vegetables.

Tell whether each situation is a *permutation* or *combination*. Then solve.

11. How many ways can 7 potted plants be arranged on a window sill?
12. A sand bucket contains 12 seashells. How many ways can you choose 3 of them?
13. Find the value of 4!.
14. Find the odds of rolling a number greater than 4 if a ten-sided die is rolled.
15. Four number cubes are tossed. What is the probability that all of them land on four?

A card is drawn from the cards shown. Find the probability of each outcome.

16. P(5 or even)
17. P(even or 1)
18. P(odd or even)
19. P(2 or greater than 5)

FCAT Practice

20. **STANDARDIZED TEST PRACTICE** Refer to the box-and-whisker plot shown. What percent of the daily high temperatures range from 70° to 95°?

Ⓐ 25% Ⓑ 50%
Ⓒ 75% Ⓓ 100%

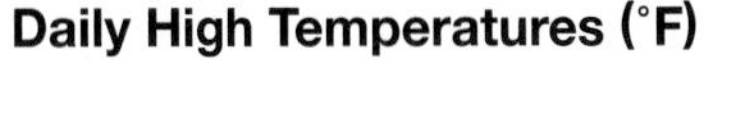

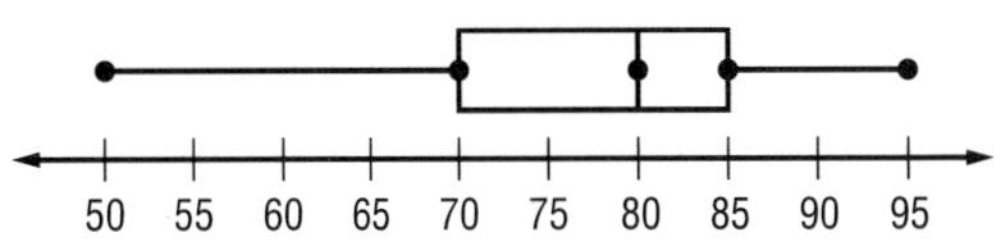

Chapter 12

Standardized Test Practice

FCAT Practice

Part 1 Multiple Choice

Record your answers on the answer sheet provided by your teacher or on a sheet of paper.

1. A package of 20 computer disks costs \$18.40. How much does each individual disk cost? (Prerequisite Skill, p. 715)

Ⓐ \$0.46 Ⓑ \$0.92
Ⓒ \$1.60 Ⓓ \$1.84

2. The point system for a basketball contest is shown below.

basket made	basket missed
gain 3 points	lose 3 points

Suppose Shantelle made 8 baskets and missed 4 baskets, then what was her total score? (Lesson 1-2)

Ⓐ 4 points Ⓑ 12 points
Ⓒ 24 points Ⓓ 32 points

3. An office has 45 light fixtures. Each fixture uses two light bulbs and each bulb costs \$0.89. Which expression could be used to find the total cost of replacing all of the bulbs in the office? (Lesson 1-2)

Ⓐ 45(0.89) Ⓑ 2(45 + 0.89)
Ⓒ $\frac{45}{2}$(0.89) Ⓓ 2(45)(0.89)

4. What is the value of m in the equation $\frac{5}{8} + m = \frac{3}{4}$? (Lesson 5-9)

Ⓐ $\frac{1}{2}$ Ⓑ $\frac{2}{5}$
Ⓒ $\frac{3}{8}$ Ⓓ $\frac{1}{8}$

5. Rita makes \$6.80 per hour. If she gets a 5% raise, what will be her new hourly rate? (Lessons 6-5 and 6-7)

Ⓐ \$0.34 Ⓑ \$3.40
Ⓒ \$7.14 Ⓓ \$10.20

6. Juliet recorded the distance and the time she walked every day.

Day	Distance (mi)	Time (min)
1	2	28
2	3	42
3	4	?

What is the best estimate of how many minutes it will take her to walk 4 miles? (Lesson 8-1)

Ⓐ 48 min Ⓑ 52 min
Ⓒ 56 min Ⓓ 64 min

7. A line passes through the points at (4, 0) and (8, 8). Which of the following points also lies on the line? (Lesson 8-7)

Ⓐ (2, −2) Ⓑ (6, 4)
Ⓒ (6, 6) Ⓓ (10, 6)

8. What is the surface area of the cube? (Lesson 11-4)

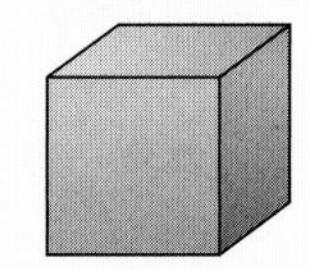

Volume: 27 in^3

Ⓐ 9 in^2 Ⓑ 36 in^2
Ⓒ 54 in^2 Ⓓ 324 in^2

9. How many different four-digit numbers can be formed using the digits 5, 6, 7, and 8 if each digit is used only once? (Lesson 12-7)

Ⓐ 26 Ⓑ 24
Ⓒ 12 Ⓓ 10

10. A bag contains 4 red marbles, 3 blue marbles, and 2 white marbles. One marble is chosen without replacement. Then another marble is chosen. What is the probability that the first marble is red and the second marble is blue? (Lesson 12-9)

Ⓐ $\frac{7}{72}$ Ⓑ $\frac{4}{27}$
Ⓒ $\frac{1}{6}$ Ⓓ $\frac{4}{9}$

The Princeton Review Test-Taking Tip

Questions 1–10 Eliminate the answer choices you know to be wrong. Then take your best guess from the choices that remain. If you can eliminate at least one answer choice, it is better to answer a question than to leave it blank.

FCAT Practice

Part 2 Short Response/Grid In

Record your answers on the answer sheet provided by your teacher or on a sheet of paper.

11. What is the value of $(0.3)^4$? (Lesson 4-2)

12. Sixteen pounds of ground beef will be divided into patties measuring one-quarter pound each. How many patties can be made? (Lesson 5-6)

13. Suppose the segment shown is translated 3 units to the left. What are the coordinates of the endpoints of the resulting segment? (Lesson 10-3)

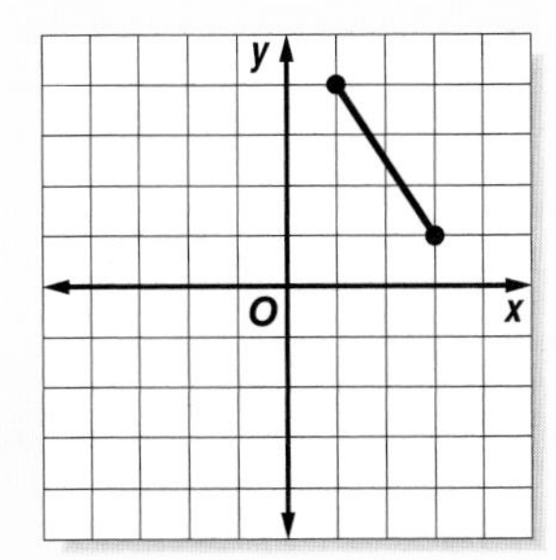

14. What is the area of the trapezoid? (Lesson 10-5)

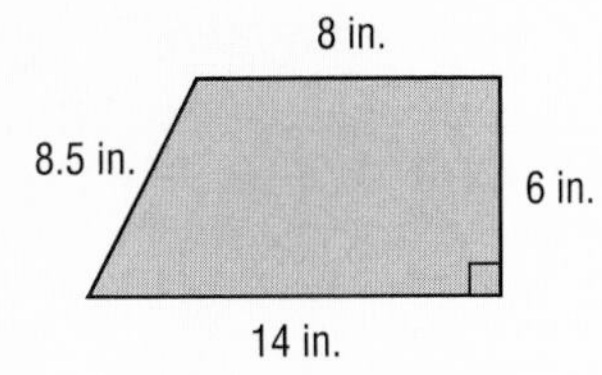

15. Use the formula $V = \frac{1}{3}\pi r^2 h$ to find the volume of a cone with a radius of 9 feet and a height of 12 feet. Round to the nearest tenth. (Lesson 11-3)

16. A can of soup is 12 cm high and has a diameter of 8 cm. A rectangular label is being designed for this can of soup. If the label will cover the surface of the can except for its top and bottom, what is the width and length of the label, to the nearest centimeter? (Lesson 11-4)

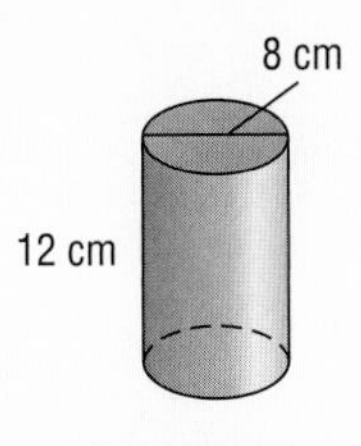

For Exercises 17 and 18, use the following box-and-whisker plot.

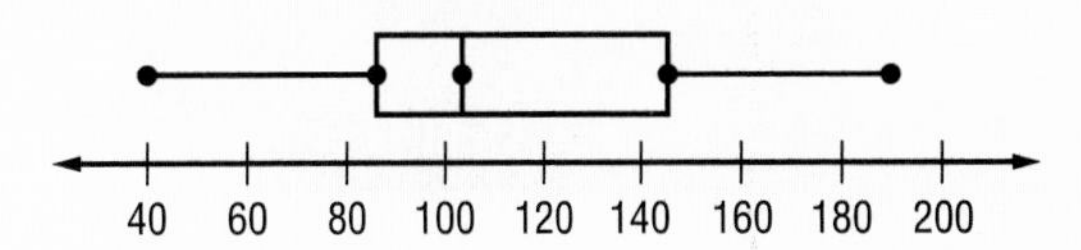

17. What is the median price of the scooters? (Lesson 12-3)

18. About what percent of these scooters cost less than $90? (Lesson 12-3)

19. Only two of the five school newspaper editors can represent the school at the state awards banquet. How many different combinations of two editors can be selected to go to the banquet? (Lesson 12-7)

FCAT Practice

Part 3 Open Ended

Record your answers on a sheet of paper. Show your work.

20. The scores on a mathematics test are given in the frequency table. (Lesson 12-4)

Score	Tally	Frequency
60–69	I	1
70–79	IIII	4
80–89	~~IIII~~	5
90–99	II	2

a. Display the data in a histogram.

b. Which interval contains the greatest number of test scores?

21. If you order a "surprise pizza" special at a certain restaurant, the restaurant chooses two toppings at random. The available toppings are pepperoni, sausage, onions, green peppers, mushrooms, and black olives. (Lessons 12-6, 12-7, and 12-9)

a. List all of the possible "surprise pizzas."

b. How many different "surprise pizzas" are available at this restaurant?

c. If you order a "surprise pizza," what is the probability that it will have pepperoni or sausage on it?

Chapter 13

Polynomials and Nonlinear Functions

What You'll Learn

- **Lesson 13-1** Identify and classify polynomials.
- **Lessons 13-2 through 13-4** Add, subtract, and multiply polynomials.
- **Lesson 13-5** Determine whether functions are linear or nonlinear.
- **Lesson 13-6** Explore different representations of quadratic and cubic functions.

Key Vocabulary

- **polynomial** (p. 669)
- **degree** (p. 670)
- **nonlinear function** (p. 687)
- **quadratic function** (p. 688)
- **cubic function** (p. 688)

Why It's Important

You have studied situations that can be modeled by linear functions. Many real-life situations, however, are not linear. These can be modeled using nonlinear functions. *You will use a nonlinear function in Lesson 13-6 to determine how far a skydiver falls in 4.5 seconds.*

Getting Started

Prerequisite Skills To be successful in this chapter, you'll need to master these skills and be able to apply them in problem-solving situations. Review these skills before beginning Chapter 13.

For Lesson 13-1 **Monomials**

Determine the number of monomials in each expression. *(For review, see Lesson 4-1.)*

1. $2x^3$ **2.** $a + 4$ **3.** $8s - 5t$

4. $x^2 + 3x - 1$ **5.** $\frac{1}{t}$ **6.** $9x^3 + 6x^2 + 8x - 7$

For Lesson 13-4 **Distributive Property**

Use the Distributive Property to write each expression as an equivalent algebraic expression. *(For review, see Lesson 3-1.)*

7. $5(a + 4)$ **8.** $2(3y - 8)$ **9.** $-4(1 + 8n)$

10. $6(x + 2y)$ **11.** $(9b - 9c)3$ **12.** $5(q - 2r + 3s)$

For Lesson 13-5 **Linear Functions**

Determine whether each equation is linear. *(For review, see Lesson 8-2.)*

13. $y = x - 2$ **14.** $y = x^2$ **15.** $y = -\frac{1}{2}x$

FOLDABLES™
Study Organizer

Make this Foldable to help you organize information about polynomials and nonlinear functions. Begin with a sheet of 11" × 17" paper.

Step 1 Fold

Fold the short sides toward the middle.

Step 2 Fold Again

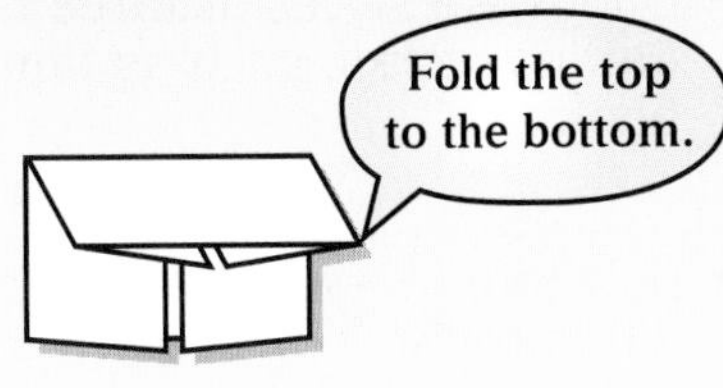
Fold the top to the bottom.

Step 3 Cut

Open. Cut along the second fold to make four tabs.

Step 4 Label

+ − × Functions

Label each of the tabs as shown.

Reading and Writing As you read and study the chapter, write examples of each concept under each tab.

Reading Mathematics

Prefixes and Polynomials

You can determine the meaning of many words used in mathematics if you know what the prefixes mean. In Lesson 4-1, you learned that the prefix *mono* means one and that a monomial is an algebraic expression with one term.

Monomials	Not Monomials
5	$x + y$
$2x$	$8n^2 - n + 1$
y^3	$a^3 + 4a^2 + a - 6$

The words in the table below are used in mathematics and in everyday life. They contain the prefixes *bi*, *tri*, and *poly*.

Prefix	Words	
bi	• bisect – to divide into two congruent parts • biannual – occurring twice a year • bicycle – a vehicle with two wheels	X Y Z **bisect**
tri	• triangle – a figure with three sides • triathlon – an athletic contest with three phases • trilogy – a series of three related literary works, such as films or books	A C B **triangle**
poly	• polyhedron – a solid with many flat surfaces • polychrome – having many colors • polygon – a figure with many sides	**polyhedron**

Reading to Learn

1. How are the words in each group of the table related?
2. What do the prefixes *bi*, *tri*, and *poly* mean?
3. Write the definition of *binomial*, *trinomial*, and *polynomial*.
4. Give an example of a binomial, a trinomial, and a polynomial.
5. **RESEARCH** Use the Internet or a dictionary to make a list of other words that have the prefixes *bi*, *tri*, and *poly*. Give the definition of each word.

13-1 Polynomials

What You'll Learn

- Identify and classify polynomials.
- Find the degree of a polynomial.

Vocabulary

- polynomial
- binomial
- trinomial
- degree

How are polynomials used to approximate real-world data?

Heat index is a way to describe how hot it feels outside with the temperature and humidity combined. Some examples are shown below.

	Temperature (°F)		
Humidity (%)	80	90	100
40	79	93	110
45	80	95	115
50	81	96	120

Heat Index

To calculate heat index, meteorologists use an expression similar to the one below. In this expression, x is the percent humidity, and y is the temperature.

$$-42 + 2x + 10y - 0.2xy - 0.007x^2 - 0.05y^2 + 0.001x^2y + 0.009xy^2 - 0.000002x^2y^2$$

a. How many terms are in the expression for the heat index?

b. What separates the terms of the expression?

CLASSIFY POLYNOMIALS Recall that a *monomial* is a number, a variable, or a product of numbers and/or variables. An algebraic expression that contains one or more monomials is called a **polynomial**. In a polynomial, there are no terms with variables in the denominator and no terms with variables under a radical sign.

Study Tip

Classifying Polynomials
Be sure expressions are written in simplest form.

- $x + x$ is the same as $2x$, so the expression is a monomial.
- $\sqrt{25}$ is the same as 5, so the expression is a monomial.

A polynomial with two terms is called a **binomial**, and a polynomial with three terms is called a **trinomial**.

Polynomial	Number of Terms	Examples
monomial	1	$4, x, 2y^3$
binomial	2	$x + 1, a - 5b, c^2 + d$
trinomial	3	$a + b + c, x^2 + 2x + 1$

The terms in a binomial or a trinomial may be added or subtracted.

Example 1 Classify Polynomials

Determine whether each expression is a polynomial. If it is, classify it as a *monomial, binomial,* or *trinomial.*

a. $2x^3 + 5x + 7$

This is a polynomial because it is the sum of three monomials. There are three terms, so it is a trinomial.

b. $t - \frac{1}{t^2}$

The expression is *not* a polynomial because $\frac{1}{t^2}$ has a variable in the denominator.

Concept Check Is $0.5x + 10$ a polynomial? Explain.

DEGREES OF POLYNOMIALS The **degree** of a monomial is the sum of the exponents of its variables. The degree of a nonzero constant such as 6 or 10 is 0. The constant 0 has no degree.

Example 2 Degree of a Monomial

Find the degree of each monomial.

a. $5a$

The variable a has degree 1, so the degree of $5a$ is 1.

b. $-3x^2y$

x^2 has degree 2 and y has degree 1. The degree of $-3x^2y$ is $2 + 1$ or 3.

Study Tip

Degrees
The degree of a is 1 because $a = a^1$.

A polynomial also has a degree. The degree of a polynomial is the same as that of the term with the greatest degree.

Example 3 Degree of a Polynomial

Find the degree of each polynomial.

a. $x^2 + 3x - 2$

term	degree
x^2	2
$3x$	1
2	0

The greatest degree is 2. So the degree of $x^2 + 3x - 2$ is 2.

b. $a^2 + ab^2 + b^4$

term	degree
a^2	2
ab^2	1 + 2 or 3
b^4	4

The greatest degree is 4. So the degree of $a^2 + ab^2 + b^4$ is 4.

Career Choices

Ecologist
An ecologist studies the relationships between organisms and their environment.

Online Research
For more information about a career as an ecologist, visit: www.pre-alg.com/careers

Example 4 Degree of a Real-World Polynomial

ECOLOGY **In the early 1900s, the deer population of the Kaibab Plateau in Arizona was affected by hunters and by the food supply. The population from 1905 to 1930 can be approximated by the polynomial $-0.13x^5 + 3.13x^4 + 4000$, where x is the number of years since 1900. Find the degree of the polynomial.**

$$\underbrace{-0.13x^5}_{\text{degree 5}} + \underbrace{3.13x^4}_{\text{degree 4}} + \underbrace{4000}_{\text{degree 0}}$$

So, $-0.13x^5 + 3.13x^4 + 4000$ has degree 5.

Concept Check Find the degree of the polynomial at the beginning of the lesson.

Check for Understanding

Concept Check

1. **Explain** how to find the degree of a monomial and the degree of a polynomial.
2. **OPEN ENDED** Write three binomial expressions. Explain why they are binomials.

3. **FIND THE ERROR** Carlos and Tanisha are finding the degree of $5x + y^2$.

Carlos	Tanisha
$5x$ has degree 1. y^2 has degree 2. $5x + y^2$ has degree 1 + 2 or 3.	$5x$ has degree 1. y^2 has degree 2. $5x + y^2$ has degree 2.

Who is correct? Explain your reasoning.

Guided Practice

Determine whether each expression is a polynomial. If it is, classify it as a *monomial, binomial,* or *trinomial.*

4. -7
5. $\frac{d}{2}$
6. $\frac{1}{x} - x$
7. $a^5 + a^3$
8. $y^2 - 4$
9. $x^2 + xy^2 - y^2$

Find the degree of each polynomial.

10. $4b^2$
11. 121
12. $8x^3y^2$
13. $3x + 5$
14. $r^3 + 7r$
15. $d^2 + c^4$

Application

GEOMETRY For Exercises 16 and 17, refer to the square at the right with a side length of *x* units.

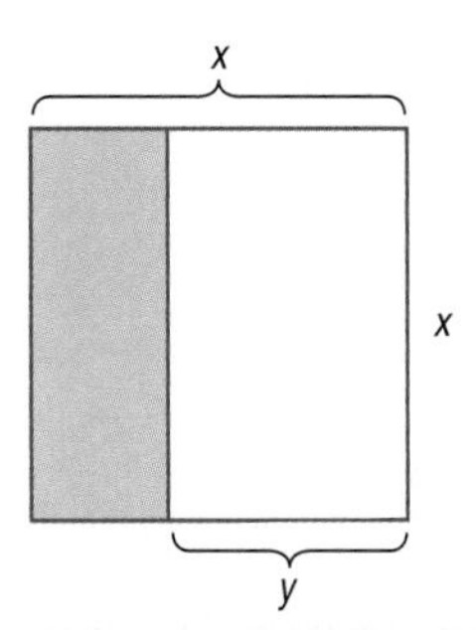

16. Write a polynomial expression for the area of the small blue rectangle.
17. What is the degree of the polynomial you wrote in Exercise 16?

Practice and Apply

Homework Help

For Exercises	See Examples
18–29	1
30–41	2, 3
44–46	4

Extra Practice

See page 755.

Determine whether each expression is a polynomial. If it is, classify it as a *monomial, binomial,* or *trinomial.*

18. 16
19. $x^2 - 7x$
20. $11a^2 + 4$
21. $-\frac{1}{3}w^2$
22. $\sqrt{15c}$
23. $8 - \frac{2}{k}$
24. $r^4 + r^2s^2$
25. $12 - n + n^4$
26. $ab^2 + 3a - b^2$
27. $\sqrt{y} + y$
28. $\frac{ab}{c} - c$
29. $x^2 - \frac{1}{2}x + \frac{1}{3}$

Find the degree of each polynomial.

30. 3
31. 56
32. ab
33. $12c^3$
34. xyz^2
35. $9s^4t$
36. $2 - 8n$
37. $g^5 + 5h$
38. $x^2 + 3x + 2$
39. $4y^3 + 6y^2 - 5y - 1$
40. $d^2 + c^4d^2$
41. $x^3 - x^2y^3 + 8$

Tell whether each statement is *always, sometimes,* or *never* true. Explain.

42. A trinomial has a degree of 3.
43. An integer is a monomial.

44. **MEDICINE** Doctors can study a patient's heart by injecting dye in a vein near the heart. In a normal heart, the amount of dye in the bloodstream after t seconds is given by $-0.006t^4 + 0.140t^3 - 0.53t^2 + 1.79t$. Find the degree of the polynomial.

LANDSCAPING For Exercises 45 and 46, use the information below and the diagram at the right. Lee wants to plant flowers along the perimeter of his vegetable garden.

45. Write a polynomial that represents the perimeter of the garden in feet.

46. What is the degree of the polynomial?

47. RESEARCH Suppose your grandparents deposited \$100 in your savings account each year on your birthday. On your fifth birthday, there would have been approximately $100x^4 + 100x^3 + 100x^2 + 100x + 100$ dollars, where x is the annual interest rate plus 1. Research the current interest rate at your family's bank. Using that interest rate, how much money would you have on your next birthday?

48. CRITICAL THINKING Find the degree of $a^{x+3} + x^{x-2}b^3 + b^{x+2}$.

49. WRITING IN MATH Answer the question that was posed at the beginning of the lesson.

How are polynomials used to approximate real-world data?

Include the following in your answer:

- a description of how the value of heat index is found, and
- an explanation of why a linear equation cannot be used to approximate the heat index data.

50. Choose the expression that is *not* a binomial.

Ⓐ $x^2 - 1$ Ⓑ $a + b$ Ⓒ $m^3 + n^3$ Ⓓ $7x + 2x$

51. State the degree of $4x^3 + xy - y^2$.

Ⓐ 1 Ⓑ 2 Ⓒ 3 Ⓓ 4

Maintain Your Skills

Mixed Review

A number cube is rolled. Determine whether each event is *mutually exclusive* or *inclusive*. Then find the probability. *(Lesson 12-9)*

52. P(odd or greater than 3)

53. P(5 or even)

54. A number cube is rolled. Find the odds that the number is greater than 2. *(Lesson 12-8)*

Find the volume of each solid. If necessary, round to the nearest tenth. *(Lesson 11-3)*

55.

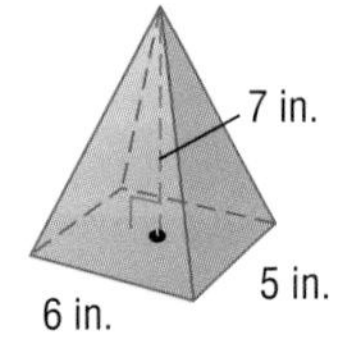

56.

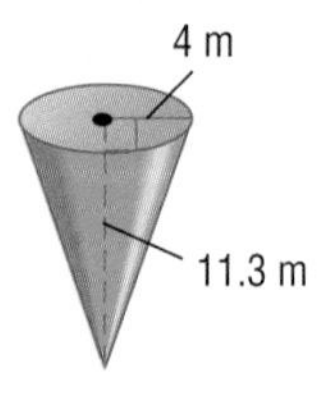

Getting Ready for the Next Lesson

PREREQUISITE SKILL Rewrite each expression using parentheses so that the terms having variables of the same power are grouped together. *(To review **properties of addition**, see Lesson 1-4.)*

57. $(x + 4) + 2x$

58. $3x^2 - 1 + x^2$

59. $(6n + 2) + (3n + 5)$

60. $(a + 2b) + (3a + b)$

61. $(s + t) + (5s - 3t)$

62. $(x^2 + 4x) + (7x^2 - 3x)$

Algebra Activity

A Follow-Up of Lesson 13-1

Modeling Polynomials with Algebra Tiles

In a set of algebra tiles, 1 represents the integer 1, x represents the variable x, and x^2 represents x^2. Red tiles are used to represent -1, $-x$, and $-x^2$.

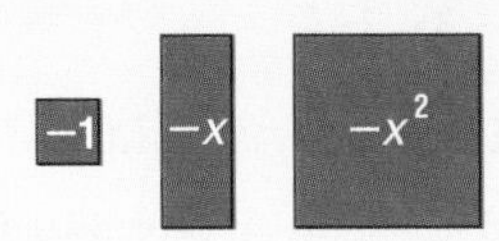

You can use these tiles to model monomials.

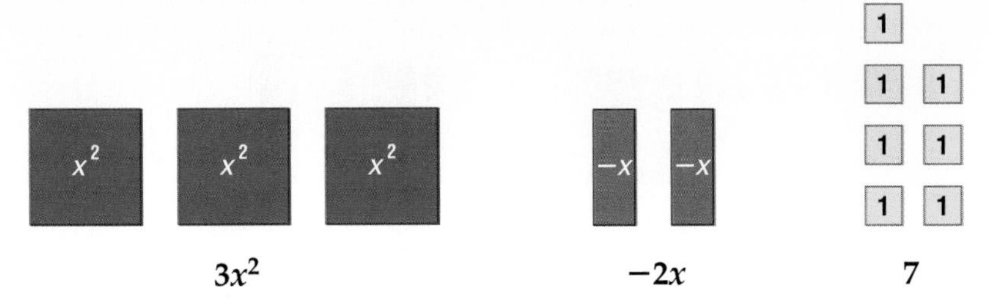

$3x^2$ $\quad$ $-2x$ $\quad$ 7

You can also use algebra tiles to model polynomials. The polynomial $2x^2 - 3x + 4$ is modeled below.

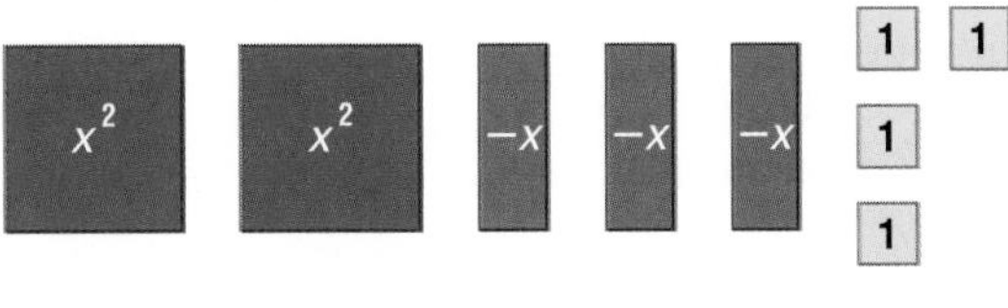

$2x^2 - 3x + 4$

Model and Analyze

Use algebra tiles to model each polynomial.

1. $-3x^2$ **2.** $5x + 3$ **3.** $4x^2 - x$ **4.** $2x^2 + 2x - 3$

5. Explain how you can tell whether an expression is a monomial, binomial, or trinomial by looking at the algebra tiles.

6. Name the polynomial modeled below.

7. Explain how you would find the degree of a polynomial using algebra tiles.

13-2 Adding Polynomials

What You'll Learn

- Add polynomials.

Sunshine State Standards
MA.A.3.3.1-3, MA.D.2.3.1-6, MA.D.2.3.1-7, MA.D.2.3.2-1

How can you use algebra tiles to add polynomials?

Consider the polynomials $2x^2 - 3x + 4$ and $-x^2 + x - 2$ modeled below.

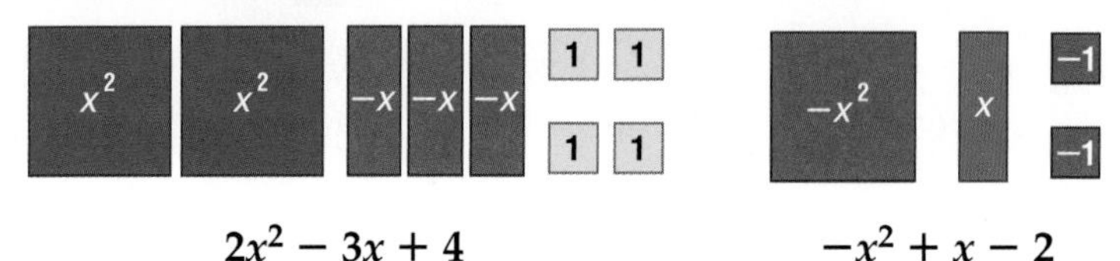

$2x^2 - 3x + 4$ $\qquad$ $-x^2 + x - 2$

Follow these steps to add the polynomials.

Step 1 Combine the tiles that have the same shape.

Step 2 When a positive tile is paired with a negative tile that is the same shape, the result is called a *zero pair*. Remove any zero pairs.

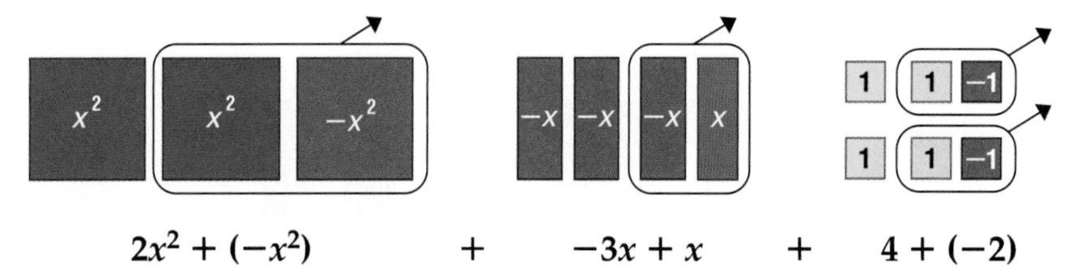

$2x^2 + (-x^2) \quad + \quad -3x + x \quad + \quad 4 + (-2)$

a. Write the polynomial for the tiles that remain.

b. Find the sum of $x^2 + 4x + 2$ and $7x^2 - 2x + 3$ by using algebra tiles.

c. **Compare and contrast** finding the sums of polynomials with finding the sum of integers.

Study Tip

Algebra Tiles
Tiles that are the same shape and size represent like terms.

ADD POLYNOMIALS Monomials that contain the same variables to the same power are *like terms*. Terms that differ only by their coefficient are called like terms.

Like Terms	Unlike Terms
$2x$ and $7x$	$-6a$ and $7b$
$-x^2y$ and $5x^2y$	$4ab^2$ and $4a^2b$

You can add polynomials by combining like terms.

Example 1 Add Polynomials

Find each sum.

a. $(3x + 5) + (2x + 1)$

Method 1 Add vertically.

$$\begin{array}{rl} 3x + 5 & \\ (+)\ 2x + 1 & \text{Align like terms.} \\ \hline 5x + 6 & \text{Add.} \end{array}$$

Method 2 Add horizontally.

$(3x + 5) + (2x + 1)$
$= (3x + 2x) + (5 + 1)$ Associative and Commutative Properties
$= 5x + 6$

The sum is $5x + 6$.

Study Tip

Negative Signs

- When a monomial has a negative sign, the coefficient is a negative number.
  $-4x \rightarrow$ coefficient is -4.
  $-b \rightarrow$ coefficient is -1.
- When a term in a polynomial is subtracted, "add its opposite" by making the coefficient negative.
  $x - 2y \rightarrow x + (-2y)$

b. $(2x^2 + x - 7) + (x^2 + 3x + 5)$

Method 1

$$\begin{array}{rl} 2x^2 + x - 7 & \\ (+)\ \ x^2 + 3x + 5 & \text{Align like terms.} \\ \hline 3x^2 + 4x - 2 & \text{Add.} \end{array}$$

Method 2

$(2x^2 + x - 7) + (x^2 + 3x + 5)$ Write the expression.

$= (2x^2 + x^2) + (x + 3x) + (-7 + 5)$ Group like terms.

$= 3x^2 + 4x - 2$ Simplify.

The sum is $3x^2 + 4x - 2$.

c. $(9c^2 + 4c) + (-6c + 8)$

$(9c^2 + 4c) + (-6c + 8)$ Write the expression.

$= 9c^2 + (4c - 6c) + 8$ Group like terms.

$= 9c^2 - 2c + 8$ Simplify.

The sum is $9c^2 - 2c + 8$.

d. $(x^2 + xy + 2y^2) + (6x^2 - y^2)$

$$\begin{array}{r} x^2 + xy + 2y^2 \\ (+)\ 6x^2 - \ \ y^2 \\ \hline 7x^2 + xy + \ \ y^2 \end{array}$$

Leave a space because there is no other term like *xy*.

The sum is $7x^2 + xy + y^2$.

Concept Check Name the like terms in $b^2 + 5b - ab + 9b^2$.

Polynomials are often used to represent measures of geometric figures.

Example 2 Use Polynomials to Solve a Problem

GEOMETRY The lengths of the sides of golden rectangles are in the ratio 1:1.62. So, the length of a golden rectangle is approximately 1.62 times greater than the width.

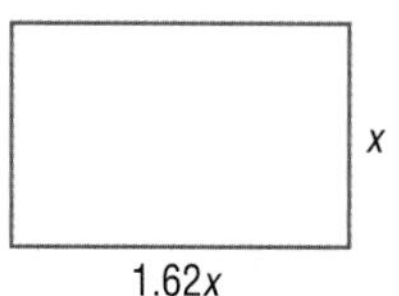

a. Find a formula for the perimeter of a golden rectangle.

$P = 2\ell + 2w$ Formula for the perimeter of a rectangle

$P = 2(1.62x) + 2x$ Replace ℓ with $1.62x$ and w with x.

$P = 3.24x + 2x$ or $5.24x$ Simplify.

A formula for the perimeter of a golden rectangle is $P = 5.24x$, where x is the measure of the width.

b. Find the length and the perimeter of a golden rectangle if its width is 8.3 centimeters.

length $= 1.62x$ Length of a golden rectangle

$= 1.62(8.3)$ or 13.446 Replace x with 8.3 and simplify.

perimeter $= 5.24x$ Perimeter of a golden rectangle

$= 5.24(8.3)$ or 43.492 Replace x with 8.3 and simplify.

The length of the golden rectangle is 13.446 centimeters, and the perimeter is 43.492 centimeters.

More About. . .

Geometry

The ancient Greeks often incorporated the golden ratio into their art and architecture.

Source: www.mcn.net

Check for Understanding

Concept Check

1. **Name** the like terms in $(x^2 + 5x + 2) + (2x^2 - 4x + 7)$.

2. **OPEN ENDED** Write two binomials that share only one pair of like terms.

3. **FIND THE ERROR** Hai says that $7xyz$ and $2zyx$ are like terms. Devin says they are not. Who is correct? Explain your reasoning.

Guided Practice

Find each sum.

4. $\begin{array}{r} 4x + 5 \\ (+)\ -x - 3 \\ \hline \end{array}$

5. $\begin{array}{r} 3a^2 - 9a + 6 \\ (+)4a^2 \qquad -2 \\ \hline \end{array}$

6. $(x + 3) + (2x + 5)$

7. $(13x - 7y) + 3y$

8. $(2x^2 + 5x) + (9 - 7x)$

9. $(3x^2 - 2x + 1) + (x^2 + 5x - 3)$

Application

10. **GEOMETRY** Find the perimeter of the figure at the right.

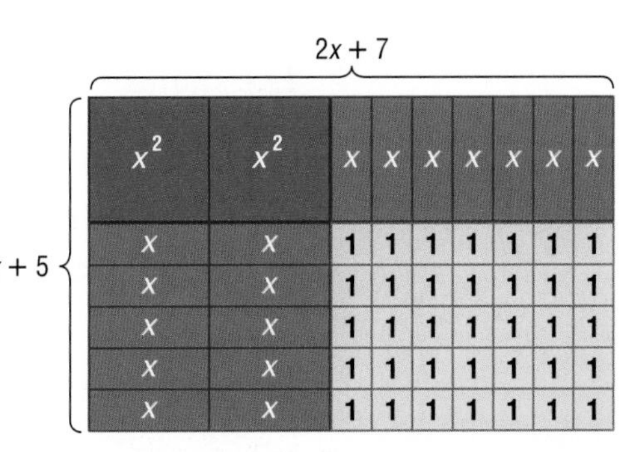

Practice and Apply

Homework Help

For Exercises	See Examples
11–22	1
23–27	2

Extra Practice

See page 756.

Find each sum.

11. $\begin{array}{r} -5x + 4 \\ (+)\ 8x - 1 \\ \hline \end{array}$

12. $\begin{array}{r} 7b - 5 \\ (+)-9b + 8 \\ \hline \end{array}$

13. $\begin{array}{r} 10x^2 + 5xy + 7y^2 \\ (+)\quad x^2 \qquad\quad -3y^2 \\ \hline \end{array}$

14. $\begin{array}{r} 4a^3 + a^2 + 8a - 8 \\ (+)\qquad 2a^2 \qquad + 6 \\ \hline \end{array}$

15. $(3x + 9) + (x + 5)$

16. $(4x + 3) + (x - 1)$

17. $(6y - 5r) + (2y + 7r)$

18. $(8m - 2n) + (3m + n)$

19. $(x^2 + y) + (4x^2 + xy)$

20. $(3a^2 + b^2) + (3a + b^2)$

21. $(5x^2 + 6x + 4) + (2x^2 + 3x + 1)$

22. $(-2x^2 + x - 5) + (x^2 - 3x + 2)$

Find each sum. Then evaluate if $a = -3$, $b = 4$, and $c = 2$.

23. $(3a + 5b) + (2a - 9b)$

24. $(a^2 + 7b^2) + (5 - 3b^2) + (2a^2 - 7)$

25. $(3a + 5b - 4c) + (2a - 3b + 7c) + (-a + 4b - 2c)$

GEOMETRY **For Exercises 26–28, refer to the triangle.**

26. Find the sum of the measures of the angles.

27. The sum of the measures of the angles in any triangle is 180°. Find the value of x.

28. Find the measure of each angle.

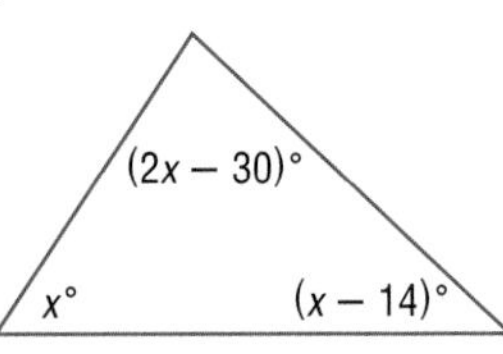

FINANCE **For Exercises 29–31, refer to the information below.**
Jason and Will both work at the same supermarket and are paid the same hourly rate. At the end of the week, Jason's paycheck showed that he worked 23 hours and had $12 deducted for taxes. Will worked 19 hours during the same week and had $10 deducted for taxes. Let x represent the hourly pay.

29. Write a polynomial expression to represent Jason's pay for the week.

30. Write a polynomial expression to represent Will's pay for the week.

31. Write a polynomial expression to represent the total weekly pay for Jason and Will.

32. CRITICAL THINKING In the figure at the right, x^2 is the area of the larger square, and y^2 is the area of each of the two smaller squares. What is the perimeter of the whole rectangle? Explain.

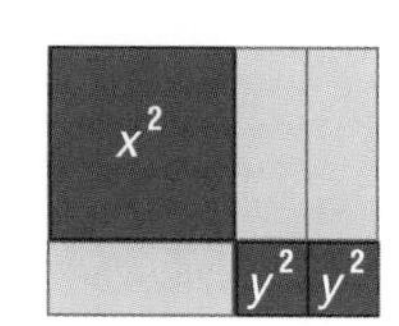

33. WRITING IN MATH Answer the question that was posed at the beginning of the lesson.

How can you use algebra tiles to add polynomials?

Include the following in your answer:

- a description of algebra tiles that represent like terms, and
- an explanation of how zero pairs are used in adding polynomials.

34. Choose the pair of terms that are *not* like terms.

(A) $6cd$, $12cd$ (B) $\frac{x}{2}$, $5x$ (C) a^2, b^2 (D) x^2y, yx^2

35. What is the sum of $11x + 2y$ and $x - 5y$?

(A) $10x - 3y$ (B) $12x - 3y$ (C) $12x + 3y$ (D) $12x - 5y$

Maintain Your Skills

Mixed Review **Find the degree of each polynomial.** *(Lesson 13-1)*

36. a^3b **37.** $3x - 5y + z^2$ **38.** $c^2 - 7c^3y^4$

A card is drawn from a standard deck of 52 playing cards. Find each probability. *(Lesson 12-9)*

39. P(2 or jack) **40.** P(10 or red) **41.** P(ace or black 7)

42. Determine whether the prisms are similar. Explain. *(Lesson 11-6)*

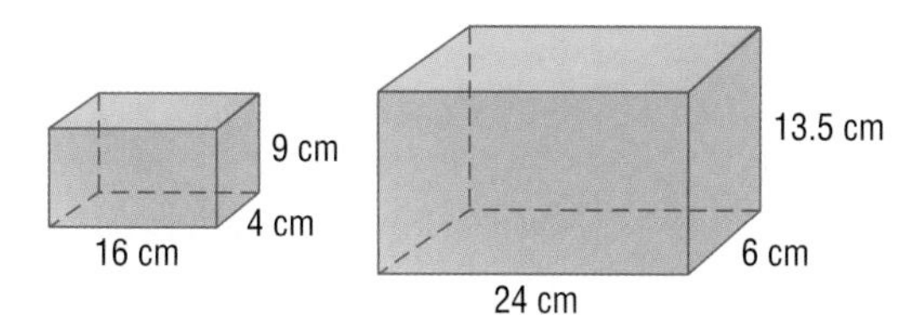

Getting Ready for the Next Lesson **PREREQUISITE SKILL** **Rewrite each expression as an addition expression by using the additive inverse.** *(To review **additive inverse**, see Lesson 2-3.)*

43. $15c - 26$ **44.** $x^2 - 7$ **45.** $1 - 2x$

46. $6b - 3a^2$ **47.** $(n + rt) - r^2$ **48.** $(s + t) - 2s$

13-3 Subtracting Polynomials

Sunshine State Standards
MA.A.3.3.1-3, MA.D.2.3.1-7, MA.D.2.3.2-1

What You'll Learn

- Subtract polynomials.

How is subtracting polynomials similar to subtracting measurements?

At the North Pole, buoy stations drift with the ice in the Arctic Ocean. The table shows the latitudes of two North Pole buoys in April, 2000.

Station	Latitude
1	89° 35.4'N = 89 degrees 35.4 minutes
5	85° 27.3'N = 85 degrees 27.3 minutes

a. What is the difference in degrees and the difference in minutes between the two stations?

b. Explain how you can find the difference in latitude between any two locations, given the degrees and minutes.

c. The longitude of Station 1 is 162° 16′ 36″ and the longitude of Station 5 is 68° 8′ 2″. Find the difference in longitude between the two stations.

Reading Math

Symbols
The symbol " in 68° 8' 2" is read as *seconds*.

SUBTRACT POLYNOMIALS When you subtract measurements, you subtract like units. Consider the subtraction of latitude measurements shown below.

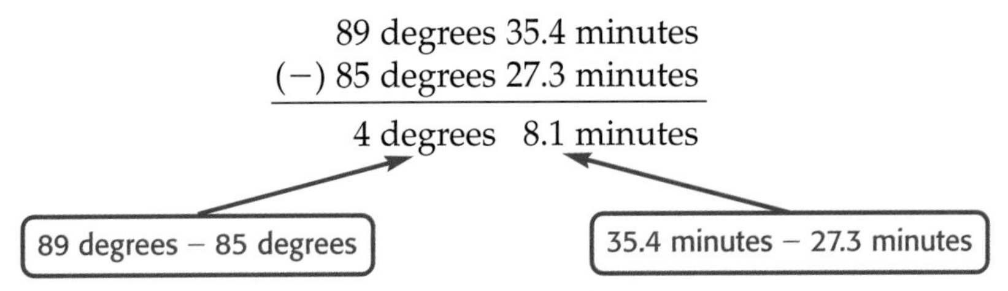

89 degrees 35.4 minutes
(−) 85 degrees 27.3 minutes
4 degrees 8.1 minutes
89 degrees − 85 degrees
35.4 minutes − 27.3 minutes

Similarly, when you subtract polynomials, you subtract like terms.

$$\begin{array}{r} 5x^2 + 14x - 9 \\ (-)\ x^2 + 8x + 2 \\ \hline 4x^2 + 6x - 11 \end{array}$$

$5x^2 - 1x^2 = 4x^2$

$14x - 8x = 6x$

$-9 - 2 = -11$

Example 1 Subtract Polynomials

Find each difference.

a. $(5x + 9) - (3x + 6)$

$$\begin{array}{r} 5x + 9 \\ (-)\ 3x + 6 \\ \hline 2x + 3 \end{array}$$

Align like terms. Subtract.

The difference is $2x + 3$.

b. $(4a^2 + 7a + 4) - (3a^2 + 2)$

$$\begin{array}{r} 4a^2 + 7a + 4 \\ (-)\ 3a^2 \qquad + 2 \\ \hline a^2 + 7a + 2 \end{array}$$

Align like terms. Subtract.

The difference is $a^2 + 7a + 2$.

Recall that you can subtract a rational number by adding its *additive inverse.*

$10 - 8 = 10 + (-8)$ The additive inverse of 8 is -8.

You can also subtract a polynomial by adding its additive inverse. To find the additive inverse of a polynomial, multiply the entire polynomial by -1.

Polynomial	Multiply by -1	Additive Inverse
t	$-1(t)$	$-t$
$x + 3$	$-1(x + 3)$	$-x - 3$
$-a^2 + b^2 - c$	$-1(-a^2 + b^2 - c)$	$a^2 - b^2 + c$

Example 2 Subtract Using the Additive Inverse

Find each difference.

a. $(3x + 8) - (5x + 1)$

The additive inverse of $5x + 1$ is $(-1)(5x + 1)$ or $-5x - 1$.

$(3x + 8) - (5x + 1)$

$= (3x + 8) + (-5x - 1)$ To subtract $(5x + 1)$, add $(-5x - 1)$.

$= (3x - 5x) + (8 - 1)$ Group the like terms.

$= -2x + 7$ Simplify.

The difference is $-2x + 7$.

b. $(4x^2 + y^2) - (-3xy + y^2)$

The additive inverse of $-3xy + y^2$ is $(-1)(-3xy + y^2)$ or $3xy - y^2$. Align the like terms and add the additive inverse.

$$\begin{array}{lr} & 4x^2 \qquad + y^2 \\ (-) & -3xy + y^2 \\ \hline \end{array} \rightarrow \begin{array}{lr} & 4x^2 \qquad + y^2 \\ (+) & 3xy - y^2 \\ \hline & 4x^2 + 3xy + 0 \end{array}$$

The difference is $4x^2 + 3xy$.

Study Tip

Zeros

It can be helpful to add zeros as placeholders when a term in one polynomial does not have a corresponding like term in another polynomial.

$$\begin{array}{lr} & 4x^2 + 0xy + y^2 \\ (+) & 0x^2 + 3xy - y^2 \\ \hline \end{array}$$

Concept Check What is the additive inverse of $a^2 + 9a - 1$?

Example 3 Subtract Polynomials to Solve a Problem

SHIPPING **The cost for shipping a package that weighs x pounds from Dallas to Chicago is shown in the table at the right. How much more does the Atlas Service charge for shipping the package?**

Shipping Company	Cost ($)
Atlas Service	$4x + 280$
Bell Service	$3x + 125$

difference in cost = cost of Atlas Service − cost of Bell Service

$= (4x + 280) - (3x + 125)$ Substitution

$= (4x + 280) + (-3x - 125)$ Add additive inverse.

$= (4x - 3x) + (280 - 125)$ Group like terms.

$= x + 155$ Simplify.

The Atlas Service charges $x + 155$ dollars more for shipping a package that weighs x pounds.

Check for Understanding

Concept Check

1. **Describe** how subtraction and addition of polynomials are related.
2. **OPEN ENDED** Write two polynomials whose difference is $x^2 + 2x - 4$.

Guided Practice

Find each difference.

3. $\begin{array}{r} r^2 + 5r \\ (-)\ r^2 + \ \ r \\ \hline \end{array}$

4. $\begin{array}{r} 3x^2 + 5x + 4 \\ (-)\ x^2 \qquad - 1 \\ \hline \end{array}$

5. $(9x + 5) - (4x + 3)$
6. $(2x + 4) - (-x + 5)$
7. $(3x^2 + x) - (8 - 2x)$
8. $(6a^2 - 3a + 9) - (7a^2 + 5a - 1)$

Application

9. **GEOMETRY** The perimeter of the isosceles trapezoid shown is $16x + 1$ units. Find the length of the missing base of the trapezoid.

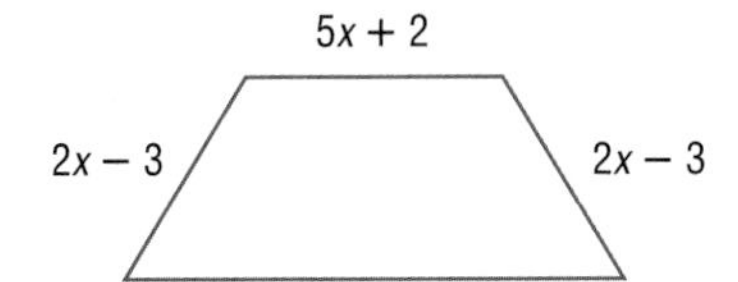

Practice and Apply

Homework Help

For Exercises	See Examples
10–15	1
16–25	2
26, 27	3

Extra Practice

See page 756.

Find each difference.

10. $\begin{array}{r} 8k + 9 \\ (-)\ k + 2 \\ \hline \end{array}$

11. $\begin{array}{r} -n^2 + 1n \\ (-)\ \ n^2 - 5n \\ \hline \end{array}$

12. $\begin{array}{r} 5a^2 + 9a - 12 \\ (-)\ -3a^2 + 5a - \ \ 7 \\ \hline \end{array}$

13. $\begin{array}{r} 6y^2 - 5y + 3 \\ (-)\ 5y^2 + 2y - 7 \\ \hline \end{array}$

14. $\begin{array}{r} 5x^2 - 4xy \qquad \\ (-)\qquad - 3xy + 2y^2 \\ \hline \end{array}$

15. $\begin{array}{r} 9w^2 \qquad + 7 \\ (-)-6w^2 + 2w - 3 \\ \hline \end{array}$

16. $(3x + 4) - (x + 2)$
17. $(7x + 5) - (3x + 2)$
18. $(2y + 5) - (y + 8)$
19. $(3t - 2) - (5t - 4)$
20. $(2x + 3y) - (x - y)$
21. $(a^2 + 6b^2) - (-2a^2 + 4b^2)$
22. $(x^2 + 6x) - (3x^2 + 7)$
23. $(9n^2 - 8) - (n + 4)$
24. $(6x^2 + 3x + 9) - (2x^2 + 8x + 1)$
25. $(3x^2 - 5xy + 7y^2) - (x^2 - 3xy + 4y^2)$

26. **GEOMETRY** Alyssa plans to trim a picture to fit into a frame. The area of the picture is $2x^2 + 11x + 12$ square units, but the area inside the frame is only $2x^2 + 5x + 2$ square units. How much of the picture will Alyssa have to trim so that it will fit into the frame?

27. **TEMPERATURE** The highest recorded temperature in North Carolina occurred in 1983. The lowest recorded temperature in North Carolina occurred two years later. The difference between these two record temperatures is 68°F more than the sum of the temperatures. Write an equation to represent this situation. Then find the record low temperature in North Carolina.

More About. . .

Temperature

The highest temperature ever officially recorded in the United States was 134°F on July 10, 1913, in Death Valley, California.

Source: www.infoplease.com

28. **CRITICAL THINKING** Suppose A and B represent polynomials. If $A + B = 3x^2 + 2x - 2$ and $A - B = -x^2 + 4x - 8$, find A and B.

29. **WRITING IN MATH** Answer the question that was posed at the beginning of the lesson.

How is subtracting polynomials similar to subtracting measurements?

Include the following in your answer:

- a comparison between subtracting measurements with two parts and subtracting polynomials with two terms, and
- an example of a subtraction problem involving measurements that have two parts, and an explanation of how to find the difference.

30. What is $(5x - 7) - (3x - 4)$?

Ⓐ $2x - 3$ Ⓑ $2x + 3$ Ⓒ $2x - 11$ Ⓓ $2x + 11$

31. Write the additive inverse of $-4h^2 - hk - k^2$.

Ⓐ $4h^2 - hk - k^2$ Ⓑ $4h^2 + hk + k^2$

Ⓒ $-4h^2 + hk + k^2$ Ⓓ $-4h^2 + hk - k^2$

Maintain Your Skills

Mixed Review

Find each sum. *(Lesson 13-2)*

32. $(2x - 3) + (x - 1)$
33. $(11x + 2y) + (x - 5y)$
34. $(5x^2 - 7x + 9) + (3x^2 + 4x - 6)$
35. $(4t - t^2) + (8t + 2)$

Determine whether each expression is a polynomial. If it is, classify it as a *monomial, binomial,* or *trinomial.* *(Lesson 13-1)*

36. $\frac{1}{5a^2}$
37. $x^2 + 9$
38. $c^2 - d^3 + cd$

39. Make a stem-and-leaf plot for the set of data shown below. *(Lesson 12-1)*
72, 64, 68, 66, 70, 89, 91, 54, 59, 71, 71, 85

Getting Ready for the Next Lesson

PREREQUISITE SKILL Simplify each expression.
*(To review **multiplying monomials**, see Lesson 4-2.)*

40. $x(3x)$
41. $(2y)(4y)$
42. $(t^2)(6t)$
43. $(4m)(m^2)$
44. $(w^2)(-3w)$
45. $(2r^2)(5r^3)$

Practice Quiz 1

Lessons 13-1 through 13-3

Find the degree of each polynomial. *(Lesson 13-1)*

1. cd^3
2. $a - 4a^2$
3. $x^2y + 7x^2 - 21$

Find each sum or difference. *(Lessons 13-2 and 13-3)*

4. $(2x - 8) + (x - 7)$
5. $(4x + 5) - (2x + 3)$
6. $(5d^2 - 3) - (2d^2 - 7)$
7. $(3r + 6s) + (5r - 9s)$
8. $(x^2 + 4x + 2) + (7x^2 - 2x + 3)$
9. $(9x - 4y) - (12x - 9y)$

10. **GEOMETRY** The perimeter of the triangle is $8x + 3y$ centimeters. Find the length of the third side. *(Lessons 13-2 and 13-3)*

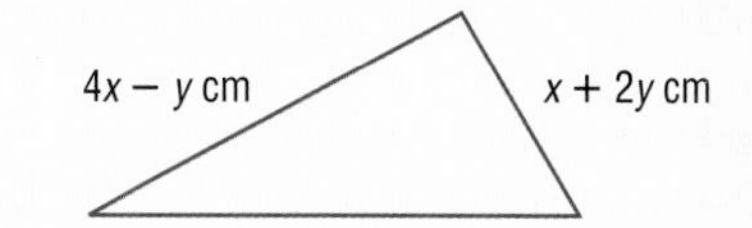

Algebra Activity

A Preview of Lesson 13-4

Modeling Multiplication

Recall that algebra tiles are named based on their area. The area of each tile is the product of the width and length.

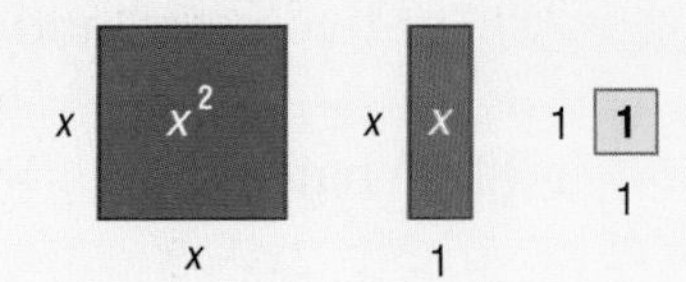

These algebra tiles can be placed together to form a rectangle whose length and width each represent a polynomial. The area of the rectangle is the product of the polynomials.

Use algebra tiles to find $x(x + 2)$.

Step 1 Make a rectangle with a width of x and a length of $x + 2$. Use algebra tiles to mark off the dimensions on a product mat.

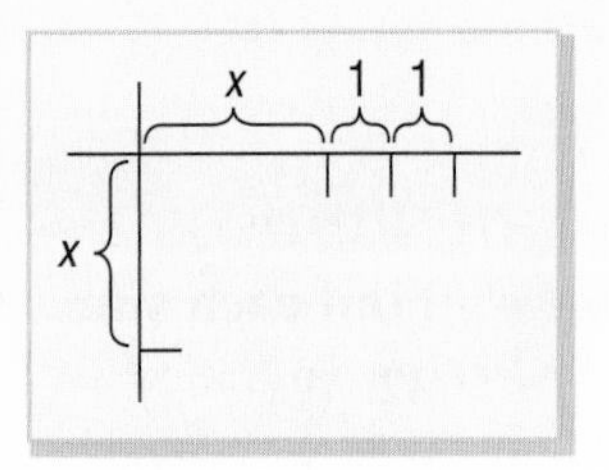

Step 2 Using the marks as a guide, fill in the rectangle with algebra tiles.

Step 3 The area of the rectangle is $x^2 + x + x$. In simplest form, the area is $x^2 + 2x$. Therefore, $x(x + 2) = x^2 + 2x$.

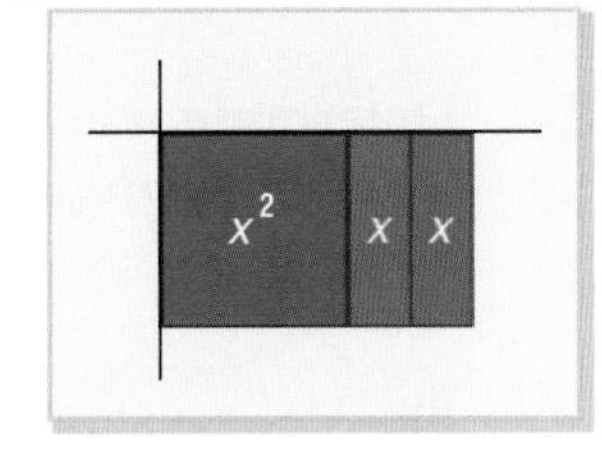

Model and Analyze

Use algebra tiles to determine whether each statement is *true* or *false*.

1. $x(x + 1) = x^2 + 1$
2. $x(2x + 3) = 2x^2 + 3x$
3. $(x + 2)2x = 2x^2 + 4x$
4. $2x(3x + 1) = 6x^2 + x$

Find each product using algebra tiles.

5. $x(x + 5)$
6. $(2x + 1)x$
7. $(2x + 4)2x$
8. $3x(2x + 1)$
9. There is a square garden plot that measures x feet on a side.
   a. Suppose you double the length of the plot and increase the width by 3 feet. Write two expressions for the area of the new plot.
   b. If the original plot was 10 feet on a side, what is the area of the new plot?

Extend the Activity

10. Write a multiplication sentence that is represented by the model at the right.

x^2	x^2	x	x	x
x	x	1	1	1
x	x	1	1	1
x	x	1	1	1

13-4 Multiplying a Polynomial by a Monomial

Sunshine State Standards
MA.D.2.3.2-1, MA.D.2.3.2-2

What You'll Learn

- Multiply a polynomial by a monomial.

How is the Distributive Property used to multiply a polynomial by a monomial?

The Grande Arche office building in Paris, France, looks like a hollowed-out prism, as shown in the photo at the right.

a. Write an expression that represents the area of the rectangular region outlined on the photo.

b. Recall that $2(4 + 1) = 2(4) + 2(1)$ by the Distributive Property. Use this property to simplify the expression you wrote in part **a**.

c. The Grande Arche is approximately w feet deep. Explain how you can write a polynomial to represent the volume of the hollowed-out region of the building. Then write the polynomial.

MULTIPLY A POLYNOMIAL AND A MONOMIAL You can model the multiplication of a polynomial and a monomial by using algebra tiles.

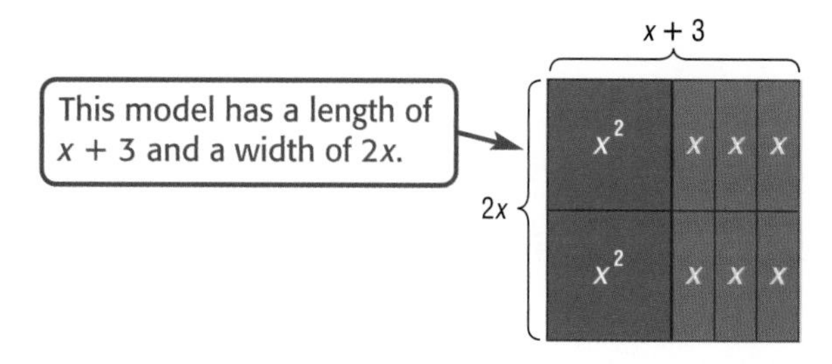

Study Tip

Look Back
To review the **Distributive Property**, see Lesson 3-1.

The model shows the product of $2x$ and $x + 3$. The rectangular arrangement contains 2 x^2 tiles and 6 x tiles. So, the product of $2x$ and $x + 3$ is $2x^2 + 6x$. In general, the Distributive Property can be used to multiply a polynomial and a monomial.

Example 1 Products of a Monomial and a Polynomial

Find each product.

a. $4(5x + 1)$

$4(5x + 1) = 4(5x) + 4(1)$ Distributive Property

$= 20x + 4$ Simplify.

b. $(2x - 6)(3x)$

$(2x - 6)(3x) = 2x(3x) - 6(3x)$ Distributive Property

$= 6x^2 - 18x$ Simplify.

Example 2 Product of a Monomial and a Polynomial

Find $3a(a^2 + 2ab - 4b^2)$.

$3a(a^2 + 2ab - 4b^2)$

$= 3a(a^2) + 3a(2ab) - 3a(4b^2)$ Distributive Property

$= 3a^3 + 6a^2b - 12ab^2$ Simplify.

Concept Check What is the product of x^2 and $x + 1$?

Sometimes problems can be solved by simplifying polynomial expressions.

Example 3 Use a Polynomial to Solve a Problem

Reading Math

Verbal Problems
When reading a verbal problem such as *30 meters longer than 6 times its width*, it is often helpful to make a drawing.

POOLS The world's largest swimming pool is the Orthlieb Pool in Casablanca, Morocco. It is 30 meters longer than 6 times its width. If the perimeter of the pool is 1110 meters, what are the dimensions of the pool?

Explore You know the perimeter of the pool. You want to find the dimensions of the pool.

Plan Let w represent the width of the pool. Then $6w + 30$ represents the length. Write an equation.

Perimeter	equals	twice	the sum of the length and width.
P	$=$	2	$(\ell + w)$

Study Tip

Look Back
To review **equations with grouping symbols**, see Lesson 3-1.

Solve

$P = 2(\ell + w)$	Write the equation.
$1110 = 2(6w + 30 + w)$	Replace P with 1110 and ℓ with $6w + 30$.
$1110 = 2(7w + 30)$	Combine like terms.
$1110 = 14w + 60$	Distributive Property
$1050 = 14w$	Subtract 60 from each side.
$75 = w$	Divide each side by 14.

The width is 75 meters, and the length is $6w + 30$ or 480 meters.

Examine Check the reasonableness of the results by estimating.

$P = 2(\ell + w)$	Formula for perimeter of a rectangle
$P \approx 2(500 + 80)$	Round 480 to 500 and 75 to 80.
$P \approx 2(580)$ or about 1160	

Since 1160 is close to 1110, the answer is reasonable.

Check for Understanding

Concept Check

1. **Determine** whether the following statement is *true* or *false*.
   If you change the order in which you multiply a polynomial and a monomial, the product will be different.
   Explain your reasoning or give a counterexample.
2. **Explain** the steps you would take to find the product of $x^3 - 7$ and $4x$.
3. **OPEN ENDED** Write a monomial and a polynomial, each having a degree no greater than 1. Then find their product.

Guided Practice **Find each product.**

4. $(5y - 4)3$
5. $a(a + 4)$
6. $t(7t + 8)$
7. $(3x - 7)4x$
8. $a(2a + b)$
9. $-5(3x^2 - 7x + 9)$

Application 10. **TENNIS** The perimeter of a tennis court is 228 feet. The length of the court is 6 feet more than twice the width. What are the dimensions of the tennis court?

Practice and Apply

Homework Help

For Exercises	See Examples
11–26	1, 2
27–30	3

Extra Practice
See page 756.

Find each product.

11. $7(2n + 5)$
12. $(1 + 4b)6$
13. $t(t - 9)$
14. $(x + 5)x$
15. $-a(7a + 6)$
16. $y(3 + 2y)$
17. $4n(10 + 2n)$
18. $-3x(6x - 4)$
19. $3y(y^2 - 2)$
20. $ab(a^2 + 7)$
21. $5x(x + y)$
22. $4m(m^2 - m)$
23. $7(-2x^2 + 5x - 11)$
24. $-3y(6 - 9y + 4y^2)$
25. $4c(c^3 + 7c - 10)$
26. $6x^2(-2x^3 + 8x + 1)$

Solve each equation.

27. $30 = 6(-2w + 3)$
28. $-3(2a - 12) = 3a - 45$

29. **BASKETBALL** The dimensions of high school basketball courts are different than the dimensions of college basketball courts, as shown in the table. Use the information in the table to find the length and width of each court.

	Basketball Courts	
Measure	High School (ft)	College (ft)
Perimeter	268	288
Width	w	w
Length	$2w - 16$	$(2w - 16) + 10$

More About. . .

Basketball

Basketball originated in Springfield, Massachusetts, in 1891. The first game was played with a soccer ball and two peach baskets used as goals.

Source: www.infoplease.com

30. **BOXES** A box large enough to hold 43,000 liters of water was made from one large sheet of cardboard.

   a. Write a polynomial that represents the area of the cardboard used to make the box. Assume the top and bottom of the box are the same. (*Hint*: $(2x + 2y)(6x - 2y) = 12x^2 + 8xy - 4y^2$)

   b. If x is 1.2 meters and y is 0.1 meter, what is the total amount of cardboard in square meters used to make the box?

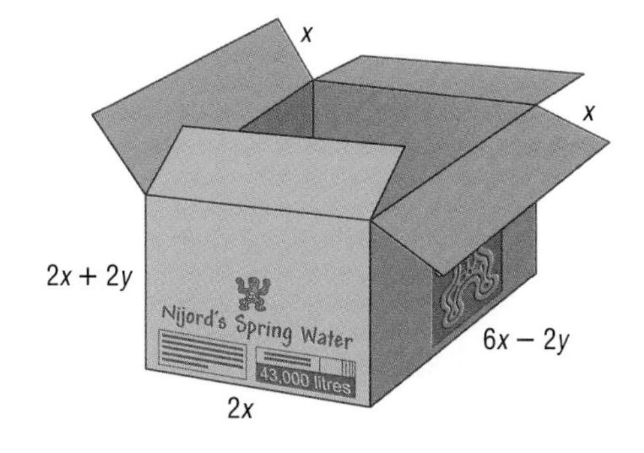

31. **CRITICAL THINKING** You have seen how algebra tiles can be used to connect multiplying a polynomial by a monomial and the Distributive Property. Draw a model and write a sentence to show how to multiply two binomials: $(a + b)(c + d)$.

32. WRITING IN MATH Answer the question that was posed at the beginning of the lesson.

How is the Distributive Property used to multiply a polynomial by a monomial?

Include the following in your answer:

- a description of the Distributive Property, and
- an example showing the steps used to multiply a polynomial and a monomial.

33. What is the product of $2x$ and $x - 8$?

(A) $2x - 8$ (B) $2x^2 - 8$ (C) $2x^2 - 16$ (D) $2x^2 - 16x$

34. The area of the rectangle is 252 square centimeters. Find the length of the longer side.

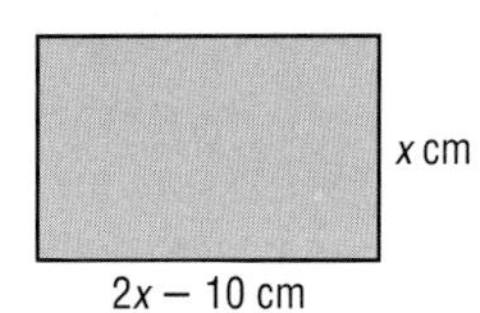

(A) 18 cm (B) 16 cm

(C) 14 cm (D) 10 cm

Maintain Your Skills

Mixed Review **Find each sum or difference.** *(Lessons 13-2 and 13-3)*

35. $(2x - 1) + 5x$

36. $(9a + 3a^2) + (a + 4)$

37. $(y^2 + 6y + 2) + (3y^2 - 8y + 12)$

38. $(4x - 7) - (2x + 2)$

39. $(9x + 8y) - (x - 3y)$

40. $(13n^2 + 6n + 5) - (6n^2 + 5)$

41. **STATISTICS** Describe two ways that a graph of sales of several brands of cereal could be misleading. *(Lesson 12-5)*

State whether each transformation of the triangles is a *reflection*, *translation*, or *rotation*. *(Lesson 10-3)*

42.

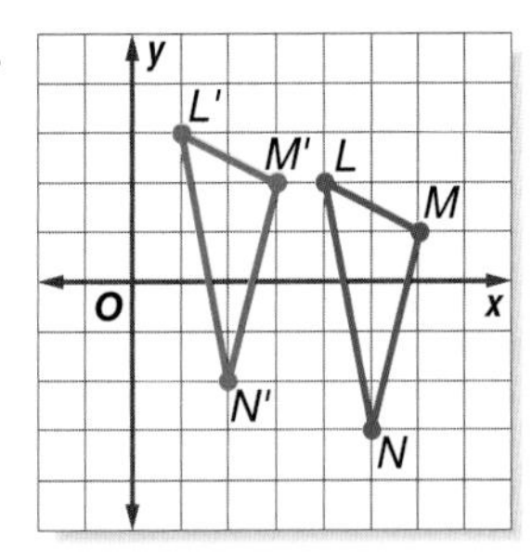

43.

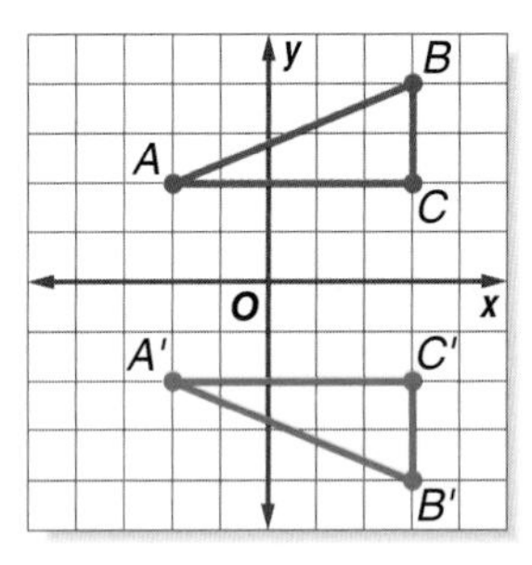

Getting Ready for the Next Lesson PREREQUISITE SKILL **Complete each table to find the coordinates of four points through which the graph of each function passes.**
*(To review **using tables to find ordered pair solutions**, see Lesson 8-2.)*

44. $y = 4x$

x	4x	(x, y)
0		
1		
2		
3		

45. $y = 2x^2 - 3$

x	$2x^2 - 3$	(x, y)
0		
1		
2		
3		

46. $y = x^3 + 1$

x	$x^3 + 1$	(x, y)
0		
1		
2		
3		

13-5 Linear and Nonlinear Functions

Sunshine State Standards
MA.D.1.3.1-1, MA.D.1.3.1-3, MA.D.1.3.2-1

Vocabulary

- nonlinear function
- quadratic function
- cubic function

What You'll Learn

- Determine whether a function is linear or nonlinear.

How can you determine whether a function is linear?

The sum of the lengths of three sides of a new deck is 40 feet. Suppose x represents the width of the deck. Then the length of the deck is $40 - 2x$.

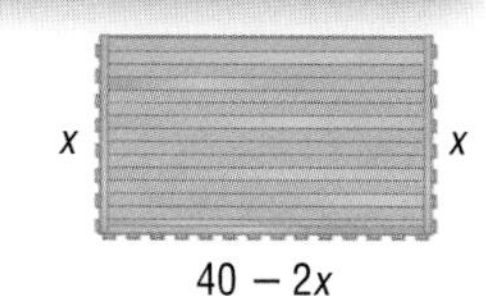

a. Write an expression to represent the area of the deck.

b. Find the area of the deck for widths of 6, 8, 10, 12, and 14 feet.

c. Graph the points whose ordered pairs are (width, area). Do the points fall along a straight line? Explain.

NONLINEAR FUNCTIONS In Lesson 8-2, you learned that linear functions have graphs that are straight lines. These graphs represent constant rates of change. **Nonlinear functions** do not have constant rates of change. Therefore, their graphs are *not* straight lines.

Example 1 Identify Functions Using Graphs

Determine whether each graph represents a *linear* or *nonlinear* function. Explain.

a.

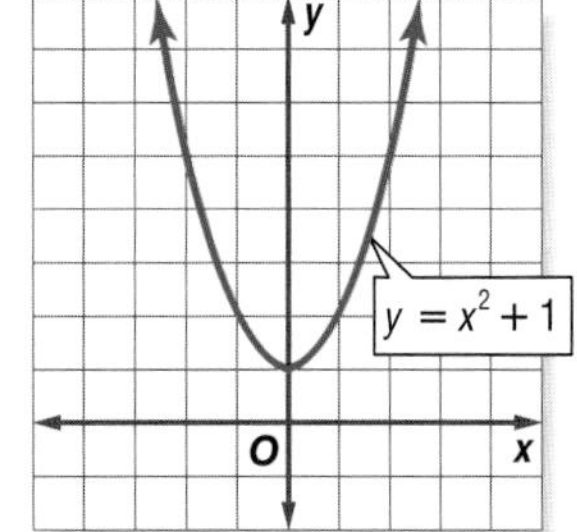

The graph is a curve, not a straight line, so it represents a nonlinear function.

b.

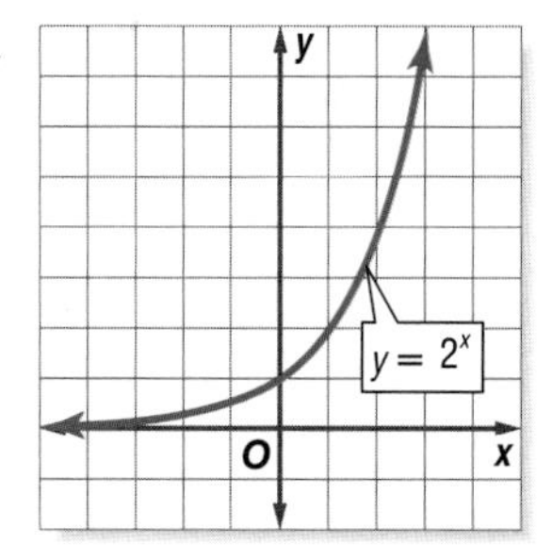

This graph is also a curve, so it represents a nonlinear function.

Recall that the equation for a linear function can be written in the form $y = mx + b$, where m represents the constant rate of change. Therefore, you can determine whether a function is linear by looking at its equation.

Example 2 Identify Functions Using Equations

Determine whether each equation represents a *linear* or *nonlinear* function.

a. $y = 10x$

This is linear because it can be written as $y = 10x + 0$.

b. $y = \frac{3}{x}$

This is nonlinear because x is in the denominator and the equation cannot be written in the form $y = mx + b$.

The tables represent the functions in Example 2. Compare the rates of change.

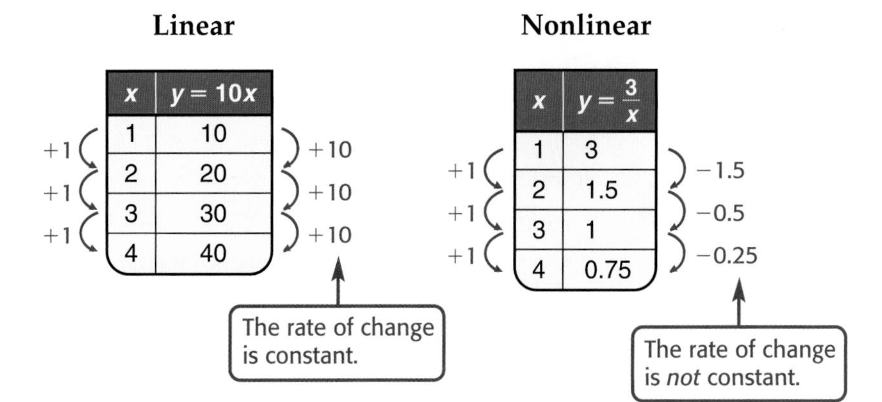

Linear

x	$y = 10x$
1	10
2	20
3	30
4	40

Nonlinear

x	$y = \frac{3}{x}$
1	3
2	1.5
3	1
4	0.75

A nonlinear function does not increase or decrease at the same rate. You can check this by using a table.

Example 3 *Identify Functions Using Tables*

Determine whether each table represents a *linear* or *nonlinear* function.

a.

x	y
10	120
15	100
20	80
25	60

(+5, +5, +5; −20, −20, −20)

As x increases by 5, y decreases by 20. So this is a linear function.

b.

x	y
2	4
4	16
6	36
8	64

(+2, +2, +2; +12, +20, +28)

As x increases by 2, y increases by a greater amount each time. So this is a nonlinear function.

Some nonlinear functions are given special names.

Key Concept — *Quadratic and Cubic Functions*

A **quadratic function** is a function that can be described by an equation of the form $y = ax^2 + bx + c$, where $a \neq 0$.

A **cubic function** is a function that can be described by an equation of the form $y = ax^3 + bx^2 + cx + d$, where $a \neq 0$.

Reading Math

Cubic

Cubic means three-dimensional. A cubic function has a variable to the third power.

Examples of these and other nonlinear functions are shown below.

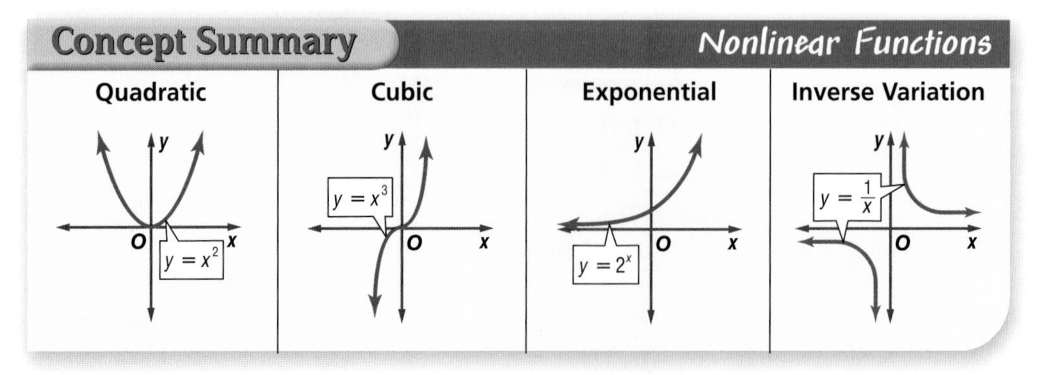

FCAT Practice
Standardized Test Practice

Example 4 Describe a Linear Function

Multiple-Choice Test Item

Which rule describes a linear function?

(A) $y = 7x^3 + 2$ (B) $y = (x - 1)5x$ (C) $4x + 3y = 12$ (D) $-2x^2 + 6y = 8$

Read the Test Item

A rule describes a relationship between variables. A rule that can be written in the form $y = mx + b$ describes a relationship that is linear.

Solve the Test Item

- $y = 7x^3 + 2 \rightarrow$ cubic equation — The variable has an exponent of 3.
  $-2x^2 + 6y = 8 \rightarrow$ quadratic equation — The variable has an exponent of 2.

  You can eliminate choices A and D.
- $y = (x - 1)5x$
  $y = 5x^2 - 5x$ This is a quadratic equation. Eliminate choice B.

The answer is C.

CHECK $4x + 3y = 12 \rightarrow y = -\frac{4}{3}x + 4$

This equation is in the form $y = mx + b$. ✓

The Princeton Review

Test-Taking Tip

Find out if there is a penalty for incorrect answers. If there is no penalty, making an educated guess can only increase your score, or at worst, leave your score the same.

Check for Understanding

Concept Check

1. **Describe** two methods for determining whether a function is linear.
2. **Explain** whether a company would prefer profits that showed linear growth or exponential growth.
3. **OPEN ENDED** Use newspapers, magazines, or the Internet to find real-life examples of nonlinear situations.

Guided Practice

Determine whether each graph, equation, or table represents a *linear* or *nonlinear* function. Explain.

4. 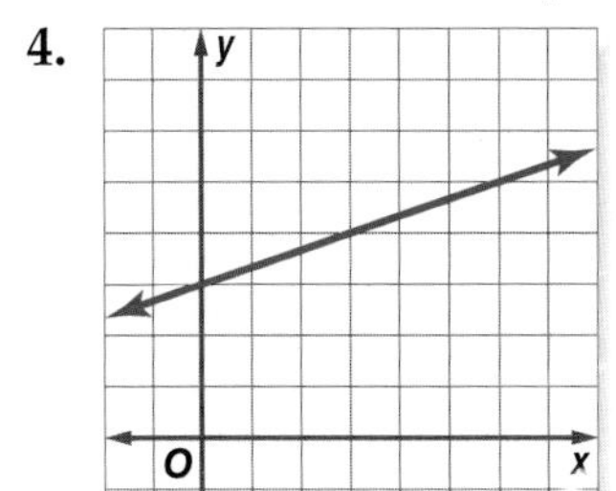

5. 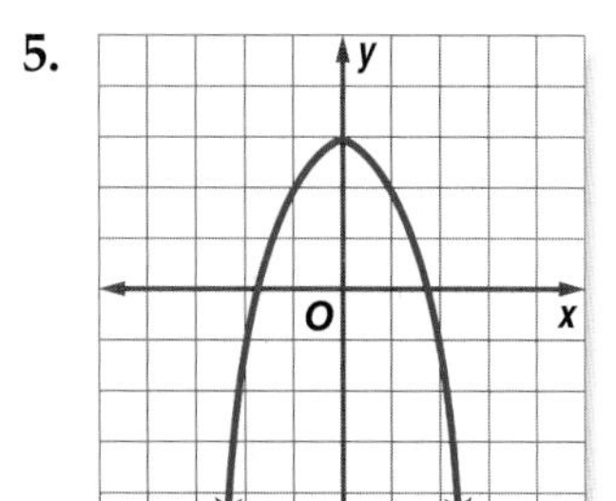

6. $y = \frac{x}{5}$

7. $xy = 12$

8. 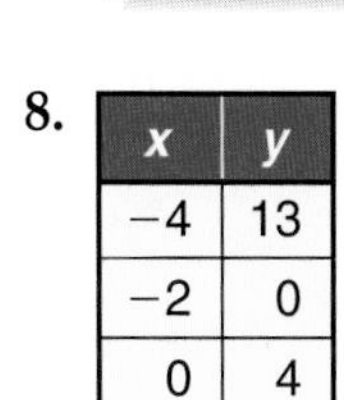

x	y
−4	13
−2	0
0	4
2	0

9. 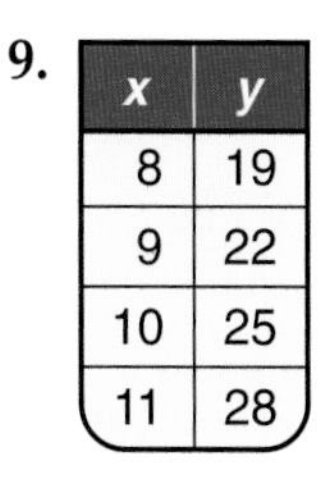

x	y
8	19
9	22
10	25
11	28

FCAT Practice
Standardized Test Practice

10. Which rule describes a nonlinear function?

(A) $x + y = 100$ (B) $y = \frac{8}{x}$ (C) $9 = 11x - y$ (D) $x = y$

www.pre-alg.com/extra_examples/fcat

Practice and Apply

Homework Help

For Exercises	See Examples
11–16, 27	1
17–22	2
23–26, 29	3

Extra Practice

See page 757.

Determine whether each graph, equation, or table represents a *linear* or *nonlinear* function. Explain.

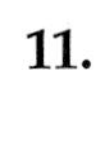

11. 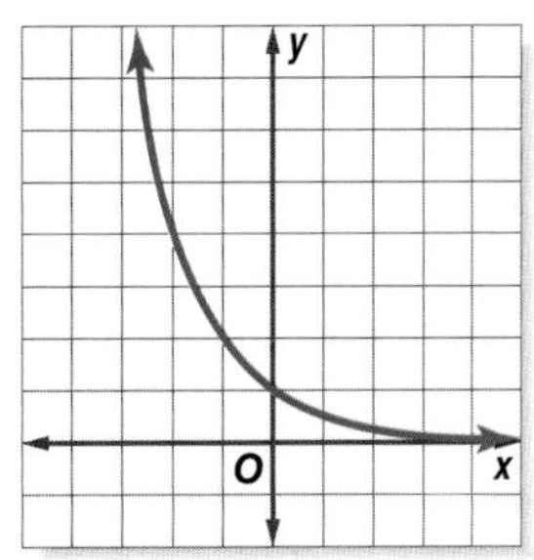

12. 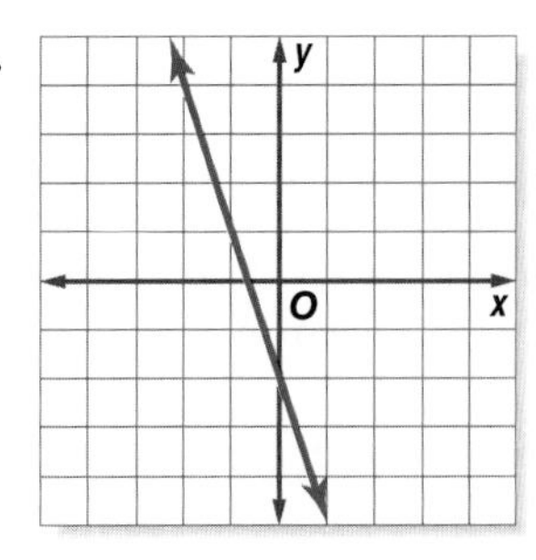

13. 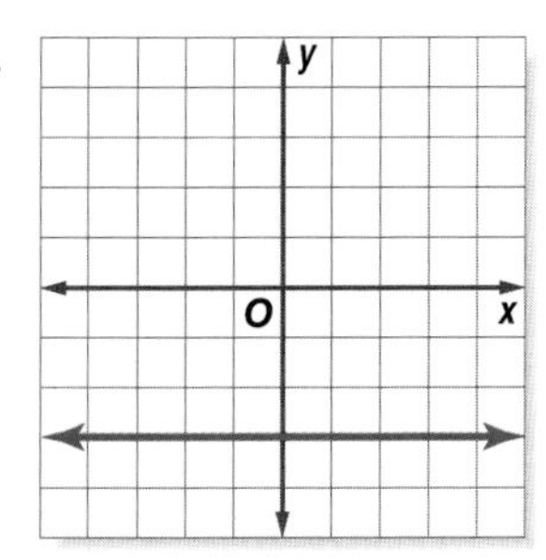

14. 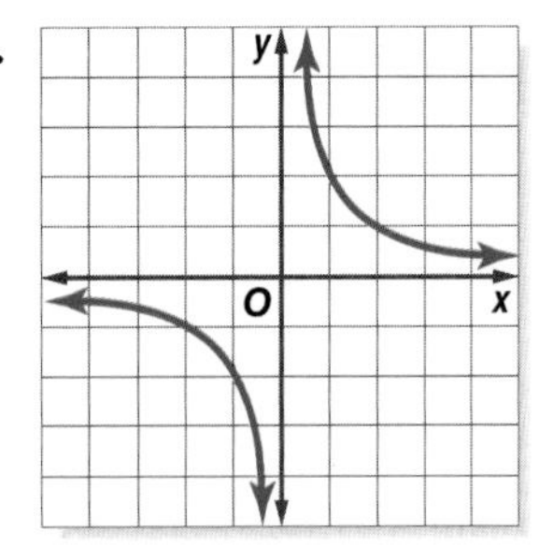

15. 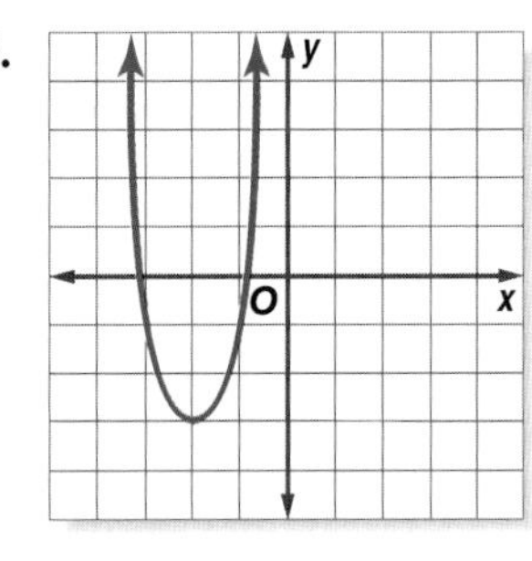

16. 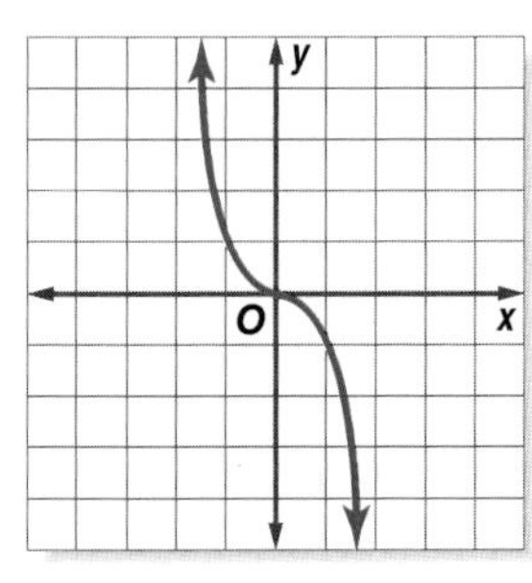

WebQuest

The trend in farm income can be modeled with a nonlinear function. Visit www.pre-alg.com/webquest to continue work on your WebQuest project.

17. $y = 0.9x$

18. $y = x^3 + 2$

19. $y = \frac{3x}{4}$

20. $2x + 3y = 12$

21. $y = 4^x$

22. $xy = -6$

23.

x	y
9	−2
11	−8
13	−14
15	−20

24.

x	y
4	1
5	4
6	9
7	16

25.

x	y
−4	12
−2	0
0	4
2	0

26.

x	y
−10	20
−9	18
−8	16
−7	14

27. TECHNOLOGY The graph shows the increase of trademark applications for internet-related products or services. Would you describe this growth as linear or nonlinear? Explain.

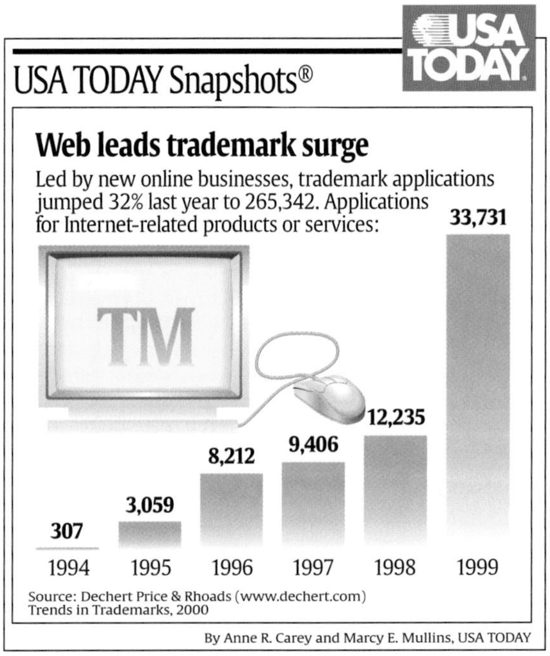

Online Research Data Update Is the growth of the Internet itself linear or nonlinear? Visit www.pre-alg.com/data_update to learn more.

USA TODAY Snapshots®

Web leads trademark surge

Led by new online businesses, trademark applications jumped 32% last year to 265,342. Applications for Internet-related products or services:

1994	1995	1996	1997	1998	1999
307	3,059	8,212	9,406	12,235	33,731

Source: Dechert Price & Rhoads (www.dechert.com) Trends in Trademarks, 2000

By Anne R. Carey and Marcy E. Mullins, USA TODAY

28. CRITICAL THINKING Are all graphs of straight lines linear functions? Explain.

More About. . .

Patents

Abraham Lincoln is the only U.S. President to hold a patent. He invented a device to lift boats over shallow waters.

Source: www.historyplace.com

29. PATENTS The table shows the years in which the first six million patents were issued. Is the number of patents issued a linear function of time? Explain.

Year	Number of Patents Issued
1911	1 million
1936	2 million
1961	3 million
1976	4 million
1991	5 million
1999	6 million

Source: *New York Times*

30. WRITING IN MATH Answer the question that was posed at the beginning of the lesson.

How can you determine whether a function is linear?

Include the following in your answer:

- a list of ways in which a function can be represented, and
- an explanation of how each representation can be used to identify the function as linear or nonlinear.

FCAT Practice

Standardized Test Practice

31. Which equation represents a linear function?

Ⓐ $y = \frac{1}{2}x$ Ⓑ $3xy = 12$ Ⓒ $x^2 - 1 = y$ Ⓓ $y = x(x + 4)$

32. Determine which general rule represents a nonlinear function if $a > 1$.

Ⓐ $y = ax$ Ⓑ $y = \frac{x}{a}$ Ⓒ $y = a^x$ Ⓓ $y = a + x$

Maintain Your Skills

Mixed Review

Find each product. *(Lesson 13-4)*

33. $t(4 + 9t)$ **34.** $5n(-1 + 3n)$ **35.** $(a - 2b)ab$

Find each difference. *(Lesson 13-3)*

36. $(2x + 7) - (x - 1)$ **37.** $(4x + y) - (5x + y)$ **38.** $(6a - a^2) - (8a + 3)$

39. GEOMETRY Classify a 65° angle as *acute, obtuse, right,* or *straight.* *(Lesson 9-3)*

Getting Ready for the Next Lesson

PREREQUISITE SKILL Use a table to graph each line.

*(To review **graphing equations**, see Lesson 8-3.)*

40. $y = -x$ **41.** $y = x - 4$ **42.** $y = 2x + 2$ **43.** $y = -\frac{1}{2}x + 3$

Practice Quiz 2

Lessons 13-4 and 13-5

Find each product. *(Lesson 13-4)*

1. $c(2c^2 - 8)$ **2.** $(4x + 2)3x$ **3.** $a^2(5 + a + 2a^2)$

Determine whether each equation represents a *linear* or *nonlinear* function. Explain. *(Lesson 13-5)*

4. $y = 9x$ **5.** $y = 0.25x^3$

13-6 Graphing Quadratic and Cubic Functions

What You'll Learn

- Graph quadratic functions.
- Graph cubic functions.

How are functions, formulas, tables, and graphs related?

You can find the area of a square A by squaring the length of a side s. This relationship can be represented in different ways.

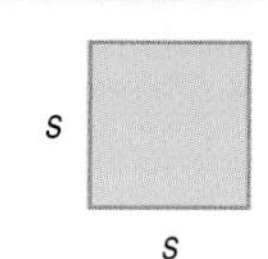

Equation

Area equals length of a side squared.

$$A = s^2$$

Table

s	s^2	(s, A)
0	$0^2 = 0$	(0, 0)
1	$1^2 = 1$	(1, 1)
2	$2^2 = 4$	(2, 4)

Graph

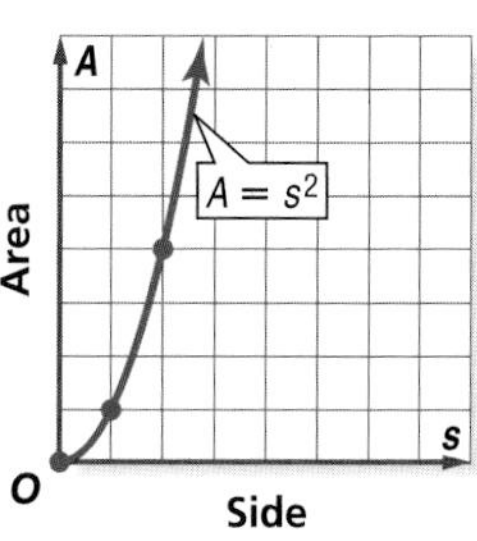

a. The volume of cube V equals the cube of the length of an edge a. Write a formula to represent the volume of a cube as a function of edge length.

b. Graph the volume as a function of edge length. (*Hint*: Use values of a like 0, 0.5, 1, 1.5, 2, and so on.)

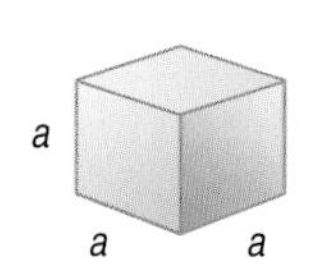

QUADRATIC FUNCTIONS In Lesson 13-5, you saw that functions can be represented using graphs, equations, and tables. This allows you to graph quadratic functions such as $A = s^2$ using an equation or a table of values.

Example 1 Graph Quadratic Functions

Graph each function.

a. $y = 2x^2$

Make a table of values, plot the ordered pairs, and connect the points with a curve.

x	$2x^2$	(x, y)
−1.5	$2(-1.5)^2 = 4.5$	(−1.5, 4.5)
−1	$2(-1)^2 = 2$	(−1, 2)
0	$2(0)^2 = 0$	(0, 0)
1	$2(1)^2 = 2$	(1, 2)
1.5	$2(1.5)^2 = 4.5$	(1.5, 4.5)

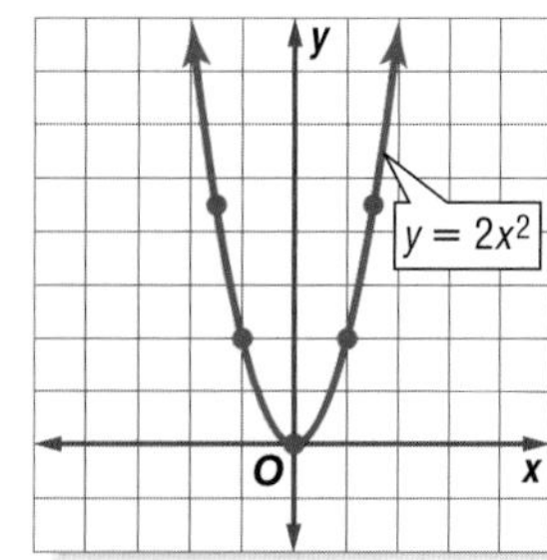

Study Tip

Graphing
It is often helpful to substitute decimal values of x in order to graph points that are closer together.

b. $y = x^2 - 1$

x	$x^2 - 1$	(x, y)
−2	$(-2)^2 - 1 = 3$	(−2, 3)
−1	$(-1)^2 - 1 = 0$	(−1, 0)
0	$(0)^2 - 1 = -1$	(0, −1)
1	$(1)^2 - 1 = 0$	(1, 0)
2	$(2)^2 - 1 = 3$	(2, 3)

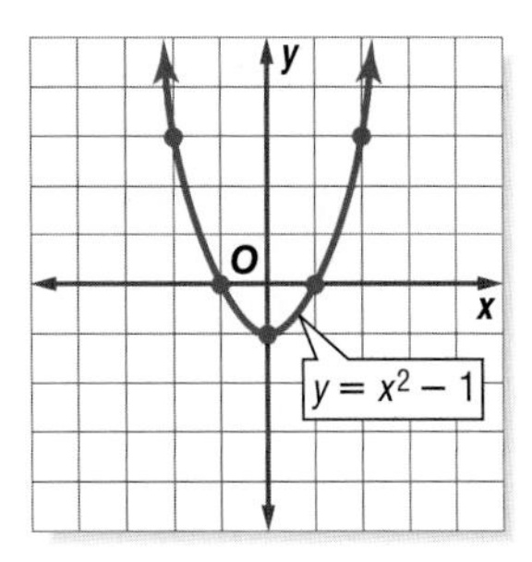

c. $y = -x^2 + 3$

x	$-x^2 + 3$	(x, y)
−2	$-(-2)^2 + 3 = -1$	(−2, −1)
−1	$-(-1)^2 + 3 = 2$	(−1, 2)
0	$-(0)^2 + 3 = 3$	(0, 3)
1	$-(1)^2 + 3 = 2$	(1, 2)
2	$-(2)^2 + 3 = -1$	(2, −1)

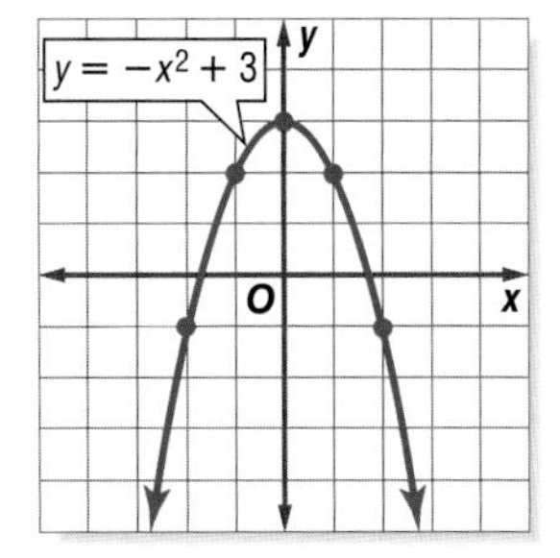

You can also write a rule from a verbal description of a function, and then graph.

Example 2 Use a Function to Solve a Problem

SKYDIVING The distance in feet that a skydiver falls is equal to sixteen times the time squared, with the time given in seconds. Graph this function and estimate how far he will fall in 4.5 seconds.

More About...

Skydiving

Skydivers fall at a speed of 110–120 miles per hour.

Source: www.aviation.about.com

Words Distance is equal to sixteen times the time squared.

Variables Let d = the distance in feet and t = the time in seconds.

	Distance	is equal to	sixteen	times	the time squared.
Equation	d	$=$	16	$\cdot$	t^2

The equation is $d = 16t^2$. Since the variable t has an exponent of 2, this function is nonlinear. Now graph $d = 16t^2$. Since time cannot be negative, use only positive values of t.

t	$d = 16t^2$	(t, d)
0	$16(0)^2 = 0$	(0, 0)
1	$16(1)^2 = 16$	(1, 16)
2	$16(2)^2 = 64$	(2, 64)
3	$16(3)^2 = 144$	(3, 144)
4	$16(4)^2 = 256$	(4, 256)
5	$16(5)^2 = 400$	(5, 400)
6	$16(6)^2 = 576$	(6, 576)

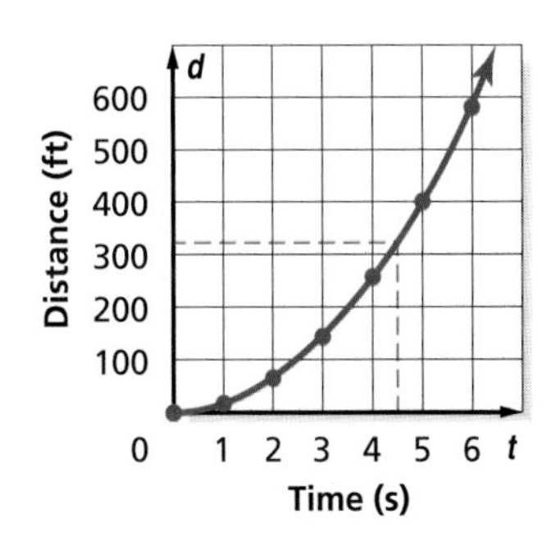

By looking at the graph, we find that in 4.5 seconds, the skydiver will fall approximately 320 feet. You could find the exact distance by substituting 4.5 for t in the equation $d = 16t^2$.

CUBIC FUNCTIONS You can also graph cubic functions such as the formula for the volume of a cube by making a table of values.

Example 3 Graph Cubic Functions

Graph each function.

a. $y = x^3$

x	$y = x^3$	(x, y)
−1.5	$(-1.5)^3 \approx -3.4$	(−1.5, −3.4)
−1	$(-1)^3 = -1$	(−1, −1)
0	$(0)^3 = 0$	(0, 0)
1	$(1)^3 = 1$	(1, 1)
1.5	$(1.5)^3 \approx 3.4$	(1.5, 3.4)

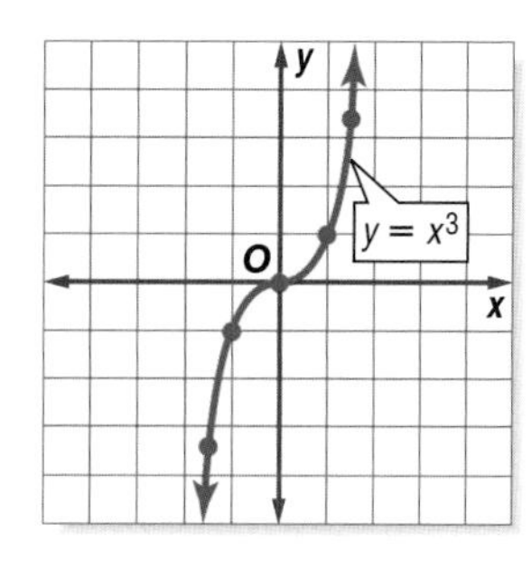

b. $y = x^3 - 1$

x	$y = x^3 - 1$	(x, y)
−1.5	$(-1.5)^3 - 1 \approx -4.4$	(−1.5, −4.4)
−1	$(-1)^3 - 1 = -2$	(−1, −2)
0	$(0)^3 - 1 = -1$	(0, −1)
1	$(1)^3 - 1 = 0$	(1, 0)
1.5	$(1.5)^3 - 1 \approx 2.4$	(1.5, 2.4)

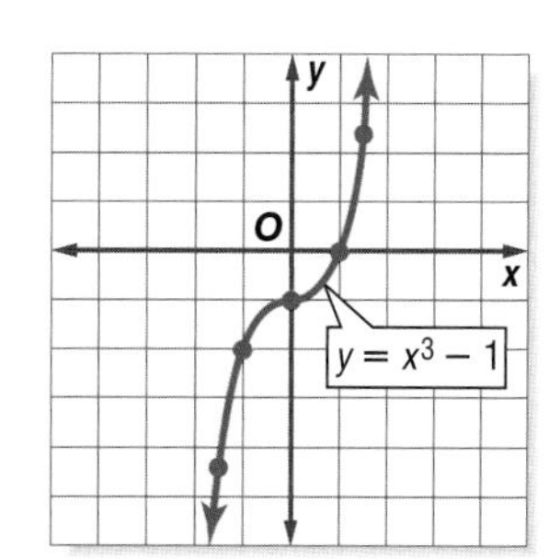

Check for Understanding

Concept Check

1. **Describe** one difference between the graph of $y = nx^2$ and the graph of $y = nx^3$ for any rational number n.
2. **Explain** how to determine whether a function is quadratic.
3. **OPEN ENDED** Write a quadratic function and explain how to graph it.

Guided Practice

Graph each function.

4. $y = x^2$
5. $y = -2x^2$
6. $y = x^2 + 1$
7. $y = -x^3$
8. $y = 0.5x^3$
9. $y = x^3 - 2$

Application

10. **GEOMETRY** A cube has edges measuring a units.
    a. Write a quadratic equation for the surface area S of the cube.
    b. Graph the surface area as a function of a. (*Hint*: Use values of a like 0, 0.5, 1, 1.5, 2, and so on.)

Practice and Apply

Homework Help

For Exercises	See Examples
11–22	1, 3
31–33	2

Extra Practice

See page 757.

Graph each function.

11. $y = 3x^2$
12. $y = 0.5x^2$
13. $y = -x^2$
14. $y = 3x^3$
15. $y = -2x^3$
16. $y = -0.5x^2$
17. $y = 2x^3$
18. $y = 0.1x^3$
19. $y = x^3 + 1$
20. $y = x^2 - 3$
21. $y = \frac{1}{2}x^2 + 1$
22. $y = \frac{1}{3}x^3 + 2$

23. Graph $y = x^2 - 4$ and $y = -4x^3$. Are these equations functions? Explain.

24. Graph $y = x^2$ and $y = x^3$ in the first quadrant on the same coordinate plane. Explain which graph shows faster growth.

The *maximum point* of a graph is the point with the greatest y value coordinate. The *minimum point* is the point with the least y value coordinate. Find the coordinates of each point.

25. the maximum point of the graph of $y = -x^2 + 7$

26. the minimum point of the graph of $y = x^2 - 6$

Graph each pair of equations on the same coordinate plane. Describe their similarities and differences.

27. $y = x^2$
$y = 3x^2$

28. $y = 0.5x^3$
$y = 2x^3$

29. $y = 2x^2$
$y = -2x^2$

30. $y = x^3$
$y = x^3 - 3$

CONSTRUCTION **For Exercises 31–33, use the information below and the figure at the right.**

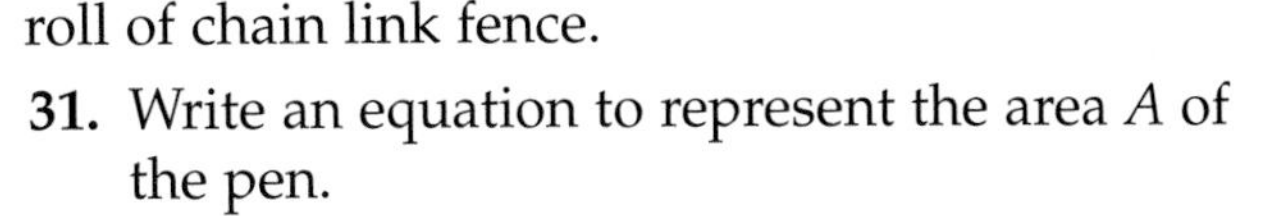

A dog trainer is building a dog pen with a 100-foot roll of chain link fence.

31. Write an equation to represent the area A of the pen.

32. Graph the equation you wrote in Exercise 31.

33. What should the dimensions of the dog pen be to enclose the maximum area inside the fence? (*Hint*: Find the coordinates of the maximum point of the graph.)

GEOMETRY **Write a function for each of the following. Then graph the function in the first quadrant.**

34. the volume V of a rectangular prism as a function of a fixed height of 2 units and a square base of varying lengths s

35. the volume V of a cylinder as a function of a fixed height of 0.2 unit and radius r

36. CRITICAL THINKING Describe how you can find real number solutions of the quadratic equation $ax^2 + bx + c = 0$ from the graph of the quadratic function $y = ax^2 + bx + c$.

37. WRITING IN MATH Answer the question that was posed at the beginning of the lesson.

How are functions, formulas, tables, and graphs related?

Include the following in your answer:

- an explanation of how to make a graph by using a rule, and
- an explanation of how to write a rule by using a graph.

38. Which equation represents the graph at the right?

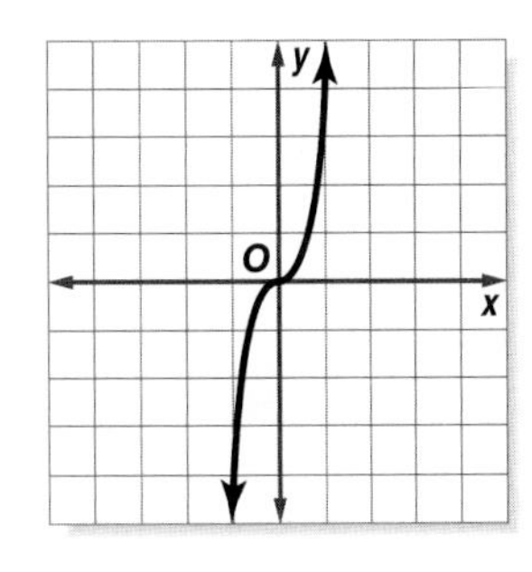

Ⓐ $y = 4x^2$

Ⓑ $y = -4x^2$

Ⓒ $y = 4x^3$

Ⓓ $y = -4x^3$

39. For a certain frozen pizza, as the cost goes from $2 to $4, the demand can be modeled by the formula $y = -10x^2 + 60x + 180$, where x represents the cost and y represents the number of pizzas sold. Estimate the cost that will result in the greatest demand.

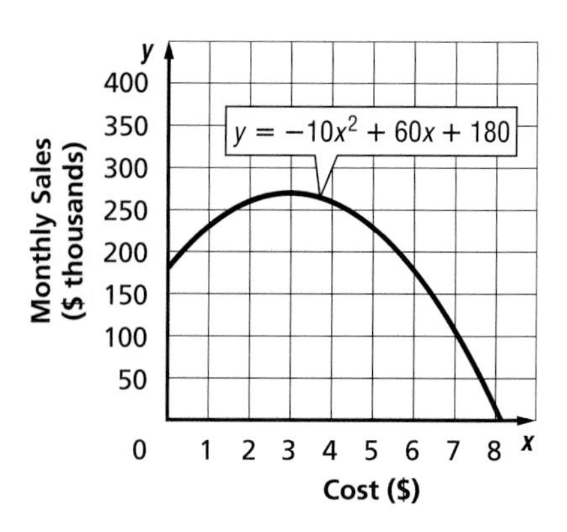

Ⓐ $0 Ⓑ $2
Ⓒ $3 Ⓓ $8

Extending the Lesson **Just as you can estimate the area of irregular figures, you can also estimate the area under a curve that is graphed on the coordinate plane. Estimate the shaded area under each curve to the nearest square unit.**

40.

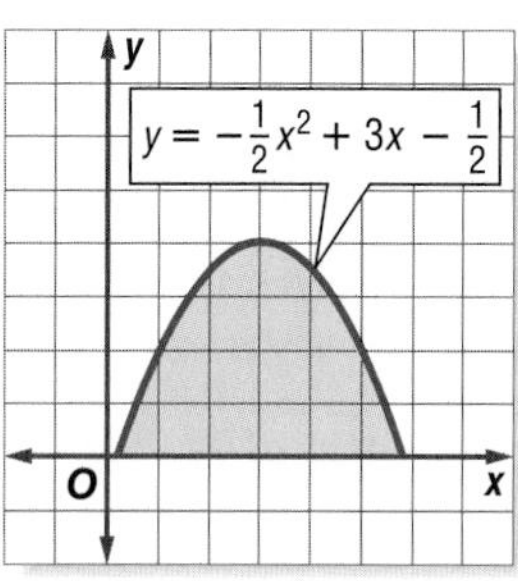

41.

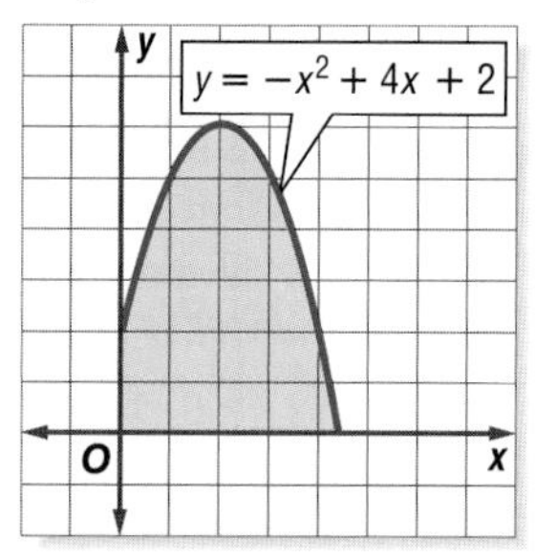

Maintain Your Skills

Mixed Review 42. **SCIENCE** The graph shows how water vapor pressure increases as the temperature increases. Is this relationship linear or nonlinear? Explain. *(Lesson 13-5)*

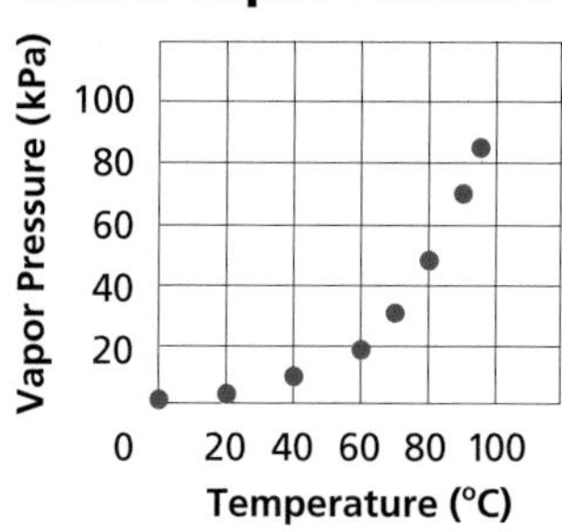

Find each product. *(Lesson 13-4)*

43. $(2x - 4)5$ 44. $n(n + 6)$ 45. $3y(8 - 7y)$

Write an equation in slope-intercept form for the line passing through each pair of points. *(Lesson 8-7)*

46. (3, 6) and (0, 9) 47. (2, 5) and (−1, −7) 48. (−4, −3) and (8, 6)

Family Farms

It is time to complete your project. Use the information and data you have gathered to prepare a Web page about farming or ranching in the United States. Be sure to include at least five graphs or tables that show statistics about farming or ranching and at least one scatter plot that shows a farming or ranching statistic over time, from which you can make predictions.

www.pre-alg.com/webquest

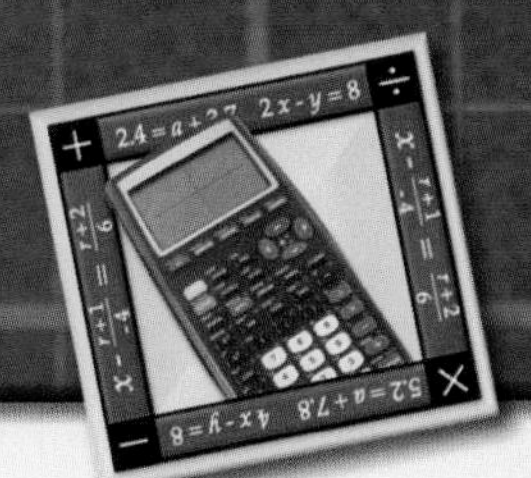

Graphing Calculator Investigation

A Follow-Up of Lesson 13-6

Families of Quadratic Functions

A quadratic function can be described by an equation of the form $ax^2 + bx + c$, where $a \neq 0$. The graph of a quadratic function is called a **parabola**. Recall that families of linear graphs share the same slope or y-intercept. Similarly, families of parabolas share the same maximum or minimum point, or have the same shape.

Graph $y = x^2$ and $y = x^2 + 4$ on the same screen and describe how they are related.

Step 1 *Enter the function $y = x^2$.*

- Enter $y = x^2$ as Y1.

KEYSTROKES: [Y=] [X,T,θ,n] [x^2] [ENTER]

Step 2 *Enter the function $y = x^2 + 4$.*

- Enter $y = x^2 + 4$ as Y2.

KEYSTROKES: [Y=] [X,T,θ,n] [x^2] [+] 4 [ENTER]

Step 3 *Graph both quadratic functions on the same screen.*

- Display the graph.

KEYSTROKES: [ZOOM] 6

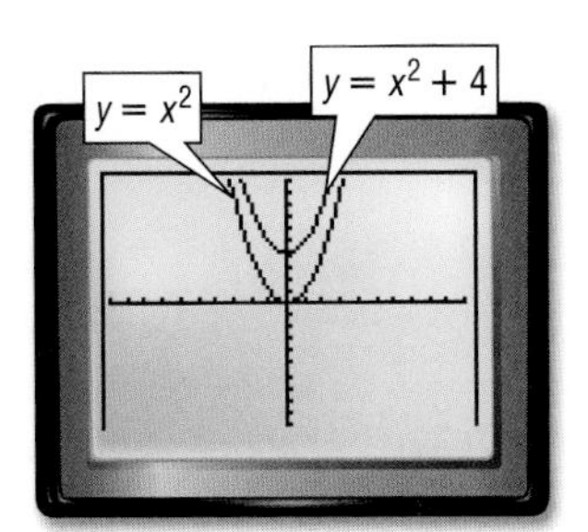

The first function graphed is Y1 or $y = x^2$. The second is Y2 or $y = x^2 + 4$. Press [TRACE] and move along each function by using the right and left arrow keys. Move from one function to another by using the up and down arrow keys.

The graphs are similar in that they are both parabolas. However, the graph of $y = x^2$ has its vertex at (0, 0), whereas the graph of $y = x^2 + 4$ has its vertex at (0, 4).

Exercises

1. Graph $y = x^2$, $y = x^2 - 5$, and $y = x^2 - 3$ on the same screen and draw the parabolas on grid paper. Compare and contrast the three parabolas.
2. **Make a conjecture** about how adding or subtracting a constant c affects the graph of a quadratic function.
3. The three parabolas at the right are graphed in the standard viewing window and have the same shape as the graph of $y = x^2$. Write an equation for each, beginning with the lowest parabola.

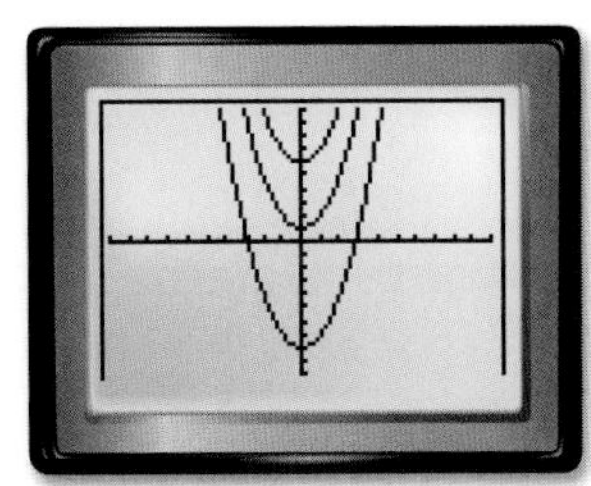

4. Clear all functions from the [Y=] menu. Enter $y = 0.4x^2$ as Y1, $y = x^2$ as Y2, and $y = 3x^2$ as Y3. Graph the functions in the standard viewing window on the same screen. Then draw the graphs on the same coodinate grid. How does the shape of the parabola change as the coefficient of x^2 increases?

www.pre-alg.com/other_calculator_keystrokes

Chapter 13

Study Guide and Review

Vocabulary and Concept Check

binomial (p. 669)
cubic function (p. 688)
degree (p. 670)
nonlinear function (p. 687)
polynomial (p. 669)
quadratic function (p. 688)
trinomial (p. 669)

Choose the correct term to complete each sentence.

1. A (binomial, trinomial) is the sum or difference of three monomials.
2. Monomials that contain the same variables with the same (power, sign) are like terms.
3. The function $y = 2x^3$ is an example of a (cubic, quadratic) function.
4. The equation $y = x^2 + 5x + 1$ is an example of a (cubic, quadratic) function.
5. x^2 and $4x^2$ are examples of (binomials, like terms).
6. The equation $y = 4x^3 + x^2 + 2$ is an example of a (quadratic, cubic) function.
7. The graph of a quadratic function is a (straight line, curve).
8. To multiply a polynomial and a monomial, use the (Distributive, Commutative) Property.

Lesson-by-Lesson Review

13-1 Polynomials

See pages 669–672.

Concept Summary

- A polynomial is an algebraic expression that contains one or more monomials.
- A binomial has two terms and a trinomial has three terms.
- The degree of a monomial is the sum of the exponents of its variables.

State whether $x^3 - 2xy$ is a *monomial, binomial,* or *trinomial*. Then find the degree.

The expression is the difference of two monomials. So it is a binomial. x^3 has degree 3, and $-2xy$ has degree $1 + 1$ or 2. So, the degree of $x^3 - 2xy$ is 3.

Exercises **Determine whether each expression is a polynomial. If it is, classify it as a *monomial, binomial,* or *trinomial*.** *See Example 1 on page 669.*

9. $c^2 + 3$ **10.** -5 **11.** $4t^4$ **12.** $\frac{6}{a} + b$
13. $3x^2 + 4x - 2$ **14.** $x + y$ **15.** $\sqrt{n}$ **16.** $1 + 3x + 5x^2$

Find the degree of each polynomial. *See Examples 2 and 3 on page 670.*

17. $2x$ **18.** $5xy$ **19.** $3a^2b$ **20.** $n^2 - 4$
21. $x^6 + y^6$ **22.** $2xy + 6yz^2$ **23.** $2x^5 + 9x + 1$ **24.** $x^2 + xy^2 - y^4$

13-2 Adding Polynomials

See pages 674–677.

Concept Summary

- To add polynomials, add like terms.

Example **Find $(5x^2 - 8x + 2) + (x^2 + 6x)$.**

$$\begin{array}{rl} 5x^2 - 8x + 2 & \\ (+)\ \ x^2 + 6x & \text{Align like terms.} \\ \hline 6x^2 - 2x + 2 & \text{Add.} \end{array}$$

The sum is $6x^2 - 2x + 2$.

Exercises **Find each sum.** *See Example 1 on pages 674 and 675.*

25. $\begin{array}{r} 3b + 8 \\ (+)\ 5b - 5 \\ \hline \end{array}$

26. $\begin{array}{r} 2x^2 + 3x - 4 \\ (+)\ 6x^2 - \ \ x + 5 \\ \hline \end{array}$

27. $\begin{array}{r} 4y^2 + 2y + 3 \\ (+)\ y^2 \qquad - 7 \\ \hline \end{array}$

28. $(9m - 3n) + (10m + 4n)$

29. $(-3y^2 + 2) + (4y^2 - y - 3)$

13-3 Subtracting Polynomials

See pages 678–681.

Concept Summary

- To subtract polynomials, subtract like terms or add the additive inverse.

Example **Find $(4x^2 + 7x + 4) - (x^2 + 2x + 1)$.**

$$\begin{array}{rl} 4x^2 + 7x + 4 & \\ (-)\ \ x^2 + 2x + 1 & \text{Align like terms.} \\ \hline 3x^2 + 5x + 3 & \text{Subtract.} \end{array}$$

The difference is $3x^2 + 5x + 3$.

Exercises **Find each difference.** *See Examples 1 and 2 on pages 678 and 679.*

30. $\begin{array}{r} a^2 + 15 \\ (-) 3a^2 - 10 \\ \hline \end{array}$

31. $\begin{array}{r} 4x^2 - 2x + 3 \\ (-)\ \ x^2 + 2x - 4 \\ \hline \end{array}$

32. $\begin{array}{r} 18y^2 + 3y - 1 \\ (-)\ 2y^2 + \qquad + 6 \\ \hline \end{array}$

33. $(x + 8) - (2x + 7)$

34. $(3n^2 + 7) - (n^2 - n + 4)$

13-4 Multiplying a Polynomial by a Monomial

See pages 683–686.

Concept Summary

- To multiply a polynomial and a monomial, use the Distributive Property.

Example **Find $-3x(x + 8y)$.**

$$\begin{aligned} -3x(x + 8y) &= -3x(x) + (-3x)(8y) && \text{Distributive Property} \\ &= -3x^2 - 24xy && \text{Simplify.} \end{aligned}$$

Exercises **Find each product.** *See Example 1 on page 683.*

35. $5(4t - 2)$

36. $(2x + 3y)7$

37. $k(6k + 3)$

38. $4d(2d - 5)$

39. $-2a(9 - a^2)$

40. $6(2x^2 + xy + 3y^2)$

Chapter 13 For More ...
• Extra Practice, see pages 755–757.
• Mixed Problem Solving, see page 770.

13-5 Linear and Nonlinear Functions

See pages 687–691.

Concept Summary

- Nonlinear functions do not have constant rates of change.

Example **Determine whether each graph, equation, or table represents a *linear* or *nonlinear* function. Explain.**

a.

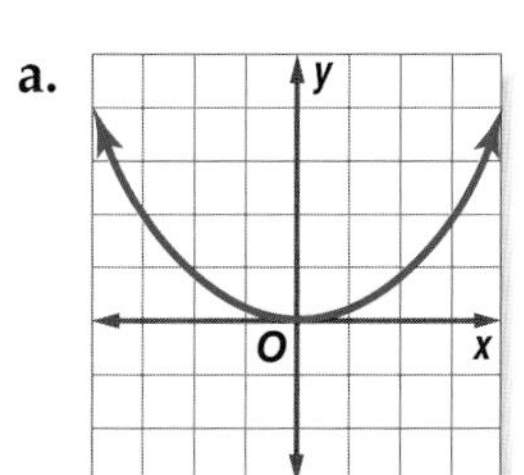

Nonlinear; graph is not a straight line.

b. $y = x + 12$

Linear; equation can be written as $y = mx + b$.

c.

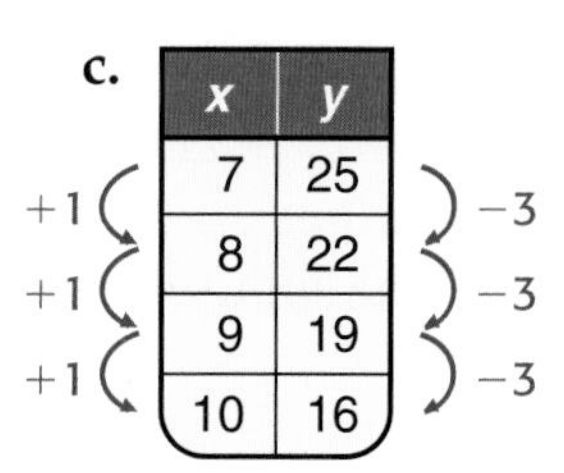

x	y
7	25
8	22
9	19
10	16

Linear; rate of change is constant.

Exercises **Determine whether each graph, equation, or table represents a *linear* or *nonlinear* function. Explain.** *See Examples 1–3 on pages 687 and 688.*

41.

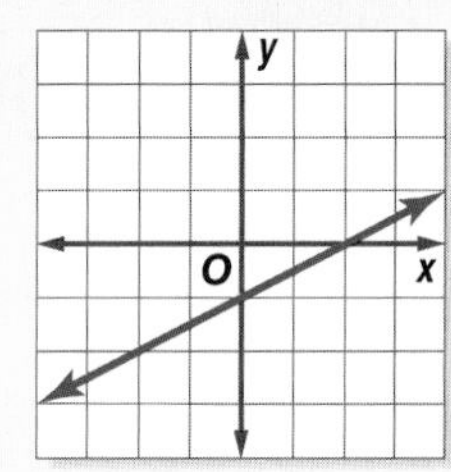

42. $y = \frac{x}{2}$

43.

x	y
−6	3
−4	4
−2	6
0	9

13-6 Graphing Quadratic and Cubic Functions

See pages 692–696.

Concept Summary

- Quadratic and cubic functions can be graphed by plotting points.

Example **Graph $y = -x^2 + 3$.**

x	$y = -x^2 + 3$	(x, y)
−2	$-(-2)^2 + 3 = -1$	(−2, −1)
−1	$-(-1)^2 + 3 = 2$	(−1, 2)
0	$-(0)^2 + 3 = 3$	(0, 3)
1	$-(1)^2 + 3 = 2$	(1, 2)
2	$-(2)^2 + 3 = -1$	(2, −1)

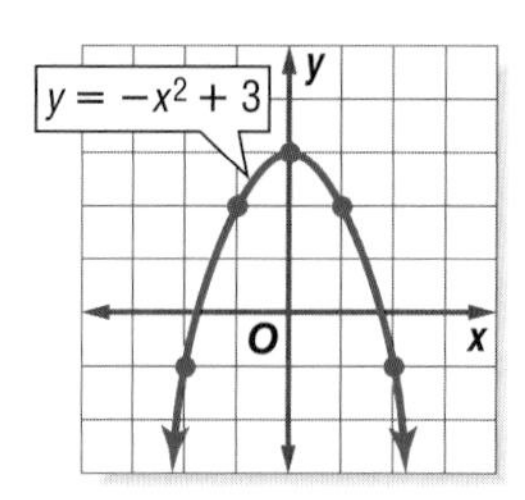

Exercises **Graph each function.** *See Examples 1 and 3 on pages 692–694.*

44. $y = x^2 + 2$
45. $y = -3x^2$
46. $y = x^3 - 2$
47. $y = x^3 + 1$
48. $y = -x^3$
49. $y = 2x^2 + 4$

Chapter 13 Practice Test

Vocabulary and Concepts

1. **Define** polynomial.
2. **Explain** how the degree of a monomial is found.
3. **OPEN ENDED** Draw the graph of a linear and a nonlinear function.

Skills and Applications

Determine whether each expression is a polynomial. If it is, classify it as a *monomial, binomial,* or *trinomial.*

4. $3x^3 - 2x + 7$
5. $6 + \frac{5}{m}$
6. $\frac{3}{5}p^4$

Find the degree of each polynomial.

7. $5ab^3$
8. $w^5 - 3w^3y^4 + 1$

Find each sum or difference.

9. $(5y + 8) + (-2y + 3)$
10. $(5a - 2b) + (-4a + 5b)$
11. $(-3m^3 + 5m - 9) + (7m^3 - 2m^2 + 4)$
12. $(6p + 5) - (3p - 8)$
13. $(5w - 3x) - (6w + 4x)$
14. $(-2s^2 + 4s - 7) - (6s^2 - 7s - 9)$

Find each product.

15. $x(3x - 5)$
16. $-5a(a^2 - b^2)$
17. $6p(-2p^2 + 3p - 4)$

Determine whether each graph, equation, or table represents a *linear* or *nonlinear* function. Explain.

18. 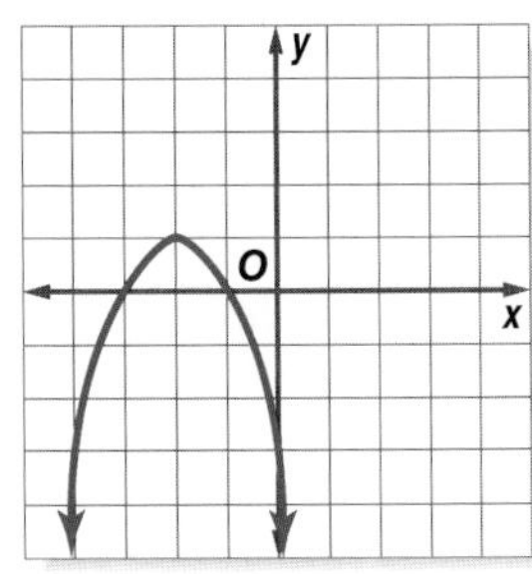

19. $5x - 6y = 2$

20.

x	y
1	10
3	7
5	3
7	−2

Graph each function.

21. $y = 2x^2$
22. $y = \frac{1}{2}x^3$
23. $y = -x^2 + 3$

24. **GEOMETRY** Refer to the rectangle.

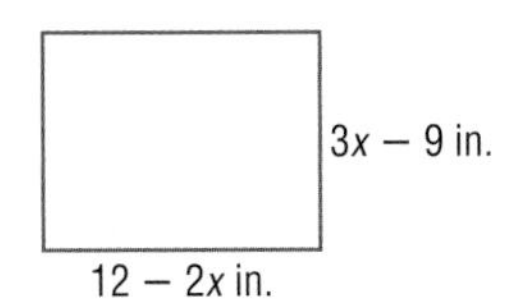

a. Write an expression for the perimeter of the rectangle.

b. Find the value of x if the perimeter is 14 inches.

FCAT Practice

25. **STANDARDIZED TEST PRACTICE** The length of a garden is equal to 5 less than four times its width. The perimeter of the garden is 40 feet. Find the length of the garden.

Ⓐ 1 ft Ⓑ 5 ft Ⓒ 10 ft Ⓓ 15 ft

Chapter 13 Standardized Test Practice

FCAT Practice

Part 1 Multiple Choice

Record your answers on the answer sheet provided by your teacher or on a sheet of paper.

1. If n represents a positive number, which of these expressions is equivalent to $n + n + n$? (Lesson 1-3)

Ⓐ n^3 Ⓑ $3n$

Ⓒ $n + 3$ Ⓓ $3(n + 1)$

2. Connor sold 4 fewer tickets to the band concert than Miguel sold. Kylie sold 3 times as many tickets as Connor. If the number of tickets Miguel sold is represented by m, which of these expressions represents the number of tickets that Kylie sold? (Lesson 3-6)

Ⓐ $m - 4$

Ⓑ $4 - 3m$

Ⓒ $3m - 4$

Ⓓ $3(m - 4)$

3. Melissa's family calculated that they drove an average of 400 miles per day during their three-day trip. They drove 460 miles on the first day and 360 miles on the second day. How many miles did they drive on the third day? (Lesson 5-7)

Ⓐ 340 Ⓑ 380

Ⓒ 410 Ⓓ 420

4. What is the ratio of the length of a side of a square to its perimeter? (Lesson 6-1)

Ⓐ $\frac{1}{16}$ Ⓑ $\frac{1}{4}$

Ⓒ $\frac{1}{3}$ Ⓓ $\frac{1}{2}$

The Princeton Review **Test-Taking Tip**

Question 2 If you have time at the end of a test, go back to check your calculations and answers. If the test allows you to use a calculator, use it to check your calculations.

5. The table shows values of x and y, where x is proportional to y. What are the missing values, S and T? (Lesson 6-3)

x	3	9	S
y	5	T	35

Ⓐ $S = 36$ and $T = 3$

Ⓑ $S = 21$ and $T = 15$

Ⓒ $S = 15$ and $T = 21$

Ⓓ $S = 3$ and $T = 36$

6. In the figure at the right, lines ℓ and m are parallel. Choose two angles whose measures have a sum of 180°. (Lesson 10-2)

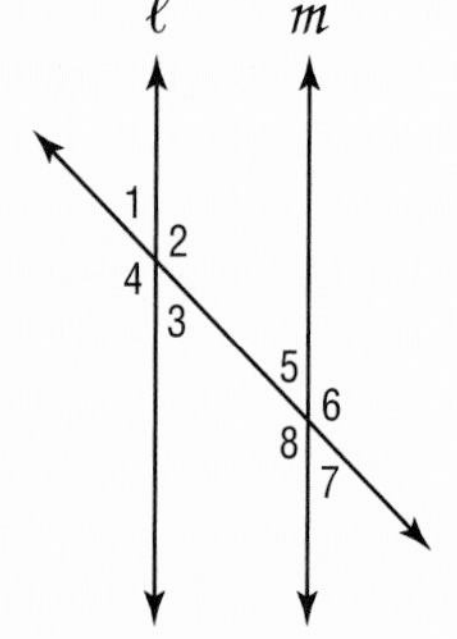

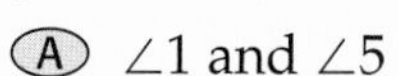

Ⓐ $\angle 1$ and $\angle 5$

Ⓑ $\angle 2$ and $\angle 8$

Ⓒ $\angle 2$ and $\angle 5$

Ⓓ $\angle 4$ and $\angle 8$

7. The point represented by coordinates $(4, -6)$ is reflected across the x-axis. What are the coordinates of the image? (Lesson 10-3)

Ⓐ $(-6, 4)$ Ⓑ $(-4, -6)$

Ⓒ $(-4, 6)$ Ⓓ $(4, 6)$

8. If $2x^2 - 3x + 7$ is subtracted from $4x^2 + 6x - 3$, what is the difference? (Lesson 13-3)

Ⓐ $2x^2 + 9x - 10$

Ⓑ $2x^2 + 3x + 4$

Ⓒ $-2x^2 + 3x + 4$

Ⓓ $-2x^2 - 9x + 10$

9. Which function includes all of the ordered pairs in the table? It may help you to sketch a graph of the points. (Lesson 13-6)

x	−2	−1	1	2	3
y	4	2	−2	−4	−6

Ⓐ $y = -x^2$ Ⓑ $y = -2x$

Ⓒ $y = -x + 2$ Ⓓ $y = x^2$

Aligned and verified by The Princeton Review

FCAT Practice

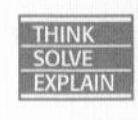

Part 2 Short Response/Grid In

Record your answers on the answer sheet provided by your teacher or on a sheet of paper.

10. The table shows the number of sandwiches sold during twenty lunchtimes. What is the mode? (Lesson 5-8)

Number of Sandwiches Sold				
9	8	10	14	12
16	9	7	10	11
11	8	9	8	7
12	14	8	9	9

11. One machine makes plastic containers at a rate of 360 containers per hour. A newer machine makes the same containers at a rate of 10 containers per minute. If both machines are run for four hours, how many containers will they make? (Lesson 6-1)

12. What percent of 275 is 165? (Lesson 6-5)

13. Mrs. Rosales can spend \$8200 on equipment for the computer lab. Each computer costs \$850 and each printer costs \$325. Mrs. Rosales buys 8 computers. Write an inequality that can be used to find p, the number of printers she could buy. (Lesson 7-6)

14. What is the y-intercept of the graph shown at the right? Each square represents 1 unit. (Lesson 8-6)

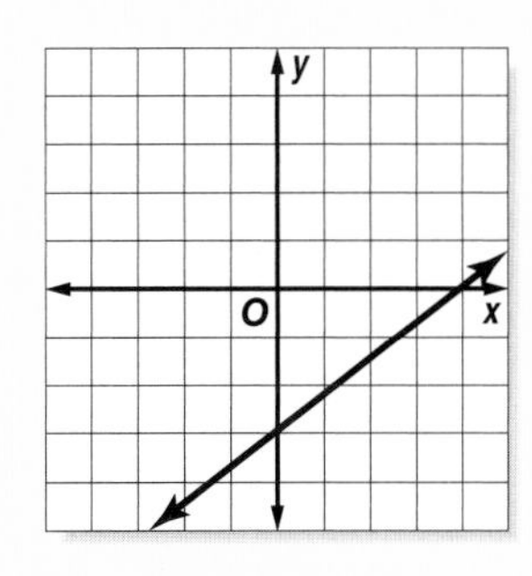

15. The area of a triangle is $\frac{1}{2}(b \times h)$. What is the area of the kite? (Lesson 10-5)

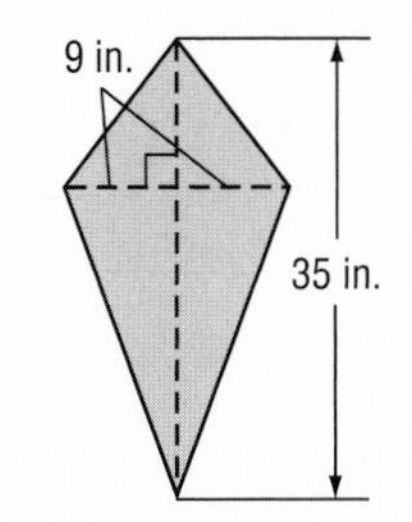

16. Brooke wants to fill her new aquarium two-thirds full of water. The aquarium dimensions are 20 inches by 20 inches by $8\frac{1}{2}$ inches. What volume of water, in cubic inches, is needed? (Lesson 11-2)

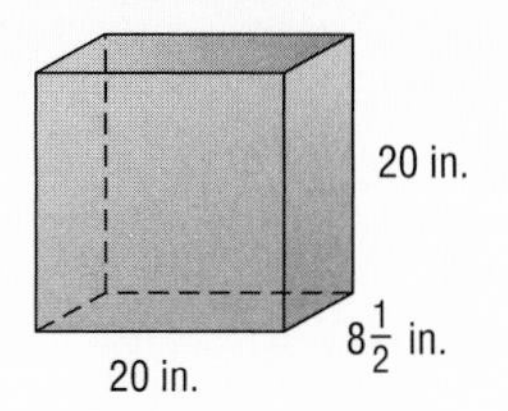

17. Let $s = 3x^2 - 2x - 1$ and $t = -2x^2 + x + 2$. Find $s + t$. (Lesson 13-2)

18. The perimeter of a soccer field is 1040 feet. The length of the field is 40 feet more than 2 times the width. What is the length of the field? (Lesson 13-4)

FCAT Practice

Part 3 Open Ended

Record your answers on a sheet of paper. Show your work.

19. An artist created a sculpture using five cylindrical posts. Each post has a diameter of 12 inches. The heights of the posts are 6 feet, 5 feet, 4 feet, 3 feet, and 2 feet. (Lesson 11-2)

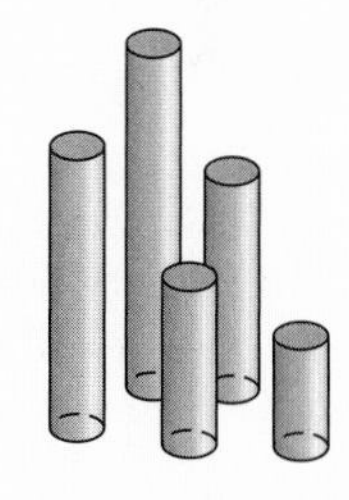

a. What is the total volume of all five posts? $V = \pi r^2 h$ is the formula for the volume of a cylinder. Use $\pi = 3.14$.

b. The posts are made of a material whose density is 12 pounds per cubic foot. How much does the sculpture weigh?

20. Refer to the table below. (Lesson 13-6)

a. Graph the ordered pairs in the table as coordinate points.

b. Sketch a line or curve through the points.

c. Write a quadratic function that includes all of the ordered pairs in the table.

x	y
−2	0
−1	−3
0	−4
1	−3
2	0

Student Handbook

Skills

Prerequisite Skills
Problem-Solving Strategy: Solve a Simpler Problem **706**
Problem-Solving Strategy: Work Backward **707**
Problem-Solving Strategy: Make a Table or List **708**
Problem-Solving Strategy: Guess and Check **709**
Comparing and Ordering Decimals **710**
Rounding Decimals **711**
Estimating Sums and Differences of Decimals **712**
Adding and Subtracting Decimals **713**
Estimating Products and Quotients of Decimals **714**
Multiplying and Dividing Decimals **715**
Estimating Sums and Differences of Fractions and Mixed Numbers **716**
Estimating Products and Quotients of Fractions and Mixed Numbers **717**
Converting Measurements within the Metric System **718**
Converting Measurements within the Customary System **720**
Displaying Data on Graphs **722**

Extra Practice **724**

Mixed Problem Solving **758**

Reference

English-Spanish Glossary **R1**

Selected Answers **R19**

Photo Credits **R51**

Index **R52**

Symbols and Properties, Formulas and Measures .... **Inside Back Cover**

Prerequisite Skills

Problem-Solving Strategy: Solve a Simpler Problem

One of the strategies you can use to solve a problem is to **solve a simpler problem**. To use this strategy, first solve a simpler or more familiar case of the problem. Then use the same concepts and relationships to solve the original problem.

Example 1 **Find the sum of the numbers 1 through 500.**

Consider a simpler problem. Find the sum of the numbers 1 through 10. Notice that you can group the addends into partial sums as shown below.

$1 + 2 + 3 + 4 + 5 + 6 + 7 + 8 + 9 + 10 = 55$

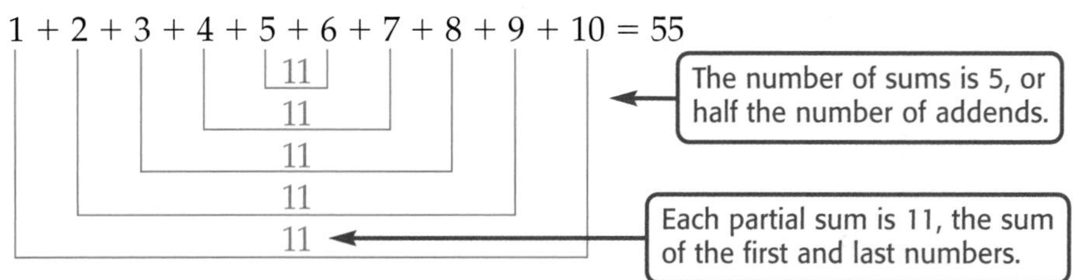
11
11
11
11
11
The number of sums is 5, or half the number of addends.
Each partial sum is 11, the sum of the first and last numbers.

The sum is 5×11 or 55.

Use the same concepts to find the sum of the numbers 1 through 500.

$$1 + 2 + 3 + \ldots + 499 + 500 = 250 \times 501 = 125{,}250$$

Multiply half the number of addends, 250, by the sum of the first and last numbers, 501.

A similar problem-solving strategy is to use subgoals.

Example 2 **Two workers can make two chairs in two days. How many chairs can 8 workers working at the same rate make in 20 days?**

First find how many chairs each worker can make in two days. Divide 2 chairs by 2 workers. → $2 \div 2 = 1$

So, each worker can make 1 chair in 2 days. To find how many chairs each worker can make in 20 days, divide 20 by 2. → $20 \div 2 = 10$

Now find how many chairs 8 workers can make by multiplying 8 by 10. → $8 \times 10 = 80$

So, 8 workers can make 80 chairs in 20 days.

Solve each problem by first solving a simpler problem.

1. Find the sum of the numbers 1 through 1000.
2. Find the number of squares of any size in the game board shown at the right.
3. How many links are needed to join 30 pieces of chain into one long chain?

Solve each problem by using subgoals.

4. Three people can pick six baskets of apples in one hour. How many baskets of apples can 12 people pick in one-half hour?
5. A shirt shop has 112 orders for T-shirt designs. Three designers can make 12 shirts in 2 hours. How many designers are needed to complete the orders in 8 hours?

Problem-Solving Strategy: Work Backward

In most problems, a set of conditions or facts is given and an end result must be found. However, some problems start with the result and ask for something that happened earlier. The strategy of **working backward** can be used to solve problems like this. To use this strategy, start with the end result and *undo* each step.

Example **Paco spent half of the money he had this morning on lunch. After lunch, he loaned his friend a dollar. Now he has $1.50. How much money did Paco start with?**

Start with the end result, $1.50, and work backward to find the amount Paco started with.

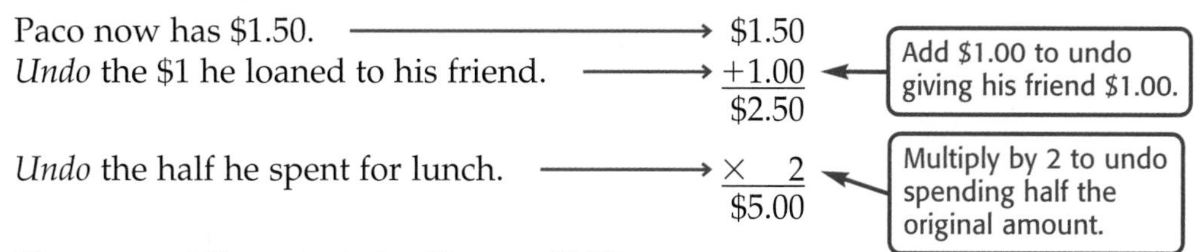

Paco now has $1.50. → $1.50
Undo the $1 he loaned to his friend. → +1.00
$2.50
Undo the half he spent for lunch. → × 2
$5.00
Add $1.00 to undo giving his friend $1.00.
Multiply by 2 to undo spending half the original amount.

The amount Paco started with was $5.00.

CHECK Paco started with $5.00. If he spent half of that, or $2.50, on lunch and loaned his friend $1.00, he would have $1.50 left. This matches the amount stated in the problem, so the solution is correct.

Solve each problem by working backward.

1. Katie used half of her allowance to buy a ticket to the class play. Then she spent $1.75 for an ice cream cone. Now she has $2.25 left. How much is her allowance?
2. Michele put $15 of her paycheck in savings. Then she spent one-half of what was left on clothes. She paid $24 for a concert ticket and later spent one-half of what was then left on a book. When she got home, she had $14 left. What was the amount of Michele's paycheck?
3. A certain number is multiplied by 3, and then 5 is added to the result. The final answer is 41. What is the number?
4. Mr. and Mrs. Jackson each own an equal number of shares of a stock. Mr. Jackson sells one-third of his shares for $2700. What was the total value of Mr. and Mrs. Jackson's stock before the sale?
5. A certain bacteria doubles its population every 12 hours. After 3 full days, there are 1600 bacteria in a culture. How many bacteria were there at the beginning of the first day?
6. Masao had some pieces of bubble gum. He gave one-fourth of the gum to Bob. Bob then gave half of his gum to Lisa. Lisa gave a third of her gum to Maria. If Maria has 3 pieces of gum, how many pieces of gum did Masao have in the beginning?
7. To catch a 7:30 A.M. bus, Carla needs 30 minutes to get dressed, 30 minutes for breakfast, and 15 minutes to walk to the bus stop. What time should she wake up?
8. Justin rented three times as many DVDs as Cole last month. Cole rented four fewer than Maria, but four more than Paloma. Maria rented 10 DVDs. How many DVDs did each person rent?

Problem-Solving Strategy: Make a Table or List

One strategy for solving problems is to **make a table**. A table allows you to organize information in an understandable way.

Example 1 **A fruit machine accepts dollars, and each piece of fruit costs 65 cents. If the machine gives only nickels, dimes, and quarters, what combinations of those coins are possible as change for a dollar?**

The machine will give back $1.00 − $0.65 or 35 cents in change in a combination of nickels, dimes, and quarters.

Make a table showing different combinations of nickels, dimes, and quarters that total 35 cents. Organize the table by starting with the combinations that include the most quarters.

quarters	dimes	nickels
1	1	0
1	0	2
0	3	1
0	2	3
0	1	5
0	0	7

The total for each combination of the coins is 35 cents. There are 6 combinations possible.

A similar strategy is to **list possibilities**. When you make a list, use an organized approach so you do not leave out important items.

Example 2 **How many ways can you receive change for a quarter if at least one coin is a dime?**

List the possibilities. Start with the ways that use the fewest number of coins.

1. dime, dime, nickel
2. dime, dime, 5 pennies
3. dime, nickel, nickel, nickel
4. dime, nickel, nickel, 5 pennies
5. dime, nickel, 10 pennies
6. dime, 15 pennies

There are 6 possibilities.

Solve each problem by making a table or list.

1. How many ways can you make change for a half-dollar using only nickels, dimes, and quarters?
2. A number cube has faces numbered 1 to 6. If a red and a blue cube are tossed and the faces landing up are added, how many ways can you roll a sum less than 8?
3. A penny, a nickel, a dime, and a quarter are in a purse. How many amounts of money are possible if you grab two coins at random?
4. If the sides of a rectangular garden are whole numbers and the area of the garden is 48 square feet, how many combinations of side lengths are possible?
5. Malcolm had 55 football cards. He traded 8 cards for 5 from Damon. He traded 6 more for 4 from Ines and 5 for 3 from Christopher. Finally, he traded 12 cards for 9 from Sam. How many cards does Malcolm have now?

Problem-Solving Strategy: Guess and Check

To solve some problems, you can make a reasonable guess and then check it in the problem. You can then use the results to improve your guess until you find the solution. This strategy is called **guess and check**.

Example **The product of two consecutive even integers is 1088. What are the integers?**

The product is close to 1000.
Make a guess. Let's try 24 and 26. ⟶ $24 \times 26 = 624$ ⟵ This product is too low.

Adjust the guess upward.
Try 30 and 32. ⟶ $30 \times 32 = 960$ ⟵ This product is still too low.

Adjust the guess upward again.
Try 34 and 36. ⟶ $34 \times 36 = 1224$ ⟵ This product is too high.

Try between 30 and 34.
Try 32 and 34. ⟶ $32 \times 34 = 1088$ ⟵ This is the correct product.

The integers are 32 and 34.

Use the guess-and-check strategy to solve each problem.

1. The product of two consecutive odd integers is 783. What are the integers?
2. Paula is three times as old as Courtney. Four years from now she will be just two times as old as Courtney. How old are Paula and Courtney now?
3. The product of a number and its next two consecutive whole numbers is 120. What is the number?
4. Stamps for postcards cost \$0.21, and stamps for first-class letters cost \$0.34. Diego wants to send postcards and letters to 10 friends. If he has \$2.75 for stamps, how many postcards and how many letters can he send?
5. Each hand in the human body has 27 bones. There are 6 more bones in the fingers than in the wrist. There are 3 fewer bones in the palm than in the wrist. How many bones are in each part of the hand?
6. The Science Club sold candy bars and soft pretzels to raise money for an animal shelter. They raised a total of \$62.75. They made 25¢ profit on each candy bar and 30¢ profit on each pretzel sold. How many of each did they sell?
7. Luis has the same number of quarters, dimes, and nickels. In all he has \$4 in change. How many of each coin does he have?
8. Kelsey sold tickets to the school musical. She had 12 bills worth \$175 for the tickets she sold. If all the money was in \$5 bills, \$10 bills, and \$20 bills, how many of each bill did she have?
9. You can buy standard-sized postcards in packages of 5 and large-sized postcards in packages of 3. How many packages of each should you buy if you need exactly 16 postcards?

Comparing and Ordering Decimals

To determine which of two decimals is greater, you can compare the digits in each place-value position, or you can use a number line.

Method 1 Use place value.

Line up the decimal points of the two numbers. Starting at the left, compare the digits in each place-value position. In the first position where the digits are different, the decimal with the greater digit is the greater decimal.

Method 2 Use a number line.

Graph each number on a number line. On a number line, numbers to the right are greater than numbers to the left.

Example 1 **Which is greater, 4.35 or 4.8?**

Method 1 Use place value.

4.35 Line up the decimal points.
4.8 The digits in the tenths place are not the same.

8 tenths $>$ 3 tenths, so 4.8 $>$ 4.35.

Method 2 Use a number line.

Compare the decimals on a number line.

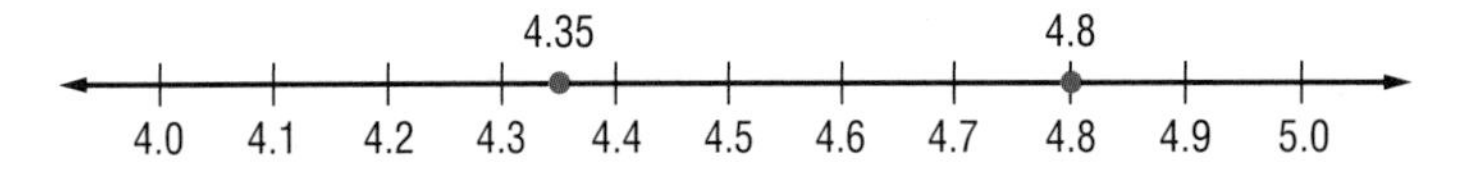

4.8 is to the right of 4.35. So 4.8 $>$ 4.35.

Example 2 **Order 0.8, 1.52, and 1.01 from least to greatest.**

0.8 is less than both 1.52 and 1.01.
1.01 is less than 1.52.

Thus, the order from least to greatest is 0.8, 1.01, 1.52.

Replace each ● with $<$ or $>$ to make a true sentence.

1. 4.05 ● 4.45
2. 2.26 ● 2.28
3. 3.005 ● 3.05
4. 8.7 ● 82.1
5. 6.2 ● 6.008
6. 15.601 ● 16.9
7. 1.9 ● 1.96
8. 8.9 ● 7.99
9. 0.66 ● 0.582
10. 7.14 ● 7.2
11. 0.048 ● 0.11
12. 10.1 ● 1.01
13. 32.1 ● 3.215
14. 1.098 ● 2
15. 9.1 ● 9.005
16. 16.8 ● 16.791
17. 0.943 ● 0.4991
18. 0.117 ● 0.95

Order each set of decimals from least to greatest.

19. {0.2, 0.01, 0.6}
20. {1.2, 2.4, 0.04, 2.2}
21. {3.5, 0.6, 2.06, 0.28}
22. {0.8, 0.07, 1.001, 0.392}
23. {7.06, 7.026, 7.061, 7.009, 7.1}
24. {0.82, 0.98, 0.103, 0.625, 0.809}

Rounding Decimals

Rewriting a number to a certain place value is called **rounding**. Look at the digit to the right of the place being rounded.

- If the digit to the right is *less than or equal to* 4, the digit being rounded stays the same.
- If the digit to the right is *greater than or equal to* 5, the digit being rounded increases by one.

The place-value chart below shows how to round 3.81 to the nearest one (or whole number).

Tens	Ones	Tenths	Hundredths	Thousandths
	3.	8	1	

- 3 is in the ones place
- 8 is to the right of 3
- $8 > 5$

So, 3.81 rounded to the nearest one (or whole number) is 4.

Example 1 **Round each number to the nearest one (or whole number).**

a. 8.3
8.3 rounds to 8.

b. 9.6
9.6 rounds to 10.

Example 2 **Round each number to the nearest tenth.**

a. 16.08
16.08 rounds to 16.1.

b. 29.54
29.54 rounds to 29.5.

Example 3 **Round each number to the nearest hundredth.**

a. 50.345
50.345 rounds to 50.35.

b. 19.998
19.998 rounds to 20.00.

Round each number to the nearest whole number.

1. 3.2
2. 64.8
3. 50.57
4. 16.08
5. 41.29
6. 38.726
7. 74.455
8. 86.299
9. 79.603

Round each number to the nearest tenth.

10. 16.57
11. 1.05
12. 43.827
13. 53.865
14. 80.349
15. 24.731
16. 49.5463
17. 131.9884
18. 68.3553

Round each number to the nearest hundredth.

19. 62.624
20. 44.138
21. 85.5639
22. 105.3582
23. 99.9862
24. 24.8715
25. 458.7625
26. 206.6244
27. 153.2965

Round each number to the nearest dollar.

28. $40.29
29. $72.50
30. $36.82

Estimating Sums and Differences of Decimals

Estimation is often used to provide a quick and easy answer when an exact answer is not necessary. It is also an excellent way to quickly see if your answer is reasonable or not.

Example 1 **Estimate each sum or difference to the nearest whole number.**

a. 16.9 + 5.4

$$\begin{array}{r} 16.9 \\ +\ 5.4 \\ \hline \end{array} \rightarrow \begin{array}{r} 17 \\ +\ 5 \\ \hline 22 \end{array}$$

Round to the nearest whole number.

b. 200.35 − 174.82

$$\begin{array}{r} 200.35 \\ -174.82 \\ \hline \end{array} \rightarrow \begin{array}{r} 200 \\ -175 \\ \hline 25 \end{array}$$

Round to the nearest whole number.

You can also use rounding to estimate answers involving money.

Example 2 **Estimate each sum or difference to the nearest dollar.**

a. \$67.07 + \$52.64 + \$0.85

$$\begin{array}{r} \$67.07 \\ 52.64 \\ +\ \ 0.85 \\ \hline \end{array} \rightarrow \begin{array}{r} \$67.00 \\ 53.00 \\ +\ \ 1.00 \\ \hline \$121.00 \end{array}$$

Round to the nearest dollar.

b. \$89.42 − \$8.94

$$\begin{array}{r} \$89.42 \\ -\ \ 8.94 \\ \hline \end{array} \rightarrow \begin{array}{r} \$89.00 \\ -\ \ 9.00 \\ \hline \$80.00 \end{array}$$

Round to the nearest dollar.

Estimate each sum or difference to the nearest whole number.

1. 12.5 + 44.8
2. 8.6 + 11.9
3. 34.32 + 19.51
4. 15.9 + 20.32
5. 32 − 29.75
6. 125.8 − 22.4
7. 159.7 − 124.8
8. 8.890 + 15.98
9. 0.7 + 1.663
10. 52.4 − 21.01
11. 26.55 − 10
12. 2.79 + 5.9 + 0.02
13. 42.1 + 16.25 + 8.96
14. 209.5 − 110
15. 18 − 12.49

Estimate each sum or difference to the nearest dollar.

16. \$6.89 + \$1.20
17. \$5.72 + \$4.35
18. \$1.68 − \$0.99
19. \$5.00 − \$2.56
20. \$20.00 − \$15.34
21. \$12.86 + \$3.33
22. \$4.99 + \$3.29
23. \$50.00 − \$39.89
24. \$92.30 − \$40.00
25. \$16.39 − \$11.80
26. \$84.99 + \$5.52
27. \$132.62 − \$45.81
28. \$20.19 + \$3.60 + \$5.08
29. \$4.80 + \$7.65 + \$2.59
30. \$325.44 + \$125.10

31. Annual precipitation in Seattle, Washington, is about 37.19 inches. The city of Spokane receives only about 16.49 inches annually. About how much more precipitation does Seattle receive than Spokane?

32. The Adventure Club holds monthly aluminum can recycling drives. During the last three months, they collected \$45.45, \$45.19, and \$44.95 from the drives. About how much did the club collect altogether?

Adding and Subtracting Decimals

To add or subtract decimals, write the numbers in a column and line up the decimal points. Then add or subtract as with whole numbers, and bring down the decimal point.

Example 1 **Find each sum or difference.**

a. 8.2 + 3.4

$$\begin{array}{r} 8.2 \\ +\ 3.4 \\ \hline 11.6 \end{array}$$

Line up the decimal points. Then add.

b. 36.98 − 15.22

$$\begin{array}{r} 36.98 \\ -\ 15.22 \\ \hline 21.76 \end{array}$$

Line up the decimal points. Then subtract.

In some cases, you may want to *annex*, or place zeros at the end of the decimals, to help align the columns. Then add or subtract.

Example 2 **Find each sum or difference.**

a. 21.43 + 5.2

$$\begin{array}{r} 21.43 \\ +\ 5.2 \\ \hline \end{array} \rightarrow \begin{array}{r} 21.43 \\ +\ 5.20 \\ \hline 26.63 \end{array}$$

Annex a zero to align the columns.

b. 7 − 1.75

$$\begin{array}{r} 7 \\ -\ 1.75 \\ \hline \end{array} \rightarrow \begin{array}{r} \overset{6}{\cancel{7}}.\overset{9}{\cancel{0}}\overset{1}{0} \\ -\ 1.75 \\ \hline 5.25 \end{array}$$

Annex two zeros to align the columns.

Find each sum or difference.

1. $\begin{array}{r} 42.3 \\ +\ 0.81 \\ \hline \end{array}$

2. $\begin{array}{r} 5.86 \\ -\ 1.51 \\ \hline \end{array}$

3. $\begin{array}{r} 13 \\ -\ 0.324 \\ \hline \end{array}$

4. 2.3 + 1.1

5. 11.5 + 4.2

6. 9.5 − 8.3

7. 24.8 − 3.6

8. 3.57 − 2.17

9. 7.43 − 5.34

10. 6.40 + 7.36

11. 15.20 + 0.16

12. 7.97 − 4.29

13. 8.70 + 0.64

14. 56.88 − 12.35

15. 4.192 + 1.255

16. 14.6 + 20.81

17. 5.2 − 3.01

18. 1.9 − 1.65

19. 6.38 − 1.1

20. 4.86 − 0.3

21. 9.43 + 1.8

22. 70.3 + 7.03

23. 0.5 + 1.674

24. 25 − 8.3

25. 18 − 12.31

26. 2.85 + 23.6

27. 0.8 + 9.612

28. 6.8 + 5.09 + 0.03

29. 0.5 + 2.41 + 6.7

30. 0.563 + 5.8 + 6.89

31. 41.30 + 0.28 + 6.15

32. 4.52 + 0.167 + 12.9

33. 23.4 + 9.865 + 18.26

34. Find the sum of 27.38 and 6.8.

35. Add $26.59, $1.80, and $13.

36. Find the difference of 42.05 and 11.621.

37. How much more than $102.90 is $115?

38. Karen plans to buy a softball for $6.50, a softball glove for $37.99, and sliders for $13.79. Find the cost of these items before tax is added.

Estimating Products and Quotients of Decimals

You can use rounding to estimate products and quotients of decimals.

Example 1 **Estimate each product or quotient to the nearest whole number.**

a. 3.8×2.1

$3.8 \times 2.1 \rightarrow 4 \times 2 = 8$ Round 3.8 to 4 and round 2.1 to 2.

3.8×2.1 is about 8.

b. $16.45 \div 3.92$

$16.45 \div 3.92 \rightarrow 16 \div 4 = 4$ Round 16.45 to 16 and round 3.92 to 4.

$16.45 \div 3.92$ is about 4.

You can use mental math and compatible numbers to estimate products and quotients of decimals. **Compatible numbers** are rounded so it is easy to compute with them mentally.

Example 2 **Estimate each product or quotient to the nearest whole number.**

a. 7×98.24

$7 \times 98.24 \rightarrow 7 \times 100 = 700$ Even though 98.24 rounds to 98, 100 is a compatible number because it is easy to mentally compute 7×100.

7×98.24 is about 700.

b. $47.5 \div 5.23$

$47.5 \div 5.23 \rightarrow 48 \div 6 = 8$ Even though 5.23 rounds to 5, 6 is a compatible number because 48 is divisible by 6.

$47.5 \div 5.23$ is about 8.

Rewrite each expression using rounding and compatible numbers. Then estimate each product or quotient.

1. 9.2×4.89
2. 6.75×5.25
3. $12.19 \div 3.8$
4. $39.79 \div 4.61$
5. 11.2×6.25
6. $15.2 \div 2.7$
7. $47.2 \div 5.1$
8. $16.53 \div 8.36$
9. $4.32(107.6)$
10. 26×10.9
11. $73.2 \div 6.99$
12. $19.1(21.60)$

Estimate each product or quotient.

13. 4.6×8.3
14. 5.12×5.9
15. $7.5 \div 4.2$
16. $9.27 \div 3.31$
17. $19.8(2.6)$
18. $41.75 \div 6$
19. $36.24 \div 8.7$
20. 5.85×7.55
21. $8.1 \div 2.2$
22. $7.9(9.12)$
23. $6.1 \div 2.1$
24. 9×96.42
25. 13×9.1
26. $10.1 \div 4.7$
27. $28.6(5)$
28. $21 \div 7.6$
29. $81 \div 10.5$
30. $52.7 \div 5.3$
31. 47.74×2
32. 204.5×3
33. $41.79 \div 7.23$

34. The speed of the spine-tailed swift has been measured at 106.25 miles per hour. At that rate, about how far can it travel in 1.8 hours?

Multiplying and Dividing Decimals

To multiply decimals, multiply as with whole numbers. Then add the total number of decimal places in the factors. Place the same number of decimal places in the product, counting from right to left.

Example 1 **Find each product.**

a. 6.3(2.1)

```
   6.3   ← 1 decimal place
 × 2.1   ← 1 decimal place
    63
  12 6
 13.23   ← 2 decimal places
```

The product is 13.23.

b. 9.47(0.5)

```
  9.47   ← 2 decimal places
 × 0.5   ← 1 decimal place
 4.735   ← 3 decimal places
```

The product is 4.735.

To divide decimals, move the decimal point in the divisor to the right, and then move the decimal point in the dividend the same number of places. Align the decimal point in the quotient with the decimal point in the dividend.

Example 2 **Find each quotient.**

a. 1.20 ÷ 0.8

```
       1.5
 0.8)1.2 0     Move each decimal
     8         point right 1 place.
     4 0
     4 0
       0
```

The quotient is 1.5.

b. 32 ÷ 0.25

```
          128
 0.25)32.00    Move each decimal
      25       point right 2 places.
       70
       50
       200
       200
         0
```

The quotient is 128.

Find each product or quotient.

1. 1.2(3)
2. 8(3.4)
3. 0.2 × 7.2
4. 1.4(6.1)
5. 0.63 ÷ 0.9
6. 8.4 ÷ 0.4
7. 0.06 × 3
8. 42 ÷ 0.8
9. 3.9(8.2)
10. 0.2(3.1)
11. 27 ÷ 0.3
12. 64 ÷ 0.4
13. 0.4 ÷ 2
14. 14.4 ÷ 0.16
15. 15.6 × 38
16. 0.51 ÷ 0.03
17. 5.7(3.8)
18. 7.07(4)
19. 1.25 × 12
20. 62.9 ÷ 100
21. 6.5(0.13)
22. 14.9(0.56)
23. 0.384 ÷ 1.2
24. 4.2 ÷ 1.05
25. 25.9 ÷ 2.8
26. 0.47 × 3.01
27. 1.01(6.2)
28. 9 ÷ 0.375
29. 50 ÷ 0.25
30. 500 ÷ 3.2
31. 0.001(7.09)
32. 6.32 × 0.81
33. 2.92 ÷ 0.002

34. Find the product of 13.6 and 9.15.

35. What is the quotient of 72.05 and 0.11?

36. If one United States dollar can be exchanged for 128.46 Spanish pesetas, how many pesetas would you receive for $50?

Estimating Sums and Differences of Fractions and Mixed Numbers

You can use rounding to estimate sums and differences of fractions and mixed numbers. To estimate the sum or difference of proper fractions, round each fraction to 0, $\frac{1}{2}$, or 1.

Example 1 **Estimate each sum or difference.**

a. $\frac{5}{8} + \frac{9}{10}$

$\frac{5}{8} + \frac{9}{10} \rightarrow \frac{1}{2} + 1 = 1\frac{1}{2}$

The sum of $\frac{5}{8}$ and $\frac{9}{10}$ is about $1\frac{1}{2}$.

b. $\frac{5}{6} - \frac{3}{8}$

$\frac{5}{6} - \frac{3}{8} \rightarrow 1 - \frac{1}{2} = \frac{1}{2}$

$\frac{5}{6} - \frac{3}{8}$ is about $\frac{1}{2}$.

To estimate the sum or difference of mixed numbers, round each mixed number to the nearest whole number or to the nearest $\frac{1}{2}$.

Example 2 **Estimate each sum or difference.**

a. $3\frac{3}{8} + 15\frac{15}{16}$

$3\frac{3}{8} + 15\frac{15}{16} \rightarrow 3\frac{1}{2} + 16 = 19\frac{1}{2}$

The sum of $3\frac{3}{8}$ and $15\frac{15}{16}$ is about $19\frac{1}{2}$.

b. $10\frac{3}{4} - 4\frac{1}{6}$

$10\frac{3}{4} - 4\frac{1}{6} \rightarrow 11 - 4 = 7$

$10\frac{3}{4} - 4\frac{1}{6}$ is about 7.

Round each fraction to 0, $\frac{1}{2}$, or 1.

1. $\frac{9}{10}$
2. $\frac{1}{8}$
3. $\frac{13}{25}$
4. $\frac{3}{14}$
5. $\frac{9}{15}$
6. $\frac{78}{81}$

Estimate each sum or difference.

7. $\frac{8}{9} + \frac{1}{4}$
8. $\frac{14}{15} + \frac{5}{6}$
9. $\frac{47}{90} + \frac{3}{24}$
10. $\frac{11}{12} + \frac{4}{9}$
11. $\frac{15}{16} + 9\frac{3}{4}$
12. $1\frac{5}{12} + \frac{7}{18}$
13. $5\frac{10}{11} + \frac{3}{5}$
14. $21\frac{8}{9} + 6\frac{4}{25}$
15. $32\frac{3}{56} + 18\frac{2}{75}$
16. $\frac{4}{5} - \frac{1}{10}$
17. $\frac{7}{9} - \frac{13}{18}$
18. $\frac{9}{10} - \frac{3}{8}$
19. $5\frac{1}{5} - 2\frac{3}{4}$
20. $8\frac{3}{5} - 2\frac{1}{8}$
21. $16\frac{34}{35} - 3\frac{1}{6}$
22. $35\frac{7}{8} - 4\frac{1}{2}$
23. $15\frac{4}{9} + 13\frac{9}{11}$
24. $140\frac{4}{5} - 120\frac{2}{15}$

25. About how much longer than $\frac{5}{6}$ minute is $4\frac{1}{2}$ minutes?

26. Estimate the sum $3\frac{3}{10} + 2\frac{4}{5} + 3\frac{1}{3}$.

27. About how much more is $19\frac{3}{4}$ inches than $10\frac{7}{8}$ inches?

28. Estimate the sum of $7\frac{1}{3}$, $6\frac{4}{5}$, $6\frac{3}{4}$, $7\frac{1}{10}$, and $6\frac{15}{16}$.

29. A board that is $63\frac{5}{8}$ inches long is about how much longer than a board that is $62\frac{1}{4}$ inches long?

Estimating Products and Quotients of Fractions and Mixed Numbers

You can estimate products and quotients of fractions and mixed numbers using rounding and compatible numbers. Compatible numbers are rounded to make it easy to compute with them mentally.

Example **Estimate each product or quotient.**

a. $\frac{5}{16} \times 30$

$\frac{5}{16} \times 30 \rightarrow \frac{1}{3} \times 30$

$\frac{5}{16}$ is close to $\frac{5}{15}$ and $\frac{5}{15} = \frac{1}{3}$.

$\frac{1}{3}$ and 30 are compatible numbers.

Think: $\frac{1}{3} \times 30 = 10$

$\frac{5}{16} \times 30$ is about 10.

b. $9\frac{7}{8} \div 5$

$9\frac{7}{8} \div 5 \rightarrow 10 \div 5 = 2$ Round $9\frac{7}{8}$ to 10. 10 and 5 are compatible numbers.

$9\frac{7}{8} \div 5$ is about 2.

Estimate each product or quotient.

1. $\frac{1}{4} \cdot 11$

2. $\frac{1}{3}(20)$

3. $\frac{1}{3} \times 14$

4. $\frac{1}{4}(15)$

5. $\frac{7}{15} \times 120$

6. $\frac{11}{20}(62)$

7. $\frac{31}{40} \cdot 100$

8. $\frac{6}{13} \times 150$

9. $\frac{1}{5}(44)$

10. $1\frac{5}{6} \cdot 30$

11. $2\frac{1}{4} \cdot 22$

12. $4\frac{4}{5} \times 24$

13. $5\frac{7}{8} \div 2$

14. $8\frac{1}{4} \div 4$

15. $14\frac{6}{7} \div 3$

16. $50 \div 4\frac{7}{8}$

17. $61 \div 2\frac{4}{5}$

18. $148 \div 3\frac{1}{4}$

19. $79 \div 1\frac{9}{10}$

20. $75 \div 2\frac{11}{16}$

21. $88 \div 2\frac{1}{8}$

22. Kim needs $3\frac{1}{2}$ batches of cookies. If one recipe calls for $2\frac{1}{4}$ cups of flour, about how many cups of flour are needed?

23. Mario wants to place photographs of people in one vertical row on a poster board that is $17\frac{1}{2}$ inches long. If each photograph is $2\frac{3}{4}$ inches long, about how many photographs can Mario place on the poster board?

24. A basketball hoop has a diameter of $18\frac{1}{2}$ inches. Estimate the circumference of the hoop. (*Hint:* To estimate the circumference of a circle, multiply the diameter by 3.)

ing Measurements within the Metric System

units of length in the metric system are defined in terms of the meter (m). he diagram below shows the relationships between some common metric units.

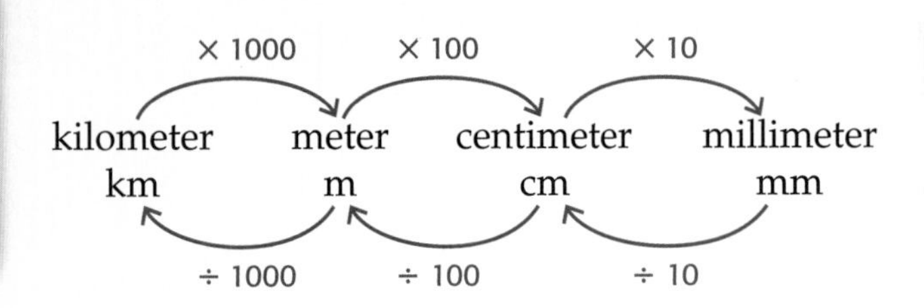

Comparing Metric and Customary Units of Length
1 mm ≈ 0.04 inch (height of a comma)
1 cm ≈ 0.4 inch (half the width of a penny)
1 m ≈ 1.1 yards (width of a doorway)
1 km ≈ 0.6 mile (length of a city block)

- To convert from larger units to smaller units, multiply.
- To convert from smaller units to larger units, divide.

There will be a greater number of smaller units than larger units.

Converting From Larger Units to Smaller Units	Converting From Smaller Units to Larger Units
1 km = 1 × 1000 = 1000 m	1 mm = 1 ÷ 10 = 0.1 cm
1 m = 1 × 100 = 100 cm	1 cm = 1 ÷ 100 = 0.01 m
1 cm = 1 × 10 = 10 mm	1 m = 1 ÷ 1000 = 0.001 km

There will be fewer larger units than smaller units.

Example 1 **Complete each sentence.**

a. 3 km = __?__ m

$3 \times 1000 = 3000$
3 km = 3000 m

To convert from kilometers to meters, multiply by 1000.

b. 9.75 cm = __?__ mm

$9.75 \times 10 = 97.5$
9.75 cm = 97.5 mm

To convert from centimeters to millimeters, multiply by 10.

c. 42 mm = __?__ cm

$42 \div 10 = 4.2$
42 mm = 4.2 cm

To convert from millimeters to centimeters, divide by 10.

The basic unit of capacity in the metric system is the liter (L). A liter and milliliter (mL) are related in a manner similar to meter and millimeter.

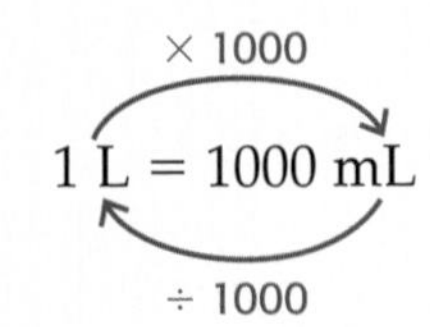

Comparing Metric and Customary Units of Capacity
1 mL ≈ 0.03 ounce (drop of water)
1 L ≈ 1 quart (bottle of ketchup)

Example 2 **Complete each sentence.**

a. 2.5 L = __?__ mL

$2.5 \times 1000 = 2500$
2.5 L = 2500 mL

To convert from larger units to smaller units, multiply.

b. 860 mL = __?__ L

$860 \div 1000 = 0.86$
860 mL = 0.86 L

To convert from smaller units to larger units, divide.

The *mass* of an object is the amount of matter that it contains. The basic unit of mass in the metric system is the kilogram (kg). Kilogram, gram (g), and milligram (mg) are related in a manner similar to kilometer, meter, and millimeter.

Comparing Metric and Customary Units of Mass
1 g ≈ 0.04 ounce (one raisin)
1 kg ≈ 2.2 pounds (textbook)

1 kg = 1000 g   1 g = 1000 mg

Example 3 **Complete each sentence.**

a. 3400 mg = __?__ g

3400 ÷ 1000 = 3.4
3400 mg = 3.4 g

To convert from smaller units to larger units, divide.

b. 74.2 kg = __?__ g

74.2 × 1000 = 74,200
74.2 kg = 74,200 g

To convert from larger units to smaller units, multiply.

State which metric unit you would probably use to measure each item.

1. amount of water in a pitcher
2. distance between two cities
3. thickness of a coin
4. amount of water in a medicine dropper
5. length of a textbook
6. mass of a pencil
7. length of a football field
8. width of a quarter
9. thickness of a pencil
10. gas in the tank of a car
11. vanilla used in a cookie recipe
12. mass of a table tennis ball
13. bag of sugar
14. mass of a horse

Complete each sentence.

15. 5 km = __?__ m
16. 3.5 cm = __?__ mm
17. 6 L = __?__ mL
18. 370 mL = __?__ L
19. 20 mm = __?__ cm
20. 4000 g = __?__ kg
21. 18 cm = __?__ mm
22. 0.75 L = __?__ mL
23. 935 cm = __?__ m
24. 210 mm = __?__ cm
25. 65 g = __?__ kg
26. 2 m = __?__ cm
27. 52.9 kg = __?__ g
28. 800 m = __?__ km
29. 9.05 kg = __?__ g
30. 0.62 km = __?__ m
31. 1250 mL = __?__ L
32. 20,000 mg = __?__ g
33. 3100 m = __?__ km
34. 2.6 m = __?__ cm
35. 36 mg = __?__ g
36. 7 mm = __?__ cm
37. 0.085 L = __?__ mL
38. 125.9 g = __?__ kg

39. The mass of a sample of rocks is 1.56 kilograms. How many grams are in 1.56 kilograms?
40. How many milliliters are in 0.09 liter?
41. Runners often participate in races that are 10 kilometers long. How many meters are in 10 kilometers?
42. How many centimeters are in 0.58 meter?
43. A can holds 355 milliliters of soft drink. How many liters is this?

Converting Measurements within the Customary System

The units of length in the customary system are inch, foot, yard, and mile. The table at the right shows the relationships among these units.

Customary Units of Length
1 foot (ft) = 12 inches (in.)
1 yard (yd) = 3 feet
1 mile (mi) = 5280 feet

- To convert from larger units to smaller units, multiply.
- To convert from smaller units to larger units, divide.

Larger Units → **Smaller Units**

5 ft $= 5 \times 12$ $= 60$ in.
4 yd $= 4 \times 3$ $= 12$ ft

There will be a greater number of smaller units than larger units.

Smaller Units → **Larger Units**

24 in. $= 24 \div 12 = 2$ ft
15 ft $= 15 \div 3$ $= 5$ yd

There will be fewer larger units than smaller units.

Example 1 **Complete each sentence.**

a. 6 yd = __?__ ft

$6 \times 3 = 18$
6 yd = 18 ft

To convert from yards to feet, multiply by 3.

b. 1.5 mi = __?__ ft

$1.5 \times 5280 = 7920$
1.5 mi = 7920 ft

To convert from miles to feet, multiply by 5280.

c. 120 in. = __?__ ft

$120 \div 12 = 10$
120 in. = 10 ft

To convert from inches to feet, divide by 12.

The units of weight in the customary system are ounce, pound, and ton. The table at the right shows the relationships among these units.

Customary Units of Weight
1 pound (lb) = 16 ounces (oz)
1 ton (T) = 2000 pounds

- To convert from larger units to smaller units, multiply.
- To convert from smaller units to larger units, divide.

Larger Units → **Smaller Units**

3 T $= 3 \times 2000 = 6000$ lb
2 lb $= 2 \times 16$ $= 32$ oz

Smaller Units → **Larger Units**

48 oz $= 48 \div 16$ $= 3$ lb
4000 lb $= 4000 \div 2000 = 2$ T

Example 2 **Complete each sentence.**

a. 120 oz = __?__ lb

$120 \div 16 = 7.5$
120 oz = 7.5 lb

To convert from smaller units to larger units, divide.

b. 4 T = __?__ lb

$4 \times 2000 = 8000$
4 T = 8000 lb

To convert from larger units to smaller units, multiply.

Capacity is the amount of liquid or dry substance a container can hold. Customary units of capacity are fluid ounce, cup, pint, quart, and gallon. The relationships among these units are shown in the table.

Customary Units of Capacity
1 cup (c) = 8 fluid ounces (fl oz)
1 pint (pt) = 2 cups
1 quart (qt) = 2 pints
1 gallon (gal) = 4 quarts

As with units of length and units of weight, to convert from larger units to smaller units, multiply. To convert from smaller units to larger units, divide.

Prerequisite Skills

Example 3 **Complete each sentence.**

a. 3 gal = ___?___ qt

$3 \times 4 = 12$ larger unit → smaller unit
3 gal = 12 qt

b. 2 c = ___?___ fl oz

$2 \times 8 = 16$ larger unit → smaller unit
2 c = 16 fl oz

c. 12 pt = ___?___ qt

$12 \div 2 = 6$ smaller unit → larger unit
12 pt = 6 qt

d. 8 c = ___?___ qt

$8 \div 2 = 4$ First, convert cups to pints.
8 c = 4 pt

$4 \div 2 = 2$ Next, convert pints to quarts.
4 pt = 2 qt
So, 8 c = 2 qt.

Complete each sentence.

1. 5 ft = ___?___ in.
2. 2 gal = ___?___ qt
3. 96 oz = ___?___ lb
4. 2 T = ___?___ lb
5. 9 ft = ___?___ yd
6. 6 c = ___?___ pt
7. 2 mi = ___?___ ft
8. 72 in. = ___?___ ft
9. 3 lb = ___?___ oz
10. 7 yd = ___?___ ft
11. 32 fl oz = ___?___ c
12. 15,840 ft = ___?___ mi
13. 2 qt = ___?___ pt
14. 5 pt = ___?___ c
15. 16 qt = ___?___ gal
16. 3000 lb = ___?___ T
17. 6 pt = ___?___ qt
18. 8 pt = ___?___ c
19. 14 pt = ___?___ qt
20. 8 yd = ___?___ ft
21. 5 gal = ___?___ qt
22. 36 qt = ___?___ gal
23. 5 c = ___?___ fl oz
24. 120 in. = ___?___ ft
25. 30 in. = ___?___ ft
26. 6.5 lb = ___?___ oz
27. 12 oz = ___?___ lb

Solve each problem by breaking it into simpler parts.

28. How many inches are in a yard?
29. How many ounces are in a ton?
30. How many cups are in a gallon?

Displaying Data in Graphs

Statistics involves collecting, analyzing, and presenting information. The information that is collected is called data. Displaying **data** in graphs makes it easier to visualize the data.

- **Bar graphs** are used to compare the frequency of data. The bar graph below compares the amounts of recycled materials.

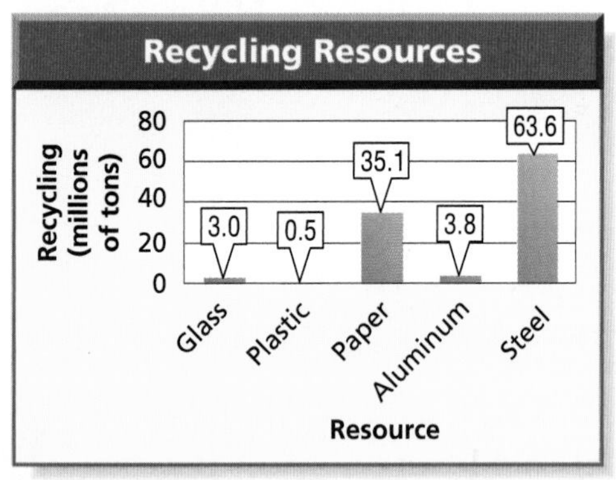

Source: Bureau of Mines

- **Double bar graphs** compare two sets of data. The double bar graph below shows movie preferences for men and women.

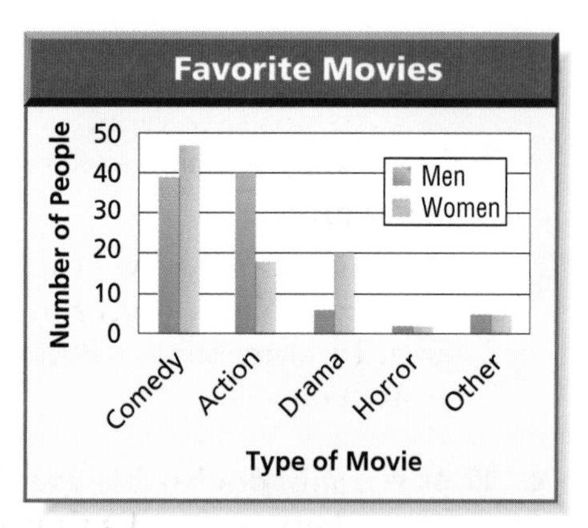

- **Line graphs** usually show how values change over a period of time. The line graph at the right shows the results of the women's Olympic high jump event from 1972 to 2000.

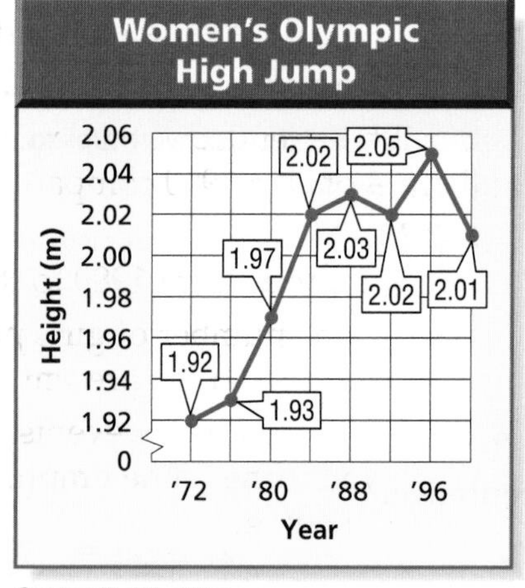

Source: *The World Almanac*

- **Double line graphs**, like double bar graphs, show two sets of data. The double line graph below compares the number of boys and the number of girls participating in high school athletics.

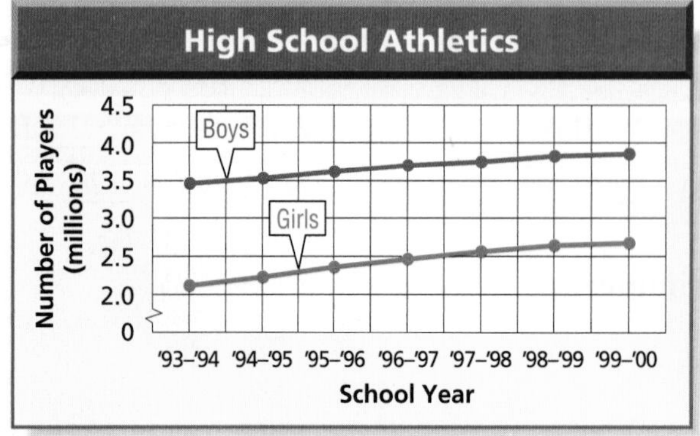

Source: Based on National Federation of State High School statistics

- **Circle graphs** show how parts are related to the whole. The circle graph at the right shows how electricity is generated in the United States.

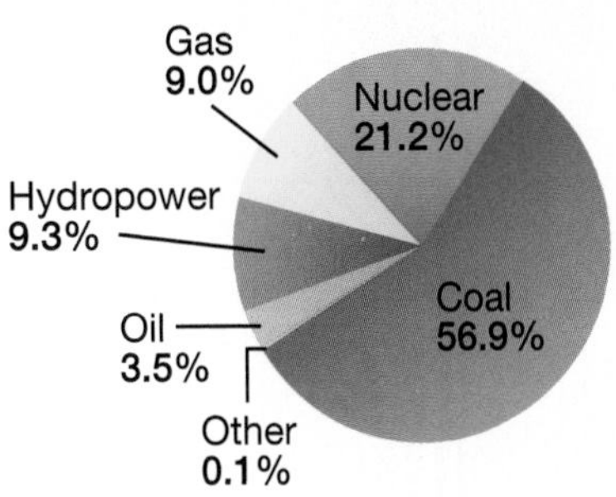

Example **A newspaper wants to display the high temperature of the past week. Should they use a line graph, circle graph, or double bar graph?**

Since the data would show how values change over a period of time, a line graph would give the reader a clear picture of what temperatures were and the changes in temperature.

Determine whether a bar graph, double bar graph, line graph, double line graph, or circle graph is the best way to display each of the following sets of data. Explain your reasoning.

1. the number of people who have different kinds of pets
2. the percent of students in class who have 0, 1, 2, 3, or more than 3 siblings
3. the number of teens who attended art museums, symphony concerts, rock concerts, and athletic events in 1990 compared to the number who attended the same events this year
4. the minimum wage every year from 1980 to the present
5. the number of boys and the number of girls participating in volunteer programs each year from 1995 to the present
6. The table below shows the number of events at recent Olympic games. Would the data be best displayed using a line graph, circle graph, or double bar graph? Explain your reasoning.

Olympic Year	1968	1972	1976	1980	1984	1988	1992	1996	2000
Number of Events	172	196	199	200	223	237	257	271	300

The graph at the right represents the age of Internet users.

7. Explain how the graph is useful in displaying the data.
8. Describe any advantages or disadvantages of using a different type of graph to display the data.

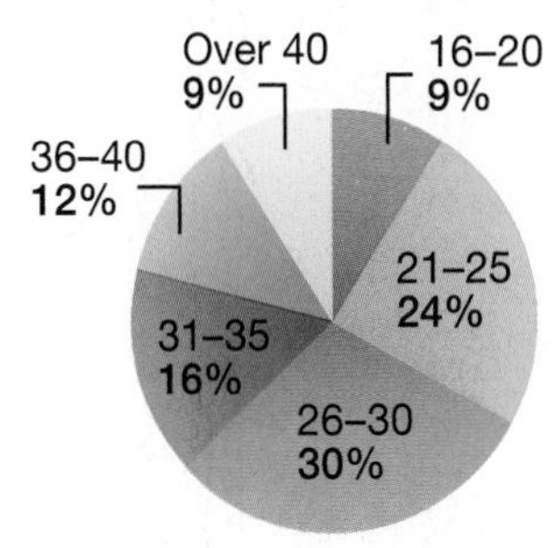

Extra Practice

Lesson 1-1

(pages 6–10)

Solve.

Weight not over (ounces)	Rate
0.5	$0.50
1.0	$0.95
1.5	$1.34
2.0	$1.73
2.5	$2.12
3.0	$2.51
3.5	$2.90
4.0	$3.29

1. **POSTAL SERVICE** The U.S. Postal Service offers air mail service to other countries. The rates for International Air Mail letters and packages are shown in the table at the right. Determine the air mail rate for a package that weighs 5.5 ounces.
   a. Write the *Explore* step. What do you know and what do you need to find?
   b. Write the *Plan* step. What strategy will you use? What do you estimate the answer to be?
   c. *Solve* the problem using your plan. What is your answer?
   d. *Examine* your solution. Is it reasonable? Does it answer the question?
2. **POSTAL SERVICE** In 1995, the state of Florida celebrated the 150th anniversary of its statehood. The U.S. Postal Service issued a stamp, the first to bear the 32-cent price, to honor the occasion. Ninety million of the commemorative stamps were issued. About how much postage did the stamps represent?
   a. Which method of computation do you think is most appropriate for this problem? Justify your choice.
   b. Solve the problem using the four-step plan. Be sure to examine your solution.

Find the next term in each list.

3. 3, 8, 13, 18, 23, …
4. 32, 29, 26, 23, 20, …
5. 6, 7, 9, 12, 16, …

Lesson 1-2

(pages 12–16)

Find the value of each expression.

1. $8 + 7 + 12 \div 4$
2. $20 \div 4 - 5 + 12$
3. $(25 \cdot 3) + (10 \cdot 3)$
4. $36 \div 6 + 7 - 6$
5. $30 \cdot (6 - 4)$
6. $(40 \cdot 2) - (6 \cdot 11)$
7. $\frac{86 - 11}{11 + 4}$
8. $\frac{12 + 84}{11 + 13}$
9. $\frac{5 \cdot 5 + 5}{5 \cdot 5 - 15}$
10. $(19 - 8)4$
11. $75 - 5(2 \cdot 6)$
12. $81 \div 27 \times 6 - 2$
13. Find the value of *thirty-two divided by the product of four and two.*

Write a numerical phrase for each verbal phrase.

14. three increased by nine
15. fifteen divided by three
16. six less than ten

Lesson 1-3

(pages 17–21)

ALGEBRA Evaluate each expression if $a = 2$, $b = 4$, and $c = 3$.

1. $ba - ac$
2. $4b + a \cdot a$
3. $11 \cdot c - ab$
4. $4b - (a + c)$
5. $7(a + b) - c$
6. $8a + 8b$
7. $\frac{8(a + b)}{4c}$
8. $36 - 12c$
9. $\frac{9(b + a)}{c - 1}$
10. $abc - bc$
11. $28 - bc + a$
12. $a(b - c)$

ALGEBRA Translate each phrase into an algebraic expression.

13. nine more than a
14. eleven less than k
15. three times p
16. the product of some number and five
17. twice Shelly's score decreased by 18
18. the quotient of 16 and n

Lesson 1-4

(pages 23–27)

Name the property shown by each statement.

1. $1 \cdot 4 = 4$
2. $6 + (b + 2) = (6 + b) + 2$
3. $9(6n) = (9 \cdot 6)n$
4. $8t \cdot 0 = 0 \cdot 8t$
5. $0(13n) = 0$
6. $7 + t = t + 7$

Find each sum or product mentally.

7. $6 + 8 + 14$
8. $5 \cdot 18 \cdot 2$
9. $0(13 \cdot 6)$
10. $8 + 4 + 12 + 16$
11. $8 \cdot 20 \cdot 10$
12. $4 \cdot 14 \cdot 5$

ALGEBRA **Simplify each expression.**

13. $(12 + x) + 9$
14. $2 \cdot (6 \cdot x)$
15. $(5 \cdot m) \cdot 3$

Lesson 1-5

(pages 28–32)

ALGEBRA **Find the solution of each equation from the list given.**

1. $16 - f = 11$; 3, 5, 7
2. $9 = \frac{72}{m}$; 8, 9, 11
3. $4b + 1 = 17$; 3, 4, 5
4. $17 + r = 25$; 6, 7, 8
5. $9 = 7n - 12$; 3, 5, 7
6. $67 = 98 - q$; 21, 26, 31

ALGEBRA **Solve each equation mentally.**

7. $13 - u = 7$
8. $23 = w + 6$
9. $88 + y = 96$
10. $9z = 45$
11. $88 = 11d$
12. $5t = 0$
13. $13g = 39$
14. $\frac{x}{2} = 8$
15. $\frac{84}{h} = 12$

ALGEBRA **Define a variable. Then write an equation and solve.**

16. The sum of a number and 8 is 14.
17. Twelve less than a number is 50.
18. The product of a number and ten is seventy.
19. A number divided by three is nine.

Lesson 1-6

(pages 33–38)

Use the grid at the right to name the point for each ordered pair.

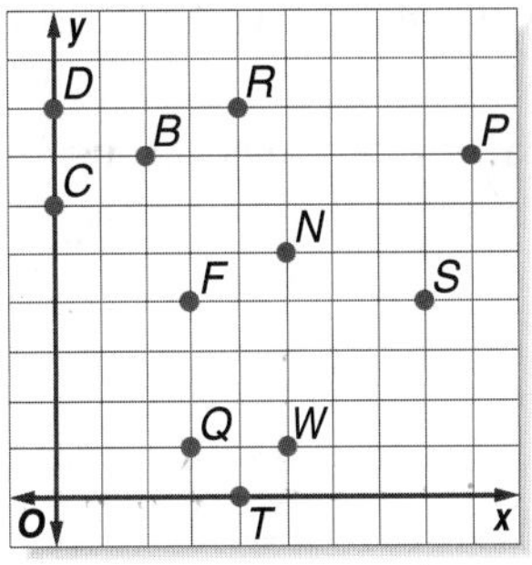

1. (9, 7)
2. (5, 5)
3. (3, 1)
4. (2, 7)
5. (8, 4)
6. (4, 0)

Refer to the coordinate system shown at the right. Write the ordered pair that names each point.

7. *R*
8. *P*
9. *W*
10. *C*
11. *D*
12. *F*

Express each relation as a table and as a graph. Then determine the domain and range.

13. {(3, 6), (4, 9), (5, 1)}
14. {(2, 1), (4, 4), (6, 7), (4, 3)}

Lesson 1-7

(pages 40–44)

Determine whether a scatter plot of the data for the following might show a *positive, negative,* or *no* relationship. Explain your answer.

1. speed of airplane and miles traveled in three hours
2. weight and shoe size
3. outside temperature and heating bill

GAMES For Exercises 4–6, use the following information. The number of pieces in a jigsaw puzzle and the number of minutes required for a person to complete it is shown below.

Number of Pieces	100	60	500	750	1000	800	75
Time (min)	35	20	175	315	395	270	25

4. Make a scatter plot of the data.
5. Does the scatter plot show any relationship? If so, is it positive or negative? Explain your reasoning.
6. Suppose Dave purchases a puzzle having 650 pieces. Predict the length of time it will take him to complete the puzzle.

Lesson 2-1

(pages 56–61)

Replace each ● with <, >, or = to make a true sentence.

1. $-4 \bullet -8$
2. $-6 \bullet 3$
3. $0 \bullet -5$
4. $-12 \bullet -9$
5. $12 \bullet -25$
6. $3 \bullet -7$
7. $0 \bullet -2$
8. $-15 \bullet 12$
9. $5 \bullet -7$
10. $|6| \bullet -2$
11. $-2 \bullet |-3|$
12. $|-7| \bullet |-4|$

Order the integers in each set from least to greatest.

13. $\{-1, 2, -5\}$
14. $\{0, -2, 8, 5, -9\}$
15. $\{100, -34, -86, 21, 0\}$
16. $\{-1, 16, -43, 8, 27, -40\}$
17. $\{0, -23, 75, -15, 24\}$
18. $\{-6, 6, -5, 18\}$

Evaluate each expression.

19. $|-3| + |9|$
20. $|-18| - |5|$
21. $|12 + 7|$
22. $-|6|$
23. $|-8| + |4|$
24. $-|-20|$
25. $|15 - 12|$
26. $|8 + 9|$
27. $-|4| \cdot |-5|$
28. $|-6| \cdot |8|$
29. $-|12| \cdot |9|$
30. $-||-16| + |-22||$

Lesson 2-2

(pages 64–68)

Find each sum.

1. $5 + (-6)$
2. $-17 + 24$
3. $15 + (-29)$
4. $-6 + 13$
5. $50 + (-14)$
6. $-21 + (-4)$
7. $30 + (-7)$
8. $(-3) + (-10)$
9. $-15 + 26$
10. $-17 + 4 + -2$
11. $50 + (-16) + (-11)$
12. $-17 + 8 + (-14)$
13. $-11 + 15 + -6$
14. $23 + (-64)$
15. $-1 + 14 + (-13)$
16. $33 + -18 + 7$
17. $-75 + (-13)$
18. $26 + 14 + (-71)$
19. $8 + (-9) + (-1)$
20. $-16 + (-12) + 13$
21. $35 + (-60)$
22. $12 + -20 + 16$
23. $100 + (-54) + (-17)$
24. $11 + (-22) + (-33)$

Lesson 2-3

(pages 70–74)

Find each difference.

1. $8 - 17$
2. $-15 - 3$
3. $10 - 21$
4. $20 - (-5)$
5. $5 - (-9)$
6. $-12 - (-7)$
7. $-19 - (-6)$
8. $-16 - (-23)$
9. $-56 - 32$
10. $-49 - (-52)$
11. $-6 - 9 - (-7)$
12. $-6 - (-10) - 7$
13. $17 - 33$
14. $-21 - 19$
15. $12 - (-24)$
16. $-35 - (-18)$
17. $-54 - 27$
18. $32 - (-18)$
19. $-26 - (-41)$
20. $99 - (-1)$
21. $-12 - (-25)$
22. $18 - (-43)$
23. $-66 - 13$
24. $54 - 100$

ALGEBRA **Evaluate each expression if $x = 6$, $y = -8$, $z = -3$, and $w = 4$.**

25. $y - z$
26. $3 - z$
27. $y - 5$
28. $x - y$
29. $14 - y - x$
30. $6 + x - z$
31. $y + z + w$
32. $w - z + 11$

Lesson 2-4

(pages 75–79)

Find each product.

1. $-4(2)$
2. $-8(-5)$
3. $13(-4)$
4. $-5 \cdot 6 \cdot 10$
5. $-6(-2)(-14)$
6. $18(-3)(6)$
7. $4(-10)(-3)$
8. $-9(3)(2)$
9. $12(-8)$

ALGEBRA **Simplify each expression.**

10. $-3 \cdot 5x$
11. $7(-8m)$
12. $-10(-3k)$
13. $-4y(-8z)$
14. $(-2r)(-3s)$
15. $6(-2m)(3n)$

ALGEBRA **Evaluate each expression.**

16. $-6t$, if $t = 15$
17. $7p$, if $p = -9$
18. $-4k$, if $k = -16$
19. aw, if $a = 0$ and $w = -72$
20. dk, if $d = -12$ and $k = 11$
21. st, if $s = -8$ and $t = -10$
22. $3hp$, if $h = 9$ and $p = -3$
23. $-5bc$, if $b = -6$ and $c = 2$
24. $-4wx$, if $w = -1$ and $x = -8$

Lesson 2-5

(pages 80–84)

Find each quotient.

1. $-36 \div 9$
2. $112 \div (-8)$
3. $-72 \div 2$
4. $-26 \div (-13)$
5. $-144 \div 6$
6. $-180 \div (-10)$
7. $304 \div (-8)$
8. $-216 \div (-9)$
9. $80 \div (-5)$
10. $-105 \div 15$
11. $120 \div (-30)$
12. $-200 \div (-8)$
13. $42 \div (-6)$
14. $144 \div (-12)$
15. $-360 \div 9$
16. $-84 \div -6$
17. $125 \div (-5)$
18. $180 \div (-15)$
19. $-400 \div 20$
20. $72 \div (-9)$
21. $-156 \div (-2)$

ALGEBRA **Evaluate each expression if $x = -5$, $y = -3$, $z = 2$, and $w = 7$.**

22. $25 \div x$
23. $-42 \div w$
24. $3 \div y$
25. $2x \div z$
26. $-3x \div y$
27. $x \div (-1)$
28. $xyz \div 10$
29. $yz \div 2$
30. $\frac{3y}{-3}$
31. $\frac{6 - y}{y}$
32. $\frac{w}{-7}$
33. $\frac{w - x}{y}$

Lesson 2-6

(pages 85–89)

Name the point for each ordered pair graphed at the right.

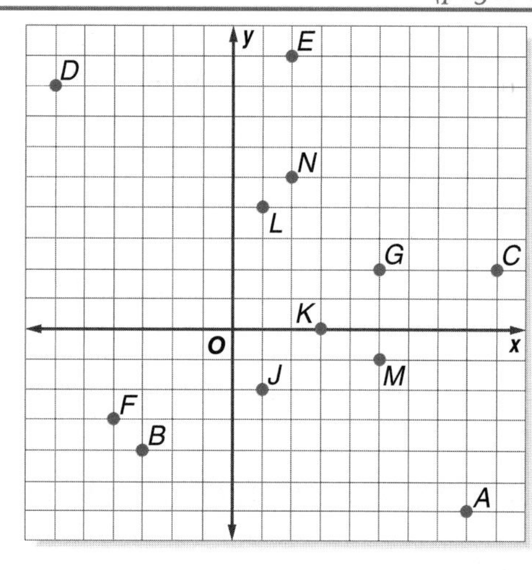

1. $(-6, 8)$
2. $(1, -2)$
3. $(9, 2)$
4. $(1, 4)$
5. $(-3, -4)$
6. $(2, 5)$
7. $(3, 0)$
8. $(5, -1)$

Graph and label each point on a coordinate plane. Name the quadrant in which each point is located.

9. $H(-2, -5)$
10. $P(1, 5)$
11. $R(-3, 1)$
12. $M(4, -2)$
13. $K(-4, 5)$
14. $G(3, -5)$

Extra Practice

Lesson 3-1

(pages 98–102)

Use the Distributive Property to write each expression as an equivalent expression. Then evaluate it.

1. $2(4 + 5)$
2. $4(5 + 3)$
3. $3(7 - 6)$
4. $(2 + 5)9$
5. $(10 - 4)3$
6. $-6(1 + 3)$

ALGEBRA **Use the Distributive Property to write each expression as an equivalent algebraic expression.**

7. $3(m + 4)$
8. $(y + 7)5$
9. $-6(x + 3)$
10. $(p - 4)5$
11. $-3(s - 9)$
12. $5(x + y)$
13. $b(c + 3d)$
14. $(a - b)(-5)$
15. $-6(v - 3w)$
16. $5(x + 12)$
17. $(m - 6)(4)$
18. $-2(a - b)$
19. $(8 - m)(-3)$
20. $8(p - 3q)$
21. $(2x + 3y)(4)$
22. $-5(9 - z)$
23. $21(k - 3)$
24. $(7 - 2h)(-3)$

sson 3-2

(pages 103–107)

Ide ify the like terms in each expression.

1. 3 $4x + x$
2. $5n + 2 - 3n$
3. $6 + 1 + 7y$
4. $2c +$ $+ 8d$
5. $3a - 9 + b$
6. $2 + 6k + 7 - 5k$

ALGEBRA **Si plify each expression.**

7. $8k + 2k + 7$
8. $3 + 2b + b$
9. $t + 2t$
10. $9(3 + 2x)$
11. $4(y + 2) - 2$
12. $(6 + 3e)4$
13. $4 + 9c + 3(c + 2)$
14. $5(7 + 2s) + 3(s + 4)$
15. $9(f + 2) + 14f$
16. $5a - 9a$
17. $-6 + 4x + 9 - 2x$
18. $6a + 11 + (-15) + 9a$
19. $2(8w - 7)$
20. $3(2d + 5) + 4d$
21. $2 + 4p - 6(p - 2)$
22. $-3(b + 4)$
23. $-6 + 3s + 11 - 5s$
24. $3(x - 5) + 7(x + 2)$
25. $3q - r + q + 6r$
26. $8(r + 1) + 7$
27. $3p - 2(p + 6q)$
28. $a + 2b + 4a$
29. $9x - 12 + 12$
30. $1 + g + 5g - 2$

Lesson 3-3

(pages 110–114)

ALGEBRA **Solve each equation. Check your solution.**

1. $y + 49 = 26$
2. $d + 31 = -24$
3. $q - 8 = 16$
4. $x - 16 = 32$
5. $40 = a + 12$
6. $b + 12 = -1$
7. $21 = u + 6$
8. $-52 = p + 5$
9. $-14 = 5 - g$
10. $121 = k + (-12)$
11. $-234 = m - 94$
12. $110 = x + 25$
13. $f - 7 = 84$
14. $y - 864 = 652$
15. $475 + z = -18$
16. $x + 12 = -9$
17. $15 - h = 11$
18. $16 = p + 21$
19. $-13 + t = -2$
20. $86 = x + 43$
21. $y - 11 = -14$

ALGEBRA **Write and solve an equation to find each number.**

22. The sum of −6 and a number is 8.
23. When 3 is subtracted from a number, the result is −5.
24. When 7 is added to a number, the result is −9.
25. When a number is decreased by 8, the result is 5.

Lesson 3-4

(pages 115–119)

ALGEBRA **Solve each equation. Check your solution.**

1. $-y = -32$
2. $7r = -56$
3. $\frac{t}{-3} = 12$
4. $4 = \frac{s}{-14}$
5. $\frac{b}{47} = -2$
6. $64 = -4n$
7. $-144 = 12q$
8. $\frac{r}{11} = -12$
9. $-5g = -385$
10. $-16x = -176$
11. $-21 = \frac{y}{-4}$
12. $-372 = 31k$
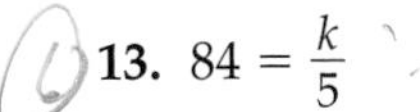
13. $84 = \frac{k}{5}$
14. $-b = 19$
15. $\frac{v}{112} = -9$
16. $-3x = -27$
17. $\frac{p}{-12} = 4$
18. $5q = -100$
19. $\frac{d}{11} = -8$
20. $-9n = -45$
21. $125 = -25z$

ALGEBRA **Write and solve an equation for each sentence.**

22. The product of 8 and a number is −40.
23. The quotient of a number and −3 is 27.
24. When 6 is multiplied by a number, the result is −24.

Lesson 3-5

(pages 120–124)

ALGEBRA **Solve each equation. Check your solution.**

1. $3t - 13 = 2$
2. $-8j - 7 = 57$
3. $9d - 5 = 4$
4. $6 - 3w = -27$
5. $\frac{k}{6} + 8 = 12$
6. $-4 = \frac{q}{8} - 19$
7. $15 - \frac{n}{7} = 13$
8. $44 = -4 + 8p$
9. $21 - h = -32$
10. $-19 = 11b - (-3)$
11. $6 = 20 + \frac{x}{3}$
12. $9 + 3a = -3$
13. $2x - 8 = 10$
14. $\frac{m}{4} - 6 = 10$
15. $-12 + 3p = 3$
16. $-18 = 6a - 6$
17. $\frac{t}{-3} + 11 = 23$
18. $3 + 2v = 11$
19. $16 = \frac{k}{3} - 11$
20. $-6g - 12 = -60$
21. $15 - 4c = -21$

Lesson 3-6

(pages 126–130)

ALGEBRA Write and solve an equation for each sentence.

1. Five less than three times a number is 13.
2. The product of 2 and a number is increased by 9. The result is 17.
3. Ten more than four times a number is 46.
4. The quotient of a number and −8, less 5 is −2.
5. Three more than two times a number is 11.
6. The quotient of a number and six, increased by 2 is −5.
7. The product of −3 and a number, decreased by 9 is 27.

Lesson 3-7

(pages 131–136)

ALGEBRA Solve by replacing the variables with the given values.

1. $d = rt$, if $d = 366$ and $t = 3$.
2. $S = (n - 2) \cdot 180$, if $n = 8$.
3. $A = bh$, if $A = 36$ and $h = 12$.
4. $P = 4s$, if $P = 108$.
5. $V = \ell wh$, if $\ell = 27$, $w = 5$, and $h = 2$.
6. $h = 69 + 2F$, if $F = 42$.

Find the perimeter and area of each rectangle.

7. a rectangle 23 centimeters long and 9 centimeters wide
8. a 16-foot by 14-foot rectangle
9. a rectangle with a length of 31 meters and a width of 3 meters
10. a square with sides 7 meters long

Find the missing dimension of each rectangle.

	Length	Width	Area	Perimeter
11.	9 ft		126 ft^2	46 ft
12.		18 in.	108 in^2	48 in.
13.	13 yd		273 yd^2	68 yd
14.		12 cm	168 cm^2	52 cm
15.		3 m	162 m^2	114 m

16. The perimeter of a rectangle is 50 meters. Its width is 10 meters. Find the length.
17. The area of a rectangle is 96 square inches. Its length is 12 inches. Find the width.

Lesson 4-1

(pages 148–152)

Use divisibility rules to determine whether each number is divisible by 2, 3, 5, 6, or 10.

1. 98
2. 243
3. 800
4. 252
5. 105
6. 210
7. 225
8. 180

List all the factors of each number.

9. 77
10. 42
11. 81
12. 132

ALGEBRA Determine whether each expression is a monomial. Explain why or why not.

13. $-6h$
14. $9 - v$
15. g
16. $3n + 9$
17. 112
18. $2(x + 9)$

Lesson 4-2

(pages 153–157)

ALGEBRA **Write each expression using exponents.**

1. $8 \cdot 8 \cdot 8 \cdot 8$
2. 9
3. $(-6)(-6)(-6)(-6)(-6)$
4. $(y \cdot y \cdot y) \cdot (y \cdot y \cdot y \cdot y)$
5. $a \cdot b \cdot b$
6. $4 \cdot 4 \cdot 4 \cdot 4 \cdot x \cdot x \cdot x \cdot y$
7. $3q \cdot 3q \cdot 3q \cdot 3q \cdot 3q \cdot 3q$
8. $\underbrace{n \cdot n \cdot n \cdot \ldots \cdot n}_{17\ factors}$
9. $(x + y)(x + y)$

Express each number in expanded form.

10. 56
11. 231
12. 4075

ALGEBRA **Evaluate each expression if $m = 3$, $n = 2$, and $p = -4$.**

13. $3m^2$
14. $n^0 + m$
15. 7^4
16. -5^3
17. p^3
18. $2(m - p)^2$
19. $-2n^3 + m$
20. $m - p^2$
21. $(m + n + p)^3$
22. $5p - m^2$
23. $(n + p)^4$
24. $(m - n)^8$

Lesson 4-3

(pages 159–163)

Determine whether each number is *prime* or *composite*.

1. 57
2. 369
3. 116
4. 125
5. 83
6. 99
7. 91
8. 79

Write the prime factorization of each number. Use exponents for repeated factors.

9. 21
10. 44
11. 51
12. 65
13. 30
14. 28
15. 117
16. 88
17. 54
18. 32
19. 300
20. 210

ALGEBRA **Factor each number or monomial completely.**

21. 40
22. $630a$
23. 187
24. 310
25. 510
26. 1589
27. $-18ab^2$
28. $-117x^3$
29. $105j^2k^5$

Lesson 4-4

(pages 164–168)

ALGEBRA **Find the GCF of each set of numbers or monomials.**

1. 27, 45
2. 30, 12
3. 16, 40, 28
4. 18, 17, 15
5. 112, 216
6. 120, 245
7. $84k$, $108k^2$
8. $135ab$, $171b$
9. $185fg$, $74f^2g$
10. $44m$, $60n$
11. $90gh$, $225k$
12. 8, $28h$
13. $16w$, $28w^3$
14. $24a$, $30ab$, $66a^2$
15. $13z$, $39yz$, $52y$

ALGEBRA **Factor each expression.**

16. $3m + 12$
17. $5x + 15$
18. $4 + 8b$
19. $7x + 21$
20. $2a + 100$
21. $42 - 14b$
22. $5f - 25$
23. $11p - 66$
24. $7y - 21$
25. $48 + 12s$
26. $18 - 2w$
27. $24k + 96$
28. $2y + 14$
29. $42 - 7b$
30. $13w + 39$

Lesson 4-5

(pages 169–173)

Write each fraction in simplest form. If the fraction is already in simplest form, write *simplified*.

1. $\frac{3}{54}$
2. $\frac{3}{16}$
3. $\frac{6}{58}$
4. $\frac{15}{55}$
5. $\frac{10}{90}$
6. $\frac{20}{49}$
7. $\frac{8}{20}$
8. $\frac{99}{9}$
9. $\frac{18}{54}$
10. $\frac{21}{64}$
11. $\frac{40}{76}$
12. $\frac{49}{56}$
13. $\frac{22}{66}$
14. $\frac{42}{49}$
15. $\frac{110}{200}$
16. $\frac{b}{b^4}$
17. $\frac{16p}{24p}$
18. $\frac{21x^2y}{81y}$
19. $\frac{32d^2}{6d}$
20. $\frac{72ab}{8b}$
21. $\frac{120z^3x}{18zx}$
22. Fourteen inches is what part of 1 yard?
23. Nine hours is what part of one day?

Lesson 4-6

(pages 175–179)

ALGEBRA **Find each product or quotient. Express using exponents.**

1. $r^4 \cdot r^2$
2. $\frac{2^9}{2^3}$
3. $\frac{b^{18}}{b^5}$
4. $12^3 \cdot 12^8$
5. $x \cdot x^9$
6. $(2s^6)(4s^2)$
7. $w^3 \cdot w^4 \cdot w^2$
8. $(-2)^2(-2)^5(-2)$
9. $\frac{4^7}{4^6}$
10. $3(f^{17})(f^2)$
11. $(5k)^2 \cdot k^7$
12. $\frac{6m^8}{3m^2}$
13. $(3x^4)(-6x)$
14. $(4k^4)(-3k)^3$
15. $\left(\frac{42}{-6}\right)\left(\frac{g^{10}}{g^3}\right)$

Lesson 4-7

(pages 181–185)

ALGEBRA **Write each expression using a positive exponent.**

1. y^{-9}
2. m^{-4}
3. 5^{-3}
4. 2^{-7}
5. 6^{-3}
6. a^{-11}

Write each fraction as an expression using a negative exponent other than −1.

7. $\frac{1}{p^4}$
8. $\frac{1}{b^9}$
9. $\frac{1}{5^3}$
10. $\frac{1}{7^4}$
11. $\frac{1}{15^2}$
12. $\frac{1}{25}$
13. $\frac{1}{c^7}$
14. $\frac{1}{64}$

Write each decimal using a negative exponent.

15. 0.01
16. 0.00001
17. 0.0001
18. 0.001
19. 0.1
20. 0.000001

Evaluate each expression if $x = 3$ and $y = -2$.

21. x^{-2}
22. 9^y
23. y^{-3}
24. x^{-3}
25. y^{-4}
26. $(xy)^{-2}$

Lesson 4-8

(pages 186–190)

Express each number in scientific notation.

1. 9040
2. 0.015
3. 6,180,000
4. 27,210,000
5. 0.00004637
6. 0.00546
7. 500,300,100
8. -0.0000032

Express each number in standard form.

9. -9.5×10^{-3}
10. 8.245×10^{-4}
11. 8.2×10^{4}
12. -9.102040×10^{2}
13. 4.02×10^{3}
14. 1.6×10^{-2}
15. 2.41023×10^{6}
16. 4.21×10^{-5}

Lesson 5-1

(pages 200–204)

Write each fraction or mixed number as a decimal. Use a bar to show a repeating decimal.

1. $\frac{6}{10}$
2. $\frac{4}{25}$
3. $-\frac{1}{8}$
4. $1\frac{3}{4}$
5. $\frac{5}{6}$
6. $\frac{9}{20}$
7. $-4\frac{7}{12}$
8. $\frac{8}{11}$
9. $3\frac{4}{18}$
10. $-\frac{3}{16}$
11. $8\frac{36}{44}$
12. $\frac{6}{15}$

Replace each ● with <, >, or = to make a true sentence.

13. $\frac{7}{8}$ ● $\frac{5}{6}$
14. 0.04 ● $\frac{5}{9}$
15. $\frac{1}{3}$ ● $\frac{2}{7}$
16. $\frac{3}{5}$ ● $\frac{12}{20}$
17. $\frac{1}{2}$ ● 0.75
18. 0.3 ● $\frac{1}{3}$
19. $\frac{2}{3}$ ● 0.64
20. $\frac{2}{20}$ ● 0.10
21. $0.\overline{5}$ ● $\frac{5}{9}$
22. $2.\overline{1}$ ● $2\frac{1}{10}$
23. $3\frac{7}{8}$ ● 3.78
24. $-\frac{6}{7}$ ● $-\frac{5}{6}$

Lesson 5-2

(pages 205–209)

Write each number as a fraction.

1. $3\frac{4}{5}$
2. $-1\frac{2}{9}$
3. 15
4. $2\frac{3}{8}$
5. -13
6. $2\frac{6}{7}$
7. 36
8. $-1\frac{3}{5}$

Write each decimal as a fraction or mixed number in simplest form.

9. 0.6
10. 0.05
11. 0.38
12. 4.12
13. 0.375
14. -3.24
15. 0.222...
16. $-0.\overline{4}$

Identify all sets to which each number belongs.

17. $-4\frac{2}{5}$
18. 6
19. $3\frac{1}{3}$
20. -10
21. 5.9
22. $-\frac{3}{1}$
23. $\frac{16}{8}$
24. 7.02002000...

Lesson 5-3

(pages 210–214)

Find each product. Write in simplest form.

1. $\frac{2}{5} \cdot \frac{3}{16}$
2. $3\frac{1}{4} \cdot \frac{2}{11}$
3. $\frac{3}{5}\left(-\frac{5}{12}\right)$
4. $\frac{5}{8} \cdot \frac{2}{3}$
5. $-\frac{9}{10} \cdot \frac{5}{24}$
6. $\frac{1}{7} \cdot \frac{21}{22}$
7. $\frac{4}{5} \cdot \frac{1}{8}$
8. $2\frac{2}{6} \cdot 6\frac{2}{7}$
9. $2\left(-\frac{7}{12}\right)$
10. $1\frac{3}{7}\left(-9\frac{4}{5}\right)$
11. $-\frac{6}{7}\left(-\frac{6}{7}\right)$
12. $\frac{6c}{10} \cdot \frac{2}{c}$
13. $\frac{p^3}{4} \cdot \frac{12}{p}$
14. $\frac{ab}{9} \cdot \frac{3}{b^2}$
15. $\frac{4x}{3y} \cdot \frac{12y^4}{x^2}$

MEASUREMENT Complete.

16. __?__ inches = $\frac{5}{12}$ yard
17. __?__ minutes = $\frac{1}{5}$ hour
18. $\frac{3}{4}$ pound = __?__ ounces
19. $\frac{7}{8}$ day = __?__ hours

Lesson 5-4

(pages 215–219)

Find the multiplicative inverse of each number.

1. $\frac{4}{7}$
2. $-\frac{5}{9}$
3. $\frac{1}{4}$
4. $5\frac{3}{8}$
5. 6
6. −18
7. $\frac{7}{10}$
8. 2.35

Find each quotient. Write in simplest form.

9. $\frac{4}{5} \div \frac{2}{5}$
10. $-\frac{1}{3} \div \frac{6}{7}$
11. $\frac{4}{9} \div \frac{1}{5}$
12. $\frac{2}{3} \div \frac{1}{9}$
13. $\frac{4}{5} \div \left(-\frac{8}{15}\right)$
14. $\frac{1}{12} \div \frac{3}{4}$
15. $\frac{3}{4} \div \frac{15}{16}$
16. $16 \div 1\frac{7}{8}$
17. $2\frac{1}{6} \div \left(-1\frac{1}{5}\right)$
18. $-11 \div 3\frac{1}{7}$
19. $\frac{8}{45} \div \frac{10}{27}$
20. $-22 \div \left(-5\frac{1}{2}\right)$
21. $\frac{w}{5} \div \frac{w}{35}$
22. $\frac{ab}{12} \div \frac{b}{16}$
23. $\frac{21y}{8x^2} \div \frac{7y}{16x}$

Lesson 5-5

(pages 220–224)

Find each sum or difference. Write in simplest form.

1. $\frac{2}{7} + \frac{3}{7}$
2. $\frac{8}{15} - \frac{4}{15}$
3. $\frac{3}{7} + \frac{4}{7}$
4. $-\frac{8}{9} + \frac{1}{9}$
5. $\frac{5}{6} - \frac{1}{6}$
6. $\frac{7}{12} - \frac{5}{12}$
7. $\frac{5}{12} + \frac{11}{12}$
8. $-\frac{3}{14} - \frac{5}{14}$
9. $3\frac{1}{4} + \left(-\frac{3}{4}\right)$
10. $\frac{3}{8} - \left(-1\frac{1}{8}\right)$
11. $4\frac{9}{10} - 1\frac{1}{10}$
12. $-5\frac{3}{5} + \left(-2\frac{1}{5}\right)$

ALGEBRA Find each sum or difference. Write in simplest form.

13. $\frac{n}{5} + \frac{3n}{5}$
14. $\frac{15}{k} - \frac{8}{k}, k \neq 0$
15. $12\frac{7}{8}s - 7\frac{3}{8}s$
16. $-6\frac{4}{9}t - 3\frac{2}{9}t$
17. $6\frac{1}{4}g + \left(-6\frac{3}{4}g\right)$
18. $7\frac{2}{5}n - \left(-4\frac{2}{5}n\right)$

Lesson 5-6

(pages 226–230)

Find the least common multiple (LCM) of each set of numbers or monomials.

1. 30, 18
2. 4, 16
3. $3m$, 12
4. $6a$, $17a^5$
5. 2, 5, 7
6. $9x^2y$, $12xy^3$

Find the least common denominator (LCD) of each pair of fractions.

7. $\frac{2}{5}, \frac{6}{25}$
8. $\frac{3}{12}, \frac{4}{5}$
9. $\frac{4}{6}, \frac{7}{9}$
10. $\frac{5}{9}, \frac{7}{12}$
11. $\frac{1}{4}, \frac{5}{6}$
12. $\frac{11}{20}, \frac{3}{8}$
13. $\frac{3}{10p}, \frac{7}{5p^3}$
14. $\frac{1}{a^2}, \frac{2}{3a^4}$

Replace each ● with <, >, or = to make a true statement.

15. $\frac{2}{3}$ ● $\frac{3}{4}$
16. $-\frac{5}{8}$ ● $-\frac{3}{5}$
17. $\frac{4}{6}$ ● $\frac{7}{12}$
18. $\frac{11}{18}$ ● $\frac{33}{54}$
19. $\frac{4}{19}$ ● $\frac{8}{38}$
20. $\frac{9}{15}$ ● $\frac{1}{2}$

Lesson 5-7

(pages 232–236)

Find each sum or difference. Write in simplest form.

1. $\frac{1}{5} + \frac{2}{7}$
2. $\frac{4}{5} + \frac{7}{9}$
3. $\frac{1}{9} - \frac{7}{12}$
4. $\frac{8}{11} - \frac{4}{5}$
5. $\frac{7}{12} - \left(-\frac{4}{11}\right)$
6. $-\frac{9}{14} + \frac{15}{16}$
7. $-\frac{3}{8} + \left(-1\frac{5}{12}\right)$
8. $-\frac{2}{15} - 3\frac{1}{5}$
9. $-5\frac{1}{3} + \left(-\frac{1}{6}\right)$
10. $3\frac{2}{5} + 2\frac{4}{7}$
11. $-4\frac{1}{8} + 2\frac{5}{9}$
12. $-3\frac{3}{7} - 5\frac{1}{14}$
13. $11\frac{3}{5} - \left(-6\frac{5}{8}\right)$
14. $\frac{5}{14} + \frac{2}{21}$
15. $2\frac{1}{7} - 3\frac{1}{3}$

Lesson 5-8

(pages 238–242)

Find the mean, median, and mode for each set of data. If necessary, round to the nearest tenth.

1. 82, 79, 93, 91, 95
2. 88, 85, 76, 94, 85, 97
3. 23, 32, 19, 27, 41, 21, 26, 32, 23
4. 7.4, 8.3, 6.1, 5.4, 6.8, 7.1, 8.0, 9.2
5. 0.57, 12.81, 12.6, 0.96, 6.1, 14.3, 4.1, 12.81, 0.96
6. 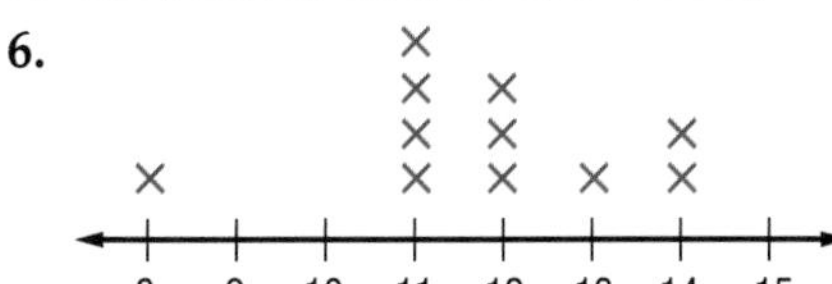

7. 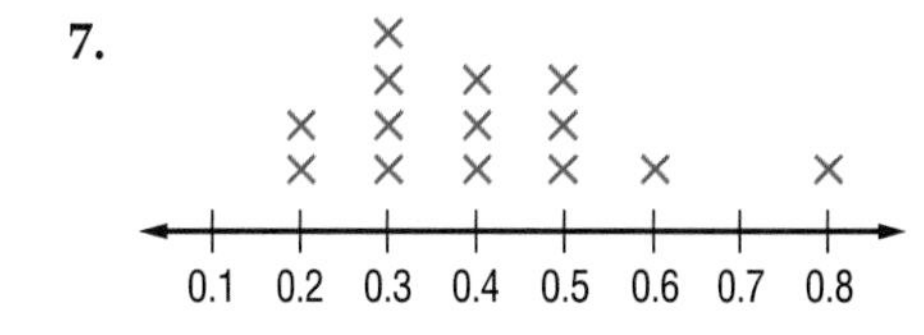

8. **POPULATION** The population of the Canadian provinces and territories in 2000 is shown in the table. Find the mean, median, and mode of the data. If necessary, round to the nearest tenth.

Province/Territory	Population (thousands)	Province/Territory	Population (thousands)
Newfoundland	538.8	Saskatchewan	1023.6
Prince Edward Island	138.9	Northwest Territories	42.1
Nova Scotia	941.0	Alberta	2997.2
New Brunswick	756.6	Yukon	30.7
Quebec	7372.4	British Columbia	4063.8
Ontario	11,669.3	Nunavut	27.7
Manitoba	1147.9		

Lesson 5-9

(pages 244–248)

ALGEBRA Solve each equation. Check your solution.

1. $a - 4.86 = 7.2$
2. $n + 6.98 = 10.3$
3. $87.64 = f - (-8.5)$
4. $x - \frac{2}{5} = -\frac{8}{15}$
5. $3\frac{3}{4} + m = 6\frac{5}{8}$
6. $4\frac{1}{6} = r + 6\frac{1}{4}$
7. $7\frac{1}{3} = c - \frac{4}{5}$
8. $-4.62 = h + (-9.4)$
9. $w - 1\frac{1}{5} = \frac{2}{9}$
10. $\frac{2}{3}w = \frac{1}{6}$
11. $6 = -\frac{3}{4}x$
12. $-0.5m = -10$
13. $-\frac{1}{9}t = 7$
14. $5\frac{2}{3} = y - \frac{1}{8}$
15. $-14.8 = -7.1 + t$

Extra Practice

Lesson 5-10

(pages 249–252)

State whether each sequence is *arithmetic*, *geometric*, or *neither*. If it is arithmetic or geometric, state the common difference or common ratio and write the next three terms of the sequence.

1. 3.5, 4.3, 5.1, …
2. 125, 75, 45, …
3. 5, 10, 20, …
4. 2401, 49, 7, …
5. $\frac{1}{2}, \frac{5}{6}, 1\frac{1}{6}$
6. $-\frac{4}{5}, 2, -5, 12\frac{1}{2}, \ldots$
7. $\frac{1}{4}, \frac{1}{2}, 1, 2, \ldots$
8. 23, 18, 13, …
9. 45, 43, 39, 33, …
10. 2, 4, 8, 16, …
11. 100, 75, 50, …
12. $\frac{1}{5}, 1, 5, 25, \ldots$

Lesson 6-1

(pages 264–268)

Express each ratio as a fraction in simplest form.

1. 15 vans out of 40 vehicles
2. 6 pens to 14 pencils
3. 12 dolls out of 18 toys
4. 8 red crayons out of 36 crayons
5. 18 boys out of 45 students
6. 30 birds to 6 birds
7. 98 ants to 14 ladybugs
8. 140 dogs to 12 cats
9. 321 pennies to 96 dimes
10. 3 cups to 3 quarts

Express each ratio as a unit rate. Round to the nearest tenth, if necessary.

11. 343.8 miles on 9 gallons
12. $7.95 for 5 pounds
13. $52 for 8 tickets
14. $43.92 for 4 CDs
15. 450 miles in 8 hours
16. $3.96 for 12 cans of pop
17. $3.84 for 64 ounces
18. 200 yards in 32.3 seconds

19. **MONEY** Which costs more per notebook, a 4-pack of notebooks for $3.98 or a 5-pack of notebooks for $4.99? Explain.

20. **ANIMALS** A cheetah can run 70 miles in 1 hour. How many feet is this per second? Round to the nearest whole number.

Lesson 6-2

(pages 270–274)

ALGEBRA **Solve each proportion.**

1. $\frac{7}{k} = \frac{49}{63}$
2. $\frac{s}{4.8} = \frac{30.6}{28.8}$
3. $\frac{6}{11} = \frac{19.2}{g}$
4. $\frac{8}{13} = \frac{b}{65}$
5. $\frac{x}{12} = \frac{26}{24}$
6. $\frac{21}{p} = \frac{3}{9}$
7. $\frac{6.5}{8} = \frac{w}{20}$
8. $\frac{10}{4.21} = \frac{7}{y}$

Write a proportion that could be used to solve for each variable. Then solve.

9. 6 plums at $1
   10 plums at *d*
10. 8 gallons at $9.36
    f gallons at $17.55
11. 3 packages at $53.67
    7 packages at *m*
12. 10 cards at $7.50
    p cards at $18
13. 12 cookies at $3.00
    16 cookies at *s*
14. 6 toy cars at $4.50
    c toy cars at $6.75

Lesson 6-3

(pages 276–280)

On a set of architectural drawings for a school, the scale is $\frac{1}{2}$ inch = 4 feet. Find the actual length of each room.

	Room	Drawing Distance
1.	Classroom	5 inches
2.	Principal's Office	1.75 inches
3.	Library	$7\frac{1}{2}$ inches
4.	Cafeteria	$9\frac{1}{4}$ inches
5.	Gymnasium	12.2 inches
6.	Nurse's Office	1.3 inches

Lesson 6-4

(pages 281–285)

Express each decimal or fraction as a percent. Round to the nearest tenth percent, if necessary.

1. 0.42
2. 0.06
3. 1.35
4. 0.001
5. 0.99
6. 3.6
7. 0.8
8. 0.0052
9. 0.00009
10. $\frac{17}{50}$
11. $\frac{9}{25}$
12. $\frac{12}{8}$
13. $\frac{7}{40}$
14. $\frac{11}{33}$
15. $\frac{36}{27}$

Express each percent as a fraction or mixed number in simplest form and as a decimal.

16. 32%
17. 15%
18. $88\frac{1}{2}$%
19. 250%
20. 21%
21. 64%
22. 25%
23. 131%
24. 72.5%
25. $66\frac{2}{3}$%
26. 0.06%
27. 315%

Extra Practice

Lesson 6-5

(pages 288–292)

Use the percent proportion to solve each problem. Round to the nearest tenth.

1. What is 81% of 134?
2. 52.08 is 21% of what number?
3. 11.18 is what percent of 86?
4. What is 120% of 312?
5. 140 is what percent of 400?
6. 430.2 is 60% of what number?
7. 32 is what percent of 80?
8. What is 15% of 125?
9. 22 is what percent of 110?
10. 9.4 is 40% of what number?
11. What is 41.5% of 95?
12. 17.92 is what percent of 112?

13. **FOOD** If 28 of the 50 soup cans on a shelf are chicken noodle soup, what percent of the cans are chicken noodle soup?

14. **SCHOOL** Of the students in a classroom, 60% are boys. If there are 20 students, how many are boys?

Lesson 6-6

(pages 293–297)

Find the percent of each number mentally.

1. 40% of 60
2. 25% of 72
3. 50% of 96
4. $33\frac{1}{3}$% of 24
5. 150% of 42
6. $37\frac{1}{2}$% of 80
7. 200% of 125
8. $66\frac{2}{3}$% of 45

Estimate. Explain which method you used to estimate.

9. 60% of 49
10. 19% of 41
11. 82% of 60
12. 125% of 81
13. $\frac{1}{2}$% of 502
14. 31% of 19

Lesson 6-7

(pages 298–302)

Solve each problem using an equation.

1. 9.28 is what percent of 58?
2. What number is 43% of 110?
3. 80% of what number is 90?
4. What number is 61% of 524?
5. 126 is what percent of 90?
6. 52% of what number is 109.2?
7. 62% of what number is 29.76?
8. 54 is what percent of 90?
9. Find 78% of 125.
10. What is 0.2% of 12?
11. 66% of what number is 49.5?
12. 36.45 is what percent of 81?

Find the discount to the nearest cent.

13. $35 skirt, 20% off
14. $108 lamp, 25% off

Find the interest to the nearest cent.

15. $1585 at 6% for 5 years
16. $2934 at 5.75% for $3\frac{1}{2}$ years

17. **BOOKS** A dictionary is on sale at a 15% discount. Find the sale price of the dictionary if it normally sells for $29.99.

Lesson 6-8

(pages 304–308)

State whether each change is a *percent of increase* or a *percent of decrease*. Then find the percent of change. Round to the nearest tenth, if necessary.

1. from \$56 to \$42
2. from \$26 to \$29.64
3. from \$22 to \$37.18
4. from \$137.50 to \$85.25
5. from \$455 to \$955.50
6. from \$3 to \$15
7. from \$750.75 to \$765.51
8. from \$953 to \$476.50
9. from \$101.25 to \$379.69
10. from \$836 to \$842.27
11. from \$18 to \$24
12. from \$250 to \$100
13. from \$107.50 to \$92
14. from \$365 to \$394.20

15. **BASEBALL CARDS** A baseball card collection contains 340 baseball cards. What is the percent of change if 25 cards are removed from the collection?

Lesson 6-9

(pages 310–314)

There are 4 blue marbles, 6 red marbles, 3 green marbles, and 2 yellow marbles in a bag. Suppose you select one marble at random. Find the probability of each outcome. Express each probability as a fraction and as a percent. Round to the nearest percent.

1. P(green)
2. P(blue)
3. P(red)
4. P(yellow)
5. P(not green)
6. P(white)
7. P(blue or red)
8. P(not yellow)
9. P(neither red nor green)
10. P(red or yellow)
11. P(not orange)
12. P(neither blue nor yellow)
13. P(not red)
14. P(not green or yellow)

15. Suppose two number cubes are rolled. What is the probability of rolling a sum greater than 8?

16. **COOKIES** A sample from a package of assorted cookies revealed that 20% of the cookies were sugar cookies. Suppose there are 45 cookies in the package. How many can be expected to be sugar cookies?

Lesson 7-1

(pages 330–333)

ALGEBRA Solve each equation. Check your solution.

1. $-7h - 5 = 4 - 4h$
2. $5t - 8 = 3t + 12$
3. $m + 2m + 1 = 7$
4. $2y + 5 = 6y + 25$
5. $3z - 1 = 23 - 3z$
6. $5a - 5 = 7a - 19$
7. $5x + 12 = 3x - 6$
8. $3x - 5 = 7x + 7$
9. $5c + 9 = 8c$
10. $3p = 4 - 9p$
11. $6z + 5 = 4z - 7$
12. $2a + 4.2 = 3a - 1.6$
13. $3.21 - 7y = 10y - 1.89$
14. $1.9s + 6 = 3.1 - s$
15. $12b - 5 = 3b$
16. $9 + 11a = -5a + 21$
17. $6 - x = -5$
18. $2.8 - 3w = 4.6 - w$
19. $2.9y + 1.7 = 3.5 + 2.3y$
20. $2.85a - 7 = 12.85a - 2$

Extra Practice

Lesson 7-2

(pages 334–338)

ALGEBRA Solve each equation. Check your solution.

1. $6(m - 2) = 12$
2. $4(x - 3) = 4$
3. $5(2d + 4) = 35$
4. $w + 6 = 2(w - 6)$
5. $3(b + 1) = 4b - 1$
6. $7w - 6 = 3(w + 6)$
7. $4(k - 6) = 6(k + 2)$
8. $3x - 0.8 = 3x + 4$
9. $\frac{5}{9}g + 8 = \frac{1}{6}g + 1$
10. $\frac{s - 3}{7} = \frac{s + 5}{9}$

11. **ALGEBRA** Find the solution of $3(3x + 4) - 2 = 9x + 10$.

12. **NUMBER THEORY** Four times the sum of three consecutive integers is 48.
    a. Write an equation that could be used to find the integers.
    b. What are the integers?

Lesson 7-3

(pages 340–344)

ALGEBRA For the given value, state whether each inequality is *true* or *false*.

1. $5 \geq 2t - 12; t = 11$
2. $7 + n < 25; n = 4$
3. $6r - 18 > 0; r = 3$
4. $3n + 2 < 26; n = 3$
5. $h - 19 < 13; h = 28$
6. $20m \geq 10; m = 0$

ALGEBRA Graph each inequality on a number line.

7. $b \geq 4$
8. $x < -2$
9. $y > 2$
10. $m \leq 0$
11. $p > -1$
12. $q \geq -3$

ALGEBRA Write an inequality for each sentence.

13. At least 295 students attend Greenville Elementary School.
14. An electric bill increased by $15 is now more than $80.
15. If 8 times a number is decreased by 2, the result is less than 15.
16. Citizens who are 18 years of age or older can vote.
17. One dozen jumbo eggs must weigh at least 30 ounces.
18. A healthful breakfast cereal should contain no more than 5 grams of sugar.

Lesson 7-4

(pages 345–349)

ALGEBRA Solve each inequality and check your solution.

1. $m + 9 < 14$
2. $k + (-5) < -12$
3. $-15 < v - 1$
4. $-7 + f \geq 47$
5. $r > -15 - 8$
6. $18 \geq s - (-4)$
7. $38 < r - (-6)$
8. $z - 9 \leq -11$
9. $-16 + c \geq 1$
10. $d + 1.4 < 6.8$
11. $-3 + x > 11.9$
12. $-0.2 \geq 0.3 + y$
13. $h + 5.7 > 21.3$
14. $t - 8.5 > -4.2$
15. $-13.2 > w - 4.87$
16. $a + \frac{5}{12} \geq \frac{7}{18}$
17. $7\frac{1}{2} < n - \left(-\frac{7}{8}\right)$
18. $\frac{2}{3} \leq a - \frac{5}{6}$
19. $-7.42 \leq d - 5.9$

Lesson 7-5
(pages 350–354)

ALGEBRA Solve each inequality and check your solution.

1. $6p < 78$
2. $\frac{m}{-3} > 24$
3. $-18 < 3b$
4. $-5k \geq 125$
5. $-75 > \frac{a}{5}$
6. $\frac{w}{6} < -5$
7. $8 < \frac{2}{3}c$
8. $\frac{m}{1.3} \geq 0.5$
9. $0.4y > -2$
10. $-\frac{1}{2}d \leq -5\frac{1}{2}$
11. $\frac{2}{7}t < 4$
12. $\frac{1}{5}m \geq 4\frac{3}{5}$
13. $\frac{y}{-13} > -20$
14. $14t < 266$
15. $\frac{g}{-25} \geq 8$

16. The product of a number and -4 is greater than or equal to -20. What is the number?

Lesson 7-6
(pages 355–359)

ALGEBRA Solve each inequality and check your solution.

1. $2m + 1 < 9$
2. $-3k - 4 \leq -22$
3. $-2 > 10 - 2x$
4. $-6a + 2 \geq 14$
5. $3y + 2 < -7$
6. $\frac{d}{4} + 3 \geq -11$
7. $\frac{x}{3} - 5 < 6$
8. $-5g + 6 < 3g + 26$
9. $-3(m - 2) > 12$
10. $\frac{r}{5} - 6 \leq 3$
11. $\frac{3(n + 1)}{7} \geq \frac{n + 4}{5}$
12. $\frac{n + 10}{-3} \leq 6$

13. Five plus three times a number is less than the difference of two times the same number and 4. What is the number?

Lesson 8-1
(pages 369–373)

Determine whether each relation is a function. Explain.

1. $\{(3, 6), (35, 64), (1, 1), (21, 7)\}$
2. $\{(32, 24), (27, 24), (36, 24), (45, 24)\}$
3. $\{(2, 9), (3, 18), (4, 27), (2, 36)\}$
4. $\left\{\left(\frac{1}{2}, 3\right), \left(\frac{1}{4}, 5\right), \left(\frac{1}{6}, 7\right), \left(\frac{1}{8}, 9\right), \left(\frac{1}{10}, 11\right)\right\}$
5. $\{(1, 0), (1, 9), (1, 18)\}$
6. $\{(5, 5), (6, 6), (7, 7), (8, 7)\}$

7.

x	y
8	8
15	8
22	51
29	22

8.

x	y
−2	4
−1	5
0	6
−1	7
−2	8

9.

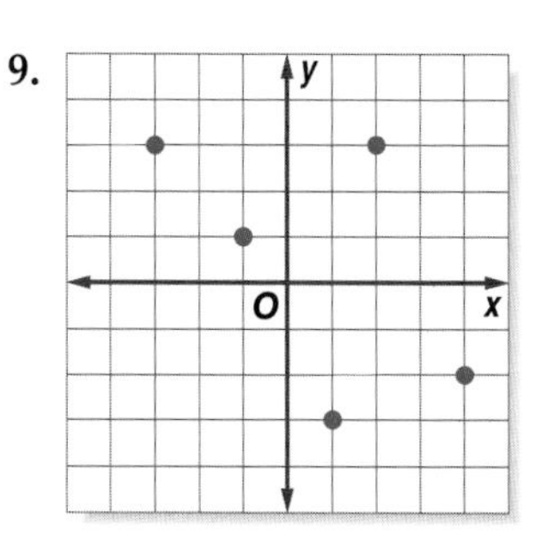

10.

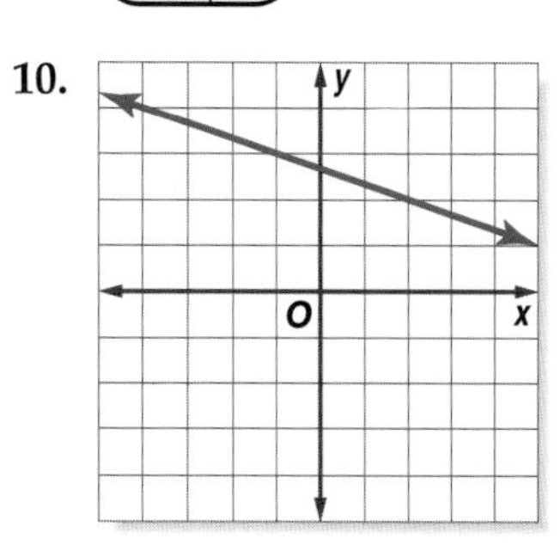

Lesson 8-2

(pages 375–379)

Find four solutions of each equation. Write the solutions as ordered pairs.

1. $x = 4$
2. $y = 0$
3. $x + y = 2$
4. $y = 2x - 6$
5. $x - y = 5$
6. $3x - y = 8$
7. $y = \frac{1}{2}x - 3$
8. $y = \frac{1}{3}x + 1$
9. $2x + y = -2$
10. $2x + 3y = 12$
11. $x + 2y = -4$
12. $2x - 4y = 8$

ALGEBRA **Graph each equation by plotting ordered pairs.**

13. $y = x + 4$
14. $y = 4x$
15. $x + y = 3$
16. $y = x - 3$
17. $y = -2x + 5$
18. $2x + y = 6$

Lesson 8-3

(pages 381–385)

Find the x-intercept and the y-intercept for the graph of each equation.

1. $y = x + 7$
2. $y = 3x + 12$
3. $4x + 3y = 24$
4. $y = 8 - 2x$
5. $-5x + y = -10$
6. $y = \frac{2}{3}x - 7$

ALGEBRA **Graph each equation using the x- and y-intercepts.**

7. $2x + y = 6$
8. $-4x + y = 8$
9. $3x - 3y = -12$
10. $x + 2y = -4$
11. $y = -x - 6$
12. $y = -1$

Lesson 8-4

(pages 387–391)

Find the slope of each line.

1.

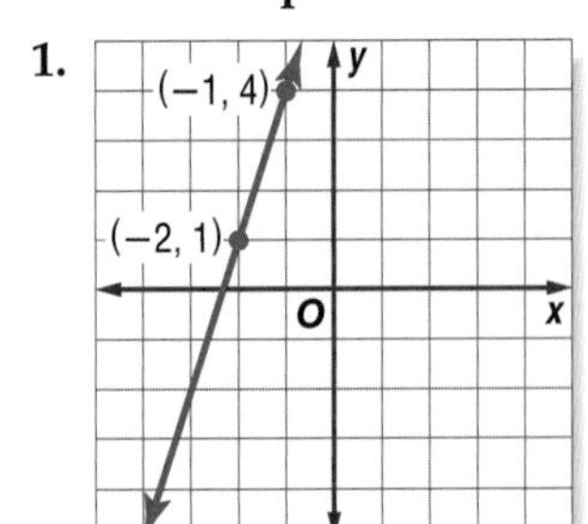

2.

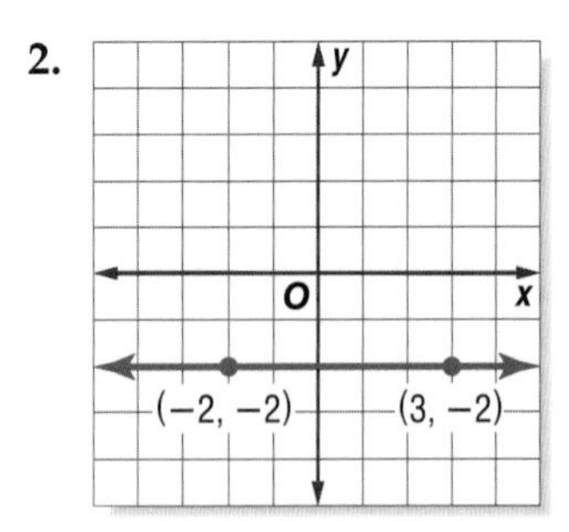

Find the slope of the line that passes through each pair of points.

3. $P(3, 8)$, $Q(4, -3)$
4. $D(4, 5)$, $E(-3, -9)$
5. $L(-1, 2)$, $M(0, 5)$
6. $J(6, 2)$, $K(6, -4)$
7. $B(8, -3)$, $C(-4, 1)$
8. $D(1, 5)$, $E(3, 10)$
9. $H(7, 2)$, $I(-2, -2)$
10. $K(2, -4)$, $L(5, -19)$
11. $G(5, 6)$, $H(7, 6)$
12. $A(-6, -3)$, $B(-9, 4)$
13. $P(-1, -6)$, $Q(-5, -10)$
14. $B(5, 9)$, $C(-4, -5)$

Lesson 8-5

(pages 393–397)

ALGEBRA **Find the rate of change for each linear function.**

1.

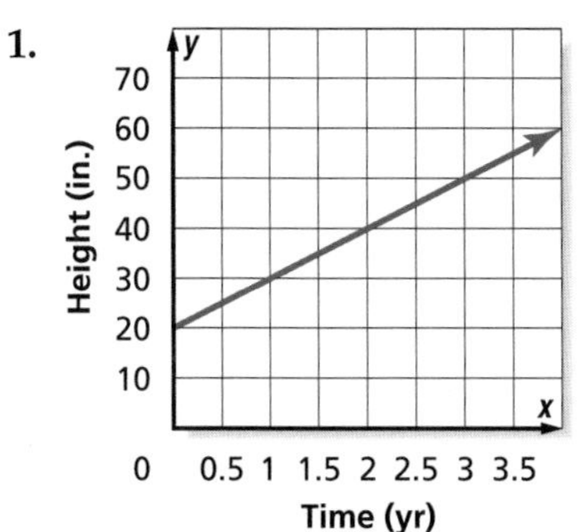

2.

y
35
30
25
20
15
10
5
0
5 10 15 20 25 30 35
x
Distance (ft)
Time (s)

3.

Cookies Purchased	x	0	1	2	3
Balance ($)	y	6	5.6	5.2	4.8

Suppose y varies directly with x. Write an equation relating x and y.

4. $y = 12$ when $x = -3$

5. $y = 45$ when $x = 15$

6. $y = 8$ when $x = 18$

7. $y = 7.6$ when $x = 4$

Lesson 8-6

(pages 398–401)

State the slope and the y-intercept for the graph of each equation.

1. $y = x + 9$

2. $y = 2x - 5$

3. $y = -6x$

4. $y = \frac{3}{2}x$

5. $y = \frac{1}{3}x + 8$

6. $x + 2y = 12$

Graph each equation using the slope and y-intercept.

7. $y = 3x - 2$

8. $x - 3y = 9$

9. $y = \frac{1}{2}x + 4$

10. $y = -\frac{2}{3}x - 1$

11. $x - y = -4$

12. $2x + 4y = -4$

13. $y = x + 5$

14. $3x + y = 9$

Lesson 8-7

(pages 404–408)

ALGEBRA **Write an equation in slope-intercept form for each line.**

1. slope = 3, y-intercept = −4

2. slope = $\frac{3}{4}$, y-intercept = 1

3. slope = −7, y-intercept = −2

4. slope = $\frac{5}{8}$, y-intercept = 9

5. slope = $-\frac{1}{2}$, y-intercept = 0

6. slope = 0, y-intercept = −6

ALGEBRA **Write an equation in slope-intercept form for the line passing through each pair of points.**

7. (4, 7) and (0, 3)

8. (3, −6) and (−1, 2)

9. (8, 7) and (0, 0)

10. (1, 4) and (3, −6)

11. (−2, 5) and (3, 9)

12. (3, −1) and (5, −1)

Extra Practice

Lesson 8-8

(pages 409–413)

TECHNOLOGY For Exercises 1–3, use the table that shows the percent of U.S. households owning more than one television set.

Year	1955	1960	1965	1970	1975	1980	1985	1990	1995	2000
Percent	4	12	22	35	43	50	57	65	71	76

1. Make a scatter plot and draw a best-fit line.
2. Write an equation in slope-intercept form for the best-fit line.
3. Use the equation to predict what percent of U.S. households will own more than one television set in 2010.

Extra Practice

Lesson 8-9

(pages 414–418)

Solve each system of equations by graphing.

1. $x + 2y = 6$
   $y = -0.5x + 3$
2. $y = -2$
   $4x + 3y = 2$
3. $y = x + 4$
   $y = -2x + 4$
4. $y = 4 + \frac{2}{3}x$
   $2x = 3y$
5. $y = x - 2$
   $y = -\frac{1}{3}x + 2$
6. $y = \frac{1}{2}x + 6$
   $2x + y = 1$

ALGEBRA Solve each system of equations by substitution.

7. $x + y = 4$
   $y = 2$
8. $2x + y = 8$
   $x = 2$
9. $y = x - 7$
   $x = 7$
10. $x = 3$
    $y = 4$
11. $y = 1$
    $2y + x = 1$
12. $y = 2x + 3$
    $y = 5$

Lesson 8-10

(pages 419–422)

Graph each inequality.

1. $y > 2x - 2$
2. $y \geq x$
3. $y < 1$
4. $x + y \leq -1$
5. $y + 3x \leq 0$
6. $x < -3$
7. $2x + 3y \geq 12$
8. $-2x + y > -1$
9. $y \geq -4$

WORK For Exercises 10–12, use the following information.
Seth can tutor students and volunteer at a soup kitchen no more than 9 evenings per month.

10. Write an inequality to represent this situation.
11. Graph the inequality.
12. Use the graph to determine how many days each month that Seth could tutor and volunteer. List two possibilities.

Lesson 9-1

(pages 436–440)

Find each square root, if possible.

1. $\sqrt{36}$
2. $-\sqrt{81}$
3. $\sqrt{\frac{1}{4}}$
4. $-\sqrt{144}$
5. $\sqrt{-25}$
6. $\sqrt{1.96}$
7. $\sqrt{-100}$
8. $-\sqrt{0.49}$
9. $\sqrt{400}$

Use a calculator to find each square root to the nearest tenth.

10. $\sqrt{21}$
11. $\sqrt{99}$
12. $-\sqrt{60}$
13. $\sqrt{124}$
14. $-\sqrt{350}$
15. $\sqrt{18.6}$
16. $-\sqrt{42}$
17. $-\sqrt{84.2}$
18. $\sqrt{182}$

Estimate each square root to the nearest whole number. Do not use a calculator.

19. $\sqrt{21}$
20. $-\sqrt{85}$
21. $\sqrt{7.3}$
22. $\sqrt{1.99}$
23. $-\sqrt{62}$
24. $\sqrt{74.1}$
25. $\sqrt{810}$
26. $-\sqrt{88.8}$
27. $\sqrt{1000}$

Lesson 9-2

(pages 441–445)

Name all of the sets of numbers to which each real number belongs. Let N = natural numbers, W = whole numbers, Z = integers, Q = rational numbers, and I = irrational numbers.

1. 15
2. 0
3. $\frac{3}{8}$
4. 0.666...
5. 1.75
6. $\sqrt{2}$
7. 5.14726...
8. $-\sqrt{36}$
9. 0.3535...

Replace each ● with <, >, or = to make a true statement.

10. $3\frac{3}{4}$ ● $\sqrt{15}$
11. $-\sqrt{41}$ ● -6.8
12. 5.2 ● $\sqrt{27.04}$
13. $-\sqrt{110}$ ● -10.5

ALGEBRA **Solve each equation. Round to the nearest tenth, if necessary.**

14. $x^2 = 14$
15. $y^2 = 25$
16. $34 = p^2$
17. $55 = h^2$
18. $225 = k^2$
19. $324 = m^2$
20. $d^2 = 441$
21. $r^2 = 25{,}000$
22. $10{,}000 = x^2$

Lesson 9-3

(pages 447–451)

Use a protractor to find the measure of each angle. Then classify each angle as *acute*, *right*, or *obtuse*.

1. $m\angle PQW$
2. $m\angle VQW$
3. $m\angle TQW$
4. $m\angle SQW$
5. $m\angle SQR$
6. $m\angle VQR$

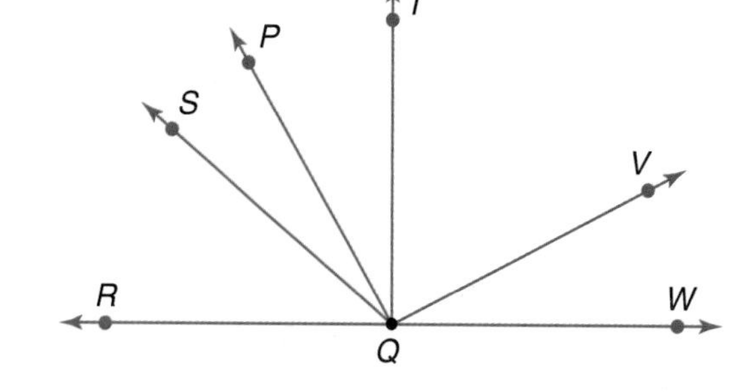

Use a protractor to draw an angle having each measurement. Then classify each angle as *acute*, *obtuse*, *right*, or *straight*.

7. 35°
8. 115°
9. 90°
10. 160°
11. 180°
12. 18°

Extra Practice

Lesson 9-4

(pages 453–457)

Find the value of x in each triangle. Then classify each triangle as *acute, right,* or *obtuse*.

1.

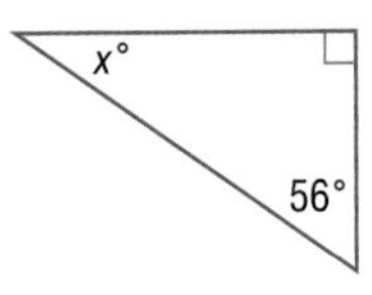

2.

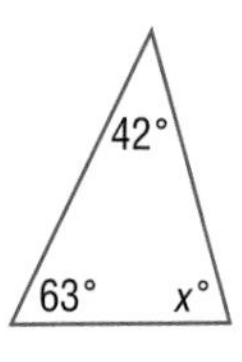

3.

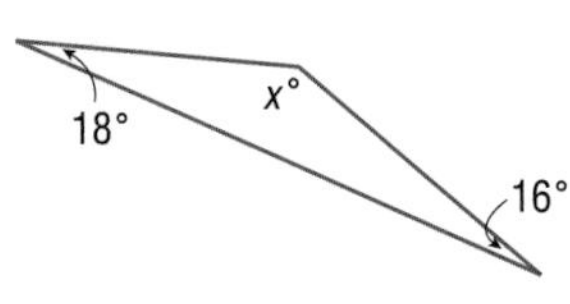

4.

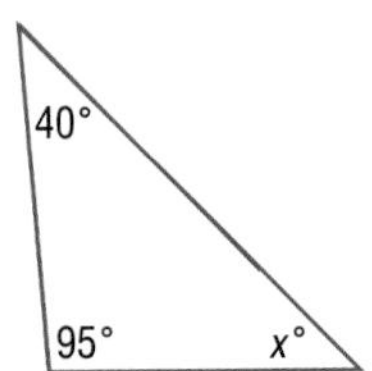

5.

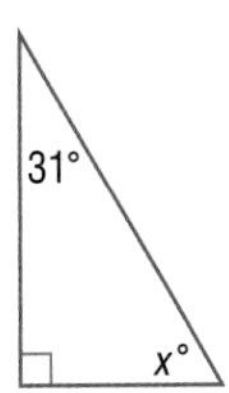

6.

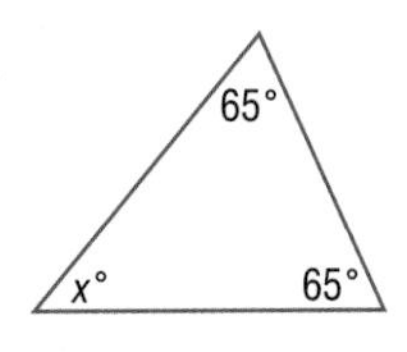

7. **ALGEBRA** The measure of the angles of a triangle are in the ratio 1:2:3. What is the measure of each angle?

8. **ALGEBRA** Determine the measures of the angles of $\triangle ABC$ if the measures of the angles of a triangle are in the ratio 1:1:2.

9. **ALGEBRA** Suppose the measures of the angles of a triangle are in the ratio 1:9:26. What is the measure of each angle?

Lesson 9-5

(pages 460–464)

Find the length of the hypotenuse in each right triangle. Round to the nearest tenth, if necessary.

1.

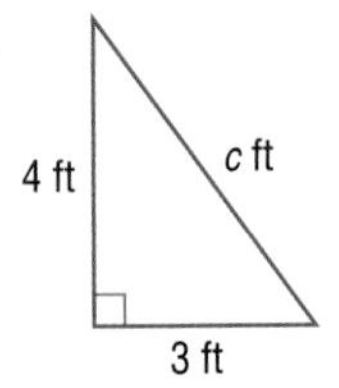

2.

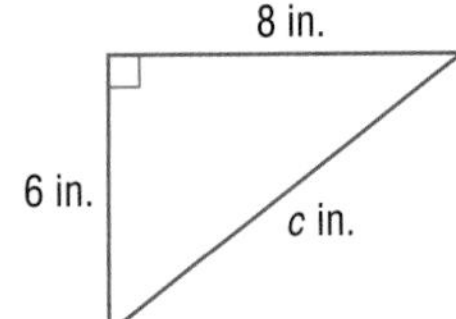

3.

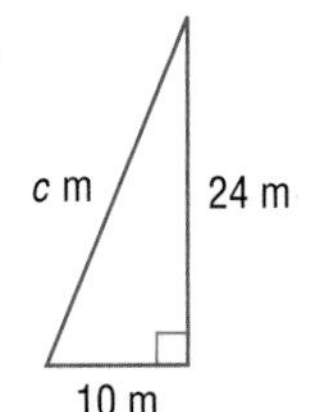

If c is the measurement of the hypotenuse, find each missing measure. Round to the nearest tenth, if necessary.

4. $a = 7$ m, $b = 24$ m
5. $a = 18$ in., $c = 30$ in.
6. $b = 10$ ft, $c = 20$ ft
7. $a = 3$ cm, $c = 9$ cm
8. $b = 8$ m, $c = 32$ m
9. $a = 32$ yd, $c = 65$ yd

Lesson 9-6

(pages 466–470)

Find the distance between each pair of points. Round to the nearest tenth, if necessary.

1. $A(2, 6), B(-4, 2)$
2. $C(-3, 9), D(2, 4)$
3. $E(6, -4), F(1, -6)$
4. $G(0, -1), H(9, -1)$
5. $I(-8, -3), J(2, 2)$
6. $K(3, 0), L(-7, -2)$

The coordinates of the endpoints of a segment are given. Find the coordinates of the midpoint of each segment.

7. $M(3, 5), N(7, 1)$
8. $O(-6, 2), P(0, 8)$
9. $Q(4, -9), R(-2, 7)$
10. $S(13, -1), T(-5, -3)$

Lesson 9-7
(pages 471–475)

In Exercises 1–4, the triangles are similar. Write a proportion to find each missing measure. Then find the value of x.

1. 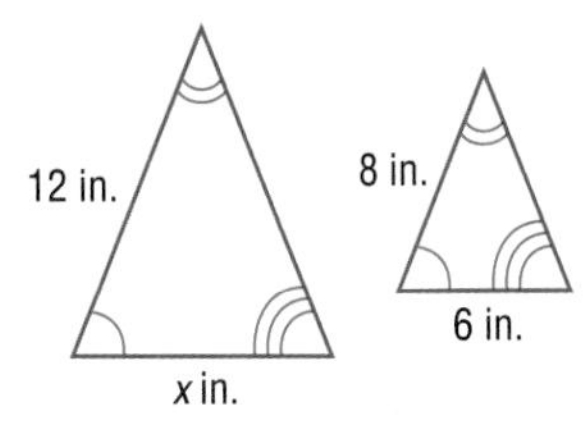

2. 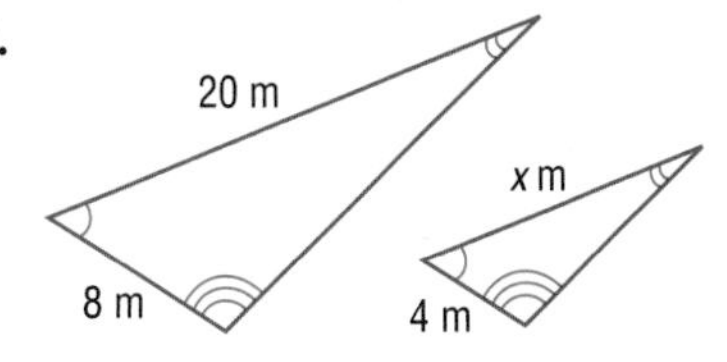

3. 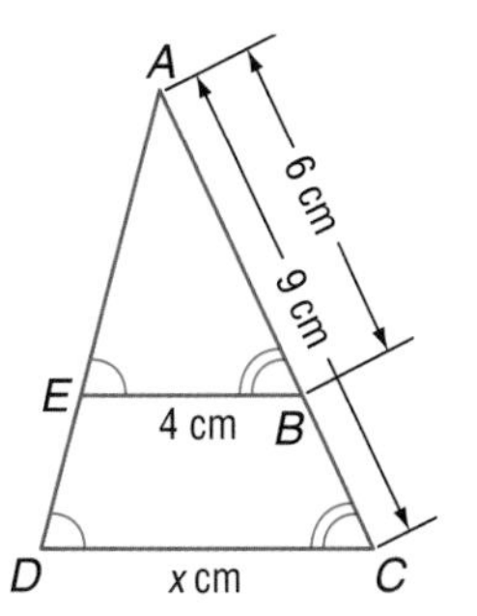

4. 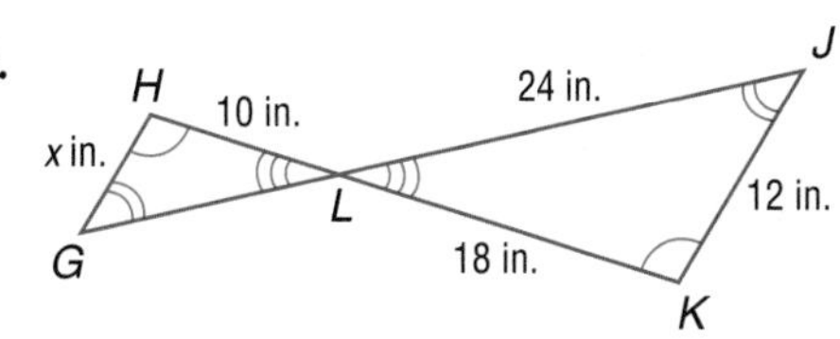

Lesson 9-8
(pages 477–481)

Find each sine, cosine, or tangent. Round to four decimal places, if necessary.

1. sin A
2. sin B
3. tan F
4. sin E
5. cos A
6. tan A

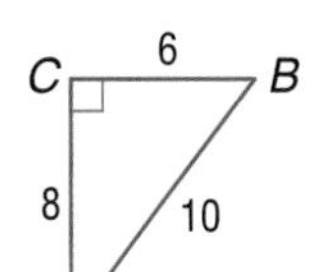

Use a calculator to find each value to the nearest ten thousandth.

7. sin 21°
8. tan 83°
9. cos 45°
10. tan 10°
11. sin 72°
12. cos 3°

Lesson 10-1
(pages 490–497)

In the figure at the right, ℓ is parallel to m and p is a transversal. If the measure of angle 2 is 38°, find the measure of each angle.

1. $\angle 1$
2. $\angle 4$
3. $\angle 3$
4. $\angle 6$
5. $\angle 5$
6. $\angle 8$

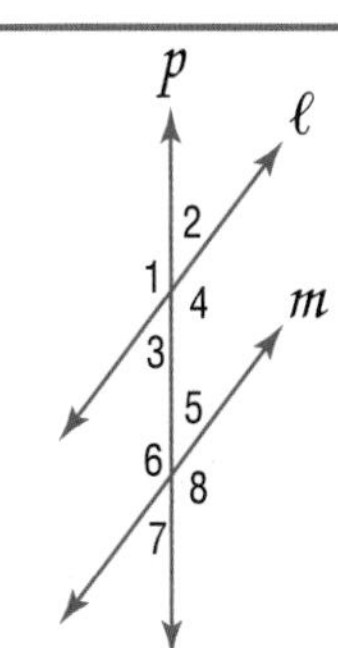

Find the value of x in each figure.

7. 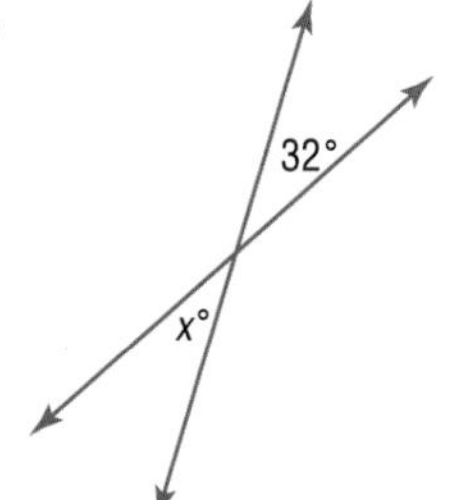

8. 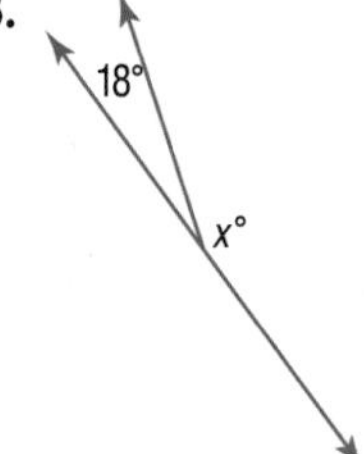

9. 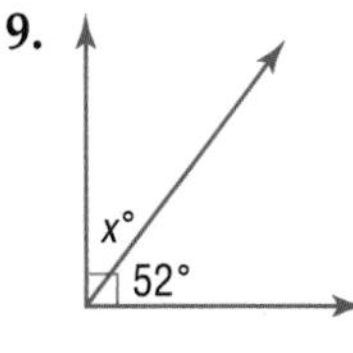

10. 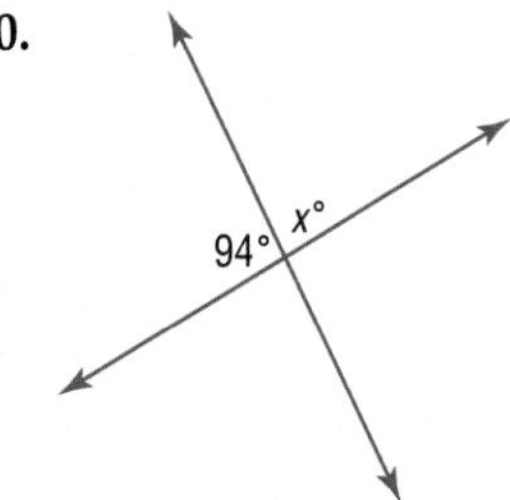

Lesson 10-2
(pages 500–504)

For each pair of congruent triangles, name the corresponding parts. Then complete the congruence statement.

1. 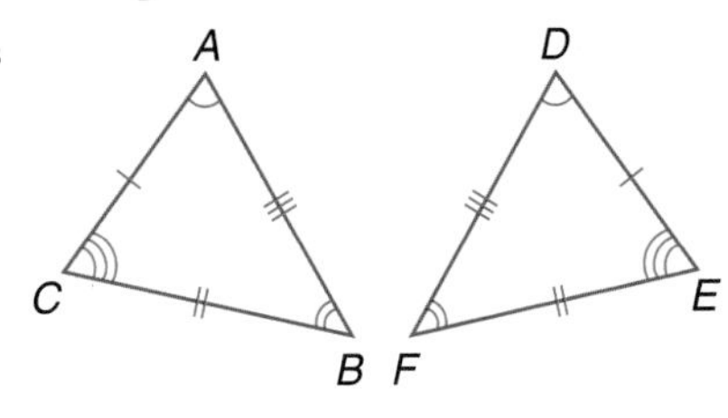

$\triangle ABC \cong \triangle$ __?__

2. 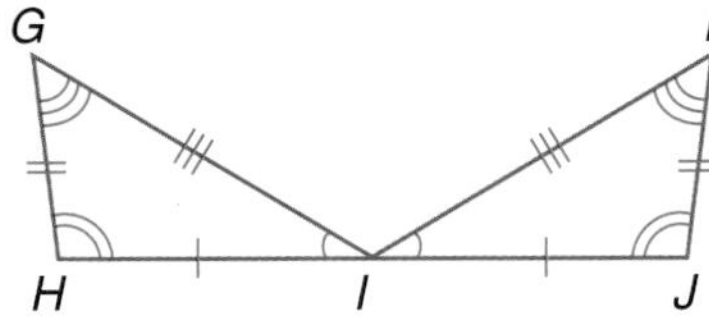

$\triangle GHI \cong \triangle$ __?__

Complete each congruence statement if $\triangle JKL \cong \triangle DGW$.

3. $\angle K \cong$ __?__
4. $\overline{WG} \cong$ __?__
5. $\angle D \cong$ __?__
6. $\overline{KL} \cong$ __?__
7. $\overline{DG} \cong$ __?__
8. $\angle W \cong$ __?__

Lesson 10-3
(pages 506–511)

Find the coordinates of the vertices of each figure after the given translation. Then graph the translation image.

1. (2, −1)

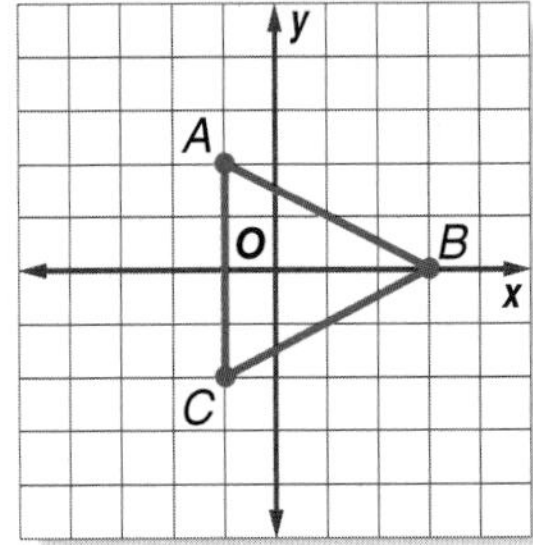

2. (−3, −2)

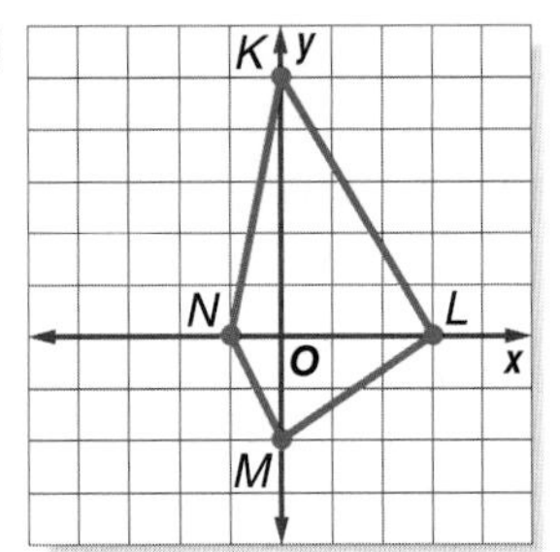

Find the coordinates of the vertices of each figure after a reflection over the given axis. Then graph the reflection image.

3. x-axis

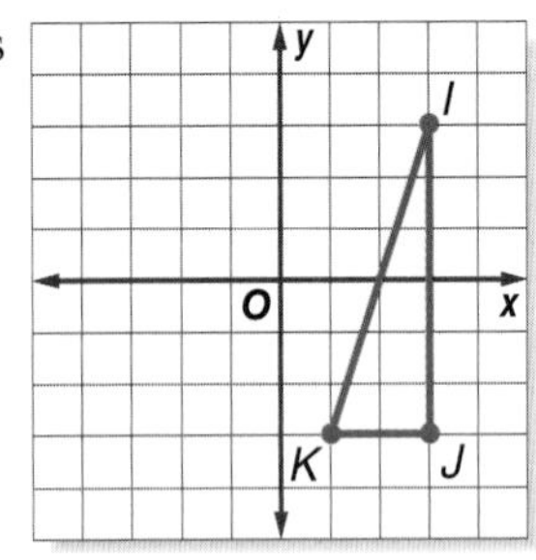

4. y-axis

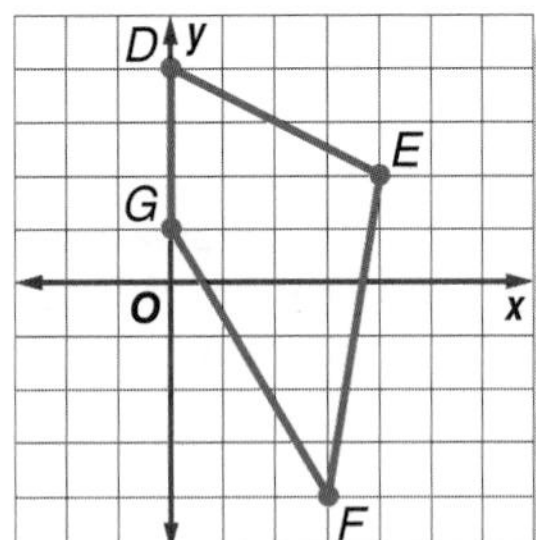

Lesson 10-4
(pages 513–517)

ALGEBRA **Find the value of x. Then find the missing angle measure.**

1. 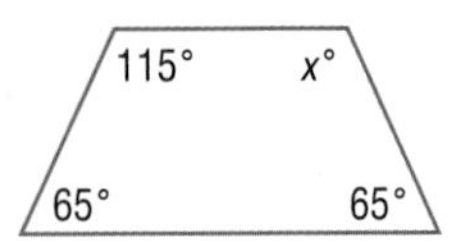

2. 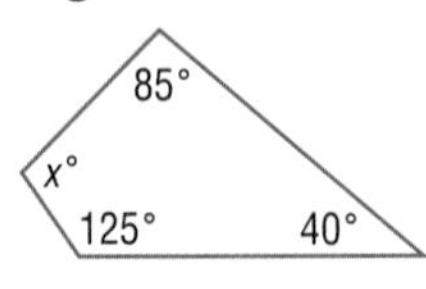

3. 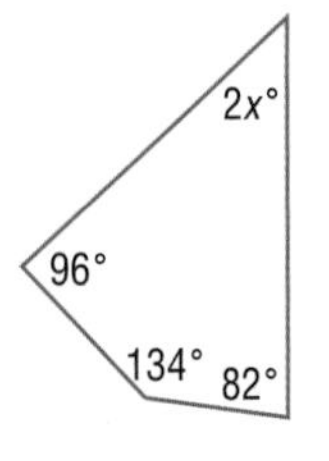

4. 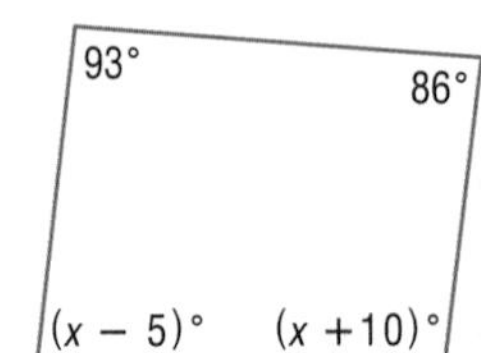

Lesson 10-5

(pages 520–525)

Find the area of each figure.

1. 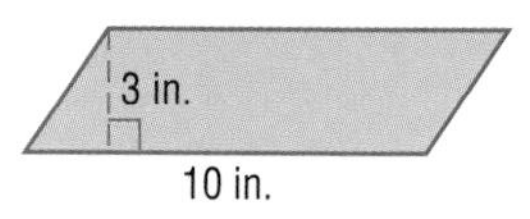

2. 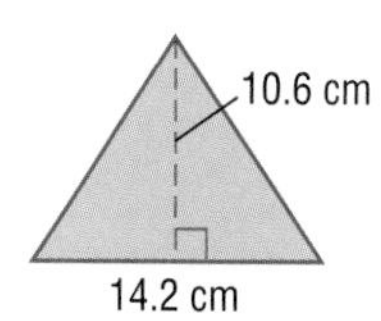

3. 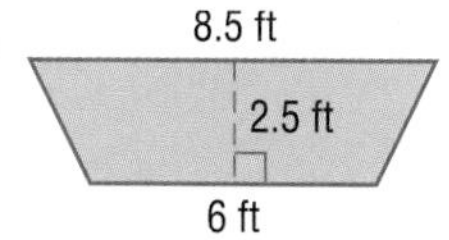

4. 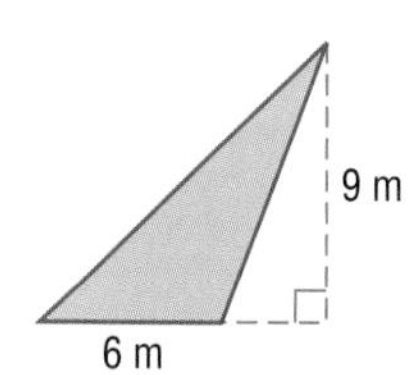

5. What is the height of a parallelogram with a base of 3.4 inches and an area of 32.3 inches?

6. The bases of a trapezoid measure 8 meters and 12 meters. Find the measure of the height if the trapezoid has an area of 70 square meters.

Lesson 10-6

(pages 527–531)

Classify each polygon. Then determine whether it appears to be regular or not regular.

1. 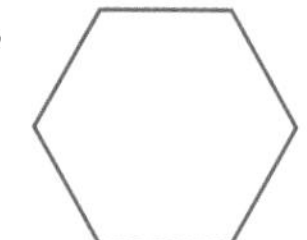

2. 

Find the sum of the measures of the interior angles of each polygon.

3. decagon
4. pentagon
5. nonagon
6. hexagon
7. octagon
8. 15-gon

Lesson 10-7

(pages 533–538)

Find the circumference and area of each circle. Round to the nearest tenth.

1. 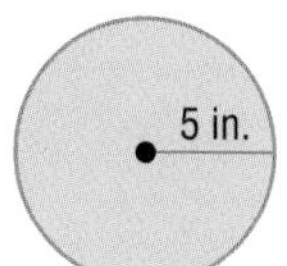

2. 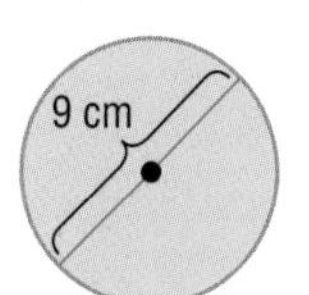

3. 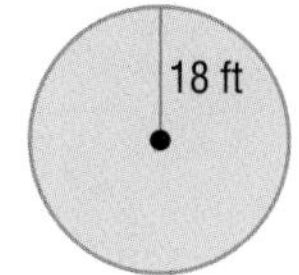

4. 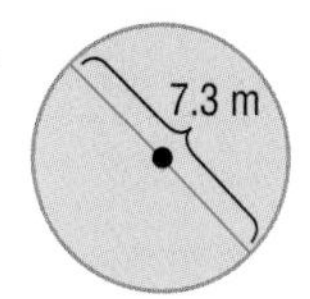

5. The radius is 8.2 feet.
6. The diameter is 1.3 yd.
7. The diameter is 5.2 yd.
8. The radius is 4.8 cm.

9. Find the diameter of a circle if its circumference is 18.5 feet. Round to the nearest tenth.

10. A circle has an area of 62.9 square inches. What is the radius of the circle? Round to the nearest tenth.

Lesson 10-8
(pages 539–543)

Find the area of each figure. Round to the nearest tenth.

1.
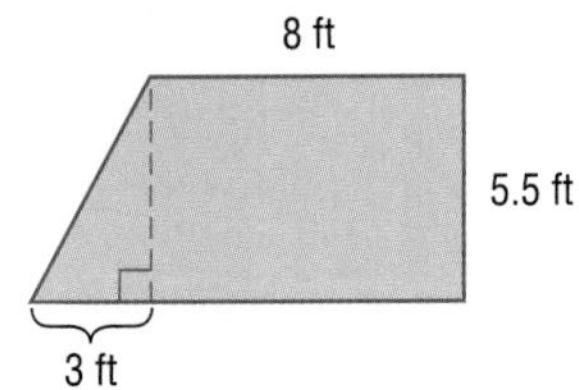

2.
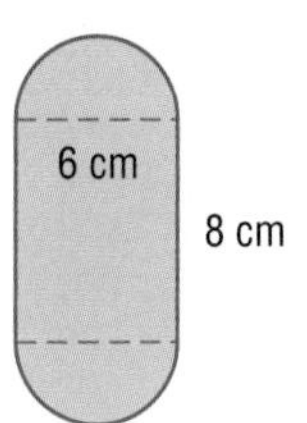

3.
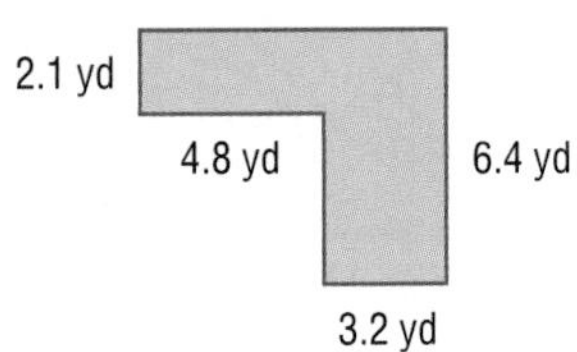

4.
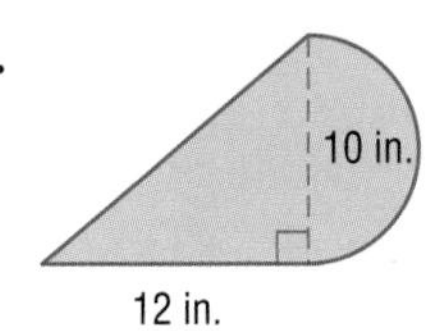

Lesson 11-1
(pages 556–561)

Identify each solid. Name the bases, faces, edges, and vertices.

1.
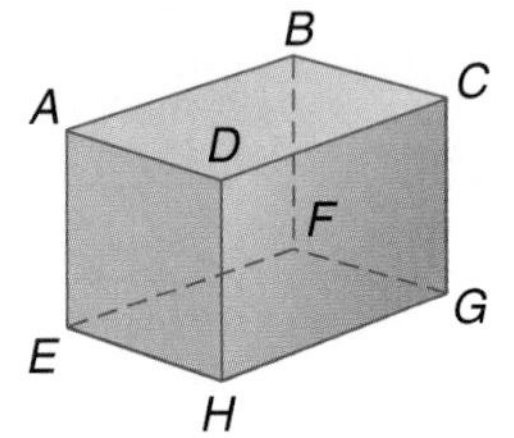

2.
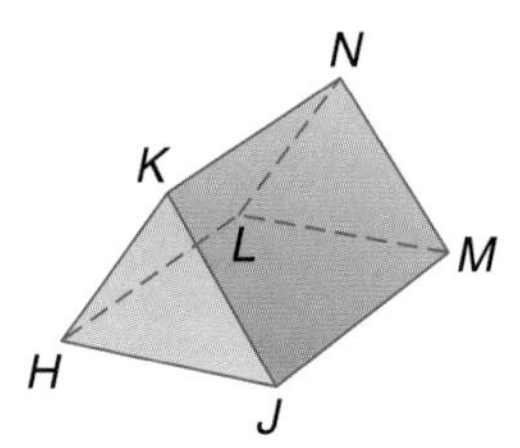

For Exercises 3–5, use the figure in Exercise 2.

3. State whether $\overline{HK}$ and $\overline{KN}$ are *parallel*, *skew*, or *intersecting*.
4. Name a segment that is skew to $\overline{JM}$.
5. Identify two planes that appear to be parallel.

Lesson 11-2
(pages 563–567)

Find the volume of each solid. If necessary, round to the nearest tenth.

1.
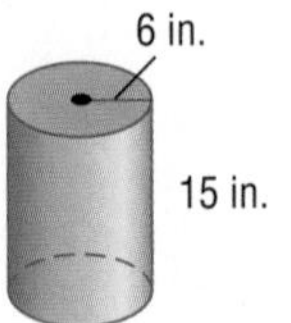

2.
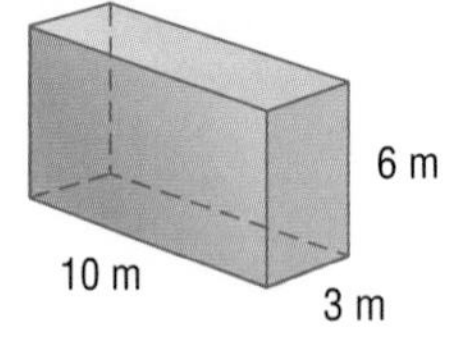

3.
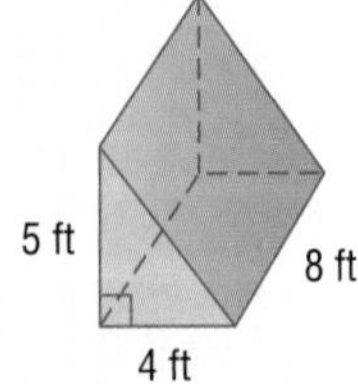

4.
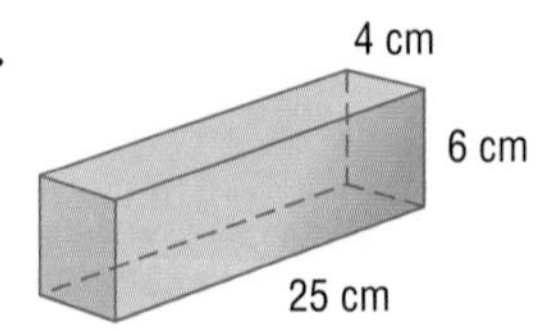

5. rectangular prism: length $2\frac{1}{2}$ yd, width 7 yd, height 12 yd
6. cylinder: diameter 9.2 mm, height 16 mm
7. triangular prism: base of triangle 3.1 cm, altitude of triangle 1.7 cm, height of prism 5.0 cm
8. Find the height of a rectangular prism with a length of 13 inches, width of 5 inches, and volume of 292.5 cubic inches.

Lesson 11-3
(pages 568–572)

Find the volume of each solid. If necessary, round to the nearest tenth.

1. 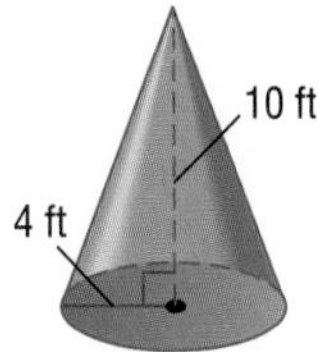

2. 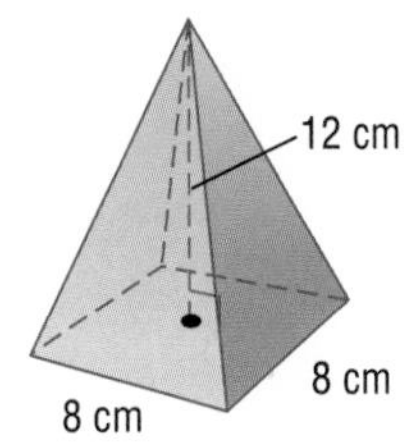

3. cone: diameter 10 yd, height 7 yd
4. rectangular pyramid: length 6 in., width 6 in., height 9 in.
5. square pyramid: length $3\frac{1}{4}$ ft, height 12 ft
6. cone: radius 3.6 cm, height 20 cm
7. hexagonal pyramid: base area 185 m^2, height 7 m

Lesson 11-4
(pages 573–577)

Find the surface area of each solid. If necessary, round to the nearest tenth.

1. 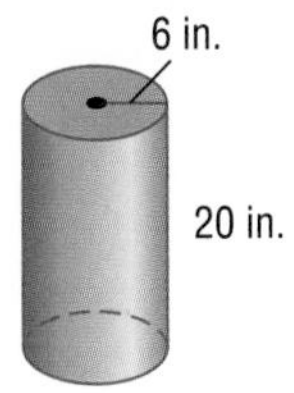

2. 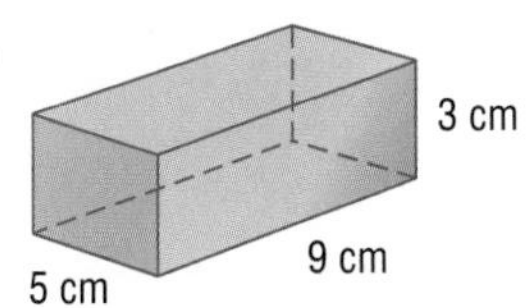

3. cube: side length 6 ft
4. cylinder: diameter 8 m, height 12 m
5. cylinder: radius 2.5 cm, height 5 cm
6. cube: side length 4.9 m
7. rectangular prism: length 7.6 mm, width 8.4 mm, height 7.0 mm
8. triangular prism: right triangle 3 in. by 4 in. by 5 in., height of prism 10 in.

Lesson 11-5
(pages 578–582)

Find the surface area of each solid. If necessary, round to the nearest tenth.

1. 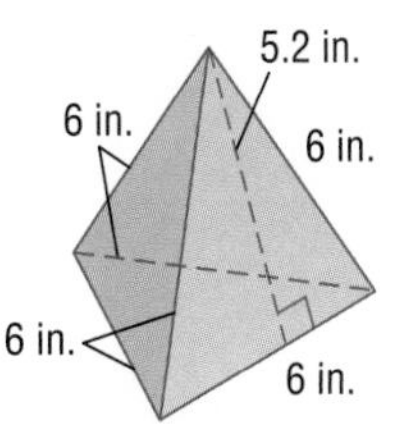

2. 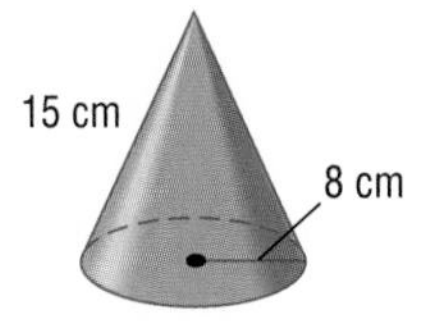

3. 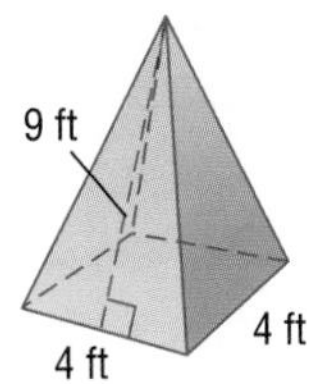

4. 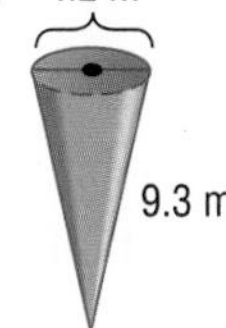

5. square pyramid: base side length 1.8 mm, slant height 3.0 mm
6. cone: radius 4 in., slant height 7 in.
7. cone: diameter 15.2 cm, slant height 12.3 cm

Lesson 11-6

(pages 584–588)

Determine whether each pair of solids is similar.

1. 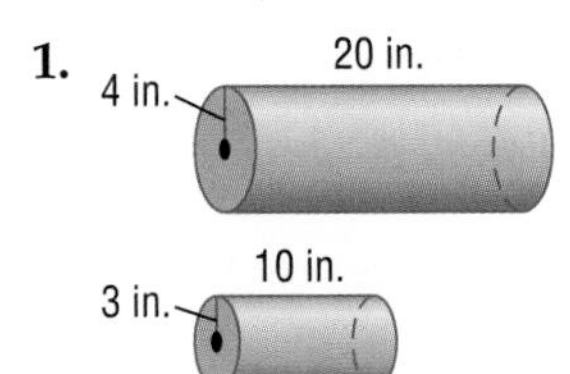

2. 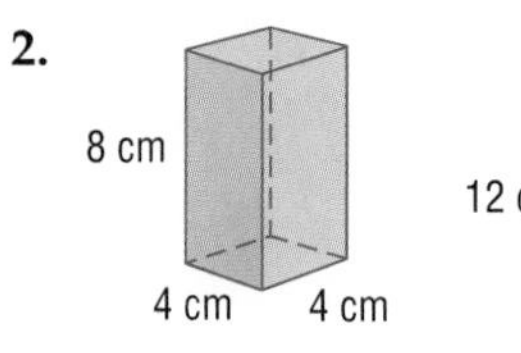

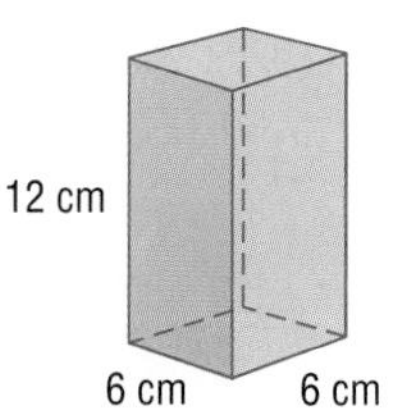

Find the missing measure of each pair of similar solids.

3. 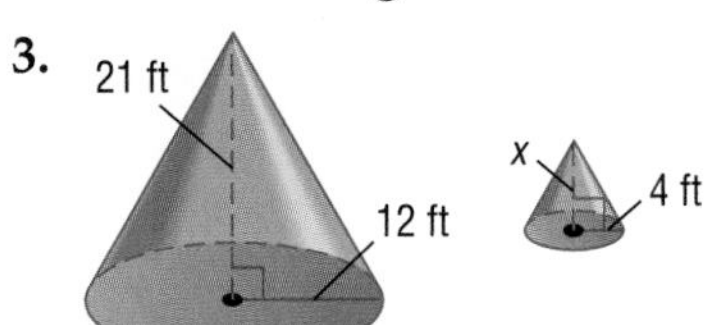

4. 10 yd, 4 yd, 4 yd
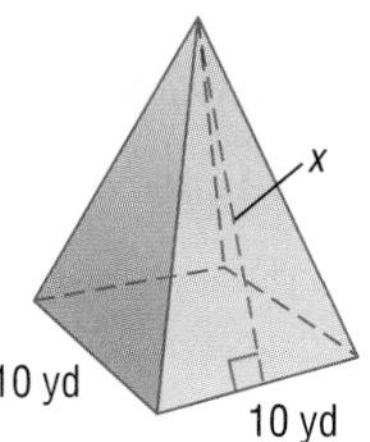

Determine whether each pair of solids is *sometimes*, *always*, or *never* similar. Explain.

5. two cylinders
6. a square pyramid and a triangular pyramid

Lesson 11-7

(pages 590–594)

Determine the number of significant digits in each measure.

1. 625 ft
2. 30 g
3. 0.24 mm
4. 36.83 L
5. 6.0 in.
6. 3900 ft
7. 4.007 cm
8. 0.0105 m
9. 0.550 g

Calculate. Round to the correct number of significant digits.

10. 32 yd + 16.9 yd
11. 14.36 in. − 9.4 in.
12. 20.86 cm − 0.375 cm
13. 9.600 m + 4.271 m
14. 8 ft · 6.2 ft
15. 7.50 m · 3.01 m
16. 41.61 in. + 18.4 in. − 3.65 in. + 7.371 in.

Lesson 12-1

(pages 606–611)

Display each set of data in a stem-and-leaf plot.

1. 37, 44, 32, 53, 61, 59, 49, 69
2. 3, 26, 35, 8, 21, 24, 30, 39, 35, 5, 38
3. 15.7, 7.4, 0.6, 0.5, 15.3, 7.9, 7.3
4. 172, 198, 181, 182, 193, 171, 179, 186, 181
5. 55, 62, 81, 75, 71, 69, 74, 80, 67
6. 121, 142, 98, 106, 111, 125, 132, 109, 117, 126
7. 17, 54, 37, 86, 24, 69, 77, 92, 21
8. 7.3, 6.1, 8.9, 6.7, 8.2, 5.4, 9.3, 10.2, 5.9, 7.5, 8.3

For Exercises 9–11, use the stem-and-leaf plot shown at the right.

9. What is the greatest value?
10. In which interval do most of the values occur?
11. What is the median value?

Stem	Leaf
7	2 2 3 5 9
8	0 1 1 4 6 6 8 9
9	3 4 8

9|4 = 94

Lesson 12-2

(pages 612–616)

Find the range and interquartile range for each set of data.

1. {44, 37, 23, 35, 61, 95, 49, 96}
2. {30, 62, 35, 80, 12, 24, 30, 39, 53, 38}
3. {7.15, 4.7, 6, 5.3, 30.1, 9.19, 3.2}
4. {271, 891, 181, 193, 711, 791, 861, 818}
5.

Stem	Leaf
2	0 1 1 2 4 7 9
3	3 3 6 8 8 8
4	2 4 5 7 9 9
5	2 9

3 | 6 = 36

6.

Stem	Leaf
4	0 2 2 3 4 5 6 6 7 8
5	1 2 5 5 5 9
6	4 7 8 8
7	0 0 1 4 9 9 9
8	1 7 9
9	0 0 1 3 5

8 | 7 = 87

Lesson 12-3

(pages 617–621)

Draw a box-and-whisker plot for each set of data.

1. 32, 54, 88, 17, 29, 73, 65, 52, 99, 103, 43, 13, 8, 59, 40, 37, 23
2. 42, 23, 31, 27, 32, 48, 37, 25, 19, 26, 30, 41, 32, 29
3. 124, 327, 215, 278, 109, 225, 186, 134, 251, 308, 179
4. 126, 432, 578, 312, 367, 400, 275, 315, 437, 299, 480, 365, 278

VOLLEYBALL **For Exercises 5–7, use the box-and-whisker plot shown.**

Heights (in.) of Players on Volleyball Team

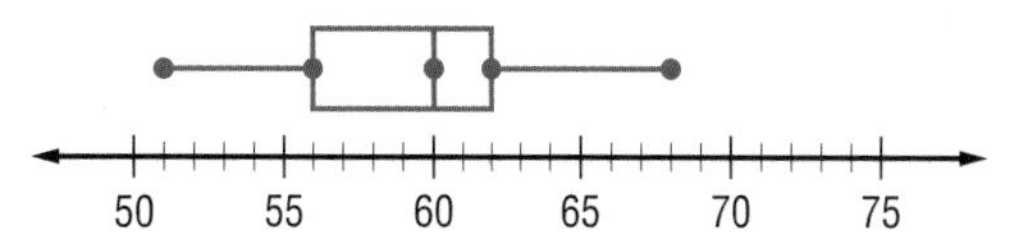

5. What is the height of the tallest player?
6. What percent of the players are between 56 and 68 inches tall?
7. Explain what the length of the box-and-whisker plot tells us about the data.

Lesson 12-4

(pages 623–628)

Display each set of data in a histogram.

1.

Weekly Exercise Time		
Time (h)	Tally	Frequency
0–2	𝍸 III	8
3–5	IIII	4
6–8	II	2
9–11	III	3

2.

Weekly Grocery Bill		
Amount ($)	Tally	Frequency
0–49	𝍸 I	6
50–99	𝍸 𝍸 II	12
100–149	𝍸 III	8
150–199	IIII	4
200–249	II	2

3.

Daily High Temperatures in August		
Temperature (°F)	Tally	Frequency
60–69	II	2
70–79	𝍸 𝍸	10
80–89	𝍸 I	6
90–99	III	3

4.

Score on Math Test		
Score	Tally	Frequency
50–59	II	2
60–69	I	1
70–79	𝍸 III	8
80–89	𝍸 𝍸 IIII	14
90–99		0

Lesson 12-5
(pages 630–633)

MONEY **For Exercises 1–2, refer to the graphs below.**

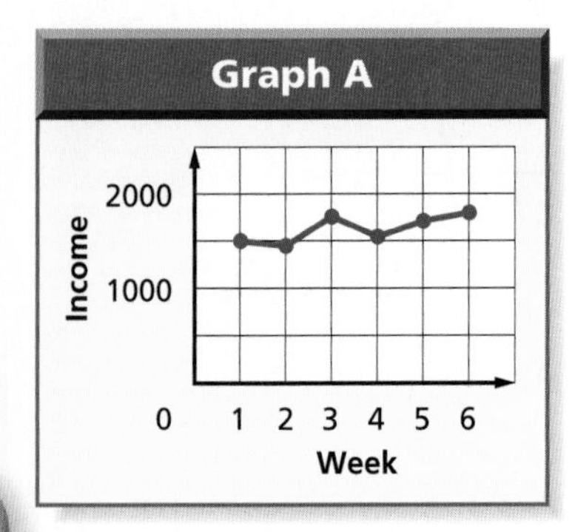

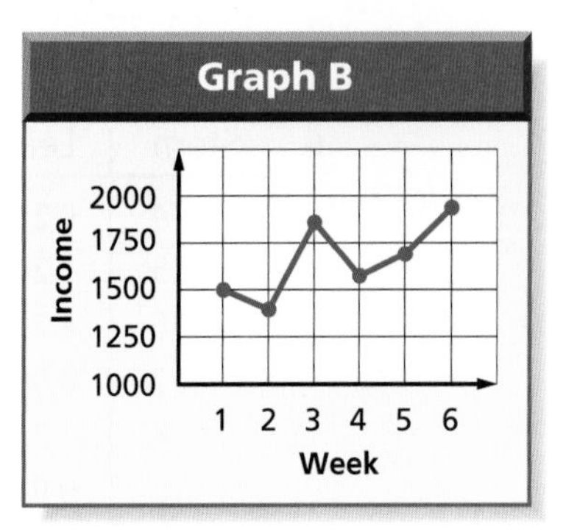

Extra Practice

1. Explain why the graphs look different.
2. Which graph appears to show that the income has been fairly consistent? Explain your reasoning.

Lesson 12-6
(pages 635–639)

Find the number of possible outcomes for each situation.

1. Engagement rings come in silver, gold, and white gold. The diamond can weigh $\frac{1}{2}$ karat, $\frac{1}{3}$ karat, or $\frac{1}{4}$ karat. The diamond can have 4 possible shapes.
2. A dress can be long, tea-length, knee-length, or mini. It comes in 2 colors and the dress can be worn on or off the shoulders.
3. The first digit of a 7 digit phone number is a 2. The last digit is a 3.
4. A chair can be a rocker, recliner, swivel, or straight back. It is available in fabric, vinyl, or leather.

Find the probability of each event.

5. Three coins are tossed. What is the probability of three tails?
6. Two six-sided number cubes are rolled. What is the probability of getting an odd sum?
7. A ten-sided die is rolled and a coin is tossed. Find the probability of the coin landing on tails and the die landing on a number greater than 3.

Lesson 12-7
(pages 641–645)

Tell whether each situation is a *permutation* or a *combination*. Then solve.

1. Seven people are running for four seats on student council. How many ways can the students be elected?
2. How many ways can the letters of the word ISLAND be arranged?
3. How many ways can five candles be arranged in three candlesticks?
4. How many ways can six students line up for a race?
5. How many ways can you select three books from a shelf containing 12 books?
6. **GEOMETRY** Determine the number of line segments that can be drawn between any two vertices of a pentagon.

Lesson 12-8

(pages 646–649)

Find the odds of each outcome if the spinner below is spun.

1. blue
2. a color with less than 5 letters
3. a color that begins with a consonant
4. red, yellow, or blue

Find the odds of each outcome if a 10-sided die is rolled.

5. number less than 7
6. odd number
7. composite number
8. number divisible by 3

Lesson 12-9

(pages 651–655)

A deck of Euchre cards consists of 4 nines, 4 tens, 4 jacks, 4 queens, 4 kings, and 4 aces. Suppose one card is selected and not replaced. Find the probability of each outcome.

1. 3 nines in a row
2. a black jack and a red queen
3. a nine of clubs, a black king, and a red ace
4. 4 face cards in a row

A number from 6 to 19 is drawn. Find the probability.

5. P(13 or even)
6. P(13 or less than 7)
7. P(even or odd)
8. P(14 or greater than 20)
9. P(even or less than 10)
10. P(odd or greater than 10)

Lesson 13-1

(pages 669–672)

Determine whether each expression is a polynomial. If it is, classify it as a *monomial*, *binomial*, or *trinomial*.

1. $3x^2 + 5$
2. $\frac{6}{x} + 9x$
3. $\frac{2}{3}p^4$
4. $-6x^2 + 3x - 5$
5. $\sqrt{w} - 6$
6. $16 - 3m + m^3$
7. $\frac{d}{15}$
8. $t^2 - 2$
9. $\frac{x}{y} + z$

Find the degree of each polynomial.

10. 38
11. $4b + 9$
12. cd
13. $4x$
14. $a^2 - 6$
15. $11r + 5s$
16. x^2y
17. $n^2 - n$
18. $6a^2b^2$
19. $3y^2 - 2$
20. $9cd^3 - 5$
21. $-5p^3 + 8q^2$
22. $w^2 + 2x - 3y^3 - 7z$
23. $\frac{x^3}{6} - x$
24. $-17n^2p - 11np^3$

Lesson 13-2

(pages 674–677)

Find each sum.

1. $$\begin{array}{r} -6m + 7 \\ \underline{(+)\ 9m - 2} \end{array}$$

2. $$\begin{array}{r} 12y - 4 \\ \underline{(+)\ -8y + 9} \end{array}$$

3. $$\begin{array}{r} 5x + y \\ \underline{(+)\ 9x - 2y} \end{array}$$

4. $$\begin{array}{r} 7c^2 - 10c + 5 \\ \underline{(+)\ 4c^2 - 4c - 8} \end{array}$$

5. $$\begin{array}{r} 2a^2 + 5ab + 6b^2 \\ \underline{(+)\ 3a^2 \qquad\quad - b^2} \end{array}$$

6. $$\begin{array}{r} 3d^3 + 2d^2 + 6d - 4 \\ \underline{(+)\quad -4d^2 \qquad - 3} \end{array}$$

7. $(3a + 4) + (a + 2)$
8. $(8m - 3) + (4m + 1)$
9. $(5x - 3y) + (2x - y)$
10. $(8p^2 - 2p + 3) + (-3p^2 - 2)$
11. $(-11r^2 + 3s) + (5r^2 - s)$
12. $(3a^2 + 5a + 1) + (2a^2 - 3a - 6)$

Find each sum. Then evaluate if $m = -2$, $n = 4$, and $p = 3$.

13. $(3m - 5n) + (-6m + 8n)$
14. $(m^2 + 2p^2) + (-4m^2 - 6p^2)$
15. $(-2m + 3n + 4p) + (5m - 6n - 8p)$

Lesson 13-3

(pages 678–681)

Find each difference.

1. $$\begin{array}{r} 2a + 7 \\ \underline{(-)\ a + 3} \end{array}$$

2. $$\begin{array}{r} -3k^2 + 6k \\ \underline{(-)\ 4k^2 + k} \end{array}$$

3. $$\begin{array}{r} 6x^2 - 4x + 11 \\ \underline{(-)\ 5x^2 + 5x - 4} \end{array}$$

4. $$\begin{array}{r} 9r^2 \qquad + 1 \\ \underline{(-)\ 2r^2 + 3r - 7} \end{array}$$

5. $$\begin{array}{r} 8n^2 + 3mn - 9 \\ \underline{(-)\ 4n^2 + 2mn \quad} \end{array}$$

6. $$\begin{array}{r} -5b^2 - 2ab \qquad \\ \underline{(-)\qquad - 10ab + 6a^2} \end{array}$$

7. $(3n + 2) - (n + 1)$
8. $(-3c + 2d) - (7c - 6d)$
9. $(4x^2 + 1) - (3x^2 - 4)$
10. $(5a - 4b) - (-a + b)$
11. $(-12a + 9b) - (3a - 7b)$
12. $(3w^3 + 5w - 6) - (5w^3 - 2w + 5)$
13. $(2t^2 - 5) - (t + 8)$
14. $(x^2 + xy - 9y^2) - (3x^2 - xy + 3y^2)$

Lesson 13-4

(pages 683–686)

Find each product.

1. $2(3a - 7)$
2. $(8c + 1)4$
3. $n(5n + 6)$
4. $t(2 - t)$
5. $(3k - 5)k$
6. $(a + b)a$
7. $4n(5n - 3)$
8. $-3x(4 - x)$
9. $6m(-m^2 + 3)$
10. $5(3x - 2)$
11. $(2p + 9)8$
12. $m(3m - 4)$
13. $-2w(6 - w)$
14. $ab(a + b)$
15. $7t(-3t + 4w)$
16. $4x(2x + y)$
17. $(c^2 - 3d)2c$
18. $-5z(z^2 - 9z)$
19. $-5x(2x^2 - 3x + 1)$
20. $7r(r^2 - 3r + 7)$
21. $-3az(2z^2 + 4az + a^2)$

Lesson 13-5

(pages 687–691)

Determine whether each graph, equation, or table represents a *linear* or a *nonlinear* function. Explain.

1. 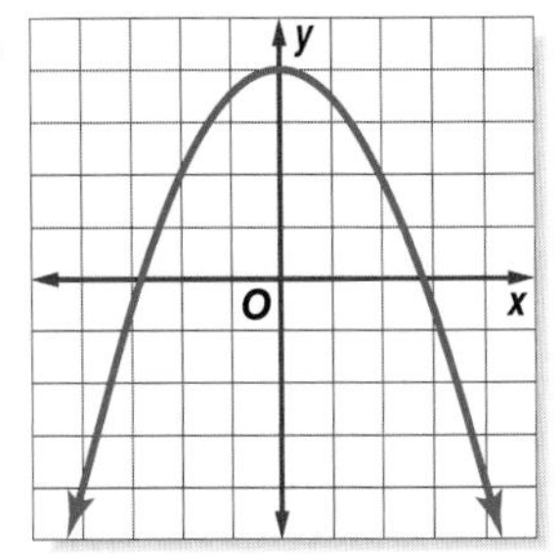

2. 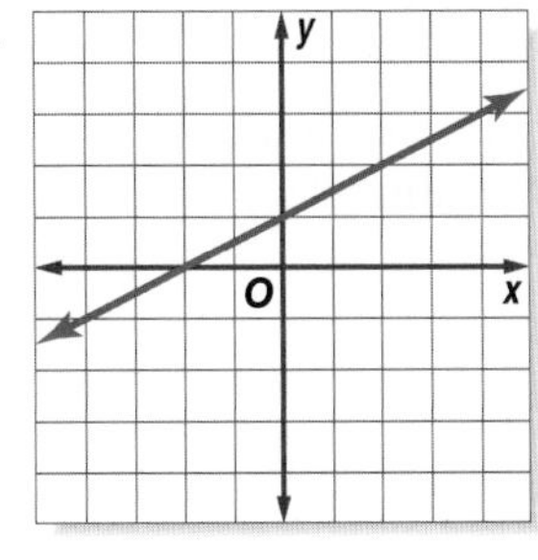

3. 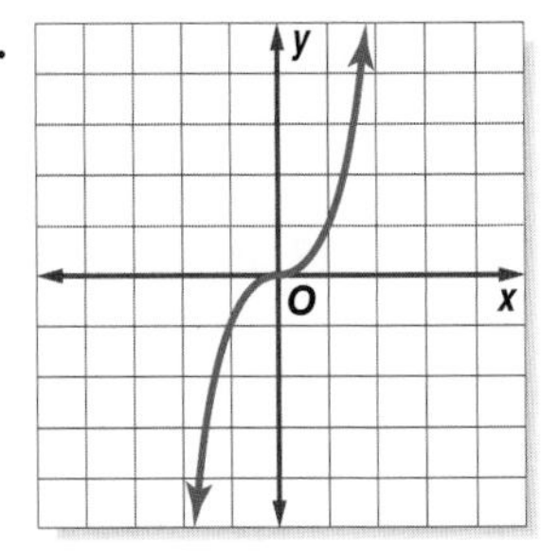

4. 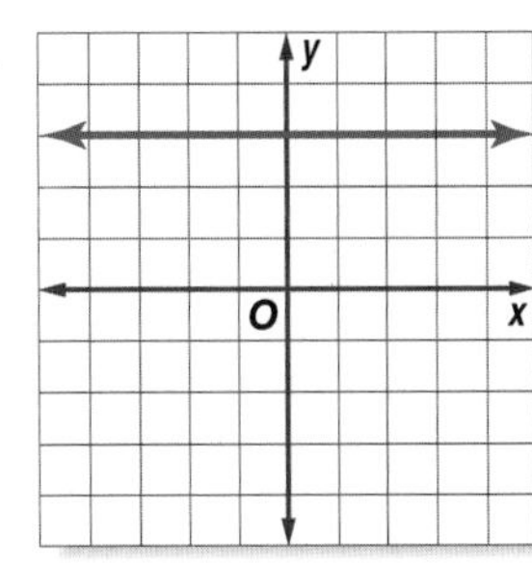

5. $y = -3x$

6. $y = 2x^3 - 5$

7. $-2x + 5y = 10$

8. $x = 7y$

9. $y = (-2)^x$

10. $y = \frac{6}{x}$

11.

x	y
2	5
4	7
6	9
8	11

12.

x	y
5	7
10	13
15	19
20	25

Lesson 13-6

(pages 692–696)

Graph each function.

1. $y = 3x^2$
2. $y = -2x^2$
3. $y = \frac{1}{2}x^2$
4. $y = x^3$
5. $y = 0.3x^3$
6. $y = x^3 - 2$
7. $y = x^2 + 4$
8. $y = \frac{1}{3}x^2 - 3$
9. $y = -0.5x^2 + 1$

Mixed Problem Solving

Chapter 1 The Tools of Algebra (pages 4–53)

1. **PATTERNS** How many cubes are in the tenth figure if the pattern below continues? *(Lesson 1-1)*

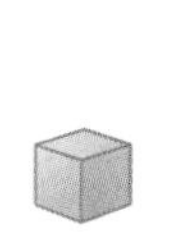
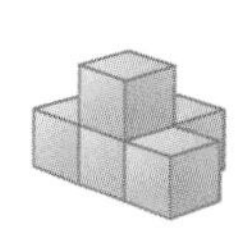

figure 1 figure 2 figure 3

2. **SPACE EXPLORATION** On one flight, the space shuttle *Endeavour* traveled 6.9 million miles and circled Earth 262 times. About how many miles did the shuttle travel on each trip around Earth? *(Lesson 1-1)*

3. **TREES** A conservation group collects seeds from trees at historic homes, grows them into saplings, and sells them to the public. Each sapling costs $35, and $7 is added to each order for shipping and handling. Write and then evaluate an expression for the total cost of one order of six saplings. *(Lesson 1-2)*

4. **SALES** For a school fund-raiser, Sophia sold 15 white chocolate hearts at $4.25 each, 36 milk chocolate hearts at $3.75 each, and 22 milk chocolate assortments at $7.45 each. How much money did Sophia raise? *(Lesson 1-2)*

SPACE For Exercises 5 and 6, use the following information.
Objects weigh six times more on Earth than they do on the moon because the force of gravity is greater. *(Lesson 1-3)*

5. Write an expression for the weight of an object on Earth if its weight on the moon is x.

6. A scientific instrument weighs 34 pounds on the moon. How much does the instrument weigh on Earth?

VOLLEYBALL For Exercises 7 and 8, use the following information.
A volleyball net is 3 feet 3 inches tall. The bottom of the net is to be set 4 feet 8 inches from the floor. *(Lesson 1-4)*

7. Write an expression for the distance from the floor to the top of the net.

8. Find the distance from the floor to the top of the net.

9. **NEWSPAPERS** Nick sold 86 newspapers on Monday, 79 on Tuesday, 68 on Wednesday, and 83 on Friday. How many newspapers did Nick sell on Thursday if he sold a total of 391 in the five days? *(Lesson 1-5)*

10. **FOOD** Kristin is planning to buy twice as many blueberry bagels as plain bagels for a staff meeting. Write a relation to show the different possibilities. *(Lesson 1-6)*

GEOLOGY For Exercises 11 and 12, use the following information.
The underground temperature of rocks in degrees Celsius is estimated by the expression $35x + 20$, where x is the depth in kilometers. *(Lesson 1-6)*

11. Make a list of ordered pairs in which the x-coordinate represents the depth and the y-coordinate represents the temperature for depths of 0, 2, and 4 kilometers.

12. Graph the ordered pairs.

13. **EMPLOYMENT** The scatter plot shows the years of experience and salaries of twenty people. Do the data show a *positive*, *negative*, or *no* relationship? Explain. *(Lesson 1-7)*

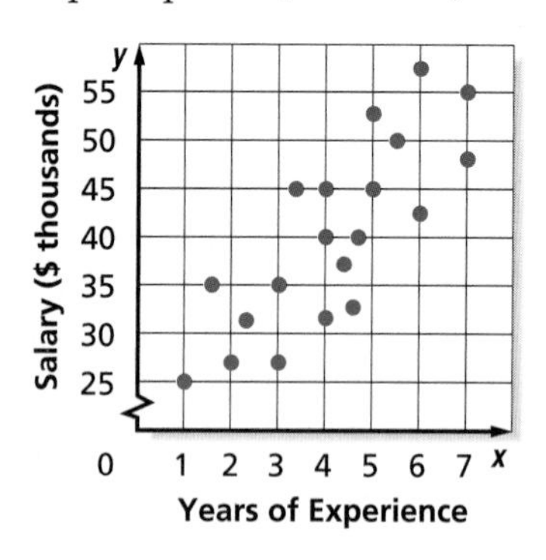

14. **BIRDS** The table shows the average lengths and widths of five bird eggs.

Bird	Length (cm)	Width (cm)
Canadian goose	8.6	5.8
robin	1.9	1.5
turtledove	3.1	2.3
hummingbird	1.0	1.0
raven	5.0	3.3

Source: *Animals as Our Companions*

Make a scatter plot of the data and predict the width of an egg 6 centimeters long. *(Lesson 1-7)*

1. **ASTRONOMY** Mars is about 228 million kilometers from the Sun. Earth is about 150 million kilometers from the Sun. Write two inequalities that compare the two distances. *(Lesson 2-1)*

2. **GAMES** In a popular television game show, one contestant finished the regular round with a score of -200, and another contestant finished with a score of -500. Write two inequalities that compare their scores. *(Lesson 2-1)*

3. **MONEY** Tino had \$250 in his checking account at the beginning of April. During the month he wrote checks in the amounts of \$72, \$37, and \$119. He also made one deposit of \$45. Find Tino's account balance at the end of April. *(Lesson 2-2)*

4. **ASTRONOMY** At noon, the average temperature on the moon is 112°C. During the night, the average temperature drops 252°C. What is the average temperature of the moon's surface during the night? *(Lesson 2-2)*

5. **SUBMARINES** The research submarine *Alvin* is located at 1500 meters below sea level. It descends another 1250 meters to the ocean floor. How far below sea level is the ocean floor? *(Lesson 2-3)*

6. **METEOROLOGY** Windchill factor is an estimate of the cooling effect the wind has on a person in cold weather. If the outside temperature is 10°F and the wind makes it feel like -25°F, what is the difference between the actual temperature and how cold it feels? *(Lesson 2-3)*

7. **GEOGRAPHY** The highest point in Africa is Mount Kilimanjaro. Its altitude is 5895 meters. The lowest point on the continent is Lake Assal. Its altitude is -155 meters. Find the difference between these altitudes. *(Lesson 2-3)*

8. **GEOLOGY** In December, 1994, geologists found that the Bering Glacier had come to a stop. The glacier had been retreating at a rate of about 2 feet per day. If the retreat resumes at the old rate, what integer represents how far the glacier will have advanced after 28 days? *(Lesson 2-4)*

9. **SPORTS** The Wildcat football team was penalized the same amount of yardage four times during the third quarter. The total of the four penalties was 60 yards. If -60 represents a loss of 60 yards, write a division sentence to represent this situation. Then express the number of yards of each penalty as an integer. *(Lesson 2-5)*

10. **AEROSPACE** To simulate space travel, NASA's Lewis Research Center in Cleveland, Ohio, uses a 430-foot shaft. If the free fall of an object in the shaft takes 5 seconds to travel the -430 feet, on average how far does the object travel in each second? *(Lesson 2-5)*

11. **MAPS** A map of a city can be created by placing the following buildings at the given coordinates: City Hall (1, 2), High School (-3, 6), Fire Department (4, -2), Recreation Center (0, 3). Draw and label the map. *(Lesson 2-6)*

GEOMETRY For Exercises 12 and 13, use the following information.
A vertex of a polygon is a point where two sides of the polygon intersect.

12. Identify the coordinates of the vertices in the triangle below.

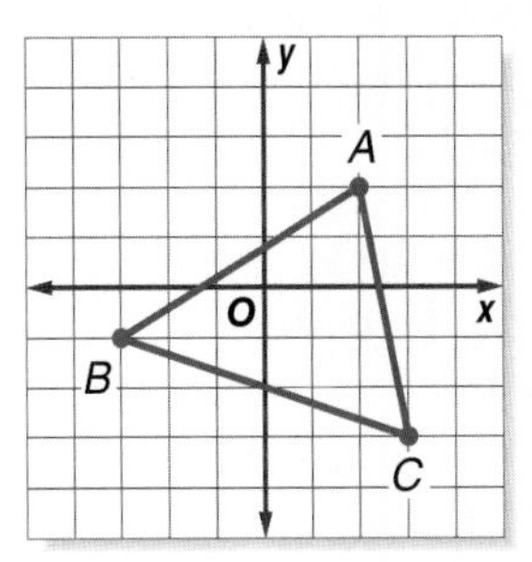

13. Add 2 to each x-coordinate. Graph the new ordered pairs. Describe how the position of the new triangle relates to the original triangle. *(Lesson 2-6)*

Mixed Problem Solving

1. **BUSINESS** A local newspaper can be ordered for delivery on weekdays or Sundays. A weekday paper is 35¢, and the Sunday edition is $1.50. The Stadlers ordered delivery of the weekday papers. The month of March had 23 weekdays and April had 20. How much should the carrier charge the Stadlers for those two months? *(Lesson 3-1)*

SHOPPING For Exercises 2 and 3, use the following information.
One pair of jeans costs $23, and one T-shirt costs $15. *(Lesson 3-1)*

2. Write two equivalent expressions for the total cost of 3 pairs of jeans and 3 T-shirts.
3. Find the total cost.

4. **ENTERTAINMENT** Kyung bought 3 CDs that each cost x dollars, 2 tapes that each cost $10, and a video that cost $14. Write an expression in simplest form that represents the total amount that Kyung spent. *(Lesson 3-2)*

TRANSPORTATION For Exercises 5 and 6, use the following information.
A minivan is rated for maximum carrying capacity of 900 pounds.

5. If the luggage weighs 100 pounds, what is the maximum weight allowable for passengers?
6. What is the maximum average weight allowable for each of 5 passengers? *(Lesson 3-3)*

7. **GEOMETRY** The perimeter of any square is 4 times the length of one of its sides. If the perimeter of a square is 72 centimeters, what is the length of each side of the square? *(Lesson 3-4)*

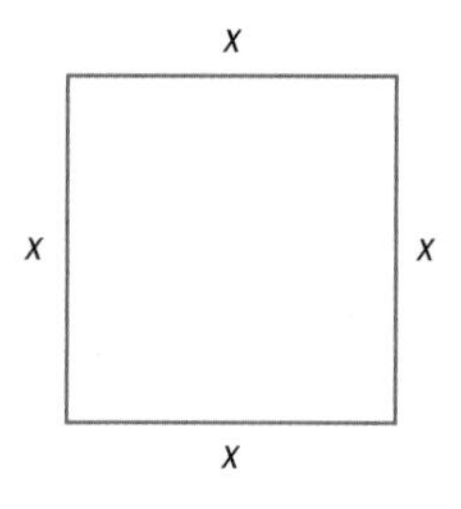

8. **PURCHASING** Mr. Rockwell bought a television set. The price was $362. He paid $75 down and will pay the balance in 7 equal payments. How much is each payment? *(Lesson 3-5)*

9. **SPORTS** Marcie paid $75 to join a tennis club for the summer. She will also pay $10 for each hour that she plays. If Marcie has budgeted $225 to play tennis this summer, how many hours can she play tennis? *(Lesson 3-5)*

FENCING For Exercises 10 and 11, use the following information.
Wanda uses 130 feet of fence to enclose a rectangular flower garden. She also used the 50-foot wall of her house as one side of the garden. What is the width of the garden? *(Lesson 3-6)*

10. Write an equation that represents this situation.
11. Solve the equation to find the width of the garden.

WORKING For Exercises 12 and 13, use the following information.
Kate worked a 40-hour week and was paid $410. This amount included a $50 bonus. *(Lesson 3-6)*

12. Write an equation that represents this situation.
13. What was Kate paid per hour?

14. **GEOMETRY** The perimeter of the triangle below is 27 yards.

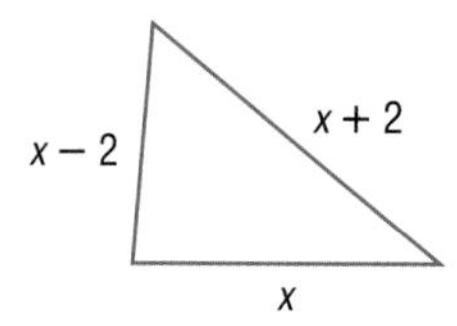

Find the lengths of the sides of the triangle. *(Lesson 3-7)*

15. **TRAVEL** The Flynn family plans to drive 600 miles for their summer vacation. The speed limit on the highways they plan to use is 55 miles per hour. If they do not exceed the speed limit, how many hours of driving should it take them? *(Lesson 3-7)*

SCIENCE For Exercises 16 and 17, use the following information.
Acceleration is the rate at which velocity is changing with respect to time. To find the acceleration, find the change in velocity by subtracting the starting velocity, s from the final velocity, f. Then divide by the time, t. *(Lesson 3-7)*

16. Write the formula for acceleration, a.
17. A motorcycle goes from 2 m/s to 14 m/s in 6 seconds. Find its acceleration.

1. **LUNCHTIME** A group of 136 sixth graders needs to be seated in the cafeteria for lunch. If all of the tables need to be full, should the school use tables that seat 6, 8, or 10 students each? *(Lesson 4-1)*

2. **PATTERNS** In a pattern, the number of colored tiles used in row x is 3^x. Find the number of tiles used in rows 4, 5, and 6 of the pattern. *(Lesson 4-2)*

3. **ELECTRICITY** The amount of power lost in watts P can be found by using the formula $P = I^2R$, where I is current in amps, and R is resistance in ohms. The resistance of the wire leading from the source of power to a home is 2 ohms. If an electric stove causes a current of 41 amps to flow through the wire, find the power lost from the wire powering the stove. *(Lesson 4-2)*

4. **CODES** Prime numbers are used to code and decode information. Suppose two prime numbers p and q are chosen so that $n = pq$. Then the key to the code is n. Find p and q if $n = 1073$. *(Lesson 4-3)*

5. **INTERIOR DESIGN** Mrs. Garcia has two different fabrics to make square pillows for her living room. One fabric is 48 inches wide, and the other fabric is 60 inches wide. How long should each side of the pillows be if they are all the same size and no fabric is wasted? *(Lesson 4-4)*

6. **ECONOMICS** The graph below shows how each dollar spent by the Federal Government is used.

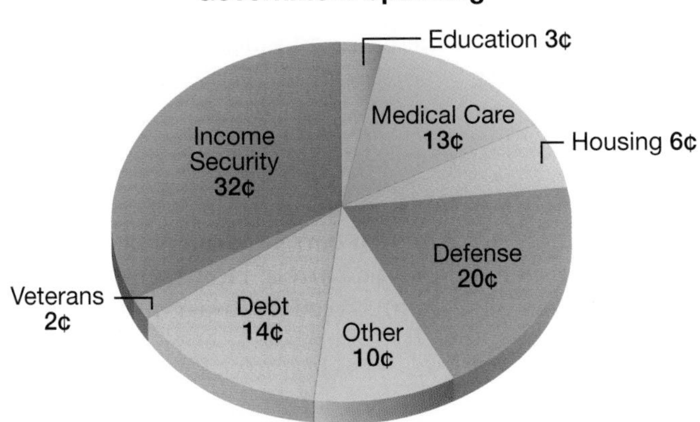

Source: The Tax Foundation

Write a fraction in simplest form comparing the amount spent on housing assistance and the total amount spent. *(Lesson 4-5)*

7. **BASKETBALL** Sydney made 8 out of 14 free throws in her last basketball game. Write her success as a fraction in simplest form. *(Lesson 4-5)*

8. **TRANSPORTATION** Cameron spends 18 minutes traveling to work. What fraction of the day is this? *(Lesson 4-5)*

9. **EARTHQUAKES** The table below describes different earthquake intensities.

Earthquake	Richter Scale	Intensity
A	8	10^7
B	4	10^3

Find $10^7 \div 10^3$ to determine how much more intense Earthquake A was than Earthquake B. *(Lesson 4-6)*

ASTRONOMY For Exercises 10 and 11, use the following information.

Any two objects in space have an attraction that can be calculated by using a formula that includes the universal gravitational constant, 6.67×10^{-11} Nm^2/kg^2. (N is newtons.) *(Lesson 4-7)*

10. Write the universal gravitational constant using positive exponents.

11. Write the constant as a decimal.

BIOLOGY For Exercises 12 and 13, use the following information.

Deoxyribonucleic acid, or DNA, contains the genetic code of an organism. The length of a DNA strand is about 10^{-7} meter. *(Lesson 4-7)*

12. Write the length of a DNA strand using positive exponents.

13. Write the length of a DNA strand as a decimal.

14. **SCIENCE** Atoms are extremely small particles about two millionths of an inch in diameter. Write this measure in standard form and in scientific notation. *(Lesson 4-8)*

15. **BUSINESS** A large corporation estimates its yearly revenue at $\$4.72 \times 10^8$. Write this number in standard form. *(Lesson 4-8)*

1. **FURNITURE** A shelf $16\frac{5}{8}$ inches wide is to be placed in a space that is $16\frac{3}{4}$ inches wide. Will the shelf fit in the space? Explain. *(Lesson 5-1)*

2. **MEASUREMENT** A piece of metal is 0.025 inch thick. What fraction of an inch is this? *(Lesson 5-2)*

3. **HEALTH** You can stay in the Sun 15 times longer than usual without burning by applying SPF number 15. If you usually burn after $\frac{1}{4}$ hour in the Sun, how long could you stay in the Sun using SPF 15 lotion? *(Lesson 5-3)*

4. **MONEY** A dollar bill remains in circulation about $1\frac{1}{4}$ years. A coin lasts about $22\frac{1}{2}$ times longer. How long is a coin in circulation? *(Lesson 5-3)*

5. **FOOD** If each guest at a party eats two-thirds of a small pizza, how many guests would finish 12 small pizzas? *(Lesson 5-4)*

6. **PUBLISHING** A magazine page is 8 inches wide. The articles are printed in three columns with $\frac{1}{4}$ inch of space in between and $\frac{3}{8}$-inch margins on each side, as shown below.

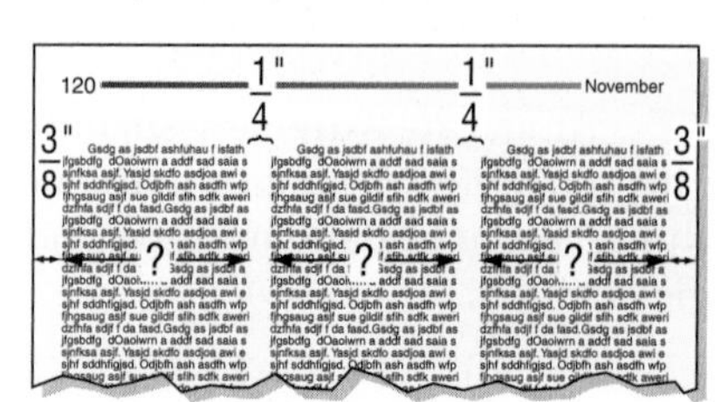

How wide should an author set the columns on her computer so that they are the same width as in the magazine? *(Lesson 5-4)*

7. **REMODELING** In their basement, the Jacksons installed $\frac{3}{8}$-inch thick paneling over a layer of dry wall that is $\frac{5}{8}$ inch thick. How thick are the wall coverings? *(Lesson 5-5)*

8. **COLLEGE** In a college dormitory, $\frac{3}{8}$ of the residents are from Ohio, and $\frac{2}{5}$ of the residents are from New York. Which state has a greater representation? *(Lesson 5-6)*

9. **NUTRITION** A survey found that $\frac{1}{6}$ of American households bought bottled water in 2000. Only $\frac{1}{17}$ of American households bought bottled water in 1993. What fraction of the population bought bottled water in 2000 that did not in 1993? *(Lesson 5-7)*

10. **EMPLOYMENT** The table below shows the earnings per woman for every $100 earned by a man in the same occupation for two years. *(Lesson 5-8)*

Occupation	Earnings ($)	
	Year 1	Year 2
Nurse	99.50	104.70
Teacher	88.60	90.30
Police Officer	91.20	94.20
Food Service	102.50	105.60
Postal Clerk	93.40	94.60

Find the mean, median, and mode of the earnings for each year.

11. **OIL PRODUCTION** Texas and Alaska produced a total of 1372.2 million barrels of oil. Alaska produced 684.0 million barrels. How many barrels of oil were produced in Texas? *(Lesson 5-9)*

12. **ON-LINE SERVICE** The cost of using an Internet service provider for 5, 6, 7, and 8 hours is given by the sequence $9.95, $12.90, $15.85, and $18.80, respectively. Is the cost an arithmetic or geometric sequence? Explain. *(Lesson 5-10)*

1. **SHOPPING** Best buys in grocery stores are generally found by comparing unit rates such as cents per ounce. Which bag of nachos shown in the table at the right is the better buy? *(Lesson 6-1)*

Size	Price
16-oz	$2.49
32-oz	$3.69

2. **COOKING** A recipe that makes 72 cookies calls for $4\frac{1}{2}$ cups of flour. How many cups of flour would be needed to make 48 cookies? *(Lesson 6-2)*

3. **FERRIS WHEEL** In a scale model of a Ferris wheel, the diameter of the wheel is 5 inches. If the actual height of the wheel is 55 feet, what is the scale of the model? *(Lesson 6-3)*

4. **BUSINESS** An executive of a marshmallow company said that marshmallows are 80% air. What fraction of a marshmallow is air? *(Lesson 6-4)*

5. **HEALTH** Doctors estimate that 3 babies out of every 1000 are likely to get a cold during their first month. What percent is this? *(Lesson 6-4)*

6. **NUTRITION** Refer to the nutritional label from a bag of pretzels shown below.

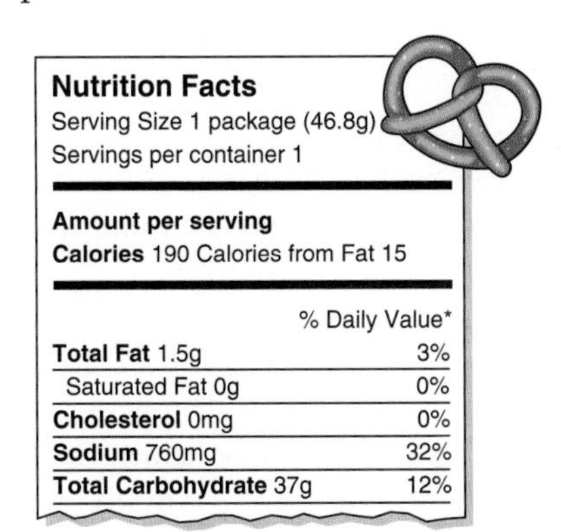

Nutrition Facts
Serving Size 1 package (46.8g)
Servings per container 1
Amount per serving
Calories 190 Calories from Fat 15
% Daily Value*
Total Fat 1.5g 3%
Saturated Fat 0g 0%
Cholesterol 0mg 0%
Sodium 760mg 32%
Total Carbohydrate 37g 12%

The 760 milligrams of sodium (salt) in one serving is 32% of the recommended daily value. What is the total recommended daily value of sodium? *(Lesson 6-5)*

7. **FAST FOOD** A certain hamburger has 560 Calories, and 288 of these are from fat. About what percent of the Calories are from fat? *(Lesson 6-6)*

8. **MONEY** If Simone wants to leave a tip of about 15% on a dinner check of $23.85, how much should she leave? *(Lesson 6-6)*

9. **BUSINESS** Many car dealers offer special interest rates as incentives to attract buyers. How much interest would a person pay for the first month of a $5500 car loan if the monthly interest rate is 0.24%? *(Lesson 6-7)*

10. **PETS** Hedgehogs are becoming so popular as pets that some breeders have reported a 250% increase in sales in recent years. If a breeder sold 50 hedgehogs one year before the increase, how many should he or she expect to sell a year from now? *(Lesson 6-8)*

11. **BRAND NAMES** The graph below shows the results of a survey.

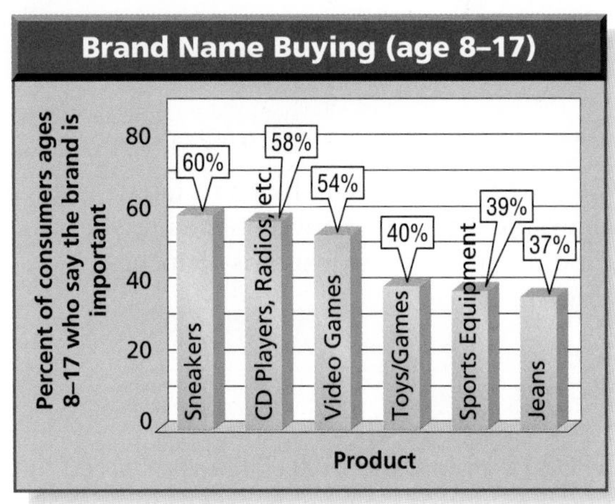

Source: International Mass Retail Association

How many of a class of 423 ninth grade students would you expect to say that they consider brand name when buying jeans? *(Lesson 6-9)*

12. **CANDY** In a small bag of colored chocolate candies, there are 15 green, 23 red, and 18 yellow candies. What is the probability of selecting a red candy if one is taken from the bag at random? *(Lesson 6-9)*

Mixed Problem Solving

CENSUS For Exercises 1–3, use the following information.
The table below shows the 2000 populations and the average rates of change in population in the 1990s for Buffalo, New York, and Corpus Christi, Texas. Suppose the population of each city continued to increase or decrease at these rates. *(Lesson 7-1)*

City	Population in 2000	Yearly Rate of Change
Buffalo, NY	293,000	−3500
Corpus Christi, TX	277,000	+2000

1. Write an expression for the population of Buffalo after x years.

2. Write an expression for the population of Corpus Christi after x years.

3. In how many years would the population of the two cities be the same?

4. **INTERNET** One Internet provider charges \$19.95 a month plus \$0.21 per minute, and a second provider charges \$24.95 a month plus \$0.16 per minute. For how many minutes is the cost of the plans the same? *(Lesson 7-1)*

5. **GEOMETRY** The length of a rectangle is three times the difference between its width and two. Find the width if the length is 15 inches. *(Lesson 7-2)*

6. **SPORTS** More than 100,000 fans attended the opening football game of the season. Write an inequality for the number of people who attended. *(Lesson 7-3)*

7. **SCHOOL** Julie has math and English homework tonight. She has no more than 90 minutes to spend on her homework. Suppose Julie spends 35 minutes completing her math homework. Write and solve an inequality to find how much time she can spend on her English homework. *(Lesson 7-4)*

8. **SAVINGS** Curtis is saving money to buy a new mountain bike. The bikes that he likes start at \$375, and he has already saved \$285. Write and solve an inequality to find the amount he must still save. *(Lesson 7-4)*

9. **STATISTICS** The Boston Marathon had more than 2,600,000 spectators along its 26-mile route. Write and solve an inequality to find the average number of spectators per mile. *(Lesson 7-5)*

10. **GROCERY SHOPPING** Mrs. Hiroshi spends at least twice as much on her weekly grocery shopping as she did one year ago. Last year, she spent \$54 each week. How much is Mrs. Hiroshi now spending each week on groceries? *(Lesson 7-5)*

11. **GEOMETRY** An *acute angle* has a measure less than 90°. If the measure of an acute angle is $2x$, write and solve an inequality to find the possible values of x. *(Lesson 7-5)*

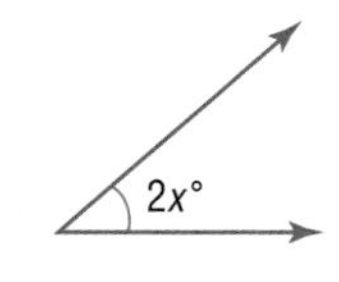

12. **SHOPPING** Luis plans to spend at most \$85 on jeans and shirts. He bought 2 shirts for \$15.30 each. How much can he spend on jeans? *(Lesson 7-6)*

13. **CAR SALES** A car salesperson receives a monthly salary of \$1000 plus a 3% commission on every car sold. For what amount of monthly sales will the salesperson earn more than \$2500? *(Lesson 7-6)*

14. **SCHOOL** Dave has earned scores of 73, 85, 91, and 82 on the first four of five math tests for the grading period. He would like to finish the grading period with a test average of at least 82. What is the minimum score Dave needs to earn on the fifth test in order to achieve his goal? *(Lesson 7-6)*

SHIPPING RATES For Exercises 1–3, use the following information.
The shipping costs for mail-order merchandise are given in the table below. *(Lesson 8-1)*

Total Price of Merchandise	Shipping Cost
\$0–\$30.00	\$4.25
\$30.01–\$70.00	\$5.75
\$70.01 and over	\$6.95

1. What is the shipping cost of merchandise totaling \$75?
2. For what price of merchandise is the shipping cost \$5.75?
3. Does the table represent a function? Explain.

PHYSICS For Exercises 4 and 5, use the following information.
As a thunderstorm approaches, you see lightning as it occurs, but you hear the accompanying thunder a short time afterward. The distance y in miles that sound travels in x seconds is given by $y = 0.21x$. *(Lesson 8-2)*

4. Find three ordered pairs that relate x and y.
5. How far away is lightning when thunder is heard 2.5 seconds after the lightning is seen?

AVIATION For Exercises 6 and 7, use the following information.
The steady descent of a jetliner is represented by the equation $a = 24{,}000 - 1500t$, where t is the time in minutes and a is the altitude in feet. *(Lesson 8-3)*

6. Name the x-intercept of the graph of the equation.
7. What does the x-intercept represent?

8. **KITES** Drew is flying a kite in the park. The kite is a horizontal distance of 20 feet from Drew's position and a vertical distance of 70 feet. Find the slope of the kite string. *(Lesson 8-4)*

9. **FUEL** The cost of gasoline varies directly as the number of gallons bought. If it costs \$27.80 to fill a 20-gallon tank, what would it cost to fill a 12-gallon tank? *(Lesson 8-5)*

10. **BUSINESS** A company's monthly cost y is given by $y = 1500 + 12x$, where x represents the number of items produced. State the slope and y-intercept of the graph of the equation and describe what they represent. *(Lesson 8-6)*

CAR RENTAL For Exercises 11 and 12, use the following information.
It costs \$59 per day plus \$0.12 per mile driven to rent a minivan. *(Lesson 8-7)*

11. Write an equation in slope-intercept form that shows the cost y for renting a minivan for one day and driving x miles.
12. Find the daily rental cost if 30 miles are driven.

13. **NUTRITION** The graph below shows energy bar sales in the United States during the month of October.

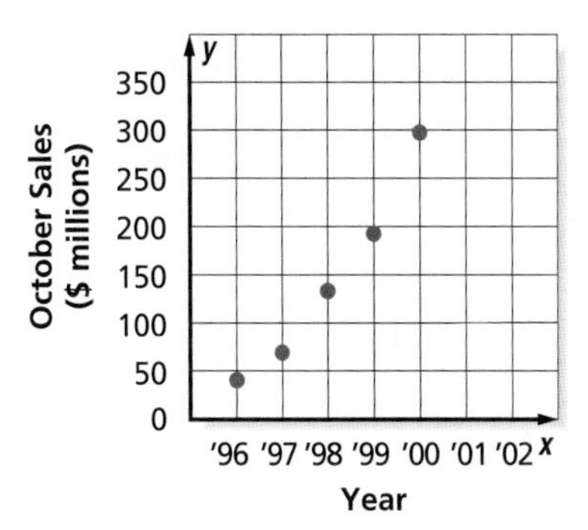

Use the graph to predict energy bar sales during October 2002. *(Lesson 8-8)*

14. **SCHOOL CONCERT** Tickets for the fall concert cost \$3 for students and \$5 for nonstudents. If a total of 140 tickets were sold and \$590 was collected, how many of each type of ticket was sold? *(Lesson 8-9)*

FINANCE For Exercises 15–17, use the following information.
Silvina earns \$3.50 per hour for weeding the garden and \$5.00 per hour for mowing the lawn. Suppose she wants to earn at least \$35 this week. *(Lesson 8-10)*

15. Write an inequality to represent this situation. Let x represent the number of hours she weeds and let y represent the number of hours she mows.
16. Graph the inequality.
17. Determine two possible ways that she can earn at least \$35 this week.

1. **CONSTRUCTION** A banquet facility must allow at least 4 square feet for each person on the dance floor. Reston's Hotel is adding a square dance floor that will be large enough for 100 people. How long should it be on each side? *(Lesson 9-1)*

2. **PHYSICS** The time t in seconds that it takes an object to fall d feet can be estimated by using $d = 0.5gt^2$. In this formula, g is acceleration due to gravity, 32 ft/s^2. If a ball is dropped from the top of a 55-foot building, how long does it take to hit the ground? *(Lesson 9-2)*

BOOKS For Exercises 3 and 4, use the following information.
The graph shows the results of a survey, which asked which types of books people buy the most. *(Lesson 9-3)*

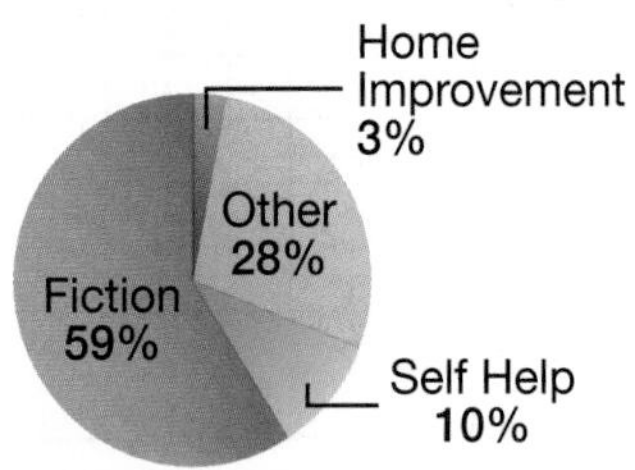

Source: *USA TODAY*

3. Classify the angle labeled Self Help as *acute*, *obtuse*, *right*, or *straight*.

4. Find the measure of the angle labeled Fiction.

5. **AVIATION** Airplane flight paths can be described using angles and compass directions. The path of a particular airplane is described as 36° west of north. Draw a diagram that represents this path. *(Lesson 9-3)*

6. **UTILITIES** A support cable is sometimes attached to give a utility pole stability. If the cable makes an angle of 65° with the ground, what is the measure of the angle formed by the cable and the pole? *(Lesson 9-4)*

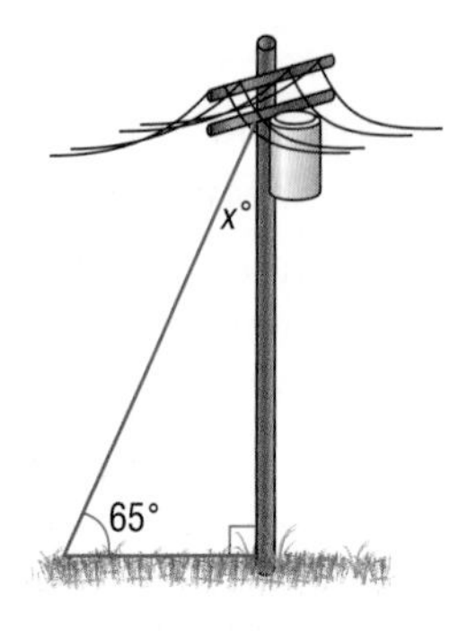

7. **BASEBALL** A baseball diamond is actually a square with 90 feet between the bases. What is the distance between home plate and second base? *(Lesson 9-5)*

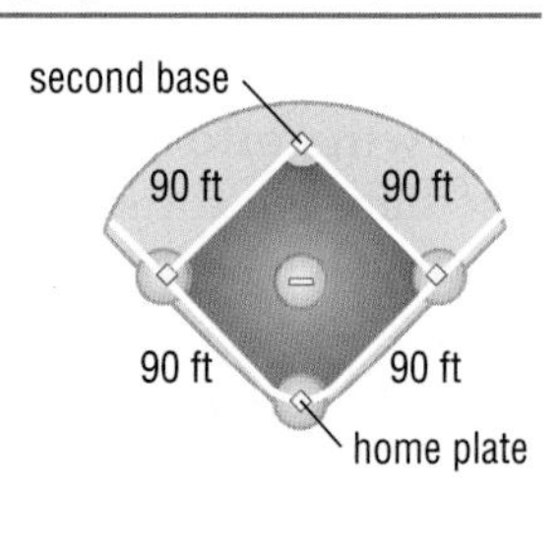

8. **SAILING** A rope from the top of a sailboat mast is attached to a point 6 feet from the base of the mast. If the rope is 24 feet long, how high is the mast? *(Lesson 9-5)*

9. **TRAVEL** Matt's home is at (−4, 9) on the map. His friend Carlos' home is at (6, 3) on the same map. They want to meet halfway between their two homes. What are the coordinates on the map where they should plan to meet? *(Lesson 9-6)*

10. **HISTORY** The largest known pyramid is Khufu's pyramid. At a certain time of day, a yardstick casts a shadow 1.5 feet long, and the pyramid casts a shadow 241 feet long. Use shadow reckoning to find the height of the pyramid. *(Lesson 9-7)*

11. **SURVEYING** A surveyor needs to find the distance across a river and draws the sketch shown below.

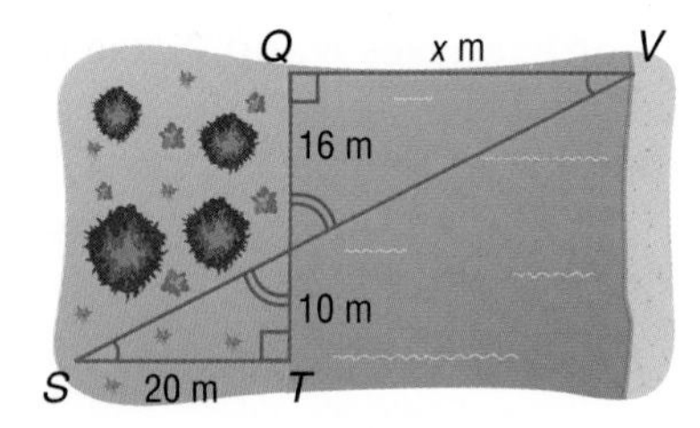

Find the distance across the river. *(Lesson 9-7)*

12. **RECREATION** Maxine is flying a kite on a 75-yard string. The string is making a 45° angle with the ground. How high above the ground is the kite? *(Lesson 9-8)*

13. **MAINTENANCE** A 15-foot ladder is propped against a house. The angle it forms with the ground is 60°. To the nearest foot, how far up the side of the house does the ladder reach? *(Lesson 9-8)*

1. **TRANSPORTATION** The angle at the corner where two streets intersect is 125°. If a bus cannot make a turn at an angle of less than 70°, can bus service be provided on a route that includes turning that corner in both directions? Explain. *(Lesson 10-1)*

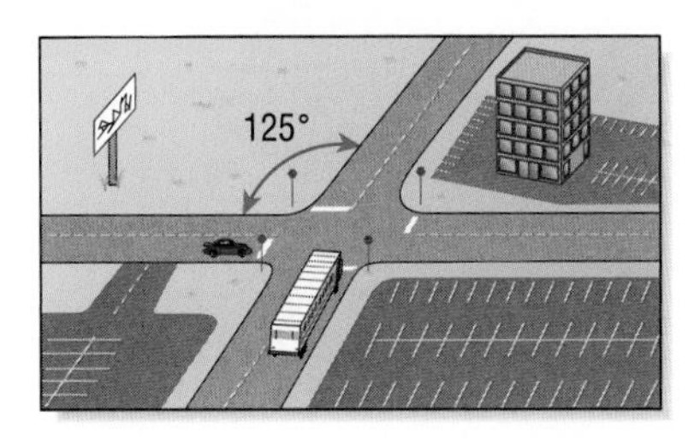

BRIDGES For Exercises 2 and 3, use the following information.
The figure below shows part of the support structure of a bridge. Name a triangle that seems to be congruent to each triangle below. *(Lesson 10-2)*

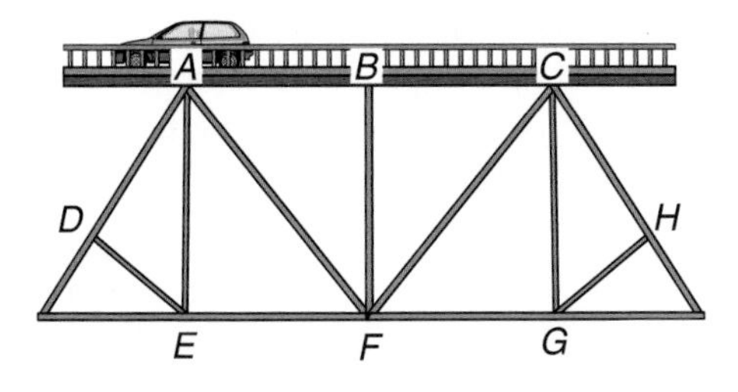

2. $\triangle AFB$

3. $\triangle CHG$

4. **MOVING** A historic house in the shape of a rectangle has coordinates $A(-3, 5)$, $B(4, 5)$, $C(4, -3)$, and $D(-3, -3)$ on a map. The house is going to be moved to a new site 3 units east and two units north. Find the coordinates of the house once it reaches the new site. *(Lesson 10-3)*

5. **SHAPES** Name three items in your room that are quadrilaterals. Classify the shapes. *(Lesson 10-4)*

6. **GEOGRAPHY** The state of Indiana is shaped almost like a trapezoid. Estimate the area of the state. *(Lesson 10-5)*

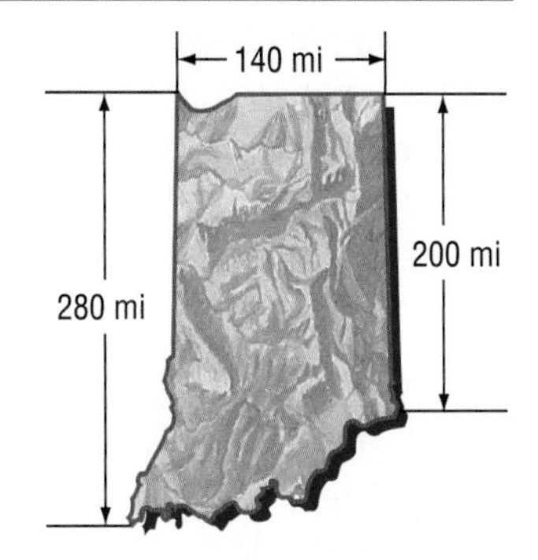

SIGNS For Exercises 7 and 8, use the following information.
Part of a driver's license exam includes identifying road signs by color and by shape. Identify the shape of each road sign pictured below. *(Lesson 10-6)*

7. 

8. 

MANUFACTURING For Exercises 9 and 10, use the following information.
Some cafeteria trays are designed so that four people can place their trays around a square table without bumping corners, as shown below. The top and bottom of the tray are parallel. *(Lesson 10-6)*

9. What is the shape of the tray?

10. Find the measure of each angle of the tray so that the trays will fit side-to-side around the table.

11. **PUBLIC SAFETY** A tornado warning system can be heard for a 2-mile radius. Find the area that will benefit from the warning. *(Lesson 10-7)*

12. **CITY PLANNING** The circular region inside the streets at DuPont Circle in Washington, D.C., is 250 feet across. What is the area of the region? *(Lesson 10-7)*

13. **GEOMETRY** Find the area of a figure that is formed using a rectangle having width equal to 8 feet and length equal to 5 feet and a half circle with a diameter of 6 feet. *(Lesson 10-8)*

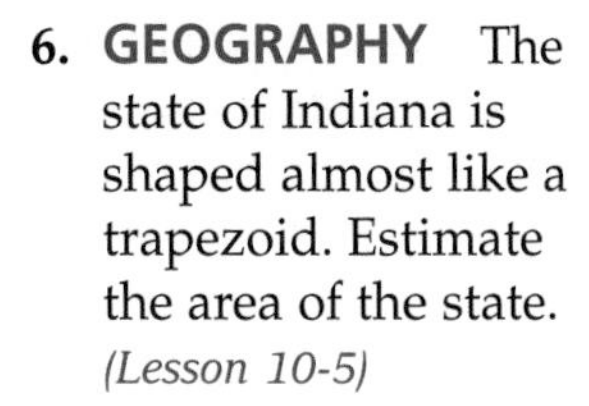

1. **PRESENTS** Mateo received a gift wrapped in the shape of a rectangular pyramid. How many faces, edges, and vertices are on the gift box? *(Lesson 11-1)*

2. **PET CARE** Tina has an old fish tank in the shape of a circular cylinder. The tank is 2 feet in diameter and 6 feet high. How many cubic feet of water does it hold? Round to the nearest cubic foot. *(Lesson 11-2)*

3. **CHEMISTRY** A quartz crystal is a hexagonal prism. It has a base area of 1.41 square centimeters and a volume of 4.64 cubic centimeters. What is its height? If necessary, round to the nearest hundredth. *(Lesson 11-2)*

4. **BAKING** A rectangular cake pan is 30 centimeters by 21 centimeters by 5 centimeters. A round cake pan has a diameter of 21 centimeters and a height of 4 centimeters. Which holds more batter, the rectangular pan or two round pans? *(Lesson 11-2)*

5. **MONUMENTS** The top of the Washington Monument is a square pyramid 54 feet high and 34 feet long on each side. What is the volume of this top part of the monument? *(Lesson 11-3)*

6. **MANUFACTURING** A carton of canned fruit holds 24 cans. Each can has a diameter of 7.6 centimeters and a height of 10.8 centimeters. Approximately how much paper is needed to make the labels for the 24 cans? If necessary, round to the nearest tenth. *(Lesson 11-4)*

7. **CAMPING** How much canvas was used to make the A-frame tent shown below? (*Hint:* Be sure to include the floor of the tent.) *(Lesson 11-4)*

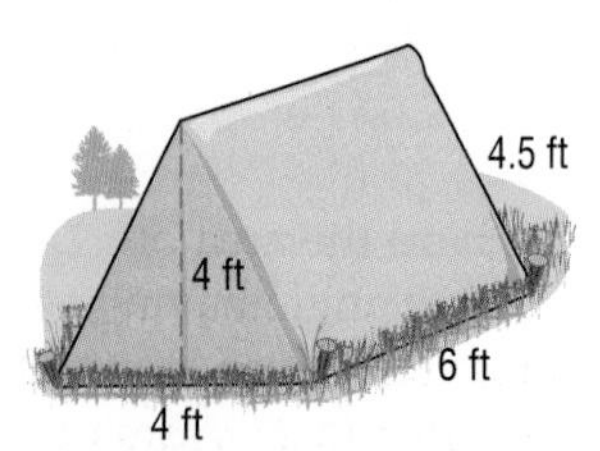

8. **HISTORY** The Pyramid of Cestius is a monument in Rome. It is a square pyramid with the dimensions shown below.

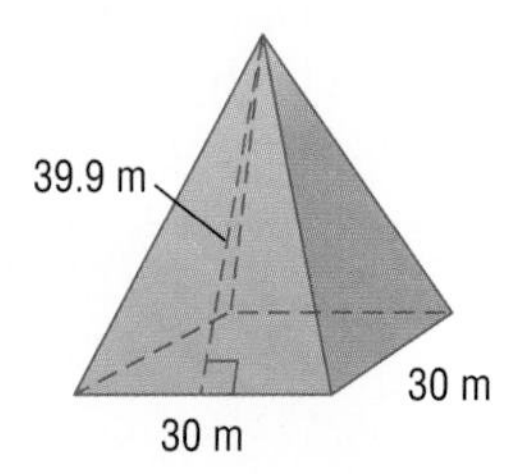

What is its lateral area? If necessary, round to the nearest tenth. *(Lesson 11-5)*

9. **TEPEES** The largest tepee in the United States is in the shape of a cone with a diameter of 42 feet and a slant height of about 47.9 feet. How much canvas was used for the cover of the tepee? If necessary, round to the nearest tenth. *(Lesson 11-5)*

10. **SHIPPING** Are the two packing tubes shown below similar solids? *(Lesson 11-6)*

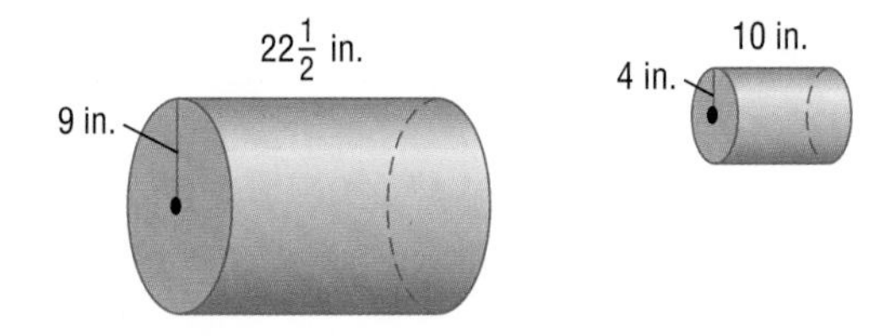

11. **MODELS** A miniature greenhouse is a rectangular prism with a volume of 16 cubic feet. The scale factor of this greenhouse to a larger greenhouse of the same shape is $\frac{1}{4}$. What is the volume of the larger greenhouse? $\left(\textit{Hint:}\text{ scale factor} = \frac{a}{b}, \text{ ratio of volumes} = \frac{a^3}{b^3}\right)$ *(Lesson 11-6)*

12. **DECORATING** Alicia is wallpapering a wall that is 8.25 feet high and 23.7 feet wide. What is the area of the wall? Round to the correct number of significant digits. *(Lesson 11-7)*

1. **ARCHITECTURE** *The World Almanac* lists fifteen tall buildings in New Orleans, Louisiana. The number of floors in each of these buildings is listed below.

51	53	45	39	36
47	42	33	32	31
33	28	28	25	23

Make a stem-and-leaf plot of the data. *(Lesson 12-1)*

2. **WORLD CULTURES** Many North American Indians hold conferences called *powwows*, to celebrate their culture and heritage through various ceremonies and dances. The ages of participants and observers in a Menominee Indian powwow are shown in the chart below.

Participants	20, 18, 12, 13, 14, 72, 65, 23, 25, 43, 67, 35, 68, 13, 56
Observers	43, 55, 70, 63, 15, 41, 9, 42, 75, 25, 16, 18, 51, 80, 75, 39, 23, 55, 50, 54, 60, 43

Find the range and interquartile range for each group. *(Lesson 12-2)*

3. **CONSUMERISM** The average retail price for one gallon of unleaded gasoline at a certain station are shown in the table below.

Year	1	2	3	4	5
Price ($)	1.21	1.20	0.92	0.95	0.95
Year	6	7	8	9	10
Price ($)	1.02	1.16	1.14	1.13	1.11

Make a box-and-whisker plot of the data. *(Lesson 12-3)*

4. **HOMEWORK** The frequency table below shows the amount of time students spend doing homework each week.

Weekly Homework Time		
Number of Hours	Tally	Frequency
0–3	IIII	5
4–7	~~IIII~~ ~~IIII~~ IIII	14
8–11	~~IIII~~ ~~IIII~~ ~~IIII~~ III	18
12–15	~~IIII~~ ~~IIII~~ I	11

Display the data in a histogram. *(Lesson 12-4)*

5. **ENTERTAINMENT** The graph below displays data about movie attendance.

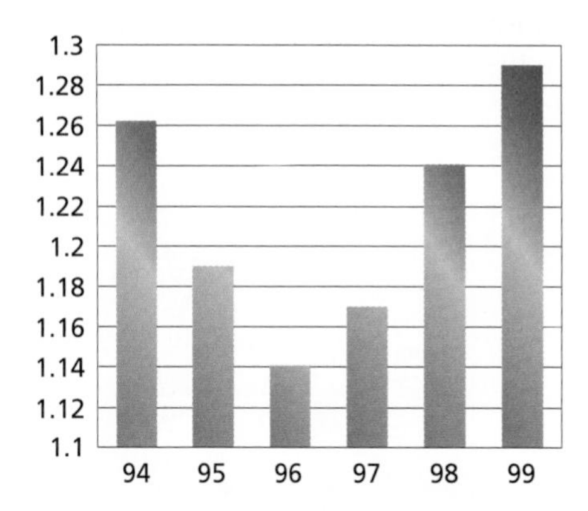

Tell why the graph appears to be misleading. *(Lesson 12-5)*

6. **BUSINESS** The Yogurt Oasis advertises that there are 1512 ways to enjoy a one-topping sundae. They offer six flavors of frozen yogurt, six different serving sizes, and several different toppings. How many toppings do they offer? *(Lesson 12-6)*

7. **VOLLEYBALL** How many different 6-player starting squads can be formed from a volleyball team of 15 players? *(Lesson 12-7)*

8. **TELEVISION** The odds in favor of a person in North America appearing on television sometime in their lifetime is 1:3. If there are 32 students in your class, predict how many will appear on television. *(Lesson 12-8)*

9. **BUSINESS** An auto dealer finds that of the cars coming in for service, 70% need a tune up and 50% need a new air filter. What is the probability that a car brought in for service needs both a tune up and a new air filter? *(Lesson 12-9)*

10. **ECONOMICS** Thirty-one percent of minimum-wage workers are between 16 and 19 years old. Twenty-two percent of the minimum-wage workers are between 20 and 24 years old. If a person who makes minimum wage is selected at random, what is the probability that he or she will be between 16 and 24 years old? *(Lesson 12-9)*

ARCHITECTURE For Exercises 1 and 2, use the following information.
The polynomial $2xy + 2y^2 + 2yz$ represents the total area of the first floor shown in the plan below. *(Lesson 13-1)*

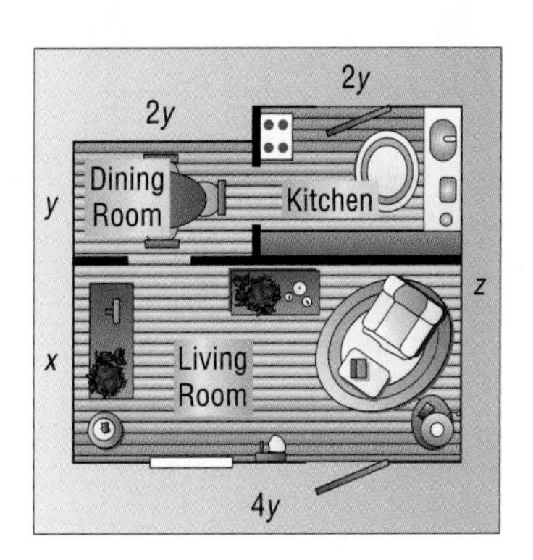

1. Find the degree of the polynomial.

2. Find an expression to represent the area of the living room. Then classify the expression as a *monomial*, *binomial*, or *trinomial*.

3. **CONSTRUCTION** A standard unit of measurement for a window is the *united inch*. You can find the united inches of a window by adding the length and width of the window. If the length of a window is $3x - 5$ inches and the width is $x + 7$ inches, what is the size of the window in united inches? *(Lesson 13-2)*

4. **GEOMETRY** The perimeter of the triangle below is $4x + 4$ centimeters. Find the length of the hypotenuse of the triangle. *(Lesson 13-3)*

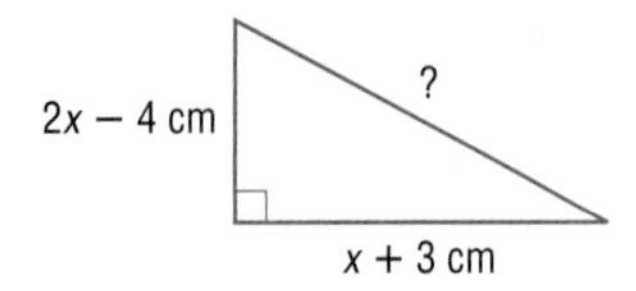

5. **GEOMETRY** Find the area of the shaded region. Write in simplest form. *(Lesson 13-4)*

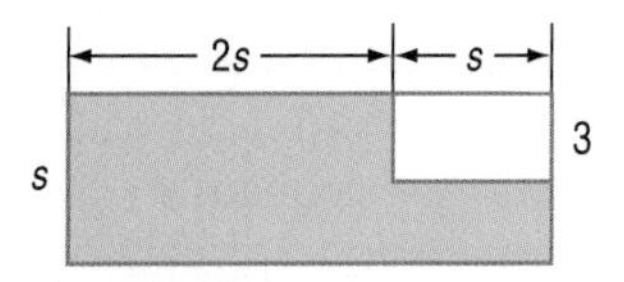

MANUFACTURING For Exercises 6 and 7, use the following information.
The figure below shows a pattern for a cardboard box before it has been cut and folded. *(Lesson 13-4)*

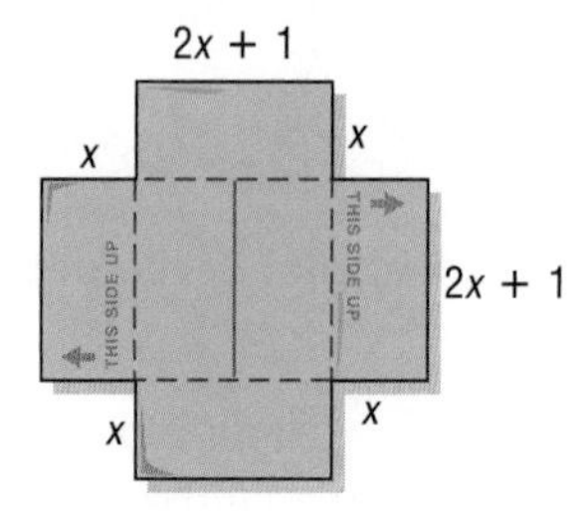

6. Find the area of each rectangular region and add to find a formula for the number of square inches of cardboard needed.

7. Find the surface area if x is 2.5 inches.

8. **PRODUCTION** The XYZ Production Company states that the cost y of producing x items is given by the equation $y = 2500 + 3.2x$. Does this equation represent a *linear* or *nonlinear* function? *(Lesson 13-5)*

9. **INTERNET** The graph below shows the increase in electronic mailboxes in the United States.

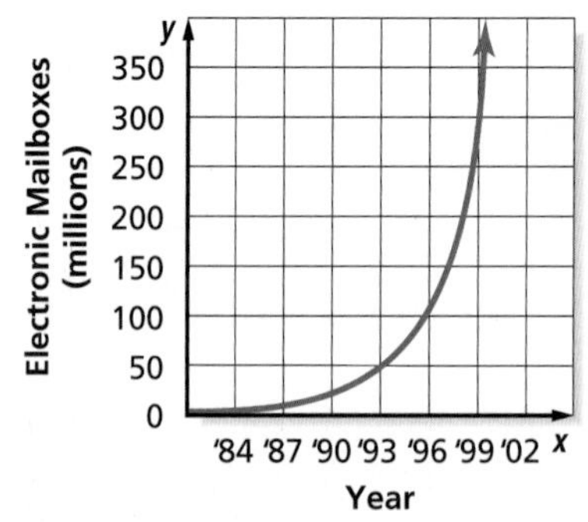

Does this graph represent a *linear* or *nonlinear* function? Explain. *(Lesson 13-5)*

10. **POPULATION** The population growth of a particular species of insect is given by the equation $y = 2x^3$, where x represents time elapsed in days and y represents the population size. Graph this equation. *(Lesson 13-6)*

Glossary/Glosario

Cómo usar el glosario en español:
1. Busca el término en inglés que desees encontrar.
2. El término en español, junto con la definición, se encuentran en la columna de la derecha.

English / Español

A

absolute value (58) The distance a number is from zero on the number line.

accuracy (589) The degree of conformity of a measurement with the true value.

acute angle (449) An angle with a measure greater than 0° and less than 90°.

acute triangle (454) A triangle that has three acute angles.

adjacent angles (493) Two angles that have the same vertex, share a common side, and do *not* overlap.

algebraic expression (17) An expression that contains sums and/or products of variables and numbers.

algebraic fraction (170, 211) A fraction with one or more variables in the numerator or denominator.

alternate exterior angles (492) Nonadjacent exterior angles found on opposite sides of the transversal. In the figure below, ∠1 and ∠7, ∠2 and ∠8 are alternate exterior angles.

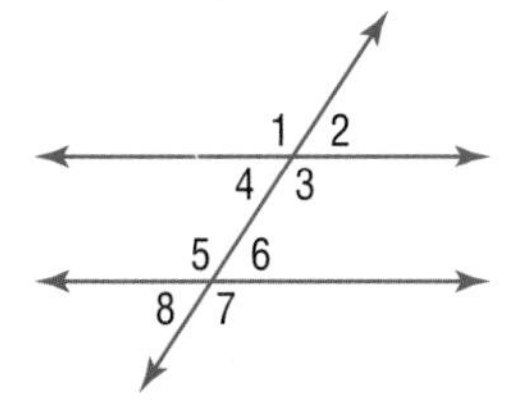

alternate interior angles (492) Nonadjacent interior angles found on opposite sides of the transversal. In the figure above, ∠4 and ∠6, ∠3 and ∠5 are alternate interior angles.

altitude (520) A line segment that is perpendicular to the base of a figure with endpoints on the base and the side opposite the base.

angle (447) Two rays with a common endpoint form an angle. The rays and vertex are used to name an angle. The angle below is ∠*ABC*.

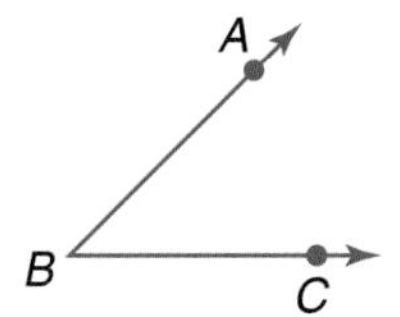

area (132) The measure of the surface enclosed by a geometric figure.

valor absoluto Distancia que un número dista de cero en la recta numérica.

exactitud Grado de conformidad de una medida con el valor verdadero.

ángulo agudo Ángulo con una medida mayor que 0° y menor que 90°.

triángulo acutángulo Triángulo que posee tres ángulos agudos.

ángulos adyacentes Dos ángulos que poseen el mismo vértice, comparten un lado y *no* se traslapan.

expresión algebraica Expresión que contiene sumas y/o productos de números y variables.

fracción algebraica Fracción con una o más variables en el numerador o denominador.

ángulos alternos externos Ángulos exteriores no adyacentes que se encuentran en lados opuestos de una transversal. En la siguiente figura, ∠1 y ∠7, ∠2 y ∠8 son ángulos alternos externos.

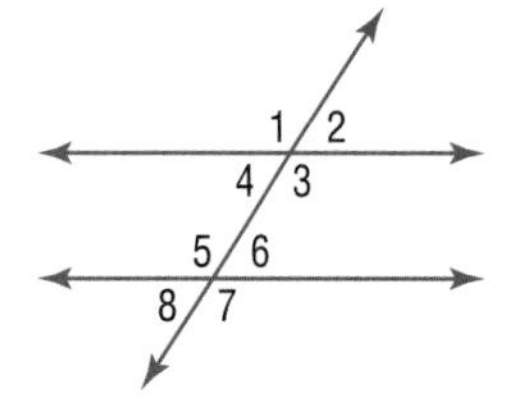

ángulos alternos internos Ángulos interiores no adyacentes que se encuentran en lados opuestos de una transversal. En la figura anterior, ∠4 y ∠6, ∠3 y ∠5 son ángulos alternos internos.

altura Segmento de recta perpendicular a la base de una figura y cuyos extremos yacen en la base y en el lado opuesto de la base.

ángulo Dos rayos con un punto común forman un ángulo. Los rayos y el vértice se usan para identificar el ángulo. El siguiente ángulo es ∠*ABC*.

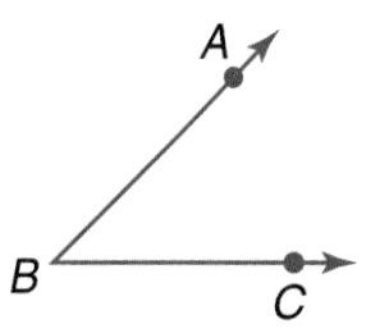

área Medida de la superficie que encierra una figura geométrica.

arithmetic sequence (249) A sequence in which the difference between any two consecutive terms is the same.

sucesión aritmética Sucesión en que la diferencia entre dos términos consecutivos cualesquiera es siempre la misma.

average (82) The sum of data divided by the number of items in the data set, also called the mean.

promedio Suma de los datos dividida entre el número de elementos en el conjunto de datos. También llamado media.

B

back-to-back stem-and-leaf plot (607) Used to compare two sets of data. The leaves for one set of data are on one side of the stem and the leaves for the other set of data are on the other side.

diagrama de tallo y hojas consecutivo Se usa para comparar dos conjuntos de datos. Las hojas de uno de los conjuntos de datos aparecen en un lado del tallo y las del otro al otro lado de éste.

bar graph (722) A graphic form using bars to make comparisons of statistics.

gráfica de barras Tipo de gráfica que usa barras para comparar estadísticas.

bar notation (201) In repeating decimals the line or bar placed over the digits that repeat. For example, $2.\overline{63}$ indicates the digits 63 repeat.

notación de barra En decimales periódicos, la línea o barra que se escribe encima de los dígitos que se repiten. Por ejemplo, en $2.\overline{63}$ la barra encima del 63 indica que los dígitos 63 se repiten.

base (153) In 2^4, the base is 2. The base is used as a factor as many times as given by the exponent (2). That is, $2^4 = 2 \times 2 \times 2 \times 2$.

base En 2^4, la base es 2. La base se usa como factor las veces que indique el exponente (2). Es decir, $2^4 = 2 \times 2 \times 2 \times 2$.

base (288) In a percent proportion, the whole quantity, or the number to which the part is being compared.

$$\frac{\text{part}}{\text{base}} = \frac{\text{percent}}{100}$$

base En una proporción porcentual, toda la cantidad o número al que se compara la parte.

$$\frac{\text{parte}}{\text{base}} = \frac{\text{por ciento}}{100}$$

base (520, 521, 522) The base of a parallelogram or a triangle is any side of the figure. The bases of a trapezoid are the parallel sides.

base La base de un paralelogramo o de un triángulo es cualquier lado de la figura. Las bases de un trapecio son los lados paralelos.

base (557) The bases of a prism are any two parallel congruent faces.

base Las bases de un prisma son cualquier par de caras paralelas y congruentes.

best-fit line (409) On a scatter plot, a line drawn that is very close to most of the data points. The line that best fits the data.

recta de ajuste óptimo En una gráfica de dispersión, una recta que está muy cercana a la mayoría de los puntos de datos. La recta que mejor se ajusta a los datos.

binomial (669) A polynomial with exactly two terms.

binomio Polinomio con exactamente dos términos.

boundary (419) A line that separates a graph into half planes.

frontera Recta que divide una gráfica en semiplanos.

box-and-whisker plot (617) A diagram that divides a set of data into four parts using the median and quartiles. A box is drawn around the quartile values and whiskers extend from each quartile to the extreme data points.

diagrama de caja y patillas Diagrama que divide un conjunto de datos en cuatro partes usando la mediana y los cuartiles. Se dibuja una caja alrededor de los cuartiles y se extienden patillas de cada uno de ellos a los valores extremos.

C

center (533) The given point from which all points on the circle are the same distance.

centro Punto dado del cual equidistan todos los puntos de un círculo.

circle (533) The set of all points in a plane that are the same distance from a given point called the center.

círculo Conjunto de todos los puntos del plano que están a la misma distancia de un punto dado del plano llamado centro.

circle graph (723) A type of statistical graph used to compare parts of a whole.

gráfica circular Tipo de gráfica estadística que se usa para comparar las partes de un todo.

circumference (533) The distance around a circle.

circunferencia Longitud del contorno de un círculo.

coefficient (103) The numerical part of a term that contains a variable.

coeficiente Parte numérica de un término que contiene una variable.

combination (642) An arrangement or listing in which order is not important.

combinación Arreglo o lista en que el orden no es importante.

common difference (249) The difference between any two consecutive terms in an arithmetic sequence.

diferencia común Diferencia entre dos términos consecutivos cualesquiera de una sucesión aritmética.

common multiples (226) Multiples that are shared by two or more numbers. For example, some common multiples of 4 and 6 are 0, 12, and 24.

múltiplos comunes Múltiplos compartidos por dos o más números. Por ejemplo, algunos múltiplos comunes de 4 y 6 son 0, 12 y 24.

common ratio (250) The ratio between any two consecutive terms in a geometric sequence.

razón común Razón entre dos términos consecutivos cualesquiera de una sucesión geométrica.

compatible numbers (714) Numbers that have been rounded so when the numbers are divided by each other, the remainder is zero.

números compatibles Números redondeados de modo que cuando se dividen, el residuo es cero.

complementary (493) Two angles are complementary if the sum of their measures is 90°.

complementarios Dos ángulos son complementarios si la suma de sus medidas es 90°.

composite number (159) A whole number that has more than two factors.

número compuesto Número entero que posee más de dos factores.

compound event (650) Two or more simple events.

evento compuesto Dos o más eventos simples.

cone (569) A three-dimensional figure with one circular base. A curved surface connects the base and vertex.

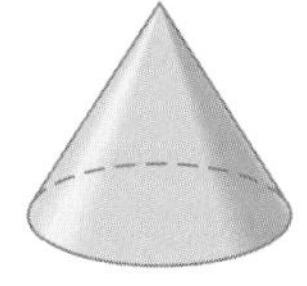

cono Figura tridimensional con una base circular, la cual posee una superficie curva que une la base con el vértice.

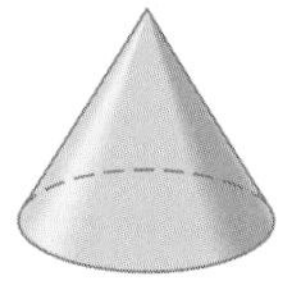

congruent (455, 493, 500) Line segments that have the same length, or angles that have the same measure, or figures that have the same size and shape.

congruentes Segmentos de recta que tienen la misma longitud o ángulos que tienen la misma medida o figuras que poseen la misma forma y tamaño.

conjecture (7) An educated guess.

conjetura Suposición informada.

constant (103) A term without a variable.

constante Término sin variables.

constant of variation (394) The slope, or rate of change, in the equation $y = kx$, represented by k.

constante de variación La pendiente, o tasa de cambio, en la ecuación $y = kx$, representada por k.

converse (462) The statement formed by reversing the phrases after *if* and *then* in an if-then statement.

recíproca Un enunciado que se forma intercambiando los enunciados que vienen a continuación de *si-entonces* en un enunciado *si-entonces*.

coordinate (57) A number that corresponds with a point on a number line.

coordenada Número que corresponde a un punto en la recta numérica.

coordinate plane (33) Another name for the coordinate system.

plano de coordenadas Otro nombre para el sistema de coordenadas.

coordinate system (33) A coordinate system is formed by the intersection of two number lines that meet at right angles at their zero points, also called a coordinate plane.

sistema de coordenadas Un sistema de coordenadas se forma de la intersección de dos rectas numéricas perpendiculares que se intersecan en sus puntos cero. También llamado plano de coordenadas.

corresponding angles (492) Angles that have the same position on two different parallel lines cut by a transversal. In the figure, $\angle 1$ and $\angle 5$, $\angle 2$ and $\angle 6$, $\angle 3$ and $\angle 7$, $\angle 4$ and $\angle 8$ are corresponding angles.

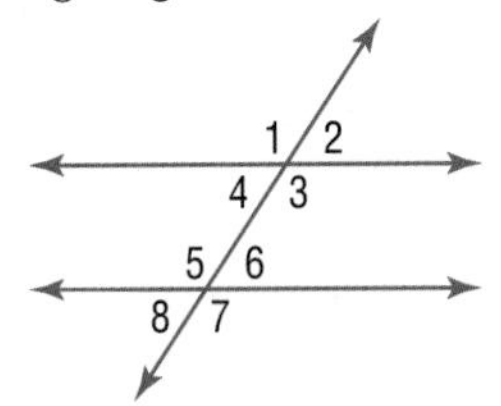

ángulos correspondientes Ángulos que tienen la misma posición en dos rectas paralelas distintas cortadas por una transversal. En la figura, $\angle 1$ y $\angle 5$, $\angle 2$ y $\angle 6$, $\angle 3$ y $\angle 7$, $\angle 4$ y $\angle 8$ son ángulos correspondientes.

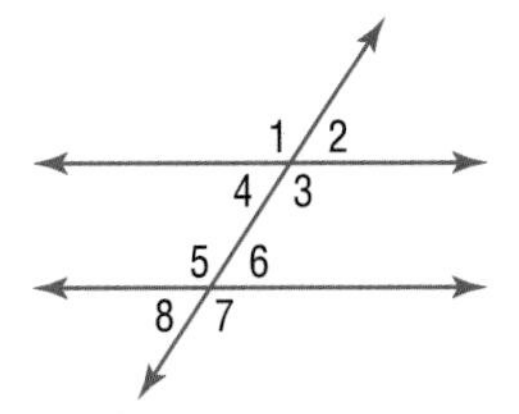

corresponding parts (500) Parts of congruent or similar figures that match.

partes correspondientes Partes de figuras congruentes o semejantes que se corresponden mutuamente.

cosine (477) If $\triangle ABC$ is a right triangle and A is an acute angle,

$$\text{cosine } \angle A = \frac{\text{measure of the leg adjacent to } \angle A}{\text{measure of the hypotenuse}}$$

coseno Si $\triangle ABC$ es un ángulo rectángulo y A es uno de sus ángulos agudos,

$$\text{coseno } \angle A = \frac{\text{medida del cateto adyacente a } \angle A}{\text{medida de la hipotenusa}}$$

counterexample (25) An example that shows a conjecture is not true.

contraejemplo Ejemplo que muestra que una conjetura no es verdadera.

cross products (270) If $\frac{a}{c} = \frac{b}{d}$, then $ad = bc$. If $ad = bc$, then $\frac{a}{c} = \frac{b}{d}$.

productos cruzados Si $\frac{a}{c} = \frac{b}{d}$, entonces $ad = bc$. Si $ad = bc$, entonces $\frac{a}{c} = \frac{b}{d}$.

cubic function (688) A function that can be described by an equation of the form $y = ax^3 + bx^2 + cx + d$, where $a \neq 0$.

función cúbica Función que puede describirse por una ecuación de la forma $y = ax^3 + bx^2 + cx + d$, donde $a \neq 0$.

cylinder (565) A solid that has two parallel, congruent bases (usually circular) connected with a curved side.

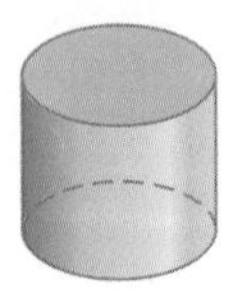

cilindro Sólido que posee dos bases congruentes y paralelas (por lo general circulares) unidas por un lado curvo.

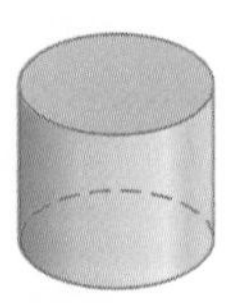

D

deductive reasoning (25) The process of using facts, properties, or rules to justify reasoning or reach valid conclusions.

razonamiento deductivo Proceso de usar hechos, propiedades o reglas para justificar un razonamiento o para sacar conclusiones válidas.

defining a variable (18) Choosing a variable and a quantity for the variable to represent in an equation.

definir una variable Seleccionar una variable y una cantidad para la variable psts representst en una ecuación.

degree (448) The most common unit of measure for angles.

grado La unidad de medida angular más común.

degree (670) The sum of the exponents of the variables of a monomial.

grado Suma de los exponentes de las variables de un monomio.

dependent events (651) Two or more events in which the outcome of one event does affect the outcome of the other event(s).

eventos dependientes Dos o más eventos en que el resultado de uno de ellos afecta el resultado del otro o de los otros eventos.

diagonal (528) A line segment that joins two nonconsecutive vertices of a polygon.

diagonal Segmento de recta que une dos vértices no consecutivos de un polígono.

diameter (533) The distance across a circle through its center.

diámetro Distancia de un lado a otro de un círculo medida a través de su centro.

dilation (512) A transformation that alters the size of a figure but not its shape.

dilatación Transformación que altera el tamaño de una figura, pero no su forma.

dimensional analysis (212) The process of including units of measurement when computing.

análisis dimensional Proceso que incorpora las unidades de medida al hacer cálculos.

direct variation (394) A special type of linear equation that describes rate of change. A relationship such that as x increases in value, y increases or decreases at a constant rate.

variación directa Tipo especial de ecuación lineal que describe tasas de cambio. Relación en que a medida que x aumenta de valor, y aumenta o disminuye a una tasa constante .

discount (299) The amount by which the regular price of an item is reduced.

descuento Cantidad por la que se reduce el precio normal de un artículo.

Distance Formula (466) The distance between two points, with coordinates (x_1, y_1) and (x_2, y_2), is given by $d = \sqrt{(x_2 - x_1)^2 + (y_2 - y_1)^2}$.

Fórmula de la distancia La distancia entre dos puntos, con coordenadas (x_1, y_1) y (x_2, y_2), se calcula con $d = \sqrt{(x_2 - x_1)^2 + (y_2 - y_1)^2}$.

divisible (148) A number is divisible by another if, upon division, the remainder is zero.

divisible Un número es divisible entre otro si, al dividirlos, el residuo es cero.

domain (35) The domain of a relation is the set of all x-coordinates from each pair.

dominio El dominio de una relación es el conjunto de coordenadas x de todos los pares .

E

edge (556) Where two planes intersect in a line.

arista Recta en donde se intersecan dos planos.

empty set (336) A set with no elements shown by the symbol { } or ∅.

conjunto vacío Conjunto que carece de elementos y que se denota con el símbolo { } o ∅.

equation (28) A mathematical sentence that contains an equals sign (=).

ecuación Enunciado matemático que contiene el signo de igualdad (=).

equilateral triangle (455) A triangle with all sides congruent.

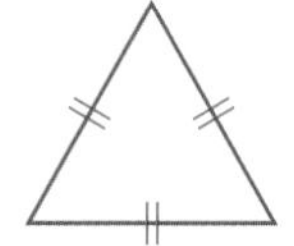

triángulo equilátero Un triángulo cuyos lados son todos congruentes.

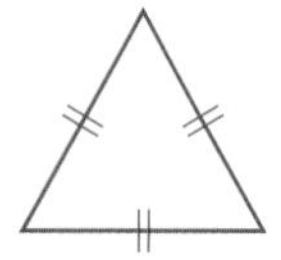

equivalent equations (111) Two or more equations with the same solution. For example, $x + 4 = 7$ and $x = 3$ are equivalent equations.

ecuaciones equivalentes Dos o más ecuaciones con las mismas soluciones. Por ejemplo, $x + 4 = 7$ y $x = 3$ son ecuaciones equivalentes.

equivalent expressions (98) Expressions that have the same value.

expresiones equivalentes Expresiones que tienen el mismo valor.

evaluate (12) Find the numerical value of an expression.

evaluar Calcular el valor numérico de una expresión.

expanded form (154) A number expressed using place value to write the value of each digit in the number.

forma desarrollada Número que se escribe usando el valor de posición para indicar el valor de cada dígito de un número.

experimental probability (311) What actually occurs in a probability experiment.

probabilidad experimental Lo que realmente sucede en un experimento probabilístico.

exponent (153) In 2^4, the exponent is 4. The exponent tells how many times the base, 2, is used as a factor. So, $2^4 = 2 \times 2 \times 2 \times 2$.

exponente En 2^4, el exponente es 4. El exponente indica cuántas veces se usa la base, 2, como factor. Así, $2^4 = 2 \times 2 \times 2 \times 2$.

exterior angles (492, 531) Four of the angles formed by the transversal and two parallel lines. Exterior angles lie outside the two parallel lines.

ángulos exteriores Cuatro de los ángulos formados por una transversal y dos rectas paralelas. Los ángulos exteriores yacen fuera de las dos rectas paralelas.

F

face (556) A flat surface, the side or base of a prism.

cara Superficie plana, el lado o la base de un prisma.

factorial (642) The expression *n factorial* ($n!$) is the product of all counting numbers beginning with n and counting backward to 1.

factorial La expresión *n factorial* ($n!$) es el producto de todos los números naturales, comenzando con n y contando al revés hasta llegar al 1.

factors (148) Two or more numbers that are multiplied to form a product.

factores Dos o más números que se multiplican para formar un producto.

factor tree (160) A way to find the prime factorization of a number. The factors branch out from the previous factors until all the factors are prime numbers.

árbol de factores Forma de encontrar la factorización prima de un número. Los factores se ramifican de los factores anteriores hasta que todos los factores son números primos.

formula (131) An equation that shows a relationship among certain quantities.

fórmula Ecuación que muestra la relación entre ciertas cantidades.

frequency table (623) A chart that indicates the number of values in each interval.

tabla de frecuencias Tabla que indica el número de valores en cada intervalo.

function (369) A function is a special relation in which each element of the domain is paired with exactly one element in the range.

función Una función es una relación especial en que a cada elemento del dominio le corresponde un único elemento del rango.

Fundamental Counting Principle (636) If event M can occur in m ways and is followed by event N that can occur in n ways, then the event M followed by event N can occur in $m \cdot n$ ways.

Principio fundamental de contar Si el evento M puede ocurrir de m maneras y lo sigue un evento N que puede ocurrir de n maneras, entonces el evento M seguido del evento N puede ocurrir de $m \cdot n$ maneras.

G

geometric sequence (250) A sequence in which the ratio between any two consecutive terms is the same.

sucesión geométrica Sucesión en que la razón entre dos términos consecutivos cualesquiera es siempre la misma.

graph (34) A dot at the point that corresponds to an ordered pair on a coordinate plane.

gráfica Marca puntual en el punto que corresponde a un par ordenado en un plano de coordenadas.

greatest common factor (GCF) (164) The greatest number that is a factor of two or more numbers.

máximo común divisor (MCD) El número mayor que es factor de dos o más números.

greatest possible error (594) One-half the precision unit, used to describe the actual measure.

error máximo posible La mitad de la unidad de precisión. Se usa para describir la medida exacta.

H

half plane (419) The region that contains the solution for an inequality.

semiplano Región que contiene la solución de una desigualdad.

histogram (623) A histogram uses bars to display numerical data that have been organized into equal intervals.

histograma Un histograma usa barras para exhibir datos numéricos que han sido organizados en intervalos iguales.

hypotenuse (460) The side opposite the right angle in a right triangle.

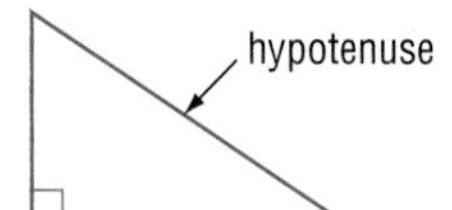

hipotenusa Lado opuesto al ángulo recto en un triángulo rectángulo.

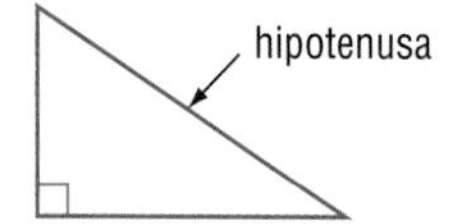

I

identity (336) An equation that is true for every value of the variable.

identidad Ecuación que es verdadera para cada valor de la variable.

independent events (650) Two or more events in which the outcome of one event does *not* influence the outcome of the other event(s).

eventos independientes Dos o más eventos en que el resultado de uno de ellos *no* afecta el resultado del otro o de los otros eventos .

indirect measurement (472) Using the properties of similar triangles to find measurements that are difficult to measure directly.

medición indirecta Uso de las propiedades de triángulos semejantes para hacer mediciones que son difíciles de realizar directamente.

inductive reasoning (7) Reasoning based on a pattern of examples or past events.

razonamiento inductivo Rezonamiento basada en un patrón de ejemplos o de sucesos pasados.

inequality (57, 340) A mathematical sentence that contains $<, >, \neq, \leq$, or $\geq$.

desigualdad Enunciado matemático que contiene $<, >, \neq, \leq$ o $\geq$.

integers (56) The whole numbers and their opposites. $\ldots, -3, -2, -1, 0, 1, 2, 3, \ldots$

enteros Los números enteros y sus opuestos. $\ldots, -3, -2, -1, 0, 1, 2, 3, \ldots$

interior angles (492, 528) Four of the angles formed by the transversal and two parallel lines. Interior angles lie between the two parallel lines.

ángulos interiores Cuatro de los ángulos formados por una transversal y dos rectas paralelas. Los ángulos interiores yacen entre las dos rectas paralelas.

interquartile range (613) The range of the middle half of a set of data. It is the difference between the upper quartile and the lower quartile.

amplitud intercuartílica Amplitud de la mitad central de un conjunto de datos. Es la diferencia entre el cuartil superior y el inferior.

inverse operations (110) Operations that undo each other, such as addition and subtraction.

operaciones inversas Operaciones que se anulan mutuamente, como la adición y la sustracción.

irrational number (441) A number that cannot be expressed as $\frac{a}{b}$, where a and b are integers and b does not equal 0.

número irracional Número que no puede escribirse como $\frac{a}{b}$, donde a y b son enteros y b no es igual a 0.

isosceles triangle (455) A triangle that has at least two congruent sides.

triángulo isósceles Triángulo que posee por lo menos dos lados congruentes.

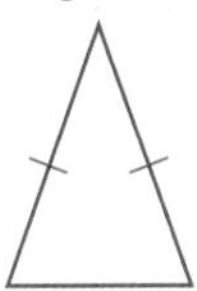

L

lateral area (578) The sum of the areas of the lateral faces of a solid.

área lateral Suma de las àreas de las caras laterales de un sólido.

lateral faces (578) The lateral faces of a prism, cylinder, pyramid, or cone are all the surface of the figure except the base or bases.

caras laterales Las caras laterales de un prisma, cilindro, pirámide o cono son todas las superficies de la figura, excluyendo la base o las bases.

least common denominator (LCD) (227) The least common multiple of the denominators of two or more fractions.

mínimo común denominador (mcd) El mínimo común múltiplo de los denominadores de dos o más fracciones.

least common multiple (LCM) (226) The least of the nonzero common multiples of two or more numbers. The LCM of 4 and 6 is 12.

mínimo común múltiplo (mcm) El menor de los múltiplos comunes no nulos de dos o más números. El MCM de 4 y 6 es 12.

leaves (606) In a stem-and-leaf plot, the next greatest place value of the data after the stem forms the leaves.

hojas En un diagrama de tallo y hojas, las hojas las forma el segundo valor de posición mayor después del tallo.

legs (460) The sides that are adjacent to the right angle of a right triangle.

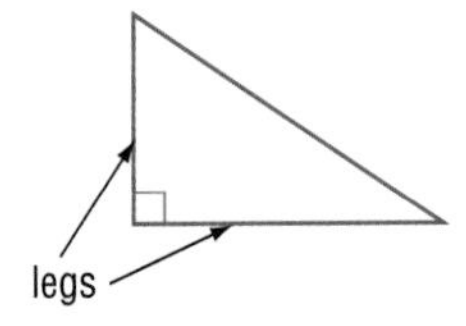

catetos Lados adyacentes al ángulo recto de un triángulo rectángulo.

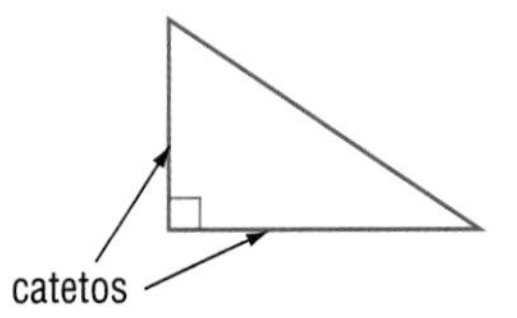

like terms (103, 674) Expressions that contain the same variables to the same power, such as $2n$ and $5n$ or $6xy^2$ and $4xy^2$.

términos semejantes Expresiones que tienen las mismas variables elevadas a los mismos exponentes, como $2n$ y $5n$ ó $6xy^2$ y $4xy^2$.

line (447) A never-ending straight path.

recta Trayectoria rectilínea interminable.

line graph (722) A type of statistical graph used to show how values change over a period of time.

gráfica lineal Tipo de gráfica estadística que se usa para mostrar cómo cambian los valores durante un período de tiempo.

line of symmetry (506) Each half of a figure is a mirror image of the other half when a line of symmetry is drawn.

eje de simetría Cuando se traza un eje de simetría, cada mitad de una figura es una imagen especular de la otra mitad.

line segment (453) Part of a line containing two endpoints and all the points between them.

segmento de recta Parte de una recta que contiene dos extremos y todos los puntos entre éstos.

linear equation (375) An equation in which the variables appear in separate terms and neither variable contains an exponent other than 1. The graph of a linear equation is a straight line.

ecuación lineal Ecuación en que las variables aparecen en términos separados y en la cual ninguna de ellas tiene un exponente distinto de 1. La gráfica de una ecuación lineal es una recta.

lower quartile (613) The median of the lower half of a set of data, indicated by LQ.

cuartil inferior Mediana de la mitad inferior de un conjunto de datos, se denota con CI.

M

mean (82, 238) The sum of data divided by the number of items in the data set, also called the average.

media Suma de los datos dividida entre el número de elementos en el conjunto de datos. También llamada promedio.

measures of central tendency (238) For a list of numerical data, numbers that can represent the whole set of data.

medidas de tendencia central Números que pueden representar todo el conjunto de datos en una lista de datos numéricos.

measures of variation (612) Used to describe the distribution of statistical data.

medidas de variación Se usan para describir la distribución de datos estadísticos.

median (238) In a set of data, the middle number of the ordered data, or the mean of the two middle numbers.

mediana En un conjunto de datos, el número central de los datos ordenados numéricamente o la media de los dos números centrales.

midpoint (468) On a line segment, the point that is halfway between the endpoints.

punto medio En un segmento de recta, el punto que equidista de ambos extremos.

Midpoint Formula (468) On a coordinate plane, the coordinates of the midpoint of a segment whose endpoints have coordinates at (x_1, y_1) and (x_2, y_2) are given by $\left(\frac{x_1 + x_2}{2}, \frac{y_1 + y_2}{2}\right)$.

Fórmula del punto medio En el plano de coordenadas, el punto medio de un segmento cuyos extremos son (x_1, y_1) y (x_2, y_2) se calcula con la fórmula $\left(\frac{x_1 + x_2}{2}, \frac{y_1 + y_2}{2}\right)$.

mixed number (200) The indicated sum of a whole number and a fraction. For example, $3\frac{1}{2}$.

número mixto Suma de un entero y una fracción. Por ejemplo, $3\frac{1}{2}$.

mode (238) The number or numbers that occurs most often in a set of data.

moda Número o números de un conjunto de datos que aparecen más frecuentemente.

monomial (150) An expression that is a number, a variable, or a product of numbers and/or variables.

monomio Expresión que es un número, una variable y/o un producto de números y variables.

multiple (226) The product of a number and a whole number.

múltiplo Producto de un número por un número entero.

mutually exclusive events (652) Two or more events that cannot happen at the same time.

eventos mutuamente exclusivos Dos o más eventos que no pueden ocurrir simultáneamente.

Glossary/Glosario

N

negative number (56) A number less than zero.

número negativo Número menor que cero.

nonlinear function (687) A function with a graph that is not a straight line.

función no lineal Función cuya gráfica no es una recta.

null set (336) A set with no elements shown by the symbol { } or ∅.

conjunto vacío Conjunto que carece de elementos y que se denota con el símbolo { } o ∅.

numerical expression (12) A combination of numbers and operations such as addition, subtraction, multiplication, and division.

expresión numérica Combinación de números y operaciones, como adición, sustracción, multiplicación y división.

O

obtuse angle (449) An angle with a measure greater than 90° but less than 180°.

ángulo obtuso Ángulo que mide más de 90°, pero menos de 180°.

obtuse triangle (454) A triangle with one obtuse angle.

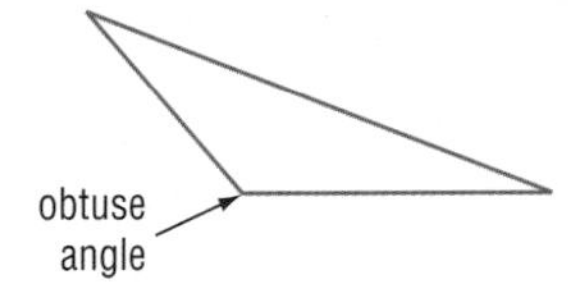

triángulo obtusángulo Triángulo que posee un ángulo obtuso.

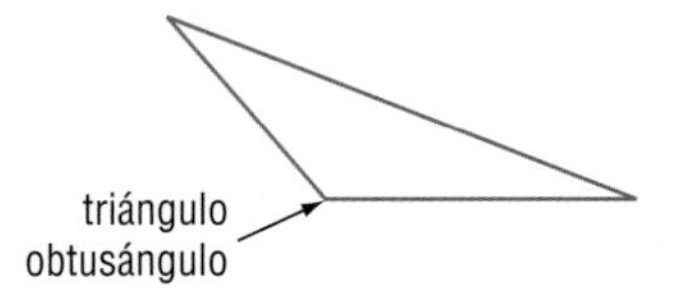

odds (646) A way to describe the chance of an event occurring.

posibilidades Una manera de describir la oportunidad de que ocurra un evento.

open sentence (28) An equation that contains a variable.

enunciado abierto Ecuación que contiene una variable.

opposites (66) Two numbers with the same absolute value but different signs.

opuestos Dos números que tienen el mismo valor absoluto, pero que tienen distintos signos.

ordered pair (33) A pair of numbers used to locate any point on a coordinate plane.

par ordenado Par de números que se usa para ubicar cualquier punto en un plano de coordenadas.

order of operations (12)

1. Simplify the expressions inside grouping symbols.
2. Evaluate all powers.
3. Do all multiplications and/or divisions from left to right.
4. Do all additions and/or subtractions from left to right.

orden de las operaciones

1. Reduce las expresiones dentro de símbolos de agrupamiento.
2. Evalúa todas las potencias.
3. Ejecuta todas las multiplicaciones y/o divisiones de izquierda a derecha.
4. Ejecuta todas las adiciones y/o sustracciones de izquierda a derecha.

origin (33) The point at which the number lines intersect in a coordinate system.

origen Punto de intersección de las rectas numéricas de un sistema de coordenadas.

outcomes (310) Possible results of a probability event.

resultado Resultados posibles de un experimento probabilístico.

outliers (621) Data that are more than 1.5 times the interquartile range from the quartiles.

valores atípicos Datos que distan de los cuartiles más de 1.5 veces la amplitud intercuartílica.

P

parallel lines (492) Two lines in the same plane that do not intersect.

rectas paralelas Dos rectas en el mismo plano que no se intersecan.

parallelogram (514) A quadrilateral with opposite sides parallel and congruent.

paralelogramo Cuadrilátero con lados opuestos congruentes y paralelos.

part (288) In a percent proportion, the number being compared to the whole quantity.

parte En una proporción porcentual, el número que se compara con la cantidad total.

percent (281) A ratio that compares a number to 100.

por ciento Razón que compara un número con 100.

percent equation (298) An equivalent form of percent proportion, where % is written as a decimal.

$$\text{Part} = \text{Percent} \times \text{Base}$$

ecuación porcentual Forma equivalente a la proporción porcentual en que el % se escribe como decimal.

$$\text{Parte} = \text{Por ciento} \times \text{Base}$$

percent of change (304) The ratio of the increase or decrease of an amount to the original amount.

porcentaje de cambio Razón del aumento o disminución de una cantidad a la cantidad original.

percent of decrease (306) The ratio of an amount of decrease to the previous amount, expressed as a percent. A negative percent of change.

porcentaje de disminución Razón de la cantidad de disminución a la cantidad original, escrita como por ciento. Un por ciento de cambio negativo.

percent of increase (305) The ratio of an amount of increase to the original amount, expressed as a percent.

porcentaje de aumento Razón de la cantidad de aumento a la cantidad original, escrita como por ciento.

percent proportion (288)

$\frac{\text{part}}{\text{base}} = \frac{\text{percent}}{100}$ or $\frac{a}{b} = \frac{p}{100}$

proporción porcentual

$\frac{\text{parte}}{\text{base}} = \frac{\text{por ciento}}{100}$ o $\frac{a}{b} = \frac{p}{100}$

perfect squares (436) Rational numbers whose square roots are whole numbers. 25 is a perfect square because $\sqrt{25} = 5$.

cuadrados perfectos Números racionales cuyas raíces cuadradas son números racionales. 25 es un cuadrado perfecto porque $\sqrt{25} = 5$.

perimeter (132) The distance around a geometric figure.

perímetro Longitud alrededor de una figura geométrica.

period (201) In a repeating decimal, the digit or digits that repeats. The period of $0.\overline{6}$ is 6.

período En un decimal periódico, el dígito o dígitos que se repiten. El período de $0.\overline{6}$ es 6.

permutation (641) An arrangement or listing in which order is important.

permutación Arreglo o lista en que el orden es importante.

perpendicular lines (494) Lines that intersect to form a right angle.

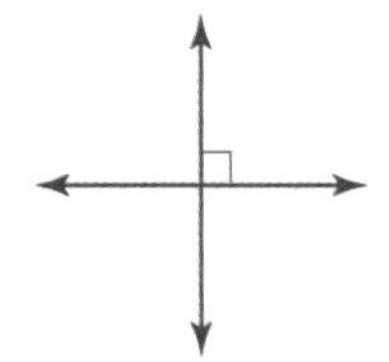

rectas perpendiculares Rectas que se intersecan formando un ángulo recto.

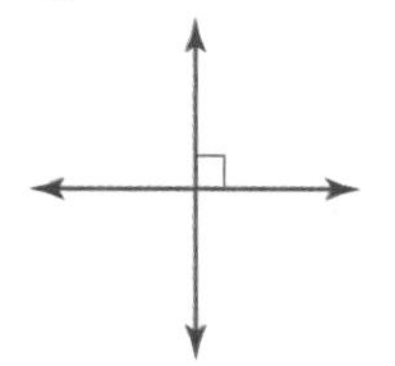

pi, π (533) The ratio of the circumference of a circle to the diameter of the circle. Approximations for π are 3.14 and $\frac{22}{7}$.

pi, π Razón de la circunferencia de un círculo al diámetro del mismo. 3.14 y $\frac{22}{7}$ son aproximaciones de π.

plane (556) A two-dimensional flat surface that extends in all directions and contains at least three noncollinear points.

plano Superficie plana bidimensional que se extiende en todas direcciones y que contiene por lo menos tres puntos no colineales.

point (447) A specific location in space with no size or shape.

punto Ubicación específica en el espacio sin tamaño o forma.

polygon (527) A simple closed figure in a plane formed by three or more line segments.

polígono Figura simple y cerrada en el plano formada por tres o más segmentos de recta.

polyhedron (556) A solid with flat surfaces that are polygons.

poliedro Sólido con superficies planas que son polígonos.

polynomial (669) An algebraic expression that contains the sums and/or products of one or more monomials.

polinomio Expresión algebraica que contiene sumas y/o productos de uno o más monomios.

power (153) A number that is expressed using an exponent.

potencia Número que puede escribirse usando un exponente.

precision (590) The exactness to which a measurement is made.

precisión Exactitud con que se realiza una medida.

prime factorization (160) A composite number expressed as a product of prime factors. For example, the prime factorization of 63 is $3 \times 3 \times 7$.

factorización prima Número compuesto escrito como producto de factores primos. Por ejemplo, la factorización prima de 63 es $3 \times 3 \times 7$.

prime number (159) A whole number that has exactly two factors, 1 and itself.

principle (300) The amount of money in an account.

prism (557) A polyhedron that has two parallel, congruent bases in the shape of polygons.

rectangular prism triangular prism

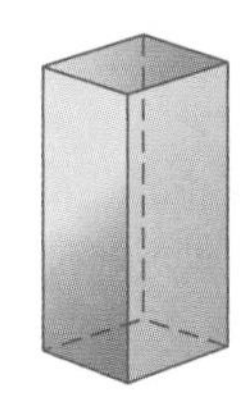
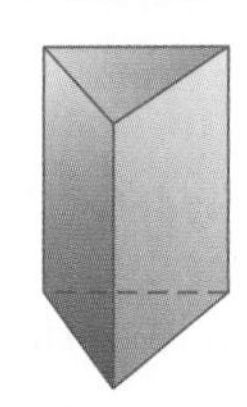

probability (310) The ratio of the number of ways a certain event can occur to the number of possible outcomes.

$$P(\text{event}) = \frac{\text{number of favorable outcomes}}{\text{number of possible outcomes}}$$

properties (23) Statements that are true for any numbers.

proportion (270) A statement of equality of two or more ratios.

protractor (448) An instrument used to measure angles.

pyramid (557) A polyhedron that has a polygon for a base and triangles for sides.

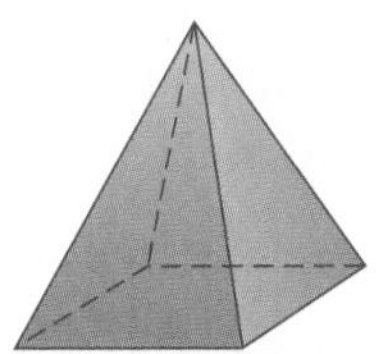

Pythagorean Theorem (460) If a triangle is a right triangle, then the square of the length of the hypotenuse is equal to the sum of the squares of the lengths of the legs or $c^2 = a^2 + b^2$.

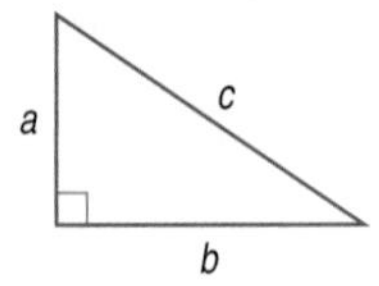

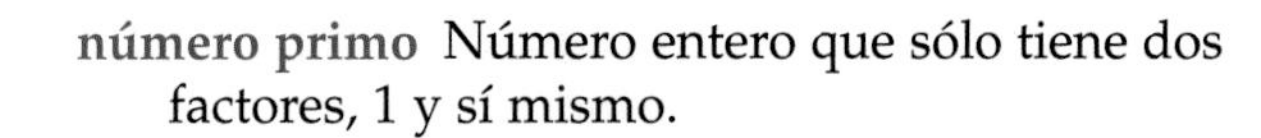

número primo Número entero que sólo tiene dos factores, 1 y sí mismo.

capital Cantidad de dinero en una cuenta.

prisma Poliedro que posee dos bases congruentes y paralelas en forma de polígonos.

prisma rectangular prisma triangular

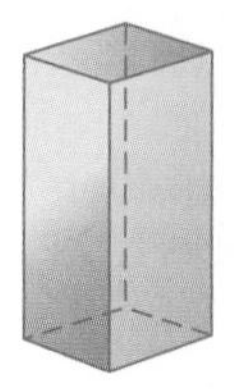
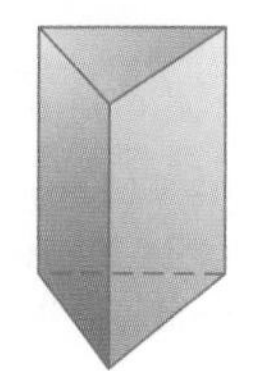

probabilidad La razón del número de maneras en que puede ocurrir el evento al número de resultados posibles.

$$P(\text{evento}) = \frac{\text{número de resultados favorables}}{\text{número de resultados posibles}}$$

propiedades Enunciados que son verdaderos para cualquier número.

proporción Enunciado de la igualdad de dos o más razones.

transportador Instrumento que se usa para medir ángulos.

pirámide Poliedro cuya base es un polígono y cuyos lados son triángulos.

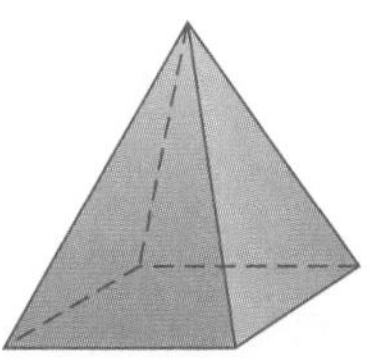

Teorema de Pitágoras Si un triángulo es rectángulo, entonces el cuadrado de la longitud de la hipotenusa es igual a la suma de los cuadrados de las longitudes de los catetos, o $c^2 = a^2 + b^2$.

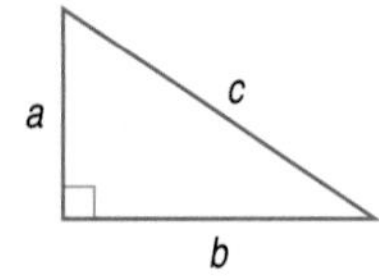

Q

quadrants (86) The four regions into which the x-axis and y-axis separate the coordinate plane.

quadratic function (688) A function that can be described by an equation of the form $y = ax^2 + bx + c$, where $a \neq 0$.

quadrilateral (513) A closed figure with four sides and four vertices, including squares, rectangles, and trapezoids.

cuadrantes Las cuatro regiones en que los ejes x y y dividen el plano de coordenadas.

función cuadrática Función que puede describirse por una ecuación de la forma $y = ax^2 + bx + c$, donde $a \neq 0$.

cuadrilátero Figura cerrada de cuatro lados y cuatro vértices, incluyendo cuadrados, rectángulos y trapecios.

quartiles (613) The values that divide a set of data into four equal parts.

cuartiles Valores que dividen un conjunto de datos en cuatro partes iguales.

R

radical sign (436) The symbol $\sqrt{}$ used to indicate a nonnegative square root.

signo radical El símbolo $\sqrt{}$ que se usa para indicar la raíz cuadrada no negativa.

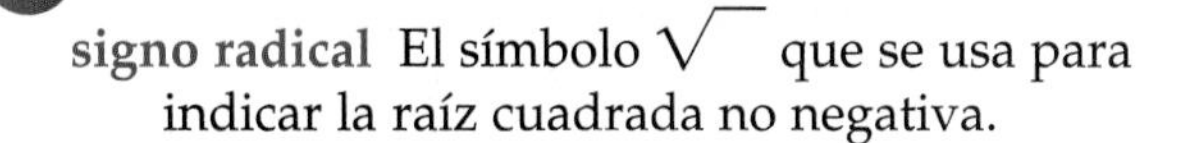

radius (533) The distance from the center to any point on the circle.

radio Distancia del centro a cualquier punto de un círculo.

range (35) The range of a relation is the set of all y-coordinates from each ordered pair.

rango El rango de una relación es el conjunto de coordenadas y de todos los pares.

range (612) A measure of variation that is the difference between the least and greatest values in a set of data.

amplitud Medida de variación que es la diferencia entre los valores máximo y mínimo de un conjunto de datos.

rate (265) A ratio of two measurements having different units.

tasa Razón de dos medidas que tienen unidades distintas.

rate of change (393) A change in one quantity with respect to another quantity.

tasa de cambio Cambio de una cantidad con respecto a otra.

ratio (264) A comparison of two numbers by division. The ratio of 2 to 4 can be stated as 2 out of 4, 2 to 4, 2:4, or $\frac{2}{4}$.

razón Comparación de dos números mediante división. La razón de 2 a 4 puede escribirse como 2 de cada 4, 2 a 4, 2:4 ó $\frac{2}{4}$.

rational number (205) A number that can be written as a fraction in the form $\frac{a}{b}$, where a and b are integers and $b \neq 0$.

número racional Número que puede escribirse como una fracción de la forma $\frac{a}{b}$ donde a y b son enteros y $b \neq 0$.

ray (447) A part of a line that extends indefinitely in one direction.

rayo Parte de una recta que se extiende indefinidamente en una dirección.

real numbers (441) The set of rational numbers together with the set of irrational numbers.

números reales El conjunto de los números racionales junto con el de números irracionales.

reciprocal (215) Another name for a multiplicative inverse.

recíproco Otro nombre del inverso multiplicativo.

rectangle (514) A parallelogram with four right angles.

rectángulo Paralelogramo con cuatro ángulos rectos.

reflection (506) A transformation where a figure is flipped over a line. Also called a flip.

reflexión Transformación en que una figura se voltea a través de una recta.

regular polygon (529) A polygon having all sides congruent and all angles congruent.

polígono regular Polígono cuyos lados son todos congruentes y cuyos ángulos son también todos congruentes.

relation (35) A set of ordered pairs.

relación Conjunto de pares ordenados.

repeating decimal (201) A decimal whose digits repeat in groups of one or more. Examples are 0.181818… and 0.8333… .

decimal periódico Decimal cuyos dígitos se repiten en grupos de uno o más. 0.181818… y 0.8333… son ejemplos de este tipo de decimales.

rhombus (514) A parallelogram with four congruent sides.

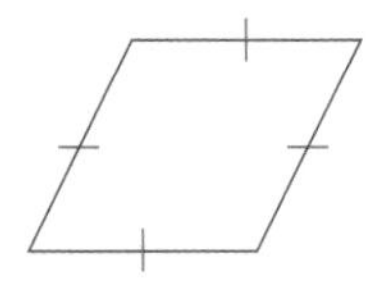

rombo Paralelogramo con cuatro lados congruentes.

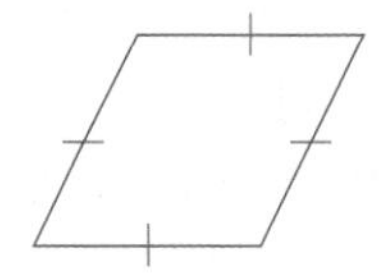

right angle (449) An angle that measures 90°.

ángulo recto Ángulo que mide 90°.

right triangle (454) A triangle with one right angle.

triángulo rectángulo Triángulo que tiene un ángulo recto.

rotation (506) A transformation where a figure is turned around a fixed point. Also called a turn.

rotación Transformación en que una figura se hace girar alrededor de un punto fijo. También se llama vuelta.

S

sample (309) A subgroup or subset of a population used to represent the whole population.

muestra Subgrupo o subconjunto de una población que se usa para representarla.

sample space (311) The set of all possible outcomes.

espacio muestral Conjunto de todos los resultados posibles.

scale (276) The relationship between the measurements on a drawing or model and the measurements of the real object.

escala Relación entre las medidas de un dibujo o modelo y las medidas de la figura verdadera.

scale drawing (276) A drawing that is used to represent an object that is too large or too small to be drawn at actual size.

dibujo a escala Dibujo que se usa para representar una figura que es demasiado grande o pequeña como para ser dibujada de tamaño natural.

scale factor (277) The ratio of a length on a scale drawing or model to the corresponding length on the real object.

factor de escala Razón de la longitud en un dibujo a escala o modelo a la longitud correspondiente en la figura verdadera.

scale model (276) A model used to represent an object that is too large or too small to be built at actual size.

modelo a escala Modelo que se usa para representar una figura que es demasiado grande o pequeña como para ser construida de tamaño natural.

scalene triangle (455) A triangle with no congruent sides.

triángulo escaleno Triángulo que no tiene lados congruentes.

scatter plot (40) A graph that shows the relationship between two sets of data.

gráfica de dispersión Gráfica en que se muestra la relación entre dos conjuntos de datos.

scientific notation (186) A number in scientific notation is expressed as $a \times 10^n$, where $1 \leq a < 10$ and n is an integer. For example, $5{,}000{,}000 = 5.0 \times 10^6$.

notación científica Un número en notación científica se escribe como $a \times 10^n$, donde $1 \leq a < 10$ y n es un entero. Por ejemplo, $5{,}000{,}000 = 5.0 \times 10^6$.

sequence (249) An ordered list of numbers, such as, 0, 1, 2, 3, or 2, 4, 6, 8.

sucesión Lista ordenada de números, como 0, 1, 2, 3 ó 2, 4, 6, 8.

sides (447) The two rays that make up an angle.

lados Los dos rayos que forman un ángulo.

significant digits (590) The digits recorded from measurement, indicating the precision of the measurement.

dígitos significativos Los dígitos de una medición que indican la precisión de la medición.

similar solids (584) Solids that have the same shape but not necessarily the same size.

sólidos semejantes Sólidos que tienen la misma forma, pero no necesariamente el mismo tamaño.

similar triangles (471) Triangles that have the same shape but not necessarily the same size.

triángulos semejantes Triángulos que tienen la misma forma, pero no necesariamente el mismo tamaño.

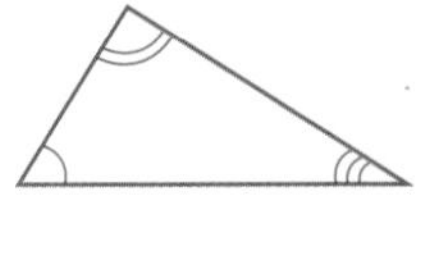
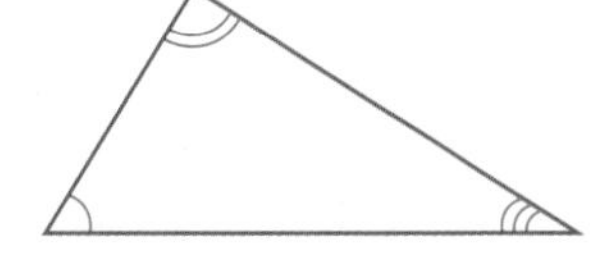
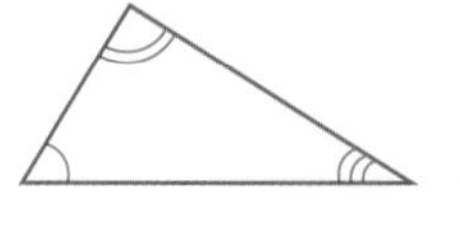
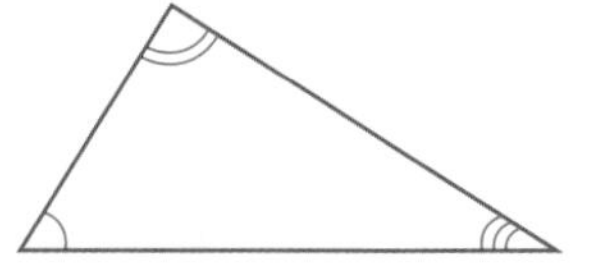

simple event (310) One outcome or a collection of outcomes.

evento simple Resultado o colección de resultados.

simple interest (300) The amount of money paid or earned for the use of money.

$I = prt$ (Interest = principal × rate × time)

interés simple Cantidad que se paga o que se gana por usar el dinero.

$I = crt$ (Interés = capital × rédito × tiempo)

simplest form (104) An algebraic expression in simplest form has no like terms and no parentheses.

forma reducida Una expresión algebraica reducida no tiene ni términos semejantes ni paréntesis.

simplest form (169) A fraction is in simplest form when the GCF of the numerator and the denominator is 1.

forma reducida Una fracción está reducida si el MCD de su numerador y denominador es 1.

simplify (25) To write an expression in a simpler form.

reducir Escribir una expresión en forma más simple.

simulation (656) The process of acting out a situation to see possible outcomes.

simulación Proceso de representación de una situación para averiguar los resultados posibles.

sine (477) If $\triangle ABC$ is a right triangle and A is an acute angle,

$$\text{sine } \angle A = \frac{\text{measure of the leg opposite } \angle A}{\text{measure of the hypotenuse}}.$$

seno Si $\triangle ABC$ es un triángulo rectángulo y A es un ángulo agudo,

$$\text{seno } \angle A = \frac{\text{medida del cateto opuesto a } \angle A}{\text{medida de la hipotenusa}}.$$

skew lines (558) Lines that are neither intersecting nor parallel. Skew lines lie in different planes.

rectas alabeadas Rectas que no se intersecan y que no son paralelas. Las rectas alabeadas yacen en distintos planos.

slant height (578) The length of the altitude of a lateral face of a regular pyramid.

altura oblicua En una pirámide regular, la longitud de la altura de una cara lateral.

slope (387) The ratio of the rise, or vertical change, to the run, or horizontal change. The slope describes the steepness of a line.

$$\text{slope} = \frac{\text{rise}}{\text{run}}$$

pendiente Razón de la elevación o cambio vertical al desplazamiento o cambio horizontal. La pendiente describe la inclinación de una recta.

$$\text{pendiente} = \frac{\text{elevación}}{\text{desplazamiento}}$$

slope-intercept form (398) A linear equation in the form $y = mx + b$, where m is the slope and b is the y-intercept.

forma pendiente-intersección Una ecuación lineal de la forma $y = mx + b$, donde m es la pendiente y b es la intersección y.

solution (28) A value for the variable that makes an equation true. For $x + 7 = 19$, the solution is 12.

solución Valosss y que posee cuatro ángulos rectos.

solving the equation (28) The process of finding a solution to an equation.

resolver la ecuación (28) Proceso de hallar una solución a una ecuación.

solving a right triangle (461) Using the Pythagorean Theorem to find the length of the third side of a right triangle, if the lengths of the other two sides are known.

resolver un triángulo rectángulo Uso del Teorema de Pitágoras para hallar la longitud de un tercer lado de un triángulo rectángulo, si se conocen las longitudes de los otros dos lados.

square (514) A parallelogram with all sides congruent and four right angles.

cuadrado Paralelogramo cuyos lados son todos congruentes y que posee cuatro ángulos rectos.

square root (436) One of the two equal factors of a number. The square root of 25 is 5 since $5^2 = 25$.

raíz cuadrada Uno de los dos factores iguales de un número. Una raíz cuadrada de 25 es 5 porque $5^2 = 25$.

standard form (154) A number is in standard form when it does not contain exponents. The standard form for seven hundred thirty-nine is 739.

forma estándar Un número está en forma estándar si no contiene exponentes. Por ejemplo, la forma estándar de setecientos treinta y nueve es 739.

stem-and-leaf plot (606) A system used to condense a set of data where the greatest place value of the data forms the stem and the next greatest place value forms the leaves.

diagrama de tallo y hojas Sistema que se usa para condensar un conjunto de datos, en que el valor de posición máximo de los datos forma el tallo y el segundo valor de posición máximo forma las hojas.

stems (606) The greatest place value common to all the data values is used for the stem of a stem-and-leaf plot.

tallos Máximo valor de posición común a todos los datos que se usa como el tallo en un diagrama de tallo y hojas.

straight angle (449) An angle with a measure equal to 180°.

ángulo llano Ángulo que mide 180°.

supplementary (494) Two angles are supplementary if the sum of their measures is 180°.

suplementarios Dos ángulos son suplementarios si sus medidas suman 180°.

surface area (573) The sum of the areas of all the surfaces (faces) of a 3-dimensional figure.

área de superficie Suma de las áreas de todas las superficies (caras) de una figura tridimensional.

system of equations (414) A set of equations with the same variables. The solution of the system is the ordered pair that is a solution for all of the equations.

sistema de ecuaciones Conjunto de ecuaciones con las mismas variables. La solución del sistema es el par ordenado que resuelve ambas ecuaciones.

T

tangent (477) If $\triangle ABC$ is a right triangle and A is an acute angle,

$$\text{tangent } \angle A = \frac{\text{measure of the leg opposite } \angle A}{\text{measure of the leg adjacent to } \angle A}$$

tangente Si $\triangle ABC$ es un triángulo rectángulo y A es un ángulo agudo,

$$\text{tangente } \angle A = \frac{\text{medida del cateto opuesto al } \angle A}{\text{medida del cateto adyacente a } \angle A}$$

term (103) When plus or minus signs separate an algebraic expression into parts, each part is a term.

término Cada una de las partes de una expresión algebraica separadas por los signos de adición o sustracción.

term (249) Each number within a sequence is called a term.

término Cada número de una sucesión se llama término.

terminating decimal (200) A decimal whose digits end. Every terminating decimal can be written as a fraction with a denominator of 10, 100, 1000, and so on.

decimal terminal Decimal cuyos dígitos terminan. Todo decimal terminal puede escribirse como una fracción con un denominador de 10, 100, 1000, etc.

theoretical probability (311) What should occur in a probability experiment.

probabilidad teórica Lo que debería ocurrir en un experimento probabilístico.

transformation (506) A movement of a geometric figure.

transformación Desplazamiento de una figura geométrica.

translation (506) A transformation where a figure is slid from one position to another without being turned. Also called a slide.

translación Transformación en que una figura se desliza sin girar, de una posición a otra. También se llama deslizamiento.

transversal (492) A line that intersects two parallel lines to form eight angles.

transversal Recta que interseca dos rectas paralelas formando ocho ángulos.

trapezoid (514) A quadrilateral with exactly one pair of parallel sides.

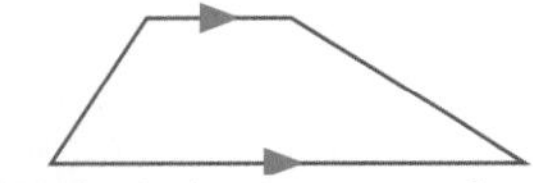

trapecio Cuadrilátero con sólo un par de lados paralelos.

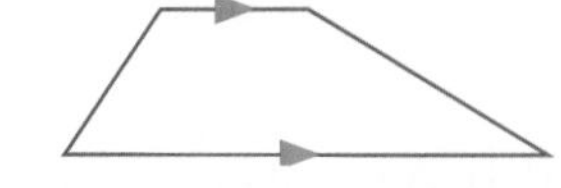

tree diagram (635) A diagram used to show the total number of possible outcomes.

diagrama de árbol Diagrama que se usa para mostrar el número total de resultados posibles.

triangle (453) A polygon having three sides.

triángulo Polígono de tres lados.

trigonometric ratio (477) A ratio of the lengths of two sides of a right triangle. The tangent, sine, and cosine ratios are three trigonometric ratios.

razón trigonométrica Razón de las longitudes de dos lados de un triángulo rectángulo. La tangente, el seno y el coseno son tres razones trigonométricas.

trigonometry (477) The study of the properties of triangles. Trigonometry means *angle measurement*.

trigonometría Estudio de las propiedades de los triángulos. La palabra significa *medida de ángulos*.

trinomial (669) A polynomial with three terms.

trinomio Polinomio de tres términos.

two-step equation (120) An equation that contains two operations.

ecuación de dos pasos Ecuación que contiene dos operaciones.

U

unit rate (265) A rate simplified so that it has a denominator of 1.

tasa unitaria Tasa reducida que tiene denominador igual a 1.

upper quartile (613) The median of the upper half of a set of data, indicated by UQ.

cuartil superior Mediana de la mitad superior de un conjunto de datos, denotada por CS.

V

variable (17) A placeholder for any value.

variable Marcador de posición para cualquier valor.

Venn diagram (164) A diagram that is used to show the relationships among sets of numbers or objects by using overlapping circles in a rectangle.

diagrama de Venn Diagrama que se usa para mostrar las relaciones entre conjuntos de números o elementos mediante círculos, que pueden traslaparse, dentro de un rectángulo.

vertex (447) The common endpoint of the rays forming an angle.

vértice Extremo común de los dos rayos que forman un ángulo.

vertex (453) A vertex of a polygon is a point where two sides of the polygon intersect.

vértice El vértice de un polígono es un punto en que se intersecan dos lados del mismo.

vertex (556) Where three or more planes intersect in a point.

vértice Punto en que se intersecan tres o más planos.

vertical angles (493) Two pairs of opposite angles formed by two intersecting lines. The angles formed are congruent. In the figure, the vertical angles are $\angle 1$ and $\angle 3$, $\angle 2$ and $\angle 4$.

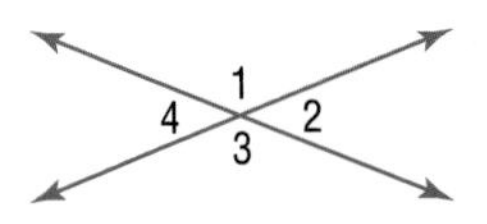

ángulos opuestos por el vértice Dos pares de ángulos opuestos formados por dos rectas que se intersecan. Los ángulos que resultan son congruentes. En la figura, los ángulos opuestos por el vértice son $\angle 1$ y $\angle 3$, $\angle 2$ y $\angle 4$.

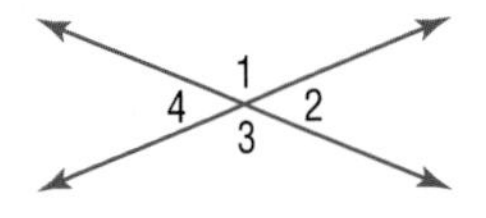

vertical line test (370) If any vertical line drawn on the graph of a relation passes through no more than one point on the graph for each value of x in the domain, then the relation is a function.

prueba de la recta vertical Si todas las rectas verticales trazadas en la gráfica de una relación no pasan por más un punto para cada valor de x en el dominio, entonces la relación es una función.

volume (563) The measure of space occupied by a solid region.

volumen Medida del espacio que ocupa un sólido.

X

***x*-axis** (33) The horizontal number line which helps to form the coordinate system.

eje *x* Recta numérica horizontal que forma parte de un sistema de coordenadas.

x-coordinate (33) The first number of an ordered pair.

coordenada x El primer número de un par ordenado.

x-intercept (381) The x-coordinate of a point where a graph crosses the x-axis.

intersección x La coordenada x de un punto en que una gráfica interseca el eje x.

Y

y-axis (33) The vertical number line which helps to form the coordinate system.

eje y Recta numérica vertical que forma parte de un sistema de coordenadas.

y-coordinate (33) The second number of an ordered pair.

coordenada y El segundo número en un par ordenado.

y-intercept (381) The y-coordinate of a point where a graph crosses the y-axis.

intersección y La coordenada y de un punto en que una gráfica interseca el eje y.

Selected Answers

Chapter 1 The Tools of Algebra

Page 5 Chapter 1 Getting Started
1. 14.8 **3.** 3.1 **5.** 2.95 **7.** 3.55 **9.** 7.88 **11.** Sample answer: 1200 **13.** Sample answer: 120 **15.** Sample answer: 20,000 **17.** Sample answer: 220 **19.** Sample answer: 14 **21.** Sample answer: \$5 **23.** Sample answer: 120 **25.** Sample answer: 4 **27.** Sample answer: 10

Pages 9–10 Lesson 1-1
1. when an exact answer is not needed **3.** 1:33 P.M. **5.** 17 **7.** 3072 **9.** 178 beats per min **11.** 17 **13.** 25 **15.** 34 **17.** 27 **19.**

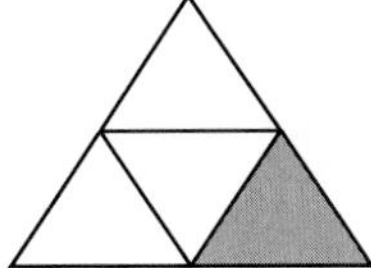

21. Since \$68 + \$15 + \$20 + \$16 = \$119, Ryan does not have enough money for the ski trip. **23.** about 5 h **25.** Sample answer: about 21,800 transplants **27a.** There are more even products. Since any even number multiplied by any number is even, and only an odd number multiplied by an odd number is odd, there are more even products in the table. There are about 3 times as many evens as odds. **27b.** Yes; in the addition table, there is only one more even number than odd. **29.** B **31.** 3 **33.** 35 **35.** 109

Pages 14–16 Lesson 1-2
1. Sample answer: $(8 - 3) \cdot 2$ **3.** Emily; she followed the order of operations and divided first. **5.** ÷; 20 **7.** −; 66 **9.** ×; 15 **11.** 12 − 9 **13.** 4 **15.** 25 **17.** 38 **19.** 2 **21.** 55 **23.** 24 **25.** 64 **27.** 180 **29.** 50 **31.** 6 − 3 **33.** 9 × 5 **35.** 24 ÷ 6 **37.** 3 × \$6 **39.** $(4 \times 2) + (2 \times 13)$ **41.** $(3 \times 57) + (2 \times 12)$ **43.** $61 - (15 + 3) = 43$ **45.** $56 \div (2 + 6) - 4 = 3$ **47.** $(50 \times 25) + (7 \times 24) + (4 \times 22) + (3 \times 16)$ **49.** 0-07-825200-8 **51.** Sample answer: $111 - (1 + 1 + 1) \times (11 + 1)$ **53.** C **55.** 64 **57.** 28 **59.** \$275 **61.** Sample answer: about 26 compact cars **63.** 126 **65.** 563

Pages 19–21 Lesson 1-3
1. Sample answer: $7n$ and $3x - 1$; $2 + 3$ and 3×8 **3.** Sample answer: $4 \times c \times d$ **5.** 6 **7.** 17 **9.** $g - 5$ **11.** $7 + n \div 8$ **13.** 11 **15.** 38 **17.** 2 **19.** 9 **21.** 27 **23.** 56 **25.** 53 **27.** 44 **29.** 32 **31.** 71°F **33.** $s + \$200$ **35.** $h - 6$ **37.** $5q - 4$ **39.** $n \div 6 + 9$ **41.** $17 - 4w$ **43.** 10 **45.** $x + 3$ **47.** $p - 4$ **49.** $s = c + m - d$ **51.** 1 **53.** D **55.** 7 **57.** 9 **59.** 36 **61.** 22

Page 21 Practice Quiz 1
1. 14 **3.** 79 **5.** 22

Pages 26–27 Lesson 1-4
1. Sample answer: $3 \cdot 4 = 4 \cdot 3$ **3.** Kimberly; the Associative Property only holds true if all numbers are added or all numbers are multiplied, not a combination of the two. **5.** Additive Identity **7.** 28 **9.** 45 **11.** $n + 13$ **13.** \$42; To find the total cost, add the three costs together. Since the order in which the costs are added does not matter, the Commutative Property of Addition holds true and makes the addition easier. By adding 4 and 26, the result is 30, and 30 + 12 is 42. **15.** Multiplicative Identity **17.** Commutative Property of Multiplication **19.** Associative Property of Addition **21.** Additive Identity **23.** Associative Property of Multiplication **25.** Commutative Property of Addition **27.** 55 **29.** 40 **31.** 990 **33.** 0 **35.** false; $(100 \div 10) \div 2 \neq 100 \div (10 \div 2)$ **37.** false; $9 - 3 \neq 3 - 9$ **39.** $m + 12$ **41.** a + 27 **43.** $12y$ **45.** $48c$ **47.** $75s$
49. There are many real-life situations in which the order in which things are completed does not matter. Answers should include the following.

- Reading the sports page and then the comics, or reading the comics and then the sports page. No matter the order, both parts of the newspaper will be read.
- When washing clothes, you would add the detergent and then wash the clothes, not wash the clothes and then add the detergent. Order matters.

51. B **53.** 36 **55.** $w - 12$ **57.** 35 **59.** 15, 21 **61.** 296 **63.** 1050 **65.** 7493

Pages 30–32 Lesson 1-5
1. Sample answer: $b + 7 = 12$ and $8 - h = 3$ **3.** 6 **5.** 5 **7.** 6 **9.** Symmetric **11.** Let n = the number; $n + 8 = 23$; 15 **13.** C **15.** 11 **17.** 12 **19.** 5 **21.** 15 **23.** 9 **25.** always **27.** 15 **29.** 0 **31.** 15 **33.** 17 **35.** 11 **37.** 9 **39.** 3 **41.** 4 **43.** Let h = the number; $h - 10 = 27$; 37 **45.** Let w = the number; $9 + w = 36$; 27 **47.** Let x = the number; $3x = 45$; 15 **49.** \$8 **51.** Symmetric Property of Equality **53.** Symmetric Property of Equality **55.** 3 **57.** Sample answer: Once the variable(s) are replaced in the open sentence, the order of operations is used to find the value of the expression. Answers should include the following.

- To evaluate an expression, replace the variable(s) with the given values, and then find the value of the expression.
- To solve an open sentence, find the value of the variable that makes the sentence true.

59. B **61.** $23 + d$ **63.** $10 - n$ **65.** 11 **67.** 42 **69.** 18 **71.** 50 **73.** 90

Page 32 Practice Quiz 2
1. Identity (×) **3.** $24h$ **5.** 8

Pages 36–38 Lesson 1-6
1. Sample answer: (3, 5); the x-coordinate is 3 and the y-coordinate is 5. **3.** The domain of a relation is the set of x-coordinates. The range is the set of y-coordinates.
5.

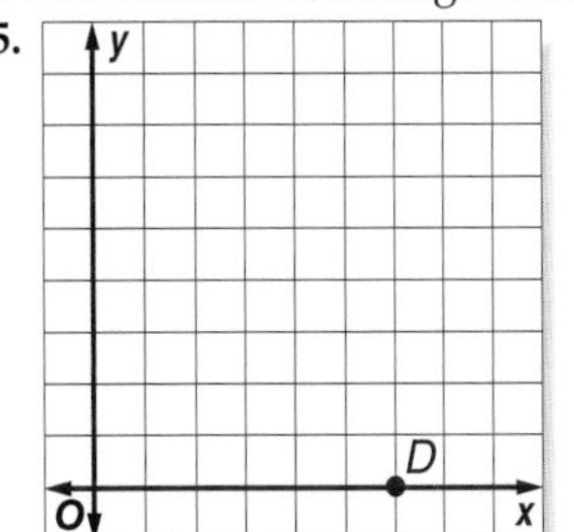

7. (6, 5)

9.

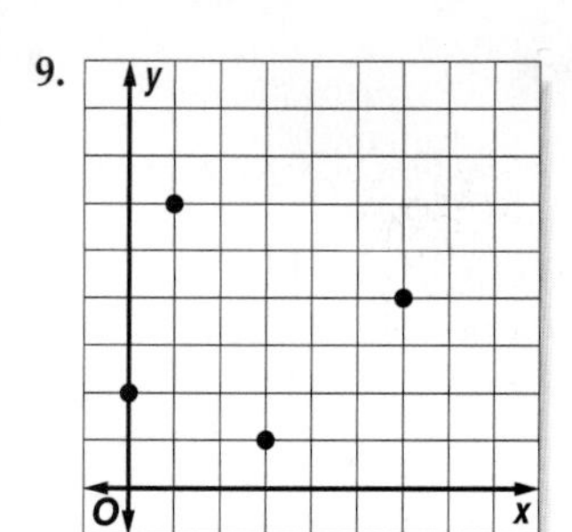

x	y
1	6
6	4
0	2
3	1

domain = {1, 6, 0, 3};
range = {6, 4, 2, 1}

11.

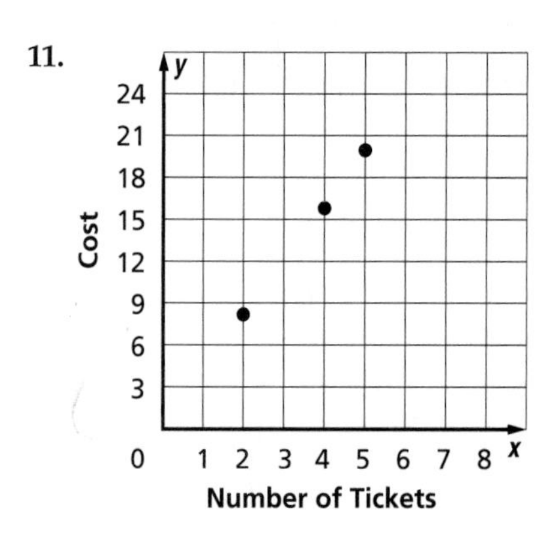

13.

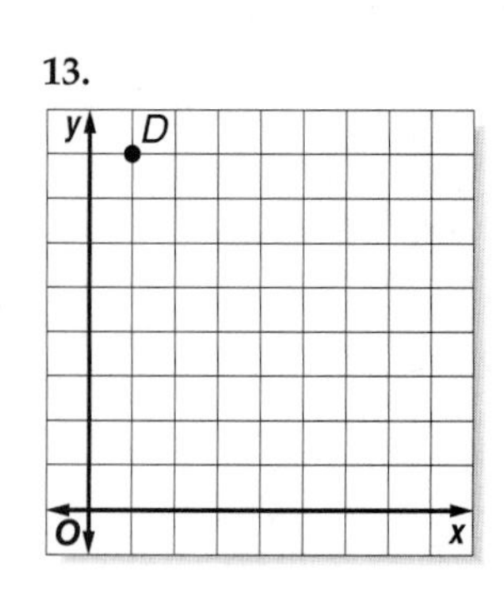

15.

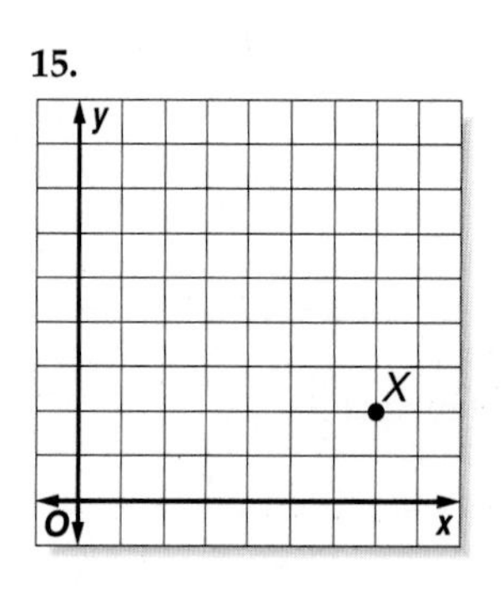

17.

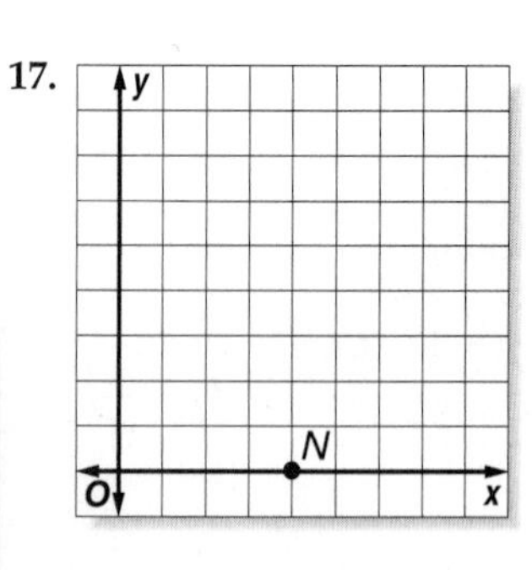

19. (7, 3) **21.** (6, 6)
23. (3, 4) **25.** on the x-axis; on the y-axis

27. **Speed of a House Mouse**

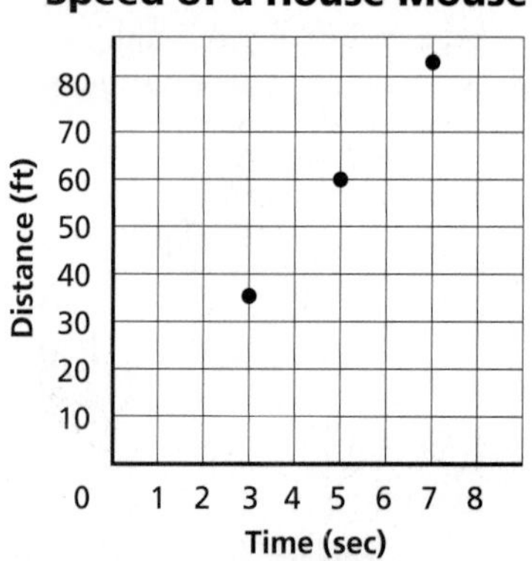

29. **Science Experiment**

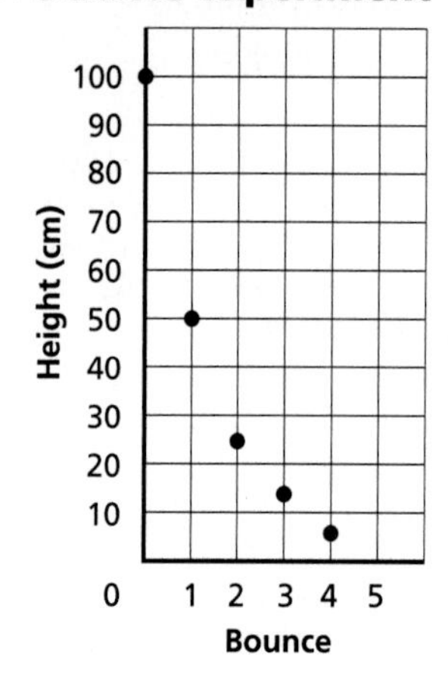

31.

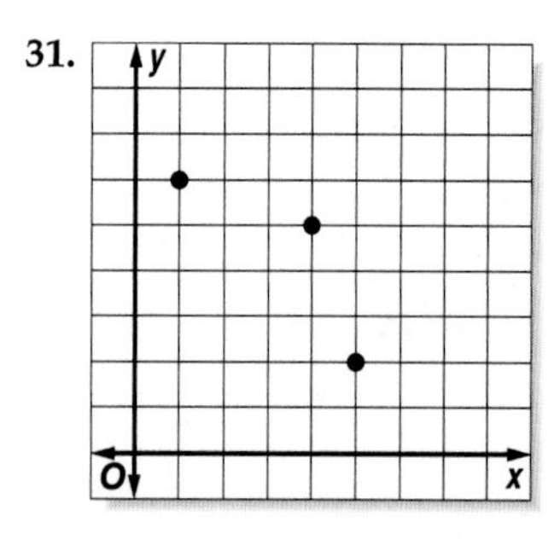

x	y
4	5
5	2
1	6

domain = {4, 5, 1};
range = {5, 2, 6}

33.

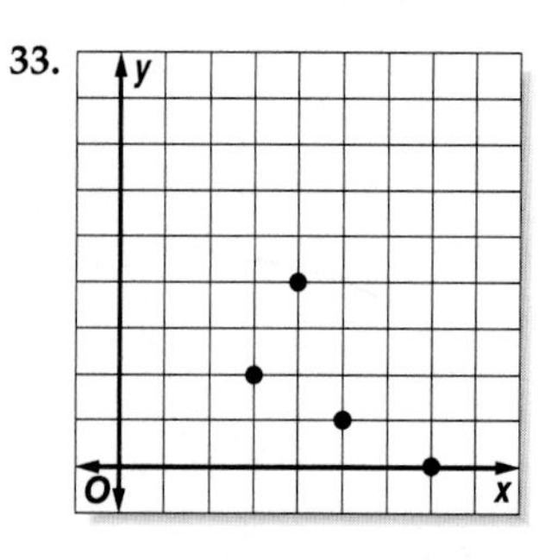

x	y
7	0
3	2
4	4
5	1

domain = {7, 3, 4, 5};
range = {0, 2, 4, 1}

35.

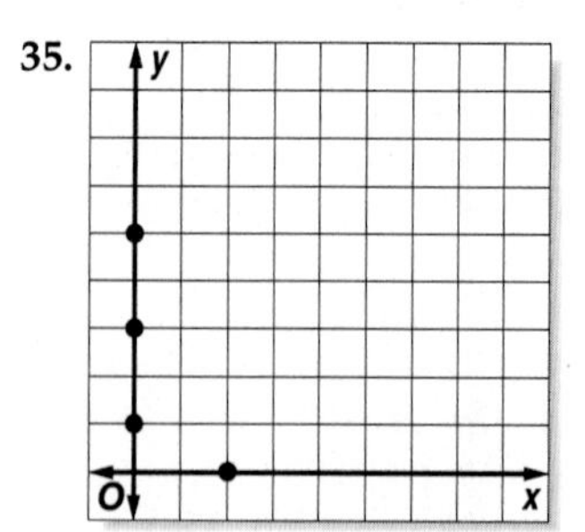

x	y
0	1
0	3
0	5
2	0

domain = {0, 2};
range = {1, 3, 5, 0}

37. (0, 14.7), (1, 10.2), (2, 6.4), (3, 4.3), (4, 2.7), (5, 1.6) **39.** domain = {0, 1, 2, 3, 4, 5}; range = {14.7, 10.2, 6.4, 4.3, 2.7, 1.6} **41.** {(0, 100), (1, 95), (2, 90), (3, 85), (4, 80), (5, 75)} **43.** about 93°C; about 96°C **45.** Ordered pairs can be used to graph real-life data by expressing the data as ordered pairs and then graphing the ordered pairs. Answers should include the following.

- The x- and y-coordinate of an ordered pair specifies the point on the graph.
- longitude and latitude lines.

47. D **49a.**

49b. triangle **49c.** (4, 2), (4, 8), (10, 2) **49d.** triangle **49e.** The figures have the same shape but not the same size.

51. 6 **53.** Multiplicative Identity **55.** 7 **57.** $10 \cdot 30$ **59.** 12 **61.** 28 **63.** 7 **65.** 9

Pages 42–44 Lesson 1-7

1. Sample answer: make predictions, draw conclusions, spot trends **3.** negative, positive, and none **5.** No; hair color is not related to height. **7.** Since the points appear to be random, there is no relationship. **9.** The number of songs on a CD usually does not affect the cost of the CD; no. **11.** As speed increases, distance traveled increases; positive. **13.** The size of a television screen and the number of channels it receives are not related; no. **15.** The number decreases.

17.

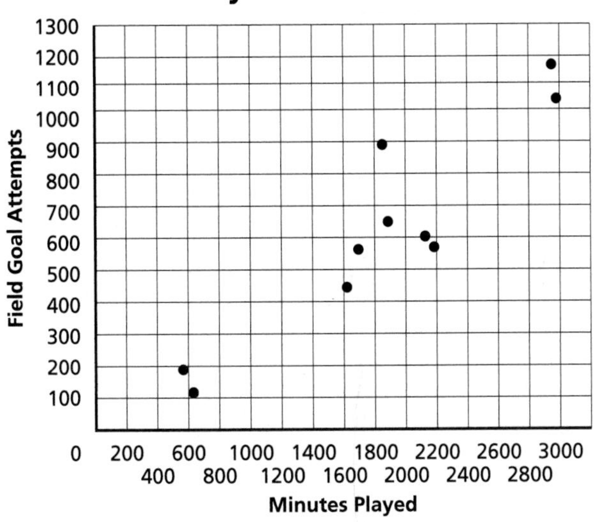

19. about 800 **21.** Sample answer: Yes; as more emphasis is placed on standardized tests, students will become more comfortable taking the tests, and the scores will increase. **23.** C

25.

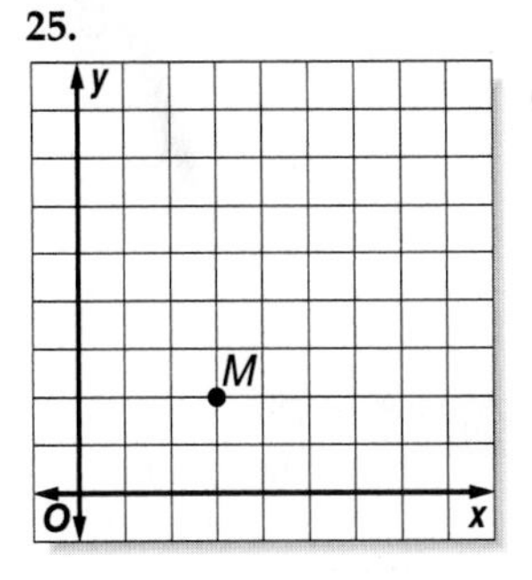

27.

29. (0, 4) **31.** domain = {0, 4, 2, 6}; range = {9, 8, 3, 1} **33.** 7 **35.** $b + 18$ **37.** 31

Pages 47–50 Chapter 1 Study Guide and Review

1. d **3.** e **5.** c **7.** 20 **9.** 22 **11.** 22 **13.** 16 **15.** 12 **17.** 14 **19.** 10 **21.** 25 **23.** Commutative Property of Addition **25.** Multiplicative Property of Zero **27.** 10 **29.** 17 **31.** 6

33.

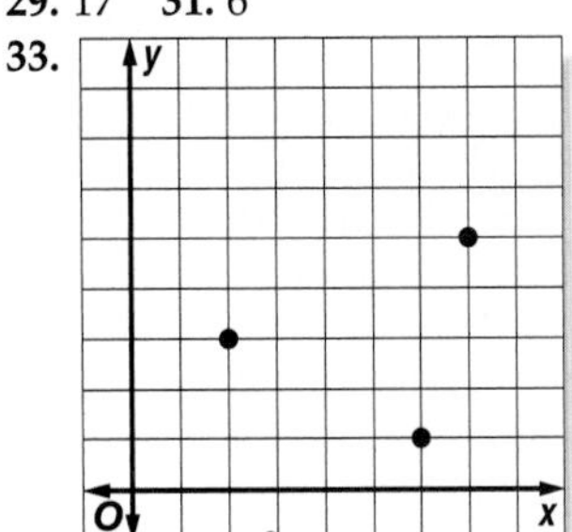

x	y
2	3
6	1
7	5

domain: {2, 6, 7}; range: {3, 1, 5}

35. Positive; as the height increases, the circumference increases.

Chapter 2 Integers

Page 55 Chapter 2 Getting Started

1. 22 **3.** 42 **5.** 8 **7.** 4 **9.** T **11.** V **13.** Q

Pages 59–61 Lesson 2-1

1. Draw a number line. Draw a dot at −4. **3.** The absolute value of a number is its distance from 0 on a number line.

5. +15

8 9 10 11 12 13 14 15 16

7. $-4 < 2$; $2 > -4$ **9.** < **11.** > **13.** 10 **15.** 21 **17.** 3

19. −54, −52, −45, −37, −36, −34, −27, −27, −2

21. −6

−8 −7 −6 −5 −4 −3 −2 −1 0

23. +9

5 6 7 8 9 10 11 12 13

25. −5

−8 −7 −6 −5 −4 −3 −2 −1 0

27.

−3 −2 −1 0 1 2 3 4 5

29.

−8 −6 −4 −2 0 2 4 6 8

31. $-5 > -10$; $-10 < -5$ **33.** $248 < 425$; $425 > 248$ **35.** $212 > 32$; $32 < 212$ **37.** > **39.** > **41.** > **43.** > **45.** {−15, −4, −2, −1} **47.** {−60, −57, 38, 98, 188} **49.** 46 **51.** −5 **53.** 7 **55.** 2 **57.** 9 **59.** −20 **61.** 4 **63.** 40 **65.** 3

67.

−54

−90 −70 −50 −30 −10

69. $-54 > -70$ **71.** 7 **73.** Sometimes; if A and B are both positive, both negative, or one is 0, it is true. If one number is negative and the other is positive, it is false. **75.** B **77.** Positive; as height increases, so does arm length.

79.

x	y
3	2
3	4
2	1
2	4

{(3, 2), (3, 4), (2, 1), (2, 4)}

81. Commutative Property of Multiplication **83.** Commutative Property of Multiplication **85.** 388 **87.** 17 **89.** 1049

Selected Answers

Pages 67–68 Lesson 2-2

1a. Negative; both addends are negative. **1b.** Positive; $|12| > |-2|$. **1c.** Negative; $|-11| > |9|$. **1d.** Positive; both addends are positive. **3.** −6 **5.** 5 **7.** 3 **9.** 4 **11.** $4 + (-5) = -1$ **13.** −7 **15.** −11 **17.** −16 **19.** −21 **21.** −66 **23.** 2 **25.** −2 **27.** 6 **29.** −26 **31.** 21 **33.** −2 **35.** −6 **37.** −3 **39.** 0 **41.** −5 **43.** 8 **45.** 40 **47.** +107,680

49. To add integers on a number line, start at 0. Move right to show positive integers and left to show negative integers. Answers should include the following.

- Sample answer:

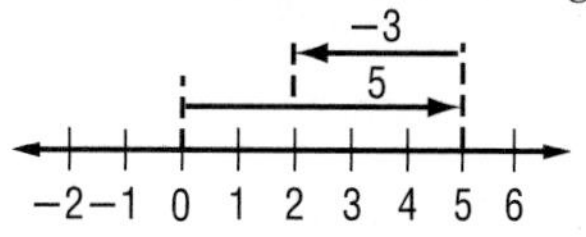

- Sample answer:

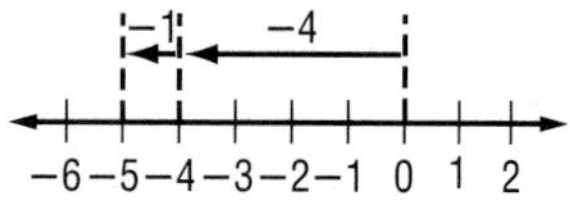

51. D **53.** {−12, −9, −8, 0, 3, 14} **55.** no relationship **57.** 6 **59.** 20 **61.** 25 **63.** 42 **65.** 65

Pages 72–74 Lesson 2-3

1. Sample answer: 5, −5; −9, 9 **3.** −3 **5.** 9 **7.** −2 **9.** 20 **11.** 21 **13.** Utah, Washington, Wisconsin, or Wyoming **15.** −1 **17.** −3 **19.** −9 **21.** −12 **23.** 10 **25.** 12 **27.** 3 **29.** −9 **31.** −14 **33.** −28 **35.** 239 **37.** 1300 **39.** 14,776 ft **41.** 24 **43.** −36 **45.** −9 **47.** −20 **49.** −23 **51.** −17 **53.** −10,822 **55a.** False; $3 - 4 \neq 4 - 3$ **55b.** False; $(5 - 2) - 1 \neq 5 - (2 - 1)$ **57.** A **59.** −2450 **61.** 3 **63.** 7 **65.** $\frac{x}{5}$ **67.** $\frac{86}{b}$ **69.** 20 **71.** 75 **73.** 120

Page 74 Practice Quiz 1

1. −80, −70, −69 **3.** 6 **5.** −7 **7.** 32 **9.** −9

Pages 77–79 Lesson 2-4

1. $3(-5) = -15$ **3.** Sample answer: $(-4)(9)(2)$ **5.** −40 **7.** 28 **9.** −540 **11.** $-21y$ **13.** 120 **15.** A **17.** −42 **19.** −72 **21.** −70 **23.** 128 **25.** 45 **27.** 130 **29.** −308 **31.** 528 **33.** −1344 **35.** −56°F **37.** $-96y$ **39.** $-55b$ **41.** $108mn$ **43.** $-135xy$ **45.** $-88bc$ **47.** $-90jk$ **49.** −99 **51.** 80 **53.** −216 **55.** 248 ft **57a.** True; $3(-5) = -5(3)$ **57b.** True; $-2(3 \cdot 5) = (-2 \cdot 3)(5)$ **59.** B **61.** −16 **63.** 4 **65.** 10 **67.** 126°F **69.** −14 **71.** (6, 2) **73.** (1, 5) **75.** (5, 5) **77.** 480 **79.** 550 **81.** 6 **83.** 15 **85.** 4

Pages 83–84 Lesson 2-5

1. Sample answer: $-16 \div 4 = -4$ **3.** 11 **5.** −3 **7.** −10 **9.** −13 **11.** 0 **13.** 9 **15.** 8 **17.** 10 **19.** −50 **21.** −11 **23.** −11 **25.** 19 **27.** −12 **29.** −13 **31.** 16 **33.** 49 points **35.** 61 **37.** Sample answer: $x = -144$; $y = 12$; $z = -12$

39. When the signs of the integers are the same, both a product and a quotient are positive; when the signs are different, the product and quotient are negative. Answers should include the following.

- Sample answer: $4 \cdot (-6) = -24$ and $-24 \div 4 = -6$; $-3 \cdot 2 = -6$ and $-6 \div (-3) = 2$
- Sample answers: same sign: $-30 \div (-5) = 6$, $30 \div 5 = 6$; different signs: $-24 \div 8 = -3$, $24 \div (-8) = -3$

41. B **43.** −39 **45.** $-50cd$ **47.** B **49.** D

Page 84 Practice Quiz 2

1. −84 **3.** −126 **5.** −31 **7.** −25 **9.** $-20xy$

Pages 87–89 Lesson 2-6

1. Sample answer: (3, 6) represents a point 3 units to the right and 6 units up from the origin. (6, 3) represents a point 6 units to the right and 3 units up from the origin. **3.** Keisha; a point in Quadrant I has two positive coordinates. Interchanging the coordinates will still result in two positive coordinates, and the point will be in Quadrant I. **5.** (1, 3) **7.** (5, −4) **9.** II **11.** III **13.** (−2, 4) **15.** (4, −2) **17.** (2, 2) **19.** (0, −2) **21.** (−3, −5)

23–34.

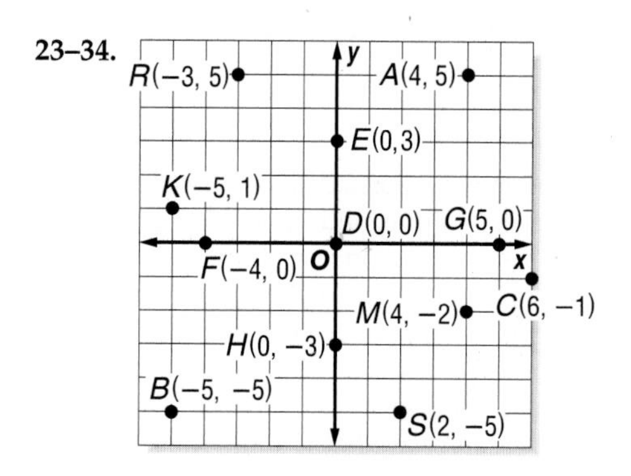

23. I **25.** IV **27.** IV **29.** none **31.** none **33.** none

35. Sample answer:

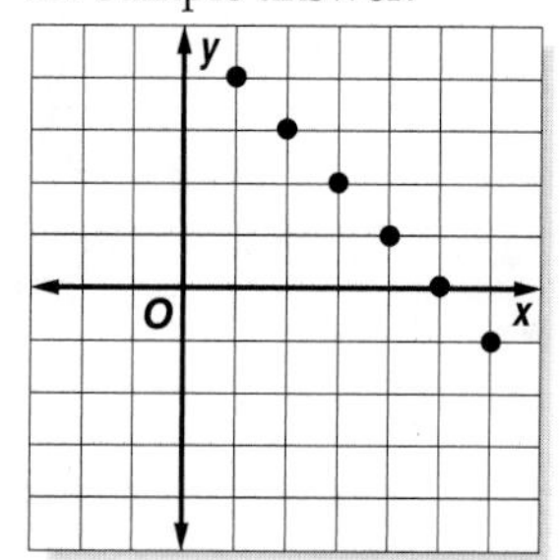

x	y
1	4
2	3
3	2
4	1
5	0
6	−1

The points are along a line slanting down to the right, crossing the y-axis at 5 and the x-axis at 5.

37. Sample answer:

x	y
−2	−4
−1	−2
0	0
1	2
2	4
3	6

The points are along a line slanting up, through the origin.

39. Sample answer:

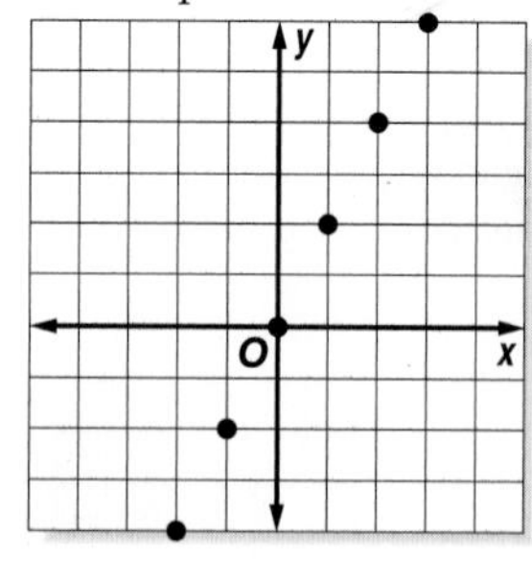

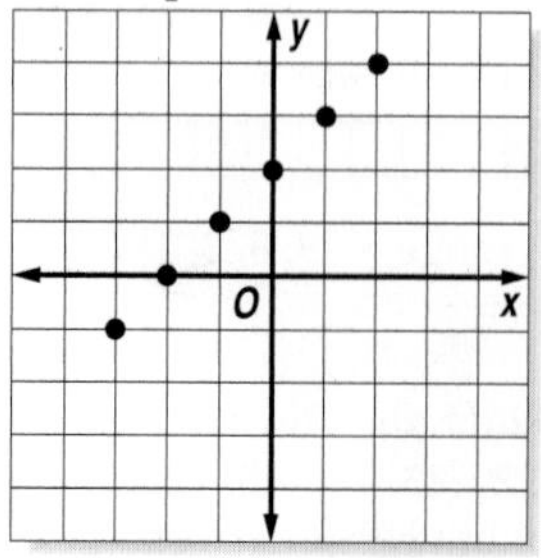

x	y
−3	−1
−2	0
−1	1
0	2
1	3
2	4

The points are along a line slanting up, crossing the y-axis at 2 and the x-axis at −2.

41. 5-point star

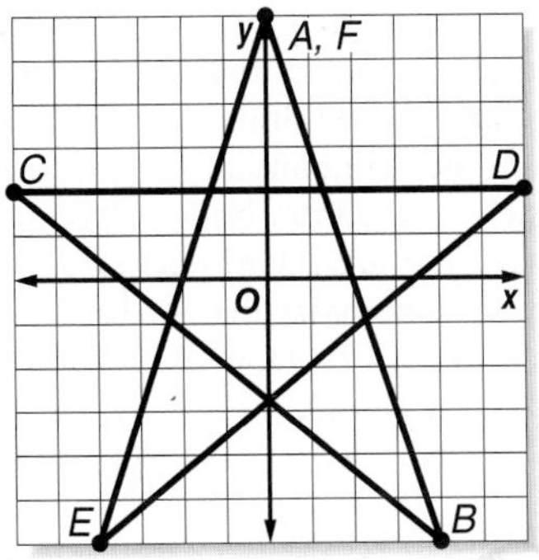

43. Sample answer:

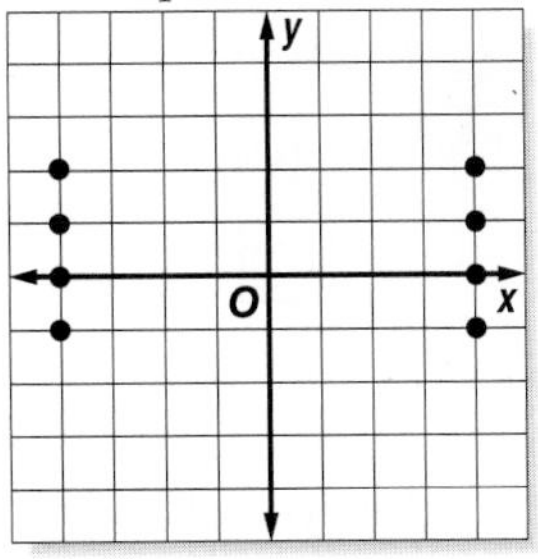

The graph can include any integer pairs where $x > 3$ or $x < -3$.

45. The new triangle is twice the size of the original triangle, and is moved to the right and up.

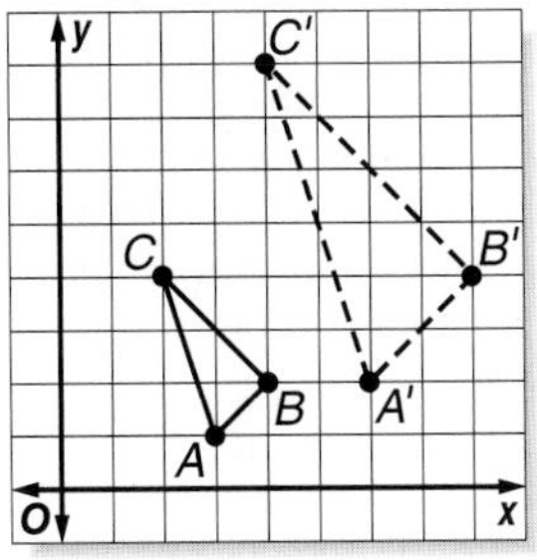

47. The new triangle is translated right 2 units and up 2 units; it is the same size as the original triangle.

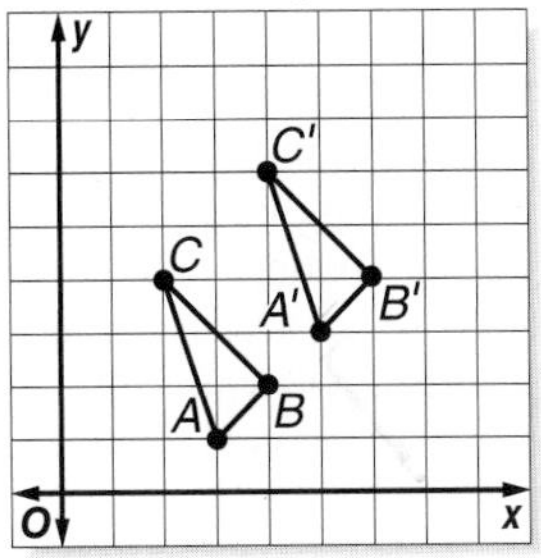

51. Sample answer:

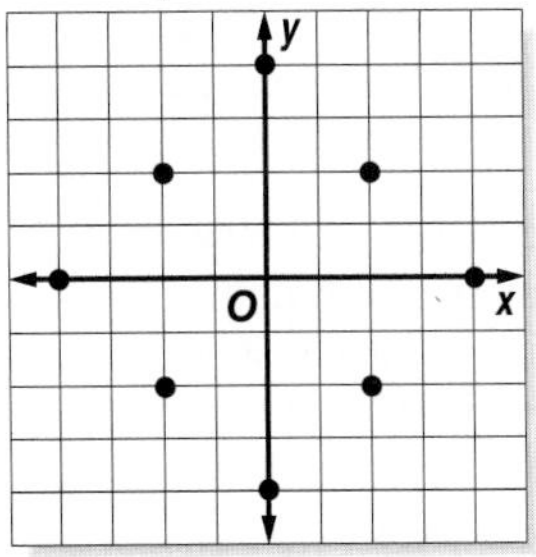

The points lie outside a rhombus defined by (0, 4), (4, 0), (0, −4), and (−4, 0).

53. D **55.** −3 **57.** 8 **59.** −96 **61.** 13°F **63.** $24h$ **65.** $45b$ **67.** 0

Pages 90–92 Chapter 2 Study Guide and Review

1. negative number **3.** coordinate **5.** integers **7.** inequality **9.** = **11.** > **13.** 25 **15.** 22 **17.** −5 **19.** −4 **21.** −10 **23.** −8 **25.** 5 **27.** −4 **29.** 9 **31.** −66 **33.** 48 **35.** 7 **37.** −4 **39.** 2

40–43.

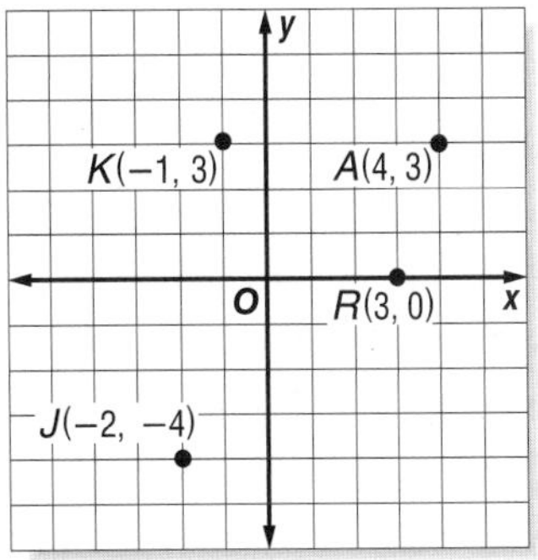

41. III **43.** None

Chapter 3 Equations

Page 97 Chapter 3 Getting Started

1. −6 **3.** 10 **5.** 5 + (−7) **7.** −5 + (−9) **9.** −3 **11.** 0 **13.** $5 + 2n$ **15.** $n - 3$

Pages 100–102 Lesson 3-1

1. Sample answer: 2(3 + 4) = 2 · 3 + 2 · 4 **3.** 5 · 7 + 5 · 8, 75 **5.** 2 · 6 + 4 · 6, 36 **7.** $3n + 6$ **9.** $-6x + 30$ **11.** \$56.25 **13.** 5 · 7 + 5 · 3, 50 **15.** 4 · 3 + 3 · 3, 21 **17.** 8 · 2 + 8 · 2, 32 **19.** 6 · 8 + 6(−5), 18 **21.** −3 · 9 + (−3)(−2), −21 **23.** 10(−5) − 3(−5), −35 **25.** 12(\$15 + \$10 + \$8), 12(\$15) + 12(\$10) + 12(\$8); \$396 **27.** $5y + 30$ **29.** $7y + 56$ **31.** $10y + 20$ **33.** $10 + 5x$ **35.** $9m - 18$ **37.** $15s - 45$ **39.** $12x - 36$ **41.** $2w - 20$ **43.** $-5a - 50$ **45.** $-5w + 40$ **47.** $-5a + 30$ **49.** $3a + 3b$ **51.** \$488.75 **53.** No; 3 + (4 · 5) = 23, (3 + 4)(3 + 5) = 56 **55.** C **57.** 8(20 + 3) = 184 **59.** 16(10 + 1) = 176 **61.** 9(100 + 3) = 927 **63.** 12(1000 + 4) = 12,048 **65.** 6 **67.** 4 **69.** 21, 25, 29 **71.** 80, 160, 320 **73.** −8 + (−4) **75.** 3 + (−9) **77.** −7 + (−10)

Pages 105–107 Lesson 3-2

1. terms that contain the same variable or are constants **3.** Koko; $5x + x = 6x$, not $5x$. **5.** terms: $2m$, $-1n$, $6m$; like terms: $2m$, $6m$; coefficients: 2, −1, 6: constant: none **7.** $8a$ **9.** $7c + 12$ **11.** $9y$ **13.** $-3y - 16$ **15.** $4x + 12y$ **17.** terms: 3, $7x$, $3x$, x; like terms: $7x$, $3x$, x; coefficients: 7, 3, 1; constant: 3 **19.** terms $2a$, 5c, $-1a$, $6a$; like terms: $2a$, $-1a$, $6a$; coefficients: 2, 5, −1, 6; constant: none **21.** terms: $6m$, $-2n$, 7; like terms: none; coefficients: 6, −2; constant: 7 **23.** $7x$ **25.** $11y$ **27.** $7a + 3$ **29.** $7y + 9$ **31.** $2x$ **33.** $-y$ **35.** $-4x + 8$ **37.** $8y$ **39.** $-x + 12$ **41.** $5b + 6$ **43.** $-4a - 6$ **45.** −8 **47.** $16m + 2n$ **49.** $-9c + 2d$ **51.** $3s + 80$ **53.** $5d - 2$ **55.** $6x + 2$ **57a.** Distributive Property **57b.** Commutative Property **57c.** Substitution Property of Equality **57d.** Distributive Property **59.** C **61.** $-2y - 16$ **63.** III **65.** 17 **67.** 2 **69.** −11 **71.** −5 **73.** −13

Page 107 Practice Quiz 1

1. $6x + 12$ **3.** $7y - 4$ **5.** $2m + 15$

Pages 113–114 Lesson 3-3

1. Addition Property of Equality **3.** 11 **5.** −4 **7.** 55

9.

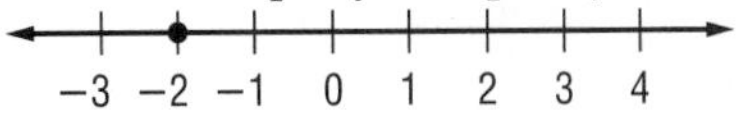

11. C **13.** 13 **15.** −8 **17.** −15 **19.** −1 **21.** 15 **23.** 24 **25.** 36 **27.** −4 **29.** −31 **31.** 118 **33.** $n + 9 = -2$; −11 **35.** $n - 3 = -6$; −3

37.

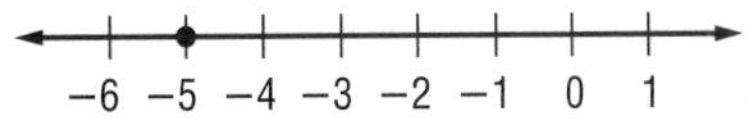

Selected Answers

39. 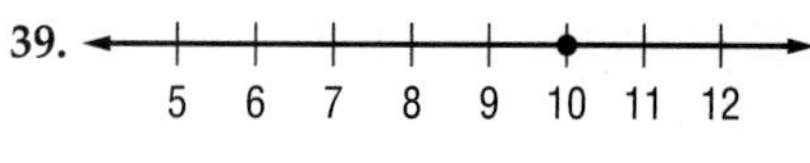

41. −4 −3 −2 −1 0 1 2 3

43. $12 = x - 20$; 32 **47.** 17 million **49.** When you solve an equation, you perform the same operation on each side so that the two sides remain equal. Answers should include the following.
- In an equation, both sides are equal. In a balance scale, the weight of the items on both sides are equal.
- The Addition and Subtraction Properties of Equality allow you to add or subtract the same number from each side of an equation. The two sides of the equation remain equal.

51. C **53.** $3t + 12$ **55.** $-4z + 4$ **57.** $-4m - 1$
59. Additive Inverse Property **61.** 84 **63.** −25 **65.** −9
67. −4

Pages 117–119 Lesson 3-4
1. Multiplication Property of Equality **3.** Sample answer: $-5x = -20$ **5.** −5 **7.** 27 **9.** 66 **11.** 7 **13.** −8 **15.** 8 **17.** 24 **19.** 14 **21.** −33 **23.** 9 **25.** −43 **27.** −135 **29.** 130 **31.** 29 **33.** −168 **35.** $6x = -42$; −7
37. $\frac{x}{-4} = 8$; −32

39. −9 −8 −7 −6 −5 −4 −3 −2

41. −3 −2 −1 0 1 2 3 4

43. 30 31 32 33 34 35 36 37

45. $12{,}000 = 5x$; 2400 mi^2 **47.** $6p = 24$, 4 painters
49a. True; one pyramid balances two cubes, so this is the same as adding one cube to each side. **49b.** True; one pyramid and one cube balance three cubes, which balance one cylinder. **49c.** False; one cylinder and one pyramid balance five cubes. **51.** B **53.** 13 **55.** −28 **57.** $7y + 6$ **59.** −36 **61.** 10 **63.** −2 **65.** 2 **67.** −19 **69.** −27

Pages 122–124 Lesson 3-5
1. You undo the operations in reverse order. **3.** 8 **5.** −2 **7.** −40 **9.** −4 **11.** 10 **13.** 2 **15.** 4 **17.** 8 **19.** 13 **21.** 3 **23.** 28 **25.** 64 **27.** 65 **29.** 21 **31.** 11 **33.** 30 **35.** 33 **37.** −13 **39.** 5 **41.** 5 **43.** -2 **45.** 10 **47.** 3 h **49.** 131 bikes **51.** $5x - 2 = 8$ **53.** C **55.** −7 **57.** −2 **59.** −2 **61.** $-5y - 15$ **63.** $-9y + 36$ **65.** $-8r + 40$ **67.** (2, −3) **69.** (−3, −4) **71.** $x \div 15$ **73.** $2x + 10$

Pages 128–130 Lesson 3-6
1. is, equals, is equal to **3.** Ben; *Three less than* means that three is subtracted *from* a number. **5.** $2n - 4 = -2$, 1 **7.** $2x + 5 = 37$, 21 yr **9.** $3n + 20 = -4$, −8 **11.** $10n - 8 = 82$, 9 **13.** $\frac{n}{-4} - 8 = -42$, 136 **15.** $3n - 8 = -2$, 2 **17.** $17 - 2n = 5$, 6 **19.** $4n + 3n + 5 = 47$, 6 **21.** $8 - 5x = -7$, 3 h **23.** $2x + 2 = 12$. 5 million people **25.** Sample answer: By 2020, Texas is expected to have 10 thousand more people age 85 or older than New York will have. Together, they are expected to have 846 thousand people age 85 or older. Find the expected number of people age 85 or older in New York by 2020. **27.** Two-step equations can be used when you start with a certain amount and increase or decrease at a certain rate. Answers should include the following.
- You've been running 15 minutes each day as part of a fitness program. You plan to increase your time by 5 minutes each week. After how many weeks do you plan to run 30 minutes each day? ($5w + 15 = 30$, 3 weeks)
- You are three years older than your sister is. Together the sum of your ages is 21. How old is your sister? ($2x + 3 = 21$, 9 years old)

29. D **31.** 4 **33.** 5 **35.** −11 **37.** −4 **39.** −5 **41.** 6

Page 130 Practice Quiz 2
1. −13 **3.** −18 **5.** $3n + 20 = 32$, 4

Pages 133–136 Lesson 3-7
1. $d = rt$ **3.** Sample answer:

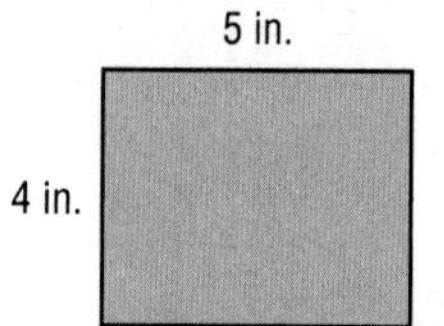

5. 34 km, 30 km^2 **7.** 4 in. **9.** 8 h **11.** 15 mph **13.** 54 cm, 162 cm^2 **15.** 136 in., 900 in^2 **17.** 48 m, 144 m^2 **19.** 20 m, 25 m^2 **21.** 11 yd **23.** 5 m **25.** 39 ft **27.** 19 yd **29.** 390 yd, 9000 yd^2 **31.** $d = 2r$ **33.** 4300 ft^2 **35.** ≈ 23.5 mph
37. **39.**

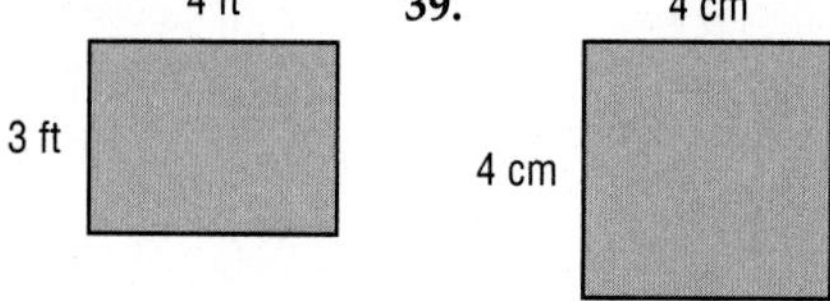

41. Sometimes; a 3-inch by 4-inch rectangle has a perimeter of 14 inches and an area of 12 square inches; a 6-inch by 8-inch rectangle has a perimeter of 28 inches and an area of 48 square inches. **43.** Formulas are important in math and science because they summarize the relationships among quantities. Answers should include the following.
- Sample answer: The formula to find acceleration is $a = \frac{v_f - v_i}{t}$ where v_f is the final velocity and v_i is the initial velocity.
- You can find the acceleration of an automobile with this formula.

45. C **47.** −9 **49.** −8 **51.** $-4x + 9$ **53a.** {(1870, 14), (1881, 600), (1910, 1000), (2000, 1500)} **53b.** domain: {1870, 1881, 1910, 2000}, range: {14, 600, 1000, 1500}

Pages 138–140 Chapter 3 Study Guide and Review
1. like terms **3.** Multiplication Property of Equality **5.** Distributive Property **7.** coordinate **9.** constant **11.** $3h + 18$ **13.** $-5k - 5$ **15.** $9t - 45$ **17.** $-2b + 8$ **19.** $9a$ **21.** $-2n - 8$ **23.** 3 **25.** 5 **27.** 8 **29.** −18 **31.** 1 **33.** −6 **35.** $2n + 3 = 53$; 25 **37.** 34 ft, 72 ft^2

Chapter 4 Factors and Fractions

Page 147 Chapter 4 Getting Started
1. $2x + 2$ **3.** $-2k - 16$ **5.** $12c + 24$ **7.** $7a + 7b$ **9.** 14 **11.** 28 **13.** 63 **15.** 30 **17.** 45 **19.** 78 **21.** 0.39 **23.** 0.005

Pages 150–152 Lesson 4-1
1. Use the rules for divisibility to determine whether 18,450 is divisible by both 2 and 3. If it is, then the number is also divisible by 6 and there is no remainder. **3a.** Sample

answer: 102 **3b.** Sample answer: 1035 **3c.** Sample answer: 343 **5.** 2 **7.** 2, 5, 10 **9.** 1, 2, 4, 5, 8, 10, 16, 20, 40, 80 **11.** yes; a number **13.** No; two terms are added. **15.** 2000, 2004, and 2032 are leap years. **17.** 3, 5 **19.** 2, 3, 5, 6, 10 **21.** 2 **23.** 5 **25.** 2, 3, 6 **27.** 2, 5, 10 **29.** 1, 2, 3, 6, 19, 38, 57, 114 **31.** 1, 5, 13, 65 **33.** 1, 2, 4, 31, 62, 124 **35.** 1, 3, 5, 9, 15, 27, 45, 135 **37.** yes; a number **39.** No; one term is subtracted from another term. **41.** No; two terms are added. **43.** No; one term is subtracted from another term. **45.** yes; the product of a number and a variable **47.** yes; the product of numbers and variables **49.** 6 ways; 1×72, 2×36, 3×24, 4×18, 6×12, 8×9 **51.** Alternating rows of a flag contain 6 stars and 5 stars, respectively. Fifty is not divisible by a number that would make the arrangement of stars in an appropriate-sized rectangle. **53.** Never; a number that has 10 as a factor is divisible by $2 \cdot 5$, so it is always divisible by 5. **55a.** 24 cases **55b.** 36 bags **55c.** Sample answer: 12 cases, 18 bags; 14 cases, 15 bags; 16 cases, 12 bags **57.** The side lengths or dimensions of a rectangle are factors of the number that is the area of the rectangle. Answers should include the following.

- A rectangle with dimensions and area labeled; for example, a 4×5 rectangle would have length 5 units, width 4 units, and area 20 square units.
- Factors are numbers that are multiplied to form a product. The dimensions of a rectangle are factor pairs of the area since they are multiplied to form the area.

59. C **61.** 34 in., 60 in^2 **63.** $5n - 2 = 3$; 1 **65.** 2 **67.** 64 **69.** -27 **71.** 2304

Pages 155–157 Lesson 4-2

1. Sample answer: 2^5, x^5 **3.** When n is even, $1^n = (-1)^n = 1$. When n is odd, $1^n = 1$ and $(-1)^n = -1$. **5.** 7^2 **7.** $(2 \times 10^3) + (6 \times 10^2) + (9 \times 10^1) + (5 \times 10^0)$ **9.** -11 **11.** 13^2 **13.** 6^1 **15.** $(-8)^4$ **17.** $(-t)^3$ **19.** m^4 **21.** $2x^2y^2$ **23.** $9(p + 1)^2$ **25.** $(8 \times 10^2) + (0 \times 10^1) + (3 \times 10^0)$ **27.** $(2 \times 10^4) + (3 \times 10^3) + (7 \times 10^2) + (8 \times 10^1) + (1 \times 10^0)$ **29.** 1000 **31.** -32 **33.** 81 **35.** -54 **37.** 13 **39.** 9 **41.** 243 **43.** $81 = 9^2$ or 3^4, $64 = 8^2$ or 4^3 or 2^6 **45.** $(-8)^3$; $(-8)(-8)(-8)$; -512 **47.** Always; the product of two negative numbers is always positive. **49.** $2^1, 2^2, 2^3, 2^4, 2^5$ **51.** After 10 folds, the noodles are $5(2^{10}) = 5(1024)$ or 5120 feet long, which is slightly less than a mile. So, after 11 folds the length of the noodles will be greater than a mile. **53.** = **55.** $6 \cdot 3^2$ cm^2 **57.** No; the surface area is multiplied by 4. The volume is multiplied by 8.
59. As the capacity of computer memory increases, the factors of 2 in the number of megabytes increases. Answers should include the following.

- Computer data are measured in small units that are based on factors of 2.
- In describing the amount of memory in modern computers, it would be impractical to list all the factors of 2. Using exponents is a more efficient way to describe and compare computer data.

61. B **63.** 2, 5, 10 **65.** 150 mph **67.** 4 **69.** $3y + 8$ **71.** 1, 5 **73.** 1, 2, 4, 8, 16 **75.** 1, 5, 7, 35

Pages 161–163 Lesson 4-3

1. A prime number has exactly two factors: 1 and itself. A composite number has more than two factors.
3. Francisca; 4 is not prime. **5.** prime **7.** $2 \cdot 3^2$ **9.** $2 \cdot 5^2$ **11.** $5 \cdot a \cdot a \cdot b$ **13.** 3 and 5, 5 and 7, 11 and 13, 17 and 19, 29 and 31, 41 and 43 **15.** composite **17.** composite **19.** composite **21.** prime **23.** 3^4 **25.** $3^2 \cdot 7$ **27.** $2^2 \cdot 5^2$ **29.** $2 \cdot 5 \cdot 11$ **31.** $3 \cdot 3 \cdot t \cdot t$ **33.** $-1 \cdot 5 \cdot 5 \cdot z \cdot z \cdot z$ **35.** $-1 \cdot 2 \cdot 19 \cdot m \cdot n \cdot p$ **37.** $3 \cdot 7 \cdot g \cdot h \cdot h \cdot h$ **39.** $2 \cdot 2 \cdot 2 \cdot 2 \cdot 2 \cdot 2 \cdot n \cdot n \cdot n$ **41.** $-1 \cdot 2 \cdot 2 \cdot 2 \cdot 3 \cdot 5 \cdot r \cdot r \cdot s \cdot t \cdot t \cdot t$ **43.** Sample answer: $-25x$
45. The number of rectangles that can be modeled to represent a number indicate whether the number is prime or composite. Answers should include the following.

- If a number is prime, then only one rectangle can be drawn to represent the number. If a number is composite, then more than one rectangle can be drawn to represent the number.
- If a model has a length or width of 1, then the number may be prime or composite. If a model does not have a length or width of 1, then the number must be composite.

47. C **49.** $(-5)^3h^2k$ **51.** yes **53.** no **55.** -9 **57.** -28 **59.** $5x - 35$ **61.** $10a + 60$ **63.** $72 - 8y$

Page 163 Practice Quiz 1

1. 3, 5 **3.** none **5.** 37 **7.** $7 \cdot 11 \cdot x$ **9.** $-1 \cdot 23 \cdot n \cdot n \cdot n$

Pages 166–168 Lesson 4-4

1. Sample answer: Find the prime factorization of each number. Multiply the factors that are common to both.
3. Jack; the common prime factors of the expressions are 2 and 11, so the GCF is $2 \cdot 11$ or 22. **5.** 3 **7.** 14 **9.** 36 **11.** $14n$ **13.** $3(n + 3)$ **15.** $5(3 + 4x)$ **17.** 4 **19.** 8 **21.** 10 **23.** 9 **25.** 8 **27.** 5 **29.** 4 **31.** 3 **33.** 4 **35.** $4x$ **37.** $2s$ **39.** $14b$ **41.** $4n$ **43.** Sample answer: $2x$, $6x^2$ **45.** $3(r + 4)$ **47.** $3(2 + y)$ **49.** $7(2 + 3c)$ **51.** $4(y - 4)$
53a. 7; Sample answer:

7 14 21 28 35
↓ ↓ ↓ ↓ ↓
7(1) 7(2) 7(3) 7(4) 7(5)

The terms increase by a factor of 7.
53b. 42, 49 **55a.** 6-in. squares **55b.** 20 tiles **57.** Yes; the GCF of 2 and 8 is 2. **59.** D **61.** yes **63.** no **65.** yes **67.** $3 \cdot 3 \cdot n$ **69.** $-1 \cdot 5 \cdot j \cdot k$ **71.** 92 **73.** -6 **75.** -10 **77.** 36 **79.** 24 **81.** 1000

Pages 171–173 Lesson 4-5

1. The GCF of the numerator and denominator is 1. **3.** $\frac{1}{7}$ **5.** simplified **7.** $\frac{16}{17}$ **9.** $\frac{a}{2}$ **11.** simplified **13.** B **15.** $\frac{5}{6}$ **17.** $\frac{2}{9}$ **19.** $\frac{9}{22}$ **21.** simplified **23.** $\frac{1}{3}$ **25.** $\frac{5}{12}$ **27.** $\frac{19}{20}$ **29.** $\frac{3}{92}$ **31.** $\frac{y^2}{1}$ or y^2 **33.** $\frac{20}{21}$ **35.** $\frac{1}{8t}$ **37.** $\frac{7z^2}{4}$ **39.** simplified **41.** $\frac{gh}{3}$ **43.** $\frac{24}{25}$ **45a.** yes; $\frac{330}{440} = \frac{3}{4}$ **45b.** No; $\frac{294}{349}$ cannot be simplified. **45c.** yes; $\frac{264}{528} = \frac{1}{2}$ **47.** $\frac{31}{50}$ **49.** $\frac{7}{10}$
51. Fractions represent parts of a whole. So, measurements that contain parts of units can be represented using fractions. Answers should include the following.

- Measurements can be given as parts of a whole because smaller units make up larger units. For example, inches make up feet.
- Twelve inches equals 1 foot. So, 3 inches equals $\frac{3}{12}$ or $\frac{1}{4}$ foot.

53. A **55.** 2 **57.** 5 **59.** composite **61.** prime **63.** 27 **65.** $(6 \cdot 7)(k^3)$ **67.** $(3 \cdot -5)(x^4 \cdot x^2)$

Pages 177–179 Lesson 4-6

1. Neither; the factors have different bases. **3.** Sample answer: $5 \cdot 5^2 = 5^3$ **5.** a^6 **7.** $-12x^5$ **9.** 10^2 **11.** a^4 **13.** 3^5 **15.** d^{10} **17.** n^9 **19.** 9^9 **21.** $18y^5$ **23.** $8a^3b^{10}$ **25.** 5^3 **27.** b^3 **29.** m^{12} **31.** $(-2)^1$ or -2 **33.** n^6 **35.** k^2m **37.** 9^7 **39.** 7^9 **41.** 10^2 or 100 times **43.** 3 **45.** 2 times **47.** 8 **49.** 5

51. Each level on the Richter scale is 10 times greater than the previous level. So, powers of 10 can be used to compare earthquake magnitudes. Answers should include the following.

- On the Richter scale, each whole-number increase represents a 10-fold increase in the magnitude of seismic waves.
- An earthquake of magnitude 7 is 10^5 times greater than an earthquake of magnitude 2 because $10^7 \div 10^2 = 10^{7-2}$ or 10^5.

53. B **55.** simplified **57.** $\frac{3x}{2y}$ **59.** 2 **61.** a **63.** Positive; as the high temperature increases, the amount of electricity that is used also increases. **65.** $-\frac{1}{10}$ **67.** $-\frac{1}{20}$ **69.** $\frac{1}{64}$

Pages 183–185 Lesson 4-7

1. To get each successive power, divide the previous power by 3. Therefore, $3^0 = 3 \div 3$ or 1. **3.** $\frac{1}{5^2}$ **5.** $\frac{1}{t^6}$ **7.** 3^{-4} **9.** 7^{-2} **11.** $\frac{1}{32}$ **13.** 10^{-3} **15.** $\frac{1}{5^3}$ **17.** $\frac{1}{(-3)^3}$ **19.** $\frac{1}{10^4}$ **21.** $\frac{1}{a^{10}}$ **23.** $\frac{1}{q^4}$ **25.** x^2 **27.** $\frac{1}{5^4}$; 0.0016 **29.** 5^{-5} **31.** 13^{-2} **33.** 9^{-2} **35.** 2^{-4} or 4^{-2} **37.** 10^{-2} or 100^{-1} **39.** 10^{-5} **41.** $-\frac{1}{128}$ **43.** $\frac{1}{729}$ **45.** 128 times **47.** x^{-5} **49.** x^{-3}

51. a^3b^{-2} or $\frac{a^3}{b^2}$

53. Yes; $(x^3)^{-2} = \frac{1}{(x^3)^2}$

$= \frac{1}{x^3 \cdot x^3}$ or $\frac{1}{x^6}$

$(x^{-2})^3 = (x^{-2})(x^{-2})(x^{-2})$

$= x^{-6}$ or $\frac{1}{x^6}$

55. C **57.** 9×10^{-1} **59.** $(1 \times 10^{-1}) + (7 \times 10^{-2}) + (3 \times 10^{-3})$ **61.** 3^7 **63.** 5^3 **65.** $8y + 48$ **67.** $5n - 15$ **69.** 720 **71.** 40.5 **73.** 0.0005

Page 185 Practice Quiz 2

1. 5 **3.** $2a$ **5.** $\frac{2}{5}$ **7.** $-2n^7$ **9.** $\frac{1}{b^6}$

Pages 188–190 Lesson 4-8

1. Sample answer: Numbers that are greater than 1 can be expressed as the product of a factor and a positive power of 10. So, these numbers are written in scientific notation using positive exponents. Numbers between 0 and 1 cannot be expressed as the product of a factor and a whole number power of 10, so they are written in scientific notation using negative exponents. **3.** 0.000308 **5.** 849,500 **7.** 6.97×10^5 **9.** 1.0×10^{-3} **11.** Mars, Venus, Earth **13.** 57,200 **15.** 0.005689 **17.** 0.0901 **19.** 2505 **21.** 2.0×10^6 **23.** 6.0×10^{-3} **25.** 5.0×10^7 **27.** 5.894×10^6 **29.** 4.25×10^{-4} **31.** 6.25×10^6 **33.** 7.53×10^{-7} **35.** 2.3×10^5 **37.** 5000 **39.** Arctic, Indian, Atlantic, Pacific **41.** 6.1×10^{-5}, 0.0061, 6.1×10^{-2}, 6100, 6.1×10^4 **43.** 48,396 **45.** 2.52×10^5; 252,000 **47.** Bezymianny; Santa Maria; Agung; Mount St. Helens tied with Hekla 1947; Hekla, 1970; Ngauruhoe **49.** 3.14 **51.** B **53.** $\frac{1}{81}$ **55.** $\frac{1}{49}$ **57.** $15a^4$ **59.** $c + \$2.50$

Pages 191–194 Chapter 4 Study Guide and Review

1. true **3.** true **5.** true **7.** true **9.** 3 **11.** 5 **13.** 2, 3, 6 **15.** 2, 5, 10 **17.** 27 **19.** 25 **21.** 90 **23.** 112 **25.** $3^2 \cdot 5$ **27.** $2^2 \cdot 17$ **29.** $7 \cdot 7 \cdot k$ **31.** $2 \cdot 13 \cdot p \cdot p \cdot p$ **33.** 6 **35.** n **37.** $2(t + 10)$ **39.** $2(15 + 2n)$ **41.** $\frac{3}{5}$ **43.** $\frac{10}{17}$ **45.** simplified **47.** $\frac{5c^2}{8b}$ **49.** c^4 **51.** r^2 **53.** $\frac{1}{7^2}$ **55.** $\frac{1}{b^4}$ **57.** $\frac{1}{(-4)^3}$ **59.** 0.0029 **61.** 70,450 **63.** 8.0×10^{-3} **65.** 4.571×10^7

Chapter 5 Rational Numbers

Page 199 Chapter 5 Getting Started

1. 0.6 **3.** 34 **5.** 0.2 **7.** −75 **9.** −1.7 **11.** $\frac{3}{5}$ **13.** $\frac{18}{25}$ **15.** 6 **17.** −13 **19.** 15 **21.** 9

Pages 202–204 Lesson 5-1

1. Sample answer: write the fractions as decimals and then compare. **3.** Sample answer: $0.\overline{14}$ **5.** 2.08 **7.** $0.2\overline{6}$ **9.** < **11.** > **13.** 0.2 **15.** 0.32 **17.** 7.3 **19.** 5.125 **21.** $0.\overline{1}$ **23.** $-0.\overline{45}$ **25.** $0.1\overline{6}$ **27.** 0.3125 **29.** 0.83 **31.** $\frac{7}{9}$, 0.8, $\frac{7}{8}$ **33.** < **35.** < **37.** > **39.** = **41.** > **43.** > **45.** Sample answer: 0.7 and $0.\overline{7}$; $\frac{1}{6} = 0.1\overline{6}$ and $\frac{8}{9} = 0.\overline{8}$; 0.7 and $0.\overline{7}$ are both greater than $0.1\overline{6}$ and less than $0.\overline{8}$. **47.** This is greater than those who chose English in the survey because $\frac{1}{7} \approx 0.14$, and $0.14 > 0.13$. **49.** All coins were made with a fraction of silver that was contained in a silver dollar. Answers should include the following.

- A quarter had one-fourth the amount of silver as a silver dollar, a dime had one-tenth the amount, and a nickel had one-twentieth the amount.
- It is easier to perform arithmetic operations using decimals rather than using fractions.

51. D **53.** 7.7×10^{-2} **55.** 9.25×10^5 **57.** $\frac{1}{(-2)^7}$ **59.** $\frac{1}{y^3}$ **61.** 29 **63.** −32 **65.** 56 **67.** $\frac{1}{13}$ **69.** $\frac{2}{3}$ **71.** $\frac{4}{7}$ **73.** $\frac{1}{8}$

Pages 207–209 Lesson 5-2

1. any number that can be written as a fraction **3.** $-\frac{7}{3}$ **5.** $\frac{4}{5}$ **7.** $-\frac{7}{9}$ **9.** I, Q **11.** $\frac{39}{1{,}000{,}000}$ **13.** $-\frac{11}{7}$ **15.** $\frac{60}{1}$ **17.** $\frac{9}{100}$ **19.** $1\frac{17}{25}$ **21.** $8\frac{1}{250}$ **23.** $-\frac{1}{3}$ **25.** $5\frac{2}{3}$ **27.** $2\frac{25}{99}$ **29.** $\frac{3}{50}$ **31.** $\frac{59}{200}$ **33.** $\frac{4}{25}$ **35.** I, Q **37.** N, W, I, Q **39.** Q **41.** not rational **43.** $200\frac{19}{100}$ **45.** Sometimes; $\frac{1}{2}$ and 2 are both rational numbers, but only 2 is an integer. **47.** $\frac{1}{1250}$ in. **49.** Yes; $2\frac{3}{8} = 2.375$ and $2.375 > 2.37$. **51.** The set of rational numbers includes the set of natural numbers, whole numbers, and integers. In the same way, natural numbers are part of the set of whole numbers and the set of whole numbers is part of the set of integers. Answers should include the following.

- The number 5 belongs to the set of natural numbers, whole numbers integers, and rational numbers.
- The number $\frac{1}{2}$ belongs only to the set of rational numbers.

53. C **55.** −7.8 **57.** $2.\overline{5}$ **59.** 3,050,000 **61.** 0.01681 **63.** $(4 \times 10^2) + (8 \times 10^1) + (3 \times 10^0)$ **65.** 24 cm; 27 cm²

67. $8 \cdot 2 + 1 \cdot 2$ **69.** $7x + 28$ **71.** Sample answer: $-5 \cdot 4 = -20$ **73.** Sample answer: $7 \cdot 2 = 14$ **75.** Sample answer: $16 \cdot 2 = 32$

Pages 212–214 Lesson 5-3

1. Sample answer: $\frac{1}{2}, \frac{1}{3}$ **3.** $\frac{3}{20}$ **5.** $\frac{5}{9}$ **7.** $\frac{13}{22}$ **9.** $\frac{6}{7}$ **11.** $\frac{8}{t}$ **13.** $\frac{12}{49}$ **15.** $-\frac{1}{40}$ **17.** $\frac{8}{45}$ **19.** $\frac{1}{3}$ **21.** $-\frac{1}{4}$ **23.** $\frac{3}{8}$ **25.** $1\frac{1}{6}$ **27.** $3\frac{1}{3}$ **29.** $14\frac{2}{3}$ **31.** 10 **33.** $-4\frac{4}{9}$ **35.** 6 **37.** 27 **39.** 27 **41.** $\frac{8c}{11}$ **43.** $\frac{xz^2}{3}$ **45.** $\frac{9}{25}$ **47.** $\frac{11}{60}$ **49.** 12.7 **51.** 10.257 **53a.** Sample answer: $\frac{8}{4} \times \frac{3}{5}$ **53b.** Sample answer: $\frac{3}{4} \times \frac{5}{6}$ **55.** A **57.** $\frac{5}{8}$ **59.** $\frac{13}{32}$ **61.** $-\frac{1}{5}$ **63.** $\frac{7}{9}$ **65.** $0.1\overline{6}$ **67.** -4.875 **69.** $8n$ **71.** $2t$ **73.** 9

Pages 217–219 Lesson 5-4

1. Dividing by a fraction is the same as multiplying by its reciprocal. **3.** $\frac{5}{4}$ **5.** $\frac{8}{25}$ **7.** $\frac{4}{5}$ **9.** $1\frac{7}{15}$ **11.** $-1\frac{29}{36}$ **13.** $\frac{3a}{2}$ **15.** 6 boards **17.** $-\frac{5}{1}$ or -5 **19.** $\frac{1}{24}$ **21.** $-\frac{9}{29}$ **23.** $\frac{8}{9}$ **25.** $-\frac{15}{22}$ **27.** $\frac{15}{16}$ **29.** -1 **31.** $1\frac{1}{2}$ **33.** -10 **35.** $1\frac{1}{3}$ **37.** 2 **39.** $-6\frac{1}{4}$ **41.** $\frac{4}{3}$ **43.** $\frac{5}{6r}$ **45.** $\frac{16}{t^5}$ **47.** 6 ribbons **49.** $\frac{81}{256}$ **51.** 8 days **53.** Dividing by a fraction is the same as multiplying by its reciprocal. Answers should include the following.

- For example, a model of two circles, each divided into four sections, represents $2 \div \frac{1}{4}$. Since there are 8 sections, $2 \div \frac{1}{4} = 8$.
- Division of fractions and multiplication of fractions are inverse operations. So, $2 \div \frac{1}{4}$ equals $2 \cdot 4$ or 8.

55. C **57.** $\frac{5}{24}$ **59.** $-\frac{10}{21}$ **61.** Q **63.** not rational **65.** 7 **67.** $1\frac{1}{7}$ **69.** $6\frac{1}{4}$ **71.** $3\frac{2}{3}$ **73.** $1\frac{2}{3}$

Pages 222–224 Lesson 5-5

1.

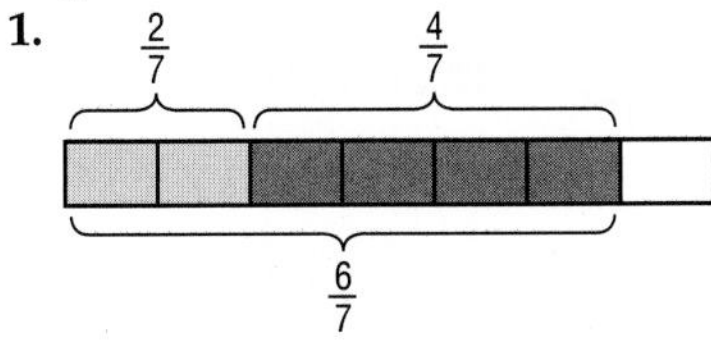

3. Kayla; Ethan incorrectly left out the negative sign on the first term. **5.** $\frac{4}{7}$ **7.** $-\frac{3}{4}$ **9.** $8\frac{1}{2}$ **11.** $\frac{8r}{11}$ **13.** $-\frac{1}{3x}$ **15.** $\frac{3}{5}$ **17.** $\frac{2}{3}$ **19.** $-\frac{1}{2}$ **21.** $-1\frac{1}{2}$ **23.** $-\frac{2}{9}$ **25.** $11\frac{4}{5}$ **27.** $9\frac{1}{3}$ **29.** $\frac{3}{4}$ **31.** $4\frac{6}{7}$ **33.** $8\frac{1}{8}$ **35.** $2\frac{3}{5}$ **37.** $\frac{1}{5}$ **39.** $\frac{5x}{8}$ **41.** $\frac{3}{m}$ **43.** $2\frac{3}{7}c$ **45.** $5\frac{3}{8}$ in. **47.** 38 ft **49.** When you use a ruler or a tape measure, measurements are usually a fraction of an inch. Answers should include the following.

- The marks on a ruler represent $\frac{1}{16}$ of an inch, $\frac{1}{8}$ of an inch, $\frac{1}{4}$ of an inch, and $\frac{1}{2}$ of an inch.
- Fractional measures are used in sewing and construction.

51. A **53.** $-1\frac{7}{8}$ **55.** $\frac{3}{10}$ **57.** $1\frac{1}{3}$ **59.** 42 cm; 90 cm^2 **61.** $5^2 \cdot 7$ **63.** $2^2 \cdot 3 \cdot n$ **65.** $2 \cdot 3 \cdot 7 \cdot a^2 \cdot b$

Page 224 Practice Quiz 1

1. 0.16 **3.** 3.125 **5.** $\frac{3}{25}$ **7.** $\frac{2}{27}$ **9.** $\frac{5}{12}$

Pages 228–230 Lesson 5-6

1. The LCM involves the common multiples of a set of numbers; the LCD is the LCM of the denominators of two or more fractions. **3.** 24 **5.** 70 **7.** 48 **9.** 8 **11.** $100x$ **13.** = **15.** front gear: 5; back gear: 13 **17.** 60 **19.** 48 **21.** 84 **23.** 100 **25.** 96 **27.** 84 **29.** 630 **31.** $112a^2b$ **33.** $75n^4$ **35.** 15 **37.** 35 **39.** 24 **41.** $16c^2d$ **43.** 60 s **45.** $<$ **47.** $>$ **49.** = **51.** $>$ **53.** amphibians **55.** 12 and 18 **57.** Never; sample answer: 5 and 6 do not contain any factors in common, and the LCM of 5 and 6 is 30. **59a.** If two numbers are relatively prime, then their LCM is the product of the two numbers. For example, the LCM of 4 and 5 is $2^2 \cdot 5$ or 20; the LCM of 6 and 25 is $2 \cdot 3 \cdot 5^2$ or 150. **59b.** Always; the LCM contains all of the factors of both numbers. Therefore, it must contain any common factors. **61.** B **63.** $\frac{1}{2}$ **65.** $1\frac{1}{7}$ **67.** 3 **69.** $\frac{ad}{5}$ **71.** $7 + 2n = 11$; 2 **73.** -20 **75.** Sample answer: $0 + 1 = 1$ **77.** Sample answer: $1 + 2 = 3$ **79.** Sample answer: $8 + 7 = 15$

Pages 234–236 Lesson 5-7

1. Find the least common denominator. **3.** José; he finds a common denominator by multiplying the denominators. Daniel incorrectly adds the numerators and the denominators of unlike fractions. **5.** $\frac{2}{9}$ **7.** $-\frac{5}{6}$ **9.** $-4\frac{1}{4}$ **11.** $\frac{9}{10}$ **13.** $\frac{5}{28}$ **15.** $1\frac{1}{16}$ **17.** $\frac{1}{8}$ **19.** $-\frac{1}{8}$ **21.** $1\frac{1}{15}$ **23.** $-7\frac{5}{9}$ **25.** $-11\frac{7}{9}$ **27.** $-14\frac{5}{8}$ **29.** $3\frac{11}{36}$ **31.** $5\frac{3}{5}$ lb **33.** $8\frac{3}{4}$ in. **35.** Sample answer: Fill the $\frac{1}{2}$-cup. From the $\frac{1}{2}$-cup, fill the $\frac{1}{3}$-cup. $\frac{1}{6}$ cup will be left in the $\frac{1}{2}$-cup because $\frac{1}{2} - \frac{1}{3} = \frac{1}{6}$. **37.** Find the LCM of the denominators. Then rename the fractions as like fractions with the LCM as the denominators. Answers should include the following.

- For example, the LCM of 4 and 6 is 12. So, $\frac{1}{4} + \frac{5}{6} = \frac{3}{12} + \frac{10}{12} = \frac{13}{12}$ or $1\frac{1}{12}$.
- Writing the prime factorization of the denominators is the first step in finding the LCM of the denominators, which is the LCD. Then the fractions can be added or subtracted.

39. B **41.** Sample answer: $\frac{1}{2} + \frac{1}{10}$ **43.** 36 **45.** $6n^3$ **47.** $9\frac{1}{2}$ **49.** $2\frac{4}{5}$ **51.** $1\frac{4}{5}$ **53.** 27 **55.** -9

Pages 241–242 Lesson 5-8

1. Mean; when an extreme value is added to the other data, it can raise or lower the sum, and therefore the mean. **3.** 12.4; 8; none **5.** 3.6; 3.5; 4 **7.** Sample answer: 13 could be an extreme value because it is 12 less than the next value. It lowers the mean by 2.1. **9.** 98 **11.** 8.9; 8; 8 **13.** 7.6; 7.5; 7.1 and 7.4 **15.** 4.3; 4.2; 4.1 and 4.2 **17.** Sample answer: the median, 95, or the modes, 95 and 97, because most students scored higher than mean, which is 91. **19.** Sample answer: The median home price would be useful because it is not affected by the cost of the very expensive homes. The cost of half the homes in the county would be greater than the median cost and half would be less. **21.** C **23.** $9\frac{5}{6}$ **25.** $-3\frac{7}{8}$ **27.** $>$ **29.** 18 **31.** -5 **33.** 7.9 **35.** 3

Pages 246–248 Lesson 5-9

1. Subtraction Property of Equality **3.** Ling; dividing 0.3 by 3 does not isolate the variable on one side. **5.** 18.7 **7.** $1\frac{13}{30}$ **9.** $-2\frac{1}{12}$ **11.** –90 **13.** 29.15 in. **15.** -2.4 **17.** 12.24 **19.** 5.9 **21.** $-\frac{31}{36}$ **23.** $6\frac{7}{9}$ **25.** $-\frac{2}{15}$ **27.** $-2\frac{3}{10}$ **29.** 4 **31.** 5 **33.** -32 **35.** $1\frac{1}{4}$ **37.** $5\frac{3}{4}$ **39.** $2\frac{7}{8}$ **41.** $13\frac{1}{2}$ in. by $21\frac{1}{2}$ in. **43.** \$16.66 **45.** $2\frac{1}{4}$ ft **47.** Equations with fractions can be written to represent the number of vibrations per second for different notes. To solve, multiply each side of the equation by the reciprocal of the fraction. Answers should include the following.

- For example, if n vibrations per second produce middle C, then $\frac{5}{4}n$ vibrations per second produce the note E above middle C.
- The equation $\frac{5}{3}n = 440$ represents the number of vibrations per second to produce middle C. To solve, multiply each side by the reciprocal of $\frac{5}{3}$, $\frac{3}{5}$.

49. D **51.** 13; 12; 11 and 12 **53.** 70.8; 66; 60 **55.** $\frac{17}{24}$ **57.** $\frac{17}{36}$ **59.** $7\frac{11}{15}$ **61.** $\frac{3}{8}$ **63.** -6 **65.** -6 **67.** 5 **69.** $-\frac{1}{3}$

Page 248 Practice Quiz 2

1. 72 **3.** 10 **5.** $\frac{13}{16}$ **7.** $1\frac{2}{3}$ **9.** 32.4; 30.5; 29

Pages 251–252 Lesson 5-10

1. Arithmetic sequences have a common difference and the terms can be found by adding or subtracting. Geometric sequences have a common ratio and the terms can be found by multiplying or dividing. **3.** A; 4; 19, 23, 27 **5.** neither **7.** A; $-\frac{1}{3}$; $\frac{5}{3}$, $\frac{4}{3}$, 1 **9.** \$11,027.36 **11.** A; 11; 38, 49, 60 **13.** G; 3; 162, 486, 1458 **15.** A; -3; 13, 10, 7 **17.** G; $-\frac{1}{5}$; $-\frac{1}{125}$, $\frac{1}{625}$, $-\frac{1}{3125}$ **19.** A; $\frac{1}{6}$; $\frac{2}{3}$, $\frac{5}{6}$, 1 **21.** A; -0.5, 2.5, 2, 1.5 **23.** neither **25.** G; $-\frac{1}{3}$; $\frac{2}{9}$, $-\frac{2}{27}$, $\frac{2}{81}$ **27.** G; $\frac{1}{2}$; $\frac{1}{32}$, $\frac{1}{64}$, $\frac{1}{128}$ **29a.** Arithmetic; the common difference is \$3. **29b.** \$48 **31.** Find the pattern, continue the sequence, and use the new values to make predictions. Answers should include the following.

- The difference between any two consecutive terms in an arithmetic sequence is the common difference. To find the next value in such a sequence, add the common difference to the last term. The ratio of any two consecutive terms in a geometric sequence is the common ratio. So, to find the next value in such a sequence, multiply the last term by the common ratio.
- Sequences occurring in nature include geysers spouting every few minutes, the arrangement of geese in migration patterns, and ocean tides.

33. A **35.** 56 **37.** -5.28 **39.** 10 **41.** 3.3; 3.3; 3.6 **43.** $102\frac{7}{8}$ in. **45.** b^3 **47.** 6 **49.** 15

Pages 254–258 Chapter 5 Study Guide and Review

1. rational **3.** algebraic fraction **5.** LCD **7.** reciprocal **9.** arithmetic **11.** 0.45 **13.** $-0.4\overline{6}$ **15.** $6.\overline{36}$ **17.** $\frac{3}{5}$ **19.** $\frac{1}{8}$ **21.** $4\frac{11}{25}$ **23.** $\frac{5}{9}$ **25.** $1\frac{7}{9}$ **27.** $3\frac{4}{11}$ **29.** $-\frac{7}{27}$ **31.** $6\frac{2}{5}$ **33.** $1\frac{1}{3}$ **35.** -3 **37.** $9\frac{4}{9}$ **39.** x **41.** -6 **43.** $-3\frac{3}{4}$ **45.** $-4\frac{1}{5}$ **47.** $\frac{4}{21x}$ **49.** $\frac{5}{9}$ **51.** $1\frac{1}{3}$ **53.** $2\frac{3}{5}$ **55.** $\frac{4}{3x}$ **57.** 120 **59.** $21c^2$ **61.** $<$ **63.** $>$ **65.** $1\frac{2}{3}$ **67.** $2\frac{5}{6}$ **69.** $1\frac{11}{15}$ **71.** $3\frac{1}{6}$ **73.** 6.0; 6.5; 3.6 and 7.2 **75.** 3.25 **77.** $-2\frac{1}{2}$ **79.** G; 3; 81, 243, 729 **81.** neither

Chapter 6 Ratio, Proportion, and Percent

Page 263 Chapter 6 Getting Started

1. 24 **3.** 10,560 **5.** 480 **7.** 4000 **9.** 6 **11.** 580 **13.** 50 **15.** 15,000 **17.** 48.8 **19.** 13.44 **21.** 0.18 **23.** 3.04 **25.** $\frac{1}{3}$ **27.** simplified **29.** simplified **31.** $\frac{3}{4}$

Pages 266–268 Lesson 6-1

1. Sample answer:

3. Sample answer: \$12 per person **5.** $\frac{5}{8}$ **7.** $\frac{1}{3}$ **9.** 0.75 inch/hour **11.** 24.2 miles/gallon **13.** 576 **15.** No; the ratio will be 2 to 3. **17.** $\frac{2}{5}$ **19.** $\frac{3}{7}$ **21.** $\frac{9}{32{,}000}$ **23.** $\frac{1760}{1}$ **25.** $\frac{17}{118}$ **27.** $\frac{3}{1}$ **29.** 0.10 cents/pencil **31.** 4.5 m/sec **33.** 7.8 ft/h **35.** 39 pages/week **37.** The 6-pack of soda costs \$0.37 per can. The 12-pack of soda costs \$0.35 per can. So, the 12-pack is less expensive. **39.** 26.4 **41.** 19.2 **43.** 72 **45.** 105 **47.** about 579 mi/h **49.** \$9 **51.** C **53b.** The ratios should be close in value. **53c.** Sample answer: Pyramid of Khufu in Giza, Egypt; The Taj Mahal in India; The Lincoln Memorial in Washington, D.C. **55.** arithmetic; 0.3; 13.3, 13.6, 13.9 **57.** $-\frac{1}{12}$ **59.** $13\frac{1}{3}$ **61.** 5.2×10^7 **63.** 3.8×10^{-2} **65.** 30 **67.** 40 **69.** 13

Pages 272–274 Lesson 6-2

1. A statement of equality of two ratios. **3.** Yes **5.** 15 **7.** 4.2 **9.** yes **11.** no **13.** yes **15.** 4 **17.** 20 **19.** 15 **21.** 1.4 **23.** 7.5 **25.** 0.94 **27.** 0.8 **29.** 15 **31.** 26 **33.** $\frac{12}{3} = \frac{72}{m}$; 18 **35.** $\frac{5}{6.25} = \frac{d}{8.75}$; 7 **37.** about 360.8 ft **39.** \$22.47 **41.** chocolate pieces: 3 c; peanuts: $1\frac{1}{2}$ c **43.** $\frac{a}{c} = \frac{b}{d}$, $\frac{b}{a} = \frac{d}{c}$, or $\frac{c}{a} = \frac{d}{b}$ **45.** C **47.** 57.3 mph **49.** 4 **51.** $\frac{4}{7y}$ **53.** 102 **55.** 6.5

Pages 278–280 Lesson 6-3

1. 1 unit 3 units **3.** 40 mi **5.** 1 in. = 5 ft

7.

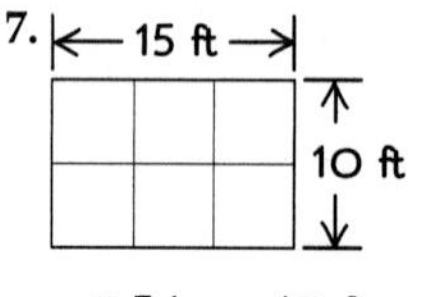

9. 12 ft **11.** 24.6 ft **13.** 11.4 ft **15.** $49\frac{1}{2}$ ft **17.** $\frac{2}{3}$ **19.** 1 cm = 0.25 cm **21.** A scale factor less than 1 means that the drawing or model is drawn smaller than actual size. A scale factor of 1 means that the drawing or model is drawn actual size. A

scale factor greater than 1 means the drawing or model is drawn larger than actual size. **23.** B **25a.** $\frac{1}{2}$ **25b.** $\frac{1}{4}$
25c. The perimeter of a 3-inch by 5-inch rectangle is 16 inches. The area is 15 square inches. For a 6-inch by 10-inch rectangle, the perimeter should be double the 3-inch by 5-inch rectangle. That is, 16 × 2 or 32 inches. The area should be 4 times the area of the 3-inch by 5-inch rectangle. That is, 4 × 15 or 60 square inches. Since the perimeter of the 6-inch by 10-inch rectangle is 32 and its area is 60, the conjecture is true. **27.** 3.5 **29.** 21.6 **31.** $6\frac{1}{12}$
33. $18t^5$ **35.** $\frac{4m^3}{3}$ **37.** $\frac{1}{20}$ **39.** $\frac{2}{5}$ **41.** $\frac{39}{50}$ **43.** $\frac{41}{50}$

Pages 283–285 Lesson 6-4
1. Sample answer: write an equivalent fraction with a denominator of 100 or express the fraction as a decimal and then express the decimal as a percent. A fraction is greater than 100% if it is greater than 1. It is less than 1% if it is less than $\frac{1}{100}$. **3.** $\frac{3}{10}$, 0.3 **5.** $1\frac{1}{4}$, 1.25 **7.** $1\frac{7}{20}$, 1.35 **9.** 45%
11. 0.8% **13.** 133.3% **15.** daily newspaper **17.** $\frac{22}{25}$, 0.88
19. $\frac{7}{8}$, 0.875 **21.** $3\frac{1}{2}$, 3.5 **23.** $\frac{61}{100}$, 0.61 **25.** $2\frac{23}{100}$, 2.23
27. $\frac{53}{10{,}000}$, 0.0053 **29.** 9% **31.** 270% **33.** 0.06% **35.** 22.5%
37. 175% **39.** 1.7% **41.** $\frac{37}{1000}$ **43.** 0.45 **45.** 19%
47. 61%, $\frac{2}{3}$, 0.69 **49.** 0.4 **51.** There are only two possibilities that satisfy the conditions, $\frac{1}{4}$ and $\frac{2}{5}$.
53. Percents are related to fractions and decimals because they can be expressed as them. Answers should include the following.

•

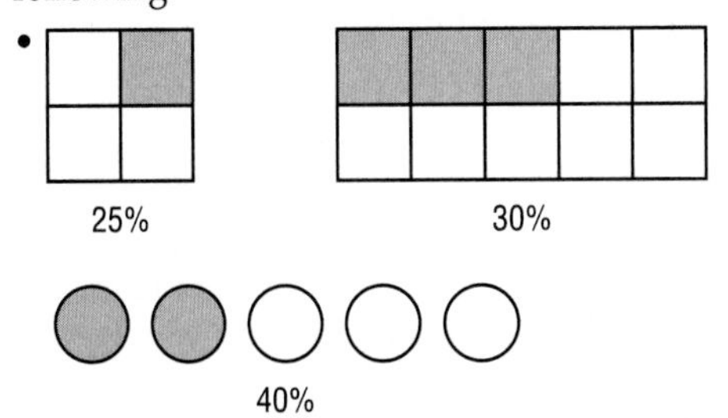

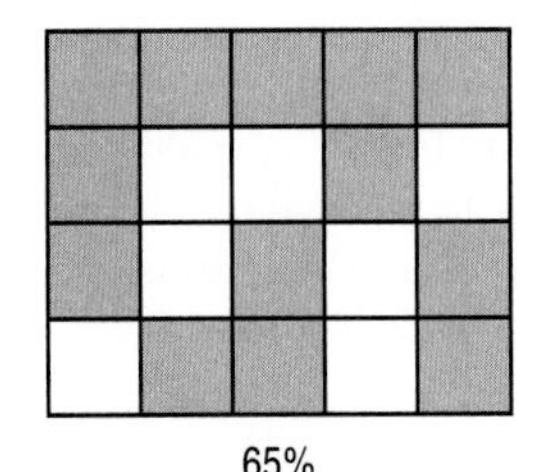

• $25\% = \frac{1}{4} = 0.25$; $30\% = \frac{3}{10} = 0.3$; $40\% = \frac{2}{5} = 0.4$;
$65\% = \frac{13}{20} = 0.65$ **55.** A **57.** $\frac{1}{12}$ **59.** $\frac{11}{21}$ **61.** $1\frac{3}{13}$
63. composite **65.** composite **67.** 800 **69.** 0.94 **71.** 320

Pages 291–292 Lesson 6-5
1. $\frac{\text{number correct}}{50} = \frac{\%}{100}$ **3.** 40% **5.** 104 **7.** 60% **9.** 45%
11. 50 **13.** 77 **15.** 36% **17.** 112 **19.** 0.2% **21.** 16%
23. about 70% **25.** 26.8 lbs
27. In real-world situations, percents are important because they show how something compares to the whole. Answers should include the following.

- For example, the outer layer of the new state quarters is an alloy of 3 parts copper to 1 part nickel.
- Thus, there are 4 parts to the outer layer (copper, copper, copper, nickel). The outer layer is $\frac{3}{4}$ or 75% copper and $\frac{1}{4}$ or 25% nickel.

29. $\frac{21}{50}$ **31.** $1\frac{1}{5}$ **33.** $\frac{7}{9}$ **35.** $3\frac{1}{2}$ **37.** 8 **39.** 6 **41.** 25

Page 292 Practice Quiz 1
1. \$0.14 per can **3.** 1 ft = 3069 ft **5.** 19.5

Pages 295–297 Lesson 6-6
1. 18% is about 20% or $\frac{1}{5}$. 216 is about 220. $\frac{1}{5}$ of 220 is 44. So, 18% of 216 is about 44. **5.** Sample answer: 13
7. Sample answer: 72 **9.** Sample answer: 8; fraction method: $\frac{1}{3} \times 24$ or 8 **11.** Sample answer: 21; meaning of percent method: 152% means about 150 for every 100 or about 15 for every 10. 14 has 1 tens. 1 × 15 = 15. So, 152% of 14 is about 21. **13.** Sample answer: 14 **15.** Sample answer: 33 **17.** Sample answer: 49 **19.** Sample answer: 4
21. Sample answer: 90 **23.** Sample answer: 375
25. Sample answer: $\frac{1}{4} \times 8$ or 2 billion **27.** Sample answer: 27; fraction method: $\frac{3}{10} \times 90$ or 27 **29.** Sample answer: 36; fraction method: $\frac{2}{5} \times 90$ or 36 **31.** Sample answer: 2; 1% method: Since 1% of 806 is about 8, $\frac{1}{4}$% of 806 is about $\frac{1}{4}$ of 8 or 2. **33.** Sample answer: 78; meaning of percent: 127% means about 130 for every 100 or about 13 for every 10. 64 has 6 tens. 6 × 13 = 78. So, 127% of 64 is about 78.
35. Sample answer: 450; meaning of percent: 295% means about 300 for every 100 or about 30 for every 10. 145 has one 100 and about 5 tens. (300 × 1) + (30 × 5) = 300 + 150 or 450. **37.** Sample answer: Pluto and Mars, or Neptune and Jupiter **39.** Sample answer: $\frac{1}{3} \times 90{,}000$ or 30,000 miles
41. about 10% **43.** 7 to 8 **45.** C **47.** 21 **49.** Maine: 31,813 sq mi; New Hampshire: 8238 sq mi; West Virginia: 18,779 sq mi; Vermont: 7279 sq mi; Alabama: 35,071 sq mi
51. 160% **53.** 0.77 **55.** 4.21 **57.** 8.9 **59.** −21 **61.** 0.5
63. 0.25 **65.** 0.07

Pages 300–302 Lesson 6-7
1. Use the percent equation in any situation where the rate and base are known. **3.** I = interest; p = principal; r = annual interest rate; t = time in years **5.** 50 **7.** 3%
9. \$1680 **11.** 1.5 years **13.** 95% **15.** 50 **17.** 63 **19.** 37.5%
21. 14.52 **23.** 218% **25.** 0.9% **27.** 13.2 **29.** \$25.49
31. \$17 **33.** \$18.50 **35.** \$1113.75 **37.** \$244.76 **39.** 16
41. \$1.20 **43.** If you know two of the three values, you can use the percent proportion to solve for the missing value. Answers should include the following.

- To find the amount of tax on an item, you can use the percent proportion or the percent equation.
- For example, the following methods can be used to find 6% tax on \$24.99.

Method 1: Percent Proportion
$$\frac{x}{24.99} = \frac{6}{100}$$
Method 2: Percent Equation
$$n = 0.06(24.99)$$
Using either method, $x = 1.50$. The amount of tax is \$1.50.
45. B **47.** Sample answer: 250; meaning of percent method: 126% means about 125 for every 100 and 12.5 for every 10.

198 has about 2 one-hundreds. $125 \times 2 = 250$. So, 126% of 198 is about 250. **49.** 121 **51.** 41 **53.** 38 cm **55.** $8w - 24$ **57.** 89% **59.** 156% **61.** 22.4%

Pages 306–308 Lesson 6-8

1. If the amount increases, it is a percent of increase. If the amount decreases, it is a percent of decrease. **3.** Mark; he divided the difference of the new amount and the original amount by the original amount. **5.** −60%; D **7.** 10.1%; I **9.** A **11.** 170%; I **13.** −12%; D **15.** −5.4%; D **17.** 164%; I **19.** 10.1% **21.** −14.3% **23.** 150%

25. The amount by which a rectangle is increased or decreased can be represented by a percent. Answers should include the following.

- If the size of the new rectangle is greater than the size of the original rectangle, the percent of increase is greater than 100%.
- Original Rectangle

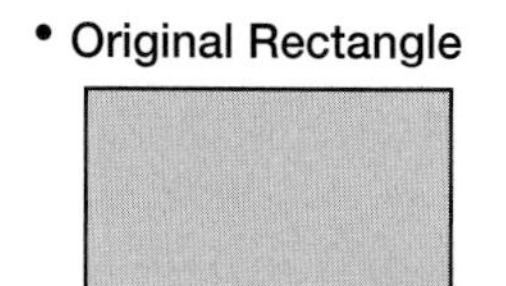

Less than 100%

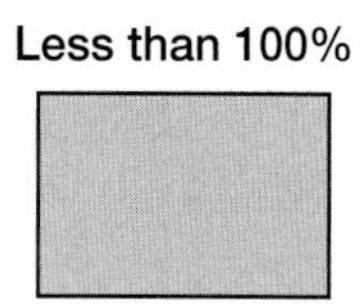

Greater than 100%

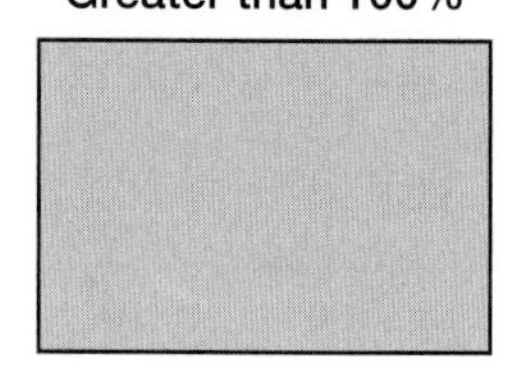

27. D **29.** \$149.85 **31.** \$13.46 **33.** Sample answer: 63; fraction method: $\frac{9}{10} \times 70$ or 63 **35.** integer, rational **37.** rational **39.** 20% **41.** $83\frac{1}{3}\%$

Page 308 Practice Quiz 2

1. Sample answer: 28; fraction method: $\frac{2}{5} \times 70$ or 28 **3.** \$14.50 **5.** 158%

Pages 312–314 Lesson 6-9

1. The event will not happen.
3. Sample answer: Spinning the spinner shown and having it land on 4.

5. $\frac{1}{2}$; 50% **7.** $\frac{2}{5}$; 40% **9.** 1; 100% **11.** 75 **13.** $\frac{1}{4}$; 25% **15.** $\frac{1}{2}$; 50% **17.** 0; 0% **19.** $\frac{7}{8}$; 87.5% **21.** $\frac{2}{9}$; 22.2% **23.** $\frac{11}{18}$; 61.1% **25.** $\frac{1}{3}$; $33\frac{1}{3}\%$ **27.** 1; 100% **29.** $\frac{1}{6}$ **31.** $\frac{1}{16}$ **33.** $\frac{1}{4}$ **35.** 0.05; 5% **37.** 444

39. Once the probability or likeliness of something happening is known, then you can use the probability to make a prediction. For example, in football, if you know the number of field goals a player has made in the past, you can use the information to predict the number of field goals he/she will make in upcoming games. Answers should include the following.

- E: $\frac{12}{100}$ or 12%; A, I: $\frac{9}{100}$ or 9%; O: $\frac{8}{100}$ or 8%; N, R, T: $\frac{6}{100}$ or 6%; D, L, S, U: $\frac{4}{100}$ or 4%; G: $\frac{3}{100}$ or 3%; B, C, F, H, M, P, V, W, Y, blank: $\frac{2}{100}$ or 2%; J, K, Q, X, Z: $\frac{1}{100}$ or 1%

41. 146.9% **43.** 23.9 **45.** $2x^5$ **47.** $\frac{18n^2}{7}$

Pages 316–320 Chapter 6 Study Guide and Review

1. proportion **3.** scale factor **5.** experimental probability **7.** $\frac{3}{4}$ **9.** $\frac{1}{4}$ **11.** $\frac{20}{9}$ **13.** 30 **15.** 0.9 **17.** 45.6 ft **19.** $\frac{7}{20}$; 0.35 **21.** $\frac{2}{25}$; 0.08 **23.** $1\frac{1}{5}$; 1.2 **25.** $\frac{5}{8}$; 0.625 **27.** 24% **29.** 45.2% **31.** 40% **33.** 7.5% **35.** 40% **37.** 25 **39.** 50 **41.** 43 **43.** 9 **45.** 8 **47.** Sample answer: 16; fraction method: $\frac{1}{2} \times 32$ or 16

49. Sample answer: 10; fraction method: $\frac{1}{5} \times 50$ or 10

51. Sample answer: 1; 1% method: Since 1% of 304 is about 3, $\frac{1}{3}\%$ of 304 is about $\frac{1}{3}$ of 3 or 1. **53.** 48% **55.** 94.5 **57.** 200 **59.** D; −70% **61.** I; 86.2% **63.** $\frac{1}{3}$ **65.** $\frac{2}{3}$ **67.** 0

Chapter 7 Equations and Inequalities

Page 327 Chapter 7 Getting Started

1. 4 **3.** 24 **5.** −44 **7.** 11 **9.** −17 **11.** 16 **13.** −14 **15.** −15 **17.** 72 **19.** −5 **21.** 1

Pages 332–333 Lesson 7-1

1. Subtraction Property of Equality **3.** −8 **5.** 3 **7.** 0.3 **9.** 75 miles **11.** 13 **13.** −7 **15.** −0.5 **17.** −6 **19.** $\frac{4}{3}$ **21.** 3 **23.** 4.2 **25.** 0.3 **27.** 3.4 **29.** $3y - 14 = y$; 7 **31.** −16 **33.** 72 **35.** 70 min **37.** 29 **39.** A **41.** −40° **43.** 12.5% **45.** 6.48 **47.** 70 **49.** $6a + 27$ **51.** $2.4c + 28$ **53.** $\frac{1}{2}n - \frac{9}{2}$

Pages 336–338 Lesson 7-2

1. Multiply 4 times $(x - 1)$. Subtract $2x$ from each side. Add 4 to each side. Divide each side by 2. **3.** 11 **5.** 13 **7.** ∅ **9.** $\ell = 7$ ft; $w = 3$ ft; $A = 21$ ft^2 **11.** 6 **13.** 2.5 **15.** 18 **17.** 3 **19.** 35 **21.** ∅ **23.** all numbers **25.** 6 **27.** all numbers **29.** 0 **31.** 70 yd by 150 yd **33.** w: 3 ft; ℓ: 13 ft; A: 39 ft^2 **35.** triangle: 7, 8, 9; rectangle, 4, 8; perimeter: 24 **37.** 2 gal

39. Many equations include grouping symbols. You must use the Distributive Property to correctly solve the equation. Answers should include the following.

- The Distributive Property states that $a(b + c) = ab + ac$.
- You use the Distributive Property to remove the grouping symbols when you are solving equations.

41. D **43.** 0.4 **45.** 40% **47.** 0.375 **49.** 3.24 **51.** −7 **53.** −24 **55.** −8

Page 338 Practice Quiz 1

1. $2x = 5x - 150$; 50 **3.** $3.\overline{1}$ **5.** all numbers

Pages 342–344 Lesson 7-3

1. An inequality represents all numbers greater or less than a given number. A number line graph can represent all those numbers. **3.** $n + 14 \geq 25$ **5.** true

7.

−2 −1 0 1 2 3 4 5 6

9.

1 2 3 4 5 6 7 8 9

11. $x \geq -20$ **13.** $f > 18{,}000$ **15.** $86 \geq 2w$ **17.** true **19.** false **21.** true

23. −1 0 1 2 3 4 5 6 7

25. 4 5 6 7 8 9 10 11 12

27. 6 7 8 9 10 11 12 13 14

29. −2 −1 0 1 2 3 4 5 6

31. −6 −5 −4 −3 −2 −1 0 1 2

33. −8 −7 −6 −5 −4 −3 −2 −1 0

35. $x > 13$ **37.** $x \leq -3$ **39.** $x \geq -32$ **41.** $m \geq 6.4$
43. $b + 14{,}600 > 30{,}000$; $b > 15{,}400$ **45.** Symmetric: If $a < b$, then $b < a$; not true. Sample counterexample: $4 < 5$, but $5 \not< 4$. Transitive: If $a < b$ and $b < c$, then $a < c$; true.
47. B

49a. −3 −2 −1 0 1 2 3 4 5

49b. −3 −2 −1 0 1 2 3 4 5

51. 5 **53.** $4n - 6 = 3n + 2$; 8 **55.** G: −16, 32, −64 **57.** 13
59. −7 **61.** 19.8

Pages 347–349 Lesson 7-4

1. Use addition to undo subtraction; use subtraction to undo addition. **3.** Sample answer: Joanna works part-time at a clothing store. Part-time workers must work fewer than 30 hours a week. If Joanna has already worked 8 hours this week, how many hours can she still work? **5.** $y \geq -7$
7. $a > 11$ **9.** $t > -3$

11. $x \leq 10$ 7 8 9 10 11 12 13 14 15

13. $p < 2$ **15.** $b \leq -22$ **17.** $y \leq 5$ **19.** $r \leq 7$ **21.** $j \leq -4$
23. $w < 6$ **25.** $y \leq 1.4$ **27.** $f \geq 5.4$ **29.** $b < 3\frac{1}{4}$

31. $n < 5$ 1 2 3 4 5 6 7 8 9

33. $p > 2$ 1 2 3 4 5 6 7 8 9

35. $x \leq -5$ −7 −6 −5 −4 −3 −2 −1 0 1

37. $m \geq 27$ 23 25 27 29 31

39. $b < 1\frac{3}{4}$ 0 $\frac{1}{2}$ 1 $1\frac{1}{2}$ 2

41. $s \leq 4\frac{2}{3}$ 3 4 5

43. 980 lb **45.** $42 + x \geq 74$; $x \geq 32$; at least 32 mph
47. Always; subtracting x gives $-1 < 0$, which is always true. **49.** C **51.** F **53.** F **55.** 3 cm, 9 cm **57.** 8 + 32
59. $5x - 17.5$ **61.** −3 **63.** −15 **65.** −2 **67.** −4 **69.** −48

Pages 353–354 Lesson 7-5

1. Multiply each side by −12 and reverse the inequality symbol. **3.** Tamika is correct. She divided each side of the inequality by 9. Since 9 is a positive number, she did not reverse the inequality symbol.

5. $x \geq -2$ −4 −3 −2 −1 0 1 2 3 4

7. $a > 50$ 10 30 50 70 90

9. $m \geq -8.4$ −9 −8.6 −8.2 −7.8 −7.4

11. $y \geq -72$ −80 −76 −72 −68 −64

13. C

15. $y > 9$ 5 6 7 8 9 10 11 12 13

17. $b \geq -3$ −4 −3 −2 −1 0 1 2 3 4

19. $t \leq 5$ 1 2 3 4 5 6 7 8 9

21. $h \leq 98$ 90 94 98 102 106

23. $z \geq 3$ 1 2 3 4 5 6 7 8 9

25. $w \leq 2$ −4 −3 −2 −1 0 1 2 3 4

27. $r > 4$ 1 2 3 4 5 6 7 8 9

29. $k > -18$ −20 −18 −16 −14 −12

31. $t \geq -16$ −20 −18 −16 −14 −12

33. $n \leq 4$ 1 2 3 4 5 6 7 8 9

35. $y > 5.7$ 5 5.2 5.4 5.6 5.8

37. $b < 72$ 70 74 78 82 86

39a. $40m \geq 2000$ **39b.** at least 50 min **41.** Inequalities can be used to compare the weights of objects on different planets. Answers should include the following.

- Comparing the weight of an astronaut in a space suit on Mars to the same astronaut on the moon: $113 > 50$.
- If you multiply or divide the astronaut's weight by the same number, the inequality comparing the weights on different planets would still be true.

43. 40 **45.** $c \leq -20$ **47.** $2n \leq 14$ **49.** $-\frac{5}{21}$ **51.** $\frac{2a}{c}$ **53.** 4
55. −9 **57.** 32

Page 354 Practice Quiz 2

1.
−4 −3 −2 −1 0 1 2 3 4

3. $a \leq 40$ **5.** $n \geq 124$ **7.** $r < -225$ **9.** $g \leq 7$

Pages 357–359 Lesson 7-6

1. Check the solution by replacing the variable with a number in the solution. If the inequality is true, the solution checks. **3.** Jerome is correct. By the Distributive Property $2(2y + 3) = 4y + 6$ not $4y + 3$.

5. $n > 2$
1 2 3 4 5 6 7 8 9

7. $c > 1$
−4 −3 −2 −1 0 1 2 3 4

9. $x \geq 8$
1 2 3 4 5 6 7 8 9

11. $d < -9\frac{1}{2}$
−10 −9 −8 −7 −6

13. $x > 8$
1 2 3 4 5 6 7 8 9

15. $c > -4$
−4 −3 −2 −1 0 1 2 3 4

17. $x > 3$
−4 −3 −2 −1 0 1 2 3 4

19. $k \geq -5$
−7 −6 −5 −4 −3 −2 −1 0 1

21. $n < -5$
−7 −6 −5 −4 −3 −2 −1 0 1

23. $b \leq 2$
−4 −3 −2 −1 0 1 2 3 4

25. $y \geq 30$
10 30 50 70 90

27. $c < -4$
−10 −8 −6 −4 −2

29. $n > 7$

31. $n \leq 4.5$
1 2 3 4 5 6 7 8 9

33. $t \leq -1$
−4 −3 −2 −1 0 1 2 3 4

35. $c < -4$
−10 −8 −6 −4 −2

37. $0.55c + 0.35 \leq 2$; 3 candy bars **39.** $2s + 10 \geq 40$, 15 subscriptions **41.** $x < 200$; Sample explanation: The inequality finds at what mileage Able's charge is greater than Baker's charge. **43.** more than 100 minutes **45.** $k < 3$ and $k > -3$, or $k = \{-2, -1, 0, 1, 2\}$ **47.** D **49.** $-5 < x < 1$ **51.** $y \geq 37$ **53.** $n \leq 5$ **55.** $a < -18.6$ **57.** 0.5% **59.** $16.\overline{6}$ mpg **61.** \$1.25 an issue **63.** 6 m

Pages 360–362 Chapter 7 Study Guide and Review

1. true **3.** false; identity **5.** false; inequality **7.** true **9.** false; is greater than or equal to **11.** 3 **13.** 4 **15.** 1 **17.** −1 **19.** −1.2 **21.** 9 **23.** 16 **25.** all numbers **27.** true

29. $b \geq 17$
14 16 18 20 22 24

31. $t < 3\frac{1}{2}$
0 1 2 3 4 5

33. $k \leq 5.1$
−1 0 1 2 3 4 5 6 7 8 9

35. $y \leq -7$
−12 −10 −8 −6 −4

37. $a \geq -2\frac{2}{5}$
−4 −2 0 2 4 6

39. $n \leq 4$ **41.** $t > -12$ **43.** $b > -4\frac{1}{4}$

Chapter 8 Functions and Graphing

Page 367 Chapter 8 Getting Started

1.

x	y
0	4
−3	3

domain = {0, −3}; range = {4, 3}

3.

x	y
6	8
7	10
8	12

domain = {6, 7, 8}; range = {8, 10, 12}

5.

x	y
−8	5
7	−1
6	1
1	−2

domain = {−8, 7, 6, 1}; range = {5, −1, 1, −2}

7. E **9.** A **11.** F **13.** true **15.** true **17.** true

Pages 371–373 Lesson 8-1

1. Sample answer: a set of ordered pairs: {(1, 2), (4, 3), (−2, −1), (−3, 3)}

a table:

x	y
1	2
4	3
−2	−1
−3	3

a graph:

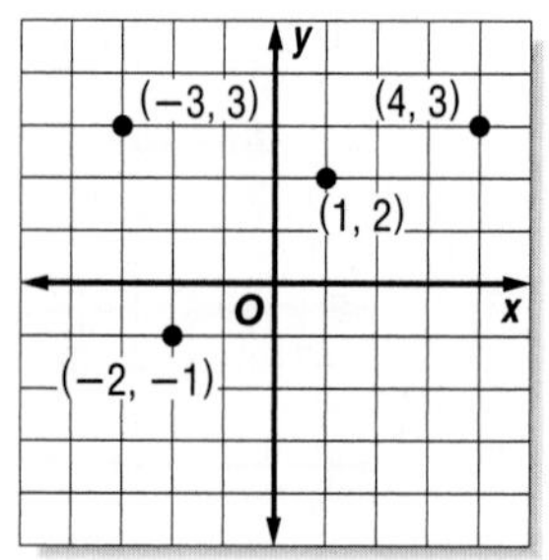

3. Sample answer: This graph does not represent a function because when x equals 1, there are two y values, 0 and 2.

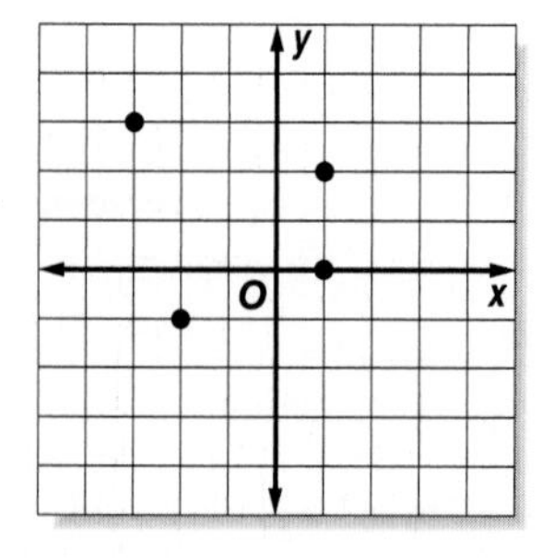

5. Yes; each x value is paired with only one y value. **7.** No; 5 is paired with 4 and 14. **9.** Yes; any vertical line passes through no more than one point of the graph. **11.** As wind speed increases, the windchill temperature decreases.
13. Yes; each x value is paired with only one y value.
15. No; 5 in the domain is paired with -4 and -1 in the range. **17.** No; -2 in the domain is paired with 5 and 1 in the range. **19.** Yes; each x value is paired with only one y value. **21.** No; a vertical line passes through more than one point. **23.** Yes; any vertical line passes through no more than one point of the graph. **25.** Generally, as the years progress, the number of farms decreases. An exception is in the year 2000. **27.** Generally, as the years progress, the size of farms increases. An exception is in the year 2000. **29.** Generally, as foot length increases, height increases. **31.** Sometimes; a relation that has a member of the domain paired with more than one member in the range is not a function.
33. For a given wind speed, there is only one windchill temperature for each actual temperature. So, the relationship between actual temperatures and windchill temperatures is a function. Answers should include the following.

- For a given wind speed, as the actual temperature increases, the windchill temperature increases.
- Since the relationship between actual temperatures and windchill temperatures is a function, there cannot be two different windchill temperatures for the same actual temperature when the wind speed remains the same.

35. A **37.** $a < -21$ **39.** $x < 6$ **41.** $\varnothing$ **43.** 18 **45.** 75 **47.** 13 **49.** 5 **51.** 0

Pages 377–379 Lesson 8-2

1. Sample answer: Infinitely many values can be substituted for x, or the domain.

3.

x	x + 5	y
−3	−3 + 5	2
−1	−1 + 5	4
0	0 + 5	5
1	1 + 5	6

$(-3, 2), (-1, 4), (0, 5), (1, 6)$

5. Sample answer: $(-1, -4), (0, 0), (1, 4), (2, 8)$ **7.** Sample answer: $(-1, 1), (0, 6), (1, 11), (2, 16)$

9.

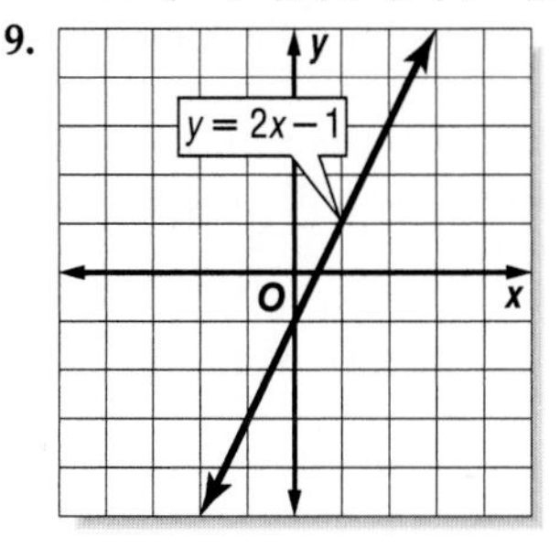

11. Sample answer: (1, 186,000) means that light travels 186,000 miles in 1 second. (2, 372,000) means that light travels 372,000 miles in 2 seconds.

13.

x	2x + 6	y
−4	2(−4) + 6	−2
0	2(0) + 6	6
2	2(2) + 6	10
4	2(4) + 6	14

$(-4, -2), (0, 6), (2, 10), (4, 14)$

15. Sample answer: $(-1, -8), (0, -7), (1, -6), (2, -5)$
17. Sample answer: $(-1, 5), (0, 0), (1, -5), (2, -10)$
19. Sample answer: $(-1, -2), (0, 1), (1, 4), (2, 7)$
21. Sample answer: $(-1, -5), (0, -6), (1, -7), (2, -8)$
23. Sample answer: $(-1, -13), (0, -10), (1, -7), (2, -4)$
25. Sample answer: $(-1, 0), (-1, 1), (-1, 2), (-1, 3)$
27. 6.2 mi **29.** Quadrant I; a person cannot have a negative age or heart rate.

31.

y = x + 5

33.

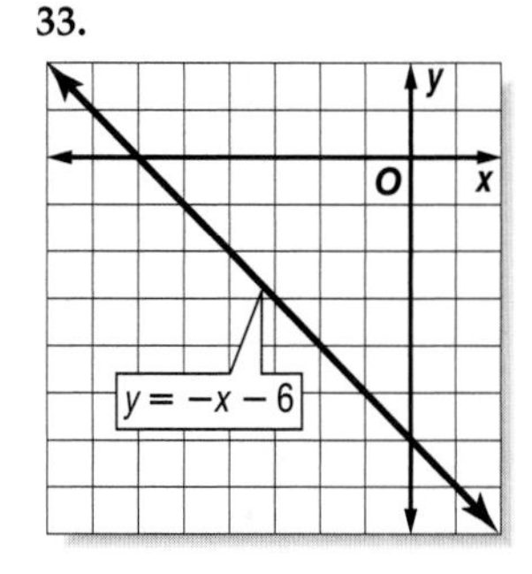

35.

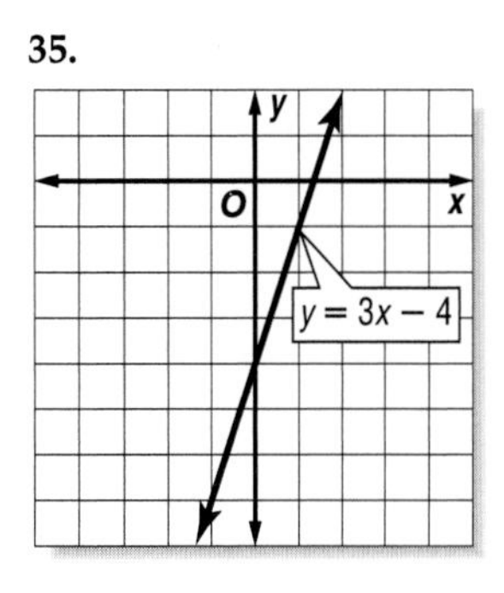

37.

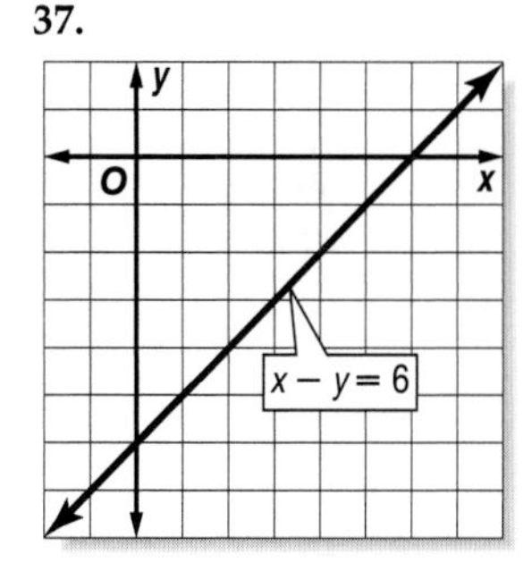

39.

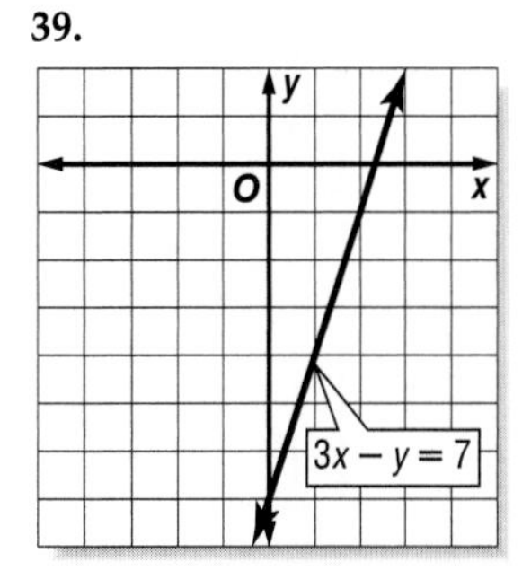

41.

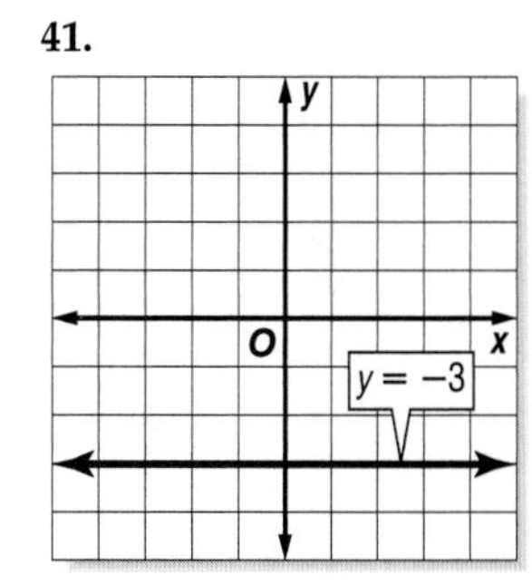

43.

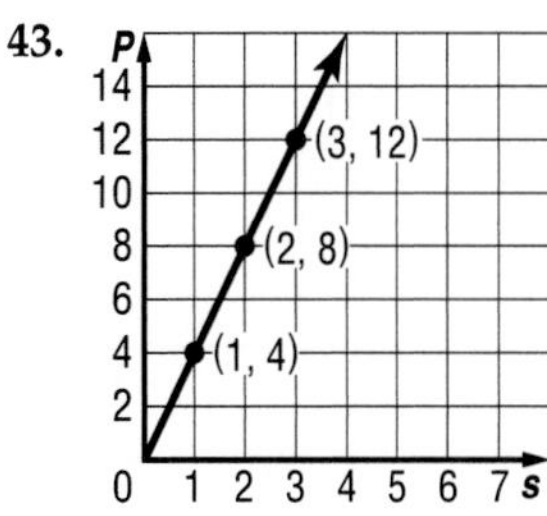

45. Yes; the points lie on a straight line. **47.** Yes; the points lie on a straight line. **49.** No; the exponent of x is not 1.
51. Sample answer: In the first table, as the x values increase by 1, the y values increase by 2. In the second table, as the x values increase by 1, the y values do not change by a constant amount. **53.** B **55.** Yes; each x value is paired with only one y value. **57.** No; 11 in the domain is paired with 8 and 21 in the range.
59. $x < 4$;

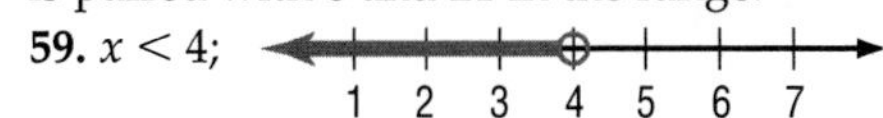

61. $\frac{6}{7}$ **63.** 3 **65.** 4

Selected Answers

Pages 384–385 Lesson 8-3

1. To find the x-intercept, let $y = 0$ and solve for x. To find the y-intercept, let $x = 0$ and solve for y. **3.** -1; -3 **5.** -4; 4 **7.** 3; 2

9.

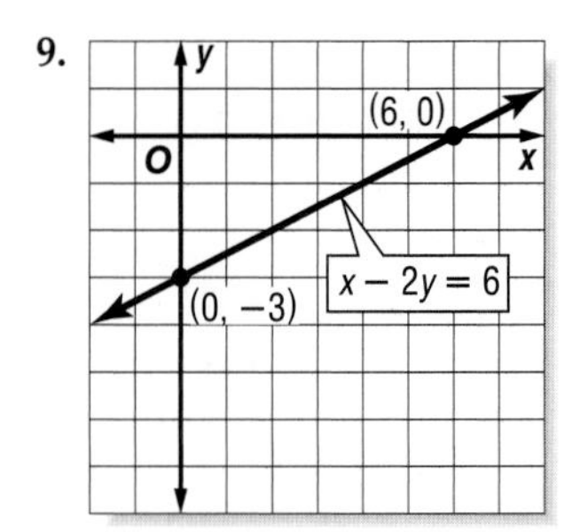

11.

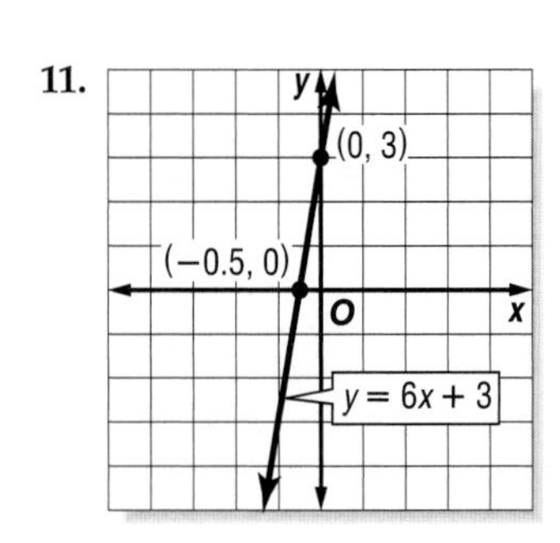

The y-intercept 3 represents the base fee of $3.

13. 1; 1 **15.** none; -5 **17.** -5; 5 **19.** none; -4 **21.** 8; -4 **23.** 5; 4

25.

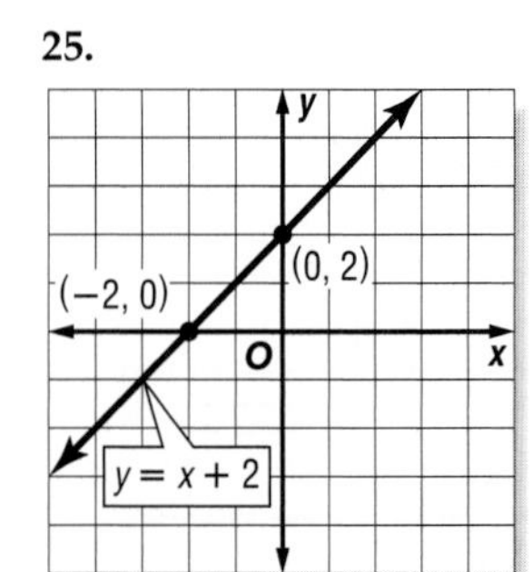

27.

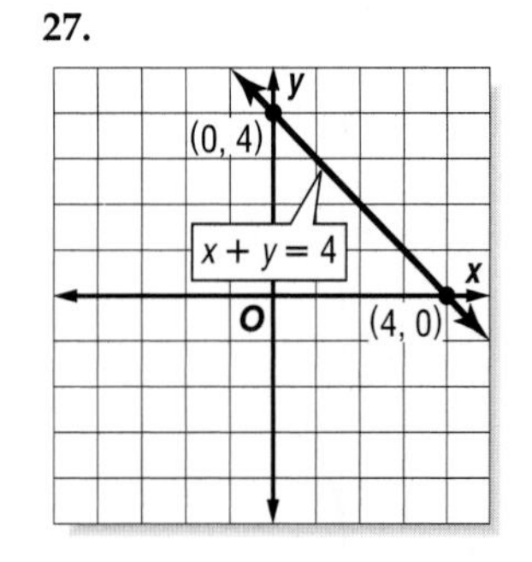

29.

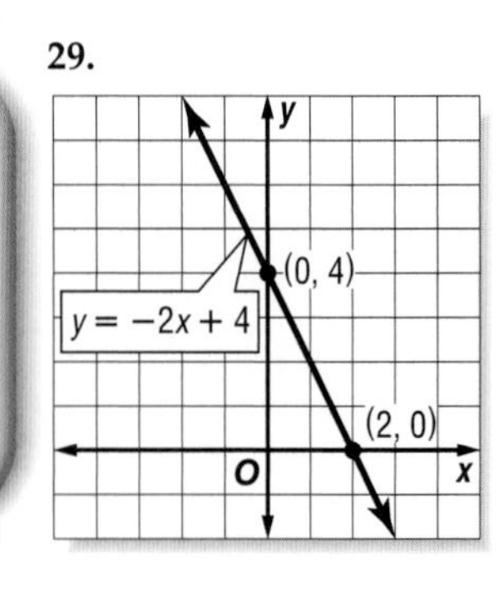

31.

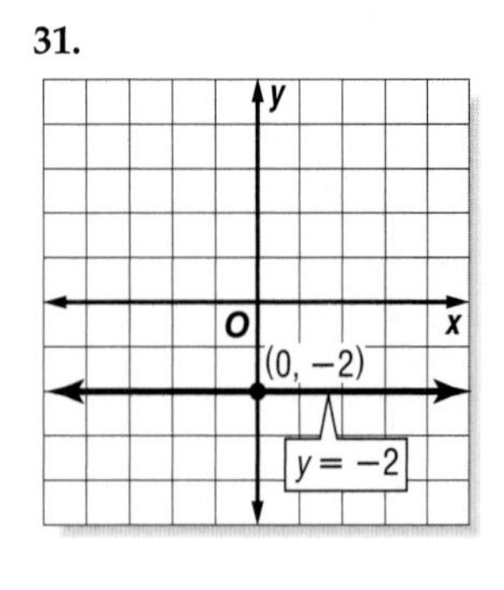

33.

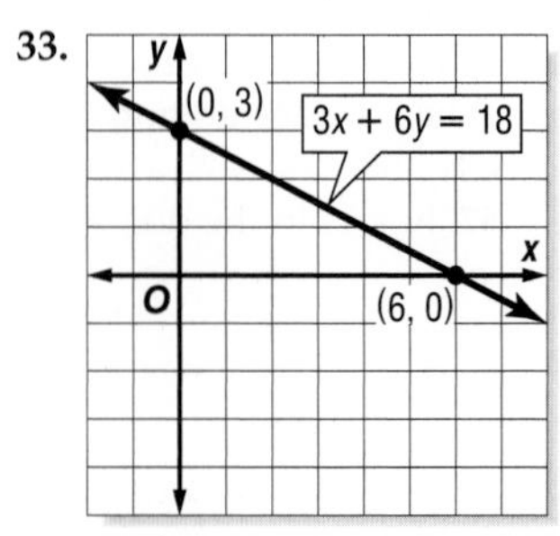

35.

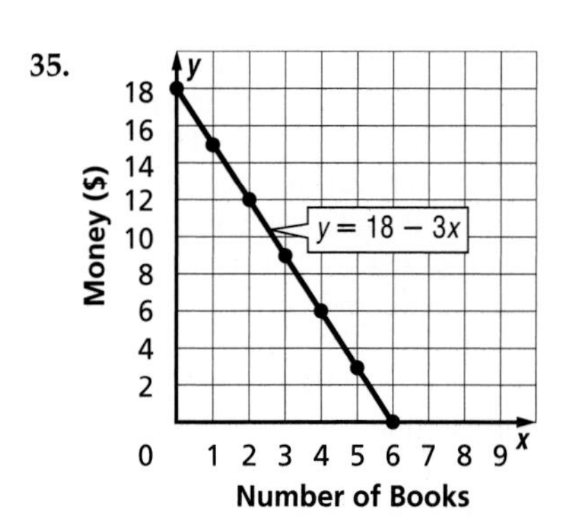

The x-intercept 6 represents the number of books that she can buy with no money left over. The y-intercept 18 represents the money she has before she buys any books. **37.** The x- and y-intercept are both 0. Therefore, the line passes through the origin. Since two points are needed to graph a line, $y = 2x$ cannot be graphed using only the intercepts.

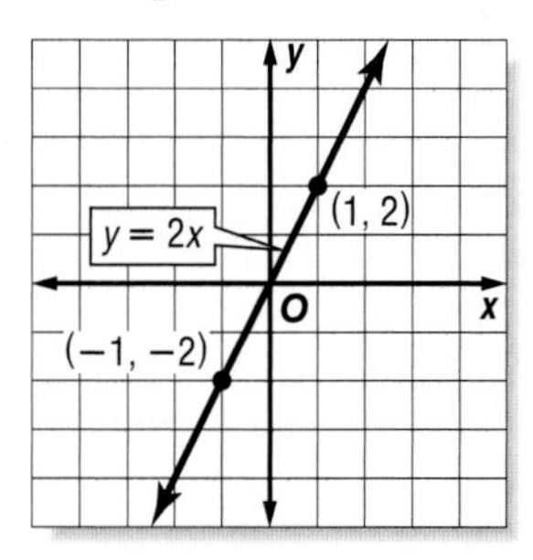

39. B **41.** Sample answer: $(-1, 5)$, $(0, 7)$, $(1, 9)$, $(2, 11)$ **43.** Sample answer: $(-1, 1)$, $(0, 5)$, $(1, 9)$, $(2, 13)$ **45.** no **47.** $n > 12$ **49.** 2.8% **51.** -16 **53.** 25

Pages 389–391 Lesson 8-4

1. Sample answer: horizontals do not rise, so slope $= \frac{\text{rise}}{\text{run}} = \frac{0}{\text{run}}$ or 0. **3.** Mike; Chloe should have subtracted -2 from 11 in the denominator. **5.** $-\frac{2}{3}$ **7.** 2 **9.** 0 **11.** B **13.** 4 **15.** $-\frac{3}{4}$ **17.** 0 **19.** 1 **21.** $-\frac{3}{2}$ **23.** undefined **25.** $-\frac{3}{4}$ **27.** It decreased; negative slope.

29. Slope can be used to describe the steepness of roller coaster hills. Answers should include the following.

- Slope is the steepness of a line or incline. It is the ratio of the rise to the run.
- An increase in rise with no change in run makes a roller coaster hill steeper. An increase in run with no change in rise makes a roller coaster less steep.

31. A **33.** 2; 6 **35.** Sample answer: $(-1, 3)$, $(0, 5)$, $(1, 7)$, $(2, 9)$ **37.** Sample answer: $(-1, 8)$, $(0, 7)$, $(1, 6)$, $(2, 5)$ **39.** $y = -2x$ **41.** $y = \frac{1}{3}x$

Pages 395–397 Lesson 8-5

1. The slope is 60, the rate of change is 60 units for every 1 unit, and the constant of variation is 60. **3.** Justin; any linear function, including direct variations, has a rate of change. **5.** increase of $12 per hour **7.** $y = 6x$ **9.** increase of 12 in./ft **11.** decrease of 2°F/min

13. Sample answer: The population of wild condors decreased from 1966 to 1990. Then the population increased from 1990 to 1996. The population of condors in captivity increased slowly from 1966 to 1982, then increased more rapidly from 1982 to 1996.

	Rate of Change (number per year)	
Interval	Condors in the Wild	Condors in Captivity
1966–1982	−2.25	0.1875
1982–1990	−3	4.625
1990–1992	0	8
1992–1994	1.5	14.5
1994–1996	12.5	3.5

15. $y = -5x$ **17.** $y = 0.75x$ **19.** $y = 2.54x$ **21.** A line representing the relationship between time and distance has a slope that is equal to the speed. Answers should include the following.

- Sample drawing:

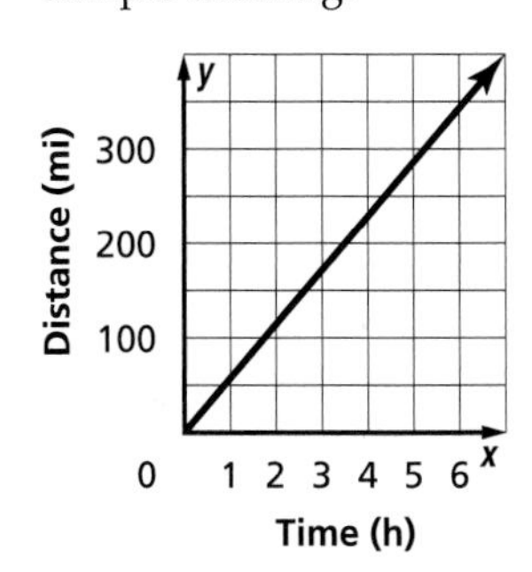

- As speed increases, the slope of the graph becomes steeper.

23. B **25.** 2

27.

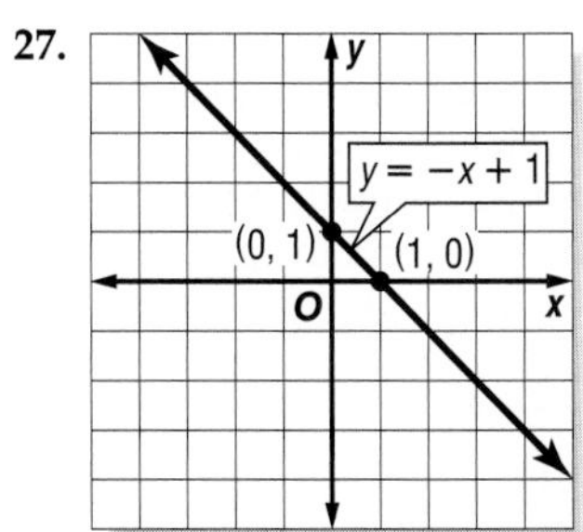

29. 14 **31.** $y = 1 - 3x$

Page 397 Practice Quiz 1

1. No; 1 is paired with 2 and −3.

3.

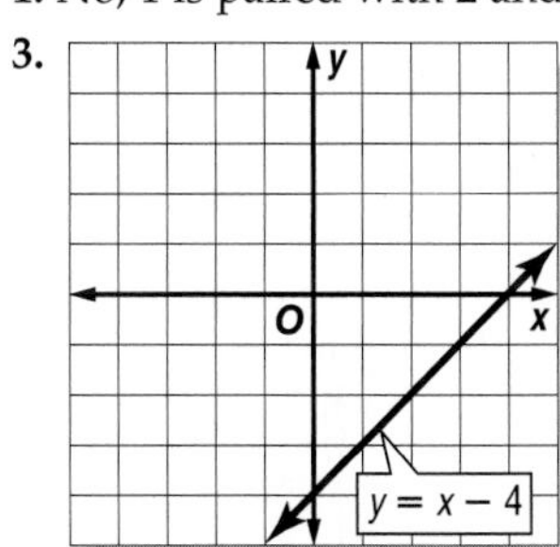

5. −9; 9 **7.** 5; −4 **9.** −2

Pages 400–401 Lesson 8-6

1. a **3.** Alex; the equation in slope-intercept form is $y = -\frac{1}{2}x + 4$. **5.** −1; 0

7.

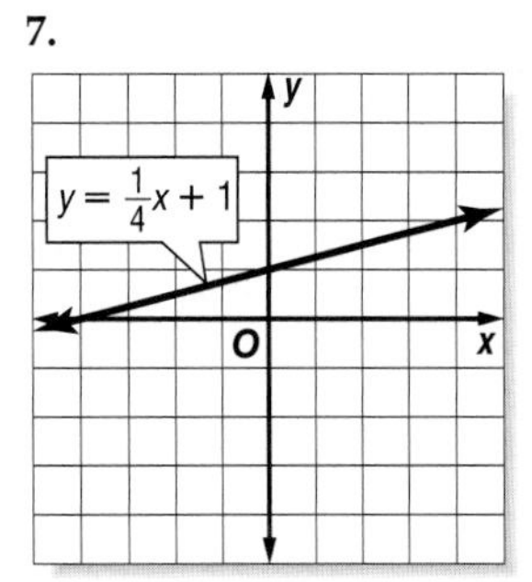

9.

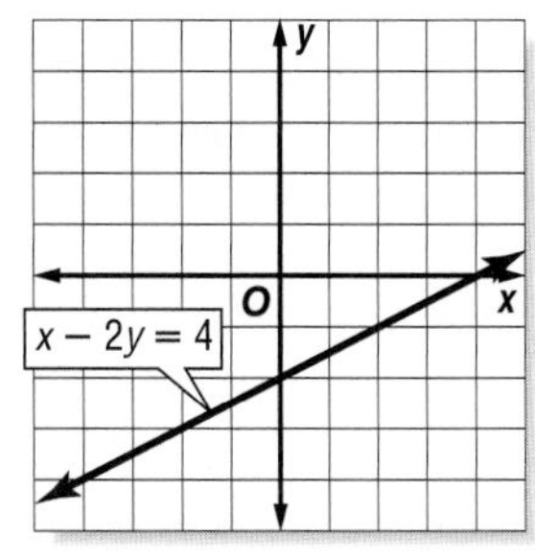

11. The y-intercept 25 represents the charge for a basic cake. Slope 1.5 represents the cost per additional slice.

13. 2; −4 **15.** −2; −3 **17.** 0; 4

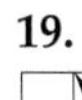

19.

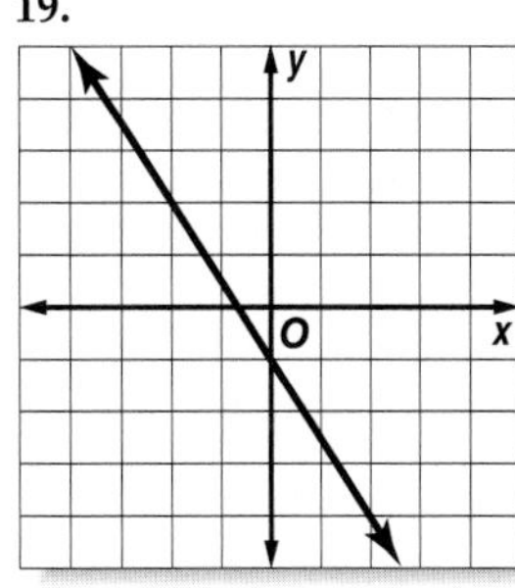

21.

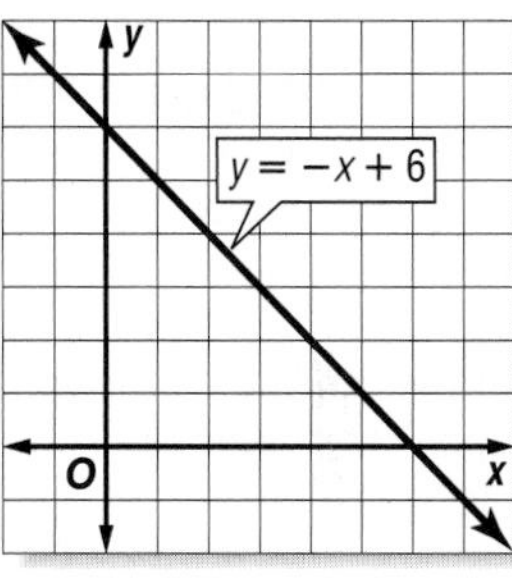

23.

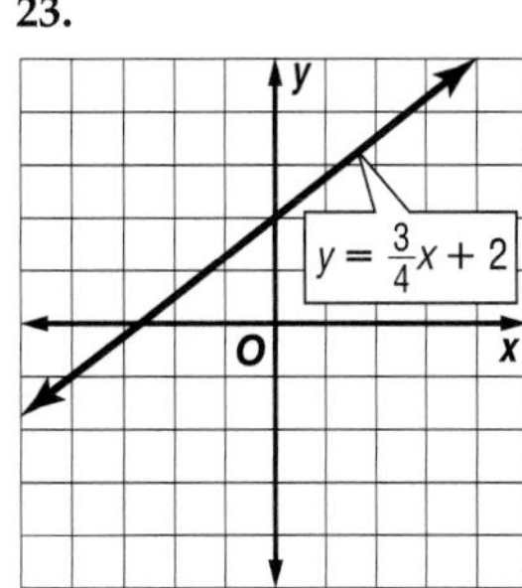

25.

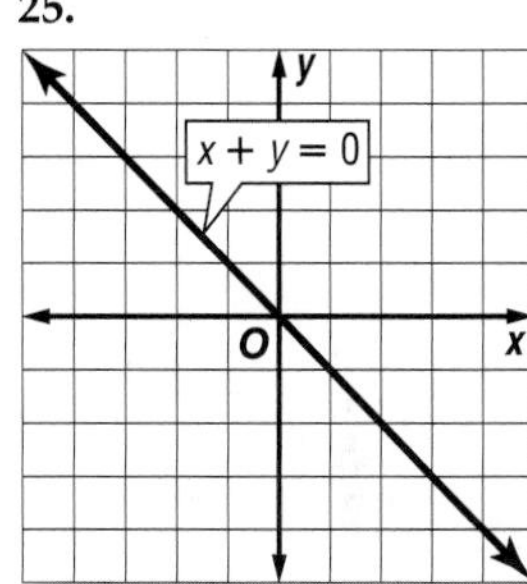

27.

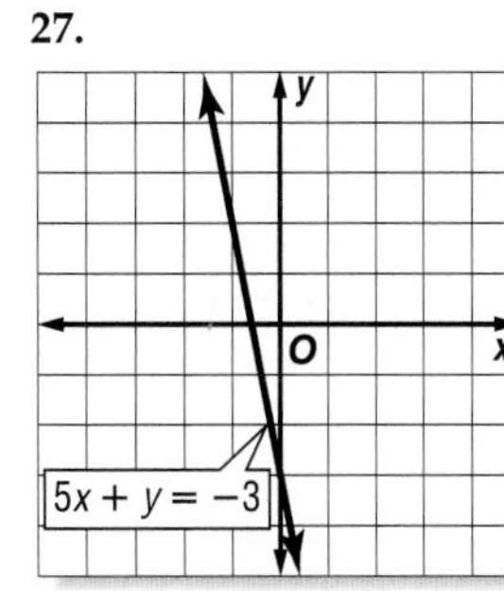

29.

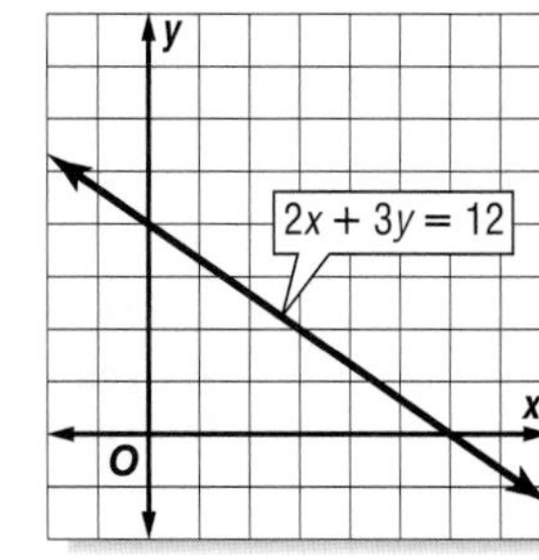

31.

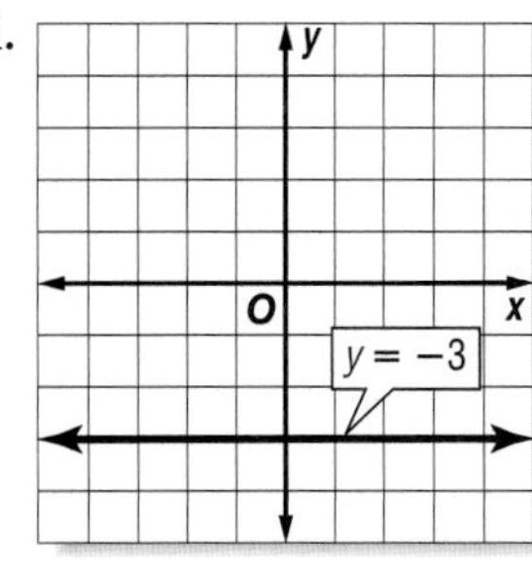

33. slope = −50, the descent in feet per minute; y-intercept = 300, initial altitude **35.** $-\frac{b}{m}$. Replace y with 0

Selected Answers

in $y = mx + b$ and solve for x. **37.** B **39.** $y = -4x$ **41.** 2 **43.** 4 **45.** $6x = 2x + 28$; 7 **47.** $-\frac{3}{5}$ **49.** 29 **51.** 11

Pages 407–408 Lesson 8-7

1. Sample answer: Find the y-intercept b and another point on the line. Use the points to determine the slope m. Then substitute these values in $y = mx + b$ and write the equation.

3. $y = \frac{1}{2}x + 1$ **5.** $y = -2x + 3$ **7.** $y = \frac{1}{2}x + 1$ **9.** $y = \frac{3}{4}x - 1$ **11.** $y = 2x + 6$ **13.** $y = 5$ **15.** $y = -\frac{1}{3}x + 8$ **17.** $y = 2x + 3$ **19.** $y = -2.5$ **21.** $y = -\frac{1}{2}x$ **23.** $y = x + 1$ **25.** $y = -x$ **27.** $y = 7$ **29.** $y = 4x - 3$ **31.** $y = 1088x$; The speed of sound is 1088 feet per second.

33a. $d = 0.5(0.7c \times 1.06)$ or $d = 0.371c$ **33b.** $18.55 **35.** C **37.** 6; 7 **39.** 3; −2 **41.** positive

Pages 410–413 Lesson 8-8

1. Sample answer: Use a ruler to extend the line so that it passes through the x value for which you want to predict. Locate the x value on the line and determine the corresponding y value. Or, write an equation for the best-fit line and substitute the desired value of x to find the corresponding value of y.

3. Sample answer:

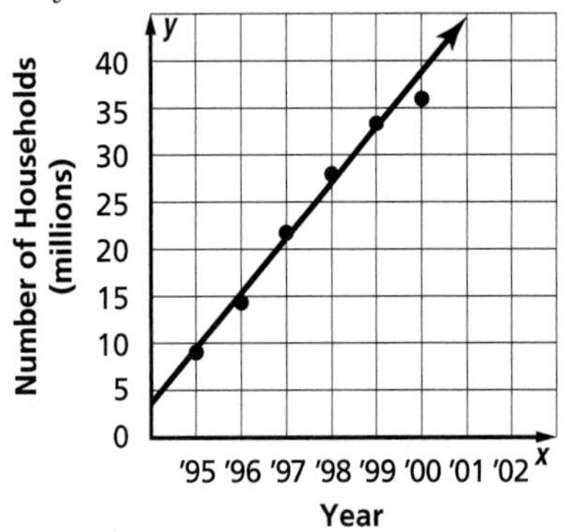

5. $y = 20x + 320$

7. Sample answer:

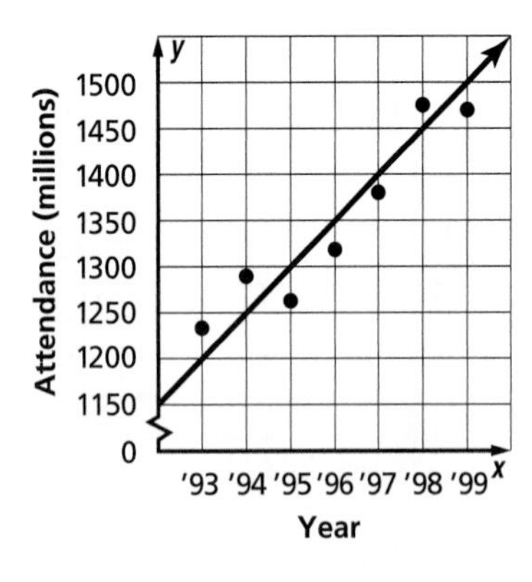

9. Sample answer:

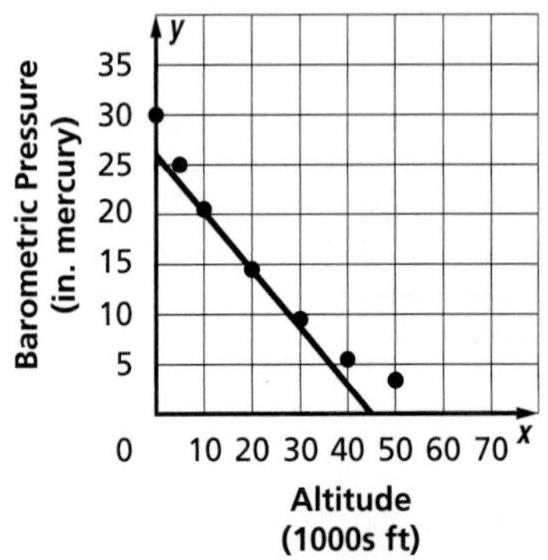

11. No, the equation gives a negative value for barometric pressure, which is not possible. Also, the data in the scatter plot do not appear to be linear. **13.** Sample answer: 238 in. **15.** As latitude increases, temperature decreases. **17.** Sample answer: 61.3°F **19.** The data describing the life expectancy for past generations can be displayed using a scatter plot. Then a line is drawn as close to as many of the points as possible. Then the line can be extended and used to predict the life expectancy for future generations. Answers should include the following.

- A best-fit line is a line drawn as close to as many of the data points as possible.
- Although the points may not be exactly linear, a best-fit line can be used to approximate the data set.

21. A **23.** $y = -2x + 2$ **25.** $y = -4$

27.

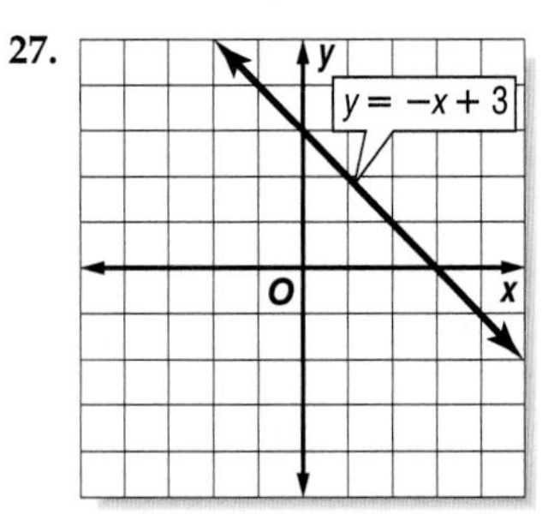

29. $n \leq 7$ **31.** $d > -\frac{3}{2}$ **33.** 2 **35.** 4.5 **37.** 3 **39.** −1 **41.** −2

Pages 416–418 Lesson 8-9

1. Sample answer: A group of two or more equations form a system of equations. The solution is the ordered pair that satisfies all the equations in the system. If the equations are graphed, the solution is the coordinates of the point where the graphs intersect. The system has infinitely many solutions if the equations are the same and the graphs coincide. **3.** (−5, −4)

5. no solution

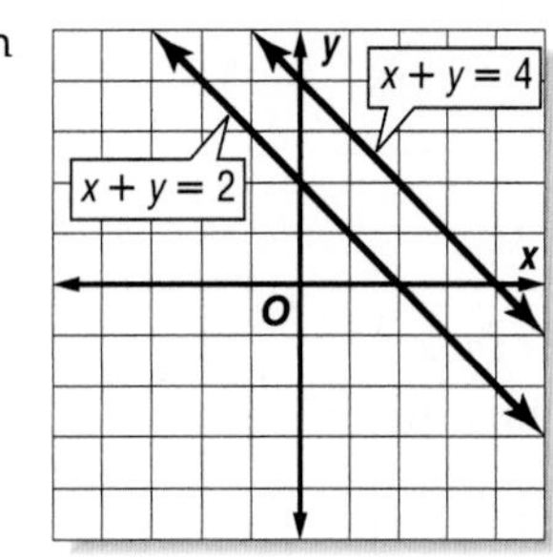

7. (2, 6) **9.** (3, −2) **11.** infinitely many

13. (3, 3)

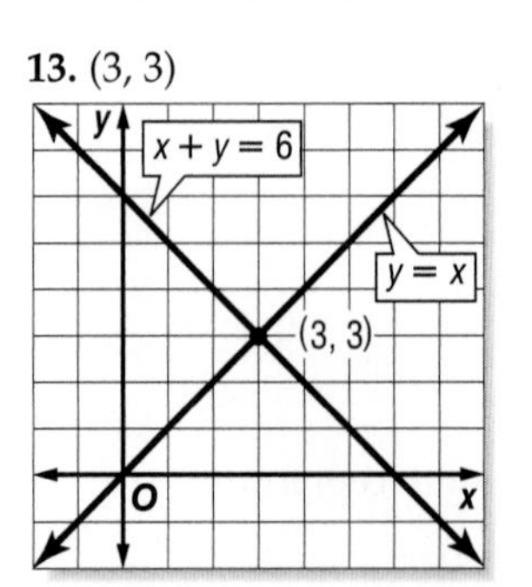

15. infinitely many

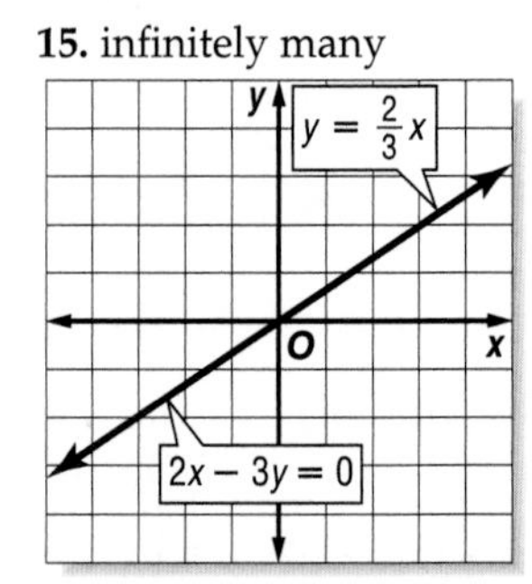

Selected Answers

17. (−2, 4)

19. (−2, 0) **21.** (−4, 10) **23.** (3, 6) **25.** (6, 11); an order of 6 pounds will cost the same, $11.00, for both sites.

27a. Sample answer: No; since slope represents rate, the slopes of the graphs are the same. Therefore, the runners will never meet.

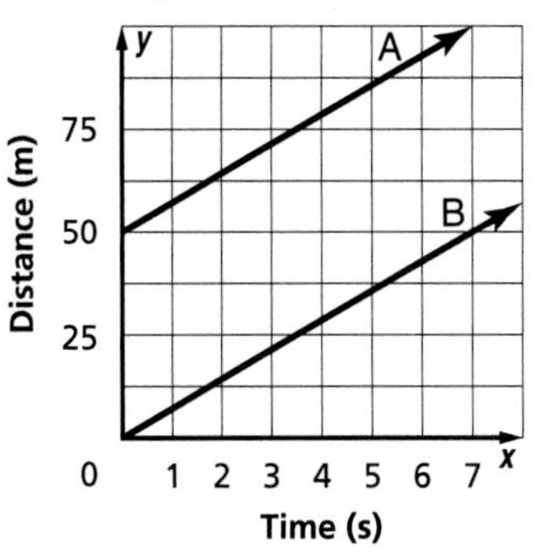

27b. Sample answer: The line representing runner B has a steeper slope than the line representing runner A. The intersection point represents the time and distance at which runner B catches up to runner A.

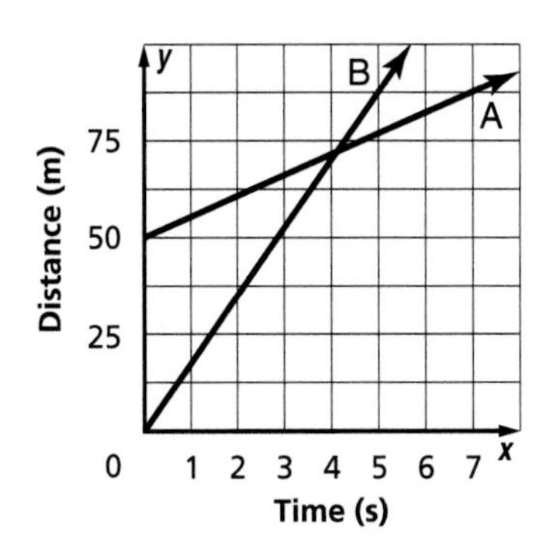

29. C **31.** positive slope **33.** $y = -3x$ **35.** yes **37.** no **39.** yes

Page 418 Practice Quiz 2

1. −1; 8 **3.** $-\frac{1}{2}$; 3 **5.** $y = 1$

7. Sample answer:

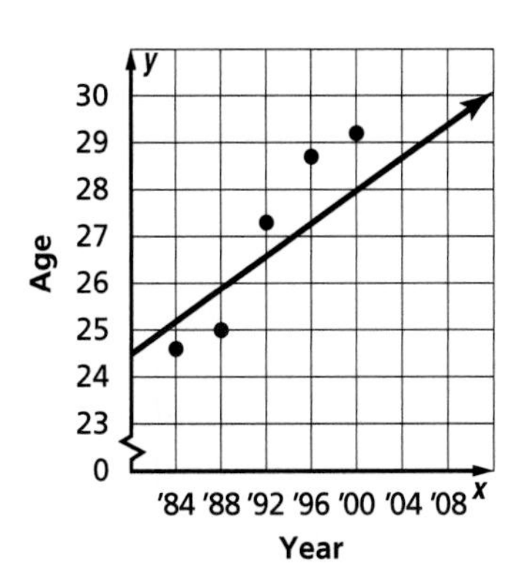

9. (0, 5)

Pages 421–422 Lesson 8-10

1. $y < -x + 3$ **3a.** Sample answer: (0, −1), (1, 0), (2, 1)
3b. Sample answer: (0, 0), (1, 1), (2, 2)

5.

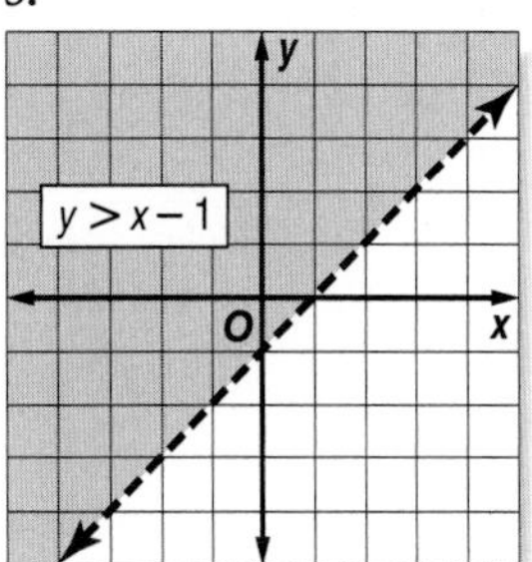

7.

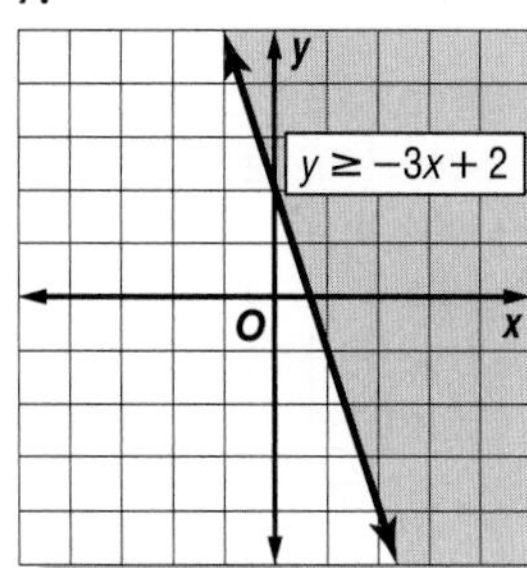

9.

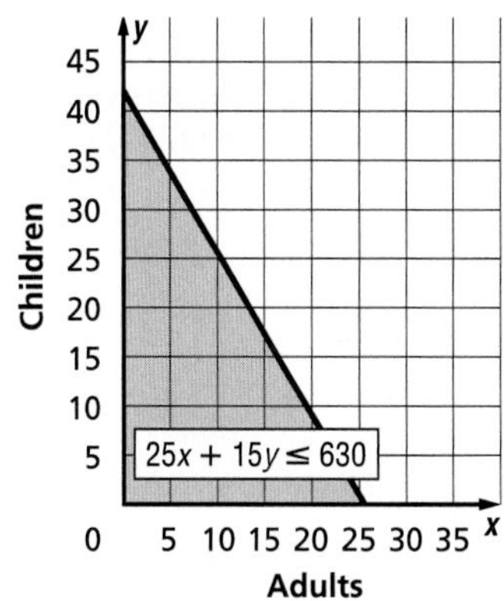

Sample answer:
5 adults, 30 children;
10 adults, 22 children;
15 adults, 15 children

11.

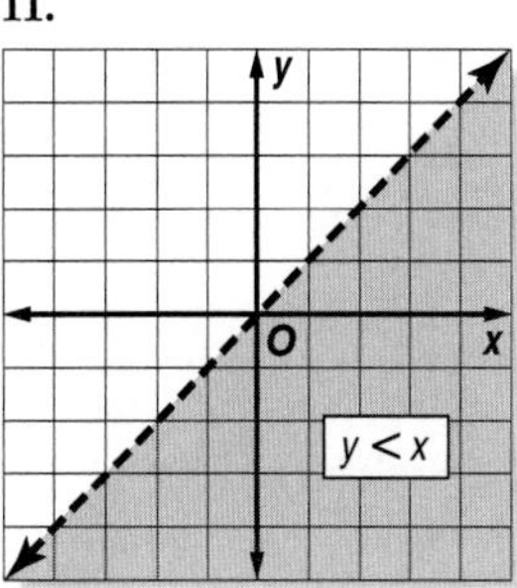

13.

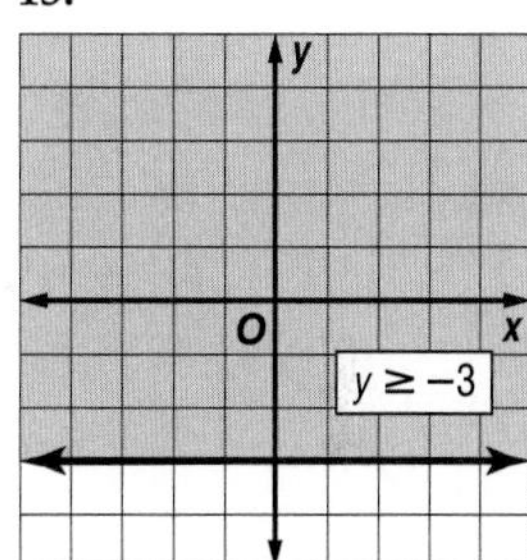

15.

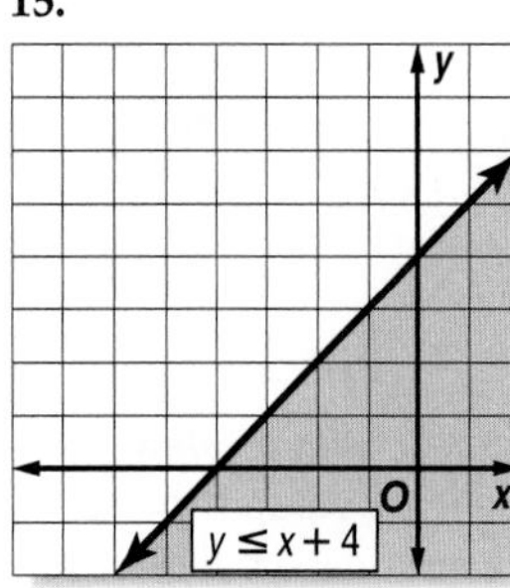

17.

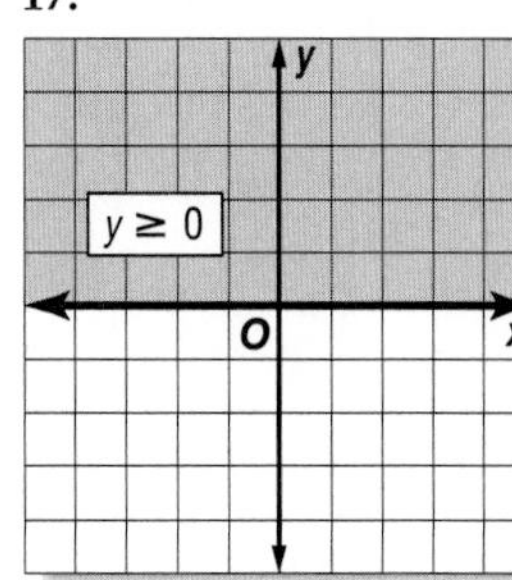

19.

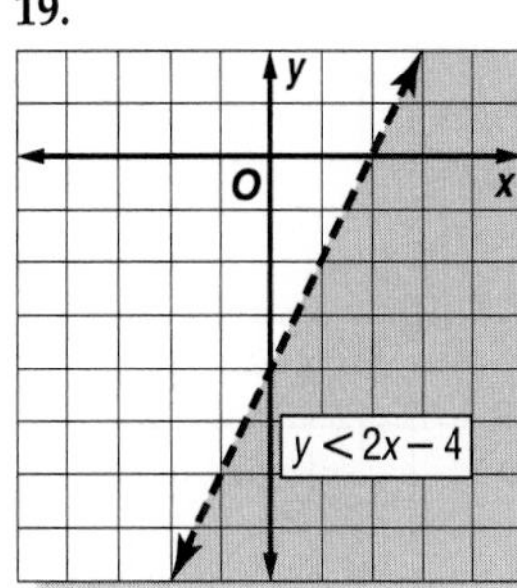

21.

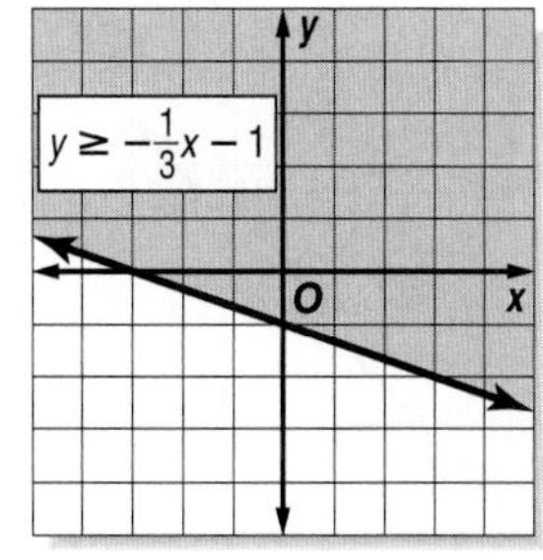

23.

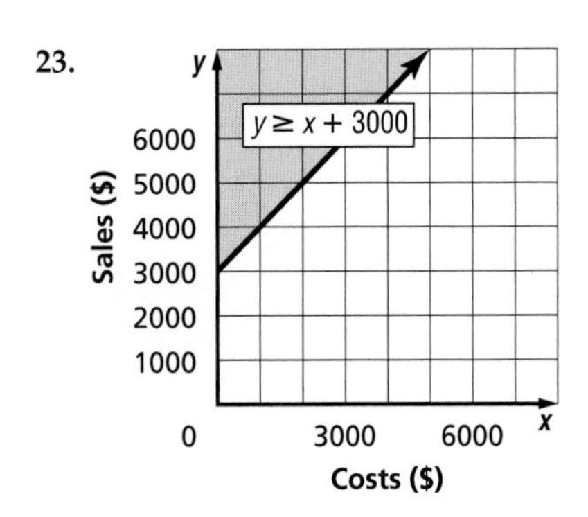

25. No; the number of sales and the costs cannot be negative. **27.** $10x + 25y \leq 1440$ **29.** Sample answer: 25 small, 45 large; 50 small, 30 large; 100 small, 10 large

31.

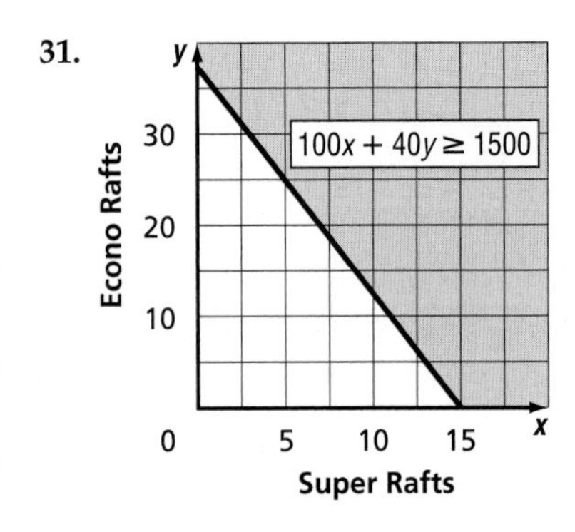

33a. $y > x - 1, y \leq -\frac{3}{2}x + 2$
33b. Sample answers: (0, 2), (−1, 2), (−3, −2) **35.** C

37. (0, 3)

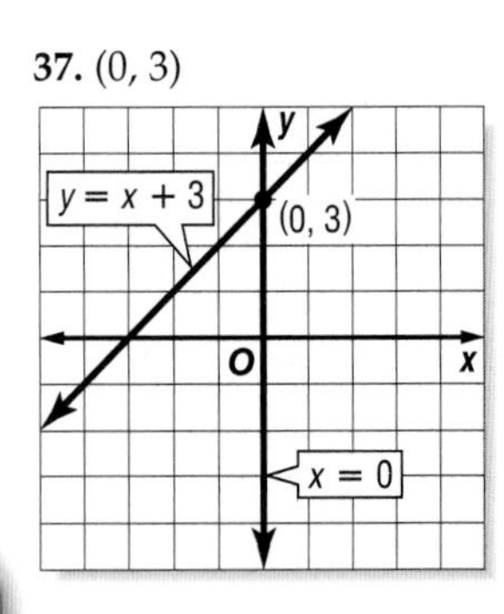

39. (2, −5)

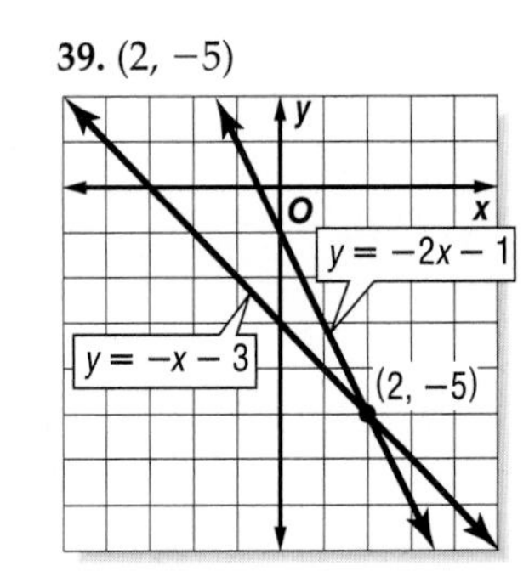

41. 0.4 **43.** $-0.\overline{5}$

Pages 424–428 Chapter 8 Study Guide and Review

1. d **3.** h **5.** a **7.** f **9.** c **11.** Yes; each x value is paired with only one y value. **13.** Yes; each x value is paired with only one y value.

15.

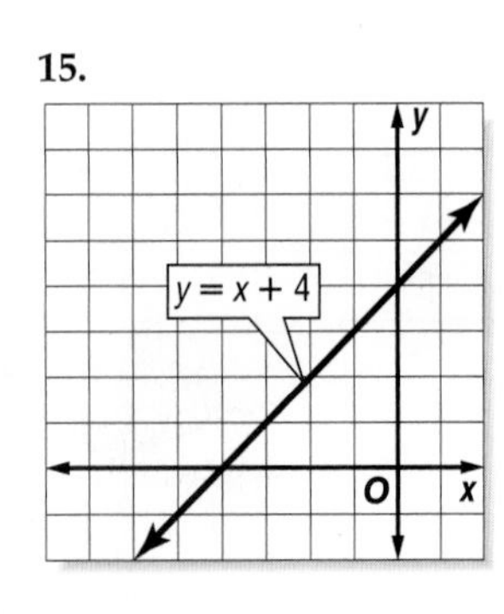

17.

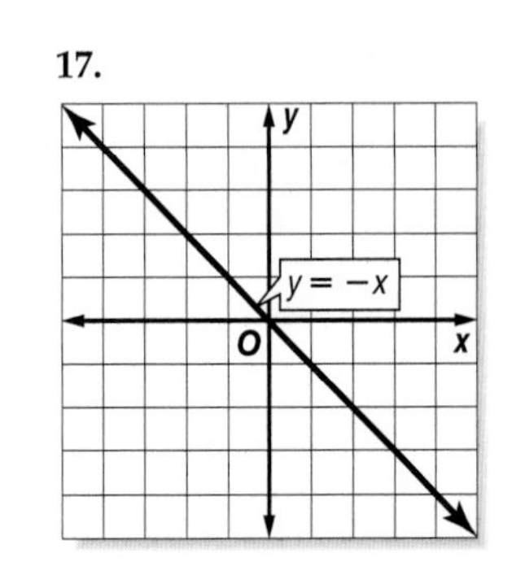

19.

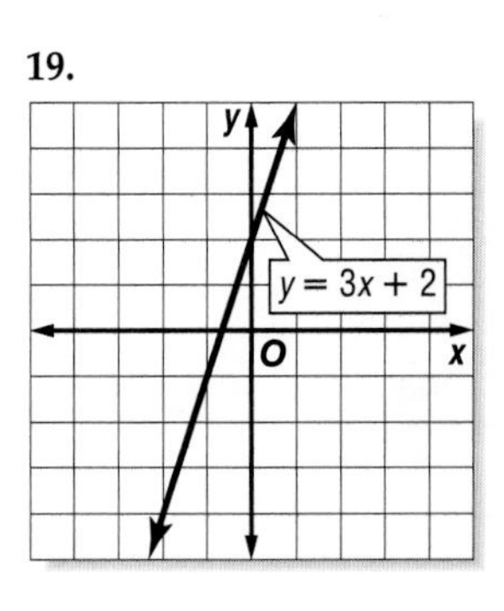

21.

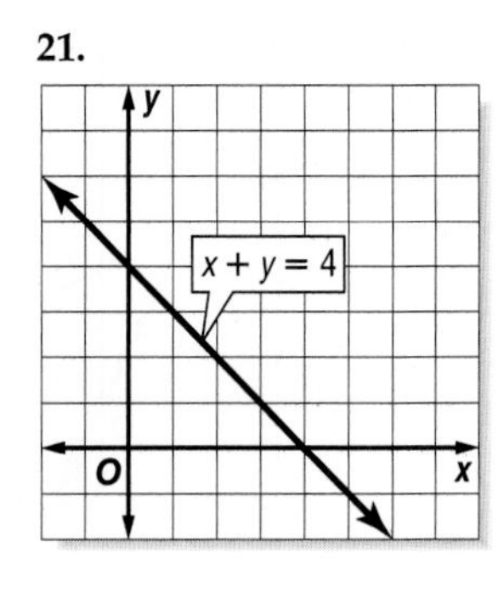

23.

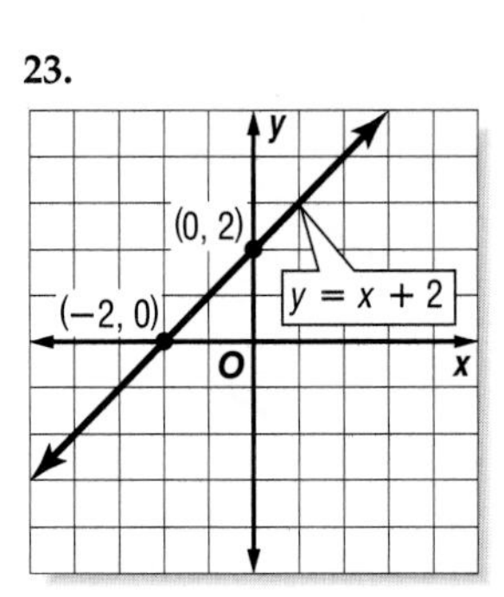

25.

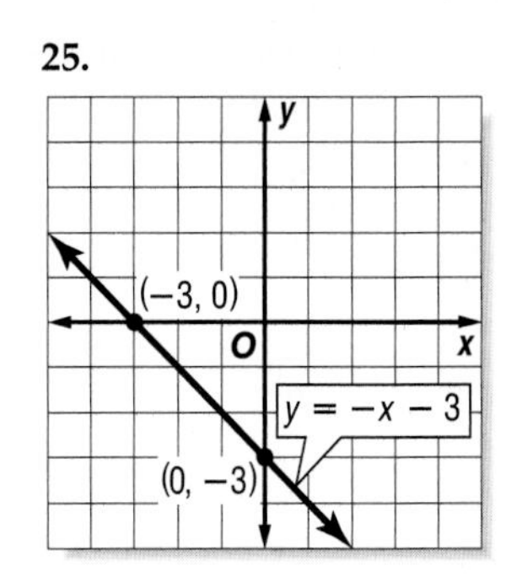

27.

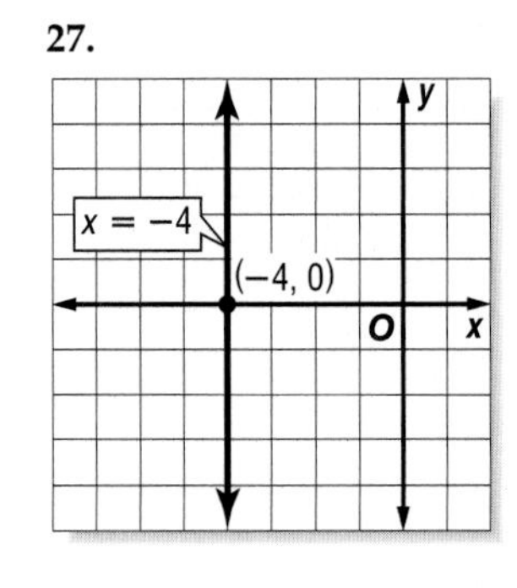

29.

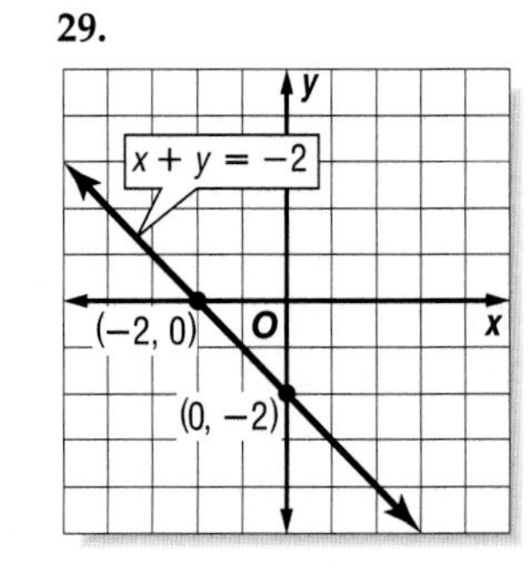

31. 1 **33.** $\frac{7}{8}$ **35.** undefined **37.** increase of 8 m/s

39. **41.**

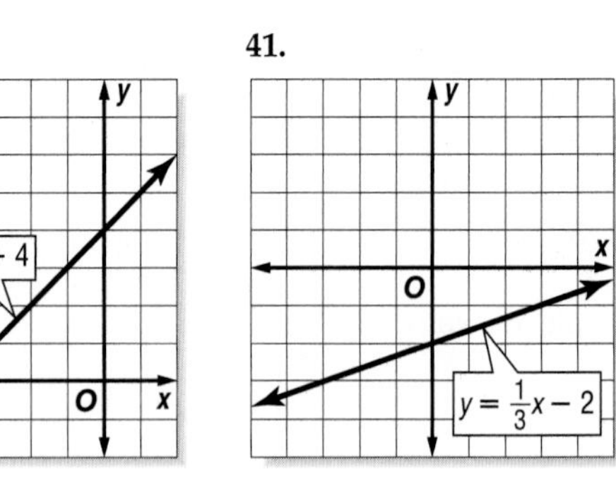

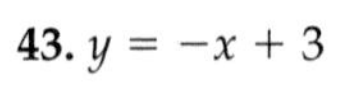

43. $y = -x + 3$

45. Sample answer:

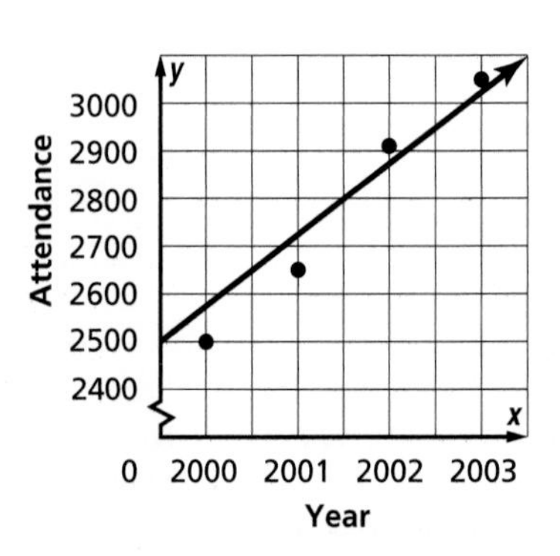

Selected Answers

47. (3, 3)

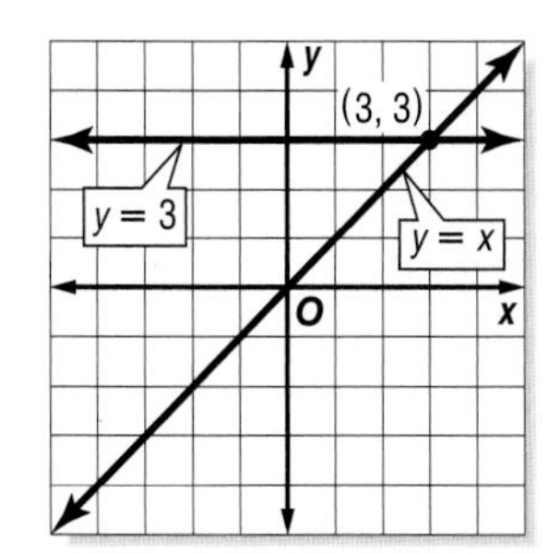

49. no solution

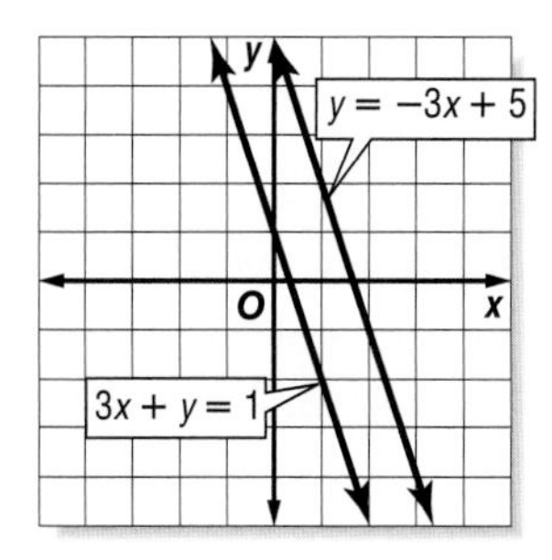

51. (4, 4)

53.

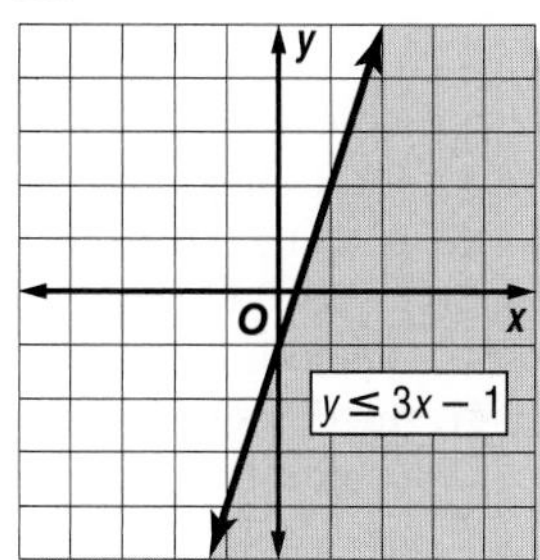

55.

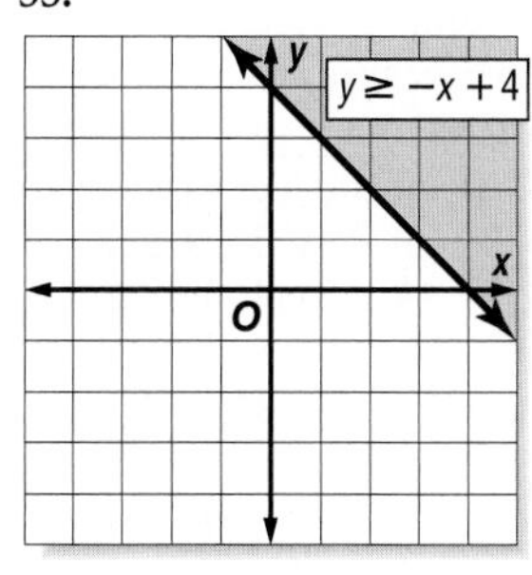

Chapter 9 Real Numbers and Right Triangles

Page 435 Chapter 9 Getting Started

1. $<$ **3.** $<$ **5.** $>$ **7.** $<$ **9.** 8 **11.** 60 **13.** 9 **15.** 11 **17.** 8 **19.** 34 **21.** 97

Pages 438–440 Lesson 9-1

1. A positive number squared results in a positive number, and a negative number squared results in a positive number. **3.** Sample answer: $-\sqrt{1.69}$

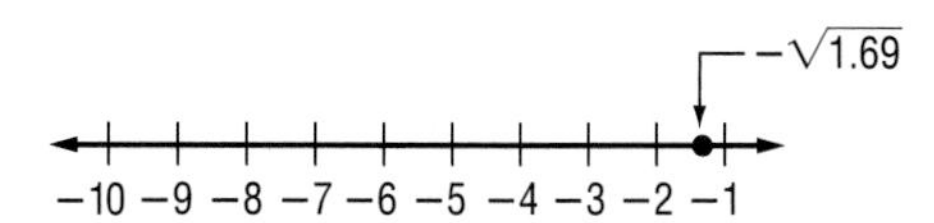

5. −8 **7.** 3.9 **9.** 8 **11.** 44.9 mi **13.** 6 **15.** −5 **17.** not possible **19.** 14 **21.** 18, −18 **23.** 1.5 **25.** 7.5 **27.** −9.3 **29.** 15.8 **31.** −1.7 **33.** 70.7 **35.** 9 **37.** −7 **39.** 14 **41.** −17 **43.** −2 **45.** 6 **47.** 14.8 mi **49.** 9; Since $64 < 65 < 81$, $\sqrt{64} < \sqrt{65} < \sqrt{81}$. Thus it follows that $8 < \sqrt{65} < 9$. So, 9 is greater than $\sqrt{65}$. **51a.** 10.4 in. **51b.** 14.2 cm **51c.** 8.4 m **53.** 10 **55.** 2 **57.** Sample answer: A number that has a rational square root will have an ending digit of 0, 1, 4, 5, 6, or 9. The last digit is the ending digit in one of the squares from 1–100. There are just six ending digits. **59.** D **61.** Sample answer: addition and subtraction **63.** a

65.

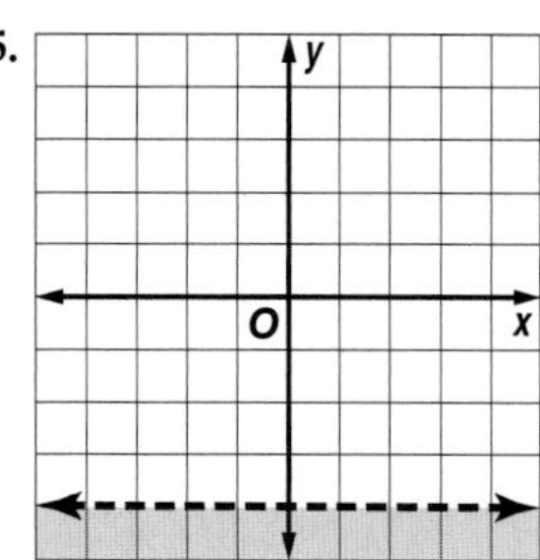

67. (5, 7) **69.** (−4, −3) **71.** $\frac{2}{3}$ **73.** 9 **75.** It can be written as a fraction. **77.** It can be written as $\frac{3}{4}$. **79.** It can be written as $\frac{6}{1}$.

Pages 443–445 Lesson 9-2

1. Whereas rational numbers can be expressed in the form $\frac{a}{b}$, where a and b are integers and b does not equal 0, irrational numbers cannot. **3.** N, W, Z, Q **5.** Q **7.** $<$ **9.** $\frac{10}{3}$, $3\frac{3}{5}$, $\sqrt{13}$, $3.\overline{7}$ **11.** 8.6, −8.6 **13.** N, W, Z, Q **15.** Q **17.** Q **19.** I **21.** I **23.** Z, Q **25.** Q **27.** Z, Q **29.** always **31.** always **33.** $>$ **35.** $<$ **37.** $=$ **39.** $\frac{6}{5}$, $\sqrt{4}$, $2.\overline{1}$, $5\frac{1}{4}$ **41.** $-10\frac{1}{2}$, $-\sqrt{105}$, −10, −1.05 **43.** Sample answer: $\sqrt{4}$ and $\sqrt{49}$ **45.** 7, −7 **47.** 4.7, −4.7 **49.** 12, −12 **51.** 11.3, −11.3 **53.** 1.3, −1.3 **55.** 1.9, −1.9 **57.** 12 or −12 **59.** 4 or −4 **61.** 36

63. If a square has an area that is not a perfect square, the lengths of the sides will be irrational. Answers should include the following.

- Area = 56 in^2 Area = 64 in^2

65. B **67.** 2621.2 ft^2 **69.** 7 **71.** 3 **73.** Sample answer: (0, 5), (1, 6), (−6, 2) **75.** $x > 5$ **77.** \$0.53/cupcake

79. 3:00

81.

10:00

83.

7:45

Pages 449–451 Lesson 9-3

1. N; $\overrightarrow{NM}$, $\overrightarrow{NP}$; ∠1, ∠MNP, ∠PNM, ∠N **3.** 20°; acute **5.** 70°; acute **7.** acute (55°) **9.** straight angle

11. 15°; acute **13.** 85°; acute **15.** 135°; obtuse **17.** 60°; acute **19.** 90°

21. acute (40°)

23. acute (65°)

25.

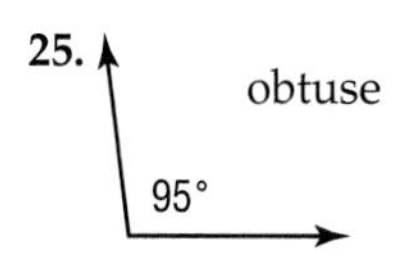

27. obtuse (155°)

29. 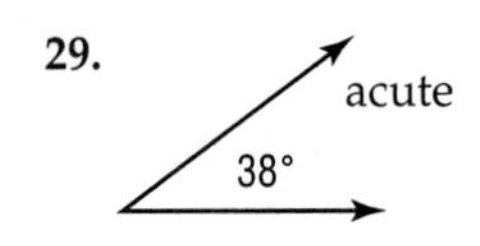

31a. obtuse **31b.** right **31c.** acute **33.** Intense: about 68°; Moderate: about 104°; Light: 72°; No standard routine: about 47°; Don't exercise regularly: about 68° **35.** 24 **37.** C **39.** Z, Q **41.** I **43.** 4 **45.** $x - 7 \geq 22$ **47.** 99 **49.** 60 **51.** 18

Page 451 Practice Quiz 2

1. 6 **3.** 8.2, −8.2 **5.** obtuse

Pages 455–457 Lesson 9-4

1. Whereas an isosceles triangle has at least two sides congruent, an equilateral triangle has three sides congruent.
3. Sample answer: 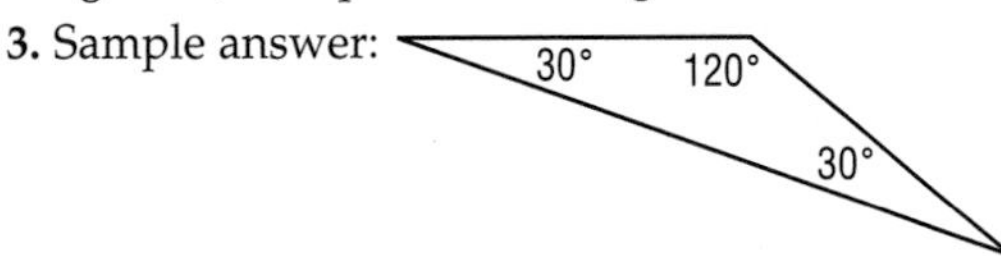

5. 90; right **7.** right isosceles **9.** acute isosceles **11.** 27; right **13.** 112; obtuse **15.** 90; right **17.** 20°, 60°, 100° **19.** acute equilateral **21.** right scalene **23.** acute scalene **25.** always **27.** obtuse **29.** acute **31.** not possible **33.** not possible **35.** 45°; 50°; 85° **37.** 10, 15, 21 **39.** B **41.** 180°; straight **43.** 145°; obtuse **45.** 9, −9 **47.** 9.2, −9.2 **49.** 144 **51.** 324 **53.** 729

Pages 462–464 Lesson 9-5

1. Sample answer: 8, 15, 17 **3.** 25 **5.** 15 **7.** no **9.** B **11.** 26 **13.** 12.5 **15.** 18.9 **17.** 23 in. **19.** 12 **21.** 12.1 **23.** 37.3 **25.** 28 **27.** 33.5 **29.** yes **31.** no **33.** no **35.** 32 in.
37. There is enough information to find the lengths of the legs. Since the right triangle is an isosceles right triangle, we know that the lengths of the legs are equal. So, in the Pythagorean Theorem, we can say that $a = b$. In addition, we know that $c = 8$.

$c^2 = a^2 + b^2$	*Pythagorean Theorem*
$8^2 = a^2 + a^2$	*$a = b$ and $c = 8$*
$8^2 = 2a^2$	*Add a^2 and a^2.*
$64 = 2a^2$	*Evaluate 8^2.*
$\frac{64}{2} = \frac{2a^2}{2}$	*Divide each side by 2.*
$32 = a^2$	*Simplify.*
$\sqrt{32} = a$	*Take the square root of each side.*

39. D **41.** 17 units **45.** 90; right **49.** $n \geq -9$ **51.** 5

Pages 468–470 Lesson 9-6

1. the point halfway between two endpoints
3. Sample answer:

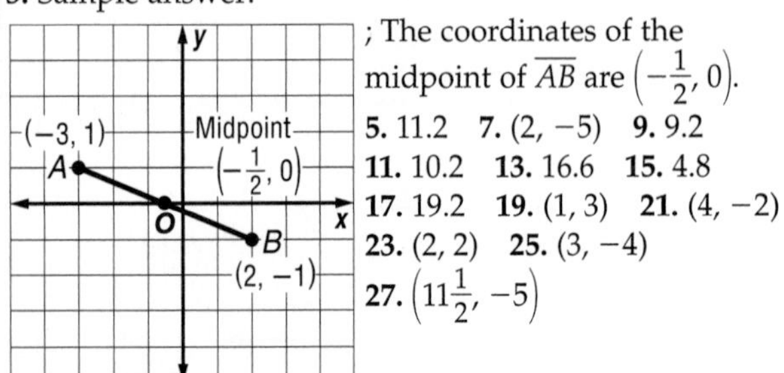

; The coordinates of the midpoint of $\overline{AB}$ are $\left(-\frac{1}{2}, 0\right)$. **5.** 11.2 **7.** (2, −5) **9.** 9.2 **11.** 10.2 **13.** 16.6 **15.** 4.8 **17.** 19.2 **19.** (1, 3) **21.** (4, −2) **23.** (2, 2) **25.** (3, −4) **27.** $\left(11\frac{1}{2}, -5\right)$

29. Yes; $\overline{PM}$ and $\overline{MN}$ have equal measures. **31.** (−2, 3) **33.** C **35.** 9.2 **37.** 35.1 **39.** 2, 1 **41.** –4, –6 **43.** 28 **45.** 21 **47.** 59

Page 470 Practice Quiz 2

1. obtuse isosceles **3.** Yes; $34^2 = 30^2 + 16^2$ **5.** (−1, 5)

Pages 473–475 Lesson 9-7

1. Sample answer:

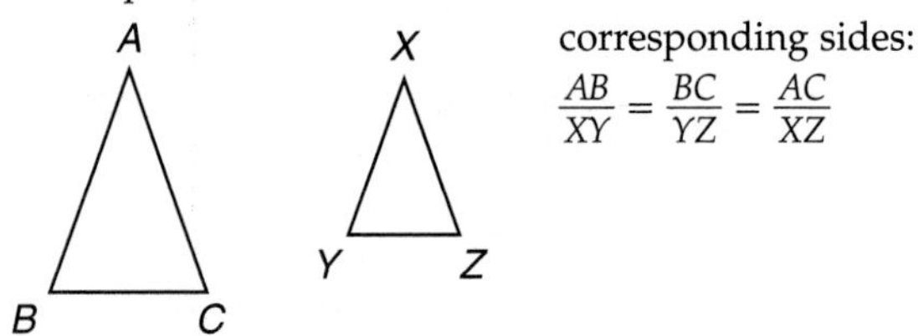

corresponding sides: $\frac{AB}{XY} = \frac{BC}{YZ} = \frac{AC}{XZ}$

3. $\frac{x}{15} = \frac{6}{9}$; 10 **5.** 6 km **7.** $\frac{x}{4} = \frac{10}{5}$; 8 **9.** $\frac{x}{18} = \frac{20}{15}$; 24 **11.** $\frac{x}{3} = \frac{7}{4.2}$; 5 **13.** always **15.** 68 yd **17.** 6 ft **19.** Sample answer: 4 units by 6 units by 8 units and 1 unit, 1.5 units, 2 units **21.** A **23.** 3.6 **25.** 7.1 **27.** 77.4 **29.** $\frac{1}{4}$ **31.** $1\frac{1}{2}$ **33.** 0.7143 **35.** 0.225

Pages 479–481 Lesson 9-8

1. The sine ratio compares the measure of the leg opposite the angle to the measure of the hypotenuse. The cosine ratio compares the measure of the leg adjacent to the angle to the measure of the hypotenuse. The tangent ratio compares the measure of the leg opposite the angle to the measure of the leg adjacent to the angle. **3.** 0.8824 **5.** 1.875 **7.** 0.9455 **9.** 25.6 **11.** 7.8 **13.** 0.9231 **15.** 0.7241 **17.** 2.4 **19.** 0.8944; 0.4472; 2.0 **21.** 0.1045 **23.** 0.8572 **25.** 0.2126 **27.** 6.8 **29.** 24.9 **31.** 97.6 **33.** about 21 m **35.** about 44 m **37.** 24.0 **39.** 40.4 **41.** sin 45° = cos 45°; sin 60° = cos 30°; sin 30° = cos 60° **43.** To find heights of buildings. Answers should include the following.

- If two of the three measures of the sides of a right triangle are known, you can use the Pythagorean Theorem to find the measure of the third side. If the measure of an acute angle and the length of one side of a right triangle are known, trigonometric ratios can be used to find the missing measures.
- 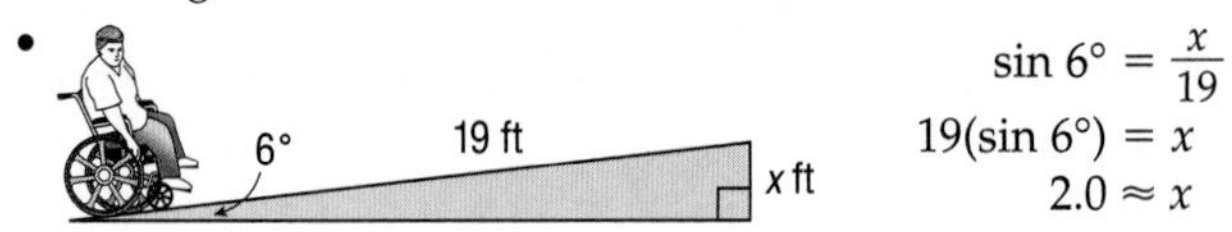

$\sin 6° = \frac{x}{19}$
$19(\sin 6°) = x$
$2.0 \approx x$

The height of the ramp is about 2 feet.
45. C **47.** $\left(1\frac{1}{2}, \frac{1}{2}\right)$ **49.** (−3, −4) **51.** $a < -6$

Pages 483–486 Chapter 9 Study Guide and Review

1. real numbers **3.** obtuse **5.** hypotenuse **7.** trigonometric ratio **9.** 6 **11.** –9 **13.** not possible **15.** 9, −9 **17.** 1.2, −1.2 **19.** 90°; right **21.** 35°; acute **23.** obtuse isosceles **25.** 13.7 **27.** 15.9 **29.** 9.4 **31.** (3, 5) **33.** (0, −2) **35.** $\frac{AB}{HJ} = \frac{BC}{JK}$; 10 **37.** 1.3333

Chapter 10 Two-Dimensional Figures

Page 491 Chapter 10 Getting Started

1. 44 **3.** 51 **5.** 28 **7.** 39 **9.** 44 **11.** 71.8 **13.** 12.5 **15.** 10.7 **17.** $10\frac{1}{6}$ **19.** $3\frac{7}{8}$ **21.** $5\frac{3}{8}$ **23.** $6\frac{2}{9}$

Pages 495–497 Lesson 10-1

1. Complementary angles have a sum of 90° and supplementary angles have a sum of 180°. **3.** 56° **5.** 124° **7.** 28 **9.** D **11.** 127° **13.** 53° **15.** 127° **17.** 148 **19.** 175

21. 9 **23.** 41° **25.** 73° **27.** 112° **29.** 5 **31.** 9 **33.** They are supplementary. **35.** A **37.** Sample answer: Both graphs intersect to form right angles. **39.** Sample answer: The slopes of the graphs of perpendicular lines are negative reciprocals of each other. **41.** 0.8910 **43.** 7.5 **45.** $-9m$

47. 45° **49.** 145°

Pages 502–504 Lesson 10-2

1. They have the same size and shape. **3.** $\angle J \cong \angle C$, $\angle K \cong \angle B$, $\angle M \cong \angle G$, $\overline{KM} \cong \overline{BG}$, $\overline{MJ} \cong \overline{GC}$, $\overline{KJ} \cong \overline{BC}$; $\triangle BGC$ **5.** $\angle M$ **7.** $\overline{MN}$ **9.** 15 ft **11.** $\angle K \cong \angle N$, $\angle J \cong \angle P$, $\angle M \cong \angle M$, $\overline{KJ} \cong \overline{NP}$, $\overline{JM} \cong \overline{PM}$, $\overline{KM} \cong \overline{NM}$; $\triangle NPM$ **13.** $\angle Z \cong \angle S$, $\angle W \cong \angle T$, $\angle Y \cong \angle R$, $\overline{ZW} \cong \overline{ST}$, $\overline{WY} \cong \overline{TR}$, $\overline{ZY} \cong \overline{SR}$; $\triangle STR$ **15.** sometimes **17.** 30° **19.** $\angle F$ **21.** $\angle A$ **23.** $\overline{HF}$ **25.** $\overline{DC}$ **27.** 20 **29.** 6 **31.** Sample answer: $\triangle ZQN$ **33.** Sample answer: $\triangle WQO$ **35.** Congruent triangles can be found on objects in nature like leaves and animals. Answers should include the following.

- Congruent triangles are triangles with the same angle measures and the same side lengths.
- Sample answer: A bird's wings when extended are an example of congruent angles. Another example of congruent angles would be the wings of a butterfly.

37. D **39.** 0.1219 **41.** 1.1585 **43.** -6

45–49.

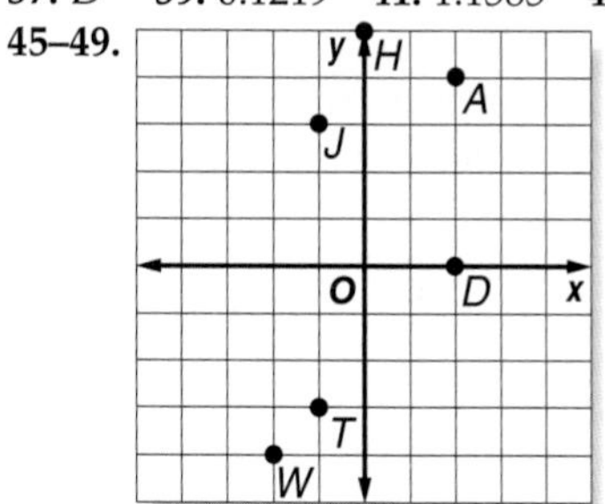

Pages 509–511 Lesson 10-3

1. The figure is moved 5 units to the right and 2 units down.

3.

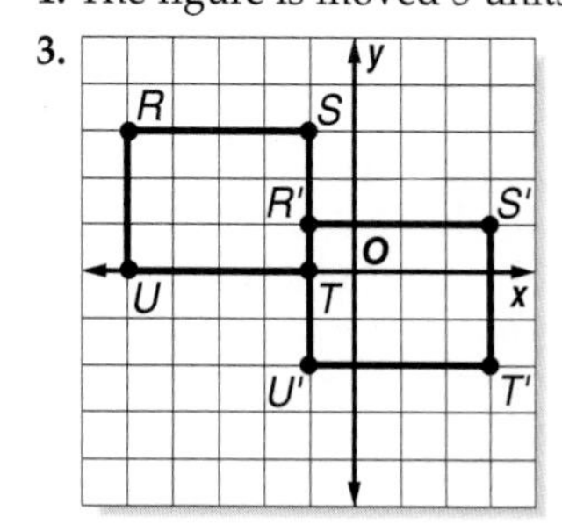

5.

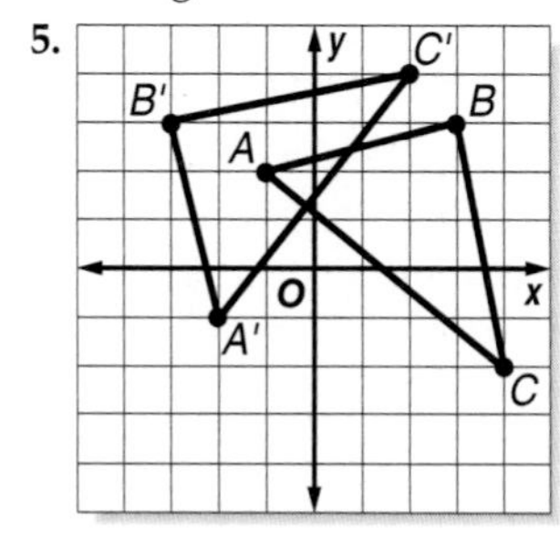

7. $A'(-2, 5)$, $B'(1, 3)$, $C'(-2, 0)$;

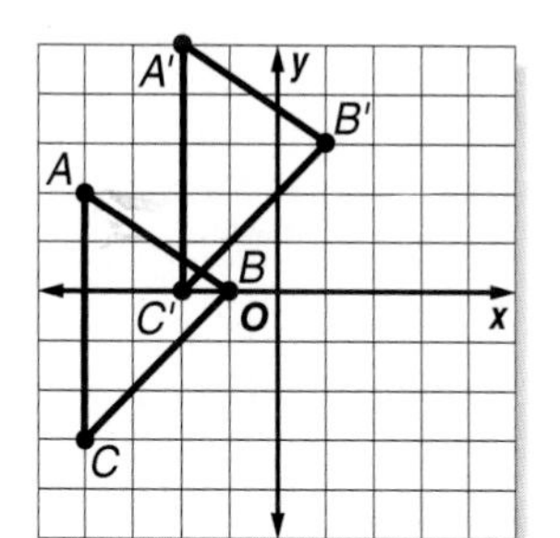

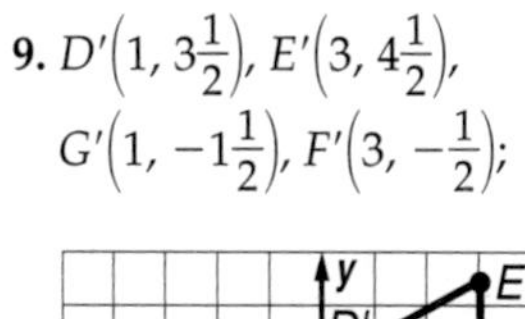

9. $D'\left(1, 3\frac{1}{2}\right)$, $E'\left(3, 4\frac{1}{2}\right)$, $G'\left(1, -1\frac{1}{2}\right)$, $F'\left(3, -\frac{1}{2}\right)$;

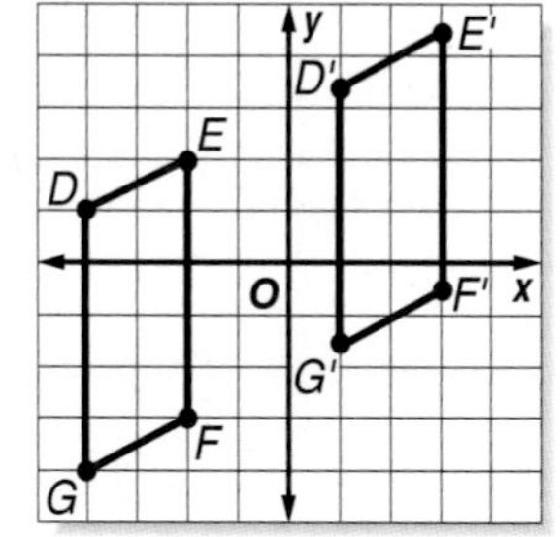

11. $R'(-4, -3)$, $S'(0, -3)$, $T'(-4, 1)$;

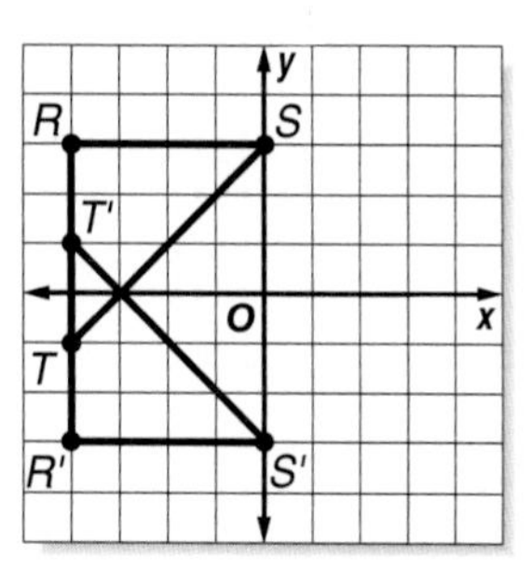

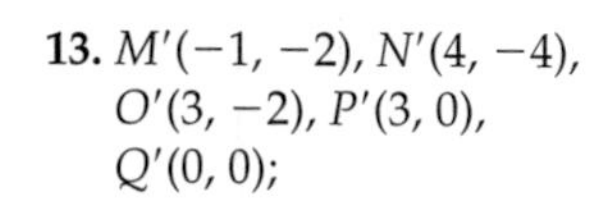

13. $M'(-1, -2)$, $N'(4, -4)$, $O'(3, -2)$, $P'(3, 0)$, $Q'(0, 0)$;

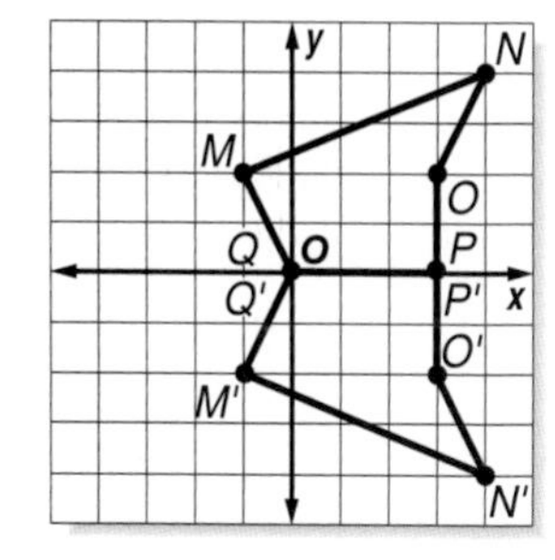

15.

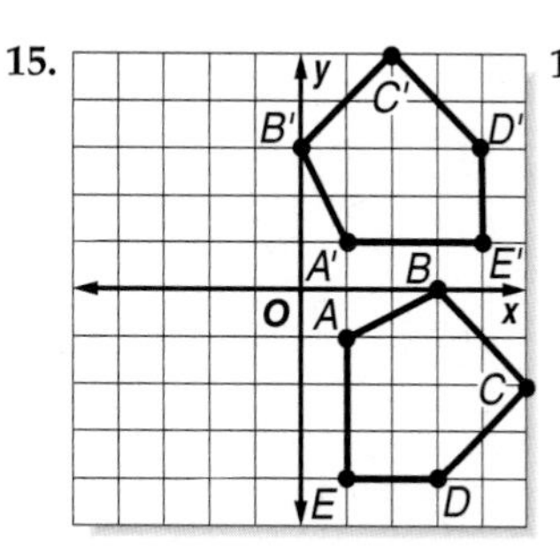

17.

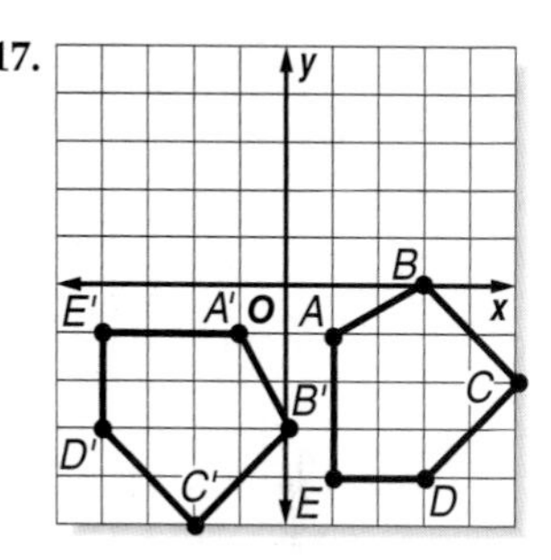

19. translation

21. Sample answer:

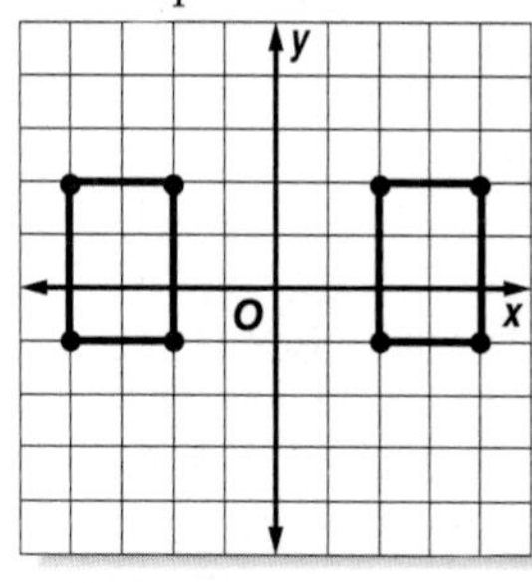

; The image of the figure's reflection is the same as the image of its translation to the right 6 units.

23. reflection **25.** Since many of the movements used in recreational activities involve rotating, sliding, and flipping, they are examples of transformations. Answers should include the following.

- a translation is a slide, a reflection is a mirror-image, and a rotation is a turn.
- sample answer: the swing represents a rotation, the scooter represents a translation, and the skateboard represents reflection.

27. C **29.** $\overline{DF}$ **31.** 110°

33. $x \geq 9.6$; 7.0 8.0 9.0 10.0

35. 113 **37.** 47 **39.** 106

Pages 515–517 Lesson 10-4

1. Sample answer: A textbook is an example of a quadrilateral; the tiles in a shuffleboard scoring region are examples of parallelograms; a "dead end" road sign is an example of a rhombus; and a floppy disk is an example of a square. **3.** 110; 110° **5.** rectangle **7.** trapezoids, parallelograms **9.** 102; 102° **11.** 60; 60°; 60°; 120° **13.** 70; 70°; 80°; 120° **15.** Sample answer: A chessboard; it is a square because it is a parallelogram with 4 congruent sides and 4 right angles. **17.** square **19.** parallelogram **21.** rhombus **23.** always **25.** sometimes **27.** quadrilateral

29a. Yes; a rhombus is equilateral but may not be equiangular.

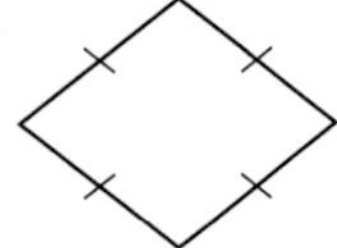

29b. Yes; a rectangle is equiangular but may not be equilateral.

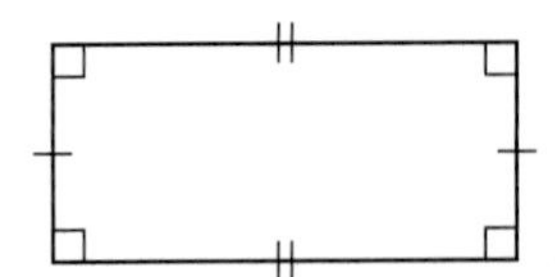

31. B **33.**

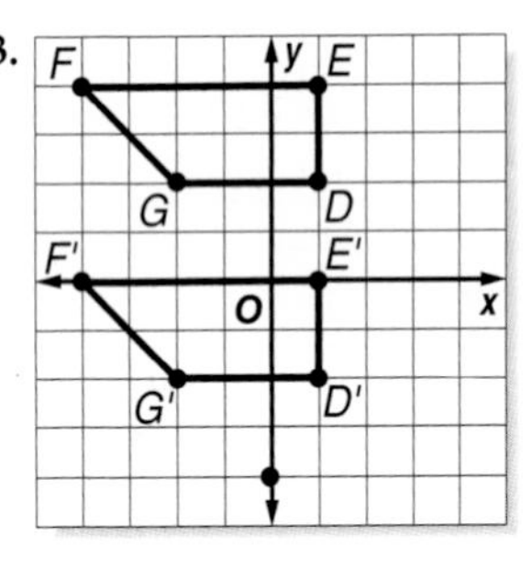

35. $\overline{MA}$ **37.** 14.4 **39.** 28.52

Page 517 Practice Quiz 1

1. 35° **3.** $Q'(3, -3)$, $R'(5, -6)$, $S'(7, -3)$ **5.** 60; 60°; 120°

Pages 523–525 Lesson 10-5

1. Sample answer:

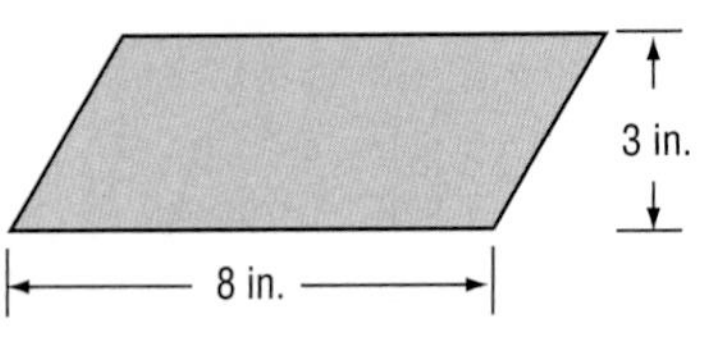

3. 8 ft^2 **5.** 60 m^2 **7.** 11 m^2 **9.** 90 cm^2 **11.** 43.99 cm^2 **13.** 28 in^2 **15.** 22.8 yd^2 **17.** 36.75 m^2 **19.** about 95,284 mi^2 **21.** 147 km^2 **23.** 57 ft^2 **25.** 5 bags **27.** 16 in. **29.** 24.5%
31. The area of a parallelogram is found by multiplying the base and the height of the parallelogram. The area of a rectangle is found by multiplying the length and the width of the rectangle. Since in a parallelogram, the base is the length of the parallelogram, and the height is the width of the parallelogram, both areas are found by multiplying the length and the width. Answers should include the following.

- Parallelograms and rectangles are similar in that they are quadrilaterals with opposite sides parallel and opposite sides congruent. They are different in that rectangles always have 4 right angles.
- 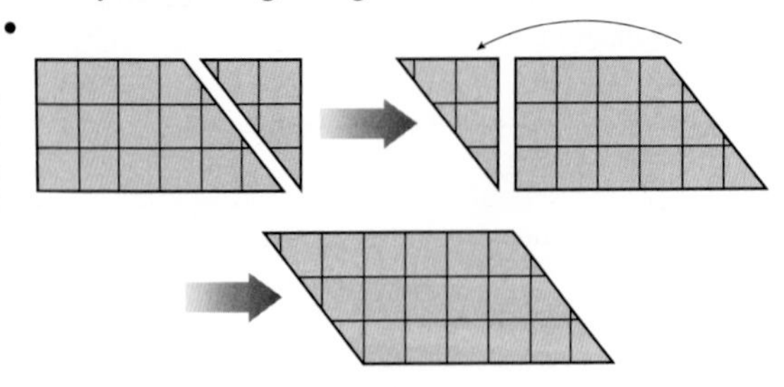

33. B **35.** 24; 24°; 96° **37.** 30 **39.** 26 **41.** 900 **43.** 1260

Pages 529–531 Lesson 10-6

1.

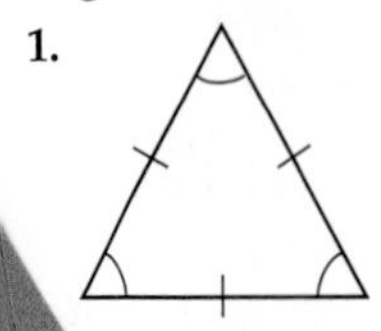

3. The number of triangles is 2 less than the number of sides. **5.** octagon; regular **7.** 128.6° **9.** hexagon; regular **11.** nonagon; not regular **13.** decagon, not regular **15.** 540° **17.** 1440°
19. 2880° **21.** Sample answer: hexagon, triangle, decagon, quadrilateral, and pentagon **23.** 140° **25.** 135° **27.** 150° **29.** octagons, squares **31.** 180 cm **33.** 58.5 in.
35. In tessellations, polygons are fit together to create a pattern such that there are no gaps or spaces. Answers should include the following.

- Sample answer: 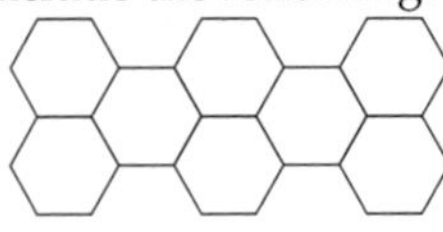
- Sample answer: 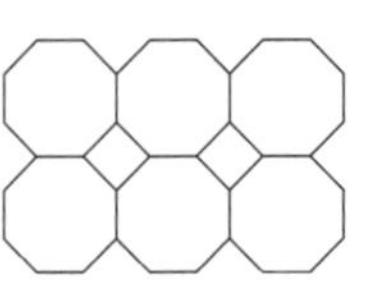

37. B **39.** 120° **41.** 60° **43.** 30° **45.** 18 cm^2 **47.** quadrilateral **49.** $16.4x$ **51.** 33.9 **53.** 18.1

Pages 536–538 Lesson 10-7

1. Multiply 2 times π times the radius. **3.** Mark; since the diameter of the circle is 7 units, its radius is 3.5 units. Thus, the area of the circle is $\pi \cdot (3.5)^2$ or 38.5 square units.
5. 50.3 m; 201.1 m^2 **7.** 8.2 km; 5.3 km^2 **9.** about 9.1 mi^2 **11.** 40.8 in.; 132.7 in^2 **13.** 131.9 km; 1385.4 km^2 **15.** 79.8 m; 506.7 m^2 **17.** 22.9 cm; 41.9 cm^2 **19.** 96.6 in; 742.6 in^2 **21.** 1.5 m **23.** 13 cm **25.** about 3979 mi **27.** d **29.** a **31.** about 26 times **33.** are safe: 25.4 in^2; are unsafe: 24.2 in^2; don't know: 14.0 in^2 **35.** 4 **37.** C **39.** 125° **41.** 235° **43.** 125° **45.** 62.5° **47.** Sample answer: $\overline{SP}$, $\overline{SQ}$, $\overline{RP}$ **49.** 144° **51.** 38 m^2 **53.** $x < -7$ **55.** 150.5

Page 538 Practice Quiz 2

1. 112.86 cm^2 **3.** 32 m^2 **5.** 29.5 in.; 69.4 in^2

Pages 541–543 Lesson 10-8

1. Sample answer:

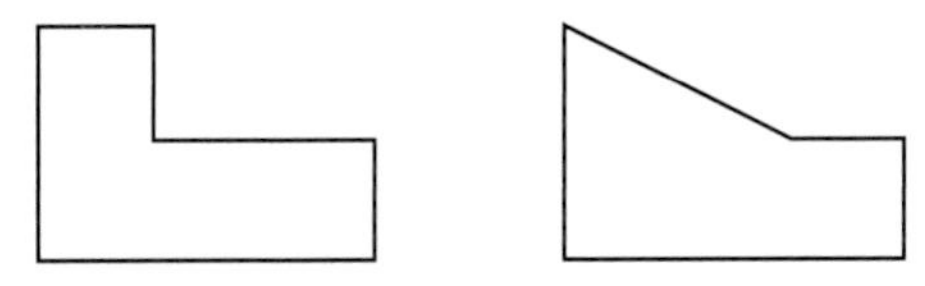

3. 49.5 yd^2 **5.** 2 **7.** 72 ft^2 **9.** 56.1 cm^2 **11.** 18.3 in^2 **13.** 45 ft^2 **15.** 86.8 yd^2 **17.** 85 units^2 **19.** 6963.5 yd^2
21. Sample answer: Separate the area into a rectangle and a trapezoid. **23.** 68,679 mi^2 **25.** You can use polygons to find the area of an irregular figure by finding the area of each individual polygon and then finding the total area of the irregular figure. Answers should include the following.

- 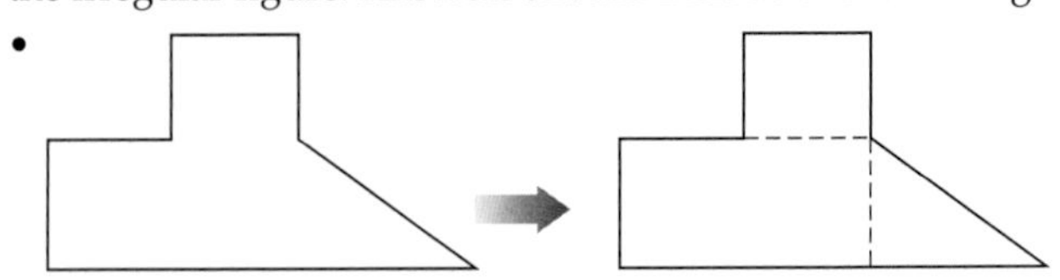
- To find the area of the irregular figure shown above, the figure can be separated into two rectangles and a triangle.

27. C **29.** 44.0 cm, 153.9 cm^2 **31.** 540° **33.** 1080° **35.** $-24xy$

Pages 544–548 Chapter 10 Study Guide and Review

1. supplementary **3.** corresponding angles **5.** equilateral **7.** 109° **9.** 71° **11.** $\angle H$ **13.** $\overline{RS}$ **15.** $\overline{GF}$

17. 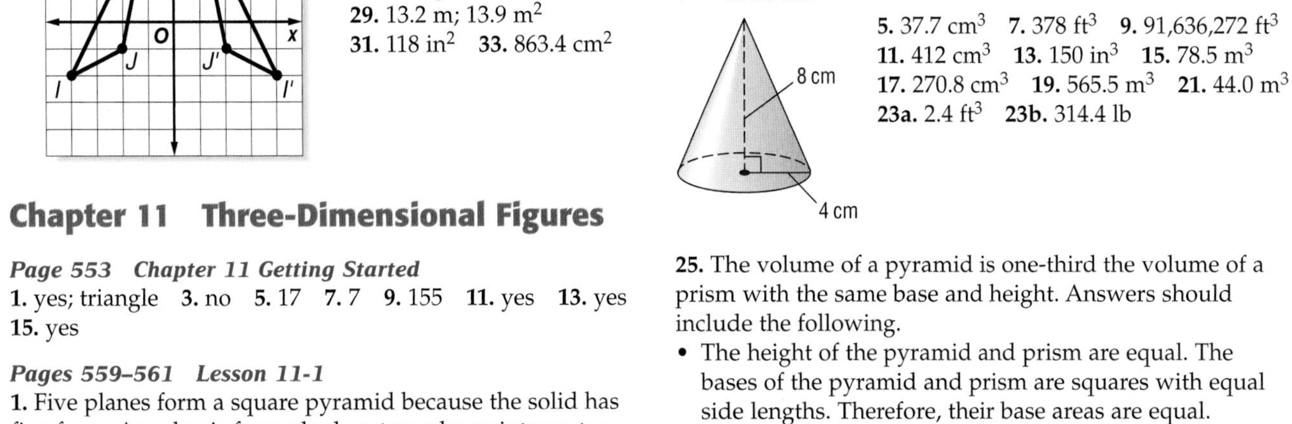

19. 112; 112° **21.** 70; 70°; 70° **23.** 11 yd^2
25. hexagon; 720°
27. decagon; 1440°
29. 13.2 m; 13.9 m^2
31. 118 in^2 **33.** 863.4 cm^2

Chapter 11 Three-Dimensional Figures

Page 553 Chapter 11 Getting Started
1. yes; triangle **3.** no **5.** 17 **7.** 7 **9.** 155 **11.** yes **13.** yes **15.** yes

Pages 559–561 Lesson 11-1
1. Five planes form a square pyramid because the solid has five faces. An edge is formed when two planes intersect; a vertex is formed when three or four planes intersect.
3. triangular pyramid; any one of the following faces can be considered a base: *RST, QRS, QST, QRT*; $\overline{QR}, \overline{QS}, \overline{QT}, \overline{RT}, \overline{RS}, \overline{ST}$; *Q, R, S, T* **5.** intersecting

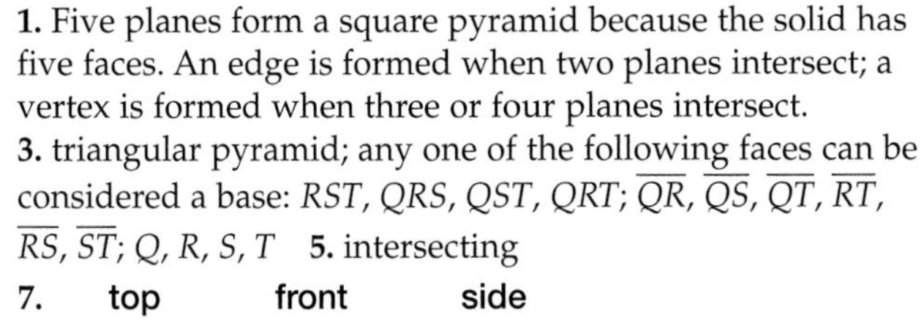

7. 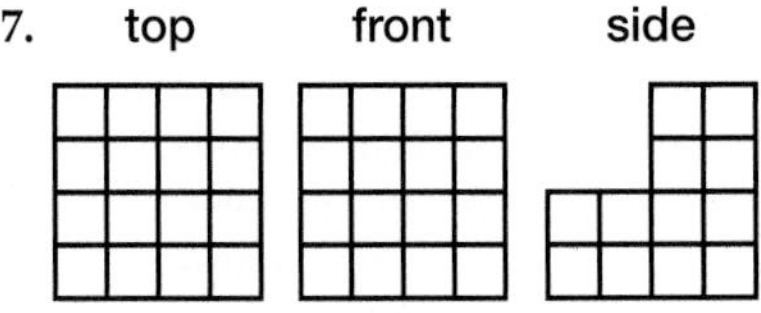

9. rectangular prism; *LMNP, QRST* or *LPTQ, MNSR* or *PNST, LMRQ*; *LMNP, QRST, LPTQ, MNSR, PNST, LMRQ*; $\overline{LM}, \overline{MN}, \overline{NP}, \overline{PL}, \overline{LQ}, \overline{MR}, \overline{NS}, \overline{PT}, \overline{QR}, \overline{RS}, \overline{ST}, \overline{TQ}$; *L, M, N, P, Q, R, S, T* **11.** triangular pyramid; any one of the following faces can be considered a base: *WXY, WYZ, WZX, XYZ*; $\overline{WX}, \overline{WY}, \overline{WZ}, \overline{XZ}, \overline{XY}, \overline{YZ}$; *W, X, Y, Z* **13.** $\overline{WR}$
15. skew **17.** top **19.** rectangular and pentagonal prisms
23. Never; only three or more planes can intersect in a single point. **25.** Sometimes; three planes may intersect in a line. Or, the planes may be parallel and not intersect at all.
27. Two-dimensional figures form three-dimensional figures. Answers should include the following.
- Two-dimensional figures have length and width and therefore lie in a single plane. Three-dimensional figures have length, width, and depth.
- Two-dimensional figures form the faces of three-dimensional figures.

29. D **31.** 37.7 cm; 113.1 cm^2 **33.** 0.5000 **35.** $c < 8$
37. $n > -11$ **39.** $y < 1\frac{3}{4}$ **41.** 14 in^2 **43.** 6.5 cm^2

Pages 565–567 Lesson 11-2
1. Sample answer: Finding how much sand is needed to fill a child's rectangular sandbox; measure the length, width, and height, and then multiply to find the volume.
3. 183.6 cm^3 **5.** 1608.5 ft^3 **7.** 282.7 yd^3 **9.** 198.9 ft
11. 512 cm^3 **13.** 748 in^3 **15.** 88.0 ft^3 **17.** 225 mm^3
19. 18.6 m^3 **21.** 6.2 m **23.** 1728 **25.** 1,000,000 **27.** 11.5 in.
29. The volume will be greater if the height is $8\frac{1}{2}$ inches. By using the formula for circumference, you can find the radius and volume of each cylinder. If the height is $8\frac{1}{2}$ inches, the volume is 86.5 in^3; if the height is 11 inches, the volume is 67.7 in^3. **31.** C **33.** Sample answer: $\overline{QT}$ and $\overline{YZ}$ **35.** $x > -8$ **37.** 25 **39.** 4 **41.** 28

Pages 570–572 Lesson 11-3
1. The base is a circle.
3. Sample answer:
$V = 134.0\ cm^3$

5. 37.7 cm^3 **7.** 378 ft^3 **9.** 91,636,272 ft^3
11. 412 cm^3 **13.** 150 in^3 **15.** 78.5 m^3
17. 270.8 cm^3 **19.** 565.5 m^3 **21.** 44.0 m^3
23a. 2.4 ft^3 **23b.** 314.4 lb

25. The volume of a pyramid is one-third the volume of a prism with the same base and height. Answers should include the following.
- The height of the pyramid and prism are equal. The bases of the pyramid and prism are squares with equal side lengths. Therefore, their base areas are equal.
- The formula for the volume of a pyramid is one-third times the formula for the volume of a prism.

27. B **29.** 904.8 cm^3 **31.** 107.5 mm^3 **33.** 10.1 in^3
35. 7.8 **37.** 36

Page 572 Practice Quiz 1
1. triangular prism **3.** 12.6 cm^3 **5.** ≈11,545,353 ft^3

Pages 575–577 Lesson 11-4
1. Sample answer: The surface of a solid is two-dimensional. Its area is the sum of the face areas, which are given in square units. **3.** 150 ft^2 **5.** 571.8 in^2 **7.** 140.7 mm^2
9. 282 in^2 **11.** 264 m^2 **13.** 1885.0 ft^2 **15.** 294 ft^2
17. 628.3 in^2 **19.** 264.0 in^2 **21.** 754 ft^2 **23.** The surface area is 4 times greater. **25.** D **27.** square or rectangle
29. rectangle **31.** 565.5 in^3 **33.** 15.21 **35.** 1.3 **37.** $3\frac{1}{4}$

Pages 580–582 Lesson 11-5
1. Slant height is the altitude of a triangular face of a pyramid; it is the length from the vertex of a cone to the edge of its base. Height of a pyramid or cone is the altitude of the whole solid. **3.** Sample answer: An architect might use the formulas to calculate the amount of materials needed for parts of a structure. **5.** 42.9 m^2 **7.** 336 ft^2
9. 96.3 in^2 **11.** 62.4 in^2 **13.** 339.3 in^2 **15.** 311.9 m^2
17. 506.6 mm^2 **19.** Style 8M has 29 in^2 more plastic.
21. 339 ft^2; 4 squares **23.** Many building materials are priced and purchased by square footage. Architects use surface area when designing buildings. Answers should include the following.
- Surface area is used in covering building exteriors and in designing interiors.
- It is important to know surface areas so the amounts and costs of building materials can be estimated.

25. C **27.** 5541.8 mm^2 **29.** 7 ft^2 **31.** 25.1 in^3 **33.** infinitely many solutions **35.** 3 **37.** 3 **39.** 3.15

Pages 586–588 Lesson 11-6
1. Sample answer: The cones are similar because the ratios comparing their radii and slant heights are equal: $\frac{2}{3} = \frac{5}{7.5}$.

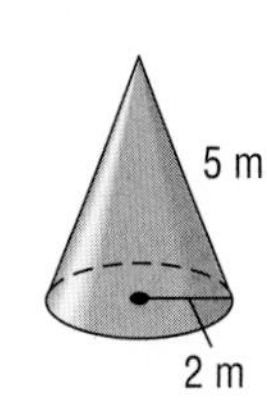

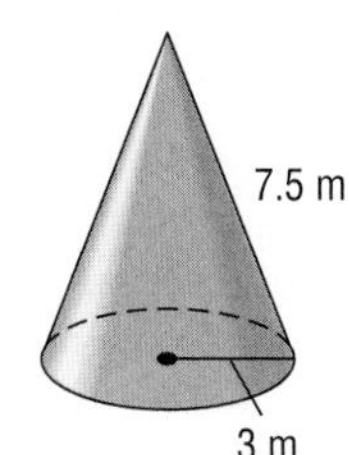

Selected Answers

3. no **5.** $x = 33\frac{1}{3}$ ft **7.** 525 m **9.** 2,976,750 m^3 **11.** yes **13.** yes **15.** $x = 7$ m, $y = 18$ m **17.** Sometimes; bases must be the same polygon and the corresponding side lengths must be proportional. **19.** always; same shape, diameters or radii are proportional **21.** 200^3 or 8,000,000 times greater **23.** If the ratios of corresponding linear dimensions are equal, the solids are similar. Answers should include the following.

- For example, if the ratio comparing the heights of two cylinders equals the ratio comparing their radii, then the cylinders are similar.
- A cone and a prism are not similar.

25. C **27.** 184 ft^2 **29.** 65° **31.** $-\frac{5}{9}$ **33.** 19.3 **35.** 6.0 **37.** 9.4

Page 588 Practice Quiz 2

1. 166 mm^2 **3.** 9.4 m^2 **5.** Yes; their corresponding dimensions are proportional.

Pages 592–594 Lesson 11-7

1. Josh; the first two 0s are not significant because they are placeholders for the decimal point. The 0 between 2 and 5 is significant because it is between two significant digits and shows the actual value in the thousandths place. **3.** Sample answer: 0.012 or 2500 **5.** 3 **7.** 3 **9.** 7.1 L **11.** 544.5 in^2 **13.** $\frac{1}{32}$ in. **15.** 3 **17.** 2 **19.** 1 **21.** 4 **23.** 45 in. **25.** 2 ft^2 **27.** 84.47 m **29.** 49.5 cm **31.** 26 m^2 **33.** b; The precision unit of ruler a is $\frac{1}{8}$ inch. The precision unit of ruler b is $\frac{1}{16}$ inch. **35.** No, the numbers are estimated to the nearest 0.1 million. **37.** 12,400,000; 3 **39.** Not necessarily, because the actual size of the 2.0 mm wrench can range from 1.95 mm to 2.05 mm and the actual size of the 2 mm bolt can range from 1.5 mm to 2.5 mm. So, the bolts could be larger than the corresponding wrenches. **41.** A **43.** 0.05 mi; between 132.75 mi and 132.85 mi **45.** 26.5 cm^2 **47.** 14.8 m^2

Pages 595–598 Chapter 11 Study Guide and Review

1. false; lateral area **3.** false; slant height **5.** false; cylinder **7.** true **9.** rectangular prism; *QRST*, *UVWX* or *QTXU*, *RSWV* or *QRVU*, *TSWX*; *QRST*, *UVWX*, *QTXU*, *RSWV*, *QRVU*, *TSWX*; $\overline{QR}$, $\overline{RS}$, $\overline{ST}$, $\overline{TZ}$, $\overline{UV}$, $\overline{VW}$, $\overline{WX}$, $\overline{XU}$, $\overline{QU}$, $\overline{TX}$, $\overline{SW}$, $\overline{RV}$; *Q*, *R*, *S*, *T*, *U*, *V*, *W*, *X* **11.** rectangular pyramid; *YNPZ*; *AYZ*, *AZP*, *ANP*, *ANY*, *YNPZ*; $\overline{AY}$, $\overline{AZ}$, $\overline{AP}$, $\overline{AN}$, $\overline{YZ}$, $\overline{ZP}$, $\overline{PN}$, $\overline{NY}$; *A*, *Y*, *Z*, *P*, *N* **13.** 0.8 mm^3 **15.** 4 ft^3 **17.** 1721.9 cm^3 **19.** 548.7 mm^2 **21.** 45 in^2 **23.** 37.1 in^2 **25.** yes **27.** 19.0 cm **29.** 0.9 m **31.** 973 ft^2

Chapter 12 More Statistics and Probability

Page 605 Chapter 12 Getting Started

1. 16; 15; none **3.** 5.3, 5.8, 6.5 **5.** $\frac{1}{3}$ **7.** $\frac{1}{2}$ **9.** $\frac{1}{2}$ **11.** $\frac{1}{2}$ **13.** $\frac{1}{8}$ **15.** $\frac{1}{6}$ **17.** $\frac{1}{6}$

Pages 608–611 Lesson 12-1

1. Sample answer: The age of the youngest President at the time of his inauguration was 42 and the age of the oldest President to be inaugurated was 69. However, most of the Presidents were 50 to 59 years old at the time of their inauguration.

3.

Stem	Leaf
0	6 7
1	2 5 5
2	0
3	5
4	0 1 *2|0 = 20*

5. 50, 99 **7.** Sample answer: The lowest score was 50. The highest score was 99. Most of the scores were in the 70–79 interval. **9.** Chicken; whereas chicken sandwiches have 8–20 grams of fat, burgers have 10–36 grams of fat.

11.

Stem	Leaf
1	3
2	6 8
3	4 9
4	
5	2 3
6	2 2 7
7	7 9
8	4 *7|7 = 77*

13.

Stem	Leaf
0	6 6 8
1	0 1 4 6 8
2	0 2 5
3	2 7
4	
5	0 *3|7 = 37*

15.

Stem	Leaf
40	0 0 5
41	
42	5
43	
44	
45	0 6
46	
47	
48	0 4 7
49	
50	
51	0
52	0
⋮	
62	5
⋮	
76	4
77	
78	9 *76|4 = 764*

17. never **19.** 35% **21.** Sample answer: In the most populated U.S. cities, about 27 to 35% of the people exercise daily. **23.** 5 **25.** The average number of games won by the teams in the Big East Conference is less than the average number of games won by the teams in the Big Ten Conference. **29a.** Sample answer: It would be easier to find the median in a stem-and-leaf plot because the data are arranged in order from least to greatest. **29b.** Sample answer: It would be easier to find the mean in a table because once you find the sum, it may be easier to count the number of items in a table. **29c.** Sample answer: It would be easier to find the mode in a stem-and-leaf plot because the value or values that occur most often are grouped together. **31.** C **33.** no **35.** 62.8 ft; 314.2 ft^2 **37.** 247% **39.** 0.65% **41.** 57.1% **43.** 1.6% **45.** 6 **47.** 1.1

Pages 614–616 Lesson 12-2

1. The range describes how the entire set of data is distributed, while the interquartile range describes how the middle half of the data is distributed. **3.** Sample answer: {8, 9, 13, 25, 26, 26, 26, 27, 28, 30, 35, 40} **5.** 27; 6 **7.** 24 h **9.** 77; 39 **11.** 50; 10 **13.** 25; 15.5 **15.** 22 **17.** Feb.: 13.5; July: 8 **19.** National League: 42, 44, 48, 38, 10; American League: 39, 44, 49, 39, 10 **21a.** Sample answer: {43, 49, 50, 50, 58, 60, 60, 66, 70, 70, 71, 78} **21b.** Sample answer: {15, 18, 20, 20, 44, 60, 60, 64, 70, 70, 75, 79} **21c.** Sample answer: The first set of data has a smaller interquartile range, thus the data in the first set are more tightly clustered around the median and the data in the second are more spread out over the range. **23.** C

25.

Stem	Leaf
0	9
1	2 4 5 8
2	1 7
3	7 *3|7 = \$37*

27. 461.8 cm^3 **29.** 1.6 ft **31.** {0.2, 0.3, 0.6, 0.8, 1.2, 1.4, 1.5} **33.** {9.8, 9.9, 10.5, 10.9, 11.2, 11.4}

Pages 619–621 Lesson 12-3

1. lower quartile, least value; upper quartile, greatest value

3. Sample answer: {28, 30, 52, 68, 90, 92}

5.

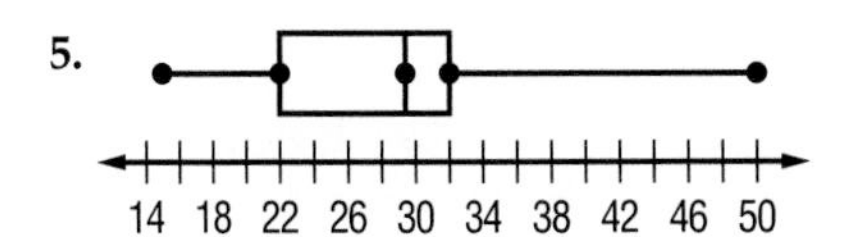

7. Sample answer: The length of the box-and-whisker plot shows that the winning times of the men's marathons are not concentrated around a certain time. **9.** The most fuel-efficient SUV and the least fuel-efficient sedan both average 22 miles per gallon.

11.

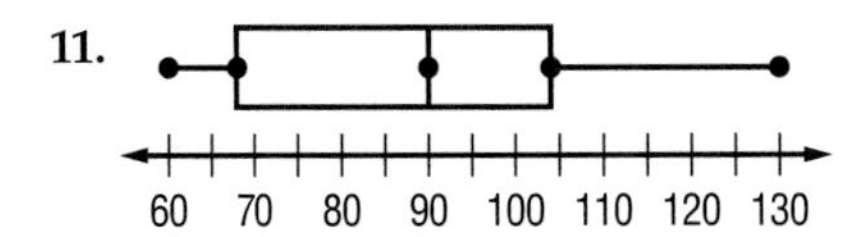

13.

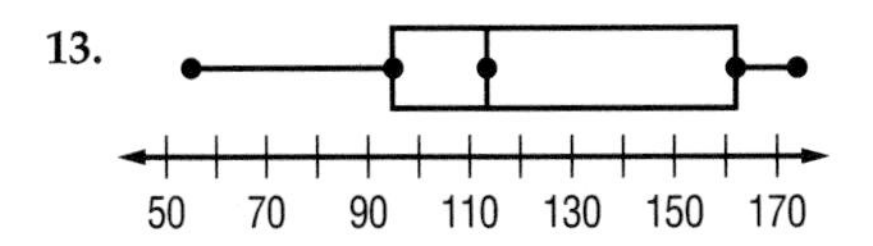

15.

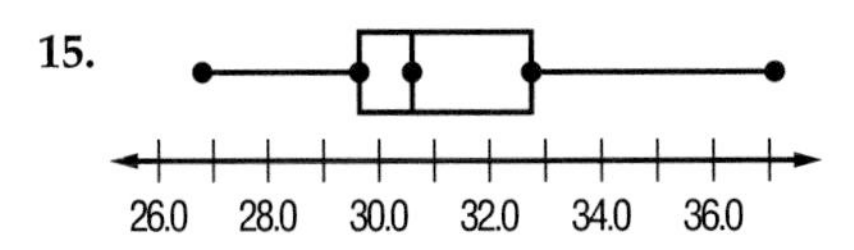

17. 50% **19.** Sample answer: The least number of games won for NFC is 3 and the least number of games won for the AFC is 1. The most number of games won for the NFC is 12 and the most number of games won for the AFC is 13. In addition, for both conferences, the median number of games won is about 9. **21.** A box-and-whisker plot would clearly display any upper and lower extreme temperatures and the median temperature. Answers should include the following.

•

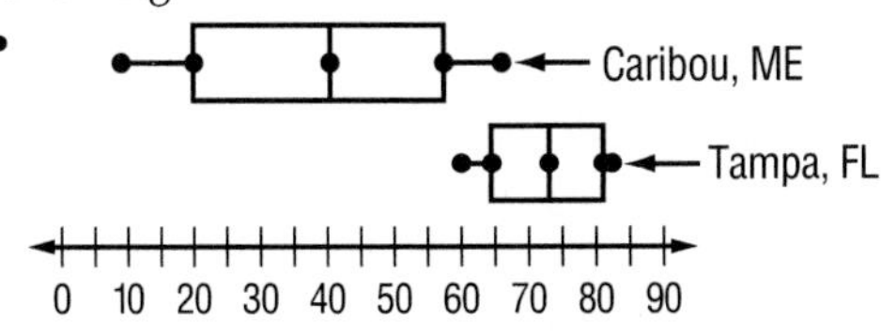

- Sample answer: Tampa has a median temperature of 73 and Caribou has a median temperature of 40.5. Whereas the highest average temperature for Tampa is 82, the highest average temperature for Caribou is 66.
- Sample answer: You can easily see how the temperatures vary.

23. C **25.** 4

27.

Stem	Leaf
1	4
2	1 4 7
3	9
4	8 2|4 = 24

29. 14

Pages 625–628 Lesson 12-4

1. Sample answer: Because the intervals are continuous.

3.

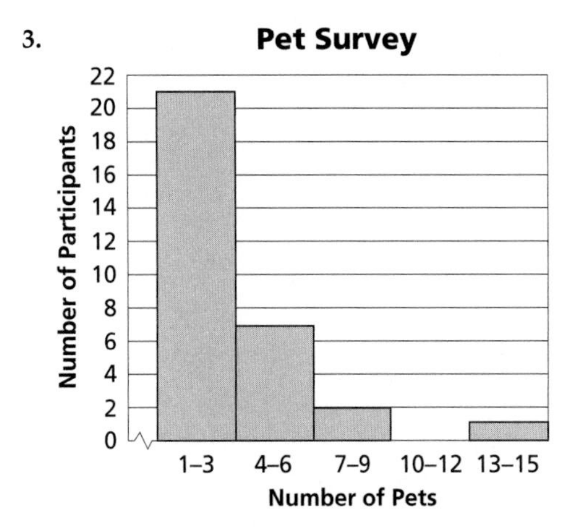

5. the number of states that have a certain number of roller coasters **7.** The numbers between 10 and 34 are omitted. **9.** 2 or more national monuments

11.

Weekly Study Time

Number of Students

0 1 2 3 4 5 6 7 8 9 10 11 12

0–3 4–6 7–9 10–12 13–15

Time (hr)

13.

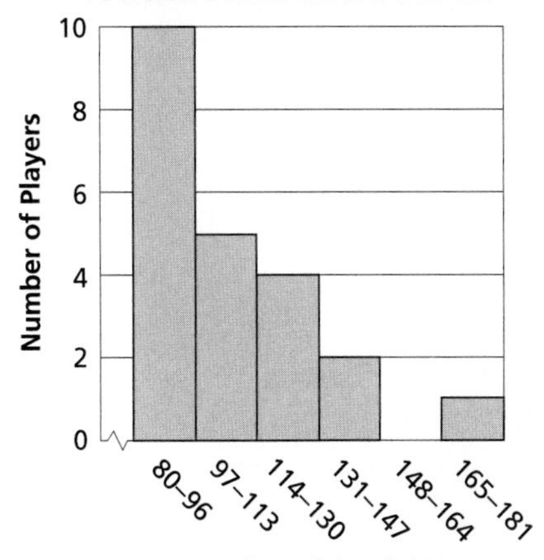

15. 13 **17.** 37.5% **19.** Dallas **21.** true **23.** false **25.** true **27.** A **29.** absolute frequency: 2, 9, 17, 12, 7, 2; relative frequency: $\frac{2}{49}, \frac{9}{49}, \frac{17}{49}, \frac{12}{49}, \frac{1}{7}, \frac{2}{49}$; cumulative frequency: 2, 11, 28, 40, 47, 49

31.

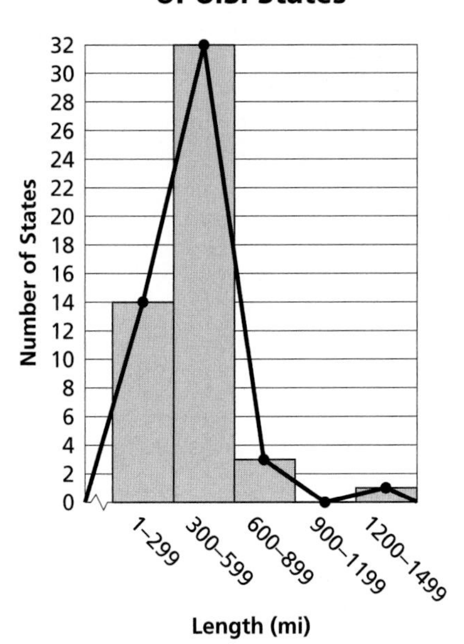

33. 44; 15 **35.** Sample answer: Brand B costs one-third more than brand C.

Page 628 Practice Quiz 1

1.

Stem	Leaf
1	6 8 9 9
2	0 2 6 8 8 8
3	0 2
4	0 2 3
5	5
6	
7	
8	6 9 $5\|5 = 55$

3.

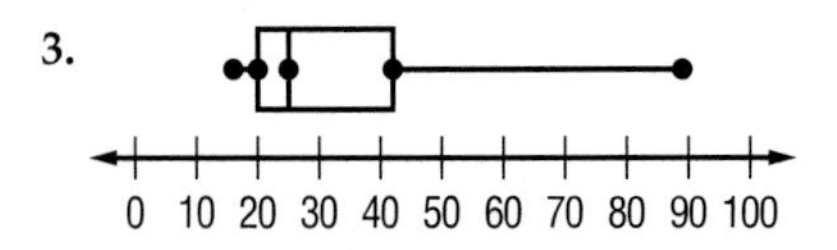

5.

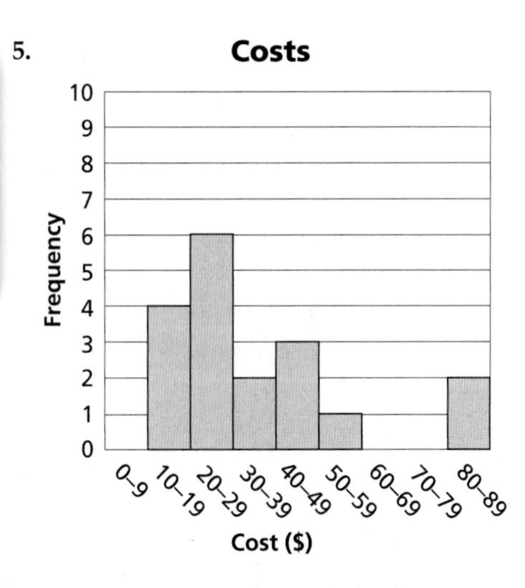

Pages 631–633 Lesson 12-5

1. Sample answer: inconsistent vertical scale and break in vertical scale **3.** Graph A has a break in the vertical scale. **5.** From the vertical scale, you can see that the number of area codes in 1999 is about 1.5 times the number of area codes in 1996. The graph is misleading because the drawing of the phone for 1999 is about 3 times the size of the phone for 1996. Also, there is a break in the vertical scale. **7.** Graph B **9.** Graph B; the vertical scale used makes the decrease in the unemployment rates appear more drastic.

11a. Sample answer:

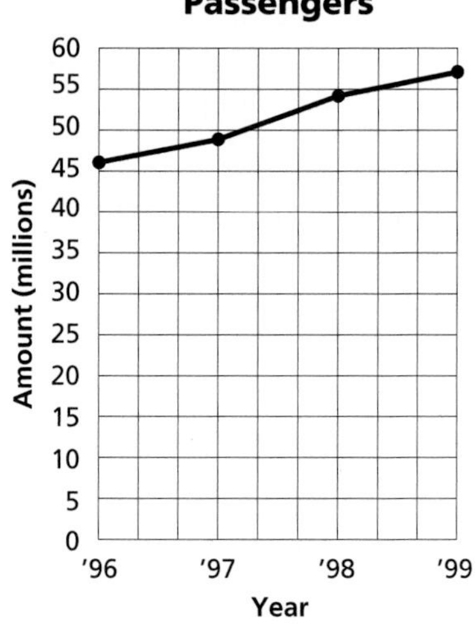

11b. Sample answer:

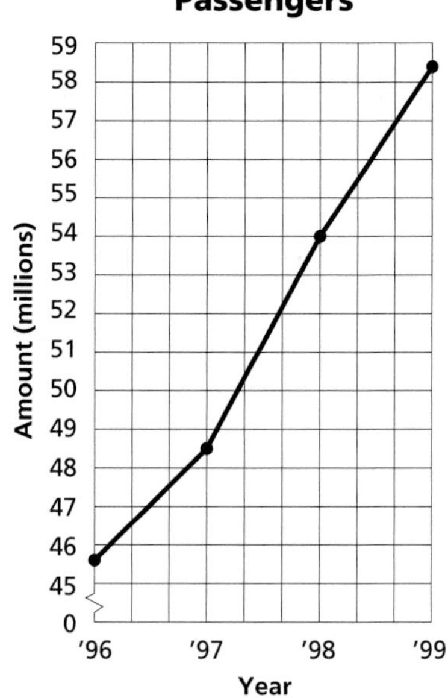

13. B

15.

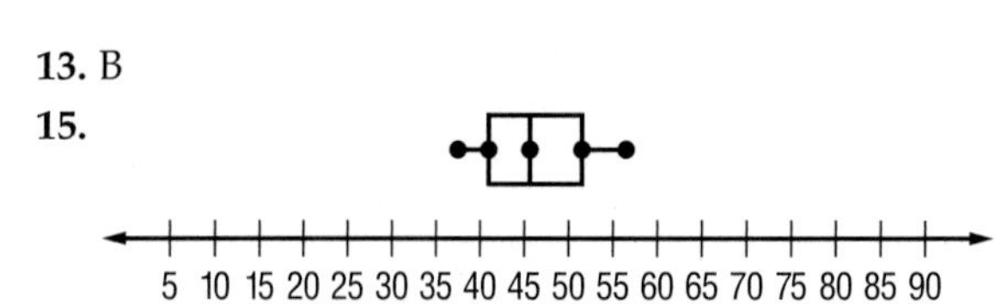

17. 20.3 m^2 **19.** $\frac{1}{6}$ or $16\frac{2}{3}\%$ **21.** $\frac{5}{6}$ or $83\frac{1}{3}\%$

23. $\frac{5}{6}$ or $83.\overline{3}\%$

Pages 637–639 Lesson 12-6

1. Sample answer: Both methods find the number of outcomes. Using the Fundamental Counting Principle is faster and uses less space; using a tree diagram shows what each outcome is. **3.** First find the number of outcomes possible. The possible outcomes are taco-chicken, taco-beef, taco-bean, burrito-chicken, burrito-beef, and burrito-bean, for a total of 6 outcomes. Two of the outcomes have chicken filling, so the probability of chicken filling is $\frac{2}{6}$ or $\frac{1}{3}$. **5.** $\frac{1}{16}$ **7.** $\frac{1}{4}$ **9.** 9

11.

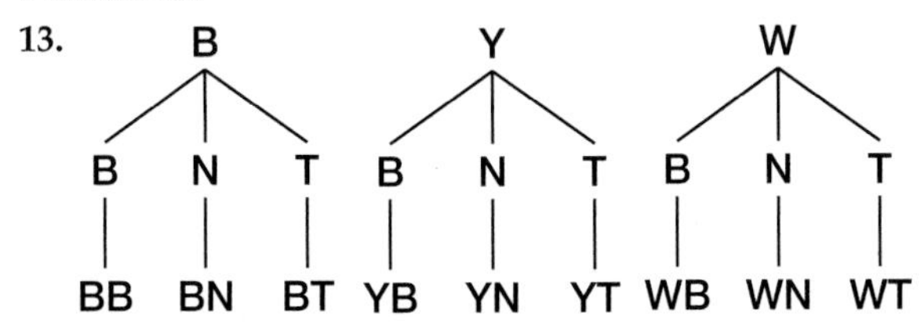

8 outcomes

13.

B Y W

B N T B N T B N T

BB BN BT YB YN YT WB WN WT

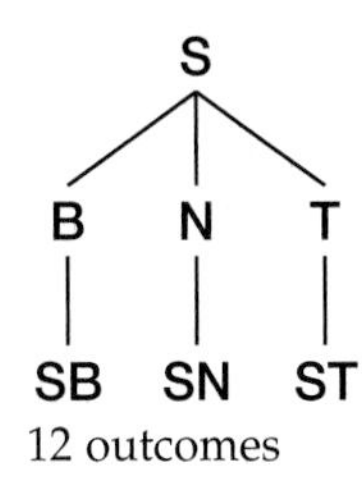

12 outcomes

15. 36 outcomes **17.** 24 outcomes **19.** 1024 outcomes **21.** $\frac{5}{18}$ **23.** $\frac{1}{100{,}000}$ **25.** Either; the probability of rolling either odd or even is one-half. **27.** Sample answer: any two letters followed by any four digits **29.** D **31.** Sample answer: Vertical scale that does not start at zero. **33.** 11–20

35.

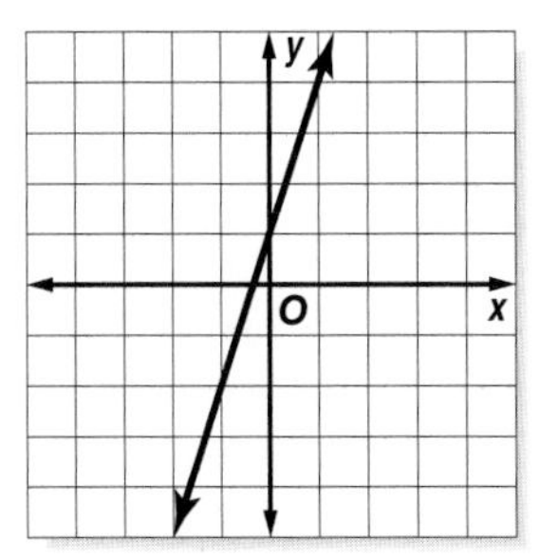

37.

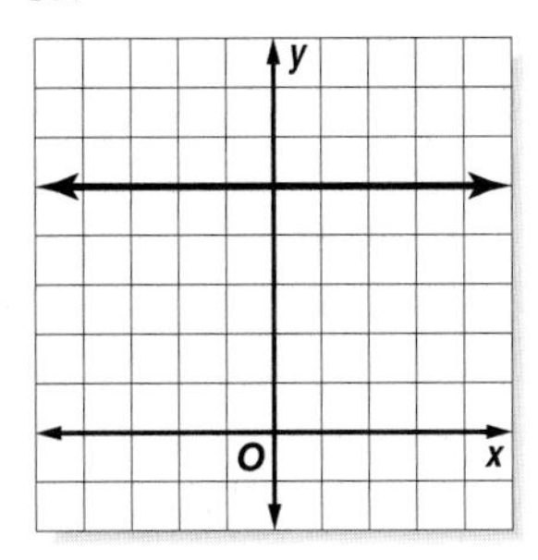

39. 15 **41.** 28

Pages 643–645 Lesson 12-7

1. Sample answer: How many 3-digit numbers can be made from the digits 1, 2, 3, and 4 if no digit is repeated? **3.** Sarah; five CDs from a collection of 30 is a combination because order is not important. **5.** P; 120 ways **7.** C; 84 ways **9.** 220 pizzas **11.** P; 24 flags **13.** C; 105 ways **15.** P; 1320 ways **17.** C; 364 ways **19.** 40,320 **21.** 39,916,800 **23.** 36 handshakes **25.** 3003 ways **27.** 10 combinations **29.** $\frac{2}{5}$ **31.** When order is not important, duplicate arrangements are not included in the number of arrangements. Answers should include the following.

- When order was not important there were half as many pairs.
- Order is important when you arrange things in a line; order is not important when you choose a group of things.

33. C **35.** Yes; the vertical axis does not include zero. **37.** Q **39.** I **41.** 1:8 **43.** 11:2

Page 645 Practice Quiz 2

1. Sample answer: Bars are different widths. **3.** 256 outcomes **5.** 15 segments

Pages 648–649 Lesson 12-8

1. Write a ratio comparing the ways the event can occur to the ways the event cannot occur. **3.** Hoshi; the probability of rolling a 2 is 1 out of 6. The odds of rolling a 2 are 1 in 5. **5.** 1:2 **7.** 5:1 **9.** 1:2 **11.** 2:1 **13.** 1:1 **15.** 1:3 **17.** 5:7 **19.** 31:5 **21.** 1:17 **23.** 5:4 **25.** 9:4 **27.** $\frac{1}{8192}$ **29.** 47:53 **31.** No; 512 is the total number of possible outcomes, so it could not be a number in the odds of winning. **33.** A **35.** 6 ways

37. 3 4 5 6 7 8 9 10 ; $a \geq 6$

39. −6 −5 −4 −3 −2 −1 0 1 2 ; $y > -3$

41. $\frac{1}{3}$ **43.** $\frac{1}{56}$

Pages 653–655 Lesson 12-9

1. Independent and dependent events are similar because both are a connection of two or more simple events and the probability of the compound event is found by multiplying the probabilities of each simple event. They are different because the second event in a dependent event is influenced by the outcome of the first event. Therefore, the probability of the second event used in calculating the probability of the compound event is dependent on the outcome of the first event. **3.** The result of the 1st event must be taken into account. **5.** $\frac{1}{12}$ **7.** $\frac{3}{14}$ **9.** $\frac{3}{4}$ **11.** $\frac{1}{30}$ **13.** $\frac{1}{5}$ **15.** $\frac{1}{91}$ **17.** $\frac{10}{91}$ **19.** $\frac{5}{364}$ **21.** $\frac{5}{8}$ **23.** $\frac{3}{7}$ **25.** 1 **27.** $\frac{2}{9}$ **29.** 0.16 **31a.** Sample answer: 3 red, 2 white, and 4 blue **31b.** Sample answer: The numerator must be 24. Any combination of 3, 2, and 4 will have a probability of $\frac{1}{21}$. **33.** B **35.** $\frac{2}{3}$ **37.** 3:10 **39.** 17,576,000 **41.** 65° **43.** 65° **45.** 90° **47.** 40.6 **49.** 16

Pages 658–662 Chapter 12 Study Guide and Review

1. b **3.** c **5.** d **7.**

Stem	Leaf
4	5 5
5	0 0 5 5
6	0 0 0 0 0
7	5 5 5
8	5 5 $7 \mid 5 = 75¢$

9. 23; 7 **11.** 22; 10 **13.** 80°F **15.** 50% **17.** Graph A **19.** 16 outcomes **21.** C; 364 ways **23.** C; 21 ways **25.** 3:5 **27.** 1:3 **29.** $\frac{1}{30}$

Chapter 13 Polynomials and Nonlinear Functions

Page 667 Chapter 13 Getting Started

1. 1 **3.** 2 **5.** 0 **7.** $5a + 20$ **9.** $-4 - 32n$ **11.** $27b - 27c$ **13.** yes **15.** yes

Pages 670–672   Lesson 13-1

1. The degree of a monomial is the sum of the exponents of its variables. The degree of a polynomial is the same as the degree of the term with the greatest degree. **3.** Tanisha; the degree of a binomial is the degree of the term with the greater degree. **5.** yes; monomial **7.** yes; binomial **9.** yes; trinomial **11.** 0 **13.** 1 **15.** 4 **17.** 2 **19.** yes; binomial **21.** yes; monomial **23.** no **25.** yes; trinomial **27.** no **29.** yes; trinomial **31.** 0 **33.** 3 **35.** 5 **37.** 5 **39.** 3 **41.** 5 **43.** Always; any number is a monomial. **45.** $2x + 2y + z + xy$ **49.** Polynomials approximate real-world data by using variables to represent quantities that are related. Answers should include the following.

- Heat index is found by using a polynomial in which one variable represents the percent humidity and another variable represents the temperature.
- Heat index cannot be approximated using a linear equation because the values do not change at a constant rate.

51. C **53.** mutually exclusive; $\frac{2}{3}$ **55.** 70 in^3 **57.** $(x + 2x) + 4$ **59.** $(6n + 3n) + (2 + 5)$ **61.** $(s + 5s) + (t - 3t)$

Pages 676–677   Lesson 13-2

1. x^2 and $2x^2$; $5x$ and $-4x$; 2 and 7 **3.** Hai; the terms have the same variables in a different order. **5.** $7a^2 - 9a + 4$ **7.** $13x - 4y$ **9.** $4x^2 + 3x - 2$ **11.** $3x + 3$ **13.** $11x^2 + 5xy + 4y^2$ **15.** $4x + 14$ **17.** $8y + 2r$ **19.** $5x^2 + xy + y$ **21.** $7x^2 + 9x + 5$ **23.** $5a - 4b; -31$ **25.** $4a + 6b + c; 14$ **27.** 56 **29.** $23x - 12$ **31.** $42x - 22$ **33.** Use algebra tiles to model each polynomial and combine the tiles that have the same size and shape. Answers should include the following.

- Algebra tiles that represent like terms have the same size and shape.
- When adding polynomials, a red tile and a white tile that have the same size and shape are zero pairs and may be removed. The result is the sum of the polynomials.

35. B **37.** 2 **39.** $\frac{2}{13}$ **41.** $\frac{3}{26}$ **43.** $15c + (-26)$ **45.** $1 + (-2x)$ **47.** $(n + rt) + (-r^2)$

Pages 680–681   Lesson 13-3

1. Subtracting one polynomial from another is the same as adding the additive inverse. **3.** $4r$ **5.** $5x + 2$ **7.** $3x^2 + 3x - 8$ **9.** $7x + 5$ units **11.** $-2n^2 + 6n$ **13.** $y^2 - 7y + 10$ **15.** $15w^2 - 2w + 10$ **17.** $4x + 3$ **19.** $-2t + 2$ **21.** $3a^2 + 2b^2$ **23.** $9n^2 - n - 12$ **25.** $2x^2 - 2xy + 3y^2$ **27.** $x - y = 68 + (x + y)$; -34°F **29.** In subtracting polynomials and in subtracting measurements, like parts are subtracted. Answers should include the following.

- To subtract measurements with two or more units, subtract the like units. To subtract polynomials with two or more terms, subtract the like terms.
- For example, to subtract 1 foot 5 inches from 3 feet 8 inches, subtract the feet 3 − 1 and subtract the inches 8 − 5. The difference is 2 feet 3 inches.

31. B **33.** $12x - 3y$ **35.** $-t^2 + 12t + 2$ **37.** yes; binomial

39.

Stem	Leaf
5	4 9
6	4 6 8
7	0 1 1 2
8	5 9
9	1 5|4 = 54

41. $8y^2$ **43.** $4m^3$ **45.** $10r^5$

Page 681   Practice Quiz 1

1. 4 **3.** 3 **5.** $2x + 2$ **7.** $8r - 3s$ **9.** $-3x + 5y$

Pages 684–686   Lesson 13-4

1. False; the order in which numbers or terms are multiplied does not change the product, by the Commutative Property of Multiplication; $x(2x + 3) = 2x^2 + 3x$ and $(2x + 3)x = 2x^2 + 3x$. **3.** Sample answer: $2x(x + 1) = 2x^2 + 2x$ **5.** $a^2 + 4a$ **7.** $12x^2 - 28x$ **9.** $-15x^2 + 35x - 45$ **11.** $14n + 35$ **13.** $t^2 - 9t$ **15.** $-7a^2 - 6a$ **17.** $40n + 8n^2$ **19.** $3y^3 - 6y$ **21.** $5x^2 + 5xy$ **23.** $-14x^2 + 35x - 77$ **25.** $4c^4 + 28c^2 - 40c$ **27.** –1 **29.** high school, 84 ft by 50 ft; college, 94 ft by 50 ft

31.

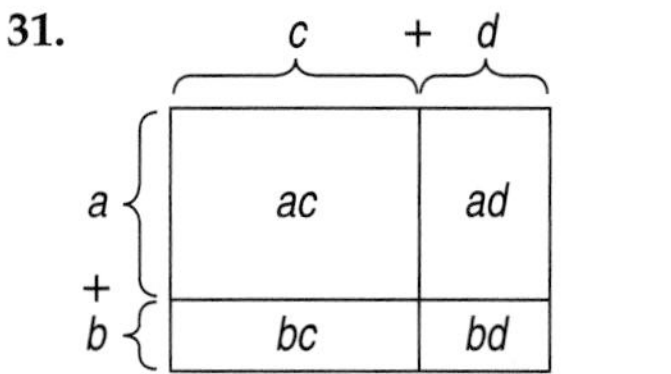

$(a + b)(c + d) = ac + ad + bc + bd$

33. D **35.** $7x - 1$ **37.** $4y^2 - 2y + 14$ **39.** $8x + 11y$ **41.** Sample answer: The scales are labeled inconsistently or the bars on a bar graph are different widths. **43.** reflection

45.

x	$2x^2 - 3$	(x, y)
0	−3	(0, −3)
1	−1	(1, −1)
2	5	(2, 5)
3	15	(3, 15)

Pages 689–691   Lesson 13-5

1. Sample answer: Determine whether an equation can be written in the form $y = mx + b$ or look for a constant rate of change in a table of values. **3.** Sample answer: population growth **5.** Nonlinear; graph is a curve. **7.** Nonlinear; equation cannot be written as $y = mx + b$. **9.** Linear; rate of change is constant. **11.** Nonlinear; graph is a curve. **13.** Linear; graph is a straight line. **15.** Nonlinear; graph is a curve. **17.** Linear; equation can be written as $y = 0.9x + 0$. **19.** Linear; equation can be written as $y = \frac{3}{4}x + 0$. **21.** Nonlinear; equation cannot be written as $y = mx + b$. **23.** Linear; rate of change is constant. **25.** Nonlinear; rate of change is not constant. **27.** Nonlinear; the points (year, applications) would lie on a curved line, not on a straight line. Or, the rate of change is not constant. **29.** No, the difference between the years varies, so the change is not constant. **31.** A **33.** $4t + 9t^2$ **35.** $a^2b - 2ab^2$ **37.** $-x$ **39.** acute

41.

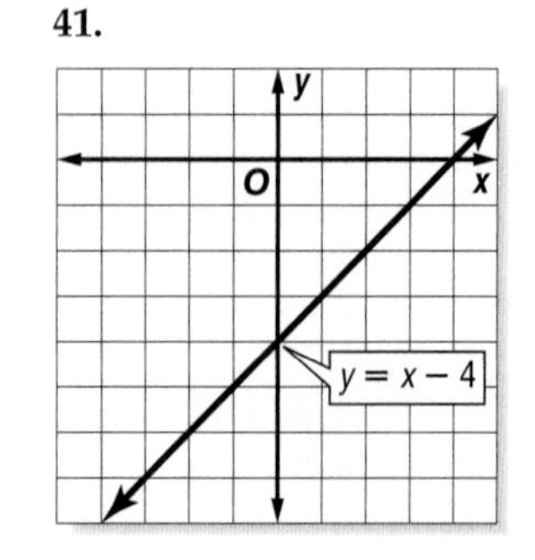

43.

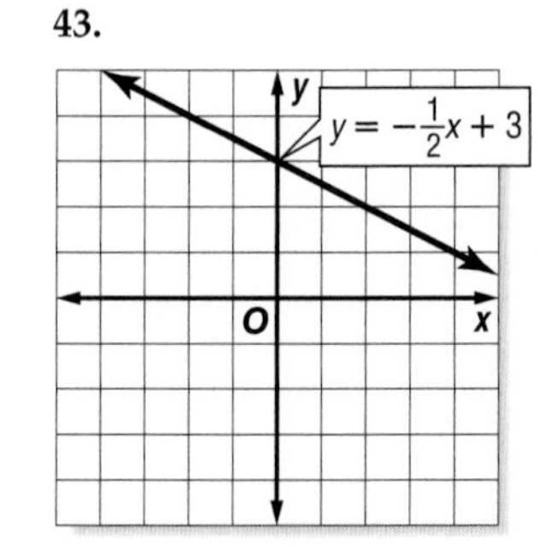

Page 691 Practice Quiz 2
1. $2c^3 - 8c$ **3.** $5a^2 + a^3 + 2a^4$ **5.** Nonlinear; equation cannot be written as $y = mx + b$.

Pages 694–696 Lesson 13-6
1. Sample answer: The graph of $y = nx^2$ has line symmetry and the graph of $y = nx^3$ does not. **3.** Sample answer: $y = x^2 + 3$; make a table of values and plot the points.

5.
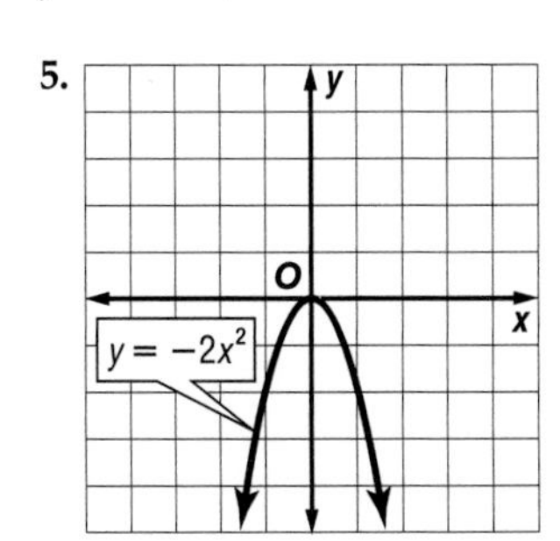

7.
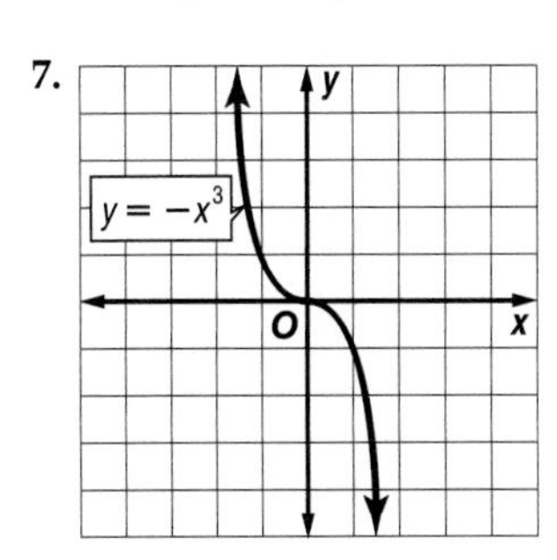

9.
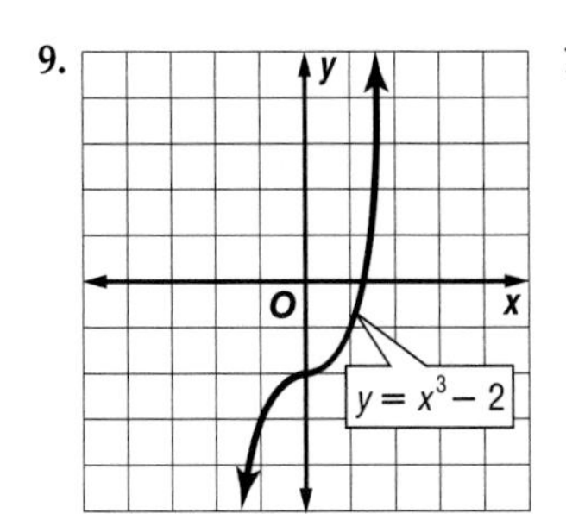

11.
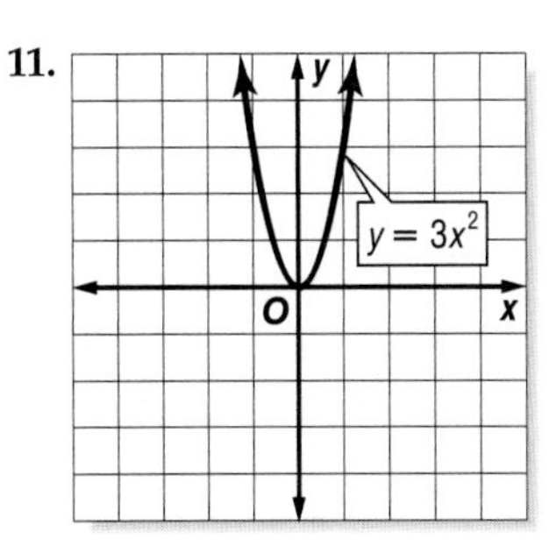

13.
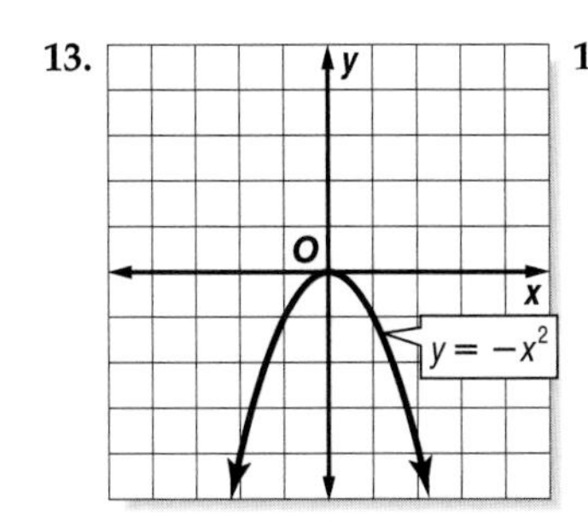

15.
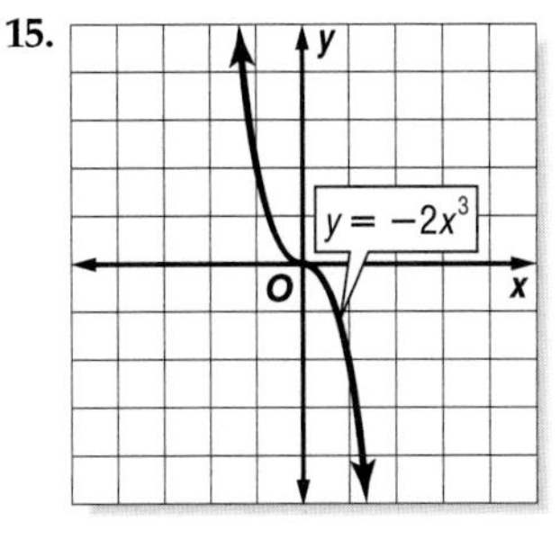

17.
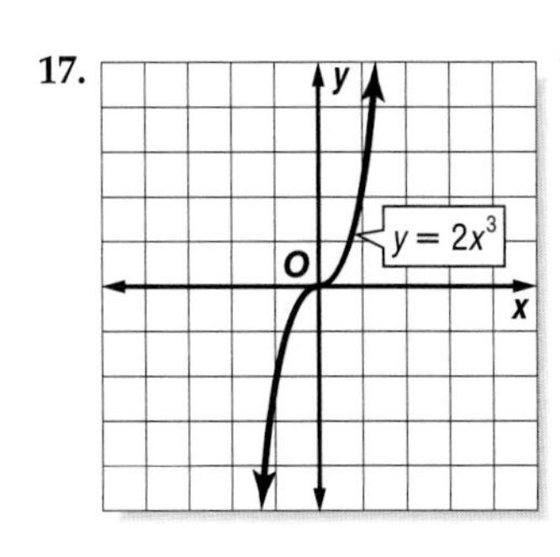

19.
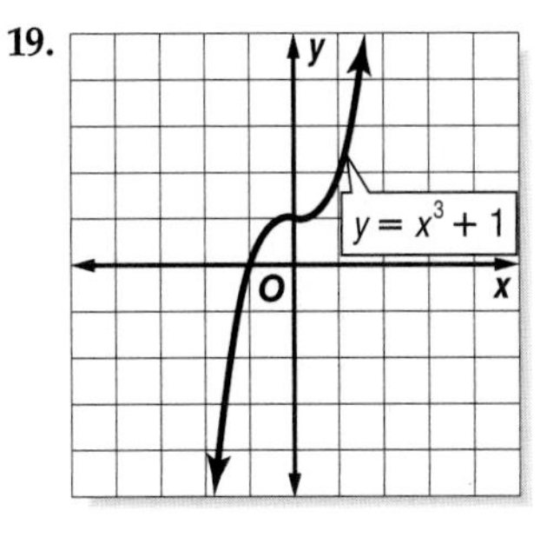

21.
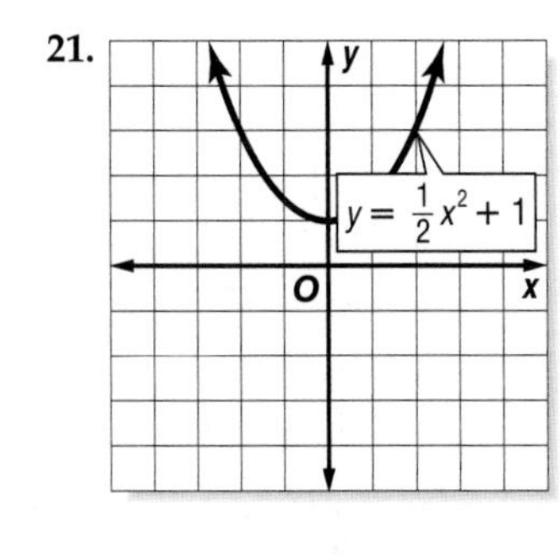

23.
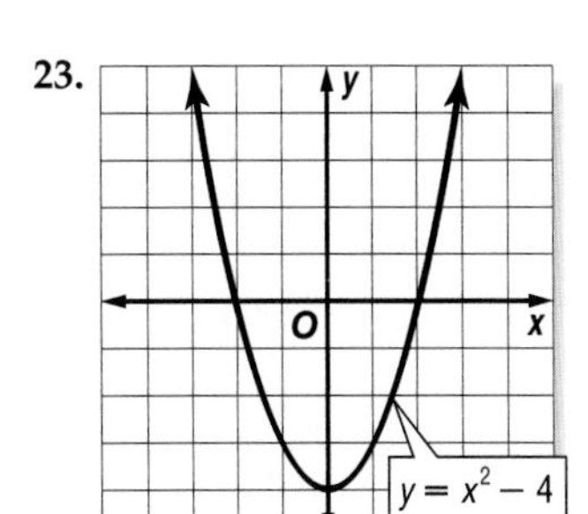

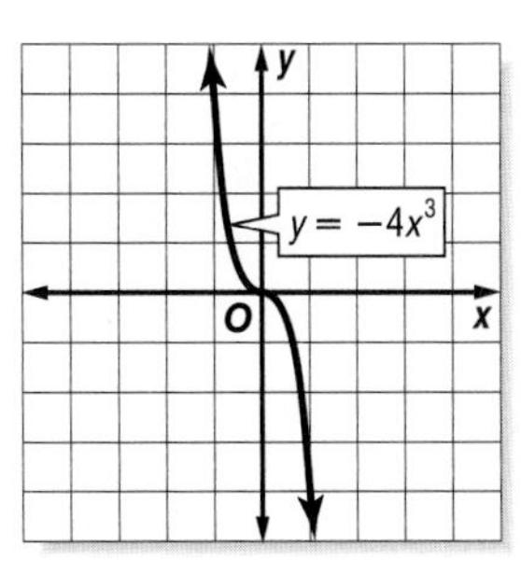

Both equations are functions because every value of x is paired with a unique value of y.

25. (0, 7)

27.
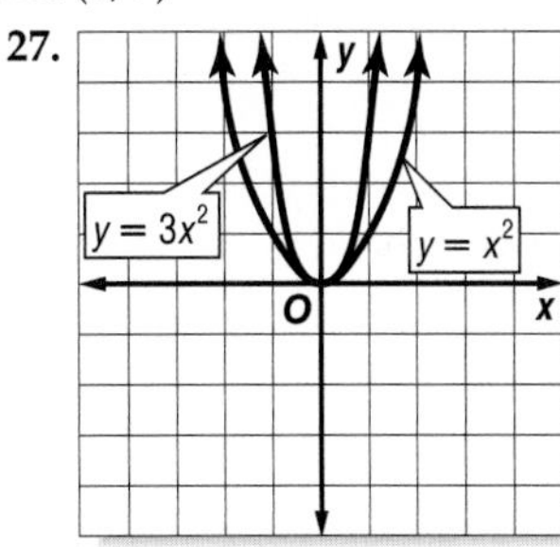

Similar shape; $y = 3x^2$ is more narrow.

29.
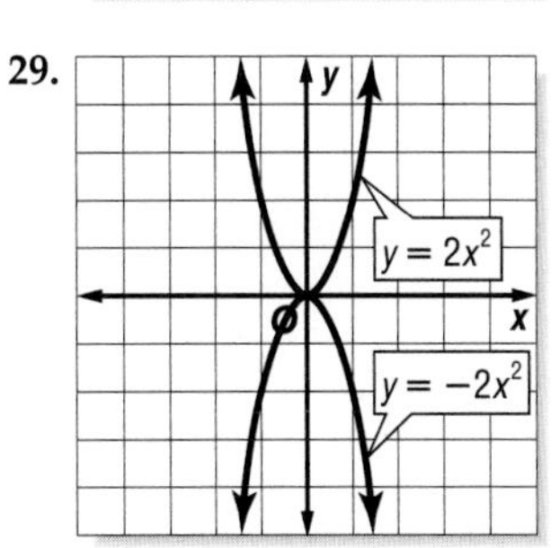

Same shape; $y = -2x^2$ is $y = 2x^2$ reflected over the x-axis.

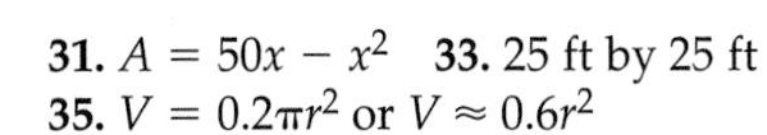
31. $A = 50x - x^2$ **33.** 25 ft by 25 ft
35. $V = 0.2\pi r^2$ or $V \approx 0.6r^2$

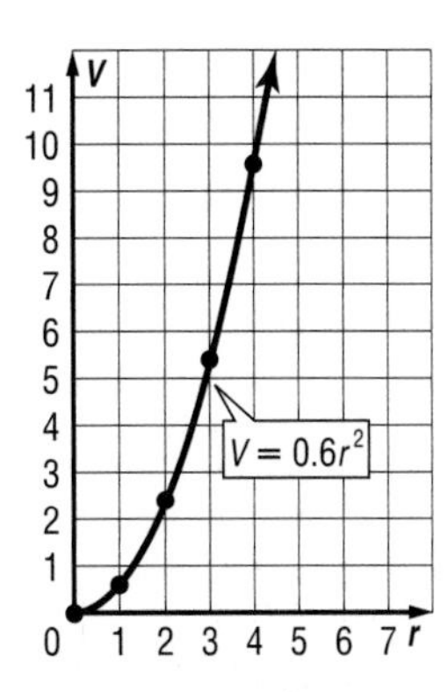

37. Formulas, tables, and graphs are interchangeable ways to represent functions. Answers should include the following.
- To make a graph, use a rule to make a table of values. Then plot the points and connect them to make a graph.
- To write a rule, find points that lie on a graph and make a table of values using the coordinates. Look for a pattern and write a rule that describes the pattern.

39. C **41.** Sample answer: 18 units^2 **43.** $10x - 20$
45. $24y - 21y^2$ **47.** $y = 4x - 3$

Pages 698–700 Chapter 13 Study Guide and Review
1. trinomial **3.** cubic **5.** like terms **7.** curve **9.** yes; binomial **11.** yes; monomial **13.** yes; trinomial **15.** no

17. 1 **19.** 3 **21.** 6 **23.** 5 **25.** $8b + 3$ **27.** $5y^2 + 2y - 4$ **29.** $y^2 - y - 1$ **31.** $3x^2 - 4x + 7$ **33.** $-x + 1$ **35.** $20t - 10$ **37.** $6k^2 + 3k$ **39.** $-18a + 2a^3$ **41.** Linear; graph is a straight line. **43.** Nonlinear; rate of change is not constant.

45.

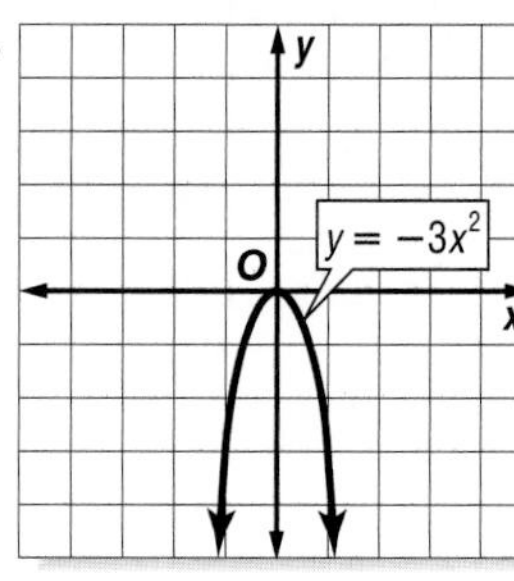

47.

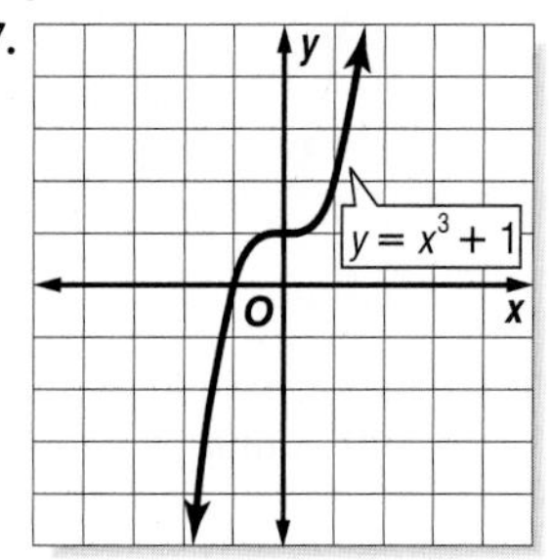

49.

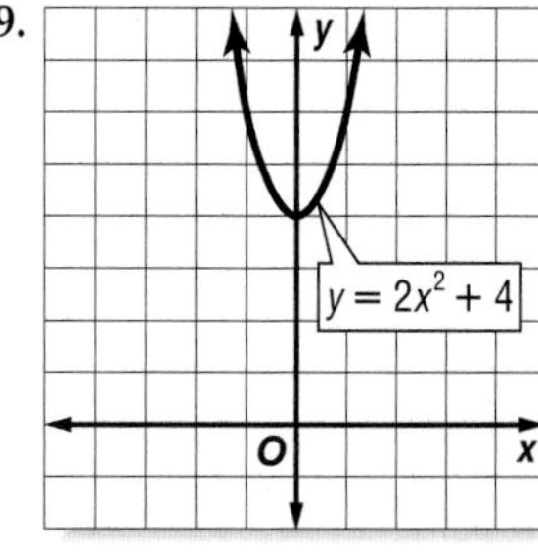

Selected Answers

Photo Credits

Cover Alan Schein/CORBIS Stock Market; **ix** AP Photo/Jackson Laizure; **x** Roy Ooms/Masterfile; **xi** David Young-Wolff/PhotoEdit; **xii** Robert Fried/Stock Boston; **xiii** Kay Chernush/Getty Images; **xiv** Tony Demin/International Stock; **xv** Douglas Faulkner/Photo Researchers; **xvi** Chris Cole/DUOMO; **xvii** Adam Jones/Photo Researchers; **xviii** SuperStock; **xix** Georgette Douwma/Getty Images; **xx** Matt Meadows; **xxi** Antonio M. Rosario/Getty Images; **1** Laura Sifferlin; **2–3** Ulf Sjostedt/Getty Images, (bkgd)SuperStock; **4–5** Keren Su/CORBIS; **10** Doug Martin; **15** AP/Wide World Photos; **19** TempSport/CORBIS; **27** Ken Eward/Photo Researchers; **31** Erik Dreyer/Getty Images; **35** Michael Boys/CORBIS; **37** Rhoda Sidney/Stock Boston; **42** Galen Rowell/CORBIS; **54–55** Jeff Hunter/Getty Images; **57** AFP/CORBIS; **66** NASA; **73** Doug Martin; **78** Mike Mazzaschi/Stock Boston; **83** Roy Ooms/Masterfile; **96–97** Jan Butchofsky-Houser/CORBIS; **99** AP/Wide World Photos; **105** Kit Kittle/CORBIS; **106** Mary Kate Denny/PhotoEdit; **112** file photo; **116** Stuart Westmorland/CORBIS; **123** Index Stock/Eric Kamp; **127** David Young-Wolff/PhotoEdit; **129** Dwayne Newton/PhotoEdit; **131** PhotoDisc; **134** AFP/CORBIS; **144–145** David Young-Wolff/Getty Images, (bkgd)Jeffrey Myers/Stock Boston/PictureQuest; **146–147** Thomas Mangelsen/Minden Pictures; **149** Elizabeth Whiting & Associates/CORBIS; **151** Aaron Haupt; **156** Christophe Loviny/CORBIS; **165** AP/Wide World Photos; **167** Chris Hellier/CORBIS; **172** Geoff Butler; **178** Laura Sifferlin; **180** Latent Image; **182** Ralph A. Clevenger/CORBIS; **184** VCG/Getty Images; **189** Robert Fried/Stock Boston; **190** Roger Ressmeyer/CORBIS; **198–199** SuperStock; **203** Darryl Torckler/Getty Images; **208** Lester Lefkowitz/CORBIS Stock Market; **214** JPL/NASA; **217** Tony Anderson/Getty Images; **219** David Barnes/CORBIS Stock Market; **220** John Cancalosi/Stock Boston; **223** Tony Freeman/PhotoEdit; **229** NASA; **231 237** Geoff Butler; **239** Mark M. Lawrence/CORBIS Stock Market; **247** Andy Caulfield/Getty Images; **251** Tom Szuba/Masterfile; **253** (t)Kay Chernush/Getty Images, (b)Geoff Butler; **262–263** SuperStock; **267** Chuck Pefley/Stock Boston; **272** Louisville Slugger Museum, Louisville, KY; **275** Matt Meadows; **276** (t)Landscape plans provided by Greenscapes Landscape Architects and Contractors, Columbus, OH, (bl)Matt Meadows, (br)Ron Kimball/Ron Kimball Stock; **277** Kunio Owaki/CORBIS Stock Market; **278** Jeff Greenberg/Southern Stock/PictureQuest; **284** Gary White Photography/StockFood; **288** Geoff Butler; **296** Frans Lanting/Minden Pictures; **299** Tony Demin/International Stock; **306** Michael Paras/International Stock; **324–325** Mark Richards/PhotoEdit, (bkgd)Marvin E. Newman/Getty Images; **326–327** Michael Melford/Getty Images; **331** SW Productions/Index Stock/Picturequest; **332** Flash! Light/Stock Boston; **337** Matt Meadows; **344** Bob Daemmrich/Stock Boston; **347** Bob Daemmrich; **348** Douglas Faulkner/Photo Researchers; **354** DUOMO/CORBIS; **356** SuperStock; **358** Doug Martin; **366–367** Allsport; **370** Peter/Stef Lamberti/Getty Images; **378** Simon Bottomley/Getty Images; **383** David Keaton/CORBIS Stock Market; **386** Matt Meadows; **387** Tony Freeman/PhotoEdit; **392** Doug Martin; **395** NASA/JSC; **396** Tom McHugh/Photo Researchers; **399** Cris Haigh/Getty Images; **401** DUOMO; **408** Gail Shumway/Getty Images; **412** Michael Steele/Allsport; **417** Chris Cole/DUOMO; **422** Roy Ooms/Masterfile; **432–433** Index Stock/David Ball, (bkgd)David Zaitz/Photonica; **434–435** Walter Bibikow/Getty Images; **438** Kunio Owaki/CORBIS Stock Market; **444** Kennan Ward/CORBIS Stock Market; **456** Garry Gay/ImageBank; **463** Steven E. Sutton/DUOMO; **472** Geoff Butler; **474** Adam Jones/Photo Researchers; **480** David Young-Wolff/PhotoEdit; **490–491** J. David Andrews/Masterfile; **500** SuperStock; **503** Chuck Szymanski/International Stock; **511** Pete Saloutos/CORBIS Stock Market; **513** Richard Hamilton Smith/CORBIS; **515** (l)Pat LaCroix/Getty Images, (r)Irene Rica Pereira. Untitled, 1951. Oil on board, 101.5 × 61 cm (40 × 24 in.). Solomon R. Guggenheim Museum, New York, NY. Gift of Jerome B. Lurie, 1981; **516** Elizabeth Murray. Museum of Modern Art, New York, NY. Acquired through the Bernhill Fund. Gift of Agnes Gund; **524** SuperStock; **529** Papilio/CORBIS; **530** Roy Lichtenstein. *Modern Painting with Clef*, 1967. Oil on synthetic polymer and pencil on canvas. 252.4 × 458.2 ($100\frac{1}{8} \times 180\frac{3}{8}$ in.)Hirshhorn Museum and Sculpture Garden, Smithsonian Institution; **537** Charles E. Rotkin/CORBIS; **543** Aaron Haupt; **552–553** Rohan/Getty Images; **556** EyeWire, Yann Arthus-Bertrand/CORBIS; **558** Walter Bibikow/Getty Images; **560** (t)SHOE 8/31/94 ©Tribune Media Services. All Rights Reserved. Reprinted by permission, (bl)Roberto Otero/Black Star/PictureQuest, (br)Hermitage Museum, St. Petersburg, Russia/SuperStock; **564** Georgette Douwma/Getty Images; **571** Greg Johnston/International Stock; **575** Bob Mullenix; **578** SuperStock; **579** Bob Daemmrich; **581** (l)John Elk III/Stock Boston, (r)John Elk III/Stock Boston; **583** Matt Meadows; **584** Aaron Haupt; **586** NASA; **593** Kunio Owaki/CORBIS Stock Market; **602–603** Buddy Mays/CORBIS; **604–605** Larry Fisher/Masterfile; **607 610** CORBIS; **614** Eastcott/Momatiuk/Getty Images; **620 632** Bettmann/CORBIS; **634** David Young-Wolff/PhotoEdit; **638** Michael Zito/SportsChrome USA; **640** Aaron Haupt; **644** Mark Burnett; **649** Aaron Haupt; **651** Matt Meadows; **654** CORBIS; **656** Aaron Haupt; **666–667** (bkgd)Digital Stock, David Madison Sports Images; **670** Raymond Gehman/CORBIS; **675** Antonio M. Rosario/Getty Images; **680** Bill Ross/CORBIS; **683** Dallas & John Heaton/CORBIS; **685** Bettmann/CORBIS; **691** Mercury Archives/Getty Images; **693** Photri/Tom Sanders/CORBIS Stock Market.

Index

Absolute frequencies, 627

Absolute value, 58, 59, 61, 63, 64, 65, 66, 91
algebraic expressions, 58
integers, 90

Activity. *See* Algebra Activity; Geometry Activity

Acute angles, 449, 450, 451, 457, 484, 745

Acute triangles, 454, 456, 746

Addends, 63, 64, 66

Addition, 11, 16, 726, 734, 735, 756
algebraic fractions, 221
Associative Property, 24, 26, 49, 63, 66, 130, 725
Commutative Property, 23, 26, 32, 49, 63, 66, 107, 114, 725
decimals, 5, 538, 712, 713
fractions, 220, 233, 716
integers, 62–63, 64–66, 74, 82, 84, 91, 97, 107, 199, 327
like fractions, 220–224, 256
measurements, 591–592
mixed numbers, 220, 233, 491, 716
polynomials, 674–677, 699
properties, 672
solving equations, 111–112, 113, 139, 244
solving inequalities, 345–349
unlike fractions, 232–236, 257

Addition expressions, 97, 102, 677

Addition Property of Equality, 111, 360

Addition table, 63

Additive identity, 24, 48, 49, 66

Additive inverse, 66, 67, 70, 71, 72, 91, 112, 114, 677, 679, 681, 699

Adjacent angles, 476, 493, 495

Algebra Activity
Analyzing Data, 237
Area and Geoboards, 518–519
Base 2, 158
Capture-recapture, 275
Constructions, 498–499
Dilations, 512
Equations with Variables on Each Side, 328–329
Fibonacci Sequence, 253
Graphing Irrational Numbers, 465
Half-Life Simulation, 180
Input and Output, 367
It's All Downhill, 386
Juniper Green, 231
Modeling Multiplication, 682
Modeling Polynomials with Algebra Tiles, 673
Probability and Pascal's Triangle, 640
Pythagorean Theorem, 458–459
Ratios in Right Triangles, 476
Scatter Plots, 39
Simulations, 656–657
Slope and Rate of Change, 392
Solving Equations using Algebra Tiles, 108–109
Symmetry, 505
Taking a Survey, 309
Tessellation, 532
Using a Percent Model, 286–287

Algebra Connection, 25, 35, 58, 77, 150, 161, 166, 170, 176, 181, 211, 217, 221, 227

Algebraic equations, solving, 30, 31, 118, 130, 136, 152, 157, 163, 219, 230, 235, 242, 246, 248, 252, 328–329, 330–333, 338, 344, 349, 443, 444, 457, 484, 685

Algebraic expressions, 17, 19, 69, 99, 667
absolute value, 58
evaluating, 17, 19, 20, 27, 32, 38, 44, 48, 51, 55, 59, 60, 68, 72, 73, 74, 77, 78, 79, 82, 83, 84, 89, 93, 118, 147, 155, 156, 168, 170, 182, 185, 190, 213, 218, 222, 223, 235, 338, 373
factoring, 166, 167, 193
finding value, 27, 32, 48, 74, 457
identify parts, 104
simplifying, 25, 26, 27, 32, 44, 48, 51, 77, 78, 84, 89, 100, 103–107, 118, 139, 147, 157, 213, 224, 439
translating verbal phrases into, 18, 19, 20, 32, 74, 105, 724
writing, 20, 97, 124, 125

Algebraic fractions, 211
addition, 221
division, 217
lowest common denominator, 227
multiplication, 211
simplifying, 169–173, 185, 193

Algebraic relationships, graphing, 87

Algebra tiles. *See also* Modeling
adding integers, 62–63
algebraic expressions, 103, 106
Distributive Property, 99
equations with variables on each side, 328–329
multiplication, 682
multiplication, polynomials by monomial, 683
polynomials, 673, 674, 677
properties of equality, 120, 123
representing expressions, 103
solving equations, 108–109
solving two-step equations, 120

Alternate exterior angles, 492, 493

Alternate interior angles, 492, 493

Altitude, 520

Angles, 447–451, 484
acute, 449, 450, 451, 457, 484, 745
adjacent, 476, 493, 495
alternate exterior, 492, 493
alternate interior, 492, 493
central, 452, 538
complementary, 493, 494, 495, 496, 517, 544
congruent, 493
corresponding, 474, 492, 493, 500, 544
of elevation, 480
exterior, 492, 531
inscribed, 538
interior, 492, 528, 530, 538, 543
measurement, 447, 448
measures, 448, 454
obtuse, 449, 450, 451, 457, 484, 745
quadrilateral, 514
right, 449, 450, 451, 457, 484, 745
straight, 447, 449, 450, 451, 457, 484, 745
supplementary, 494, 495, 496
vertical, 493, 495

Applications. *See also* Cross-Curriculum Connections; More About
accounting, 67
aerospace, 759
ages, 128
aircraft, 172
air pressure, 37
animals, 172, 184, 229, 266, 306, 619, 639

aquariums, 576
architecture, 277, 479, 558, 579, 580, 581, 587, 626, 769, 770
art, 463, 509, 515, 516, 530
astronauts, 606
astronomy, 759, 761
auto racing, 229
aviation, 765, 766
baking, 768
ballooning, 134
banking, 300, 301
baseball, 450, 615–616, 766
baseball cards, 739
basketball, 83, 241, 337, 761
batteries, 567
bicycles, 536
bicycling, 135
birds, 291, 758
birthdays, 106
blood types, 655
books, 291, 766
boxes, 685
brand names, 763
breakfast, 202
bridges, 767
buildings, 480, 559
business, 16, 21, 73, 123, 190, 247, 301, 307, 384, 399, 400, 421, 760, 761, 763, 765, 769
calculators, 301
calendars, 151
camping, 768
candy, 10, 358, 763
carpentry, 167, 218, 390
car rental, 332, 358, 765
cars, 251
car sales, 764
catering, 385
census, 764
chores, 290, 291
city planning, 767
codes, 761
college, 762
comics, 560
communication, 10, 185, 631
community service, 134
computers, 177, 203, 638
construction, 589, 695, 766, 770
consumerism, 769
cooking, 218, 247, 516, 763
crafts, 421, 576
cruises, 107
currency, 273
cycling, 228
decorating, 768
design, 277, 279–280, 321
driving, 313
earnings, 353
earthquakes, 177, 761
ecology, 670
economics, 761, 769
education, 618
elections, 113
electricity, 761
employment, 758, 762
energy, 83, 179
engineering, 566
entertainment, 36, 141, 391, 411, 415, 421, 760, 769
environment, 289
farming, 372
fast food, 763
fencing, 760
ferris wheel, 763
finance, 677, 765
fish, 291
fitness, 450
flags, 523
floods, 78
flooring, 444
flowers, 644
food, 21, 48, 219, 284, 296, 301, 312, 314, 608, 626, 643, 758, 762
food costs, 397
food service, 129, 637
football, 64, 67
forestry, 297
fuel, 765
fund-raising, 358, 581
furniture, 762
games, 516, 638, 653, 726, 759
gardening, 15, 213, 223, 429
gardens, 279
golf, 67
grilling, 235
grocery shopping, 764
handshakes, 644
health, 9, 248, 610, 762, 763
height, 31, 141
highway maintenance, 570
hiking, 358
home improvement, 540
home repair, 252, 390
homework, 343, 769
hourly pay, 240
houses, 234
insects, 279
interior design, 761
Internet, 764, 770
kites, 765
landmarks, 473
landscape design, 540
landscaping, 135, 502, 672
lawn care, 524
license plates, 644
light bulbs, 136
literature, 163
lunchtime, 761
machinery, 209
magazines, 267
maintenance, 766
manufacturing, 208, 767, 768, 770
maps, 88, 292, 472, 473, 759
masonry, 592
measurement, 118, 171, 172, 183, 189, 207, 213, 222, 372, 378, 593, 762
measurement error, 593
media, 283
medicine, 10, 122, 184, 671
metals, 566
meteorology, 128, 157, 246, 759
microwaves, 567
military, 133
mining, 572
mirrors, 511
models, 768
money, 9, 10, 51, 100, 105, 152, 173, 295, 296, 357, 385, 409, 754, 759, 762, 763
monuments, 537, 768
movies, 101, 632
moving, 767
music, 151, 536
Native Americans, 129
newspapers, 758
nutrition, 239, 341, 762, 763, 765
oceanography, 74
oil production, 762
olympics, 128, 273, 619, 624
online service, 762
paint, 274
painting, 118
parades, 167
parks, 474
patterns, 758, 761
pet care, 768
pets, 73, 763
phone calling cards, 123
phone services, 358
photography, 272, 273
picnics, 407
pools, 123, 577, 684
population, 68, 114, 129, 302, 307, 735, 770
postal service, 7, 724
presents, 768
presidents, 629
pressure, 411
production, 770
public safety, 767
publishing, 15, 235, 246, 762
purchasing, 760
quilts, 504
racing, 449
reading, 631
real estate, 301, 525
recreation, 479, 766
remodeling, 762
rides, 475
roller coasters, 279, 439, 625
running, 135
safety, 343, 480, 495, 496

sailing, 766
salaries, 242
sales, 21, 121, 135, 358, 363, 418, 758
savings, 303, 347
school, 42, 107, 185, 203, 213, 307, 358, 413, 420, 537, 608, 631, 764
school colors, 642
school concert, 765
school supplies, 106
sewing, 218, 223, 235, 259
shadows, 473, 480
shapes, 767
shipping, 679, 768
shipping rates, 765
shoes, 283
shopping, 26, 101, 106, 265, 267, 300, 363, 376, 760, 763, 764
signs, 767
skyscrapers, 438
snacks, 274
soccer, 353
sound, 155
space, 187, 188, 189, 296, 758
space exploration, 758
space shuttle, 16, 19
space travel, 212
spending, 411
sports, 93, 101, 238, 301, 515, 542, 759, 760, 764
statistics, 83, 686, 764
statues, 279
submarines, 759
surveying, 766
surveys, 243
swimming, 410, 641
technology, 393, 411, 690, 744
telephone rates, 251
telephones, 229
television, 463, 769
temperature, 78
tennis, 685
tepees, 768
tessellations, 529
tests, 241
test scores, 40
time, 449
towers, 502
toys, 117
traffic laws, 614
trains, 247
transportation, 10, 14, 204, 348, 760, 761, 767
travel, 8, 15, 16, 134, 156, 212, 219, 259, 267, 594, 620, 630, 632, 760, 766
trees, 534, 758
trucks, 202
utilities, 766
vacations, 241, 625
volleyball, 753, 758, 769
voting, 235
walkways, 542
water parks, 623
weather, 59, 60, 71, 72, 73, 83, 89, 93, 113, 248, 307, 333, 348, 371, 608, 615–616
weather records, 79
white house, 208
wildlife, 129
working, 760
world cultures, 769
write a problem, 130

Approximations, 594, 672

Area, 132, 137, 518–519. *See also* Surface area
circles, 533–538, 548, 749
irregular figures, 539–543, 548
parallelograms, 520, 547
rectangles, 132, 133, 134, 135, 140, 141, 152, 214, 224, 349, 671, 730, 767
square, 523
trapezoids, 521, 522–523, 547
triangles, 521, 522, 523, 547

Arithmetic sequences, 249–252, 258, 259, 268, 344, 736

Assessment
Multiple Choice, 52, 94, 142, 196, 260, 322, 364, 430, 488, 550, 600, 664, 702
Open Ended, 53, 95, 143, 197, 261, 323, 365, 431, 489, 551, 601, 665, 703
Practice Chapter Test, 51, 93, 141, 195, 259, 321, 363, 429, 487, 549, 599, 663, 701
Practice Quiz, 21, 32, 74, 84, 107, 130, 163, 185, 224, 248, 292, 308, 338, 354, 397, 418, 451, 470, 517, 538, 572, 588, 628, 645, 681, 691
Prerequisite Skills, 5, 10, 16, 21, 27, 32, 38, 55, 61, 68, 74, 79, 84, 97, 102, 107, 114, 119, 124, 130, 147, 152, 157, 163, 168, 173, 179, 185, 199, 204, 209, 213, 219, 224, 230, 236, 242, 248, 263, 268, 274, 280, 285, 292, 297, 302, 308, 327, 333, 338, 344, 349, 354, 367, 373, 379, 385, 391, 397, 401, 408, 413, 418, 435, 440, 445, 451, 457, 464, 470, 475, 491, 497, 504, 511, 517, 525, 531, 538, 553, 561, 567, 572, 577, 582, 588, 605, 611, 616, 621, 628, 633, 639, 645, 649, 667, 672, 677, 681, 686, 691, 706–723
Short Response/Grid In, 53, 95, 124, 143, 197, 261, 323, 354, 365, 431, 489, 517, 531, 537, 551, 601, 665, 703
Standardized Test Practice, 10, 16, 21, 29, 30, 32, 39, 44, 51, 61, 68, 74, 76, 77, 79, 84, 89, 93, 102, 107, 112, 113, 114, 119, 124, 130, 136, 141, 152, 157, 163, 168, 171, 173, 179, 184, 190, 204, 209, 213, 219, 224, 230, 236, 240, 241, 242, 247, 252, 259, 268, 274, 280, 285, 292, 297, 302, 305, 306, 308, 314, 321, 333, 338, 344, 349, 351, 353, 354, 359, 363, 373, 379, 385, 389, 390, 391, 397, 401, 408, 413, 418, 422, 429, 440, 445, 451, 457, 461, 462, 464, 470, 475, 481, 487, 494, 495, 497, 504, 511, 517, 525, 531, 537, 543, 549, 561, 564, 566, 567, 572, 577, 582, 588, 594, 599, 611, 616, 621, 627, 633, 639, 645, 647, 648, 649, 655, 663, 672, 677, 681, 686, 689, 691, 695, 701
Test-Taking Tips, 29, 52, 76, 95, 112, 143, 171, 196, 240, 305, 322, 351, 364, 389, 430, 461, 488, 494, 550, 564, 600, 647, 664, 689, 702

Associative Property of Multiplication, 24, 26, 49, 51, 76, 78, 114, 522, 725

Average (mean), 82, 92

B

Backsolving, 29, 112

Back-to-back stem-and-leaf plots, 607–608, 609, 611

Bar graphs, 722–723
double, 722–723
misleading, 631

Bar notation, 201

Base 2 numbers, 158

Base 10 numbers, 158

Bases, 153, 290, 299, 520, 557, 559, 570, 595, 599

Best-fit lines, 409–413, 427, 429

Bias, 634

Bilateral symmetry, 505

Binary numbers, 158

Binomials, 669, 671, 676, 681, 698, 755

Box-and-whisker plots, 617–621, 622, 633, 659, 753
double, 619
drawing, 619, 620
interpretation, 618–619

Brackets [ ], as grouping symbol, 12

Cake method, 161

Calculator skills, 437, 438, 439, 478, 479, 480, 497, 504, 531, 534, 561. *See also* Graphing Calculator Investigation

Career Choices
- architect, 579
- artist, 515
- biologist, 42
- business owner, 399
- carpenter, 223
- cartographer, 472
- civil engineer, 480
- ecologist, 670
- interior designer, 278
- marine biologist, 348
- meteorologist, 129
- musician, 172
- real estate agent, 358
- statistician, 649
- veterinarian, 73

Center, circles, 533

Central angles, 452, 538

Challenge. *See* Critical Thinking: Extend the Activity; Extending the Lesson

Change, rate of, 393–397, 426

Checking reasonableness, 82, 586, 684

Checking solutions, 99, 111, 113, 116, 117, 118, 121, 122, 123, 124, 130, 136, 141, 152, 157, 163, 176, 201, 204, 212, 217, 230, 242, 245, 246, 248, 252, 258, 297, 330, 331, 332, 334, 335, 336, 338, 346, 347, 351, 352, 353, 354, 355, 356, 359, 360, 373, 377, 379, 382, 406, 413, 414, 420, 437, 438, 454, 514, 561, 588, 689, 729, 736, 739, 740, 741

Chords, 538

Circle graphs, 451, 452

Circles
- area, 535–538, 548, 749
- center, 533
- circumference, 533–538, 548, 749
- diameter, 533, 536, 749
- pi, 208, 533, 534
- radius, 533, 536, 749

Circumference
- circles, 533–538, 548, 749

Closure Property, 27

Coefficients, 103, 104, 105, 106
- negative, 122, 161

Combinations, 642–643, 644, 661, 754
- notation, 643

Commission, 301

Common Misconceptions, 58, 132, 154, 175, 355, 420, 449, 467, 507, 557, 591, 617. *See also* Find the Error

Common multiples, 226

Common ratios, 250, 251, 258

Commulative, 23

Communication
- choose, 592
- classify, 529
- compare and contrast, 228, 251, 312, 455, 549, 637, 643, 653, 674
- define, 19, 36, 59, 105, 207, 272, 300, 468, 523, 614
- describe, 202, 234, 283, 371, 389, 395, 429, 515, 559, 580, 589, 599, 653, 680, 689, 694
- determine, 150, 272, 684
- display, 608
- draw, 222, 266, 549
- estimate, 295
- explain, 59, 83, 87, 117, 122, 133, 141, 150, 155, 161, 166, 171, 177, 188, 202, 217, 241, 266, 295, 300, 306, 342, 347, 377, 384, 407, 410, 416, 421, 438, 443, 468, 495, 502, 529, 570, 575, 580, 586, 614, 619, 625, 637, 648, 670, 684, 689, 694
- find, 228, 241, 540
- identify, 207, 608
- list, 42, 51, 128, 336, 353, 421, 438, 570
- name, 36, 42, 246, 332, 449, 455, 631, 676
- replace, 228
- state, 67, 77, 117, 141, 177, 251, 400, 429
- tell, 9, 14, 26, 30, 113, 312, 529, 535, 619
- write, 19, 51, 77, 133, 207, 357, 421, 509, 540

Commutative Property of Addition, 23, 26, 32, 49, 63, 66, 107, 114, 725

Commutative Property of Multiplication, 23, 26, 49, 61, 75, 78, 522, 725

Compare and order
- decimals, 710
- fractions, 228
- integers, 57, 59
- numbers in scientific notation, 188
- percents, 283, 284
- real numbers, 442

Complementary angles, 493, 494, 495, 496, 517, 544

Complex solids, volume, 564

Composite numbers, 159, 161, 162, 192, 229, 285, 731

Compound events, 650–655, 662

Compound interest, 303

Computation, choosing method, 8

Concept Summary, 41, 47, 48, 49, 50, 82, 90, 91, 92, 138, 139, 140, 148, 154, 191, 192, 193, 194, 207, 254, 255, 256, 257, 258, 289, 293, 298, 316, 317, 318, 319, 320, 341, 360, 361, 362, 377, 416, 424, 425, 426, 427, 428, 442, 483, 484, 485, 486, 495, 544, 545, 546, 547, 548, 595, 596, 597, 598, 658, 659, 660, 661, 662, 688, 698, 699, 700

Cones, 569
- diameter, 571
- radius, 571
- surface area, 578–582, 597
- volume, 568–572, 596, 616

Congruence statements, 501, 502, 503, 511, 517, 545, 549

Congruent angles, 493

Conjectures, 7, 12, 22, 25, 26, 32, 38, 39, 63, 137, 155, 162, 175, 180, 204, 215, 253, 275, 303, 368, 386, 392, 393, 445, 453, 476, 512, 563, 583, 584, 640, 697

Constants, 103, 104, 105, 106
- proportionality, 394
- variation, 394, 395

Constructions, 498–499

Converting measurements, 397, 566

Coordinate plane, 45, 57, 87, 88, 89, 92, 93, 124, 728
- reflections, 508
- rotations, 509
- transformations, 506–511, 545–546
- translations, 507

Coordinates, 69, 468, 486

Index

Coordinate system, 33–34, 36, 38, 39, 44, 50, 51, 79, 85–89, 92, 367, 475, 504, 725

Corresponding parts, 471, 500, 501, 502, 503

Cosine, 477–481, 486, 747

Counterexamples, 25, 26, 27, 73, 208, 444, 510, 684

Counting outcomes, 661, 635, 639

Critical Thinking, 10, 16, 21, 27, 32, 37, 44, 60, 61, 68, 73, 78, 84, 89, 102, 107, 114, 119, 124, 130, 135, 152, 157, 163, 168, 173, 179, 184, 190, 204, 209, 213, 219, 223, 230, 236, 242, 247, 251, 268, 274, 280, 284, 285, 292, 297, 302, 307, 314, 333, 338, 344, 348, 354, 358, 373, 379, 385, 391, 397, 401, 408, 412, 417, 422, 440, 445, 450, 457, 464, 469, 475, 481, 497, 504, 511, 517, 525, 531, 537, 543, 560, 567, 572, 577, 581, 587, 593, 610, 616, 621, 627, 633, 639, 644, 649, 654, 672, 677, 680, 685, 690, 695

Cross-Curriculum Connections. *See also* Applications; More About
- biology, 42, 156, 761
- chemistry, 68, 178, 768
- earth science, 235, 383
- geography, 60, 73, 208, 284, 296, 332, 524, 543, 617, 759, 767
- geology, 571, 758, 759
- history, 151, 156, 168, 537, 570, 587, 766, 768
- life science, 178, 291
- physical science, 184, 190, 251, 396
- physics, 765, 766
- science, 20, 27, 36, 37, 45, 131, 291, 378, 438, 536, 571, 614, 696, 760, 761

Cross products, 270, 271, 272

Cubes, 584
- surface area, 157, 694
- volume, 157

Cubic functions, 688
- graphing, 692–696, 700

Cubic units, 575

Cumulative frequencies, 627

Customary system, 118
- converting measurements, 720–721, 734

Cylinders
- surface area, 573–577, 597
- volume, 563–567, 596

D

Data
- analysis, 39, 180, 237, 240, 253, 275, 386, 392, 562, 583, 640, 656, 657
- box-and-whisker plots, 633, 659, 753
- collecting, 39, 180, 237, 253, 275, 369, 386, 392, 562, 583, 640, 656, 657
- comparisons, 269
- display, 606, 722–723
- frequency tables, 627
- histograms, 623–628, 629, 633, 660, 753
- line plot, 239
- mean, 82, 92, 238–239, 240, 248, 252, 258, 605, 618, 735
- median, 238–239, 240, 248, 252, 258, 605, 611, 615, 618, 735
- mode, 238–239, 240, 248, 252, 258, 605, 735
- modeling, 657
- patterns, 22
- scatter plots, 39, 40–42, 43, 45–46, 50, 61, 68, 107, 408, 410, 411, 412, 422, 427, 429, 726
- stem-and-leaf plots, 606–611, 658, 681, 752

Decimal numbers, 158

Decimals, 206, 281–285, 318, 435
- addition, 5, 538, 712, 713
- comparison with fractions, 202
- division, 242, 714, 715
- estimating, 5
- multiplication, 263, 491, 714, 715
- with negative exponents, 183
- ordering, 616, 710
- repeating, 201, 202, 203, 206, 214, 224, 254, 338, 733
- rounding, 588, 711
- subtraction, 5, 712, 713
- terminating, 200, 203, 206
- writing as fractions or mixed numbers, 207, 208, 224, 254, 422, 733
- writing as percents, 282–283, 297, 302
- writing fractions as, 200–209, 214, 338, 475
- writing mixed numbers as, 209, 214

Decision making. *See* Critical Thinking; Problem solving

Deductive reasoning, 25, 107

Degrees, 448

Density, 566

Dependent events, 652

Diagonals, 528, 558

Diameter
- circles, 533, 536
- cones, 571

Differences, 13, 21, 61, 70, 71, 72, 73, 74, 84, 91, 93, 118, 230, 235, 242, 256, 257, 327, 349, 605, 712, 716, 727, 734, 735, 756. *See also* Subtraction

Dilations, 512

Dimensional analysis, 212, 213, 217, 266, 267, 280

Direct variation, 394

Discounts, 299, 308, 738

Distance, 746

Distance Formula, 131, 466–470, 485–486

Distributive Property, 98–102, 104, 107, 138, 147, 163, 164, 166, 185, 193, 209, 333, 334, 338, 349, 361, 667, 683, 686, 699, 728

Divisibility Rules, 148, 149, 150, 151, 157, 191, 730

Division, 11
- decimals, 242, 714, 715
- integers, 80–84, 92, 114, 199, 327
- negative numbers, 351–352
- notation, 13
- positive numbers, 350–351
- rational numbers, 215–219, 256
- solving equations, 115–119, 139, 245, 435
- solving inequalities, 350–354

Division expressions, 117

Division Property of Equality, 115

Division sentence, 80

Divisor, 69

Domain, 35, 36, 37, 38, 44, 50, 51, 136, 367, 725

Double bar graphs, 722–723

Double box-and-whisker plots, 619

Double line graphs, 722–723

E

Edges, 556, 557, 559, 595, 599

Endpoints, 468, 469, 481, 486, 527, 746

Enrichment. *See* Critical Thinking; Extend the Activity; Extending the Lesson

Equals sign, 126

Equations, 28, 49, 96–143, 126, 326. *See also* Algebraic equations; Linear equations; Percent equations; Systems of equations; Two-step equations
graphing, 691
identifying functions, 687–688
negative coefficients, 122
solving, 28–29, 49, 68, 113, 118, 121, 122, 123, 124, 141, 157, 443, 491, 511, 588, 725, 729, 730, 736, 739, 740, 744, 745
addition, 111–112, 139
division, 115–119, 139, 435
with grouping symbols, 334–338, 361
by multiplication, 115–119, 139
rational numbers, 244–248, 258
subtraction, 110–111, 139
variables on each side, 330–333, 360
solving systems, 414–418, 428
solving two-step, 120–122, 140, 327, 354, 451
translating verbal sentences into, 30, 125
writing inequalities, 340–342
writing linear, 404–408, 427
writing two-step, 126–130

Equilateral figures, 517

Equivalent equations, 104, 111

Equivalent expressions, 98

Error Analysis. *See* Find the Error; Common Misconceptions

Estimation, 5, 9, 121, 209, 220, 230, 233, 294, 295, 296, 297, 298, 302, 308, 319, 321, 397, 437, 438, 439, 445, 456, 564, 567, 586, 712, 714, 716, 717, 738, 745

Evaluating expressions. *See* Expressions

Even numbers, 148

Events
compound, 650–655, 662
dependent, 652
inclusive, 672
independent, 650–651
mutually exclusive, 652, 672
simple, 310

Expanded form, 154, 156, 185, 192

Experimental probability, 311

Exponents, 153–157, 190, 192, 252, 435, 464, 731, 732
negative, 181–185, 187, 194
positive, 182, 183, 185, 187, 194

Expressions, 12–13. *See also* Addition expressions; Algebraic expressions; Binomial expressions; Division expressions; Equivalent expressions; Numerical expressions; Subtraction expressions; Verbal expressions
evaluating, 12, 107, 154, 155, 724, 726, 727, 731, 732
with exponents, 154, 155
factoring, 731
finding value, 14, 21, 724
simplifying, 104, 114, 497, 525, 725, 727, 728
translating, into words, 11

Extend the Activity, 39, 158, 253, 368, 386, 392, 640, 656, 657, 682

Extending the Lesson, 32, 102, 168, 185, 213, 236, 252, 268, 280, 333, 344, 359, 392, 440, 445, 464, 497, 531, 538, 572, 577, 582, 594, 621, 627, 655, 696

Extra Practice, 724–757

Extreme value, 239

Faces, 557, 559, 595, 599

Factor expressions, 166

Factorial notation, 642

Factoring, 157

Factor monomials, 161

Factors, 69, 79, 148–152, 151, 191, 730

Factor tree, 160

Families of quadratic functions, 697

Family of graphs, 402–403

Find the Error, 14, 26, 72, 87, 100, 105, 128, 161, 166, 212, 222, 234, 246, 278, 291, 306, 347, 353, 357, 389, 395, 400, 462, 479, 536, 565, 592, 643, 648, 671, 676. *See also* Common Misconceptions

Foldables™ Study Organizer, iv, 1, 5, 55, 97, 147, 199, 263, 327, 367, 435, 491, 553, 605, 667

Formulas, 131–133, 140, 300
distance, 466–470, 485–486
midpoint, 466–470, 485–486

Four-step problem-solving plan, 5, 6–8, 9, 10, 47–48, 51

Fraction bars as grouping symbol, 12, 13, 224, 264

Fractions, 281, 318. *See also* Algebraic fractions
addition, 220, 233, 716
comparison, 202, 228
computations, 605
division, 216, 717
multiplication, 567, 649, 717
percents as, 281–282, 283
simplifying, 171, 172, 173, 179, 185, 199, 732
subtraction, 221, 716
unit, 236
writing decimals, 200–209, 214, 224, 254, 338, 422, 475, 733
writing rational numbers, 205
writing ratios, 264–265
writing repeating decimals, 206
writing simplest form, 263
writing terminating decimals, 206

Frequencies
absolute, 627
cumulative, 627
relative, 627

Frequency polygon, 628

Frequency tables, comparison to histograms, 627

Functional notation, 380

Functions, 369–373, 424, 537, 741
graphing, 370, 757
quadratic and cubic, 692–696, 700
linear, 687–691, 700
nonlinear, 687–691, 700

Fundamental Counting Principle, 636, 661

Geometric sequences, 249–252, 258, 259, 268, 344, 736

Geometry
angles, 447–451, 484
acute, 449, 450, 451, 457, 484, 745
adjacent, 476, 493, 495
alternate exterior, 492, 493
alternate interior, 492, 493
central, 452, 538
complementary, 493, 494, 495, 496, 517, 544
congruent, 493

Index

corresponding, 474, 492, 493, 500, 544
of elevation, 480
exterior, 492, 531
inscribed, 538
interior, 492, 528, 530, 538, 543
measures, 447, 448, 454
obtuse, 449, 450, 451, 457, 484, 745
quadrilateral, 514
right, 449, 450, 451, 457, 484, 745
straight, 447, 449, 450, 451, 457, 484, 745
supplementary, 494, 495, 496
vertical, 493, 495
area
circles, 533–538, 548, 561, 749
irregular figures, 539–543, 548
parallelograms, 520–521, 547
rectangles, 132, 133, 134, 135, 140, 141, 152, 214, 224, 349, 671, 730, 767
shaded region, 770
squares, 439, 523
triangles, 445, 561
circumference of circle, 561, 749
diameter of circle, 533, 749
inequality involving acute angle, 764
isosceles triangle, 469
length of hypotenuse in right triangles, 462, 470
length of rectangle, 764
length of sides of golden rectangles, 675
line segments, 473, 644, 645
obtuse angles, 691
patterns, 9
perimeter, 469
equilateral triangles, 394
geometric figures, 106
golden rectangles, 675
isosceles trapezoid, 680
isosceles triangles, 680
rectangles, 132, 133, 134, 135, 140, 141, 152, 224, 302, 335, 336, 349, 359, 363, 385, 417, 676, 701, 730
squares, 378, 394, 440, 463, 760
triangles, 467, 503, 591, 681, 760, 770
pi, 208, 533, 534
radius of circle, 443, 533, 536, 749
ratio of width to length, 266
rectangular prism, 464
right angles, 464, 691
scalene triangle, 469
straight angles, 691
sum of angles of triangle, 676
surface area, 594, 751
cubes, 157, 694
prisms and cylinders, 573–577, 597
pyramids and cones, 578–582, 597
three-dimensional figures, 552–601
triangles, 38, 453–457, 468, 485, 528
acute, 454, 456, 746
angle measures, 453
area, 521, 522, 523, 547
congruent, 500–504, 545, 748
equilateral, 455
isosceles, 455
obtuse, 454, 456, 746
obtuse isosceles, 455, 470, 485
right, 454, 456, 460, 461, 462, 463, 464, 481, 485, 746
right scalene, 470, 485
scalene, 455
similar, 471–475, 486
vertice, 88
vertices, 88, 447, 453, 527, 545–546, 556, 557, 559, 595, 599
volume, 750, 751
complex solid, 564
cones, 568–572, 596, 616
cubes, 157
prisms and cylinders, 563–567, 596
pyramids and cones, 568–572, 596
rectangular prisms, 695

Geometry Activity
Building Three-Dimensional Figures, 554–555
Similar solids, 583
Volume, 562

Golden ratio, 253, 268

Golden rectangle, 268

Graphing
algebraic relationships, 87
cubic functions, 692–696, 700
equations, 691
functions, 370, 757
inequalities, 342, 343, 353, 354, 361, 362, 379, 419–422, 440, 740, 744
linear equations, 376–377
quadratic functions, 692–696, 700
solutions of equations, 111, 113, 118, 428, 744
systems of equations, 414, 416, 417, 422, 428
systems of inequalities, 346, 348

Graphing Calculator Investigation
Box-and-Whisker Plots, 622
Families of Graphs, 402–403
Families of Quadratic Functions, 697
Finding Angles of a Right Triangle, 482
Function Tables, 374
Graphing Inequalities, 423
Histograms, 629
Mean and Median, 243
Probability Simulation, 315
Scatter Plots, 45–46

Graph pairs, 88

Graph points, 55, 85, 86, 87, 92, 93

Graphs
bar, 722–723
circle, 723
data display, 722–723
identifying functions, 687
line, 722–723
relations as, 35, 36, 37, 50
writing equations, 405

Greatest common factor (GCF), 164–168, 179, 193, 199, 214, 252, 731

Greatest possible error, 594

Grid In. *See* Assessment.

Grouping symbols, 48
brackets, 12
fraction bars, 12, 13
inequalities, 356
parentheses, 12, 15, 75
solving equations, 334–338, 361

Guess and Check, 709

Half-Life Simulation, 180

Heptagon, 528

Hexagon, 528

Hexagonal prism, 572

Hexagonal pyramid, 571

Histograms, 623–628, 629, 633, 660, 753
comparison to frequency tables, 627
interpretation, 624

Homework Help, 9, 14, 20, 26, 31, 36, 43, 59, 67, 73, 78, 83, 88, 101, 106, 113, 118, 123, 129, 134, 151, 155, 162, 167, 172, 178, 183, 189, 203, 208, 213, 218, 222, 229, 235, 241, 246, 251, 267, 279, 284, 291,

296, 301, 307, 313, 332, 337, 343, 348, 353, 357, 371, 378, 384, 390, 396, 400, 407, 411, 417, 421, 439, 444, 450, 456, 463, 469, 474, 480, 496, 503, 510, 516, 524, 530, 542, 559, 566, 571, 576, 580, 587, 593, 609, 615, 620, 626, 632, 638, 644, 648, 653, 671, 676, 680, 685, 690, 694

Horizontal lines, 383

Hypotenuse, 460, 463, 470, 475, 477, 481, 485, 487, 746

Identity Property, 32, 63

Inclusive events, 672

Independent events, 650–651

Indirect measurement, 471–475, 486

Inductive reasoning, 7, 25, 71

Inequalities, 326, 340–344, 361, 367, 464
- graphing, 342, 343, 353, 354, 361, 362, 379, 419–422, 428, 440, 740, 744
- solving, 354, 373, 379, 385, 445, 481, 561, 567, 649, 740, 741
  - addition, 345–349, 361
  - division, 350–354, 362
  - multiplication, 350–354, 362
  - multi-step, 355–359, 362
  - subtraction, 345–349, 361
- symbols, 57
- truth, 341
- writing, 740

Integers, 69, 451, 487, 745
- absolute value, 90
- addition, 62–63, 64–66, 74, 82, 84, 91, 97, 107, 199, 327
- comparing and ordering, 56, 57, 59, 60, 68
- division, 80–84, 92, 114, 199, 327
- modeling real-world situations, 61
- multiplication, 75–79, 82, 84, 91–92, 97, 152, 199, 327
- number line, 56, 57, 59, 68
- problem solving, 66
- for real-world situations, 56–57
- subtraction, 70–72, 74, 82, 84, 91, 118, 199, 327, 385
- writing as fractions, 205

Intercepts, graphing linear equations using, 381–385, 425

Interest, 299, 738

Internet Connections
- www.pre-alg.com/careers, 42, 73, 129, 172, 223, 278, 339, 348, 358, 399, 472, 480, 515, 579, 649, 670
- www.pre-alg.com/chapter_test, 51, 93, 141, 259, 321, 363, 429, 487, 549, 599, 663, 701
- www.pre-alg.com/data_update, 43, 68, 114, 119, 157, 190, 229, 267, 348, 396, 439, 543, 587, 649, 690
- www.pre-alg.com/extra_examples, 7, 13, 19, 25, 35, 41, 57, 65, 71, 77, 81, 87, 99, 105, 111, 117, 121, 127, 133, 149, 155, 161, 165, 171, 177, 187, 201, 207, 211, 217, 221, 227, 233, 239, 245, 249, 265, 271, 277, 283, 289, 295, 299, 305, 311, 331, 335, 341, 347, 351, 355, 369, 377, 383, 389, 395, 399, 405, 409, 415, 419–422, 437, 443, 449, 455, 461, 467, 473, 479, 493, 501, 507, 515, 521, 529, 535, 541, 557, 565, 569, 575, 579, 585, 591, 607, 613, 619, 625, 631, 637, 643, 647, 651, 669, 675, 679, 683, 689, 693
- www.pre-alg.com/other_calculator_keystrokes, 45, 243, 315, 402, 423, 482, 629, 697
- www.pre-alg.com/self_check_quiz, 9, 15, 21, 27, 31, 37, 43, 59, 67, 79, 83, 89, 101, 107, 113, 119, 123, 129, 135, 151, 157, 163, 167, 173, 179, 183, 189, 203, 209, 213, 219, 223, 229, 235, 241, 247, 251, 267, 273, 279, 285, 291, 297, 301, 307, 313, 333, 337, 343, 349, 353, 357, 371, 379, 385, 391, 397, 401, 407, 411, 417, 421, 439, 445, 451, 457, 463, 469, 475, 481, 497, 503, 511, 517, 525, 531, 537, 543, 559, 567, 571, 577, 581, 587, 593, 609, 615, 621, 627, 633, 639, 645, 649, 653, 671, 677, 681, 685, 691, 695
- www.pre-alg.com/standardized_test, 53, 95, 143, 197, 261, 322, 365, 430, 489, 551, 601, 665, 703
- www.pre-alg.com/usa_today, 8, 289, 290, 312, 610, 654
- www.pre-alg.com/vocabulary_review, 47, 90, 138, 191, 316, 360, 483, 595, 658, 698
- www.pre-alg.com/webquest.com, 43, 79, 135, 136, 145, 173, 242, 301, 314, 325, 333, 412, 422, 433, 481, 542, 571, 594, 603, 626, 690, 696

Interquartile range, 613, 614, 615, 628, 659, 753

Intersecting lines, 493, 560

Inverse operations, 110, 121, 258, 373, 440

Inverse Property of Multiplication, 215

Investigations. *See* Algebra Activity; Geometry Activity; Graphing Calculator Investigation; Spreadsheet Investigation; WebQuest

Irrational numbers, 206, 441, 443, 444, 451, 484, 487, 745

Irregular figures, area, 539–543, 548

Isosceles triangles
- obtuse, 455, 470, 485

Key Concepts, 18, 23, 29, 58, 64, 65, 66, 70, 75, 76, 80, 81, 111, 115, 117, 132, 175, 176, 181, 186, 210, 215, 216, 221, 232, 233, 238, 270, 271, 282, 288, 310, 345, 350, 352, 376, 382, 388, 394, 436, 441, 449, 453, 454, 455, 460, 466, 468, 472, 477, 492, 493, 500, 507, 508, 514, 520, 521, 522, 528, 533, 557, 563, 565, 568, 569, 573, 575, 579, 585, 636, 646, 650, 652, 688

Keystrokes. *See* Graphing Calculator Investigation; Internet Connections

Lateral area, 578

Lateral faces, 578

Least common denominator (LCD), 227, 228, 229, 234, 236, 252, 735

Least common multiple (LCM), 226–230, 228, 236, 257, 735

Leaves, 606

Like fractions
- addition, 220–224, 256
- subtraction, 220–224, 256

Like terms, 103, 104, 105, 106, 122, 674, 676, 677, 728

Linear equations, 408, 447, 667
- graphing, 376–377
- with intercepts, 381–385, 425
- two variables, 375–379, 425
- writing, 404–408, 427

Index

Linear functions, 687–691, 700, 701, 757

Line graphs, 722–723
double, 722–723

Line of symmetry, 506, 507

Line plot, 239

Lines
best-fit, 409–413, 427, 429
horizontal, 383
intersecting, 493, 560
parallel, 492–497, 495, 497, 544, 560
perpendicular, 494, 495
skew, 558, 560, 567
vertical, 383

Line segments, 453, 644, 645

Line symmetry, 505

Lists, 708

Logical Reasoning. *See* Critical Thinking

Look Back, 18, 71, 76, 100, 111, 166, 176, 210, 244, 264, 266, 293, 330, 334, 335, 369, 442, 460, 466, 497, 523, 535, 540, 570, 584, 607, 635, 637, 652, 683, 684

Major arc, 538

Make a Conjecture. *See* Conjectures

Make a table or list, 708

Markup, 301

Maximum point, 695

Mean, 82, 92, 238–239, 248, 252, 258, 605, 618, 735

Measurements, 213
addition, 592
converting, 168, 263, 272, 718–721, 734
customary, 118, 720–721, 734
metric, 718–719
multiplication, 592

Measures of central tendency, 238–242, 258, 605, 735

Measures of variation, 612–616, 659

Median, 238–239, 248, 252, 258, 605, 611, 615, 618, 735

Mental Math, 25, 26, 29, 30, 31, 38, 44, 49, 51, 66, 74, 79, 102, 104, 122, 127, 150, 160, 201, 282, 525, 725, 738

Metric system, converting measurements, 718–719

Midpoint, 468, 486, 746

Midpoint Formula, 466–470, 485–486

Minimum point, 695

Minor arc, 538

Mixed numbers, 200–201
addition, 220, 233, 491, 716
division, 216, 717
multiplication, 211, 717
subtraction, 221, 234, 716
writing as decimals, 200, 201, 202, 214, 422
writing as fractions, 205, 209
writing decimals as, 207–208, 224, 733

Mode, 238–239, 248, 252, 258, 605, 735

Modeling, 276–280, 317, 464
adding integers, 62–63, 222
algebraic expressions, 103, 106
area
circles, 535
parallelograms, 520–521
rectangles, 132
trapezoids, 522
triangles, 521
area and geoboards, 518–519
circumference of circles, 533
congruent triangles, 500
constructions, 498–499
direct variation, 394, 395
distance formula, 466
Distributive Property, 99
equations with variables on each side, 328–329
integers on number line, 65
irrational numbers, 465
manipulatives, 62–63
midpoint formula, 468
monomials, 673
multiplication, 682
polynomial by monomial, 683
percent, 286–287
polyhedrons, 557
polynomials, 673, 677
prime numbers, 162
properties of equality, 123
Pythagorean Theorem, 458–459, 460
rational numbers, 207, 210, 215
ratios of similar solids, 585
real-world situations, 61
right triangles, 476
similar triangles, 472
simulations, 656–657
slope, 388
solving equations, 108–109
solving two-step equations, 120
surface area
cones, 579
cylinders, 575
rectangular prisms, 573
three-dimensional figures, 554–555, 560
triangles, 453
trigonometric ratios, 477
volume, 562
cones, 569
cylinders, 565
prisms, 563, 565
pyramids, 568

Monomials, 148–152, 161, 163, 191, 195, 667, 668, 669, 671, 681, 698, 730, 731, 755
degrees, 670
division, 175–179, 176, 194
factoring, 161, 162, 192, 195
greatest common factor (GCF), 166, 167, 193
identify, 150
least common multiple (LCM), 227
multiplication, 175–179, 194, 681
powers, 179

More About. *See also* Applications; Cross-Curriculum Connections
airports, 247
amusement parks, 99, 644
animals, 203, 474
aquariums, 564
architecture, 277, 503
art, 560
astronomy, 66
attractions, 272
aviation, 112
baby-sitting, 106
backpacking, 356
baseball cards, 105
basketball, 610, 685
billiards, 456
calling cards, 123
candy, 10
cellular phones, 332
cheerleading, 217
decorating, 337
endangered species, 396
energy, 83
food, 284
football, 15
fruit drinks, 575
games, 511, 651
golf, 57

gymnastics, 463
hang gliding, 401
health, 378
hurricanes, 239
Internet, 654
jobs, 632
movie industry, 31
oceans, 189
oranges, 593
parks, 116
patients, 691
planets, 229
plants, 35
pole vaulting, 412
pools, 395
population, 267
presidents, 607
rafting, 422
ranching, 118
recycling, 208
scooters, 127
scuba diving, 370
skateboards, 299, 638
skydiving, 693
snow, 529
snowboards, 417
soccer, 19, 134
sound, 408
space, 586
sports, 344, 620
state fairs, 347
stock market, 306
swimming, 354
technology, 162
temperature, 680
tides, 78
track and field, 165
travel, 219
videos, 331
water, 182
weather, 444
weddings, 149

Multiple Choice. *See* Assessment

Multiple representations, 18, 23, 29, 58, 64, 65, 66, 70, 75, 76, 80, 81, 98, 110, 111, 115, 117, 132, 175, 176, 186, 207, 210, 215, 216, 221, 225, 232, 233, 238, 270, 288, 298, 310, 345, 350, 352, 377, 388, 394, 436, 453, 460, 466, 468, 472, 477, 500, 514, 522, 533, 535, 563, 565, 568, 569, 573, 575, 579, 613, 636, 650, 652

Multiplication, 11, 756
algebraic fractions, 211
decimals, 263, 491, 714, 715
fractions, 567, 649
integers, 75–79, 82, 84, 91–92, 97, 152, 199, 327
measurements, 592
mixed numbers, 211
monomials, 176, 681
negative fractions, 211
negative numbers, 351–352
notation, 13
polynomials by monomials, 683–686, 699
positive numbers, 350–351
rational numbers, 256, 553, 577
solving equations, 115–119, 139, 245
solving inequalities, 350–354

Multiplication Properties
Associative, 24, 26, 49, 522
Commutative, 23, 26, 49, 75, 522
of Zero, 24, 49, 61, 725

Multiplicative identity, 24, 38, 49, 725

Multiplicative inverse, 215–216, 218, 504, 734

Multi-step inequalities, solving, 355–359, 362

Mutually exclusive events, 652, 672

Natural numbers, 205, 444, 451, 487, 745

Negative coefficients, 122, 161

Negative exponents, 181–185, 187, 194

Negative fractions, multiplication, 211

Negative integers, 64

Negative numbers, 56
division, 351–352
multiplication, 351–352

Negative slope, 388

Nonlinear functions, 687–691, 700, 701, 757

Nonscientific calculators, 12

Number line, 56, 57, 113, 740
absolute value, 58
addition, 64, 65, 68
comparison and ordering of integers, 56, 57
integers, 56, 57, 59
multiplication, 75, 77
subtraction, 70

Numbers, 12–13, 48
base 2, 158
base 10, 158
binary, 158
comparison, 283
composite, 159, 161, 162, 192, 229, 285, 731
decimal, 158
even, 148
factoring, 149
irrational, 206, 441, 443, 444, 451, 484, 487, 745
mixed, 200–201
natural, 205, 444, 451, 487, 745
negative, 56, 351–352
odd, 148
positive, 56, 350–351
prime, 159, 161, 162, 192, 440, 731
properties, 24
rational, 198–261, 205–209, 210–214, 215–219, 244–248, 255, 256, 258, 440, 441, 443, 444, 451, 487, 553, 577, 745
real, 441, 745
triangular, 457
whole, 27, 84, 205, 216, 444, 451

Number theory, 156, 162, 740

Numerical expressions, 12, 14
evaluating, 13, 17, 21, 22, 58, 60, 72, 74
finding value, 14
translating, 11, 13, 14, 15, 38, 51
writing for verbal phase, 724

Octagonal prism, 566

Octagons, 528

Odd numbers, 148

Odds, 646–649, 662, 755
notation, 647

Online Research. *See also* Internet Connections; Research
career choices, 42, 73, 129, 172, 223, 278, 348, 358, 399, 472, 480, 515, 579, 649, 670
data update, 43, 68, 114, 119, 157, 190, 229, 267, 348, 396, 439, 543, 587, 625, 690

Open circles, 346

Open Ended, 9, 14, 19, 26, 30, 36, 42, 59, 67, 72, 77, 83, 87, 100, 105, 113, 117, 128, 133, 155, 158, 161, 162, 166, 171, 177, 183, 188, 202, 212, 217, 222, 228, 234, 241, 246, 266, 272, 278, 283, 291, 295, 300, 306, 312, 332, 336, 342, 347, 353, 357, 371, 377, 384, 389, 395, 400, 407, 410, 416, 421, 429, 438, 443, 449,

Index

Index

455, 462, 468, 473, 479, 487, 495, 502, 509, 515, 523, 529, 535, 540, 559, 565, 570, 575, 580, 586, 592, 599, 608, 614, 619, 625, 631, 637, 643, 648, 653, 670, 676, 680, 684, 689, 694. *See also* Assessment

Open sentences, 28, 31, 32
solving, 28

Operations
integers, 82
inverse, 110

Opposites, 66, 476

Ordered pairs, 33–35, 36, 37, 39, 44, 50, 51, 55, 61, 79, 84, 85, 86, 87, 88, 124, 136, 367, 377, 378, 391, 422, 425, 429, 725, 728, 742
functions, 369–370

Ordered pair solutions, tables, 375, 378, 686

Ordering decimals, 616

Ordering integers, 59, 68, 87, 88, 726

Order of operations, 12–13, 14, 16, 24, 48, 51, 147, 154, 192, 401, 464, 525

Origin, 33, 34

Outcomes, 310

Outliers, 621

Parabola, 697

Parallel lines, 495, 560
angles and, 492–497, 497, 544

Parallelogram, 514, 515, 546
altitude, 520
area, 520–521, 547
base, 520

Parentheses ( ), 15, 75
as grouping symbol, 12, 15
solving equations, 334–335

Pascal's triangle, 640

Patterns, 9, 16, 47, 48, 55, 74, 84, 102, 167

Pentagon, 528

Percent equations, 298–302, 319, 373

Percent-fraction equivalents, 293

Percent of change, 304–308, 320

Percent of decrease, 306, 320, 739

Percent of increase, 305, 306, 320, 739

Percent proportion, 288–292, 297, 318, 321, 738

Percents, 281–308, 318, 359, 525, 737, 738
applying, 289, 290, 299–308
of change, 304–308
comparing, 283
compound interest, 303
decimals as, 282–283
equations, 298–302
estimating with, 294–295
finding mentally, 293
finding percent, 288, 289, 299
finding the base, 290, 299
finding the part, 290, 298
fractions as, 282
greater than 100%, 283
less than 1%, 283
modeling, 281, 286, 287
and predictions, 312
and probability, 310
proportion, 288–292
simple interest, 299, 300
writing as decimals, 282, 298
writing as fractions, 281–282, 283, 288, 475

Perfect squares, 436, 438

Perimeter, 137
polygon, 530
rectangle, 132, 134, 140
rectangles, 132, 133, 134, 135, 140, 141, 152, 224, 302, 335, 336, 349, 359, 363, 385, 417, 676, 701, 730

Permutations, 641–645, 661, 754

Perpendicular lines, 494, 495

Perspective, 554–555

Pi, 208, 533, 534

Plane, 556

Points, 447
writing equations given two, 406

Polygons, 527–531, 547, 553
classification, 527–528, 530, 549, 749
frequency, 628
interior angles, 528
measures of angles, 528–529, 530
regular, 529

Polyhedron, 556

Polynomials, 669–672, 698, 755
addition, 674–677, 699
classification, 669, 681, 698, 701
degrees, 670, 671, 677, 681, 698, 701
multiplication, 683–686, 699
monomials, 683–686, 699
subtraction, 678–681, 699

Positive exponents, 182, 183, 185, 187, 194

Positive integer, 64

Positive numbers, 56
division, 350–351
multiplication, 350–351

Positive slope, 388, 391

Powers, 153–157, 174, 192, 194

Practice Chapter Test. *See* Assessment

Practice Quiz. *See* Assessment

Precision, 590–594, 598

Precision unit, 590, 591, 592, 593

Predictions, 310–314, 320, 395
best-fit lines, 409, 410, 412, 413
equations, 410
scatter plots, 42
sequences, 251

Prerequisite Skills. *See also* Assessment
Adding and Subtracting Decimals, 713
Comparing and Ordering Decimals, 710
Converting Measurements within the Customary System, 720–721
Converting Measurements within the Metric System, 718–719
Displaying Data in Graphs, 722–723
Estimating Products and Quotients of Decimals, 714
Estimating Products and Quotients of Fractions and Mixed Numbers, 717
Estimating Sums and Differences of Decimals, 712
Estimating Sums and Differences of Fractions and Mixed Numbers, 716
Getting Ready for the Next Lesson, 10, 16, 21, 27, 32, 38, 61, 68, 74, 79, 84, 102, 107, 114, 119, 124, 130, 152, 157, 163, 168, 173, 179, 185, 204, 209, 213, 219, 224, 230, 236, 242, 248, 268, 274, 280, 285, 292, 297, 302, 308, 333, 338, 344, 349, 354, 373, 379, 385, 391, 397, 401, 408, 413, 418, 440, 445,

451, 457, 464, 470, 475, 497, 504, 511, 517, 525, 531, 538, 561, 567, 572, 577, 582, 588, 611, 616, 621, 628, 633, 639, 645, 649, 672, 677, 681, 686, 691
Getting Started, 5, 55, 97, 147, 199, 263, 327, 367, 435, 491, 553, 605, 667
Multiplying and Dividing Decimals, 715
Problem-Solving Strategy
Guess and Check, 709
Make a Table or List, 708
Solve a Simpler Problem, 706
Work Backward, 707, 709, 712, 713, 714, 715, 716, 717, 719
Rounding Decimals, 711

Prime, 285
relatively, 168
twin, 162

Prime factorization, 159–162, 192, 224, 236, 731

Prime factors, 227, 230

Prime numbers, 159, 161, 162, 192, 440, 731

Prisms, 557, 595
height, 564
hexagonal, 572
octagonal, 566
rectangular, 557, 560, 563, 565, 566, 575, 576, 582, 597
surface area, 573–577, 597
triangular, 557, 562, 564, 566, 572, 587
volume, 563–567, 596

Probability, 310–314, 320, 333, 338, 604–605, 672, 677, 739, 754, 755
compound events, 650–655, 662
dependent events, 652
event, 636–637
experimental, 311
mutually exclusive events, 652
notation, 650
outcomes, 310
simple, 605
simple events, 310
theoretical, 311
two independent events, 650–651

Probability simulation, 315

Problem solving
adding polynomials, 675
angles, 449
area, 523
circumference, 534
combinations, 643
comparing fractions in, 202
direct variation, 395
distance formula, 467
distributive property, 99
dividing powers, 177
divisibility rules, 149
equations, 116, 125, 405
four-step plan, 5, 6–8, 9, 10, 47–48, 51
fractions, 234
functions, 693
graphing equations, 399
graphing inequalities, 420
guess and check, 709
inequalities, 347
integers, 66, 76
integers in, 76
making a table or list, 708
mixed, 758–770
polynomials, 675, 684
proportions, 271
scientific notation, 187
solving simpler problem, 706
square roots, 438
subtracting integers, 71
subtracting polynomials, 679
trigonometric ratios, 479
volume, 570
work backward, 122, 707, 709, 712, 713, 714, 715, 716, 717, 719

Product of Powers rule, 184

Products, 27, 69, 74, 78, 79, 84, 91–92, 93, 147, 175, 176, 177, 178, 185, 224, 252, 255, 280, 285, 327, 567, 577, 605, 715, 725, 727, 732, 756. *See also* Multiplication
estimating, 714, 717
signs, 79

Projects. *See* WebQuest

Properties of Equality, 23–25, 29, 30, 31, 49
Substitution, 18
Symmetric, 29, 30, 31
Transitive, 29, 30, 31

Proportions, 262, 270–274, 280, 285, 317, 413, 440, 470, 473, 474, 553, 582, 599, 677, 747
identifying, 271
property, 271
solving, 271, 272, 737

Protractor, 448, 449, 450, 457, 464, 484, 497, 512, 745

Pyramids, 557, 595
hexagonal, 571
rectangular, 557, 570, 572, 577
square, 557, 571, 581, 582, 585
surface area, 578–582, 597
triangular, 557
volume, 568–572, 596

Pythagorean Theorem, 460–464, 469, 485
converse, 462

Pythagorean triple, 459

Quadrants, 86, 87, 92

Quadratic functions, 688
families, 697
graphing, 692–696, 700

Quadrilaterals, 513–517, 528, 546
angles of, 514
classification, 514–515, 516, 531
naming, 513

Quantitative relationships, 96

Quartiles, 613

Quotient of Powers rule, 184

Quotients, 13, 38, 69, 79, 83, 84, 89, 92, 93, 177, 178, 185, 218, 224, 230, 242, 252, 256, 274, 280, 327, 714, 715, 717, 727, 732, 734. *See also* Division

Radical sign, 436, 437

Radius
circles, 443, 533, 536, 749
cones, 571

Range, 35, 36, 37, 44, 50, 51, 136, 367, 612–616, 618, 628, 659, 725, 753

Rate of change, 392, 393–397, 426, 743

Rates, 264–268, 316
converting, 266, 280

Rational numbers, 198–261, 440, 441, 443, 444, 451, 487, 745
division, 215–219, 256
identifying and classifying, 206
multiplication, 210–214, 255, 553, 577
solving equations, 244–248, 258
writing as fractions, 205–206

Ratios, 264–268, 273, 316, 321, 359, 386, 445, 454, 481, 553, 645, 736
common, 250, 251, 258
comparison, 269
writing as fractions, 264–265

Reading and Writing, 5, 55, 97, 147, 199, 263, 327, 367, 435, 491, 553, 605, 667

Reading Math, 17, 23, 24, 29, 56, 57, 64, 75, 80, 88, 98, 103, 148, 149, 150, 159, 177, 200, 205, 206, 281, 300, 311, 341, 370, 381, 383, 437, 448, 453, 471, 472, 477, 493, 500, 508, 624, 641, 642, 643, 647, 650, 678, 684, 688

Reading Mathematics
Dealing with Bias, 634
Factors and Multiples, 225
Language of Functions, 380
Learning Geometry Vocabulary, 446
Learning Mathematics Prefixes, 526
Learning Mathematics Vocabulary, 69
Making Comparisons, 269
Meanings of At Most and At Least, 339
Powers, 174
Precision and Accuracy, 589
Prefixes and Polynomials, 668
Translating Expressions into Words, 11
Translating Verbal Problems into Equations, 125

Real numbers, 441, 745

Real number system, 441–445, 484

Reasonableness, 7, 586

Reasoning. *See also* Critical Thinking
deductive, 25
inductive, 7, 25, 71

Reciprocals, 215, 247

Rectangles, 514, 546
area, 132, 133, 134, 135, 140, 141, 152, 214, 224, 349, 671, 730, 767
perimeter, 132, 133, 134, 135, 140, 141, 152, 224, 302, 335, 336, 349, 359, 363, 385, 417, 676, 701, 730

Rectangular prisms, 557, 560, 563, 565, 566, 575, 576, 582, 597

Rectangular pyramid, 557, 570, 572, 577

Reflections, 506, 508, 509, 510, 512, 686, 748

Regular polygon, 529

Regular tessellations, 527

Relations, 35, 50, 367
as tables and graphs, 35, 36, 37, 50

Relationships, types of, 41, 42, 43, 51 50, 68

Relative frequencies, 627

Relatively prime, 168

Repeating decimals, 201, 202, 203, 206, 214, 224, 254, 338, 733

Replacement set, 102

Research, 43, 46, 83, 113, 151, 225, 253, 268, 291, 308, 380, 446, 526, 530, 543, 560, 583, 589, 610, 668, 672. *See also* Online Research

Review
Lesson-by-Lesson, 47–53, 90–92, 138, 191–194, 254–258, 316, 360–362, 424–428, 483–486, 544–548, 595, 658–662, 698
Mixed, 16, 21, 27, 32, 44, 61, 68, 74, 79, 84, 89, 102, 107, 114, 119, 124, 130, 136, 152, 157, 163, 168, 173, 179, 185, 190, 204, 209, 213, 219, 224, 230, 236, 242, 248, 252, 268, 274, 280, 285, 292, 297, 302, 308, 314, 333, 338, 344, 349, 354, 359, 373, 379, 385, 391, 397, 401, 408, 413, 418, 422, 440, 445, 451, 457, 464, 470, 475, 481, 497, 504, 511, 517, 525, 538, 543, 561, 567, 572, 577, 582, 588, 594, 611, 616, 621, 628, 633, 639, 649, 655, 672, 677, 681, 686, 691, 696

Rhombus, 514, 546

Rotational symmetry, 505

Rotations, 506, 507, 508, 509, 510, 512, 532, 686
coordinate plane, 509

Rounding, 9, 10, 201, 242, 283, 284, 291, 314, 443, 445, 457, 462, 463, 468, 469, 475, 478, 480, 481, 482, 484, 486, 487, 491, 531, 534, 535, 536, 543, 548, 549, 565, 566, 568, 569, 570, 571, 572, 577, 580, 581, 582, 588, 592, 593, 596, 597, 598, 599, 605, 611, 672, 711, 735, 736, 737, 738, 745, 746, 747, 749, 750, 751, 752

S

Sample space, 311

Scale, 276, 631, 737

Scale drawings, 276–280, 317

Scale factor, 277, 285, 583, 587

Scale model, 276

Scalene triangles
right, 470, 485

Scatter plots, 39, 40–42, 43, 45–46, 50, 61, 68, 107, 408, 410, 411, 412, 422, 427, 429, 726
constructing, 40
interpreting, 41
predictions, 42
trends, 44

Scientific calculators, 12, 16

Scientific notation, 186–190, 188, 194, 195, 204, 268, 733

Segment measure, 472

Sequences
arithmetic, 249–252, 258, 259, 268, 344, 736
Fibonacci, 253
geometric, 249–252, 258, 259, 344, 736

Sets, 733
replacement, 102

Shadow reckoning, 473

Short Response. *See* Assessment

Side lengths of rectangles, 152

Sides of the angle, 447

Sierpinski's triangle, 471

Significant digits, 590–594, 598, 599, 752

Similar figures, 471–475, 486

Similar solids, 599

Simple interest, 300

Simplest form, 104

Simulation
probability, 315

Simulations, 656–657
half-life, 180

Sine, 477–481, 486, 747

Skew lines, 558, 560, 567

Slant height, 578

Slides, 506

Slope, 387–391, 395, 397, 398, 400, 401, 404, 408, 412, 413, 418, 426, 427, 429, 470, 639, 742, 743
negative, 388
positive, 388, 391
undefined, 389
zero, 388

Slope-intercept form, 398–401, 401, 406, 407, 413, 418, 427, 696, 742, 743

Solids, 556, 750, 751, 752
similar, 584–588, 598

Solution, 28

Solving Equations. *See* Equations

Spreadsheet Investigation
Circle Graphs and Spreadsheets, 452
Compound Interest, 303
Expressions and Spreadsheets, 22
Perimeter and Area, 137

Spreadsheets, 22

Square pyramid, 557, 571, 581, 582, 585

Square roots, 436–440, 451, 483, 628, 745
solving equations, 443

Squares, 436–440, 483, 514, 546
area, 439, 523

Square units, 575

Standard form, 154, 188, 189, 194, 195, 209, 733

Standardized Test Practice. *See* Assessment

Statistics, 418, 604–665, *See also* Data
bias, 634
misleading, 630–633, 661

Stem-and-leaf plots, 606–611, 658, 681, 752
back-to-back, 607–608, 609, 611
interpretation, 607

Stems, 606

Straight line, 102

Study organizer. *See* Foldables™ Study Organizer

Study Tips, 6, 7, 12, 13, 18, 25, 28, 34, 41, 58, 60, 65, 70, 71, 76, 82, 86
algebra tiles, 674
alternate strategy, 406
alternative methods, 128, 170, 220, 250, 265, 335, 522
altitudes, 521
angles, 447
Base, 289
box-and-whisker plots, 618
break in scale, 624
calculating with Pi, 534
calculator, 187
checking equations, 111, 405
checking reasonableness of results, 494
checking solutions, 121, 346, 376
check your work, 443
choosing a method, 164
choosing points, 388
choosing x values, 375
circles and rectangles, 574
classifying polynomials, 669
common misconceptions, 58, 132, 154, 175, 355, 420, 449, 467, 507, 557, 591, 617
commutative, 160
cones, 569
congruence statements, 501
cross products, 271
cylinders, 565
degrees, 670
diagonals, 558
different forms, 398
dimensions, 556
dividing by a whole number, 216
division expressions, 117
equiangular, 454
equivalent expressions, 104
estimation, 211, 288, 409
exponents, 155
expression, 126
finding percents, 295
first power, 153
fractions, 283
graphing, 692
graphing shortcuts, 382
growth of functions, 688
height of pyramid, 568
hypotenuse, 460
inequalities, 341, 345
irregular figures, 539
labels and scales, 631
Look Back, 18, 71, 76, 100, 111, 166, 176, 210, 244, 264, 266, 293, 330, 334, 335, 369, 442, 460, 466, 472, 497, 523, 535, 540, 570, 584, 635, 637, 652, 683, 684
mean, median, mode, 238
measures of volume, 563
mental computation, 122
mental math, 127, 160, 201, 282
multiplying more than two factors, 636
naming quadrilaterals, 513
negative fractions, 211
negative number, 352
negative signs, 233, 675
negative slopes, 399
n-gon, 528
notation, 507
parallel lines, 492
percents, 294
pi, 533, 565
plotting points, 377
positive and negative exponents, 187
positive number, 350
powers of ten, 186
precision, 590, 592
precision units, 591
prime factors, 226
probability, 310
properties, 270
Reading Math, 215, 455
reciprocals, 245
relatively prime, 230
remainders, 161
repeating decimals, 206
rolling number cubes, 646
rounding, 478
scale factors, 277, 585
slopes, 394
slopes/intercepts, 415
solutions, 421
statistics, 613, 630
substitution, 467
trends, 372
triangular pyramids, 581
use a table, 405
use a venn diagram, 169
writing prime factors, 165
zeros, 679

Substitution, 107, 379, 413, 440
solving equations, 744
solving systems of equations, 416, 417, 418, 428

Substitution Property of Equality, 18

Subtraction, 11, 328, 727, 734, 735, 756
decimals, 5, 712, 713
fractions, 221, 716
integers, 70–72, 74, 82, 84, 91, 118, 199, 327, 385
like fractions, 220–224, 256
mixed numbers, 221, 234, 716
polynomials, 678–681, 699
simplifying expressions, 100
solving equations, 110–111, 113, 139, 244
solving inequalities, 345–349
unlike fractions, 232–236, 257

Subtraction expressions, 100, 102, 104

Subtraction Property of Equality, 110, 111, 360

Sums, 61, 64, 67, 74, 79, 91, 93, 107, 230, 235, 242, 256, 257, 327, 605, 712, 716, 725, 726, 734, 735, 756. *See also* Addition

Surface area, 594, 751
cubes, 157, 694
prisms and cylinders, 573–577, 597
pyramids and cones, 578–582, 597

Symbols. *See also* Grouping symbols
bar notation, 201
break in scale, 624
congruent, 493
equality, 28
equals sign, 126
inequality, 28, 57, 59, 90, 93
≈ (is about equal to), 437
negative signs, 675
radical sign, 436
variable, 17

Symmetric Property of Equality, 29, 30, 31

Symmetry, 505
bilateral, 505
line, 505
rotational, 505
turn, 505

Systems of equations
graphing, 422
solving, 414–418, 428, 440, 582

Systems of inequalities, 422
graphing, 346, 348

Tables, 708
functions, 369–370
identifying functions, 688
ordered pair solutions, 375, 378, 686
relations, 35, 36, 37, 50
writing equations, 406

Tangent, 747

Tangent ratios, 477–481, 479, 486

Term, 103

Terminating decimals, 200, 203, 206

Tessellations, 527, 529, 530, 531, 532

Test preparation. *See* Assessment

Test-Taking Tips. *See* Assessment

Theoretical probability, 311

Three-dimensional figures, 552–601
building, 554–555
identifying, 556–561
precision and significant digits, 589, 590–594, 598
similar solids, 583, 584–588, 598
surface area
prisms and cylinders, 573–577, 597
pyramids and cones, 578–583, 597
volume
prisms and cylinders, 563–567, 596
pyramids and cones, 568–572, 596

Transformations, 512, 686
coordinate plane, 506–511, 545–546

Transitive Property of Equality, 29, 30, 31

Translations, 506, 507, 509, 510, 512, 686, 748

Transversal, 492–493, 495, 496, 544, 549, 747

Trapezoids, 514, 515, 546
area, 521, 522–523, 547

Tree diagram, 635, 637, 638, 645

Triangles, 38, 453–457, 485, 528
acute, 454, 456, 746
angle measures, 453
area, 521, 522, 523, 547
congruent, 500–504, 545, 748
equilateral, 455
isosceles, 455
obtuse, 454, 456, 746
obtuse isosceles, 455, 470, 485
right, 454, 456, 460, 461, 462, 463, 464, 481, 485, 746
right scalene, 470, 485
scalene, 455
similar, 471–475, 486
vertice, 88

Triangular numbers, 457

Triangular prisms, 557, 562, 564, 566, 572, 587

Triangular pyramid, 557

Trigonometric ratios, 477–481, 486

Trigonometry terms, 477

Trinomials, 669, 671, 681, 698, 755

Turn symmetry, 505

Twin primes, 162

Two-dimensional figures, 490–551, 561

Two-step equations, 120
solve, 327
solving, 120–122, 140, 327, 354, 451
writing, 126–130, 140

Unit fraction, 236

Unit rates, 265–268, 359, 445, 736

Unlike fractions
addition, 232–236, 257
subtraction, 232–236, 257

Unlike terms, 674, 677

USA TODAY Snapshots, iv, 8, 16, 43, 60, 101, 145, 156, 203, 213, 242, 289, 290, 312, 325, 343, 433, 450, 537, 593, 603, 610, 649, 654, 690

Variables, 17, 18, 19, 28, 30, 31, 48, 49, 725
defining, 18
linear equations in two, 375–379, 425
showing relationships, 17, 21
solving equations, 328–329, 330–333, 360, 397

Venn diagram, 164, 441

Verbal expressions
translating expressions into, 11, 38, 124, 125
translating into expressions, 13, 14, 15, 18, 19, 20, 30, 32, 51, 74, 105
writing numerical phrases, 724

Verbal problems, two-step, 127

Vertical lines, 383

Vertical line test, 370

Vertices, 88, 447, 453, 527, 545–546, 556, 557, 559, 595, 599

Volume, 750, 751
complex solid, 564
cones, 568–572, 596, 616
prisms and cylinders, 563–567, 596
pyramids and cones, 568–572, 596

WebQuest, 43, 79, 135, 136, 145, 173, 242, 301, 314, 325, 333, 412, 422, 433, 481, 542, 571, 594, 603, 626, 690, 696

Whole numbers, 27, 84, 205, 216, 444, 451

Work backward, 122, 240, 362, 707, 709, 712, 713, 714, 715, 716, 717, 719

Writing in Math, 10, 16, 27, 32, 37, 44, 61, 68, 74, 79, 84, 89, 101, 106, 114, 119, 123, 130, 136, 152, 157, 162, 168, 173, 179, 184, 190, 204, 213, 219, 223, 236, 242, 247, 251, 268, 274, 280, 285, 292, 297, 302, 307, 314, 333, 338, 344, 349, 354, 359, 373, 379, 385, 391, 397, 401, 408, 412, 418, 422, 440, 445, 451, 457, 464, 469, 475, 481, 497, 504, 511, 517, 525, 531, 537, 543, 561, 567, 571, 577, 582, 588, 594, 611, 616, 621, 627, 633, 639, 645, 649, 654, 672, 677, 681, 686, 691, 695

***x*-axis,** 33, 36, 40, 86, 92, 508

***x*-coordinate,** 33, 34, 35, 38, 50, 85, 86, 508

***x*-intercept,** 381–385, 391, 397, 425, 429, 742

***y*-axis,** 33, 36, 40, 86, 92, 508, 509

***y*-coordinate,** 33, 34, 35, 36, 38, 50, 85, 86, 508

***y*-intercept,** 381–385, 391, 397, 398–399, 400, 401, 404, 406, 408, 410, 412, 413, 418, 425, 427, 429, 470, 639, 742, 743

Z

Zero, 56
  Multiplicative Property, 24, 49, 61, 725

Zero pair, 62

Zero slope, 388

Index

FCAT Mathematics Reference Sheet

Area

Triangle $A = \frac{1}{2}bh$

Rectangle $A = \ell w$

Trapezoid $A = \frac{1}{2}h(b_1 + b_2)$

Parallelogram $A = bh$

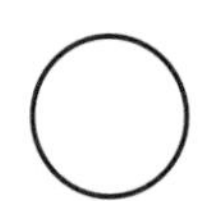

Circle $A = \pi r^2$

KEY	
b = base	d = diameter
h = height	r = radius
ℓ = length	A = area
w = width	C = circumference
$S.A.$ = surface area	V = volume
Use 3.14 or $\frac{22}{7}$ for π	

In a polygon, the sum of the measures of the interior angles is equal to $180(n - 2)$, where n represents the number of sides.

Circumference

$C = \pi d = 2\pi r$

Pythagorean theorem $c^2 = a^2 + b^2$

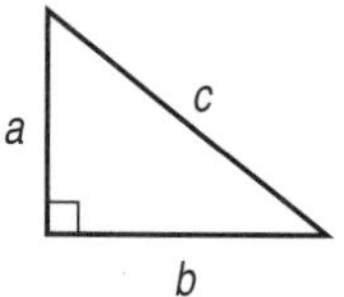

		Volume	Total Surface Area
	Right Circular Cylinder	$V = \pi r^2 h$	$S.A. = 2\pi rh + 2\pi r^2$
	Rectangular Solid	$V = \ell wh$	$S.A. = 2(\ell w) + 2(hw) + 2(\ell h)$

Conversions

1 yard = 3 feet = 36 inches
1 mile = 1,760 yards = 5,280 feet
1 acre = 43,560 square feet
1 hour = 60 minutes
1 minute = 60 seconds

1 cup = 8 fluid ounces
1 pint = 2 cups
1 quart = 2 pints
1 gallon = 4 quarts

1 liter = 1000 milliliters = 1000 cubic centimeters
1 meter = 100 centimeters = 1000 millimeters
1 kilometer = 1000 meters
1 gram = 1000 milligrams
1 kilogram = 1000 grams

1 pound = 16 ounces
1 ton = 2,000 pounds

Metric numbers with four digits are presented without a comma (e.g., 9960 kilometers). For metric numbers greater than four digits, a space is used instead of a comma (e.g., 12 500 liters).